Test Yourself

Take a chapter quiz on the *Essentials of The Living World* website. Each quiz is specially constructed to test your comprehension of key concepts. Feedback on your responses helps you gauge your mastery of the material.

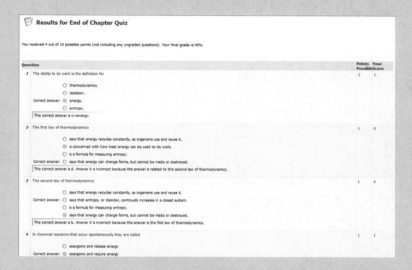

Access to Premium Learning Materials

The *Essentials of The Living World* website is your portal to exclusive study tools such as McGraw-Hill's animations, Virtual Labs, and ScienCentral videos.

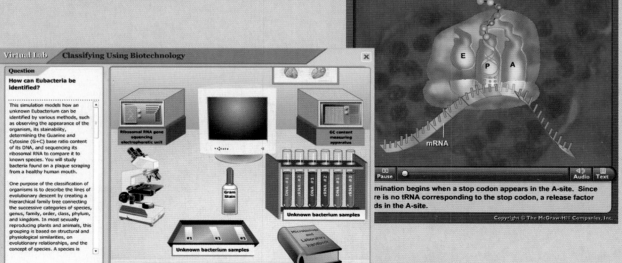

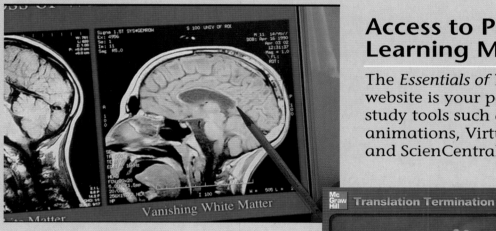

Second Edition

Essentials of
The Living World

George B. Johnson

Washington University

Jonathan B. Losos

Harvard University

Illustration Authors

William C. Ober, M.D.

and

Claire W. Garrison, R.N.

 Higher Education

Boston   Burr Ridge, IL   Dubuque, IA   New York   San Francisco   St. Louis
Bangkok   Bogotá   Caracas   Kuala Lumpur   Lisbon   London   Madrid   Mexico City
Milan   Montreal   New Delhi   Santiago   Seoul   Singapore   Sydney   Taipei   Toronto

Higher Education

ESSENTIALS OF THE LIVING WORLD, SECOND EDITION

Published by McGraw-Hill, a business unit of The McGraw-Hill Companies, Inc., 1221 Avenue of the Americas, New York, NY 10020. Copyright © 2008 by The McGraw-Hill Companies, Inc. All rights reserved. No part of this publication may be reproduced or distributed in any form or by any means, or stored in a database or retrieval system, without the prior written consent of The McGraw-Hill Companies, Inc., including, but not limited to, in any network or other electronic storage or transmission, or broadcast for distance learning.

Some ancillaries, including electronic and print components, may not be available to customers outside the United States.

 This book is printed on recycled, acid-free paper containing 10% postconsumer waste.

Printed in China
2 3 4 5 6 7 8 9 0 SDB/SDB 0 9 8 7

ISBN 978–0–07–352542–6
MHID 0–07–352542–1

Publisher: Janice Roerig-Blong
Sponsoring Editor: Thomas C. Lyon
Developmental Editor: Darlene M. Schueller
Director of Development: Kristine Tibbetts
Marketing Manager: Tamara Maury
Senior Project Manager: Sheila M. Frank
Senior Production Supervisor: Kara Kudronowicz
Senior Media Project Manager: Jodi K. Banowetz
Senior Media Producer: Eric A. Weber
Senior Coordinator of Freelance Design: Michelle D. Whitaker
Cover/Interior Designer: Christopher Reese
(USE) Cover Image: © Steve Bloom/stevebloom.com
Senior Photo Research Coordinator: Lori Hancock
Photo Research: Emily Tietz
Supplement Producer: Melissa M. Leick
Compositor: Electronic Publishing Services Inc., NYC
Typeface: 10.5/12 Times Roman
Printer: Shen Zhen Donnelley Printing Co., Ltd.

The credits section for this book begins on page C-1 and is considered an extension of the copyright page.

Library of Congress Cataloging-in-Publication Data

Johnson, George B. (George Brooks), 1942–
    Essentials of the living world / George B. Johnson. — 2nd ed.
      p. cm.
    Includes index.
    ISBN 978–0–07–352542–6 — ISBN 0–07–352542–1 (hard copy : alk. paper)
    1. Biology — Textbooks. I. Title.

QH308.2.J6199   2008
570 — dc22   2006025425
                                                         CIP

www.mhhe.com

Brief Contents

Contents

Part Seven   Plant Life

Boxed Readings

Applications Directory

Biology is having an enormous impact on modern society, affecting our lives often and deeply. To aid you in quickly turning to where a particular topic of current interest is found, this "Applications Directory" presents a list of these topics in alphabetical order, so that you can easily discover the page on which a topic is located—think of it as an index pre-sorted for topics that impact our lives.

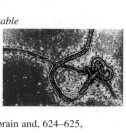

Preface

Just as species evolve, so do textbooks. With this second edition, *Essentials of the Living World* takes an important new step, as Professor Jonathan Losos of Harvard University joins the author team. Working together, we have designed this new edition of *Essentials* to teach the concepts of biology to undergraduate students like those we teach, students who would probably not become biology majors. The first edition was written by the older of us, Professor George Johnson of Washington University, the white-haired guy below, who for seven editions had been a principle author of *Biology,* a majors biology text. The challenge we faced in writing this new edition of *Essentials* was to present the key ideas of biology, the same principles taught in a majors course, to nonmajors in a clear and engaging way, without technical jargon or needless complication. With each passing year our students, and those of every biology instructor, are more impacted by biology, and this is true whether they major in science or English, art or business. With this rush of change it has become increasingly clear to us that every student must have the opportunity to understand how science works, and how the many changes science is creating can be expected to affect their lives and the future of our planet. We believe deeply that an appreciation of science at work has become an essential element of every college student's education, of citizenship in a world for which today's college students will soon be responsible.

Many of the changes we have made in the content of this second edition reflect the severe impact that human activity is having on our planet. The addition of Jonathan Losos as a co-author improves our ability as an author team to convey to students using *Essentials* the nature of these changes. Jonathan, who teaches the evolution course at Harvard University, is both a field ecologist and a frequent collaborator of DNA lab scientists. His knowledge of ecology and molecular biology adds a dimension to the author team that any text would desire, expert hands-on knowledge right at the focus of current change in biology and its impact on today's world. We have been faculty colleagues, and have worked together on *Biology* for the last three editions, so our union in this edition of *Essentials* has been fun as well as useful, like two kids finding a new playground.

Essentials has evolved in other important ways. While many of the changes we have made to this edition reflect the major changes going on in biology in the last few years, we have also found new ways to make the text easier for instructors to use. The most obvious of these is that while only slightly longer, *Essentials* now has seven more chapters. Only one of these seven chapters is a completely new one (Chapter 22, *Behavior and the Environment).* The others break six long chapters, which instructors found over-stuffed with information into twelve shorter, more manageable chapters. This new organization allows each topic its own focus, with twice the pedagogy afforded by the mega-chapters of the previous edition. The overlong chapters, worthy attempts to stress the connections between important topics, are now extinct in *Essentials,* dinosaurs of the evolving textbook world.

What's New?

Overview

- New advances in science presented
- Illustration integrated into body of text
- Table of Contents reorganized
- New *Inquiry & Analysis* features added

The first edition of *Essentials* featured a lean presentation of biology that focused on presenting the essential concepts in a clear and interesting way. That approach has not changed in the second edition. *Essentials* remains first and foremost a teaching tool, devoted to explaining to nonmajors the key ideas of biology and how they relate to everyday life.

This second edition of *Essentials* is new in two significant ways. First, there have been major advances in the biological sciences since the first edition, advances that affect biology in important ways. Second, the many instructors using *Essentials* have suggested several improvements in content and presentation, requesting new ways of relating what a student is reading to the illustrations that appear on the page, suggesting some chapters were overlong, and asking for better ways to convey the *process* of science to students.

New Advances in Science Presented. The primary purpose of any revision is to update the science, so that the text presents the current state of biology. A textbook is not a newspaper, and cannot incorporate new findings as soon as they are announced, lest errors creep in. Any scientist knows that results and ideas are in a constant state of flux, and that new findings suffer a harsh trial by fire after they are announced, tried, and tested in many other labs before being widely accepted. When new findings are verified, they can sometimes cause a sea change in how we think about a key idea or regard a particular theory. Sometimes whole new avenues of inquiry open, changing biology in fundamental and important ways. At least four such major advances have occurred in biology since the first edition of *Essentials,* each introduced briefly on the facing page.

Illustrations Integrated into Body of Text. Many instructors tell us that their students are visual learners, the illustrations key to helping them understand concepts, but that all too often a text's illustrations seem peripheral to the discussions in the text. In response to these instructors, the second edition of *Essentials* actively integrates illustrations into text discussions. The text includes a description of the figure that accompanies it, pointing out aspects that are key about the figure. A student reading the text can't help but follow through the figure with the description provided in the text. For complex ideas, there are numbered links between the text and the figure. The numbers clearly link the point under discussion in the text with the appropriate place in the figure, allowing a student to quickly refer to the area of the figure under discussion.

Table of Contents Reorganized. As noted earlier, many instructors told us they prefer the text be presented in a larger number of shorter chapters, so that the concepts don't arrive in a flood that students have trouble absorbing so quickly. In response, we have split each of six long chapters of the first edition of *Essentials* into two, so that where there were six, there are now twelve. This eases the pace of the text, providing added pedagogy for key topics.

New *Inquiry & Analysis* Feature Added. Biology students, and in particular nonmajors, enter a biology course with little knowledge or skill in dealing with experimental data, so key to understanding biology. To help students become familiar with the challenge of analyzing and interpreting data, every chapter of the second edition of *Essentials* ends with an analysis of experimental data related to that chapter's content. Some of these *Inquiry & Analysis* features are full-page presentations, others one-half page. All of them provide a short description of an actual experiment, a graph presenting data collected during that experiment, and then a series of questions to guide the students in interpreting the data. The questions help the student understand how the data set is presented, why it is presented in this way, and how the graph is best interpreted to draw the proper conclusions about the experimental outcome. Concepts that help a student interpret the data (for example what a log scale is and why it is used) are set off in color, so that the student can easily identify helpful information in the body of the text. To support these new *Inquiry & Analysis* features, a two-page discussion about data presentation and analysis has been added to Chapter One. This section introduces the student to the concept of variables, and teaches the student how to present data in various types of graphs. This discussion targets under-prepared students, helping them to better understand the presentation of data in graphs and charts. This both prepares them for the *Inquiry & Analysis* features, and allows them to gain more knowledge from the data presented throughout the text.

Advances in Science

Small RNAs Take a Lead Role in Molecular Research

For decades, DNA has been the focus of molecular biology, but findings in the last few years have put RNA center stage. While biologists are accustomed to viewing RNA as a copy of DNA used to guide a cell's production of proteins, it comes as a great surprise that small RNA molecules like the one you see below are also active players in gene regulation. Apparently most animals and plants regulate the expression of their genes at two levels: 1. proteins  act to promote or inhibit the "reading" of genes by the enzyme called RNA polymerase that produces the messenger RNA copy used to make proteins; 2. small RNAs inhibit the use of particular messenger RNAs by binding to those that match the small RNA so that they cannot be read properly—imagine trying to sit in a seat occupied by someone else. This new level of gene control appears very widespread, and has already led to major advances in gene technology and medical treatment. At present, small RNAs are perhaps the most active research area in molecular biology. Learn more on page 236.

Global Warming Gets Real

Global warming has been a reality to science for many years, researchers concluding that recent decades were the warmest in the last thousand years. Industry and government in the United States have been slow to accept this conclusion, however, attacking "hockey stick" graphs (like the one below showing little change in global temperatures for many centuries, followed by a rapid rise in the most recent century) as statistically flawed. This objection was put to rest for good in 2006 by a comprehensive study by the National Academy of Sciences. Temperature measurements, variation in ancient tree rings, temperature measures in deep holes in the earth, all lead to the same conclusion—the warming in the last 25 years exceeds any peaks since 1600. Global warming is real, and demands our attention. Learn more on pages 465 and 480.

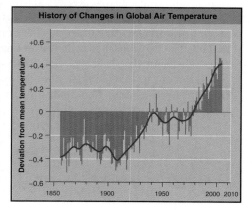

History of Changes in Global Air Temperature

Evolution Under Fire

The teaching of evolution in America's public schools, long controversial, has been under sharp recent attack by proponents of "intelligent design" (the proposal that life is too complex to have resulted from evolution without intelligent help). The sequencing of the genomes (all the DNA of an organism) of many vertebrates in the last few years has finally allowed a direct test of Darwin's proposition—did vertebrates evolve or not? The numbers in the diagram below (or see page 304) are the percent of DNA letters that are the same as the human genome. As you can see, the longer ago a vertebrate diverged, the more differences that have accumulated between its DNA and the human sequence. This settles the issue once and for all—vertebrates evolved. Learn more on page 280.

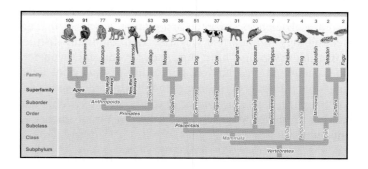

The Emergence of Bird Flu

In the last two years the world has been alarmed by the possibility of a lethal influenza pandemic. Influenza, or "flu," is a respiratory illness caused by a virus. The virus, common in birds in Asia, can be lethal in humans. The 1918 Spanish flu pandemic killed about 2% of those infected, but because the virus spread easily from person to person, enough people caught it to kill between 40 million and 100 million worldwide in two years. In 2004 a new strain of flu emerged, infecting people handling Asian ducks and chickens. Of 180 people infected in the following two years, 100 died—a mortality rate of 54%! Alarmed researchers reinvestigating the 1918 flu virus found it too had jumped directly from birds to humans. In both instances, the virus is able to replicate in humans and cause death, and in both cases the bird virus is so different from familiar flu that humans have virtually no immunity. The difference between the 1918 flu virus and the 2004 bird flu is that today's virus is not efficiently transmitted between humans. Researchers warn us, however, that relatively few mutations would be required to effect this change in infectivity. The situation is being watched closely as scientists debate the best course of action. Learn more on page 325.

Updates and Additions

Every chapter of this new edition of *Essentials* has been carefully reworked to incorporate new research findings and to better integrate the illustrations with the text.

Chapter 1 Added a 2-page section, "How Scientists Analyze and Present Experimental Results." This discussion will help underprepared students understand the presentation of data in graphs and charts throughout the text.

Chapter 2 The ecology section in this chapter was revised, making it more of an overview of the study of ecology. The specific topic of population growth was dropped in this chapter (leaving the topic for later in the text) and replaced with the general concept of ecosystems, communities, niches, competition, predation, and symbiosis.

Chapter 3 This chapter now covers just basic chemistry. The discussion of biomolecules has been moved to a separate chapter. It expands the discussion of the medical uses of radioactive isotopes, clarifies the discussion of carbon-14 dating, expands the discussion of electronegativity with regards to water molecules and polarity, clarifies the discussion of the involvement of hydrogen ions in determining pH, adds a new boxed feature Today's Biology "Acid Rain," which shows an application of the pH discussion, and expands the discussion of buffers, specifically buffering systems in the blood.

Chapter 4 A separate chapter is now devoted to the topic of biomolecules. It improves the organization and presentation of the protein structure discussion, adds art within the running text to help clarify nucleotide and nucleic acid structure, adds A Closer Look boxed reading "Discovering the Structure of DNA" to point out how knowledge of the structure of nucleotides contributed to working out the structure of DNA, and expands the discussion of lipids with more examples.

Chapter 5 Discussions are expanded of cell size with the example of multicellular organisms, and of cholesterol as a component of the lipid bilayer; the introduction to eukaryotic cell structure is expanded, and the process of osmosis clarified.

Chapter 6 A discussion of biochemical pathways has been added, and the discussion of redox reactions expanded.

Chapter 7 A separate chapter is now devoted to photosynthesis. The information on cellular respiration has been moved to a separate chapter. This chapter expands the discussion of C3 photosynthesis, adding how ADP and NADP$^+$ recycles, and also expands the discussion of photorespiration.

Chapter 8 A separate chapter is now devoted to cellular respiration which expands the introduction to cellular respiration to include discussions of NADH and FADH2.

Chapter 9 A separate chapter is now devoted to mitosis. The information on meiosis has been moved to a separate chapter. This chapter has been revised to clarify the difference between homologous chromosomes and sister chromatids. The section "Curing Cancer" is moved into a boxed reading so instructors have more flexibility in choosing to cover it.

Chapter 10 A separate chapter is now devoted to meiosis. The discussion of reduction division is clarified, and a new figure more clearly shows independent assortment of chromosomes.

Chapter 11 The discussion of gene linkage is expanded, and a new figure is added supporting the discussion.

Chapter 12 A separate chapter is now devoted to DNA structure, replication, and mutations. It revises the figure on point mutation, clarifying that the mutation occurs in the DNA.

Chapter 13 A separate chapter is now devoted to gene expression, including transcription, translation, and gene regulation.

Chapter 14 A new feature "Can modified genes escape from GM crops?" experimentally evaluates the potential of GM plants to donate their modified genes to neighboring plants. The discussion of whether development is reversible or irreversible is expanded, including a new figure showing isolation and transfer of nucleus in a cloning experiment. A new section, "Progress with Reproductive Cloning" includes a figure showing a timeline of successful cloning experiments. The discussion of problems with reproductive cloning is expanded, including a new figure on genomic imprinting.

Chapter 15 The discussion of the fossil evidence for evolution is expanded, including a new box feature Today's Biology "Darwin and Moby Dick." Expands the discussion of the molecular evidence, using new genomic data to directly test Darwin. A new section, Section 15.3 Evolution's Critics, is added. This is a five-page discussion addressing the criticisms of evolution put forward by critics, and evaluating their proposition of intelligent design as an alternative to evolution. The chapter is revised to clarify the discussion of Hardy-Weinberg, and expand the discussions of genetic drift and nonrandom mating with examples. A new boxed reading Author's Corner "Are Bird-Killing Cats Nature's Way of Making Better Birds?" explores backyard natural selection. The discussion of "Problems with the Biological Species Concept" is clarified. The discussion of the "Pace of Evolution" is combined here in one place (in the first edition, this was discussed in two different places).

Chapter 16 The description of cladistics is clarified. The cat family tree (figure 16.7) is revised based on new molecular research. The organization of the chapter is revised to discuss the domain Bacteria first, and then Archaea.

Chapter 18 The discussion of all life cycles has been expanded and clarified, particularly the discussion of fruits in relation to the angiosperm life cycle.

Chapter 19 The old introduction is replaced with a new, more visual presentation overviewing the general features of animals. An entire new section, Section 19.2, discusses Five Key Transitions in Body Plan.

Chapter 20 The discussion of food chains has been expanded to elaborate more fully on the role of decomposers. The discussion of ecological pyramids has been reworked and clarified.

Chapter 21 The introduction to communities has been revised because students will have been introduced to the basic concept of communities in chapter 2 and are now prepared for a more comprehensive treatment. The chapter expands the discussion of predator-prey interactions.

Chapter 22 The treatment of migration has been reworked, with a clarified discussion of the experiment on migratory behavior of starlings.

Chapter 23 Updated information is presented on the ozone hole, and on global warming. A discussion is added on alternative energy sources (solar, wind, geothermal, nuclear, and hydrogen gas). The discussion of curbing population growth is significantly expanded, and a new section is added on population pyramids and consumption of resources.

Chapter 25 A separate chapter is now devoted to the circulatory system, the respiratory system material having moved to its own chapter. In the introduction, how blood circulates through the body is clarified.

Chapter 26 A separate chapter is now devoted to the respiratory system. The description of countercurrent flow in the gills is clarified, and a discussion of surfactants is added. The discussion of CO_2 transport and buffering in the blood is expanded and clarified.

Chapter 27 A separate chapter is now devoted to the digestive system, homeostasis and the urinary system moving to a separate chapter. The traditional food pyramid is replaced by the "new" food pyramid. The boxed reading on Diabetes and Obesity is updated, and the discussion of diversity seen in the digestive systems of different types of animals is expanded.

Chapter 28 A separate chapter is now devoted to homeostasis and the urinary system. The discussion of negative feedback loops is expanded, including a new figure, figure 28.1. A new boxed reading explains "How Hormones Control Your Kidney's Functions."

Chapter 29 A discussion of memory T cells and their role in the secondary immune response is added, along with an extended treatment of memory B cells. The discussion of the overall immune response is expanded and clarified, tying together the cellular and humoral responses. The discussion of flu immunization is expanded, and the data on HIV/AIDS are updated.

Chapter 31 The discussion of the thyroid is expanded to include hyperthyroidism.

Chapter 32 The description of the placenta and the discussion of allometric growth are clarified.

Chapter 33 A new figure shows vascular cambium and secondary growth. The discussion of water and carbohydrate transport in plants is clarified.

Chapter 34 New figures show various methods of asexual reproduction in plants. The discussion of germination is expanded. The discussion on the chemical reactions in photoperiodism is clarified.

Learning Tools

Focusing on the Essential Ideas of Biology

Biology is at its core a set of ideas, and if students can master these essential concepts, then the rest of biology comes easily to them. *Essentials* was written to focus on concepts rather then terminology and information, to teach how things work and why things happen the way they do, rather than merely naming parts or giving descriptive information. In this way we intended to make *Essentials* a more effective learning tool.

1. **Process Boxes.** However clearly it is written, there is no way a text can avoid the fact that some processes like photosynthesis and the Krebs cycle are complex. To aid in a student's learning of complex ideas, we have prepared special "This is how it works" Process Boxes for some four dozen essential processes that students encounter in introductory biology. Each of these process boxes walks the student through a complex process, one step at a time, so that the central idea is not lost in the details.

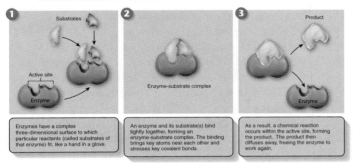

Enzymes have a complex three-dimensional surface to which particular reactants (called substrates of that enzyme) fit, like a hand in a glove.

An enzyme and its substrate(s) bind tightly together, forming an enzyme-substrate complex. The binding brings key atoms near each other and stresses key covalent bonds.

As a result, a chemical reaction occurs within the active site, forming the product. The product then diffuses away, freeing the enzyme to work again.

2. **Integration of Illustrations Into Text.** The use of numbered steps in *Essentials* to link different aspects of an illustration to those parts of the text discussing them puts illustrations to work, actively integrating them into a student's reading. Increasingly, today's students are visual learners, and this linkage amplifies the effectiveness of a student's reading. For examples of this method of integrating illustrations into the text, see Chapter 7, page 127 (left), in which the text discusses how two photosystems interact during photosynthesis. The numbers link the points under discussion with the appropriate aspect of the illustration. As another example, Chapter 5's pages 90–91 link the discussion of the operation of the cell's

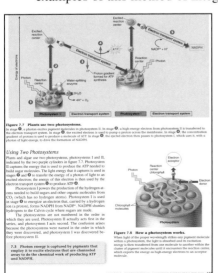

Figure 7.7 Plants use two photosystems.

Using Two Photosystems

Figure 7.8 How a photosystem works.

endomembrane system to aspects of the artwork presented on page 91. The student does not examine the figure as a separate activity, but rather as an active part of reading the discussion.

3. **Interactive Learning.** Perhaps the most difficult challenge facing nonmajors students taking a science course for the first time is assessment—although students don't yet understand a point, they may think that they do, not knowing enough to perceive their ignorance. To aid students in meeting this challenge, *Essentials* has placed on its ARIS website a series of interactive assessments. Students can take quizzes that probe what they understand, pinpointing any material that they have not yet mastered and pointing them to the appropriate learning module in each instance.

Arranging This Text for Effective Learning

This text is organized to teach the key ideas of biology effectively to today's nonmajors students. As college instructors ourselves, in *Essentials* we put to work what we have learned in a combined 40 years of teaching biology to freshmen—key strategies we found to be successful in our own classrooms.

1. **Connecting the Essential Concepts.** We have written *Essentials* as a series of learning modules presented on one or two pages. This format makes it easier for students to understand how the essential concepts within the chapter fit together conceptually. It also allows instructors to customize the text to their courses, as they can clearly point out which modules will be included in lecture.

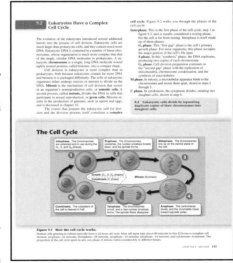

2. **Relating Essential Concepts to Evolution.** It is no accident that *Essentials* begins with a chapter on evolution and ecology. These ideas, central to biology, provide the student a framework within which to explore the world of the cell and gene function, which occupy the initial third of the text. Students learn about cells and genes much more readily when they are presented in an evolutionary context, as biology rather than as molecular machinery. The fact that within each mitochondrion is a tiny circle of DNA makes perfect sense when mitochondria are

understood as relicts descended from a bacterium engulfed long ago by an ancestral eukaryotic cell.

Because of evolution's central role in biology, it is important that students be prepared to deal with the long-standing controversy over the teaching of evolution in public schools. Chapter 15 presents the student with a detailed overview of the controversy, evaluating the arguments advanced against evolution objectively on their scientific merit.

3. **Relating Essential Concepts to Everyday Life.** One of the principle roles of nonmajors biology courses is to create educated citizens. In writing *Essentials* we have endeavored to relate what the student is learning to the biology each student ought to know to live as an informed citizen in the twenty-first century. Students also engage much more actively in the course when they can see how what they are studying relates to their own everyday lives.

Throughout the text, *Essentials* presents full-page boxed readings written by the authors that make connections between a chapter's contents and the everyday world: Today's Biology essays examine important new advances; The Scientific Process essays focus on how scientific analysis is carried out; A Closer Look essays allow a more detailed examination of interesting points; and Author's Corner essays take a more personal view (the author's) of how science relates to our everyday lives.

Much of the current impact of biology on everyday life centers on progress in DNA research and cell biology. This "new biology" is only beginning to impact nonmajors courses, but it is of great interest to students. Genetically modified foods, cloning of genetically identical animals, the potential of using stem cells to repair damaged tissues—all these topics appear in the news daily, and students need a way to link what they are learning in biology to them. Chapter 14, The New Biology, provides a brief roadmap.

Helping Students Learn the Process of Science

A principle goal of a nonmajors biology course is to educate students about science—what science is, and how it is done. In every chapter of *Essentials,* the student sees science in action. Hypotheses are advanced and their predictions tested. We have made two additions in this edition to help students learn how science is done.

1. **Data Analysis and Presentation.** Many students taking a course in nonmajors biology have little or no background in the sciences. Much of the difficulty they have with the course in many cases simply reflects a lack of experience with handling numbers. Data are at the heart of science, and no

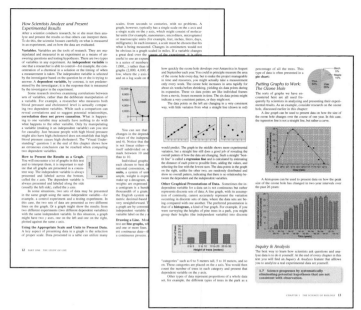

student can navigate a biology course successfully without being able to read a graph. To aid such students, we have included in Chapter 1 (pages 12–13) a quick overview of how scientists analyze and present experimental results.

2. ***Inquiry & Analysis* Features.** Perhaps the most important thing a student can learn about how science is done has little to do with equipment and laboratories. Science at its heart is a process of objectively asking and answering questions. To aid students in gaining some appreciation of this, each chapter of *Essentials* ends with a full-page or half-page feature *Inquiry & Analysis* that poses a particular scientific hypothesis and sorts out how to evaluate it with an experiment, collecting real data. The student is then challenged to analyze the data and reach a conclusion about the validity of

the hypothesis. This process is the nuts-and-bolts of the process of science, and by mastering it a student goes a long way toward learning how a scientist thinks. The answers to *Inquiry & Analysis* questions are available to the instructor on the ARIS website.

For the Instructor

McGraw-Hill offers the instructor powerful tools to support the second edition of *Essentials of The Living World.*

ARIS for Essentials of The Living World
aris.mhhe.com

McGraw-Hill's ARIS—Assessment, Review, and Instruction System—for *Essentials of The Living World* is a complete, online tutorial, electronic homework, and course management system. Instructors can create and share course materials and assignments with colleagues with a few clicks of the mouse. All PowerPoint lectures, assignments, quizzes, and tutorials are directly tied to text-specific materials. Instructors can edit questions, import their own content, and create announcements and due dates for assignments. ARIS has automatic grading and reporting of easy-to-assign homework, quizzing, and testing. All student activity within ARIS is automatically recorded and available to the instructor through a fully integrated grade book that can be downloaded to Excel. Contact your McGraw-Hill sales representative for more information on getting started with ARIS.

ARIS Presentation Center

Build instructional materials wherever, whenever, and however you want!
ARIS Presentation Center is an online digital library containing assets such as photos, artwork, animations, PowerPoints, and other media types that can be used to create customized lectures, visually enhanced tests and quizzes, compelling course websites, or attractive printed support materials.

Access to your book, access to all books!
The Presentation Center library includes thousands of assets from many McGraw-Hill titles. This ever-growing resource gives instructors the power to utilize assets specific to an adopted textbook as well as content from all other books in the library.

Nothing could be easier!
Accessed from the instructor side of your textbook's ARIS website, Presentation Center's dynamic search engine allows you to explore by discipline, course, textbook chapter, asset type, or keyword. Simply browse, select, and download the files you need to build engaging course materials. All assets are copyright McGraw-Hill Higher Education but can be used by instructors for classroom purposes.

Art
Full-color digital files of all illustrations in the book can be readily incorporated into lecture presentations, exams, or custom-made classroom materials.

Photos
Digital files of all photographs from the book can be reproduced for multiple classroom uses.

Tables
Tables from the text are available in electronic format.

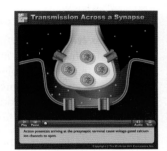

Animations
Over 200 full-color animations that support the topics discussed in the text are provided. Instructors are able to harness the visual impact of processes in motion by importing these files into classroom presentations or online course materials.

PowerPoint Lecture Outlines
Ready-made presentations that combine art and lecture notes are provided for each of the chapters in the text. These outlines can be used as they are or can be tailored to reflect your preferred lecture topics and sequences.

PowerPoint Image Slides
For instructors who prefer to create individualized lectures, illustrations, photos, and tables are preinserted by chapter into PowerPoint slides for convenience.

Instructor's Testing and Resource CD-ROM

This cross-platform CD-ROM contains the Instructor's Manual and Test Bank, available in both Word and PDF formats. The Instructor's Manual contains lecture outlines, learning objectives, key terms, lecture suggestions and enrichment tips, and critical-thinking questions. The Test Bank provides questions that can be used for homework assignments or the preparation of exams. The computerized Test Bank allows the user to quickly create customized exams. This user-friendly program allows you to search for questions by topic or format; edit existing questions or add new ones; and scramble questions and answer keys for multiple versions of the same test.

Transparencies

A set of full-color acetate transparencies is available to supplement classroom lectures. These transparencies include key figures from the textbook for classroom projection, and they are available to adopters.

eInstruction Classroom Performance System (CPS)

Wireless technology brings interactivity into the classroom or lecture hall. Instructors and students receive immediate feedback through wireless response pads that are easy to use and engage students. eInstruction can be used by instructors to:

- Take attendance
- Administer quizzes and exams
- Create a lecture with intermittent questions
- Manage lectures and student comprehension through use of the CPS grade book
- Integrate interactivity into their PowerPoint presentations

Course Delivery Systems

With help from our partners WebCT, Blackboard, eCollege, and other course management systems, professors can take complete control over their course content. Course cartridges containing content from the ARIS textbook website, online testing, and powerful student tracking features are readily available for use within these platforms.

McGraw-Hill: Biology Digitized Video Clips

McGraw-Hill is pleased to offer adopting instructors a new presentation tool—digitized biology video clips on DVD! Licensed from some of the highest quality science video producers in the world, these brief segments range from about five seconds to just under three minutes in length and cover all areas of general biology from cells to ecosystems. Engaging and informative, McGraw-Hill's Biology Digitized Video Clips will help capture students' interest while illustrating key biological concepts and processes such as mitosis, how cilia and flagella work, and how some plants have evolved into carnivores.

ScienCentral Videos

McGraw-Hill is pleased to announce an exciting new partnership with ScienCentral, Inc., to provide brief biology news videos for use in lecture or for student study and assessment purposes. A complete set of ScienCentral videos are located within this text's ARIS course management system, and each video includes a learning objective and quiz questions. These active learning tools enhance a biology course by engaging students in real-life issues and applications such as developing new cancer treatments and understanding how methamphetamine damages the brain. ScienCentral, Inc., funded in part by grants from the National Science Foundation, produces science and technology content for television, video, and the Web.

Virtual Labs

A complete set of over 20 online introductory biology laboratory exercises are available on this text's ARIS course management system. Each exercise is self-contained with all instructions and assessment for students. For use either stand-alone or as a supplement to a traditional lab course, this interactive learning tool will help students understand key biological concepts and processes.

For the Student

ARIS for Essentials of The Living World

aris.mhhe.com

McGraw-Hill's ARIS—Assessment, Review, and Instruction System—for *Essentials of The Living World* offers access to a vast array of premium online content to fortify the learning experience.

- **Text-Specific Study Tools**—The ARIS site features quizzes, study tools, and other features tailored to coincide with each chapter of the text.

- **Course Assignments and Announcements**—Students of instructors choosing to utilize McGraw-Hill's ARIS tools for course administration will receive a course code to log into their specific course for assignments.

Student Interactive CD-ROM

This interactive CD-ROM is an indispensable resource for studying topics covered in the text. It includes chapter outlines, chapter-based quizzes, animations of complex processes, flashcards, PowerPoint lecture outlines, and PowerPoint slides of art and photos found in the textbook. All of the materials are organized chapter-by-chapter. Direct links of the text's ARIS website are also provided.

Student Study Guide

This valuable student resource contains activities and questions to help reinforce chapter concepts. The guide provides concept outlines, key terms, and sample quizzes.

Acknowledgments

Every author knows that he or she labors on the shoulders of many others; the text you see is the result of hard work by an army of "behind-the-scenes" editors, spelling and grammar checkers, photo researchers, and artists that perform their magic on our manuscript; and an even larger army of production managers and staff that then transform this manuscript into a bound book. We cannot thank them all. Tom Lyon and Darlene Schueller were our editorial team, with whom we worked every day. Publisher Janice Roerig-Blong solved the many management problems her authors inadvertently created in their excess of enthusiasm, and provided valuable advice and support. Marty Lange, the editor-in-chief, oversaw all of this with humor and consistent support. Sheila Frank spearheaded our production team, which worked miracles with a very tight schedule. The photo program was carried out by Lori Hancock, who did a super job. Michelle Whitaker did a great job with the design. The book was produced by Electronic Publishing Services Inc., where Erin Daniel oversaw the art program, and another team headed by Jim Hill went to work to prepare the book for printing.

Our long-time, off-site developmental editors and right arms Liz Sievers and Megan Berdelman have again played an invaluable role in overseeing every detail of a complex revision, getting the work out despite husbands falling off roofs and babies being born. Their intelligence and perseverance continue to play a major role in the quality of this book.

The marketing of this new edition was planned and supervised by Tamara Maury, a battle-wise general not afraid to fight in the trenches alongside the many able sales reps that present our book to instructors.

Lastly but not least, we would like to extend a special thanks to Michael Lange, New Projects editor at McGraw-Hill, for his continued strong support of this project. It is still a new project in our hearts.

George Johnson
Jonathan Losos

Reviewers

Both of us have authored other texts, and all of our writing efforts have taught us the great value of reviewers in improving our texts. Scientific colleagues from around the country have provided numerous suggestions on how to improve the content of this second edition of *Essentials,* and many instructors and students using the first edition have suggested ways to clarify explanations, improve presentations, and expand on important topics. Indeed, the enthusiasm of their help was a bit overwhelming, and led to a revision far more extensive than we had planned. The instructors listed below provided detailed comments. We have tried to listen carefully to all of you. Every one of you has our thanks.

Jonathan Akin
Northwestern State University of Louisiana

Chander P. Arora
Los Angeles Valley College

Neil R. Baker
The Ohio State University

Mathew J. Bateman
Chadron State College

Jane B. Beiswenger
University of Wyoming

Joseph S. Bettencourt
Marist College

Cindy Birkner
Webber Twp High School / Rend Lake College

Mark Bolyard
Southern Illinois University Edwardsville

Bradford Boyer
Suffolk County Community College

Linda Kay Brandt
Henry Ford Community College

Edward M. Brecker
Palm Beach Community College

Adriene L. Brown
Tacoma Community College

Paul J. Bybee
Utah Valley State College

Beth B. Campbell
Itawamba Community College

Charlotte Carter
Stillman College

Thomas T. Chen
Santa Monica College

Thomas Chubb
Villanova University

Genevieve C. Chung
Broward Community College

Jeffrey Scott Coker
Elon University

James Crowder
Brookdale Community College

Judy M. Dacus
Cedar Valley College

Michael Dennis
Montana State University, Billings

Darren Divine
Community College of Southern Nevada

Jeff E. Engel
Western Illinois University

Steven E. Fields
Winthrop University

Jose Fierro
Florida Community College-Kent

Michelle A. Fisher
Three Rivers Community College

Katherine Foreman
Moraine Valley Community College

Michelle L. Fulton
Allan Hancock College

Uma Garimella
University of Central Arkansas

Sandra Gibbons
Moraine Valley Community College

Dalton R. Gossett
Louisiana State University-Shreveport

Carla J. Guthridge
Cameron University

Bob Harms
St. Louis Community College

Thomas E. Hemmerly
Middle Tennessee State University

Keith R. Hench
Kirkwood Community College

Juliana G. Hinton
McNeese State University

Mario Hollomon
Texas Southern University

Thomas Horvath
SUNY College at Oneonta

Pavla Hoyer
Los Angeles Valley College

Carol B. Johnson
Texas A&M University

Daniel Kainer
Montgomery College

Ragupathy Kannan
University of Arkansas-Fort Smith

Arnold J. Karpoff
University of Louisville

Dennis J. Kitz
Southern Illinois University Edwardsville

Michael E. Kovach
Baldwin-Wallace College

Kipp C. Kruse
Eastern Illinois University

Kim Lackey
Stillman College

Thomas G. Lammers
University of Wisconsin Oshkosh

Vic Landrum
Washburn University

Brenda G. Leicht
University of Iowa

Eric C. Lovely
Arkansas Tech University

Steven P. Lynch
Louisiana State University Shreveport

Janice B. Lynn
Auburn University at Montgomery

Susan Meiers
Western Illinois University

Timothy Metz
Campbell University

Daryl G. Miller
Broward Community College

Sheila Gibbs Miracle
Southeast Kentucky Community and Technical College

Jeanne Mitchell
Truman State University

Susan E. Mounce
Eastern Illinois University

Scott Murton
Lake Michigan College

Archana Nair
Tomball College

Rod Nelson
University of Arkansas - Fort Smith

Nathan Opolot Okia
Auburn University Montgomery

Joseph A.J. Orkwiszewski
Villanova University

Murali T. Panen
Luzerne County Community College

Mary Phillips
Tulsa Community College

David Pindel
Corning Community College

Michael Stuart Plotkin
University of California, Davis

Paul R. Ramsey
Louisiana Tech University

Jill D. Reid
Virginia Commonwealth University

Darryl Ritter
Okaloosa-Walton College

Michael L. Rutledge
Middle Tennessee State University

Kim Cleary Sadler
Middle Tennessee State University

Sangha Saha
Harold Washington College

Mark Sandheinrich
University of Wisconsin - La Crosse

Robert J. Schodorf
Lake Michigan College

Brian W. Schwartz
Columbus State University

A. K. M. Shahjahan
Baton Rouge Community College

Rick L. Simonson
University of Nebraska at Kearney

Kerri M. Skinner
University of Nebraska at Kearney

Marc Smith
Sinclair Community College

Linda D. Smith-Staton
Pellissippi State Technical Community College

J. Carolyn Sorrels
Itawamba Community College

Jim Stegge
Rochester Community and Technical College

Alicia Steinhardt
Hartnell Community College West Valley Community College

Judith Stewart
Community College of Southern Nevada

Season R. Thomson
Germanna Community College

Cheryl Vanier
University of Nevada Las Vegas

Manuel Varela
Eastern New Mexico University

Thomas V. Vogel
Western Illinois University

Marva Volk
Tulsa Community College, Southeast Campus

Eric Wada
American River College

Michael Wenzel
California State University, Sacramento

Robert R. Wise
University of Wisconsin Oshkosh

Brenda Reams Woodard
Northwestern State University of Louisiana

Lan Xu
South Dakota State University

Tony Yates
Seminole State College

Gregory Zagursky
Radford University

Michelle Zurawski
Moraine Valley Community College

Symposium Participants

Sarah M. Bales
Moraine Valley Community College

Genevieve C. Chung
Broward Community College

Susan L. Decker
North Central Texas College

Bud Donahou
Northwest Mississippi Community College

Bob Harms
St. Louis Community College

Clark L. Ovrebo
University of Central Oklahoma

Forrest Payne
University of Arkansas at Little Rock

Darryl Ritter
Okaloosa-Walton College

Kim Cleary Sadler
Middle Tennessee State University

J. Carolyn Sorrels
Itawamba Community College

Thomas V. Vogel
Western Illinois University

Brian Wainscott
Community College of Southern Nevada

Robin R. Whitekiller
University of Central Arkansas

Guided Tour

Accurate and Engaging Illustrations

The brilliant and vivid illustrations in *Essentials of The Living World* bring the study of biology to life! All of the figures have been carefully rendered to convey realistic, three-dimensional detail. Each drawing has been precisely labeled to coordinate with the text discussions.

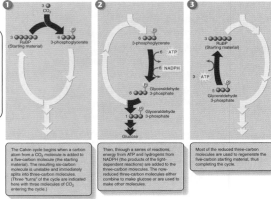

Process Boxes

Process Boxes break down complex processes into a series of smaller steps, allowing the student to track the key occurrences and learn them as they go.

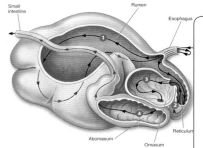

Step-By-Step Format

Essentials of The Living World presents concepts in an easy-to-follow stepwise format. Numbered steps within figures trace the sequence of events, and brief explanations describe what is happening in each step. Combining process descriptions with artwork creates a self-contained snapshot that summarizes concepts in a convenient and consistent format.

Engaging Figure Presentations

Figures are arranged in cohesive layouts that emphasize relationships among figure parts. For example, sharing labels between photos and drawings allow for easy comparison of structure appearance between the two mediums.

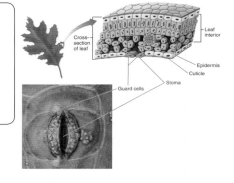

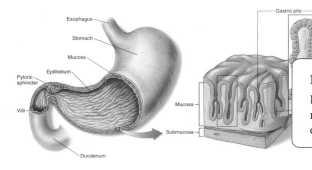

Multi-Level Perspective

Illustrations depicting complex structures connect the macroscopic and microscopic views to help students connect the two levels.

Combination Art

Drawings of structures are often paired with micrographs to enhance visualization.

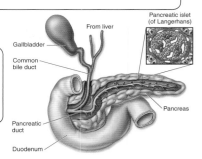

Illustrated Tables

The inclusion of figures within many of the tables in the text now makes it easier for the student to understand the table information at a glance, and helps remind them of key structures or processes.

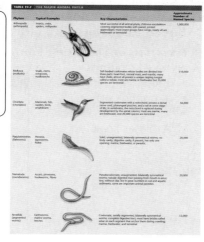

Systematic Pedagogy

Essentials of The Living World structures each chapter around a consistent framework of pedagogic devices. Whatever the subject matter of a chapter, students can develop a consistent learning strategy.

Chapter-at-a-Glance

Each chapter begins with an outline that gives the student an overview of the content contained within the chapter. Reviewing the outline before reading the chapter will help them focus their attention on the major concepts they should take away from the chapter. A chapter outline provides a quick overview of the chapter contents and organization.

Opening Vignette

The vignette is designed to pique students' interest and help them recognize the application and relevance of the topics presented in each chapter.

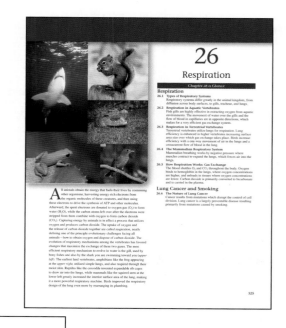

Modular Format

Each page or two-page spread in ***Essentials of The Living World*** is organized as an independent module with its own numbered heading and a highlighted summary. The **Numbered Headings** reinforce the Chapter-at-a-Glance outline while the **Section Summary** reviews the main ideas presented in the module. This system organizes the information in the chapter within a clear conceptual framework, which in turn helps the student learn and retain the material.

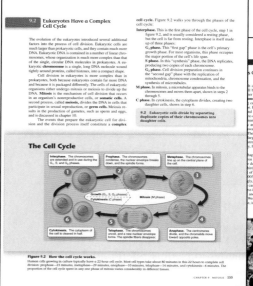

Boxed Readings

Essentials of The Living World features four types of boxed readings:

- Today's Biology
- Author's Corner
- The Scientific Process
- A Closer Look

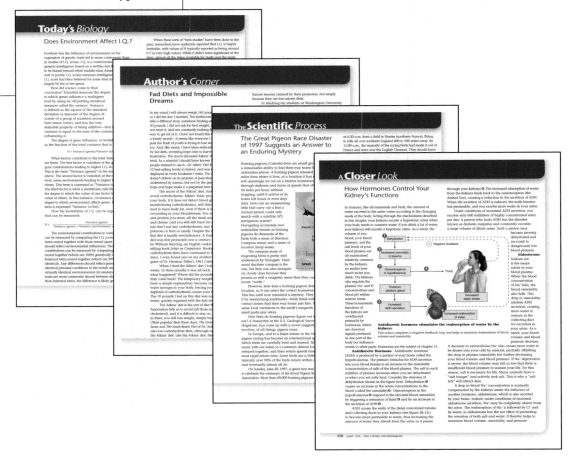

Inquiry & Analysis

The *Inquiry & Analysis* features provide a short description of an actual experiment, a graph presenting data collected during the experiment, and a series of questions to assist in interpreting the data.

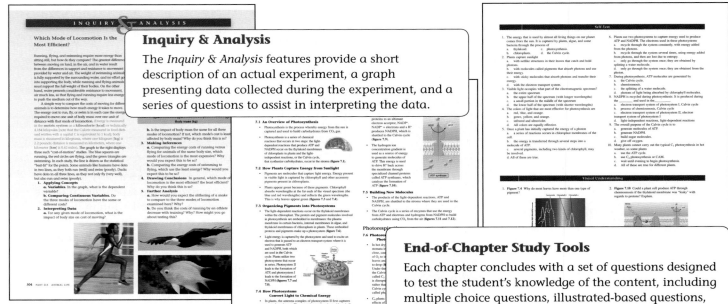

End-of-Chapter Study Tools

Each chapter concludes with a set of questions designed to test the student's knowledge of the content, including multiple choice questions, illustrated-based questions, and application questions. Answers to the multiple-choice questions can be found in Appendix A, while extended answers for all of the questions are found on the ARIS site at **aris.mhhe.com**.

Organismal Level. At the organismal level, in the second section of figure 1.4, cells are organized into three levels of complexity.

1. **Tissues.** The most basic level is that of **tissues,** which are groups of similar cells that act as a functional unit. Nerve tissue is one kind of tissue, composed of cells called neurons that are specialized to carry electrical signals from one place to another in the body.
2. **Organs.** Tissues, in turn, are grouped into **organs,** which are body structures composed of several different tissues grouped together in a structural and functional unit. Your brain is an organ composed of nerve cells and a variety of connective tissues that form protective coverings and distribute blood.
3. **Organ Systems.** At the third level of organization, organs are grouped into **organ systems.** The nervous system, for example, consists of sensory organs, the brain and spinal cord, and neurons that convey signals to and from them.

Populational Level. Organisms are further organized into several hierarchical levels within the living world, as you can see in the third section of figure 1.4.

1. **Population.** The most basic of these is the **population,** which is a group of organisms of the same species living in the same place. A flock of geese living together on a pond is a population.

POPULATIONAL LEVEL

Population

Species

Community

2. **Species.** All the populations of a particular kind of organism together form a **species,** its members similar in appearance and able to interbreed. All Canada geese, whether found in Canada, Minnesota, or Missouri, are basically the same, members of the species *Branta canadensis.*
3. **Community.** At a higher level of biological organization, a **community** consists of all the populations of different species living together in one place. Geese, for example, may share their pond with ducks, fish, grasses, and many kinds of insects. All interact in a single pond community.
4. **Ecosystem.** At the highest tier of biological organization, a biological community and the soil and water within which it lives together constitute an ecological system, or **ecosystem.**

Emergent Properties

At each higher level in the living hierarchy, novel properties emerge, properties that were not present at the simpler level of organization. These **emergent properties** result from the way in which components interact, and often cannot be guessed just by looking at the parts themselves. You have the same array of cell types as a giraffe, for example. Examining a collection of its individual cells gives little clue of what your body is like.

The emergent properties of life are not magical or supernatural. They are the natural consequence of the hierarchy or structural organization which is the hallmark of life. Water, which makes up 50–75% of your body's weight, and ice are both made of H_2O molecules, but one is liquid and the other solid because the H_2O molecules in ice are more organized.

Functional properties emerge from more complex organization. Metabolism is an emergent property of life. The chemical reactions within a cell arise from interactions between molecules that are orchestrated by the orderly environment of the cell's interior. Consciousness is an emergent property of the brain that results from the interactions of many neurons in different parts of the brain.

> **1.3** Cells, multicellular organisms, and ecological systems each are organized in a hierarchy of increased complexity. Life's hierarchical organization is responsible for the emergent properties that characterize so many aspects of the living world.

Ecosystem

1.4 Biological Themes

Just as every house is organized into thematic areas such as bedroom, kitchen, and bathroom, so the living world is organized by major *themes,* such as how energy flows within the living world from one part to another. As you study biology in this text, five general themes will emerge repeatedly, themes that serve to both unify and explain biology as a science (table 1.1):

1. evolution;
2. the flow of energy;
3. cooperation;
4. structure determines function;
5. homeostasis.

Evolution

Evolution is genetic change in a species over time. Charles Darwin was an English naturalist who, in 1859, proposed the idea that this change is a result of a process called **natural selection.** Simply stated, those organisms whose characteristics make them better able to survive the challenges of their environment live to reproduce, passing their favorable characteristics on to their offspring. Darwin was thoroughly familiar with variation in domesticated animals (in addition to many nondomesticated organisms), and he knew that varieties of pigeons could be selected by breeders to exhibit exaggerated characteristics, a process called **artificial selection.** You can see some of these extreme-looking pigeons pictured in table 1.1 under the heading "evolution." We now know that the characteristics selected are passed on through generations because DNA is transmitted from parent to offspring. Darwin visualized how selection in nature could be similar to that which had produced the different varieties of pigeons. Thus, the many forms of life we see about us on earth today, and the way we ourselves are constructed and function, reflect a long history of natural selection. Evolution will be explored in more detail in chapters 2 and 15.

The Flow of Energy

All organisms require energy to carry out the activities of living—to build bodies and do work and think thoughts. All of the energy used by most organisms comes from the sun and is passed in one direction through ecosystems. The simplest way to understand the flow of energy through the living world is to look at who uses it. The first stage of energy's journey is its capture by green plants, algae, and some bacteria by the process of photosynthesis. This process uses energy from the sun to synthesize sugars that photosynthetic organisms like plants store in their bodies. Plants then serve as a source of life-driving energy for animals that eat them. Other animals, like the eagle in table 1.1, may then eat the plant eaters. At each stage, some energy is used for the processes of living, some is transferred, and much is lost, primarily as heat. The flow of energy is a key factor in shaping ecosystems, affecting how many and what kinds of animals live in a community.

Cooperation

The ants cooperating in the upper right photo in table 1.1 protect the plant on which they live from predators and shading by other plants, while this plant returns the favor by providing the ants with nutrients (the yellow structures at the tips of the leaves). This type of cooperation between different kinds of organisms has played a critical role in the evolution of life on earth. For example, organisms of two different species that live in direct contact, like the ants and the plant on which they live, form a type of relationship called **symbiosis.** Animal cells possess organelles that are the descendants of symbiotic bacteria, and symbiotic fungi helped plants first invade land from the sea. The coevolution of flowering plants and insects—where changes in flowers influenced insect evolution and in turn, changes in insects influenced flower evolution—has been responsible for much of life's great diversity.

Structure Determines Function

One of the most obvious lessons of biology is that biological structures are very well suited to their functions. You will see this at every level of organization: within cells, the shape of the proteins called enzymes that cells use to carry out chemical reactions are precisely suited to match the chemicals the enzymes must manipulate. Within the many kinds of organisms in the living world, body structures seem carefully designed to carry out their functions—the long tongue with which a moth sucks nectar from a deep flower is one example. The superb fit of structure to function in the living world is no accident. Life has existed on earth for over 2 billion years, a long time for evolution to favor changes that better suit organisms to meet the challenges of living. It should come as no surprise to you that after all this honing and adjustment, biological structures carry out their functions well.

Homeostasis

The high degree of specialization we see among complex organisms is only possible because these organisms act to maintain a relatively stable internal environment, a process introduced earlier called homeostasis. Without this constancy, many of the complex interactions that need to take place within organisms would be impossible, just as a city cannot function without rules to maintain order. Maintaining homeostasis in a body as complex as yours requires a great deal of signaling back-and-forth between cells.

As already stated, you will encounter these biological themes repeatedly in this text. But just as a budding architect must learn more than the parts of buildings, so your study of biology should teach you more than a list of themes, concepts, and parts of organisms. Biology is a dynamic science that will affect your life in many ways, and that lesson is one of the most important you will learn. It is also an awful lot of fun.

> **1.4 The five general themes of biology are (1) evolution, (2) the flow of energy, (3) cooperation, (4) structure determines function, and (5) homeostasis.**

TABLE 1.1   BIOLOGICAL THEMES

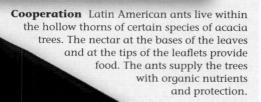

Cooperation Latin American ants live within the hollow thorns of certain species of acacia trees. The nectar at the bases of the leaves and at the tips of the leaflets provide food. The ants supply the trees with organic nutrients and protection.

Evolution Charles Darwin's studies of artificial selection in pigeons provided key evidence that selection could produce the sorts of changes predicted by his theory of evolution. The differences that have been obtained by artificial selection of the wild European rock pigeon (*top*) and such domestic races as the red fantail (*middle*) and the fairy swallow (*bottom*), with its fantastic tufts of feathers around its feet, are indeed so great that the birds probably would, if wild, be classified in different major groups.

The Flow of Energy Energy passes from the sun to plants to plant-eating animals to animal-eating animals, such as this eagle.

Homeostasis Homeostasis often involves water balance to maintain proper blood chemistry. All complex organisms need water—some, like this hippo, luxuriate in it. Others, like the kangaroo rat that lives in arid conditions where water is scarce, obtain water from food and never actually drink.

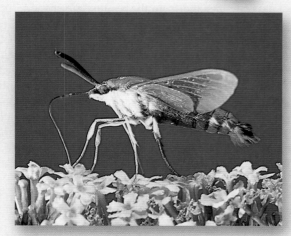

Structure Determines Function With its long tongue, the hummingbird clear-wing moth is able to reach the nectar deep within these flowers.

1.5 How Scientists Think

Deductive Reasoning

Science is a process of investigation, using observation, experimentation, and reasoning. Not all investigations are scientific. For example, when you want to know how to get to Chicago from St. Louis, you do not conduct a scientific investigation—instead, you look at a map to determine a route. In other investigations, you make individual decisions by applying a "guide" of accepted general principles. This is called **deductive reasoning.** Deductive reasoning, using general principles to explain specific observations, is the reasoning of mathematics, philosophy, politics, and ethics; deductive reasoning is also the way a computer works. All of us rely on deductive reasoning to make everyday decisions—like whether you need to slow down while driving along a city street in figure 1.5. We use general principles as the basis for examining and evaluating these decisions.

Inductive Reasoning

Where do general principles come from? Religious and ethical principles often have a religious foundation; political principles reflect social systems. Some general principles, however, are not derived from religion or politics but from observation of the physical world around us. If you drop an apple, it will fall, whether or not you wish it to and despite any laws you may pass forbidding it to do so. Science is devoted to discovering the general principles that govern the operation of the physical world.

How do scientists discover such general principles? Scientists are, above all, observers: they look at the world to understand how it works. It is from observations that scientists determine the principles that govern our physical world.

This way of discovering general principles by careful examination of specific cases is called **inductive reasoning.** Inductive reasoning first became popular about 400 years ago, when Isaac Newton, Francis Bacon, and others began to conduct experiments and from the results infer general principles about how the world operates. The experiments were sometimes quite simple. Newton's consisted simply of releasing an apple from his hand and watching it fall to the ground. This simple observation is the stuff of science. From a host of particular observations, each no more complicated than the falling of an apple, Newton inferred a general principle—that all objects fall toward the center of the earth. This principle was a possible explanation, or **hypothesis,** about how the world works. You also make observations and formulate general principles based on your observations, like forming a general principle about the timing of traffic lights in figure 1.5. Like Newton, scientists work by forming and testing hypotheses, and observations are the materials on which they build them.

> **1.5** Science uses inductive reasoning to infer general principles from detailed observation.

DEDUCTIVE REASONING

An Accepted General Principle
When traffic lights along city streets are "timed" to change at the time interval it takes traffic to pass between them, the result will be a smooth flow of traffic.

DEDUCTIVE REASONING

Using a General Principle to Make Everyday Decisions

Traveling at the speed limit, you approach each intersection anticipating that the red light will turn green as you reach the intersection.

INDUCTIVE REASONING

Observations of Specific Events

Driving down the street at the speed limit, you observe that the red traffic light turns green just as you approach the intersection.

Maintaining the same speed, you observe the same event at the next several intersections: the traffic lights turn green just as you approach the intersections. When you speed up, however, the light doesn't change until after you reach the intersection.

INDUCTIVE REASONING

Formation of a General Principle
You conclude that the traffic lights along this street are "timed" to change in the time it takes your car, traveling at the speed limit, to traverse the distance between them.

Figure 1.5 Deductive and inductive reasoning.
A deduction is a conclusion drawn from general principles. An inference is a conclusion drawn from specific observations. In this hypothetical example, a driver who assumes that the traffic signals are timed can use deductive reasoning to expect that the traffic lights will change predictably at intersections. In contrast, a driver who is not aware of the general control and programming of traffic signals can use inductive reasoning to determine that the traffic lights are timed as the driver encounters similar timing of signals at several intersections.

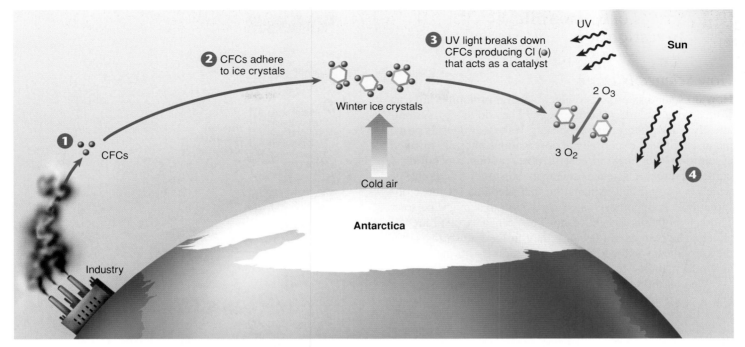

Figure 1.6 How CFCs attack and destroy ozone.
CFCs are stable chemicals that accumulate in the atmosphere as a by-product of industrial society ❶. In the intense cold of the Antarctic, these CFCs adhere to tiny ice crystals in the upper atmosphere ❷. UV light causes the breakdown of CFCs, producing Cl. Cl acts as a catalyst, converting O_3 into O_2 ❸. As a result, more harmful UV radiation reaches the earth's surface ❹.

1.6 Science in Action: A Case Study

In 1985 Joseph Farman, a British earth scientist working in Antarctica, made an unexpected discovery. Analyzing the Antarctic sky, he found far less ozone (O_3, a form of oxygen gas) than should be there—a 30% drop from a reading recorded five years earlier in the Antarctic!

At first it was argued that this thinning of the ozone (soon dubbed the "ozone hole") was an as-yet-unexplained weather phenomenon. Evidence soon mounted, however, implicating synthetic chemicals as the culprit. Detailed analysis of chemicals in the Antarctic atmosphere revealed a surprisingly high concentration of chlorine, a chemical known to destroy ozone. The source of the chlorine was a class of chemicals called **chlorofluorocarbons (CFCs)**. CFCs (the purple balls ❶ in figure 1.6) have been manufactured in large amounts since they were invented in the 1920s, largely for use as coolants in air conditioners, propellants in aerosols, and foaming agents in making Styrofoam. CFCs were widely regarded as harmless because they are chemically unreactive under normal conditions. But in the atmosphere over Antarctica, CFCs condense onto tiny ice crystals ❷; in the spring, the CFCs break down and produce chlorine, which acts as a catalyst, attacking and destroying ozone, turning it into oxygen gas without the chlorine being used up ❸.

The thinning of the ozone layer in the upper atmosphere 25 to 40 kilometers above the surface of the earth is a serious matter. The ozone layer protects life from the harmful ultraviolet (UV) rays from the sun that bombard the earth continuously. Like invisible sunglasses, the ozone layer filters out these dangerous rays. So when ozone is converted to oxygen gas, the UV rays are able to pass through to the earth ❹. When UV rays damage the DNA in skin cells, it can lead to skin cancer. It is estimated that every 1% drop in the atmospheric ozone concentration leads to a 6% increase in skin cancers.

The world currently produces less than 200,000 tons of CFCs annually, down from 1986 levels of 1.1 million tons. As scientific observations have become widely known, governments have rushed to correct the situation. By 1990, worldwide agreements to phase out production of CFCs by the end of the century had been signed. Production of CFCs declined by 86% in the following ten years.

Nonetheless, most of the CFCs manufactured since they were invented are still in use in air conditioners and aerosols and have not yet reached the atmosphere. As these CFCs move slowly upward through the atmosphere, the problem can be expected to continue. Ozone depletion is still producing major ozone holes over the Antarctic.

But the worldwide reduction in CFC production is having a major impact. The period of maximum ozone depletion will peak in the next few years, and researchers' models predict that after that the situation should gradually improve, and that the ozone layer will recover by the middle of the 21st century. Clearly, global environmental problems can be solved by concerted action.

> **1.6 Industrially produced CFCs catalytically destroy ozone in the upper atmosphere.**

Stages of a Scientific Investigation

How Science Is Done

How do scientists establish which general principles are true from among the many that might be? They do this by systematically testing alternative proposals. If these proposals prove inconsistent with experimental observations, they are rejected as untrue. After making careful observations concerning a particular area of science, scientists construct a hypothesis, which is a suggested explanation that accounts for those observations. A hypothesis is a proposition that might be true. Those hypotheses that have not yet been disproved are retained. They are useful because they fit the known facts, but they are always subject to future rejection if—in the light of new information—they are found to be incorrect.

We call the test of a hypothesis an experiment. Suppose that a room appears dark to you. To understand why it appears dark, you propose several hypotheses. The first might be, "The room appears dark because the light switch is turned off." An alternative hypothesis might be, "The room appears dark because the lightbulb is burned out." And yet another alternative hypothesis might be, "I am going blind." To evaluate these hypotheses, you would conduct an experiment designed to eliminate one or more of the hypotheses. For example, you might reverse the position of the light switch. If you do so and the light does not come on, you have disproved the first hypothesis. Something other than the setting of the light switch must be the reason for the darkness. Note that a test such as this does not prove that any of the other hypotheses are true; it merely demonstrates that one of them is not. A successful experiment is one in which one or more of the alternative hypotheses is demonstrated to be inconsistent with the results and is thus rejected.

As you proceed through this text, you will encounter a great deal of information, often accompanied by explanations. These explanations are hypotheses that have withstood the test of experiment. Many will continue to do so; others will be revised as new observations are made. Biology, like all science, is in a constant state of change, with new ideas appearing and replacing old ones.

The Scientific Process

Joseph Farman, who first reported the ozone hole, is a practicing scientist, and what he was doing in Antarctica was science. Science is a particular way of investigating the world, of forming general rules about why things happen by observing particular situations. A scientist like Farman is an observer, someone who looks at the world in order to understand how it works.

Scientific investigations can be said to have six stages as illustrated in figure 1.7: ❶ observing what is going on; ❷ forming a set of hypotheses; ❸ making predictions; ❹ testing them and ❺ carrying out controls, until one or more of the hypotheses have been eliminated; and ❻ forming conclusions based on the remaining hypothesis.

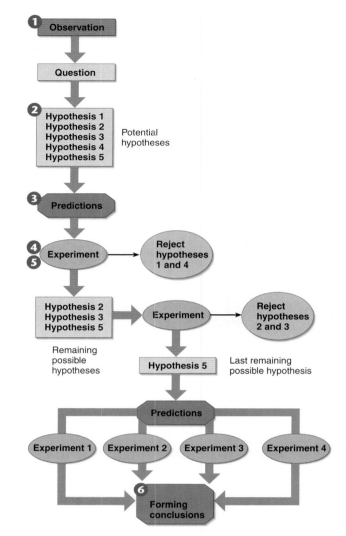

Figure 1.7   The scientific process.
This diagram illustrates the stages of a scientific investigation. First, observations are made. Then a number of potential explanations (hypotheses) are suggested in response to an observation. Experiments are conducted to eliminate any hypotheses. Next, predictions are made based on the remaining hypotheses, and further experiments (including control experiments) are carried out in an attempt to eliminate one or more of the hypotheses. Finally, any hypothesis that is not eliminated is retained. If it is validated by numerous experiments and stands the test of time, a hypothesis may eventually become a theory.

1. **Observation.** The key to any successful scientific investigation is careful **observation.** Farman and other scientists had studied the skies over the Antarctic for many years, noting a thousand details about temperature, light, and levels of chemicals. You can see an example in figure 1.8, where the black and purple represent the lowest levels of ozone that the scientists recorded. Had these scientists not kept careful records of what they observed, Farman might not have noticed that ozone levels were dropping.

2. **Hypothesis.** When the unexpected drop in ozone was reported, environmental scientists made a guess about what was destroying the ozone—that perhaps

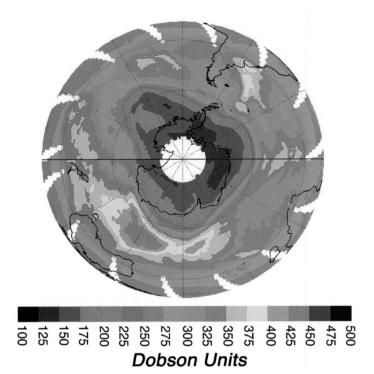

Figure 1.8 The ozone hole.

The swirling colors represent different concentrations of ozone over the South Pole as viewed from a satellite on September 15, 2001. As you can easily see, there is an "ozone hole" (the *purple* areas) over Antarctica covering an area about the size of the United States. (The color *white* indicates areas where no data were available.)

the culprit was CFCs. Of course, this was not a guess in the true sense; scientists had some working knowledge of CFC and what it might be doing in the upper atmosphere. We call such a guess a hypothesis. A hypothesis is a guess that might be true. What the scientists guessed was that chlorine from CFCs released into the atmosphere was reacting chemically with ozone over the Antarctic, converting ozone (O_3) into oxygen gas (O_2) and in the process removing the ozone shield from our earth's atmosphere. Often, scientists will form **alternative hypotheses** if they have more than one guess about what they observe. In this case, there were several other hypotheses advanced to explain the ozone hole. One suggestion explained it as the result of convection. A hypothesis was proposed that the seeming depletion of ozone was in fact a normal consequence of the spinning of the earth; the ozone spun away from the polar regions much as water spins away from the center as a clothes washer moves through its spin cycle. Another hypothesis was that the ozone hole was a transient phenomenon, due perhaps to sunspots, and would soon disappear.

3. **Predictions.** If the CFC hypothesis is correct, then several consequences can reasonably be expected. We call these expected consequences **predictions.** A prediction is what you expect to happen if a hypothesis is true. The CFC hypothesis predicts that if CFCs are responsible for producing the ozone hole, then it should be possible to detect CFCs in the upper Antarctic atmosphere as well as the chlorine released from CFCs that attack the ozone.

4. **Testing.** Scientists set out to test the CFC hypothesis by attempting to verify some of its predictions. We call the test of a hypothesis an **experiment.** To test the hypothesis, atmospheric samples were collected from the stratosphere over 6 miles up by a high-altitude balloon. Analysis of the samples revealed CFCs, as predicted. Were the CFCs interacting with the ozone? The samples contained free chlorine and fluorine, confirming the breakdown of CFC molecules. The results of the experiment thus support the hypothesis.

5. **Controls.** Events in the upper atmosphere can be influenced by many factors. We call each factor that might influence a process a **variable.** To evaluate alternative hypotheses about one variable, all the other variables must be kept constant so that we do not get misled or confused by these other influences. This is done by carrying out two experiments in parallel: in the first experimental test, we alter one variable in a known way to test a particular hypothesis; in the second, called a **control experiment,** we do *not* alter that variable. In all other respects, the two experiments are the same. To further test the CFC hypothesis, scientists carried out control experiments in which the key variable was the amount of CFCs in the atmosphere. Working in laboratories, scientists reconstructed the atmospheric conditions, solar bombardment, and extreme temperatures found in the sky far above the Antarctic. If the ozone levels fell without addition of CFCs to the chamber, then CFCs could not be what was attacking the ozone, and the CFC hypothesis must be wrong. Carefully monitoring the chamber, however, scientists detected no drop in ozone levels in the absence of CFCs. The result of the control was thus consistent with the predictions of the hypothesis.

6. **Conclusion.** A hypothesis that has been tested and not rejected is tentatively accepted. The hypothesis that CFCs released into the atmosphere are destroying the earth's protective ozone shield is now supported by a great deal of experimental evidence and is widely accepted. While other factors have also been implicated in ozone depletion, destruction by CFCs is clearly the dominant phenomenon. A collection of related hypotheses that have been tested many times and not rejected is called a **theory.** A theory indicates a higher degree of certainty; however, in science, nothing is "certain." The theory of the ozone shield—that ozone in the upper atmosphere shields the earth's surface from harmful UV rays by absorbing them—is supported by a wealth of observation and experimentation and is widely accepted. The explanation for the destruction of this shield is still at the hypothesis stage.

How Scientists Analyze and Present Experimental Results

After a scientist conducts research, he or she must then analyze and present the results so that others can interpret them. To do this, the scientist focuses carefully on what is measured in an experiment, and on how the data are evaluated.

Variables. Variables are the tools of research. They are manipulated and measured in an experiment as a means of answering questions and testing hypotheses. There are two types of variables in any experiment. An **independent variable** is one that a researcher is able to control—for example, the concentration of a chemical in a solution or the timing of when a measurement is taken. The independent variable is selected by the investigator based on the question he or she is trying to answer. A **dependent variable,** by contrast, is not predetermined by the investigator; it is the response that is measured by the investigator in the experiment.

Some research involves examining correlations between sets of variables, rather than the deliberate manipulation of a variable. For example, a researcher who measures both blood pressure and cholesterol level is actually comparing two dependent variables. While such a comparison can reveal correlations and so suggest potential relationships, **correlation does not prove causation**. What is happening to one variable may actually have nothing to do with what happens to the other variable. Only by manipulating a variable (making it an independent variable) can you test for causality. Just because people with high blood pressure might also have high cholesterol does not establish that high blood pressure *causes* high cholesterol. The "Visual Understanding" question 1 at the end of this chapter shows how an erroneous conclusion can be reached when comparing two dependent variables.

How to Present the Results as a Graph. You will encounter a lot of graphs in this text and to interpret them, it is important to realize that all graphs are presented in a consistent way. The independent variable is always presented and labeled across the bottom, called the *x axis*. The dependent variable is always presented and labeled along the side (usually the left side), called the *y axis*.

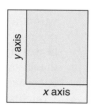

In some situations, two sets of data may be presented in the same graph using the same independent variable—for example, a control experiment and a testing experiment. In this case, the two sets of data are presented as two different lines on the graph. Or a graph might show the results from two different experiments (two different dependent variables) with the same independent variable. In this situation, a graph might have two *y* axes, one on the left and one on the right, plotted against the same *x* axis.

Using the Appropriate Scale and Units to Present Data. A key aspect of presenting data in a graph is the selection of proper scale. Data presented in a table can utilize many scales, from seconds to centuries, with no problems. A graph, however, typically has a single scale on the *x* axis and a single scale on the *y* axis, which might consist of molecular units (for example, nanometers, microliters, micrograms) or macroscopic units (for example, feet, inches, liters, days, milligrams). In each instance, a scale must be chosen that fits what is being measured. Changes in centimeters would not be obvious in a graph scaled in miles. If a variable changes a great deal over the course of the experiment, it is often useful to use an expanding scale. A **log** or **logarithmic scale** is a series of numbers plotted as powers of 10 (1, 10, 100, 1,000,...) rather than in the linear progression seen on most graphs (2,000, 4,000, 6,000...). Consider the two graphs below, where the *y* axis is plotted on a linear scale on the left and on a log scale on the right.

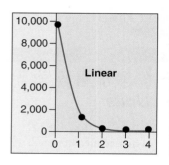 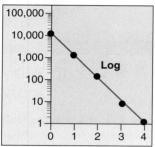

You can see that the log scale more clearly displays changes in the dependent variable (the *y* axis) for the upper values of the independent variable (the *x* axis, values 2, 3, and 4). Notice that the interval *between* each *y* axis number is not linear either—the interval between each number is itself subdivided on a log scale. Thus, 50 (the fourth tick mark between 10 and 100) is plotted much closer to 100 than to 10.

Individual graphs use different units of measurement, each chosen to best display the experimental data. By international convention, scientific data are presented in **metric units,** a system of units expressed as powers of 10. For example, weight is expressed in units called *grams*. Ten grams make up a decagram, and 1,000 grams is a kilogram. Smaller weights are expressed as a portion of a gram—for example, a centigram is a hundredth of a gram, and a milligram is a thousandth of a gram. The metric system was chosen over the English system used in the United States because the metric decimal-based system simplifies calculations and is very straightforward. The units of measurement employed in a graph are by convention indicated in parentheses next to the independent variable label on the *x* axis and the dependent variable label on the *y* axis.

Drawing a Line. Most of the graphs that you will find in this text are **line graphs,** which are graphs composed of data points and one or more lines. Line graphs are typically used to present *continuous data*—that is, data that are discrete samples of a continuous process. An example might be data measuring how quickly the ozone hole develops over Antarctica in August and September each year. You could in principle measure the area of the ozone hole every day, but to make the project manageable in time and resources, you might actually take a measurement only every week. The ozone hole increases in area rapidly for about six weeks before shrinking, yielding six data points during its expansion. These six data points are like individual frames from a movie, frozen moments in time. The six data points might indicate a very consistent pattern, or they might not.

The data points on the left are changing in a very consistent way, with little variation from what a straight line (drawn in red)

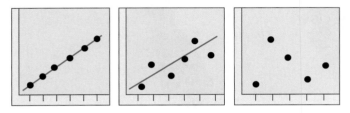

would predict. The graph in the middle shows more experimental variation, but a straight line still does a good job of revealing the overall pattern of how the data are changing. Such a straight "best-fit line" is called a **regression line** and is calculated by estimating the distance of each point to possible lines, adding the values, and selecting the line with the lowest sum. The data points in the graph on the right, unlike the other two, are randomly distributed and show no overall pattern, indicating that there is no relationship between the dependent and the independent variables.

Other Graphical Presentations of Data. Sometimes the independent variable for a data set is not continuous but rather represents discrete sets of data. A line graph, with its assumption of continuity, cannot accurately represent the variation occurring in discrete sets of data, where the data sets are being compared with one another. The preferred presentation is that of a **histogram,** a kind of bar graph. For example, if you were surveying the heights of pine trees in a park, you might group their heights (the independent variable) into discrete

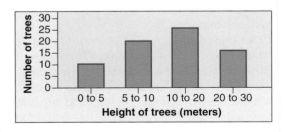

"categories" such as 0 to 5 meters tall, 5 to 10 meters, and so on. These categories are placed on the *x* axis. You would then count the number of trees in each category and present that dependent variable on the *y* axis.

Other types of data represent proportions of a whole data set, for example, the different types of trees in the park as a

percentage of all the trees. This type of data is often presented in a **pie chart:**

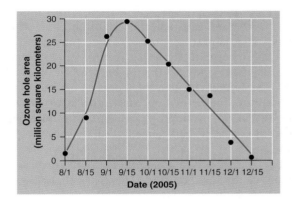

Putting Graphs to Work: The Ozone Hole

The sorts of graphs we have encountered here are all used frequently by scientists in analyzing and presenting their experimental results. As an example, consider research on the ozone hole, discussed earlier in this chapter:

A *line graph* can be used to present data on how the size of the ozone hole changes over the course of one year. In this case, the regression line is not a straight line, but rather a curve.

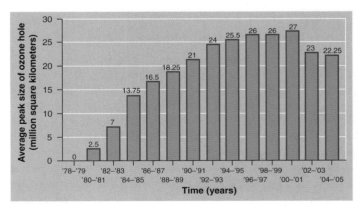

A *histogram* can be used to present data on how the peak size of the ozone hole has changed in two-year intervals over the past 26 years:

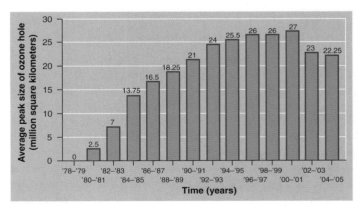

Inquiry & Analysis

The best way to learn how scientists ask questions and analyze data is to do it yourself. At the end of every chapter in this text you will find an *Inquiry & Analysis* feature that alllows you to analylze a real experimental data set yourself.

1.7 Science progresses by systematically eliminating potential hypotheses that are not consistent with observation.

A theory is a unifying explanation for a broad range of observations. Thus we speak of the theory of gravity, the theory of evolution, and the theory of the atom. Theories are the solid ground of science, that of which we are the most certain. There is no absolute truth in science, however, only varying degrees of uncertainty. The possibility always remains that future evidence will cause a theory to be revised. A scientist's acceptance of a theory is always provisional. For example, in another scientist's experiment, evidence that is inconsistent with a theory may be revealed. As information is shared throughout the scientific community, previous hypotheses and theories may be modified, and scientists may formulate new ideas.

Very active areas of science are often alive with controversy, as scientists grope with new and challenging ideas. This uncertainty is not a sign of poor science but rather of the push and pull that is the heart of the scientific process. The hypothesis that the world's climate is growing warmer due to humanity's excessive production of carbon dioxide (CO_2), for example, has been quite controversial, although the weight of evidence has increasingly supported the hypothesis.

The word theory is thus used very differently by scientists than by the general public. To a scientist, a theory represents that of which he or she is most certain; to the general public, the word theory implies a *lack* of knowledge or a guess. How often have you heard someone say, "It's only a theory!"? As you can imagine, confusion often results. In this text the word theory will always be used in its scientific sense, in reference to a generally accepted scientific principle.

The Scientific "Method"

It was once fashionable to claim that scientific progress is the result of applying a series of steps called the **scientific method;** that is, a series of logical "either/or" predictions tested by experiments to reject one alternative. The assumption was that trial-and-error testing would inevitably lead one through the maze of uncertainty that always slows scientific progress. If this were indeed true, a computer would make a good scientist—but science is not done this way! If you ask successful scientists like Farman how they do their work, you will discover that without exception they design their experiments with a pretty fair idea of how they will come out. Environmental scientists understood the chemistry of chlorine and ozone when they formulated the CFC hypothesis, and they could imagine how the chlorine in CFCs would attack ozone molecules. A hypothesis that a successful scientist tests is not just any hypothesis. Rather, it is a "hunch" or educated guess in which the scientist integrates all that he or she knows. The scientist also allows his or her imagination full play, in an attempt to get a sense of what *might* be true. It is because insight and imagination play such a large role in scientific progress that some scientists are so much better at science than others

Figure 1.9   Nobel Prize winner.
Sherwood Rowland, along with Mario Molina and Paul Crutzen, won the 1995 Nobel Prize in Chemistry for discovering how CFCs act to catalytically break down atmospheric ozone in the stratosphere, the chemistry responsible for the "ozone hole" over the Antarctic.

(figure 1.9)—just as Beethoven and Mozart stand out among composers.

The Limitations of Science

Scientific study is limited to organisms and processes that we are able to observe and measure. Supernatural and religious phenomena are beyond the realm of scientific analysis because they cannot be scientifically studied, analyzed, or explained. Supernatural explanations can be used to explain any result, and cannot be disproven by experiment or observation. Scientists in their work are limited to objective interpretations of observable phenomena.

It is also important to recognize that there are practical limits to what science can accomplish. While scientific study has revolutionized our world, it cannot be relied upon to solve all problems. For example, we cannot pollute the environment and squander its resources today, in the blind hope that somehow science will make it all right sometime in the future. Nor can science restore an extinct species. Science identifies solutions to problems when solutions exist, but it cannot invent solutions when they don't.

> **1.8   A scientist does not follow a fixed method to form hypotheses but relies also on judgment and intuition.**

Author's Corner

Where Are All My Socks Going?

All my life, for as far back as I can remember, I have been losing socks. Not pairs of socks, mind you, but single socks. I first became aware of this peculiar phenomenon when as a young man I went away to college. When Thanksgiving rolled around that first year, I brought an enormous duffle bag of laundry home. My mother, instead of braining me, dumped the lot into the washer and dryer, and so discovered what I had not noticed—that few of my socks matched anymore.

That was over forty years ago, but it might as well have been yesterday. All my life, I have continued to lose socks. This last Christmas I threw out a sock drawer full of socks that didn't match, and took advantage of sales to buy a dozen pairs of brand-new ones. Last week, when I did a body count, three of the new pairs had lost a sock!

Enough. I set out to solve the mystery of the missing socks. How? The way Sherlock Holmes would have, scientifically. Holmes worked by eliminating those possibilities that he found not to be true. A scientist calls possibilities "hypotheses" and, like Sherlock, rejects those that do not fit the facts. Sherlock tells us that when only one possibility remains unrejected, then—however unlikely—it must be true.

Hypothesis 1: It's the socks. I have four pairs of socks bought as Christmas gifts but forgotten until recently. Deep in my sock drawer, they have remained undisturbed for five months. If socks disappear because of some intrinsic property (say the manufacturer has somehow designed them to disappear to generate new sales), then I could expect at least one of these undisturbed ones to have left the scene by now. However, when I looked, all four pairs were complete. Undisturbed socks don't disappear. Thus I reject the hypothesis that the problem is caused by the socks themselves.

Hypothesis 2: Transformation, a fanciful suggestion by science fiction writer Avram Davidson in his 1958 story "Or All the Seas with Oysters" that I cannot get out of the quirky corner of my mind. I discard the socks I have worn each evening in a laundry basket in my closet. Over many years, I have noticed a tendency for socks I have placed in the closet to disappear. Over that same long period, as my socks are disappearing, there is something in my closet that seems to multiply—COAT HANGERS! Socks are larval coat hangers! To test this outlandish hypothesis, I had only to move the laundry basket out of the closet. Several months later, I was still losing socks, so this hypothesis is rejected.

Hypothesis 3: Static cling. The missing single socks may have been hiding within the sleeves of sweat shirts or jackets, inside trouser legs, or curled up within seldom-worn garments. Rubbing around in the dryer, socks can garner quite a bit of static electricity, easily enough to cause them to cling to other garments. Socks adhering to the outside of a shirt or pant leg are soon dislodged, but ones that find themselves within a sleeve, leg, or fold may simply stay there, not "lost" so much as misplaced. However, after a

diligent search, I did not run across any previously lost socks hiding in the sleeves of my winter garments or other seldom-worn items, so I reject this hypothesis.

Hypothesis 4: I lose my socks going to or from the laundry. Perhaps in handling the socks from laundry basket to the washer/dryer and back to my sock drawer, a sock is occasionally lost. To test this hypothesis, I have pawed through the laundry coming into the washer. No single socks. Perhaps the socks are lost after doing the laundry, during folding or transport from laundry to sock drawer. If so, there should be no single socks coming out of the dryer. But there are! The singletons are first detected among the dry laundry, before folding. Thus I eliminate the hypothesis that the problem arises from mishandling the laundry. It seems the problem is in the laundry room.

Hypothesis 5: I lose them during washing. Perhaps the washing machine is somehow "eating" my socks. I looked in the washing machine to see if a sock could get trapped inside, or chewed up by the machine, but I can see no possibility. The clothes slosh around in a closed metal container with water passing in and out through little holes no wider than a pencil. No sock could slip through such a hole. There is a thin gap between the rotating cylinder and the top of the washer through which an errant sock might escape, but my socks are too bulky for this route. So I eliminate the hypothesis that the washing machine is the culprit.

Hypothesis 6: I lose them during drying. Perhaps somewhere in the drying process socks are being lost. I stuck my head in our clothes dryer to see if I could see any socks, and I couldn't. However, as I look, I can see a place a sock could go—behind the drying wheel! A clothes dryer is basically a great big turning cylinder with dry air blowing through the middle. The edges of the turning cylinder don't push hard against the side of the machine. Just maybe, every once in a while, a sock might get pulled through, sucked into the back of the machine.

To test this hypothesis, I should take the back of the dryer off and look inside to see if it is stuffed with my missing socks. My wife, knowing my mechanical abilities, is not in favor of this test. Thus, until our dryer dies and I can take it apart, I shall not be able to reject hypothesis 6. Lacking any other likely hypothesis, I take Sherlock Holmes' advice and tentatively conclude that the dryer is the culprit.

1.9 Four Theories Unify Biology as a Science

The Cell Theory: Organization of Life

As was stated at the beginning of this chapter, all organisms are composed of cells, life's basic units. Cells were discovered by Robert Hooke in England in 1665. Hooke was using one of the first microscopes, one that magnified 30 times. Looking through a thin slice of cork, he observed many tiny chambers, which reminded him of monks' cells in a monastery. Not long after that, the Dutch scientist Anton van Leeuwenhoek used microscopes capable of magnifying 300 times, and discovered an amazing world of single-celled life in a drop of pond water like you see in figure 1.10. He called the bacterial and protist cells he saw "wee animalcules." However, it took almost two centuries before biologists fully understood their significance. In 1839, the German biologists Matthias Schleiden and Theodor Schwann, summarizing a large number of observations by themselves and others, concluded that all living organisms consist of cells. Their conclusion forms the basis of what has come to be known as the **cell theory.** Later, biologists added the idea that all cells come from other cells. The cell theory, one of the basic ideas in biology, is the foundation for understanding the reproduction and growth of all organisms. The nature of cells and how they function is discussed in detail in chapter 5.

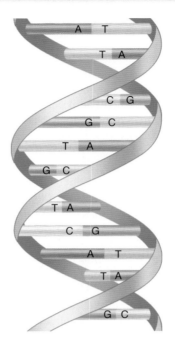

Figure 1.11 Genes are made of DNA.

Winding around each other like the rails of a spiral staircase, the two strands of a DNA molecule make a double helix. Because of its size and shape, the nucleotide represented by the letter A can only pair with the nucleotide represented by the letter T, and likewise G can only pair with C.

Figure 1.10 Life in a drop of pond water.

All organisms are composed of cells. Some organisms, including these protists, are single-celled, while others, such as plants, animals, and fungi, consist of many cells.

The Gene Theory: Molecular Basis of Inheritance

Even the simplest cell is incredibly complex, more intricate than a computer. The information that specifies what a cell is like—its detailed plan—is encoded in a long cablelike molecule called **DNA (deoxyribonucleic acid).** Researchers James Watson and Francis Crick discovered in 1953 that each DNA molecule is formed from two long chains of building blocks, called nucleotides, wound around each other. You can see in figure 1.11 that the two chains face each other, like two lines of people holding hands. The chains contain information in the same way this sentence does—as a sequence of letters. There are four different nucleotides in DNA (symbolized as A, T, C, and G in the figure), and the sequence in which they occur encodes the information. Specific sequences of several hundred to many thousand nucleotides make up a *gene,* a discrete unit of hereditary information. A gene might encode a particular protein, or a different kind of unique molecule called RNA, or a gene might act to regulate other genes. All organisms on earth encode their genes in strands of DNA. This prevalence of DNA lead to the development of the **gene theory.** Illustrated in figure 1.12, the gene theory states that the proteins and RNA molecules encoded by an organism's genes determine what it will be like. The entire set of DNA instructions that specifies a cell is called its **genome.** The sequence of the human genome, 3 billion nucleotides long, was decoded in 2001, a triumph of scientific investigation. How genes function is the subject of chapter 13. In chapter 14 we explore how detailed knowledge of genes is revolutionizing biology and having an impact on the lives of all of us.

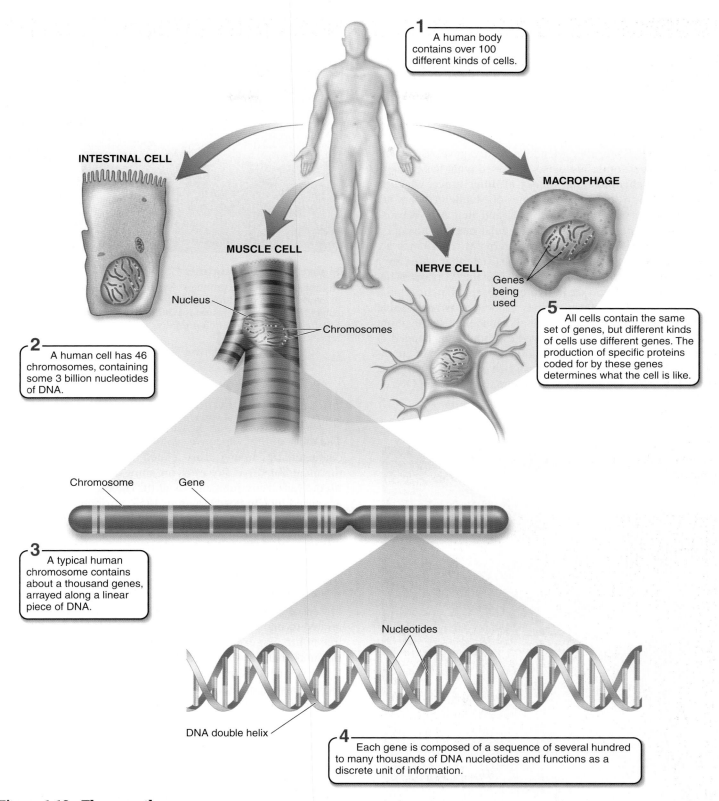

1 A human body contains over 100 different kinds of cells.

INTESTINAL CELL

MACROPHAGE

MUSCLE CELL

NERVE CELL

Nucleus

Chromosomes

Genes being used

2 A human cell has 46 chromosomes, containing some 3 billion nucleotides of DNA.

5 All cells contain the same set of genes, but different kinds of cells use different genes. The production of specific proteins coded for by these genes determines what the cell is like.

Chromosome   Gene

3 A typical human chromosome contains about a thousand genes, arrayed along a linear piece of DNA.

Nucleotides

DNA double helix

4 Each gene is composed of a sequence of several hundred to many thousands of DNA nucleotides and functions as a discrete unit of information.

Figure 1.12   The gene theory.

The gene theory states that what an organism is like is determined in large measure by its genes. Here you see how the many kinds of cells in the body of each of us are determined by which genes are used in making each particular kind of cell.

The Theory of Heredity: Unity of Life

The storage of hereditary information in genes composed of DNA is common to all living things. The **theory of heredity** first advanced by Gregor Mendel in 1865 states that the genes of an organism are inherited as discrete units. A triumph of experimental science developed long before genes and DNA were understood, Mendel's theory of heredity is the subject of chapter 11. Soon after Mendel's theory gave rise to the field of genetics, other biologists proposed what has come to be called the **chromosomal theory of inheritance,** which in its simplest form states that the genes of Mendel's theory are physically located on chromosomes, and that it is because chromosomes are parceled out in a regular manner during reproduction that Mendel's regular patterns of inheritance are seen. In modern terms, the two theories state that genes are a component of a cell's chromosomes (like the 23 pairs of human chromosomes you see in figure 1.13), and that the regular duplication of these chromosomes in meiosis is responsible for the pattern of inheritance we call Mendelian segregation. Sometimes a character is conserved essentially unchanged in a long line of descent, reflecting a fundamental role in the biology of the organism, one not easily changed once adopted. Other characters might be modified due to changes in DNA.

Figure 1.14 Prosimians.
Humans and apes are primates. Among the earliest primates to evolve were prosimians, small nocturnal (night-active) insect eaters. These lemurs, prosimians native to Madagascar, show the features characteristic of all primates: grasping fingers and toes and binocular vision.

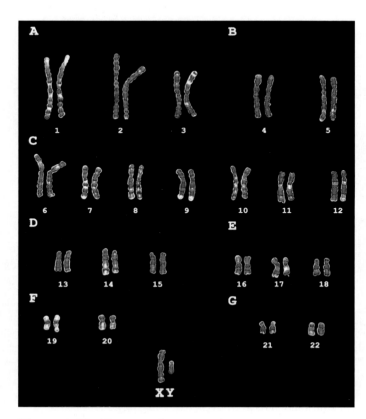

Figure 1.13 Human chromosomes.
The chromosomal theory of inheritance states that genes are located on chromosomes. This human karyotype (an ordering of chromosomes) shows banding patterns on chromosomes that represent clusters of genes.

The Theory of Evolution: Diversity of Life

The unity of life, which we see in the retention of certain key characteristics among many related life-forms, contrasts with the incredible diversity of living things that have evolved to fill the varied environments of earth. The **theory of evolution,** advanced by Charles Darwin in 1859, attributes the diversity of the living world to natural selection. Those organisms best able to respond to the challenges of living will leave more offspring, he argued, and thus become more common. It is because the world offers diverse opportunities that it contains so many different life-forms. An essential component of Darwin's theory is, in his own words, that evolution is a process of "descent with modification," that all living organisms are related to one another in a common tree of descent, a family tree of life. For example, the first primate was a small arboreal mammal that lived some 65 million years ago. About 40 million years ago, the primates split into two groups: one group, prosimians (figure 1.14), has changed little since then, while the other, monkeys, apes, and our family line, has continued to evolve. Today scientists can decipher each of all the thousands of genes (the genome) of an organism. By comparing genomes of different organisms, researchers can literally reconstruct the tree of life (figure 1.15). The organisms at the base of the tree are more ancient life-forms, having evolved earlier in the history of life on earth. The higher branches indicate other organisms that evolved later.

As you learned in section 1.1, biologists divide all living organisms into six kingdoms, based on similar overall characteristics. In recent years, biologists have added a classification level above kingdoms, based on fundamental differences in cell structure. The six kingdoms are each now assigned into one of three great groups called *domains:* Bacteria, Archaea,

Figure 1.15 The tree of life.
Biologists who compare genomes have reached a broad consensus about the major branches of the tree of life. The tree illustrated here was proposed in 2003. It shows that crocodiles seem to be more closely related to birds than to other reptiles. Fungi are thought to share a common ancestor with animals because both have a single whiplike flagellum on their reproductive cells. Thus, although they first appeared before plants, fungi seem to be more closely related to animals than plants.

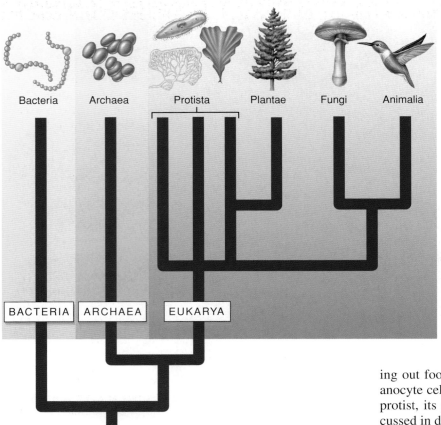

Figure 1.16 The three domains of life.
Biologists categorize all living things into three overarching groups called domains: Bacteria, Archaea, and Eukarya. Domain Bacteria contains the kingdom Bacteria, and domain Archaea contains the kingdom Archaea. Domain Eukarya is composed of four more kingdoms: Protista, Plantae, Fungi, and Animalia.

and Eukarya. Bacteria (the yellow zone in figure 1.16) and Archaea (the pink zone in figure 1.16) each consist of one kingdom of prokaryotes (single-celled organisms with little internal structure). Four more kingdoms composed of organisms with more complexly-organized cells are placed within the domain Eukarya, the eukaryotes (the purple zone in figure 1.16).

The simplest and most ancient of the Eukarya is the Kingdom Protista, mostly composed of tiny unicellular organisms. The protists are the most diverse of the four eukaryotic kingdoms, and gave rise to the other three. Kingdom Plantae, the land plants, arose from a kind of photosynthetic protist called a green algae. Kingdom Fungi, the mushrooms and yeasts, arose from a protist that has not yet been clearly identified. Kingdom Animalia, of which we are a member, is thought to have arisen from a kind of protist called a choanoflagellate. The simplest animals, sponges, have special flagellated cells called choanocytes that line the body interior. By waving the choanocyte flagella, the sponge draws water into its interior through pores, filtering out food particles from the water as it passes. Each choanocyte cell of a sponge closely resembles a choanoflagellate protist, its presumed ancestor. The kingdoms of life are discussed in detail in chapter 16.

1.9 The theories uniting biology state that cellular organisms store hereditary information in DNA. Sometimes DNA alterations occur, which when preserved result in evolutionary change. Today's biological diversity is the product of a long evolutionary journey.

INQUIRY & ANALYSIS

Is the Size of the Ozone Hole Increasing?

Since 1980 scientists have measured the size of the ozone hole centered over Antarctica. The peak size each year is presented on the graph to the right.

1. **Applying Concepts** In the graph, what is the dependent variable? Explain.

2. **Making Inferences** Is the size of the ozone hole increasing from 1980 to 1990? from 1995 to 2005?

3. **Drawing Conclusions** Does the graph support the conclusion that the size of the ozone hole over Antarctica is increasing? Explain. What might account for the difference you observe between 1980–90 and 1995–2005?

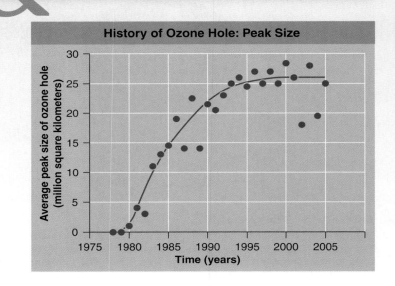

Biology and the Living World

1.1 The Diversity of Life

- Biology is the study of life. Although all living organisms share common characteristics, they are also diverse and are therefore categorized into six groups called kingdoms.

- The six kingdoms are Archaea, Bacteria, Protista, Fungi, Plantae, and Animalia (**figure 1.1**).

1.2 Properties of Life

- All living organisms share five basic properties: cellular organization, all living organisms are composed of cells; metabolism, all living organisms use energy; homeostasis, all living organisms maintain stable internal conditions; growth and reproduction, all living organisms grow and reproduce; heredity, all living organisms possess genetic information  that determines how each organism looks and functions, and this information is passed on to future generations.

1.3 The Organization of Life

- Living organisms exhibit increasing levels of complexity within their cells, within their bodies, and within ecosystems (**figure 1.4**).

- Novel properties that appear in each level of the hierarchy of living organisms are called emergent properties. These properties are the natural consequences of ever more complex structural organization.

1.4 Biological Themes

- Five themes emerge from the study of biology: evolution, the flow of energy, cooperation, structure determines function, and homeostasis. These themes are used to examine the similarities and differences among organisms (**table 1.1**).

The Scientific Process

1.5 How Scientists Think

- Scientists use reasoning when examining the world. Deductive reasoning is the process of using general principles to explain individual observations. Inductive reasoning is the process of using specific observations to formulate general principles (**figure 1.5**).

1.6 Science in Action: A Case Study

- The scientific investigation of the "ozone hole" revealed that industrially produced CFCs were responsible for the thinning of the ozone layer in the earth's atmosphere (**figure 1.6**).

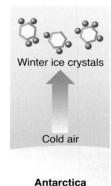

Winter ice crystals

Cold air

Antarctica

1.7 Stages of a Scientific Investigation

- Scientific investigations involve using observations to formulate hypotheses. Hypotheses are possible explanations of these observations that can be used in forming predictions that can be tested experimentally. Some hypotheses are rejected based on experimentation, while others are tentatively accepted.

- Scientific investigations use a series of six stages, called the scientific process, to study a scientific question. These stages are observations, forming hypotheses, making predictions, testing, establishing controls, and drawing conclusions (**figure 1.7**).

1.8 Theory and Certainty

- Hypotheses that hold up to testing over time are sometimes combined into statements called theories. Theories carry a higher degree of certainty, although no theory in science is absolute.

- The process of science was once viewed as a series of "either/or" predictions that were tested experimentally. This process, referred to as the scientific "method," did not take into account the importance of insight and imagination that are necessary to good scientific investigations.

Core Ideas of Biology

1.9 Four Theories Unify Biology as a Science

- There are four unifying themes in the study of biology: cell theory, gene theory, the theory of heredity, and the theory of evolution.

- The cell theory states that all living organisms are composed of cells, which grow and reproduce to form other cells (**figure 1.10**).

- The gene theory states that long molecules in the cell, called DNA, encode instructions for producing cellular components. These instructions, organized into discrete units called genes, determine how an organism looks and functions (**figure 1.12**).

- The theory of heredity states that the genes of an organism are passed as discrete units from parent to offspring through a process of heredity (**figure 1.13**).

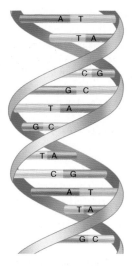

- The theory of evolution states that modifications in genes that are passed from parent to offspring result in changes in future generations. These changes lead to greater diversity among organisms over time, ultimately leading to the formation of new groups of organisms (**figure 1.15**).

- Organisms are organized into kingdoms based on similar characteristics. The organisms within a kingdom show similarities but exhibit differences from those of other kingdoms. The kingdoms are further organized into one of three major groups called domains based on their cellular characteristics. The three domains are Bacteria, Archaea, and Eukarya (**figure 1.16**).

1. Biologists categorize all living things based on related characteristics into large groups, called
   a. kingdoms.
   c. courses.
   b. planets.
   d. territories.
2. Living things can be distinguished from nonliving things because they have
   a. complexity.
   c. cellular organization.
   b. movement.
   d. response to a stimulus.
3. Living things are organized. Choose the answer that illustrates this complexity, and is arranged from smallest to largest.
   a. cell, atom, molecule, tissue, organelle, organ, organ system, organism, population, species, community, ecosystem
   b. atom, molecule, organelle, cell, tissue, organ, organ system, organism, population, species, community, ecosystem
   c. atom, molecule, organelle, cell, tissue, organ, organ system, organism, community, population, species, ecosystem
   d. atom, molecule, cell wall, cell, organ, organelle, organism, species, population, community, ecosystem
4. At each higher level in the hierarchy of living things, properties occur that were not present at the simpler levels. These properties are referred to as
   a. novelistic properties.
   c. incremental properties.
   b. complex properties.
   d. emergent properties.
5. The five general biological themes include
   a. evolution, energy flow, competition, structure determines function, and homeostasis.
   b. evolution, energy flow, cooperation, structure determines function, and homeostasis.
   c. evolution, growth, competition, structure determines function, and homeostasis.
   d. evolution, growth, cooperation, structure determines function, and homeostasis.
6. When you are trying to understand something new, you begin by observation, and then put the observations together in a logical fashion to form a general principle. This method is called
   a. inductive reasoning.
   c. theory production.
   b. rule enhancement.
   d. deductive reasoning.
7. When trying to figure out explanations for observations, you usually construct a series of possible hypotheses. Then you make predictions of what will happen if each hypothesis is true, and
   a. test each hypothesis, using appropriate controls, to determine which hypothesis is true.
   b. test each hypothesis, using appropriate controls, to rule out as many as possible.
   c. use logic to determine which hypothesis is most likely true.
   d. use logic to determine which hypotheses are most likely false.
8. Which of the following statements is correct regarding a hypothesis?
   a. After sufficient testing, you can conclude that it is true.
   b. If it explains the observations, it doesn't need to be tested.
   c. After sufficient testing, you can accept it as probable, being aware that it may be revised or rejected in the future.
   d. You never have any degree of certainty that it is true; there are too many variables.
9. Cell theory states that
   a. all organisms have cell walls and all cell walls come from other cells.
   b. all cellular organisms undergo sexual reproduction.
   c. all living organisms use cells for energy, either their own or they ingest cells of other organisms.
   d. all living organisms consist of cells, and all cells come from other cells.
10. The gene theory states that all the information that specifies what a cell is and what it does
    a. is different for each cell type in the organism.
    b. is passed down, unchanged, from parents to offspring.
    c. is contained in a long molecule called DNA.
    d. is contained in the body's nucleus.

1. For over two centuries global temperatures have been warming, and over this same period of time the number of pirate ship attacks has steadily decreased. Does this graph support the conclusion that the number of pirate attacks has decreased because of warmer temperatures? Explain.

Number of pirate ship attacks (approximate)

2,000 — 1720 1700
1,000 — 1770 1800 1840 1880 1930 1960 1985
0 —
13.0   14.0   15.0
Global mean temperature °C

2. **Figure 1.5** You notice that on cloudy days people often carry umbrellas, folded or in a case. You also note that when umbrellas are open there are many car accidents. You conclude that open umbrellas cause car accidents. Referring back to figure 1.5, explain the type of reasoning used to reach this conclusion, and why it can sometimes be a problem.

1. You are the biologist in a group of scientists who have traveled to a distant star system and landed on a planet. You see an astounding array of shapes and forms. You have three days to take samples of living things before returning to earth. How do you decide what is alive?

2. St. John's wort is an herb that has been used for hundreds of years as a remedy for mild depression. How might a modern-day scientist research its effectiveness?

2
Evolution and Ecology

These four finches live on the Galápagos Islands, a cluster of volcanic islands far out to sea off the coast of South America. All descendants of a single ancestral migrant, blown to the islands from the mainland long ago, the Galápagos finches gave Darwin valuable clues about how natural selection shapes the evolution of species. The two upper finches are ground finches, their different beaks adapting them to eat different-sized seeds. The finch on the left consumes smaller, slender seeds. The stouter beak of the finch on the right enables it to crack open larger, drier seeds. On the lower left is a woodpecker finch, a kind of tree finch that carries around a cactus spine, which it uses to probe for insects in deep crevices. On the lower right is a warbler finch, which like its namesake eats crawling insects. Each of these species utilizes food resources differently. Evolution and ecology tell us different but related things about the diversity of the living world: ecological interactions generate the selective pressures that shape the evolution of groups like Darwin's finches.

The great diversity of life on earth—ranging from bacteria to elephants and roses—is the result of a long process of **evolution,** the change that occurs in organisms' characteristics through time. In 1859, the English naturalist Charles Darwin (1809–82; figure 2.1) first suggested an explanation for why evolution occurs, a process he called **natural selection.** Biologists soon became convinced Darwin was right and now consider evolution one of the central concepts of the science of biology. A second key concept, and one that closely relates to evolution, is that of **ecology,** how organisms live in their environment. It has been said that evolution is the consequence of ecology over time. Ecology is of increasing concern to all of us, as a growing human population places ever-greater stress on our planet. In this chapter, we introduce these two key related concepts, evolution and ecology, to provide a foundation as you begin to explore the living world. While these concepts are presented here with broad strokes, highlighting only the key ideas, both will be revisited in much more detail later in the text (evolution in chapters 15–19 and ecology in chapters 20–23).

2.1 Darwin's Voyage on HMS *Beagle*

The theory of evolution proposes that a population can change over time, sometimes forming a new species, which is a population or group of populations that possess similar characteristics and can interbreed and produce fertile offspring. This famous theory provides a good example of how a scientist develops a hypothesis—in this case, a hypothesis of how evolution occurs—and how, after much testing, the hypothesis is eventually accepted as a theory.

Charles Robert Darwin was an English naturalist who, after 30 years of study and observation, wrote one of the most famous and influential books of all time. This book, *On the Origin of Species by Means of Natural Selection, or The Preservation of Favoured Races in the Struggle for Life,* created a sensation when it was published, and the ideas Darwin expressed in it have played a central role in the development of human thought ever since.

In Darwin's time, most people believed that the various kinds of organisms and their individual structures resulted from direct actions of the Creator. Species were thought to be specially created and unchangeable over the course of time. In contrast to these views, a number of earlier philosophers had presented the view that living things must have changed during the history of life on earth. Darwin proposed a concept he called natural selection as a coherent, logical explanation for this process. Darwin's book, as its title indicates, presented a conclusion that differed sharply from conventional wisdom. Although his theory did not challenge the existence of a Divine Creator, Darwin argued that this Creator did not simply create things and then leave them forever unchanged. Instead, Darwin's God expressed Himself through the operation of natural laws that produced change over time—evolution.

Figure 2.1 The theory of evolution by natural selection was proposed by Charles Darwin.
This rediscovered photograph appears to be the last ever taken of the great biologist. It was taken in 1881, the year before Darwin died.

The story of Darwin and his theory begins in 1831, when he was 22 years old. The small British naval vessel HMS *Beagle* that you see in figure 2.2 was about to set sail on a five-year navigational mapping expedition around the coasts of South America. The red arrows in figure 2.3 indicates the route taken by the HMS *Beagle.* The young (26-year-old) captain of HMS *Beagle,* unable by British naval tradition to have social contact with his crew, and anticipating a voyage that would last many years, wanted a gentleman companion, someone to talk to. Indeed, the *Beagle*'s previous skipper had broken down and shot himself to death after three solitary years away from home.

On the recommendation of one of his professors at Cambridge University, Darwin, son of a wealthy doctor and very much a gentleman, was selected to serve as the captain's companion, primarily to share his table at mealtime during every shipboard dinner of the long voyage. Darwin paid his own expenses, and even brought along a manservant.

Darwin took on the role of ship's naturalist (the official naturalist, a man named Robert McKormick, left the ship before the first year was out). During this long voyage, Darwin had the chance to study a wide variety of plants and animals on continents and islands and in distant seas. He was able to

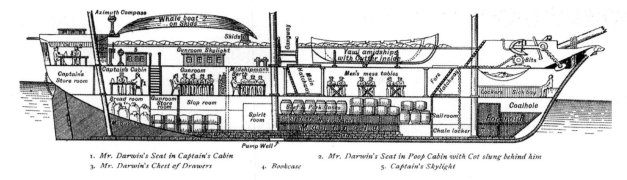

Figure 2.2   Cross section of HMS *Beagle*.

HMS *Beagle*, a 10-gun brig of 242 tons, only 90 feet in length, had a crew of 74 people! After he first saw the ship, Darwin wrote to his college professor Henslow: "The absolute want of room is an evil that nothing can surmount."

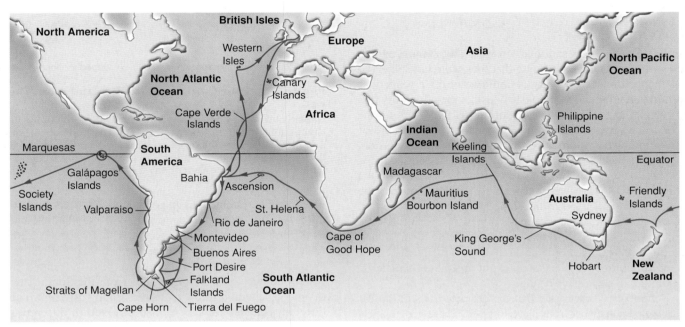

Figure 2.3   The five-year voyage of HMS *Beagle*.

Although the ship sailed around the world, most of the time was spent exploring the coasts and coastal islands of South America, such as the Galápagos Islands. Darwin's studies of the animals of these islands played a key role in the eventual development of his theory of evolution by means of natural selection.

explore the biological richness of the tropical forests, examine the extraordinary fossils of huge extinct mammals in Patagonia at the southern tip of South America, and observe the remarkable series of related but distinct forms of life on the **Galápagos Islands.** Such an opportunity clearly played an important role in the development of his thoughts about the nature of life on earth.

When Darwin returned from the voyage at the age of 27, he began a long period of study and contemplation. During the next 10 years, he published important books on several different subjects, including the formation of oceanic islands from coral reefs and the geology of South America. He also devoted eight years of study to barnacles, a group of small marine animals with shells that inhabit rocks and pilings, eventually writing a four-volume work on their classification and natural history. In 1842, Darwin and his family moved out of London to a country home at Down, in the county of Kent. In these pleasant surroundings, Darwin lived, studied, and wrote for the next 40 years.

2.1   Darwin was the first to propose natural selection as the mechanism of evolution that produced the diversity of life on earth.

Darwin's Evidence

One of the obstacles that had blocked the acceptance of any theory of evolution in Darwin's day was the incorrect notion, widely believed at that time, that the earth was only a few thousand years old. The discovery of thick layers of rocks, evidences of extensive and prolonged erosion, and the increasing numbers of diverse and unfamiliar fossils discovered during Darwin's time made this assertion seem less and less likely. The great geologist Charles Lyell (1797–1875), whose *Principles of Geology* (1830) Darwin read eagerly as he sailed on HMS *Beagle,* outlined for the first time the story of an ancient world of plants and animals in flux. In this world, species were constantly becoming extinct while others were emerging. It was this world that Darwin sought to explain.

What Darwin Saw

When HMS *Beagle* set sail, Darwin was fully convinced that species were immutable, meaning that they were not subject to being changed. Indeed, it was not until two or three years after his return that he began to seriously consider the possibility that they could change. Nevertheless, during his five years on the ship, Darwin observed a number of phenomena that were of central importance to him in reaching his ultimate conclusion. For example, in the rich fossil beds of southern South America, he observed fossils of the extinct armadillo shown on the right in figure 2.4. They were surprisingly similar in form to the armadillos that still lived in the same area, shown on the left. Why would similar living and fossil organisms be in the same area unless the earlier form had given rise to the other? Later, Darwin's observations would be strengthened by the discovery of other examples of fossils that show intermediate characteristics, pointing to successive change.

Repeatedly, Darwin saw that the characteristics of similar species varied somewhat from place to place. These geographical patterns suggested to him that organismal lineages change gradually as individuals move into

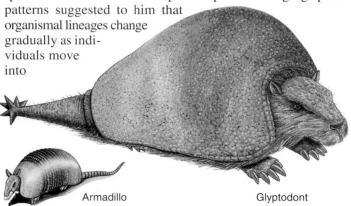

Figure 2.4 Fossil evidence of evolution.

The now-extinct glyptodont was a large 2,000-kilogram South American armadillo (about the size of a small car), much larger than the modern armadillo, which weighs an average of about 4.5 kilograms and is about the size of a house cat. The similarity of fossils such as the glyptodonts to living organisms found in the same regions suggested to Darwin that evolution had taken place.

Large ground finch (seeds)

Cactus finch (cactus fruits and flowers)

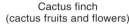

Vegetarian finch (buds)

Woodpecker finch (insects)

Figure 2.5   Four Galápagos finches and what they eat.

Darwin observed 14 different species of finches on the Galápagos Islands, differing mainly in their beaks and feeding habits. These four finches eat very different food items, and Darwin surmised that the very different shapes of their beaks represented evolutionary adaptations improving their ability to do so.

new habitats. On the Galápagos Islands, 900 kilometers (540 miles) off the coast of Ecuador, Darwin encountered a variety of different finches on the islands. The 14 species, although related, differed slightly in appearance. Darwin felt it most reasonable to assume all these birds had descended from a common ancestor blown by winds from the South American mainland several million years ago. Eating different foods, on different islands, the species had changed in different ways, most notably in the size of their beaks. The larger beak of the ground finch in the upper left of figure 2.5 is better suited to crack open the large seeds it eats. As the generations descended from the common ancestor, these ground finches changed and adapted, what Darwin referred to as "descent with modification"—evolution.

In a more general sense, Darwin was struck by the fact that the plants and animals on these relatively young volcanic islands resembled those on the nearby coast of South America. If each one of these plants and animals had been created independently and simply placed on the Galápagos Islands, why didn't they resemble the plants and animals of islands with similar climates, such as those off the coast of Africa, for example? Why did they resemble those of the adjacent South American coast instead?

> **2.2   The fossils and patterns of life that Darwin observed on the voyage of HMS *Beagle* eventually convinced him that evolution had taken place.**

The Theory of Natural Selection

It is one thing to observe the results of evolution but quite another to understand how it happens. Darwin's great achievement lies in his formulation of the hypothesis that evolution occurs because of natural selection.

Darwin and Malthus

Of key importance to the development of Darwin's insight was his study of Thomas Malthus's *Essay on the Principle of Population* (1798). In his book, Malthus pointed out that populations of plants and animals (including human beings) tend to increase geometrically, while the ability of humans to increase their food supply increases only arithmetically. A geometric progression is one in which the elements increase by a constant factor; the blue line in figure 2.6 shows the progression 2, 6, 18, 54, . . . and each number is three times the preceding one. An arithmetic progression, in contrast, is one in which the elements increase by a constant difference; the red line shows the progression 2, 4, 6, 8, . . . and each number is two greater than the preceding one.

Because populations increase geometrically, virtually any kind of animal or plant, if it could reproduce unchecked, would cover the entire surface of the world within a surprisingly short time. Instead, populations of species remain fairly constant year after year, because death limits population numbers. Malthus's conclusion provided the key ingredient that was necessary for Darwin to develop the hypothesis that evolution occurs by natural selection.

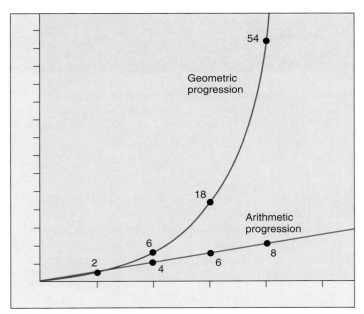

Figure 2.6 Geometric and arithmetic progressions.
An arithmetic progression increases by a constant difference (for example, units of 1 or 2 or 3), while a geometric progression increases by a constant factor (for example, by 2 or by 3 or by 4). Malthus contended that the human growth curve was geometric, but the human food production curve was only arithmetic. Can you see the problems this difference would cause?

Natural Selection

Sparked by Malthus's ideas, Darwin saw that although every organism has the potential to produce more offspring than can survive, only a limited number actually do survive and produce further offspring. Combining this observation with what he had seen on the voyage of HMS *Beagle,* as well as with his own experiences in breeding domestic animals, Darwin made an important association (figure 2.7): those individuals that possess physical, behavioral, or other attributes that help them live in their environment are more likely to survive than those that do not have these characteristics. By surviving, they gain the opportunity to pass on their favorable characteristics to their offspring. As the frequency of these characteristics increases in the population, the nature of the population as a whole will gradually change. Darwin called this process *selection.* The driving force he identified has often been referred to as survival of the fittest. However, this is not to say the biggest or the strongest always survive. These characteristics may be favorable in one environment but less favorable in another. The organisms that are "best suited" to their particular environment survive more often, and therefore produce more offspring than others in the population, and in this sense are the "fittest."

Darwin was thoroughly familiar with variation in domesticated animals and began *On the Origin of Species* with a detailed discussion of pigeon breeding. He knew that breeders selected certain varieties of pigeons and other animals, such as dogs, to produce certain characteristics, a process Darwin called **artificial selection.** Once this had been done, the animals would breed true for the characteristics that had been selected. Darwin had also observed that the differences purposely developed between domesticated races or breeds were often greater than those that separated wild species. Domestic pigeon breeds, for example, show much greater variety than

> *"Can we doubt . . . that individuals having any advantage, however slight, over others, would have the best chance of surviving and procreating their kind? On the other hand, we may feel sure that any variation in the least degree injurious would be rigidly destroyed. This preservation of favorable variations, I call Natural Selection."*

Figure 2.7 An excerpt from Charles Darwin's *On the Origin of Species.*

all of the hundreds of wild species of pigeons found throughout the world. Such relationships suggested to Darwin that evolutionary change could occur in nature too. Surely if pigeon breeders could foster such variation by "artificial selection," nature through environmental pressures could do the same, playing the breeder's role in selecting the next generation—a process Darwin called **natural selection.**

Darwin's theory provides a simple and direct explanation of biological diversity, or why animals are different in different places—because habitats differ in their requirements and opportunities, the organisms with characteristics favored locally by natural selection will tend to vary in different places.

Darwin Drafts His Argument

Darwin drafted the overall argument for evolution by natural selection in a preliminary manuscript in 1842. After showing the manuscript to a few of his closest scientific friends, however, Darwin put it in a drawer and for 16 years turned to other research. No one knows for sure why Darwin did not publish his initial manuscript—it is very thorough and outlines his ideas in detail. Some historians have suggested that Darwin was wary of igniting public, and even private, criticism of his evolutionary ideas—there could have been little doubt in his mind that his theory of evolution by natural selection would spark controversy. Others have proposed that Darwin was simply refining his theory, although there is little evidence he altered his initial manuscript in all that time.

Wallace Has the Same Idea

The stimulus that finally brought Darwin's theory into print was an essay he received in 1858. A young English naturalist named Alfred Russel Wallace (1823–1913) sent the essay to Darwin from Malaysia; it concisely set forth the theory of evolution by means of natural selection, a theory Wallace had developed independently of Darwin. Like Darwin, Wallace had been greatly influenced by Malthus's 1798 book. Colleagues of Wallace, knowing of Darwin's work, encouraged him to communicate with Darwin. After receiving Wallace's essay, Darwin arranged for a joint presentation of their ideas at a seminar in London. Darwin then completed his own book, expanding the 1842 manuscript that he had written so long ago, and submitted it for publication.

Publication of Darwin's Theory

Darwin's book appeared in November 1859 and caused an immediate sensation. Although people had long accepted that humans closely resembled apes in many characteristics, the possibility that there might be a direct evolutionary relationship was unacceptable to many. Darwin did not actually discuss this idea in his book, but it followed directly from the principles he outlined. In a subsequent book, *The Descent of Man,* Darwin presented the argument directly, building a powerful case that humans and living apes have common ancestors. Many people were deeply disturbed with the suggestion that human beings were descended from the same ancestor as

Figure 2.8 Darwin greets his monkey ancestor.
In his time, Darwin was often portrayed unsympathetically, as in this drawing from an 1874 publication.

apes, and Darwin's book on evolution caused him to become a victim of the satirists of his day—the cartoon in figure 2.8 is a vivid example. Darwin's arguments for the theory of evolution by natural selection were so compelling, however, that his views were almost completely accepted within the intellectual community of Great Britain after the 1860s.

> **2.3** The fact that populations do not really expand geometrically implies that nature acts to limit population numbers. The traits of organisms that survive to produce more offspring will be more common in future generations—a process Darwin called natural selection.

2.4 The Beaks of Darwin's Finches

Darwin's Galápagos finches played a key role in his argument for evolution by natural selection. He collected 31 specimens of finch from three islands when he visited the Galápagos Islands in 1835. Darwin, not an expert on birds, had trouble identifying the specimens. He believed by examining their beaks that his collection contained wrens, "gross-beaks," and blackbirds.

The Importance of the Beak

Upon Darwin's return to England, ornithologist John Gould examined the finches. Gould recognized that Darwin's collection was in fact a closely related group of distinct species, all similar to one another except for their beaks. In all, 14 species are now recognized, 13 from the Galápagos and one from far-distant Cocos Island. The two ground finches with the larger beaks in figure 2.9 feed on seeds that they crush in their beaks, whereas those with narrower beaks eat insects, including the warbler finch (named for its resemblance to a mainland bird). Other species include fruit and bud eaters, and species that feed on cactus fruits and the insects they attract; some populations of the sharp-beaked ground finch even include "vampires" that creep up on seabirds and use their sharp beaks to drink their blood. Perhaps most remarkable are the tool users, like the woodpecker finch you see in the upper left of the figure, that picks up a twig, cactus spine, or leaf stalk,

Figure 2.10 A gene shapes the beaks of Darwin's finches.

A cell signalling molecule called "bone morphogenic protein 4" (BMP4) has been shown by DNA researchers to tailor the shape of the beak in Darwin's finches.

trims it into shape with its beak, and then pokes it into dead branches to pry out grubs.

The differences in the beaks of Darwin's finches are due to differences in the genes of the birds. When biologists compare the DNA of large ground finches (with stout beaks for cracking seeds) to the DNA of cactus finches (with slender, pointed beaks for retrieving nectar), the only growth factor gene that is different in the DNA of the two species is BMP4 (figure 2.10). The difference is in how the gene is used. The large ground finches, with larger beaks, make more BMP4 protein at an earlier stage than do the cactus finches.

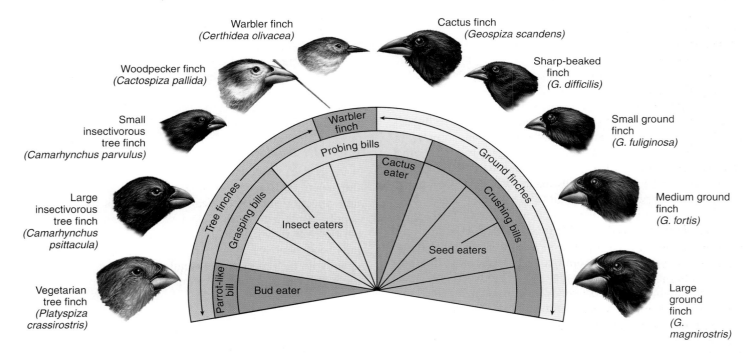

Figure 2.9 A diversity of finches on a single island.

Ten species of Darwin's finches from Isla Santa Cruz, one of the Galápagos Islands. The ten species show differences in beaks and feeding habits. These differences presumably arose when the finches arrived and encountered habitats lacking small birds. Scientists concluded that all of these birds derived from a single common ancestor.

The correspondence between the beaks of the 14 finch species and their food source immediately suggested to Darwin that evolution had shaped them:

"Seeing this gradation and diversity of structure in one small, intimately related group of birds, one might really fancy that from an original paucity of birds in this archipelago, one species has been taken and modified for different ends."

Was Darwin Wrong?

If Darwin's suggestion that the beak of an ancestral finch had been "modified for different ends" is correct, then it ought to be possible to see the different species of finches acting out their evolutionary roles, each using its beak to acquire its particular food specialty. The four species that crush seeds within their beaks, for example, should feed on different seeds, with those with stouter beaks specializing on harder-to-crush seeds.

Many biologists visited the Galápagos after Darwin, but it was 100 years before any tried this key test of his hypothesis. When the great naturalist David Lack finally set out to do this in 1938, observing the birds closely for a full five months, his observations seemed to contradict Darwin's proposal! Lack often observed many different species of finch feeding together on the same seeds. His data indicated that the stout-beaked species and the slender-beaked species were feeding on the very same array of seeds.

We now know that it was Lack's misfortune to study the birds during a wet year, when food was plentiful. The size of the finch's beak is of little importance in such flush times; slender and stout beaks work equally well to gather the abundant tender small seeds. Later work revealed a very different picture during dry years, when few seeds are available.

A Closer Look

Starting in 1973, Peter and Rosemary Grant of Princeton University and generations of their students have studied the medium ground finch, *Geospiza fortis,* on a tiny island in the center of the Galápagos called Daphne Major. These finches feed preferentially on small tender seeds, abundantly available in wet years. The birds resort to larger, drier seeds that are harder to crush when small seeds are hard to find. Such lean times come during periods of dry weather, when plants produce few seeds, large or small.

By carefully measuring the beak shape of many birds every year, the Grants were able to assemble for the first time a detailed portrait of evolution in action. The Grants found that beak depth changed from one year to the next in a predictable fashion. During droughts, plants produced few seeds, and all available small seeds quickly were eaten, leaving large seeds as the major remaining source of food. As a result, birds with large beaks survived better, because they were better able to break open these large seeds. Consequently, the average beak depth of birds in the population increased the next year because this next generation included offspring of the large-

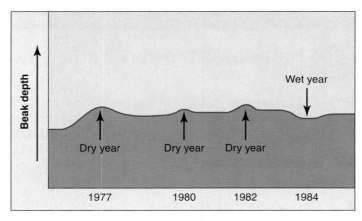

Figure 2.11 Evidence that natural selection alters beak size in *Geospiza fortis*.

In dry years, when only large, tough seeds were available, the mean beak size increased. In wet years, when many small seeds were available, smaller beaks became more common.

beaked birds that survived. The offspring of the surviving "dry year" birds had larger beaks, an evolutionary response which led to the peaks you see in the graph in figure 2.11. The reason they are peaks and not plateaus is that the average beak size decreased again when wet seasons returned because the larger beak size was no longer more favorable when seeds were plentiful and so smaller-beaked birds survived to reproduce.

Could these changes in beak dimension reflect the action of natural selection? An alternative possibility might be that the changes in beak depth do not reflect changes in gene frequencies but rather are simply a response to diet, with poorly fed birds having stouter beaks. To rule out this possibility, the Grants measured the relation of parent beak size to offspring beak size, examining many broods over several years. The depth of the beak was passed down faithfully from one generation to the next, suggesting the differences in beak size indeed reflected gene differences.

Support for Darwin

If the year-to-year changes in beak depth can be predicted by the pattern of dry years, then Darwin was right after all—natural selection influences beak size based on available food supply. Birds with stout beaks have an advantage during dry periods, for they can break the large, dry seeds that are the only food available. When small seeds become plentiful once again with the return of wet weather, a smaller beak proves a more efficient tool for harvesting smaller seeds.

2.4 In Darwin's finches, natural selection adjusts the shape of the beak in response to the nature of the food supply, adjustments that are occurring even today.

Evolution Repeats Itself in Caribbean Lizards

Darwin would have been puzzled at the average American's reluctance to accept his theory of evolution. The evidence supporting Darwin's theory is clear, and every year more supporting evidence accumulates.

There is a sticky point in Darwin's argument, however. If evolution is indeed guided by natural selection, as Darwin claims, then two environments that are similar should select in the same way—similar habitats should select for the same sorts of critters, all else being equal.

Is Darwin right? Do two communities of animals living in similar habitats evolve to be the same? Does evolution repeat itself at the community level?

This is not an easy question to answer, simply because it's difficult to find an array of similar but independent habitats to compare.

But not impossible. A team of researchers led by biology professor Jonathan Losos of Harvard University (an author of this text) has spent the last several years studying lizards of the genus *Anolis* (commonly called "anoles"), which live on large Caribbean islands. He has focused on Puerto Rico, Cuba, Haiti, and Jamaica. All four islands are inhabited by a diverse array of anole lizards (there are 57 species on Cuba alone), and all four islands have quite similar habitats and vegetation.

Unlike rats and cockroaches, which are generalists and much the same wherever you find them, anole lizards are specialists. In Puerto Rico, for example, one slender anole species with a long tail lives only in the grass. On narrow twigs at the base of trees you find a different species, also slender, but with stubby legs. On the higher branches of the tree a third species is found, of stocky build and long legs. High up in the leafy canopy of the tree lives a fourth giant green species.

Do the four Caribbean islands have similar lizard communities? Yes. If you go to Cuba, to Haiti, or to Jamaica, you can find on each island a species that looks nearly identical to each of the specialists on Puerto Rico, living in the same type of habitat and behaving in much the same manner.

Does this striking similarity of anole communities on the four islands indicate that the "Darwin experiment" has given the same result four times running?

The striking similarity of anole communities living on the four islands might be explained two different ways:

Hypothesis A Lizards migrated between the islands. A specialist anole like the one that lives in grass may have evolved only once, but then traveled to the other islands, perhaps on floating driftwood. If this is true, the similarity of communities is not the result of evolution repeating itself, but just a matter of specialists finding their way to the habitats they prefer.

Hypothesis B Lizards evolved in parallel on the four islands. The anole communities on the four islands may have evolved their similarity independently, evolution taking the same course again and again.

Working with Allan Larson of Washington University and Todd Jackman (now at Villanova University), Losos was able to choose between these two hypotheses by looking at the DNA of the lizards. The team compared several genes from more than 50 anole species. Points of similarity allowed them to construct a "phylogenetic tree," a family tree that showed who was related to who.

If hypothesis A is correct, then all the leaf specialists should be closely related to one another, whatever island they live on. The same would be expected for the four twig species, and also for the branch and canopy species.

On the other hand, if hypothesis B is correct, then a leaf specialist on one island should be more closely related to the other lizards on the same island, regardless of their specialty, than to a leaf specialist on another island.

Has evolution repeated itself? Yes. The DNA data are clear-cut: specialist species on one island are not closely related to the same specialists elsewhere, and are closely related to other anoles inhabiting the same island. Hypothesis B is supported by the data. The four lizard communities evolved independently to be similar to one another.

The Losos research team has gone on to examine the functional consequences of Caribbean anole specializations, to see if natural selection can reasonably explain how each species has evolved. Why do some anole species have long legs, for example, while others have short stubby ones? These studies, involving both field and laboratory experiments, are science at its very best, insightful and fun. The rich picture of lizard evolution that is emerging would have delighted Darwin.

How Natural Selection Produces Diversity

Darwin believed that each Galápagos finch species had adapted to the particular foods and other conditions on the particular island it inhabited. Because the islands presented different opportunities, a cluster of species resulted. Presumably, the ancestor of Darwin's finches reached these newly formed islands before other land birds, so that when it arrived, all of the niches where birds occur on the mainland were unoccupied. A *niche* is what a biologist calls the way a species makes a living—the biological (that is, other organisms) and physical (climate, food, shelter, etc.) conditions with which an organism interacts as it attempts to survive and reproduce. As the new arrivals to the Galápagos moved into vacant niches and adopted new lifestyles, they were subjected to diverse sets of selective pressures. Under these circumstances, the ancestral finches rapidly split into a series of populations, some of which evolved into separate species.

The phenomenon by which a cluster of species change, as they occupy a series of different habitats within a region, is called *adaptive radiation*. Figure 2.12 shows how the 14 species of Darwin's finches on the Galápagos Islands and Cocos Island are thought to have evolved. The ancestral population, indicated by the base of the brackets, migrated to the islands about 2 million years ago and underwent adaptive radiation giving rise to the 14 different species. Such species clusters are often particularly impressive on island groups, in series of lakes, or in other sharply discontinuous habitats.

The descendants of the original finches that reached the Galápagos Islands now occupy many different kinds of habitats on the islands. The 14 species that inhabit the Galápagos Islands and Cocos Island occupy four types of niches:

1. **Ground finches.** There are six species of *Geospiza* ground finches. Most of the ground finches feed on seeds. The size of their beaks is related to the size of the seeds they eat. Some of the ground finches feed primarily on cactus flowers and fruits and have longer, larger, more pointed beaks.
2. **Tree finches.** There are five species of insect-eating tree finches. Four species have beaks that are suitable for feeding on insects. The woodpecker finch has a chisel-like beak. This unique bird carries around a twig or a cactus spine, which it uses to probe for insects in deep crevices.
3. **Vegetarian finch.** The very heavy beak of this bud-eating bird is used to wrench buds from branches.
4. **Warbler finches.** These unusual birds play the same ecological role in the Galápagos woods that warblers play on the mainland, searching continually over the leaves and branches for insects. They have a slender, warblerlike beak.

2.5   Darwin's finches, all derived from one similar mainland species, have radiated widely on the Galápagos Islands, filling unoccupied niches in a variety of ways.

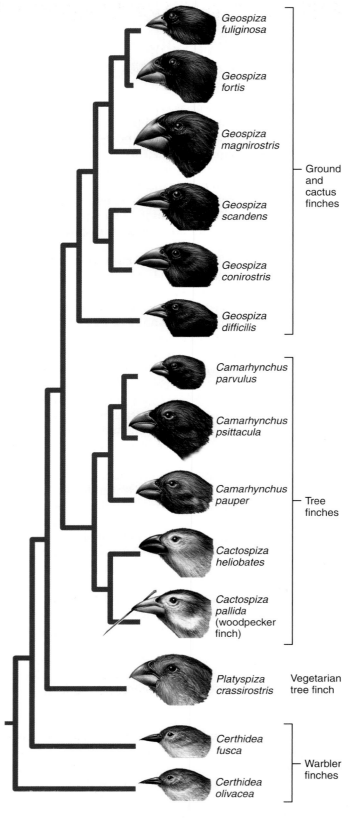

Figure 2.12   An evolutionary tree of Darwin's finches.

This family tree was constructed by comparing DNA of the 14 species. Their position at the base of the finch tree suggests that warbler finches were among the first adaptive types to evolve in the Galápagos.

Figure 2.20 Resource partitioning among lizard species.
Species of *Anolis* lizards in the Caribbean partition their tree habitats in a variety of ways. Some species of anoles occupy the canopy of trees (a), others use twigs on the periphery (b), and still others are found at the base of the trunk (c). In addition, some use grassy areas in the open (d). This same pattern of resource partitioning has evolved independently on different Caribbean islands.

2.10 How Species Evolve to Occupy Different Niches Within an Ecosystem

Each organism in an ecosystem confronts the challenge of survival in a different way. As we have just discussed, the niche an organism occupies is the sum total of all the ways it uses the resources of its environment, and may be described in terms of space utilization, food consumption, temperature range, appropriate conditions for mating, requirements for moisture, and other factors. *Niche* is not synonymous with **habitat,** the place where an organism lives. *Habitat* is a place, and *niche* is a pattern of living. Many species can share a habitat, but as we shall see, no two species can long occupy exactly the same niche.

Resource Partitioning

Competition is the struggle of two organisms to use the same resource when there is not enough of the resource to satisfy both. When two species compete for the same resource, the species that utilizes the resource more efficiently will eventually outcompete the other in that location and drive it to extinction there. Ecologists call this *the principle of competitive exclusion:* no two species with the same niche can coexist. Persistent competition between two species is rare in nature. Either one species drives the other to extinction, or natural selection favors changes that reduce the competition between them. In resource partitioning, species that live in the same geographical area avoid competition by living in different portions of the habitat or by using different food or other resources. A clear example of this is the *Anolis* lizards you see in figure 2.20. The rather exotic blue species in the upper left of the figure lives high in the tree, where it doesn't need to compete for food and space with the dark brown species in the upper right that lives on the tree's trunk.

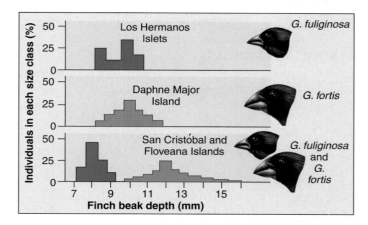

Figure 2.21 Character displacement.
These two species of Galápagos finches (genus *Geospiza*) have beaks of similar sizes when living apart, but different sizes when living together.
Data from E. J. Heske, et al. Ecology, 1994.

The changes that evolve in two species to reduce niche overlap—that is, to lessen the degree to which they compete for the same resources—are called **character displacements.** Character displacement can be seen clearly among Darwin's finches. The two Galápagos finches in figure 2.21 have beaks of similar size when each is living on an island where the other does not occur. On islands where they are found living together, the two species have evolved beaks of different sizes, one adapted to larger seeds, the other to smaller ones. In essence, the two finches have subdivided the food niche, creating two new smaller niches. By partitioning the available food resources, the two species have avoided direct competition with each other, and so are able to live together in the same habitat.

> **2.10 Species that live together partition available resources, reducing competition between them.**

2.11 Predation

Species that live together in a community interact in many ways, one of which is to eat one another. **Predation** is the consuming of one organism by another. The organism doing the eating is called the *predator*, while the organism being eaten is the *prey*. Predation includes everything from a leopard capturing and eating an antelope to a whale grazing on millions of microscopic ocean plankton. Nor is predation limited to eating animals. Locusts eating the leaves of plants are predators too.

Predator/Prey Cycles

Why doesn't a predator simply exterminate its prey, and then become extinct itself, having nothing left to eat. This is just what happens in laboratory experiments. However, if refuges are provided for the prey, its population drops to low levels but not to extinction. Low prey population levels then provide inadequate food for the predators, causing the predator population to decrease. When this occurs, the prey population can recover. In this way, predator and prey populations cycle in their abundance. Many small mammals like the lemming in figure 2.22 undergo population cycles. Some of them are true predator-prey cycles, while others are the result of cyclic climatic factors.

Defense Mechanisms

Animals defend themselves against predators with chemical defenses, warning coloration, and camouflage. The green frog you see in figure 2.23 is a poison dart frog from Latin America. The mucus that covers its skin contains a very toxic poison—a few micrograms will kill a person, a very effective chemical defense. Its bright green color is another defense, advertising its toxicity and so warning any potential predator not to attempt eating it. How does a predator know not to eat bright green frogs? By learning. One attempt is rarely repeated.

Plants have evolved many mechanisms to defend themselves from their predators, called herbivores. The most obvious are thorns, spines, and prickles. Chemical defenses are even more crucial, and are widespread in plants. Mustard oils, which give the sharp pungent taste to mustard, capers, cabbage, and horseradish, are toxic to many groups of insects.

Coevolution

Sometimes, predators and prey undergo long-term reciprocal evolutionary adjustments, a process termed **coevolution**. For example, some plants, like the mustard plant, produce oils that are toxic to many caterpillars. The cabbage butterfly caterpillar has evolved the ability to break down mustard oils and can feed on these plants. Like having the only key to a grocery store, the cabbage butterfly in figure 2.24 is able to use a new resource without competing with other herbivores for it. Responding with another evolutionary adjustment, the butterfly sense organs have evolved to detect mustard oils.

> **2.11 Members of a community often consume one another. As a result, prey species often evolve defensive adaptations.**

Figure 2.22 Lemmings are rodents, eaten by predatory mammals and birds.

The populations of lemmings and their predators decrease and increase in a cyclic manner as both populations are affected by the population sizes of each other.

Figure 2.23 Warning coloration serves as a defense mechanism in Dendrobatidae.

The skin coloration of this poison dart frog warns potential predators of the toxic mucus covering the skin, and predators keep their distance.

Figure 2.24 Coevolution between mustard plant and white cabbage butterfly (Pieris rapae).

The mustard plant evolved a defense mechanism of toxic oils that deter predators. The caterpillar of this white cabbage butterfly coevolved a mechanism to break down the mustard oil, allowing it to feed on the plants.

2.12 Symbiosis

Not all interactions among the organisms of a community involve winners and losers, as both competition and predation do. Other relationships are more cooperative. **Symbiosis** is an interaction in which two or more kinds of organisms interact in cooperative, more-or-less permanent relationships. The major kinds of symbiotic relationships include (1) **commensalism,** in which one species benefits from the relationship while the other neither benefits nor is harmed; (2) **mutualism,** in which both participating species benefit; and (3) **parasitism,** in which one species benefits but the other is harmed.

Commensalism

Commensalism is a symbiotic relationship that benefits one species and neither hurts nor helps the other. The birds you see in figure 2.25 are oxpeckers, busily pecking ticks and other insects off of an impala for their dinner. In this symbiotic relationship, the oxpeckers receive a clear benefit in the form of nutrition. If the removal of the ticks benefits the impala, then the relationship is mutually beneficial and the relationship would be considered a form of mutualism. However, there is no evidence that this is so. In this instance, as in most examples of commensalism, it is difficult to be certain whether the partner receives a benefit or not.

Mutualism

Mutualism is a symbiotic relationship in which both species benefit. The pistol shrimp patrolling the surface of the coral in figure 2.26 is defending its homestead from sea stars, which prey on coral. When it encounters a sea star on the coral, the shrimp attacks, pinching the sea star's spines and tube feet and making loud snapping sounds with its enlarged pincers. The loud popping sounds, which have given the shrimp its name, are so intense they stun small fish. Protection from being eaten by sea stars certainly provides a benefit to the coral. The shrimp also clearly benefit, obtaining food and shelter from the coral. Because both parties benefit, their relationship is an example of mutualism.

Parasitism

Parasitism is a symbiotic relationship which benefits one species at the expense of the other. Typically the parasite is much smaller than its host, and remains closely associated with it. Parasitism is sometimes considered a special kind of predator-prey relationship in which the predator is much smaller than the prey, but unlike a predator the parasite typically does not kill its host. Parasites are very common among animals. The head louse you see in figure 2.27, for example, is one of two sucking lice that parasitize humans. Eggs like the one you see in the photograph are cemented to hair follicles.

2.12   In symbiosis, two species interact closely, with at least one species benefiting.

Figure 2.25   The oxpeckers and impala relationship is an example of commensalism.

Figure 2.26   The pistol shrimp defends the coral, which he calls home—an example of mutualism.

Figure 2.27   The head louse, shown here with its egg, feeds on its host and is an example of parasitism.

Does the Presence of One Species Limit the Population Size of Others?

Implicit in Darwin's theory of evolution is the idea that species in nature compete for limiting resources. Does this really happen? Some of the best evidence of competition between species comes from experimental field studies, studies conducted not in the laboratory but out in natural populations. By setting up experiments in which two species occur either alone or together, scientists can determine whether the presence of one species has a negative impact on the size of the population of the other species. This experiment concerns a variety of seed-eating rodents that occur in North American deserts. In 1988, researchers set up a series of 50-meter × 50-meter enclosures to investigate the effect of kangaroo rats on smaller seed-eating rodents. Kangaroo rats were removed from half of the enclosures, but not from the other enclosures. The walls of all the enclosures had holes that allowed rodents to come and go, but in plots without kangaroo rats the holes were too small to allow the kangaroo rats to enter.

The graph to the right displays data collected over the course of the next three years as researchers monitored the number of the smaller rodents present in the enclosures. To estimate the population sizes, researchers determined how many small rodents could be captured in a fixed interval. Data were collected for each enclosure immediately after the kangaroo rats were removed in 1988, and at three-month intervals thereafter. The graph presents the relative population size—that is, the total number of captures averaged over the number of enclosures (an **average** is the numerical mean value, calculated by adding a list of values and then dividing this sum by the number of items in the list. For example, if a total of 30 rats were captured from 3 enclosures, the average would be 10 rats). As you can see, the two kinds of enclosures do not contain the same number of small rodents.

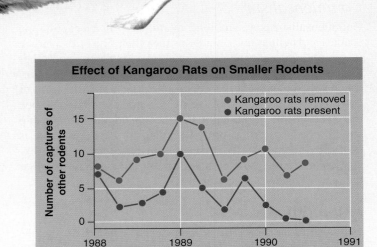

Effect of Kangaroo Rats on Smaller Rodents

● Kangaroo rats removed
● Kangaroo rats present

1. **Applying Concepts**
   a. Variable. In the graph, what is the dependent variable?
   b. Relative Magnitude. Which of the two kinds of enclosures maintains the highest population of small rodents? Does it have kangaroo rats or have they been removed?
2. **Interpreting Data**
   a. What is the average number of small rodents in each of the two plots immediately after kangaroo rats were removed? after one year? after two?
   b. At what point is the difference between the two kinds of enclosures the greatest?
3. **Making Inferences**
   a. What precisely is the observed impact of kangaroo rats on the population size of small rodents?
   b. Examine the magnitude of the difference between the number of small rodents in the two plots. Is there a trend?
4. **Drawing Conclusions**
   Do these results support the hypothesis that kangaroo rats compete with other small rodents to limit their population sizes?
5. **Further Analysis**
   a. Can you think of any cause other than competition that would explain these results? Suggest an experiment that could potentially eliminate or confirm this alternative.
   b. Do the populations of the two kinds of enclosures change in synchrony (that is, grow and shrink at the same times) over the course of a year? If so, why might this happen? How would you test this hypothesis?

Evolution

2.1 Darwin's Voyage on the HMS *Beagle*

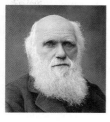

- Darwin's voyage on the HMS *Beagle* allowed him the opportunity to study a large variety of animal and plant species across the globe. Years after his voyage, he published the book *On the Origin of Species,* in which he proposed natural selection as the mechanism that underlies the process of evolution.

- The theory of evolution through natural selection is overwhelmingly accepted by scientists but is viewed as controversial by some in the general public.

2.2 Darwin's Evidence

- Darwin observed fossils in South America of extinct species that resembled living species. On the Galápagos Islands, Darwin observed finches that differed slightly in appearance between islands but resembled finches found on the South American mainland. These observations laid the groundwork for his proposal of evolution through natural selection (**figures 2.4** and **2.5**).

2.3 The Theory of Natural Selection

- Key to Darwin's hypothesis of evolution by natural selection was the observation by Malthus that the food supply limits population growth. A population, be it plants, animals, etc., grows only as large as that which can live off of the available food (**figure 2.6**).

- Using Malthus's observations and his own observations, Darwin proposed that individuals that are better suited to their environments survive to produce offspring, gaining the opportunity to pass their characteristics on to future generations. Therefore, future generations may be different from ancestral populations, a process known as evolution.

- Spurred by a similar proposal put forth years later by Alfred Wallace, Darwin published his book in 1859.

Darwin's Finches: Evolution in Action

2.4 The Beaks of Darwin's Finches

- By observing the different sizes and shapes of beaks in the closely related finches of the Galápagos Islands and correlating the beaks with the types of food consumed by the different birds, Darwin concluded that the birds' beaks were modified from an ancestral species based on the food available, each suited to its food supply (**figure 2.9**).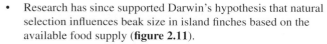

- Research has since supported Darwin's hypothesis that natural selection influences beak size in island finches based on the available food supply (**figure 2.11**).

2.5 How Natural Selection Produces Diversity

- The 14 species of finches found on the islands off the coast of South America evolved from a mainland species that adapted to different niches, a process called adaptive radiation (**figure 2.12**).

Ecology

2.6 What Is Ecology?

- Ecology is the study of how organisms interact with each other and with their physical environment. There are five levels of ecological organization: populations, communities, ecosystems, biomes, and the biosphere.

2.7 A Closer Look at Ecosystems

- An ecosystem is a physical environment that contains a community of various organisms. These organisms interact with each other and with their physical environment, extracting energy and raw materials from it (**figures 2.14** and **2.15**). Similar ecosystems found throughout the world are called biomes.

2.8 Communities

- The array of organisms that live together in an area is called a community. Different species in a community compete and cooperate with each other to make the community stable. Communities are often identified by the dominant species (**figure 2.16**).

2.9 The Niche and Competition

- A habitat is the place where an organism lives while its niche is the way it uses resources such as food and space. Limited resources lead to competition between organisms of different species, called interspecific competition, and between organisms of the same species, called intraspecific competition (**figures 2.18** and **2.19**).

2.10 How Species Evolve to Occupy Different Niches Within an Ecosystem

- Resource partitioning is the process whereby two competitive species coexist by utilizing different portions of the habitat or different resources such as food. The varying beak sizes seen in Darwin's finches suggest that the bird species evolved in part due to resource partitioning (**figures 2.20** and **2.21**).

2.11 Predation

- Interactions where one organism consumes another is called predatory/prey relationships. The hunted is called the prey and the hunter is the predator. The population sizes of predator and prey often oscillate in response to each other (**figure 2.22**).

- Species of prey may evolved defense mechanisms to protect themselves and this in turn can lead to coevolution in the characteristics of the predator species (**figures 2.23** and **2.24**).

2.12 Symbiosis

- In symbiosis, two or more species live together in closely linked relationships. Commensalism is where one organism benefits and the other neither benefits nor is harmed (**figure 2.25**). Mutualism involves cooperation between species where they both benefit (**figure 2.26**). In parasitism, one organism serves as a host to another organism such that the host is harmed (**figure 2.27**).

1. The theory of evolution states that
   a. once a species is formed, it remains stable over time, and populations do not change.
   b. populations always change over time, forming new species.
   c. populations can change over time, sometimes forming new species.
   d. populations that change over time usually go extinct, such as the dinosaurs.

2. A key observation made by Darwin was
   a. a species always looked almost exactly the same wherever he saw it; traits were preserved.
   b. the characteristics of a species varied in different places; there were geographic patterns.
   c. the characteristics of a species varied in different places; it was unpredictable.
   d. everywhere he went there were wildly different and unrelated organisms.

3. Darwin was greatly influenced by Thomas Malthus, who pointed out
   a. populations increase geometrically.
   b. populations increase arithmetically.
   c. populations are capable of geometric increase, yet remain at constant levels.
   d. the food supply usually increases faster than the population that depends on it.

4. Darwin proposed that individuals with traits that help them live in their immediate environment are more likely to survive and reproduce than individuals without those traits. He called this
   a. natural selection.
   b. the principle of population growth.
   c. the theory of evolution.
   d. Malthusian growth.

5. A great deal of research has been done on Darwin's finches over the last 70 years. The research
   a. seems to often contradict Darwin's original ideas.
   b. seems to agree with Darwin's original ideas.
   c. does not show any clear patterns that support or refute Darwin's original ideas.
   d. sometimes supports his ideas and sometimes does not.

6. The way a species makes its living—that is, the biological and physical conditions in which it exists—is called its
   a. population.   c. community.
   b. territory.   d. niche.

7. In the levels of ecological organization, the lowest level, composed of individuals of a single species who live near each other, share the same resources, and can potentially interbreed is called a
   a. population.   c. ecosystem.
   b. community.   d. biome.

8. Within an ecosystem
   a. materials flow through once and are lost, while energy cycles and recycles.
   b. materials and energy both flow through once and are lost.
   c. materials and energy both cycle and recycle.
   d. energy flows through once and is lost, while materials cycle and recycle.

9. A major principle of ecology states that no two species can occupy exactly the same niche. One will utilize resources more efficiently than the other and will drive the second species to extinction. This is the principle of
   a. competition.
   b. natural selection.
   c. community organization.
   d. competitive exclusion.

10. The following are examples of symbiosis except
    a. mutualism.
    b. predation.
    c. parasitism.
    d. commensalism.

1. **Figure 2.15** Carbon cycles through the ecosystem between the atmosphere, organisms (such as producers, herbivores, and carnivores), and decomposers. This natural cycle maintains a somewhat constant level of carbon in the atmosphere. The burning of fossil fuels releases carbon into the cycle that has been trapped inside the earth. How does this affect the cycle and what problems can this cause?

2. **Figure 2.20** Using Darwin's reasoning, explain how these four species of lizards, all closely related, came to be separate species on a Caribbean island.

1. Many evolutionary pathways seem to have moved from larger to smaller organisms over time, such as the glyptodont and the armadillo, the mammoth and the elephant. Yet other organisms, such as the tiny eohippus and the horse, have increased in size over time. Explain how this can occur.

2. *Chthamalus* is a small barnacle that lives on rocky shores. It can survive on rocks that are underwater, or on rocks that are exposed during low tide. *Balanus* is a larger barnacle that can outcompete *Chthamalus* if they both want the same spot. *Balanus* can only live on rocks that are seldom exposed to air. Explain what could happen when they are both present.

3

The Chemistry of Life

These trees in the Blue Ridge Mountains of North Carolina have
been seriously damaged by acid rain. The death of this forest
must have seemed a calamity to the animals that live there. A
porcupine knows no chemistry, has no way to comprehend what has
happened, or why. Later in this chapter, you will explore what causes
acid rain and snow, and how the acid has killed forests like this one. A
famous conservation saying is that "you cannot save what you don't
understand." In order to understand acid rain, you must first come to
understand some simpler things, the nuts and bolts that underlie what
happens in nature. All living things—in fact, everything you can see
in the picture above—are made of tiny particles called atoms, linked
together in assemblies called molecules. This is where we will have to
start, if we want to understand things like what happened to this forest.
Then, with molecules under our belt, we will need to get more specific
and consider the nature of rain. What is rain and snow made of? Water.
We will need to take a very careful look at water. When we do, we
will see that when some chemicals are added to water, a chemically
active mixture called an acid results. Acid rain is water containing such
chemicals. Understanding this gives us the mental tool we need to attack
the problem of what happened to this forest and how to stop it. In just
this way, chemistry underlies much of what you will learn in biology.

3.1 Atoms

Organisms are chemical machines, and to understand them we must learn a little chemistry. Any substance in the universe that has mass and occupies space is defined as **matter.** All matter is composed of extremely small particles called **atoms.** An atom is the smallest particle into which a substance can be divided and still retain its chemical properties.

Every atom has the same basic structure you see in figure 3.1. At the center of every atom is a small, very dense nucleus formed of two types of subatomic particles, **protons** (purple balls) and **neutrons** (pink balls). Whizzing around the core is an orbiting cloud of a third kind of subatomic particle, the **electron** (depicted on the left by a yellow cloud and on the right by yellow balls on concentric rings). Neutrons have no electrical charge, whereas protons have a positive charge and electrons a negative one. In each atom, there is an orbiting electron for every proton in the nucleus. The electron's negative charge balances the proton's positive charge. The atom is said to be electrically neutral.

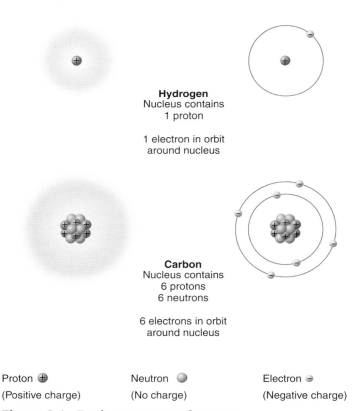

Hydrogen
Nucleus contains
1 proton

1 electron in orbit
around nucleus

Carbon
Nucleus contains
6 protons
6 neutrons

6 electrons in orbit
around nucleus

Proton ⊕   Neutron ◯   Electron ⊖
(Positive charge)   (No charge)   (Negative charge)

Figure 3.1   Basic structure of atoms.

All atoms have a nucleus consisting of protons and neutrons, except hydrogen, the smallest atom, which has only one proton and no neutrons in its nucleus. Carbon, for example, has six protons and six neutrons in its nucleus. Electrons spin around the nucleus in orbitals a far distance away from the nucleus. The electrons determine how atoms react with each other.

An atom is typically described by the number of protons in its nucleus or by the overall mass of the atom. The terms *mass* and *weight* are often used interchangeably, but they have slightly different meanings. Mass refers to the amount of a substance, whereas weight refers to the force gravity exerts on a substance. Hence, an object has the same mass whether it is on the earth or the moon, but its weight will be greater on the earth, because the earth's gravitational force is greater than the moon's. For example, an astronaut weighing 180 pounds on earth will weigh about 30 pounds on the moon. He didn't lose any significant mass during his flight to the moon, there is just less gravitational pull on his mass.

The number of protons in the nucleus of an atom is called the **atomic number.** For example, the atomic number of carbon is 6 because it has six protons. Atoms with the same atomic number (that is, the same number of protons) have the same chemical properties and are said to belong to the same **element.** Formally speaking, an element is any substance that cannot be broken down into any other substance by ordinary chemical means.

Neutrons are similar to protons in mass, and the number of protons and neutrons in the nucleus of an atom is called the **mass number.** A carbon atom that has six protons and six neutrons has a mass number of 12. The mass of atoms and subatomic particles is measured in units called *daltons*. A proton's mass is approximately 1 dalton (actually 1.007 daltons), as is a neutron's mass (1.009 daltons). In contrast, an electron's mass is only 1/1,840 of a dalton, so its contribution to the overall mass of an atom is negligible. The atomic numbers and mass numbers of some of the most common elements on earth are shown in table 3.1. Although precise measurements of mass numbers are often presented, it is also common to round the mass number to an integer value.

Electrons Determine What Atoms Are Like

Electrons have very little mass (only about 1/1,840 the mass of a proton). Of all the mass contributing to your weight, the portion that is contributed by electrons is less than the mass of your eyelashes. And yet electrons determine the chemical behavior of atoms because they are the parts of atoms that come close enough to each other in nature to interact. Almost all the volume of an atom is empty space. Protons and neutrons lie at the core of this space, while orbiting electrons are very far from the nucleus. If the nucleus of an atom were the size of an apple, the orbit of the nearest electron would be more than a mile out!

Electrons Carry Energy

Because electrons are negatively charged, they are attracted to the positively charged nucleus, but they also repel the negative charges of each other. It takes work to keep them in orbit, just as it takes work to hold an apple in your hand when gravity is pulling the apple down toward the ground. The apple in your hand is said to possess **energy,** the ability to do work, because of its position—if you were to release it, the apple would fall. Similarly, electrons have energy of position, called *potential*

TABLE 3.1	ELEMENTS COMMON IN LIVING ORGANISMS		
Element	Symbol	Atomic Number	Mass Number
Hydrogen	H	1	1.008
Carbon	C	6	12.011
Nitrogen	N	7	14.007
Oxygen	O	8	15.999
Sodium	Na	11	22.989
Magnesium	Mg	12	24.305
Phosphorus	P	15	30.974
Sulfur	S	16	32.064
Chlorine	Cl	17	35.453
Potassium	K	19	39.098
Calcium	Ca	20	40.080
Iron	Fe	26	55.847
Iodine	I	53	126.904

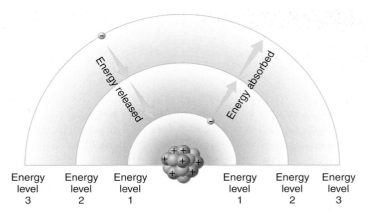

Figure 3.2 The electrons of atoms possess potential energy.
Electrons circulate rapidly around the nucleus in paths called orbitals. Energy level 1 is the lowest potential energy level because it is closest to the nucleus. When an electron absorbs energy, it moves from level 1 to the next higher energy level (level 2). When an electron loses energy, it falls to a lower energy level closer to the nucleus.

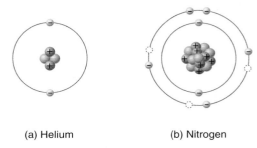

(a) Helium   (b) Nitrogen

Figure 3.3 Electrons in electron shells.
(a) An atom of helium has two protons, two neutrons, and two electrons. The electrons fill the one orbital in its one electron shell. (b) An atom of nitrogen has seven protons, seven neutrons, and seven electrons. Two electrons fill the orbital in the innermost electron shell, and five electrons occupy orbitals in the second electron shell. The orbitals in the second electron shell can hold up to eight electrons; therefore there are three vacancies in the outer electron shell of a nitrogen atom.

energy. It takes work to oppose the attraction of the nucleus, so moving the electron farther out to a more distant shell, as shown by the set of arrows on the right side of figure 3.2, requires an input of energy and results in an electron with greater potential energy. Moving an electron in toward the nucleus has the opposite effect (the set of arrows on the left side); energy is released, and the electron has less potential energy. Consider again an apple held in your hand. If you carry the apple up to a second-story window, it has a greater potential energy when you drop it, compared to when it is dropped at ground level. Similarly, if you lower the apple until it is six inches from the ground, it has less potential energy. Cells use the potential energy of atoms to drive chemical reactions, as we will discuss in chapter 6.

While the energy levels of an atom are often visualized as well-defined circular orbits around a central nucleus as shown in figure 3.1, such a simple picture is not realistic. These energy levels, called *electron shells,* often consist of complex three-dimensional shapes, and the exact location of an individual electron at any given time is impossible to specify. However, some locations are more probable than others, and it is often possible to say where an electron is *most likely* to be located. The volume of space around a nucleus where an electron is most likely to be found is called the **orbital** of that electron.

Each electron shell has a specific number of orbitals, and each orbital can hold up to two electrons. The first shell in any atom contains one orbital. Helium, shown in figure 3.3*a,* has one electron shell with one orbital that corresponds to the lowest energy level. The orbital contains two electrons, shown above and below the nucleus. In atoms with more than one electron shell, the second shell contains four orbitals and holds up to eight electrons. Nitrogen, shown in figure 3.3*b,* has two electron shells, shown as two concentric rings; the first one is completely filled with two electrons, but three of the four orbitals in the second electron shell are not filled

because nitrogen's second shell contains only five electrons (openings in orbitals are indicated with dotted circles). In atoms with more than two electron shells, subsequent shells also contain up to four orbitals and a maximum of eight electrons. Atoms with incomplete electron orbitals tend to be more reactive because they lose, gain, or share electrons in order to fill their outermost electron shell. This losing, gaining, or sharing of electrons is the basis for chemical reactions in which chemical bonds form between atoms. Chemical bonds will be discussed later in this chapter.

> **3.1** Atoms, the smallest particles into which a substance can be divided, are composed of electrons orbiting a nucleus composed of protons and neutrons. Electrons determine the chemical behavior of atoms.

Ions and Isotopes

Ions

Sometimes an atom may gain or lose an electron from its outer shell. Atoms in which the number of electrons does not equal the number of protons because they have gained or lost one or more electrons are called **ions.** All ions are electrically charged. An atom of sodium (on the left in figure 3.4), for example, becomes a positively charged ion, called a *cation* (on the right), when it loses an electron, because one proton in the nucleus is left with an unbalanced charge (11 positively charged protons and only 10 negatively charged electrons). Negatively charged ions, called *anions*, can also form, when an atom gains an electron from another atom.

Isotopes

The number of neutrons in an atom of a particular element can vary without changing the chemical properties of the element. Atoms that have the same number of protons but different numbers of neutrons are called **isotopes.** Isotopes of an atom have the same atomic number but differ in their mass number. Most elements in nature exist as mixtures of different isotopes. For example, there are three isotopes of the element carbon, all of which possess six protons (the purple balls in figure 3.5). The most common isotope of carbon (99% of all carbon) has six neutrons (the pink balls). Because its mass number is 12 (six protons plus six neutrons), it is referred to as carbon-12 (on the left). The isotope carbon-14 (on the right) is rare (1 in 1 trillion atoms of carbon) and unstable, such that its nucleus tends to break up into particles with lower atomic numbers, a process called **radioactive decay.** Radioactive isotopes are used in medicine and in dating fossils.

Medical Uses of Radioactive Isotopes

When most people hear the word "radioactive" they picture atomic bombs exploding into mushroom clouds and the devastation that results. While it is true that the radiation emitted from radioactive isotopes can damage cells of the human body, it is also true that isotopes can be used in many medical procedures. Short-lived isotopes, those that decay fairly rapidly and produce harmless products, are commonly used as tracers in the body. A **tracer** is a radioactive substance that is taken up and used by the body. Emissions from the radioactive isotope tracer are detected using special laboratory equipment, and can reveal key diagnostic information about the functioning of the body. For example, PET and PET/CT (positron emission tomography/computerized tomography) imaging procedures can be used to identify a cancerous area in the body. First, a radioactive tracer is injected into the body. This tracer is taken up by all cells, but it is taken up in larger amounts in cells with higher metabolic activities, such as cancer cells. Images are then taken of the body, and areas emitting greater amounts of the tracer can be seen. For example, in the images in figure 3.6, the radioactive-emitting cancer site appears as a black area on the left and a yellow glowing area on the right. There are many other uses of radioactive isotopes in medicine both in detection and treatment of disorders.

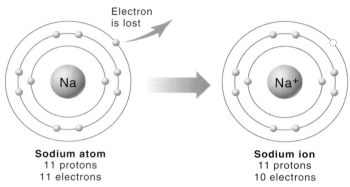

Sodium atom
11 protons
11 electrons

Sodium ion
11 protons
10 electrons

Figure 3.4 Making a sodium ion.

An electrically neutral sodium atom has 11 protons and 11 electrons. Sodium ions bear one positive charge when they ionize and lose one electron. Sodium ions have 11 protons and only 10 electrons.

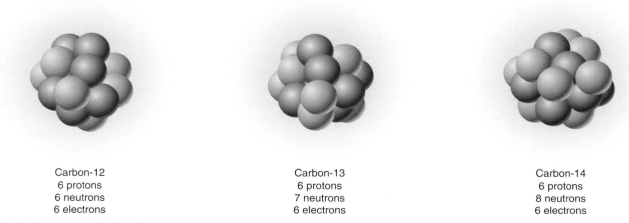

Carbon-12
6 protons
6 neutrons
6 electrons

Carbon-13
6 protons
7 neutrons
6 electrons

Carbon-14
6 protons
8 neutrons
6 electrons

Figure 3.5 Isotopes of the element carbon.

The three most abundant isotopes of carbon are carbon-12, carbon-13, and carbon-14. The yellow "clouds" in the diagrams represent the orbiting electrons, whose numbers are the same for all three isotopes. Protons are shown in purple, and neutrons are shown in pink.

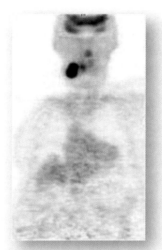
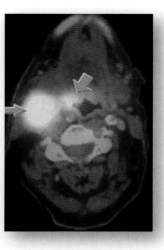

Figure 3.6 Using a radioactive tracer to identify cancer.

In certain medical imaging procedures, the patient ingests a radioactive tracer that is absorbed in greater amounts by cells with higher metabolic activities, such as cancer cells. The tracer emits radioactivity that is detected using PET and PET/CT equipment. A cancerous area in the neck is seen in these two images, as a dark black area on the left and a bright yellow glowing area on the right.

Dating Fossils

Fossils are created when the remains, footprints, or other traces of organisms become buried in sand or sediment. Over time, the calcium in bone and other hard tissues becomes mineralized as the sediment is converted to rock. A fossil is any record of prehistoric life—generally taken to mean older than 10,000 years. By dating the rocks in which fossils occur, biologists can get a very good idea of how old the fossils are. Rocks are usually dated by measuring the degree of radioactive decay of certain radioactive atoms among rock-forming minerals. A radioactive atom is one whose nucleus is unstable and eventually flies apart, creating more stable atoms of another element. Because the rate of decay of a radioactive element (the percent of atoms that undergo decay in a minute) is constant, scientists can use the amount of radioactive decay to date fossils. The older the fossil, the greater the fraction of its radioactive atoms that have decayed.

A widely employed method of dating fossils less than 50,000 years old is the carbon-14 (^{14}C) **radioisotopic dating** method illustrated in figure 3.7. Most carbon atoms have a mass number of 12 (^{12}C). However, a tiny but fixed proportion of the carbon atoms in the atmosphere consists of carbon atoms with a mass number of 14 (^{14}C). This isotope of carbon is created by the bombardment of nitrogen-14 atoms with cosmic rays. This proportion of ^{14}C (designated A in the figure) is captured by plants in photosynthesis, and is the proportion present in the carbon molecules of the animal's body that eats the plants, in this case a rabbit. After the plant or animal dies, it no longer accumulates any more carbon, and the ^{14}C present at the time of death gradually decays over time back to nitrogen-14 (^{14}N). The amount of ^{14}C (A) decreases while the amount of ^{12}C stays the same. Scientists can determine how long ago an organism died by measuring

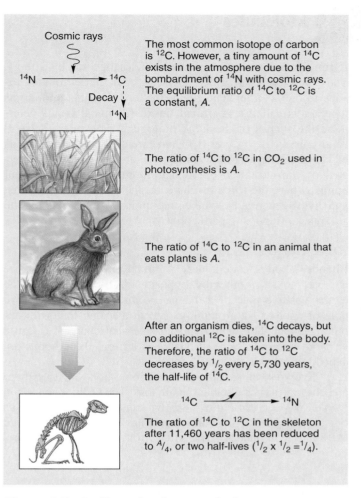

Figure 3.7 Radioactive isotope dating.

This diagram illustrates radioactive dating using carbon-14, a short-lived isotope.

the ratio of ^{14}C to ^{12}C in its remains or in the fossilized rock. Over time, the ratio of ^{14}C to ^{12}C decreases. It takes 5,730 years for half of the ^{14}C ($1/2A$ or $A/2$) present in the sample to be converted to ^{14}N by this process. This length of time is called the **half-life** of the isotope. Because the half-life of an isotope is a constant that never changes, the extent of radioactive decay allows you to date a sample. Thus a sample that had a quarter of its original proportion of ^{14}C remaining ($1/4A$ or $A/4$) would be approximately 11,460 years old (two half-lives—5,730 years for half of the ^{14}C to decay to a level of $A/2$ and another 5,730 years of the remaining ^{14}C to decay to a level of $A/4$).

For fossils older than 50,000 years, there is too little ^{14}C remaining to measure precisely, and scientists instead examine the decay of potassium-40 (^{40}K) into argon-40 (^{40}Ar), which has a half-life of 1.3 billion years.

3.2 Isotopes of an element differ in the number of neutrons they contain, but all have the same chemical properties.

Molecules

A **molecule** is a group of atoms held together by energy. The energy acts as "glue," ensuring that the various atoms stick to one another. The energy or force holding two atoms together is called a **chemical bond.** Chemical bonds determine the shapes of the large biological molecules that will be discussed in chapter 4. There are three principal kinds of chemical bonds: ionic bonds, where the force is generated by the attraction of oppositely charged ions; covalent bonds, where the force results from the sharing of electrons; and hydrogen bonds, where the force is generated by the attraction of opposite partial electrical charges.

Ionic Bonds

Chemical bonds called **ionic bonds** form when atoms are attracted to each other by opposite electrical charges. Just as the positive pole of a magnet is attracted to the negative pole of another, so an atom can form a strong link with another atom if they have opposite electrical charges. Because an atom with an electrical charge is an ion, these bonds are called ionic bonds.

Everyday table salt is built of ionic bonds. The sodium and chlorine atoms of table salt are ions. The sodium you see in the yellow panels of figure 3.8a gives up the sole electron in its outermost shell (the shell underneath has eight) and chlorine, in the light green panels, gains an electron to complete its outer-most shell. Recall from section 3.1 that an atom is more stable when its outermost electron shell is filled (with 2 electrons in the innermost shell or 8 electrons in shells that are farther out from the nucleus). To achieve this stability, an atom will give up or accept electrons from another atom. As a result of this electron hopping, sodium atoms in table salt are positive sodium ions and chlorine atoms are negative chloride ions. Because each ion is attracted electrically to all surrounding ions of opposite charge, this causes the formation of an elaborate matrix of sodium and chloride ionic bonds—a crystal. The sodium chloride crystal shown in figure 3.8b reveals an organized structure of alternating sodium (yellow) and chloride (light green) ions. That is why table salt is composed of tiny crystals and is not a powder.

The two key properties of ionic bonds that make them form crystals are that they are strong (although not as strong as covalent bonds) and that they are *not* directional. A charged atom is attracted to the electrical field contributed by all near-by atoms of opposite charge. Ionic bonds do not play an important part in most biological molecules because of this lack of directionality. Complex, stable shapes require the more specific associations made possible by directional bonds.

Covalent Bonds

Strong chemical bonds called **covalent bonds** form between two atoms when they share electrons. Most of the atoms in your body are linked to other atoms by covalent bonds. Why do atoms in molecules share electrons? Remember, all atoms seek to fill up their outermost shell of orbiting electrons, which in all atoms (except tiny hydrogen and helium) takes eight electrons. For example, an atom with six outer shell electrons seeks to share them with an atom that has two outer shell electrons or with two atoms that have single outer shell electrons. If only one pair of electrons is shared between two atoms, the

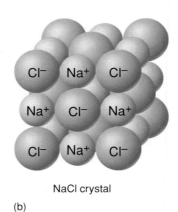

Sodium atom	Chlorine atom

+ Sodium ion	− Chloride ion

(a)

NaCl crystal

(b)

Figure 3.8 The formation of the ionic bond in table salt.

(a) When a sodium atom donates an electron to a chlorine atom, the sodium atom, lacking that electron, becomes a positively charged sodium ion. The chlorine atom, having gained an extra electron, becomes a negatively charged chloride ion. (b) Sodium chloride forms a highly regular lattice of alternating sodium ions and chloride ions. You are familiar with these crystals as everyday table salt.

covalent bond is called a *single covalent bond.* If two pairs of electrons are shared, it is called a *double covalent bond;* and if three pairs of electrons are shared it is called a *triple covalent bond.* Water (H_2O), for example, is a molecule in which oxygen that has six outer electrons, shown in the outer shell in figure 3.9*a*, forms single covalent bonds with two hydrogens that have one outer electron each, indicated by the arrow in the figure. Because the carbon atom has four electrons in its outermost shell, carbon can form as many as four covalent bonds in its attempt to fully populate its outermost shell of electrons. Because there are many ways four covalent bonds can form, carbon atoms participate in many different kinds of molecules.

The two key properties of covalent bonds that make them ideal for their molecule-building role in living systems are that (1) they are strong, involving the sharing of lots of energy; and (2) they are very directional—bonds form between two specific atoms, rather than a generalized attraction of one atom for its neighbors.

Hydrogen Bonds

Weak chemical bonds of a very special sort called **hydrogen bonds** play a key role in biology. To understand them we need to look at covalent bonds again briefly. When a covalent bond forms between two atoms, one nucleus may be much better at attracting the shared electrons than the other, an aspect of the atom called its *electronegativity.* In water, for example, the shared electrons are much more strongly attracted to the oxygen atom than to the hydrogen atoms; oxygen has a higher electronegativity. When this happens, shared electrons spend more time in the vicinity of the oxygen atom, which as a result becomes somewhat negative in charge; they spend less time in the vicinity of the hydrogens, and these become somewhat positive in charge. The space-filling model of water in figure 3.9*b* shows the charge distribution in the water molecule. The charges are not full electrical charges like ions possess but rather tiny *partial charges,* signified by the Greek letter delta (δ). What you end up with is a sort of molecular magnet, with positive and negative ends, or "poles." Molecules like this are said to be *polar.* Molecules that don't exhibit a large difference in electronegativities of its atoms, like carbon-hydrogen bonds, are called *nonpolar* molecules. Polar molecules, like water, can also be held near each other with a special type of bond, called a **hydrogen bond**. Hydrogen bonds occur when the positive end of one **polar molecule** is attracted to the negative end of another, like two magnets drawn to each other. Figure 3.10 shows how water molecules are held together with hydrogen bonds (indicated by the dashed lines) with partially negative oxygens, in red, being attracted to the partially positive hydrogens, in blue.

Two key properties of hydrogen bonds cause them to play an important role in biological molecules, the molecules found in living organisms. First, they are weak and so are not effective over long distances like more powerful covalent and ionic bonds. Second, as a result of their weakness, hydrogen bonds are highly directional—polar molecules must be very close for the weak attraction to be effective. Hydrogen bonds are too weak to actually form stable molecules by themselves. In-

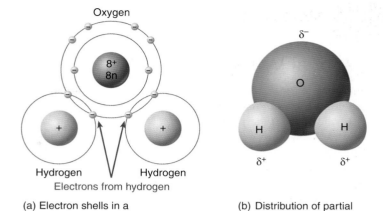

(a) Electron shells in a water molecule

(b) Distribution of partial charges in a water molecule

Figure 3.9   Water molecules contain two covalent bonds.
Each water molecule is composed of one oxygen atom and two hydrogen atoms. (a) The oxygen atom shares one electron with each participating hydrogen atom. (b) A three-dimensional representation of a water molecule showing the partial charges of the molecule.

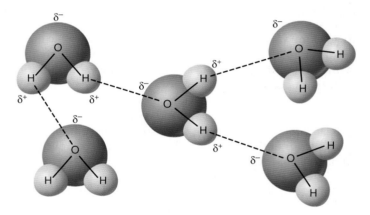

Figure 3.10   Hydrogen bonding in water molecules.
Hydrogen bonds, indicated by the dashed lines, hold water molecules together. The partial positive charge of the hydrogen atom of one molecule is attracted to the partial negative charge of the oxygen atom of another molecule. This attraction of partial charges holds water molecules together.

stead, they act like Velcro, forming a tight bond by the additive effects of *many* weak interactions. Hydrogen bonds stabilize the shapes of many important biological molecules by causing certain parts of a molecule to be attracted to other parts. Later in chapter 4, we will discuss the role of hydrogen bonding in maintaining the structures of large biological molecules such as proteins and DNA.

3.3   Molecules are atoms linked together by chemical bonds. Ionic bonds form when electrically charged ions attract each other. Most biological molecules are held together by covalent bonds, which are strong linkages created by the sharing of electrons. Hydrogen bonds are important but weaker bonds.

Author's *Corner*

How Tropical Lizards Climb Vertical Walls

Science is most fun when it tickles your imagination. This is particularly true when you see something you know just can't be true. A few years ago, my wife, Barbara, and I were on a tropical vacation, and I was lying on the bed when a little lizard walked up the wall beside me and across the ceiling, stopping right over my head and looking down at me.

This was no special effect, no trick with mirrors. I was seeing it with my own eyes, real as day, right there above me. The lizard, a green gecko about the size of a toothbrush, stood upside down on the ceiling and seemed to laugh at me for several minutes before trotting over to the far wall and down.

How did my gecko perform this gripping feat? Investigators have puzzled over the adhesive properties of geckos for decades. What force prevented gravity from dropping the gecko on my nose?

The most reasonable hypothesis seemed suction— salamanders' feet form suction cups that let them climb walls, so maybe geckos' do too. The way to test this is to see if the feet adhere in a vacuum, with no air to create suction. Salamander feet don't adhere, but gecko feet do. It's not suction.

How about friction? Cockroaches climb using tiny hooks that grapple onto irregularities in the surface, much as rock climbers use crampons. Geckos, however, happily run up walls of smooth polished glass that no cockroach can climb. It's not friction.

Electrostatic attraction? Clothes in a dryer stick together because of electrical charges created by their rubbing together. You can stop this by adding a "static remover" that is itself heavily ionized. But a gecko's feet still adhere in ionized air. It's not electrostatic attraction.

Could it be glue? Many insects use adhesive secretions from glands in their feet to aid climbing. But there are no gland cells in gecko feet, no secreted chemical. It's not glue.

There was one tantalizing clue, however, the kind that experimenters love. Gecko feet seem to get stickier on surfaces with highly ordered molecules. This suggests that geckos are tapping directly into the molecular structure of the surfaces they walk on!

Tracking down this clue, Robert Full of the University of California, Berkeley, and his research team took a closer look at gecko feet. Geckos have rows of tiny hairs on the bottoms of their feet, like the bristles of a toothbrush. There are about half a million of these hairs on each foot, pointed toward the heel.

When you look at these hairs under a microscope, the end of each hair is divided into between 400 and 1,000 fine projections, the projections sticking out from the tip like tiny stiff brushes.

When a gecko takes a step, it drives the sole of its foot into the surface and pushes it backward. This shoves the forest of tips directly against the surface. The atoms of each gecko tip become closely engaged with the atoms of the surface, and *that* is the force that defies gravity. When two atoms approach each other very closely—closer than the diameter of an atom—a subtle nuclear attraction called "**Van der Waals forces**" comes into play. These forces are individually very weak, but when lots of atoms add their little bits, the sum can add up to quite a lot.

Full and his team used microelectrical mechanical sensors (originally designed to be used with atomic force microscopes) to measure the force exerted by a single hair removed from a gecko's foot. It was 200 micronewtons, a tiny force but stupendous for a single hair. Enough to hold up an ant. A million hairs could support a small child. My little gecko, ceiling-walking with 2 million of them, could have carried an 80-pound backpack—talk about being overengineered!

If they stick that well, how do geckos ever come unstuck? For a gecko's foot to stick, each hair projection must butt up squarely against the surface, so the hair's individual atoms can come into play. Tipped past a critical angle—30 degrees—the attractive forces between hair and surface atoms weaken to nothing. The trick is to tip the foot hairs until the projections let go. Geckos release their feet by curling up each toe and peeling it off, sort of like undoing Velcro.

Now I can laugh with my little gecko friend, should I see it again, for I know its secret.

3.4 Hydrogen Bonds Give Water Unique Properties

Three-fourths of the earth's surface is covered by liquid water. About two-thirds of your body is water, and you cannot exist long without it. All other organisms also require water. It is no accident that tropical rain forests are bursting with life, whereas dry deserts seem almost lifeless except after rain. The chemistry of life, then, is water chemistry.

Water has a simple atomic structure, an oxygen atom linked to two hydrogen atoms by single covalent bonds. The chemical formula for water is thus H_2O. It is because the oxygen atom attracts the shared electrons more strongly than the hydrogen atoms that water is a *polar molecule* and so can form *hydrogen bonds*. Water's ability to form hydrogen bonds is responsible for much of the organization of living chemistry, from membrane structure to how proteins fold.

The weak hydrogen bonds that form between a hydrogen atom of one water molecule and the oxygen atom of another produce a lattice of hydrogen bonds within liquid water. Each of these bonds is individually very weak and short-lived—a single bond lasts only 1/100,000,000,000 of a second. However, like the grains of sand on a beach, the cumulative effect of large numbers of these bonds is enormous and is responsible for many of the important physical properties of water (table 3.2).

Heat Storage

The temperature of any substance is a measure of how rapidly its individual molecules are moving. Because of the many hydrogen bonds that water molecules form with one another, a large input of thermal energy is required to disrupt the organization of liquid water and raise its temperature. Because of this, water heats up more slowly than almost any other compound and holds its temperature longer. That is a major reason why your body is able to maintain a relatively constant internal temperature.

Ice Formation

If the temperature is low enough, very few hydrogen bonds break in water. Instead, the lattice of these bonds assumes a crystal-like structure, forming a solid we call ice. Interestingly, ice is less dense than water—that is why icebergs and ice cubes float. Why is ice less dense? This is best understood by comparing the molecular structures of water and ice that you see in figure 3.11. At temperatures above freezing, water molecules in figure 3.11a move around each other with hydrogen bonds breaking and forming. As temperatures drop, the movement of water molecules decreases, allowing hydrogen bonds to stabilize, holding individual molecules farther apart, as in figure 3.11b, making the ice structure less dense.

TABLE 3.2	THE PROPERTIES OF WATER
Property	**Explanation**
Heat storage	Hydrogen bonds require considerable heat before they break, minimizing temperature changes.
Ice formation	Water molecules in an ice crystal are spaced relatively far apart because of hydrogen bonding.
High heat of vaporization	Many hydrogen bonds must be broken for water to evaporate.
Cohesion	Hydrogen bonds hold molecules of water together.
High polarity	Water molecules are attracted to ions and polar compounds.

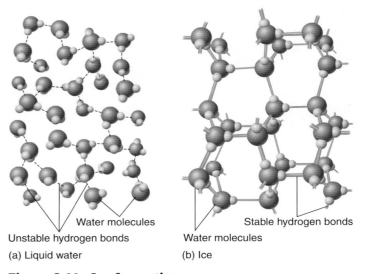

Water molecules

Unstable hydrogen bonds

(a) Liquid water

Stable hydrogen bonds

Water molecules

(b) Ice

Figure 3.11 Ice formation.
When water (a) cools below 0°C, it forms a regular crystal structure (b) that floats. The individual water molecules are spaced apart and held in position by hydrogen bonds.

High Heat of Vaporization

If the temperature is high enough, many hydrogen bonds break in water, with the result that the liquid is changed into vapor. A considerable amount of heat energy is required to do this—every gram of water that evaporates from your skin removes 2,452 joules of heat from your body, which is equal to the energy released by lowering the temperature of 586 grams of water 1°C. That is why sweating cools you off; as the sweat evaporates (vaporizes) it takes energy with it, in the form of heat, cooling the body.

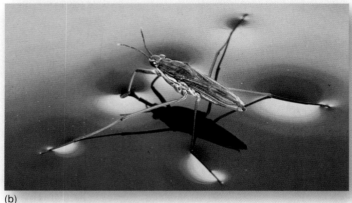

(a)

(b)

Figure 3.12   Cohesion.

(a) Cohesion allows water molecules to stick together and form droplets. (b) Surface tension is a property derived from cohesion—that is, water has a "strong" surface due to the force of its hydrogen bonds. Some insects, such as this water strider, literally walk on water.

Cohesion

Because water molecules are very polar, they are attracted to other polar molecules—hydrogen bonds bind polar molecules to each other. When the other polar molecule is another water molecule, the attraction is called **cohesion.** The surface tension of water is created by cohesion. Surface tension is the force that causes water to bead, like on the spider web in figure 3.12, or supports the weight of the water strider. When the other polar molecule is a different substance, the attraction is called **adhesion.** Capillary action—such as water moving up into a paper towel—is created by adhesion. Water clings to any substance, such as paper fibers, with which it can form hydrogen bonds. Adhesion is why things get "wet" when they are dipped in water and why waxy substances do not—they are composed of nonpolar molecules that don't form hydrogen bonds with water molecules.

High Polarity

Water molecules in solution always tend to form the maximum number of hydrogen bonds possible. Polar molecules form hydrogen bonds and are attracted to water molecules. Polar

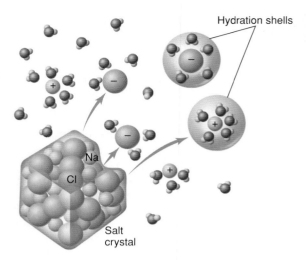

Hydration shells

Na

Cl

Salt crystal

Figure 3.13   How salt dissolves in water.

Salt is soluble in water because the partial charges on water molecules are attracted to the charged sodium and chloride ions. The water molecules surround the ions, forming what are called hydration shells. When all of the ions have been separated from the crystal, the salt is said to be dissolved.

molecules are called **hydrophilic** (from the Greek *hydros,* water, and *philic,* loving), or water-loving, molecules. Water molecules gather closely around any molecule that exhibits an electrical charge, whether a full charge (ion) or partial charge (polar molecule). When a salt crystal dissolves in water as you see happening in figure 3.13, what really happens is that individual ions break off from the crystal and become surrounded by water molecules. The blue hydrogen atoms of water are attracted to the negative charge of the chloride ions and the red oxygen atoms are attracted to the positive charge of the sodium ions. Water molecules orient around each ion like a swarm of bees attracted to honey, and this shell of water molecules, called a *hydration shell,* prevents the ions from reassociating with the crystal. Similar shells of water form around all polar molecules, and polar molecules that dissolve in water in this way are said to be **soluble** in water.

Nonpolar molecules like oil do not form hydrogen bonds and are not water-soluble. When nonpolar molecules are placed in water, the water molecules shy away, instead forming hydrogen bonds with other water molecules. The nonpolar molecules are forced into association with one another, crowded together to minimize their disruption of the hydrogen bonding of water. It seems almost as if the nonpolar compounds shrink from contact with water, and for this reason they are called **hydrophobic** (from the Greek *hydros,* water, and *phobos,* fearing). Many biological structures are shaped by such hydrophobic forces, as will be discussed in chapter 4.

> **3.4   Water molecules form a network of hydrogen bonds in liquid and dissolve other polar molecules. Many of the key properties of water arise because it takes considerable energy to break liquid water's many hydrogen bonds.**

3.5 Water Ionizes

The covalent bonds within a water molecule sometimes break spontaneously. When it happens, one of the protons (hydrogen atom nuclei) dissociates from the molecule. Because the dissociated proton lacks the negatively charged electron it was sharing in the covalent bond with oxygen, its own positive charge is no longer counterbalanced, and it becomes a positively charged ion, **hydrogen ion** (H^+). The rest of the dissociated water molecule, which has retained the shared electron from the covalent bond, is negatively charged and forms a **hydroxide ion** (OH^-). This process of spontaneous ion formation is called **ionization.** It can be represented by a simple chemical equation, in which the chemical formulas for water and the two ions are written down, with an arrow showing the direction of the dissociation:

$$H_2O \longleftrightarrow OH^- + H^+$$
water   hydroxide   hydrogen
                    ion   ion

Because covalent bonds are strong, spontaneous ionization is not common. In a liter of water, only roughly 1 molecule out of each 550 million is ionized at any instant in time, corresponding to 1/10,000,000 (that is, 10^{-7}) of a mole of hydrogen ions. (A mole is a measurement of weight. One mole of any object is the weight of 6.022×10^{23} units of that object.) The concentration of H^+ in water can be written more easily by simply counting the number of decimal places after the digit "1" in the denominator:

$$[H^+] = \frac{1}{10,000,000}$$

pH

A more convenient way to express the hydrogen ion concentration of a solution is to use the pH scale (figure 3.14). This scale defines pH as the negative logarithm of the hydrogen ion concentration in the solution:

$$pH = -\log [H^+]$$

Since the logarithm of the hydrogen ion concentration is simply the exponent of the molar concentration of H^+, the pH equals the exponent times –1. Thus, pure water, with an $[H^+]$ of 10^{-7} mole/liter, has a pH of 7. Recall that for every hydrogen ion formed when water dissociates, a hydroxide ion is also formed, meaning that the dissociation of water produces H^+ and OH^- in equal amounts. Therefore, a pH value of 7 indicates neutrality—a balance between H^+ and OH^-—on the pH scale.

Note that the pH scale is *logarithmic,* which means that a difference of 1 on the pH scale represents a 10-fold change in hydrogen ion concentration. This means that a solution with a pH of 4 has *10 times* the concentration of H^+ present in one with a pH of 5.

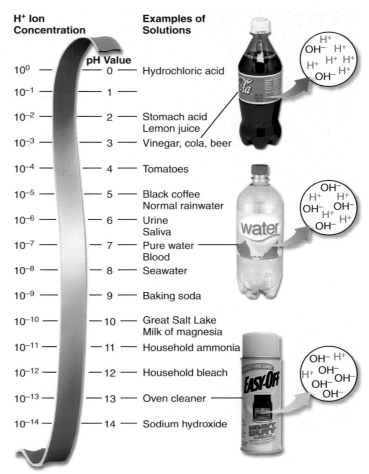

Figure 3.14   The pH scale.
A fluid is assigned a value according to the number of hydrogen ions present in a liter of that fluid. The scale is logarithmic, so that a change of only 1 means a 10-fold change in the concentration of hydrogen ions; thus lemon juice with a pH of 2 is 100 times more acidic than tomatoes with a pH of 4, and seawater is 10 times more basic than pure water.

Acids. Any substance that dissociates in water to increase the concentration of H^+ is called an **acid.** Acidic solutions have pH values below 7. The stronger an acid is, the more H^+ it produces and the lower its pH. For example, hydrochloric acid (HCl), which is abundant in your stomach, ionizes completely in water. This means that a dilution of 10^{-1} mole/liter of HCl will dissociate to form 10^{-1} mole/liter of H^+, giving the solution a pH of 1. The pH of champagne, which bubbles because of the carbonic acid dissolved in it, is about 3.

Bases. A substance that combines with H^+ when dissolved in water is called a **base.** By combining with H^+, a base lowers the H^+ concentration in the solution. Basic (or alkaline) solutions, therefore, have pH values above 7. Very strong bases, such as sodium hydroxide (NaOH), have pH values of 12 or more.

Acid Rain

As you study biology, you will learn that hydrogen ions play many roles in the chemistry of life. When conditions become overly acidic—too many hydrogen ions—serious damage to organisms often results. One important example of this is acid precipitation, more informally called **acid rain**. Acid precipitation is just what it sounds like, the presence of acid in rain or snow. Where does the acid come from? Tall smokestacks from coal-burning power plants send smoke high into the atmosphere through these stacks, each of which is over 65 meters tall. The smoke the stacks belch out contains high concentrations of sulfur dioxide (SO_2), because the coal that the plants burn is rich in sulfur. The sulfur-rich smoke is dispersed and diluted by winds and air currents. Since the 1950s, such tall stacks have become popular in the United States and Europe—there are now over 800 of them in the United States alone.

In the 1970s, 20 years after the stacks were introduced, ecologists began to report evidence that the tall stacks were not eliminating the problems associated with the sulfur, just exporting the ill effects elsewhere. The lakes and forests of the Northeast suffered drastic drops in biodiversity, forests dying and lakes becoming devoid of life. It turned out that the SO_2 introduced into the upper atmosphere by high smokestacks combines with water vapor to produce sulfuric acid (H_2SO_4). When this water later falls back to earth as rain or snow, it carries the sulfuric acid with it. When schoolchildren measured the pH of natural rainwater as part of a nationwide project in 1989, rain and snow in the Northeast often had a pH as low as 2 and 3—more acidic than vinegar.

After accumulating in soils for over fifty years, the effects of acid rain are now only too evident. The impact of acid rain on forests first became evident in the Northeast. Some 15% of the lakes in New England have become chronically acidic and are dying biologically as their pH levels fall to below 5.0. Many of the forests of the northeastern United States and Canada have also been seriously damaged The trees in the photo above and on the first page of this chapter show the ill effects of acid precipitation. In the last decades, acid added to forest soils has caused the loss from these soils of over half the essential plant nutrients calcium and magnesium. Researchers blame excess acids for dissolving Ca^{++} and Mg^{++} ions into drainage waters much faster than weathering rocks can replenish them. Without them, trees stop growing and die.

Now, some thirty years later, acid rain effects are becoming apparent in the Southeast as well. Researchers suggest the reason for the delay is that southern soils are generally thicker than northern ones and thus able to

sponge up far more acid. But now that southern forest soils are becoming saturated, they too are beginning to die. In a third of the southeastern streams studied, fish are declining or already gone.

The solution is straightforward: capture and remove the emissions instead of releasing them into the atmosphere. Progressively tougher pollution laws over the past three decades have reduced U.S. emissions of sulfur dioxide by about 40% from its 1973 peak of 28.8 metric tons a year. Despite this significant progress, much remains to be done. Unless levels are cut further, researchers predict forests may not recover for centuries.

An informed public will be essential. While textbook treatments have in the past tended to minimize the impact of this issue on students ("the vast majority of North American forests are not suffering substantially from acid precipitation"), it is important that we face the issue squarely and support continued efforts to address this serious problem.

Buffers

The pH inside almost all living cells, and in the fluid surrounding cells in multicellular organisms, is fairly close to 7. The many proteins that govern metabolism are all extremely sensitive to pH, and slight alterations in pH can cause the molecules to take on different shapes that disrupt their activities. For this reason, it is important that a cell maintain a constant pH level. The pH of your blood, for example, is 7.4, and you would survive only a few minutes if it were to fall to 7.0 or rise to 7.8.

Yet the chemical reactions of life constantly produce acids and bases within cells. Furthermore, many animals eat substances that are acidic or basic; Coca-Cola, for example, is acidic, and egg white is basic. What keeps an organism's pH constant? Cells contain chemical substances called buffers that minimize changes in concentrations of H^+ and OH^-.

A **buffer** is a substance that takes up or releases hydrogen ions into solution as the hydrogen ion concentration of the solution changes. Hydrogen ions are donated to the solution when their concentration falls and taken from the solution when their concentration rises. The graph in figure 3.15 shows how buffers work. The blue line indicates changes in pH. As a base is added to the solution, the H^+ concentration falls and the pH should rise sharply, but by contributing H^+ to the solution the buffer works to keep the pH within a range, called the buffering range (the darker blue bar). Only when the buffering capacity is exceeded does the pH begin to rise. What sort of substance will act in this way? Within organisms, most buffers consist of pairs of substances, one an acid and the other a base.

The key buffer in human blood is an acid-base pair consisting of *carbonic acid* (acid) and *bicarbonate* (base). These two substances interact in a pair of reversible reactions. First, carbon dioxide (CO_2) and H_2O join to form carbonic acid (H_2CO_3) (step **2** in figure 3.16), which in a second reaction dissociates to yield bicarbonate ion (HCO_3^-) and H^+ (step

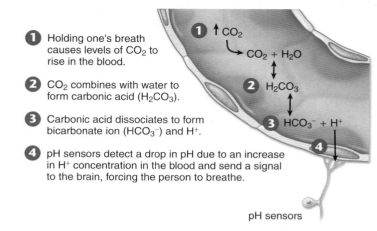

1 Holding one's breath causes levels of CO_2 to rise in the blood.

2 CO_2 combines with water to form carbonic acid (H_2CO_3).

3 Carbonic acid dissociates to form bicarbonate ion (HCO_3^-) and H^+.

4 pH sensors detect a drop in pH due to an increase in H^+ concentration in the blood and send a signal to the brain, forcing the person to breathe.

pH sensors

Figure 3.16   Holding your breath.
CO_2 accumulates in the blood when a person holds his/her breath. CO_2 combines with water, forming carbonic acid. Carbonic acid dissociates into bicarbonate and H^+, which acts to lower the pH. The drop in pH is detected by sensors that stimulate the brain, causing the person to breathe.

3). If some acid or other substance adds H^+ to the blood, the HCO_3^- acts as a base and removes the excess H^+ by forming H_2CO_3. Similarly, if a basic substance removes H^+ from the blood, H_2CO_3 dissociates, releasing more H^+ into the blood. The forward and reverse reactions that interconvert H_2CO_3 and HCO_3^- thus stabilize the blood's pH.

For example, when you breathe in, your body takes up oxygen from the air and when you breathe out, your body releases carbon dioxide. When you hold your breath, CO_2 accumulates in your blood and drives the chemical reactions in figure 3.16 producing carbonic acid. Can you hold your breath indefinitely? No, but not for the reason you might think. It is not lack of oxygen that forces you to breathe, but too much carbon dioxide. If you try and hold your breath for very long, CO_2 accumulates in the blood, as shown in step **1** in figure 3.16, triggering the formation of carbonic acid in step **2** that dissociates into bicarbonate ion and H^+ in step **3**. This increase in H^+ causes the blood to become more acidic. If the pH in the blood drops too low, pH sensors that are located in some of the major blood vessels of the body detect the change (step **4**) and send signals to the brain. These signals, along with other sensory processes, stimulate the area of the brain that controls respiration, causing it to increase the rate of breathing. Hyperventilating, breathing very quickly, has the opposite effect, lowering the levels of CO_2 in the blood. That is why you are told to breathe into a paper bag when you are hyperventilating, to increase your intake of CO_2.

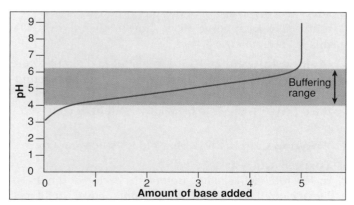

Figure 3.15   Buffers minimize changes in pH.
Adding a base to a solution neutralizes some of the acid present and so raises the pH. Thus, as the curve moves to the right, reflecting more and more base, it also rises to higher pH values. What a buffer does is to make the curve rise or fall very slowly over a portion of the pH scale, called the "buffering range" of that buffer.

3.5 A tiny fraction of water molecules spontaneously ionize at any moment, forming H^+ and OH^-. The pH of a solution is a measure of its H^+ concentration. Low pH values indicate high H^+ concentrations (acidic solutions), and high pH values indicate low H^+ concentrations (basic solutions).

Using Radioactive Decay to Date the Iceman

In the fall of 1991, sticking out of the melting snow on the crest of a high pass near the mountainous border between Italy and Austria, two Austrian hikers found a corpse. Right away it was clear the body was very old, frozen in an icy trench where he had sought shelter long ago and only now released as the ice melted. In the years since this startling find, scientists have learned a great deal about the dead man, who they named Ötzi. They know his age, his health, the shoes and clothing he wore, what he ate, and that he died from an arrow that ripped through his back. Its tip is still embedded in the back of his left shoulder. From the distribution of chemicals in his teeth and bones, we know he lived his life within 60 kilometers of where he died.

How old was this Iceman? Scientists answered this key question by measuring the degree of decay of the short-lived carbon isotope ^{14}C in Ötzi's body. This procedure is discussed earlier in this chapter (see figure 3.7). The graph to the right displays the radioactive decay curve of the carbon isotope carbon-14 (^{14}C); it takes 5,730 years for half of the ^{14}C present in a sample to decay to nitrogen-14 (^{14}N). When Ötzi's carbon isotopes were analyzed, researchers determined that the ratio of ^{14}C to ^{12}C (a **ratio** is the size of one variable relative to another), also written as the fraction $^{14}C/^{12}C$, in Ötzi's body is 0.435 of the fraction found in tissues of a person who has recently died.

1. **Applying Concepts**
   a. Variable. In the graph, what is the dependent variable?
   b. Proportion. What proportion (a **proportion** is the size of a variable relative to the whole) of the ^{14}C present in Ötzi's body when he died is still there today?

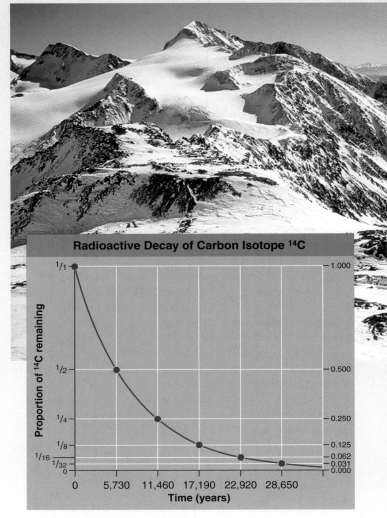

Radioactive Decay of Carbon Isotope ^{14}C

2. **Interpreting Data** Plot this proportion on the ^{14}C radioactive decay curve above. How many half-lives does this point represent?

3. **Making Inferences** If Ötzi were indeed a recent corpse, made to look old by the harsh weather conditions found on the high mountain pass, what would you expect the ratio of ^{14}C to ^{12}C to be, relative to that in your own body?

4. **Drawing Conclusions** How old is the Iceman Ötzi?

5. **Further Analysis**
   a. The radioactive iodine isotope ^{131}I decays at a half-life of eight days. Plotted on the graph above, would its radioactive decay curve be above or below that of ^{14}C?
   b. Scientists often employ the radioactive decay of isotope potassium-40 (^{40}K) into argon-40 (^{40}Ar) to date old material. ^{40}K has a half-life of 1.3 billion years. Would it be a better or poorer isotope than ^{14}C to use in dating Ötzi?

Some Simple Chemistry

3.1 Atoms

- An atom is the smallest particle that retains the chemical properties of its substance. Atoms contain a core nucleus of protons and neutrons with electrons orbiting around the nucleus (**figure 3.1**).

- The number of protons in an atom is called its atomic number. The mass that is contributed by the protons and neutrons is called the atom's mass number. All atoms that have the same atomic number are said to be the same element.

- Protons are positively charged particles, neutron particles carry no charge, and electrons are negatively charged particles that orbit around the nucleus in electron shells. Electrons also carry energy based on their distance from the nucleus and determine the chemical behavior of the atom (**figure 3.2**).

- Most electron shells hold up to eight electrons and atoms will undergo chemical reactions in order to fill the outermost electron shell, either by gaining, losing, or sharing electrons (**figure 3.3**).

3.2 Ions and Isotopes

- Ions are atoms that have gained one or more electrons (negative ions called anions) or lost one or more electrons (positive ions called cations) (**figure 3.4**).

- Isotopes are atoms that have the same number of protons but differing numbers of neutrons (**figure 3.5**). Some isotopes are unstable and have applications in medicine (**figure 3.6**) and dating fossils (**figure 3.7**).

3.3 Molecules

- Molecules form when atoms are held together with energy. The force holding atoms together is called a chemical bond. There are three main types of chemical bonds.

- Ionic bonds form when ions of opposite charge are attracted to each other (**figure 3.8**).

- Covalent bonds form when two atoms share electrons (**figure 3.9**).

- Hydrogen bonds form between polar molecules. The atoms of polar molecules are held together by covalent bonds in which the shared electrons are unevenly distributed around their nuclei, giving them a slightly positive end and a slightly negative end. Hydrogen bonds form when the positive end of one molecule is attracted to the negative end of another (**figure 3.10**).

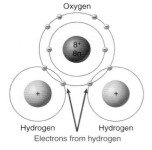

Oxygen

8+
8n

Hydrogen   Hydrogen
Electrons from hydrogen

Water: Cradle of Life

3.4 Hydrogen Bonds Give Water Unique Properties

- Water molecules are polar molecules that form hydrogen bonds with each other and with other polar molecules. Many of the physical properties of water are attributed to this hydrogen bonding.

- Water molecules that are held together through hydrogen bonding are more difficult to separate, and as a result a significant amount of thermal energy is needed to pull the molecules apart. For this reason, water heats up slowly and holds it temperature longer.

- The hydrogen bonds that hold water molecules together become more stable at lower temperatures and as a result, they lock water molecules into place in solid crystal structures called ice (**figure 3.11**).

- Water molecules held together require a significant input of thermal energy to vaporize the liquid into a gas. This high heat of vaporization is used by our bodies to cool themselves.

- Because water molecules are polar molecules, they will form hydrogen bonds with other polar molecules. If the other polar molecules are water molecules, the process is called cohesion. If the other polar molecules are some other substance, the process is called adhesion.

- When water molecules form hydrogen bonds with other polar molecules, water molecules will tend to surround other polar molecules, forming a barrier around them called a hydration shell. Polar molecules are said to be hydrophilic and are water-soluble (**figure 3.13**). Nonpolar molecules do not form hydrogen bonds with water molecules and will cluster together when placed in water. They are said to be hydrophobic and are water-insoluble.

3.5 Water Ionizes

- Water molecules dissociate forming negatively charged hydroxide ions (OH^-) and positively charged hydrogen ions (H^+). This property of water is significant because the concentration of hydrogen ions in a solution determines its pH.

- A solution with a higher hydrogen ion concentration is an acid with specific chemical properties, and a solution with a lower hydrogen ion concentration is a base with different chemical properties (**figure 3.14**).

- Substances called buffers control changes in pH by taking up or releasing H^+ into the solution and control pH within a range called the buffer range (**figure 3.15**).

- A buffer that functions in the human body is an acid-base pair consisting of carbonic acid and bicarbonate. This buffering process involves a series of two reversible reactions, one which generates hydrogen ions that reduces the pH and the reverse reaction that takes up hydrogen ions from solution that increases pH (**figure 3.16**).

- This carbonic acid-bicarbonate buffering system works in the blood to regulate breathing (**figure 3.16**).

1. The smallest particle into which a substance can be divided and still retain all of its chemical properties is
   a. matter.
   c. a molecule.
   b. an atom.
   d. mass.

2. An atom that has gained or lost one or more electrons is
   a. an isotope.
   c. an ion.
   b. a neutron.
   d. radioactive.

3. Atoms are held together by a force called a bond. The three types of bonds are
   a. positive, negative, and neutral.
   b. ionic, doric, and corinthian.
   c. magnetic, electric, and radioactive.
   d. ionic, covalent, and hydrogen.

4. Carbon has four electrons in its outer electron shell, therefore
   a. it has a completely filled outer electron shell.
   b. it can form four single covalent bonds.
   c. it does not react with any other atom.
   d. it has a positive charge.

5. The partial separation of charge in the water molecule
   a. results from the electrons' greater attraction to the oxygen atom.
   b. results from oxygen's higher electronegativity.
   c. indicates that the water molecule is a polar molecule.
   d. All of these are correct.

6. Water has some very unusual properties. These properties occur because of the
   a. hydrogen bonds between the individual water molecules.
   b. covalent bonds between the individual water molecules.
   c. hydrogen bonds within each individual water molecule.
   d. ionic bonds between the individual water molecules.

7. Which of the following properties are somehow related to the need for significant thermal energy to break hydrogen bonds?
   a. cohesion and adhesion
   b. hydrophobic and hydrophilic
   c. heat storage and heat of vaporization
   d. ice formation and high polarity

8. The attraction of water molecules to other water molecules is called
   a. cohesion.
   b. capillary action.
   c. solubility.
   d. adhesion.

9. Water sometimes ionizes, a single molecule breaking apart into an hydrogen ion and a hydroxide ion. Other materials may dissociate in water, resulting in an increase of either (1) hydrogen ions or (2) hydroxide ions in the solution. We call the results
   a. (1) acids and (2) bases.
   b. (1) bases and (2) acids.
   c. (1) neutral solutions and (2) neutronic solutions.
   d. (1) hydrogen solutions and (2) hydroxide solutions.

10. A buffer acts in all but which of the following ways?
    a. to absorb H^+ from the solution
    b. to maintain pH levels within a range
    c. to keep pH from ever changing
    d. to release H^+ into the solution

Visual Understanding

1. **Figure 3.7** Based on the figure, explain why it is difficult to use carbon-14 dating on things that are older than about 50,000 years.

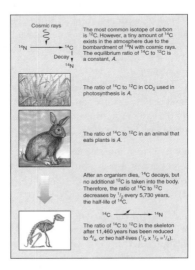

2. **Figure 3.9a** This figure shows the oxygen atom forming covalent bonds with two hydrogen atoms. The carbon atom, like oxygen, has two electrons in its innermost shell, but only four electrons in its outermost shell. Using this water molecule as a guide, draw a diagram showing how carbon forms covalent bonds with four hydrogen atoms in a methane (CH_4) molecule.

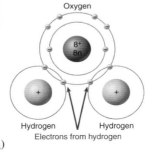

3. **Figure 3.12** How can these things—drops of water clinging to a web and insects walking on water—occur?

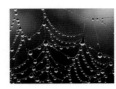

Challenge Questions

1. What happens if an atom gains or loses electrons? What happens if an atom gains or loses neutrons? What happens if an atom gains or loses protons?

2. You are on a 10-day backpacking trip with a small group of friends. Yvonne has washed out a set of water bottles with bleach. Before she can rinse them, Carlos, hot and thirsty, picks one up and drinks it, drops it, and begins to choke. What is the problem, and what can you do for him?

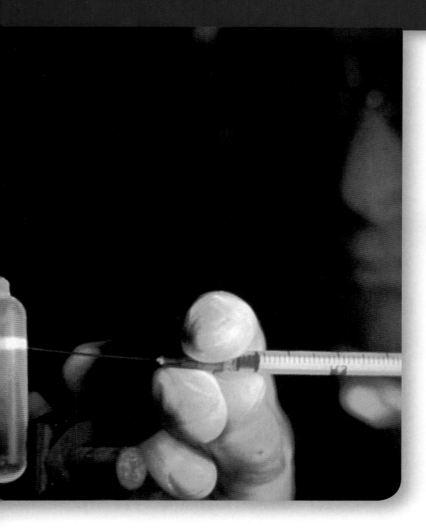

4

Molecules of Life

Using a syringe, this researcher is gently removing a glowing band of DNA, to be used in an experiment studying heredity. DNA, the carrier of an organism's genes, is one of several kinds of very large molecules found in all organisms. A molecule is a collection of tiny atoms linked together. Incredibly, every atom in your body was at one time part of a star. On earth, atoms are the basic chemical elements. Only a few are found in any significant numbers in living things. An essential atom for life is carbon, which can assemble into DNA and other very large molecules. Interacting with water, these long carbon chains twist about each other, or fold up into compact masses. Much of the chemistry that goes on in organisms, determining what each individual is like, depends on the actions of large folded molecules called proteins. By promoting particular chemical reactions, proteins trigger the production of structural materials like carbohydrates and energy storage molecules like lipids. Because DNA encodes the information needed to assemble each protein present in an organism, it is the library of life.

4.1 Polymers Are Built of Monomers

An **organic molecule** is a molecule formed by living organisms that consists of a carbon-based core with special groups attached. These groups of atoms have special chemical properties and are referred to as *functional groups.* Functional groups tend to act as units during chemical reactions and to confer specific chemical properties on the molecules that possess them. Five principal functional groups are listed in figure 4.1; the last column indicates the types of organic molecules that contain these functional groups.

The bodies of organisms contain thousands of different kinds of organic molecules, but much of the body is made of just four kinds: *proteins, nucleic acids, carbohydrates,* and *lipids.* Called **macromolecules** because they can be very large, these four are the building materials of cells, the "bricks and mortar" that make up the bodies of cells and the machinery that runs within them.

The body's macromolecules are assembled by sticking smaller bits, called **monomers,** together, much as a train is built by linking rail-cars together. A molecule built up of long chains of similar subunits is called a **polymer**. Table 4.1 lists the monomers in the first column that make up the polymers that are the basis for many cellular structures.

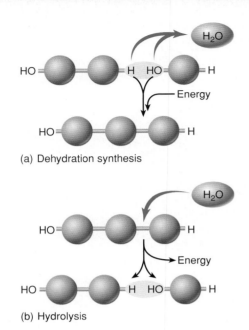

(a) Dehydration synthesis

(b) Hydrolysis

Figure 4.2   Dehydration and hydrolysis.
(a) Biological molecules are formed by linking subunits. The covalent bond between subunits is formed in dehydration synthesis, a process during which a water molecule is eliminated. (b) Breaking such a bond requires the addition of a water molecule, a reaction called hydrolysis.

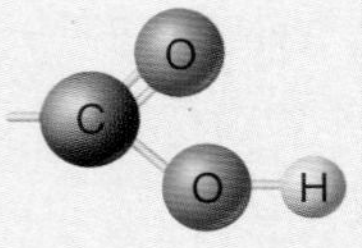

Group	Structural Formula	Ball-and-Stick Model	Found In
Hydroxyl	—OH	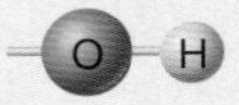	Carbohydrates
Carbonyl	C=O		Lipids
Carboxyl	—C with O and OH		Proteins
Amino	—N with H and H		Proteins
Phosphate	$-O-P-O^-$ with O^- and O		DNA, ATP

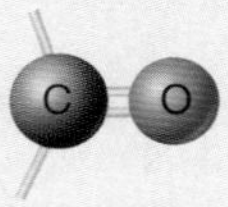

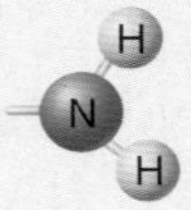

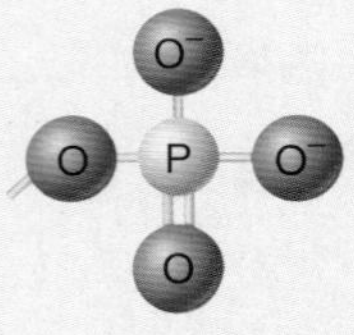

Figure 4.1 Five principal functional groups.
Most chemical reactions that occur within organisms involve transferring a functional group from one molecule to another or breaking a carbon-carbon bond.

Making (and Breaking) Macromolecules

The four different kinds of macromolecules (proteins, nucleic acids, carbohydrates, and lipids) all put their subunits together in the same way: a covalent bond is formed between two subunits in which a hydroxyl group (OH) is removed from one subunit and a hydrogen (H) is removed from the other. This process that is illustrated in figure 4.2*a*, is called **dehydration synthesis** because, in effect, the removal of the OH and H groups (highlighted by the blue oval) constitutes removal of a molecule of water—the word *dehydration* means "taking away water." This process requires the help of a special class of proteins called **enzymes** to facilitate the positioning of the molecules so that the correct chemical bonds are stressed and broken. The process of tearing down a molecule, such as the protein or fat contained in food that is consumed, is essentially the reverse of dehydration synthesis: instead of removing a water molecule, one is added. When a water molecule comes in, as shown in figure 4.2*b*, a hydrogen becomes attached to one subunit (on the left side of the bond) and a hydroxyl to another (on the right side), and the covalent bond is broken. The breaking up of a polymer in this way is called **hydrolysis**.

> **4.1   Macromolecules are formed by linking subunits together into long chains, removing a water molecule as each link is formed.**

TABLE 4.1 MACROMOLECULES

Monomer	Polymer	Cellular structure
Amino Acid	Polypeptide	Intermediate filament

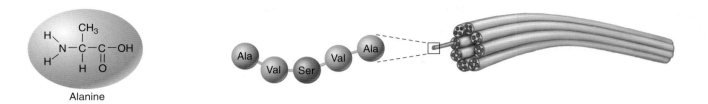

Alanine

Nucleotide	DNA strand	Chromosome

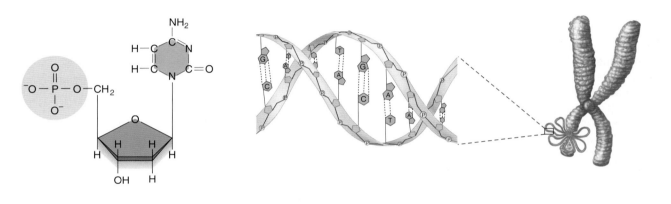

Monosaccharide	Starch	Starch grains in a chloroplast

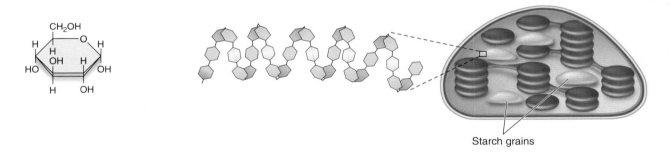

Starch grains

Fatty acid	Fat molecule	Adipose cells with fat droplets

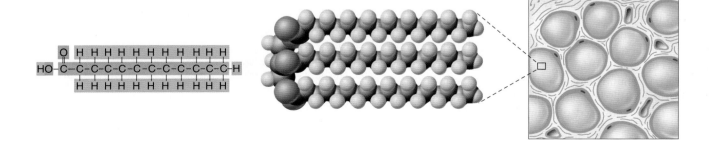

4.2 Proteins

Complex macromolecules called **proteins** are a major group of biological macromolecules within the bodies of organisms. Perhaps the most important proteins are *enzymes,* which have the key role in cells of lowering the energy required to initiate particular chemical reactions. Other proteins play structural roles. Cartilage, bones, and tendons all contain a structural protein called collagen. Keratin, another structural protein, forms the horns of a rhinoceros and the feathers of a bird. Still other proteins act as chemical messengers within the brain and throughout the body. Table 4.2 presents you with an overview of the wide-ranging functions of proteins.

Amino Acids

Despite their diverse functions, all proteins have the same basic structure: a long polymer chain made of subunits called amino acids. **Amino acids** are small molecules with a simple basic structure: a central carbon atom to which an amino group (–NH₂), a carboxyl group (–COOH), a hydrogen atom (H), and a functional group, designated "R," are bonded. There are 20 common kinds of amino acids that differ from one another by the identity of their functional R group.

The 20 amino acids are classified into four general groups with representative amino acids shown in figure 4.3 (their R groups are highlighted in white). Six of the amino acids are nonpolar, differing chiefly in size—the most bulky contain ring structures (like phenylalanine in the upper left), and amino acids containing them are called *aromatic.* Another six are polar but uncharged (like asparagine in the upper right), and these differ from one another in the strength of their polarity. Five more are polar and are capable of ionizing to a charged form (like aspartic acid in the lower left). The remaining three possess special chemical groups (like the white highlighted area of proline in the lower right) that are important in forming links between protein chains or in forming kinks in their shapes. The polarity of the R groups is important to the proper folding of the protein into its functional shape, which is discussed later.

Protein Structure

An individual protein is made by linking specific amino acids together in a particular order, just as a word is made by linking a specific sequence of letters of the alphabet together in a particular order. The covalent bond linking two amino acids together is called a **peptide bond** and forms by dehydration synthesis. Recall from section 4.1 that in a dehydration synthesis, water is formed as a by-product of the reaction. You can see in figure 4.4 that a water molecule is released as the polypeptide chain forms.

Long chains of amino acids linked by peptide bonds are called **polypeptides.** The hemoglobin proteins in your red blood cells are each composed of four polypeptide chains.

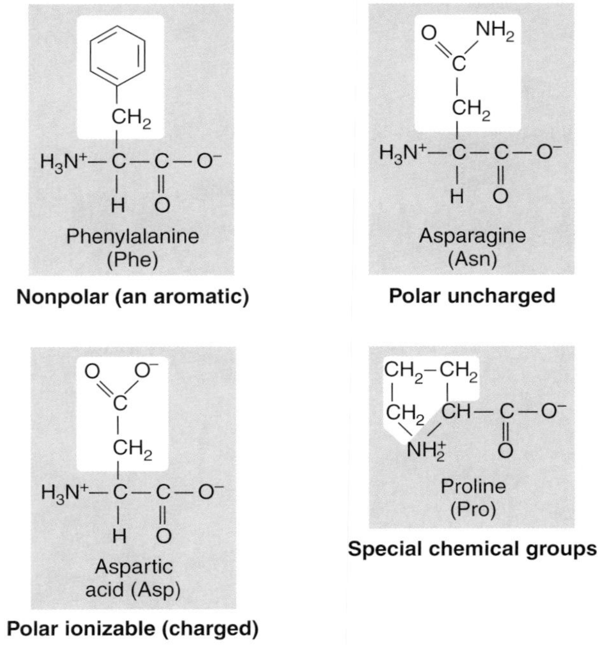

Figure 4.3   Examples of amino acids.
There are four general groups of amino acids that differ in their functional groups (highlighted in *white*).

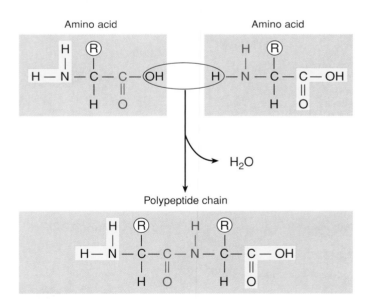

Figure 4.4   The formation of a peptide bond.
Every amino acid has the same basic structure, with an amino group (–NH₂) at one end and a carboxyl group (–COOH) at the other. The only variable is the functional, or "R," group. Amino acids are linked by dehydration synthesis to form peptide bonds. Chains of amino acids linked in this way are called polypeptides and are the basic structural components of proteins.

TABLE 4.2	THE MANY FUNCTIONS OF PROTEINS		
Function	**Class of Protein**	**Examples**	**Examples of Use**
Enzyme catalysis	Enzymes	Hydrolytic enzymes	Cleave macromolecules
		Proteases	Break down proteins
		Polymerases	Produce nucleic acids
		Kinases	Phosphorylate sugars and proteins
Defense	Immunoglobulins	Antibodies	Mark foreign proteins for elimination
	Toxins	Snake venom	Block nerve function
	Cell surface antigens	MHC proteins	"Self" recognition
Transport	Circulating transporters	Hemoglobin	Carries O_2 and CO_2 in blood
		Myoglobin	Carries O_2 and CO_2 in muscle
		Cytochromes	Electron transport
	Membrane transporters	Sodium-potassium pump	Excitable membranes
		Proton pump	Chemiosmosis
		Glucose transporter	Transport sugar into cells
Support	Fibers	Collagen	Forms cartilage
		Keratin	Forms hair, nails
		Fibrin	Forms blood clots
Motion	Muscle	Actin	Contraction of muscle fibers
		Myosin	Contraction of muscle fibers
Regulation	Osmotic proteins	Serum albumin	Maintains osmotic concentration of blood
	Gene regulators	*lac* repressor	Regulates transcription
	Hormones	Insulin	Controls blood glucose levels
		Vasopressin	Increases water retention by kidneys
		Oxytocin	Regulates uterine contractions and milk let down
Storage	Ion binding	Ferritin	Stores iron, especially in spleen
		Casein	Stores ions in milk
		Calmodulin	Binds calcium ions

The hemoglobin functions as a carrier of oxygen from your lungs to the cells of your body.

Some proteins form long, thin fibers, whereas others are globular, with the long strands of polypeptides coiled up and folded back on themselves or intertwined with other polypeptides. The shape of a protein is very important because it determines the protein's function. If we picture a polypeptide as a long strand similar to a reed, a protein might be the basket woven from it. Importantly, the sequence of amino acids in a polypeptide determines the protein's structure. There are four general levels of protein structure, primary, secondary, tertiary, and quaternary (figure 4.5); all are ultimately determined by the sequence of amino acids.

Primary Structure. The sequence of amino acids of a polypeptide chain is termed the polypeptide's **primary structure.** The amino acids are linked together by peptide bonds, forming long chains like the "beaded strand" at the top of figure 4.5. The primary structure of a protein, the sequence of its amino acids, determines all other levels of protein structure.

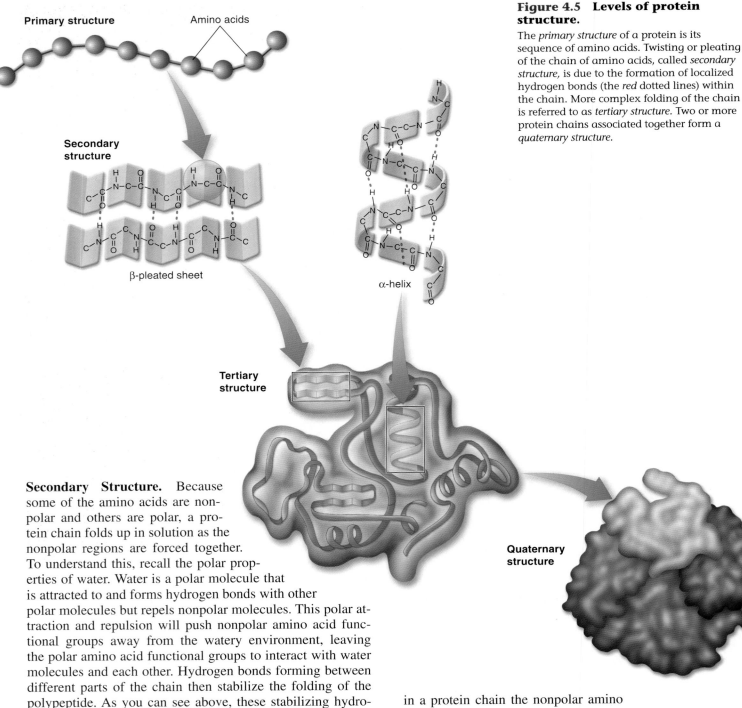

Primary structure

Amino acids

Secondary structure

β-pleated sheet

α-helix

Tertiary structure

Quaternary structure

Figure 4.5 Levels of protein structure.

The *primary structure* of a protein is its sequence of amino acids. Twisting or pleating of the chain of amino acids, called *secondary structure,* is due to the formation of localized hydrogen bonds (the *red* dotted lines) within the chain. More complex folding of the chain is referred to as *tertiary structure.* Two or more protein chains associated together form a *quaternary structure.*

Secondary Structure. Because some of the amino acids are nonpolar and others are polar, a protein chain folds up in solution as the nonpolar regions are forced together. To understand this, recall the polar properties of water. Water is a polar molecule that is attracted to and forms hydrogen bonds with other polar molecules but repels nonpolar molecules. This polar attraction and repulsion will push nonpolar amino acid functional groups away from the watery environment, leaving the polar amino acid functional groups to interact with water molecules and each other. Hydrogen bonds forming between different parts of the chain then stabilize the folding of the polypeptide. As you can see above, these stabilizing hydrogen bonds, indicated by red dotted lines, do not involve the R groups themselves, but rather the polypeptide backbone. This initial folding is called the **secondary structure** of a protein. Hydrogen bonding within this secondary structure can fold the polypeptide into coils, called α-helices, and sheets, called β-pleated sheets.

Tertiary Structure. The final three-dimensional shape, or **tertiary structure,** of the protein, folded and twisted in the case of a globular molecule, is determined by exactly where

in a protein chain the nonpolar amino acids occur. Again, the repulsion of the nonpolar amino acids by water will force these amino acids toward the interior of the globular protein, leaving the polar amino acids exposed to the outside.

Quaternary Structure. When a protein is composed of more than one polypeptide chain, the spatial arrangement of the several component chains is called the **quaternary structure** of the protein, like the four subunits that make up the quaternary structure of the protein in figure 4.5.

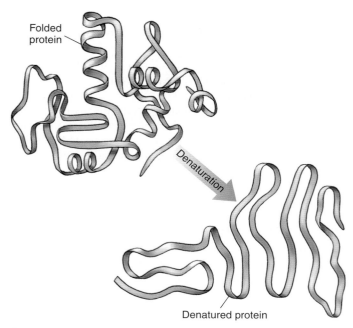

Folded protein

Denaturation

Denatured protein

Figure 4.6   Protein denaturation.
Changes in a protein's environment, such as variations in temperature or pH, can cause a protein to unfold and lose its shape in a process called denaturation. In this denatured state, proteins are biologically inactive.

How Proteins Fold into Their Functional Shape

The polar nature of the watery environment in the cell influences how the polypeptide folds into the functional protein. The protein in figure 4.6 is folded in such a way that allows it to carry out its function. If the polar nature of the protein's environment changes by either increasing temperature or lowering pH, both of which alter hydrogen bonding, the protein may unfold, as in the lower right of the figure. When this happens the protein is said to be **denatured.** When the polar nature of the solvent is reestablished, some proteins may spontaneously refold.

Many structural proteins form long cables that have architectural roles in cells, providing strength and determining shape. Some of these structural proteins are no doubt familiar to you as pictured in figure 4.7. The globular proteins called *enzymes* have three-dimensional shapes with grooves or depressions that precisely fit a particular sugar or other chemical (like the red molecule binding to the enzyme in figure 4.8); once in the groove, the chemical is encouraged to undergo a reaction—often, one of its chemical bonds is stressed as the chemical is bent by the enzyme, like a foot in a flexing shoe. This process of enhancing chemical reactions is called **catalysis,** and proteins are the catalytic agents of cells, determining what chemical processes take place and where and when.

(a)

(b)

(c)

(d)

Figure 4.7   Structural proteins.
This class of proteins has a wide variety of functions. (a) Fibrin: traps red blood cells in forming a blood clot; (b) silk: forms a spider's web; (c) keratin: in human hair; (d) keratin: in a peacock's feather.

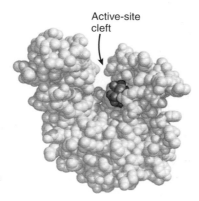

Active-site cleft

Figure 4.8   Globular protein: an enzyme.
This enzyme (*blue*) has a deep groove that binds a specific chemical (*red*) at a site on the enzyme called the active site. The bound molecule then undergoes a chemical reaction catalyzed by the enzyme.

Chaperone Proteins

How does a protein fold into a specific shape? As just discussed, nonpolar amino acids play a key role. Until recently, investigators thought that newly made proteins fold spontaneously as hydrophobic interactions with water shove nonpolar amino acids into the protein interior. We now know this is too simple a view. Protein chains can fold in so many different ways that trial and error would simply take too long. In addition, as the open chain folds its way toward its final form, nonpolar "sticky" interior portions are exposed during intermediate stages. If these intermediate forms are placed in a test tube in the same protein environment that occurs in a cell, they stick to other unwanted protein partners, forming a gluey mess.

How do cells avoid this? A vital clue came in studies of unusual mutations (changes in DNA) that prevented viruses from replicating in bacterial cells—it turned out the virus proteins could not fold properly! Further study revealed that normal cells contain special proteins called **chaperone proteins** that help new proteins fold correctly. When the bacterial gene encoding its chaperone protein is disabled by mutation, the bacteria die, clogged with lumps of incorrectly folded proteins. Fully 30% of the bacteria's proteins fail to fold to the right shape.

Molecular biologists have now identified more than 17 kinds of proteins that act as molecular chaperones. Many are heat shock proteins, produced in greater amounts if a cell is exposed to elevated temperature; high temperatures cause proteins to unfold, and heat shock chaperone proteins help the cell's proteins refold.

To understand how a chaperone works, examine figure 4.9 closely. The misfolded protein (the purple wormlike struc-ture) enters inside the chaperone. There, in a way not clearly understood, the visiting protein is induced to unfold, and then refold again, before it leaves. You can see in the third panel of the diagram the protein has unfolded into a long polypeptide chain. In the fourth panel, the polypeptide chain has then refolded into a different shape. The chaperone protein has in this way "rescued" a protein that was caught in a wrongly folded state, and given it another chance to fold correctly. To demonstrate this rescue capability, investigators "fed" a deliberately misfolded protein malate dehydrogenase to chaperone proteins; the malate dehydrogenase was rescued, refolding to its active shape.

Protein Folding and Disease

There are tantalizing suggestions that chaperone protein deficiencies may play a role in certain diseases by failing to facilitate the intricate folding of key proteins. Cystic fibrosis is a hereditary disorder in which a mutation disables a protein that plays a vital part in moving ions across cell membranes. In at least some cases, the vital membrane protein appears to have the correct amino acid sequence, but fails to fold to its final form. It has also been speculated that chaperone deficiency may be a cause of the protein clumping in brain cells that produces the amyloid plaques characteristic of Alzheimer's disease.

> **4.2** Proteins are chains of amino acids that fold into complex shapes. The sequence of its amino acid determines a protein's function. Chaperone proteins help newly produced proteins to fold properly.

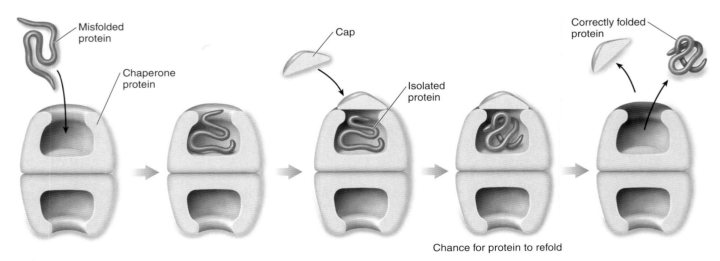

Chance for protein to refold

Figure 4.9 How one type of chaperone protein works.
This barrel-shaped chaperone protein is a heat shock protein, produced in elevated amounts at high temperatures. An incorrectly folded protein enters one chamber of the barrel, and a cap seals the chamber and confines the protein. The isolated protein is now prevented from aggregating with other misfolded proteins, and it has a chance to refold properly. After a short time period, the protein is ejected, folded or unfolded, and the cycle can repeat itself.

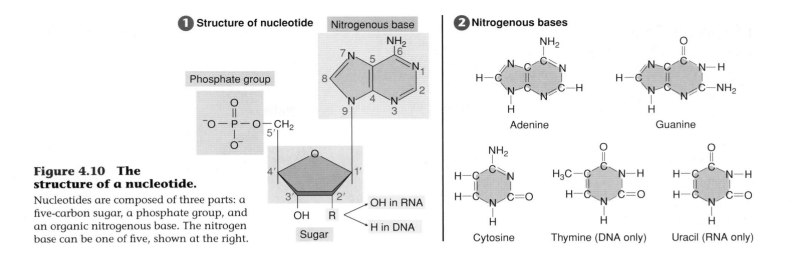

① Structure of nucleotide

Phosphate group

Nitrogenous base

Figure 4.10 The structure of a nucleotide.
Nucleotides are composed of three parts: a five-carbon sugar, a phosphate group, and an organic nitrogenous base. The nitrogen base can be one of five, shown at the right.

OH in RNA
H in DNA

Sugar

② Nitrogenous bases

Adenine

Guanine

Cytosine

Thymine (DNA only)

Uracil (RNA only)

4.3 Nucleic Acids

Very long polymers called nucleic acids serve as the information storage devices of cells, just as CDs or hard drives store the information that computers use. Nucleic acids are long polymers of repeating subunits called nucleotides. Each nucleotide is a complex organic molecule composed of three parts shown in figure 4.10 **①**: a five-carbon sugar (in blue), a phosphate group (in yellow, PO_4), and an organic nitrogen-containing base (in orange).

In the formation of a nucleic acid, the individual sugars are linked in a line by the phosphate groups in very long **polynucleotide chains:**

How does the long, chainlike structure of a nucleic acid permit it to store the information necessary to specify what a human being is like? If nucleic acids were simply a monotonous repeating polymer, it could not encode the message of life. Imagine trying to write a story using only the letter *E* and no spaces or punctuation. All you could ever say is "EEEEEEE. . . ." You need more than one letter to write—the English alphabet uses 26 letters. Nucleic acids can encode information because they contain more than one kind of nucleotide. There are five different kinds of nucleotides: two larger ones that contain the nitrogenous

bases adenine and guanine (shown in the top row of figure 4.10 **②**), and three smaller ones (in the bottom row) that contain the nitrogenous bases cytosine, thymine, and uracil. Nucleic acids encode information by varying the identity of the nucleotide at each position in the polymer.

DNA and RNA

Nucleic acids come in two varieties, **deoxyribonucleic acid (DNA)** and **ribonucleic acid (RNA),** both polymers of nucleotides with some differences. RNA is similar to DNA, but with two major chemical differences. First, RNA molecules contain ribose sugars in which four of the five carbons bond to a hydroxyl group (—OH). In DNA, one of the hydroxyl groups is replaced with a hydrogen atom (this is the carbon labeled 2′ in figure 4.10 **①**). Second, RNA molecules do not contain the thymine nucleotide; it uses uracil instead. Structurally, RNA is also different. RNA is a long, single strand of nucleotides and is used by cells in making proteins using genetic instructions encoded within DNA. The sequence of nucleotides in DNA determines the order of amino acids in the primary structure of the protein. DNA consists of *two* nucleotide strands wound around each other in a **double helix,** like strands of a pearl necklace twisted together. You can see this difference in structure by comparing the blue double-stranded DNA molecule in figure 4.11 with the green single-stranded RNA molecule.

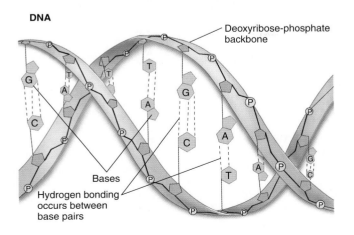

DNA

Deoxyribose-phosphate backbone

Bases

Hydrogen bonding occurs between base pairs

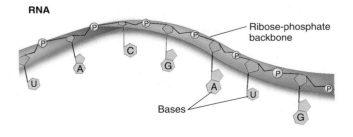

RNA

Ribose-phosphate backbone

Bases

Figure 4.11 How DNA structure differs from RNA.
DNA contains two nucleic acid strands wrapped around each other, while RNA is single-stranded.

The Double Helix

Why is DNA a *double* helix? When scientists looked carefully at the structure of the DNA double helix, they found that the bases of each chain point inward toward the other (like the DNA double strand shown in figure 4.12). The bases of the two chains are linked in the middle of the molecule by hydrogen bonds (the dotted lines between the two strands), like two columns of people holding hands across. The key to understanding why DNA is a double helix is revealed by looking at the bases: *only two base pairs are possible.* The distance between the two strands is consistent, this suggests that two big bases cannot pair together—the combination is simply too bulky to fit; similarly, two little ones cannot, as they pinch the helix inward too much. To form a double helix, it is necessary to pair a big base with a little one. *In every DNA double helix, adenine (A) pairs with thymine (T) and guanine (G) pairs with cytosine (C).* The reason A doesn't pair with C and G doesn't pair with T is that these base pairs cannot form proper hydrogen bonds—the electron-sharing atoms are not pointed at each other.

A and C cannot properly align to form hydrogen bonds.

G and T cannot properly align to form hydrogen bonds.

A and T can align to form two hydrogen bonds.

G and C can align to form three hydrogen bonds.

The simple A–T, G–C pairs within the DNA double helix allow the cell to copy the information in a very simple way. It just unzips the helix and adds the matching bases to each strand! That is the great advantage of a double helix—it actually contains two copies of the information, one the mirror image of the other. If the sequence of one chain is ATTGCAT, the sequence of its partner in the double helix *must* be TAACGTA. The fidelity with which hereditary information is passed from one generation to the next is a direct result of this simple double-entry bookkeeping, which makes accurate copying of the genetic message possible.

4.3 Nucleic acids like DNA are long chains of the nucleotides A, T, G, and C. The sequence of the nucleotides specifies the amino acid sequence of proteins.

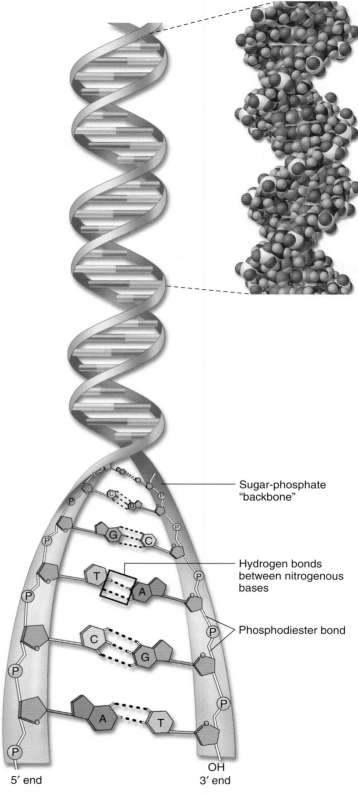

Sugar-phosphate "backbone"

Hydrogen bonds between nitrogenous bases

Phosphodiester bond

5′ end

3′ end
OH

Figure 4.12 The DNA double helix.

The DNA molecule is composed of two nucleotide chains twisted together to form a double helix. The two chains of the double helix are joined by hydrogen bonds between A–T and G–C base pairs. The section of DNA on the right is a space-filling model of DNA, where atoms are indicated by colored balls.

A Closer Look

Discovering the Structure of DNA

By the middle of the last century, biologists were increasingly sure that DNA was the molecule that stored the hereditary information, but investigators were puzzled over how such a seemingly simple molecule could carry out such a complex function.

A key observation was made by chemist Erwin Chargaff shortly after the end of the Second World War. He noted that in DNA molecules, the amount of adenine, A, always equals the amount of thymine, T, and the amount of guanine, G, always equals the amount of cytosine, C. This observation (A=T, G=C), known as Chargaff's rule, strongly suggested that DNA has a regular structure, but did not reveal what it was.

The significance of the regularities pointed out by Chargaff were not immediately obvious, but they became clear when a young British chemist, Rosalind Franklin, carried out an X-ray diffraction analysis of DNA. In X-ray diffraction, a molecule is bombarded with a beam of X rays. When individual rays encounter atoms, their path is bent or diffracted, and the diffraction pattern is recorded on photographic film. The pattern that resulted using DNA resembled the ripples created by tossing a rock into a smooth lake. When carefully analyzed, a molecule's pattern can reveal information about the three-dimensional structure of the molecule.

X-ray diffraction works best on substances that can be prepared as perfectly regular crystalline arrays. However at the time of Franklin's analysis, it was impossible to obtain true crystals of natural DNA, so she had to use DNA in the form of fibers. Franklin worked in the same laboratory as Oxford biochemist Maurice Wilkins, who was able to prepare more uniformly oriented DNA fibers than anyone had previously. Using these fibers, Franklin succeeded in obtaining crude diffraction information on natural DNA. The diffraction patterns she obtained seemed to suggest that the DNA molecule had the shape of a coiled spring or corkscrew, a form called a helix.

Learning informally of Franklin's results before they were published in 1953, James Watson and Francis Crick, two young investigators at Cambridge University, quickly worked out a likely structure for the DNA molecule, which we now know was substantially correct. The key to their understanding the structure of DNA was Watson and Crick's insight that each DNA molecule is actually made up of two chains of nucleotides that are intertwined—a double helix.

Backbone. In Watson and Crick's historic 1953 model (in the photograph, Watson is peering at the model as Crick points), each DNA molecule is composed of two complementary polynucleotide strands that form a double helix, with the bases extending into the interior of the helix. An analogy that is often made is to a spiral staircase where the two strands of the double helix are the handrails on the staircase.

Complementarity. What holds the two strands together? Watson and Crick proposed that the bases from opposite strands can form hydrogen bonds with each other to join the two complementary strands. Although each individual base pair is of low energy, the sum of many base pairs has enough energy that the molecule is very stable. To return to our spiral staircase analogy, where the backbone is the handrails, the base pairs are the stairs themselves.

Because of differences in size and position of particular atoms, only two hydrogen bonding pairs are possible in such a double helix: adenine (A) can form hydrogen bonds with thymine (T), and guanine (G) can form hydrogen bonds with cytosine (C). The Watson-Crick model thus in a very direct and simple way explained what had until then been one of the great mysteries of DNA, Chargaff's observation that adenine and thymine always occur in the same proportions in any DNA molecule, as do guanine and cytosine.

At the heart of the Watson-Crick model of DNA is a seemingly simple concept with some profound implications. Because only two base pairs are possible, if we know the sequence of one strand, we automatically know the sequence of the other strand; wherever there is an A in one strand, there must be a T in the other, and wherever there is a G in one strand, there must be a C in the other. This concept of a mirror-image relationship is called *complementary.* You see the importance: if more than two base pairs were possible, then the sequence of one DNA strand would not allow us to know for sure the sequence of the other. It is this fundamental insight that has made Watson and Crick's discovery one of the most profound of the twentieth century.

Watson and Crick continued on in the area of DNA research, but Franklin's career was cut short by her untimely death due to cancer at the age of 37.

4.4 Carbohydrates

Polymers called **carbohydrates** make up the structural framework of cells and play a critical role in energy storage. A carbohydrate is any molecule that contains carbon, hydrogen, and oxygen in the ratio 1:2:1. Some carbohydrates are simple, small monomers or dimers and are called **simple carbohydrates.** Others are long polymers and are called **complex carbohydrates.** Because they contain many carbon-hydrogen (C–H) bonds, carbohydrates are well-suited for energy storage. Such C–H bonds are the ones most often broken by organisms to obtain energy. Table 4.3 on the facing page shows some examples.

Simple Carbohydrates

The simplest carbohydrates are the *simple sugars* or **monosaccharides** (from the Greek *monos,* single, and *saccharon,* sweet). These molecules consist of one subunit. For example, glucose, the sugar that carries energy to the cells of your body,

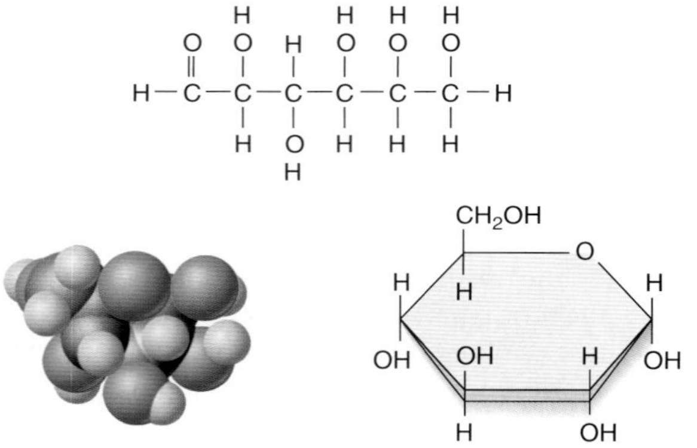

Figure 4.13 The structure of glucose.
Glucose is a monosaccharide and consists of a linear six-carbon molecule that forms a ring when added to water. This illustration shows three ways glucose can be represented diagrammatically.

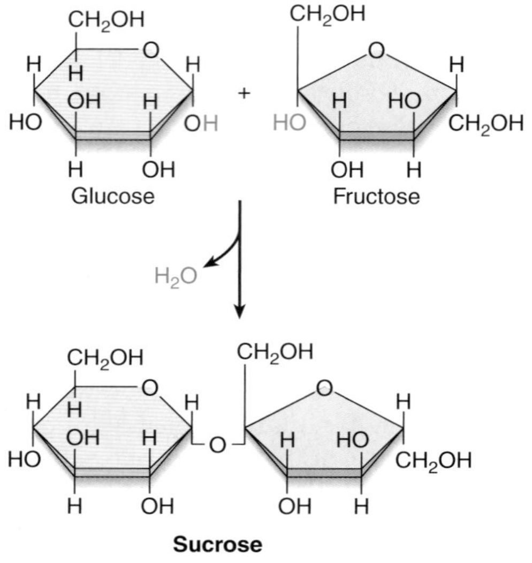

Figure 4.14 Formation of sucrose.
The disaccharide sucrose is formed from glucose and fructose in a dehydration reaction.

is made of six carbons and has the chemical formula $C_6H_{12}O_6$. A molecule of glucose is pictured in several ways in figure 4.13. The long chain of carbon atoms at the top of the figure is its formal chemical structure. When placed in water, the chain folds into the ring structure shown on the lower right. The individual atoms are depicted in the "3-D" space-filling model you see in the lower left. Another type of simple carbohydrate is a **disaccharide,** which forms when two monosaccharides link together through a dehydration reaction. In figure 4.14 you can see how the disaccharide sucrose (table sugar) is made by linking two six-carbon sugars together, a glucose (orange) and a fructose (green).

Complex Carbohydrates

Organisms store their metabolic energy by converting sugars, which are soluble, into insoluble forms that can be deposited in specific storage areas in the body. This trick is achieved by linking the sugars together into long polymer chains called **polysaccharides.** Plants and animals store energy in polysaccharides formed from glucose. The glucose polysaccharide that plants use to store energy is called **starch**—that is why potatoes are referred to as "starchy" food. In animals, energy is stored in **glycogen,** a highly insoluble macromolecule formed of glucose polysaccharides that are very long and highly branched. Plants and animals also use glucose chains as building materials, linking the subunits together in different orientations not recognized by most enzymes. These structural polysaccharides are chitin, in animals, and **cellulose,** in plants. The cellulose deposited in the cell walls of the plant cells, like those wrapping around the caption in figure 4.15 below, cannot be digested by humans and makes up the fiber in our diets.

> **4.4 Carbohydrates are molecules made of C, H, and O atoms. As sugars they store energy in C–H bonds.**

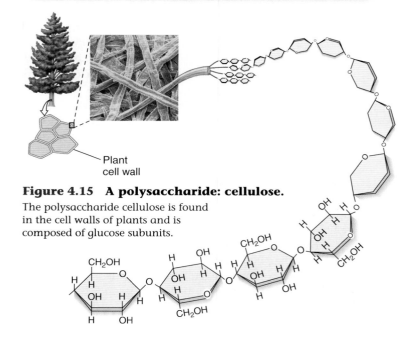

Figure 4.15 A polysaccharide: cellulose.
The polysaccharide cellulose is found in the cell walls of plants and is composed of glucose subunits.

TABLE 4.3   CARBOHYDRATES AND THEIR FUNCTION

Carbohydrate	Example	Description

Transport Disaccharides

Lactose

Glucose is transported within some organisms as a disaccharide. In this form, it is less readily metabolized because the normal glucose-utilizing enzymes of the organism cannot break the bond linking the two monosaccharide subunits. One type of disaccharide is called lactose. Many mammals supply energy to their young in the form of lactose.

Storage Polysaccharides

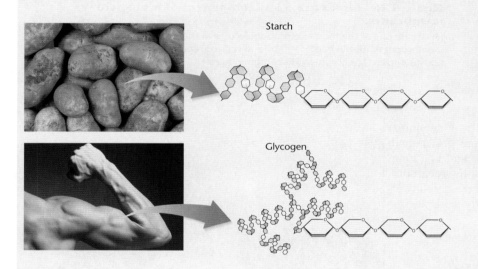

Starch

Organisms store energy in long chains of glucose molecules called polysaccharides. The chains tend to coil up in water, making them insoluble and ideal for storage. The storage polysaccharides found in plants are called starches, which can be branched or unbranched.

Glycogen

In animals, glucose is stored as glycogen. Glycogen is similar to starch in that it consists of long chains of glucose that coil up in water and are insoluble. But glycogen chains are much longer and highly branched.

Structural Polysaccharides

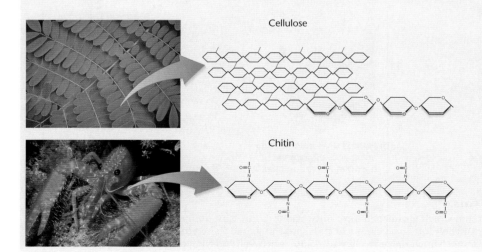

Cellulose

Cellulose is a structural polysaccharide found in the cell walls of plants; its glucose subunits are joined in a way that ca nnot be broken down readily. Cleavage of the links between the glucose subunits in cellulose requires an enzyme most organisms lack. Some animals, such as cows, are able to digest cellulose by means of bacteria and protists they harbor in their digestive tract, which provide the necessary enzymes.

Chitin

Chitin is a type of structural polysaccharide found in the external skeletons of many invertebrates, including insects and crustaceans, and in the cell walls of fungi. Chitin is a modified form of cellulose with a nitrogen group added to the glucose units. When cross-linked by proteins, it forms a tough, resistant surface material.

4.5 Lipids

For long-term storage, organisms usually convert glucose into fats, another kind of storage molecule that contains more energy-rich C–H bonds than carbohydrates. Fats and all other biological molecules that are not soluble in water but soluble in oil are called **lipids.** Lipids are insoluble in water not because they are long chains like starches but rather because they are nonpolar. In water, fat molecules cluster together because they cannot form hydrogen bonds with water molecules. This is why oil forms into a layer on top of water when the two substances are mixed. This is also what drives the formation of membranes that surround cells. A section of a cell membrane is pictured in figure 4.16. Lipids called phospholipids make up the two layers of the membrane (the right side is peeled apart), while another lipid, cholesterol, can be seen as the yellow structures embedded within the layers of the membrane. Membranes are discussed in more detail in chapter 5.

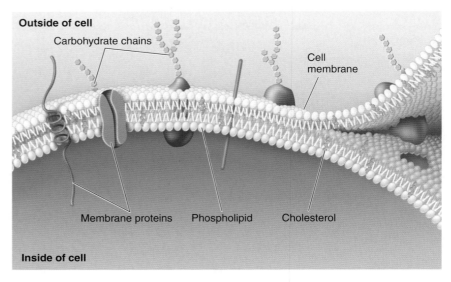

Figure 4.16   Lipids are a key component of biological membranes.

Lipids are one of the most common molecules in the human body, because the membranes of all the body's 100 trillion cells are composed largely of lipids called phospholipids. Membranes also contain cholesterol, another type of lipid.

Fats

Fat molecules are lipids composed of two kinds of subunits: fatty acids (the gray boxed structures in figure 4.17a) and glycerol (the orange boxed structure). A **fatty acid** is a long chain of carbon and hydrogen atoms (called a hydrocarbon) ending in a carboxyl (—COOH) group. The three carbons of glycerol form the backbone to which three fatty acids are at-

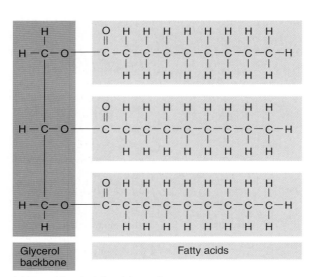

(a) Fat molecule (triacylglycerol)

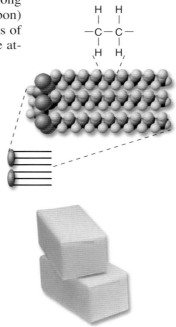

(b) Hard fat (saturated): Fatty acids with single bonds between all carbon pairs

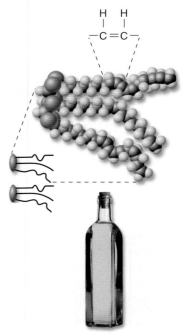

(c) Oil (unsaturated): Fatty acids that contain double bonds between one or more pairs of carbon atoms

Figure 4.17   Saturated and unsaturated fats.

(a) Fat molecules each contain a three-carbon glycerol to which is attached three fatty acid tails. (b) Most animal fats are "saturated" (every carbon atom carries the maximum load of hydrogens). Their fatty acid chains fit closely together, and these triacylglycerols form immobile arrays called hard fats. (c) Most plant fats are unsaturated, which prevents close association between triacylglycerols and produces oils.

of cells. In 1838, botanist Matthias Schleiden made a careful study of plant tissues and developed the first statement of the cell theory. He stated that all plants "are aggregates of fully individualized, independent, separate beings, namely the cells themselves." In 1839, Theodor Schwann reported that all animal tissues also consist of individual cells.

The idea that all organisms are composed of cells is called the **cell theory.** In its modern form, the cell theory includes three principles:

1. All organisms are composed of one or more cells, within which the processes of life occur.
2. Cells are the smallest living things. Nothing smaller than a cell is considered alive.
3. Cells arise only by division of a previously existing cell. Although life likely evolved spontaneously in the environment of the early earth, biologists have concluded that no additional cells are originating spontaneously at present. Rather, life on earth represents a continuous line of descent from those early cells.

Most Cells Are Very Small

Cells are not all the same size. Individual marine alga cells, for example, can be up to 5 centimeters long—as long as your little finger. In contrast, the cells of your body are typically from 5 to 20 micrometers (μm) in diameter, too small to see with the naked eye. It would take anywhere from 100 to 400 human cells to span the diameter of the head of a pin. The cells of bacteria are even smaller than yours, only a few micrometers thick.

Why Aren't Cells Larger?

Why are most cells so tiny? Most cells are small because larger cells do not function as efficiently. In the center of every cell is a command center that must issue orders to all parts of the cell, directing the synthesis of certain enzymes, the entry of ions and molecules from the exterior, and the assembly of new cell parts. These orders must pass from the core to all parts of the cell, and it takes them a very long time to reach the periphery of a large cell. For this reason, an organism made up of relatively small cells is at an advantage over one composed of larger cells.

Another reason cells are not larger is the advantage of having a greater **surface-to-volume ratio.** As cell size increases, volume grows much more rapidly than surface area. For a round cell, surface area increases as the square of the diameter, whereas volume increases as the cube. To visualize this, consider the two single cells in figure 5.2. The large cell to the right is 10 times bigger than the small cell, but while its surface area is 100 times greater (10^2), its volume is 1,000 (10^3) times the volume of the smaller cell. A cell's surface provides the interior's only opportunity to interact with the environment with substances passing into and out of the cell across its surface, and large cells have far less surface for each unit of volume than do small ones.

Some larger cells, however, function quite efficiently in part because they have structural features that increase surface area. Cells in the nervous system, for example, called neurons,

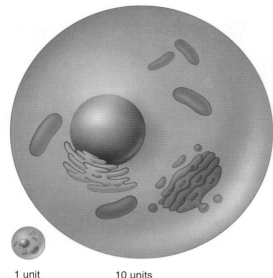

Cell radius (r)	1 unit	10 units
Surface area ($4\pi r^2$)	12.57 units^2	1,257 units^2
Volume ($\frac{4}{3}\pi r^3$)	4.189 units^3	4,189 units^3

Figure 5.2 Surface-to-volume ratio.
As a cell gets larger, its volume increases at a faster rate than its surface area. If the cell radius increases by 10 times, the surface area increases by 100 times, but the volume increases by 1,000 times. A cell's surface area must be large enough to meet the needs of its volume.

are long slender cells, some extending more than a meter in length. These cells efficiently interact with their environment because although they are long, they are thin, some less than 1 micrometer in diameter, and so their interior regions are not far from the surface at any given point.

Another structural feature that increases the surface area of a cell are small "fingerlike" projections called microvilli. The cells that line the small intestines of the human digestive system are covered with microvilli that dramatically increase the surface area of the cells.

With few exceptions however, cells don't usually grow much larger than 50 micrometers (a micrometer is one-millionth of a meter). For organisms to get much larger, they are usually composed of many cells. By grouping together many smaller cells, these multicellular organisms vastly increase their total surface-to-volume ratio.

An Overview of Cell Structure

All cells are surrounded by a delicate membrane, called a *plasma membrane*, that controls the permeability of the cell to water and dissolved substances. A semifluid matrix called *cytoplasm* fills the interior of the cell. It used to be thought that the cytoplasm was uniform, like Jell-O, but we now know that it is highly organized. Your cells, for example, have an internal framework that both gives the cell its shape and positions components and materials within its interior. In the following sections, we explore the membranes that encase all living cells and then examine in detail their interiors.

Visualizing Cells

How many cells are big enough to see with the unaided eye? Other than egg cells, not many are (figure 5.3). Most are less than 50 micrometers in diameter, far smaller than the period at the end of this sentence.

The Resolution Problem. How do we study cells if they are too small to see? The key is to understand why we can't see them. The reason we can't see such small objects is the limited resolution of the human eye. **Resolution** is defined as the minimum distance two points can be apart and still be distinguished as two separated points. On the visibility scale in figure 5.3 below, you can see that the limit of resolution of the human eye (the blue bar at the bottom) is about 100 micrometers. This limit occurs because when two objects are closer together than about 100 micrometers, the light reflected from each strikes the same "detector" cell at the rear of the eye. Only when the objects are farther apart than 100 micrometers will the light from each strike different cells, allowing your eye to resolve them as two objects rather than one.

Microscopes. One way to increase resolution is to increase magnification, so that small objects appear larger. Robert Hooke and Antony van Leeuwenhoek used glass lenses to magnify small cells and cause them to appear larger than the 100-micrometer limit imposed by the human eye. The glass lens adds additional focusing power. Because the glass lens makes the object appear closer, the image on the back of the eye is bigger than it would be without the lens.

Modern *light microscopes* use two magnifying lenses (and a variety of correcting lenses) to achieve very high magnification and clarity. The first lens focuses the image of the object on the second lens, which magnifies it again and focuses it on the back of the eye. Microscopes that magnify in stages using several lenses are called **compound microscopes.** They can resolve structures that are separated by more than 200 nanometers (nm). The six entries in the upper portion of table 5.1 are images viewed through various types of light microscopes.

Increasing Resolution. Light microscopes, even compound ones, are not powerful enough to resolve many structures within cells. For example, a membrane is only 5 nanometers thick. Why not just add another magnifying stage to the microscope and so increase its resolving power? Because when two objects are closer than a few hundred nanometers, the light beams reflecting from the two images start to overlap. The only way two light beams can get closer together and still be resolved is if their wavelengths are shorter.

One way to avoid overlap is by using a beam of electrons rather than a beam of light. Electrons have a much shorter wavelength, and a microscope employing electron beams has 1,000 times the resolving power of a light microscope. **Transmission electron microscopes (TEM),** so called because the electrons used to visualize the specimens are transmitted through the material, are capable of resolving objects only 0.2 nanometer apart—just twice the diameter of a hydrogen atom! The entry on the left under electron microscopes in table 5.1 is an example of an image captured using TEM.

A second kind of electron microscope, the **scanning electron microscope (SEM),** beams the electrons onto the surface of the specimen. The electrons reflected back from the surface of the specimen, together with other electrons that the specimen itself emits as a result of the bombardment, are amplified and transmitted to a screen, where the image can be viewed and photographed. Scanning electron microscopy yields striking three-dimensional images and has improved our understanding of many biological and physical phenomena. The entry on the right in table 5.1 under electron microscopes is an SEM image.

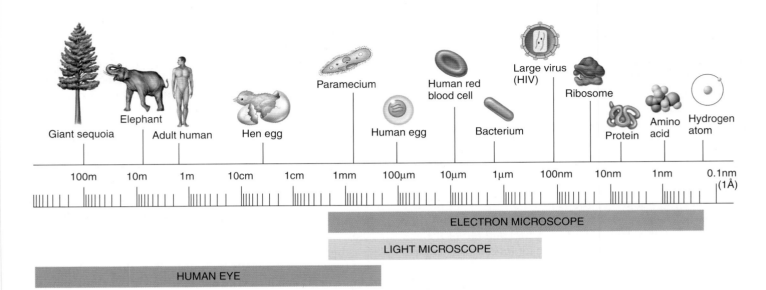

Figure 5.3 A scale of visibility.
Most cells are microscopic in size, although vertebrate eggs are typically large enough to be seen with the unaided eye. Prokaryotic cells are generally 1 to 2 micrometers (μm) across.

TABLE 5.1   TYPES OF MICROSCOPES

Light Microscopes

Bright-field microscope: Light is simply transmitted through a specimen in culture, giving little contrast. Staining specimens improves contrast but requires that cells be fixed (not alive), which can cause distortion or alteration of components.

28.36 µm

Dark-field microscope: Light is directed at an angle toward the specimen; a condenser lens transmits only light reflected off the specimen. The field is dark, and the specimen is light against this dark background.

67.74 µm

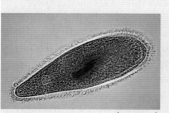

Phase-contrast microscope: Components of the microscope bring light waves out of phase, which produces differences in contrast and brightness when the light waves recombine.

32.81 µm

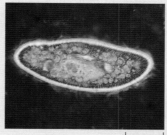

Differential-interference-contrast microscope: Out-of-phase light waves to produce differences in contrast are combined with two beams of light travelling close together, which create even more contrast, especially at the edges of structures.

26.6 µm

Fluorescence microscope: A set of filters transmits only light that is emitted by fluorescently stained molecules or tissues.

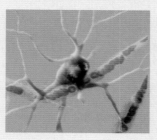

Confocal microscope: Light from a laser is focused to a point and scanned across the specimen in two directions. Clear images of one plane of the specimen are produced, while other planes of the specimen are excluded and do not blur the image. Fluorescent dyes and false coloring enhances the image.

Electron Microscopes

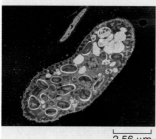

Transmission electron microscope: A beam of electrons is passed through the specimen. Electrons that pass through are used to form an image. Areas of the specimen that scatter electrons appear dark. False coloring enhances the image.

2.56 µm

Scanning electron microscope: An electron beam is scanned across the surface of the specimen, and electrons are knocked off the surface. Thus, the surface topography of the specimen determines the contrast and the content of the image. False coloring enhances the image.

6.76 µm

Visualizing Cell Structure by Staining Specific Molecules. A powerful tool for the analysis of cell structure has been the use of stains that bind to specific molecular targets. This approach has been used in the analysis of tissue samples, or histology, for many years and has been improved dramatically with the use of antibodies that bind to very specific molecular structures. This process, called immunocytochemistry, uses antibodies generated in animals such as rabbits or mice. When these animals are injected with specific proteins, they will produce antibodies that specifically bind to the injected protein, which can be purified from their blood. These purified antibodies can then be chemically bonded to enzymes, stains, or fluorescent molecules that glow when exposed to specific wavelengths of light. When cells are washed in a solution containing the antibodies, they bind to cellular structures that contain the target molecule and can be seen with light microscopy. The image produced using fluorescence microscopy in table 5.1 shows the cytoskeleton made of cablelike structures inside the cell. This approach has been used extensively in the analysis of cell structure and function.

5.1 All living things are composed of one or more cells, each a small volume of cytoplasm surrounded by a cell membrane. Most cells and their components are so small they can only be viewed using microscopes.

5.2 The Plasma Membrane

Encasing all living cells is a delicate sheet of molecules called the **plasma membrane.** It would take more than 10,000 of these molecular sheets, which are about 5 nanometers thick, piled on top of one another to equal the thickness of this sheet of paper. However, the sheets are not simple in structure, like a soap bubble's skin. Rather, they are made up of a diverse collection of proteins floating within a lipid framework like small boats bobbing on the surface of a pond. Regardless of the kind of cell they enclose, all plasma membranes have the same basic structure of proteins embedded in a sheet of lipids, called the **fluid mosaic model.**

The lipid layer that forms the foundation of a plasma membrane is composed of modified fat molecules called **phospholipids.** A phospholipid molecule can be thought of as a polar head with two nonpolar tails attached to it. The head of a phospholipid molecule has a phosphate chemical group linked to it—the yellow sphere in figure 5.4*a*—making it extremely polar (and thus water-soluble). The other end of the phospholipid molecule is composed of two long fatty acid chains. Recall from chapter 4 that fatty acids are long chains of carbon atoms with attached hydrogen atoms. The carbon atoms are the gray spheres you see in figure 5.4*a*. The fatty acid tails are strongly nonpolar and thus water-insoluble.

Imagine what happens when a collection of phospholipid molecules is placed in water. A structure called a **lipid bilayer** forms spontaneously. How can this happen? The long nonpolar tails of the phospholipid molecules are pushed away by the water molecules that surround them, shouldered aside as the water molecules seek partners that can form hydrogen bonds. After much shoving and jostling, every phospholipid molecule ends up with its polar head facing water and its nonpolar tail facing away from water. The phospholipid molecules form a *double* layer, called a bilayer. The shaded blue areas in figure 5.5 represent watery environments inside and outside the plasma membrane that push the nonpolar tails to the

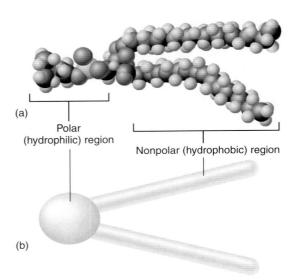

(a)

Polar (hydrophilic) region

Nonpolar (hydrophobic) region

(b)

Figure 5.4   Phospholipid structure.
One end of a phospholipid molecule is polar and the other is nonpolar. (a) The molecular structure is shown with colored spheres representing individual atoms (*black* for carbon, *blue* for hydrogen, *red* for oxygen, and *yellow* for phosphorus). (b) The phospholipid is often depicted diagrammatically as a ball with two tails.

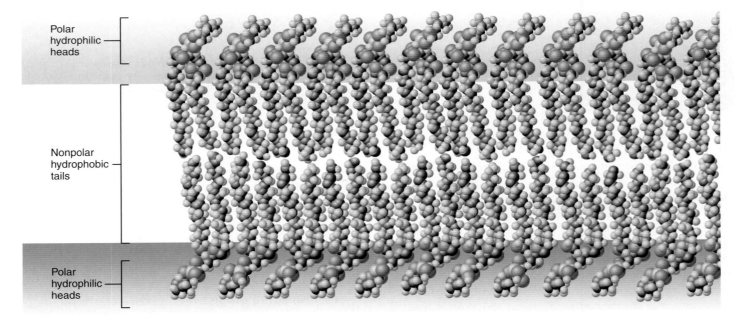

Polar hydrophilic heads

Nonpolar hydrophobic tails

Polar hydrophilic heads

Figure 5.5   The lipid bilayer.
The basic structure of every plasma membrane is a double layer of lipids. This diagram illustrates how phospholipids aggregate to form a bilayer with a nonpolar interior when placed in water.

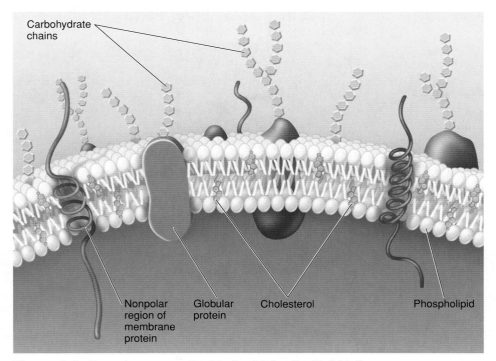

Figure 5.6 Proteins are embedded within the lipid bilayer.
A variety of proteins protrude through the lipid bilayer. Membrane proteins function as channels, receptors, and cell surface markers. Carbohydrate chains are often bound to these proteins and to phospholipids in the membrane. These chains serve as distinctive identification tags, unique to particular types of cells.

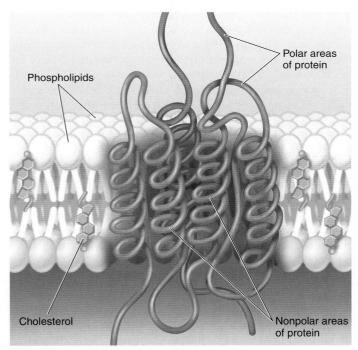

Figure 5.7 Nonpolar regions lock proteins into membranes.
A spiral helix of nonpolar amino acids (*red*) extends across the nonpolar lipid interior, while polar (*purple*) portions of the protein protrude out from the bilayer. The protein cannot move in or out because such a movement would drag polar segments of the protein into the nonpolar interior of the membrane.

interior of the bilayer. Because there are two layers with the tails facing each other, no tails are ever in contact with water. Because the interior of a lipid bilayer is completely nonpolar, it repels any water-soluble molecules that attempt to pass through it, just as a layer of oil stops the passage of a drop of water (that's why ducks do not get wet).

Cholesterol, another nonpolar lipid molecule, resides in the interior portion of the bilayer. Cholesterol is a multiringed molecule that affects the fluid nature of the membrane. Although cholesterol is important in maintaining the integrity of the plasma membrane, it can accumulate in blood vessels, forming plaques that lead to cardiovascular disease.

Proteins Within the Membrane

The second major component of every biological membrane is a collection of **membrane proteins** that float within the lipid bilayer. As you can see in figure 5.6, some proteins (the purple structures) pass through the lipid bilayer, providing channels through which molecules and information pass. While some membrane proteins are fixed into position, others move about freely.

Many membrane proteins project up from the surface of the plasma membrane like buoys, often with carbohydrate chains (the orange chains in figure 5.6) or lipids attached to their tips like flags. These **cell surface proteins** act as markers to identify particular types of cells, or as beacons to bind specific hormones or proteins to the cell.

Protein channels that extend all the way across the bilayer provide passageways for ions and polar molecules like water so they can pass into and out of the cell. How do these **transmembrane proteins** manage to span the membrane, rather than just floating on the surface in the way that a drop of water floats on oil? The part of the protein that actually traverses the lipid bilayer is a specially constructed spiral helix of nonpolar amino acids—the red coiled areas of the transmembrane protein you see in figure 5.7. Water responds to these nonpolar amino acids much as it does to nonpolar lipid chains, and as a result the helical spiral is held within the lipid interior of the bilayer, anchored there by the strong tendency of water to avoid contact with these nonpolar amino acids.

> **5.2** All cells are encased within a delicate lipid bilayer sheet, the plasma membrane, within which are embedded a variety of proteins that act as markers or channels through the membrane.

Today's *Biology*

Membrane Defects Can Cause Disease

The year 1993 marked an important milestone in the treatment of human disease. That year the first attempt was made to cure **cystic fibrosis** (**CF**), a deadly genetic disorder, by transferring healthy genes into sick individuals. Cystic fibrosis is a fatal disease in which the body cells of affected individuals secrete a thick mucus that clogs the airways of the lungs. The cystic fibrosis patient in the photograph is breathing into a Vitalograph, a device that measures lung function. These same secretions block the ducts of the pancreas and liver so that the few patients who do not die of lung disease die of liver failure. Cystic fibrosis is usually thought of as a children's disease because until recently few affected individuals lived long enough to become adults. Even today half die before their mid-twenties. There is no known cure.

Cystic fibrosis results from a defect in a single gene that is passed down from parent to child. It is the most common fatal genetic disease of Caucasians. One in 20 individuals possesses at least one copy of the defective gene. Most of these individuals are not afflicted with the disease; only those children who inherit a copy of the defective gene from each parent succumb to cystic fibrosis—about 1 in 2,500 infants.

Cystic fibrosis has proven difficult to study. Many organs are affected, and until recently it was impossible to identify the nature of the defective gene responsible for the disease. In 1985 the first clear clue was obtained. An investigator, Paul Quinton, seized on a commonly observed characteristic of cystic fibrosis patients, that their sweat is abnormally salty, and performed the following experiment. He isolated a sweat duct from a small piece of skin and placed it in a solution of salt (NaCl) that was three times as concentrated as the NaCl inside the duct. He then monitored the movement of ions. Diffusion tends to drive both the sodium (Na^+) and the chloride (Cl^-) ions into the duct because of the higher outer ion concentrations. In skin isolated from normal individuals, Na^+ and Cl^- both entered the duct, as expected. In skin isolated from cystic fibrosis individuals, however, only Na^+ entered the duct—no Cl^- entered. For the first time, the molecular nature of cystic fibrosis became clear. Water accompanies chloride, and was not entering the ducts because chloride was not, creating thick mucus. Cystic fibrosis is a defect in a plasma membrane protein called CFTR (cystic fibrosis *t*ransmembrane *c*onductance *r*egulator) that normally regulates passage of Cl^- into and out of the body's cells.

The defective *cf* gene was isolated in 1987, and its position on a particular human chromosome (chromosome 7) was pinpointed in 1989. Interestingly, many cystic fibrosis patients produce a CFTR protein with a normal amino acid sequence. The *cf* mutation in these cases appears to interfere with how the CFTR protein folds, preventing it from folding into a functional shape.

Soon after the *cf* gene was isolated, experiments were begun to see if it would be possible to cure cystic fibrosis by gene therapy—that is, by transferring healthy *cf* genes into the cells with defective ones. In 1990 a working *cf* gene was successfully transferred into human lung cells growing in tissue culture, using adenovirus, a cold virus, to carry the gene into the cells. The CFTR-defective cells were "cured," becoming able to transport chloride ions across their plasma membranes. Then in 1991 a team of researchers successfully transferred a normal human *cf* gene into the lung cells of a living animal—a rat. The *cf* gene was first inserted into the adenovirus genome because adenovirus is a cold virus and easily infects lung cells. The treated virus was then inhaled by the rat. Carried piggyback, the *cf* gene entered the rat lung cells and began producing the normal human CFTR protein within these cells!

These results were very encouraging, and at first the future for all cystic fibrosis patients seemed bright. Clinical tests using adenovirus to introduce healthy *cf* genes into cystic fibrosis patients were begun with much fanfare in 1993.

They were not successful. As described in detail in chapter 14, there were insurmountable problems with the adenovirus being used to transport the *cf* gene into cystic fibrosis patients. The difficult and frustrating challenge that cystic fibrosis researchers had faced was not over. Research into clinical problems is often a time-consuming and frustrating enterprise, never more so than in this case. Recently, as chapter 14 recounts, new ways of introducing the healthy *cf* gene have been tried with better results. The long, slow journey toward a cure has taught us not to leap to the assumption that a cure is now at hand, but the steady persistence of researchers has taken us a long way, and again the future for cystic fibrosis patients seems bright.

5.3 Prokaryotic Cells

There are two major kinds of cells: prokaryotes and eukaryotes. **Prokaryotes** have a relatively uniform cytoplasm that is not subdivided by interior membranes into separate compartments. They do not, for example, have special membrane-bounded compartments, called *organelles,* or a *nucleus* (a membrane-bounded compartment that holds the hereditary information). All bacteria and archaea are prokaryotes; all other organisms are eukaryotes.

Prokaryotes are the simplest cellular organisms. Over 5,000 species are recognized, but doubtless many times that number actually exist and have not yet been described. Although these species are diverse in form, their organization is fundamentally similar: small cells typically about 1 to 10 micrometers thick; enclosed like all cells by a plasma membrane, but with no distinct interior compartments. Outside of almost all bacteria is a *cell wall,* a framework of carbohydrates cross-linked into a rigid structure. In some bacteria another layer called the *capsule* encloses the cell wall. Bacterial cells are single-celled organisms. They assume many shapes, like the sausage or spiral shapes shown in figure 5.8*a* and *b.* They can also adhere in chains and masses like the spherical cells in figure 5.8*c,* but in these cases the individual cells remain functionally separate from one another.

If you were able to magnify your vision and peer into a prokaryotic cell, you would be struck by its simple organization. The entire interior of the cell, the blue cytoplasm in figure 5.9, is one unit, with little or no internal support structure (the rigid wall, the purple layer surrounding the cell, supports the cell's shape) and no internal compartments bounded by membranes. Scattered throughout the cytoplasm of prokaryotic cells are small structures called *ribosomes,* the small spherical structures you see inside the cell in figure 5.9. Ribosomes are the sites where proteins are made, but they are not considered organelles because they lack a membrane boundary. The DNA is found in a region of the cytoplasm called the *nucleoid region.* Although the DNA is localized in this region of the cytoplasm, it is not considered a nucleus because, as you can see in figure 5.9, the nucleoid region and its associated DNA are not enclosed within an internal membrane.

Some prokaryotes use a *flagellum* (plural, *flagella*) to move. Flagella are long, threadlike structures, made of protein fibers that project from the surface of a cell. They are used in locomotion and feeding. There may be none, one, or more per cell depending on the species. Bacteria can swim at speeds up to 20 cell diameters per second, rotating their flagella like screws.

Pili (singular, **pilus**) are short flagella (only several micrometers long, and about 7.5 to 10 nanometers thick) that occur on the cells of some prokaryotes. Pili help the prokaryotic cells attach to appropriate substrates and aid in the exchange of genetic information between cells.

> **5.3** Prokaryotic cells lack a nucleus and do not have an extensive system of interior membranes.

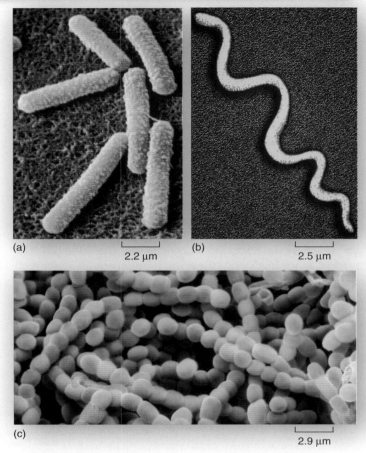

(a) ⊢ 2.2 µm ⊣ (b) ⊢ 2.5 µm ⊣

(c) ⊢ 2.9 µm ⊣

Figure 5.8 Bacterial cells have different shapes.

(a) *Bacillus* is a rod-shaped bacterium. (b) *Treponema* is a coil-shaped bacterium; rotation of internal filaments produces a cork-screw movement. (c) *Streptomyces* is a more or less spherical bacterium in which the individuals adhere in chains.

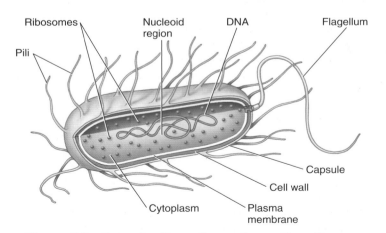

Figure 5.9 Organization of a prokaryotic cell.

Prokaryotic cells lack internal compartments. Not all prokaryotic cells have a flagellum or a capsule like the one illustrated here, but all do have a nucleoid region, ribosomes, a plasma membrane, cytoplasm, and a cell wall.

5.4 Eukaryotic Cells

For the first 1 billion years of life on earth, all organisms were prokaryotes, cells with very simple interiors. Then, about 1.5 billion years ago, a new kind of cell appeared for the first time, the eukaryotic cell. Eukaryotic cells are much larger and profoundly different from prokaryotic cells, with a complex interior organization. All cells alive today except bacteria and archaea are of this new kind.

Figures 5.10 and 5.11 present cross-sectional diagrams of idealized animal and plant cells. As you can see, the interior of a eukaryotic cell is much more complex than the prokaryotic cell you encountered in figure 5.9. The **plasma membrane** ❶ encases a semifluid matrix called the **cytoplasm** ❷, which contains within it the nucleus and various cell structures called organelles. An **organelle** is a specialized structure within which particular cell processes occur. Each organelle, such as a mitochondrion ❸, has a specific function in the eukaryotic cell. The organelles are anchored at specific locations in the cytoplasm by an interior scaffold of protein fibers, the **cytoskeleton** ❹.

One of the organelles is very visible when these cells are examined with a microscope, filling the center of the cell like the pit of a peach. Seeing it, the English botanist Robert Brown in 1831 called it the **nucleus** ❺ (plural, *nuclei*), from the Latin word for "kernel." Inside the nucleus, the DNA is wound tightly around proteins and packaged into compact units called chromosomes. It is the nucleus that gives **eukaryotes** their name, from the Greek words *eu,* true, and *karyon,* nut; by way of contrast, the earlier-evolving bacteria and archaea are called prokaryotes ("before the nut").

If you examine the organelles in figures 5.10 and 5.11, you can see that most of them form separate compartments within the cytoplasm, bounded by their own membranes. *The hallmark of the eukaryotic cell is this compartmentalization.* This internal compartmentalization is achieved by an extensive **endomembrane system** ❻ that weaves through the cell interior, providing extensive surface area for many membrane-associated cell processes to occur.

Vesicles ❼ (small membrane-bounded sacs that store and transport materials) form in the cell either by budding off of the endomembrane system or by the incorporation of lipids and protein in the cytoplasm. These many closed-off compartments allow different processes to proceed simultaneously without interfering with one another, just as rooms do in a house. Thus the

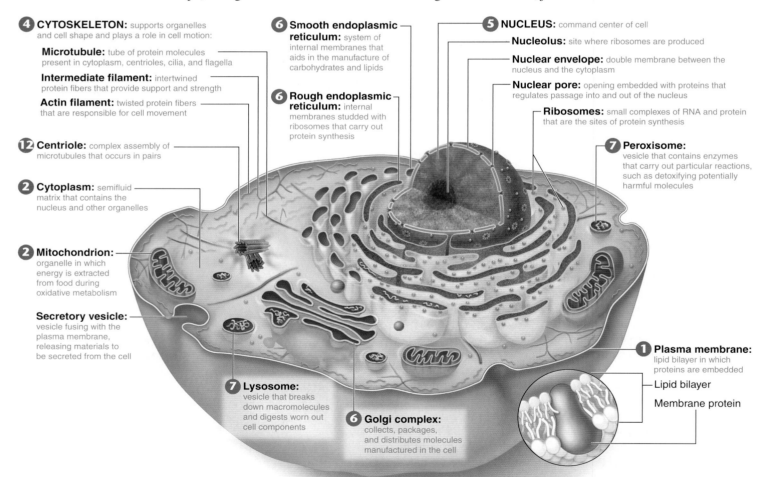

❹ CYTOSKELETON: supports organelles and cell shape and plays a role in cell motion:

Microtubule: tube of protein molecules present in cytoplasm, centrioles, cilia, and flagella

Intermediate filament: intertwined protein fibers that provide support and strength

Actin filament: twisted protein fibers that are responsible for cell movement

⓬ Centriole: complex assembly of microtubules that occurs in pairs

❷ Cytoplasm: semifluid matrix that contains the nucleus and other organelles

❷ Mitochondrion: organelle in which energy is extracted from food during oxidative metabolism

Secretory vesicle: vesicle fusing with the plasma membrane, releasing materials to be secreted from the cell

❻ Smooth endoplasmic reticulum: system of internal membranes that aids in the manufacture of carbohydrates and lipids

❻ Rough endoplasmic reticulum: internal membranes studded with ribosomes that carry out protein synthesis

❼ Lysosome: vesicle that breaks down macromolecules and digests worn out cell components

❻ Golgi complex: collects, packages, and distributes molecules manufactured in the cell

❺ NUCLEUS: command center of cell

Nucleolus: site where ribosomes are produced

Nuclear envelope: double membrane between the nucleus and the cytoplasm

Nuclear pore: opening embedded with proteins that regulates passage into and out of the nucleus

Ribosomes: small complexes of RNA and protein that are the sites of protein synthesis

❼ Peroxisome: vesicle that contains enzymes that carry out particular reactions, such as detoxifying potentially harmful molecules

❶ Plasma membrane: lipid bilayer in which proteins are embedded

Lipid bilayer

Membrane protein

Figure 5.10 Structure of an animal cell.

In this generalized diagram of an animal cell, the plasma membrane encases the cell, which contains the cytoskeleton and various cell organelles and interior structures suspended in a semifluid matrix called the cytoplasm. Some kinds of animal cells possess fingerlike projections called microvilli. Other types of eukaryotic cells, for example many protist cells, may possess flagella, which aid in movement, or cilia, which can have many different functions.

5.9 Outside the Plasma Membrane

Cell Walls Offer Protection and Support

Plants, fungi, and many protists cells share a characteristic with bacteria that is not shared with animal cells—that is, they have **cell walls,** which protect and support their cells. Eukaryotic cell walls are chemically and structurally different from bacterial cell walls. In plants, cell walls are composed of fibers of the polysaccharide cellulose, while in fungi they are composed of chitin. The **primary walls** of plant cells are laid down when the cell is still growing. These are the thinner, outer walls of the cells shown below in figure 5.24. Between the walls of adjacent cells is a sticky substance called the **middle lamella,** which glues the cells together. Some plant cells produce strong **secondary walls,** which are deposited inside the primary walls. As you can see in the photo, the secondary cell walls are very thick compared to the primary walls and therefore are not deposited until the cell has finished increasing in size.

An Extracellular Matrix Surrounds Animal Cells

As we discussed, many types of eukaryotic cells possess a cell wall exterior to the plasma membrane. The wall acts to protect the cell, maintain its shape, and prevent excessive water uptake. Animal cells are the great exception, lacking the cell walls that encase the cells of plants, fungi, and most protists. Animal cells secrete an elaborate mixture of **glycoproteins** (proteins with short chains of sugars attached to them) into the space around them, forming the **extracellular matrix** (**ECM**), which performs a function different than cell walls.

The fibrous protein collagen, the same protein in fingernails and hair, is abundant in the ECM. Figure 5.25 shows how these fibers of collagen and another fibrous protein, elastin, are embedded within a complex web of other glycoproteins called proteoglycans, which form a protective layer over the cell surface.

The ECM is attached to the plasma membrane by a third kind of glycoprotein, **fibronectin.** As you can see in the figure, fibronectin molecules bind not only to ECM glycoproteins but also to proteins called **integrins,** which are an integral part of the plasma membrane. Integrins extend into the cytoplasm, where they are attached to the microfilaments of the cytoskeleton. Linking ECM and cytoskeleton, integrins allow the ECM to influence cell behavior in important ways, altering gene expression and cell migration patterns by a combination of mechanical and chemical signaling pathways. In this way, the ECM can help coordinate the behavior of all the cells in a particular tissue.

> **5.9** Plant and protist cells encase themselves within a strong cell wall. In animal cells, which lack a cell wall, the cytoskeleton is linked by integrin proteins to a web of glycoproteins called the extracellular matrix.

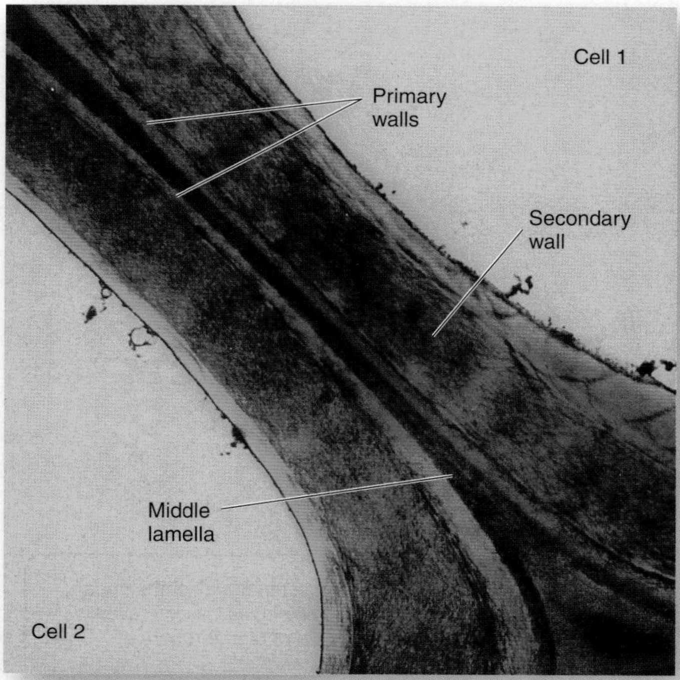

Figure 5.24   Cell walls in plants.

Plant cell walls are thick, strong, and rigid. Primary cell walls are laid down when the cell is young. Thicker secondary cell walls may be added later when the cell is fully grown. The middle lamella lies between the walls of adjacent cells and glues the cells together.

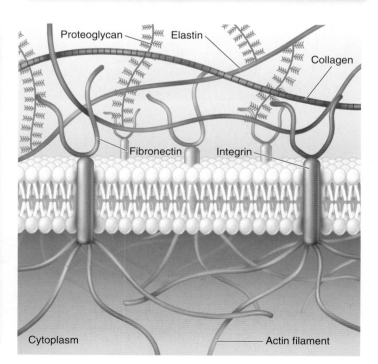

Figure 5.25   The extracellular matrix.

Animal cells are surrounded by an extracellular matrix composed of various glycoproteins that give the cells support, strength, and resilience.

5.10 Diffusion and Osmosis

For cells to survive, food particles, water, and other materials must pass into the cell, and waste materials must be eliminated. All of this moving back and forth across the cell's plasma membrane occurs in one of three ways: (1) water and other substances diffuse through the membrane, (2) food particles and sometimes liquids are engulfed by the membrane folding around them, or (3) proteins in the membrane act as doors that admit certain molecules only. First we will examine diffusion.

Diffusion

How a molecule moves—just where it goes—is totally random, like shaking marbles in a cup, so if two kinds of molecules are added together, they soon mix. The random motion of molecules always tends to produce uniform mixtures when a substance moves from regions where its concentration is high to regions where its concentration is lower (that is, *down* the **concentration gradient**). How does a molecule "know" in what direction to move? It doesn't—molecules don't "know" anything. A molecule is equally likely to move in any direction and is constantly changing course in random ways. There are simply more molecules able to move from where they are common than from where they are scarce. This mixing process is called **diffusion.** Diffusion is the net movement of molecules down a concentration gradient toward regions of lower concentration (that is, where there are relatively fewer of them) as a result of random motion. For example, the lump of sugar dropped into a beaker of water in figure 5.26 will break apart into individual sugar molecules that will move about randomly. However, they will tend to travel away from the area of high concentration (the sugar cube) to an area of lower concentration (the rest of the beaker). Eventually, the substance will achieve a state of **equilibrium,** where there is no net movement toward any particular direction (as shown in panel 4). The individual molecules of the substance are still in motion, but there is no net change.

Osmosis

Diffusion allows molecules like oxygen, carbon dioxide, and nonpolar lipids to cross the plasma membrane. Ions and polar molecules, by contrast, cannot cross the very nonpolar environment found in the lipid core of the membrane bilayer. However, the movement of water molecules, which are very polar, is not blocked—water diffuses freely across the plasma membrane. How is this possible? Water molecules pass

Diffusion

1 Lump of sugar

A lump of sugar is dropped into a beaker of water.

2 Sugar molecule

Sugar molecules begin to break off from the lump.

3

More and more sugar molecules move away and randomly bounce around.

4

Eventually, all of the sugar molecules become evenly distributed throughout the water.

Figure 5.26 How diffusion works.
Diffusion is the mixing process that spreads molecules through the cell interior. To see how diffusion works, visualize a simple experiment in which a lump of sugar is dropped into a beaker of water.

Osmosis

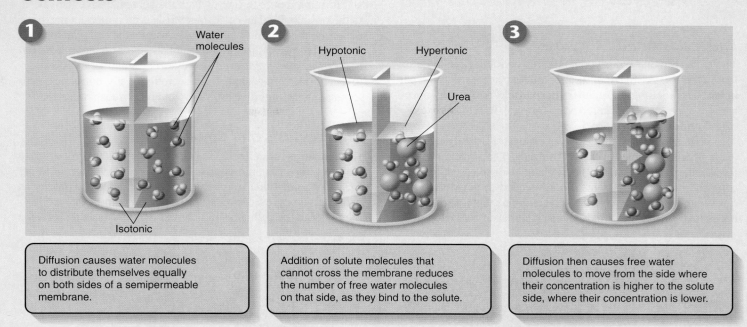

Figure 5.27   How osmosis works.
Osmosis is the net movement of water across a membrane toward the side with less "free" water.

through small channels, called **aquaporins,** that traverse the membrane. These water channels are very selective, even blocking the passage of protons (hydrogen ions), which are smaller than water molecules. This selectivity is due to a cluster of positively charged amino acids that line the pore and repel protons, which are also positively charged.

As in diffusion, water passes into and out of a cell down its concentration gradient, a process called **osmosis.** However, the movement of water into and out of a cell is dependent upon the concentration of other substances in solution. To understand how water moves into and out of a cell, let's focus on the water molecules already present inside a cell. What are they doing? Many of them are interacting with the sugars, proteins, and other polar molecules inside. Remember, water is very polar itself and readily interacts with other polar molecules. These social water molecules are not randomly moving about as they were outside; instead, they remain clustered around the polar molecules they are interacting with. As a result, while water molecules keep coming into the cell by random motion, they don't randomly come out again. The simple experiment in figure 5.27 illustrates what happens. Think of the right side of the beaker as the inside of a cell, and the left side is a watery environment. When the polar molecule urea is present in the cell, water molecules cluster around each urea molecule and are no longer able to pass through the membrane to the "outside." In effect, the polar solute has reduced the number of free water molecules. Because the "outside" of the cell (on the left) has more unbound water molecules, water moves by diffusion into the cell (from the left to the right).

The concentration of *all* molecules dissolved in a solution (the **solutes**) is called the osmotic concentration of the solution. If two solutions have unequal osmotic concentrations, the solution with the higher solute concentration, like the right side of the beaker above, is said to be **hypertonic** (Greek *hyper,* more than), and the solution with the lower one, like the left side of the beaker, is **hypotonic** (Greek *hypo,* less than). If the osmotic concentrations of the two solutions are equal, the solutions are **isotonic** (Greek *iso,* the same).

Movement of water into a cell by osmosis creates pressure, called **osmotic pressure,** which can cause a cell to swell and burst. Most animal cells cannot withstand osmotic pressure unless their plasma membranes are braced to resist the swelling. If placed in pure water, they soon burst like overinflated balloons. That is why the cells of so many kinds of organisms have cell walls to stiffen their exteriors. In fact, this osmotic pressure, called *turgor pressure* in plants, is important for plant cells to maintain their shape. Without adequate water inside the cells, the plants wilt. In animals, the fluids bathing the cells have as many polar molecules dissolved in them as the cells do, so the problem doesn't arise.

> **5.10 Random movements of molecules cause them to mix uniformly in solution, a process called diffusion. Water associated with polar solutes is not free to diffuse, and there is a net movement of water across a membrane toward the side with less "free" water, a process called osmosis.**

Bulk Passage into and out of Cells

Endocytosis and Exocytosis

The cells of many eukaryotes take in food and liquids by extending their plasma membranes outward toward food particles. The membrane engulfs the particle and forms a vesicle—a membrane-bordered sac—around it. This process is called **endocytosis** (figure 5.28).

The reverse of endocytosis is **exocytosis,** the discharge of material from vesicles at the cell surface. The vesicle in figure 5.29 contains a substance to be discharged, or released, from the cell. The purple particles remain suspended in the vesicle as it fuses with the cell membrane. The membrane that forms the vesicle is made of phospholipids and as it comes in contact with the plasma membrane, the phospholipids of both membranes interact, forming a pore through which the contents leave the vesicle to the outside. In plant cells, exocytosis is an important means of exporting the materials needed to construct the cell wall through the plasma membrane. Among protists, the discharge of a contractile vacuole is a form of exocytosis. In animal cells, exocytosis provides a mechanism for secreting many hormones, neurotransmitters, digestive enzymes, and other substances.

Phagocytosis and Pinocytosis

If the material the cell takes in is particulate (made up of discrete particles), such as an organism, like the red bacterium in figure 5.28*a,* or some other fragment of organic matter, the process is called **phagocytosis** (Greek *phagein,* to eat, and *cytos,* cell). If the material the cell takes in is liquid or substances dissolved a

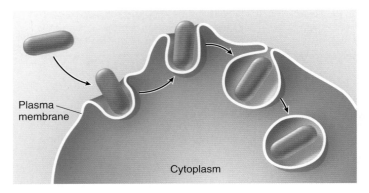

(a) Phagocytosis

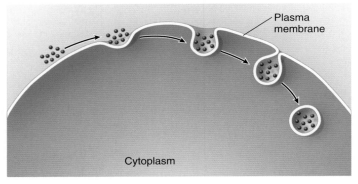

(b) Pinocytosis

Figure 5.28 Endocytosis.
Endocytosis is the process of engulfing material by folding the plasma membrane around it, forming a vesicle. (a) When the material is an organism or some other relatively large fragment of organic matter, the process is called phagocytosis. (b) When the material is a liquid, the process is called pinocytosis.

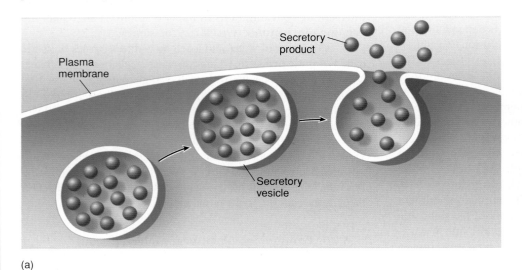

(a)

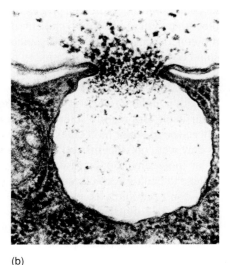

(b)

Figure 5.29 Exocytosis.
Exocytosis is the discharge of material from vesicles at the cell surface. (a) Proteins and other molecules are secreted from cells in small pockets called secretory vesicles, whose membranes fuse with the plasma membrane, thereby allowing the secretory vesicles to release their contents to the cell surface. (b) In the photomicrograph, you can see exocytosis taking place explosively.

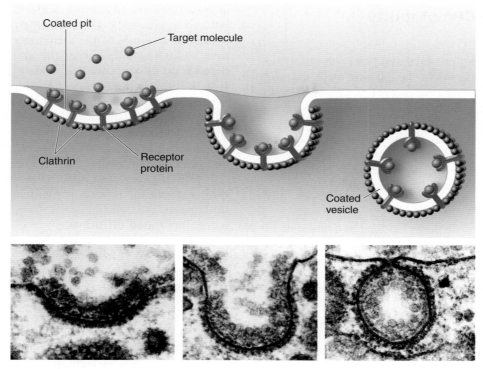

Figure 5.30 Receptor-mediated endocytosis.
Cells that undergo receptor-mediated endocytosis have pits coated with the protein clathrin that initiate endocytosis when target molecules bind to receptor proteins in the plasma membrane. In the photomicrographs, a coated pit appears in the plasma membrane of a developing egg cell, covered with a layer of proteins (80,000×). When an appropriate collection of molecules gathers in the coated pit, the pit deepens, and eventually seals off to form a vesicle.

liquid like the small particles in figure 5.28*b*, it is called **pinocytosis** (Greek *pinein,* to drink). Pinocytosis is common among animal cells. Mammalian egg cells, for example, "nurse" from surrounding cells; the nearby cells secrete nutrients that the maturing egg cell takes up by pinocytosis. Virtually all eukaryotic cells constantly carry out these kinds of endocytosis, trapping particles and extracellular fluid in vesicles and ingesting them. Endocytosis rates vary from one cell type to another. They can be surprisingly high: some types of white blood cells ingest 25% of their cell volume each hour!

Receptor-Mediated Endocytosis

Specific molecules are often transported into eukaryotic cells through **receptor-mediated endocytosis,** illustrated in figure 5.30. Molecules to be transported into the cell, the red balls in the figure, first bind to specific receptors in the plasma membrane. The transport process is specific to only molecules that have a shape that fits snugly into the receptor. The plasma membrane of a particular kind of cell contains a characteristic battery of receptor types, each for a different kind of molecule.

The portion of the receptor molecule inside the membrane is trapped in an indented pit coated with the protein clathrin, visible in the photos as well as in the drawing above. The pits act like molecular mousetraps, closing over to form an internal vesicle when the right molecule enters the pit. The

trigger that releases the trap is the binding of the properly fitted target molecule to a receptor embedded in the membrane of the pit. When binding occurs, the cell reacts by initiating endocytosis. The process is highly specific and very fast.

One type of molecule that is taken up by receptor-mediated endocytosis is called low-density lipoprotein (LDL). The LDL molecules bring cholesterol into the cell where it can be incorporated into membranes. Cholesterol plays a key role in determining the stiffness of the body's membranes. In the human genetic disease called hypercholesterolemia, the receptors lack tails and so are never caught in the clathrin-coated pits and, thus, are never taken up by the cells. The cholesterol stays in the bloodstream of affected individuals, coating their arteries and leading to heart attacks.

It is important to understand that receptor-mediated endocytosis in itself does not bring substances directly into the cytoplasm of a cell. The material taken in is still separated from the cytoplasm by the membrane of the vesicle.

5.11 The plasma membrane can engulf materials by endocytosis, folding the membrane around the material to encase it within a vesicle. Exocytosis is essentially this process in reverse, expelling substances using vesicles.

Selective Permeability

From the point of view of efficiency, the problem with endocytosis is that it is expensive to carry out—the cell must make and move a lot of membrane. Also, endocytosis is not picky—in pinocytosis particularly, engulfing liquid does not allow the cell to choose which molecules come in. Cells solve this problem by using proteins in the plasma membrane as channels to pass molecules into and out of the cell. Because each kind of channel allows passage of only a certain kind of molecule, the cell can control what enters and leaves, an ability called **selective permeability.**

Selective Diffusion

Some channels act like open doors. As long as a molecule fits the channel, it is free to pass through in either direction. Diffusion tends to equalize the concentration of such molecules on both sides of the membrane, with the molecules moving toward the side where they are scarcest. This mechanism of transport is called **selective diffusion.** One class of selectively open channels consists of ion channels, which are pores that span the membrane. Ions that fit the pore can diffuse through it in either direction. Such ion channels play an essential role in signaling by the nervous system.

Facilitated Diffusion

Most diffusion occurs through use of a special carrier protein. This protein binds only certain kinds of molecules, such as a particular sugar, amino acid, or ion (shown as the red balls in figure 5.31). The molecule physically binds to the carrier on one side of the membrane and is released to the other side. The direction of the molecule's net movement depends on its concentration gradient across the membrane. If the concentration is greater outside the cell, the molecule is more likely to bind to the carrier on the extracellular side of the membrane, as shown in panel 1, and be released on the cytoplasmic side, as in panel 3. If the concentration of the molecule is greater inside the cell, the net movement will be from inside to outside. Thus the net movement always occurs from high concentration to low, just as it does in simple diffusion, but the process is facilitated by the carriers. For this reason, this mechanism of transport is given a special name, **facilitated diffusion.**

A characteristic feature of transport by carrier proteins is that its rate can be saturated. If the concentration of a substance is progressively increased, the rate of transport of the substance increases up to a certain point and then levels off. There are a limited number of carrier proteins in the membrane, and when the concentration of the transported substance is raised high enough, all the carriers will be in use. The transport system is then said to be "saturated." When an investigator wishes to know if a particular substance is being transported across a membrane by a carrier system, or is diffusing across, he or she conducts experiments to see if the transport system can be saturated. If it can be saturated, it is carrier-mediated; if it cannot be saturated, it is not.

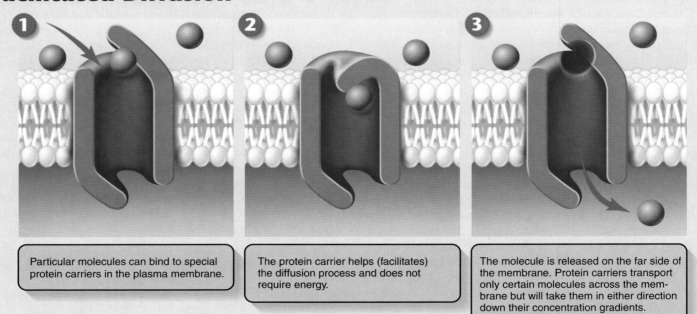

Facilitated Diffusion

1 Particular molecules can bind to special protein carriers in the plasma membrane.

2 The protein carrier helps (facilitates) the diffusion process and does not require energy.

3 The molecule is released on the far side of the membrane. Protein carriers transport only certain molecules across the membrane but will take them in either direction down their concentration gradients.

Figure 5.31   How facilitated diffusion works.

Active Transport

Other channels through the plasma membrane are closed doors. These channels open only when energy is provided. They are designed to enable the cell to maintain high or low concentrations of certain molecules, much more or less than exists outside the cell. If the doors were open, the molecules would simply flood in or out by facilitated diffusion. Instead, like motor-driven turnstiles, the channels operate only when energy is provided, and they move a certain substance only in one direction (*up* its concentration gradient). The operation of these one-way, energy-requiring channels results in **active transport,** the movement of molecules across a membrane to a region of higher concentration by the expenditure of energy.

You might think that the plasma membrane possesses all sorts of active transport channels for the transport of important sugars, amino acids, and other molecules, but in fact, almost all of the active transport in cells is carried out by only two kinds of channels, the sodium-potassium pump and the proton pump.

The Sodium-Potassium Pump. The first of these, the **sodium-potassium (Na+-K+) pump,** expends metabolic energy to actively pump sodium ions (Na+) in one direction, out of cells, and potassium ions (K+) in one direction, into cells. More than one-third of all the energy expended by your body's cells is spent driving Na+-K+ pump channels. This energy is derived from *adenosine triphosphate (ATP),* a molecule we will learn about in chapter 6. The transportation of two different ions in opposite directions happens because energy causes a change in the shape of the membrane protein carrier. Figure 5.32 walks you through one cycle of the pump. When the carrier is open to the inside of the cell, as in panel 1, Na+ binds to three binding sites on the carrier. Energy, supplied by the breaking of an ATP, as in panel 2, causes a change in the shape of the carrier such that it shifts to open to the outside. Once open to the outside of the cell, the Na+ ions leave the carrier and two K+ ions bind to the carrier, as in panel 3. The shape of the protein changes back, as in panel 4, and the K+ ions are released to the interior of the cell. Each channel can move over 300 sodium ions per second when working full tilt. As a result of all this pumping, there are far fewer sodium ions in the cell. This concentration gradient, paid for by the expenditure of considerable metabolic energy in the form of ATP molecules, is exploited by your cells in many ways. Two of the most important are (1) the conduction of signals along nerve cells (discussed in detail in chapter 30) and (2) the pulling of valuable molecules such as sugars and amino acids into the cell *against* their concentration gradient!

We will focus for a moment on this second process. The plasma membranes of many cells are studded with facilitated diffusion channels, which offer a path for sodium

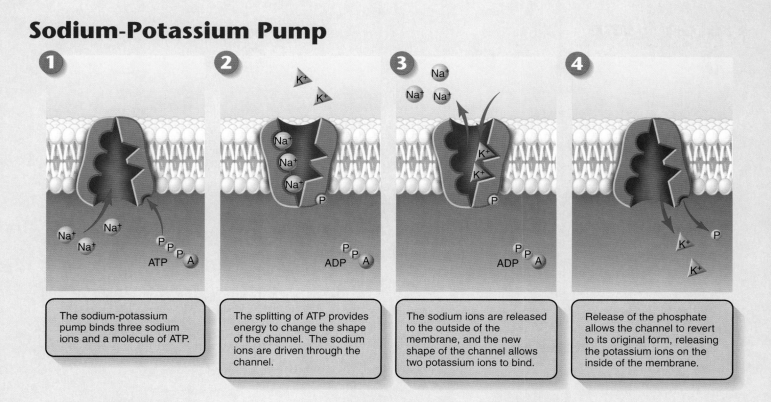

Sodium-Potassium Pump

1 The sodium-potassium pump binds three sodium ions and a molecule of ATP.

2 The splitting of ATP provides energy to change the shape of the channel. The sodium ions are driven through the channel.

3 The sodium ions are released to the outside of the membrane, and the new shape of the channel allows two potassium ions to bind.

4 Release of the phosphate allows the channel to revert to its original form, releasing the potassium ions on the inside of the membrane.

Figure 5.32   How the sodium-potassium pump works.

ions that have been pumped out by the Na⁺-K⁺ pump to diffuse back in. There is a catch, however; these channels require that the sodium ions have a partner in order to pass through—like a dancing party where only couples are admitted through the door. These special channels won't let sodium ions across unless another molecule tags along, crossing hand in hand with the sodium ion. In some cases the partner molecule is a sugar, as shown in figure 5.33, in others an amino acid or other molecule. Because so many sodium ions are trying to get back in, this diffusion pressure drags in the partner molecules as well, even if they are already in high concentration within the cell. In this way, sugars and other actively transported molecules enter the cell—via special **coupled channels.** Their movement is in fact a form of facilitated diffusion driven by the active transport of sodium ions.

The Proton Pump. The second major active transport channel is the **proton pump.** The orange channel in figure 5.34 is a proton pump, a complex channel that expends metabolic energy to pump protons across membranes. Just as in the sodium-potassium pump, this creates a diffusion pressure that tends to drive protons back across again, but in this case the only channels open to them are not coupled channels but rather channels that make ATP, the purple channel in the figure. This pump is the key to cell metabolism, which is the way cells convert photosynthetic energy

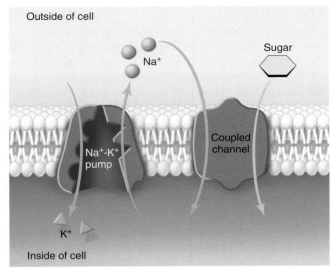

Figure 5.33   A coupled channel.

The active transport of a sugar molecule into a cell typically takes place in two stages—facilitated diffusion of the sugar coupled to active transport of sodium ions. The sodium-potassium pump keeps the Na⁺ concentration higher outside the cell than inside. For Na⁺ to diffuse back in through the coupled channel requires the simultaneous transport of a sugar molecule as well. Because the concentration gradient for Na⁺ is steeper than the opposing gradient for sugar, Na⁺ and sugar move into the cell.

Proton Pump

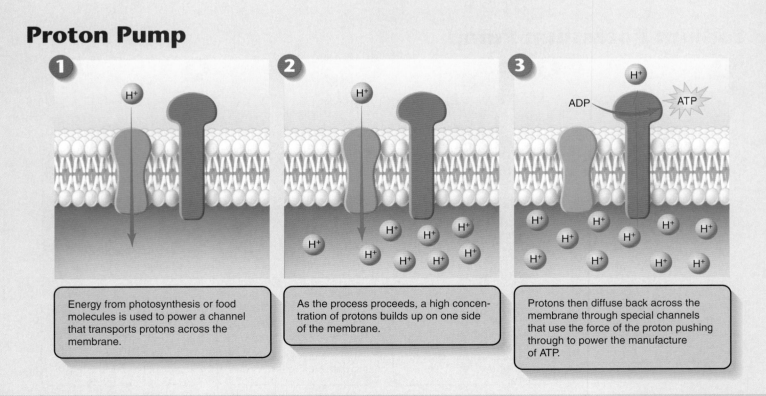

1 Energy from photosynthesis or food molecules is used to power a channel that transports protons across the membrane.

2 As the process proceeds, a high concentration of protons builds up on one side of the membrane.

3 Protons then diffuse back across the membrane through special channels that use the force of the proton pushing through to power the manufacture of ATP.

Figure 5.34   How the proton pump works.

TABLE 5.3   MECHANISMS FOR TRANSPORT ACROSS CELL MEMBRANES

Process	Passage Through Membrane	How it Works	Example
Passive Processes			
Diffusion			
Direct		Random molecular motion produces net migration of molecules toward region of lower concentration.	Movement of oxygen into cells
Protein channel		Polar molecules pass through a protein channel.	Movement of ions in or out of cell
Facilitated Diffusion			
Protein carrier		Molecule binds to carrier protein in membrane and is transported across; net movement is toward region of lower concentration.	Movement of glucose into cells
Osmosis			
Aquaporins		Diffusion of water across selectively permeable membrane.	Movement of water into cells placed in a hypotonic solution
Active Processes			
Endocytosis			
Membrane vesicle			
Phagocytosis		Particle is engulfed by membrane, which folds around it and forms a vesicle.	Ingestion of bacteria by white blood cells
Pinocytosis		Fluid droplets are engulfed by membrane, which forms vesicles around them.	"Nursing" of human egg cells
Receptor-mediated endocytosis		Endocytosis is triggered by a specific receptor.	Cholesterol uptake
Exocytosis			
Membrane vesicle		Vesicles fuse with plasma membrane and eject contents.	Secretion of mucus
Active Transport			
Protein carrier			
Na^+-K^+ pump		Carrier expends energy to transport a substance across a membrane against its concentration gradient.	Movement of Na^+ and K^+ against their concentration gradients
Coupled transport		Molecules are transported across a membrane against their concentration gradients by the cotransport of another substance down its concentration gradient.	Coupled uptake of glucose into cells against its concentration gradient
Proton pump		Protons are pumped across membranes against their concentration gradient. The proton gradient then drives the formation of ATP through another channel.	Proton pump in chemiosmosis

or chemical energy from food into ATP. Its activity is referred to as **chemiosmosis.** We discuss chemiosmosis at greater length in chapters 7 and 8. Table 5.3 summarizes the mechanisms for transport across plasma membranes that we have discussed.

5.12 Cells are selectively permeable, admitting only certain molecules. Facilitated diffusion is selective transport across a membrane in the direction of lower concentration. Active transport is energy-driven transport across a membrane toward a region of higher concentration.

What Limits the Rate of a Cell's Glucose Uptake?

Although the concentration of the sugar glucose within a human red blood cell (less than 0.5 mM [mM is a **measure of concentration,** which means millimolar, or one-thousandth of a mole per liter]) is much lower than in blood plasma (5mM), very little can diffuse into the cell directly across the lipid bilayer because glucose molecules are large and polar, and so cannot cross the highly nonpolar interior of the lipid bilayer. The blue line on the bottom of the graph to the right is the calculated curve for the rate of glucose uptake if it enters a cell solely by simple diffusion across a lipid bilayer membrane (the **rate** of a molecular process is simply the speed at which it occurs, in this case, measured as the number of molecules diffused per unit volume [ml] per unit time [hour]). The actual transport rate into red blood cells is much higher, because the membranes of red blood cells contain facilitated diffusion channels for the transport of glucose molecules across the membrane. These glucose transport channels account for some 2% of the total protein in a red blood cell's plasma membrane. As you can see in the diagram below, a channel protein is composed of several segments of polar amino acids, looping back and forth to create a passage spanning the membrane. In the glucose channel, 12 such polar segments form a polar "door" through which glucose molecules can diffuse.

The red line on the graph to the upper right presents an experiment measuring the rate of glucose uptake by red blood cells via facilitated diffusion as a function of how much glucose is present in the surrounding extracellular fluid. The rate of glucose uptake (measured as micromoles per milliliter per hour) is plotted against the extracellular concentration of glucose.

1. **Applying Concepts**
   a. **Variable.** In this study, which is the dependent variable?
   b. **Rate.** Is the rate of glucose uptake affected by the extracellular glucose concentration? How?
   c. What would be the predicted rate of glucose uptake in blood plasma? [Hint: What is the concentration of glucose in blood plasma mentioned above?]

2. **Interpreting Data**
   a. What is the difference in the uptake rate between 1 mM and 4 mM? between 10 mM and 14 mM?

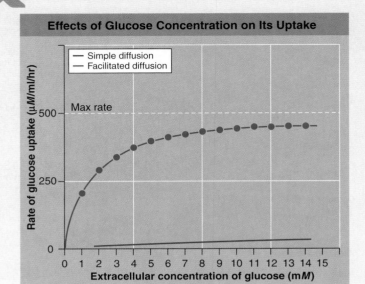

Effects of Glucose Concentration on Its Uptake

— Simple diffusion
— Facilitated diffusion

Max rate

y-axis: Rate of glucose uptake (μM/ml/hr)
x-axis: Extracellular concentration of glucose (mM)

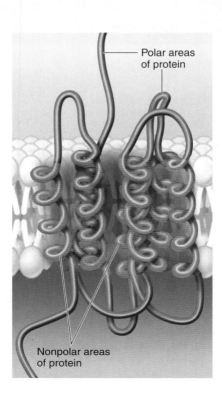

Polar areas of protein

Nonpolar areas of protein

b. What general statement can be made regarding the effect of increasing glucose levels in the blood on the uptake of glucose by red blood cells?

3. **Making Inferences** If the external glucose concentration was increased to 30 mM, what would you predict would be the rate of glucose uptake? to 100 mM?

4. **Drawing Conclusions**
   a. Why does the rate of glucose uptake by red blood cells in this experiment never exceed 500 μM/ml/hr?
   b. At what glucose concentration are half of the transport channels occupied?

5. **Further Analysis**
   a. The concentration of glucose within red blood cells is not significantly higher when extracellular glucose concentrations are 14 mM than when they are 1 mM, because enzymes within the cells quickly convert any arriving glucose molecules into other metabolites. Predict what the above glucose uptake curve would look like if the experiment was repeated in the presence of chemicals that completely inhibited these enzymes.
   b. Cells in other tissues of the human body take up glucose more rapidly than 500 micromoles per milliliter per hour. How would you predict the plasma membranes of these cells to be different from those of red blood cells?

6.2   The Laws of Thermodynamics

Running, thinking, singing, reading these words—all activities of living organisms involve changes in energy. A set of universal laws we call the laws of thermodynamics govern these and all other energy changes in the universe.

The First Law of Thermodynamics

The first of these universal laws, the **first law of thermodynamics,** concerns the amount of energy in the universe. It states that energy can change from one state to another (from potential to kinetic, for example) but it can never be destroyed, nor can new energy be made. The total amount of energy in the universe remains constant.

A lion eating a giraffe is in the process of acquiring energy. Rather than creating new energy or capturing the energy in sunlight, the lion is merely transferring some of the potential energy stored in the giraffe's tissues to its own body (just as the giraffe obtained the potential energy stored in the plants it ate while it was alive). Within any living organism, this chemical potential energy can be shifted to other molecules and stored in chemical bonds, or it can be converted into kinetic energy, or into other forms of energy such as light or electrical energy. During each conversion, some of the energy dissipates into the environment as **heat energy,** a measure of the random motions of molecules (and, hence, a measure of one form of kinetic energy). Energy continuously flows through the biological world in one direction, with new energy from the sun constantly entering the system to replace the energy dissipated as heat.

Heat can be harnessed to do work only when there is a heat gradient—that is, a temperature difference between two areas. This is how a steam engine functions. In old steam locomotives like you see in figure 6.2, heat was used to move the wheels. First, a boiler (not shown) heats up water to create steam. The steam is then pumped into the cylinder of the steam engine, where it moves the piston to the right. The moving of this piston then does the work of the steam engine by moving a lever that turns the wheel. Cells are too small to maintain significant internal temperature differences, so heat

Disorder happens "spontaneously"

Organization requires energy

Figure 6.3   Entropy in action.
As time elapses, a teenager's room becomes more disorganized. It takes energy to clean it up.

energy is incapable of doing the work of cells. Thus, although the total amount of energy in the universe remains constant, the energy available to do useful work in a cell decreases, as progressively more of it dissipates as heat.

The Second Law of Thermodynamics

The **second law of thermodynamics** concerns this transformation of potential energy into heat, or random molecular motion. It states that the disorder in a closed system like the universe is continuously increasing. Put simply, disorder is more likely than order. For example, it is much more likely that a column of bricks will tumble over than that a pile of bricks will arrange themselves spontaneously to form a column. In general, energy transformations proceed spontaneously to convert matter from a more ordered, less stable form, to a less ordered, more stable form. Without an input of energy from the teenager (or a parent), the ordered room in figure 6.3 falls into disorder.

Entropy

Entropy is a measure of the degree of disorder of a system, so the second law of thermodynamics can also be stated simply as "entropy increases." When the universe formed 10 to 20 billion years ago, it held all the potential energy it will ever have. It has become progressively more disordered ever since, with every energy exchange increasing the entropy of the universe.

> **6.2**   The first law of thermodynamics states that energy cannot be created or destroyed; it can only undergo conversion from one form to another. The second law states that disorder (entropy) in the universe tends to increase. Life converts energy from the sun to other forms of energy that drive life processes; the energy is never lost, but as it is used, more and more of it is converted to heat, the energy of random molecular motion.

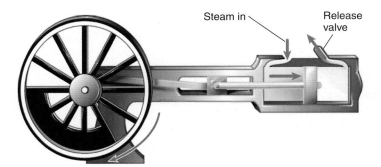

Steam in
Release valve

Figure 6.2   A steam engine.
In a steam engine, heat is used to produce steam. The expanding steam pushes against a piston that causes the wheel to turn.

6.3 | Chemical Reactions

In a chemical reaction, the molecules that you start with are called **reactants,** or sometimes **substrates,** whereas the molecules that you end up with after the reaction is over are called the **products** of the reaction. Not all chemical reactions are equally likely to occur. Just as a boulder is more likely to roll downhill than uphill, so a reaction is more likely to occur if it releases energy than if it needs to have energy supplied. Consider how the chemical reaction proceeds in the first panel of figure 6.4. Like when rolling a boulder uphill, energy needs to be supplied. This is because the product of the reaction contains more energy than the reactant. This type of chemical reaction, called **endergonic,** does not occur spontaneously. By contrast, an **exergonic** reaction, shown in the second panel, tends to occur spontaneously because the product has less energy than the reactant, like a boulder that has rolled downhill.

Activation Energy

If all chemical reactions that release energy tend to occur spontaneously, it is fair to ask, "Why haven't all exergonic reactions occurred already?" Clearly they have not. If you ignite gasoline, it burns with a release of energy. So why doesn't all the gasoline in all the automobiles in the world just burn up right now? It doesn't because the burning of gasoline, and almost all other chemical reactions, requires an input of energy to get it started—a kick in the pants such as a match or spark plug. Even in exergonic reactions where the product contains or stores less energy than the reactants, it is first necessary to break existing chemical bonds in the reactants, and this takes energy. The extra energy required to destabilize existing chemical bonds and so initiate a chemical reaction is called **activation energy,** indicated by brackets in figure 6.4*b* and *c.* You must first nudge a boulder out of the hole it sits in before it can roll downhill. Activation energy is simply a chemical nudge.

Catalysis

Just as all the gasoline in the world doesn't burn up right now, so all the exergonic reactions in your cells don't spontaneously happen either. Each reaction waits for something to nudge it along, to supply it with sufficient activation energy to get going. One way to make an exergonic reaction more likely to happen is to lower the necessary activation energy. Like digging away the ground below your boulder, lowering activation energy reduces the nudge needed to get things started. The process of lowering the activation energy of a reaction is called **catalysis.** Catalysis cannot make an endergonic reaction occur spontaneously—you cannot avoid the need to supply energy—but it can make a reaction, endergonic or exergonic, proceed much faster. Compare the activation energy levels (the red arched arrows) in the second and third panels below: the catalyzed reaction has a lower barrier to overcome.

> **6.3   Chemical reactions occur when the covalent bonds linking atoms together are formed or broken. It takes energy to initiate chemical reactions.**

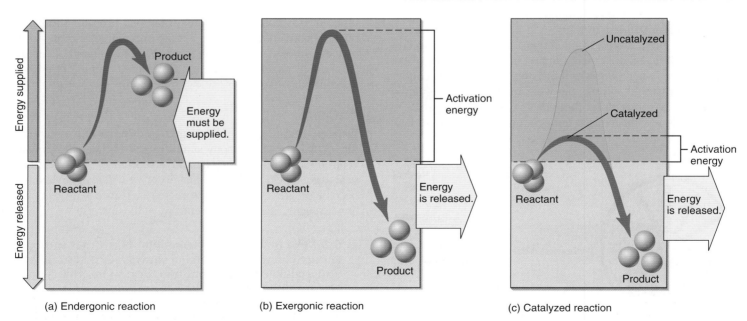

(a) Endergonic reaction   (b) Exergonic reaction   (c) Catalyzed reaction

Figure 6.4   Chemical reactions and catalysis.
(a) The products of endergonic reactions contain more energy than the reactants. (b) The products of exergonic reactions contain less energy than the reactants, but exergonic reactions do not necessarily proceed rapidly because it takes energy to get them going. The "hill" in this energy diagram represents energy that must be supplied to destabilize existing chemical bonds. (c) Catalyzed reactions occur faster because the amount of activation energy required to initiate the reaction—the height of the energy hill that must be overcome—is lowered.

6.4   How Enzymes Work

Proteins called **enzymes** are the catalysts used by cells to touch off particular chemical reactions. By controlling which enzymes are present, and when they are active, cells are able to control what happens within themselves, just as a conductor controls the music an orchestra produces by dictating which instruments play when.

An enzyme works by binding to a specific molecule and stressing the bonds of that molecule in such a way as to make a particular reaction more likely. The key to this activity is the shape of the enzyme. An enzyme is specific for a particular reactant, or substrate, because the enzyme surface provides a mold that very closely fits the shape of the desired reactant. For example, the blue-colored lysozyme enzyme in figure 6.5 is contoured to fit a specific sugar molecule (the yellow reactant). Other molecules that fit less perfectly simply don't adhere to the enzyme's surface. The site on the enzyme surface where the reactant fits is called the **active site.** The site on the reactant that binds to an enzyme is called the **binding site.** Proteins are not rigid. The binding of the reactant induces the enzyme to change its shape slightly. In figure 6.5b, the edges of the lysozyme now hug the sugar molecule, leading to an "induced fit" between the enzyme and its reactant, like a glove molding to a hand.

An enzyme lowers the activation energy of a particular reaction. In the case of lysozyme, an enzyme found in human

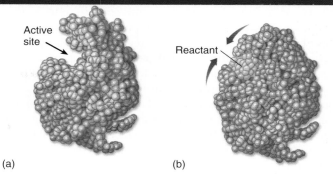

(a)   (b)

Figure 6.5   Enzyme shape determines its activity.
(a) A groove runs through the lysozyme enzyme (*blue* in this diagram) that fits the shape of the reactant (in this case, a chain of sugars). (b) When such a chain of sugars, indicated in *yellow,* slides into the groove, it induces the protein to change its shape slightly and embrace the substrate more intimately. This induced fit causes a chemical bond between two sugar molecules within the chain to break.

tears, the enzyme has an antibacterial function, encouraging the breaking of a particular chemical bond in molecules that make up the cell wall of bacteria. The enzyme weakens the bond by drawing away some of its electrons. Alternatively, an enzyme may encourage the formation of a link between two reactants, like the blue and red colored molecules in figure 6.6, by holding them near each other. Regardless of the type of reaction, the enzyme is not affected by the chemical reaction and is available to be used again.

How Enzymes Work

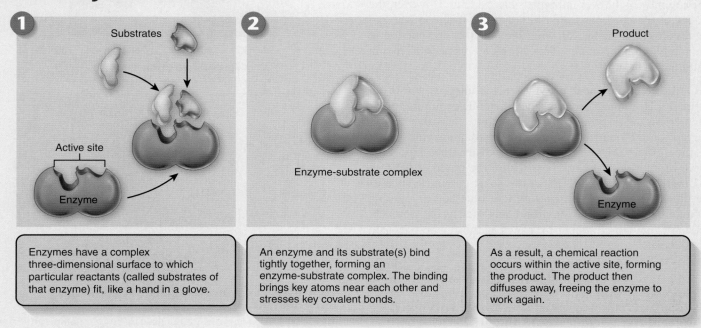

1 Enzymes have a complex three-dimensional surface to which particular reactants (called substrates of that enzyme) fit, like a hand in a glove.

2 An enzyme and its substrate(s) bind tightly together, forming an enzyme-substrate complex. The binding brings key atoms near each other and stresses key covalent bonds.

3 As a result, a chemical reaction occurs within the active site, forming the product. The product then diffuses away, freeing the enzyme to work again.

Figure 6.6   How enzymes work.

Biochemical Pathways

Every organism contains thousands of different kinds of enzymes that together catalyze a bewildering variety of reactions. Often several of these reactions occur in a fixed sequence called a **biochemical pathway,** the product of one reaction becoming the substrate for the next. You can see in the biochemical pathway shown in figure 6.7 how the initial substrate is altered by enzyme 1 so that it now fits into the active site of another enzyme, becoming the substrate for enzyme 2, and so on until the final product is produced. Because these reactions occur in sequence, the enzymes involved are often positioned near each other in the cell. For example, the four enzymes involved in this biochemical pathway are all embedded in the membrane near each other. The close proximity of the enzymes allows the reactions of the biochemical pathway to proceed faster. Biochemical pathways are the organizational units of metabolism. We will discuss them at length in chapters 7 and 8.

Factors Affecting Enzyme Activity

Temperature and pH can have a major influence on the action of enzymes. Enzyme activity is affected by any change in condition that alters the enzyme's three-dimensional shape.

Temperature. When the temperature increases, the bonds that determine enzyme shape are too weak to hold the enzyme's peptide chains in the proper position and the enzyme denatures. As a result, enzymes function best within an optimum temperature range, which is relatively narrow for most human enzymes. In the human body, enzymes work best at temperatures near normal body temperature of 37°C, as shown by the brown curve in figure 6.8a. Also notice that the rates of enzyme reactions tend to drop quickly at higher temperatures, when the enzyme proteins begin to unfold. This is why an extremely high fever in humans can be fatal. However, the shapes of the enzymes found in hotsprings bacteria (the red curve) are more stable, allowing the enzymes to function at much higher temperatures. This allows the bacteria to live in water that is near 70°C.

pH. In addition, most enzymes also function within an optimal pH range, because the shape-determining polar interactions of enzyme proteins are quite sensitive to hydrogen ion (H$^+$) concentration. Most human enzymes, such as the protein-degrading enzyme trypsin (the dark blue curve in figure 6.8b) work best within the range of pH 6 to 8. Blood has a pH of 7.4. However, some enzymes, such as the digestive enzyme pepsin (the light blue curve) are able to function in very acidic environments such as the stomach, but can't function at higher pHs such as those where trypsin works best.

> **6.4** Enzymes are proteins that catalyze chemical reactions within cells. Sometimes enzymes are organized into biochemical pathways. Enzymes are sensitive to temperature and pH, because both of these variables influence protein shape.

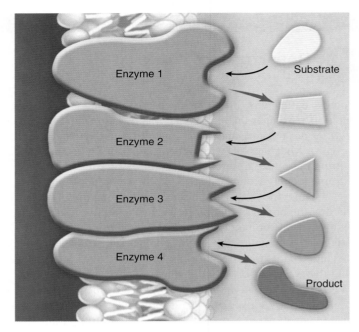

Figure 6.7 A biochemical pathway.
The original substrate is acted on by enzyme 1, changing the substrate to a new form recognized by enzyme 2. Each enzyme in the pathway acts on the product of the previous stage.

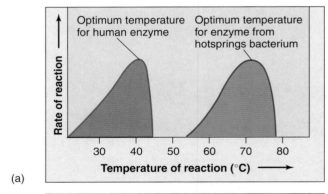

(a)

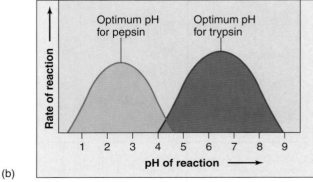

(b)

Figure 6.8 Enzymes are sensitive to their environment.
The activity of an enzyme is influenced by both (a) temperature and (b) pH. Most human enzymes work best at temperatures of about 40°C and within a pH range of 6 to 8.

Allosteric Enzyme Regulation

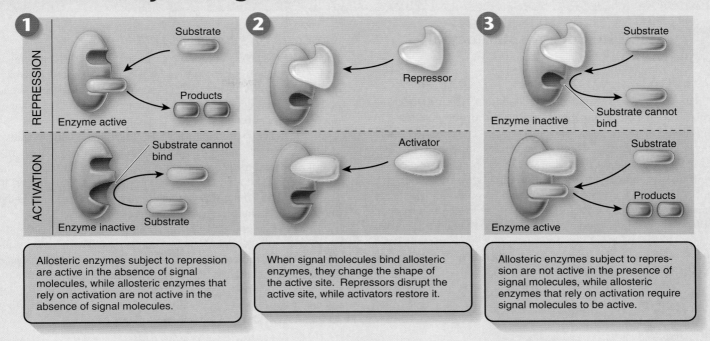

① REPRESSION

Substrate → Products
Enzyme active

① ACTIVATION

Substrate cannot bind
Enzyme inactive — Substrate

Allosteric enzymes subject to repression are active in the absence of signal molecules, while allosteric enzymes that rely on activation are not active in the absence of signal molecules.

②

Repressor

Activator

When signal molecules bind allosteric enzymes, they change the shape of the active site. Repressors disrupt the active site, while activators restore it.

③

Substrate
Enzyme inactive — Substrate cannot bind

Substrate → Products
Enzyme active

Allosteric enzymes subject to repression are not active in the presence of signal molecules, while allosteric enzymes that rely on activation require signal molecules to be active.

Figure 6.9 How cells control enzymes.

6.5 How Cells Regulate Enzymes

Because an enzyme must have a precise shape to work correctly, it is possible for the cell to control when an enzyme is active by altering its shape. Many enzymes have shapes that can be altered by the binding to their surfaces of "signal" molecules. Such enzymes are called *allosteric* (Latin, other shape). Enzymes can be inhibited or activated by the binding of signal molecules. For example, the upper tan panels of figure 6.9 show an enzyme that is inhibited. The binding of a signal molecule, called a **repressor** (the yellow molecule in panel 2), alters the shape of the enzyme's active site such that it cannot bind the substrate (the red molecule). In other cases, the enzyme may not be able to bind the reactants *unless* the signal molecule is bound to the enzyme. The lower set of pink panels shows a signal molecule serving as an **activator.** The red substrate cannot bind to the enzyme's active site unless the activator (the yellow molecule) is in place, altering the shape of the active site. The site where the signal molecule binds to the enzyme surface is called the **allosteric site.**

Enzymes are often regulated by a mechanism called **feedback inhibition,** where the product of the reaction acts as the repressor. Feedback inhibition can occur in two ways: *competitive inhibitors* and *noncompetitive inhibitors*. The blue molecule in figure 6.10a functions as a competitive inhibitor, blocking the active site so that the substrate cannot bind. The yellow molecule in figure 6.10b functions as a noncompetitive inhibitor. It binds to an allosteric site, changing the shape of the enzyme such that it is unable to bind to the substrate, like in the upper panels in figure 6.9.

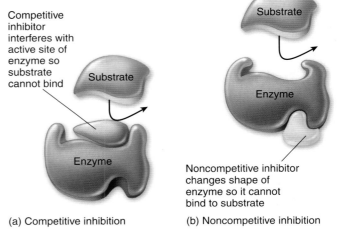

Competitive inhibitor interferes with active site of enzyme so substrate cannot bind

Substrate

Enzyme

(a) Competitive inhibition

Substrate

Enzyme

Noncompetitive inhibitor changes shape of enzyme so it cannot bind to substrate

(b) Noncompetitive inhibition

Figure 6.10 How enzymes can be inhibited.

(a) In competitive inhibition, the inhibitor interferes with the active site of the enzyme. (b) In noncompetitive inhibition, the inhibitor binds to the enzyme at a place away from the active site, effecting a conformational change in the enzyme so that it can no longer bind to its substrate. In feedback inhibition, the inhibitor molecule is the product of the reaction.

Many drugs and antibiotics work by inhibiting enzymes. Statin drugs like Lipitor lower cholesterol by inhibiting a key enzyme cells use to make cholesterol. The antibiotic penicillin inhibits an enzyme bacteria use in making cell walls. As humans lack this enzyme, we are not harmed by the drug.

> **6.5 An enzyme's activity can be affected by signal molecules that bind to them, changing their shape.**

6.6 ATP: The Energy Currency of the Cell

Cells use energy to do all those things that require work, but how does the cell use energy from the sun or the potential energy stored in molecules to power its activities? The sun's radiant energy and the energy stored in molecules are energy sources, but like money that is invested in stocks and bonds or real estate, these energy sources cannot be used directly to run a cell, any more than money invested in stocks can be used to buy a candy bar at the store. To be useful, the energy from the sun or food molecules must first be converted to a source of energy that a cell can use, like someone converting stocks and bonds to ready cash. The "cash" molecule in the body is **adenosine triphosphate (ATP).** ATP is the energy currency of the cell.

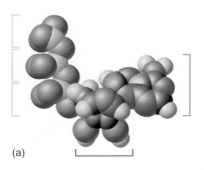

(a)

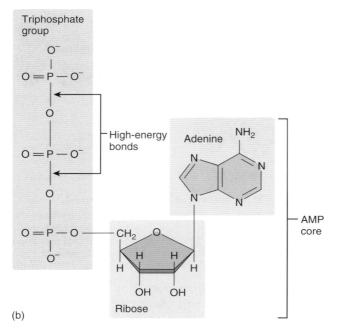

(b)

Figure 6.11   The parts of an ATP molecule.
The model (a) and structural diagram (b) both show that ATP consists of three phosphate groups attached to a ribose (five-carbon sugar) molecule. The ribose molecule is also attached to an adenine molecule (also one of the nitrogenous bases of DNA and RNA). When the endmost phosphate group is split off from the ATP molecule, considerable energy is released.

Structure of the ATP Molecule

Each ATP molecule is composed of the three parts shown in figure 6.11: (1) a sugar (colored blue) serves as the backbone to which the other two parts are attached, (2) adenine (colored peach) is one of the four nitrogenous bases in DNA and RNA, and (3) a chain of three phosphates (colored yellow) contain high-energy bonds.

As you can see in the figure, the phosphates carry negative electrical charges, and so it takes considerable chemical energy to hold the line of three phosphates next to one another at the end of ATP. Like a coiled spring, the phosphates are poised to push apart. It is for this reason that the chemical bonds linking the phosphates are such chemically reactive bonds.

When the endmost phosphate is broken off an ATP molecule, a sizable packet of energy is released. The reaction converts ATP to adenosine diphosphate, ADP. The second phosphate group can also be removed, yielding additional energy and leaving adenosine monophosphate (AMP). Most energy exchanges in cells involve cleavage of only the outermost bond, converting ATP into ADP and P_i, inorganic phosphate:

$$\text{ATP} \longrightarrow \text{ADP} + P_i + \text{energy}$$

Exergonic reactions require activation energy, and endergonic reactions require the input of even more energy, and so these reactions in the cell are usually coupled with the breaking of the phosphate bond in ATP, called *coupled reactions.* Because almost all chemical reactions in cells require less energy than is released by this reaction, ATP is able to power many of the cell's activities, producing heat as a by-product. Table 6.1 introduces you to some of the key cellular activities powered by the breakdown of ATP.

Cells convert energy from the sun or food into molecules of ATP. ATP is then broken down within cells, liberating energy that the cells use to perform life functions. ATP therefore cycles in the cell. The ATP-ADP cycle is shown in figure 6.12. Some cells convert energy from the sun into molecules of ATP through the process of **photosynthesis,** the subject of chapter 7. This ATP is then used to manufacture sugar molecules, converting the energy from ATP into potential energy stored in the bonds that hold the atoms together. All cells convert the potential energy found in

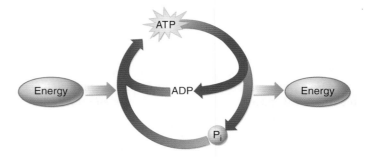

Figure 6.12   The ATP-ADP cycle.
In mitochondria and chloroplasts, chemical or the sun's energy is harnessed to form ATP from ADP and inorganic phosphate. When ATP is used to drive the living activities of cells, the molecule is cleaved back to ADP and inorganic phosphate, which are then available to form new ATP molecules.

TABLE 6.1 HOW CELLS USE ATP ENERGY TO POWER CELLULAR WORK

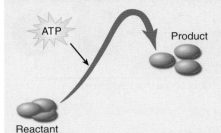

Biosynthesis

Cells use the energy released from the exergonic hydrolysis of ATP to drive endergonic reactions like those of protein synthesis, an approach called energy coupling.

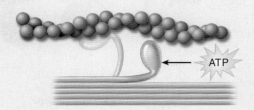

Contraction

In muscle cells, filaments of protein repeatedly slide past each other to achieve contraction of the cell. An input of ATP is required for the filaments to reset and slide again.

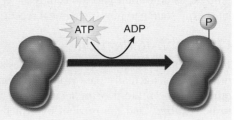

Chemical Activation

Proteins can become activated when a high-energy phosphate from ATP attaches to the protein, activating it. Other types of molecules can also become phosphorylated by transfer of a phosphate from ATP.

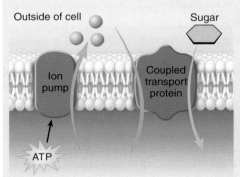

Importing Metabolites

Metabolite molecules such as amino acids and sugars can be transported into cells against their concentration gradients by coupling the intake of the metabolite to the inward movement of an ion moving down its concentration gradient, this ion gradient being established using ATP.

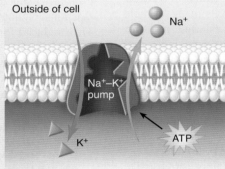

Active Transport: Na⁺–K⁺ Pump

Most animal cells maintain a low internal concentration of Na^+ relative to their surroundings, and a high internal concentration of K^+. This is achieved using a protein called the sodium-potassium pump, which actively pumps Na^+ out of the cell and K^+ in, using energy from ATP.

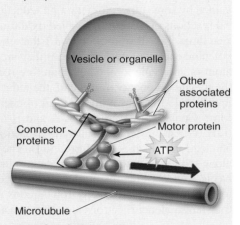

Cytoplasmic Transport

Within a cell's cytoplasm, vesicles or organelles can be dragged along microtubular tracks using molecular motor proteins, which are attached to the vesicle or organelle with connector proteins. The motor proteins use ATP to power their movement.

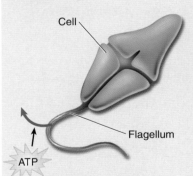

Flagellar Movements

Microtubules within flagella slide past each other to produce flagellar movements. ATP powers the sliding of the microtubules.

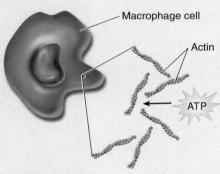

Cell Crawling

Actin filaments in a cell's cytoskeleton continually assemble and disassemble to achieve changes in cell shape and to allow cells to crawl over substrates or engulf materials. The dynamic character of actin is controlled by ATP molecules bound to actin filaments.

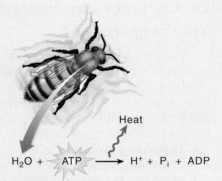

Heat Production

The hydrolysis of the ATP molecule releases heat. Reactions that hydrolyze ATP often take place in mitochondria or in contracting muscle cells and may be coupled to other reactions. The heat generated by these reactions can be used to maintain an organism's temperature.

food molecules into ATP through **cellular respiration,** the subject of chapter 8. The potential energy of food, in the form of glucose, is converted into molecules of ATP used by the cell as an energy source to power chemical reactions.

Oxidation-Reduction

In many of the reactions of photosynthesis and cellular respiration, electrons actually pass from one atom or molecule to another. When an atom or molecule loses an electron, it is said to be *oxidized,* and the process by which this occurs is called **oxidation.** The name reflects the fact that in biological systems, oxygen, which attracts electrons strongly, is the most common electron acceptor. Conversely, when an atom or molecule gains an electron, it is said to be *reduced,* and the process is called **reduction.** Oxidation and reduction always take place together, because every electron that is lost by an atom through oxidation is gained by some other atom through reduction. Therefore, chemical reactions of this sort are called **oxidation-reduction (redox) reactions.** Recall from section 3.1 that electrons maintain their potential energy of position even if they are transferred from one atom or molecule to another; therefore redox reactions involve the transfer of energy. In general, the reduced form of an organic molecule thus has a higher level of energy than the oxidized form. Consider two molecules A and B that you see in figure 6.13. When they undergo a redox reaction, molecule A loses an electron, which is transferred to B. The energy is transferred along with the electron such that the product A˙ (colored yellow) has less energy than the product B˙ (colored red). It is important to note, however, that the electron in redox reactions isn't usually transferred alone. Electrons are often transferred with hydrogen.

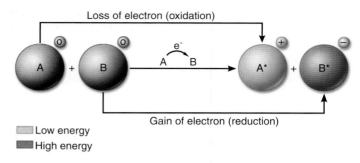

Loss of electron (oxidation)

Gain of electron (reduction)

- ▢ Low energy
- ▨ High energy

Figure 6.13 Redox reactions.
Oxidation is the loss of an electron; reduction is the gain of an electron. Here the charges of molecules A and B are shown in small circles to the upper right of each molecule. Molecule A loses energy as it loses an electron, while molecule B gains energy as it gains an electron.

For example, in chapter 7 you will encounter an electron carrier called $NADP^+$, and in chapter 8 you will encounter another electron carrier called NAD^+. Both of these molecules undergo redox reactions in photosynthesis and cellular respiration, respectively. Their reduced forms are NADPH and NADH because the electrons that reduce these carriers are transferred along with hydrogens. Oxidation-reduction reactions play a key role in the flow of energy through biological systems because the electrons that pass from one atom to another carry energy with them.

> **6.6 Cells store chemical energy in ATP and use ATP to drive chemical reactions. Energy is often transferred with electrons. Oxidation is the loss of an electron; reduction is the gain of one.**

INQUIRY & ANALYSIS

Do Enzymes Physically Attach to Their Substrates?

In order to determine if an enzyme actually binds to the chemical it catalyzes (its "substrate"), chemist Victor Henri in 1903 carried out the experiment whose results you see in the graph, measuring the reaction rate (*V*) at different substrate concentrations (*S*).

1. **Applying Concepts** In the graph, what is the dependent variable? Explain.
2. **Making Inferences** As *S* increases, *V* increases by smaller and smaller amounts. Is there a maximum reaction rate? Indicate it on the graph. What does this saturation effect tell you?
3. **Drawing Conclusions** Does this result provide support for the hypothesis that an enzyme binds physically to its substrate? Explain. If the hypothesis were incorrect, what would you expect the graph to look like?

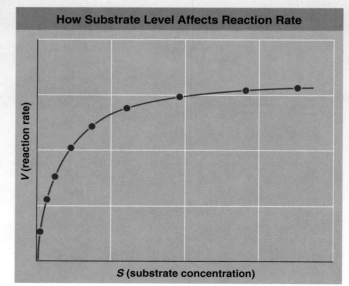

How Substrate Level Affects Reaction Rate

V (reaction rate)

S (substrate concentration)

Cells and Energy

6.1 The Flow of Energy in Living Things

- Energy is the ability to do work. Energy exists in two states: kinetic energy and potential energy.

- Kinetic energy is the energy of motion. Potential energy is stored energy, which exists in objects that aren't in motion but have the capacity to move (**figure 6.1**). All of the work carried out by living things involves the transformation of potential energy into kinetic energy.

- Energy flows from the sun to the earth, where it is trapped by photosynthetic organisms and stored in carbohydrates as potential energy. This energy is transferred during chemical reactions.

6.2 The Laws of Thermodynamics

- The laws of thermodynamics describe changes in energy in our universe. The first law of thermodynamics explains that energy can never be created or destroyed, only changed from one state to another. The total amount of energy in the universe remains constant.

- The second law of thermodynamics explains that the conversion of potential energy into random molecular motion, or disorder, is constantly increasing. Entropy, which is a measure of disorder in a system, is constantly increasing such that disorder is more likely than order. Energy must be used to maintain order (**figure 6.3**).

Cell Chemistry

6.3 Chemical Reactions

- Chemical reactions involve the breaking or formation of covalent bonds. The starting molecules are called the reactants, and the molecules produced by the reaction are called the products.

- Some chemical reactions are more likely to occur than others. Chemical reactions in which the products contain more potential energy than the reactants are called endergonic reactions (**figure 6.4***a*). Chemical reactions that release more energy than the energy put into them are called exergonic reactions (**figure 6.4***b*).

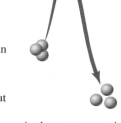

- All chemical reactions require an input of energy, even exergonic reactions that occur spontaneously. This energy required to start a reaction is called activation energy. A chemical reaction proceeds faster when its activation energy is lowered, and this occurs through a process called catalysis (**figure 6.4***c*).

Enzymes

6.4 How Enzymes Work

- Enzymes are cellular proteins that lower the activation energy of chemical reactions in the cell. Enzymes are catalysts.

- An enzyme binds the reactants, also called the substrates, of a reaction to its active site, increasing the likelihood that chemical bonds will break or form—therefore the reaction will require less energy (**figures 6.5** and **6.6**). The enzyme lowers the activation energy of the reaction. The enzyme is not affected by the reaction and can be used over and over again (**figure 6.6**).

- Sometimes enzymes work in a series of reactions called a biochemical pathway. The product of one reaction becomes the substrate for the next reaction. The enzymes that are involved in a biochemical pathway are usually located near each other in the cell (**figure 6.7**).

- In the cell, chemical reactions are regulated by controlling which enzymes are present in the cell. Other factors can also affect enzyme function such as temperature and pH, and so most enzymes have an optimal temperature and pH range (**figure 6.8**).

- Temperatures above the optimal range can disrupt the bonds that hold the enzyme in its proper shape, decreasing its ability to catalyze a chemical reaction. The bonds that hold the enzyme's shape are also affected by hydrogen ion concentrations, and so increasing or decreasing the pH (hydrogen ion concentrations) can disrupt the enzyme's function.

6.5 How Cells Regulate Enzymes

- An enzyme can be inhibited or activated in the cell as a means of regulation by temporarily altering the enzyme's shape. An enzyme can be inhibited when a molecule, called a repressor, binds to the enzyme, altering the shape of the active site so that it cannot bind the substrate. Some enzymes need to be activated, or turned on, in order to bind to their substrate. A molecule called an activator binds to the enzyme, changing the shape of the active site so that it is able to bind the substrate (**figures 6.9** and **6.10**).

- Enzymes are often regulated by feedback inhibition, a process whereby the product of the reaction functions as a repressor.

How Cells Use Energy

6.6 ATP: The Energy Currency of the Cell

- Cells require energy to do work. This cellular energy is stored in molecules of ATP (**table 6.1**). ATP contains a sugar, an adenine, and a chain of three phosphates. The three phosphates are held together with high-energy bonds that are chemically reactive (**figure 6.11**). When the endmost phosphate bond breaks, considerable energy is released. A cell uses this energy to drive other reactions in the cell (**figure 6.12**).

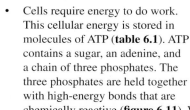

- Coupled reactions, called oxidation-reduction or redox reactions, involve the transfer of electrons from one atom or molecule to another. The atom or molecule that loses an electron is said to be oxidized and loses energy. The atom or molecule that gains the electron is said to be reduced and gains energy (**figure 6.13**).

1. The ability to do work is the definition for
   a. thermodynamics.   c. energy.
   b. radiation.   d. entropy.
2. The first law of thermodynamics
   a. says that energy recycles constantly, as organisms use and reuse it.
   b. is concerned with how heat energy can be used to do work.
   c. is a formula for measuring entropy.
   d. says that energy can change forms, but cannot be made or destroyed.
3. The second law of thermodynamics
   a. says that energy recycles constantly, as organisms use and reuse it.
   b. says that entropy, or disorder, continually increases in a closed system.
   c. is a formula for measuring entropy.
   d. says that energy can change forms, but cannot be made nor destroyed.
4. Chemical reactions that occur spontaneously are called
   a. exergonic and release energy.
   b. exergonic and their products contain more energy.
   c. endergonic and release energy.
   d. endergonic and their products contain more energy.
5. The catalysts that help an organism carry out needed chemical reactions are called
   a. hormones.   c. reactants.
   b. enzymes.   d. substrates.

6. Factors that affect the activity of an enzyme molecule include
   a. peptides and energy.
   b. substrates and reactants.
   c. temperature and pH.
   d. entropy and thermodynamics.
7. In order for an enzyme to work properly
   a. it must have a particular shape.
   b. the temperature must be within certain limits.
   c. the pH must be within certain limits.
   d. All of these are true.
8. One way a cell can stop an enzyme from working when the cell does not need more of the product is that the cell
   a. changes its internal temperature to denature the enzyme.
   b. changes its internal pH to inhibit enzyme activity.
   c. destroys the enzyme.
   d. produces repressor molecules that alter enzyme shape.
9. In competitive inhibition
   a. an enzyme molecule has to compete with other enzyme molecules for the necessary substrate.
   b. an enzyme molecule has to compete with other enzyme molecules for the necessary energy.
   c. an inhibitor molecule competes with the substrate for the same binding site on the enzyme.
   d. two different substrates compete for the same binding site on the enzyme.
10. The source of immediate, or "ready cash," energy in the body is
    a. ATP molecules.   c. enzyme molecules.
    b. DNA molecules.   d. product molecules.

Visual Understanding

1. **Figure 6.8** You eat a hamburger. Salivary amylase begins to digest the carbohydrates in the bun while you are still chewing. Pepsin works in your stomach to digest the protein, and trypsin is active in your small intestine to break the bonds between specific amino acids. How does the optimum pH for pepsin and trypsin reflect this chain of events?

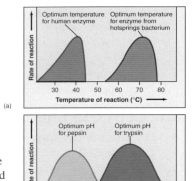

2. **Table 6.1** ATP comes primarily from the breakdown of glucose. If your blood glucose level drops, what sorts of problems can that cause?

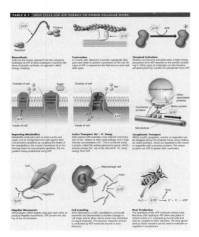

Challenge Questions

1. Living organisms on the surface of the planet and the top layer of the ocean get their energy from the photosynthetic plants and bacteria that capture the sun's energy and transform it to molecules that other organisms can use. At the bottom of some deep ocean trenches there are entire colonies of many types of organisms. What do they use for energy?
2. Why do you need enzymes to help your cells do their jobs?

3. What sorts of things can keep an enzyme from doing its job?
4. After a rain your new iron fenceposts are rusting. Explain how this is an example of a redox reaction, just like the ones that occur in your body. Now can you think of a reason why blood coming from the lungs is bright red?

7

Photosynthesis: Acquiring Energy from the Sun

I n this forest glade, you can literally see the pulse of life flowing through the organisms of an ecosystem. Sunlight beams down, a stream of energy in the form of packets of light called photons. Everywhere the light falls, there are plants: trees and shrubs and flowers and grasses, all with green leaves intercepting the energy as it rains down. In the cells of each leaf are organelles called chloroplasts that contain light-gathering pigments in their membranes. These pigments, notably the pigment chlorophyll, which makes leaves green, absorb photons of light and use the energy to strip electrons from water molecules. The chloroplasts use these electrons to reduce CO_2—that is, to add hydrogens (a hydrogen atom, you will recall, is just a proton with an associated electron)—and so make organic molecules. This process of capturing the sun's energy to build molecules is called photosynthesis—literally, using "light to build." In this chapter, we will delve into photosynthesis, tracing how light energy is captured, converted to chemical energy, and put to work assembling organic molecules. In the roots and other tissues of the plants, the opposite process is taking place. Organic molecules are being broken down in the process of cellular respiration to provide energy to power growth and cellular activities. These reactions, which take place largely in another kind of organelle called a mitochondrion, are the subject of the following chapter. Together, chloroplasts and mitochondria carry out a flow of energy driven by the power of sunlight.

7.1 An Overview of Photosynthesis

Life is powered by sunshine. All of the energy used by almost all living cells comes ultimately from the sun, captured by plants, algae, and some bacteria through the process of **photosynthesis.** Thus, life is only possible because our earth is awash in energy streaming inward from the sun. Each day, the radiant energy that reaches the earth is equal to that of about 1 million Hiroshima-sized atomic bombs. About 1% of it is captured by photosynthesis and provides the energy that drives us all. The other 99%, which is not absorbed by the photosynthetic machinery in the cells, is reflected back into space, warms the land and water, or is used in the water cycle (evaporation and precipitation).

Photosynthesis occurs within the plasma membrane in many kinds of bacteria, in the cells of algae, and within leaves (but not all cells) of plants. Recall from chapter 5 that the cells of plant leaves contain organelles called chloroplasts that actually carry out photosynthesis. You can see them as green specks within the blue mesophyll cells in the cutaway drawing of a leaf in figure 7.1. As the levels of magnification increase in the figure, you can eventually "see" right into a chloroplast. No other structures in a plant, other than chloroplasts, are able to carry out photosynthesis.

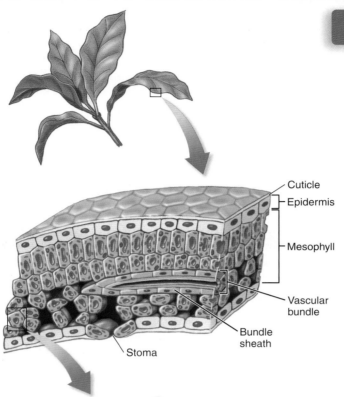

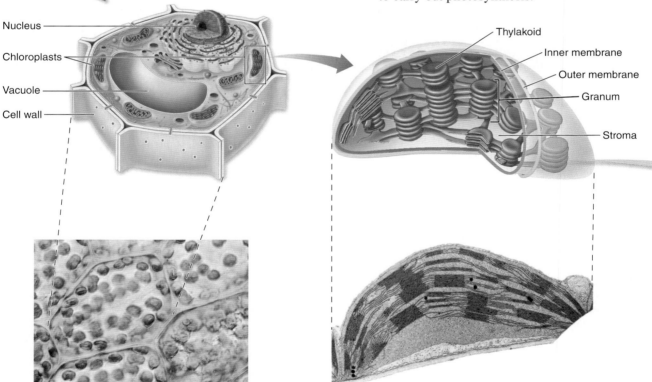

Figure 7.1   Journey into a leaf.

Plant cells within leaves contain chloroplasts, in which thylakoid membranes are stacked. Within the thylakoid, chlorophyll pigments grouped in photosystems drive the reactions of photosynthesis.

Photosynthesis takes place in three stages, all indicated in the lower right panel of figure 7.1: ❶ capturing energy from sunlight—accomplished by the photosystem; ❷ using the energy to make ATP and another key molecule, NADPH— accomplished by the light-dependent reactions; and ❸ using the ATP and NADPH to power the synthesis of carbohydrates from CO_2 in the air—accomplished by the Calvin cycle.

The first two stages take place only in the presence of light and are commonly called the **light-dependent reactions.** The third stage, the formation of organic molecules from atmospheric CO_2, is called the **Calvin cycle,** but is also referred to as the **light-independent reactions** because it doesn't require light directly. However, these reactions do depend indirectly on the light-dependent ones, as they require the ATP and NADPH produced by the light-dependent reactions; perhaps for this reason, several of the enzymes involved in the Calvin cycle are activated by light-dependent reactions.

The overall process of photosynthesis may be summarized by the following simple equation:

$$6\ CO_2 + 12\ H_2O + light \longrightarrow C_6H_{12}O_6 + 6\ H_2O + 6\ O_2$$

carbon dioxide   **water**   **energy**   **glucose**   **water**   **oxygen**

Inside the Chloroplast

The chloroplast is the site of all three stages of photosynthesis. The internal membranes of chloroplasts are organized into flattened sacs called *thylakoids,* and often numerous thylakoids are stacked on top of one another in columns called *grana,* as shown below. Surrounding the thylakoid membrane system is a semiliquid substance called *stroma.* In the membranes of thylakoids, chlorophyll pigments are grouped together in a network called a *photosystem,* which will be discussed in section 7.3.

The photosystem is the starting point of photosynthesis, acting as an antenna to capture photons (units of light energy). A lattice of proteins anchors pigment molecules in precise positions on the antenna. A pigment molecule is a molecule that absorbs light. The primary pigment molecule in most photosystems is **chlorophyll.** The photosystem lattice positions the chlorophyll molecules around one another so that when a photon of light strikes any chlorophyll molecule in the photosystem, the excitation passes from one chlorophyll to another. This is not a chemical reaction, in which an electron physically passes between atoms. Rather, it is energy that passes between chlorophylls. A crude analogy to this form of energy transfer is the initial "break" in a game of pool. If the cue ball squarely hits the point of the triangular array of 15 pool balls, the two balls at the far corners of the triangle fly off, and none of the central balls move at all. The kinetic energy is transferred through the central balls to the most distant ones.

Eventually the energy arrives at a key chlorophyll molecule that is touching a membrane-bound protein. An excited electron is then transferred from that key chlorophyll to an acceptor molecule in the membrane, which passes it in turn to a series of other proteins that put the energy to work making ATP and NADPH in the light-dependent reactions and building organic molecules in the Calvin cycle. We will discuss these stages of photosynthesis in the following sections.

7.1   Photosynthesis uses energy from sunlight to power the synthesis of organic molecules from CO_2 in the air. In plants, photosynthesis takes place in specialized compartments within chloroplasts.

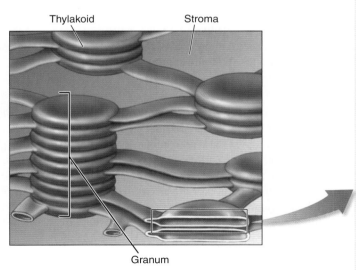

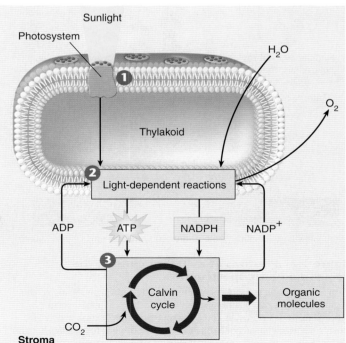

How Plants Capture Energy from Sunlight

Where is the energy in light? What is there about sunlight that a plant can use to create chemical bonds? The revolution in physics in the twentieth century taught us that light actually consists of tiny packets of energy called **photons,** which have properties both of particles and of waves. When light shines on your hand, your skin is being bombarded by a stream of these photons smashing onto its surface.

Sunlight contains photons of many energy levels, only some of which we "see." We call the full range of these photons the **electromagnetic spectrum.** As you can see in figure 7.2, some of the photons in sunlight have shorter wavelengths (toward the left side of the spectrum) and carry a great deal of energy—for example, gamma rays and ultraviolet (UV) light. Others such as radio waves carry very little energy and have longer wavelengths (hundreds to thousands of meters long). Our eyes perceive photons carrying intermediate amounts of energy as **visible light,** because the retinal pigment molecules in our eyes, which are different from chlorophyll, absorb only those photons of intermediate wavelength. Plants are even more picky, absorbing mainly blue and red light and reflecting back what is left of the visible light. To understand why plants are green, look at the

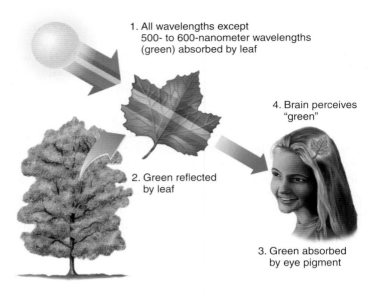

Figure 7.3 Why are plants green?
A leaf containing chlorophyll absorbs a broad range of photons—all the colors in the spectrum except for the photons around 500 to 600 nanometers. The leaf reflects these colors. These reflected wavelengths are absorbed by the visual pigments in our eyes, and our brains perceive the reflected wavelengths as "green."

green tree in figure 7.3. The full spectrum of visible light shines on the leaves of this tree, and only the green wavelengths of light are not absorbed. They are reflected off the leaf, which is why our eyes perceive leaves as green.

How can a leaf or a human eye choose which photons to absorb? The answer to this important question has to do with the nature of atoms. Remember that electrons spin in particular orbits around the atomic nucleus, at different energy levels. Atoms absorb light by boosting electrons to higher energy levels, using the energy in the photon to power the move. Boosting the electron requires just the right amount of energy, no more and no less, just as when climbing a ladder you must raise your foot just so far to climb a rung. A particular kind of atom absorbs only certain photons of light, those with the appropriate amount of energy.

Pigments

As mentioned earlier, molecules that absorb light energy are called **pigments.** When we speak of visible light, we refer to those wavelengths that the pigment within human eyes, called *retinal,* can absorb—roughly from 380 nanometers (violet) to 750 nanometers (red). Other animals use different pigments for vision and thus "see" a different portion of the electromagnetic spectrum. For example, the pigment in insect eyes absorbs at shorter wavelengths than retinal. That is why bees can see ultraviolet light, which we cannot see, but are blind to red light, which we can see.

As noted, the main pigment in plants that absorbs light is chlorophyll. Its two forms, chlorophyll *a* and chlorophyll *b,* are similar in structure, but slight differences in their chemical "side groups" produce slight differences in their absorption spectra. An absorption spectrum is a graph indicating how effectively a pigment absorbs different wavelengths of visible light. For example, chlorophyll molecules will absorb photons

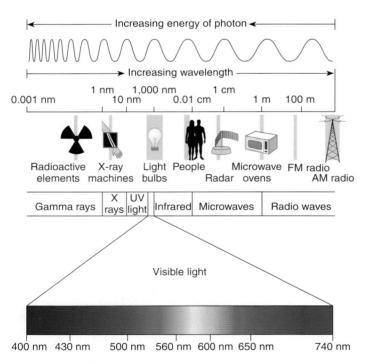

Figure 7.2 Photons of different energy: the electromagnetic spectrum.
Light is composed of packets of energy called photons. Some of the photons in light carry more energy than others. Light, a form of electromagnetic energy, is conveniently thought of as a wave. The shorter the wavelength of light, the greater the energy of its photons. Visible light represents only a small part of the electromagnetic spectrum, that with wavelengths between about 400 and 740 nanometers.

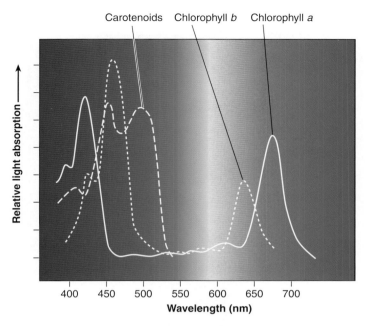

Figure 7.4 Absorption spectra of chlorophylls and carotenoids.

The peaks represent wavelengths of sunlight strongly absorbed by the two common forms of photosynthetic pigment, chlorophyll *a* and chlorophyll *b,* and by accessory pigments called carotenoids. Chlorophylls absorb predominantly violet-blue and red light, in two narrow bands of the spectrum, while they reflect the green light in the middle of the spectrum. Carotenoids absorb mostly blue and green light and reflect orange and yellow light.

at the ends of the visible spectrum, the peaks you see in figure 7.4. While chlorophyll absorbs fewer kinds of photons than our visual pigment retinal, it is much more efficient at capturing them. Chlorophyll molecules capture photons with a metal ion (magnesium) that lies at the center of a complex carbon ring. Photons excite electrons of the magnesium ion, which are then channeled away by the carbon atoms.

While chlorophyll is the primary pigment involved in photosynthesis, plants also contain other pigments called *accessory pigments* that absorb light of wavelengths not captured by chlorophyll. **Carotenoids** are a group of accessory pigments that capture violet to blue-green light. As you can see in figure 7.4, these wavelengths of light are not efficiently absorbed by chlorophyll.

Accessory pigments give color to flowers, fruits, and vegetables but are also present in leaves, their presence usually masked by chlorophyll. During the warm months, when plants are actively producing food through photosynthesis, their cells are filled with chlorophyll-containing chloroplasts that cause the leaves to appear green, like the oak leaves in figure 7.5 on the left. In the fall, the days become shorter and cooler and for many species, leaves stop their food-making processes. Their chlorophyll molecules break down and are not replaced. When this happens, the colors reflected by accessory pigments become visible. The leaves turn colors of yellow, orange, and red like the oak leaves on the right .

> **7.2 Plants use pigments like chlorophyll to capture photons of blue and red light, reflecting photons of green wavelengths.**

Oak leaf in summer

Oak leaf in autumn

Figure 7.5 Fall colors are produced by pigments such as carotenoids.

During the spring and summer, chlorophyll masks the presence in leaves of other pigments called carotenoids. Cool temperatures in the fall cause leaves of deciduous trees to cease manufacturing chlorophyll. With chlorophyll no longer present to reflect green light, the orange and yellow light reflected by carotenoids gives bright colors to the autumn leaves.

Organizing Pigments into Photosystems

The light-dependent reactions of photosynthesis occur on membranes. In most photosynthetic bacteria, the proteins involved in the light-dependent reactions are embedded within the plasma membrane. In algae, intracellular membranes contain the proteins that drive the light-dependent reactions. In plants, photosynthesis occurs in specialized organelles called chloroplasts. The chlorophyll molecules and proteins involved in the light-dependent reactions are embedded in the thylakoid membranes inside the chloroplasts. A portion of a thylakoid membrane is enlarged in figure 7.6. The chlorophyll molecules can be seen as the green spheres embedded along with accessory pigment molecules within a matrix of proteins (the purple area) within the thylakoid membrane. This complex of protein and pigment makes up the **photosystem.**

The light-dependent reactions take place in five stages, illustrated in figure 7.7, which will be discussed in detail later in this chapter:

1. **Capturing light.** In stage ❶, a photon of light of the appropriate wavelength is captured by a pigment molecule, and the excitation energy is passed from one chlorophyll molecule to another.

2. **Exciting an electron.** In stage ❷, the excitation energy is funneled to a key chlorophyll *a* molecule called the **reaction center.** The excitation energy causes the transfer of an excited electron from the reaction center to another molecule that is an electron acceptor. The reaction center replaces this "lost" electron with an electron from the breakdown of a water molecule. Oxygen is produced as a by-product of this reaction.

3. **Electron transport.** In stage ❸, the excited electron is then shuttled along a series of electron-carrier molecules embedded in the membrane. This is called the **electron transport system (ETS).** As the electron passes along the electron transport system, the energy from the electron is "siphoned" out in small amounts. This energy is used to pump hydrogen ions (protons), across the membrane, indicated by the blue arrow, eventually building up a high concentration of protons inside the thylakoid.

4. **Making ATP.** In stage ❹, the high concentration of protons can be used as an energy source to make ATP. Protons are only able to move back across the membrane via special channels, the protons flooding through them, like water through a dam. The kinetic energy that is released by the movement of protons is transferred to potential energy in the building of ATP molecules from ADP. This process, called **chemiosmosis,** makes the ATP that will be used in the Calvin cycle to make carbohydrates.

5. **Making NADPH.** The electron leaves the electron transport system and enters another photosystem where it is "reenergized" by the absorption of another photon

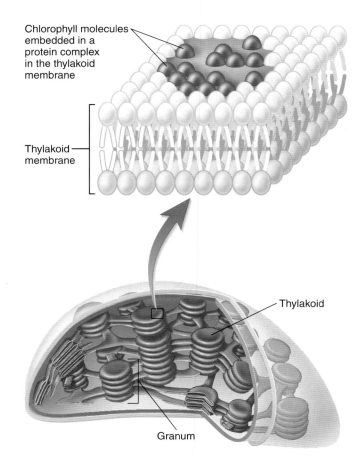

Figure 7.6 Chlorophyll embedded in a membrane.
Chlorophyll molecules are embedded in a network of proteins that hold the pigment molecules in place. The proteins are embedded within the membranes of thylakoids.

of light. In ❺, this energized electron enters another electron transport system, where it is again shuttled along a series of electron-carrier molecules. The result of this electron transport system is not the synthesis of ATP, but rather the formation of NADPH. The electron is transferred to a molecule, $NADP^+$, and a hydrogen ion that forms NADPH. This molecule is important in the synthesis of carbohydrates in the Calvin cycle.

Architecture of a Photosystem

In all but the most primitive bacteria, light is captured by photosystems. Like a magnifying glass focusing light on a precise point, a photosystem channels the excitation energy gathered by any one of its pigment molecules to a specific chlorophyll *a* molecule, the reaction center chlorophyll. For example, in figure 7.8, a chlorophyll molecule on the outer edge of the photosystem is excited by the photon, and this energy passes from one chlorophyll molecule to another, indicated by the yellow zig-zag arrow, until it reaches the reaction center molecule. This molecule then passes the energy, in the form of an excited electron, out of the photosystem to drive the synthesis of ATP and organic molecules.

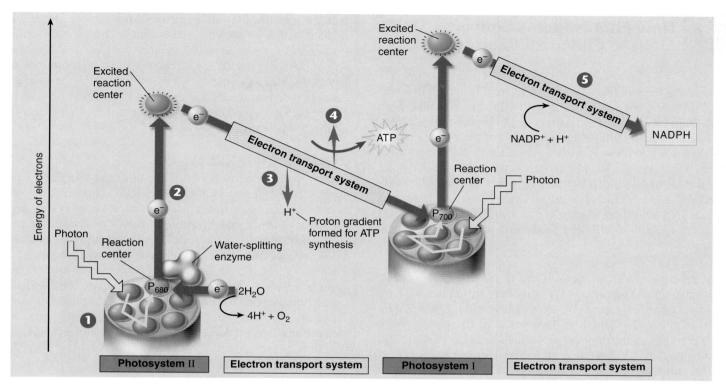

Figure 7.7 Plants use two photosystems.

In stage **❶**, a photon excites pigment molecules in photosystem II. In stage **❷**, a high-energy electron from photosystem II is transferred to the electron transport system. In stage **❸**, the excited electron is used to pump a proton across the membrane. In stage **❹**, the concentration gradient of protons is used to produce a molecule of ATP. In stage **❺**, the ejected electron then passes to photosystem I, which uses it, with a photon of light energy, to drive the formation of NADPH.

Using Two Photosystems

Plants and algae use two photosystems, photosystems I and II, indicated by the two purple cylinders in figure 7.7. Photosystem II captures the energy that is used to produce the ATP needed to build sugar molecules. The light energy that it captures is used in stages **❶** and **❷** to transfer the energy of a photon of light to an excited electron; the energy of this electron is then used by the electron transport system **❸** to produce ATP **❹**.

Photosystem I powers the production of the hydrogen atoms needed to build sugars and other organic molecules from CO_2 (which has no hydrogen atoms). Photosystem I is used in stage **❺** to energize an electron that, carried by a hydrogen ion (a proton), forms NADPH from $NADP^+$. NADPH shuttles hydrogens to the Calvin cycle where sugars are made.

The photosystems are not numbered in the order in which they are used. Photosystem II actually acts first in the series, and photosystem I acts second. The confusion arises because the photosystems were named in the order in which they were discovered, and photosystem I was discovered before photosystem II.

> **7.3 Photon energy is captured by pigments that employ it to excite electrons that are channeled away to do the chemical work of producing ATP and NADPH.**

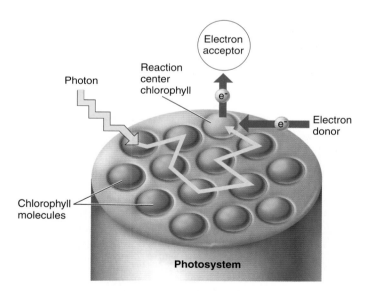

Figure 7.8 How a photosystem works.

When light of the proper wavelength strikes any pigment molecule within a photosystem, the light is absorbed and its excitation energy is then transferred from one molecule to another within the cluster of pigment molecules until it encounters the reaction center, which exports the energy as high-energy electrons to an acceptor molecule.

How Photosystems Convert Light to Chemical Energy

Plants use the two photosystems discussed in series, first one and then the other, to produce both ATP and NADPH. This two-stage process is called **noncyclic photophosphorylation,** because the path of the electrons is not a circle—the electrons ejected from the photosystems do not return to them, but rather end up in NADPH. The photosystems are replenished instead with electrons obtained by splitting water. As described earlier, photosystem II acts first. High-energy electrons generated by photosystem II are used to synthesize ATP and then passed to photosystem I to drive the production of NADPH.

Photosystem II

The reaction center of photosystem II consists of more than 10 transmembrane protein subunits that are represented by the first purple structure you see on the left in figure 7.9. The *antenna complex,* which is the portion of the photosystem that contains all the pigment molecules, consists of some 250 molecules of chlorophyll *a* and accessory pigments bound to several protein chains. The antenna complex captures energy from a photon and funnels it to a reaction center. The antenna complex you see in figure 7.9 is the same photosystem that was illustrated more diagrammatically in figure 7.7. The reaction center gives up an excited electron to a primary electron acceptor in the electron transport system. The path of the excited electron is indicated with the red arrow. After the reaction center gives up an electron to the electron transport system, there is an empty electron orbital that needs to be filled. This electron is replaced with an electron from a water molecule. In photosystem II the oxygen atoms of two

water molecules bind to a cluster of manganese atoms embedded within an enzyme and bound to the reaction center (notice the light gray water-splitting enzyme at the bottom left of photosystem II). This enzyme splits water, removing electrons one at a time to fill the holes left in the reaction center by departure of light-energized electrons. As soon as four electrons have been removed from the two water molecules, O_2 is released.

Electron Transport System

The primary electron acceptor for the light-energized electrons leaving photosystem II passes the excited electron to a series of electron-carrier molecules called the *electron transport system* (indicated by the light purple structure in figure 7.9). These proteins are embedded within the thylakoid membrane; one of them is a "proton pump" protein (discussed in chapter 5). That is, the energy of the electron is used by this protein to pump a proton from the stroma into the thylakoid space (indicated by the blue arrow through the purple electron transport system). A nearby protein in the membrane then carries the now energy-depleted electron on to photosystem I.

Making ATP: Chemiosmosis

Before progressing onto photosystem I, let's see what happens with the protons that were pumped into the thylakoid by the electron transport system. Each thylakoid is a closed compartment into which protons are pumped. Notice that figure 7.10 contains only a portion of the structures shown in figure 7.9. This is because figure 7.10 focuses your attention on the process of ATP formation, called *chemiosmosis,* which involves photosystem II, the first electron transport system, and a new structure called ATP synthase. The thylakoid membrane you see in figure 7.10 is impermeable to protons, so protons build up inside the thylakoid

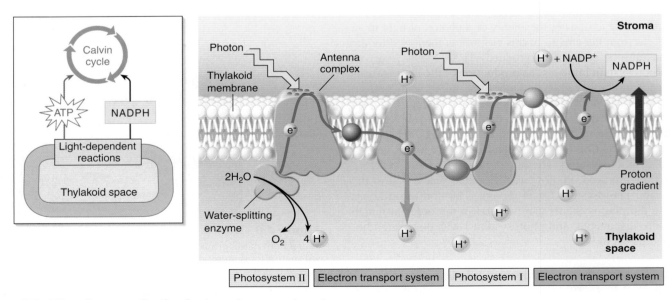

Figure 7.9   The photosynthetic electron transport system.

space, creating a very large concentration gradient. As you may recall from chapter 5, molecules in solution diffuse from areas of higher concentration to areas of lower concentration. Here protons diffuse back out of the thylakoid space, down their concentration gradient, passing through special channel proteins called *ATP synthases*. The ATP synthase is the third membrane protein on the far right side of figure 7.10, protruding like a knob out of the external surface of the thylakoid membrane. As protons pass out of the thylakoid through the ATP synthase channels, ADP is phosphorylated to ATP and released into the stroma (the fluid matrix inside the chloroplast). The stroma contains the enzymes that catalyze the light-independent reactions, where ATP is used to build molecules of sugar.

Photosystem I

Now, with ATP formed, let's return our attention to figure 7.9, with photosystem I (the second purple structure) accepting an electron from the electron transport system. The reaction center of photosystem I is a membrane complex consisting of at least 13 protein subunits. Energy is fed to it by an antenna complex consisting of 130 chlorophyll *a* and accessory pigment molecules. The electron arriving from the electron transport system has by no means lost all of its light-excited energy; almost half remains. Thus, the absorption of another photon of light energy by photosystem I boosts the electron leaving its reaction center to a very high energy level.

Making NADPH

Like photosystem II, photosystem I passes electrons to an electron transport system (the dark pink structure on the right in figure 7.9). When two of these electrons reach the end of this electron transport system, they are then donated to a molecule of $NADP^+$ to form NADPH. This reaction, which takes place on the stromal side of figure 7.9, involves an $NADP^+$, two electrons, and a proton. Because the reaction occurs on the stromal side of the membrane and involves the uptake of a proton in forming NADPH, it contributes further to the proton gradient established during photosynthetic electron transport.

Products of the Light-Dependent Reactions

The light-dependent reactions can be seen more as a stepping stone, rather than an end point of photosynthesis. All of the products of the light-dependent reactions are either waste products, such as oxygen, or are ultimately used elsewhere in the cell. The ATP and NADPH produced in the light-dependent reactions end up being passed on to the Calvin cycle in the stroma of the chloroplast. There, ATP is used to power chemical reactions that build carbohydrates. NADPH is used as the source of "reducing power," providing the hydrogens and electrons used in building carbohydrates. The next section discusses the Calvin cycle of photosynthesis.

> **7.4** **The light-dependent reactions of photosynthesis produce the ATP and NADPH needed to build organic molecules, and release O_2 as a by-product of stripping hydrogen atoms and their associated electrons from water molecules.**

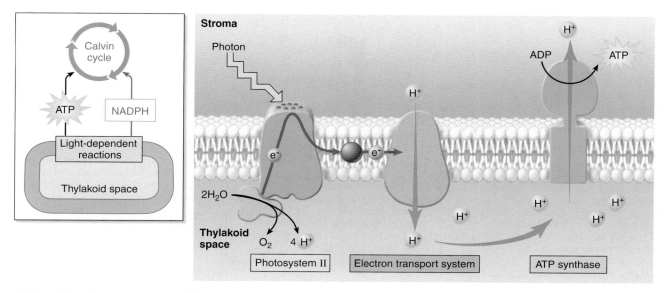

Figure 7.10   Chemiosmosis in a chloroplast.
One of the proteins of the photosynthetic electron transport system pumps protons into the interior of the thylakoid. ATP is produced on the outside surface of the membrane (stroma side), as protons diffuse back out of the thylakoid through ATP synthase channels.

Building New Molecules

The Calvin Cycle

Stated very simply, photosynthesis is a way of making organic molecules from carbon dioxide (CO_2). To build organic molecules, cells use raw materials provided by the light-dependent reactions:

1. **Energy.** ATP (provided by the ETS of photosystem II) drives the endergonic reactions.
2. **Reducing power.** NADPH (provided by the ETS of photosystem I) provides a source of hydrogens and the energetic electrons needed to bind them to carbon atoms. Recall from the discussion of redox reactions in chapter 6 that a molecule that accepts an electron is said to be reduced.

The actual assembly of new molecules employs a complex battery of enzymes in what is called the **Calvin cycle,** or **C_3 photosynthesis** (C_3 because the first molecule produced in the process is a three-carbon molecule). The process takes place in three stages, highlighted in the three panels of figure 7.11. These three stages are also indicated by different-colored pie-shaped pieces in the more detailed look at the Calvin cycle provided by figure 7.12. Both figures indicate that three turns of the cycle are needed to produce one mol-

The Calvin Cycle

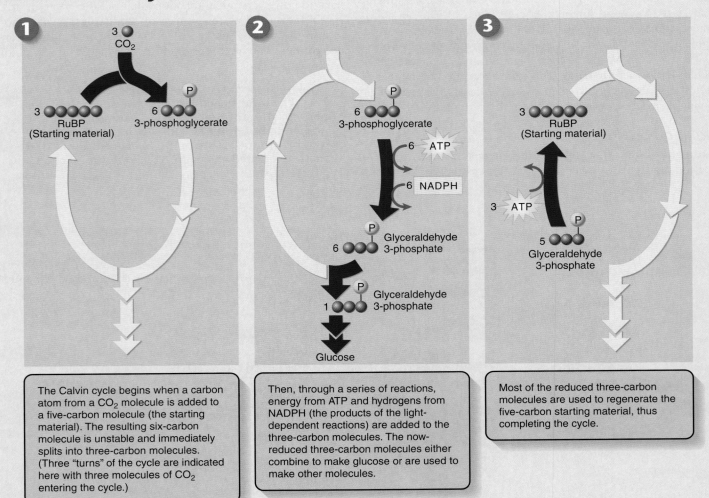

The Calvin cycle begins when a carbon atom from a CO_2 molecule is added to a five-carbon molecule (the starting material). The resulting six-carbon molecule is unstable and immediately splits into three-carbon molecules. (Three "turns" of the cycle are indicated here with three molecules of CO_2 entering the cycle.)

Then, through a series of reactions, energy from ATP and hydrogens from NADPH (the products of the light-dependent reactions) are added to the three-carbon molecules. The now-reduced three-carbon molecules either combine to make glucose or are used to make other molecules.

Most of the reduced three-carbon molecules are used to regenerate the five-carbon starting material, thus completing the cycle.

Figure 7.11 How the Calvin cycle works.
The Calvin cycle takes place in the stroma of the chloroplasts. The NADPH and the ATP that were generated by the light-dependent reactions are used in the Calvin cycle to build carbohydrate molecules. The number of carbon atoms at each stage is indicated by the number of balls. It takes six turns of the cycle to make one six-carbon molecule of glucose.

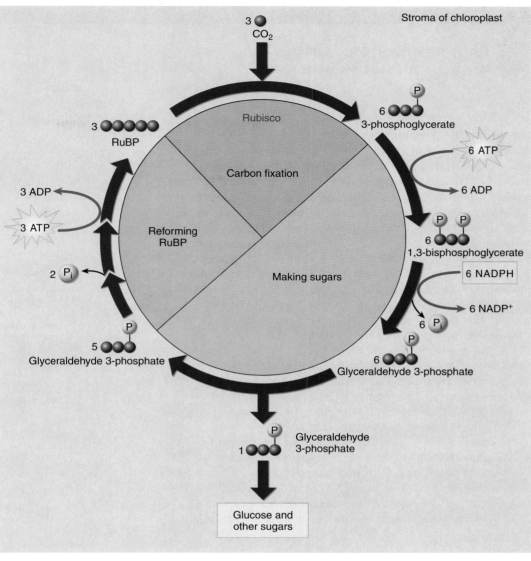

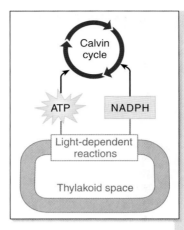

Figure 7.12
Reactions of the Calvin cycle.

For every three molecules of CO_2 that enter the cycle, one molecule of the three-carbon compound glyceraldehyde 3-phosphate (G3P) is produced. Notice that the process requires energy stored in ATP and NADPH, which are generated by the light-dependent reactions. This process occurs in the stroma of the chloroplast. The large 16-subunit enzyme that catalyzes the reaction, RuBP carboxylase, or **rubisco**, is the most abundant protein in chloroplasts and is thought to be the most abundant protein on earth.

ecule of glyceraldehyde 3-phosphate. In any *one* turn of the cycle, a carbon atom from a carbon dioxide molecule is first added to a five-carbon sugar, producing two three-carbon sugars. This process, highlighted by the dark blue arrow in panel 1 of figure 7.11 and the blue pie-shaped area in figure 7.12, is called *carbon fixation* because it attaches a carbon atom that was in a gas to an organic molecule.

Then, in a long series of reactions, the carbons are shuffled about. Eventually some of the resulting molecules are channeled off to make sugars (shown by the dark blue arrows in panel 2 of figure 7.11 and at the bottom of the cycle within the purple colored area in figure 7.12). Other molecules are used to re-form the original five-carbon sugar (the dark blue arrow in panel 3 of figure 7.11 and the dark-pink-colored area in figure 7.12), which is then available to restart the cycle. The cycle has to "turn" six times in order to form a new glucose molecule, because each turn of the cycle adds only one carbon atom from CO_2, and glucose is a six-carbon sugar.

Recycling ADP and NADP⁺

The products of the light-dependent reactions, ATP and NADPH, feed into the light-independent reactions of the Calvin cycle to make sugar molecules. To keep photosynthesis moving along, the cells must continually supply the light-dependent reactions with more ADP and $NADP^+$. This is accomplished by recycling these products from the Calvin cycle. After the phosphate bonds are broken in ATP, ADP is available for chemiosmosis. After the hydrogens and electrons are stripped from NADPH, $NADP^+$ is available to cycle back to the electron transport system of photosystem I.

> **7.5 In a series of reactions that do not directly require light, cells use ATP and NADPH provided by photosystems II and I to assemble new organic molecules.**

7.6 Photorespiration: Putting the Brakes on Photosynthesis

C_4 Photosynthesis

Many plants have trouble carrying out C_3 photosynthesis when the weather is hot. A cross section of a leaf showing how it responds to hot, arid weather is shown in figure 7.13. As temperature increases in the upper panel, plants partially close their leaf openings, called **stomata** (singular, **stoma**), to conserve water. As a result, in the lower panel you can see that CO_2 and O_2 are not able to enter and exit the leaves through these openings. The concentration of CO_2 in the leaves falls, while the concentration of O_2 in the leaves rises. Under these conditions rubisco, the enzyme that carries out the first step of the Calvin cycle, engages in **photorespiration,** where the enzyme incorporates O_2, not CO_2, into the cycle and when this occurs, CO_2 is ultimately released as a by-product. Photorespiration thus short-circuits the successful performance of the Calvin cycle.

Some plants are able to adapt to climates with higher temperatures by performing **C_4 photosynthesis.** In this process, plants such as sugarcane, corn, and many grasses are able to fix carbon using different types of cells and chemical reactions within their leaves and thereby avoiding a reduction in photosynthesis due to higher temperatures.

A cross section of a leaf from a C_4 plant is shown in figure 7.14. Examining it, you can see how these plants solve the problem of photorespiration. In the enlargement, you see two cell types: the green cell is a mesophyll cell and the tan cell is a bundle-sheath cell. In the mesophyll cell, CO_2 combines with a three-carbon molecule instead of RuBP as it did in figure 7.12,

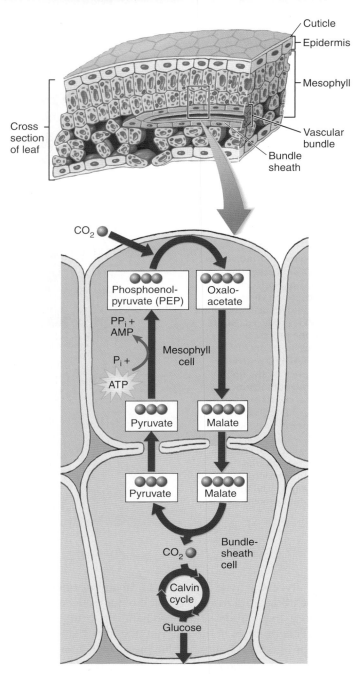

Figure 7.14   Carbon fixation in C_4 plants.
This process is called the C_4 pathway because the first molecule formed in the pathway is a four-carbon sugar, oxaloacetate. This molecule is converted into malate that is transported into bundle-sheath cells. Once there, malate undergoes a chemical reaction producing carbon dioxide. The carbon dioxide is trapped in the bundle-sheath cell, where it enters the Calvin cycle.

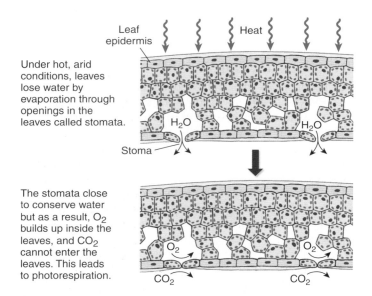

Figure 7.13   Plant response in hot weather.

The Redox Cycle

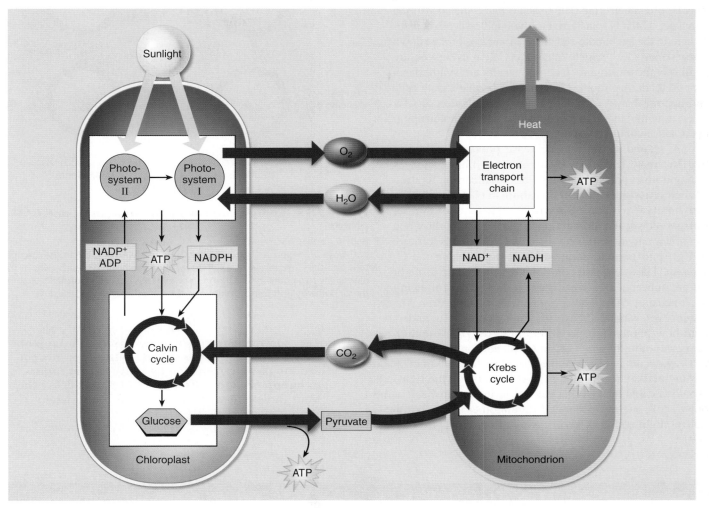

The energy-capturing metabolism of the chloroplasts studied in this chapter and the energy-utilizing metabolism of the mitochondria studied in chapter 8 are intimately related. Photosynthesis carried out by chloroplasts uses the products of cellular respiration as starting substrates and reduces carbon by adding hydrogen atoms. Cellular respiration carried out by mitochondria uses the products of photosynthesis as its starting substrates and oxidizes carbon by removing hydrogen atoms. Together, photosynthesis and cellular respiration form an oxidation-reduction cycle, as diagrammed here. Note that it is electrons that cycle between chloroplasts and mitochondria, not energy. Energy passes through the redox cycle, flowing into the cycle from the sun, passing through newly assembled molecules, and eventually flowing out of the cycle as heat.

The evolutionary history of the redox cycle can be seen in its elements. The Calvin cycle of photosynthesis uses part of the glycolytic pathway of cellular respiration, run in reverse, to produce glucose. The principal proteins involved in electron transport in chloroplasts are related to those in mitochondria, and in many cases are actually the same. Biologists believe the glycolysis stage of cellular respiration, which takes place in the cytoplasm, converting glucose to pyruvate and requiring no oxygen, to be the most ancient process in the energy cycle. The chloroplast's oxygen-generating photosynthesis is thought to have evolved next, followed later by the oxidative reactions of the mitochondria's cellular respiration.

producing a four-carbon molecule, oxaloacetate (hence the name, C_4 photosynthesis), rather than the three-carbon molecule phosphoglycerate you saw in figure 7.12. C_4 plants carry out this process in the mesophyll cells of their leaves, using a different enzyme. The oxaloacetate is then converted to malate, which is transferred to the bundle-sheath cells of the leaf. In the tan bundle-sheath cell, malate is broken down to regenerate CO_2, which enters the Calvin cycle you are familiar with from figure 7.12, and sugars are synthesized. Why go to all this trouble? Because the bundle-sheath cells are impermeable to CO_2 and so the concentration of CO_2 increases within them, so much that the rate of photorespiration is substantially lowered.

A second strategy to decrease photorespiration is used by many succulent (water-storing) plants such as cacti and pineapples. This mode of initial carbon fixation is called **crassulacean acid metabolism (CAM)** after the plant family Crassulaceae in which it was first discovered. In these plants, the stomata open during the night when it's cooler, and close during the day. CAM plants initially fix CO_2 into organic compounds at night, using the C_4 pathway. These organic compounds accumulate at night and are subsequently broken down during the following day, releasing CO_2. These high levels of CO_2 drive the Calvin cycle and decrease photorespiration. To understand how photosynthesis differs in CAM plants and C_4 plants, examine figure 7.15. In C_4 plants (on the left), the C_4 pathway occurs in mesophyll cells, while the Calvin cycle occurs in bundle-sheath cells. In CAM plants (on the right), the C_4 pathway and the Calvin cycle occur in the same cell, a mesophyll cell, but they occur at different times of the day, the C_4 cycle at night and the Calvin cycle during the day.

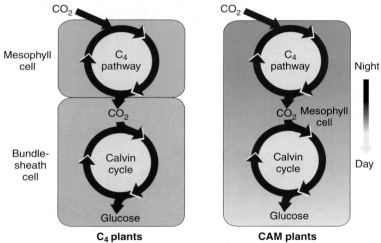

Figure 7.15 Comparing carbon fixation in C_4 and CAM plants.

Both C_4 and CAM plants utilize the C_4 and C_3 pathways. In C_4 plants, the pathways are separated spatially; the C_4 pathway takes place in the mesophyll cells and the C_3 pathway (the Calvin cycle) in the bundle-sheath cells. In CAM plants, the two pathways occur in mesophyll cells but are separated temporally; the C_4 pathway is utilized at night and the C_3 pathway during the day.

7.6 Photorespiration occurs due to a buildup of oxygen within photosynthetic cells. C_4 plants get around photorespiration by synthesizing sugars in bundle-sheath cells, and CAM plants delay the light-independent reactions until night, when stomata are open.

INQUIRY & ANALYSIS

Is Photosynthesis Faster in Brighter Light?

The graph to the right is a 1975 experiment comparing photosynthesis in C_3 and C_4 plants. Researchers found that C_3 plants reach a maximal rate of photosynthesis at moderately low levels of light, far below the maximum level of light they typically experience (indicated by the arrow), while C_4 plants do not.

1. **Applying Concepts** In the graph, what is the dependent variable?
2. **Making Inferences** Is C_3 photosynthesis faster in very bright light than in moderately bright light? C_4?
3. **Drawing Conclusions** What hypothesis would you advance to explain why the rate of C_3 photosynthesis saturates? [Hint: What organelle is actually carrying out photosynthesis? Does a plant have an unlimited number of them?]

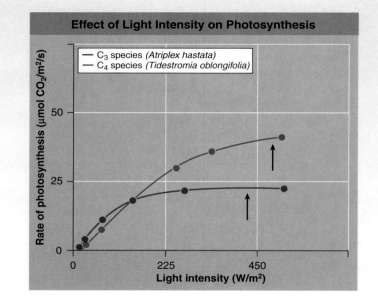

Effect of Light Intensity on Photosynthesis

— C_3 species (*Atriplex hastata*)
— C_4 species (*Tidestromia oblongifolia*)

Photosynthesis

7.1 An Overview of Photosynthesis

- Photosynthesis is the process whereby energy from the sun is captured and used to build carbohydrates from CO_2 gas.

- Photosynthesis is a series of chemical reactions that occurs in two steps: the light-dependent reactions that produce ATP and NADPH occur on the thylakoid membranes of chloroplasts in plants and the light-independent reactions, or the Calvin cycle, that synthesize carbohydrates, occur in the stroma (**figure 7.1**).

7.2 How Plants Capture Energy from Sunlight

- Pigments are molecules that capture light energy. Energy present in visible light is captured by chlorophyll and other accessory pigments present in chloroplasts.

- Plants appear green because of these pigments. Chlorophyll absorbs wavelengths at the far ends of the visual spectrum (the blue and red wavelengths) and reflects the green wavelengths. This is why leaves appear green (**figures 7.3** and **7.4**).

7.3 Organizing Pigments into Photosystems

- The light-dependent reactions occur on the thylakoid membranes within the chloroplast. The protein and pigment molecules involved in photosynthesis are embedded in membranes: the plasma membrane in certain bacteria, internal membranes in algae, and thylakoid membranes of chloroplasts in plants. These embedded proteins and pigments make up a photosystem (**figure 7.6**).

- Light energy is captured by the photosystem and used to excite an electron that is passed to an electron transport system where it is used to generate ATP and NADPH, both which are used in the Calvin cycle. Plants utilize two photosystems that occur in series. Photosystem II leads to the formation of ATP, and photosystem I leads to the formation of NADPH (**figures 7.7** and **7.8**).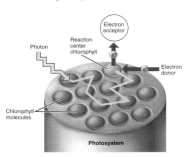

7.4 How Photosystems Convert Light to Chemical Energy

- In plants, the antenna complex of photosystem II first captures energy from the sun. The energy is transferred to a chlorophyll molecule called the reaction center, which gives up an excited electron to a series of proteins within the membrane called the electron transport system. The lost electron is replenished with an electron from a water molecule.

- The excited electron is passed from one protein to another in the electron transport system, where energy from the electron is used to operate a proton pump that pumps hydrogen ions across the membrane against a concentration gradient (**figure 7.9**).

- The electron is then transferred to a second photosystem, photosystem I, where it gets an energy boost from the capture of another photon of light. This reenergized electron is passed

along another series of proteins to an ultimate electron acceptor, NADP+. NADP+ + electrons and a H+ produces NADPH, which is shuttled to the Calvin cycle (**figure 7.9**).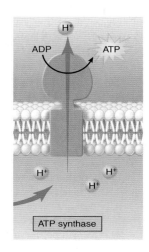

- The hydrogen ion concentration gradient is used as a source of energy to generate molecules of ATP. This energy is used to drive H+ back across the membrane through specialized channel proteins called ATP synthases, which catalyze the formation of ATP (**figure 7.10**).

7.5 Building New Molecules

- The products of the light-dependent reactions, ATP and NADPH, are shuttled to the stroma where they are used in the Calvin cycle.

- The Calvin cycle is a series of enzymes that use the energy from ATP and electrons and hydrogens from NADPH to build carbohydrates using CO_2 from the air (**figures 7.11** and **7.12**).

Photorespiration

7.6 Photorespiration: Putting the Brakes on Photosynthesis

- In hot dry weather, the stomata in the leaves close, causing the levels of O_2 to increase in the leaves and CO_2 levels to drop (**figure 7.13**). Under these conditions, the Calvin cycle, also called C_3 photosynthesis, is disrupted because O_2 rather than CO_2 enters the Calvin cycle in a process called photorespiration.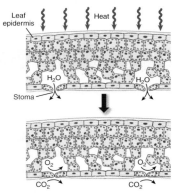

- C_4 plants reduce the effects of photorespiration by modifying the carbon-fixation step by splitting carbon fixation into two steps that take place in different cells. An alternate C_4 pathway, producing malate, is conducted in mesophyll cells. Malate is then transferred to bundle-sheath cells, where it breaks down to produce carbon dioxide that enters the Calvin cycle in the bundle-sheath cells (**figure 7.14**).

- CAM plants reduce the effects of photorespiration by fixing carbon at night. In CAM plants, carbon dioxide is processed through the C_4 pathway during the day when stomata are closed. Carbon dioxide is then released from intermediate molecules at night, where it then enters the Calvin cycle. (**figure 7.15**).

1. The energy that is used by almost all living things on our planet comes from the sun. It is captured by plants, algae, and some bacteria through the process of
   a. thylakoid.
   c. photosynthesis.
   b. chloroplasts.
   d. the Calvin cycle.
2. Plants capture sunlight
   a. with netlike structures in their leaves that catch and hold photons.
   b. with molecules called pigments that absorb photons and use their energy.
   c. with sticky molecules that absorb photons and transfer their energy.
   d. with the electron transport system.
3. Visible light occupies what part of the electromagnetic spectrum?
   a. the entire spectrum
   b. the upper half of the spectrum (with longer wavelengths)
   c. a small portion in the middle of the spectrum
   d. the lower half of the spectrum (with shorter wavelengths)
4. The colors of light that are most effective for photosynthesis are
   a. red, blue, and orange.
   b. green, yellow, and orange.
   c. infrared and ultraviolet.
   d. All colors are equally effective.
5. Once a plant has initially captured the energy of a photon
   a. a series of reactions occurs in chloroplast membranes of the cell.
   b. the energy is transferred through several steps into a molecule of ATP.
   c. several pigments, including two kinds of chlorophyll, may be involved.
   d. All of these are true.

6. Plants use two photosystems to capture energy used to produce ATP and NADPH. The electrons used in these photosystems
   a. recycle through the system constantly, with energy added from the photons.
   b. recycle through the system several times, using energy added from photons, and then are lost due to entropy.
   c. only go through the system once; they are obtained by splitting a water molecule.
   d. only go through the system once; they are obtained from the photon.
7. During photosynthesis, ATP molecules are generated by
   a. the Calvin cycle.
   b. chemiosmosis.
   c. the splitting of a water molecule.
   d. photons of light being absorbed by chlorophyll molecules.
8. NADPH is recycled during photosynthesis. It is produced during the _____ and used in the_____.
   a. electron transport system of photosystem I, Calvin cycle
   b. process of chemiosmosis, Calvin cycle
   c. electron transport system of photosystem II, electron transport system of photosystem I
   d. light-independent reactions, light-dependent reactions
9. The overall purpose of the Calvin cycle is to
   a. generate molecules of ATP.
   b. generate NADPH.
   c. build sugar molecules.
   d. give off oxygen.
10. Many plants cannot carry out the typical C_3 photosynthesis in hot weather, so some plants
   a. use the ATP cycle.
   b. use C_4 photosynthesis or CAM.
   c. wait until evening to begin photosynthesis.
   d. All of these are true for different plants.

Visual Understanding

1. **Figure 7.4** Why do most leaves have more than one type of pigment?

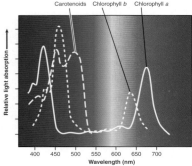

2. **Figure 7.10** Could a plant cell produce ATP through chemiosmosis if the thylakoid membrane was "leaky" with regards to protons? Explain.

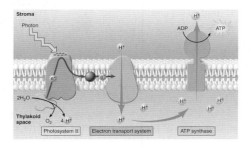

Challenge Questions

1. To reduce six molecules of carbon dioxide to glucose via photosynthesis, how many molecules of NADPH and ATP are required?
2. In theory, a plant kept in total darkness could still manufacture glucose, if it were supplied with which molecules?

3. If you were going to design a plant that would survive in the deserts of Arizona and New Mexico, how would you balance its need for CO_2 with its need to avoid water loss in the hot summer temperatures?

8

How Cells Harvest Energy from Food

A nimals such as this chipmunk depend on the energy stored in the chemical bonds of the food they eat to power their life processes. Their lives are driven by energy. All the activities this chipmunk carries out—climbing trees, chewing on acorns, seeing and smelling and hearing its surroundings, thinking the thoughts that chipmunks think—use energy. But unlike the oak tree that produces the nuts on which this chipmunk is dining, no part of the chipmunk is green. It cannot carry out photosynthesis like an oak tree, and so cannot harvest energy from the sun as the tree does. Instead, it must get its energy secondhand, by consuming organic molecules manufactured by plants. The chemical energy that the oak tree invested in making its molecules is harvested by the chipmunk in a process called cellular respiration. The same processes are used by all animals to harvest energy from molecules—and by plants too. There is no sunlight under the soil where the oak tree's roots penetrate, and like the cells of the chipmunk, these plant root cells obtain the energy to fuel their lives from cellular respiration. In this chapter, we examine cellular respiration up close. As you will see, cellular respiration and photosynthesis have much in common.

8.1 Where Is the Energy in Food?

In both plants and animals, and in fact in almost all organisms, the energy for living is obtained by breaking down the organic molecules originally produced in plants. The ATP energy and reducing power invested in building the organic molecules are retrieved by stripping away the energetic electrons and using them to make ATP. When electrons are stripped away from chemical bonds, the food molecules are being oxidized (remember, oxidation is the loss of electrons). The oxidation of foodstuffs to obtain energy is called **cellular respiration.** Do not confuse the term cellular respiration with the breathing of oxygen gas that your lungs carry out, which is called simply respiration.

The cells of plants fuel their activities with sugars and other fuel molecules, just as yours do; only the chloroplasts carry out photosynthesis. No light shines on roots below the ground, and yet root cells are just as alive as the cells in the stem and leaves. Why would plant cells manufacture organic molecules only to turn around and break them down? The organic molecules produced by plants serve two functions, to build plant tissue and to store energy for future needs. The plant breaks down "storage molecules" during cellular respiration. Nonphotosynthetic organisms eat plants, extracting energy from plant tissue and plant storage molecules in cellular respiration. Other animals, like the lion gnawing with such relish on a giraffe leg in figure 8.1, eat these animals.

In aerobic respiration, which requires oxygen, ATP is formed as electrons are harvested, transferred along an electron transport chain (similar to the electron transport system in photosynthesis), and eventually donated to oxygen gas. Eukaryotes produce the majority of their ATP from glucose in this way. Chemically, there is little difference between this oxidation of carbohydrates in a cell and the burning of wood in a fireplace. In both instances, the reactants are carbohydrates and oxygen, and the products are carbon dioxide, water, and energy:

$$C_6H_{12}O_6 + 6\ O_2 \longrightarrow 6\ CO_2 + 6\ H_2O + \textbf{energy}$$
(heat or ATP)

Cellular respiration is carried out in two stages, illustrated in figure 8.2: The first stage uses coupled reactions to make ATP. This stage, *glycolysis,* takes place in the cell's cytoplasm, (the blue area in figure 8.2). Importantly, it does not require oxygen. This ancient energy-extracting process is thought to have evolved over 2 billion years ago, when there was no oxygen in the earth's atmosphere.

The second stage, which requires oxygen, takes place within the mitochondrion (the tan sausage-shaped structure in figure 8.2). The focal point of this stage is the *Krebs cycle,* shown as the dark blue circle, a cycle of chemical reactions

Figure 8.1   Lion at lunch.
Energy that this lion extracts from its meal of giraffe will be used to power its roar, fuel its running, and build a bigger lion.

that harvests electrons from C—H chemical bonds and passes the energy-rich electrons through the electron transport chain, which uses their energy to power the production of ATP. As you see in the figure, the harvested electrons are first transferred to electron carriers, NADH and $FADH_2$, that then deliver the electrons, indicated by the long red arrow on the left, to the electron transport chain. The harvesting of electrons, a form of *oxidation,* is far more powerful than glycolysis at recovering energy from food molecules, and is how the bulk of the energy used by eukaryotic cells is extracted from food molecules.

> **8.1   Cellular respiration is the dismantling of food molecules to obtain energy. In aerobic respiration, the cell harvests energy from glucose molecules in two stages, glycolysis and oxidation. Oxygen is the final electron acceptor.**

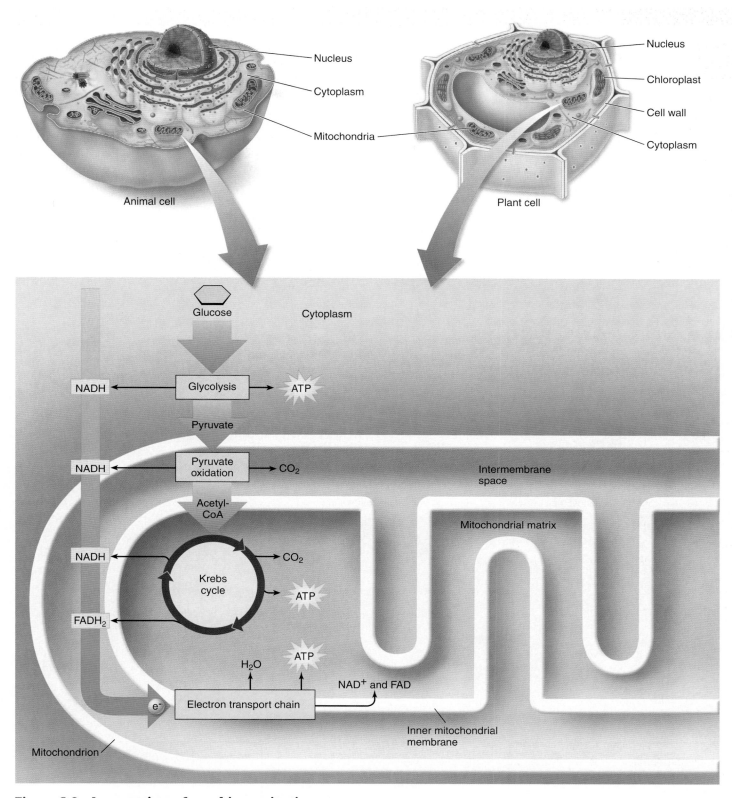

Figure 8.2 An overview of aerobic respiration.

8.2 Using Coupled Reactions to Make ATP

Glycolysis

The first stage in cellular respiration, **glycolysis** is a series of sequential biochemical reactions, a *biochemical pathway*. In 10 enzyme-catalyzed reactions, the six-carbon sugar glucose is cleaved into two three-carbon molecules called pyruvate. Figure 8.3 presents a conceptual overview of the process, while figure 8.4 provides a more detailed look at the series of 10 biochemical reactions. Where is the energy extracted? In each of two "coupled" reactions (steps 7 and 10 in figure 8.4), the breaking of a chemical bond in an exergonic reaction releases enough energy to drive the formation of an ATP molecule from ADP (an endergonic reaction). This transfer of a high-energy phosphate group from a substrate to ADP is called **substrate-level phosphorylation.** In the process, electrons and hydrogen atoms are extracted and donated to a carrier molecule called NAD^+. The NAD^+ carries the electrons as NADH to join the other electrons extracted during oxidative respiration, discussed in the following section. Only a small number of ATP molecules are made in glycolysis itself, two for each molecule of glucose, but in the absence of oxygen this is the only way organisms can get energy from food.

Glycolysis is thought to have been one of the earliest of all biochemical processes to evolve. Every living creature is capable of carrying out glycolysis. Glycolysis, although inefficient, was not discarded during the course of evolution but rather was used as the starting point for the further extraction

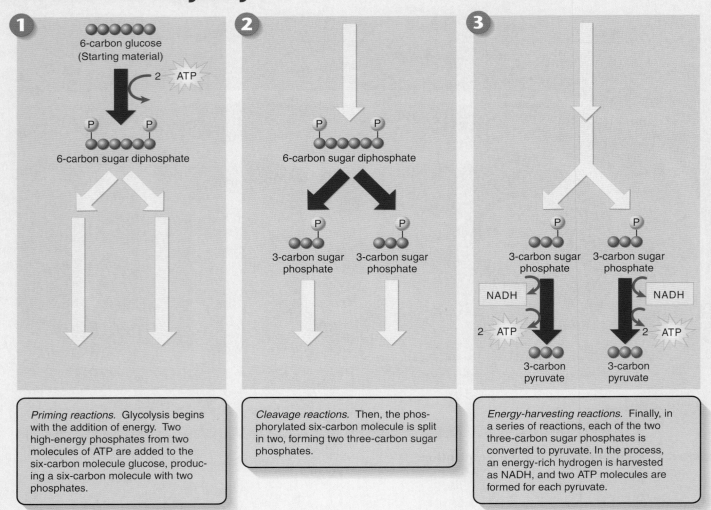

Overview of Glycolysis

1

6-carbon glucose
(Starting material)

2 ATP

P   P

6-carbon sugar diphosphate

Priming reactions. Glycolysis begins with the addition of energy. Two high-energy phosphates from two molecules of ATP are added to the six-carbon molecule glucose, producing a six-carbon molecule with two phosphates.

2

P   P

6-carbon sugar diphosphate

P   P

3-carbon sugar   3-carbon sugar
phosphate   phosphate

Cleavage reactions. Then, the phosphorylated six-carbon molecule is split in two, forming two three-carbon sugar phosphates.

3

P   P

3-carbon sugar   3-carbon sugar
phosphate   phosphate

NADH   NADH

2 ATP   2 ATP

3-carbon   3-carbon
pyruvate   pyruvate

Energy-harvesting reactions. Finally, in a series of reactions, each of the two three-carbon sugar phosphates is converted to pyruvate. In the process, an energy-rich hydrogen is harvested as NADH, and two ATP molecules are formed for each pyruvate.

Figure 8.3 How glycolysis works.

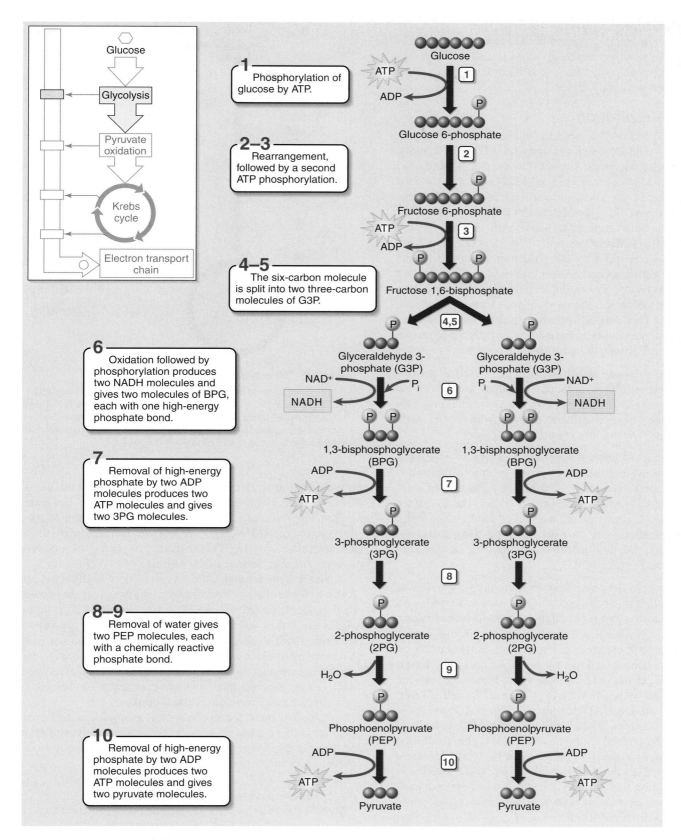

Figure 8.4 The reactions of glycolysis.
The process of glycolysis involves 10 enzyme-catalyzed reactions.

of energy by oxidation. Nature did not, so to speak, go back to the drawing board and design metabolism from scratch. Rather, new reactions, which make up what is called the *Krebs cycle,* were added onto the old, just as renovations to a house build upon what is already there.

Anaerobic Respiration

As explained in section 8.1, aerobic cellular respiration occurs in two stages—coupled reactions to make ATP in glycolysis, and oxidation reactions in the Krebs cycle and electron transport chain to make additional molecules of ATP. Oxygen is the final electron acceptor in the oxidation reactions, accepting the electrons carried by NADH. In the presence of oxygen, cells can use both stages of cellular respiration, because the required oxygen is available for the oxidation reactions. In the absence of oxygen, some organisms can still carry out oxidation reactions to make ATP by using electron acceptors other than oxygen. They are said to respire anaerobically. For example, many bacteria use sulfur, nitrate, or other inorganic compounds as the electron acceptor in place of oxygen. Other organisms use organic molecules as electron acceptors. For example, some eukaryotic cells use pyruvate, the end product of glycolysis, as an electron acceptor.

Methanogens. Among the organisms that practice anaerobic respiration are primitive archaea. A group of archaea called methanogens use CO_2 as the electron acceptor, reducing CO_2 to CH_4 (methane), using hydrogens derived from organic molecules produced by other organisms.

Sulfur Bacteria. A second anaerobic respiratory process is carried out by certain primitive bacteria. In this sulfate respiration, the bacteria derive energy from the reduction of inorganic sulfate (SO_4) to hydrogen sulfide (H_2S). The hydrogen atoms are obtained from organic molecules produced by other organisms. These bacteria thus do the same thing methanogens do, but they use SO_4 as the oxidizing (that is, electron-accepting) agent in place of CO_2.

Fermentation. In the absence of oxygen, does the pyruvate that is the product of glycolysis and the starting material for oxidative respiration just accumulate in the cytoplasm of aerobic organisms? No. It has a different fate. Recall that during glycolysis energetic electrons are extracted, carried away on protons that are contributed to a carrier molecule called NAD^+ to form NADH. In the absence of oxygen, these electrons are not used in oxidative reactions, and so soon all the cell's NAD^+ becomes converted to NADH. With no more NAD^+ available to carry away electrons, glycolysis cannot proceed. Clearly, to obtain energy from food in the absence of oxygen, a solution to this problem is needed. A home must be found for these electrons, recycling the NADH back to the NAD^+ needed in glycolysis. Adding the extracted electrons to an organic molecule, as animals and plants do when they have no oxygen to take it, is called **fermentation.**

Two types of fermentation are common among eukaryotes, illustrated in figure 8.5. Animals such as ourselves simply add the extracted electrons to pyruvate (the center arrow in the figure), forming lactate. Later, when oxygen becomes available, the process can be reversed and the electrons used

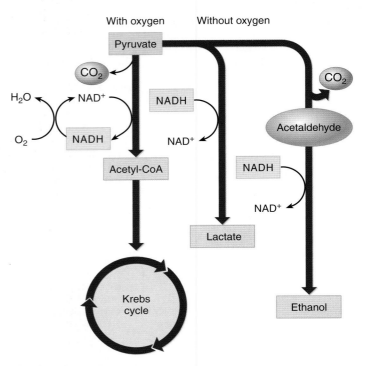

Figure 8.5 Two types of fermentation.
In the presence of oxygen, pyruvate is oxidized to acetyl-CoA and enters the Krebs cycle. In the absence of oxygen, pyruvate is reduced in a process called fermentation. When pyruvate is reduced directly, the product is lactate; when CO_2 is first removed from pyruvate and the remainder is reduced, the product is ethanol.

for energy production. This is why your arm muscles would feel tired if you were to lift this text up and down 100 times rapidly. The muscle cells use up all the oxygen, and so they start running on ATP made by glycolysis, storing the pyruvate and electrons as lactate. This so-called oxygen debt produces the tired, burning feeling in the muscle.

Single-celled fungi called yeasts adopt a different approach to fermentation. First they convert the pyruvate into another molecule called acetaldehyde (the purple oval in figure 8.5), to which they add the electron extracted during glycolysis, producing ethyl alcohol (ethanol). For centuries, humans have consumed ethyl alcohol produced by fermentation in wine and beer. Yeasts conduct this process only in the absence of oxygen, which is why wine is made in closed containers—to keep oxygen in the air away from the crushed grapes.

Although these are two common examples of fermentation, many other organisms carry out the process of fermentation, and many of these organisms are used in industrial applications, producing consumer goods.

> **8.2** In the first stage of respiration, called glycolysis, cells shuffle chemical bonds so that two coupled reactions can occur, producing ATP by substrate-level phosphorylation. When electrons that are also a product of glycolysis are not donated to oxygen, they are added to inorganic or organic molecules.

Fad Diets and Impossible Dreams

In my mind I will always weigh 165 pounds, as I did the day I married. The bathroom scale tells a different story, somehow finding another 30 pounds. I did not ask for that weight, do not want it, and am constantly looking for a way to get rid of it. I have not found this to be a lonely search—it seems like everyone I know past the flush of youth is trying to lose weight, too. And, like many, I have been seduced by fad diets, investing hope only to harvest frustration. The much discussed Atkins' diet was the fad diet I tried. As a scientist I should have known better, but so many people seemed to use it—*Dr. Atkins' Diet Revolution* is one of the 10 best-selling books in history, and was (and is) prominently displayed in every bookstore I enter. The reason this diet doesn't deliver on its promise of pain-free weight loss is well understood by science, but not by the general public. Only hope and hype make it a perpetual best seller.

The secret of the Atkins' diet, stated simply, is to avoid carbohydrates. Atkins' basic proposition is that your body, if it does not detect blood glucose (from metabolizing carbohydrates), will think it is starving and start to burn body fat, even if there is lots of fat already circulating in your bloodstream. You may eat all the fat and protein you want, all the steak and eggs and butter and cheese, and you will still burn fat and lose weight— just don't eat any carbohydrates, any bread or pasta or potatoes or fruit or candy. Despite the title of Atkins' book, this diet is hardly revolutionary. A basic low-carbohydrate diet was first promoted over a century ago in the 1860s by William Banting, an English casket maker, in his best-selling book *Letter on Corpulence*. Books promoting low-carbohydrate diets have continued to be best sellers ever since. I even found one on my mother's bookshelf, in the guise of Dr. Herman Taller's 1961 *Calories Don't Count*.

When I tried the Atkins' diet I lost 10 pounds in three weeks. In three months it was all back, and then some. So what happened? Where did the pounds go, and why did they come back? The temporary weight loss turns out to have a simple explanation: because carbohydrates act as water sponges in your body, forcing your body to become depleted of carbohydrates causes your body to lose water. The 10 pounds I lost on this diet was not fat weight but water, quickly regained with the first starchy foods I ate.

The Atkins' diet is the sort of diet the American Heart Association tells us to avoid (all those saturated fats and cholesterol), and it is difficult to stay on. If you do hang in there, you will lose weight, simply because you eat less. Other popular diets these days, *The Zone* diet of Dr. Barry Sears and *The South Beach Diet* of Dr. Arthur Agatston, are also low-carbohydrate diets, although not as extreme as the Atkins' diet. Like the Atkins' diet, they work not for the bizarre reasons claimed by their promoters, but simply because they are low-calorie diets.

In teaching my students at Washington University about fad diets, I tell them there are two basic laws that no diet can successfully violate:

1. All calories are equal.
2. (calories in) – (calories out) = fat.

The fundamental fallacy of the Atkins' diet, the Zone diet, the South Beach diet, and indeed of all fad diets, is the idea that somehow carbohydrate calories are different from fat and protein calories. This is scientific foolishness. Every calorie you eat contributes equally to your eventual weight, whether it comes from carbohydrate, fat, or protein.

To the extent these diets work at all, they do so because they obey the second law. By reducing calories in, they reduce fat. If that were all there was to it, I'd go out and buy Sears' book. Unfortunately, losing weight isn't that simple, as anyone who has seriously tried already knows. The problem is that your body will not cooperate.

If you try to lose weight by exercising and eating less, your body will attempt to compensate by metabolizing more efficiently. It has a fixed weight, what obesity researchers call a "set point," a weight to which it will keep trying to return. A few years ago, a group of researchers at Rockefeller University in New York, in a landmark study, found that if you lose weight, your metabolism slows down and becomes more efficient, burning fewer calories to do the same work—your body will do everything it can to gain the weight back! Similarly, if you gain weight, your metabolism speeds up. In this way your body uses its own natural weight control system to keep your weight at its set point. No wonder it's so hard to lose weight!

Clearly our bodies don't keep us at one weight all our adult lives. It turns out your body adjusts its fat thermostat—its set point—depending on your age, food intake and amount of physical activity. Adjustments are slow, however, and it seems to be a great deal easier to move the body's set point up than to move it down. Apparently higher levels of fat reduce the body's sensitivity to the leptin hormone that governs how efficiently we burn fat. That is why you can gain weight, despite your set point resisting the gain—your body still issues leptin alarm calls to speed metabolism, but your brain doesn't respond with as much sensitivity as it used to. Thus the fatter you get, the less effective your weight control system becomes.

This doesn't mean that we should give up and learn to love our fat. Rather, now that we are beginning to understand the biology of weight gain, we must accept the hard fact that we cannot beat the requirements of the two diet laws. The real trick is not to give up. Eat less and exercise more, and keep at it. In one year, or two, or three, your body will readjust its set point to reflect the new reality you have imposed by constant struggle. There simply isn't any easy way to lose weight.

8.3 Harvesting Electrons from Chemical Bonds

The first step of oxidative respiration in the mitochondrion is the oxidation of the three-carbon molecule called pyruvate, which is the end product of glycolysis. The cell harvests pyruvate's considerable energy in two steps: first, by oxidizing pyruvate to form acetyl-CoA, and then by oxidizing acetyl-CoA in the Krebs cycle.

Step One: Producing Acetyl-CoA

Pyruvate is oxidized in a single reaction that cleaves off one of pyruvate's three carbons. This carbon then departs as part of the CO_2 molecule shown coming off the pathway with the green arrow in figure 8.6. Pyruvate dehydrogenase, the complex of enzymes that removes CO_2 from pyruvate, is one of the largest enzymes known. It contains 60 subunits! In the course of the reaction, a hydrogen and electrons are removed from pyruvate and donated to NAD^+ to form NADH. Figure 8.7 shows how an enzyme catalyzes this reaction, bringing the substrate (pyruvate) into proximity with NAD^+. Now focus again on figure 8.6. The two-carbon fragment (called an acetyl group) that remains after removing CO_2 from pyruvate is joined to a cofactor called coenzyme A (CoA) by pyruvate dehydrogenase, forming a compound known as **acetyl-CoA.** If the cell has plentiful supplies of ATP, acetyl-CoA is funneled into fat synthesis, with its energetic electrons preserved for later needs. If the cell needs ATP now, the fragment is directed instead into ATP production through the Krebs cycle.

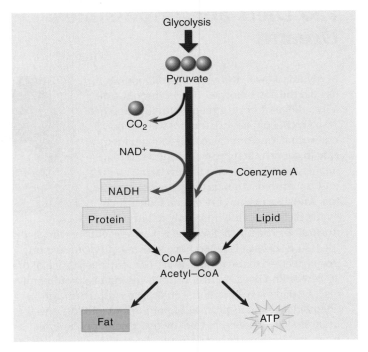

Figure 8.6   Producing acetyl-CoA.

Pyruvate, the three-carbon product of glycolysis, is oxidized to the two-carbon molecule acetyl-CoA, in the process losing one carbon atom as CO_2 and an electron (donated to NAD^+ to form NADH). Almost all the molecules you use as foodstuffs are converted to acetyl-CoA; the acetyl-CoA is then channeled into fat synthesis or into ATP production, depending on your body's needs.

Transferring Hydrogen Atoms

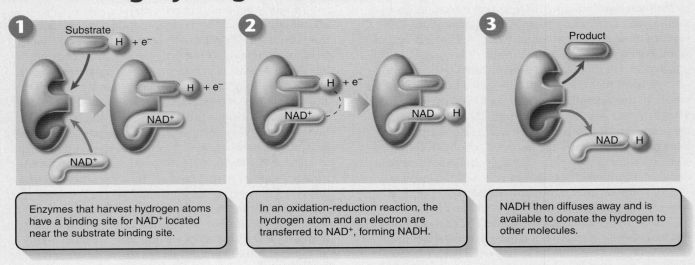

① Enzymes that harvest hydrogen atoms have a binding site for NAD^+ located near the substrate binding site.

② In an oxidation-reduction reaction, the hydrogen atom and an electron are transferred to NAD^+, forming NADH.

③ NADH then diffuses away and is available to donate the hydrogen to other molecules.

Figure 8.7   How NAD⁺ works.

Cells use NAD^+ to carry hydrogen atoms and energetic electrons from one molecule to another. NAD^+ oxidizes energy-rich molecules by acquiring their hydrogens (this proceeds 1 ⟶ 2 ⟶ 3) and then reduces other molecules by giving the hydrogens to them (this proceeds 3 ⟶ 2 ⟶ 1).

A Closer Look

Metabolic Efficiency and the Length of Food Chains

In the earth's ecosystems, the organisms that carry out photosynthesis are often consumed as food by other organisms. We call these "organism-eaters" *heterotrophs*. Humans are heterotrophs, as no human photosynthesizes.

It is thought that the first heterotrophs were ancient bacteria living in a world where photosynthesis had not yet introduced much oxygen into the oceans or atmosphere. The only mechanism they possessed to harvest chemical energy from their food was glycolysis. Neither oxygen-generating photosynthesis nor the oxidative stage of cellular respiration had evolved yet. It has been estimated that a heterotroph limited to glycolysis, as these ancient bacteria were, captures only 3.5% of the energy in the food it consumes. Hence, if such a heterotroph preserves 3.5% of the energy in the photosynthesizers it consumes, then any other heterotrophs that consume the first heterotroph will capture through glycolysis 3.5% of the energy in it, or 0.12% of the energy available in the original photosynthetic organisms. A very large base of photosynthesizers would thus be needed to support a small number of heterotrophs.

When organisms became able to extract energy from organic molecules by oxidative cellular respiration, this constraint became far less severe, because the efficiency of oxidative respiration is estimated to be about 32%. This increased efficiency results in the transmission of much more energy from one trophic level to another than does glycolysis. (A *trophic level* is a step in the movement of energy through an ecosystem.) The efficiency of oxidative cellular respiration has made possible the evolution of food chains, in which photosynthesizers are consumed by heterotrophs, which are consumed by other heterotrophs, and so on. You will read more about food chains in chapter 20.

Even with this very efficient oxidative metabolism, approximately two-thirds of the available energy is lost at each trophic level, and that puts a limit on how long a food chain can be. Most food chains, like the East African grassland ecosystem illustrated here, involve only three or rarely four trophic levels. Too much energy is lost at each transfer to allow chains to be much longer than that. For example, it would be impossible for a large human population to subsist by eating lions captured from the grasslands of East Africa; the amount of grass available there would not support enough zebras and other herbivores to maintain the number of lions needed to feed the human population. Thus, the ecological complexity of our world is fixed in a fundamental way by the chemistry of oxidative cellular respiration.

Photosynthesizers. The grass under this yellow fever tree grows actively during the hot, rainy season, capturing the energy of the sun and storing it in molecules of glucose, which are then converted into starch and stored in the grass.

Herbivores. These zebras consume the grass and transfer some of its stored energy into their own bodies.

Carnivores. The lion feeds on zebras and other animals, capturing part of their stored energy and storing it in its own body.

Scavengers. This hyena and the vultures occupy the same stage in the food chain as the lion. They are also consuming the body of the dead zebra, which has been abandoned by the lion.

Refuse utilizers. These butterflies, mostly *Precis octavia*, are feeding on the material left in the hyena's dung after the food the hyena consumed had passed through its digestive tract.

A food chain in the savannas, or open grasslands, of East Africa.
At each of these levels in the food chain, only about a third or less of the energy present is used by the recipient.

Step Two: The Krebs Cycle

The next stage in oxidative respiration is called the **Krebs cycle,** named after the man who discovered it. The Krebs cycle (not to be confused with the Calvin cycle in photosynthesis) takes place within the mitochondrion. While a complex process, it's nine reactions can be broken down into three stages, as indicated by the overview presented in figure 8.8:

Stage 1. Acetyl-CoA joins the cycle, binding to a four-carbon molecule and producing a six-carbon molecule.

Stage 2. Two carbons are removed as CO_2, their electrons donated to NAD^+, and a four-carbon molecule is left. A molecule of ATP is also produced.

Stage 3. More electrons are extracted, forming NADH and $FADH_2$; the four-carbon starting material is regenerated.

To examine the Krebs cycle in more detail, follow along the series of individual reactions illustrated in figure 8.9. The cycle starts when the two-carbon acetyl-CoA fragment produced from pyruvate is stuck onto a four-carbon sugar called oxaloacetate. Then, in rapid-fire order, a series of eight additional reactions occur (steps 2 through 9). When it is all over, two carbon atoms have been expelled as CO_2, one ATP molecule has been made in a coupled reaction, eight more energetic electrons have been harvested and taken away as NADH or on other carriers, such as $FADH_2$, which serves the same function as NADH, and we are left with the same four-carbon sugar we started with. The process of reactions is a cycle—that is, a circle of reactions. In each turn of the cycle, a new acetyl group replaces the two CO_2 molecules lost, and more electrons are extracted. Note that a single glucose molecule produces *two* turns of the cycle, one for each of the two pyruvate molecules generated by glycolysis.

In the process of cellular respiration, glucose is entirely consumed. The six-carbon glucose molecule is first cleaved into a pair of three-carbon pyruvate molecules during glycolysis. One of the carbons of each pyruvate is then lost as CO_2 in the conversion of pyruvate to acetyl-CoA, and the other two carbons are lost as CO_2 during the oxidations of the Krebs cycle. All that is left to mark the passing of the glucose molecule into six CO_2 molecules is its energy, preserved in four ATP molecules and electrons carried by 10 NADH and two $FADH_2$ carriers.

> **8.3** The end product of glycolysis, pyruvate, is oxidized to the two-carbon acetyl-CoA, yielding a pair of electrons plus CO_2. Acetyl-CoA then enters the Krebs cycle, yielding ATP, many energized electrons, and two CO_2 molecules.

Overview of the Krebs Cycle

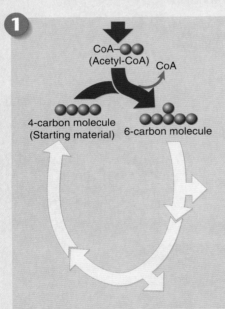

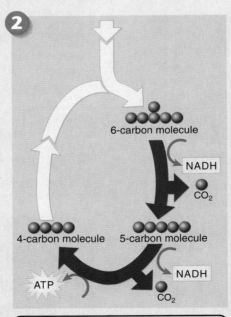

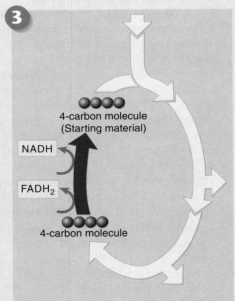

The Krebs cycle begins when a two-carbon fragment is transferred from acetyl-CoA to a four-carbon molecule (the starting material).

Then, the resulting six-carbon molecule is oxidized (a hydrogen removed to form NADH) and decarboxylated (a carbon removed to form CO_2). Next, the five-carbon molecule is oxidized and decarboxylated again, and a coupled reaction generates ATP.

Finally, the resulting four-carbon molecule is further oxidized (hydrogens removed to form $FADH_2$ and NADH). This regenerates the four-carbon starting material, completing the cycle.

Figure 8.8 How the Krebs cycle works.

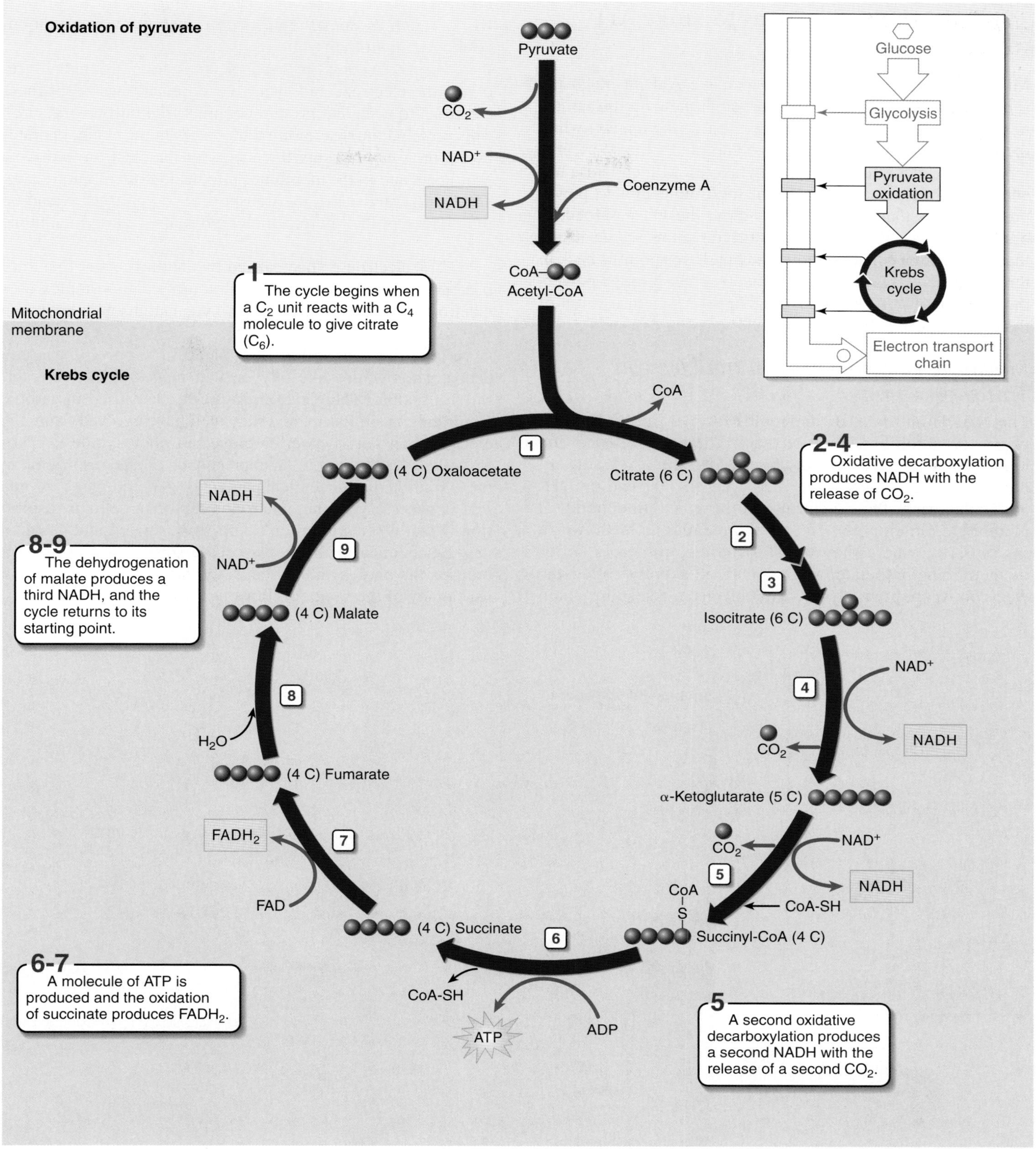

Oxidation of pyruvate

Pyruvate

CO_2

NAD^+

NADH

Coenzyme A

CoA—
Acetyl-CoA

Mitochondrial membrane

Krebs cycle

Glucose

Glycolysis

Pyruvate oxidation

Krebs cycle

Electron transport chain

1 The cycle begins when a C_2 unit reacts with a C_4 molecule to give citrate (C_6).

CoA

(4 C) Oxaloacetate

Citrate (6 C)

2-4 Oxidative decarboxylation produces NADH with the release of CO_2.

NADH

8-9 The dehydrogenation of malate produces a third NADH, and the cycle returns to its starting point.

NAD^+

(4 C) Malate

Isocitrate (6 C)

NAD^+

NADH

CO_2

H_2O

(4 C) Fumarate

α-Ketoglutarate (5 C)

CO_2

NAD^+

FADH$_2$

NADH

FAD

CoA
S

CoA-SH

(4 C) Succinate

Succinyl-CoA (4 C)

CoA-SH

6-7 A molecule of ATP is produced and the oxidation of succinate produces FADH$_2$.

ATP

ADP

5 A second oxidative decarboxylation produces a second NADH with the release of a second CO_2.

Figure 8.9   The Krebs cycle.

This series of nine enzyme-catalyzed reactions takes place within the mitochondrion.

Using the Electrons to Make ATP

Mitochondria use chemiosmosis to make ATP in much the same way that chloroplasts do, although their proton pumps transport protons *out of* an enclosed space (the matrix) while the electron transport system in chloroplasts transport protons *into* an enclosed space (the thylakoid). Mitochondria use energetic electrons extracted from food molecules to power proton pumps that drive protons across the inner mitochondrial membrane. As protons become far more scarce inside than outside, the concentration gradient drives protons back in through special ATP synthase channels. Their passage powers the production of ATP from ADP. The ATP then passes out of the mitochondrion through open ATP-passing channels.

Moving Electrons Through the Electron Transport Chain

The NADH and FADH$_2$ molecules formed during the first stages of aerobic respiration each contain electrons and hydrogens that were gained when NAD$^+$ and FAD were reduced (refer back to figure 8.2). The NADH and FADH$_2$ molecules carry their electrons to the inner mitochondrial membrane (an enlarged area of the membrane is shown in figure 8.10), where they transfer the electrons to a series of membrane-associated molecules collectively called the **electron transport chain.** The electron transport chain works much as does the electron transport system you encountered in studying photosynthesis.

A protein complex (the pink structure in figure 8.10) receives the electrons and, using a mobile carrier, passes these electrons to a second protein complex (the purple structure). This protein complex, along with others in the chain, operates as a proton pump, using the energy of the electron to drive a proton out across the membrane into the intermembrane space. The arrows indicating the transport of the protons extend up into the top of the figure, which represents the intermembrane space.

The electron is then carried by another carrier to a third protein complex (the light blue structure). This complex uses electrons such as this one to link oxygen atoms with hydrogen ions to form molecules of water.

It is the availability of a plentiful supply of electron acceptor molecules like oxygen that makes oxidative respiration possible. The electron transport chain used in aerobic respiration is similar to, and may well have evolved from, the electron transport system employed in photosynthesis. Photosynthesis is thought to have preceded cellular respiration in the evolution of biochemical pathways, generating the oxygen that is necessary as the electron acceptor in cellular respiration. Natural selection didn't start from scratch and design a new biochemical pathway for cellular respiration; instead, it built on the photosynthetic pathway that already existed, and uses many of the same reactions.

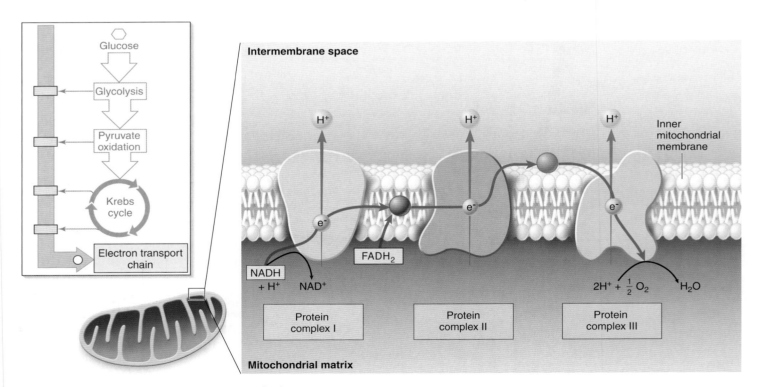

Figure 8.10 The electron transport chain.

High-energy electrons are transported (*red arrows*) along a chain of electron-carrier molecules. Three of these molecules are protein complexes that use portions of the electrons' energy to pump protons (*blue arrows*) out of the matrix and into the intermembrane space. The electrons are finally donated to oxygen to form water.

Producing ATP: Chemiosmosis

In eukaryotes, aerobic respiration takes place within the mitochondria present in virtually all cells. The internal compartment, or **matrix,** of a mitochondrion contains the enzymes that carry out the reactions of the Krebs cycle. As described earlier, the electrons harvested by oxidative respiration are passed along the electron transport chain, and the energy they release transports protons out of the matrix and into the outer compartment, sometimes called the **intermembrane space.** Proton pumps in the inner mitochondrial membrane accomplish the transport. The electrons contributed by NADH activate three of these proton pumps, and those contributed by $FADH_2$ activate two, as indicated in figure 8.10. As the proton concentration in the outer compartment rises above that in the matrix, the concentration gradient induces the protons to reenter the matrix by diffusion through special proton channels. Called ATP synthases, these channels are embedded in the inner mitochondrial membrane, as shown in figure 8.11. As the protons pass through, these channels synthesize ATP from ADP and P_i within the matrix. The ATP is then transported by facilitated diffusion out of the mitochondrion and into the cell's cytoplasm. This ATP synthesizing process is the same chemiosmosis process that you encountered in studying photosynthesis in chapter 7.

Although we have discussed electron transport and chemiosmosis as separate processes, in a cell they are integrated as shown in the overview figure 8.12. The electron transport chain uses electrons harvested in aerobic respiration (red arrows) to pump a large number of protons across the inner mitochondrial membrane (in the upper right). Their subsequent reentry into the mitochondrial matrix drives the synthesis of ATP by chemiosmosis (in the lower right).

> **8.4** The electrons harvested by oxidizing food molecules are used to power proton pumps that chemiosmotically drive the production of ATP.

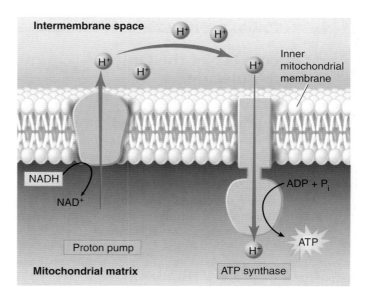

Figure 8.11   Chemiosmosis.

NADH transports high-energy electrons harvested from macromolecules to "proton pumps" that use the energy to pump protons out of the mitochondrial matrix. As a result, the concentration of protons outside the inner mitochondrial membrane rises, inducing protons to diffuse back into the matrix. Many of the protons pass through ATP synthase channels that couple the reentry of protons to the production of ATP.

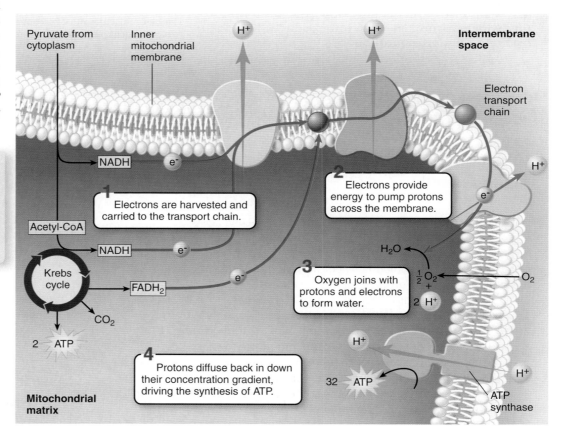

Figure 8.12   An overview of the electron transport chain and chemiosmosis.

8.5 Glucose Is Not the Only Food Molecule

We have considered in detail the fate of a molecule of glucose, a simple sugar, in cellular respiration. But how much of what you eat is sugar? As a more realistic example of the food you eat, consider the fate of a fast-food hamburger. The hamburger you eat is composed of carbohydrates, fats, protein, and many other molecules. This diverse collection of complex molecules is broken down by the process of digestion in your stomach and intestines into simpler molecules. Macromolecules are broken down by digestion into their subunits (building blocks). Recall from chapter 4 that carbohydrates are broken down into glucose, fats into fatty acids, and proteins into amino acids. These breakdown reactions produce little or no energy themselves, but prepare the way for cellular respiration—that is, glycolysis and oxidative metabolism.

We have seen what happens to the glucose. What happens to the amino acids and fatty acids? These subunits undergo chemical modifications that convert them into products that feed into cellular respiration. For example, proteins (the second set of arrows in figure 8.13) are first broken down into their individual amino acids. A series of reactions removes the nitrogen side groups and converts the rest of the amino acid into a molecule that takes part in the Krebs cycle. Thus the proteins and fats in the hamburger, like glucose, also become important sources of energy.

> **8.5** Cells also garner energy from proteins and fats, which are broken down into products that feed into cellular respiration.

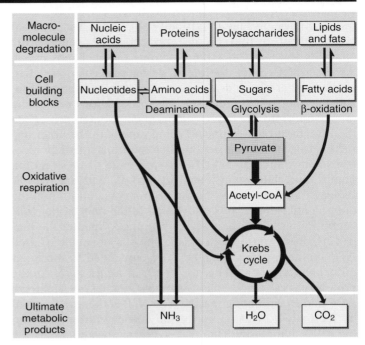

Figure 8.13 How cells obtain energy from foods.
Most organisms extract energy from organic molecules by oxidizing them. The first stage of this process, breaking down macromolecules into their subunits, yields little energy. The second stage, cellular respiration, extracts energy, primarily in the form of high-energy electrons. The subunit of many carbohydrates, glucose, readily enters glycolysis and passes through the biochemical pathways of oxidative respiration. However, the subunits of other macromolecules must be converted into products that can enter the biochemical pathways found in oxidative respiration.

INQUIRY & ANALYSIS

How Do Swimming Fish Avoid Low Blood pH?

During exercise, if oxygen is depleted, muscles use glycolysis to obtain ATP, donating the electrons to pyruvate to form lactic acid. The lactic acid is released into the blood and lowers blood pH. Fish blood has poor buffering capacity, and the experiment in the graph explores how they avoid low blood pH after vigorous exercise for up to 15 minutes.

1. **Applying Concepts** What is the dependent variable?
2. **Making Inferences** About how much of the total lactic acid is released after exercise stops? [Hint: notice the x-axis scale changes from minutes to hours, so replot all points to minutes and compare areas under the curve.]
3. **Drawing Conclusions** Is this result consistent with the hypothesis that fish maintain blood pH levels by delaying the release of lactic acid from muscles? Why might this be beneficial to the fish?

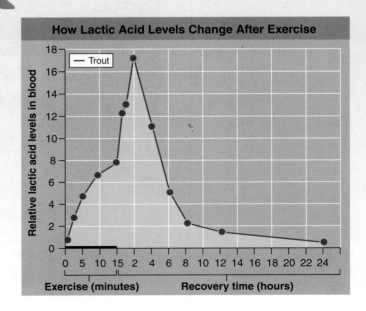

An Overview of Cellular Respiration

8.1 Where Is the Energy in Food?

- Cellular respiration is the process of harvesting energy from glucose molecules and storing it as cellular energy in ATP. Cellular respiration is carried out in two stages: glycolysis occurring in the cytoplasm and oxidation occurring in the mitochondria (**figure 8.2**).

Respiration Without Oxygen: Glycolysis

8.2 Using Coupled Reactions to Make ATP

- Glycolysis is a series of 10 chemical reactions in which glucose is broken down into two three-carbon pyruvate molecules. Two exergonic reactions are coupled with a reaction that leads to the formation of ATP. This is called substrate-level phosphorylation (**figures 8.3** and **8.4**).

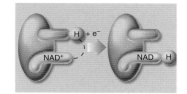

6-carbon sugar diphosphate

3-carbon sugar phosphate   3-carbon sugar phosphate

- Electrons are extracted from glucose and are donated to a carrier molecule, NAD⁺, which carries electrons and hydrogen as NADH to be used later in oxidative respiration.

- Glycolysis does not require oxygen and is therefore referred to as anaerobic respiration. There are other forms of anaerobic respiration, including fermentation (**figure 8.5**).

- During fermentation, NAD⁺ is recycled by transferring its electrons to a molecule other than oxygen. In animals, the electrons are transferred back to pyruvate and the product is lactate. In yeast, pyruvate is converted to acetaldehyde, which accepts the electrons from NADH, and the product is ethanol or ethyl alcohol.

Respiration With Oxygen: The Krebs Cycle

8.3 Harvesting Electrons from Chemical Bonds

- The two molecules of pyruvate formed in glycolysis are passed into the mitochondrion, where they are converted into two molecules of acetyl-coenzyme A (**figure 8.6**). Some of the energy in pyruvate is transferred to NAD⁺ to produce NADH, which will be used in a later step (**figure 8.7**).

- Acetyl-CoA enters a series of chemical reactions called the Krebs cycle, where one molecule of ATP is produced and energy is harvested in the form of electrons that are transferred to molecules of NAD⁺ and FAD to produce NADH and FADH₂, respectively (**figures 8.8** and **8.9**).

- The Krebs cycle makes two turns for every molecule of glucose that is oxidized.

8.4 Using the Electrons to Make ATP

- The energy stored in NADH and FADH₂ is harvested by the electron transport chain to make ATP. NADH and FADH₂ are transported to the inner mitochondrial membrane, where they transfer electrons to the electron transport chain.

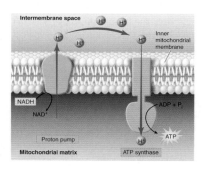

Intermembrane space

Inner mitochondrial membrane

NADH

NAD⁺

Proton pump

Mitochondrial matrix

ADP + Pᵢ

ATP

ATP synthase

- The electrons are passed along the electron transport chain, a group of proteins embedded in the inner mitochondrial membrane. The energy from the electrons drives proton pumps that pump H⁺ across the inner membrane from the matrix to the intermembrane space, creating a H⁺ concentration gradient (**figure 8.10**).

- As in photosynthesis, ATP is formed when H⁺ passes back across the membrane through ATP synthase channels, driven by the proton concentration gradient. The energy from the movement of electrons is transferred to the chemical bonds in ATP (**figure 8.11**). Thus, the energy harvested from glucose is stored in ATP (**figure 8.12**).

Other Sources of Energy

8.5 Glucose Is Not the Only Food Molecule

- Food sources other than glucose are also used in oxidative respiration. Macromolecules, such as proteins, lipids, and nucleic acids, are broken down into intermediate products that enter cellular respiration in different reaction steps (**figure 8.13**).

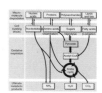

1. In animals, the energy for life is obtained by cellular respiration. This involves
   a. breaking down the organic molecules that were consumed.
   b capturing photons from plants.
   c. utilizing ATP that was produced by plants.
   d. breaking down CO_2 that was produced by plants.

2. NAD^+ is recycled during
   a. glycolysis.
   b. fermentation.
   c. the Krebs cycle.
   d. the formation of acetyl-CoA.

3. During glycolysis, ATP forms by
   a. the breakdown of pyruvate.
   b. chemiosmosis.
   c. substrate-level phosphorylation.
   d. NAD^+.

4. Which of the following processes can occur in the absence of oxygen?
   a. the Krebs cycle
   b. glycolysis
   c. chemiosmosis
   d. all of the above

5. Every living creature on this planet is capable of carrying out the rather inefficient biochemical process of glycolysis, which
   a. makes glucose, using the energy from ATP.
   b. makes ATP by splitting glucose and capturing the energy.
   c. phosphorylates ATP to make ADP, using the energy from photons.
   d. makes glucose, using oxygen and carbon dioxide and water.

6. The electrons generated from the Krebs cycle are transferred to _____ and then are shuttled to _____.
   a. NAD^+, oxygen
   b. NAD^+, electron transport chain
   c. NADH, oxygen
   d. NADH, electron transport chain

7. The final electron acceptor in lactate fermentation is
   a. pyruvate.   c. lactic acid.
   b. NAD^+ d. O_2.

8. After glycolysis, the pyruvate molecules go to the
   a. nucleus of the cell and provide energy.
   b. membranes of the cell and are broken down in the presence of CO_2 to make more ATP.
   c. mitochondria of the cell and are broken down in the presence of O_2 to make more ATP.
   d. Golgi bodies and are packaged and stored until needed.

9. The vast majority of the ATP molecules produced within a cell are produced
   a. during photosynthesis.
   b. during glycolysis.
   c. during the Krebs cycle.
   d. during the electron transport chain.

10. Cells can extract energy from foodstuffs other than glucose because
    a. proteins, fatty acids, and nucleic acids get converted to glucose and then enter oxidative respiration.
    b. each type of macromolecule has its own oxidative respiration pathway.
    c. each type of macromolecule is broken down into its subunits, which enter the oxidative respiration pathway.
    d. they can all enter the glycolytic pathway.

1. **Figure 8.5** Soft drinks are artificially carbonated, which is what causes them to fizz. Beer and sparkling wines are naturally carbonated. How does this natural carbonation occur?

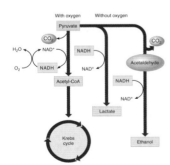

2. **Figure 8.13** Your friend Yevgeny wants to go on a low-carbohydrate diet so that he can lose some of the "baby fat" he's still carrying. He asks your advice; what do you tell him?

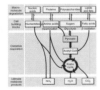

1. If cellular respiration were the stock market (you're investing ATPs and getting ATP dividends), where would you get the most return on your investment: glycolysis, the Krebs cycle, or the electron transport chain? Explain your answer.

2. How much ATP would be generated in the cells of a person who consumed a diet of pyruvate instead of glucose?

3. If you poke a hole in a mitochondrion, can it still perform oxidative respiration? Can fragments of mitochondrion perform oxidative respiration?

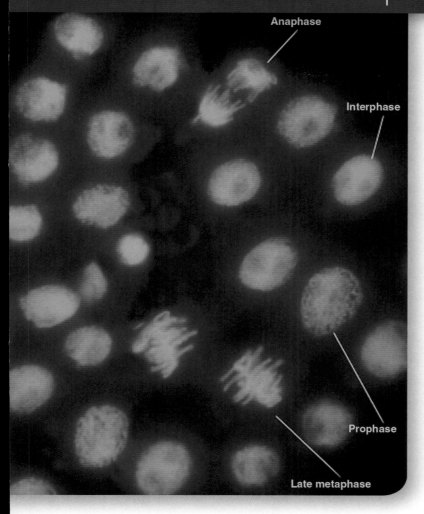

Anaphase

Interphase

Prophase

Late metaphase

9
Mitosis

The beautifully colored spheres you see above are dividing plant cells. If you look carefully, you can identify among them cells in each of the several stages of the cell division process, the subject of this chapter. The squiggly lines within individual cells are chromosomes, each tagged with a fluorescent chemical that makes it glow. Within each dividing cell, you can see that half of the chromosomes are being drawn to one of the two opposite ends of the cell, pulled by microtubules too tiny to be visible to our eyes. Some human cells divide frequently, particularly those subjected to a lot of wear and tear. The epithelial cells of your skin divide so often that your skin replaces itself every two weeks. The lining of your stomach is replaced every few days! Nerve cells, on the other hand, can live for 100 years without dividing. Cells use a battery of genes to regulate when and how frequently they divide. If some of these genes become disabled, a cell may begin to divide ceaselessly, a condition we call cancer. Exposure to DNA-damaging chemicals such as those in cigarette smoke greatly increases the chance of this sort of event occurring in the tissues exposed to the smoke, which is why smokers will more likely get lung cancer than colon cancer.

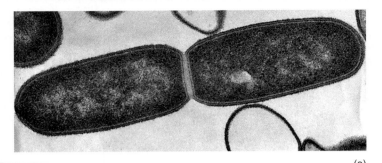

9.1 Prokaryotes Have a Simple Cell Cycle

All species reproduce, passing their hereditary information on to their offspring. In this chapter, we begin our consideration of heredity with a look at how cells reproduce. Cell division in prokaryotes takes place in two stages, which together make up a **simple cell cycle.** First the DNA is copied, and then the cell splits by a process called **binary fission.** The cell in figure 9.1*a* is undergoing binary fission.

In prokaryotes, the hereditary information—that is, the genes that specify the prokaryote—is encoded in a single circle of DNA, called a prokaryotic chromosome. Before the cell itself divides, the DNA circle makes a copy of itself, a process called *replication.* Starting at one point, the origin of replication (the point where the two strands of DNA are connected at the top of figure 9.1*b*), the double helix of DNA begins to unzip, exposing the two strands. The enlargement on the right of figure 9.1*b* shows how the DNA replicates. The purple strand is from the original DNA and the red strand is the newly formed DNA. The new double helix is formed from each naked strand by placing on each exposed nucleotide its complementary nucleotide (that is, A with T, G with C, as discussed in chapter 4). DNA replication is discussed in more detail in chapter 12. When the unzipping has gone all the way around the circle, the cell possesses two copies of its hereditary information.

When the DNA has been copied, the cell grows, resulting in elongation. The newly replicated DNA molecules are partitioned toward each end of the cell. This partitioning process involves DNA sequences near the origin of replication, and results in these sequences being attached to the membrane. When the cell reaches an appropriate size, the prokaryotic cell begins to split into two equal halves. New plasma membrane and cell wall are added at a point between where the two DNA copies are partitioned, indicated by the green divider in figure 9.1*b*. As the growing plasma membrane pushes inward, the cell is constricted in two, eventually forming two **daughter cells.** Each contains one prokaryotic chromosome and is a complete living cell in its own right.

> **9.1** Prokaryotes divide by binary fission after the DNA has replicated.

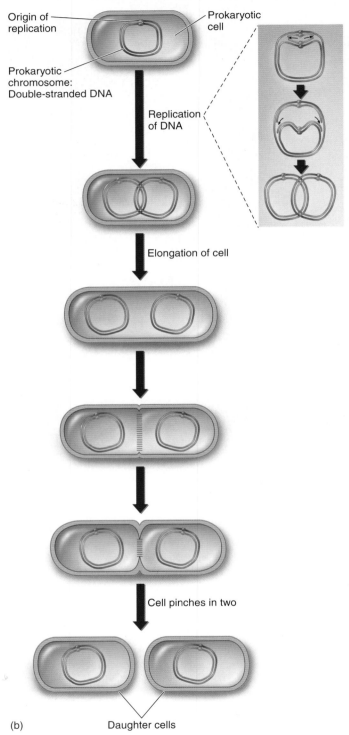

Figure 9.1 Cell division in prokaryotes.

(a) Prokaryotes divide by a process of binary fission. Here, a cell has divided in two and is about to be pinched apart by the growing plasma membrane. (b) Before the cell splits, the circular DNA molecule of a prokaryote initiates replication at a single site, called the origin of replication, moving out in both directions. When the two moving replication points meet on the far side of the molecule, its replication is complete. The cell then undergoes binary fission, where the cell divides into two daughter cells.

Eukaryotes Have a Complex Cell Cycle

The evolution of the eukaryotes introduced several additional factors into the process of cell division. Eukaryotic cells are much larger than prokaryotic cells, and they contain much more DNA. Eukaryotic DNA is contained in a number of linear chromosomes, whose organization is much more complex than that of the single, circular DNA molecules in prokaryotes. A eukaryotic **chromosome** is a single, long DNA molecule wound tightly around proteins, called histones, into a compact shape.

Cell division in eukaryotes is more complex than in prokaryotes, both because eukaryotes contain far more DNA and because it is packaged differently. The cells of eukaryotic organisms either undergo mitosis or meiosis to divide up the DNA. **Mitosis** is the mechanism of cell division that occurs in an organism's nonreproductive cells, or **somatic cells.** A second process, called **meiosis,** divides the DNA in cells that participate in sexual reproduction, or **germ cells.** Meiosis results in the production of gametes, such as sperm and eggs, and is discussed in chapter 10.

The events that prepare the eukaryotic cell for division and the division process itself constitute a **complex cell cycle.** Figure 9.2 walks you through the phases of the cell cycle:

Interphase. This is the first phase of the cell cycle, step 1 in figure 9.2, and is usually considered a resting phase, but the cell is far from resting. Interphase is itself made up of three phases:

   G_1 **phase.** This "first gap" phase is the cell's primary growth phase. For most organisms, this phase occupies the major portion of the cell's life span.

   S phase. In this "synthesis" phase, the DNA replicates, producing two copies of each chromosome.

   G_2 **phase.** Cell division preparation continues in the "second gap" phase with the replication of mitochondria, chromosome condensation, and the synthesis of microtubules.

M phase. In mitosis, a microtubular apparatus binds to the chromosomes and moves them apart, shown in steps 2 through 5.

C phase. In cytokinesis, the cytoplasm divides, creating two daughter cells, shown in step 6.

> **9.2** Eukaryotic cells divide by separating duplicate copies of their chromosomes into daughter cells.

The Cell Cycle

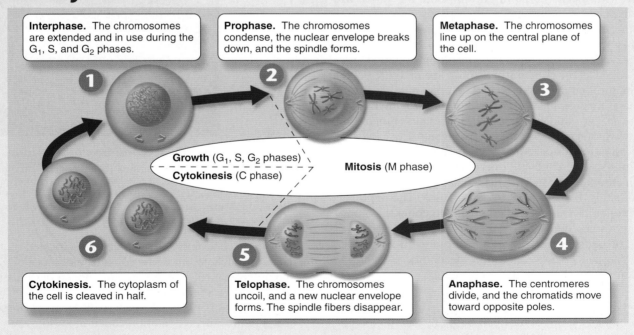

Interphase. The chromosomes are extended and in use during the G_1, S, and G_2 phases.

Prophase. The chromosomes condense, the nuclear envelope breaks down, and the spindle forms.

Metaphase. The chromosomes line up on the central plane of the cell.

Growth (G_1, S, G_2 phases)
Cytokinesis (C phase)
Mitosis (M phase)

Cytokinesis. The cytoplasm of the cell is cleaved in half.

Telophase. The chromosomes uncoil, and a new nuclear envelope forms. The spindle fibers disappear.

Anaphase. The centromeres divide, and the chromatids move toward opposite poles.

Figure 9.2 How the cell cycle works.
Human cells growing in culture typically have a 22-hour cell cycle. Most cell types take about 80 minutes in this 22 hours to complete cell division: prophase—23 minutes, metaphase—29 minutes, anaphase—10 minutes, telophase—14 minutes, and cytokinesis—4 minutes. The proportion of the cell cycle spent in any one phase of mitosis varies considerably in different tissues.

9.3　Chromosomes

Chromosomes were first observed by the German embryologist Walther Fleming in 1882, while he was examining the rapidly dividing cells of salamander larvae. When Fleming looked at the cells through what would now be a rather primitive light microscope, he saw minute threads within their nuclei that appeared to be dividing lengthwise. Fleming called their division *mitosis,* based on the Greek word *mitos,* meaning "thread."

Chromosome Number

Since their initial discovery, chromosomes have been found in the cells of all eukaryotes examined. Their number may vary enormously from one species to another. A few kinds of organisms—such as the Australian ant *Myrmecia* spp.; the plant *Haplopappus gracilis,* a relative of the sunflower that grows in North American deserts; and the fungus *Penicillium*—have only 1 pair of chromosomes, while some ferns have more than 500 pairs. Most eukaryotes have between 10 and 50 chromosomes in their body cells.

Homologous Chromosomes

Chromosomes exist in somatic cells as pairs, called **homologous chromosomes,** or **homologues.** Homologues carry information about the same traits at the same locations on each chromosome but the information can vary between homologues, which will be discussed in chapter 11. Cells that have two of each type of chromosome are called **diploid cells.** One chromosome of each pair is inherited from the mother (colored green in figure 9.3) and the other from the father (colored purple). Before cell division, each homologous chromosome replicates, resulting in two identical copies, called **sister chromatids.** You'll see in figure 9.3 that the sister chromatids remain joined together after replication at a special linkage site called the **centromere,** the knoblike structure in the middle of each chromosome. Human body cells have a total of 46 chromosomes, which are actually 23 pairs of homologous chromosomes. In their duplicated state, before mitosis, there are still only 23 pairs of chromosomes, but each chromosome has duplicated and consists of two sister chromatids, for a total of 92 chromatids. The duplicated sister chromatids can make it confusing to count the number of chromosomes in an organism, but keep in mind that the number of centromeres doesn't increase with replication, and so you can always determine the number of chromosomes simply by counting the centromeres.

The Human Karyotype

The 46 human chromosomes can be paired as homologues by comparing size, shape, location of centromeres, and so on. This arrangement of chromosomes is called a *karyotype.* An example of a human karyotype is shown in figure 9.4. You can see how the different sizes and shapes of chromosomes allow scientists to pair together the ones that are homologous. For example, chromosome 1 is much larger than chromosome

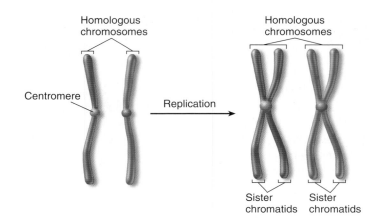

Figure 9.3　The difference between homologous chromosomes and sister chromatids.

Homologous chromosomes are a pair of the same chromosome—say, chromosome number 16. Sister chromatids are the two replicas of a single chromosome held together by the centromere after DNA replication. A duplicated chromosome looks somewhat like an X.

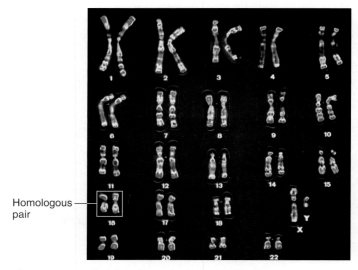

Figure 9.4　The 46 chromosomes of a human.

In this presentation, photographs of the individual chromosomes of a human male have been cut out and paired with their homologues, creating an organized display called a *karyotype.* The chromosomes are in a duplicated state, and the sister chromatids can actually be seen in many of the homologous pairs.

14, and its centromere is more centrally located on the chromosome. Each chromosome contains thousands of genes that play important roles in determining how a person's body develops and functions. For this reason, possession of all the chromosomes is essential to survival. Humans missing even one chromosome, a condition called monosomy, do not usually survive embryonic development. Nor does the human embryo develop properly with an extra copy of any one chromosome, a condition called trisomy. For all but a few of the smallest chromosomes, trisomy is fatal; even in those cases, serious problems result. We will revisit this issue of differences in chromosome number in chapter 11.

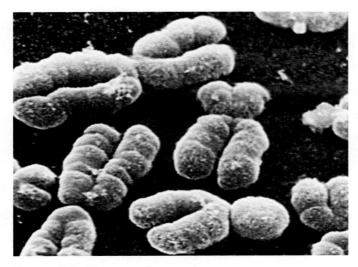

Figure 9.5 Human chromosomes.
The photograph (950×) shows human chromosomes as they appear immediately before nuclear division. Each DNA strand has already replicated, forming identical copies held together by the centromere.

Chromosome Structure

Chromosomes are composed of **chromatin,** a complex of DNA and protein; most are about 40% DNA and 60% protein. A significant amount of RNA is also associated with chromosomes because chromosomes are the sites of RNA synthesis. The DNA of a chromosome is one very long, double-stranded fiber that extends unbroken through the entire length of the chromosome. A typical human chromosome contains about 140 million (1.4×10^8) nucleotides in its DNA. Furthermore, if the strand of DNA from a single chromosome were laid out in a straight line, it would be about 5 centimeters (2 inches) long. The amount of information in one human chromosome would fill about 2,000 printed books of 1,000 pages each! Fitting such a strand into a nucleus is like cramming a string the length of a football field into a baseball—and that's only 1

of 46 chromosomes! In the cell, however, the DNA is coiled, allowing it to fit into a much smaller space than would otherwise be possible.

Chromosome Coiling

The DNA of eukaryotes is divided into several chromosomes, although the chromosomes you see in figure 9.5 hardly look like long double-stranded molecules of DNA. These chromosomes, duplicated as sister chromatids, are formed into the shape we see here by winding and twisting the long DNA strands into a much more compact form. Winding up DNA presents an interesting challenge. Because the phosphate groups of DNA molecules have negative charges, it is impossible to just tightly wind up DNA because all the negative charges would simply repel one another. In figure 9.6 you can see how the cell solves this problem. The DNA doesn't wind around itself. Instead, as you can see on the right, the DNA helix wraps around proteins with positive charges called **histones** (the pink balls). The positive charges of the histones counteract the negative charges of the DNA, so that the complex has no net charge. Every 200 nucleotides, the DNA duplex is coiled around a core of eight histone proteins, forming a complex known as a **nucleosome.** The nucleosomes, which resemble beads on a string in figure 9.6, are further coiled into a solenoid. This solenoid is then organized into looped domains. The final organization of the chromosome is not known, but it appears to involve further radial looping into rosettes around a preexisting scaffolding of protein. These are the flower-shaped structures in the figure. This complex of DNA and histone proteins, coiled tightly, forms a compact chromosome.

> **9.3** All eukaryotic cells store their hereditary information in chromosomes, but different kinds of organisms use very different numbers of chromosomes to store this information. Coiling of the DNA into chromosomes allows it to fit in the nucleus.

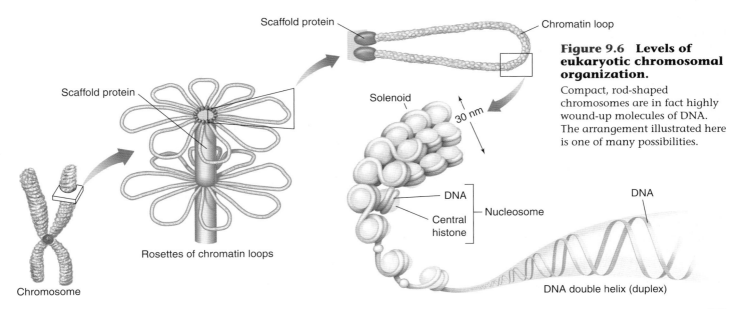

Figure 9.6 Levels of eukaryotic chromosomal organization.

Compact, rod-shaped chromosomes are in fact highly wound-up molecules of DNA. The arrangement illustrated here is one of many possibilities.

Scaffold protein · Chromatin loop

Solenoid

30 nm

DNA

Central histone — Nucleosome

DNA

Scaffold protein

Rosettes of chromatin loops

Chromosome

DNA double helix (duplex)

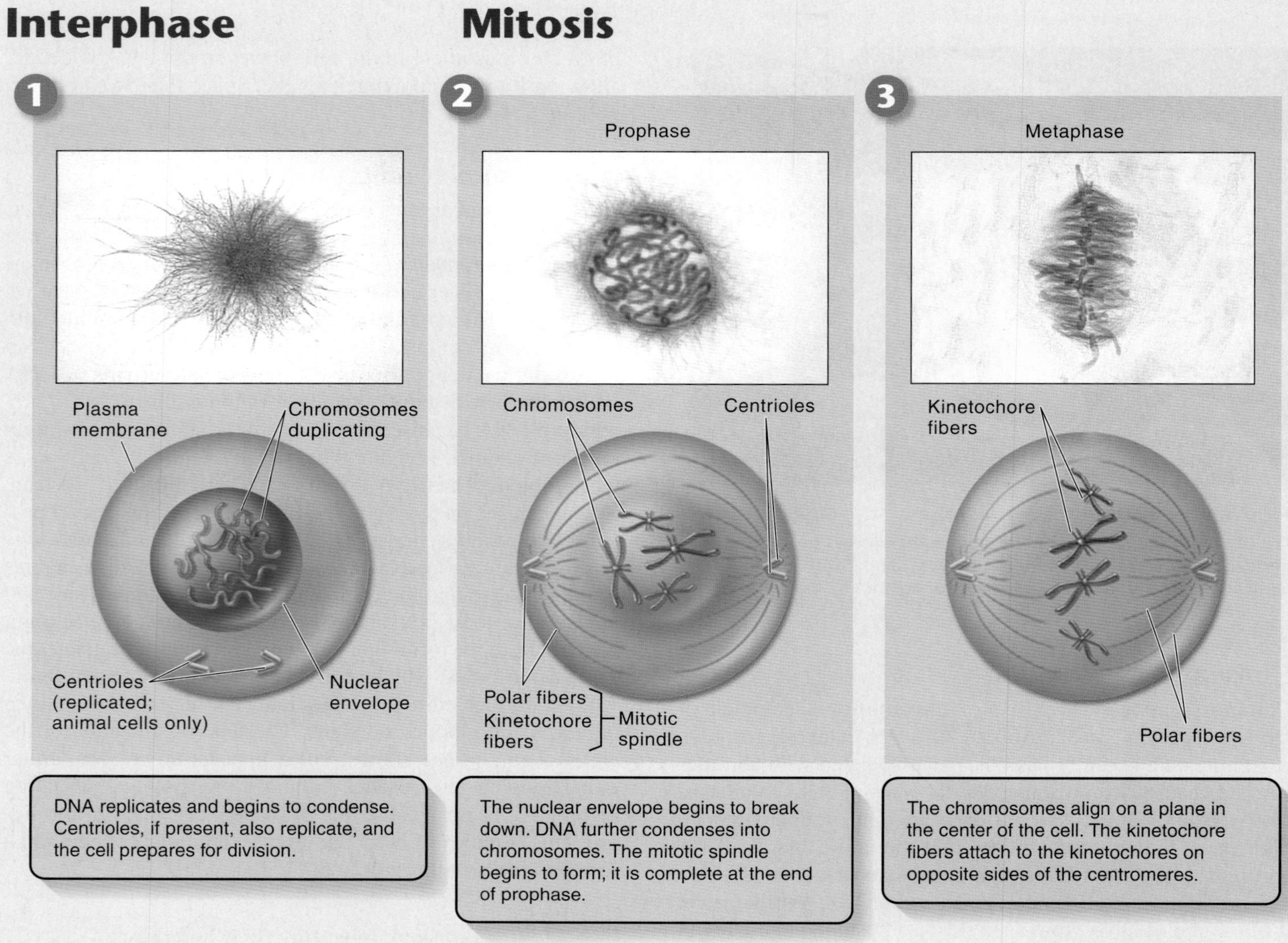

Interphase   Mitosis

①

Plasma membrane

Chromosomes duplicating

Centrioles (replicated; animal cells only)

Nuclear envelope

DNA replicates and begins to condense. Centrioles, if present, also replicate, and the cell prepares for division.

②

Prophase

Chromosomes   Centrioles

Polar fibers
Kinetochore fibers — Mitotic spindle

The nuclear envelope begins to break down. DNA further condenses into chromosomes. The mitotic spindle begins to form; it is complete at the end of prophase.

③

Metaphase

Kinetochore fibers

Polar fibers

The chromosomes align on a plane in the center of the cell. The kinetochore fibers attach to the kinetochores on opposite sides of the centromeres.

Figure 9.7 How cell division works.

Cell division in eukaryotes begins in interphase, carries through the four stages of mitosis, and ends with cytokinesis. Several features of the spindle illustrated in the drawings above appear in dividing animal cells but not in plant cells, and cannot be seen in the photographs, which are of the African blood lily *Haemanthus katharinae.* (In these exceptional photographs, the chromosomes are stained *blue* and microtubules stained *red.*)

9.4   Cell Division

Interphase

When cell division begins in interphase, chromosomes first replicate, and then begin to wind up tightly, a process called **condensation.** Sister chromatids are held together by a complex of proteins called *cohesin.* Chromosomes are not usually visible during interphase, but to clarify what is happening, they are shown in panel 1 of figure 9.7 as if they were.

Mitosis

Interphase is not a phase of mitosis, but it sets the stage for cell division. It is followed by nuclear division, called *mitosis.* Although the process of mitosis is continuous, with the stages flowing smoothly one into another, for ease of study, mitosis is traditionally subdivided into four stages: prophase, meta-

phase, anaphase, and telophase. We will be referring to the panels in figure 9.7 in the following descriptions.

Prophase: Mitosis Begins. In **prophase,** the individual condensed chromosomes, the blue structures in the photo of panel 2, first become visible with a light microscope. As the replicated chromosomes condense, the nucleolus disappears and the cell dismantles the nuclear envelope and begins to assemble the apparatus it will use to pull the replicated sister chromatids to opposite ends ("poles") of the cell. In the center of an animal cell, the pairs of centrioles separate; the two pairs of centrioles move apart toward opposite poles of the cell, forming between them as they move apart a network of protein cables called the **spindle.** In panel 2, the centrioles are positioned at the poles; the red structures in the drawing and photo are the protein cables that make up the spindle. Each cable is called a *spindle fiber* and is made of microtubules, which are long, hollow tubes of protein. Plant cells lack centrioles and instead brace the ends of the spindle toward the poles.

Cytokinesis

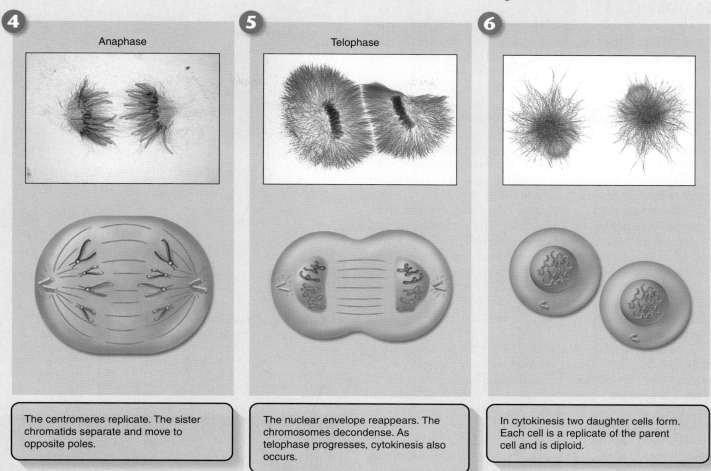

4 Anaphase

The centromeres replicate. The sister chromatids separate and move to opposite poles.

5 Telophase

The nuclear envelope reappears. The chromosomes decondense. As telophase progresses, cytokinesis also occurs.

6

In cytokinesis two daughter cells form. Each cell is a replicate of the parent cell and is diploid.

As condensation of the chromosomes continues, a second group of microtubules extends out from the poles toward the centromeres of each chromosome. Each set of microtubules continues to grow longer until it makes contact with a disk of protein, called a *kinetochore,* associated with each side of the centromere. When the process is complete, one sister chromatid of each pair is attached by microtubules to one pole and the other sister chromatid to the other pole.

Metaphase: Alignment of the Chromosomes. The second phase of mitosis, **metaphase,** begins when the chromosomes, each consisting of a pair of sister chromatids, align in the center of the cell along an imaginary plane that divides the cell in half, referred to as the equatorial plane. Panel 3 shows the chromosomes beginning to align along the equatorial plane. Microtubules attached to the kinetochores of the centromeres are fully extended back toward the opposite poles of the cell.

Anaphase: Separation of the Chromatids. In **anaphase,** enzymes cleave the cohesin link holding sister chromatids together, the kinetochores split, and the sister chromatids are freed from each other. Cell division is now simply a matter of reeling in the microtubules, dragging to the poles the sister chromatids, now referred to as daughter chromosomes. In panel 4 you see the daughter chromosomes being pulled by their centromeres, the arms of the chromosomes dangling behind. The ends of the microtubules are dismantled, one bit after another, making the tubes shorter and shorter and so drawing the chromosome attached to the far end closer and closer to the opposite poles of the cell. When they finally arrive, each pole has one complete set of chromosomes.

Telophase: Re-formation of the Nuclei. The only tasks that remain in **telophase** are the dismantling of the stage and the removal of the props. The mitotic spindle is disassembled, and a nuclear envelope forms around each set of chromosomes while they begin to uncoil, as shown in panel 5, and the nucleolus reappears.

Cytokinesis

At the end of telophase, mitosis is complete. The cell has divided its replicated chromosomes into two nuclei, which are positioned at opposite ends of the cell. Mitosis is also referred to as **karyokinesis.** You may recall from chapter 5 that the nucleus is also referred to as *karyon* (Latin for "kernel"); therefore, karyokinesis is the division of the nucleus. Toward the end of mitosis, **cytokinesis,** the division of the cytoplasm, occurs, and the cell is cleaved into roughly equal halves. Cytoplasmic organelles have already been replicated and resorted to the areas that will separate and become the daughter cells. Cytokinesis, the formation of daughter cells shown in the last panel in figure 9.7, signals the end of cell division.

In animal cells, which lack cell walls, cytokinesis is achieved by pinching the cell in two with a contracting belt of actin filaments. As contraction proceeds, a **cleavage furrow** becomes evident around the cell's circumference, where the cytoplasm is being progressively pinched inward by the decreasing diameter of the actin belt. In figure 9.8*a* you see an animal cell pinching in half during cytokinesis. Imagine the cleavage furrow deepening further, until the cell is literally pinched in two.

Plant cells have rigid walls that are far too strong to be deformed by actin filament contraction. A different approach to cytokinesis has therefore evolved in plants. Plant cells assemble membrane components in their interior, at right angles to the mitotic spindle. In figure 9.8*b*, you can see how membrane is deposited between the daughter cells by vesicles that fuse together. This expanding partition, called a **cell plate,** grows outward until it reaches the interior surface of the plasma membrane and fuses with it, at which point it has effectively divided the cell in two. Cellulose is then laid down over the new membranes, forming the cell walls of the two new cells.

Cell Death

Despite the ability to divide, no cell lives forever. The ravages of living slowly tear away at a cell's machinery. To some degree damaged parts can be replaced, but no replacement process is perfect. And sometimes the environment intervenes. If food supplies are cut off, for example, animal cells cannot obtain the energy necessary to maintain their lysosome membranes. The cells die, digested from within by their own enzymes.

During fetal development, many cells are programmed to die. In human embryos, hands and feet appear first as "paddles," but the skin cells between bones die on schedule to form the separated toes and fingers. Figure 9.9 shows a developing human hand looking like a paddle. The cells in the tissue between the bones will later die, leaving behind a set of fingers. In ducks, this cell death is not part of the developmental program, which is why ducks have webbed feet and you don't.

Human cells appear to be programmed to undergo only so many cell divisions and then die, following a plan written into the genes. In tissue culture, cell lines divide about 50 times, and then the entire population of cells dies off. Even if some of the cells are frozen for years, when they are thawed

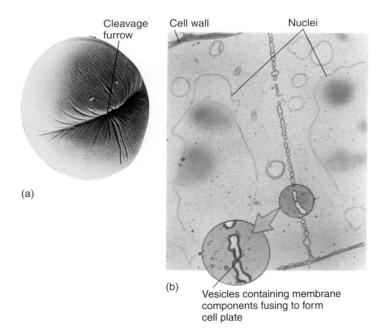

(a)

(b)

Vesicles containing membrane components fusing to form cell plate

Figure 9.8 Cytokinesis.
The division of cytoplasm that occurs after mitosis is called cytokinesis and cleaves the cell into roughly equal halves. (a) In an animal cell, such as this sea urchin egg, a cleavage furrow forms around the dividing cell. (b) In this dividing plant cell, a cell plate is forming between the two newly forming daughter cells.

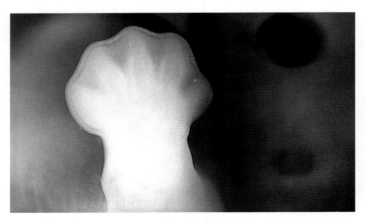

Figure 9.9   Programmed cell death.
In the human embryo, programmed cell death results in the formation of fingers and toes from paddlelike hands and feet.

they simply resume where they left off and die on schedule. Only cancer cells appear to thwart these instructions, dividing endlessly. All other cells in your body contain a hidden clock that keeps time by counting cell divisions, and when the alarm goes off the cells die.

> **9.4   The eukaryotic cell cycle starts in interphase with the condensation of replicated chromosomes; in mitosis, these chromosomes are drawn by microtubules to opposite ends of the cell; in cytokinesis, the cell is split into two daughter cells.**

9.5 Controlling the Cell Cycle

The events of the cell cycle are coordinated in much the same way in all eukaryotes. The control system human cells use first evolved among the protists over a billion years ago; today, it operates in essentially the same way in fungi as it does in humans.

The goal of controlling any cyclic process is to adjust the duration of the cycle to allow sufficient time for all events to occur. In principle, a variety of methods can achieve this goal. For example, an internal clock can be employed to allow adequate time for each phase of the cycle to be completed. This is how many organisms control their daily activity cycles. The disadvantage of using such a clock to control the cell cycle is that it is not very flexible. One way to achieve a more flexible and sensitive regulation of a cycle is simply to let the completion of each phase of the cycle trigger the beginning of the next phase, as a runner passing a baton starts the next leg in a relay race. Until recently, biologists thought this type of mechanism controlled the cell division cycle. However, we now know that eukaryotic cells employ a separate, centralized controller to regulate the process: at critical points in the cell cycle, further progress depends upon a central set of "go/no-go" switches that are regulated by feedback from the cell.

This mechanism is the same one engineers use to control many processes. For example, the furnace that heats a home in the winter typically goes through a daily heating cycle. When the daily cycle reaches the morning "turn on" checkpoint, sensors report whether the house temperature is below the set point (for example, 70°F). If it is, the thermostat triggers the furnace, which warms the house. If the house is already at least that warm, the thermostat does not start the furnace. Similarly, the cell cycle has key checkpoints where feedback signals from the cell about its size and the condition of its chromosomes can either trigger subsequent phases of the cycle or delay them to allow more time for the current phase to be completed.

Three principal checkpoints control the cell cycle in eukaryotes:

1. **Cell growth is assessed at the G_1 checkpoint.**
   Located near the end of G_1 and just before entry into S phase, the G_1 checkpoint (indicated by the lower red star in figure 9.10) makes the key decision of whether the cell should divide, delay division, or enter a resting stage. In yeasts, where researchers first studied this checkpoint, it is called START. If conditions are favorable for division, the cell begins to copy its DNA, initiating S phase. The G_1 checkpoint is where the more complex eukaryotes typically arrest the cell cycle if environmental conditions make cell division impossible or if the cell passes into an extended resting period called G_0 (figure 9.11).

2. **DNA replication is assessed at the G_2 checkpoint.**
   The second checkpoint, indicated by the red star at the

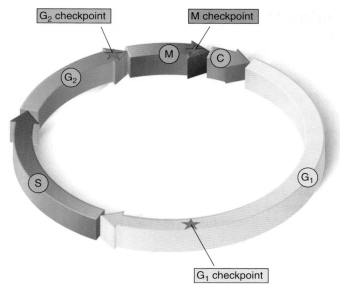

Figure 9.10 Control of the cell cycle.
Cells use a centralized control system to check whether proper conditions have been achieved before passing three key checkpoints in the cell cycle.

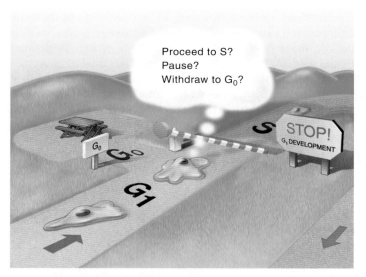

Figure 9.11 The G_1 checkpoint.
Feedback from the cell determines whether the cell cycle will proceed to the S phase, pause, or withdraw into G_0 for an extended rest period.

end of G_2 in figure 9.10, triggers the start of M phase. If this checkpoint is passed, the cell initiates the many molecular processes that signal the beginning of mitosis.

3. **Mitosis is assessed at the M checkpoint.** The third checkpoint, indicated by the red star on the arrowhead of the M phase in figure 9.10, occurs at metaphase and triggers the exit from mitosis and cytokinesis and the beginning of G_1.

> **9.5 The complex cell cycle of eukaryotes is controlled by feedback at three checkpoints.**

9.6 What Is Cancer?

Cancer is a growth disorder of cells. It starts when an apparently normal cell begins to grow in an uncontrolled way, spreading out to other parts of the body. The result is a cluster of cells, called a **tumor,** that constantly expands in size. The cluster of pink lung cells in the photo in figure 9.12 have begun to form a tumor. Benign tumors are completely enclosed by normal tissue and are said to be encapsulated. These tumors do not spread to other parts of the body and are therefore noninvasive. Malignant tumors are invasive and not encapsulated. Because they are not enclosed by normal tissue, cells are able to break away from the tumor and spread to other areas of the body. The tumor you see in figure 9.13, called a *carcinoma*, grows larger and eventually begins to shed cells that enter the bloodstream. Cells that leave a tumor and spread throughout the body, forming new tumors at distant sites, are called **metastases**.

Cancer is perhaps the most devastating and deadly disease. Of the children born in 1985, one-third will contract cancer at some time during their lives; one-fourth of the male children and one-third of the female children will someday die of cancer. Most of us have had family or friends affected by the disease. In 2005, 570,280 Americans died of cancer.

Not surprisingly, researchers are expending a great deal of effort to learn the cause of this disease. Scientists have made considerable progress in the last 30 years using molecular biological techniques, and the rough outlines of understanding are now emerging. We now know that cancer is a gene disorder of somatic tissue, in which damaged genes fail to properly control cell growth and division. The cell division cycle is regulated by a sophisticated group of proteins.

Cancer results from the damage of these genes encoding these proteins. Damage to DNA, such as damage to these genes, is called **mutation.**

Most cancers are the direct result of mutations in growth-regulating genes. There are two general classes of genes that are usually involved in cancer: proto-oncogenes and tumor-suppressor genes. Genes known as **proto-oncogenes** encode proteins that stimulate cell division. Mutations to these genes can cause cells to divide excessively. Mutated proto-oncogenes become cancer-causing genes called **oncogenes.**

The second class of cancer-causing genes are called **tumor-suppressor genes.** Cell division is normally turned off in healthy cells by proteins encoded by tumor-suppressor genes. Mutations to these genes essentially "release the brakes," allowing the cell containing the mutated gene to divide uncontrolled.

Cancer can be caused by chemicals or environmental factors such as UV rays that damage DNA, or in some instances by viruses that circumvent the cell's normal growth and division controls. Whatever the immediate cause, however, all cancers are characterized by unrestrained cell growth and division. The cell cycle never stops in a cancerous line of cells. Cancer cells are virtually immortal—until the body in which they reside dies.

> **9.6 Cancer is unrestrained cell growth and division caused by damage to genes regulating the cell division cycle.**

Figure 9.12 Lung cancer cells (300×).
These cells are from a tumor located in the alveolus (air sac) of a human lung.

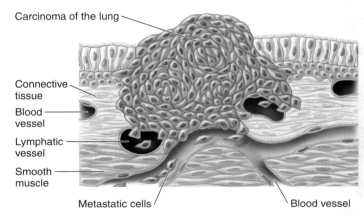

Carcinoma of the lung

Connective tissue

Blood vessel

Lymphatic vessel

Smooth muscle

Metastatic cells

Blood vessel

Figure 9.13 Portrait of a cancer.
This ball of cells is a carcinoma (cancer tumor) developing from epithelial cells that line the interior surface of a human lung. As the mass of cells grows, it invades surrounding tissues, eventually penetrating lymphatic and blood vessels, both of which are plentiful within the lung. These vessels carry metastatic cancer cells throughout the body, where they lodge and grow, forming new masses of cancerous tissue.

9.7 Cancer and Control of the Cell Cycle

Cancer results from damaged genes failing to control cell division. Researchers have identified several of these genes. One particular gene seems to be a key regulator of the cell cycle. Officially dubbed *p53* (researchers italicize the gene symbol to differentiate it from the protein), this gene plays a key role in the G_1 checkpoint of cell division. Figure 9.14 illustrates how the product of this gene, the p53 protein, monitors the integrity of DNA, checking that it is undamaged. If the p53 protein detects damaged DNA, as it does in the upper panel of this figure, it halts cell division and stimulates the activity of special enzymes to repair the damage. Once the DNA has been repaired, p53 allows cell division to continue, indicated by the upper set of arrows. In cases where the DNA cannot be repaired, p53 then directs the cell to kill itself, activating an apoptosis (cell suicide) program, indicated by the lower set of arrows.

By halting division in damaged cells, *p53* prevents the formation of tumors (even though its activities are not limited to cancer prevention). Scientists have found that *p53* is itself damaged beyond use in the majority of cancerous cells they have examined. It is precisely because *p53* is nonfunctional that these cancer cells are able to repeatedly undergo cell division without being halted at the G_1 checkpoint. The lower panel of figure 9.14 shows what happens when p53 doesn't function properly. The damaged strand is not repaired by the abnormal p53 and upon replication, leads to damaged cells. As more and more damage occurs to these cells, they become cancerous. To test this, scientists administered healthy p53 protein to rapidly dividing cancer cells in a petri dish: the cells soon ceased dividing and died. Scientists have further reported that cigarette smoke causes mutations in the *p53* gene, reinforcing the strong link between smoking and cancer described in chapter 26.

9.7 **Mutations disabling key elements of the G_1 checkpoint are associated with many cancers.**

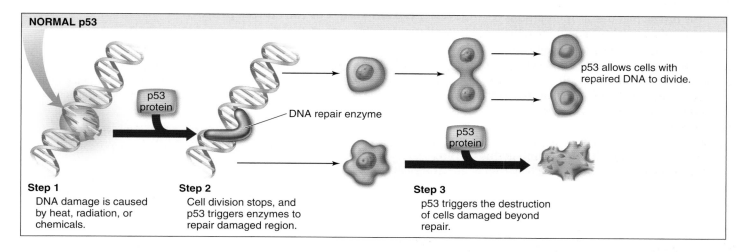

NORMAL p53

p53 allows cells with repaired DNA to divide.

p53 protein

DNA repair enzyme

p53 protein

Step 1
DNA damage is caused by heat, radiation, or chemicals.

Step 2
Cell division stops, and p53 triggers enzymes to repair damaged region.

Step 3
p53 triggers the destruction of cells damaged beyond repair.

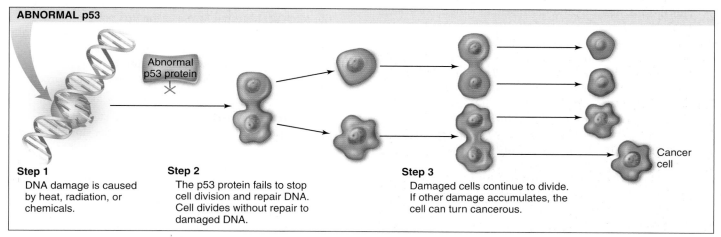

ABNORMAL p53

Abnormal p53 protein

Cancer cell

Step 1
DNA damage is caused by heat, radiation, or chemicals.

Step 2
The p53 protein fails to stop cell division and repair DNA. Cell divides without repair to damaged DNA.

Step 3
Damaged cells continue to divide. If other damage accumulates, the cell can turn cancerous.

Figure 9.14 Cell division and p53 protein.
Normal p53 protein monitors DNA, destroying cells with irreparable damage to their DNA. Abnormal p53 protein fails to stop cell division and repair DNA. As damaged cells proliferate, cancer develops.

Curing Cancer

Potential cancer therapies are being developed on many fronts. Some act to prevent the start of cancer within cells. Others act outside cancer cells, preventing tumors from growing and spreading. The figure on the right indicates targeted areas for the development of cancer treatments. The following discussion will examine each of these areas.

Preventing the Start of Cancer

Many promising cancer therapies act within potential cancer cells, focusing on different stages of the cell's "Shall I divide?" decision-making process.

1. Receiving the Signal to Divide. The first step in the decision process is receiving a "divide" signal, usually a small protein called a growth factor released from a neighboring cell. The growth factor, the red ball at #1 in the figure, is received by a protein receptor on the cell surface. Like banging on a door, its arrival signals that it's time to divide. Mutations that increase the number of receptors on the cell surface amplify the division signal and so lead to cancer. Over 20% of breast cancer tumors prove to overproduce a protein called HER2 associated with the receptor for epidermal growth factor (EGF).

Therapies directed at this stage of the decision process utilize the human immune system to attack cancer cells. Special protein molecules called *monoclonal antibodies,* created by genetic engineering, are the therapeutic agents. These monoclonal antibodies are designed to seek out and stick to HER2. Like waving a red flag, the presence of the monoclonal antibody calls down attack by the immune system on the HER2 cell. Because breast cancer cells overproduce HER2, they are killed preferentially. The biotechnology research company Genentech's recently approved monoclonal antibody, called herceptin, has given promising results in clinical tests.

Up to 70% of colon, prostate, lung, and head/neck cancers have excess copies of a related receptor, epidermal growth factor 1 (HER1). The monoclonal antibody C225, directed against HER1, has succeeded in shrinking 22% of advanced, previously incurable colon cancers in early clinical trials. Apparently blocking HER1 interferes with the ability of tumor cells to recover from chemotherapy or radiation.

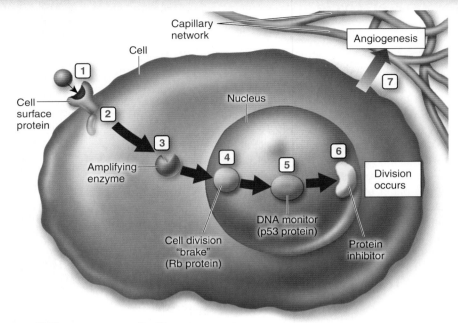

Seven different stages in the cancer process.

(1) On the cell surface, a growth factor's signal to divide is increased. (2) Just inside the cell, a protein relay switch that passes on the divide signal gets stuck in the "ON" position. (3) In the cytoplasm, enzymes that amplify the signal are amplified even more. In the nucleus, (4) a "brake" preventing DNA replication is inoperable, (5) proteins that check for damage in the DNA are inactivated, and (6) other proteins that inhibit the elongation of chromosome tips are destroyed. (7) The new tumor promotes angiogenesis, the formation of new blood vessels that promote growth.

2. Passing the Signal via a Relay Switch. The second step in the decision process is the passage of the signal into the cell's interior, the cytoplasm. This is carried out in normal cells by a protein called Ras that acts as a relay switch, #2 in the figure. When growth factor binds to a receptor like EGF, the adjacent Ras protein acts like it has been "goosed," contorting into a new shape. This new shape is chemically active, and initiates a chain of reactions that passes the "divide" signal inward toward the nucleus. Mutated forms of the Ras protein behave like a relay switch stuck in the "ON" position, continually instructing the cell to divide when it should not. Thirty percent of all cancers have a mutant form of Ras. So far, no effective therapies have been developed targeting this step.

3. Amplifying the Signal. The third step in the decision process is the amplification of the signal within the cytoplasm. Just as a TV signal needs to be amplified in order to be received at a distance, so a "divide" signal must be amplified if it is to reach the nucleus at the interior of the cell, a very long journey at a molecular scale. To get a signal all the way into the nucleus, the cell employs a sort of pony express. The "ponies" in this case are enzymes called *tyrosine kinases,* #3 in the figure. These enzymes add phosphate groups to proteins, but only at a particular amino acid, tyrosine. No other enzymes in the cell do this, so the tyrosine kinases form an elite core of signal carriers not confused by the myriad of other molecular activities going on around them.

Cells use an ingenious trick to amplify the signal as it moves toward the nucleus. Ras, when "ON," activates the initial protein kinase. This protein kinase activates other protein kinases that in their turn activate still others. The trick is that once a protein kinase enzyme is activated, it goes to work like a demon, activating hoards of others every second! And each and every one it activates behaves the same way too, activating still more, in a cascade of ever-widening effect. At each stage of the relay, the signal is amplified a thousandfold.

Mutations stimulating any of the protein kinases can dangerously increase the already amplified signal and lead to cancer. Some 15 of the cell's 32 internal tyrosine kinases have been implicated in cancer. Five percent of all cancers, for example, have a mutant hyperactive form of the protein kinase Src. The trouble begins when a mutation causes one of the tyrosine kinases to become locked into the "ON" position, sort of like a stuck doorbell that keeps ringing and ringing.

To cure the cancer, you have to find a way to shut the bell off. Each of the signal carriers presents a different problem, as you must quiet it without knocking out all the other signal pathways the cell needs. The cancer therapy drug Gleevec, a monoclonal antibody, has just the right shape to fit into a groove on the surface of the tyrosine kinase called "abl." Mutations locking abl "ON" are responsible for chronic myelogenous leukemia, a lethal form of white blood cell cancer. Gleevec totally disables abl. In clinical trials, blood counts revert to normal in more than 90% of cases.

4. Releasing the Brake. The fourth step in the decision process is the removal of the "brake" the cell uses to restrain cell division. In healthy cells this brake, a tumor-suppressor protein called Rb, blocks the activity of a protein called E2F, #4 in the figure. When free, E2F enables the cell to copy its DNA. Normal cell division is triggered to begin when Rb is inhibited, unleashing E2F. Mutations that destroy Rb release E2F from its control completely, leading to ceaseless cell division. Forty percent of all cancers have a defective form of Rb.

Therapies directed at this stage of the decision process are only now being attempted. They focus on drugs able to inhibit E2F, which should halt the growth of tumors arising from inactive Rb. Experiments in mice in which the E2F genes have been destroyed provide a model system to study such drugs, which are being actively investigated.

5. Checking That Everything Is Ready. The fifth step in the decision process is the mechanism used by the cell to ensure that its DNA is undamaged and ready to divide. This job is carried out in healthy cells by the tumor-suppressor protein p53, which inspects the integrity of the DNA, #5 in the figure. When it detects damaged or foreign DNA, p53 stops cell division and activates the cell's DNA repair systems. If the damage doesn't get repaired in a reasonable time, p53 pulls the plug, triggering events that kill the cell. In this way, mutations such as those that cause cancer are either repaired or the cells containing them eliminated. If p53 is itself destroyed by mutation, future damage accumulates unrepaired. Among this damage are mutations that lead to cancer. Fifty percent of all cancers have a disabled p53. Fully 70% to 80% of lung cancers have a mutant inactive p53—the chemical benzo[a]pyrene in cigarette smoke is a potent mutagen of p53.

6. Stepping on the Gas. Cell division starts with replication of the DNA. In healthy cells, another tumor suppressor "keeps the gas tank nearly empty" for the DNA replication process by inhibiting production of an enzyme called *telomerase*. Without this enzyme, a cell's chromosomes lose material from their tips, called *telomeres*. Every time a chromosome is copied, more tip material is lost. After some 30 divisions, so much is lost that copying is no longer possible. Cells in the tissues of an adult human have typically undergone 25 or more divisions. Cancer can't get very far with only the five remaining cell divisions, so inhibiting telomerase is a very effective natural brake on the cancer process, #6 in the figure. It is thought that almost all cancers involve a mutation that destroys the telomerase inhibitor, releasing this brake and making cancer possible. It should be possible to block cancer by reapplying this inhibition. Cancer therapies that inhibit telomerase are just beginning clinical trials.

Preventing the Spread of Cancer

7. Stopping Tumor Growth. Once a cell begins cancerous growth, it forms an expanding tumor. As the tumor grows ever-larger, it requires an increasing supply of food and nutrients, obtained from the body's blood supply. To facilitate this necessary grocery shopping, tumors leak out substances into the surrounding tissues that encourage the formation of small blood vessels, a process called angiogenesis, #7 in the figure. Chemicals that inhibit this process are called *angiogenesis inhibitors*. Two such natural angiogenesis inhibitors, angiostatin and endostatin, caused tumors to regress to microscopic size in mice, but initial human trials were disappointing.

Laboratory drugs are more promising. A monoclonal antibody drug called Avastin, targeted against a blood vessel growth promoting substance called vascular endothelial growth factor (VEGF), destroys the ability of VEGF to carry out its blood-vessel-forming job. Given to hundreds of advanced colon cancer patients in 2003 as part of a large clinical trial, Avastin improved colon cancer patients' chance of survival by 50% over chemotherapy.

Why Do Human Cells Age?

Human cells appear to have built-in life spans. In 1961 cell biologist Leonard Hayflick reported the startling result that skin cells growing in tissue culture, such as those growing in culture flasks in the photo below, will divide only a certain number of times. After about 50 population doublings cell division stops (a **doubling** is a round of cell division producing two daughter cells for each dividing cell, for example going from a population of 30 cells to 60 cells). If a cell sample is taken after 20 doublings and frozen, when thawed it resumes growth for 30 more doublings, and then stops. An explanation of the "Hayflick limit" was suggested in 1986 when researchers first glimpsed an extra length of DNA at the end of chromosomes. Dubbed telomeres, they proved to be composed of the simple DNA sequence TTAGGG, repeated nearly a thousand times. Importantly, telomeres were found to be substantially shorter in the cells of older body tissues.

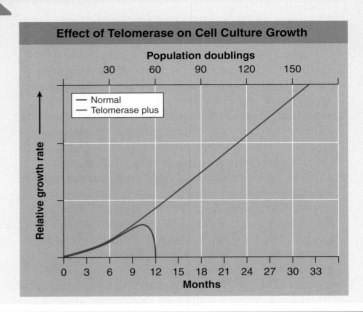

Effect of Telomerase on Cell Culture Growth

TTAGGG TTAGGG TTAGGG TTAGGG TTAGGG--------

This led to the hypothesis that a run of some 16 TTAGGGs was where the DNA replicating enzyme, called polymerase, first sat down on the DNA (16 TTAGGGs being the size of the enzyme's "footprint"), and because of being its docking spot, the polymerase was unable to copy that bit. Thus a 100-base portion of the telomere was lost by a chromosome during each doubling as DNA replicated. Eventually, after some 50 doubling cycles, each with a round of DNA replication, the telomere would be used up and there would be no place for the DNA replication enzyme to sit. The cell line would then enter senescence, no longer able to proliferate.

This hypothesis was tested in 1998. Using genetic engineering, researchers transferred into newly established human cell cultures a gene that leads to expression of an enzyme called *telomerase* that all cells possess but no body cell uses. This enzyme adds TTAGGG sequences back to the end of telomeres, in effect rebuilding the lost portions of the telomere. Laboratory cultures of cell lines with (telomerase plus) and without (normal) this gene were then monitored for many generations. The graph above displays the results.

1. **Applying Concepts**
   a. Variable. In the graph, what is the dependent variable?
   b. Comparing Continuous Processes. How do normal skin cells (blue line) differ in their growth history from telomerase plus cells with the telomerase gene (red line)?

2. **Interpreting Data**
   a. After how many doublings do the normal cells cease to divide? Is this consistent with the telomerase hypothesis?

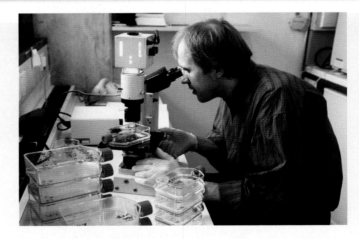

   b. After how many doublings do the telomerase plus cells cease to divide in this experiment?

3. **Making Inferences** After 9 population doublings, would the rate of cell division be different between the two cultures? after 15? Why?

4. **Drawing Conclusions** How does the addition of the telomerase gene affect the senescense of skin cells growing in culture? Does this result confirm the telomerase hypothesis this experiment had set out to test?

5. **Further Analysis**
   a. Cancer cells are thought to possess mutations disabling the cell's ability to keep the telomerase gene shut off. How would you test this hypothesis?
   b. Sperm-producing cells continue to divide throughout a male's adult life. How might this be possible? How would you test this idea?

Cell Division

9.1 Prokaryotes Have a Simple Cell Cycle

- Prokaryotic cells divide in a two-step process: DNA replication and binary fission. The DNA is a single loop of DNA called a chromosome. The DNA begins replication at a site called the origin of replication. The DNA double strand unzips and new strands form along the original strands, producing two circular chromosomes that separate to the ends of the cell. New plasma membrane and cell wall is added down the middle of the cell, splitting the cell in two, producing two daughter cells that are genetically identical to the parent cell (**figure 9.1**).

9.2 Eukaryotes Have a Complex Cell Cycle

- Eukaryotes have a more complex cell cycle. The replicated DNA is packaged into chromosomes that are distributed into daughter cells through the processes of mitosis and cytokinesis (**figure 9.2**).

9.3 Chromosomes

- Eukaryotic DNA is organized into chromosomes, and two chromosomes that carry copies of the same genes are called homologous chromosomes. Before cells divide, the DNA replicates forming two identical copies of each chromosome, called sister chromatids (**figures 9.3** and **9.4**).

- DNA in the nucleus of eukaryotic cells is packaged into compact structures called chromosomes. The DNA wraps around histone proteins, and the DNA-histone complex is tightly coiled, forming a compact chromosome (**figures 9.5** and **9.6**).

9.4 Cell Division

- When a eukaryotic cell begins to divide, it enters a stage in the cell cycle called interphase. During interphase, the DNA replicates and toward the end, the DNA begins to condense into chromosomes.

- Interphase is followed by mitosis, which consists of four phases: prophase, metaphase, anaphase, and telophase (**figure 9.7**).

- Prophase signals the beginning of mitosis. The DNA condenses into chromosomes that stay attached to their replicated partners (sister chromatids) at sites called centromeres. The nucleolus and nuclear envelope disappear, and centrioles when present begin forming the spindle. Microtubules also form from the poles and extend to the kinetochores, which are proteins located at the centromere areas of the chromosomes.

- Metaphase involves the alignment of sister chromatids along the equatorial plane. Microtubules connect the kinetochores of sister chromatids to each of the poles.

- During anaphase, the microtubules shorten, pulling the sister chromatids

apart and toward opposite poles, such that each pole receives one copy of each chromosome.

- Telophase signals the completion of nuclear division. The microtubule spindle is dismantled, the nuclear envelope and nucleolus re-form.

- Following mitosis, the cell separates into two daughter cells in a process called cytokinesis (**figure 9.8**). Cytokinesis in animal cells involves a pinching in of the cell around its equatorial plane, until the cell eventually splits into two cells. Cytokinesis in plant cells involved the formation of cell membrane and cell wall between the two poles, eventually forming two separate cells.

9.5 Controlling the Cell Cycle

- The cell cycle is controlled at three checkpoints (**figure 9.10**). At the G_1 checkpoint, before entering the S phase, the cell either initiates division or enters a period of rest called G_0 (**figure 9.11**).

- If cell division is initiated at the G_1 checkpoint, DNA replication begins and is checked at the G_2 checkpoint. Mitosis is either initiated or delayed, depending on the successful replication of DNA. The third checkpoint is toward the end of mitosis, before cytokinesis (**figure 9.10**).

Cancer and the Cell Cycle

9.6 What Is Cancer?

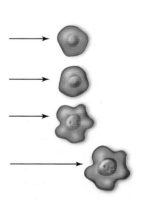

- Cancer is a growth disorder of cells, where there is a loss of control over cell division. Cells begin to divide in an uncontrolled way, forming a mass of cells called a tumor. A tumor that is encapsulated by normal tissue is called benign, but a situation in which cells can break away from the mass and spread to other tissues is called metastases. Mutations in proto-oncogenes and tumor-suppressor genes lead to uncontrolled growth—cancer (**figure 9.13**).

9.7 Cancer and Control of the Cell Cycle

- Mutations of the *p53* gene are found in many types of cancer cells. This gene plays a key role in the G_1 checkpoint, checking the condition of the DNA. If the DNA is damaged, the p53 protein will stop cell division and give the cell the opportunity to correct the damaged DNA. If the DNA cannot be repaired, the p53 protein will trigger the destruction of the cell (**figure 9.14**). When the *p53* gene is damaged by mutations, the DNA is not checked, and cells with damaged DNA can divide. Further mutations to the DNA can accumulate in the cells and result in cells that become cancerous.

1. Prokaryotes reproduce new cells by
   a. copying DNA then undergoing binary fission.
   b. splitting in half.
   c. undergoing mitosis.
   d. copying DNA then undergoing the M phase.
2. The eukaryotic cell cycle is different from prokaryotic cell division in all the following ways except
   a. the amount of DNA present in the cells.
   b. how the DNA is packaged.
   c. in the production of daughter cells.
   d. the involvement of microtubules.
3. In eukaryotes, the genetic material is found in chromosomes
   a. and the more complex the organism, the more pairs of chromosomes it has.
   b. and many organisms have only one chromosome.
   c. and most eukaryotes have between 10 and 50 pairs of chromosomes.
   d. and most eukaryotes have between 2 and 10 pairs of chromosomes.
4. Homologous chromosomes
   a. are also referred to as sister chromatids.
   b. are genetic identical.
   c. carry information about the same traits located in the same places on the chromosomes.
   d. are connected to each other at their centromeres.
5. In mitosis, when the duplicated chromosomes line up in the center of the cell, that stage is called
   a. prophase.   c. anaphase.
   b. metaphase.   d. telophase.

6. The division of the cytoplasm in the eukaryotic cell cycle is called
   a. interphase.
   b. karyokinesis.
   c. cytokinesis.
   d. binary fission.
7. The cell cycle is controlled by
   a. a series of checkpoints.
   b. an internal clock.
   c. the completion of one phase triggering the next phase.
   d. cell size—when it grows large enough the cell cycle is triggered.
8. When cell division becomes unregulated, and a cluster of cells begins to grow without regard for the normal controls, that is called
   a. a mutation.   c. metastases.
   b. cancer.   d. oncogenes.
9. Which of the following is a cancer-causing gene?
   a. a proto-oncogene
   b. a tumor-suppressor gene
   c. an oncogene
   d. the $p53$ gene
10. The normal function of the $p53$ gene in the cell is
   a. as a tumor-suppressor gene.
   b. to monitor the DNA for damage.
   c. to trigger the destruction of cells not capable of DNA repair.
   d. All of the above.

1. **Figure 9.4** This shows a set of human chromosomes. At what stage of the cell cycle are such photos taken, and why?

2. **Figure 9.6** During interphase the DNA is not visible through a microscope. Why isn't the DNA visible during interphase and why would this be the case?

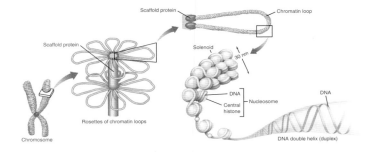

1. Why does the DNA need to change periodically from a long, double-helix chromatin molecule into a tightly wound-up chromosome? What does it do at each stage that it cannot do at the other?

2. Despite all we know about cancer today, some types of cancers are still increasing in frequency. Lung cancer in women is one of those. What reason(s) might there be for this increasing problem?

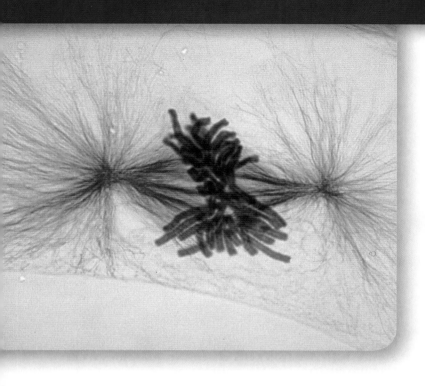

10

Meiosis

Humans, like most animals and plants, reproduce sexually. That is how you came into being: your father contributed a sperm cell which united with an egg cell from your mother to form a cell called a zygote, containing both sets of chromosomes. Dividing repeatedly by mitosis, this zygote cell eventually gave rise to your adult body, made up of an astonishing number of cells—some 100 trillion. The sperm and egg that joined to form your initial cell were the products of a special form of cell division called meiosis, the subject of this chapter. Far more intricate than mitosis, the details of meiosis are not as well understood. The basic process, however, is clear. A cell dividing by meiosis goes through two nuclear divisions, replicating the DNA before the first division but not between the two divisions. The cell you see above is a gamete-producing cell of the Oregon newt *Taricha granulosa*, a kind of salamander. The micrograph (that is, a photo taken with a microscope) captures the cell in the metaphase stage of the first meiotic division, when all the blue-stained chromosomes are lined up on the metaphase plate. Soon the red-stained spindle fibers will draw homologous chromosomes to opposite poles of the cell. When the two divisions are all over, there will be four cells, each with only half as much DNA as the initial cell. Confused? So were biologists when they first discovered meiosis. Hopefully, this chapter will make things clearer. It is important that you understand meiosis clearly, because meiosis and sexual reproduction play key roles in generating the tremendous genetic diversity that is the raw material of evolution.

10.1 Discovery of Meiosis

Only a few years after Walther Fleming's discovery of chromosomes in 1882, Belgian cytologist Pierre-Joseph van Beneden was surprised to find different numbers of chromosomes in different types of cells in the roundworm *Ascaris*. Specifically, he observed that the **gametes** (eggs and sperm) each contained two chromosomes, whereas the *somatic* (nonreproductive) cells of embryos and mature individuals each contained four.

Fertilization

From his observations, van Beneden proposed in 1887 that an egg and a sperm, each containing half the complement of chromosomes found in other cells, fuse to produce a single cell called a **zygote.** The zygote, like all of the somatic cells ultimately derived from it, contains two copies of each chromosome. The fusion of gametes to form a new cell is called **fertilization,** or **syngamy.**

Meiosis

It was clear even to early investigators that gamete formation must involve some mechanism that reduces the number of chromosomes to half the number found in other cells. If it did not, the chromosome number would double with each fertilization, and after only a few generations, the number of chromosomes in each cell would become impossibly large. For example, in just 10 generations, the 46 chromosomes present in human cells would increase to over 47,000 (46×2^{10}) chromosomes.

The number of chromosomes does not explode in this way because of a special reduction division that occurs during gamete formation, producing cells with half the normal number of chromosomes. The subsequent fusion of two of these cells ensures a consistent chromosome number from one generation to the next. This reduction division process, known as *meiosis,* is the subject of this chapter.

The Sexual Life Cycle

Meiosis and fertilization together constitute a cycle of reproduction. Two sets of chromosomes are present in the somatic cells of adult individuals, making them **diploid** cells (Greek, *di,* two), but only one set is present in the gametes, which are thus **haploid** (Greek, *haploos,* one). Figure 10.1 shows how two haploid cells, a sperm cell containing three chromosomes contributed by the father and an egg cell containing three chromosomes contributed by the mother, fuse to form a diploid zygote with six chromosomes. Reproduction that involves this alternation of meiosis and fertilization is called **sexual reproduction.** Some organisms however, reproduce by mitotic division and don't involve the fusion of gametes. Reproduction in these organisms is referred to as **asexual reproduction.** Binary fission of prokaryotes shown in chapter 9 is an example of asexual reproduction. Some organisms are able to reproduce both asexu-

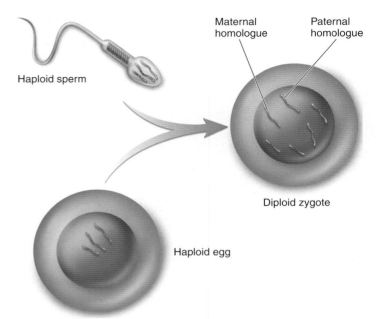

Figure 10.1   Diploid cells carry chromosomes from two parents.

A diploid cell contains two versions of each chromosome, a maternal homologue contributed by the haploid egg of the mother, and a paternal homologue contributed by the haploid sperm of the father.

Figure 10.2   Sexual and asexual reproduction.
Reproduction is not always either sexual or asexual. The strawberry reproduces both asexually (runners) and sexually (flowers).

ally and sexually. The strawberry plant pictured in figure 10.2 reproduces sexually by fertilization that occurs in their flowers. They also reproduce asexually by sending out runners, stems that grow along the ground, sending out new roots and shoots that give rise to genetically identical plants.

> **10.1** Meiosis is a process of cell division in which the number of chromosomes in certain cells is halved during gamete formation.

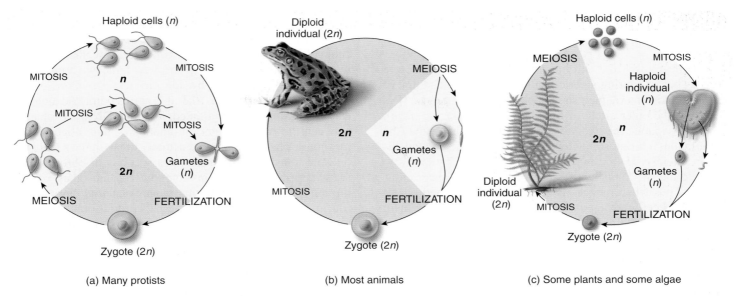

Figure 10.3 Three types of sexual life cycles.
In sexual reproduction, haploid cells or organisms alternate with diploid cells or organisms.

(a) Many protists

(b) Most animals

(c) Some plants and some algae

10.2 The Sexual Life Cycle

Somatic Tissues

The life cycles of all sexually reproducing organisms follow the same basic pattern of alternation between diploid chromosome numbers (the blue areas of the life cycles illustrated in figure 10.3) and haploid ones (the yellow areas). In most animals, fertilization results in the formation of a diploid zygote, shown in figure 10.3b, that begins to divide by mitosis. This single diploid cell eventually gives rise to all of the cells in the adult frog shown in the figure. These cells are called **somatic** cells, from the Latin word for "body." Each is genetically identical to the zygote.

In unicellular eukaryotic organisms like the protists shown in figure 10.3a, individual haploid cells function as gametes, fusing with other gamete cells. In plants like the fern shown in figure 10.3c, the haploid cells that meiosis produces divide by mitosis, forming a multicellular haploid phase, the heart-shaped structure in the figure. Some cells of this haploid phase eventually differentiate into eggs or sperm, which fuse to form a diploid zygote.

Germ-Line Tissues

In animals, the cells that will eventually undergo meiosis to produce gametes are set aside from somatic cells early in the course of development. These cells are often referred to as **germ-line** cells. Both the somatic cells and the gamete-producing germ-line cells are diploid, as indicated by blue arrows in figure 10.4. Somatic cells undergo mitosis to form genetically identical, diploid daughter cells. The germ-line cells undergo meiosis, indicated by the yellow arrows, producing haploid gametes.

> **10.2 In the sexual life cycle, there is an alternation of diploid and haploid phases.**

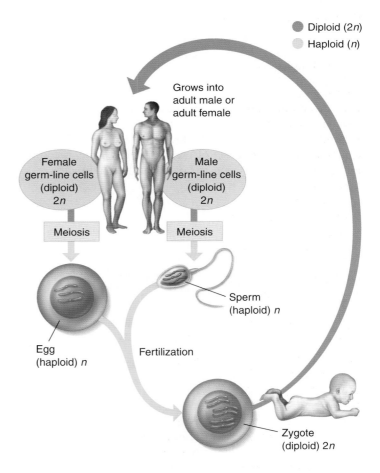

Figure 10.4 The sexual life cycle in animals.
In animals, the completion of meiosis is followed soon by fertilization. Thus, the vast majority of the life cycle is spent in the diploid stage. In this text, n stands for haploid and $2n$ stands for diploid. Germ-line cells are set aside early in development and undergo meiosis to form haploid gametes (eggs or sperm). The rest of the body cells are called somatic cells.

10.3 The Stages of Meiosis

Just as in mitosis, the chromosomes have replicated before meiosis begins, during a period called interphase. The first of the two divisions of meiosis, called **meiosis I** (meiosis I is shown in the outer circle of figure 10.7 on the facing page), serves to separate the two versions of each chromosome (the homologous chromosomes or homologues); the second division, **meiosis II** (the inner circle of figure 10.7), serves to separate the two replicas of each version, called sister chromatids. Thus when meiosis is complete, what started out as one diploid cell ends up as four haploid cells. Because there was one replication of DNA but *two* cell divisions, the process reduces the number of chromosomes by half.

Meiosis I

Meiosis I is traditionally divided into four stages:

1. **Prophase I.** The two versions of each chromosome (the two homologues) pair up and exchange segments.
2. **Metaphase I.** The chromosomes align on a central plane.
3. **Anaphase I.** One homologue with its two sister chromatids still attached moves to a pole of the cell, and the other homologue moves to the opposite pole.
4. **Telophase I.** Individual chromosomes gather together at each of the two poles.

In **prophase I,** individual chromosomes first become visible, as viewed with a light microscope, as their DNA coils more and more tightly. Because the chromosomes (DNA) have replicated before the onset of meiosis, each of the threadlike chromosomes actually consists of two sister chromatids associated along their lengths (held together by cohesin proteins in a process called *sister chromatid cohesion*) and joined at their centromeres, just as in mitosis. However, now meiosis begins to differ from mitosis. During prophase I, the two homologous chromosomes line up side by side, physically touching one another, as you see in figure 10.5. It is at this point that a process called **crossing over** is initiated, in which DNA is exchanged between the two nonsister chromatids of homologous chromosomes. The chromosomes actually break in the same place on both nonsister chromatids and sections of chromosomes are swapped between the homologous chromosomes, producing a hybrid chromosome that is part maternal chromosome (the green sections) and part paternal chromosome (the purple sections). Two elements hold the homologous chromosomes together: (1) cohesion between sister chromatids; and (2) crossovers between nonsister chromatids (homologues). Late in prophase, the nuclear envelope disperses.

In **metaphase I,** the spindle apparatus forms, but because homologues are held close together by crossovers, spindle fibers can attach to only the outward-facing kinetochore of each centromere. For each pair of homologues, the orientation on the metaphase plate is random; which homologue is oriented toward which pole is a matter of chance. Like shuffling a deck of cards, many combinations are possible—in fact, 2 raised to a power equal to the number of chromosome pairs. For example, in a hypothetical cell that has three chromosome pairs, there are eight possible orientations (2^3). Each orientation results in gametes with different combinations of parental chromosomes. This process is called **independent assortment.** The chromosomes in figure 10.6 line up along the metaphase plate but their placement—whether the maternal chromosome (the green chromosomes) is on the right or left of the plate—is completely random. However, its paternal homologue (purple) is always opposite.

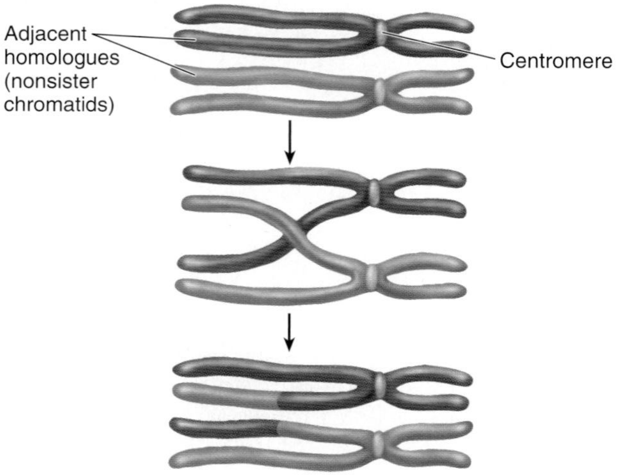

Adjacent homologues (nonsister chromatids)

Centromere

Figure 10.5 Crossing over.
In crossing over, the two homologues of each chromosome exchange portions. During the crossing over process, nonsister chromatids that are next to each other exchange chromosome arms.

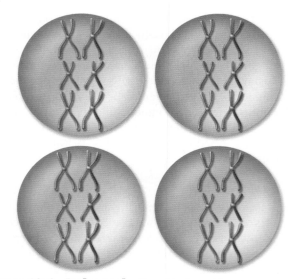

Figure 10.6 Independent assortment.
Independent assortment occurs because the orientation of chromosomes on the metaphase plate is random. Shown here are four possible orientations of chromosomes in a hypothetical cell. Each of the many possible orientations results in gametes with different combinations of parental chromosomes.

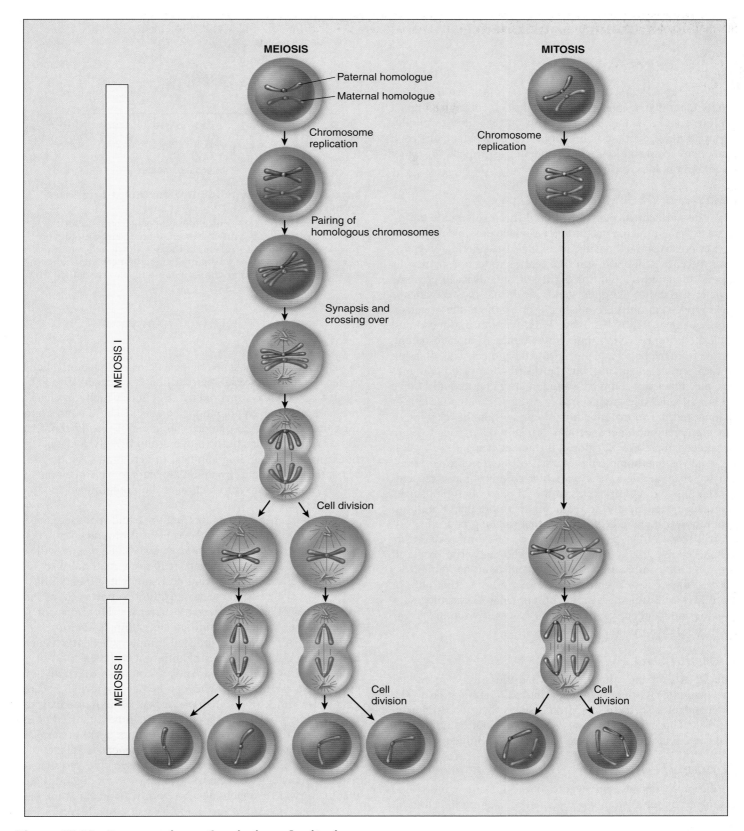

Figure 10.10 A comparison of meiosis and mitosis.

Meiosis involves two nuclear divisions with no DNA replication between them. It thus produces four daughter cells, each with half the original number of chromosomes. Crossing over occurs in prophase I of meiosis. Mitosis involves a single nuclear division after DNA replication. It thus produces two daughter cells, each containing the original number of chromosomes that are genetically identical to the parent cell.

10.5 Evolutionary Consequences of Sex

While our knowledge of how sex evolved is sketchy, it is abundantly clear that sexual reproduction has an enormous impact on how species evolve today, because of its ability to rapidly generate new genetic combinations. Three mechanisms each make key contributions: independent assortment, crossing over, and random fertilization.

Independent Assortment

The reassortment of genetic material that takes place during meiosis is the principal factor that has made possible the evolution of eukaryotic organisms, in all their bewildering diversity, over the past 1.5 billion years. Sexual reproduction represents an enormous advance in the ability of organisms to generate genetic variability. To understand, recall that most organisms have more than one chromosome. For example, the organism represented in figure 10.11 has three pairs of chromosomes, each offspring receiving three homologues from each parent, purple from the father and green from the mother. The offspring in turn produces gametes, but the distribution of homologues into the gametes is completely random. A gamete could receive all homologues that are paternal in origin, as on the far left; or it could receive all maternal homologues, as on the far right, or any combination. Independent assortment alone leads to eight possible gamete combinations. In human beings, each gamete receives one homologue of each of the 23 chromosomes, but which homologue of a particular chromosome it receives is determined randomly. Each of the 23 pairs of chromosomes migrates independently, so there are 2^{23} (more than 8 million) different possible kinds of gametes that can be produced.

To make this point to his class, one professor offers an "A" course grade to any student who can write down all the possible combinations of heads and tails (an "either/or" choice, like that of a chromosome migrating to one pole or the other) with flipping a coin 23 times (like 23 chromosomes moving independently). No student has ever won an "A," as there are over 8 million possibilities.

Crossing Over

The DNA exchange that occurs when the arms of nonsister chromatids cross over adds even more recombination to the independent assortment of chromosomes that occurs later in meiosis. Thus, the number of possible genetic combinations that can occur among gametes is virtually unlimited.

Random Fertilization

Furthermore, because the zygote that forms a new individual is created by the fusion of two gametes, each produced independently, fertilization squares the number of possible outcomes ($2^{23} \times 2^{23} = 70$ trillion).

Importance of Generating Diversity

Paradoxically, the evolutionary process is both revolutionary and conservative. It is revolutionary in that the pace of evolu-

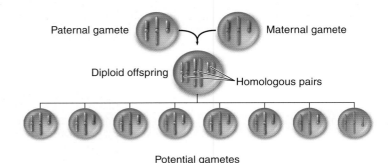

Figure 10.11 Independent assortment increases genetic variability.
Independent assortment contributes new gene combinations to the next generation because the orientation of chromosomes on the metaphase plate is random. In the cell shown here with three chromosome pairs, there are eight different gametes that can result, each with different combinations of parental chromosomes.

tionary change is quickened by genetic recombination, much of which results from sexual reproduction. It is conservative in that change is not always favored by selection, which may instead preserve existing combinations of genes. These conservative pressures appear to be greatest in some asexually reproducing organisms that do not move around freely and that live in especially demanding habitats. In vertebrates, on the other hand, the evolutionary premium appears to have been on versatility, and sexual reproduction is the predominant mode of reproduction.

Whatever the forces that led to sexual reproduction, its evolutionary consequences have been profound. No genetic process generates diversity more quickly; and, as you will see in chapter 11, genetic diversity is the raw material of evolution, the fuel that drives it and determines its potential directions. In many cases, the pace of evolution appears to increase as the level of genetic diversity increases. Programs for selecting larger stature in domesticated animals such as cattle and sheep, for example, proceed rapidly when new genetic combinations arise by crossing different varieties.

Breeding programs can only accomplish so much, however. There are limits to what can be accomplished, because genes often affect more than one aspect of an individual, setting limits on how much a character can be altered. For example, selecting for large clutch size (more eggs) in barnyard chickens eventually leads to eggs with thinner shells that break more easily. For this reason we do not have chickens that lay twice as many eggs as the best layers do now, or gigantic cattle that yield twice as much meat, or corn with an ear at the base of every leaf instead of just at the base of a few leaves.

> **10.5 Sexual reproduction increases genetic variability through independent assortment in metaphase I of meiosis, crossing over in prophase I of meiosis, and random fertilization.**

Why Sex?

Not all reproduction is sexual. In **asexual reproduction,** an individual inherits all of its chromosomes from a single parent and is, therefore, genetically identical to its parent. Prokaryotic cells reproduce asexually, undergoing binary fission to produce two daughter cells containing the same genetic information.

Most protists reproduce asexually except under conditions of stress; then they switch to sexual reproduction. Among plants and fungi, asexual reproduction is common.

In animals, asexual reproduction often involves the budding off of a localized mass of cells, which grows by mitosis to form a new individual.

Even when meiosis and the production of gametes occur, there may still be reproduction without sex. The development of an adult from an unfertilized egg, called **parthenogenesis,** is a common form of reproduction in arthropods. Among bees, for example, fertilized eggs develop into diploid females, but unfertilized eggs develop into haploid males. Parthenogenesis even occurs among the vertebrates. Some lizards, fishes, and amphibians are capable of reproducing in this way; their unfertilized eggs undergo a mitotic nuclear division without cell cleavage to produce a diploid cell, which then develops into an adult.

If reproduction can occur without sex, why does sex occur at all? This question has generated considerable discussion, particularly among evolutionary biologists. Sex is of great evolutionary advantage for populations or species, which benefit from the variability generated in meiosis by random orientation of chromosomes and by crossing over. However, evolution occurs because of changes at the level of *individual* survival and reproduction, rather than at the population level, and no obvious advantage accrues to the progeny of an individual that engages in sexual reproduction. In fact, recombination is a destructive as well as a constructive process in evolution. The segregation of chromosomes during meiosis tends to disrupt advantageous combinations of genes more often than it creates new, better adapted combinations; as a result, some of the diverse progeny produced by sexual reproduction will not be as well adapted as their parents were. In fact, the more complex the adaptation of an individual organism, the less likely that recombination will improve it, and the more likely that recombination will disrupt it. It is, therefore, a puzzle to know what a well-adapted individual gains from participating in sexual reproduction, as *all* of its progeny could maintain its successful gene combinations if

that individual simply reproduced asexually.

The Red Queen Hypothesis. One evolutionary advantage of sex may be that it allows populations to "store" forms of a trait that are currently bad but have promise for reuse at some time in the future. Because populations are constrained by a changing physical and biological environment, selection is constantly acting against such traits. But in sexual species, selection can never get rid of those variants sheltered by more dominant forms of the trait.

The evolution of most sexual species, most of the time, thus manages to keep pace with ever-changing physical and biological constraints. This "treadmill evolution" is sometimes called the "Red Queen hypothesis," after the Queen of Hearts in Lewis Carroll's *Through the Looking Glass,* who tells Alice, "Now, here, you see, it takes all the running you can do, to keep in the same place."

The DNA Repair Hypothesis. Several geneticists have suggested that sex occurs because only a diploid cell can effectively repair certain kinds of chromosome damage, particularly double-strand breaks in DNA. Both radiation and chemical events within cells can induce such breaks. As organisms became larger and longer-lived, it must have become increasingly important for them to be able to repair such damage. The synaptonemal complex, which in early stages of meiosis precisely aligns pairs of homologous chromosomes, may well have evolved originally as a mechanism for repairing double-strand damage to DNA, using the undamaged homologous chromosome as a template to repair the damaged chromosome. A transient diploid phase would have provided an opportunity for such repair. In yeast, mutations that inactivate the repair system for double-strand breaks of the chromosomes also prevent crossing over, suggesting a common mechanism for both synapsis and repair processes.

Muller's Ratchet. The geneticist Herman Muller pointed out in 1965 that asexual populations incorporate a kind of mutational ratchet mechanism—once harmful mutations arise, asexual populations have no way of eliminating them, and they accumulate over time, like turning a ratchet. Sexual populations, on the other hand, can employ recombination to generate individuals carrying fewer mutations, which selection can then favor. Sex may just be a way to keep the mutational load down.

Are New Microtubules Made When the Spindle Forms?

During interphase, before the beginning of meiosis, a relatively few long microtubules extend from the centrosome (a zone around the centrioles of animal cells where microtubules are organized) to the cell periphery. Like most microtubules, these are refreshed at a low rate with resynthesis. Late in prophase, however, a dramatic change is seen—the centrosome divides into two, and a large increase is seen in the number of microtubules radiating from each of the two daughter centrosomes. The two clusters of new microtubules are easily seen as the green fibers connecting to the two sets of purple daughter chromosomes in the micrograph of early prophase below (a **micrograph** is a photo taken through a microscope). This burst of microtubule assembly marks the beginning of the formation of the spindle characteristic of metaphase. When it first became known to cell biologists, they asked whether these were existing microtubules being repositioned in the spindle, or newly synthesized microtubules only produced just before metaphase begins.

The graph on the right above displays the results of an experiment designed to answer this question. Mammalian cells in culture (cells **in culture** are growing in the laboratory on artificial medium) were injected with microtubule subunits (tubulin) to which a fluorescent dye had been attached (a **fluorescent dye** is one that glows when exposed to ultraviolet or short-wavelength visual light). After the fluorescent subunits had become incorporated into the cells' microtubules, all the fluorescence in a small region of a cell was bleached by an intense laser beam, destroying the microtubules there. Any subsequent rebuilding of microtubules in the bleached region would have to employ the fluorescent subunits present in the cell, causing recovery of fluorescence in the bleached region. The graph at right reports this recovery as a function of time, for interphase and metaphase cells. The dotted line represents the time for 50% recovery of fluorescence ($t_{1/2}$) (that is, $t_{1/2}$ is the time required for half of the microtubules in the region to be resynthesized).

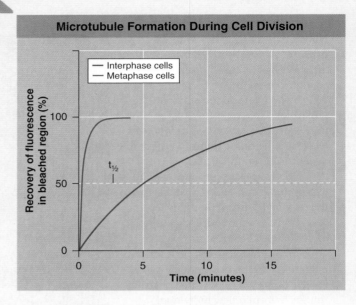

Microtubule Formation During Cell Division

Legend: Interphase cells / Metaphase cells

Y-axis: Recovery of fluorescence in bleached region (%) — 0, 50, 100
X-axis: Time (minutes) — 0, 5, 10, 15
$t_{1/2}$

1. **Applying Concepts**
   a. Variable. In the graph, what is the dependent variable?
   b. $t_{1/2}$. Are new microtubules synthesized during interphase? What is the $t_{1/2}$ of this replacement synthesis? Are new microtubules synthesized during metaphase? What is the $t_{1/2}$ of this replacement synthesis?

2. **Interpreting Data** Is there a difference in the rate at which microtubules are synthesized during interphase and metaphase? How big is the difference? What might account for it?

3. **Making Inferences**
   a. What general statement can be made regarding the relative rates of microtubule production before and during meiosis?
   b. Is there any difference in the final amount of microtubule synthesis which would occur if this experiment were to be continued for an additional 15 minutes?

4. **Drawing Conclusions** When are the microtubules of the spindle assembled?

5. **Further Analysis** The spindle breaks down after cell division is completed. Design an experiment to test whether the tubulin subunits of the spindle microtubules are recycled into other cell components, or destroyed, after meiosis.

10 µm

Meiosis

10.1 Discovery of Meiosis

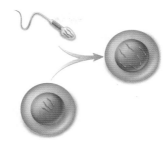

- In sexually reproducing organisms, the number of chromosomes in gametes must be halved to maintain the correct number of chromosomes in offspring (**figure 10.1**). Organisms accomplish this through a cell division process called meiosis.

- Sexual reproduction involves meiosis but some organisms also undergo asexual reproduction, which is reproducing by mitosis.

10.2 The Sexual Life Cycle

- Sexual life cycles alternate between diploid and haploid stages, but there is variation in the amount of time in the life cycle devoted to the haploid and diploid stages. Three types of sexual life cycles exist: in some types of algae the majority of the life cycle is devoted to the haploid stage, in most animals the majority of the life cycle is devoted to the diploid stage, and in some plants and algae the life cycle is split with half being devoted to a haploid stage and half to a diploid stage (**figure 10.3**).

- Germ-line cells of an organism produce gametes that contain half the number of chromosomes found in its somatic cells (**figure 10.4**).

10.3 The Stages of Meiosis

- Meiosis involves two nuclear divisions, meiosis I and meiosis II, each containing a prophase, metaphase, anaphase, and telophase. Like mitosis, the DNA replicates itself before meiosis begins, during interphase.

- Prophase I is distinguished by the exchange of genetic material between homologous chromosomes, a process called crossing over (**figure 10.5**). Homologous chromosomes align with each other along their lengths, and sections of arms of homologues are physically exchanged, changing the genetic information contained in the chromosomes.

- During metaphase I, homologous chromosomes align along the equatorial plane moved into place by microtubules called the spindle apparatus. The alignment of the chromosomes is random such that there is a shuffling of paternal and maternal chromosomes along the equatorial plane, leading to the independent assortment of chromosomes into the gametes (**figure 10.6**).

- The homologous chromosomes separate during anaphase I, being pulled apart by the spindle apparatus toward their respective poles. This differs from mitosis where sister chromatids separate in anaphase.

- In telophase I, the chromosomes cluster at the poles. This leads to the next phase of meiosis (**figure 10.7**).

- Meiosis II mirrors mitosis in that it involves the separation of sister chromatids through the phases of prophase II, metaphase II, anaphase II, and telophase II. Meiosis II differs from mitosis in that because homologous pairs were separated during meiosis I, each daughter cell has only one-half the number of chromosomes (**figure 10.8**).

Comparing Meiosis and Mitosis

10.4 How Meiosis Differs from Mitosis

- Two processes that distinguish meiosis from mitosis are crossing over through synapsis and reduction division.

- When homologous chromosomes come together during prophase I, they associate with each other along the lengths, a process called synapsis (**figure 10.9a**). This does not occur during mitosis. During synapsis, sections of homologous chromosomes are physically exchanged in a process called crossing over. Crossing over results in daughter cells that are not genetically identical with the parent cell or with each other. Mitosis results in daughter cells that are genetically identical to the parent cell and to each other.

- Meiosis also differs from mitosis in reduction division where the daughter cells contain half the number of chromosomes as the parent cell. Reduction division occurs because meiosis contains two nuclear divisions (meiosis I and meiosis II) but only one round of DNA replication (**figure 10.9b**).

- The primary reasons for the differences in meiosis and mitosis stem from the synapsis of homologous chromosomes in prophase I. Because of synapsis, the arms of homologous chromosomes are close enough to undergo crossing over. Also, the close relationship of the homologous chromosomes in synapsis block the inner kinetochores so that the spindle fibers are only able to attach to one side of the homologue, resulting of the separation of homologues in metaphase I rather then the separating of sister chromatids that occurs in mitosis and in meiosis II (**figure 10.10**).

10.5 Evolutionary Consequences of Sex

- Sexual reproduction results in the introduction of genetic variation in future generations through independent assortment, crossing over, and random fertilization (**figure 10.11**).

1. An egg and a sperm unite to form a new organism. To prevent the new organism from having twice as many chromosomes as its parents
   a. half of the chromosomes in the new organism quickly die off, leaving the correct number.
   b. half of the chromosomes from the egg, and half from the sperm, are ejected from the new cell.
   c. the large egg contains all the chromosomes, the tiny sperm only contributes some DNA.
   d. the germ cells went through meiosis; the egg and sperm only have half the parental chromosomes.
2. The diploid number of chromosomes in humans is 46. The haploid number is
   a. 138.
   b. 92.
   c. 46.
   d. 23.
3. In organisms that have sexual life cycles, there is a time when there are
   a. 1n gametes (haploid), then next there are 2n zygotes (diploid).
   b. 2n gametes (haploid), then next there are 1n zygotes (diploid).
   c. 2n gametes (diploid), then next there are 1n zygotes (haploid).
   d. 1n gametes (diploid), then next there are 2n zygotes (haploid).
4. The purpose of meiosis I is to
   a. duplicate all chromosomes.
   b. randomly separate the homologous pairs, called independent assortment.
   c. separate the duplicated sister chromatids.
   d. divide the original material into four complete diploid cells.
5. The purpose of meiosis II is to
   a. duplicate all chromosomes.
   b. randomly separate the homologous pairs, called independent assortment.
   c. separate the duplicated sister chromatids.
   d. divide the original material into four complete diploid cells.
6. During which stage of meiosis does crossing over occur?
   a. prophase I
   b. anaphase I
   c. metaphase II
   d. interphase
7. Synapsis is the process whereby
   a. homologous pairs of chromosomes separate and migrate toward a pole.
   b. homologous chromosomes exchange chromosomal material.
   c. homologous chromosomes become closely associated.
   d. the daughter cells contain half of the number of chromosomes of the parent cell.
8. The purpose of mitosis is to _____, while the purpose of meiosis is to _____.
   a. make diploid cells/make haploid cells
   b. make haploid cells/make diploid cells
   c. make cells which are either haploid or diploid/make cells which are haploid
   d. make cells which are haploid/make cells which vary in chromosome number
9. A major consequence of sex and meiosis is that each species
   a. remains pretty much the same because the chromosomes are carefully duplicated and passed to the next generation.
   b. has a lot of genetic reassortment due to processes in meiosis II.
   c. has a lot of genetic reassortment due to processes in meiosis I.
   d. has a lot of genetic reassortment due to processes in telophase II.
10. Genetic diversity is greatest in
   a. sexual reproduction.
   b. asexual reproduction.
   c. binary fission.
   d. All of the above are equal in their amount of genetic diversity.

Visual Understanding

1. **Figure 10.5** How is it that, in meiosis, you can end up with four "daughter cells" that are all genetically different from one another?

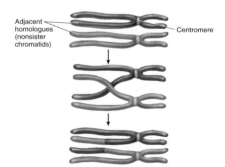

Adjacent homologues (nonsister chromatids)
Centromere

2. **Figure 10.9** Referring to the homologous chromosomes shown here during prophase I, and knowing that they stay in synapsis during metaphase I, explain why it is that sister chromatids don't separate during anaphase I as they do in mitosis.

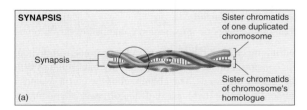

SYNAPSIS
Synapsis
Sister chromatids of one duplicated chromosome
Sister chromatids of chromosome's homologue
(a)

Challenge Questions

1. You only need one set of instructions for your body to do all the jobs it needs to carry out. So why aren't organisms simply haploid all their lives?
2. Are the germ-line cells of your body haploid or diploid? Why not the alternative?
3. An organism has 56 chromosomes in its diploid stage. Indicate how many chromosomes are present in the following, and explain your reasoning:
   a. somatic cells
   b. metaphase I
   c. metaphase II
   d. gametes

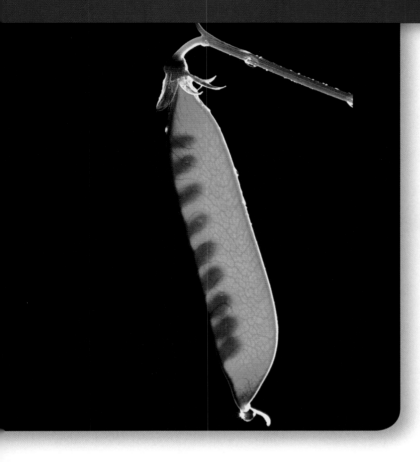

11
Foundations of Genetics

In this pea pod, you can see the shadowy outlines of seeds that will form part of the next generation of this pea plant. While the seeds appear similar to one another, the plants they produce may differ in significant ways. This is because the gametes that produced the seeds contribute chromosomes from both parents, in effect "shuffling the deck of cards" so that a progeny plant will have some characteristics from one parent and some from the other. About 150 years ago, Gregor Mendel first described this process, before anyone knew what genes, or chromosomes, were. We now understand the process of heredity in considerable detail, and can begin to devise ways of treating some of the disorders that arise in people when particular genes are damaged in germ-line tissue. In this chapter you will watch as Mendel experiments with pea plants like the one above. Unlike researchers before him, Mendel carefully counted the number of each kind of pea plant his experiments produced and looking at his results saw a beautiful simplicity. The theory he proposed to explain it has become one of the key principles of biology.

11.1 Mendel and the Garden Pea

When you were born, many things about you resembled your mother or father. This tendency for traits to be passed from parent to offspring is called **heredity. Traits** are alternative forms of a **character,** or heritable feature. How does heredity happen? Before DNA and chromosomes were discovered, this puzzle was one of the greatest mysteries of science. The key to understanding the puzzle of heredity was found in the garden of an Austrian monastery over a century ago by a monk named Gregor Mendel (figure 11.1). Mendel used the scientific process described in chapter 1 as a powerful way of analyzing the problem. Crossing pea plants with one another, Mendel made observations that allowed him to form a simple but powerful hypothesis that accurately predicted patterns of heredity—that is, how many offspring would be like one parent and how many like the other. When Mendel's rules, introduced in chapter 1 as the theory of heredity, became widely known, investigators all over the world set out to discover the physical mechanism responsible for them. They learned that hereditary traits are instructions carefully laid out in the DNA a child receives from each parent. Mendel's solution to the puzzle of heredity was the first step on this journey of understanding and one of the greatest intellectual accomplishments in the history of science.

Figure 11.1   Gregor Mendel.
The key to understanding the puzzle of heredity was solved by Mendel by cultivating pea plants in the garden of his monastery in Brunn, Austria.

Early Ideas About Heredity

Mendel was not the first person to try to understand heredity by crossing pea plants. Over 200 years earlier British farmers had performed similar crosses and obtained results similar to Mendel's. They observed that in crosses between two types— tall and short plants, say—one type would disappear in one generation, only to reappear in the next. In the 1790s, for example, the British farmer T. A. Knight crossed a variety of the garden pea that had purple flowers with one that had white flowers. All the offspring of the cross had purple flowers. If two of these offspring were crossed, however, some of *their* offspring were purple and some were white. Knight noted that the purple had a "stronger tendency" to appear than white, but he did not count the numbers of each kind of offspring.

Mendel's Experiments

Gregor Mendel was born in 1822 to peasant parents and was educated in a monastery. He became a monk and was sent to the University of Vienna to study science and mathematics. Although he aspired to become a scientist and teacher, he failed his university exams for a teaching certificate and returned to the monastery, where he spent the rest of his life, eventually becoming abbot. Upon his return, Mendel joined an informal neighborhood science club, a group of farmers and others interested in science. Under the patronage of a local nobleman, each member set out to undertake scientific investigations, which were then discussed at meetings and published in the club's own journal. Mendel undertook to repeat the classic series of crosses with pea plants done by Knight and others, but this time he intended to count the numbers of each kind of offspring in the hope that the numbers would give some hint of what was going on. Quantitative approaches to science—measuring and counting—were just becoming fashionable in Europe.

Mendel's Experimental System: The Garden Pea

Mendel chose to study the garden pea because several of its characteristics made it easy to work with:

1. Many varieties were available. Mendel selected seven pairs of lines that differed in easily distinguished traits (including the white versus purple flowers that Knight had studied 60 years earlier).
2. Mendel knew from the work of Knight and others that he could expect the infrequent version of a character to disappear in one generation and reappear in the next. He knew, in other words, that he would have something to count.
3. Pea plants are small, easy to grow, produce large numbers of offspring, and mature quickly.
4. The reproductive organs of peas are enclosed within their flowers. Figure 11.2 shows a cut-away view of the flower so that you can see the anther that holds the pollen and the carpel that holds the egg. Left alone, the flowers do not open. They simply fertilize themselves with their own pollen (male gametes). To carry out a cross, Mendel had only to pry the petals apart, reach in with a scissors, and snip off the male organs (anthers); he could then dust the female organs (the tip of the carpel) with pollen from another plant to make the cross.

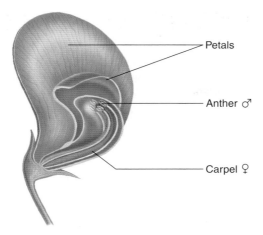

Figure 11.2 The garden pea.
Because it is easy to cultivate and because there are many distinctive varieties, the garden pea, *Pisum sativum,* was a popular choice as an experimental subject in investigations of heredity for as long as a century before Mendel's studies.

Mendel's Experimental Design

Mendel's experimental design was the same as Knight's, only Mendel counted his plants. The crosses were carried out in three steps that are presented in the three panels in figure 11.3:

1. Mendel began by letting each variety self-fertilize for several generations. This ensured that each variety was **true-breeding,** meaning that it contained no other varieties of the trait, and so would produce only offspring of the same variety when it self-pollinated. The white flower variety, for example, produced only white flowers and no purple ones in each generation. Mendel called these lines the **P generation** (P for parental).

2. Mendel then conducted his experiment: he crossed two pea varieties exhibiting alternative traits, such as white versus purple flowers in panel 2. The offspring that resulted he called the F_1 **generation** (F_1 for "first filial" generation, from the Latin word for "son" or "daughter").

3. Finally, Mendel allowed the plants produced in the crosses of step 2 to self-fertilize, and he counted the numbers of each kind of offspring that resulted in this F_2 ("second filial") **generation.** As reported by Knight and shown in panel 3, the white flower trait reappeared in the F_2 generation, although not as frequently as the purple flower trait.

> **11.1** Mendel studied heredity by crossing true-breeding garden peas that differed in easily scored alternative traits and then allowing the offspring to self-fertilize.

Mendel's Experimental Design

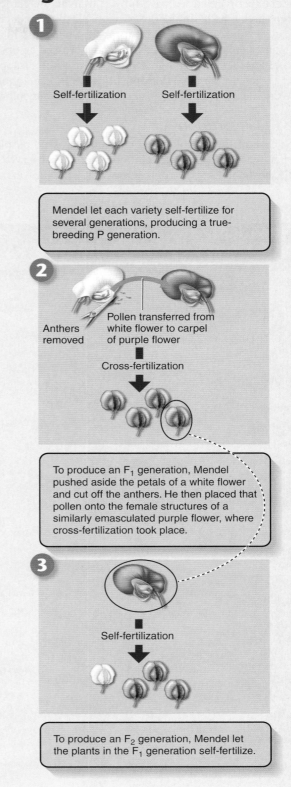

Mendel let each variety self-fertilize for several generations, producing a true-breeding P generation.

Anthers removed · Pollen transferred from white flower to carpel of purple flower

Cross-fertilization

To produce an F_1 generation, Mendel pushed aside the petals of a white flower and cut off the anthers. He then placed that pollen onto the female structures of a similarly emasculated purple flower, where cross-fertilization took place.

Self-fertilization

To produce an F_2 generation, Mendel let the plants in the F_1 generation self-fertilize.

Figure 11.3 How Mendel conducted his experiments.

11.2 What Mendel Observed

Mendel experimented with a variety of traits in the garden pea and repeatedly made similar observations. In all, Mendel examined seven pairs of contrasting traits as shown in table 11.1. For each pair of contrasting traits that Mendel crossed he observed the same result, shown in figure 11.3, where a trait disappeared in the F$_1$ generation only to reappear in the F$_2$ generation. We will examine in detail Mendel's crosses with flower color.

The F$_1$ Generation

In the case of flower color, when Mendel crossed purple and white flowers, all the F$_1$ generation plants he observed were purple; he did not see the contrasting trait, white flowers. Men-

del called the trait expressed in the F$_1$ plants **dominant** and the trait not expressed **recessive.** In this case, purple flower color was dominant and white flower color recessive. Mendel studied several other characters in addition to flower color, and for every pair of contrasting traits Mendel examined, one proved to be dominant and the other recessive. The dominant and recessive traits for each character he studied are indicated in table 11.1.

The F$_2$ Generation

After allowing individual F$_1$ plants to mature and self-fertilize, Mendel collected and planted the seeds from each plant to see what the offspring in the F$_2$ generation would look like. Mendel found (as Knight had earlier) that some F$_2$ plants exhibited white flowers, the recessive trait. The recessive trait had

TABLE 11.1	SEVEN CHARACTERS MENDEL STUDIED IN HIS EXPERIMENTS			
	Character		**F$_2$ Generation**	
	Dominant Form ×	**Recessive Form**	**Dominant: Recessive**	**Ratio**
	Purple flowers ×	White flowers	705:224	3.15:1 (3/4:1/4)
	Yellow seeds ×	Green seeds	6022:2001	3.01:1 (3/4:1/4)
	Round seeds ×	Wrinkled seeds	5474:1850	2.96:1 (3/4:1/4)
	Green pods ×	Yellow pods	428:152	2.82:1 (3/4:1/4)
	Inflated pods ×	Constricted pods	882:299	2.95:1 (3/4:1/4)
	Axial flowers ×	Terminal flowers	651:207	3.14:1 (3/4:1/4)
	Tall plants ×	Dwarf plants	787:277	2.84:1 (3/4:1/4)

Figure 11.4 Round versus wrinkled seeds.
One of the differences among varieties of pea plants that Mendel studied was the shape of the seed. In some varieties the seeds were round, whereas in others they were wrinkled.

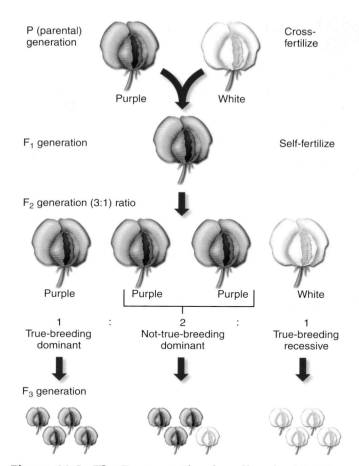

Figure 11.5 The F$_2$ generation is a disguised 1:2:1 ratio.
By allowing the F$_2$ generation to self-fertilize, Mendel found from the offspring (F$_3$) that the ratio of F$_2$ plants was one true-breeding dominant, two not-true-breeding dominant, and one true-breeding recessive.

disappeared in the F$_1$ generation, only to reappear in the F$_2$ generation. It must somehow have been present in the F$_1$ individuals but unexpressed!

At this stage Mendel instituted his radical change in experimental design. He *counted* the number of each type among the F$_2$ offspring. He believed the proportions of the F$_2$ types would provide some clue about the mechanism of heredity. In the cross between the purple-flowered F$_1$ plants, he counted a total of 929 F$_2$ individuals (see table 11.1). Of these, 705 (75.9%) had purple flowers and 224 (24.1%) had white flowers. Approximately one-fourth of the F$_2$ individuals exhibited the recessive form of the trait. Mendel carried out similar experiments with other traits, such as round versus wrinkled seeds (figure 11.4) and obtained the same result: three-fourths of the F$_2$ individuals exhibited the dominant form of the character, and one-fourth displayed the recessive form. In other words, the dominant:recessive ratio among the F$_2$ plants was always close to 3:1.

A Disguised 1:2:1 Ratio

Mendel let the F$_2$ plants self-fertilize for another generation and found that the one-fourth that were recessive were true-breeding—future generations showed nothing but the recessive trait. Thus, the white F$_2$ individuals described previously showed only white flowers in the F$_3$ generation (as shown on the right in figure 11.5). Among the three-fourths of the plants that had shown the dominant trait in the F$_2$ generation, only one-third of the individuals were true-breeding in the F$_3$ generation (as shown on the left). The others showed both traits in the F$_3$ generation (as shown in the center)—and when Mendel counted their numbers, he found the ratio of dominant to re-

cessive to again be 3:1! From these results Mendel concluded that the 3:1 ratio he had observed in the F$_2$ generation was in fact a disguised 1:2:1 ratio:

$$\begin{array}{ccc} \mathbf{1} & \mathbf{2} & \mathbf{1} \end{array}$$

true-breeding : not-true-breeding : true-breeding
dominant   dominant   recessive

> **11.2** When Mendel crossed two contrasting traits and counted the offspring in the subsequent generations, he observed that all of the offspring in the first generation exhibited one (dominant) trait, and none exhibited the other (recessive) trait. In the following generation, 25% were true-breeding for the dominant trait, 50% were not-true-breeding and appeared dominant, and 25% were true-breeding for the recessive trait.

11.3 Mendel Proposes a Theory

To explain his results, Mendel proposed a simple set of hypotheses that would faithfully predict the results he had observed. Now called Mendel's theory of heredity, it has become one of the most famous theories in the history of science. Mendel's theory is composed of five simple hypotheses:

Hypothesis 1: *Parents do not transmit traits directly to their offspring.* Rather, they transmit information about the traits, what Mendel called *merkmal* (the German word for "factor"). These factors act later, in the offspring, to produce the trait. In modern terminology, we call Mendel's factors **genes.**

Hypothesis 2: *Each parent contains two copies of the factor governing each trait.* The two copies may or may not be the same. If the two copies of the factor are the same (both encoding purple or both white flowers, for example) the individual is said to be **homozygous.** If the two copies of the factor are different (one encoding purple, the other white, for example), the individual is said to be **heterozygous.**

Hypothesis 3: *Alternative forms of a factor lead to alternative traits.* Alternative forms of a factor are called **alleles.** Mendel used lowercase letters to represent recessive alleles and uppercase letters to represent dominant ones. Thus, in the case of purple flowers, the dominant purple flower allele is represented as *P* and the recessive white flower allele is represented as *p*. In modern terms, we call the appearance of an individual, such as possessing white flowers, its **phenotype.** Appearance is determined by which alleles of the flower-color gene the plant receives from its parents, and we call those particular alleles the individual's **genotype.** Thus a pea plant might have the phenotype "white flower" and the genotype *pp*.

Hypothesis 4: *The two alleles that an individual possesses do not affect each other,* any more than two letters in a mailbox alter each other's contents. Each allele is passed on unchanged when the individual matures and produces its own gametes (egg and sperm). At the time, Mendel did not know that his factors were carried from parent to offspring on chromosomes. Figure 11.6 shows a modern view of how genes are carried on chromosomes, with homologous chromosome carrying the same genes but not necessarily the same alleles. The location of a gene on a chromosome is called its *locus* (plural loci).

Hypothesis 5: *The presence of an allele does not ensure that a trait will be expressed in the individual that carries it.* In heterozygous individuals, only the dominant allele achieves expression; the recessive allele is present but unexpressed.

These five hypotheses, taken together, constitute Mendel's model of the hereditary process. Many traits in humans exhibit dominant or recessive inheritance similar to the traits Mendel studied in peas (table 11.2).

Analyzing Mendel's Results

To analyze Mendel's results, it is important to remember that each trait is determined by the inheritance of alleles from the parents, one allele from the mother and the other from the father. These alleles, present on chromosomes, are distributed to gametes during meiosis. Each gamete receives one copy of each chromosome, and therefore one of the alleles.

Consider again Mendel's cross of purple-flowered with white-flowered plants. Like Mendel, we will assign the symbol *P* to the dominant allele, associated with the production of purple flowers, and the symbol *p* to the recessive allele,

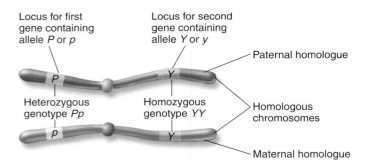

Figure 11.6   Alternative alleles of genes are located on homologous chromosomes.

TABLE 11.2	SOME DOMINANT AND RECESSIVE TRAITS IN HUMANS		
Recessive Traits	**Phenotypes**	**Dominant Traits**	**Phenotypes**
Common baldness	M-shaped hairline receding with age	Mid-digital hair	Presence of hair on middle segment of fingers
Albinism	Lack of melanin pigmentation	Brachydactyly	Short fingers
Alkaptonuria	Inability to metabolize homogenistic acid	Phenylthiocarbamide (PTC) sensitivity	Ability to taste PTC as bitter
Red-green color blindness	Inability to distinguish red and green wavelengths of light	Camptodactyly	Inability to straighten the little finger
		Polydactyly	Extra fingers and toes

associated with the production of white flowers. By convention, genetic traits are usually assigned a letter symbol referring to their more common forms, in this case "*P*" for purple flower color. The dominant allele is written in uppercase, as *P*; the recessive allele (white flower color) is assigned the same symbol in lowercase, *p*.

In this system, the genotype of an individual true-breeding for the recessive white-flowered trait would be designated *pp*. In such an individual, both copies of the allele specify the white-flowered phenotype. Similarly, the genotype of a true-breeding purple-flowered individual would be designated *PP*, and a heterozygote would be designated *Pp* (dominant allele first). Using these conventions, and denoting a cross between two strains with ×, we can symbolize Mendel's original cross as *pp* × *PP*.

Punnett Squares

The possible results from a cross between a true-breeding, white-flowered plant (*pp*) and a true-breeding, purple-flowered plant (*PP*) can be visualized with a **Punnett square.** In a Punnett square, the possible gametes of one individual are listed along the horizontal side of the square, while the possible gametes of the other individual are listed along the vertical side. The genotypes of potential offspring are represented by the cells within the square. Figure 11.7 walks you through the set-up of a Punnett square crossing two individual plants that are heterozygous for flower color (*Pp* × *Pp*). The genotypes of the parents are placed along the top and side and the genotypes of potential offspring appear in the cells.

The frequency that these genotypes occur in the offspring is usually expressed by a **probability.** For example, in a cross between a homozygous white-flowered plant (*pp*) and a homozygous purple-flowered plant (*PP*), *Pp* is the only possible genotype for all individuals in the F_1 generation as shown by the Punnett square on the left of figure 11.8. Because *P* is dominant to *p*, all individuals in the F_1 generation

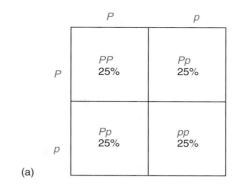

(a)

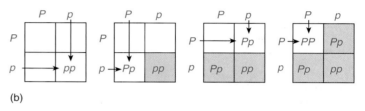

(b)

Figure 11.7 A Punnett square analysis.
(a) Each square represents 1/4 or 25% of the offspring from the cross. The squares in (b) show how the square is used to predict the genotypes of all potential offspring.

have purple flowers. When individuals from the F_1 generation are crossed, as shown by the Punnett square on the right, the probability of obtaining a homozygous dominant (*PP*) individual in the F_2 is 25% because one-fourth of the possible genotypes are *PP*. Similarly, the probability of an individual in the F_2 generation being homozygous recessive (*pp*) is 25%. Because the heterozygous genotype has two possible ways

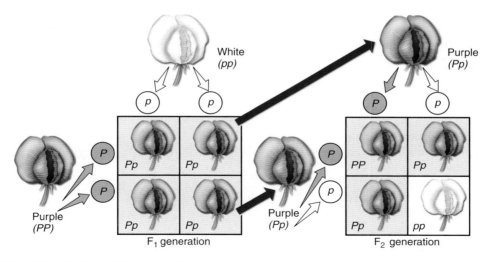

Figure 11.8 How Mendel analyzed flower color.
The only possible offspring of the first cross are *Pp* heterozygotes, purple in color. These individuals are known as the F_1 generation. When two heterozygous F_1 individuals cross, three kinds of offspring are possible: *PP* homozygotes (purple flowers); *Pp* heterozygotes (also purple flowers), which may form two ways; and *pp* homozygotes (white flowers). Among these individuals, known as the F_2 generation, the ratio of dominant phenotype to recessive phenotype is 3:1.

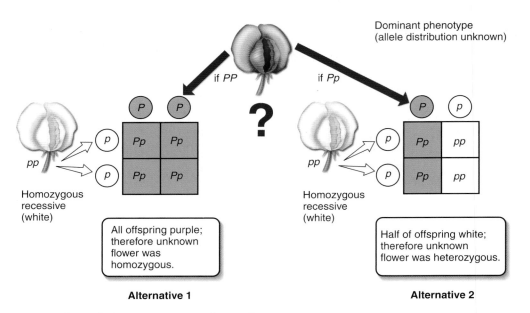

Dominant phenotype
(allele distribution unknown)

if *PP*

if *Pp*

Alternative 1

pp

Homozygous
recessive
(white)

	P	P
p	Pp	Pp
p	Pp	Pp

All offspring purple;
therefore unknown
flower was
homozygous.

?

Alternative 2

pp

Homozygous
recessive
(white)

	P	p
p	Pp	pp
p	Pp	pp

Half of offspring white;
therefore unknown
flower was heterozygous.

Figure 11.9   How Mendel used the testcross to detect heterozygotes.
To determine whether an individual exhibiting a dominant phenotype, such as purple flowers, was homozygous (*PP*) or heterozygous (*Pp*) for the dominant allele, Mendel devised the testcross. He crossed the individual in question with a known homozygous recessive (*pp*)—in this case, a plant with white flowers.

of occurring (*Pp* and *pP*), it occurs in half of the cells within the square, the probability of obtaining a heterozygous (*Pp*) individual in the F_2 is 50% (25% + 25%).

The Testcross

How did Mendel know which of the purple-flowered individuals in the F_2 generation (or the P generation) were homozygous (*PP*) and which were heterozygous (*Pp*)? It is not possible to tell simply by looking at them. For this reason, Mendel devised a simple and powerful procedure called the **testcross** to determine an individual's actual genetic composition. Consider a purple-flowered plant. It is impossible to tell whether such a plant is homozygous or heterozygous simply by looking at its phenotype. To learn its genotype, you must cross it with some other plant. What kind of cross would provide the answer? If you cross it with a homozygous dominant individual, all of the progeny will show the dominant phenotype whether the test plant is homozygous or heterozygous. It is also difficult (but not impossible) to distinguish between the two possible test plant genotypes by crossing with a heterozygous individual. However, if you cross the test plant with a homozygous recessive individual, the two possible test plant genotypes will give totally different results. To see how this works, step through a testcross of a purple-flowered plant with a white-flowered plant. Figure 11.9 shows you the two possible alternatives:

Alternative 1 (on left): unknown plant is homozygous (*PP*). *PP* × *pp:* all offspring have purple flowers (*Pp*) as shown by the four purple squares.

Alternative 2 (on right): unknown plant is heterozygous (*Pp*). *Pp* × *pp:* one-half of offspring have white flowers (*pp*) and one-half have purple flowers (*Pp*) as shown by the two white and two purple squares.

To perform his testcross, Mendel crossed heterozygous F_1 individuals back to the parent homozygous for the recessive trait. He predicted that the dominant and recessive traits would appear in a 1:1 ratio, and that is what he observed, as you can see illustrated in alternative 2 above.

For each pair of alleles he investigated, Mendel observed phenotypic F_2 ratios of 3:1 (see table 11.1) and testcross ratios very close to 1:1, just as his model predicted.

Testcrosses can also be used to determine the genotype of an individual when two genes are involved. Mendel carried out many two-gene crosses, some of which we will soon discuss. He often used testcrosses to verify the genotypes of particular dominant-appearing F_2 individuals. Thus an F_2 individual showing both dominant traits (*A_ B_*) might have any of the following genotypes: *AABB, AaBB, AABb,* or *AaBb.* By crossing dominant-appearing F_2 individuals with homozygous recessive individuals (that is, *A_ B_* × *aabb*), Mendel was able to determine if either or both of the traits bred true among the progeny and so determine the genotype of the F_2 parent.

AABB	trait A breeds true	trait B breeds true
AaBB		trait B breeds true
AABb	trait A breeds true	
AaBb		

11.3   The genes that an individual has are referred to as its genotype; the outward appearance of the individual is referred to as its phenotype. The phenotype is determined by the alleles inherited from the parents. Analyses using Punnett squares determine all possible genotypes of a particular cross.

11.4 Mendel's Laws

Mendel's First Law: Segregation

Mendel's model brilliantly predicts the results of his crosses, accounting in a neat and satisfying way for the ratios he observed. Similar patterns of heredity have since been observed in countless other organisms. Traits exhibiting this pattern of heredity are called *Mendelian traits.* Because of its overwhelming importance, Mendel's theory is often referred to as Mendel's first law, or the **law of segregation.** In modern terms, Mendel's first law states that *the two alleles of a trait separate from each other during the formation of gametes, so that half of the gametes will carry one copy and half will carry the other copy.*

Mendel's Second Law: Independent Assortment

Mendel went on to ask if the inheritance of one factor, such as flower color, influences the inheritance of other factors, such as plant height. To investigate this question, he first established a series of true-breeding lines of peas that differed from one another with respect to two of the seven pairs of characteristics he had studied. He then crossed contrasting pairs of true-breeding lines. Figure 11.10 shows an experiment in which the P generation consists of homozygous individuals with round, yellow seeds (*RRYY* in the figure) that are crossed with individuals that are homozygous for wrinkled, green seeds (*rryy*). This cross produces offspring that have round, yellow seeds and are heterozygous for both of these traits (*RrYy*). Such F₁ individuals are said to be **dihybrid.**

Mendel then allowed the dihybrid individuals to self-fertilize. If the segregation of alleles affecting seed shape and alleles affecting seed color were independent, the probability that a particular pair of seed-shape alleles would occur together with a particular pair of seed-color alleles would simply be a product of the two individual probabilities that each pair would occur separately. For example, the probability of an individual with wrinkled, green seeds appearing in the F₂ generation would be equal to the probability of an individual with wrinkled seeds (1 in 4) multiplied by the probability of an individual with green seeds (1 in 4), or 1 in 16.

In his dihybrid crosses, Mendel found that the frequency of phenotypes in the F₂ offspring closely matched the 9:3:3:1 ratio predicted by the Punnett square analysis shown in figure 11.10. He concluded that for the pairs of traits he studied, the inheritance of one trait does not influence the inheritance of the other trait, a result often referred to as Mendel's second law, or the **law of independent assortment.** We now know that this result is only valid for genes not located near one another on the same chromosome. Thus in modern terms, Mendel's second law is often stated as follows: *genes located on different chromosomes are inherited independently of one another.*

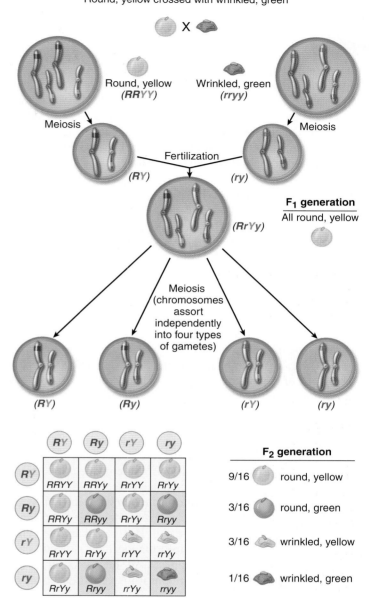

Figure 11.10 Analysis of a dihybrid cross.
This dihybrid cross shows round (*R*) versus wrinkled (*r*) seeds and yellow (*Y*) versus green (*y*) seeds. The ratio of the four possible phenotypes in the F₂ generation is predicted to be 9:3:3:1.

Mendel's paper describing his results was published in the journal of his local scientific society in 1866. Unfortunately, his paper failed to arouse much interest, and his work was forgotten. Sixteen years after his death, in 1900, several investigators independently rediscovered Mendel's pioneering paper while searching the literature in preparation for publishing their own findings, which were similar to those Mendel had quietly presented more than three decades earlier.

> **11.4** Mendel's theories of segregation and independent assortment are so well supported by experimental results that they are considered "laws."

11.5 How Genes Influence Traits

It is useful, before considering Mendelian genetics further, to gain a brief overview of how genes work. With this in mind, we will sketch, in broad strokes, a picture of how a Mendelian trait is influenced by a particular gene, how a gene can be altered by mutation, and the potential long-term evolutionary consequences of such an alteration. We will use the protein hemoglobin as our example—you can follow along on figure 11.11.

From DNA to Protein

Each cell of an individual, as shown at the bottom of figure 11.11, contains a set of DNA molecules, called its genome, that determines what the individual will be like. As you learned in chapter 4, DNA molecules are composed of two strands twisted about each other, each the mirror image of the other. Each strand is a long chain of nucleotide subunits linked together like so many pearls in a necklace. There are four kinds of nucleotides, and like an alphabet with four letters the order of nucleotides determines the message encoded in the DNA of a gene.

The human genome contains 20,000 to 25,000 genes. The DNA of the human genome is parcelled out into 23 pairs of chromosomes, each chromosome containing from 1,000 to 2,000 different genes. The bands on the chromosome indicate areas that are rich in genes. You can see on figure 11.11 that the hemoglobin gene is located on chromosome 11.

At the next level in the figure, individual genes are "read" from the chromosomal DNA by enzymes that create an RNA strand of the same sequence (except U is substituted for T). This RNA transcript of the hemoglobin (*Hb*) gene leaves the cell nucleus and acts as a work order for protein production in other parts of the cell. But, in eukaryotic cells, the RNA transcript has more information than is needed, so it is first "edited" to remove unnecessary bits before it leaves the nucleus. For example, the initial RNA gene transcript encoding the beta-subunit of the protein hemoglobin is 1,660 nucleotides long; after "editing", the resulting "messenger" RNA is 1,000 nucleotides long—you can see in the figure that the Hb mRNA is shorter than the RNA transcript of *Hb* gene.

After an RNA transcript is edited, it leaves the nucleus as messenger RNA (mRNA) and is delivered to ribosomes in the cytoplasm. Each ribosome is a tiny protein-assembly plant, and uses the sequence of the messenger RNA to determine the amino acid sequence of a particular polypeptide. In the case of beta-hemoglobin, the messenger RNA encodes a polypeptide strand of 146 amino acids.

How Proteins Determine the Phenotype

As we saw in chapter 4, polypeptide chains of amino acids that in the figure resemble beads on a string spontaneously fold in water into complex three-dimensional shapes. The beta-hemoglobin polypeptide folds into a compact mass that associates with three others to form an active hemoglobin protein molecule that is present in red blood cells. In the figure, each hemoglobin molecule binds oxygen (a process described fully in chapter 26) in the oxygen-rich environment of the lungs, and releases oxygen in the oxygen-poor environment of active tissues.

The oxygen-binding efficiency of the hemoglobin proteins in a person's bloodstream has a great deal to do with how well the body functions, particularly under conditions of strenuous physical activity, when delivery of oxygen to the body's muscles is the chief factor limiting the activity.

As a general rule, genes influence the phenotype by specifying the kind of proteins present in the body, which determines in large measure how that body functions.

How Mutation Alters Phenotype

A change in the identity of a single nucleotide within a gene, called a mutation, can have a profound effect if the change alters the identity of the amino acid encoded there. When a mutation of this sort occurs, the new version of the protein may fold differently, altering or destroying its function. For example, how well the hemoglobin protein performs its oxygen-binding duties depends a great deal on the precise shape that the protein assumes when it folds. A change in the identity of a single amino acid can have a drastic impact on that final shape. In particular, a change in the sixth amino acid of beta-hemoglobin from glutamic acid to valine causes the hemoglobin molecules to aggregate into stiff rods that deform blood cells into a sickle shape and can no longer carry oxygen efficiently. The resulting sickle-cell anemia can be fatal.

Natural Selection for Alternative Phenotypes Leads to Evolution

Because random mutations occur in all genes occasionally, populations usually contain several versions of a gene, usually all but one of them rare. Sometimes the environment changes in such a way that one of the rare versions functions better under the new conditions. When that happens, natural selection will favor the rare allele, which will then become more common. The sickle-cell version of the beta-hemoglobin gene, rare throughout most of the world, is common in Central Africa because heterozygous individuals obtain enough functional hemoglobin from their one normal allele to get along, but are resistant to malaria, a deadly disease common there, due to their other sickle-cell allele.

> **11.5** Genes determine phenotypes by specifying the amino acid sequences, and thus the functional shapes, of the proteins that carry out cell activities. Mutations, by altering protein sequence, can change a protein's function and thus alter the phenotype in evolutionarily significant ways.

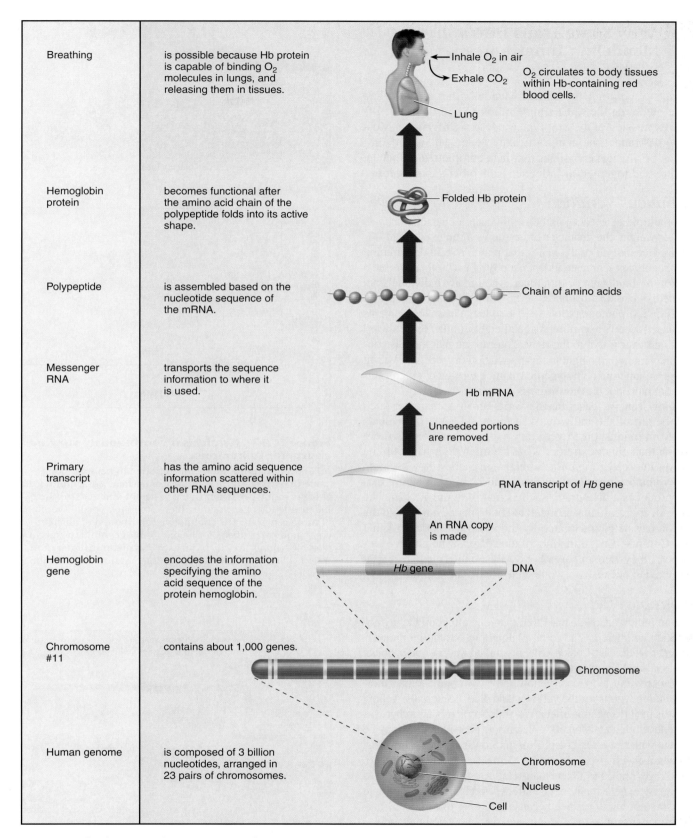

Breathing is possible because Hb protein is capable of binding O_2 molecules in lungs, and releasing them in tissues.

Inhale O_2 in air
Exhale CO_2
O_2 circulates to body tissues within Hb-containing red blood cells.
Lung

Hemoglobin protein becomes functional after the amino acid chain of the polypeptide folds into its active shape.

Folded Hb protein

Polypeptide is assembled based on the nucleotide sequence of the mRNA.

Chain of amino acids

Messenger RNA transports the sequence information to where it is used.

Hb mRNA

Unneeded portions are removed

Primary transcript has the amino acid sequence information scattered within other RNA sequences.

RNA transcript of *Hb* gene

An RNA copy is made

Hemoglobin gene encodes the information specifying the amino acid sequence of the protein hemoglobin.

Hb gene   DNA

Chromosome #11 contains about 1,000 genes.

Chromosome

Human genome is composed of 3 billion nucleotides, arranged in 23 pairs of chromosomes.

Chromosome
Nucleus
Cell

Figure 11.11   The journey from DNA to phenotype.

What an organism is like is determined in large measure by its genes. Here you see how one gene of the 20,000 to 25,000 in the human genome plays a key role in enabling you to breathe. The many steps on the journey from gene to trait are the subject of chapters 12 and 13.

11.6 Why Some Traits Don't Show Mendelian Inheritance

Scientists attempting to confirm Mendel's theory often had trouble obtaining the same simple ratios he had reported. Often the expression of the genotype is not straightforward. Most phenotypes reflect the action of many genes, and the phenotype can be affected by alleles that lack complete dominance, are expressed together, or influence each other's expression.

Continuous Variation

When multiple genes act jointly to influence a character such as height or weight, the character often shows a range of small differences. Because all of the genes that play a role in determining these phenotypes segregate independently of each other, we see a gradation in the degree of difference when many individuals are examined. A classic illustration of this sort of variation is seen in figure 11.12, a photograph of a 1914 college class. The students were placed in rows according to their heights, under 5 feet toward the left and over 6 feet to the right. You can see that there is considerable variation in height in this population of students. We call this type of inheritance **polygenic** (many genes) and we call this gradation in phenotypes **continuous variation.**

How can we describe the variation in a character such as the height of the individuals in figure 11.12*a*? Individuals range from quite short to very tall, with average heights more common than either extreme. What we often do is to group the variation into categories. Each height, in inches, is a separate phenotypic category. Plotting the numbers in each height category produces a histogram, such as that in figure 11.12*b*. The histogram approximates an idealized bell-shaped curve, and the variation can be characterized by the mean and spread of that curve. Compare this to the inheritance of plant height in Mendel's peas; they were either tall or dwarf, no intermediate height plants existed because only one gene controlled that trait.

Pleiotropic Effects

Often, an individual allele has more than one effect on the phenotype. Such an allele is said to be **pleiotropic.** When the pioneering French geneticist Lucien Cuenot studied yellow fur in mice, a dominant trait, he was unable to obtain a true-breeding yellow strain by crossing individual yellow mice with one another. Individuals homozygous for the yellow allele died, because the yellow allele was pleiotropic: one effect was yellow color, but another was a lethal developmental defect. A pleiotropic gene alteration may be dominant with respect to one phenotypic consequence (yellow fur) and recessive with respect to another (lethal developmental defect). In pleiotropy, one gene affects many characters, in marked contrast to polygeny, where many genes affect one character. Pleiotropic effects are difficult to predict, because the genes that affect a character often perform other functions we may know nothing about.

Pleiotropic effects are characteristic of many inherited disorders, such as cystic fibrosis and sickle-cell anemia, discussed later in this chapter. In these disorders, multiple symptoms can be traced back to a single gene defect. As shown in figure 11.13, cystic fibrosis patients exhibit overly

(a)

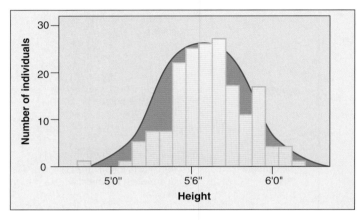

(b)

Figure 11.12   Height is a continuously varying character in humans.

(a) This photograph shows the variation in height among students of the 1914 class of the Connecticut Agricultural College. Because many genes contribute to height and tend to segregate independently of each other, there are many possible combinations of those genes. (b) The cumulative contribution of different combinations of alleles for height forms a continuous spectrum of possible heights, in which the extremes are much rarer than the intermediate values. This is quite different from the 3:1 ratio seen in Mendel's F_2 peas.

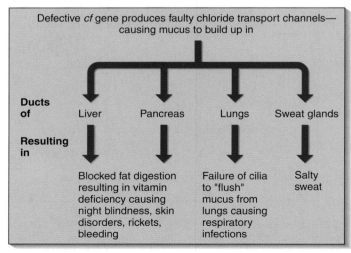

Figure 11.13   Pleiotropic effects of the cystic fibrosis gene, *cf*.

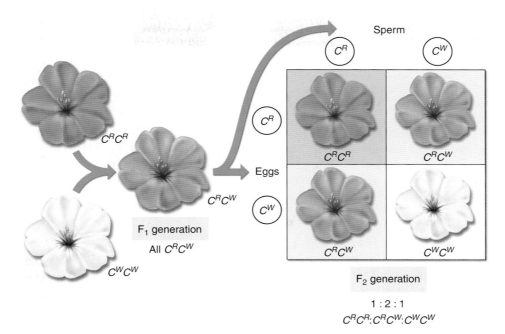

Figure 11.14   Incomplete dominance.
In a cross between a red-flowered Japanese four o'clock, genotype $C^R C^R$, and a white-flowered one ($C^W C^W$), neither allele is dominant. The heterozygous progeny have pink flowers and the genotype $C^R C^W$. If two of these heterozygotes are crossed, the phenotypes of their progeny occur in a ratio of 1:2:1 (red:pink:white).

sticky mucus, salty sweat, liver and pancreas failure, and a battery of other symptoms. All are pleiotropic effects of a single defect, a mutation in a gene that encodes a chloride ion transmembrane channel. In sickle-cell anemia, a defect in the oxygen-carrying hemoglobin molecule causes anemia, heart failure, increased susceptibility to pneumonia, kidney failure, enlargement of the spleen, and many other symptoms. It is usually difficult to deduce the nature of the primary defect from the range of its pleiotropic effects.

Incomplete Dominance

Not all alternative alleles are fully dominant or fully recessive in heterozygotes. Some pairs of alleles exhibit **incomplete dominance** and produce a heterozygous phenotype that is intermediate between those of the parents. For example, the cross of red- and white-flowered Japanese four o'clocks described in figure 11.14 produced red-, pink-, and white-flowered F_2 plants in a 1:2:1 ratio—heterozygotes are intermediate in color. Recall from the earlier discussion that with Mendel's pea plants that didn't exhibit incomplete dominance, the heterozygotes expressed the dominant phenotype.

Environmental Effects

The degree to which many alleles are expressed depends on the environment. Some alleles are heat-sensitive, for example. Traits influenced by such alleles are more sensitive to temperature or light than are the products of other alleles. The arctic foxes in figure 11.15, for example, make fur pigment only when the weather is warm. Can you see why this trait would be an advantage for the fox? Imagine a fox that didn't possess this trait and was white all year round. It would be

(a)

(b)

Figure 11.15   Environmental effects on an allele.
(a) An arctic fox in winter has a coat that is almost white, so it is difficult to see the fox against a snowy background. (b) In summer, the same fox's fur darkens to a reddish brown, so that it resembles the color of the surrounding tundra. Heat-sensitive alleles control this color change.

very visible to predators in the summer, standing out against its darker surroundings. Similarly, the *ch* allele in Himalayan rabbits and Siamese cats encodes a heat-sensitive version of tyrosinase, one of the enzymes mediating the production of melanin, a dark pigment. The ch version of the enzyme is inactivated at temperatures above about 33°C. At the surface of the main body and head, the temperature is above 33°C and the tyrosinase enzyme is inactive, while it is more active at body extremities such as the tips of the ears and tail, where the temperature is below 33°C. The dark melanin pigment this enzyme produces causes the ears, snout, feet, and tail to be black.

Epistasis

In some situations, two or more genes interact with each other, such that one gene contributes to or masks the expression of the other gene. This becomes apparent when analyzing dihybrid crosses involving these traits. Recall that when individuals heterozygous for two different genes mate (a dihybrid cross), offspring may display the dominant phenotype for both genes, either one of the genes, or for neither gene. Sometimes, however, an investigator cannot find four phenotype classes because two or more of the genotypes express the same phenotypes.

As was stated earlier, few phenotypes are the result of the action of one gene. Most traits reflect the action of many genes, some that act sequentially or jointly. **Epistasis** is an interaction between the products of two genes in which one of the genes modifies the phenotypic expression produced by the other. For example, some commercial varieties of corn, *Zea mays,* exhibit a purple pigment called anthocyanin in their seed coats, while others do not. In 1918, geneticist R. A. Emerson crossed two true-breeding corn varieties, neither exhibiting anthocyanin pigment. Surprisingly, all of the F$_1$ plants produced purple seeds.

When two of these pigment-producing F$_1$ plants were crossed to produce an F$_2$ generation, 56% were pigment producers and 44% were not. What was happening? Emerson correctly deduced that two genes were involved in producing pigment, and that the second cross had thus been a dihybrid cross like those performed by Mendel. Mendel had predicted 16 equally possible ways gametes could combine with each other, resulting in genotypes with a phenotypic ratio of 9:3:3:1 (9 + 3 + 3 + 1 = 16). How many of these were in each of the two types Emerson obtained? He multiplied the fraction that were pigment producers (0.56) by 16 to obtain 9 and multiplied the fraction that were not (0.44) by 16 to obtain 7. Thus, Emerson had a **modified ratio** of 9:7 instead of the usual 9:3:3:1 ratio. Figure 11.16 shows the results of the dihybrid cross made by Emerson. Go back and compare these results with Mendel's dihybrid cross in figure 11.10 and you can see that the F$_2$ genotypes in Emerson's results are consistent with what Mendel found, so why are the phenotypic ratios different?

Why Was Emerson's Ratio Modified? It turns out that in corn plants either one of the two genes that contribute to kernel color can block the expression of the other. One

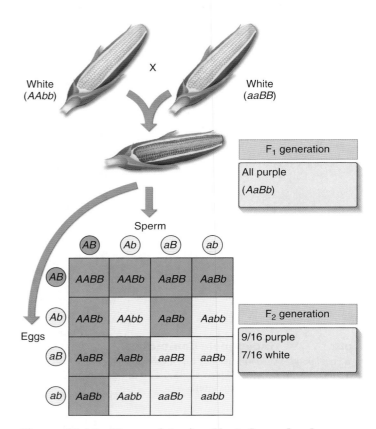

Figure 11.16 How epistasis affects kernel color.
The purple pigment found in some varieties of corn is the result of two genes. Unless a dominant allele is present at each of the two loci, no pigment is expressed.

of the genes (*B*) produces an enzyme that permits colored pigment to be produced only if a dominant allele (*BB* or *Bb*) is present. The other gene (*A*) produces an enzyme that in its dominant form (*AA* or *Aa*) allows the pigment to be deposited on the seed coat color. Thus, an individual with two recessive alleles for gene *A* (no pigment deposition) will have white seed coats even though it is able to manufacture the pigment because it possesses dominant alleles for gene *B* (purple pigment production). Similarly, an individual with dominant alleles for gene *A* (pigment can be deposited) will also have white seed coats if it has only recessive alleles for gene *B* (pigment production) and cannot manufacture the pigment.

To produce and deposit pigment, a plant must possess at least one functional copy of each enzyme gene (*A_B_*). Of the 16 genotypes predicted by random assortment, 9 contain at least one dominant allele of both genes; they produce purple progeny and are colored purple in the Punnett square in figure 11.16. The remaining 7 genotypes lack dominant alleles at either or both loci (3 + 3 + 1 = 7) and so are phenotypically the same (nonpigmented—the light-color boxes in the Punnett square), giving the phenotypic ratio of 9:7 that Emerson observed.

Today's *Biology*

Does Environment Affect I.Q.?

Nowhere has the influence of environment on the expression of genetic traits led to more controversy than in studies of I.Q. scores. I.Q. is a controversial measure of general intelligence based on a written test that many feel to be biased toward white middle-class America. However well or poorly I.Q. scores measure intelligence, a person's I.Q. score has been believed for some time to be determined largely by his or her genes.

How did science come to that conclusion? Scientists measure the degree to which genes influence a multigene trait by using an off-putting statistical measure called the *variance*. Variance is defined as the square of the standard deviation (a measure of the degree-of-scatter of a group of numbers around their mean value), and has the very desirable property of being additive—that is, the total variance is equal to the sum of the variances of the factors influencing it.

The degree of gene influence, or *heritability,* is defined as the fraction of the total variance that is genetic:

$$H = \text{Variance (genes)/Variance (total)}$$

What factors contribute to the total Variance? There are three. The first factor is variation at the gene level, some gene combinations leading to higher I.Q. scores than others. This is the term "Variance (genes)" in the simple equation above. The second factor is variation at the environmental level, some environments leading to higher I.Q. scores than others. This term is expressed as "Variance (environment)." The third factor is what a statistician calls the covariance, the degree to which the value of one factor influences the value of others. In this instance, covariance expresses the degree to which environment affects genes. The covariance term is expressed "Variance (coVar)."

Now the heritability of I.Q. can be expressed in terms that can be measured:

$$H = \frac{\text{Variance (genes)}}{\text{Variance (genes) + Variance (environment) + Variance (coVar)}}$$

The environmental contributions to variance in I.Q. can be measured by comparing the I.Q. scores of identical twins reared together with those reared apart (any differences should reflect environmental influences). The genetic contributions can be measured by comparing identical twins reared together (which are 100% genetically identical) with fraternal twins reared together (which are 50% genetically identical). Any differences should reflect genes, as twins share identical prenatal conditions in the womb and are raised in virtually identical environmental circumstances, so when traits are more commonly shared between identical twins than fraternal twins, the difference is likely genetic.

When these sorts of "twin studies" have been done in the past, researchers have uniformly reported that I.Q. is highly heritable, with values of H typically reported as being around 0.7 (a very high value). While it didn't seem significant at the time, almost all the twins available for study over the years have come from middle-class or wealthy families.

The study of I.Q. has proven controversial, because I.Q. scores are often different when social and racial groups are compared. What is one to make of the observation that I.Q. scores of poor children measure lower as a group than do scores of children of middle-class and wealthy families? This difference has led to the controversial suggestion by some that the poor are genetically inferior.

What should we make of such a harsh conclusion? To make a judgment, we need to focus for a moment on the fact that these measures of the heritability of I.Q. have all made a critical assumption, one to which population geneticists, who specialize in these sorts of things, object strongly. The assumption is that environment does not affect gene expression—that is, that the covariance term in the preceding equation is zero.

Recent studies have allowed a direct assessment of this assumption. Importantly, it proves to be flat wrong.

In November of 2003, researchers reported an analysis of twin data from a study carried out in the late 1960s. The National Collaborative Prenatal Project, funded by the National Institutes of Health, enrolled nearly 50,000 pregnant women, most of them black and quite poor, in several major U.S. cities. Researchers collected abundant data, and gave the children I.Q. tests seven years later. Although not designed to study twins, this study was so big that many twins were born, 623 births. Seven years later, 320 of these pairs were located and given I.Q. tests. This thus constitutes a huge "twin study," the first ever conducted of I.Q. among the poor.

When the data were analyzed, the results were unlike any ever reported. The heritability of I.Q. was different in different environments! Most notably, the influence of genes on I.Q. was far less in conditions of poverty, where environmental limitations seem to block the expression of genetic potential. Specifically, for families of high socioeconomic status, H = 0.72, much as reported in previous studies, but for families raised in poverty, H = 0.10, a very low value, indicating genes were making little contribution to observed I.Q. scores. The lower a child's socioeconomic status, the less impact genes had on I.Q.

These data say that the genetic contributions to I.Q. don't mean much in an impoverished environment. Clearly, improvements in the growing and learning environments of poor children can be expected to have a major impact on their I.Q. scores. Additionally, these data argue that the controversial differences reported in mean I.Q scores between racial groups may well reflect no more than poverty, and are no more inevitable.

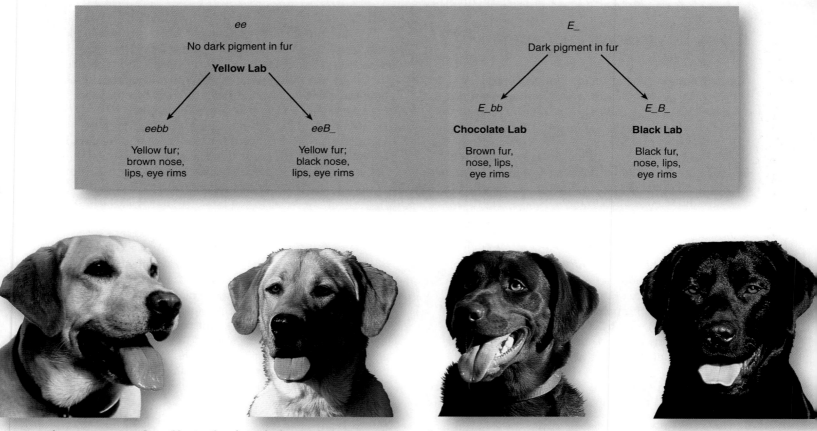

Figure 11.17 The effect of epistatic interactions on coat color in dogs.
The coat color seen in Labrador retrievers is an example of the interaction of two genes, each with two alleles. The *E* gene determines if the pigment will be deposited in the fur, and the *B* gene determines how dark the pigment will be.

Other Examples of Epistasis. In many animals, coat color is the result of epistatic interactions among genes. Coat color in Labrador retrievers, a breed of dog, is due primarily to the interaction of two genes. The *E* gene determines if dark pigment will be deposited in the fur or not. If a dog has the genotype *ee* (like the two dogs on the left in figure 11.17), no pigment will be deposited in the fur, and it will be yellow. If a dog has the genotype *EE* or *Ee* (*E_*), pigment will be deposited in the fur (like the two dogs on the right).

A second gene, the *B* gene, determines how dark the pigment will be. Dogs with the genotype *E_bb* will have brown fur and are called chocolate labs. Dogs with the genotype *E_B_* are black labs with black fur. But, even in yellow dogs, the *B* gene does have some effect. Yellow dogs with the genotype *eebb* (on the far left) will have brown pigment on their nose, lips, and eye rims, while yellow dogs with the genotype *eeB_* (the second from the left) will have black pigment in these areas. The genes for coat color in this breed have been found, and a genetic test is available to determine the coat color in a litter of puppies.

Codominance

A gene may have more than two alleles in a population, and in fact most genes possess several different alleles. Often in het-erozygotes there isn't a dominant allele; instead, the effects of both alleles are expressed. In these cases, the alleles are said to be **codominant.**

Codominance is seen in the color patterning of some animals. For example, the "roan" pattern is a coloring pattern exhibited in some varieties of horses and cattle. A roan animal expresses both white and colored hairs on at least part of its body. This intermingling of the different colored hairs creates either an overall lighter color, or patches of lighter and darker colors. The roan pattern results from a heterozygous genotype, such as produced by mating of a homozygous white and homozygous colored. Could the in-termediate color be the result of incomplete dominance? No. The heterozygote that receives a white allele and a col-ored allele does not have individual hairs that are a mix of the two colors; rather, both alleles are being expressed, with the result that the animal has some hairs that are white and some that are colored. The gray horse in figure 11.18 is exhibiting the roan pattern. It looks like it has gray hairs, but if you were able to examine its coat closely, you would see both white hairs and black hairs, giving it an overall gray color.

A human gene that exhibits more than one dominant allele is the gene that determines ABO blood type. This

Figure 11.18 Codominance in color patterning.
This roan horse (named "Rivers Top Gun" by its owners) is heterozygous for coat color. The offspring of a cross between a white homozygote and a black homozygote, it expresses both phenotypes. Some of the hairs on its body are white and some are black.

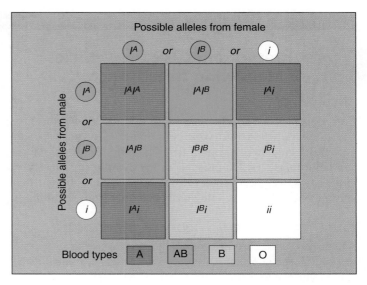

Figure 11.19 Multiple alleles controlling the ABO blood groups.
Three common alleles control the ABO blood groups. The different combinations of the three alleles result in four different blood type phenotypes: type A (either $I^A I^A$ homozygotes or $I^A i$ heterozygotes), type B (either $I^B I^B$ homozygotes or $I^B i$ heterozygotes), type AB ($I^A I^B$ heterozygotes), and type O (ii homozygotes).

gene encodes an enzyme that adds sugar molecules to lipids on the surface of red blood cells. These sugars act as recognition markers for cells in the immune system and are called cell surface antigens. The gene that encodes the enzyme, designated I, has three common alleles: I^B, whose product adds the sugar galactose; I^A, whose product adds galactosamine; and i, which codes for a protein that does not add a sugar.

Different combinations of the three I gene alleles occur in different individuals because each person may be homozygous for any allele or heterozygous for any two. An individual heterozygous for the I^A and I^B alleles produces both forms of the enzyme and adds both galactose and galactosamine to the surfaces of red blood cells. Because both alleles are expressed simultaneously in heterozygotes, the I^A and I^B alleles are codominant. Both I^A and I^B are dominant over the i allele because both I^A or I^B alleles lead to sugar addition and the i allele does not. The different combinations of the three alleles produce four different phenotypes:

1. Type A individuals add only galactosamine. They are either $I^A I^A$ homozygotes or $I^A i$ heterozygotes (the three darkest boxes in figure 11.19).
2. Type B individuals add only galactose. They are either $I^B I^B$ homozygotes or $I^B i$ heterozygotes (the three lightest-colored boxes).
3. Type AB individuals add both sugars and are $I^A I^B$ heterozygotes (the two intermediate-colored boxes).
4. Type O individuals add neither sugar and are ii homozygotes (the one white box in figure 11.19).

These four different cell surface phenotypes are called the **ABO blood groups.** A person's immune system can distinguish between these four phenotypes. If a type A individual receives a transfusion of type B blood, the recipient's immune system recognizes that the type B blood cells possess a "foreign" antigen (galactose) and attacks the donated blood cells, causing the cells to clump or agglutinate. This also happens if the donated blood is type AB. However, if the donated blood is type O, it contains no galactose or galactosamine antigens on the surfaces of its blood cells, and so elicits no immune response to these antigens. For this reason, the type O individual is often referred to as a "universal donor." In general, any individual's immune system will tolerate a transfusion of type O blood. Because neither galactose nor galactosamine is foreign to type AB individuals (whose red blood cells have both sugars), those individuals may receive any type of blood.

11.6 A variety of factors can disguise the Mendelian segregation of alleles. Among them are continuous variation, which results when many genes contribute to a trait; pleiotropic effects where one allele affects many phenotypes; incomplete dominance, which produces heterozygotes unlike either parent; environmental influences on the expression of phenotypes; and the interaction of more than one allele as seen in epistasis and codominance.

11.7 Chromosomes Are the Vehicles of Mendelian Inheritance

Chromosomes are not the only kinds of structures that segregate regularly when eukaryotic cells divide. Centrioles also divide and segregate in a regular fashion, as do the mitochondria and chloroplasts (when present) in the cytoplasm. Therefore, in the early twentieth century it was by no means obvious that chromosomes were the vehicles of hereditary information.

The Chromosomal Theory of Inheritance

A central role for chromosomes in heredity was first suggested in 1900 by the German geneticist Karl Correns, in one of the papers announcing the rediscovery of Mendel's work. Soon after, observations that similar chromosomes paired with one another during meiosis led directly to the *chromosomal theory of inheritance*, first formulated by the American Walter Sutton in 1902.

Several pieces of evidence supported Sutton's theory. One was that reproduction involves the initial union of only two cells, egg and sperm. If Mendel's model was correct, then these two gametes must make equal hereditary contributions. Sperm, however, contain little cytoplasm, suggesting that the hereditary material must reside within the nuclei of the gametes. Furthermore, while diploid individuals have two copies of each pair of homologous chromosomes, gametes have only one. This observation was consistent with Mendel's model, in which diploid individuals have two copies of each heritable gene and gametes have one. Finally, chromosomes segregate during meiosis, and each pair of homologues orients on the metaphase plate independently of every other pair. Segregation and independent assortment were two characteristics of the genes in Mendel's model.

Problems with the Chromosomal Theory

Investigators soon pointed out one problem with this theory, however. If Mendelian traits are determined by genes located on the chromosomes, and if the independent assortment of Mendelian traits reflects the independent assortment of chromosomes in meiosis, why does the number of traits that assort independently in a given kind of organism often greatly exceed the number of chromosome pairs the organism possesses? This seemed a fatal objection, and it led many early researchers to have serious reservations about Sutton's theory.

Morgan's White-Eyed Fly

The essential correctness of the chromosomal theory of heredity was demonstrated long before this paradox was resolved. A single small fly provided the confirmation. In 1910 Thomas Hunt Morgan, studying the fruit fly *Drosophila melanogaster,* detected a mutant male fly, one that differed strikingly from normal flies of the same species: its eyes were white instead of red (figure 11.20).

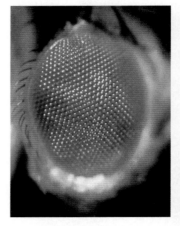

Figure 11.20   Red-eyed (wild type) and white-eyed (mutant) *Drosophila.*
The white-eye defect is hereditary, the result of a mutation in a gene located on the X chromosome. By studying this mutation, Morgan first demonstrated that genes are on chromosomes.

Morgan immediately set out to determine if this new trait would be inherited in a Mendelian fashion. He first crossed the mutant male with a normal female to see if red or white eyes were dominant. All of the F_1 progeny had red eyes, so Morgan concluded that red eye color was dominant over white. Following the experimental procedure that Mendel had established long ago, Morgan then crossed the red-eyed flies from the F_1 generation with each other. Of the 4,252 F_2 progeny Morgan examined, 782 (18%) had white eyes. Although the ratio of red eyes to white eyes in the F_2 progeny was greater than 3:1, the results of the cross nevertheless provided clear evidence that eye color segregates. However, there was something about the outcome that was strange and totally unpredicted by Mendel's theory—*all of the white-eyed F_2 flies were males!*

How could this result be explained? Perhaps it was impossible for a white-eyed female fly to exist; such individuals might not be viable for some unknown reason. To test this idea, Morgan testcrossed the female F_1 progeny with the original white-eyed male. He obtained white-eyed and red-eyed males and females in a 1:1:1:1 ratio, just as Mendelian theory predicted. Hence, a female could have white eyes. Why, then, were there no white-eyed females among the progeny of the original cross?

Sex Linkage Confirms the Chromosomal Theory

The solution to this puzzle involved sex. In *Drosophila,* the sex of an individual is determined by the number of copies of a particular chromosome, the **X chromosome,** that an individual possesses. A fly with two X chromosomes is a female, and a fly with only one X chromosome is a male. In males, the single X chromosome pairs in meiosis with a large, dissimilar partner called the **Y chromosome.** The female thus produces only X gametes, while the male produces both X and Y gametes. When fertilization involves an X sperm, the result is an XX zygote, which develops into a female; when fertilization involves a Y sperm, the result is an XY zygote, which develops into a male.

The solution to Morgan's puzzle is that the gene causing the white-eye trait in *Drosophila* resides only on the X chromosome—it is absent from the Y chromosome. (We now know that the Y chromosome in flies carries almost no functional genes.) A trait determined by a gene on the sex chromosome is said to be **sex-linked.** Knowing the white-eye trait is recessive to the red-eye trait, we can now see that Morgan's result was a natural consequence of the Mendelian assortment of chromosomes. Figure 11.21 steps you through Morgan's experiment,

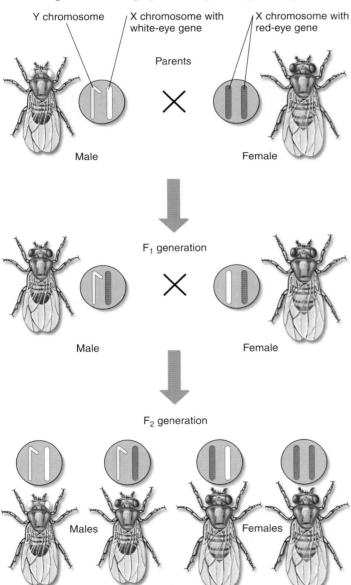

Figure 11.21   Morgan's experiment demonstrating the chromosomal basis of sex linkage.

The white-eyed mutant male fly was crossed with a normal female. The F₁ generation flies all exhibited red eyes, as expected for flies heterozygous for a recessive white-eye allele. In the F₂ generation, all of the white-eyed flies were male.

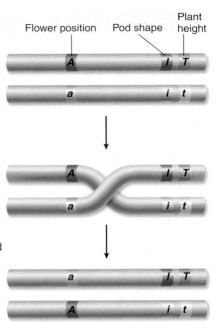

Figure 11.22 Linkage.
Genes that are located farther apart on a chromosome, like the genes for flower position (*A*) and pod shape (*I*) in Mendel's peas, will assort independently because crossing over results in recombination of these alleles. Pod shape (*I*) and plant height (*T*), however, are positioned very near each other, such that crossing over usually would not occur. These genes are said to be linked and do not undergo independent assortment.

showing both the eye color genes and the sex chromosomes. The sex chromosomes are color-coded in figure 11.21—red X chromosomes carry the red-eye gene while white X chromosomes carry the white-eye gene. The yellow hooked structure is the Y chromosome that carries no gene for eye color. In this experiment, the F₁ generation all had red eyes, while the F₂ generation contained flies with white eyes—but they were all males! This at-first-surprising result happens because the segregation of the white-eye trait has a one-to-one correspondence with the segregation of the X chromosome. In other words, the white-eye gene is on the X chromosome.

Morgan's experiment presented the first clear evidence that the genes determining Mendelian traits reside on chromosomes, just as Sutton had proposed. Now we can see that the reason Mendelian traits assort independently is because chromosomes assort independently. When Mendel observed the segregation of alternative traits in pea plants, he was observing a reflection of the meiotic segregation of the chromosomes, which contained the characters he was observing.

If genes are located on chromosomes, you might expect that two genes on the same chromosome would segregate together. However, if the two genes are located far from each other on the chromosome, like genes *A* and *I* in figure 11.22, the likelihood of crossing over occurring between them is very high, leading to independent segregation. Conversely, the closer two genes are to each other on a chromosome, like genes *I* and *T*, the less likely it is that a cross over event will occur between them. Genes that are located quite close to each other almost always segregate together, meaning that they are inherited together. The tendency of close-together genes to segregate together is called **linkage.**

11.7 Mendelian traits assort independently because they are determined by genes located on chromosomes that assort independently in meiosis.

11.8 Human Chromosomes

Each human somatic cell normally has 46 chromosomes, which in meiosis form 23 pairs. Homologous chromosomes can be identified, according to size, shape, and appearance. Of the 23 pairs of human chromosomes, 22 are perfectly matched in both males and females and are called **autosomes.** The remaining pair, the **sex chromosomes,** consist of two similar chromosomes in females and two dissimilar chromosomes in males. In humans, females are designated XX and males are XY. The genes present on the Y chromosome determine "maleness" and therefore humans who inherit the Y chromosome develop into males.

Nondisjunction

Sometimes during meiosis, sister chromatids or homologous chromosomes that paired up during metaphase remain stuck together instead of separating. The failure of chromosomes to separate correctly during either meiosis I or II is called **nondisjunction.** Nondisjunction leads to **aneuploidy,** an abnormal number of chromosomes. The nondisjunction you see in figure 11.23 occurs because the homologous pair of larger chromosomes failed to separate in anaphase I. The gametes that result from this division have unequal numbers of chromosomes. Under normal meiosis (refer back to figure 10.8), all gametes would be expected to have two chromosomes, but as you can see two of these gametes have three chromosomes, while two others have just one.

Almost all humans of the same sex have the same karyotype (refer back to figure 9.4) simply because other arrangements don't work well. Humans who have lost even one copy of an autosome (called **monosomics**) do not survive development. In all but a few cases, humans who have gained an extra autosome (called **trisomics**) also do not survive. However, five of the smallest chromosomes—those numbered 13, 15, 18, 21, and 22—can be present in humans as three copies and still allow the individual to survive for a time. The presence of an extra chromosome 13, 15, or 18 causes severe developmental defects, and infants with such a genetic makeup die within a few months. In contrast, individuals who have an extra copy of chromosome 21 or, more rarely, chromosome 22, usually survive to adulthood. In such individuals, the maturation of the skeletal system is delayed, so they generally are short and have poor muscle tone. Their mental development is also affected, and children with trisomy 21 or trisomy 22 are always mentally impaired.

Down Syndrome. The developmental defect produced by trisomy 21, an extra copy of chromosome 21 seen in the karyotype in figure 11.24, was first described in 1866 by J. Langdon Down; for this reason, it is called **Down syndrome.** About 1 in every 750 children exhibits Down syndrome, and the frequency is similar in all racial groups. It is much more common in children of older mothers. The graph in figure 11.25 shows the increasing incidence in older mothers. In mothers under 30 years old, the incidence is only about 0.6 per 1,000

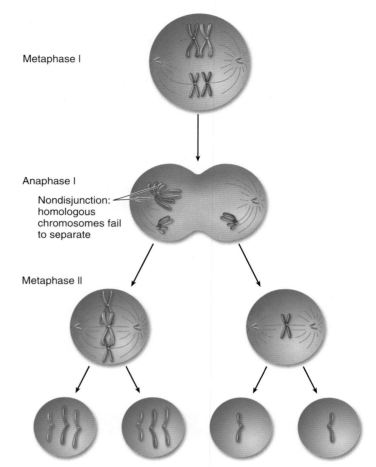

Results in four gametes: two are $n+1$ and two are $n-1$

Figure 11.23   Nondisjunction in anaphase I.
In nondisjunction that occurs during meiosis I, one pair of homologous chromosomes fails to separate in anaphase I, and the gametes that result have one too many or one too few chromosomes. Nondisjunction can also occur in meiosis II, when sister chromatids fail to separate during anaphase II.

(or 1 in 1,500 births), while in mothers 30 to 35 years old, the incidence doubles to about 1.3 per 1,000 births (or 1 in 750 births). In mothers over 45, the risk is as high as 63 per 1,000 births (or 1 in 16 births). The reason that older mothers are more prone to Down syndrome babies is that all the eggs that a woman will ever produce are present in her ovaries by the time she is born, and as she gets older they may accumulate damage that can result in nondisjunction.

Nondisjunction Involving the Sex Chromosomes

As noted, 22 of the 23 pairs of human chromosomes are perfectly matched in both males and females and are called autosomes. The remaining pair are the sex chromosomes, X and Y. In humans, as in *Drosophila* (but by no means in all diploid species), females are XX and males XY; any individual with at least one Y chromosome is male. The Y chromosome is highly condensed and bears few functional genes in most

(a)

(b)

Figure 11.24 Down syndrome.

(a) In this karyotype of a male individual with Down syndrome, the trisomy at position 21 can be clearly seen. (b) A person with Down syndrome.

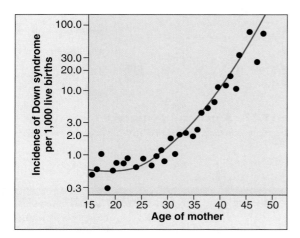

Figure 11.25 Correlation between maternal age and the incidence of Down syndrome.

As women age, the chances they will bear a child with Down syndrome increase. After a woman reaches age 35, the frequency of Down syndrome increases rapidly.

organisms. Some of the active genes the Y chromosome does possess are responsible for the features associated with "maleness." Individuals who gain or lose a sex chromosome do not generally experience the severe developmental abnormalities caused by changes in autosomes. Such individuals may reach maturity, but with somewhat abnormal features.

Nondisjunction of the X Chromosome. When X chromosomes fail to separate during meiosis, some of the gametes that are produced possess both X chromosomes and so are XX gametes; the other gametes that result from such an event have no sex chromosome and are designated "O."

Figure 11.26 shows what happens if gametes from sex chromosome nondisjunction combine with sperm. If an XX

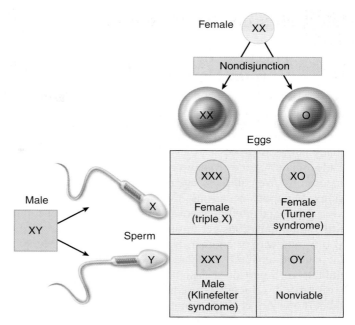

Figure 11.26 Nondisjunction of the X chromosome.

Nondisjunction of the X chromosome can produce sex chromosome aneuploidy—that is, abnormalities in the number of sex chromosomes.

gamete combines with an X gamete, the resulting XXX zygote (in the upper left above) develops into a female who is taller than average but other symptoms can vary greatly. Some are normal in most respects, others may have lower reading and verbal skills, and still others are mentally retarded. If an XX gamete combines with a Y gamete (in the lower left), the XXY zygote develops into a sterile male who has many female body characteristics and, in some cases, diminished mental capacity. This condition, called *Klinefelter syndrome,* occurs in about 1 in 500 male births.

If an O gamete fuses with a Y gamete (in the lower right), the OY zygote is nonviable and fails to develop further because humans cannot survive when they lack the genes on the X chromosome. If an O gamete fuses with an X gamete (in the upper right), the XO zygote develops into a sterile female of short stature, with a webbed neck and immature sex organs that do not undergo changes during puberty. The mental abilities of XO individuals are normal in verbal learning but lower in nonverbal/math-based problem solving. This condition, called *Turner syndrome,* occurs roughly once in every 5,000 female births.

Nondisjunction of the Y Chromosome. The Y chromosome can also fail to separate in meiosis, leading to the formation of YY gametes. When these gametes combine with X gametes, the XYY zygotes develop into fertile males of normal appearance. The frequency of the XYY genotype is about 1 per 1,000 newborn males.

11.8 Autosome loss is always lethal, and an extra autosome is with few exceptions lethal too. Additional sex chromosomes have less serious consequences, although they can lead to sterility.

11.9 The Role of Mutations in Human Heredity

The proteins encoded by most of your genes must function in a very precise fashion for you to develop properly and for the many complex processes of your body to function correctly. Unfortunately, genes sometimes sustain damage or are copied incorrectly. We call these accidental changes in genes **mutations.** Mutations occur only rarely, because your cells police your genes and attempt to correct any damage they encounter. Still, some mutations get through. Many of them are bad for you in one way or another. It is easy to see why. Mutations hit genes at random—imagine that you randomly changed the number of a part on a design of a jet fighter. Sometimes it won't matter critically—a seatbelt becomes a radio, say. But what if a key rivet in the wing becomes a roll of toilet paper? The chance of a random mutation in a gene improving the performance of its protein is about the same as that of a randomly selected part making the jet fly faster.

Most mutations are rare in human populations. Almost all result in recessive alleles, and so they are not eliminated from the population by evolutionary forces—because they are not expressed in most individuals (heterozygotes) in which they occur. Do you see why they occur mostly in heterozygotes? Because mutant alleles are rare, it is unlikely that a person carrying a copy of the mutant allele will marry someone who also carries it. Instead, he or she will typically marry someone homozygous normal, and their children would not be homozygous for the mutant allele. While most mutations are harmful to normal functions and are usually recessive, that is not to say all mutations are undesirable; some mutations can lead to enhanced function. Nor are all mutations recessive; rarely they can occur as dominant alleles.

In some cases, particular mutant alleles have become more common in human populations. In these cases, the harmful effects that they produce are called *genetic disorders.* Some of the most common genetic disorders are listed in table 11.3. To study human heredity, scientists look at the results of crosses that have already been made. They study family trees, or **pedigrees,** to identify which relatives exhibit a trait. Figure 11.27 shows a general pedigree. Females are indicated with circles and males with squares. The lines connect the parents and display the offspring. Solid shapes indicate an individual

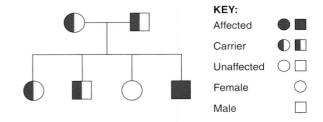

Figure 11.27   A general pedigree.
This pedigree is consistent with the inheritance of a recessive trait.

TABLE 11.3	SOME IMPORTANT GENETIC DISORDERS			
Disorder	**Symptom**	**Defect**	**Dominant/Recessive**	**Frequency Among Human Births**
Cystic fibrosis	Mucus clogs lungs, liver, and pancreas	Failure of chloride ion transport mechanism	Recessive	1/2,500 (Caucasians)
Sickle-cell anemia	Poor blood circulation	Abnormal hemoglobin molecules	Recessive	1/625 (African Americans)
Tay-Sachs disease	Deterioration of central nervous system in infancy	Defective enzyme (hexosaminidase A)	Recessive	1/3,500 (Ashkenazi Jews)
Phenylketonuria	Brain fails to develop in infancy	Defective enzyme (phenylalanine hydroxylase)	Recessive	1/12,000
Hemophilia	Blood fails to clot	Defective blood-clotting factor VIII	Sex-linked recessive	1/10,000 (Caucasian males)
Huntington's disease	Brain tissue gradually deteriorates in middle age	Production of an inhibitor of brain cell metabolism	Dominant	1/24,000
Muscular dystrophy (Duchenne)	Muscles waste away	Degradation of myelin coating of nerves stimulating muscles	Sex-linked recessive	1/3,700 (males)
Congenital hypothyroidism	Increased birth weight, puffy face, constipation, lethargy	Failure of proper thyroid development	Recessive	1/1,000 (Hispanics) 1/700 (Native Americans)
Hypercholesterolemia	Excessive cholesterol levels in blood, leading to heart disease	Abnormal form of cholesterol cell surface receptor	Dominant	1/500

affected by a disorder; half-filled shapes indicate a carrier (someone who is heterozygous); open shapes indicate an individual not affected. Then they can often determine whether the gene producing the trait is sex-linked or autosomal and whether the trait's phenotype is dominant or recessive. Frequently, they can infer which individuals are homozygous and which are heterozygous for the allele specifying the trait.

Hemophilia: A Sex-Linked Trait

Blood in a cut clots as a result of the polymerization of protein fibers circulating in the blood. A dozen proteins are involved in this process, and all must function properly for a blood clot to form. A mutation causing any of these proteins to lose their activity leads to a form of **hemophilia,** a hereditary condition in which the blood clots slowly or not at all.

Hemophilias are recessive disorders, expressed only when an individual does not possess any copy of the normal allele and so cannot produce one of the proteins necessary for clotting. Most of the genes that encode the blood-clotting proteins are on autosomes, but two (designated VIII and IX) are on the X chromosome. These two genes are sex-linked (see section 11.7): any male who inherits a mutant allele will develop hemophilia, because his other sex chromosome is a Y chromosome that lacks any alleles of those genes.

The most famous instance of hemophilia, often called the Royal hemophilia, is a sex-linked form that arose in the royal family of England. This hemophilia was caused by a mutation in gene IX that occurred in one of the parents of Queen Victoria of England (1819–1901). The pedigree in figure 11.28 shows that in the six generations since Queen Victoria, 10 of her male descendants have had hemophilia (the solid squares). The present British royal family has escaped the disorder because Queen Victoria's son, King Edward VII, did not inherit the defective allele, and all the subsequent rulers of England are his descendants. Three of Victoria's nine children did receive the defective allele, however, and they carried it by marriage into many of the other royal families of Europe.

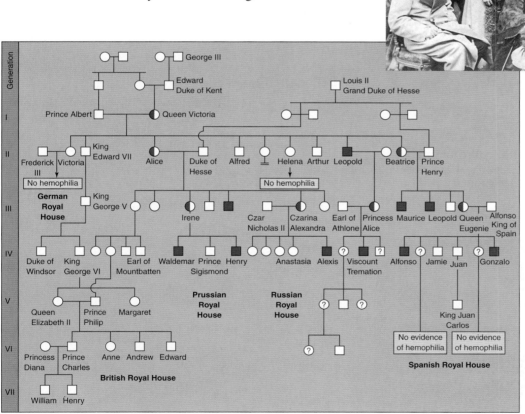

Figure 11.28 The Royal hemophilia pedigree.

Queen Victoria's daughter Alice introduced hemophilia into the Russian and Prussian royal houses, and her daughter Beatrice introduced it into the Spanish royal house. Victoria's son Leopold, himself a victim, also transmitted the disorder in a third line of descent. Half-shaded symbols represent carriers with one normal allele and one defective allele; fully shaded symbols represent affected individuals. Squares represent males; circles represent females. In this photo, Queen Victoria of England is surrounded by some of her descendants in 1894. Standing behind Victoria and wearing feathered boas are two of Victoria's granddaughters, Alice's daughter's: Princess Irene of Prussia *(right)*, and Alexandra *(left)*, who would soon become Czarina of Russia. Both Irene and Alexandra were also carriers of hemophilia.

Sickle-Cell Anemia: Recessive Trait

Sickle-cell anemia is a recessive hereditary disorder. Its inheritance is shown in the pedigree in figure 11.29, where affected individuals are homozygous, carrying two copies of the mutated gene. Affected individuals have defective molecules of hemoglobin, the protein within red blood cells that carries oxygen. Consequently, these individuals are unable to properly transport oxygen to their tissues. The defective hemoglobin molecules stick to one another, forming stiff, rodlike structures and resulting in the formation of sickle-shaped red blood cells (figure 11.30). As a result of their stiffness and irregular shape, these cells have difficulty moving through the smallest blood vessels; they tend to accumulate in those vessels and form clots. People who have large proportions of sickle-shaped red blood cells tend to have intermittent illness and a shortened life span.

The hemoglobin in the defective red blood cells differs from that in normal red blood cells in only one of hemoglobin's 574 amino acid subunits. In the defective hemoglobin, the amino acid valine replaces a glutamic acid at a single position in the protein. Interestingly, the position of the change is far from the active site of hemoglobin where the iron-bearing heme group binds oxygen. Instead, the change occurs on the outer edge of the protein. Why then is the result so catastrophic? The sickle-cell mutation puts a very nonpolar amino acid on the surface of the hemoglobin protein, creating a "sticky patch" that sticks to other such patches—nonpolar amino acids tend to associate with one another in polar environments like water. As one hemoglobin adheres to another, chains of hemoglobin molecules form.

Individuals heterozygous for the sickle-cell allele are generally indistinguishable from normal persons. However, some of their red blood cells show the sickling characteristic when they are exposed to low levels of oxygen. The allele responsible for sickle-cell anemia is particularly common among people of African descent because the sickle-cell allele is more common in Africa, as shown in the upper map of figure 11.31, where the darker green areas indicate higher incidence of the sickle-cell allele. About 9% of African Americans are heterozygous for this allele, and about 0.2% are homozygous and therefore have the disorder. In some groups of people in Africa, up to 45% of all individuals are heterozygous for this

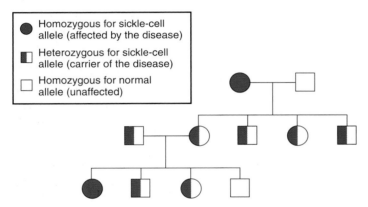

Figure 11.29 Inheritance of sickle-cell anemia.

Sickle-cell anemia is a recessive autosomal disorder. If one parent is homozygous for the recessive trait, all of the offspring will be carriers (heterozygotes) like the F_1 generation of Mendel's testcross.

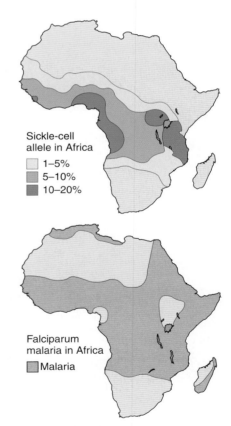

Figure 11.31 The sickle-cell allele confers resistance to malaria.

The distribution of sickle-cell anemia closely matches the occurrence of malaria in central Africa. This is not a coincidence. The sickle-cell allele, when heterozygous, confers resistance to malaria, a very serious disease.

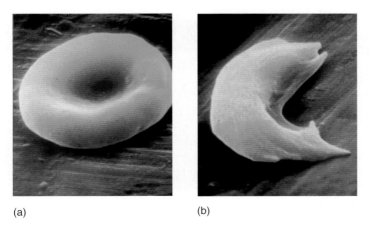

(a)   (b)

Figure 11.30 Normal and sickled red blood cells.

(a) A normal red blood cell is shaped like a flattened sphere. In individuals homozygous for the sickle-cell trait, many of the red blood cells have sickle shapes (b).

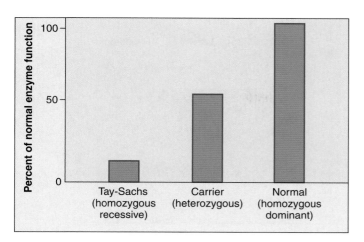

Figure 11.32 Tay-Sachs disease.
Homozygous individuals *(left bar)* typically have less than 10% of the normal level of hexosaminidase A *(right bar)*, while heterozygous individuals *(middle bar)* have about 50% of the normal level—enough to prevent deterioration of the central nervous system.

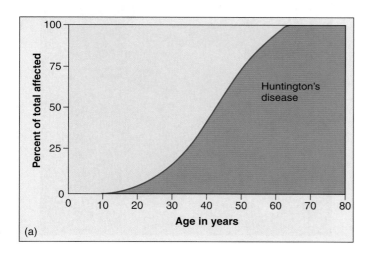

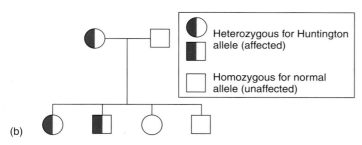

Figure 11.33 Huntington's disease is a dominant genetic disorder.
(a) Because of the late age of onset of Huntington's disease, the allele causing it persists despite being both dominant and fatal. (b) The pedigree illustrates how a dominant lethal allele can be passed from one generation to the next. Although the mother was affected, we can tell that she was heterozygous because if she were homozygous dominant, all of her children would have been affected. However, by the time she found out that she had the disease, she had probably already given birth to her children. In this way the trait passes on to the next generation even though it is fatal.

allele, and fully 6% are homozygous and express the disorder. What factors determine the high frequency of sickle-cell anemia in Africa? It turns out that heterozygosity for the sickle-cell anemia allele increases resistance to malaria, a common and serious disease in Central Africa. The lower map in figure 11.31 shows the incidence of malaria in Africa. Comparing the two maps, you can see that the area of the sickle-cell trait matches well with the incidence of malaria. The interactions of sickle-cell anemia and malaria are discussed further in chapter 15.

Tay-Sachs Disease: Recessive Trait

Tay-Sachs disease is an incurable hereditary disorder in which the brain deteriorates. Affected children appear normal at birth and usually do not develop symptoms until about the eighth month, when signs of mental deterioration appear. The children are blind within a year after birth, and they rarely live past five years of age.

The Tay-Sachs allele produces the disease by encoding a nonfunctional form of the enzyme hexosaminidase A. This enzyme breaks down *gangliosides,* a class of lipids occurring within the lysosomes of brain cells. As a result, the lysosomes fill with gangliosides, swell, and eventually burst, releasing oxidative enzymes that kill the cells. There is no known cure for this disorder.

Tay-Sachs disease is rare in most human populations, occurring in only 1 in 300,000 births in the United States. However, the disease has a high incidence among Jews of Eastern and Central Europe (Ashkenazi) and among American Jews, 90% of whom trace their ancestry to Eastern and Central Europe. In these populations, it is estimated that 1 in 28 individuals is a heterozygous carrier of the disease, and approximately 1 in 3,500 infants has the disease. Because the disease is caused by a recessive allele, most of the people who carry the defective allele do not themselves develop symptoms of the disease because, as shown by the middle bar in figure 11.32, their one normal gene produces enough enzyme activity (50%) to keep the body functioning normally.

Huntington's Disease: Dominant Trait

Not all hereditary disorders are recessive. **Huntington's disease** is a hereditary condition caused by a dominant allele that causes the progressive deterioration of brain cells. Perhaps 1 in 24,000 individuals develops the disorder. Because the allele is dominant, every individual who carries the allele expresses the disorder. Nevertheless, the disorder persists in human populations because its symptoms usually do not develop until the affected individuals are more than 30 years old, and by that time most of those individuals have already had children. Consequently, as illustrated by the pedigree in figure 11.33, the allele is often transmitted before the lethal condition develops.

11.9 Many human hereditary disorders reflect the presence of rare (and sometimes not so rare) mutations within human populations.

11.10 Genetic Counseling and Therapy

Although most genetic disorders cannot yet be cured, we are learning a great deal about them, and progress toward successful therapy is being made in many cases. However, in the absence of a cure, some parents may feel the only recourse is to try to avoid producing children with these conditions. The process of identifying parents at risk of producing children with genetic defects and of assessing the genetic state of early embryos is called **genetic counseling.** Genetic counseling can help prospective parents determine their risk of having a child with a genetic disorder and advise them on medical treatments or options if a genetic disorder is determined to exist in an unborn child.

High-Risk Pregnancies

If a genetic defect is caused by a recessive allele, how can potential parents determine the likelihood that they carry the allele? One way is through pedigree analysis, often employed as an aid in genetic counseling. By analyzing a person's pedigree, it is sometimes possible to estimate the likelihood that the person is a carrier for certain disorders. For example, if one of your relatives has been afflicted with a recessive genetic disorder such as cystic fibrosis, it is possible that you are a heterozygous carrier of the recessive allele for that disorder. When a pedigree analysis indicates that both parents of an expected child have a significant probability of being heterozygous carriers of a recessive allele responsible for a serious genetic disorder, the pregnancy is said to be a high-risk pregnancy. In such cases, there is a significant probability that the child will exhibit the clinical disorder.

Another class of high-risk pregnancies are those in which the mothers are more than 35 years old. As we have seen, the frequency of birth of infants with Down syndrome increases dramatically in the pregnancies of older women (see figure 11.25).

Genetic Screening

When a pregnancy is diagnosed as being high risk, many women elect to undergo **amniocentesis,** a procedure that permits the prenatal diagnosis of many genetic disorders. Figure 11.34 shows how an amniocentesis is performed. In the fourth month of pregnancy, a sterile hypodermic needle is inserted into the expanded uterus of the mother, and a small sample of the amniotic fluid bathing the fetus is removed. Within the fluid are free-floating cells derived from the fetus; once removed, these cells can be grown in cultures in the laboratory. During amniocentesis, the position of the needle and that of the fetus are usually observed by means of **ultrasound.** The ultrasound image in figure 11.35 clearly reveals the fetus's position in the uterus. You can see its head and a hand extending up, maybe sucking its thumb. The sound waves used in ultrasound are not harmful to mother or fetus, and they permit the person withdrawing the amniotic fluid to do so without damaging the fetus. In addition, ultrasound can be used to examine the fetus for signs of major abnormalities.

In recent years, physicians have increasingly turned to another invasive procedure for genetic screening called **chorionic villus sampling.** In this procedure, the physician removes cells from the chorion, a membranous part of the placenta that nourishes the fetus. This procedure can be used earlier in pregnancy

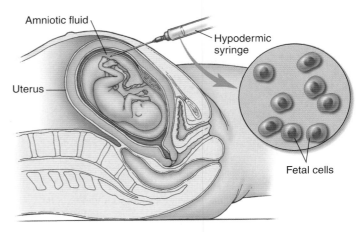

Figure 11.34   Amniocentesis.
A needle is inserted into the amniotic cavity, and a sample of amniotic fluid, containing some free cells derived from the fetus, is withdrawn into a syringe. The fetal cells are then grown in culture and their karyotype and many of their metabolic functions are examined.

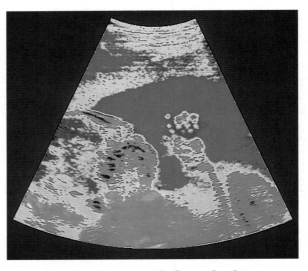

Figure 11.35   An ultrasound view of a fetus.
During the fourth month of pregnancy, when amniocentesis is normally performed, the fetus usually moves about actively. The head of the fetus (visualized in *green*) is to the *left.*

(by the eighth week) and yields results much more rapidly than does amniocentesis, but can increase the risk of miscarriage.

Genetic counselors look at three things in the cultures of cells obtained from amniocentesis or chorionic villus sampling:

1. **Chromosomal karyotype.** Analysis of the karyotype can reveal aneuploidy (extra or missing chromosomes) and gross chromosomal alterations.
2. **Enzyme activity.** In many cases, it is possible to test directly for the proper functioning of enzymes involved in genetic disorders. The lack of normal enzymatic activity signals the presence of the disorder. Thus, the lack of the enzyme responsible for breaking down phenylalanine signals PKU (phenylketonuria), the absence of the enzyme responsible for the breakdown of gangliosides indicates Tay-Sachs disease, and so forth.

3. Genetic markers. Genetic counselors can look for an association with known genetic markers. For sickle-cell anemia, Huntington's disease, and one form of muscular dystrophy (a genetic disorder characterized by weakened muscles), investigators have found other mutations on the same chromosomes that, by chance, occur at about the same place as the mutations that cause those disorders. By testing for the presence of these other mutations, a genetic counselor can identify individuals with a high probability of possessing the disorder-causing mutations. Finding such mutations in the first place is a little like searching for a needle in a haystack, but persistent efforts have proved successful in these three disorders. The associated mutations are detectable because they alter the length of the DNA segments that DNA-cleaving enzymes produce when they cut strands of DNA at particular places.

DNA Screening

The mutations that cause hereditary defects are frequently caused by alteration of a single DNA nucleotide within a key gene. Such spot differences between the version of a gene you have and the one another person has are called "single nucleotide polymorphisms," or SNPs. With the completion of the Human Genome Project (described in detail in chapter 14), researchers have begun assembling a huge database of hundreds of thousands of SNPs. Each of us differs from the standard "type sequence" in several thousand gene-altering SNPs. Screening SNPs and comparing them to known SNP databases should soon allow genetic counselors to screen each patient for copies of genes leading to hereditary disorders such as cystic fibrosis and muscular dystrophy.

Parents conceiving by *in vitro* fertilization have available a well-established screening procedure known as **preim-**

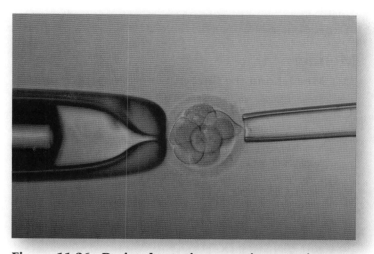

Figure 11.36 Preimplantation genetic screening.
The photograph shows a human embryo at the eight-cell stage, just before one of the eight cells is to be extracted for genetic testing by researchers.

plantation genetic screening. In this test, the egg is fertilized outside the mother, in glassware, and allowed to divide three times, until it contains eight cells. One of the eight cells is then removed from each of several such eight-cell embryos (figure 11.36) and tested for any of 150 genetic defects. The remaining seven-cell embryos are each able to develop into normal fetuses, giving the parents the choice of identifying and implanting an embryo that is disease free.

> **11.10** It has recently become possible to detect genetic defects early in pregnancy, allowing for appropriate planning by the prospective parents.

INQUIRY & ANALYSIS

How Can Two Genes on the Same Chromosome Segregate in a Cross?

The flower color and seed color genes Mendel studied in garden peas are both located on chromosome #1—and yet he observed them to segregate independently—that is, 50% recombination among alleles, the same as random chance would predict. In the graph, recombination is plotted versus physical distance between genes (100% meaning opposite ends of the chromosome).

1. **Applying Concepts** What is the dependent variable?
2. **Making Inferences** Do genes that are located farther apart on a chromosome recombine more often? To what maximum value? Explain.
3. **Drawing Conclusions** How close together do you think Mendel's flower and seed color genes are on chromosome #1?

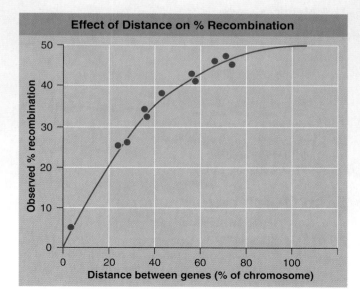

Mendel

11.1 Mendel and the Garden Pea

- Mendel studied inheritance using the garden pea (**figure 11.2**). He examined the patterns of inheritance of seven characteristics.

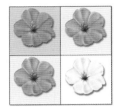

- Mendel used plants that were true-breeding for a particular characteristic as the P generation. He then crossed two P generation plants that expressed alternate forms of a characteristic. Their offspring were called the F_1 generation. He then allowed the F_1 plants to self-fertilize, giving rise to the F_2 generation (**figure 11.3**).

11.2 What Mendel Observed

- In Mendel's experiments, the F_1 generation plants all expressed the same alternative form, called the dominant trait. In the F_2 generation, 3/4 of the offspring expressed the dominant trait and 1/4 expressed the other trait, called the recessive trait. This is a 3:1 ratio (**table 11.1**). This 3:1 ratio was actually a 1:2:1 ratio— 1 true-breeding dominant, 2 not-true-breeding dominant, and 1 true-breeding recessive (**figure 11.5**).

11.3 Mendel Proposes a Theory

- Mendel's theory of heredity explains that characteristics are passed on from parent to offspring as alleles, one allele inherited from each parent (**figure 11.6**). If both of the alleles are the same (either both dominant or both recessive), the individual is homozygous for the trait. If the individual has one dominant and one recessive allele, it is heterozygous for the trait. An individual's alleles are referred to as its genotype and the expression of those alleles, what the individual looks like, is its phenotype.

- A Punnett square can be used to predict the genotypes and phenotypes of offspring of a cross (**figures 11.7** and **11.8**).

- A testcross is done to determine if a dominant individual is homozygous or heterozygous for a particular trait (**figure 11.9**).

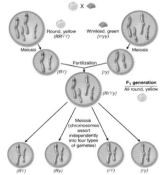

11.4 Mendel's Laws

- Mendel's law of segregation states that alleles are distributed into gametes that combine randomly to produce offspring. Mendel's law of independent assortment states that genes located on different chromosomes are inherited independently of each other (**figure 11.10**).

From Genotype to Phenotype

11.5 How Genes Influence Traits

- Genes coded in DNA determine phenotype because DNA encodes the amino acid sequences of proteins and proteins are the outward expression of genes (**figure 11.11**). Alternative forms of a gene, called alleles, result from mutations.

11.6 Why Some Traits Don't Show Mendelian Inheritance

- Not all traits follow the inheritance patterns outlined by Mendel. Continuous variation results when more than one gene contributes in a cumulative way to the phenotype (**figure 11.12**). Pleiotropic effects result when one gene influences more than one trait (**figure 11.13**). Incomplete dominance results when alternative alleles are not fully dominant or fully recessive (**figure 11.14**). The expression of some genes is influenced by environmental factors, such as temperature (**figure 11.15**). Epistasis occurs when two or more genes interact, have an additive or masking effect, resulting in different phenotypes (**figures 11.16** and **11.17**). Codominance occurs when there are more than two alleles in a population but there isn't a dominant allele—two alleles are expressed resulting in phenotypic expression of both alleles (**figures 11.18** and **11.19**).

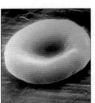

Chromosomes and Heredity

11.7 Chromosomes Are the Vehicles of Mendelian Inheritance

- Mendel explained that genes assort independently; this is because they are located on chromosomes that assort independently during meiosis. Morgan demonstrated this using an X-linked gene in fruit flies (**figure 11.21**).

11.8 Human Chromosomes

- Humans have 23 pairs of homologous chromosomes, for a total of 46 chromosomes. They have 22 pairs of autosomes and one pair of sex chromosomes. Nondisjunction occurs when homologous pairs or sister chromatids fail to separate during meiosis, resulting in gametes with too many or too few chromosomes (**figure 11.23**).

- Nondisjunction of autosomes is usually fatal, but the effects of nondisjunction of sex chromosomes is less severe (**figure 11.26**).

Human Hereditary Disorders

11.9 The Role of Mutations in Human Heredity

- Mutations can result in new alleles that can lead to genetic disorders. Pedigrees can be used to track and predict the inheritance of these disorders in families (**figure 11.27**). Disorders such as hemophilia (**figure 11.28**), sickle-cell anemia (**figures 11.29–11.31**), Tay-Sachs (**figure 11.32**), and Huntington's disease (**figure 11.33**) are all human genetic disorders.

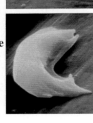

11.10 Genetic Counseling and Therapy

- Some genetic disorders can be detected during pregnancy using methods such as amniocentesis (**figure 11.34**) and chorionic villus sampling. In these procedures, fetal DNA is examined by chromosomal karyotyping, enzyme activity analysis, and searching for genetic markers associated with genetic disorders.

1. Gregor Mendel studied the garden pea plants because
   a. pea plants are small, easy to grow, grow quickly, and produce lots of flowers and seeds.
   b. he knew about studies with the garden pea that had been done for hundreds of years, and wanted to continue them, using math—counting and recording differences.
   c. he knew that there were many varieties available with distinctive characteristics.
   d. All of these.

2. Mendel examined seven characteristics, such as flower color. He crossed plants with two different forms of a character (purple flowers and white flowers). In every case the first generation of offspring (F_1) were
   a. all purple flowers.
   b. half purple flowers and half white flowers.
   c. 3/4 purple and 1/4 white flowers.
   d. all white flowers.

3. Following question 2, if Mendel then randomly mated the offspring of that first cross, or F_1 generation, the offspring in the F_2 generation were
   a. all purple flowers.
   b. half purple flowers and half white flowers.
   c. 3/4 purple and 1/4 white flowers.
   d. all white flowers.

4. Mendel then studied his results, and proposed a set of hypotheses to explain them. The basis of these hypotheses is that parents transmit
   a. traits directly to their offspring and they are expressed.
   b. some factor, or information, about traits to their offspring and it may or may not be expressed.
   c. some factor, or information, about traits to their offspring and it will always be expressed.
   d. some factor, or information, about traits to their offspring and it is expressed in every generation, perhaps in a "blended" form with information from the other parent.

5. A cross between two individuals results in a ratio of 9:3:3:1 for four possible phenotypes. This is an example of a
   a. dihybrid cross.
   b. monohybrid cross.
   c. testcross.
   d. none of these.

6. Human height shows a continuous variation from the very short to the very tall. Height is most likely controlled by
   a. epistatic genes.
   b. environmental factors.
   c. sex-linked genes.
   d. multiple genes.

7. In the human ABO blood grouping, the four basic blood types are type A, type B, type AB, and type O. The blood proteins A and B are
   a. simple dominant and recessive traits.
   b. incomplete dominant traits.
   c. codominant traits.
   d. sex-linked traits.

8. What finding finally determined that genes were carried on chromosomes?
   a. heat sensitivity of certain enzymes that determined coat color
   b. sex-linked eye color in fruit flies
   c. the finding of complete dominance
   d. establishing pedigrees

9. Nondisjunction
   a. occurs when homologous chromosomes or sister chromatids fail to separate during meiosis.
   b. may lead to Down syndrome.
   c. results in aneuploidy.
   d. All of the above.

10. If parents are concerned about their risk of producing children with serious genetic defects
    a. genetic infant design is now available.
    b. genetic correction of defects is now available.
    c. genetic modification of the parents is now available.
    d. genetic screening and prenatal diagnosis is now available.

Visual Understanding

1. **Figure 11.10** Using the four gametes shown from the F_1 generation, how many possible crosses are there? Remember that a gamete type could cross with another of the same type (such as $RY \times RY$). Draw a Punnett square for each cross, and list the ratios of genotypes and phenotypes for the F_2 generation of that cross.

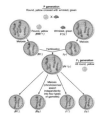

2. **Figure 11.19** Referring to this figure, use Punnett squares to illustrate whether a type A female and a type B male can have a child with type O blood.

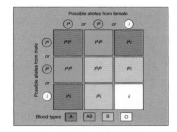

Challenge Questions

1. As Mendel struggled with understanding inheritance and formed his laws, how would the outcome have been different if he had only chosen five characteristics to study—flower position, pod shape, plant height, flower color, and seed color?

2. Kim and Su-Ling are doing fruit fly crosses in their biology class. Kim wants to test whether the female they have is homozygous or heterozygous for red eyes by mating with a red-eyed male. How should Su-Ling explain the difficulty to him?

3. Your biology class is collecting information on heredity. Michael realizes that he, along with three of his four brothers, are color blind, but his four sisters are not, and neither are his parents nor his grandparents. Can you help Michael understand what happened?

4. Kuzungu is a child orphaned by civil war in her country and raised in a group home. She has sickle-cell anemia, and type AB blood. Two couples who believe they are her grandparents ask you, a genetic counselor, to help them determine the truth. What do you suggest?

Additional Genetics Problems

These genetics problems will help you to see the far-reaching effects of Mendel's experiments. If you need help, the answers appear at www.mhhe.com/tlw5/geneticsanswers.

1. Silky feathers in chickens is a single-gene recessive trait whose effect is to produce shiny plumage.

   a. If 108 birds were raised from a cross between individuals heterozygous for this gene, how many would be expected to be silky and how many normal?

   b. If you had a normal-feathered bird, what would be the easiest cross to perform in order to determine if the bird is homozygous or heterozygous for the silky allele?

2. Among Hereford cattle there is a dominant allele called *polled;* the individuals that have this allele lack horns. Suppose you acquire a herd consisting entirely of polled cattle, and you carefully determine that no cow in the herd has horns. Some of the calves born that year, however, grow horns. You remove them from the herd and make certain that no horned adult has gotten into your pasture. Despite your efforts, more horned calves are born the next year. What is the reason for the appearance of the horned calves? If your goal is to maintain a herd consisting entirely of polled cattle, what should you do?

3. An inherited trait among humans in Norway causes affected individuals to have very wavy hair, not unlike that of a sheep. The trait, called *woolly,* is very evident when it occurs in families; no child possesses woolly hair unless at least one parent does. Imagine you are a Norwegian judge, and you have before you a woolly haired man suing his normal-haired wife for divorce because their first child has woolly hair but their second child has normal hair. The husband claims this constitutes evidence of his wife's infidelity. Do you accept his claim? Justify your decision.

4. Brachydactyly is a rare human trait that causes a shortening of the length of the fingers by a third. A review of medical records reveals that the progeny of marriages between a brachydactyl person and a normal person are approximately half brachydactylous. What proportion of offspring in matings between two brachydactylous individuals would be expected to be brachydactylous?

5. Many animals and plants bear recessive alleles for *albinism,* a condition in which homozygous individuals lack certain pigments. An albino plant, for example, lacks chlorophyll and is white, and an albino human lacks melanin. If two normally pigmented persons heterozygous for the same albinism allele marry, what proportion of their children would you expect to be albino?

6. You inherit a racehorse and decide to put him out to stud. In looking over the stud book, however, you discover that the horse's grandfather exhibited a rare disorder that causes brittle bones. The disorder is hereditary and results from homozygosity for a recessive allele. If your horse is heterozygous for the allele, it will not be possible to use him for stud because the genetic defect may be passed on. How would you determine whether your horse carries this allele?

7. Your instructor presents you with a *Drosophila* (fruit fly) with red eyes, as well as a stock of white-eyed flies and another stock of flies homozygous for the red-eye allele. You know that the presence of white eyes in *Drosophila* is caused by homozygosity for a recessive allele. How would you determine whether the single red-eyed fly was heterozygous for the white-eye allele?

8. Hemophilia is a recessive sex-linked human blood disease that leads to failure of blood to clot normally. One form of hemophilia has been traced to the royal family of England, from which it spread throughout the royal families of Europe. For the purposes of this problem, assume that it originated as a mutation either in Prince Albert or in his wife, Queen Victoria.

   a. Prince Albert did not have hemophilia. If the disease is a sex-linked recessive abnormality, how could it have originated in Prince Albert, a male, who would have been expected to exhibit sex-linked recessive traits?

   b. Alexis, the son of Czar Nicholas II of Russia and Empress Alexandra (a granddaughter of Victoria), had hemophilia, but their daughter Anastasia did not. Anastasia died, a victim of the Russian revolution, before she had any children. Can we assume that Anastasia would have been a carrier of the disease? Would your answer be different if the disease had been present in Nicholas II or in Alexandra?

9. A normally pigmented man marries an albino woman. They have three children, one of whom is an albino. What is the genotype of the father?

10. A man works in an atomic energy plant, and he is exposed daily to low-level background radiation. After several years, he has a child who has Duchenne muscular dystrophy, a recessive genetic defect caused by a mutation on the X chromosome. Neither the parents nor the grandparents have the disease. The man sues the plant, claiming that the abnormality in their child is the direct result of radiation-induced mutation of his gametes, and that the company should have protected him from this radiation. Before reaching a decision, the judge hearing the case insists on knowing the sex of the child. Which sex would be more likely to result in an award of damages, and why?

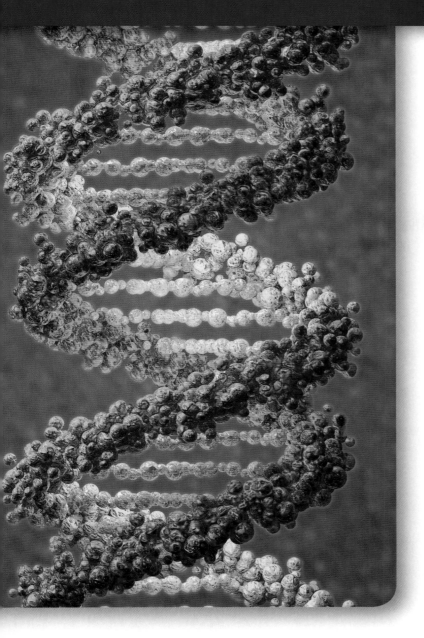

12

DNA: The Genetic Material

The realization that patterns of heredity can be explained by the segregation of chromosomes in meiosis raised a question that occupied biologists for over 50 years: What is the exact nature of the connection between hereditary traits and chromosomes? In this chapter you will examine some of the chain of experiments that have led to our current understanding of the molecular mechanisms of heredity. The experiments determining that DNA is the genetic material are among the most elegant in science. Just as in a good detective story, each conclusion has led to new questions. The intellectual path taken has not always been a straight one, the best questions not always obvious. But however erratic and lurching the course of the experimental journey, our picture of heredity has become progressively clearer, the image more sharply defined. We now understand in considerable detail how the DNA molecule copies itself, and how changes to it lead to hereditary gene mutations.

12.1 The Griffith Experiment

As we learned in chapters 9, 10, and 11, chromosomes contain genes, which, in turn, contain hereditary information. However, Mendel's work left a key question unanswered: What *is* a gene? When biologists began to examine chromosomes in their search for genes, they soon learned that chromosomes are made of two kinds of macromolecules, both of which you encountered in chapter 4: **proteins** (long chains of *amino acid subunits* linked together in a string) and **DNA** (deoxyribonucleic acid—long chains of *nucleotide* subunits linked together in a string). It was possible to imagine that either of the two was the stuff that genes are made of—information might be stored in a sequence of different amino acids, or in a sequence of different nucleotides. But which one is the stuff of genes, protein or DNA? This question was answered clearly in a variety of different experiments, all of which shared the same basic design: If you separate the DNA in an individual's chromosomes from the protein, which of the two materials is able to change another individual's genes?

In 1928, British microbiologist Frederick Griffith made a series of unexpected observations while experimenting with pathogenic (disease-causing) bacteria. Figure 12.1 takes you stepwise through his discoveries. When he infected mice with a virulent strain of *Streptococcus pneumoniae* bacteria (then known as *Pneumococcus*), the mice died of blood poisoning as you can see in panel 1. However, when he infected similar mice with a mutant strain of *S. pneumoniae* that lacked the virulent strain's polysaccharide capsule, the mice showed no ill effects, as you can see in panel 2. The capsule was apparently necessary for infection. The normal pathogenic form of this bacterium is referred to as the S form because it forms smooth colonies in a culture dish. The mutant form, which lacks an enzyme needed to manufacture the polysaccharide capsule, is called the R form because it forms rough colonies.

To determine whether the polysaccharide capsule itself had a toxic effect, Griffith injected dead bacteria of the virulent S strain into mice and as panel 3 shows, the mice remained perfectly healthy. Finally, as shown in panel 4, he injected mice with a mixture containing dead S bacteria of the virulent strain and live, capsuleless R bacteria, each of which by itself did not harm the mice. Unexpectedly, the mice developed disease symptoms and many of them died. The blood of the dead mice was found to contain high levels of live, virulent *Streptococcus* type S bacteria, which had surface proteins characteristic of the live (previously R) strain. Somehow, the information specifying the polysaccharide capsule had passed from the dead, virulent S bacteria to the live, capsuleless R bacteria in the mixture, permanently transforming the capsuleless R bacteria into the virulent S variety.

> **12.1** Hereditary information can pass from dead cells to living ones and transform them.

Transformation

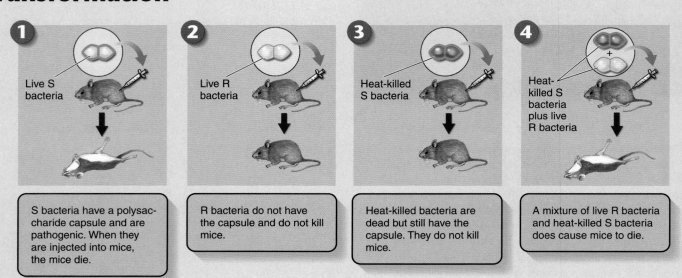

1. S bacteria have a polysaccharide capsule and are pathogenic. When they are injected into mice, the mice die.

2. R bacteria do not have the capsule and do not kill mice.

3. Heat-killed bacteria are dead but still have the capsule. They do not kill mice.

4. A mixture of live R bacteria and heat-killed S bacteria does cause mice to die.

Figure 12.1 How Griffith discovered transformation.
Transformation, the movement of a gene from one organism to another, provided some of the key evidence that DNA is the genetic material. Griffith found that extracts of dead pathogenic strains of the bacterium *Streptococcus pneumoniae* can "transform" live harmless strains into live pathogenic strains.

12.2 The Avery and Hershey-Chase Experiments

The Avery Experiments

The agent responsible for transforming *Streptococcus* went undiscovered until 1944. In a classic series of experiments, Oswald Avery and his coworkers Colin MacLeod and Maclyn McCarty characterized what they referred to as the "transforming principle." Avery and his colleagues prepared the same mixture of dead S *Streptococcus* and live R *Streptococcus* that Griffith had used, but first they removed as much of the protein as they could from their preparation of dead S *Streptococcus,* eventually achieving 99.98% purity. Despite the removal of nearly all protein from the dead S *Streptococcus,* the transforming activity was not reduced. Moreover, the properties of the transforming principle resembled those of DNA in several ways:

Same chemistry as DNA. When the purified principle was analyzed chemically, the array of elements agreed closely with DNA.

Same behavior as DNA. In an ultracentrifuge, the transforming principle migrated like DNA; in electrophoresis and other chemical and physical procedures, it also acted like DNA.

Not affected by lipid and protein extraction. Extracting the lipid and protein from the purified transforming principle did not reduce its activity.

Not destroyed by protein- or RNA-digesting enzymes. Protein-digesting enzymes did not affect the principle's activity, nor did RNA-digesting enzymes.

Destroyed by DNA-digesting enzymes. The DNA-digesting enzyme destroyed all transforming activity.

The evidence was overwhelming. They concluded that "a nucleic acid of the deoxyribose type is the fundamental unit of the transforming principle of *Pneumococcus* Type III"—in essence, that DNA is the hereditary material.

The Hershey-Chase Experiment

Avery's result was not widely appreciated at first, because most biologists still preferred to think that genes were made of proteins. In 1952, however, a simple experiment carried out by Alfred Hershey and Martha Chase was impossible to ignore. The team studied the genes of viruses that infect bacteria. These viruses attach themselves to the surface of bacterial cells and inject their genes into the interior; once inside, the genes take over the genetic machinery of the cell and order the manufacturing of hundreds of new viruses. When mature, the progeny viruses burst out to infect other cells. These bacteria-infecting viruses have a very simple structure: a core of DNA surrounded by a coat of protein.

In this experiment shown in figure 12.2, Hershey and Chase used radioactive isotopes to "label" the DNA and protein of the viruses. Radioactively tagged molecules are indicated in red in the figure. In the preparation on the

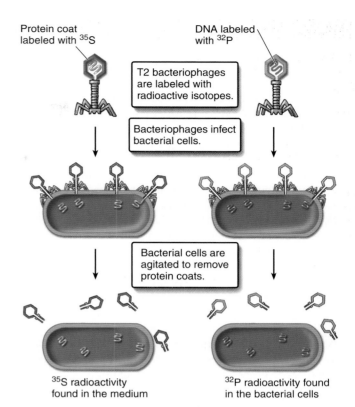

Figure 12.2   The Hershey-Chase experiment.
The experiment that convinced most biologists that DNA is the genetic material was carried out soon after World War II, when radioactive isotopes were first becoming commonly available to researchers. Hershey and Chase used different radioactive labels to "tag" and track protein and DNA. They found that when bacterial viruses inserted their genes into bacteria to guide the production of new viruses, it was DNA and not protein that was inserted. More specifically, ^{35}S radioactivity did not enter infected bacterial cells and ^{32}P radioactivity did. Clearly the virus DNA, not the virus protein, was responsible for directing the production of new viruses.

right, the viruses were grown so that their DNA contained radioactive phosphorus (^{32}P); in another preparation on the left side of the figure, the viruses were grown so that their protein coats contained radioactive sulfur (^{35}S). After the labeled viruses were allowed to infect bacteria, Hershey and Chase shook the suspensions forcefully to dislodge attacking viruses from the surface of bacteria, used a rapidly spinning centrifuge to isolate the bacteria, and then asked a very simple question: What did the viruses inject into the bacterial cells, protein or DNA? They found that the bacterial cells infected by viruses containing the ^{32}P label had labeled tracer in their interiors; cells infected by viruses containing the ^{35}S labeled tracer did not. The conclusion was clear: the genes that viruses use to specify new viruses are made of DNA and not protein.

> **12.2   Several key experiments demonstrated conclusively that DNA, not protein, is the hereditary material.**

12.3 Discovering the Structure of DNA

As it became clear that DNA was the molecule that stored the hereditary information, researchers began to question how this nucleic acid could carry out the complex function of inheritance. At the time, investigators did not know what the DNA molecule looked like.

We now know that DNA is a long, chainlike molecule made up of subunits called **nucleotides.** As you can see in figure 12.3, each nucleotide has three parts: a central sugar called deoxyribose, a phosphate (PO_4) group, and an organic base. The sugar (the lavender pentagon structure) and the phosphate group (the yellow-boxed structure) are the same in every nucleotide of DNA. However, there are four different kinds of bases: two large ones with double-ring structures, and two small ones with single rings. The large bases, called **purines,** are **A** (adenine) and **G** (guanine). The small bases, called **pyrimidines,** are **C** (cytosine) and **T** (thymine). A key observation, made by Erwin Chargaff, was that DNA molecules always had equal amounts of purines and pyrimidines. In fact, with slight variations due to imprecision of measurement, the amount of A always equals the amount of T, and the amount of G always equals the amount of C. This observation (A = T, G = C), known as **Chargaff's rule,** suggested that DNA had a regular structure.

The significance of Chargaff's rule became clear in 1953 when the British chemist Rosalind Franklin carried out an X-ray diffraction experiment. In these experiments, DNA molecules are bombarded with X-ray beams, and when individual rays encounter atoms, their paths are bent or diffracted like a thrown ball bounces off or around an object. Each atomic encounter creates a pattern on photographic film, shown in figure 12.4a, that looks like the ripples created by tossing a rock into a smooth lake. Franklin's results suggested that the DNA molecule had the shape of a coiled spring or a corkscrew, with the image in the photo from the viewpoint of looking down the center of the molecule, a form called a **helix.**

Franklin's work was shared with two researchers at Cambridge University, Francis Crick and James Watson, before it was published. Using Tinkertoy-like models of the bases, Watson and Crick deduced the true structure of DNA (figure 12.4b): The DNA molecule, shown in figure 12.4c, is a **double helix,** a winding staircase of two strands whose bases face one another. Chargaff's rule is a direct reflection of this structure—every bulky purine on one strand is paired with a slender pyrimidine on the other strand. Specifically, A (the blue bases) pairs with T (the orange bases), and G (the purple bases) pairs with C (the pink bases). Because hydrogen bonds, shown as dotted lines, can form between the **base pairs,** the molecule keeps a constant thickness.

> **12.3** The DNA molecule has two strands of nucleotides held together by hydrogen bonds between bases. The two strands wind into a double helix.

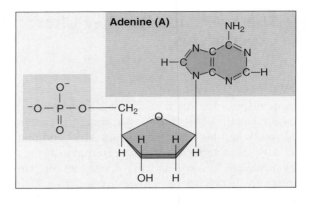

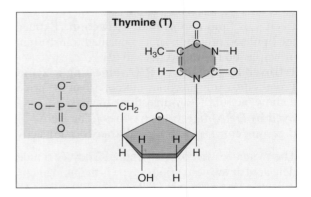

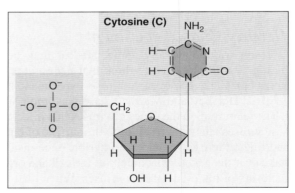

Figure 12.3   The four nucleotide subunits that make up DNA.

The nucleotide subunits of DNA are composed of three parts: a central five-carbon sugar called deoxyribose, a phosphate group, and an organic, nitrogen-containing base.

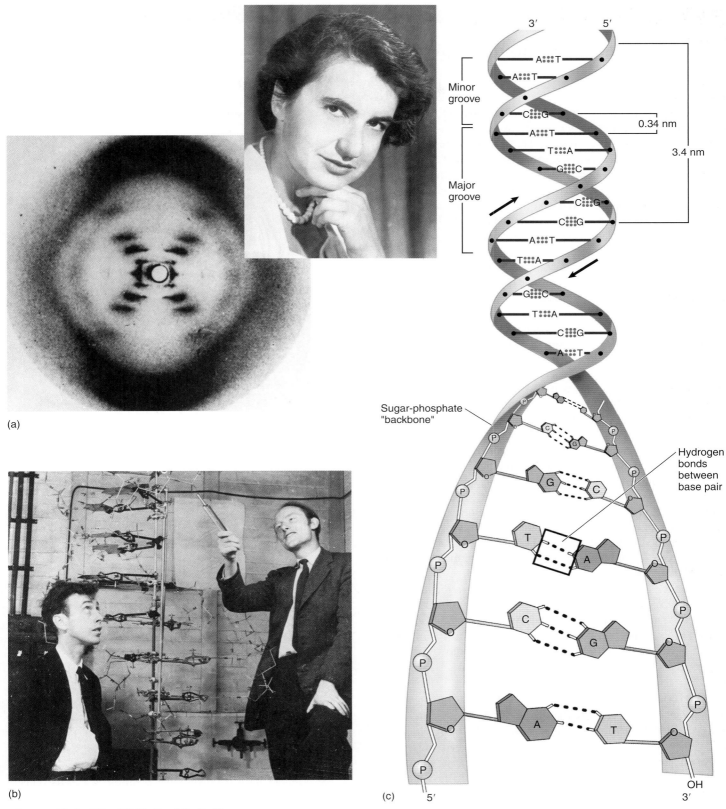

Figure 12.4 The DNA double helix.

(a) This X-ray diffraction photograph was made in 1953 by Rosalind Franklin (inset) in the laboratory of Maurice Wilkins. It suggested to Watson and Crick that the DNA molecule was a helix, like a winding staircase. (b) In 1953 Watson and Crick deduced the structure of DNA. James Watson (seated and peering up at their homemade model of the DNA molecule) was a young American postdoctoral student, and Francis Crick (pointing) was an English scientist. (c) The dimensions of the double helix were suggested by the X-ray diffraction studies. In a DNA duplex molecule, only two base pairs are possible: adenine (A) with thymine (T) and guanine (G) with cytosine (C). A G–C base pair has three hydrogen bonds; an A–T base pair has only two.

12.4 How the DNA Molecule Copies Itself

The attraction that holds the two DNA strands together is the formation of weak hydrogen bonds between the bases that face each other from the two strands. That is why A pairs with T and not C; A can only form hydrogen bonds with T. Similarly, G can form hydrogen bonds with C but not T. In the Watson-Crick model of DNA, the two strands of the double helix are said to be *complementary* to each other. One chain of the helix can have any sequence of bases, of A, T, G, and C, but this sequence completely determines that of its partner in the helix. If the sequence of one chain is ATTGCAT, the sequence of its partner in the double helix must be TAACGTA. Each chain in the helix is a complementary mirror image of the other. This **complementarity** makes it possible for the DNA molecule to copy itself during cell division in a very direct manner. But, there are three possible alternatives as to how the DNA could serve as a template for the assembly of new DNA molecules.

First, the two strands of the double helix could separate and serve as templates for the assembly of two new strands by base pairing A with T and G with C. This is what happens in figure 12.5a, with the original strand colored blue and the newly formed strands red. After replicating, the original strands rejoin, preserving the original strand of DNA and forming an entirely new strand. This is called *conservative replication.*

In the second alternative, the double helix need only "unzip" and assemble a new complementary chain along each single strand. This form of DNA replication is called *semiconservative replication,* because while the sequence of the original duplex is conserved after one round of replication, the duplex itself is not. Instead, each strand of the duplex becomes part of another duplex. You can see in figure 12.5b that the blue strand is from the original helix and the red strand is newly formed.

In the third alternative, called *dispersive replication,* the original DNA would serve as a template for the formation of new DNA strands but the new and old DNA would be dispersed among the two daughter strands. As shown in figure 12.5c, each daughter strand is made up of sections of original (blue) strands and new (red) strands.

The Meselson-Stahl Experiment

The three alternative hypotheses of DNA replication were tested in 1958 by Matthew Meselson and Franklin Stahl of the California Institute of Technology. These two scientists grew bacteria in a medium containing the heavy isotope of nitrogen, ^{15}N, which became incorporated into the bases of the bacterial DNA (the upper petri dish in figure 12.6). After several generations, samples were taken from this culture and grown in a medium containing the normal lighter isotope ^{14}N, which became incorporated into the newly replicating DNA. Bacterial samples were taken from the ^{14}N media at 20 minute intervals (❷ through ❹). DNA was extracted from all three samples and a fourth sample, ❶, that served as a control.

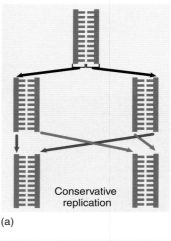
(a)

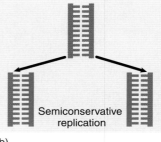
(b)

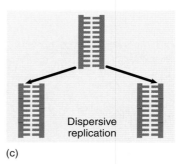
(c)

Figure 12.5 Alternative mechanisms of DNA replication.

By dissolving the DNA they had collected in a heavy salt called cesium chloride, and then spinning the solution at very high speeds in an ultracentrifuge, Meselson and Stahl were able to separate DNA strands of different densities. The centrifugal forces caused the cesium ions to migrate toward the bottom of the centrifuge tube, creating a gradient of cesium concentration, and thus a gradation of density. Each DNA strand floats or sinks in the gradient until it reaches the position where its density exactly matches the density of the cesium there. Because ^{15}N strands are denser than ^{14}N strands, they migrate farther down the tube to a denser region of cesium.

The DNA collected immediately after the transfer was all dense, as shown in test tube ❷. However, after the bacteria completed their first round of DNA replication in the ^{14}N medium, the density of their DNA had decreased to a value

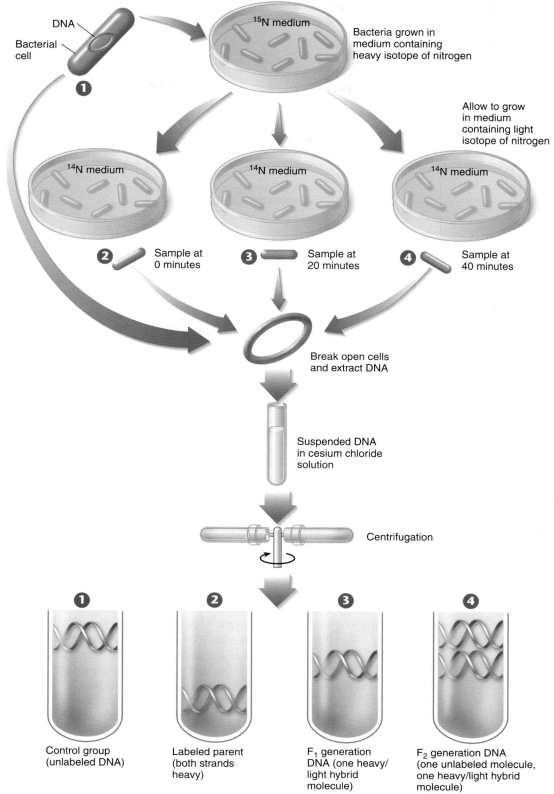

Figure 12.6 The Meselson-Stahl experiment.

Bacterial cells were grown for several generations in a medium containing a heavy isotope of nitrogen (^{15}N) and then were transferred to a new medium containing the normal lighter isotope (^{14}N). (The bacteria shown here are not drawn to scale, as tens of thousands of bacterial cells grow on even a tiny portion of a plate in culture.) At various times thereafter, samples of the bacteria were collected, and their DNA was dissolved in a solution of cesium chloride, which was spun rapidly in a centrifuge. The labeled and unlabeled DNA settled in different areas of the tube because they differed in weight. The DNA with two heavy strands settled down toward the bottom of the tube. The DNA with two light strands settled higher up in the tube. The DNA with one heavy and one light strand settled in between the other two.

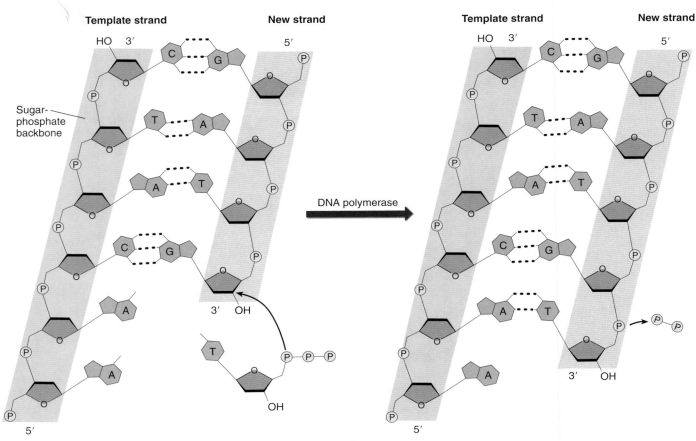

Figure 12.7　How nucleotides are added in DNA replication.
Nucleotides are added to the new growing strand of DNA by DNA polymerase. The addition of the nucleotides follows base pairing.

intermediate between ^{14}N-DNA and ^{15}N-DNA, as shown in test tube ❸. After the second round of replication, two density classes of DNA were observed, one intermediate and one equal to that of ^{14}N-DNA, as shown in test tube ❹.

Meselson and Stahl interpreted their results as follows: after the first round of replication, each daughter DNA duplex was a hybrid possessing one of the heavy strands of the parent molecule and one light strand; when this hybrid duplex replicated, it contributed one heavy strand to form another hybrid duplex and one light strand to form a light duplex. Thus, this experiment clearly ruled out conservative and dispersive DNA replication, and confirmed the prediction of the Watson-Crick model that DNA replicates in a semiconservative manner.

How DNA Copies Itself

The copying of DNA before cell division is called **DNA replication** and is overseen by an enzyme called *DNA polymerase.* An enzyme called *helicase* first unwinds the DNA double helix, then DNA polymerase reads along each single strand, the blue strand in figure 12.7, and adds the correct complementary nucleotide (A pairs with T, G with C) at each position as it moves, creating a complementary strand, the pink strand.

However, there are some limitations of the actions of DNA polymerase. First, it can only add to an existing strand; it cannot begin a strand. Another enzyme circumvents this difficulty by beginning the new strand with a section of nucleic

acids called a *primer.* These are the green segments in figure 12.8. This happens at the place where the parent DNA molecule becomes unzipped, called the **replication fork.** At the replication fork, the polymerase very actively shuttles several hundred nucleotides up one strand, building a new strand of DNA called the **leading strand,** adding on to the primer in a continuous fashion. The leading strand is the upper red strand in figure 12.8, the primer is off further to the left out of the frame of the diagram. The phosphate group of a nucleotide, called the 5′ end, attaches to the sugar end, called the 3′ end, of the nucleotide at the end of the growing strand. The directionality of building the new strand is more apparent in figure 12.7, where the phosphate group of the incoming thymine attaches to the OH group of the sugar of the guanosine. The new strand assembles in a 5′ to 3′ direction. This reveals a second limitation; DNA polymerase can only build a strand of DNA in one direction, and so it assembles the other DNA strand, called the **lagging strand,** in segments. Each lagging strand segment begins with a primer (the green segments you saw in figure 12.8), and the DNA polymerase then builds it away from the replication fork until it encounters the previous section.

Eukaryotic chromosomes each contain a single, very long molecule of DNA, one far too long to copy all the way from one end to the other with a single replication fork. Each eukaryotic chromosome is instead copied in sections of about 100,000 nucleotides, each with its own replication origin and fork.

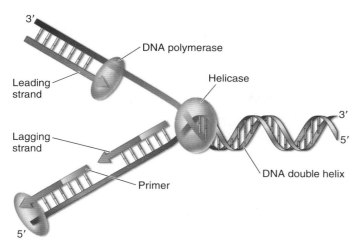

Figure 12.8   Building the leading and lagging strands.
DNA polymerase builds the leading strand as a continuous strand moving into the replication fork growing 5′ to 3′, but the lagging strand, also growing 5′ to 3′, is assembled moving away from the replication fork, in segments, each beginning with a primer.

Before the newly formed DNA molecules wind back into the double helix shape, the primers must be removed and the segments of DNA must be sealed together. Remember that the newly formed strand of DNA was assembled in sections on the lagging strand. These sections need to be covalently linked together. The enzyme that performs that function is *DNA ligase.* DNA ligase joins the ends of newly synthesized segments of DNA after the primers have been removed, resulting in one continuous strand of DNA. This process is summarized in figure 12.9, with the DNA ligase in panel 3 sealing the gaps between the DNA sections of the lagging strand.

The enormous amount of DNA that resides within the 100 trillion cells of your body represents a long series of DNA replications, starting with the DNA of a single cell—the fertilized egg. Living cells have evolved many mechanisms to avoid errors during DNA replication and to preserve the DNA from damage. These mechanisms of **DNA repair** proofread the strands of each daughter cell against one another for accuracy and correct any mistakes. But the proofreading is not perfect. If it were, no mistakes such as mutations would occur, no variation in gene sequence would result, and evolution would come to a halt, for as we discussed in chapter 11, genetic variation that alters the phenotype is the raw material on which natural selection acts and evolution occurs. We will return to the subject of mutations and evolution in chapter 15.

12.4   The basis for the great accuracy of DNA replication is complementarity. DNA's two strands are complementary mirror images of each other, so either one can be used as a template to reconstruct the other.

DNA Replication

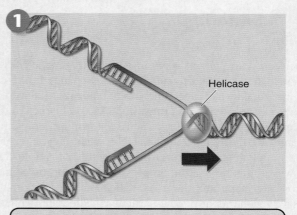

Helicase unwinds the DNA double helix for about 1,000 nucleotides.

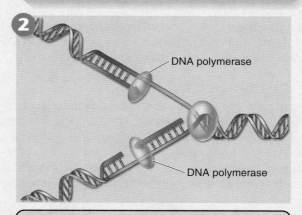

DNA polymerase assembles a complementary new strand on each old one, building the two strands in opposite directions.

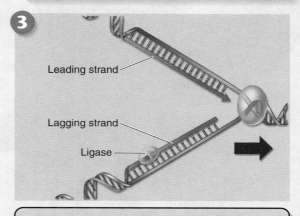

DNA ligase attaches one new strand to the previously replicated segment on the lagging strand, and helicase unwinds another segment.

Figure 12.9   How DNA replication works.

12.5 Mutation

There are two general ways in which the genetic message is altered: mutation and recombination. A change in the content of the genetic message—the base sequence of one or more genes—is referred to as a **mutation.** As you learned in the previous section, DNA copies itself by forming complementary strands along single strands of DNA when they are separated. The template strand directs the formation of the new strand. However, this replication process is not foolproof. Sometimes errors are made and these are called mutations. Some mutations alter the identity of a particular nucleotide, while others remove or add nucleotides to a gene. A change in the position of a portion of the genetic message is referred to as **recombination.** Some recombination events move a gene to a different chromosome; others alter the location of only part of a gene. The cells of eukaryotes contain an enormous amount of DNA, and the mechanisms that protect and proofread the DNA are not perfect. If they were, no variation would be generated.

Mistakes Happen

In fact, cells do make mistakes during replication, often causing a change in a cell's genetic message, or a mutation (figure 12.10). However, mutations are rare. Typically, a particular gene is altered in only one of a million gametes. If changes were common, the genetic instructions encoded in DNA would soon degrade into meaningless gibberish. Limited as it might seem,

Figure 12.10 Mutation.
Fruit flies normally have one pair of wings, extending from the thorax. This fly is a *bithorax* mutant. Because of a mutation in a gene regulating a critical stage of development, it possesses two thorax segments and thus two sets of wings.

the steady trickle of change that does occur is the very stuff of evolution. Every difference in the genetic messages that specify different organisms arose as the result of genetic change.

Kinds of Mutation

The message that DNA carries in its genes is the "instructions" of how to make proteins. As discussed in more detail in chapter 13, the sequence of nucleotides in a strand of DNA translates into the sequence of amino acids that makes up a protein. If the core message in the DNA is altered through mutation, as shown by the substitution of the adenine (in red) for the cytosine in figure 12.11, then the protein product can also be altered, sometimes to the point where it can no longer function properly. Because mutations can occur randomly in a cell's DNA, most mutations are detrimental, just as making a random change in a computer program usually worsens performance. The consequences of a detrimental mutation may be minor or catastrophic, depending on the function of the altered gene.

Mutations in Germ-Line Tissues. The effect of a mutation depends critically on the identity of the cell in which the mutation occurs. During the embryonic development of all multicellular organisms, there comes a point when cells destined to form gametes (germ-line cells) are segregated from those that will form the other cells of the body (somatic cells). Only when a mutation occurs within a germ-line cell is it passed to subsequent generations as part of the hereditary endowment of the gametes derived from that cell. Mutations in germ-line tissue are of enormous biological importance because they provide the raw material from which natural selection produces evolutionary change.

Mutations in Somatic Tissues. Change can occur only if there are new, different allele combinations available to replace the old. Mutation produces new alleles, and recombination puts the alleles together in different combinations. In animals, it is the occurrence of these two processes in germ-line tissue that is important to evolution, because mutations in somatic cells (somatic mutations) are not passed from one generation to the next. However, a somatic mutation may have drastic effects on the individual organism in which it occurs, because it is passed on to all of the cells that are descended from the original mutant cell. Thus, if a mutant lung cell divides, all cells derived from it will carry the mutation. Somatic mutations of lung cells are, as we shall see, the principal cause of lung cancer in humans.

Altering the Sequence of DNA. One category of mutational changes affects the message itself, producing alterations in the sequence of DNA nucleotides (table 12.1). If alterations involve only one or a few base pairs in the coding sequence, they are called **point mutations.** Sometimes the identity of a base changes (**base substitution**), while other times one or a few bases are added (**insertion**) or lost (**deletion**). If an insertion or deletion throws the reading of the gene message out of register, a **frame-shift mutation** results. Figure 12.11 shows a base substitution mutation that results in the change of an amino acid, from proline to threonine. This could be

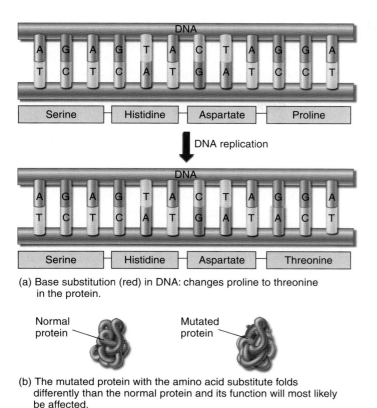

(a) Base substitution (red) in DNA: changes proline to threonine in the protein.

(b) The mutated protein with the amino acid substitute folds differently than the normal protein and its function will most likely be affected.

Figure 12.11   Base substitution mutation.

(a) Some changes in a DNA sequence can result in a change in a single amino acid. (b) This results in a mutated protein that may not function the same as the normal protein.

a minor change or catastrophic. However, suppose that this had been the deletion of a nucleotide, that the cytosine base nucleotide had been skipped during replication. This would cause a frame-shift, resulting in the rest of the polypeptide being altered after that point. Frame-shift mutations are extremely detrimental. While some point mutations arise due to spontaneous pairing errors that occur during DNA replication, others result from damage to the DNA caused by **mutagens,** usually radiation or chemicals. The latter class of mutations is of particular importance because modern industrial societies often release many chemical mutagens into the environment.

Changes in Gene Position.   Another category of mutations affects the way the genetic message is organized. In both prokaryotes and eukaryotes, individual genes may move from one place in the genome to another by **transposition** (see also page 232). When a particular gene moves to a different location, its expression or the expression of neighboring genes may be altered. In addition, large segments of chromosomes in eukaryotes may change their relative locations or undergo duplication. Such **chromosomal rearrangements** often have drastic effects on the expression of the genetic message.

The Importance of Genetic Change

All evolution begins with alterations in the genetic message: mutation creates new alleles, gene transfer and transposition alter gene location, reciprocal recombination shuffles and sorts these changes, and chromosomal rearrangement alters the organization

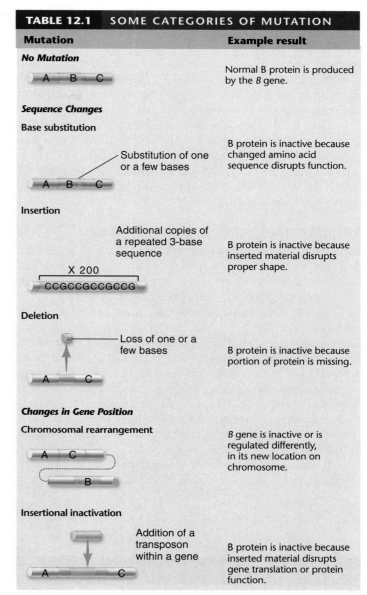

TABLE 12.1	SOME CATEGORIES OF MUTATION
Mutation	**Example result**
No Mutation	Normal B protein is produced by the *B* gene.
Sequence Changes	
Base substitution — Substitution of one or a few bases	B protein is inactive because changed amino acid sequence disrupts function.
Insertion — Additional copies of a repeated 3-base sequence	B protein is inactive because inserted material disrupts proper shape.
Deletion — Loss of one or a few bases	B protein is inactive because portion of protein is missing.
Changes in Gene Position	
Chromosomal rearrangement	*B* gene is inactive or is regulated differently, in its new location on chromosome.
Insertional inactivation — Addition of a transposon within a gene	B protein is inactive because inserted material disrupts gene translation or protein function.

of entire chromosomes. Some changes in germ-line tissue produce alterations that enable an organism to leave more offspring, and those changes tend to be preserved as the genetic endowment of future generations. An example of such a change was discussed in chapter 11, under "Environmental Effects." A genetic change occurred in the arctic fox that resulted in the production of fur pigments in warm weather but not in cold temperatures. This genetic change allowed the fox to be less conspicuous to predators, aiding its survival. This increased their ability to survive until the mating season, enabling them to leave more offspring, some of which inherited this same trait. Other changes reduce the ability of an organism to leave offspring. Those changes tend to be lost, as the organisms that carry them contribute fewer members to future generations. For example, a mutation that may cause an animal such as a zebra to lack speed will result in an early death at the jaws of a cheetah. However, the hunter is also affected by genetic mutations. A similar mutation in the cheetah will slow the animal down and result in an early death through starvation.

Evolution can be viewed as the selection of particular combinations of alleles from a pool of alternatives. The rate of evolution is ultimately limited by the rate at which these alternatives are generated. Genetic change through mutation and recombination provides the raw material for evolution.

Genetic changes in somatic cells do not pass on to off-spring, and so they have less evolutionary consequence than germ-line changes. However, changes in the genes of somatic cells can have an important and immediate impact, particularly if the gene affects development or is involved with regulation of cell proliferation.

Mutation, Smoking, and Lung Cancer

The association of particular chemicals with cancer, particularly chemicals that are potent mutagens (see chapters 9 and 26), led researchers early on to suspect that cancer might be caused, at least in part, by the action of chemicals on the body.

The hypothesis that chemicals cause cancer was first advanced over 200 years ago in 1761 by Dr. John Hill, an English physician. Hill noted unusual tumors of the nose in heavy snuff users and suggested tobacco had produced these cancers. In 1775, a London surgeon, Sir Percivall Pott, made a similar observation, noting that men who had been chimney sweeps exhibited frequent cancer of the scrotum. He suggested that soot and tars might be responsible. These and many other observations led to the hypothesis that cancer results from the action of chemicals on the body.

It was over a century before this hypothesis was directly tested. In 1915, Japanese doctor Katsusaburo Yamagiwa ap-plied extracts of coal tar to the skin of 137 rabbits every two or three days for three months. Then he waited to see what would happen. After a year, cancers appeared at the site of application in seven of the rabbits. Yamagiwa had induced cancer with the coal tar, the first direct demonstration of chemical carcinogen-esis. In the decades that followed, this approach demonstrated that many chemicals were capable of causing cancer.

These were lab studies, and many did not accept that they applied to real people. Do tars in fact induce cancer in humans? In 1949, the American physician Ernst Winder and the British epidemiologist Richard Doll independently reported that lung cancer showed a strong link to the smoking of cigarettes, which introduces tars into the lungs. Winder interviewed 684 lung cancer patients and 600 normal controls, asking whether each had ever smoked. Cancer rates were 40 times higher in heavy smokers than in nonsmokers. From these studies, it seemed likely as long as 50 years ago that tars and other chemicals in cigarette smoke induce cancer in the lungs of persistent smokers. While this suggestion was resisted by the tobacco industry, the evidence that has accumulated since these pioneering studies makes a clear case, and there is no longer any real doubt. Chemicals in cigarette smoke cause cancer.

> **12.5** Rare changes in genes, called mutations, can have significant effects on the individual when they occur in somatic tissue, but they are inherited only if they occur in germ-line tissue. Inherited changes provide the raw material for evolution. Chemicals that produce mutations in DNA, such as tars in cigarette smoke, are often potent carcinogens.

INQUIRY & ANALYSIS

Can We Observe DNA Replicate?

The photographs on the right are the original data of the Meselson and Stahl experiment illustrated in figure 12.6.

1. **Applying Concepts** What are the dark bands in these photographs? In what way, if any, do the three solutions of cesium chloride in which the bands are seen differ?

2. **Making Inferences** How do you think the researchers were able to vertically align the three photos properly? [Hint: What is the independent variable?]

3. **Drawing Conclusions** Is the location of the bands consistent with the hypothesis of semiconservative replication? Explain. What pattern would you have expected if replication had been conservative?

Migration of DNA in a Density Gradient

Photographs	Scans	Rounds of replication
		0
		1
		2

—Density→ —Density→

Genes Are Made of DNA

12.1 The Griffith Experiment

- Using *Streptococcus pneumoniae* bacteria, Griffith showed that information that controls physical characteristics can be passed from one bacterium to another, even from a dead bacterium.

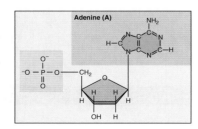

- By injecting mice with different strains of *S. pneumoniae,* Griffith determined that some strains were pathogenic, resulting in the mice's death. Bacteria of the pathogenic strains contained polysaccharide capsules (S strain), while those without the coats (R strain) were nonlethal. When he mixed dead pathogenic bacteria (S), which usually would not cause death, and live nonpathogenic bacteria (R) and injected them into mice, the mice died. The dead mice contained living S strains.

- Something passed from the dead lethal bacteria to the live nonlethal bacteria, causing them to turn deadly. Griffith did not determine whether this "transforming principle" was protein or DNA (**figure 12.1**).

12.2 The Avery and Hershey-Chase Experiments

- Avery showed that protein was not the source of this transformation. He replicated Griffith's experiment but removed all protein from the preparation. The virulent strain with its protein coat removed was still able to transform nonvirulent bacteria. This experimental result supported the hypothesis that DNA, not protein, was the transforming principle.

- Using bacterial viruses, Hershey and Chase showed that genes were carried on DNA and not proteins.

- They used two different preparations, radioactively tagging DNA in one and protein in the other. Each preparation was used to infect bacteria. When they screened the two bacterial cultures, they discovered that the infected bacteria contained radioactively tagged DNA, not protein (**figure 12.2**).

12.3 Discovering the Structure of DNA

- Once it was clear that DNA was the hereditary molecule, scientists tried to figure out its structure. The basic chemical components of DNA were determined to be nucleotides, a deoxyribose sugar attached to a phosphate group and one of four

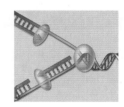

Adenine (A)

organic bases (**figure 12.3**). Chargaff observed that two sets of bases are always present in equal amounts (A equals T and C equals G).

- Watson and Crick determined that DNA is a double helix, in which A on one strand pairs with T on the other, and G similarly pairs with C. They used research by Chargaff and Franklin to come to their conclusions (**figure 12.4**).

DNA Replication

12.4 How the DNA Molecule Copies Itself

- The structure of DNA implies a method of replication, that a single strand of DNA can serve as the template for production of another strand, but there are several ways in which this could occur.

- Meselson and Stahl showed that DNA replicates semiconservatively, using each of the original strands as templates to form new strands (**figures 12.5** and **12.6**). In semiconservative replication, each new strand of DNA consists of a template strand from the parent DNA and a newly synthesized strand that is complementary to the template strand.

- Each DNA strand is copied by the actions of an enzyme called DNA polymerase.

- The two strands of the DNA molecule separate, kind of like unzipping, exposing the two single strands of DNA. DNA polymerase adds nucleotides to the new DNA strands that are complementary to the original single strand, adding on to a primer. Nucleotides are added to the growing strand in a 5′ to 3′ direction (**figure 12.7**).

- Because nucleotides can only be added onto the 3′ end of the strand, DNA copies in a continuous manner on the leading strand and in a discontinuous manner on the lagging strand. DNA segments are linked together with another enzyme called DNA ligase (**figures 12.8** and **12.9**).

- Errors can occur during the replication of DNA, which could be damaging to the cell. The cell has many mechanisms to correct damage to the DNA or mistakes made during replication. This proofreading process compares one strand against its complementary strand and corrects errors, but this system is not foolproof.

Altering the Genetic Message

12.5 Mutation

- A mutation is a change in the hereditary message (**figure 12.10**). Mutations that are passed on to offspring are the raw material on which natural selection acts to produce evolutionary changes.

- Mutations that change one or only a few nucleotides are called point mutations (**figure 12.11**). They may arise as the result of errors in pairing during DNA replication, UV radiation, or chemical mutagens. Some mutations occur through the movement of sections of DNA from one place to another, a process called transposition (**table 12.1**).

1. As a result of the experiments performed by Frederick Griffith, we found that
   a. hereditary information within a cell cannot be changed.
   b. hereditary information can be added to cells from other cells.
   c. if hereditary information is added to a cell, it will kill the organism.
   d. hereditary information in the form of proteins can be added to cells.

2. The experiment performed by Alfred Hershey and Martha Chase showed that the molecule viruses use to specify new viruses is
   a. a protein.   c. ATP.
   b. a carbohydrate.   d. DNA.

3. Erwin Chargaff, Rosalind Franklin, Francis Crick, and James Watson all worked on pieces of information relating to the
   a. structure of DNA.
   b. function of DNA.
   c. inheritance of DNA.
   d. mutations of DNA.

4. All four DNA nucleotides differ in
   a. their sizes.
   b. the number of hydrogen bonds they can form with their base pair.
   c. the type of nitrogen base.
   d. the type of sugar.

5. Which of the following lists the purine nucleotides?
   a. adenine and cytosine
   b. guanine and thymine
   c. cytosine and thymine
   d. adenine and guanine

6. If one strand of a DNA molecule has the base sequence ATTGCAT, its complementary strand will have the sequence
   a. ATTGCAT.   c. GCCATGC.
   b. TAACGTA.   d. CGGTACG.

7. Regarding the duplication of DNA, we now know that each double helix
   a. serves as a template to produce an identical double helix next to it.
   b. splits down the middle into two single helices, and each one then acts as a template to build its complement.
   c. fragments into small chunks that duplicate and reassemble.
   d. All of these are true for different types of DNA.

8. DNA polymerase can only add nucleotides to an existing chain, so _____ is required.
   a. a primer
   b. helicase
   c. a lagging strand
   d. a leading strand

9. Genetic messages can be altered in two ways:
   a. deliberately and accidentally.
   b. through the chromosome or through the protein.
   c. by mutation or by recombination.
   d. by activation or by repression.

10. Mutations can occur in
    a. germ-line tissues and be passed on to future generations.
    b. somatic tissues and be passed on to future generations.
    c. germ-line tissues and cause diseases such as lung cancer.
    d. somatic tissues and cause diseases such as cystic fibrosis.

1. **Figure 12.4c** What are some of the possible problems you see if the cytosine indicated with the red arrow is accidentally replaced with an adenine?

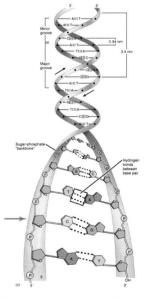

2. **Table 12.1** Your friend Gorinda wants to know if there are ever mutations that don't cause problems. What do you tell him?

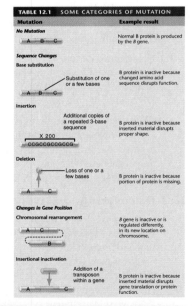

1. Based on the experiments and discoveries discussed in chapter 11 and here in chapter 12, defend the statement, attributed to Sir Isaac Newton in 1676 (though some say that Bernard of Chartres said it first, way back in about 1130!) that scientists build new ideas in science by "standing on the shoulders of giants."

2. Mutations can occur in somatic (body) cells or in germ-line cells. What problems or opportunities does each kind of cell face?

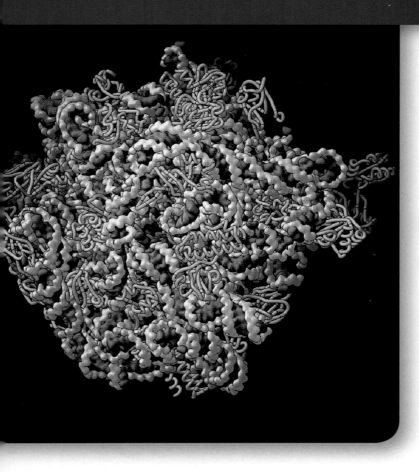

13

How Genes Work

Ribosomes, like the one you see here, are very complex cellular machines that assemble proteins, using information that has been copied from genes onto RNA molecules. The ribosomes read the gene information copied onto these messenger RNA transcripts, and use it to determine the amino acid sequence of the new protein that the ribosome is assembling. Each ribosome is made of over 50 different proteins (shown here in gold), as well as three chains of RNA composed of some 3,000 nucleotides (shown here in gray). It has been traditionally assumed that the proteins in a ribosome act as enzymes to catalyze the amino acid assembly process, with the RNA acting as a scaffold to position the proteins. In the year 2000 we learned the reverse to be true. Powerful X-ray diffraction studies revealed the complete detailed structure of a ribosome at atomic resolution. Unexpectedly, the many proteins of a ribosome are scattered over its surface like decorations on a Christmas tree. The role of these proteins seems to be to stabilize the many bends and twists of the RNA chains, the proteins acting like spot-welds between the RNA strands they touch. Importantly, there are no proteins on the inside of the ribosome where the chemistry of protein synthesis takes place—just twists of RNA. Thus, it is the ribosome's RNA, not its proteins, that catalyzes the joining together of amino acids! Clearly, our knowledge of how genes work is still increasing, often adjusting what seem to be fundamental concepts.

13.1 Transcription

The discovery that genes are made of DNA left unanswered the question of how the information in DNA is used. How does a string of nucleotides in a spiral molecule determine if you have red hair? We now know that the information in DNA is arrayed in little blocks, like entries in a dictionary, each block a gene specifying a protein. These proteins determine what a particular cell will be like.

Just as an architect protects building plans from loss or damage by keeping them safe in a central place and issuing only blueprint copies to on-site workers, so your cells protect their DNA instructions by keeping them safe within a central DNA storage area, the nucleus. The DNA never leaves the nucleus. Instead, "blueprint" copies of particular genes within the DNA instructions are sent out into the cell to direct the assembly of proteins. These working copies of genes are made of ribonucleic acid (RNA) rather than DNA. Recall that RNA is the same as DNA except that the sugars in RNA have an extra oxygen atom and T is replaced by a similar pyrimidine base called uracil, U (see figure 4.10). The path of information is thus: **DNA ⟶ RNA ⟶ protein.**

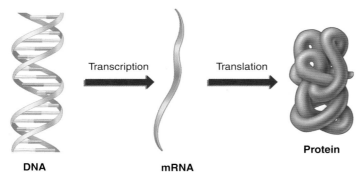

Transcription   Translation

DNA   **mRNA**   **Protein**

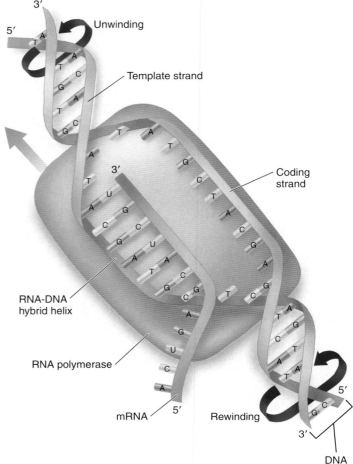

Figure 13.1 Transcription.
One of the strands of DNA functions as a template on which nucleotide building blocks are assembled into mRNA by RNA polymerase as it moves along the DNA strand.

This information path is often called the *central dogma,* because it describes the key organization used by your cells to express their genes. A cell uses three kinds of RNA in the synthesis of proteins: messenger RNA (mRNA), ribosomal RNA (rRNA), and transfer RNA (tRNA).

The use of information in DNA to direct the production of particular proteins is called **gene expression.** Gene expression occurs in two stages: in the first stage, called **transcription,** an mRNA molecule is synthesized from a gene within the DNA; in the second stage, called **translation,** this mRNA is used to direct the production of a protein.

The Transcription Process

The RNA copy of a gene used in the cell to produce a protein is called **messenger RNA (mRNA)**—it is the messenger that conveys the information from the nucleus to the cytoplasm. The copying process that makes the mRNA is called transcription—just as monks in monasteries used to make copies of manuscripts by faithfully transcribing each letter, so enzymes within the nuclei of your cells make mRNA copies of your genes by faithfully complementing each nucleotide.

In your cells, the transcriber is a large and very sophisticated protein called **RNA polymerase.** It binds to one strand of a DNA double helix at a particular site called a *promoter* and then moves along the DNA strand like a train on a track. The pink structure in figure 13.1 is the RNA polymerase and the DNA "track" guides it along. As it goes along the DNA, the polymerase pairs each nucleotide with its complementary RNA version (A with U, G with C), building an mRNA chain behind it as it moves along the DNA strand. Although shown here in a simplified way, remember that the DNA and RNA nucleotides forming hydrogen bonds between complementary base pairs actually have different chemical shapes, which is why they pair the way they do.

> **13.1 Transcription is the production of an mRNA copy of a gene by the enzyme RNA polymerase.**

13.2 Translation

The Genetic Code

The essence of Mendelian genetics is that information determining hereditary traits, traits passed from parent to child, is encoded information. The information is written within the chromosomes in blocks called genes. Genes affect Mendelian traits by directing the production of particular proteins. The essence of gene expression, of using your genes, is reading the information encoded within DNA and using that information to direct the production of the correct protein.

To correctly read a gene, a cell must translate the information encoded in DNA into the language of proteins—that is, it must convert the *order* of the gene's nucleotides into the order of amino acids in a protein, a process called **translation.** The rules that govern this translation are called the **genetic code.**

The mRNA is transcribed from the gene in a linear sequence, one nucleotide following another, beginning at the promoter. There, the RNA polymerase binds to the DNA and begins its assembly of the mRNA. Transcription ends when the RNA polymerase reaches a certain nucleotide sequence that signals it to stop.

However, the mRNA is not translated in this same way. The mRNA is "read" by a ribosome in three-nucleotide units. Each three-nucleotide sequence of the mRNA corresponds to a particular amino acid and is called a **codon.** Biologists worked out which codons correspond to which amino acids by trial-and-error experiments carried out in test tubes. In these experiments, investigators used artificial mRNAs to direct the synthesis of proteins in the tube, and then looked to see the sequence of amino acids in the newly formed proteins. An mRNA that was a string of UUUUUU . . . , for example, produced a protein that was a string of phenylalanine (Phe) amino acids, telling investigators that the codon UUU corresponded to the amino acid Phe. The entire genetic code dictionary is presented in figure 13.2. The first letters of the codon are positioned down the left side, the second across the top, and the third down the right side. To determine the amino acid encoded by a codon, say AGC, go to "A" on the left, follow the row over to the "G" column, and go down to the C on the right. As you discover, AGC encodes the amino acid serine. Because at each position of a three-letter codon, any of the four different nucleotides (U, C, A, G) may be used, there are 64 different possible three-letter codons ($4 \times 4 \times 4 = 64$) in the genetic code.

The genetic code is universal, the same in practically all organisms. GUC codes for valine in bacteria, in fruit flies, in eagles, and in your own cells. The only exception biologists have ever found to this rule is in the way in which cell organelles that contain DNA (mitochondria and chloroplasts) and a few microscopic protists read the "stop" codons. In every other instance, the same genetic code is employed by all living things.

The Genetic Code

First Letter	Second Letter								Third Letter
	U		**C**		**A**		**G**		
U	UUU	Phenylalanine	UCU	Serine	UAU	Tyrosine	UGU	Cysteine	U
	UUC		UCC		UAC		UGC		C
	UUA	Leucine	UCA		UAA	Stop	UGA	Stop	A
	UUG		UCG		UAG	Stop	UGG	Tryptophan	G
C	CUU	Leucine	CCU	Proline	CAU	Histidine	CGU	Arginine	U
	CUC		CCC		CAC		CGC		C
	CUA		CCA		CAA	Glutamine	CGA		A
	CUG		CCG		CAG		CGG		G
A	AUU	Isoleucine	ACU	Threonine	AAU	Asparagine	AGU	Serine	U
	AUC		ACC		AAC		AGC		C
	AUA		ACA		AAA	Lysine	AGA	Arginine	A
	AUG	Methionine; Start	ACG		AAG		AGG		G
G	GUU	Valine	GCU	Alanine	GAU	Aspartate	GGU	Glycine	U
	GUC		GCC		GAC		GGC		C
	GUA		GCA		GAA	Glutamate	GGA		A
	GUG		GCG		GAG		GGG		G

Figure 13.2   The genetic code (RNA codons).

A codon consists of three nucleotides read in sequence. For example, ACU codes for threonine. The first letter, A, is in the First Letter column; the second letter, C, is in the Second Letter column; and the third letter, U, is in the Third Letter column. Each of the mRNA codons is recognized by a corresponding anticodon sequence on a tRNA molecule. Most amino acids are specified by more than one codon. For example, threonine is specified by four codons, which differ only in the third nucleotide (ACU, ACC, ACA, and ACG).

Translating the RNA Message into Proteins

The final result of the transcription process is the production of an mRNA copy of a gene. Like a photocopy, the mRNA can be used without damage or wear and tear on the original. After transcription of a gene is finished, the mRNA passes out of the nucleus into the cytoplasm through pores in the nuclear membrane. There, translation of the genetic message occurs. In translation, organelles called **ribosomes** use the mRNA produced by transcription to direct the synthesis of a protein following the genetic code.

The Protein-Making Factory. Ribosomes are the protein-making factories of the cell. Each is very complex, containing over 50 different proteins and several segments of **ribosomal RNA (rRNA).** Ribosomes use mRNA, the "blueprint" copies of nuclear genes, to direct the assembly of a protein.

Ribosomes are composed of two pieces, or subunits, one nested into the other like a fist in the palm of your hand. The "fist" is the smaller of the two subunits, the pink structure in figure 13.3. Its rRNA has a short nucleotide sequence exposed on the surface of the subunit. This exposed sequence is identical to a sequence called the leader region that occurs at the beginning of all genes. Because of this, an mRNA molecule binds to the exposed rRNA of the small subunit like a fly sticking to flypaper.

The Key Role of tRNA. Directly adjacent to the exposed rRNA sequence are three small pockets or dents, called the A, P, and E sites, in the surface of the ribosome (shown in figure 13.3 and discussed shortly). These sites have just the right shape to bind yet a third kind of RNA molecule, **transfer RNA (tRNA).** It is tRNA molecules that bring amino acids to the ribosome used in making proteins. tRNA molecules are chains about 80 nucleotides long. The string of nucleotides folds back on itself, forming a 3-looped structure shown in figure 13.4a. The looped structure further folds into a compact shape shown in figure 13.4b, with a three-nucleotide sequence

at one end (the pink loop) and an amino acid attachment site on the other end (the 3′ end).

The three-nucleotide sequence, called the **anticodon,** is very important: it is the complementary sequence to 1 of the 64 codons of the genetic code! Special enzymes, called *activating enzymes,* match amino acids with their proper tRNAs, with the anticodon determining which amino acid will attach to a particular tRNA.

Because the first dent in the ribosome, called the A site (the attachment site where amino-acid-bearing tRNAs will bind) is directly adjacent to where the mRNA binds to the rRNA, three nucleotides of the mRNA are positioned directly facing the anticodon of the tRNA. Like the address on a letter, the anticodon ensures that an amino acid is delivered to its correct "address" on the mRNA where the ribosome is assembling the protein.

Making the Protein. Once an mRNA molecule has bound to the small ribosomal subunit, the other larger ribosomal subunit binds as well, forming a complete ribosome. The ribosome then begins the process of translation, illustrated in the panels of figure 13.5. Panel 1 shows how the mRNA begins to thread through the ribosome like a string passing through the hole in a doughnut. The mRNA passes through in short spurts, three nucleotides at a time, and at each burst of movement a new three-nucleotide codon on the mRNA is positioned opposite the A site in the ribosome, where a tRNA molecule first binds, as shown in panel 2.

As each new tRNA brings in an amino acid to each new codon presented at the A site, the old tRNA paired with the previous codon is passed over to the P site where peptide bonds form between the incoming amino acid and the growing peptide chain. The tRNA in the P site eventually shifts to the E site (the exit site), as shown in panel 3, and the amino acid it carried is attached to the end of a growing amino acid chain. The tRNA is then released in panel

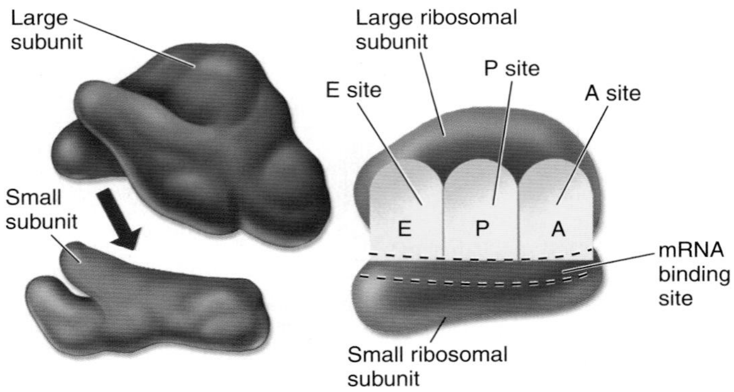

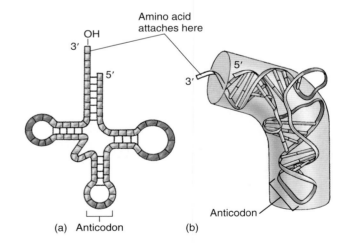

Figure 13.4   The structure of tRNA.

tRNA, like mRNA, is a long strand of nucleotides. However, unlike mRNA, hydrogen bonding occurs between its nucleotides, causing the strand to form hairpin loops, as seen in (a). The loops then fold up on each other to create the compact, three-dimensional shape seen in (b). Amino acids attach to the free, single-stranded —OH end of a tRNA molecule. A three-nucleotide sequence called the anticodon in the lower loop of tRNA interacts with a complementary codon on the mRNA.

Figure 13.3   A ribosome is composed of two subunits.

The smaller subunit fits into a depression on the surface of the larger one. The A, P, and E sites on the ribosome play key roles in protein synthesis.

Translation

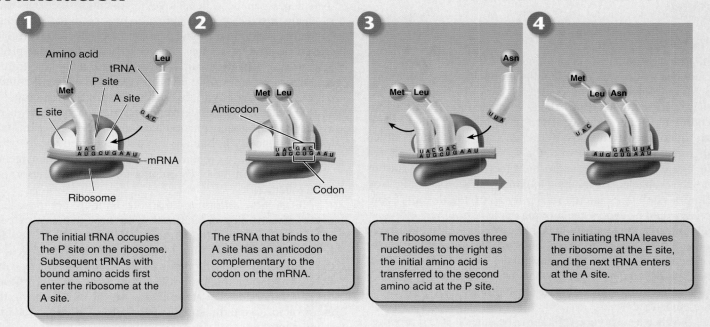

1 The initial tRNA occupies the P site on the ribosome. Subsequent tRNAs with bound amino acids first enter the ribosome at the A site.

2 The tRNA that binds to the A site has an anticodon complementary to the codon on the mRNA.

3 The ribosome moves three nucleotides to the right as the initial amino acid is transferred to the second amino acid at the P site.

4 The initiating tRNA leaves the ribosome at the E site, and the next tRNA enters at the A site.

Figure 13.5 How translation works.
The mRNA strand acts as a template for tRNA molecules. The appropriate tRNA is selected and positioned by the ribosome, which moves along the mRNA in three-nucleotide steps. tRNAs bring amino acids into the ribosome at the A site. A peptide bond is formed between the incoming amino acid and the growing polypeptide chain at the P site, and the empty tRNAs leave the ribosome at the E site.

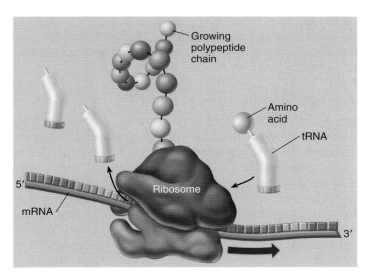

Figure 13.6 Ribosomes guide the translation process.
tRNA binds to an amino acid as determined by the anticodon sequence. Ribosomes bind the loaded tRNAs to their complementary sequences on the strand of mRNA. tRNA adds its amino acid to the growing polypeptide chain, which is released as the completed protein.

4. So as the ribosome proceeds down the mRNA, one tRNA after another is selected to match the sequence of mRNA codons. In figure 13.6 you can see the ribosome traveling along the length of the mRNA, the tRNAs bringing the amino acids into the ribosome and the growing polypeptide chain extending out from the ribosome. Translation continues until a "stop" codon is encountered, which signals the end of the protein. The ribosome complex falls apart, and the newly made protein is released into the cell.

As explained earlier, the overall flow of genetic information, the so-called "central dogma," is from DNA to mRNA to protein. For example, the protein that is being formed in figure 13.5 began with the DNA nucleotide sequence TAC-GACTTA, which is first transcribed into the mRNA sequence AUGCUGAAU. This sequence is then translated by the tRNAs into a peptide composed of the amino acids methionine—leucine—asparagine.

13.2 The genetic code dictates how a particular nucleotide sequence specifies a particular amino acid sequence. A gene is transcribed into mRNA, which is then translated into a protein. The sequence of mRNA codons dictates the corresponding sequence of amino acids in a growing protein chain.

13.3 Architecture of the Gene

Introns

In prokaryotes, a gene is an uninterrupted stretch of DNA nucleotides that is read three at a time to make a chain of amino acids. In eukaryotes, by contrast, genes are fragmented. In these more complex genes, the DNA nucleotide sequences encoding the amino acid sequence of the protein (called **exons**) are interrupted frequently by extraneous "extra stuff" called **introns.** You can see them in the segment of DNA illustrated in figure 13.7; the exons are the blue areas and the introns are the orange areas. Imagine looking at an interstate highway from a satellite. Scattered randomly along the thread of concrete would be cars, some moving in clusters, others individually; most of the road would be bare. That is what a eukaryotic gene is like: scattered exons embedded within much longer sequences of introns. In humans, only 1% to 1.5% of the genome is devoted to the exons that encode proteins, while 24% is devoted to the noncoding introns within which these exons are embedded.

When a eukaryotic cell transcribes a gene, it first produces a **primary RNA transcript** of the entire gene, shown in figure 13.7 with the exons in green and the introns in orange. Enzymes add modifications called a *5′ cap* and a *3′ poly-A tail,* which protect the RNA transcript from degradation. The primary transcript is then processed. Enzyme-RNA complexes excise out the introns and join together the exons to form the shorter mature mRNA transcript that is actually translated into protein. Notice that the mature mRNA transcript in figure 13.7 contains only exons (green segments), no introns. Because introns are excised from the RNA transcript before it is translated into protein, they do not affect the structure of the protein encoded by the gene in which they occur, despite the fact that introns represent over 90% of the nucleotide sequence of a typical human gene.

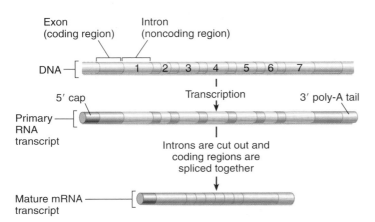

Figure 13.7 Processing eukaryotic RNA.
The gene shown here codes for a protein called ovalbumin. The ovalbumin gene and its primary transcript contain seven segments not present in the mRNA used by the ribosomes to direct the synthesis of the protein.

Why this crazy organization? It appears that many human genes can be spliced together in more than one way. In many instances, exons are not just random fragments, but rather functional modules. One exon encodes a straight stretch of protein, another a curve, yet another a flat place. Like mixing Tinkertoy parts, you can construct quite different assemblies by employing the same exons in different combinations and orders. With this sort of **alternative splicing,** the 25,000 genes of the human genome seem to encode as many as 120,000 different expressed messenger RNAs. It seems that added complexity in humans has been achieved not by gaining more gene parts (we have only about twice as many genes as a fruit fly), but rather by coming up with new ways to put them together.

Protein synthesis in eukaryotes is more complex than in prokaryotes. Figure 13.8 walks you through the entire process. Transcription (step ❶) and RNA processing (step ❷) occurs within the nucleus. In step ❸, the mRNA travels to the cytoplasm where it binds to the ribosome. In step ❹, tRNAs bind to their appropriate amino acids, that correspond to their anticodons. In steps ❺ and ❻, the tRNAs bring the amino acids to the ribosome and the mRNA is translated into a polypeptide.

Gene Families

Eukaryotic genes have other unique characteristics. For example, everything we have said in this chapter assumes that a chromosome carries one copy of a gene—one to make the enzyme that breaks down lactose, for instance, and one to encode the protein hemoglobin, which carries oxygen in your blood from lungs to tissues. However, most eukaryotic genes exist in multiple copies, clusters of almost identical sequences called **multigene families.** Multigene families may contain as few as three or as many as several hundred versions of a gene.

Transposons: Jumping Genes

Other DNA sequences are very unusual in that they are repeated hundreds of thousands of times, scattered randomly about on the chromosomes. These segments of DNA are called **transposable sequences** or **transposons.** Transposons have the remarkable ability to move about from one chromosomal location to another. Once every few thousand cell divisions, a transposon will copy itself and simply pick up and move elsewhere, jumping at random to a new location on the chromosome, taking the gene with it. Transposons appear to be molecular parasites. Fully 45% of the human genome is composed of transposable sequences.

> **13.3** The coding portions of most eukaryotic genes are embedded as exons within long sequences of noncoding introns. Many eukaryotic genes exist in multiple copies, some of which appear to have moved from one chromosomal location to another.

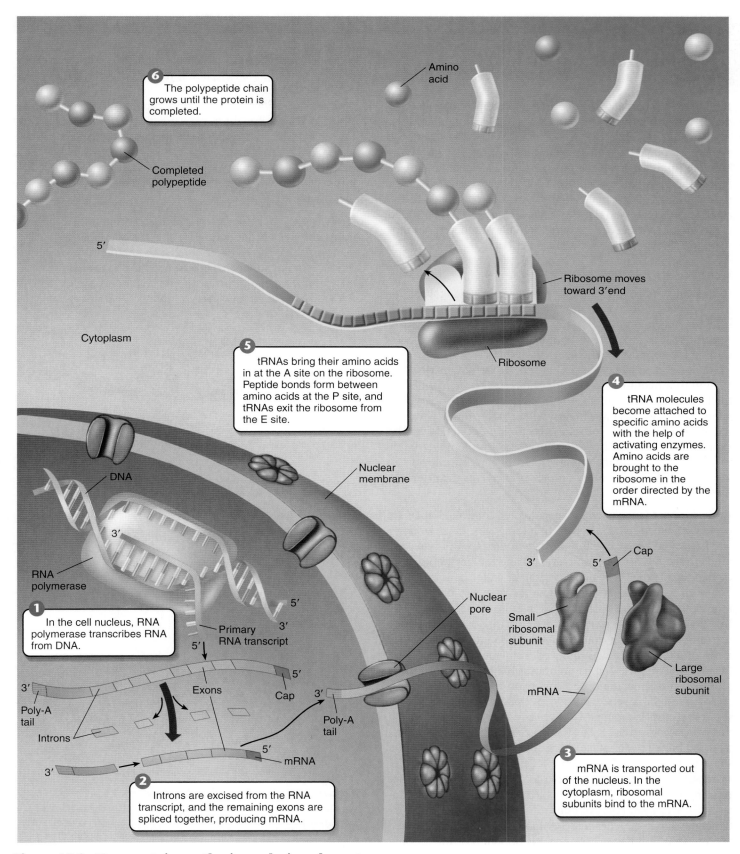

Figure 13.8 How protein synthesis works in eukaryotes.

13.4 Turning Genes Off and On

Being able to translate a gene into a protein is only part of gene expression. Every cell must also be able to regulate when particular genes are used. Imagine if every instrument in a symphony played at full volume all the time, all the horns blowing full blast and each drum beating as fast and loudly as it could! No symphony plays that way, because music is more than noise—it is the controlled expression of sound. In the same way, growth and development are due to the controlled expression of genes, each brought into play at the proper moment to achieve precise and delicate effects.

Controlling Transcription

Cells control the expression of their genes by saying *when* individual genes are to be transcribed. At the beginning of each gene are special regulatory sites that act as points of control. Specific regulatory proteins within the cell bind to these sites, turning transcription of the gene off or on.

For a gene to be transcribed, the RNA polymerase has to bind to a **promoter,** a specific sequence of nucleotides on the DNA that signals the beginning of a gene. In prokaryotes, gene expression is controlled by either blocking or allowing the RNA polymerase access to the promoter. Genes can be turned off by the binding of a **repressor,** a protein that binds to the DNA blocking the promoter. Genes can be turned on by the binding of an **activator,** a protein that makes the promoter more accessible to the RNA polymerase.

Repressors. Many genes are "negatively" controlled: they are turned off except when needed. In these genes, the regulatory site is located between the place where the RNA polymerase binds to the DNA (the promoter site) and the beginning edge of the gene. When a regulatory protein called a repressor is bound to its regulatory site, called the *operator,* its presence blocks the movement of the polymerase toward the gene. Imagine if you went to sit down to eat dinner and someone was already sitting in your chair—you could not begin your meal until this person was removed from your chair, any more than the polymerase can begin transcribing the gene until the repressor protein is removed from its promotor site.

To turn on a gene whose transcription is blocked by a repressor, all that is required is to remove the repressor. Cells do this by binding special "signal" molecules to the repressor protein; the binding causes the repressor protein to contort into a shape that doesn't fit DNA, and it falls off, removing the barrier to transcription. A specific example demonstrating how repressor proteins work is the set of genes called the *lac* operon in the bacterium *Escherichia coli.* An **operon** is a segment of DNA containing a cluster of genes that are transcribed as a unit. The *lac* operon, shown in figure 13.9, consists of both protein-encoding genes (labeled genes 1, 2, and 3, which are enzymes involved in breaking down the sugar lactose) and associated regulatory

elements—the operator (the purple segment) and promoter (the orange segment). Transcription is turned off because a repressor molecule (the purple structure straddling the DNA) binds to the operator such that RNA polymerase cannot bind to the promoter. When an *E. coli* encounters the sugar lactose, a metabolite of lactose called allolactose binds to the repressor protein and induces a twist in its shape that causes it to fall from the DNA. As you can see in the figure, RNA polymerase is no longer blocked, so it starts to transcribe the genes needed to break down the lactose to get energy.

Activators. Because RNA polymerase binds to a specific promoter site on one strand of the DNA double helix, it is necessary that the DNA double helix unzip in the vicinity of this site for the polymerase protein to be able to sit down properly. In many genes, this unzipping cannot take place without the assistance of a regulatory protein called an activator that

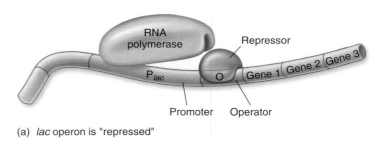

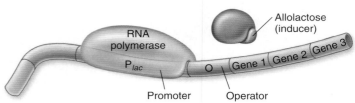

(a) *lac* operon is "repressed"

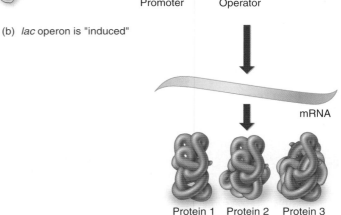

(b) *lac* operon is "induced"

Figure 13.9 How the lac operon works.

(a) The *lac* operon is shut down ("repressed") when the repressor protein is bound to the operator site. Because promoter and operator sites overlap, RNA polymerase and the repressor cannot bind at the same time. (b) The *lac* operon is transcribed ("induced") when allolactose binds to the repressor protein changing its shape so that it can no longer sit on the operator site and block polymerase binding.

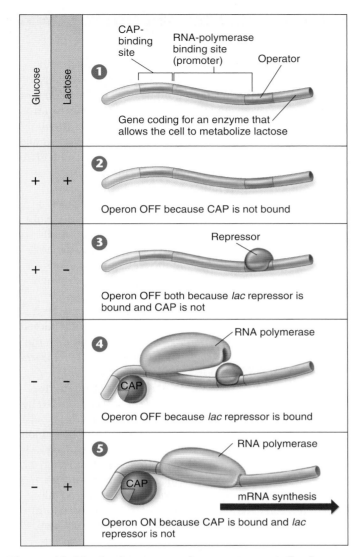

Figure 13.10 Activators and repressors at the *lac* operon.

Panel labels and content (left to right columns: Glucose, Lactose):

1. CAP-binding site / RNA-polymerase binding site (promoter) / Operator
   Gene coding for an enzyme that allows the cell to metabolize lactose

2. (Glucose +, Lactose +) Operon OFF because CAP is not bound

3. (Glucose +, Lactose −) Repressor
   Operon OFF both because *lac* repressor is bound and CAP is not

4. (Glucose −, Lactose −) RNA polymerase / CAP
   Operon OFF because *lac* repressor is bound

5. (Glucose −, Lactose +) RNA polymerase / CAP / mRNA synthesis
   Operon ON because CAP is bound and *lac* repressor is not

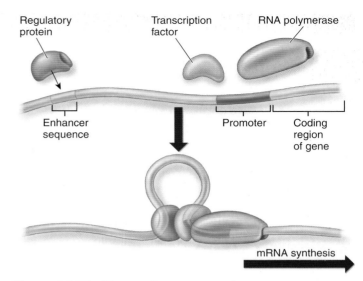

Figure 13.11 How enhancers work.
The enhancer site is located far from the gene. The binding of a regulatory protein brings the enhancer in contact with the gene.

Labels: Regulatory protein / Transcription factor / RNA polymerase / Enhancer sequence / Promoter / Coding region of gene / mRNA synthesis

binds to the DNA in this region and helps it unwind. Just as in the case of the repressor protein described previously, cells can turn genes on and off by binding "signal" molecules to the activator protein. These molecules either prevent the activator from binding to the DNA or enable it to do so.

Why bother with activators? Imagine if you had to eat every time you encountered food! Activator proteins enable a cell to cope with this sort of problem. Activators and repressors work together to control transcription. To understand how, let's consider the *lac* operon again, now shown in figure 13.10. When a bacterium encounters the sugar lactose, it may already have lots of energy in the form of glucose, as shown in panel ❷, and so does not need to break down more lactose. In such a bacterial cell, the activator protein called CAP is not able to bind to the DNA. Because RNA polymerase requires the activator to function, the *lac* operon is not expressed. Also if glucose is present and lactose is absent, not only is the acti-

vator CAP unable to bind, but also a repressor blocks the promoter, as shown in panel ❸ and in figure 13.9. In the absence of both glucose and lactose, a "low glucose" signal molecule (the green pie-shaped piece in panels ❹ and ❺) binds to CAP, and CAP is able to bind to the DNA. However, the repressor is still blocking transcription, as shown in panel ❹. Only in the absence of glucose and in the presence of lactose, the repressor is removed, the activator (CAP) is bound, and transcription proceeds, as shown in panel ❺.

Enhancers

A third level of control is exercised in eukaryotes by expanding access to complexly controlled genes. Many eukaryotic genes possess special sequences called **enhancers** that help transcription factor proteins guide RNA polymerase to the promoter at the beginning of a gene. Unlike promoters and operators, which butt right up to the start of a gene, enhancers are usually located far away from the gene.

How can an enhancer affect a promoter thousands of nucleotides away? As you can see in figure 13.11, the DNA loops around so that the enhancer (yellow) is positioned near the promoter (blue). This brings the regulatory protein attached to the enhancer (red) into direct contact with the transcription factor (green) associated with the RNA polymerase (pink) at the promoter.

This positioning of regulatory sites at a distance permits any of a large number of different regulatory sequences scattered about the DNA to influence that particular gene.

> **13.4 Cells control the expression of genes by determining when they are transcribed.** Some regulatory proteins block the binding of polymerase, and others facilitate it.

Today's *Biology*

How Small RNAs Regulate Gene Expression

Thus far we have discussed gene regulation entirely in terms of proteins that regulate the start of transcription, that is, that block or activate the "reading" of a particular gene by RNA polymerase. Within the last decade, however, it has become increasingly clear that RNA molecules can regulate the expression of genes, acting after transcription as a second level of control.

Discovery of RNA Interference

Like many important advances in science, the discovery that a class of small RNA molecules may play a major role in regulating gene expression arose unexpectedly. In 1990, plant biologists, working with petunias with deep purple flowers, introduced numerous extra copies of the flower color gene. To their surprise this did not lead to flowers with an even darker purple hue, but rather to white flowers! Somehow the introduced genes had silenced both themselves and the plant's own flower color gene.

This puzzling result began to make sense in 1998, when a simple experiment was carried out—investigators injected double-stranded RNA molecules into the nematode worm *Caenorhabditis elegans*. This resulted in the silencing of the gene whose sequence was complementary to the double-stranded RNA, and of no other gene. The investigators called this very specific gene silencing effect **RNA interference.**

What is going on here? As you will learn in chapter 17, RNA viruses (that is, viruses like the AIDS virus that use RNA rather than DNA as their hereditary material) replicate themselves through double-stranded intermediates—at a critical stage, a virus enzyme called reverse transcriptase travels along the virus RNA and assembles a complementary strand. Because the life cycle of many viruses involves a double-stranded RNA stage, RNA interference may have evolved as a cellular defense mechanism against these viruses; the evolution of this adaptation would have predated the evolutionary divergence of plants and animals. Indeed, double-stranded viral RNAs can be targeted for destruction by RNA interference machinery. Without intending to do so, the petunia and nematode researchers had stumbled across this defense, triggering it because introduced RNAs produce low levels of double-stranded copies.

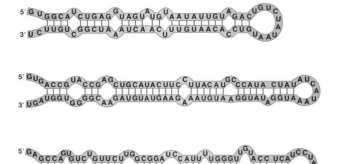

How small RNAs form double-stranded loops.

These three RNA molecules fold back to form hairpin loops because the sequences of the left and right halves are complementary, and form base pairs.

How RNA Interference Works

Investigating interference, researchers soon noted that in the process of silencing a gene, plants produced short RNA molecules (ranging in length from 21 to 28 nucleotides) that matched the gene being silenced. Researchers focusing on far larger messenger RNA (mRNA), transfer RNA (tRNA), and ribosomal RNA (rRNA) had not noticed these far smaller bits, tossing them out during experiments. These small RNAs appeared to regulate the activity of specific genes.

Soon researchers found evidence of similar small RNAs in a wide range of other organisms. In the plant *Arabidopsis thaliana,* small RNAs seemed to be involved in the regulation of genes critical to early development, while in yeasts they were identified as the agents that silence genes in tightly packed regions of the genome. In the ciliated protozoan *Tetrahymena thermophila,* the loss of major blocks of DNA during development seems guided by small RNA molecules.

The first clue of how small fragments of RNA can act to regulate gene expression emerged when researchers noted that stretches of double-stranded RNA injected into *C. elegans* can dissociate. Each single strand can then form a double-stranded RNA by folding back in a hairpin loop, like the three sections of RNA shown in the figure on this page. This occurs because the two ends of the strand have a complementary nucleotide sequence. When the RNA loops, the complementary bases form base pairings that hold the strands together much as it does the strands of a DNA duplex.

Exactly how does such a double-stranded RNA inhibit the expression of the gene from which the double-stranded RNA has been generated? The process is illustrated on the facing page. In the first stage of RNA interference, an enzyme called "dicer" recognizes long double-stranded RNA molecules and cuts them into short small RNA segments called *siRNAs (small interfering RNAs)* ❶. In the next step, the siRNAs can assemble into a ribonucleoprotein complex called *RISC (RNA Interference Silencing Complex)* ❷. RISC then unwinds the siRNA duplex, which leaves one single strand of RNA that is able to bind to mRNAs complementary to it ❸ and thus silence the genes that produced those mRNA molecules.

Once the siRNA has bound to mRNA, the silencing is achieved in one of two ways ❹: either the mRNA is inhibited by blocking its translation into protein, or the mRNA is destroyed. The choice between inhibition and destruction is thought to be governed by how closely the sequence of the siRNA matches the mRNA sequence, with destruction being the outcome for best-matched targets.

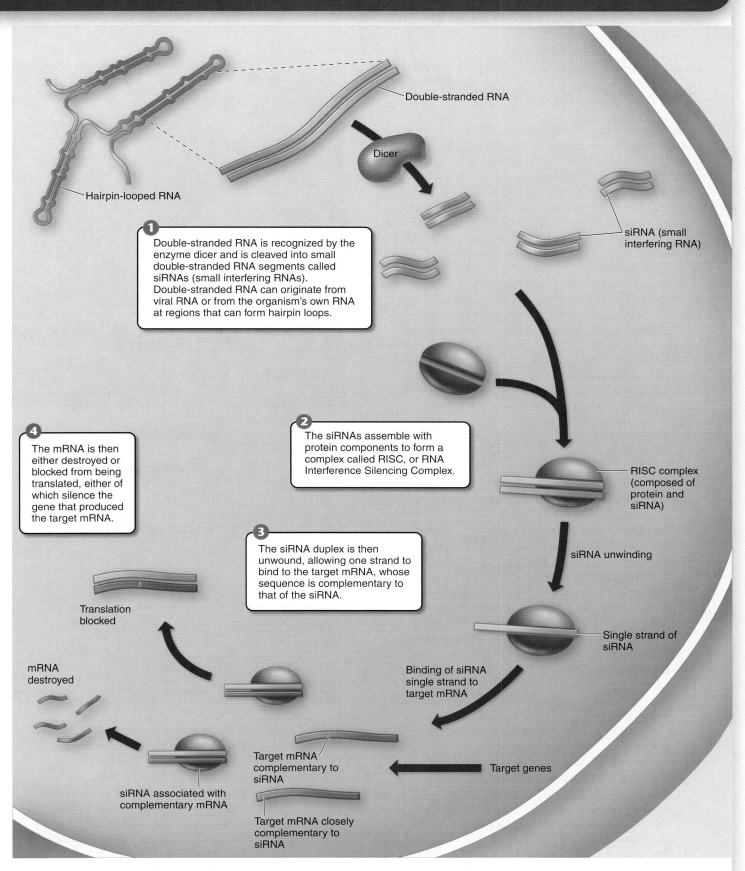

Double-stranded RNA

Dicer

Hairpin-looped RNA

siRNA (small interfering RNA)

1 Double-stranded RNA is recognized by the enzyme dicer and is cleaved into small double-stranded RNA segments called siRNAs (small interfering RNAs). Double-stranded RNA can originate from viral RNA or from the organism's own RNA at regions that can form hairpin loops.

4 The mRNA is then either destroyed or blocked from being translated, either of which silence the gene that produced the target mRNA.

2 The siRNAs assemble with protein components to form a complex called RISC, or RNA Interference Silencing Complex.

RISC complex (composed of protein and siRNA)

3 The siRNA duplex is then unwound, allowing one strand to bind to the target mRNA, whose sequence is complementary to that of the siRNA.

siRNA unwinding

Translation blocked

Single strand of siRNA

mRNA destroyed

Binding of siRNA single strand to target mRNA

siRNA associated with complementary mRNA

Target mRNA complementary to siRNA

Target genes

Target mRNA closely complementary to siRNA

A closer look at how RNA interference works.

How Does a Cell Adjust the Use of Its Genes to Fit Changing Needs?

When scientists first began to examine the molecular biology of bacteria a half-century ago, they soon noticed a curious phenomenon. Many bacterial enzymes were found to be present in the cell only when they were needed—that is, when the chemical the enzyme acted upon (its substrate) was itself present in the cell. It was as if the presence of the substrate had somehow induced the cell to start producing the enzyme that utilized it. This process was dubbed "induction" and the molecule which initiated it the *inducer*. In section 13.4 you encountered a classic example of enzyme induction, the *lac* operon.

The graph to the right below is from the 1952 paper that first asked if the *lac* gene encoding enzyme β-galactosidase (which cleaves the disaccharide sugar lactose into the two monosaccharides, glucose and galactose) was inducible by its substrate lactose. The growth of a laboratory culture of *Escherichia coli* bacteria, such as you see to the right, was monitored by measuring the amount of bacterial protein present in small samples taken from the liquid culture. The amount of the enzyme β-galactosidase was then measured in the same samples. After the culture was allowed to grow for a while, investigators added lactose to it. Later, the cells were removed and placed in fresh growth medium free of lactose, effectively removing lactose from the culture. This experiment thus poses this question: does lactose induce synthesis of β-galactosidase?

1. **Applying Concepts**
   a. Variable. In the graph, what is the dependent variable?
   b. Monitoring a Process. What is the level of β-galactosidase present before lactose is added to the growth medium? How does this level change after lactose is added? After lactose is removed? What general comparison can be made regarding the level of β-galactosidase in the presence and absence of lactose?
2. **Interpreting Data** Over the several cell generations that β-galactosidase is being induced, bacterial protein increases from 12 micrograms to 60 micrograms. How does the rate of the enzyme's synthesis compare with the rate at which overall bacterial protein is being made? Is it a constant proportion, or does its rate of synthesis change?
3. **Making Inferences** Does this experiment rule out the possibility that the sugar lactose simply activated preexisting β-galactosidase enzymes rather than causing the synthesis of new enzyme molecules?
4. **Drawing Conclusions** How does the presence of lactose affect the synthesis of the enzyme

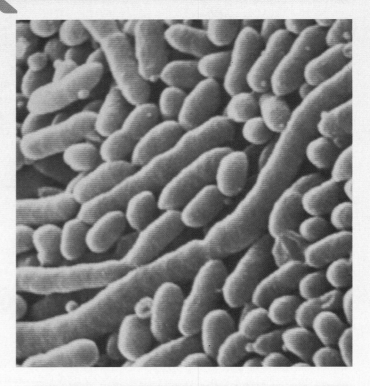

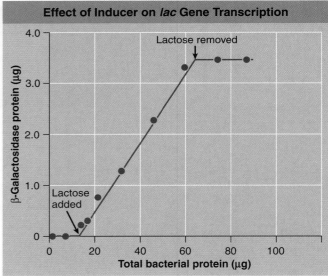

Effect of Inducer on *lac* Gene Transcription

β-galactosidase? Referring to section 13.4 of this chapter, can you suggest a mechanism that would explain this result?

5. **Further Analysis** Over the next decade, researchers obtained strains of *E. coli* bacteria with mutations inactivating the *lac* gene; such strains could not synthesize β-galactosidase even in the presence of lactose. If you had a way to introduce any gene you wished into this strain, design an experiment to confirm the conclusion you have drawn above.

From Gene to Protein

13.1 Transcription

- DNA is the storage site of genetic information in the cell. Segments of DNA called genes contain information needed to make proteins, but proteins are not synthesized directly from DNA. Instead the gene is copied into another nucleotide-based molecule, RNA. The process of gene expression, DNA to RNA to protein, is called the central dogma.

- Three types of RNA are used by the cell in gene expression: messenger RNA (mRNA), ribosomal RNA (rRNA), and transfer RNA (tRNA).

- Gene expression occurs in two stages: transcription, where an mRNA copy is made from the DNA, and translation, where the information on the mRNA is translated into a protein using rRNA and tRNA.

- DNA serves as a template on which mRNA is assembled by a protein called RNA polymerase, in a process called transcription. RNA polymerase adds complementary nucleotides onto the growing mRNA in a way that is similar to the actions of DNA polymerase (**figure 13.1**).

13.2 Translation

- The genetic information encoded in DNA is transcribed into three-nucleotide units called codons on mRNA, which correspond to particular amino acids. The rules that govern the translation of codons on the mRNA into amino acids is called the genetic code (**figure 13.2**).

- The genetic code contains 64 codons but codes for only 20 amino acids; therefore, there is duplication. In many instances, two or more different codons encode the same amino acid.

- The genetic code is universal, used by practically all organisms in the same way.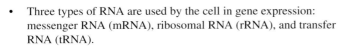

- The codons are translated from mRNA into proteins, a process called translation, using two other forms of RNA: rRNA that functions as the platform, called a ribosome, on which the amino acids are assembled into proteins, and tRNA that carry amino acids to the building protein (**figures 13.3 and 13.4**).

- A ribosome is composed of two components, a small rRNA, subunit and a large rRNA subunit. mRNA binds to the small subunit, which triggers the binding of the large subunit, forming a complete ribosome.

- A ribosome moves along mRNA, while tRNA molecules that contain anticodon sequences that are complementary to the codons on the mRNA bring amino acids to the ribosome. The amino acids add to the end of a growing polypeptide chain (**figures 13.5 and 13.6**).

13.3 Architecture of the Gene

- Eukaryotic genes contain coding regions, called exons, and noncoding regions, called introns. The introns are spliced out before translation (**figure 13.7**).

- Introns are excised from the RNA before it is translated, and so introns have no effect on the final protein produced.

- This complex organization allows eukaryotic genomes more flexibility. Exons can be spliced together in different ways, a process called alternative splicing, producing different protein products.

- The primary RNA transcript is produced in the nucleus, where it is processed, by splicing out introns. The mRNA then travels from the nucleus to the cytoplasm where it is translated into a chain of amino acids with the help of ribosomes and tRNA (**figure 13.8**).

- Some eukaryotic genes, called multigene families, are repeated on a chromosome. Genes can also move from one place to another in the genome as transposons.

Regulating Gene Expression

13.4 Turning Genes Off and On

- Most genes are regulated by controlling which genes are transcribed. Genes are turned on or shut off by attaching proteins that allow RNA polymerase access to the promoter or block the promoter, respectively.

- The *lac* operon in *E. coli* is turned off until it is needed. It is controlled by a repressor that binds to the DNA, blocking the promoter such that the RNA polymerase cannot bind. When the proteins produced by the *lac* operon genes are needed, an inducer molecule will bind to the repressor and keep it from binding to the DNA, thereby freeing the promoter so that RNA polymerase can bind (**figure 13.9**).

- The *lac* operon is also controlled by an activator that binds to the DNA, allowing the RNA polymerase to bind to the DNA. Without the activator, the RNA polymerase cannot bind to the DNA. It is only when the activator binds to the DNA and the repressor is removed from the DNA that the RNA polymerase can bind to the promoter (**figure 13.10**).

- Eukaryotic genes are controlled from distant locations called enhancers. A regulatory protein will bind to the enhancer region far from the gene. The regulatory protein, along with transcription factors, help the RNA polymerase bind to the promoter (**figure 13.11**). The enhancer is able to affect RNA polymerase binding even from a distance because the DNA forms a loop, bringing the distant enhancer region to the area of the promoter.

1. Which of the following is not a type of RNA?
   a. nRNA (nuclear RNA)
   b. mRNA (messenger RNA)
   c. rRNA (ribosomal RNA)
   d. tRNA (transfer RNA)
2. Each amino acid in a protein is specified by
   a. a gene.
   b. a promoter.
   c. an RNA molecule.
   d. a codon.
3. The three-nucleotide codon system can be arranged into _____ _____ combinations.
   a. 16
   b. 20
   c. 64
   d. 128
4. The process of obtaining a copy of the information in a gene as a strand of messenger RNA is called
   a. polymerase.
   b. expression.
   c. transcription.
   d. translation.
5. The site where RNA polymerase attaches to the DNA molecule to start the formation of an RNA molecule is called a(n)
   a. promoter.
   b. exon.
   c. intron.
   d. enhancer.

6. The process of taking the information on a strand of messenger RNA and building an amino acid chain, which will become all or part of a protein molecule, is called
   a. polymerase.   c. transcription.
   b. expression.   d. translation.
7. If an mRNA codon reads UAC, its complementary anticodon will be
   a. TUC.
   b. ATG.
   c. AUG.
   d. CAG.
8. When an mRNA leaves the cell's nucleus, it next becomes associated with
   a. proteins.
   b. a ribosome.
   c. tRNA.
   d. RNA polymerase.
9. Regarding the activity level of genes,
   a. all genes are on all the time in all cells, making the needed amino acid sequences.
   b. some genes are always off unless a promoter turns them on.
   c. some genes are always on unless a promoter turns them off.
   d. some genes are always off unless a repressor is not bound.
10. Which of the following statements is correct about prokaryotic gene expression?
    a. Prokaryotic mRNAs must have introns spliced out.
    b. Prokaryotic mRNAs contain the transcript of only one gene.
    c. Repressors block transcription by binding to the DNA.
    d. All of these statements are correct.

1. **Page 228** Assume that this figure is showing gene expression in a eukaryotic cell. What step is missing in the process?

2. **Figure 13.9** Can genes 1, 2, and 3 be transcribed? What would happen if an inducer molecule was present in the cell?

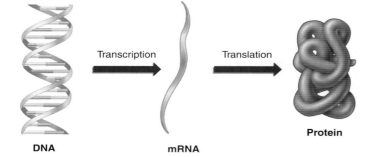

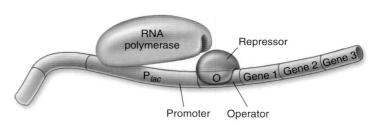

1. Compare DNA to a cookbook. The book is kept in a library and cannot be checked out (removed). Start with the letters and words in the cookbook compared with the bases and codons in DNA; end with the amino acid chain being folded into a protein, and a cake being baked.
2. What would happen if all the genes in a cell were always active?

3. The nucleotide sequence of a hypothetical gene is: TACATACTTAGTTACGTCGCCCGGAAATATC
   a. What will be the sequence on the mRNA when it is transcribed?
   b. What will be the amino acid sequence of the protein when it's translated?
   c. What would happen to the amino acid chain if the highlighted nucleotide underwent a mutation and was changed to an adenine?

14
The New Biology

This sheep, Dolly, was the first animal to be cloned from a single adult cell. The lamb you see above beside her is her offspring, normal in every respect. From Dolly we learn that genes are not lost during development. If a single adult cell can be induced to switch the proper combination of genes on and off, that one cell can develop into a normal adult individual. Embryonic stem cells are like this—poised to become any cell of the body as the embryo develops. It may be possible to replace damaged tissues with healthy tissue grown from a patient's own embryonic stem cells, as long as the disorder is not an inherited one. The approach has been used successfully in laboratory mice to cure a variety of disorders. However, its use in humans would require employing stem cells derived from the patient, and because this may involve destroying a human embryo the approach is controversial. Another approach, when the damaged tissue does result from a defective gene, is to repair rather than replace, using a virus to transfer a healthy gene into those tissues that lack it. In this chapter you will explore genomic sequencing, the application of gene technology to medicine and agriculture, reproductive cloning, stem cell tissue replacement, and gene therapy, all areas in which a revolution is reshaping biology.

14.1 Genomics

Recent years have seen an explosion of interest in comparing the entire DNA content of different organisms, a new field of biology called **genomics.** While initial successes focused on organisms with relatively small numbers of genes, researchers have recently completed the sequencing of several large eukaryotic genomes, including our own.

The full complement of genetic information of an organism—all of its genes and other DNA—is called its **genome.** The first genome to be sequenced was a very simple one: a small bacterial virus called φ-X174. Frederick Sanger, inventor of the first practical way to sequence DNA, obtained the sequence of this 5,375-nucleotide genome in 1977. This was followed by the sequencing of dozens of prokaryotic genomes. The advent of automated DNA sequencing machines in recent years has made the DNA sequencing of much larger eukaryotic genomes practical (table 14.1).

Sequencing DNA

DNA sequencing is a process that allows scientists to read each nucleotide in a strand of DNA. In sequencing DNA, a DNA fragment of unknown sequence is amplified, so there are thousands of copies of the fragment. The DNA fragments are then mixed with copies of DNA polymerase, copies of a primer (recall from chapter 12 that DNA polymerase can only add nucleotides onto an existing strand of nucleotides), a supply of the four nucleotide bases, and a supply of four different chain-terminating chemical tags. The chemical tags act as one of the four nucleotide bases in DNA synthesis, undergoing complementary base pairing. First, heat is applied to denature the double-stranded DNA fragments. The solution is then allowed to cool, allowing the primer (the lighter blue box in figure 14.1 ❶) to bind to a single strand of the DNA, and

synthesis of the complementary strand proceeds. Whenever a chemical tag is added instead of a nucleotide base, the synthesis stops, as shown in the figure. For example, the terminating red "T" was added after three normal nucleotides and synthesis stopped. Because of the relatively low concentration of the chemical tags compared with the nucleotides, a tag that binds to G on the DNA fragment, for example, will not necessarily be added to the first G site. Thus, the mixture will contain a series of double-stranded DNA fragments of different lengths, corresponding to the different distances the polymerase traveled from the primer before a chain-terminating tag was incorporated (six fragments are shown in ❶).

The series of fragments are then separated according to size by gel electrophoresis. The fragments become arrayed like the rungs of a ladder, each rung one base longer than the one below it. Compare the lengths of the fragments in ❶ and their positions on the gel in ❷, where the shortest fragment is one nucleotide long (G), and that is also the lowest rung on the gel. In automated DNA sequencing, fluorescently colored chemical tags are used to label the fragments, one color corresponding to each nucleotide. Computers read off the colors on the gel to determine the DNA sequence and display this sequence as a series of colored peaks (❸ and ❹). What made the attempt to sequence large eukaryotic genomes practical was the development in the mid-1990s of automated sequencers that perform electrophoresis of DNA fragments in capillary tubes instead of the traditional gel slabs. These systems can handle about 1,000 samples a day, with only 15 minutes of human attention. A research institute with several hundred such instruments can produce about 100 Mbp (million base pairs) every day.

> **14.1** Powerful automated DNA sequencing technology has begun to reveal the DNA sequences of entire genomes.

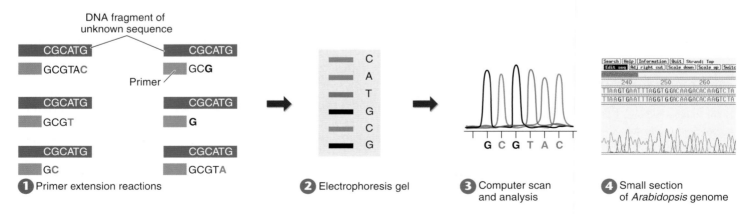

❶ Primer extension reactions ❷ Electrophoresis gel ❸ Computer scan and analysis ❹ Small section of *Arabidopsis* genome

Figure 14.1 How to sequence DNA.
❶ DNA is sequenced by adding complementary bases to a single-stranded fragment. DNA synthesis stops when a chemical tag is inserted instead of a nucleotide, resulting in different sizes of DNA fragments. ❷ The DNA fragments of varying lengths are separated by gel electrophoresis, the smaller fragments migrating farther down the gel. The boldface letters indicate the chemical tags added in step ❶ that stopped the replication process. ❸ Computers scan the gel, from smallest to largest fragments and display the DNA sequence as a series of colored peaks. ❹ Data from an automated DNA-sequencing run show the nucleotide sequence for a small section of the *Arabidopsis* (plant) genome.

TABLE 14.1	SOME EUKARYOTIC GENOMES			
Organism		Estimated Genome Size (Mbp)	Number of Genes (×1,000)	Nature of Genome
Vertebrates				
	Homo sapiens (human)	3,200	20–25	The first large genome to be sequenced; the number of transcribable genes is far less than expected; much of the genome is occupied by repeated DNA sequences.
	Pan troglodytes (chimpanzee)	2,800	20–25	There are few base substitutions between chimp and human genomes, less than 2%, but many small sequences of DNA have been lost as the two species diverged, often with significant effects.
	Mus musculus (mouse)	2,500	25	Roughly 80% of mouse genes have a functional equivalent in the human genome; importantly, large portions of the noncoding DNA of mouse and human have been conserved; overall, rodent genomes (mouse and rat) appear to be evolving more than twice as fast as primate genomes (humans and chimpanzees).
	Gallus gallus (chicken)	1,000	20–23	One-third the size of the human genome; genetic variation among domestic chickens seems much higher than in humans.
	Fugu rubripes (pufferfish)	365	35	The *Fugu* genome is only one-ninth the size of the human genome, yet it contains 10,000 more genes.
Invertebrates				
	Caenorhabditis elegans (nematode)	97	21	The fact that every cell of *C. elegans* has been identified makes its genome a particularly powerful tool in developmental biology.
	Drosophila melanogaster (fruit fly)	137	13	*Drosophila* telomere regions lack the simple repeated segments that are characteristic of most eukaryotic telomeres. About one-third of the genome consists of gene-poor centric heterochromatin.
	Anopheles gambiae (mosquito)	278	15	The extent of similarity between *Anopheles* and *Drosophila* is approximately equal to that between human and pufferfish.
Plants				
	Arabidopsis thaliana (wall cress)	115	26	*A. thaliana* was the first plant to have its genome fully sequenced. The evolution of its genome involved a whole-genome duplication followed by subsequent losses of genes and extensive local gene duplications.
	Oryza sativa (rice)	430	33–50	The rice genome contains only 13% as much DNA as the human genome, but roughly twice as many genes; like the human genome, it is rich in repetitive DNA.
Protists				
	Plasmodium falciparum (malaria parasite)	23	5	The *Plasmodium* genome has an unusually high proportion of adenine and thymine. Scarcely 5,000 genes contain the bare essentials of the eukaryotic cell.
	Dictyostelium discoideum	34	12–13	The genome for *Dictyostelium discoideum* is more closely related to animals than to plants or fungi, but it has retained more of the ancestral genome sequences than any of the other three groups of eukaryotes.
Fungi				
	Saccharomyces cerevisiae (brewer's yeast)	13	6	*S. cerevisiae* was the first eukaryotic cell to have its genome fully sequenced.

On June 26, 2000, geneticists reported that the entire human genome had been sequenced. This effort presented no small challenge, as the human genome is huge—more than 3 billion base pairs, which is the largest genome sequenced to date. To get an idea of the magnitude of the task, consider that if all 3.2 billion base pairs were written down on the pages of this book, the book would be 500,000 pages long, and it would take you about 60 years, working eight hours a day, every day, at five bases a second, to read it all.

Geography of the Genome

The preliminary report of the human genome sequence, published in 2001, estimated the number of protein-encoding genes to be 30,000. The final report, published in 2004, lowered that estimate to 20,000 to 25,000 protein-encoding genes. This is scarcely more than in nematodes (a wormlike animal) at 21,000 genes, not quite double the number in the fruit fly *Drosophila* at 13,000 genes, and but a quarter of the number that had been anticipated by scientists based on the number of unique messenger RNA (mRNA) molecules.

How can human cells contain four times as many kinds of mRNA as there are genes? Recall from chapter 13 that in a typical human gene, the sequence of DNA nucleotides that specifies a protein is broken into many bits called exons, scattered among much longer segments of nontranslated DNA called introns. Imagine this paragraph was a human gene; all the occurrences of the letter "e" could be considered exons, while the rest would be noncoding introns. Look at figure 14.2, which breaks up the human genome into different types of DNA; you see that introns make up 24% of the human genome.

When a cell uses a human gene to make a protein, it first manufactures mRNA copies of all the exons (protein-specifying fragments) of the gene, then splices the exons together. Now here's the turn of events researchers had not anticipated: the transcripts of human genes are often spliced together in different ways, called alternative splicing. As we discussed in chapter 13, each exon is actually a module; one exon may code for one part of a protein, another for a different part of a protein. When the exon transcripts are mixed in different ways, very different protein shapes can be built.

With alternative mRNA splicing, it is easy to see how 20,000 to 25,000 genes can encode four times as many proteins. The added complexity of human proteins occurs because the gene parts are put together in new ways. Great music is made from simple tunes in much the same way.

In addition to the fragmenting of genes by the scattering of exons throughout the genome, there is another interesting "organizational" aspect of the genome. Genes are not distributed evenly over the genome. The small chromosome number 19 is packed densely with genes, transcription factors, and other functional elements. The much larger chromosome numbers 4 and 8, by contrast, have few genes. On most chromosomes, vast stretches of seemingly barren DNA fill the chromosomes between scattered clusters rich in genes.

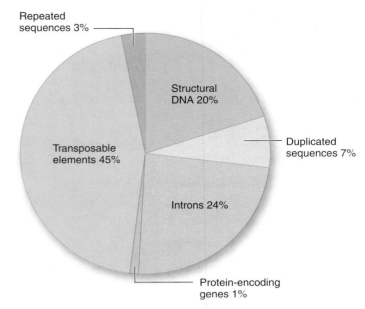

Figure 14.2   The human genome.
Very little of the human genome is devoted to protein-encoding genes. Surprisingly, most of it appears to be composed of transposable elements.

DNA That Codes for Proteins

Four different classes of protein-encoding genes are found in the human genome, differing largely in gene copy number.

Single-copy genes. Many eukaryotic genes exist as single copies at a particular location on a chromosome. Mutations in these genes produce recessive Mendelian inheritance. Silent copies inactivated by mutation, called *pseudogenes,* are as common as protein-encoding genes.

Segmental duplications. Human chromosomes contain many segmental duplications, whole blocks of genes that have been copied over from one chromosome to another. Blocks of similar genes in the same order are found throughout the genome. Chromosome 19 seems to have been the biggest borrower, with blocks of genes shared with 16 other chromosomes.

Multigene families. Many genes exist as parts of multigene families, groups of related but distinctly different genes that often occur together in a cluster. Multigene families contain from three to several dozen genes. Although they differ from each other, the genes of a multigene family are clearly related in their sequences, making it likely that they arose from a single ancestral sequence.

Tandem clusters. These groups of repeated genes consist of DNA sequences that are repeated many thousands of times, one copy following another in tandem array. By transcribing all of the copies in these tandem clusters simultaneously, a cell can rapidly obtain large amounts of the product they encode. For example, the genes encoding rRNA are present in clusters of several hundred copies.

Noncoding DNA

One of the most notable characteristics of the human genome is the startling amount of noncoding DNA it possesses. Only 1% to 1.5% of the human genome is coding DNA, devoted to genes encoding proteins. Each of your cells has about six feet of DNA stuffed into it, but of that, less than one inch is devoted to genes! Nearly 99% of the DNA in your cells seems to have little or nothing to do with the instructions that make you who you are. Table 14.2 provides an overview of the types of sequences found in the human genome. True genes are scattered about the human genome in clumps among the much larger amounts of noncoding DNA, like isolated hamlets in a desert.

There are four major types of noncoding human DNA:

Noncoding DNA within genes. As we discussed on page 244, a human gene is made up of numerous fragments of protein-encoding information (exons) embedded within a much larger matrix of noncoding DNA (introns). Together, introns make up about 24% of the human genome, and exons about 1.5%.

Structural DNA. Some regions of the chromosomes remain highly condensed, tightly coiled, and untranscribed throughout the cell cycle. Called constitutive heterochromatin, these portions—about 20% of the DNA—tend to be localized around the centromere, or located near the ends of the chromosome.

Repeated/duplicated sequences. Scattered about chromosomes are simple sequence repeats (SSRs). An SSR is a two- or three-nucleotide sequence like CA or CGG, repeated like a broken record thousands and thousands of times. SSRs make up about 3% of the human genome. An additional 7% is devoted to other sorts of duplicated sequences. Repetitive sequences with excess C and G tend to be found in the neighborhood of genes, while A- and T-rich repeats dominate the nongene deserts. The light bands on chromosome karyotypes now have an explanation—they are regions rich in GC and genes (see figure 11.24). Dark bands signal neighborhoods rich in AT and thin on genes. Chromosome 19, dense with genes, has few dark bands.

Transposable elements. Fully 45% of the human genome consists of mobile bits of DNA called transposable elements. Discovered by Barbara McClintock in 1950 (she won the Nobel Prize in Physiology or Medicine for her discovery in 1983), transposable elements are bits of DNA that are able to jump from one location on a chromosome to another—tiny molecular versions of Mexican jumping beans. They work like a "cut-and-paste" or "copy-and-paste" word processing function.

Human chromosomes contain five sorts of transposable elements. Fully 20% of the genome consists of long interspersed nuclear elements (LINEs). An ancient and very successful element, LINEs are about 6 kb (6,000 DNA bases) long, and contain all the equipment needed for transposition, including genes for a DNA-loop-nicking enzyme and a reverse transcriptase.

Nested within the genome's LINEs are over half a million copies of a parasitic element called *Alu*, composing 10% of the human genome. *Alu* is only about 300 bases long, and has no transposition machinery of its own; like a flea on a dog, *Alu* moves with the LINE it resides within. Just as a flea sometimes jumps to a different dog, so *Alu* sometimes uses the enzymes of its LINE to move to a new chromosome location. Often jumping right into genes, *Alu* transpositions cause many harmful mutations.

Three other types of transposable elements are also present in the human genome. Eight percent of the genome is devoted to long terminal repeats (LTRs). LTRs and LINEs are a type of transposon called "retrotransposons" because they involve an RNA intermediate. Three percent is devoted to DNA transposons, which copy themselves as DNA rather than RNA. And, some 4% is devoted to dead transposons, elements that have lost the signals for replication and so can no longer jump.

> **14.2 The entire 3.2-billion-base-pair human genome has been sequenced. Gene sequences vary greatly in copy number, some occurring many thousands of times, others only once. Only about 1% of the human genome is devoted to protein-encoding genes. Much of the rest is composed of transposable elements.**

TABLE 14.2	TYPES OF DNA SEQUENCES FOUND IN THE HUMAN GENOME	
Class	**Frequency**	**Description**
Protein-encoding genes	1%–1.5%	Translated exons, within some 25,000 genes scattered about the chromosomes
Introns	24%	Noncoding DNA comprising the great majority of most genes
Structural DNA	20%	Constitutive heterochromatin, localized near centromeres and telomeres
Repeated sequences	3%	Simple sequence repeats (SSRs) of a few nucleotides repeated millions of times
Duplicated sequences	7%	Duplicated sequences, other than the SSRs
Transposable elements	45%	20% long interspersed nuclear elements (LINEs), active transposons 15% other transposable elements, including long terminal repeats (LTRs) 10% the parasitic sequence (*Alu*), present in half a million copies

14.3 A Scientific Revolution

In recent years, the ability to manipulate genes and move them from one organism to another has led to great advances in medicine and agriculture. Moving genes from one organism to another is often called **genetic engineering.** Many of the gene transfers have placed eukaryotic genes into bacteria, converting the bacteria into tiny factories that produce prodigious amounts of the protein encoded by the eukaryotic gene. Other gene transfers have moved genes from one animal or plant to another.

Genetic engineering is having a major impact on medicine and agriculture (figure 14.3). Most of the insulin used to treat diabetes is now obtained from bacteria that contain a human insulin gene. In late 1990, the first transfers of genes from one human to another were carried out in attempts to correct the effects of defective genes in a rare genetic disorder called *severe combined immune deficiency syndrome.* In addition, cultivated plants and animals can be genetically engineered to resist pests, grow bigger, or grow faster.

Restriction Enzymes

The first stage in any genetic engineering experiment is to chop up the "source" DNA to get a copy of the gene you wish to transfer. This first stage is the key to successful transfer of the gene, and learning how to do it is what has led to the

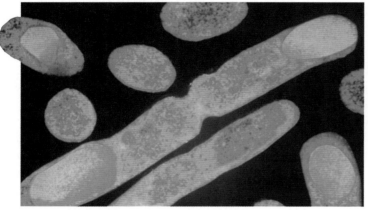

Producing insulin. The common bacteria *Escherichia coli* (*E. coli*) can be genetically engineered to contain the gene that codes for the protein insulin. The bacteria are turned into insulin-producing factories and can produce large quantities of insulin for diabetic patients. In the image above, insulin-producing sites inside genetically altered *E. coli* cells are orange.

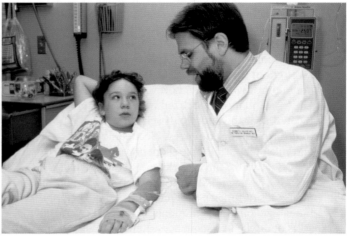

Curing disease. One of two young girls who were the first humans "cured" of a hereditary disorder by transferring into their bodies healthy versions of the gene they lacked. The transfer was successfully carried out in 1990, and the girls remain healthy.

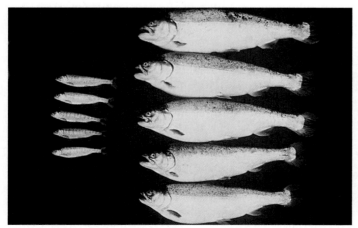

Increasing yields. The genetically engineered salmon on the *right* have shortened production cycles and are heavier than the nontransgenic salmon on the *left*.

Figure 14.3 Examples of genetic engineering.

Pest-proofing plants. The genetically engineered cotton plants on the *right* have a gene that inhibits feeding by weevils; the cotton plants on the *left* lack this gene, and produce far fewer cotton bolls.

genetic revolution. The trick is in how the DNA molecules are cut. The cutting must be done in such a way that the resulting DNA fragments have "sticky ends" that can later be inserted into another molecule of DNA.

This special form of molecular surgery is carried out by **restriction enzymes,** also called restriction endonucleases, which are special enzymes that bind to specific short sequences (typically four to six nucleotides long) on the DNA. These sequences are very unusual in that they are symmetrical—the two strands of the DNA duplex have the same nucleotide sequence, running in opposite directions! The sequence in figure 14.4 for example, is GAATTC. Try writing down the sequence of the opposite strand: it is CTTAAG—the same sequence, written backward. This sequence is recognized by the restriction enzyme *Eco*RI.

What makes the DNA fragments "sticky" is that most restriction enzymes do not make their incision in the center of the sequence; rather, the cut is made to one side. In the sequence in figure 14.4 ❶, for example, the cut is made after the first nucleotide, G/AATTC. This produces a break, with short, single strands of DNA dangling from each end. Because the two single-stranded ends are complementary in sequence, they could pair up and heal the break, with the aid of a sealing enzyme—*or they could pair with any other DNA fragment cut by the same enzyme,* because all would have the same single-stranded sticky ends. Figure 14.4 ❷ shows how DNA from another source also cut with *Eco*RI (the orange DNA) has the same sticky ends, ends which are also complementary to the original source DNA. Any gene in any organism cut by the enzyme that attacks GAATTC sequences will have the same sticky ends, and can be joined to any other with the aid of a sealing enzyme called a **ligase,** which reforms the bonds between the sugars and phosphates of DNA (figure 14.4 ❸).

Restriction enzymes are bacterial enzymes, discovered in the late 1960s by Werner Arber and Hamilton Smith (who, along with Daniel Nathans, were awarded the 1978 Nobel Prize in Physiology or Medicine). Arber had observed that bacterial viruses could infect some cells but not others. Bacteria that exhibited this "host restriction" were found to contain enzymes that could cleave foreign DNA (their own DNA being protected from the enzyme action by chemical modifications of the DNA). Any viruses that attempt to infect bacterial cells fail, because the bacterial restriction enzymes degrade the viral DNA (refer back to figure 12.2 to see how a bacterial virus infects a cell). Since their discovery, hundreds of different restriction enzymes have been identified, recognizing a wide variety of four- to six-nucleotide DNA sequences called restriction sites. Each kind of enzyme attacks only one sequence and always cuts at the same place. By trying one enzyme after another, biologists can almost always find an enzyme that cuts out the gene they seek, attacking a restriction site present by chance on both ends of the gene but not within it. Restriction enzymes are the basic tools of genetic engineering.

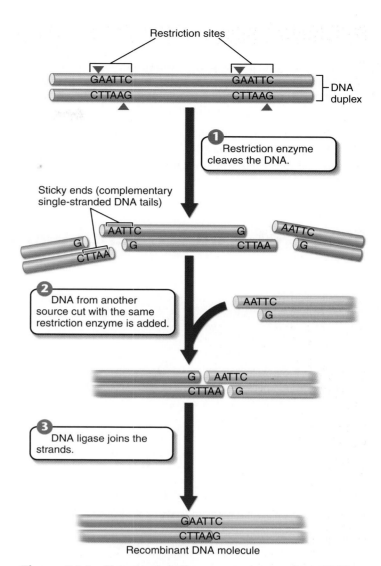

Figure 14.4 How restriction enzymes produce DNA fragments with sticky ends.

The restriction enzyme *Eco*RI always cleaves the sequence GAATTC between G and A. Because the same sequence occurs on both strands, both are cut. However, the two sequences run in opposite directions on the two strands. As a result, single-stranded tails are produced that are complementary to each other, or "sticky."

The Four Stages of a Genetic Engineering Experiment

To transfer a gene from one organism to another, three "ingredients" are required: a source of DNA that contains the gene you want to transfer, a restriction enzyme to cut the DNA, and a vehicle to carry the source DNA into the host cell. The source of DNA is often a *DNA library,* a collection of DNA fragments representing all of the DNA from an organism. The restriction enzyme selected will cut out the gene of interest from the DNA within which it occurs, and also will cut the vehicle DNA. The vehicle is a molecule of DNA that can carry the gene into a cell;

the vehicle is called a *vector*. Some of the most commonly used vectors include bacterial plasmids, a circular segment of DNA that is separate from the bacterial chromosome, and viruses.

With these ingredients in hand, a gene transfer experiment may still present unique problems, but all gene transfers share four distinct stages outlined in figure 14.5:

1. Cleaving DNA—cutting the source and vector DNA.
2. Producing recombinant DNA—placing the DNA fragments into vectors that transfer the DNA into the target cells.
3. Cloning—introducing DNA-bearing vectors into target cells and allowing the cells to reproduce.
4. Screening—selecting the particular infected cells that have received the gene of interest.

14.3 **Restriction enzymes bind to specific short sequences of DNA and cut the DNA there. This produces fragments with "sticky ends," which can be inserted into other DNA molecules.**

Stages of a Genetic Engineering Experiment

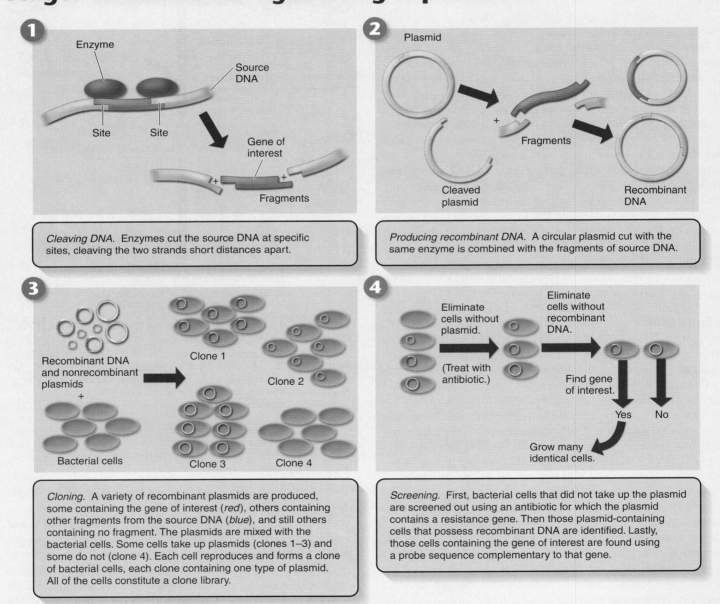

Cleaving DNA. Enzymes cut the source DNA at specific sites, cleaving the two strands short distances apart.

Producing recombinant DNA. A circular plasmid cut with the same enzyme is combined with the fragments of source DNA.

Cloning. A variety of recombinant plasmids are produced, some containing the gene of interest (*red*), others containing other fragments from the source DNA (*blue*), and still others containing no fragment. The plasmids are mixed with the bacterial cells. Some cells take up plasmids (clones 1–3) and some do not (clone 4). Each cell reproduces and forms a clone of bacterial cells, each clone containing one type of plasmid. All of the cells constitute a clone library.

Screening. First, bacterial cells that did not take up the plasmid are screened out using an antibiotic for which the plasmid contains a resistance gene. Then those plasmid-containing cells that possess recombinant DNA are identified. Lastly, those cells containing the gene of interest are found using a probe sequence complementary to that gene.

Figure 14.5   How a genetic engineering experiment works.

14.4 Genetic Engineering and Medicine

Much of the excitement about genetic engineering has focused on its potential to improve medicine—to aid in curing and preventing illness. Major advances have been made in the production of proteins used to treat illness, in the creation of new vaccines to combat infections, and in the replacement of defective genes, or gene therapy, which will be discussed later in this chapter.

Making "Magic Bullets"

Many genetic defects occur because our bodies fail to make critical proteins. *Diabetes* is such an illness. The body is unable to control levels of sugar in the blood because a critical protein, **insulin,** cannot be made. These failures can be overcome if the body can be supplied with the protein it lacks. The donated protein is in a very real sense a "magic bullet" to combat the body's inability to regulate itself.

Until recently, the principal problem with using regulatory proteins as drugs was in manufacturing the protein. Proteins that regulate the body's functions are typically present in the body in very low amounts, and this makes them difficult and expensive to obtain in quantity. With genetic engineering techniques, the problem of obtaining large amounts of rare proteins has been largely overcome. The genes encoding medically important proteins are now introduced into bacteria (table 14.3). Because the host bacteria can be grown cheaply, large amounts of the desired protein can be easily isolated. In 1982, the U.S. Food and Drug Administration approved the use of human insulin produced from genetically engineered bacteria, the first commercial product of genetic engineering.

Today hundreds of pharmaceutical companies around the world are busy producing other medically important proteins using these genetic engineering techniques. A gene added to the DNA of the mouse on the right in figure 14.6 produces human growth hormone, allowing the mouse to grow larger than its twin. The advantage of using gene engineering is clearly seen with **factor VIII,** a protein that promotes blood clotting. A deficiency in factor VIII leads to hemophilia, an inherited disorder (discussed in chapter 11), which is characterized by prolonged bleeding. For a long time, hemophiliacs received blood factor VIII that had been isolated from donated blood. Unfortunately, some of the donated blood had been infected with viruses such as HIV and hepatitis B, which were then unknowingly transmitted to those people who received blood transfusions. Today the use of genetically engineered factor VIII eliminates the risks associated with blood products obtained from other individuals.

Piggyback Vaccines

Another area of potential significance involves the use of genetic engineering to produce **subunit vaccines** against viruses such as those that cause herpes and hepatitis. Genes encoding

TABLE 14.3	GENETICALLY ENGINEERED DRUGS
Product	**Effects and Uses**
Anticoagulants	Involved in dissolving blood clots; used to treat heart attack patients
Colony-stimulating factors	Stimulate white blood cell production; used to treat infections and immune system deficiencies
Erythropoietin	Stimulates red blood cell production; used to treat anemia in individuals with kidney disorders
Factor VIII	Promotes blood clotting; used to treat hemophilia
Growth factors	Stimulate differentiation and growth of various cell types; used to aid wound healing
Human growth hormone	Used to treat dwarfism
Insulin	Involved in controlling blood sugar levels; used in treating diabetes
Interferons	Disrupt the reproduction of viruses; used to treat some cancers
Interleukins	Activate and stimulate white blood cells; used to treat wounds, HIV infections, cancer, immune deficiencies

Figure 14.6 Genetically engineered human growth hormone.

These two mice are genetically identical, but the large one has one extra gene: the gene encoding human growth hormone. The gene was added to the mouse's genome by genetic engineers and is now a stable part of the mouse's genetic endowment.

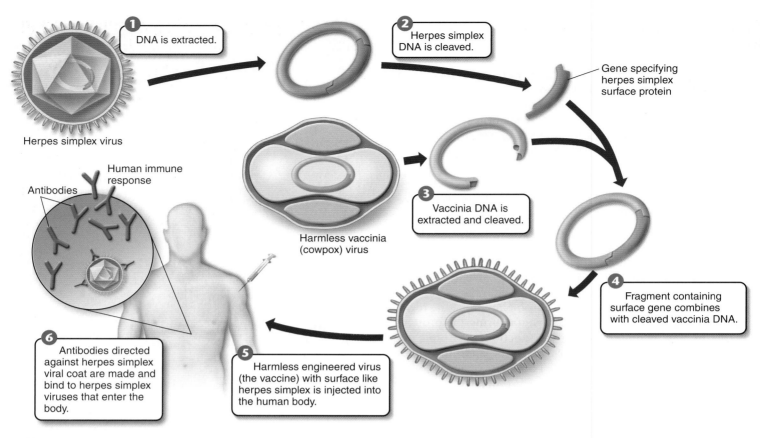

Figure 14.7   Constructing a subunit, or piggyback, vaccine for the herpes simplex virus.

part of the protein-polysaccharide coat of the herpes simplex virus or hepatitis B virus are spliced into a fragment of the vaccinia (cowpox) virus genome. The vaccinia virus, which British physician Edward Jenner used more than 200 years ago in his pioneering vaccinations against smallpox, is now used as a vector to carry the herpes or hepatitis viral coat gene into cultured mammalian cells. The steps, highlighted in figure 14.7, begin with ❶ extracting the herpes simplex viral DNA and ❷ isolating a gene that codes for a protein on the surface of the virus. The cowpox viral DNA is extracted and cleaved in step ❸ and the herpes gene is combined with the cowpox DNA in ❹. The recombinant DNA is inserted into a cowpox virus. Many copies of the recombinant virus are produced, which have the outside coat of a herpes virus. When this recombinant virus is injected into a human in step ❺, the immune system produces antibodies directed against the coat of the recombinant virus as in step ❻. It therefore develops an immunity to herpes virus. The same procedure can be done using the hepatitis virus. Vaccines produced in this way, also known as **piggyback vaccines,** are harmless because the vaccinia virus is benign, and only a small fragment of the DNA from the disease-causing virus is introduced via the recombinant virus.

The great attraction of this approach is that it does not depend upon the nature of the viral disease. In the future, similar recombinant viruses may be injected into humans to confer resistance to a wide variety of viral diseases.

In 1995, the first clinical trials began of a novel new kind of **DNA vaccine,** one that depends not on antibodies attacking injected antigens but rather on the second arm of the body's immune defense, the so-called cellular immune response, in which blood cells known as killer T cells attack infected cells. A plasmid containing a viral gene is injected and taken up by cells of the body where the gene is expressed. The infected cells are attacked and destroyed when they stick fragments of foreign proteins onto their outer surfaces that the T cells detect (the discovery by Peter Doherty and Rolf Zinkernagel that infected cells do so led to their receiving the Nobel Prize in Physiology or Medicine in 1996). With the cellular immune response activated, viruses are attacked as they enter the body. The first DNA vaccines spliced an influenza virus gene encoding an internal nucleoprotein into a plasmid, which was then injected into mice. The mice developed strong cellular immune responses to influenza. New and controversial, the approach offers great promise.

14.4   Genetic engineering has facilitated the production of medically important proteins and led to novel vaccines.

14.5 Genetic Engineering and Agriculture

One of the greatest impacts of genetic engineering on society has been the successful manipulation of the genes of crop plants to make them more resistant to disease caused by insects, to make them resistant to herbicides (chemicals that kill plants), to improve their nutritional balance and protein content, and to make them hardier, able to resist stress caused by frost, drought, and other factors.

Pest Resistance

An important effort of genetic engineers in agriculture has involved making crops resistant to insect pests without spraying with pesticides, a great saving to the environment. Consider cotton. Its fibers are a major source of raw material for clothing throughout the world, yet the plant itself can hardly survive in a field because many insects attack it. Over 40% of the chemical insecticides used today are employed to kill insects that eat cotton plants. The world's environment would greatly benefit if these thousands of tons of insecticide were not needed. Biologists are now in the process of producing cotton plants that are resistant to attack by insects.

One successful approach uses a kind of soil bacterium, *Bacillus thuringiensis* (Bt) that produces a protein that is toxic when eaten by crop pests, such as larvae (caterpillars) of butterflies. When the gene producing the Bt protein is inserted into the chromosomes of tomatoes, the plants begin to manufacture Bt protein, which makes them highly toxic to tomato hornworms (one of the most serious pests of commercial tomato crops) but the Bt protein is not harmful to humans.

Many important plant pests also attack roots. To combat these pests, genetic engineers are introducing the *Bt* gene into different kinds of bacteria, ones that colonize the roots of crop plants. Any insects eating such roots consume the bacteria and so are lethally attacked by the enzyme.

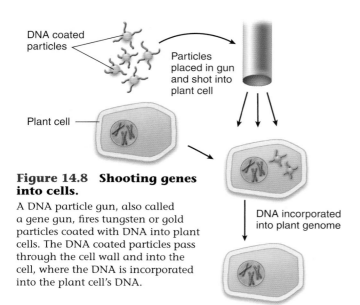

Figure 14.8 Shooting genes into cells.

A DNA particle gun, also called a gene gun, fires tungsten or gold particles coated with DNA into plant cells. The DNA coated particles pass through the cell wall and into the cell, where the DNA is incorporated into the plant cell's DNA.

DNA coated particles

Particles placed in gun and shot into plant cell

Plant cell

DNA incorporated into plant genome

Figure 14.9 Genetically engineered herbicide resistance.

All four of these petunia plants were exposed to equal doses of an herbicide. The two on *top* were genetically engineered to be resistant to glyphosate, the active ingredient in the herbicide, whereas the two dead ones on the *bottom* were not.

Herbicide Resistance

A major advance has been the creation of crop plants that are resistant to the herbicide *glyphosate*, a powerful biodegradable herbicide that kills most actively growing plants. Glyphosate is used in orchards and agricultural fields to control weeds. Growing plants need to make a lot of protein, and glyphosate stops them from making protein by destroying an enzyme necessary for the manufacture of so-called aromatic amino acids (that is, amino acids that contain a ring structure, like phenylalanine—see figure 4.3). Humans are unaffected by glyphosate because we don't make aromatic amino acids—we obtain them from plants we eat! To make crop plants resistant to this powerful plant killer, genetic engineers screened thousands of organisms until they found a species of bacteria that could make aromatic amino acids in the presence of glyphosate. They then isolated the gene encoding the resistant enzyme and successfully introduced the gene into plants. They inserted the gene into the plants using DNA particle guns, also called gene guns. You can see in figure 14.8 how a DNA particle gun works. Small tungsten or gold pellets are coated with DNA (red in the figure) that contains the gene of interest. The DNA gun literally shoots the gene into plant cells in culture where the gene is expressed. Plants that have been genetically engineered in this way are shown in figure 14.9. The two plants on top were genetically engineered to be resistant to glyphosate, the herbicide that killed the two plants at the bottom of the photo.

The creation of glyphosate-tolerant crops is of major benefit to the environment. Glyphosate is quickly broken down in the environment, which makes its use a great improvement over long-lasting chemical herbicides. Also, not having to plow to remove weeds reduces the loss of fertile topsoil to erosion.

More Nutritious Crops

In the last 10 years the cultivation of genetically modified (GM) crops of corn, cotton, soybeans, and other plants has become commonplace in the United States. In 2003, 84% of soybeans in the United States were planted with seeds genetically modified to be herbicide resistant. The result has been that less tillage was needed and as a consequence soil erosion was greatly lessened. Pest-resistant GM corn in 2003 comprised 38% of all corn planted in the United States, and pest-resistant GM cotton comprised 81% of all cotton. In both cases, the change greatly lessens the amount of chemical pesticide used in raising the crops. These benefits of soil preservation and chemical pesticide reduction, while significant, have been largely bestowed upon farmers, making their cultivation of crops cheaper and more efficient.

Like the first act of a play, these developments have served mainly to set the stage for the real action, which is only now beginning to happen. The real promise of plant genetic engineering is to produce genetically modified plants with desirable traits that directly benefit the consumer.

One recent advance, nutritionally improved "golden" rice, gives us a hint of what is to come. In developing countries large numbers of people live on simple diets that are poor sources of vitamins and minerals (what botanists called "micronutrients"). Worldwide, the two major micronutrient deficiencies are iron, which affects 1.4 billion women, 24% of the world population, and vitamin A, affecting 40 million children, 7% of the world population. The deficiencies are especially severe in developing countries where the major staple food is rice. In recent research, Swiss bioengineer Ingo Potrykus and his team at the Institute of Plant Sciences, Zurich, have gone a long way toward solving this problem. Supported by the Rockefeller Foundation and with results to be made free to developing countries, the work is a model of what plant genetic engineering can achieve.

To solve the problem of dietary iron deficiency among rice eaters, Potrykus first asked why rice is such a poor source of dietary iron. The problem, and the answer, proved to have three parts:

1. *Too little iron.* The proteins of rice endosperm have unusually low amounts of iron. To solve this problem, a ferritin gene (abbreviated as *Fe* in figure 14.10) was transferred into rice from beans. Ferritin is a protein with an extraordinarily high iron content, and so greatly increased the iron content of the rice.
2. *Inhibition of iron absorption by the intestine.* Rice contains an unusually high concentration of a chemical called phytate, which inhibits iron reabsorption in the intestine—it stops your body from taking up the iron in the rice. To solve this problem, a gene encoding an enzyme called phytase (abbreviated as *Pt*) that destroys phytate was transferred into rice from a fungus.
3. *Too little sulfur for efficient iron absorption.* The human body requires sulfur for the uptake of iron, and rice has very little of it. To solve this problem, a gene encoding a sulfur-rich metallothionin protein (abbreviated as *S*) was transferred into rice from wild rice.

To solve the problem of vitamin A deficiency, the same approach was taken. First, the problem was identified. It turns out rice only goes partway toward making beta-carotene (pro-vitamin A); there are no enzymes in rice to catalyze the last four steps. To solve the problem, genes encoding these four enzymes (abbreviated *A₁ A₂ A₃ A₄*) were added to rice from a flower, the daffodil.

The development of transgenic rice is only the first step in the battle to combat dietary deficiencies. The added nutritional value only makes up for half a person's requirements, and many years will be required to breed the genes into lines adapted to local conditions, but it is a promising start, representative of the very real promise of genetic engineering.

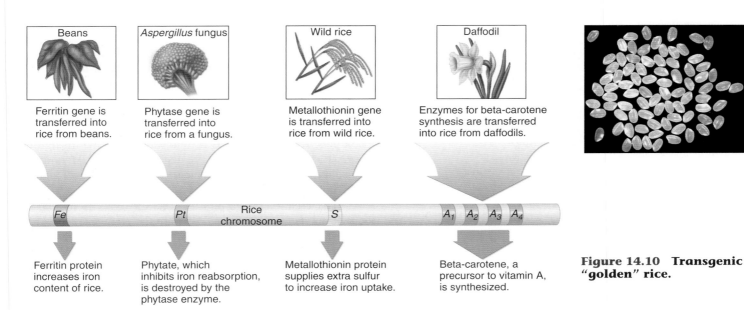

Beans	*Aspergillus* fungus	Wild rice	Daffodil
Ferritin gene is transferred into rice from beans.	Phytase gene is transferred into rice from a fungus.	Metallothionin gene is transferred into rice from wild rice.	Enzymes for beta-carotene synthesis are transferred into rice from daffodils.

Rice chromosome: Fe · Pt · S · A₁ A₂ A₃ A₄

| Ferritin protein increases iron content of rice. | Phytate, which inhibits iron reabsorption, is destroyed by the phytase enzyme. | Metallothionin protein supplies extra sulfur to increase iron uptake. | Beta-carotene, a precursor to vitamin A, is synthesized. |

Figure 14.10 Transgenic "golden" rice.

How Do We Measure the Potential Risks of Genetically Modified Crops?

The advantages afforded by genetic engineering are revolutionizing our lives. But what are the disadvantages, the potential costs and dangers, of genetic engineering? Many people, including influential activists and members of the scientific community, have expressed concern that genetic engineers are "playing God" by tampering with genetic material. Could genetically engineered products administered to plants or animals turn out to be dangerous for consumers? What kind of unforeseen impact on the ecosystem might "improved" crops have?

While the promise of genetic engineering is very much in evidence, this same genetic engineering has been the cause of considerable controversy and protest. The intense feelings generated by this dispute point to the need to understand how we measure the risks associated with the genetic engineering of plants. Two sets of risks need to be considered. The first stems from eating genetically modified foods, the other concerns potential ecological effects.

Is Eating Genetically Modified Food Dangerous? Bioengineers modify crops in two quite different ways. One class of gene modification makes the crop easier to grow; a second class of modification is intended to improve the food itself. Many consumers worry that either sort of genetically modified (GM) food may have been rendered dangerous.

The introduction of glyphosate-resistant soybeans is an example of the first class of modification, producing a GM version that is easier to grow. Is the soybean that results nutritionally different? No. The gene that confers glyphosate resistance in soybeans does so by protecting the plant's ability to manufacture aromatic amino acids. In unprotected weeds, by contrast, glyphosate blocks this manufacturing process, killing the weed. As discussed earlier in this section, humans don't make any aromatic amino acids, so glyphosate doesn't hurt us.

The real issue, of course, is whether the gene modification that renders crop plants glyphosate resistant involves introducing novel proteins with potentially dangerous consequences. Thus in early GM crops like Flavr-Savr tomatoes (genetically engineered to stay fresh longer), one of the issues raised was that the "selectable marker" (a segment of DNA that was transferred along with the genes of interest to be used in screening to ensure that the desired genes had been correctly introduced into the GM tomatoes) was still present in the modified plants, with unknown future consequences.

While new procedures eliminate this possibility in today's GM crops, a second issue raises a valid concern: Could introduced proteins like the enzyme making crops glyphosate tolerant become allergens, causing a potentially fatal immune reaction in some people? Because the potential danger of allergic reactions is quite real, every time a protein-encoding gene is introduced into a GM crop it is necessary to carry out extensive tests of the introduced protein's allergen potential.

Glyphosate-resistant corn has a single added protein, the enzyme EPSP synthetase, which catalyzes the first reaction in aromatic amino acid biosynthesis. How does this added enzyme make the plants tolerant to glyphosate? The introduced protein sequence, a version of the EPSP synthetase gene, is nearly identical to the "wild-type" EPSP synthetase in conventional corn; the only difference is two amino acids out of 444 in the protein sequence. The change alters the EPSP synthetase's shape a little, so that glyphosate no longer can inhibit it.

The key safety question for consumers is whether the change in the EPSP synthetase is capable of inducing an allergic response in humans. The EPSP synthetase in glyphosate-resistant corn has passed extensive and stringent allergenicity tests, and so glyphosate-resistant corn is approved for human consumption by the Environmental Protection Agency (EPA).

This same issue arises each time a gene is added to a crop plant. The addition of the *Bt* gene to corn offers an illuminating example of the need for careful attention to this issue. The *Cry1A* gene isolated from the soil bacterium *Bacillus thuringiensis* creates a Bt protein that binds to specific receptors in the midgut of lepidopteran insects like the European corn borer, but

Calvin and Hobbes   by Bill Watterson

is harmless to humans, wildlife, worms, and beneficial insects that can help control other pests. Thoroughly tested for allergenicity, corn with this Bt protein is approved by the EPA for human consumption.

No GM crop currently being produced in the United States contains a protein that acts as an allergen to humans. On this score, then, the risk of bioengineering to the food supply seems to be slight, so long as adequate testing of new varieties continues.

Are GM Crops Harmful to the Environment?   Those concerned about the widespread use of GM crops raise three legitimate concerns that merit careful evaluation:

1. *Harm to Other Organisms.* The first concern is the possibility of unintentional harm to other organisms. This issue is seen most clearly in the case of crops like Bt corn. Results from a small laboratory experiment suggested that pollen from Bt corn could harm larvae from the Monarch butterfly, which is, like the corn borer, a lepidopteran insect. While this preliminary report received considerable publicity, subsequent studies suggest little possibility of harm. Monarch butterflies lay their eggs on milkweed, not corn, and field management ensures that there is little if any milkweed growing in or near cornfields. In field tests few Monarch larvae are found there. Additionally, field tests confirm that corn sheds its pollen at a different time than when the Monarch larval stage occurs. Finally, the amount of Bt pollen necessary to kill a Monarch caterpillar is far more than is encountered in field tests.

   Farmers focus on the fact that GM cornfields do not need to be sprayed with chemical pesticides to control the corn borer. An estimated $9 billion in damage is caused annually by the application of pesticides in the United States, and billions of insects and other animals, including an estimated 67 million birds, are killed each year.

   In a serious attempt to learn if GM crops are a danger to other organisms, the British government undertook a comprehensive three-year "farm-scale evaluation" of the effects of GM-herbicide-resistant beet, corn, and oilseed crops on biodiversity (in this instance, numbers of kinds of insects). In the study, 60 fields each of beets, corn, and rape (oilseed) were split between conventional varieties and genetically modified herbicide-tolerant strains. In a report released in 2003, the results "reveal significant differences in the effect on biodiversity when managing genetically herbicide-resistant crops as compared to conventional varieties." The report showed that weeds are important sources of food and shelter for insects; because GM crops allow farmers to get rid of weeds more effectively, they have a greater impact on insect populations.

2. *Resistance.* All insecticides used in agriculture share the problem that pests eventually evolve resistance to them, in much the same way that bacterial populations evolve resistance to antibiotics. Use of the insecticide creates a selective pressure favoring mutations that make the pest resistant to it. This is true of chemical pesticides, and also of insecticides produced by Bt corn.

Will pests eventually become resistant to the Bt toxin, just as many have become resistant to the high levels of chemical pesticide we sprayed on crops? This is certainly a possibility, as a few species of insects have developed resistance to Bt when it was sprayed directly on crops in years past. However, despite the widespread use of Bt crops like corn, soybeans, and cotton since 1996, there are as of yet no cases of insects developing resistance to Bt plants in the field. Still, because of this possibility farmers are required to plant at least 20% non-Bt crops alongside Bt crops to provide refuges where insect populations are not under selection pressure and in this way to slow the development of resistance. It is very important that this requirement be met. If ignored, Bt-resistant insect pests may appear in future fields.

3. *Gene Flow.* How about the possibility that introduced genes will pass from GM crops to their wild or weedy relatives? This sort of gene flow happens naturally all the time, and so this is a legitimate question. For the first round of major GM crops, there is usually no potential relative around to receive the modified gene from the GM crop. There are no wild relatives of soybeans in Europe, for example. Thus there can be no gene escape from GM soybeans in Europe, any more than genes can flow from you to your pet dog or cat. For secondary crops only now being gene modified, the risks are more tangible. In 2004, Environmental Protection Agency (EPA) scientists planted eight fields in central Oregon (400 acres in all) with genetically modified creeping bentgrass (*Agrostis stolonifera*). Bentgrass is widely planted on golf course greens because it can be closely mowed. But normal bentgrass is particularly susceptible to weeds, and so requires extensive chemical spraying. The GM form of creeping bentgrass being tested by the EPA had been made resistant to the herbicide glyphosate, which would eliminate the need for most of the spraying. Tracking the spread of GM pollen from these eight fields, the EPA scientists collected seeds from potted "sentinel" bentgrass plants and from natural grasses, grew them to seedling stage, and examined the DNA of the seedling plants for the glyphosate-resistance gene. The gene was found in the potted bentgrass plants up to 13 miles away, and in the wild relative grasses nearly nine miles away from the fields! Might the spread of the glyphosate-resistance gene to surrounding grasses create hard-to-kill "superweeds," or would it be easy to use other weedkillers to control any newly resistant plants? However one interprets its findings, this study suggests that it may be difficult to control GM secondary crops from interbreeding with surrounding relatives to create new hybrids.

14.5 Genetic engineering affords great opportunities for progress in food production, although many are concerned about possible risks. On balance, the risks appear slight, and the potential benefits substantial.

Can Modified Genes Escape from GM Crops?

On the previous page of this chapter you read of a field experiment conducted in 2004 by the Environmental Protection Agency to assess the possibility that introduced genes could pass from genetically modified golf course grass to other plants. Investigators introduced a gene conferring herbicide resistance (the EPSP synthetase gene for resistance to glyphosate) into golf course bentgrass, *A. stolonifera,* and then looked to see if the gene passed from the GM grass to other plants of the same species, and also if it passed to other related species.

The map at the bottom displays the setup of this elaborate field study. A total of 178 *A. stolonifera* plants were placed outside the golf course, many of them downwind. An additional 69 bentgrass individuals were found to be already growing downwind, most of them the related species *A. gigantea*. Seeds were collected from each of these plants, and the DNA of resulting seedlings tested for the presence of the gene introduced into the GM golf course grass. In the graph, the upper red histogram (a **histogram** is a "bar graph" that sorts data into a series of discontinuous categories, the value of each bar representing the number of individuals in a category, or, as in this case, the average value of entries in that category) presents the relative frequency with which the gene was found in *A. stolonifera* plants located at various distances from the golf course. The lower blue histogram does the same for *A. gigantea* plants.

1. **Applying Concepts**
   a. Variable. In the histogram, is there a dependent variable? If so, what is it?
   b. Reading a Histogram. Does the gene conferring resistance to herbicide pass to other plants of this species, *A. stolonifera*? to individuals of the related species *A. gigantea*?
   c. What is the maximal distance over which the herbicide resistance gene is transferred to other

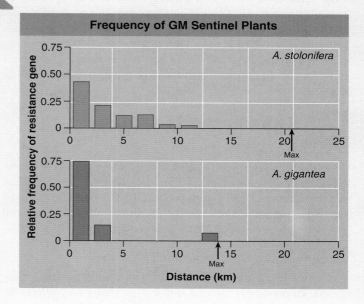

Frequency of GM Sentinel Plants

plants of this species? of the related species? What are these distances, expressed in miles?

2. **Interpreting Data**
   a. What general statement can be made about the effect of distance on the likelihood that the herbicide resistance gene will pass to another plant?
   b. Are there any significant differences in the gene flow to individuals of *A. stolonifera* and to individuals of the related species *A. gigantea*?

3. **Making Inferences** What mechanism do you propose to account for this gene flow?

4. **Drawing Conclusions** Is it fair to conclude that genetically modified traits can pass from crops to other plants? What qualifications would you place on your conclusion?

5. **Further Analysis**
   a. How would you go about testing the role of wind in mediating the gene flow observed in this study?
   b. Do you think that non-grass crops would be as likely to exhibit this sort of gene flow? How would you test this?

One of the most active and exciting areas of biology involves recently developed approaches to manipulating animal cells. In this section, you will encounter three areas where landmark progress is being made in cell technology: reproductive cloning of farm animals, stem cell research, and gene therapy. Advances in cell technology hold the promise of literally revolutionizing our lives.

14.6 Reproductive Cloning

The idea of cloning animals was first suggested in 1938 by German embryologist Hans Spemann (called the "father of modern embryology"), who proposed what he called a "fantastical experiment": remove the nucleus from an egg cell and put in its place a nucleus from another cell.

Is Development Irreversible?

Unexpectedly, early attempts to clone animals in this way failed. Experiments carried out in the 1950s and 1960s by John Gurdon were typical of many others. Using very fine pipettes (hollow glass tubes) to suck the nucleus out of a frog or toad egg, these researchers replaced the egg nucleus with a nucleus sucked out of a body cell taken from another individual (figure 14.11). If the transplanted nucleus was obtained from an adult, the egg never completed development. If instead the nucleus was obtained from an advanced embryo, the egg went on to develop into a tadpole, but most died before becoming an adult. Spemann's fantastical experiment didn't seem to work.

Nuclear transplant experiments like these were attempted without success by many investigators, until finally, in 1984, Steen Willadsen, a Danish embryologist working in Texas, suc-

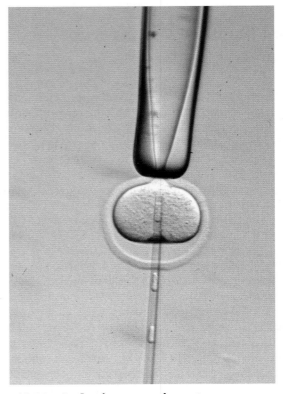

Figure 14.11 A cloning experiment.
In this photo, a nucleus is being injected from a micropipette (*bottom*) into an enucleated egg cell held in place by a pipette.

ceeded in cloning a sheep using the nucleus from a cell of an early embryo. The key to his success was in picking a cell very early in development. This exciting result was soon replicated by others in a host of other organisms, including pigs and monkeys.

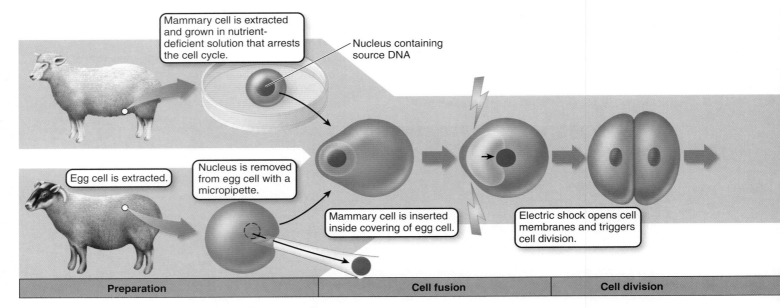

Figure 14.12 Wilmut's animal cloning experiment.
Wilmut combined a mammary cell and an egg cell (with its nucleus removed) to successfully clone a sheep.

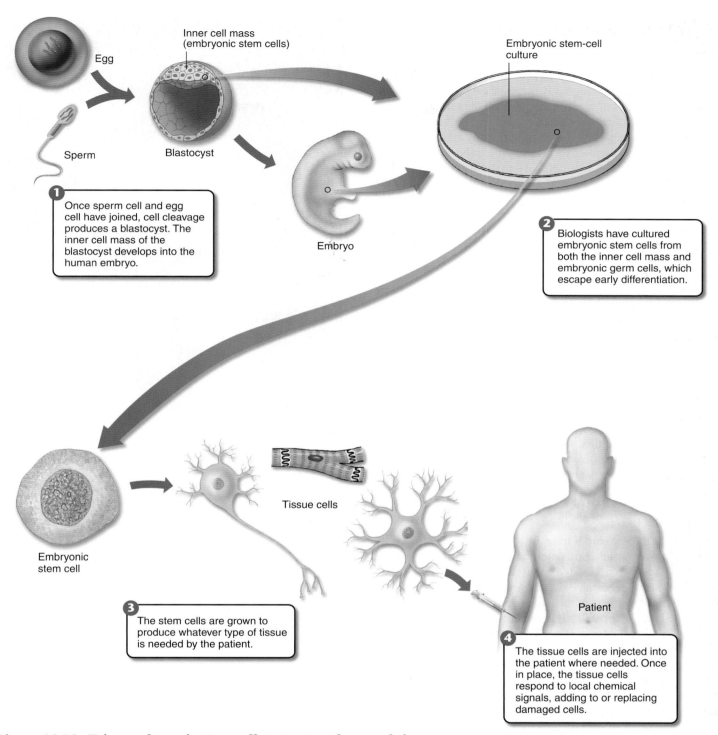

Egg

Sperm

Inner cell mass (embryonic stem cells)

Blastocyst

Embryo

Embryonic stem-cell culture

1 Once sperm cell and egg cell have joined, cell cleavage produces a blastocyst. The inner cell mass of the blastocyst develops into the human embryo.

2 Biologists have cultured embryonic stem cells from both the inner cell mass and embryonic germ cells, which escape early differentiation.

Embryonic stem cell

Tissue cells

Patient

3 The stem cells are grown to produce whatever type of tissue is needed by the patient.

4 The tissue cells are injected into the patient where needed. Once in place, the tissue cells respond to local chemical signals, adding to or replacing damaged cells.

Figure 14.16 Using embryonic stem cells to restore damaged tissue.

Embryonic stem cells can develop into any body tissue. Methods for growing the tissue and using it to repair damaged tissue in adults, such as the brain cells of multiple sclerosis patients, heart muscle, and spinal nerves, are being developed.

14.8 Therapeutic Cloning

The revolution in cell technology has been led by advances in stem cell research. By surgically transplanting embryonic stem cells, scientists can now routinely perform the remarkable feat of repairing disabled body tissues in mice. The basic strategy for repairing damaged tissues is to surgically transfer embryonic stem cells to the damaged area, where the stem cells can form healthy replacement cells. Thus embryonic stem cells transferred into the pancreas of a diabetic mouse that lacks the islet cells needed to produce insulin have been induced to become insulin-secreting islet cells (figure 14.17). The new cells produce only about 2% as much insulin as normal islet cells do, so there is still plenty to learn, but the take-home message is clear: transplanted embryonic stem cells offer a path to cure type I diabetes.

While exciting, these therapeutic uses of embryonic stem cells to cure type I diabetes, Parkinson's disease, damaged heart muscle, and injured nerve tissue were all achieved in experiments carried out using strains of mice without functioning immune systems. Why is this important? Because had these mice possessed fully functional immune systems, they almost certainly would have rejected the implanted stem cells as foreign. Humans with normal immune systems might well refuse to accept transplanted stem cells simply because they are from another individual. For such stem cell therapy to work in humans, this problem needs to be addressed and solved. The need to solve this problem has been the primary impetus behind the push to develop therapeutic cloning.

Using Cloning to Achieve Immune Acceptance

Early in 2001, a research team at the Rockefeller University reported a way around this potentially serious problem. Their solution? They first isolated skin cells from a mouse, then using the same procedure that created Dolly, they created a 120-cell embryo from them. The embryo was then destroyed, its embryonic stem cells harvested for transfer to injured tissue. This procedure, called **therapeutic cloning**, and the procedure that was used to create Dolly, called **reproductive cloning,** are contrasted in figure 14.18. You can see that steps ❶ through ❺ are essentially the same for both procedures, but the two methods proceed differently after that. In reproductive cloning, the blastocyst from step ❺ is implanted in a surrogate mother in step ❻ⓐ, developing into a baby that is genetically identical to the nucleus donor, step ❼ⓐ. In therapeutic cloning, by contrast, stem cells from the blastocyst of step ❺ are removed and grown in culture, step ❻. These stem cells develop into pancreatic islet cells in step ❼ and are injected or transplanted into the diabetic patient in step ❽, where they begin producing insulin.

Therapeutic cloning, or, more technically, **somatic cell nuclear transfer,** successfully addresses the key problem that must be solved before embryonic stem cells can be used to repair damaged human tissues, which is immune acceptance.

Figure 14.17 Embryonic stem cells growing in cell culture.
Embryonic stem cells derived from early human embryos will grow indefinitely in tissue culture. When transplanted, they can sometimes be induced to form new cells of the adult tissue into which they have been placed. This suggests exciting therapeutic uses.

Because stem cells are cloned from the body's own tissues in therapeutic cloning, they pass the immune system's "self" identity check, and the body readily accepts them.

Therapeutic Cloning Is Controversial

It is important to draw a clear distinction between therapeutic cloning and reproductive cloning. In therapeutic cloning, the cloned embryo is destroyed to obtain embryonic stem cells—whereas in reproductive cloning, the embryo develops into an adult individual. Human reproductive cloning is ethically unacceptable. The possibility of therapeutic cloning is also controversial, for fear by some that a cloned embryo might be brought to term by inserting it into a human uterus.

There is little doubt that the controversy over human therapeutic cloning will continue.

> **14.8** Therapeutic cloning involves initiating blastocyst development from a patient's tissue using nuclear transplant procedures, then using the blastocyst's embryonic stem cells to replace the patient's damaged or lost tissue.

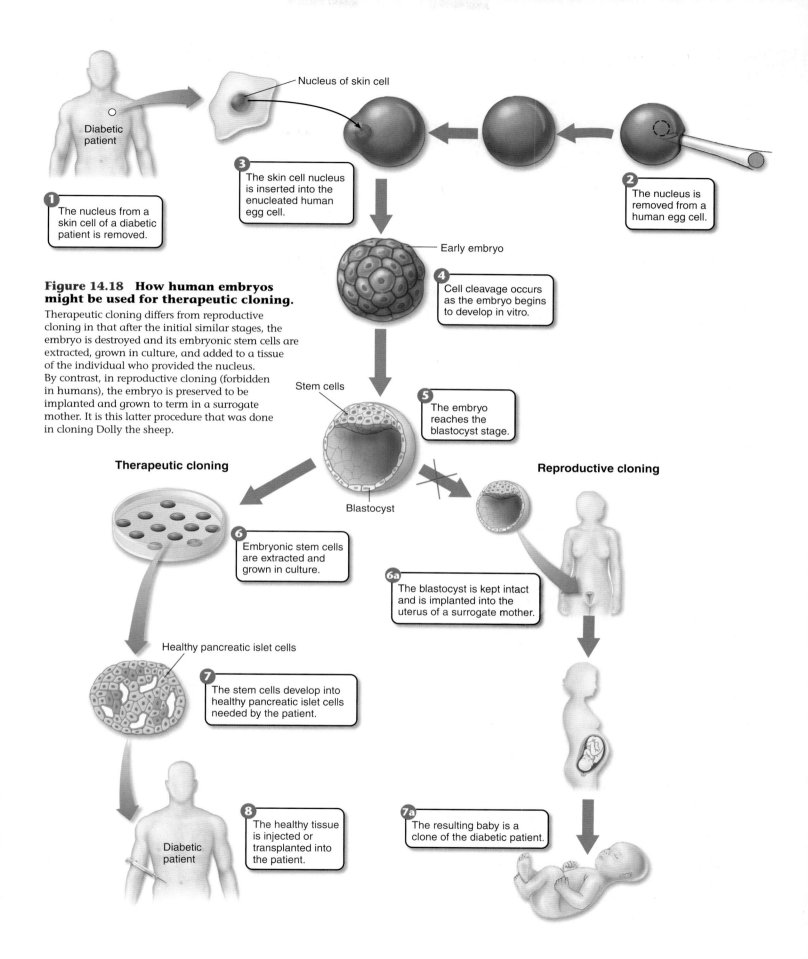

Nucleus of skin cell

1 The nucleus from a skin cell of a diabetic patient is removed.

Diabetic patient

2 The nucleus is removed from a human egg cell.

3 The skin cell nucleus is inserted into the enucleated human egg cell.

Early embryo

4 Cell cleavage occurs as the embryo begins to develop in vitro.

Figure 14.18 How human embryos might be used for therapeutic cloning.

Therapeutic cloning differs from reproductive cloning in that after the initial similar stages, the embryo is destroyed and its embryonic stem cells are extracted, grown in culture, and added to a tissue of the individual who provided the nucleus. By contrast, in reproductive cloning (forbidden in humans), the embryo is preserved to be implanted and grown to term in a surrogate mother. It is this latter procedure that was done in cloning Dolly the sheep.

Stem cells

5 The embryo reaches the blastocyst stage.

Blastocyst

Therapeutic cloning

Reproductive cloning

6 Embryonic stem cells are extracted and grown in culture.

6a The blastocyst is kept intact and is implanted into the uterus of a surrogate mother.

Healthy pancreatic islet cells

7 The stem cells develop into healthy pancreatic islet cells needed by the patient.

8 The healthy tissue is injected or transplanted into the patient.

Diabetic patient

7a The resulting baby is a clone of the diabetic patient.

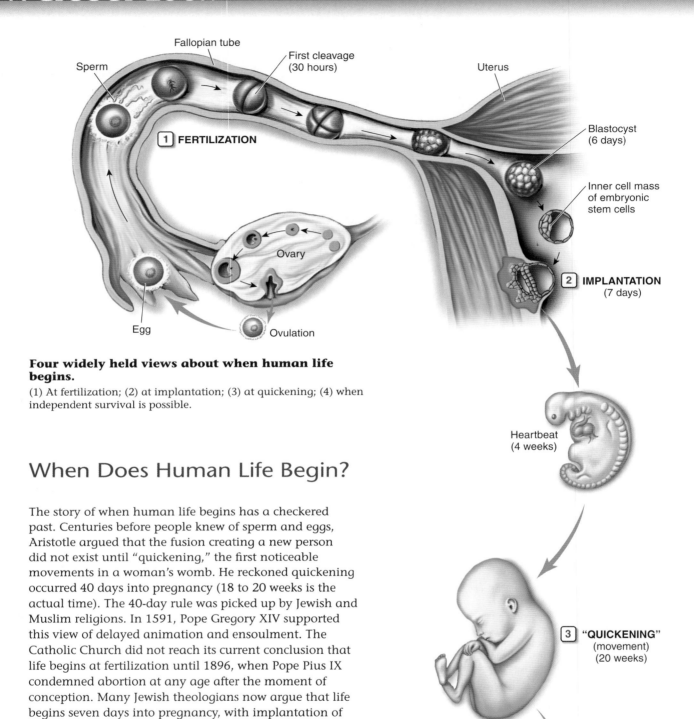

1 **FERTILIZATION**

Sperm

Fallopian tube

First cleavage
(30 hours)

Uterus

Blastocyst
(6 days)

Inner cell mass
of embryonic
stem cells

Ovary

Egg

Ovulation

2 **IMPLANTATION**
(7 days)

Heartbeat
(4 weeks)

3 **"QUICKENING"**
(movement)
(20 weeks)

4 **INDEPENDENT SURVIVAL POSSIBLE**
(third trimester)

Four widely held views about when human life begins.
(1) At fertilization; (2) at implantation; (3) at quickening; (4) when independent survival is possible.

When Does Human Life Begin?

The story of when human life begins has a checkered past. Centuries before people knew of sperm and eggs, Aristotle argued that the fusion creating a new person did not exist until "quickening," the first noticeable movements in a woman's womb. He reckoned quickening occurred 40 days into pregnancy (18 to 20 weeks is the actual time). The 40-day rule was picked up by Jewish and Muslim religions. In 1591, Pope Gregory XIV supported this view of delayed animation and ensoulment. The Catholic Church did not reach its current conclusion that life begins at fertilization until 1896, when Pope Pius IX condemned abortion at any age after the moment of conception. Many Jewish theologians now argue that life begins seven days into pregnancy, with implantation of the embryo. Gene transcription starts even later (well after stem cells are harvested), and many scientists feel human individuality cannot be said to begin until then, when the embryo starts to actually use its genes. The U.S. Supreme Court takes the position that human life begins much later, when the fetus becomes capable of independent life if separated from the mother—roughly the third trimester.

14.9 Gene Therapy

The third major advance in cell technology involves introducing "healthy" genes into cells that lack them. For decades scientists have sought to cure often-fatal genetic disorders like cystic fibrosis, muscular dystrophy, and multiple sclerosis by replacing the defective gene with a functional one.

Early Success

That such **gene transfer therapy** can work was first demonstrated in 1990. Two girls were cured of a rare blood disorder due to a defective gene for the enzyme adenosine deaminase. Scientists isolated working copies of this gene and introduced them into bone marrow cells taken from the girls. The gene-modified bone marrow cells were allowed to proliferate, then were injected back into the girls. The girls recovered and stayed healthy. For the first time, a genetic disorder was cured by gene therapy.

The Rush to Cure Cystic Fibrosis

Like hounds to a hot scent, researchers set out to apply the new approach to one of the big killers, cystic fibrosis. The defective gene, labelled *cf,* had been isolated in 1989. Five years later, in 1994, researchers successfully transferred a healthy *cf* gene into a mouse with a defective one—they in effect had cured cystic fibrosis in a mouse. They achieved this remarkable result by adding the *cf* gene to a virus that infected the lungs of the mouse, carrying the gene with it "piggyback" into the lung cells. The virus chosen as the "vector" was adenovirus, a virus that causes colds and is very infective of lung cells. To avoid any complications, the lab mice used in the experiment had their immune systems disabled.

Very encouraged by these well-publicized preliminary trials with mice, several labs set out in 1995 to attempt to cure cystic fibrosis by transferring healthy copies of the *cf* gene into human patients. Confident of success, researchers added the human *cf* gene to adenovirus, then squirted the gene-bearing virus into the lungs of cystic fibrosis patients. For eight weeks the gene therapy did seem successful, but then disaster struck. The gene-modified cells in the patients' lungs came under attack by the patients' own immune systems. The "healthy" *cf* genes were lost and with them any chance of a cure.

Problems with the Vector

Other attempts at gene therapy met with similar results, eight weeks of hope followed by failure. In retrospect, although it was not obvious then, the problem with these early attempts seems predictable. Adenovirus causes colds. Do you know anyone who has never had a cold? When you get a cold, your body produces antibodies to fight off the infection, and so all of us have antibodies directed against adenovirus. We were introducing therapeutic genes in a vector our bodies are primed to destroy.

In 1995, the newly appointed head of the National Institutes of Health (the NIH), Nobel Prize winner Harold Varmus, held a comprehensive review of human gene therapy trials. Three problems became evident in the review. (1) The adenovirus vector being used in most trials elicits a strong immune response, leading to rejection of the added gene. (2) Adenovirus infection can, in rare instances, produce a very severe immune reaction, enough to kill. If many patients are treated, such instances can be expected to occur. (3) When the adenovirus infects a cell, it inserts its DNA into the human chromosome. Unfortunately, it does so at a random location. This means that the insertion events will cause mutations—by jumping into the middle of a gene, the virus inactivates that gene. Because the spot where the adenovirus inserts is random, some of the mutations that result can be expected to cause cancer, certainly an unacceptable consequence.

Faced with these findings, Varmus called a halt to all further human clinical trials of gene therapy. "Go back to work in the laboratory," he told researchers, "until you get a vector that works."

More Promising Vectors

Within a few years, researchers had a much more promising vector. This new gene carrier is a tiny parvovirus called *adeno-associated virus* (AAV). It has only two genes and needs adenovirus to replicate. To create a vector for gene transfer, researchers remove both of the AAV genes. The shell that remains is still quite infective and can carry human genes into patients. More importantly, AAV enters human DNA far less frequently than adenovirus; thus, it is less likely to produce cancer-causing mutations. In addition, AAV does not elicit a strong immune response—cells infected with AAV are not eliminated by a patient's immune system. Finally, AAV never elicits a dangerously strong immune response, so it is safe to administer AAV to patients.

Success with the AAV Vector

In 1999, AAV successfully cured anemia in rhesus monkeys. In monkeys, humans, and other mammals, red blood cell production is stimulated by a protein called erythropoietin (EPO). People with a type of anemia caused by low red blood cell counts, like dialysis patients, get regular injections of EPO. Using AAV to carry a souped-up EPO gene into the monkeys, scientists were able to greatly elevate their red blood cell counts, curing the monkeys of anemia—and they stayed cured.

A similar experiment using AAV cured dogs of a hereditary disorder leading to retinal degeneration and blindness. These dogs had a defective gene that produced a mutant form of a protein associated with the retina of the eye and were blind. Recombinant viral DNA was made using a healthy version of the gene, shown in steps ❶ and ❷ in figure 14.19. Injection of AAV bearing the needed gene into the fluid-filled compartment behind the retina in step ❸ restored their sight in step ❹.

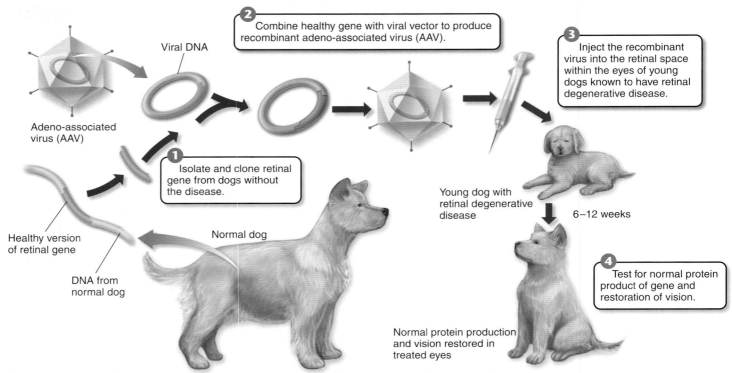

Figure 14.19 Using gene therapy to cure a retinal degenerative disease in dogs.
Researchers were able to use genes from healthy dogs to restore vision in dogs blinded by an inherited retinal degenerative disease. This disease also occurs in human infants and is caused by a defective gene that leads to early vision loss, degeneration of the retinas, and blindness. In the gene therapy experiments, genes from dogs without the disease were inserted into 3-month-old dogs that were known to carry the defective gene and that had been blind since birth. Six weeks after the treatment, the dogs' eyes were producing the normal form of the gene's protein product, and by three months, tests showed that the dogs' vision was restored.

Human clinical trials are under way again. In 2000, scientists performed the first gene therapy experiment for muscular dystrophy, injecting genes into a 35-year-old South Dakota man. He is an early traveler on what is likely to become a well-traveled therapeutic highway. Trials are also under way for cystic fibrosis, rheumatoid arthritis, hemophilia, and a wide variety of cancers. The way seems open, the possibility of progress tantalizingly close.

14.9 In principle, it should be possible to cure hereditary disorders like cystic fibrosis by transferring a healthy gene into the cells of affected tissues. Early attempts using adenovirus vectors were not often successful. New virus vectors like AAV avoid the problems of earlier vectors and offer promise of gene transfer therapy cures.

INQUIRY & ANALYSIS

Is Snuppy a Fake?

Photographs are often used to substantiate claims. The puppy in this photo is "Snuppy," the first dog to be cloned. Beside him, to the left, is the adult male dog who provided the skin cell from which Snuppy was cloned. The dog to the right of Snuppy is his surrogate mother.

1. **Making Inferences**
   a. Judging by visual similarity, which adult dog is the closer relative of Snuppy?
   b. The scientist who cloned Snuppy has been accused of faking research evidence in different experiments concerning stem cells. Given this cloud on his reputation, what evidence would you accept that Snuppy is indeed a clone?

Snuppy, with His Clone Dad & Surrogate Mom

Sequencing Entire Genomes

14.1 Genomics

- The genetic information of an organism, its genes and other DNA, is called its genome. The sequencing and study of genomes is a new area of biology called genomics.

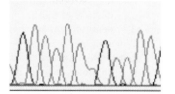

- The sequencing of entire genomes, a once long and tedious process, has been made faster and easier with automated systems (**figure 14.1**).

14.2 The Human Genome

- The human genome contains about 20,000 to 25,000 genes, not much more than other organisms (**table 14.1**) and far less than what was expected due to alternative splicing of exons.

- Genes are organized in different ways in the genome, with nearly 99% of the human genome containing noncoding segments (**figure 14.2** and **table 14.2**).

Genetic Engineering

14.3 A Scientific Revolution

- Genetic engineering is the process of moving genes from one organism to another. It is having a major impact on medicine and agriculture (**figure 14.3**).

- Restriction enzymes are a special kind of enzyme that binds to short sequences of DNA and cuts them at specific sequences. When two different molecules of DNA are cut with the same restriction enzyme, sticky ends form, allowing the segments of different DNAs to be joined (**figure 14.4**). Cleaving DNA is the first of four stages of a genetic engineering experiment. The other three stages are producing recombinant DNA, cloning the DNA in a host cell, and screening the clones for the gene of interest (**figure 14.5**).

14.4 Genetic Engineering and Medicine

- Genetic engineering has been used in many medical applications. The first application was the production of medically important proteins used to treat illnesses (**table 14.3**). Genes encoding the proteins are inserted into bacteria that become small protein-producing factories.

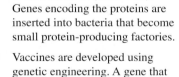

- Vaccines are developed using genetic engineering. A gene that encodes a viral protein is put into the DNA of a harmless vector that expresses the protein. The vector is injected into the body. The viral protein is produced in the body and elicits an immune response to the protein, which protects the person from an infection by that virus in the future (**figure 14.7**).

14.5 Genetic Engineering and Agriculture

- Genetic engineering has been used in manipulating crop plants to make them more cost-effective to grow or more nutritious (**figures 14.8–14.10**).

- However, GM plants are a source of controversy because of potential dangers that may result from the genetic manipulation of crop plants.

The Revolution in Cell Technology

14.6 Reproductive Cloning

- Ian Wilmut succeeded in cloning a sheep by synchronizing the cell cycles of the two cells, the nucleus donor cell, and the host cell (**figure 14.12**).

- Other animals have been successfully cloned (**figure 14.13**), but problems and complications usually arise, causing premature death. The problems with successful cloning appear to be caused by the lack of modifications that need to be made to the DNA, which turns genes "on" or "off," a process called genomic imprinting (**figure 14.14**).

14.7 Embryonic Stem Cells

- Embryonic stem cells are totipotent cells, which are cells that are able to divide and develop into any cell in the body or develop into an entire individual. These cells are present in the early embryo. Because of the totipotent nature of embryonic stem cells, they could be used to replace tissues lost or damaged due to accident or disease (**figure 14.16**).

14.8 Therapeutic Cloning

- The use of embryonic stem cells to replace damaged tissue has one major drawback, which is tissue rejection. The embryonic stem cells, harvested from an embryo, are treated as foreign cells by the patient's body. The patient's body will reject the tissue produced from the embryonic stem cell. A process called therapeutic cloning could alleviate the problems with immunological reactions.

- Therapeutic cloning is the process whereby a cell from an individual who has lost tissue function due to disease or accident is cloned, producing an embryo. Embryonic stem cells are harvested from the cloned embryo and injected into the same individual. The embryonic stem cell regrows the lost tissue without eliciting an immune response (**figure 14.18**). However, this procedure is also controversial.

14.9 Gene Therapy

- Using gene therapy, a patient with a genetic disorder is cured by replacing a defective gene with a "healthy" gene. In theory, this should work—but early attempts to cure cystic fibrosis failed because of immunological reactions to the adenovirus vector used to carry the healthy gene into the patient.

- The recent focus of gene therapy has been to identify a vector that avoids the problems encountered with the adenovirus vector. Promising results in experiments using a parvovirus called adeno-associated virus (AAV) has scientists hopeful that this vector will eliminate the problems seen with adenovirus (**figure 14.19**).

1. The total amount of DNA in an organism, including all of its genes and other DNA, is its
   a. heredity.   c. genome.
   b. genetics.   d. genomics.

2. A possible reason why humans have such a small number of genes as opposed to what was anticipated by scientists is that
   a. humans don't need more than 25,000 to function.
   b. the exons used to make a specific mRNA can be rearranged to form different proteins.
   c. the sample size used to sequence the human genome was not big enough, so the number of genes estimated could be low.
   d. the estimate will increase as scientists find out what so-called "junk DNA" actually does.

3. A protein that can cut DNA at specific DNA base sequences is called a
   a. DNase.   c. restriction enzyme.
   b. DNA ligase.   d. DNA polymerase.

4. The four steps of a genetic engineering experiment are (in order)
   a. cleaving DNA, cloning, producing recombinant DNA, and screening.
   b. cleaving DNA, producing recombinant DNA, cloning, and screening.
   c. producing recombinant DNA, cleaving DNA, screening, and cloning.
   d. screening, producing recombinant DNA, cloning, and cleaving DNA.

5. Using drugs produced by genetically engineered bacteria allows
   a. the drug to be produced in far larger amounts than in the past.
   b. humans to permanently correct the effect of a missing gene from their own systems.
   c. humans to cure cystic fibrosis.
   d. All these answers are correct.

6. Some of the advantages to using genetically modified organisms in agriculture include
   a. increased yield.
   b. unchanged nutritive value.
   c. the ease of transferring the gene to other organisms.
   d. possibility of anaphylaxis.

7. Which of the following is *not* a concern about the use of genetically modified crops?
   a. possible danger to humans after consumption
   b. insecticide resistance developing in pest species
   c. gene flow into natural relatives of GM crops
   d. harm to the crop itself from mutations

8. Genomic imprinting seems to involve
   a. protein signals that block transcription of a gene from its DNA.
   b. proteins that cause deformation of RNA polymerase.
   c. methylation or demethylation of RNA polymerase.
   d. methylation or demethylation of DNA.

9. One of the main biological problems with replacing damaged tissue through therapeutic cloning and use of embryonic stem cells is
   a. immunological rejection of the tissue by the patient.
   b. that stem cells may not target appropriate tissue.
   c. the time needed to grow sufficient amounts of tissue from stem cells.
   d. that genetic mutation of chosen stem cells may cause future problems.

10. In gene therapy, healthy genes are placed into cells with defective genes by using
    a. bacteria.
    b. micropipettes (needles).
    c. viruses.
    d. Cells are not modified genetically. Instead, healthy tissue is grown and transplanted into the patient.

1. **Figure 14.1** Can you sequence the unknown section of DNA with the DNA fragments obtained with the following DNA sequencing?

2. **Figure 14.12** Your friend Thomas wants to know why scientists can't just take the egg cell, with its own nucleus intact, and shock it to begin cell division? How do you answer him?

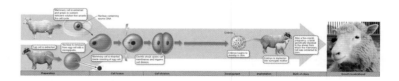

1. The goal behind therapeutic cloning is to replace tissue that is damaged due to an accident or nongenetic disease. Why wouldn't therapeutic cloning as shown in figure 14.18 work to replace tissue damage caused by genetic disorders?

2. If a person has a genetic disease such as cystic fibrosis, the hope is that we will be able to use gene therapy to cure him or her. When that happens, will the patient's future children and grandchildren also be free of the disease? Explain.

3. Much of the technology for producing GM foods is owned by multinational corporations, which seek to maintain intellectual ownership of their creations. As one example, Monsanto

Corporation requires farmers to sign contracts for glyphosate-tolerant soybeans that prevent the farmers from saving seed for replanting the next year. The company has aggressively brought suit against violators. On the one hand, companies need to be able to profit from their products, and the development costs of GM foods are enormous. Without potential profit, future GM crops will not be developed. On the other hand, in many highly populated regions of the world, people who face famine when their crops fail simply cannot afford to pay the price of seeds every year. How would you want to see this challenging issue handled?

15

Evolution and Natural Selection

This photograph shows color variants of the peppered moth, *Biston betularia,* glued to the trunk of a soot-polluted tree. Historically, the dark form of the moth was rare, but with the onset of the Industrial Revolution the dark "melanic" forms became more common. To test the hypothesis that natural selection was favoring the dark form because of industrial pollution, biologists released equal numbers of the two variants in both polluted and nonpolluted areas. In polluted areas, twice as many dark individuals survived as light ones; in unpolluted areas, the reverse occurred, with more light-colored individuals surviving. This experiment clearly demonstrated that natural selection was favoring the dark form in polluted environments. How? The most likely hypothesis was that birds were eating the moths, selecting more often those that were most visible on tree trunks. As you can see in the photo, white moths are more visible on polluted tree trunks. As a test, moths were glued to dark tree trunks to see which the birds would eat. As you might expect, they ate the most visible ones. So is camouflage the reason natural selection favored the dark moths? Probably not. Further work showed these moths don't spend their days on tree trunks. Some other effect of the pollution seems to be at work. This peppered moth story has long been held to be a classic example of evolution in action, mentioned in all texts. The evidence for natural selection is exceptionally clear, and the recent reevaluation removing predation as a potential agent of selection in this case provides an excellent example of how science works, its conclusions always subject to revision as we learn more. However, a decade ago, critics of teaching evolution in the classroom began to criticize the teaching of the peppered moth story, suggesting its investigators cheated by gluing moths to trees. Although the evidence for natural selection is very strong and does not in any way depend on these studies, the spotted moth story has disappeared from most texts. In this chapter, we will look closely at the action of natural selection, at the evidence for evolution, and at the controversy surrounding its presentation in the nation's classrooms.

15.1 Evolution: Getting from There to Here

Since the publication of *On The Origin of Species* by Charles Darwin in 1859, the idea of evolution by natural selection has played a central role in the science of biology. Although the modern concept of biological evolution has been developed considerably since Darwin, his phrase "descent with modification" still captures the essence of his proposal: all species arise from other, preexisting species. Over time, a species accumulates differences, adaptations in response to environmental challenges, such that ancestral and descendant species are not identical. It is through this concept of evolution that we are able to explain the great paradox of biology, that in life there exists both unity and diversity.

Darwin's general concept of evolution was introduced in chapter 2, but it is important—before launching into a detailed discussion of evolution and natural selection—that we first review Darwin's proposal of how natural selection, the process that leads to evolution, works. The process of natural selection can be conveniently visualized as occurring in a series of steps:

1. There is gene variation among individuals of a population.
2. This variation is often passed on to offspring, so that offspring have a tendency to have traits more like their parents than like other members of the population.
3. All populations overproduce young—only some of the offspring will survive to reproduce.
4. The likelihood that a particular individual will survive and reproduce is not random. The traits it inherits make it more or less likely that it will be able to respond to environmental challenges and survive to reproduce.
5. Those individuals with traits that aid in responding to the environment, so that they are able to reproduce, will have a better chance of adding individuals to the next generation.
6. Over time, the population changes such that the traits of the more successful reproducers become more prevalent in the population.

Microevolution Leads to Macroevolution

When the word *evolution* is mentioned, it is difficult not to conjure up images of dinosaurs roaming the earth, woolly mammoths frozen in blocks of ice, or Darwin observing a monkey. Traces of ancient life-forms, now extinct, survive as fossils that help us piece together the evolutionary story.

With such a background, we usually think of evolution in terms of changes that take place over long periods of time, changes in the kinds of animals and plants on earth as new forms arise from existing ones. This kind of evolution, called **macroevolution,** is evolutionary change on a grand scale. Macroevolution is larger, more complex changes that result in the creation of new species and higher taxonomic groups.

Much of the focus of Darwin's theory of natural selection, however, is directed not at the way in which new species form, but rather at the way that changes occur *within* a species to make that species different from its immediate ancestor. Natural selection is the process whereby some individuals in a population, those that possess certain inherited characteristics or adaptations, produce more surviving offspring than individuals lacking these characteristics. As a result, the population will gradually come to include more and more individuals with the advantageous characteristics—in other words, adaptation is the outcome of natural selection, and in this way the population evolves. Changes of this sort within populations are called **microevolution.**

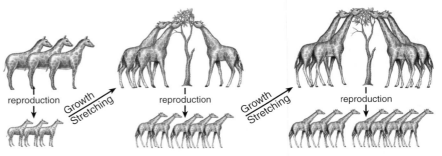

reproduction

Growth Stretching

reproduction

Growth Stretching

reproduction

Proposed ancestor of giraffes has characteristics of modern-day okapi.

The giraffe ancestor lengthened its neck by stretching to reach tree leaves, then passed the change on to offspring.

(a) **Lamarck's theory: variation is acquired.**

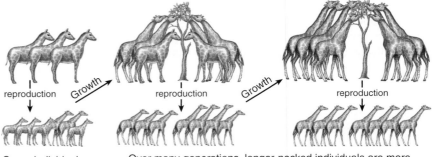

reproduction

Growth

reproduction

Growth

reproduction

Some individuals born happen to have longer necks.

Over many generations, longer-necked individuals are more successful, perhaps because they can feed on taller trees. These successful individuals have more offspring and pass the long-neck trait on to them.

(b) **Darwin's theory: variation is inherited.**

Figure 15.1 How did long necks evolve in giraffes?

(a) Lamarck proposed that the trait for a longer neck was acquired and then passed on to offspring. (b) Darwin proposed that there existed some variation in the length of the neck in the population, and over time longer necks became more prevalent in the population.

Adaptation results from microevolutionary changes that increase the likelihood of survival and reproduction of particular genetic traits in a population. In essence, Darwin's explanation of evolution is that adaptation by natural selection is responsible for evolutionary changes *within* a species (microevolution), and that the accumulation of these changes results in larger and more complex changes, leading to the development of new species and higher taxonomic groups (macroevolution).

The Key Is the Source of the Variation

Darwin did not invent the idea of evolution. Rather, he agreed with many earlier philosophers and naturalists who deduced that the many kinds of organisms around us were produced by a process of evolution. The question that had not been answered was *how* organisms evolved. What mechanism caused it?

Until Darwin, there was no consensus among biologists about the mechanism causing evolution among the earth's organisms. An idea championed by a predecessor of Darwin, the prominent biologist Jean-Baptiste Lamarck, was that evolution occurred by the inheritance of acquired characteristics. According to Lamarck, individuals passed on to offspring body and behavior changes acquired during their lives. Figure 15.1*a* illustrates Lamarck's theory. Lamarck proposed that ancestral giraffes with short necks (the left-hand side of the figure) tended to stretch their necks to feed on tree leaves, and this extension of the neck was passed on to subsequent generations, leading to the long-necked giraffe.

In Darwin's theory, by contrast, the variation is not created by experience but already exists when selection acts on it, an idea illustrated in figure 15.1*b*. Populations of ancestral giraffes contained variation—you can see individuals on the left-hand side of the figure with longer necks. Able to feed higher up on the trees, these individuals had more food and so were better able to survive and reproduce than their shorter-necked relatives. When these longer-necked animals did reproduce, they passed their long-neck trait on to their offspring. Over time, the long-neck trait became more prevalent in the population, resulting in a change in the physical characteristics of the species—evolution. Rather than individuals acquiring characteristics during their lifetimes, Darwin proposed natural selection over generations as the mechanism of evolution.

It is important to remember that natural selection can only act on the variation already present in a population. New gene variations are constantly being introduced into populations through random mutations. Of course, for the variation to be passed on to offspring, the mutation must occur in the germ cells. Mutations that only affect somatic cells (for example, mutations in lung cells that result from tobacco use and lead to lung cancer) cannot be passed on to offspring.

The Rate of Evolution

For more than a century after the publication of *On the Origin of Species,* the standard view was that evolutionary change occurred extremely slowly. Such change would be nearly imperceptible from generation to generation, but would accumulate such that, over the course of millions of years, major changes could occur. This view is termed *gradualism.*

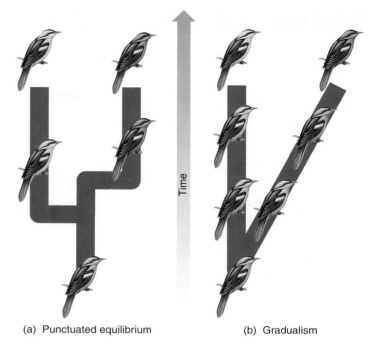

(a) Punctuated equilibrium   (b) Gradualism

Figure 15.2 Two views of the pace of macroevolution.

(a) Punctuated equilibrium surmises that species formation occurs in bursts, separated by long periods of stasis, while (b) gradualism surmises that species formation is constantly occurring by accumulating small differences.

An alternative view is that species experience long periods of little or no evolutionary change (termed *stasis*), punctuated by bursts of evolutionary change occurring over geologically short time intervals. This view, termed *punctuated equilibrium,* is contrasted with gradualism in figure 15.2. In some species, evolution appears to proceed in the spurts you see on the left in figure 15.2*a*. In times of environmental upheaval, evolutionary innovations give rise to new lines (the branch points in the figure); these lines then persist unchanged for a long time (the vertical bars), in "equilibrium," until a new spurt of evolution creates a "punctuation" in the fossil record. Some marine organisms seem to show the irregular pattern of evolutionary change the punctuated equilibrium model predicts. On the other hand, some other well-documented groups such as African mammals clearly have evolved gradually, and not in spurts, more like the pattern you see on the right in figure 15.2*b*. It appears, in fact, that gradualism and punctuated equilibrium are two ends of a continuum. Although some groups appear to have evolved solely in a gradual manner and others only in a punctuated mode, many other groups appear to show evidence of both gradual and punctuated episodes at different times in their evolutionary history.

15.1 Darwin proposed that natural selection on variants within populations leads to evolution. Viewed over long time periods, some groups evolve at a uniform rate, others in fits and starts.

The evidence that Darwin presented in *The Origin of Species* to support his theory of evolution was strong. Darwin presented both artificial selection arguments and compelling biogeographic evidence he had collected on his round-the-world voyage. You have already reviewed some of this evidence in chapter 2, section 2.2. We will now examine other lines of evidence supporting Darwin's theory.

The Fossil Record

The most direct evidence of macroevolution is found in the fossil record. **Fossils** are the preserved remains, tracks, or traces of once-living organisms. Fossils are created when organisms become buried in sediment. The calcium in bone or other hard tissue mineralizes, and the surrounding sediment eventually hardens to form rock. Most fossils are, in effect, skeletons. In the rare cases when fossils form in very fine sediment, feathers may also be preserved. When remains are frozen or become suspended in amber (fossilized plant sap), however, the entire body may be preserved. The fossils contained in layers of sedimentary rock reveal a history of life on earth.

By dating the rocks in which fossils occur, we can get an accurate idea of how old the fossils are. Rocks are dated by measuring the degree of decay of certain radioisotopes contained in the rock, a procedure you encountered in chapter 3.

Using Fossils to Test the Theory of Evolution

If the theory of evolution is correct, then the fossils we see preserved in rock should represent a history of evolutionary change. The theory makes the clear prediction that a parade of successive changes should be seen, as first one change occurs and then another. If the theory of evolution is not correct, on the other hand, then such orderly change is not expected.

To test this prediction, biologists follow a very simple procedure:

1. *Assemble a collection of fossils of a particular group of organisms.* You might for example gather together a collection of fossil titanotheres, a hoofed mammal that lived between about 50 million and 35 million years ago.
2. *Date each of the fossils.* In dating the fossils, it is important to make no reference to what the fossil is like. Imagine it as being concealed in a black box of rock, with only the box being dated.
3. *Order the fossils by their age.* Without looking in the "black boxes," place them in a series, beginning with the oldest and proceeding to the youngest.
4. *Now examine the fossils.* Do the differences between the fossils appear jumbled, or is there evidence of successive change as evolution predicts? You can judge for yourself in figure 15.3. During the 15 million years spanned by this collection of titanothere fossils, the small, bony protuberance located above the nose 50 million years ago evolved in a series of continuous changes into relatively large blunt horns.

The Fossil Record Confirms Evolution's Key Prediction

It is important not to miss the key point of the result you see illustrated in figure 15.3: evolution is an observation, not a conclusion. Because the dating of the samples is independent of what the samples are like, *successive change through time is a data statement.* While the statement that evolution is the result of natural selection is a theory advanced by Darwin, the statement that macroevolution has occurred is a factual observation.

Many other examples illustrate this clear confirmation of the key prediction of Darwin's theory. The evolution of today's large, single-hoof horse with complex molar teeth from a much smaller four-toed ancestor with much simpler molar teeth is a familiar and clearly documented instance.

While many gaps interrupted the fossil record in Darwin's era, even then scientists knew of *Archaeopteryx,* a transitional fossil between reptiles and birds. Today, the fossil record is far more complete, particularly among the vertebrates; fossils have been found linking all the major groups. The forms linking mammals to reptiles are particularly well known.

New finds continue to fill in gaps. New fossils more completely revealing our evolution from ape ancestors are found practically every year. On the facing page, you can read of how recently discovered fossils have helped complete the story of how modern whales evolved from four-legged land mammals not unlike those that graze on grass today.

Figure 15.3 Testing the theory of evolution with fossil titanotheres.

Here you see illustrated changes in a group of hoofed mammals known as titanotheres between about 50 million and 35 million years ago. During this time, the small, bony protuberance located above the nose 50 million years ago evolved into relatively large, blunt horns.

| 50 | 45 | 40 | 35 |

Millions of years ago

15.3 Evolution's Critics

Of all the major ideas of biology, evolution is perhaps the best known to the general public, because many people mistakenly believe that evolution represents a challenge to their religious beliefs. Because Darwin's theory of evolution by natural selection is the subject of often-bitter public controversy, we will examine the objections of evolution's critics in detail, to see why there is such a disconnect between science and public opinion.

History of the Controversy

An Old Conflict. Immediately after publication of *The Origin of Species,* English clergymen attacked Darwin's book as heretical; Gladstone, England's prime minister and a famous statesman, condemned it. The book was defended by Thomas Huxley and other scientists, including the foremost American botanist of the day, Asa Gray. Gradually they won over the scientific establishment, and by the turn of the century evolution was generally accepted by the world's scientific community.

The Fundamentalist Movement. By the 1920s the teaching of evolution had become frequent enough in American public schools to alarm conservative critics of evolution who saw Darwinism as a threat to their Christian beliefs. Between 1921 and 1929, fundamentalists introduced bills outlawing the teaching of evolution in 37 state legislatures. Four passed: Tennessee, Mississippi, Arkansas, and Texas.

Civil rights groups used the case of high school teacher John Scopes to challenge the Tennessee law within months of its being passed in 1925. The trial attracted national attention— you might have seen it portrayed in the film *Inherit the Wind*. Scopes, who had indeed violated the new law, lost.

After the 1920s, there were few other attempts to pass state laws preventing the teaching of evolution. Only one bill was introduced between 1930 and 1963. Why? Because Darwin's fundamentalist critics had succeeded quietly in winning their way. Textbooks published throughout the 1930s ignored evolution, new editions of texts removing the words *evolution* and *Darwin* from their indices. In the 1920–29 period, for example, the average number of words about the evolution of humans in 93 secondary school texts was 1,339; in the 1930–39 period it had dropped to 439. As late as 1950–59 it was 614. To quote biologist Ernst Mayr, "The word EVOLUTION simply disappeared from American schoolbooks."

These antievolution laws remained on the books for many years. Then in 1965 teacher Susan Epperson was dismissed for teaching evolution under the 1928 Arkansas law. In 1968 the United States Supreme Court found the Arkansas antievolution law to be unconstitutional; the 1920s laws were soon repealed.

Russian advances in space in the early 1960s created a public outcry for better American science education. New biology textbooks reintroduced evolution, and gave it renewed emphasis. The average number of words per text devoted to the evolution of man, for example, rose from 614 in the period 1950–59 to 8,977 in the period 1960–69. By the 1970s, evolution again formed the core of most biology schoolbooks.

The Scientific Creationism Movement. Again alarmed by the prevalence of evolution in public school biology classes, Darwin's critics took a new approach. It began with a proposal by the Institute for Creationism Research in 1964, which said, "Creationism is just as much a science as is evolution, and evolution is just as much a religion as is creation." This proposal has become known as *creationism science*. It was soon followed by the introduction of legislation in state legislatures mandating that "all theories of origins be accorded equal time." Creationism was represented as being as much a scientific theory as evolution, to which students had a right to be exposed.

In 1981 the state legislatures of Arkansas and Louisiana passed "equal-time" bills into law. The Louisiana equal-time law requiring "balanced treatment of creation-science and evolution-science in public schools" was struck down by the Supreme Court in 1987, which judged that creation science is not, in fact, science but rather a religious view that has no place in public science classrooms.

Local Action. In the following decades, Darwin's critics began to substitute the school board for the legislature. Unlike most European countries, which set their school curricula through a central education ministry, U.S. education is highly decentralized, with elected education boards setting science standards at the state and local level. These standards determine the content of state-wide assessment tests, and have a major impact on what is taught in classrooms.

Critics of Darwin have run successfully for seats on local and state education boards across the United States, and from these positions have begun to alter standards to lessen the impact of evolution in the classroom. Great publicity followed the removal of evolution from the Kansas state standards in 1999 and again in 2005, but in many other states the same effect has been achieved more quietly. Only 22 states today mandate the teaching of natural selection, for example. Four states fail to mention evolution at all.

Intelligent Design. In recent years, critics of Darwin have begun new attempts to combat the teaching of evolution in the schoolroom, arguing before state and local school boards that life is too complex for natural selection and so must reflect intelligent design. They go on to argue that this "theory of intelligent design (ID)" should be presented in the science classroom as an alternative to the theory of evolution.

Scientists object strongly to dubbing ID a scientific theory. The essence of science is seeking explanations in what can be observed, tested, replicated by others, and possibly falsified. Explanations that cannot be tested and potentially rejected simply aren't science. If someone invokes a nonnatural cause—a supernatural force—in their research, and you decide to test it, can you think of any way to do so such that it could be falsified? Supernatural causation is not science.

The source of sharp public controversy, intelligent design has been overwhelmingly rejected by the scientific community, which does not regard intelligent design to be science at all, but rather thinly disguised creationism, a religious view that has no place in the science classroom.

Arguments Advanced by Darwin's Critics

Critics of evolution have raised a variety of objections to Darwin's theory of evolution by natural selection:

1. **Evolution is not solidly demonstrated.** *"Evolution is just a theory,"* critics point out, as if theory meant lack of knowledge, some kind of guess. Scientists, however, use the word theory in a very different sense than the general public does (see section 1.8). Theories are the solid ground of science, supported with much experimental evidence and that of which we are most certain. Few of us doubt the theory of gravity because it is "just a theory."

2. **There are no fossil intermediates.** *"No one ever saw a fin on the way to becoming a leg,"* critics claim, pointing to the many gaps in the fossil record in Darwin's day. Since then, however, most fossil intermediates in vertebrate evolution have indeed been found. A clear line of fossils now traces the transition between whales and hoofed mammals, between reptiles and mammals, and between apes and humans. The fossil evidence of evolution between major forms is compelling.

3. **The intelligent design argument.** *"The organs of living creatures are too complex for a random process to have produced."* This classic "argument from design" was first proposed nearly 200 years ago by William Paley in his book *Natural Theology*—the existence of a clock is evidence of the existence of a clockmaker, Paley argues. Similarly, Darwin's critics argue that organs like the mammalian ear are too complex to be due to blind evolution. There must have been a designer. Biologists do not agree. The intermediates in the evolution of the mammalian ear are well documented in the fossil record. Its three tiny bones originated from the reptile lower jaw, gradually reducing in size as more advanced mammals evolved. Eventually the bones migrated into the inner ear where we now find them in present-day mammals, helping amplify sound vibrations as they pass across the middle ear. These intermediate forms were each favored by natural selection because they each had value—being able to amplify sound a little is better than not being able to amplify it at all. Complex structures like the mammalian ear evolved as a progression of slight improvements. Nor is the solution always optimal, as your own eyes attest. As you can see in the blown-up image in figure 15.9, the receptor cells are actually facing backward to the stimulus (light). No intelligent designer would design an eye backwards!

4. **Evolution violates the second law of thermodynamics.** *"A jumble of soda cans doesn't by itself jump neatly into a stack—things become more disorganized due to random events, not more organized."* Biologists point out that this argument ignores what the second law really says: disorder increases in a closed system, which the earth most certainly is not. Energy enters the biosphere from the sun, fueling life and all the processes that organize it.

5. **Proteins are too improbable.** *"Hemoglobin has 141 amino acids. The probability that the first one would be*

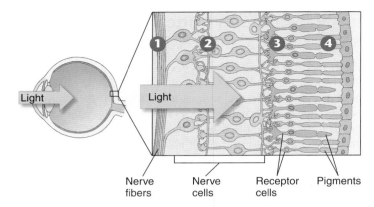

Figure 15.9 The vertebrate eye is poorly designed.
The visual pigments in a vertebrate eye that are stimulated by light are embedded in the retinal tissue, facing backward to the direction of the light. The light has to pass through nerve fibers ❶, nerve cells ❷, receptor cells ❸, before reaching the pigments ❹.

leucine is 1/20, and that all 141 would be the ones they are by chance is $(1/20)^{141}$, an impossibly rare event." You cannot use probability to argue backward. The probability that a student in a classroom has a particular birthday is 1/365; arguing this way, the probability that everyone in a class of 50 would have the birthdays they do is $(1/365)^{50}$, and yet there the class sits.

6. **Natural selection does not imply evolution.** *"No scientist has come up with an experiment where fish evolve into frogs and leap away from predators."* Is microevolution (evolution within a species) the mechanism that has produced macroevolution (evolution among species)? Most biologists that have studied the problem think so. The differences between breeds produced by artificial selection—such as Chihuahuas, dachshunds, and greyhounds—are more distinctive than differences between wild canine species. Laboratory selection experiments with insects easily create forms that cannot interbreed and thus would in nature be considered different species. Thus, production of radically different forms has indeed been observed, repeatedly. To object that evolution still does not explain really major differences, like between fish and amphibians, simply takes us back to point 2—these changes take millions of years, and are seen clearly in the fossil record.

7. **Life could not have evolved in water.** *"Because the peptide bond does not form spontaneously in water, amino acids could never have spontaneously linked together to form proteins; nor is there any chemical reason why biological proteins contain only the L-isomer and not the D-isomer."* Both of these contentions are valid, but do not require rejecting evolution. Rather, they suggest that the early evolution of life took place on a surface rather than in solution. Amino acids link up spontaneously on the surface of clays, for example, which can have a shape that selects the L-isomer. Several hypotheses about the origin of life are discussed in more detail in chapter 17.

The Irreducible Complexity Fallacy

The century-and-a-half-old "intelligent design" argument of William Paley has been recently articulated in a new molecular guise by Lehigh University biochemistry professor Michael Behe. In his 1996 book *Darwin's Black Box: The Biochemical Challenge to Evolution,* Behe argues that the intricate molecular machinery of our cells is so elaborate, our body processes so interconnected, that they cannot be explained by evolution from simpler stages in the way that Darwinists explain the evolution of the mammalian ear. The molecular machinery of the cell is "irreducibly complex." Behe defines an irreducibly complex system as "a single system composed of several well-matched, interacting parts that contribute to the basic function, wherein the removal of any one of the parts causes the system to effectively cease functioning." Each part plays a vital role. Remove just one, Behe emphasizes, and cell molecular machinery cannot function.

As an example of such an irreducibly complex system, Behe describes the series of more than a dozen blood clotting proteins that act in our body to cause blood to clot around a wound. Take out any step in the complex cascade of reactions that leads to coagulation of blood, says Behe, and your body's blood would leak out from a cut like water from a ruptured pipe. Remove a single enzyme from the complementary system that confines the clotting process to the immediate vicinity of the wound, and all your lifeblood would harden. Either condition would be fatal. The need for *all* the parts of such complex systems to work leads directly to Behe's criticism of Darwin's theory of evolution by natural selection. Behe writes that "irreducibly complex systems cannot evolve in a Darwinian fashion." Why not? Because natural selection can only act on changes that actually affect how an organism functions. If dozens of different proteins all must work correctly to clot blood, how could natural selection act to fashion any one of the individual proteins? No one protein does anything on its own, just as a portion of a watch doesn't tell time. Behe argues that, like Paley's watch, the blood clotting system must have been designed all at once, as a single functioning machine.

What's wrong with Behe's argument, as evolutionary scientists have been quick to point out, is that each part of a complex molecular machine does not evolve by itself, despite Behe's claim that it must. The several parts evolve together, in concert, precisely because evolution acts on the system, not its parts. That's the fundamental fallacy in Behe's argument. Natural selection can act on a complex system because at every stage of its evolution, the system functions. Parts that improve function are added, and, because of later changes, eventually become essential, in the same way that the second rung of a ladder becomes essential once you have added a third.

The mammalian blood clotting system, for example, has evolved in stages from much simpler systems. By comparing the amino acid sequences of the many proteins, biochemist Russell Doolittle has estimated how long it has been since each protein evolved. You can see what he has learned above in figure 15.10. The core of the vertebrate clotting system, called the "common pathway" (highlighted in blue), evolved

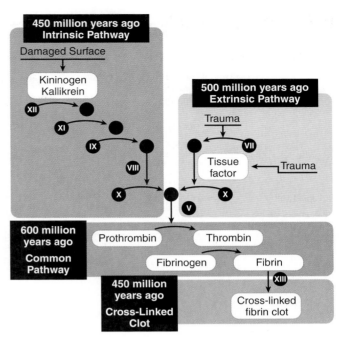

Figure 15.10 How blood clotting evolved.
The blood clotting system evolved in steps, with new proteins adding on to the preceding step.

at the dawn of the vertebrates aproximately 600 million years ago, and is found today in lampreys, the most primitive fish. As vertebrates evolved, proteins were added to the clotting system, improving its efficiency. The so-called "extrinsic pathway" (highlighted in pink), triggered by substances released from damaged tissues, was added 500 million years go. Each step in the pathway amplifies what goes before, so adding the extrinsic pathways greatly increases the amplification and thus the sensitivity of the system. Fifty million years later, a third component was added, the so-called "intrinsic pathway" (highlighted in tan). It is triggered by contact with the jagged surfaces produced by injury. Again, amplification and sensitivity were increased to ultimately end up with blood clots formed by the cross linking of fibrin (highlighted in green). At each stage as the clotting system evolved to become more complex, its overall performance came to depend on the added elements. Mammalian clotting, which utilizes all three pathways, no longer functions if any one of them is disabled. Blood clotting has become "irreducibly complex"—as the result of Darwinian evolution. Behe's claim that complex cellular and molecular processes can't be explained by Darwinism is wrong. Indeed, examination of the human genome reveals that the cluster of blood clotting genes arose through duplication of genes, with increasing amounts of change. The evolution of the blood clotting system is an observation, not a surmise. Its irreducible complexity is a fallacy.

15.3 Darwin's theory of evolution has not been accepted by religious conservatives in the United States, who object to its being taught in public school. Their criticisms are without scientific merit, but have considerable public support.

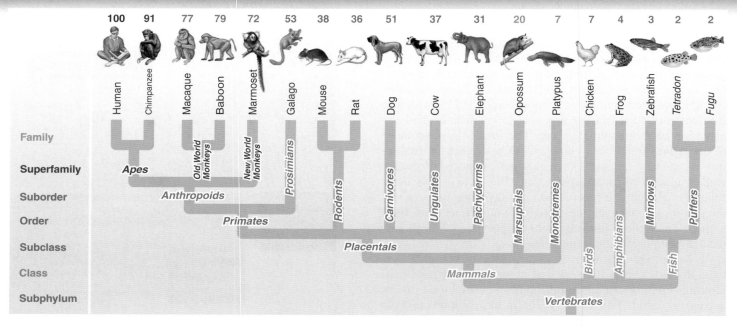

| 100 | 91 | 77 | 79 | 72 | 53 | 38 | 36 | 51 | 37 | 31 | 20 | 7 | 7 | 4 | 3 | 2 | 2 |

Human / Chimpanzee / Macaque / Baboon / Marmoset / Galago / Mouse / Rat / Dog / Cow / Elephant / Opossum / Platypus / Chicken / Frog / Zebrafish / Tetradon / Fugu

Family

Superfamily — *Apes* — *Old World Monkeys* — *New World Monkeys*

Suborder — *Anthropoids* — *Prosimians* — *Rodents* — *Carnivores* — *Ungulates* — *Pachyderms* — *Marsupials* — *Monotremes* — *Minnows* — *Puffers*

Order — *Primates*

Subclass — *Placentals* — *Birds* — *Amphibians* — *Fish*

Class — *Mammals*

Subphylum — *Vertebrates*

Putting Darwin (and Intelligent Design) To the Test

In the spring of 2006 the South Carolina Board of Education rejected a state panel's proposal to change high school standards by calling on students to critically analyze evolution. The Board stated it felt the proposal was a ploy to promote the avoidance of teaching evolution. Similar proposals to add a requirement that students critically analyze evolution had been rejected earlier in the year by the Utah and Ohio Boards of Education, and are currently under consideration in several other states.

What are we to make of this? Surely no scientist can object to critically analyzing any theory. That is what science is all about, seeking explanations for what can be observed, tested, replicated, and possibly falsified. Indeed, biologists claim that Darwin's theory of evolution has been subjected to as much critical analysis as any theory in the history of science.

So why the objection to this change in high school standards? Because many scientists and teachers, apparently including the South Carolina Board of Education, feel the change is simply intended to promote the teaching of a non-scientific alternative to evolution in classrooms.

This distinction between an assertion that can be tested and one that cannot goes to the very nature of science. Actually, nothing makes this difference more clear cut than the critical analysis so sought after by South Carolina's critics of evolution. So let's do it. Let's put Darwin to the test.

As explained earlier in the chapter, if Darwin's assertion is correct, that organisms evolved from ancestral species, then we should be able to track evolutionary changes in our DNA. The variation that we see between species reflects adaptations to environmental challenges, adaptations that result from changes in DNA. Therefore, a series of evolutionary changes should be reflected in an accumulation of genetic changes in the DNA. This hypothesis, that evolutionary changes reflect accumulated changes in DNA, leads to the following prediction: two species that are more distantly related (for example, humans and mice) should have accumulated a greater number of evolutionary differences than two species that are more closely related (say, humans and chimpanzees).

So have they? Let's compare vertebrate species to see. The "family tree" above shows how biologists believe 18 different vertebrate species are related. Apes and monkeys, because they are in the same order (primates), are considered more closely related to each other than either are to members of another order, such as mice and rats (rodents).

The wealth of genomes (a genome is all the DNA that an organism possesses) that have been sequenced since completion of the human genome project allows us to directly compare the DNA of these 18 vertebrates. To reduce the size of the task, investigators at the National Human Genome Research Institute working at the University of California, Santa Cruz, focused on 44 so-called ENCODE regions scattered around the vertebrate genomes. These regions, corresponding to 30 Mb (megabase, or million bases) or roughly 1% of the total human genome, were selected to be representative of the genome as a whole, containing protein-encoding genes as well as "junk" DNA.

For each vertebrate species, the investigators determined the similarity of its DNA to that of humans—that is, the percent of the nucleotides in that organism's 44 ENCODE regions which match those of the human genome.

You can see the result in each instance presented as a number above the picture of each organism on the vertebrate family tree. As Darwin's theory predicts, the closer the relatives, the less the genomic difference we see. The chimpanzee genome is more like the human genome (91%) than the monkey genomes are (72 to 79%). Furthermore, these five genomes, all in the same order, are more like each other than any are to those of another order, such as rodents (mouse and rat).

In general, as you proceed through the taxonomic categories of the vertebrate family tree from very distant relatives on the right (some in the same class as humans) to very close ones on the left (in the same family), you can see clearly that genomic similarity increases as taxonomic distance decreases—just as Darwin's theory predicts. The prediction of evolutionary theory is solidly confirmed.

The analysis does not have to stop here. The evolutionary history of the vertebrates is quite well known from fossils, and because many of these fossils have been independently dated using tools such as radioisotope dating, it is possible to recast the analysis in terms of concrete intervals of time, and assess directly whether or not vertebrate genomes accumulate more differences over longer periods of time as Darwin's theory predicts.

For each of the 18 vertebrates being analyzed, the graph above plots genomic similarity—how alike the DNA sequence of the vertebrate's ENCODE regions are to those of the human genome—against divergence time (that is, how many millions of years have elapsed since that vertebrate and humans shared a common ancestor in the fossil record). Thus the last common ancestor shared by chickens and humans was an early reptile called a dicynodont that lived some 250 million years ago; since then the genomes of the two species have changed so much that only 7% of their ENCODE sequences are still the same.

The result seen in the graph is striking and very clear: Over their more than 300 million year history, vertebrates have accumulated more and more genetic change in their DNA. "Descent with modification" was Darwin's

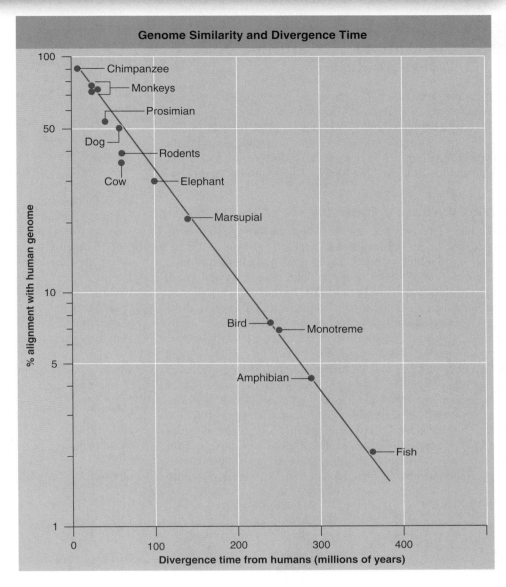

Genome Similarity and Divergence Time

definition of evolution, and that is exactly what we see in the graph. The evolution of the vertebrate genome is not a theory, but an observation.

The wealth of data made available by the human genome project has allowed a powerful test of Darwin's prediction. The conclusion to which the test leads us is that evolution is an observed fact, clearly revealed in the DNA of vertebrates.

This is the sort of critical analysis that science requires, and that the theory of evolution has again passed. Anyone suggesting that a nonscientific alternative to evolution, such as Intelligent Design, offers an alternative scientific explanation to evolution is welcome to subject it to the same sort of critical analysis you have seen employed here. Can you think of a way to do so? It is precisely because the assertion of intelligent design cannot be critically analyzed—it does not make any testable prediction—that it is not science and has no place in science classrooms.

15.4 Genetic Change Within Populations: The Hardy-Weinberg Rule

Population genetics is the study of the properties of genes in populations. Genetic variation within natural populations could not be explained by Darwin and his contemporaries. The way in which meiosis produces genetic segregation among the progeny of a hybrid had not yet been discovered. Selection, scientists then thought, should always favor an optimal form.

Genes Within Populations

From the 1920s onward, scientists began to formulate a comprehensive theory of how **alleles** (alternative forms of a gene) behave in populations, and how changes in **allele frequencies** (the proportion of alleles of a particular type in a population) lead to evolutionary change. Variation within populations puzzled many scientists; dominant alleles were believed to drive recessive alleles out of populations, with selection favoring an optimal form. The solution to the puzzle of why genetic variation persists was developed in 1908 by G. H. Hardy and W. Weinberg. Hardy and Weinberg studied the frequencies of alleles in a population's *gene pool,* which is the sum of all of the genes in a population, including all alleles in all individuals. Hardy and Weinberg pointed out that in a large population in which there is random mating, and in the absence of forces that change allele frequencies, the original genotype proportions remain constant from generation to generation. Dominant alleles do not, in fact, replace recessive ones. Because

their proportions do not change, the genotypes are said to be in **Hardy-Weinberg equilibrium.** The Hardy-Weinberg rule is viewed as a baseline to which the frequencies of alleles in a population can be compared. If the allele frequencies are not changing (they are in Hardy-Weinberg equilibrium), the population is not evolving.

Hardy and Weinberg came to this conclusion by analyzing the frequencies of alleles in successive generations. The **frequency** of something is defined as the proportion of individuals with a certain characteristic, compared to the entire population. Thus, in the population of 1,000 cats shown in figure 15.11, there are 840 black and 160 white cats. To determine the frequency of black cats, divide 840 by 1,000 (840/1,000), which is 0.84. The frequency of white cats is 160/1,000 = 0.16.

Knowing the frequency of the phenotype, one can calculate the frequency of the genotypes and alleles in the population. By convention, the frequency of the more common of two alleles (in this case *B* for the black allele) is designated by the letter *p* and that of the less common allele (*b* for the white allele) by the letter *q*. Because there are only two alleles, the sum of *p* and *q* must always equal 1 ($p + q = 1$).

In algebraic terms, the Hardy-Weinberg equilibrium is written as an equation. For a gene with two alternative alleles *B* (frequency *p*) and *b* (frequency *q*), the equation looks like this:

$$p^2 \quad + \quad 2pq \quad + \quad q^2 \quad = \quad 1$$

p^2	$2pq$	q^2	
Individuals homozygous for allele *B*	Individuals heterozygous for alleles *B* and *b*	Individuals homozygous for allele *b*	

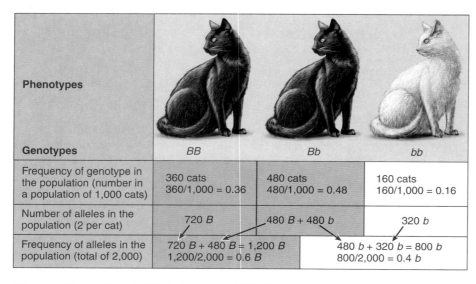

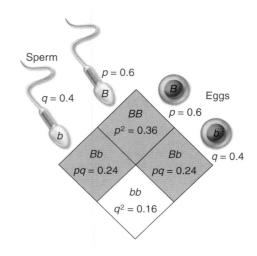

Figure 15.11 Hardy-Weinberg equilibrium.

In the absence of factors that alter them, the frequencies of gametes, genotypes, and phenotypes remain constant generation after generation. The example shown here involves a population of 1,000 cats, in which 160 are white and 840 are black. White cats are *bb,* and black cats are *BB* or *Bb.* The potential crosses in this cat population can be determined using a Punnett square analysis.

You will notice that not only does the sum of the alleles add up to 1 but so does the sum of the frequencies of genotypes.

Knowing the frequencies of the alleles in a population doesn't reveal whether the population is evolving or not. We need to look at future generations to determine this. Using the allele frequencies calculated for our population of cats, we can predict what the genotypic and phenotypic frequencies will be in future generations. The Punnett square shown in figure 15.11 is constructed with allele frequencies of 0.6 for the B allele and 0.4 for the b allele, taken from the bottom row of the chart. It might help you to consider these frequencies as percentages, with a 0.6 representing 60% of the population and 0.4 representing 40% of the population. According to the Hardy-Weinberg rule, 60% of the sperm in the population will carry the B allele (indicated as $p = 0.6$ in the Punnett square) and 40% of the sperm ($q = 0.4$) will carry the b allele. When these are crossed with eggs carrying the same allele frequencies (60% or $p = 0.6$ B allele and 40% or $q = 0.4$ b allele), the predicted genotypic frequencies can be simply calculated. The genotypic ratio for BB, in the upper quadrant, equals the frequency of B (0.6) multiplied by the frequency of B (0.6) or (0.6 x 0.6 = 0.36). So, if the population is not evolving, the genotypic ratio for BB would stay the same, and 0.36 or 36% of the cats in future generations would be homozygous dominant (BB) for coat color. Likewise, 0.48 or 48% of the cats would be heterozygous Bb (0.24 + 0.24 = 0.48), and 0.16 or 16% of the cats would be homozygous recessive bb.

Hardy-Weinberg Assumptions

The Hardy-Weinberg rule is based on certain assumptions. The equation on page 282 is true only if the following five assumptions are met:

1. The size of the population is very large or effectively infinite.
2. Individuals mate with one another at random.
3. There is no mutation.
4. There is no input of new copies of any allele from any extraneous source (such as from a nearby population) or losses of copies of alleles through emigration (individuals leaving the population).
5. All alleles are replaced equally from generation to generation (natural selection is not occurring).

How valid are the predictions made by the Hardy-Weinberg equation? For many genes, they prove to be very accurate. As an example, consider the recessive allele responsible for the serious human disease cystic fibrosis. This allele (q) is present in Caucasians in North America at a frequency of 0.022. What proportion of Caucasian North Americans, therefore, is expected to express this trait? The frequency of double-recessive individuals (q^2) is expected to be

$$q^2 = 0.022 \times 0.022 = 0.00048$$

which equals 0.48 in every 1,000 individuals or about 1 in every 2,000 individuals, very close to real estimates.

What proportion is expected to be heterozygous carriers? If the frequency of the recessive allele q is 0.022, then the frequency of the dominant allele p must be $p = 1 - q$ or:

$$p = 1 - 0.022 = 0.978$$

The frequency of heterozygous individuals ($2pq$) is thus expected to be:

$$2 \times 0.978 \times 0.022 = 0.043$$

It is estimated that 12 million individuals in the United States are carriers of the cystic fibrosis allele. In a population of 292 million people, that is a frequency of 0.041, very close to projections using the Hardy-Weinberg equation. However, if in the future the frequency of the cystic fibrosis allele in the United States were to change, this would suggest that the population is no longer following the assumptions of the Hardy-Weinberg rule. For example, if prospective parents who were carriers of the allele chose not to have children, the frequency of the allele would decrease in future generations. Mating would no longer be random, because those carrying the allele would not mate. Consider another scenario. If gene therapies were developed that were able to cure the symptoms of cystic fibrosis, patients would survive longer and would have more of an opportunity to reproduce. This would increase the frequency of the allele in future generations. An increase could also result from an influx of the allele into the population by migration, if the allele were more frequent among individuals migrating into the country.

Most human populations are large and randomly mating with respect to most traits (a few traits affecting appearance undergo strong sexual selection) and thus are similar to the ideal population envisioned by Hardy and Weinberg. For some genes, however, the observed proportion of heterozygotes does not match the value calculated from the allele frequencies. When this occurs, it indicates that something is acting on the population to alter one or more of the genotypic frequencies, whether it is selection, nonrandom mating, migration, or some other factor. Viewed in this light, Hardy-Weinberg can be viewed as a *null hypothesis*. A null hypothesis is a prediction that is made stating there will be no differences in the parameters being measured. If over several generations, the genotypic frequencies in the population do not match those predicted by the Hardy-Weinberg equation, the null hypothesis would be rejected and the assumption made that some force is acting on the population to change the frequencies of alleles. The factors that can affect the frequencies of alleles in a population are discussed in detail in the next two sections.

15.4 Mendelian inheritance does not alter allele frequencies. In a large, random-mating population that fulfills the other Hardy-Weinberg assumptions, alternative alleles B (frequency p) and b (frequency q) are expected to be present in genotypic proportions of $p^2 + 2pq + q^2 = 1$.

15.5 Agents of Evolution

Many factors can alter allele frequencies. But only five alter the proportions of homozygotes and heterozygotes enough to produce significant deviations from the proportions predicted by the Hardy-Weinberg rule: (1) mutation, (2) migration, (3) genetic drift, (4) nonrandom mating, and (5) selection (table 15.1).

Mutation

A **mutation** is a change in a nucleotide sequence in DNA. In the first panel of table 15.1, the T nucleotide undergoes mutation and is replaced with an A nucleotide. Mutation from one allele to another obviously can change the proportions of particular alleles in a population. But mutation rates are generally too low to significantly alter Hardy-Weinberg proportions of common alleles. Many genes mutate 1 to 10 times per 100,000 cell divisions. Some of these mutations are harmful, while others are neutral or, even rarer, beneficial. Also, the mutations must affect the DNA of the germ cells (egg and sperm), or the mutation will not be passed on to offspring. The mutation rate is so slow that few populations are around long enough to accumulate significant numbers of mutations. However, no matter how rare, mutation is the ultimate source of variation in a population.

Migration

Migration, defined in genetic terms as the movement of individuals between populations, can be a powerful force upsetting the genetic stability of natural populations. Migration includes movement of individuals into a population, called *immigration,* or the movement of individuals out of a population, called *emigration.* If the characteristics of the newly arrived individuals differ from those already there, and if the newly arrived individuals adapt to survive in the new area and mate successfully, then the genetic composition of the receiving population may be altered.

Sometimes migration is not obvious. Subtle movements include the drifting of gametes of plants, or of the immature stages of marine organisms, from one place to another. In the second panel in table 15.1, the bee is carrying pollen from a flower in one population to a flower in another population. By doing this, the bee may be introducing new alleles into a population. However it occurs, migration can alter the genetic characteristics of populations and prevent the maintenance of Hardy-Weinberg equilibrium. The magnitude of effects of migration is based on two factors: (1) the proportion of migrants in the population, and (2) the difference in allele frequencies between the migrants and the original population. The actual evolutionary impact of migration is difficult to assess, and depends heavily on the selective forces prevailing at the different places where the populations occur.

Genetic Drift

In small populations, the frequencies of particular alleles may be changed drastically by chance alone. In an extreme case, individual alleles of a given gene may all be represented in few individuals, and some of them may be accidentally lost

TABLE 15.1	AGENTS OF EVOLUTION	
Factor		**Description**
Mutation	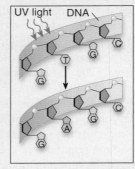	The ultimate source of variation. Individual mutations occur so rarely that mutation alone does not change allele frequency much.
Migration		A very potent agent of change. Migration acts to promote evolutionary change by enabling populations that exchange members to converge toward one another. This bee carries pollen from one population of flowers to another.
Genetic drift		Chance events may result in the loss of individuals and therefore the loss of alleles in a population. Usually occurs only in very small populations. A small number of alleles can also impact a newly formed population, such as the founder effect on an island.
Nonrandom mating	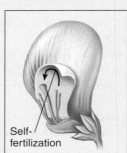 Self-fertilization	Inbreeding is the most common form of nonrandom mating. It does not alter allele frequency but decreases the proportion of heterozygotes ($2pq$). See text for explanation.
Selection	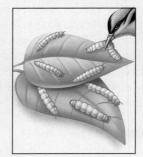	The only form that produces *adaptive* evolutionary changes. Only rapid for allele frequency greater than .01.

if those individuals fail to reproduce or die. This loss of individuals and their alleles is due to random events rather than the fitness of the individuals carrying those alleles. This is not to say that alleles are always lost with genetic drift, but allele frequencies appear to change randomly, as if the frequencies were drifting; thus, random changes in allele frequencies is known as **genetic drift.** A series of small populations that are isolated from one another may come to differ strongly as a result of genetic drift.

When one or a few individuals migrate and become the founders of a new, isolated population at some distance from their place of origin, the alleles that they carry are of special significance in the new population. Even if these alleles are rare in the source population, they will become a significant fraction of the new population's genetic endowment. Returning to the third panel in table 15.1, about half the mainland population of birds exhibited the red coloring of the migrating bird, but the red coloring will become much more common in the island population that arises from this bird. This is called the **founder effect.** As a result of the founder effect, rare alleles and combinations often become more common in new, isolated populations. The founder effect is particularly important in the evolution of organisms that occur on oceanic islands, such as the Galápagos Islands, which Darwin visited. Most of the kinds of organisms that occur in such areas were probably derived from one or a few initial founders. In a similar way, isolated human populations are often dominated by the genetic features that were characteristic of their founders, particularly if only a few individuals were involved initially (figure 15.12).

Even if organisms do not move from place to place, occasionally their populations may be drastically reduced in size. This may result from flooding, drought, earthquakes, and other natural forces or from progressive changes in the environment. The surviving individuals constitute a random genetic sample of the original population. Such a restriction in genetic variability has been termed the **bottleneck effect.** The very low levels of genetic variability seen in African cheetahs today is thought to reflect a near-extinction event in the past.

Nonrandom Mating

Individuals with certain genotypes sometimes mate with one another either more or less commonly than would be expected on a random basis, a phenomenon known as **nonrandom mating.** One type of nonrandom mating is **sexual selection,** choosing a mate often based on certain physical characteristics. Another type of nonrandom mating is inbreeding, or mating with relatives, shown by the self-fertilization of the flower in the fourth panel in table 15.1. Inbreeding increases the proportions of individuals that are homozygous because no individuals mate with any genotype but their own. As a result, inbred populations contain more homozygous individuals than predicted by the Hardy-Weinberg rule. For this reason, populations of self-fertilizing plants consist primarily of homozygous individuals, whereas outcrossing plants, which interbreed with individuals different from themselves, have a higher proportion of heterozygous individuals. Nonrandom mating alters genotype frequencies but not allele frequencies. The allele frequencies remain the same—the alleles are just distributed differently among the offspring.

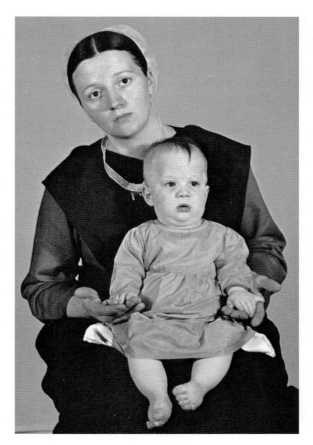

Figure 15.12 The founder effect.
This Amish woman is holding her child, who has Ellis-van Creveld syndrome. The characteristic symptoms are short limbs, dwarfed stature, and extra fingers. This disorder was introduced in the Amish community by one of its founders in the eighteenth century and persists to this day because of reproductive isolation.

Selection

As Darwin pointed out, some individuals leave behind more progeny than others, and the likelihood they will do so is affected by their inherited characteristics. The result of this process is called **selection** and was familiar even in Darwin's day to breeders of horses and farm animals. In so-called **artificial selection,** the breeder selects for the desired characteristics. For example, mating larger animals with each other produces offspring that are larger. In **natural selection,** Darwin suggested the environment plays this role, with conditions in nature determining which kinds of individuals in a population are the most fit (meaning individuals that are best suited to their environment, see section 2.3) and so affecting the proportions of genes among individuals of future populations. The environment, like the preferential feeding of the bird in the fifth panel in table 15.1, imposes the conditions that determine the results of selection and, thus, the direction of evolution.

Forms of Selection

Selection operates in natural populations of a species as skill does in a football game. In any individual game, it can be difficult to predict the winner, because chance can play an important role in the outcome; but over a long season, the teams with the most skillful players usually win the most games. In nature, those individuals best suited to their environments tend to win the evolutionary game by leaving the most offspring, although chance can play a major role in the life of any one individual. Selection is a statistical concept, just as betting is. Although you cannot predict the fate of any one individual, or any one coin toss, it is possible to predict which kind of individual will tend to become more common in populations of a species, as it is possible to predict the proportion of heads after many coin tosses.

In nature, many traits, perhaps most, are affected by more than one gene. The interactions between genes are typically complex, as you saw in chapter 11. For example, alleles of many different genes play a role in determining human height (see figure 11.12). In such cases, selection operates on all the genes, influencing most strongly those that make the greatest contribution to the phenotype. How selection changes the population depends on which genotypes are favored. Three types of natural selection have been identified: stabiliz-

ing selection, disruptive selection, and directional selection. Figure 15.13 show the results of these three types of selection on body size.

Stabilizing Selection

When selection acts to eliminate both extremes from an array of phenotypes—for example, eliminating the larger and smaller body sizes (the red-shaded areas under the curve in figure 15.13*a*)—the result is an increase in the frequency of the already common intermediate phenotype (a midsized body). This is called **stabilizing selection.** In effect, selection is operating to prevent change away from the middle range of values. In a classic study carried out after an "uncommonly severe storm of snow, rain, and sleet" on February 1, 1898, 136 starving English sparrows were collected and brought to the laboratory of H. C. Bumpus at Brown University in Providence, Rhode Island. Of these, 64 died and 72 survived. Bumpus took standard measurements on all the birds. He found that among males, the surviving birds tended to be bigger, as one might expect from the action of directional selection (discussed later). However, among females,

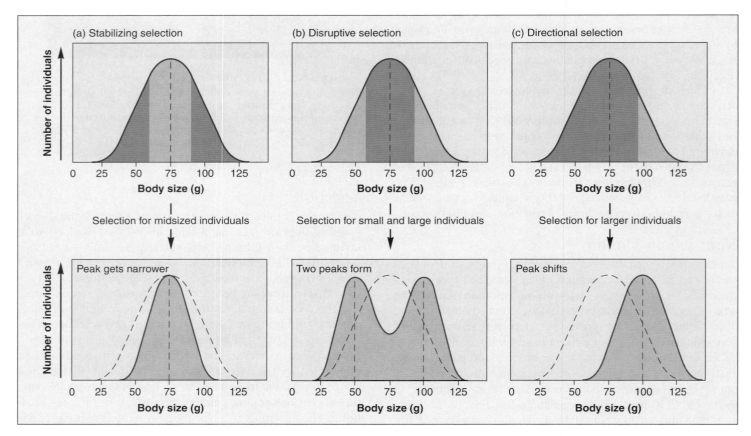

Figure 15.13   Three kinds of natural selection.
In the top panels, the *blue* areas indicate the phenotypes that are being selected for and the *red* areas are the phenotypes that are not being preferentially selected for. The bottom panels show the phenotypic results of the selection. (a) In *stabilizing selection,* individuals with midrange phenotypes are favored, with selection acting against both ends of the range of phenotypes. (b) In *disruptive selection,* individuals in the middle of the range of phenotypes of a certain trait are selected against, and the extreme forms of the trait are favored. (c) In *directional selection,* individuals concentrated toward one extreme of the array of phenotypes are favored.

the birds that survived were those that were more average in size. Among the female birds that perished were many more individuals that had extreme measurements, either very large or very small. Selection had acted most strongly against these "extreme-sized" female birds. Stabilizing selection does not change which phenotype is the most common of the population—the average-sized birds were already the most common phenotype—but rather makes it even more common by eliminating extremes. Many examples similar to Bumpus's female sparrows are known. In humans, infants with intermediate weight at birth (the blue-screened area in figure 15.14a) have the highest survival rate, with the red line indicating that infant mortality is at its lowest for babies of intermediate weight. In chickens, eggs of intermediate weight have the highest hatching success.

Disruptive Selection

In some situations, selection acts to eliminate the intermediate type, the red area under the curve in figure 15.13b, resulting in the two more extreme phenotypes becoming more common in the population. This type of selection is called **disruptive selection.** A clear example is the different beak sizes of the African black-bellied seedcracker finch *Pyrenestes ostrinus* (figure 15.14b). Populations of these birds contain individuals with large and small beaks, but very few individuals with intermediate-sized beaks. As their name implies, these birds feed on seeds, and the available seeds fall into two size categories: large and small. Only large-beaked birds, like the one on the left, can open the tough shells of large seeds, whereas birds with the smallest beaks, like the one on the right, are more adept at handling small seeds. Birds with intermediate-sized beaks are at a disadvantage with both seed types: unable to open large seeds and too clumsy to efficiently process small seeds. Consequently, selection acts to eliminate the intermediate phenotypes, in effect partitioning the population into two phenotypically distinct groups.

Directional Selection

When selection acts to eliminate one extreme from an array of phenotypes—for example, the smaller body sizes indicated by the red area under the curve in figure 15.13c—the genes determining this extreme become less frequent in the population. This form of selection is called **directional selection.** In the *Drosophila* population illustrated in figure 15.14c, flies that flew toward light, a behavior called phototropism, were eliminated from the population. The remaining flies were mated and the experiment repeated. After 20 generations of selected mating, flies exhibiting phototropism were far less frequent in the population.

> **15.5** Five evolutionary forces have the potential to significantly alter allele and genotype frequencies in populations: mutation, migration, genetic drift, nonrandom mating, and selection. Selection on traits affected by many genes can favor intermediate values, or one or both extremes.

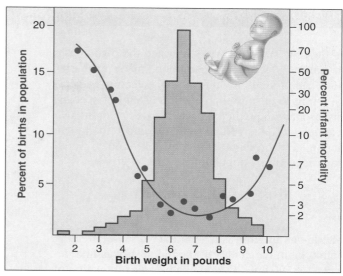

(a) Stabilizing selection for birth weight. The death rate among human babies is lowest at an intermediate birth weight between 7 and 8 pounds indicated by the *red* line. The intermediate weights are also the most common in the population, indicated by the *blue* area. Larger and smaller babies both occur less frequently and have a greater tendency to die at or near birth.

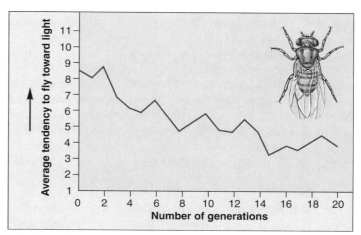

(b) Disruptive selection for large and small beaks. Differences in beak size in the black-bellied seedcracker finch of West Africa are the result of disruptive selection for two distinct food sources.

(c) Directional selection for negative phototropism in *Drosophila*. Individuals of the fly *Drosophila* were selectively bred. Flies that moved toward light were discarded, and only flies that moved away from light were used as parents for the next generation. After 20 generations, the offspring of parents that tended not to fly toward light had an ever greater tendency to avoid light.

Figure 15.14 Examples of selection.

In the time since Darwin suggested the pivotal role of natural selection in evolution, many examples have been found in which natural selection is clearly acting to change the genetic makeup of species, just as Darwin predicted. Here we will examine two examples: sickle-cell anemia (a defect in human hemoglobin proteins) and color selection in South American guppy populations.

15.6   Sickle-Cell Anemia

Sickle-cell anemia is a hereditary disease affecting hemoglobin molecules in the blood. It was first detected in 1904 in Chicago in a blood examination of an individual complaining of tiredness. You can see the original doctor's report in figure 15.15. The disorder arises as a result of a single nucleotide change in the gene encoding β-hemoglobin, one of the key proteins used by red blood cells to transport oxygen. The sickle-cell mutation changes the sixth amino acid in the β-hemoglobin chain (position B6) from glutamic acid (very polar) to valine (nonpolar). The unhappy result of this change is that the nonpolar *valine* at position B6, protruding from a corner of the hemoglobin molecule, fits nicely into a nonpolar pocket on the opposite side of another hemoglobin molecule; the nonpolar regions associate with each other. As the two-molecule unit that forms still has both a B6 valine and an opposite nonpolar pocket, other hemoglobins clump on, and long chains form as in figure 15.16a. The result is the deformed "sickle-shaped" red blood cell you see in figure 15.16b. In normal everyday hemoglobin, by contrast, the polar amino acid *glutamic acid* occurs at position B6. This polar amino acid is not attracted to the nonpolar pocket, so no hemoglobin clumping occurs, and cells are normal shaped as in figure 15.16c.

Persons homozygous for the sickle-cell genetic mutation in the β-hemoglobin gene frequently have a reduced life span. This is because the sickled form of hemoglobin does not carry oxygen atoms well, and red blood cells that are sickled do not flow smoothly through the tiny capillaries but instead jam up and block blood flow. Heterozygous individuals, who have both a defective and a normal form of the gene, make enough functional hemoglobin to keep their red blood cells healthy.

The Puzzle: Why So Common?

The disorder is now known to have originated in Central Africa, where the frequency of the sickle-cell allele is about 0.12. One in 100 people is homozygous for the defective allele and develops the fatal disorder. Sickle-cell anemia affects roughly two African Americans out of every thousand but is almost unknown among other racial groups.

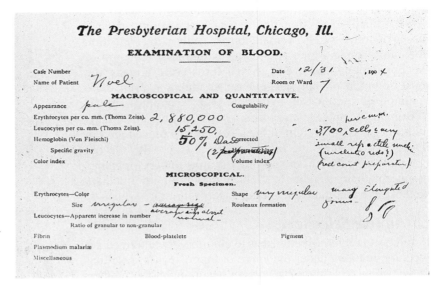

Figure 15.15   The first known sickle-cell anemia patient.
Dr. Ernest Irons's blood examination report on his patient Walter Clement Noel, December 31, 1904, described his oddly shaped red blood cells.

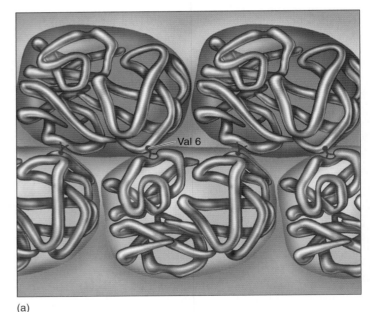

(a)

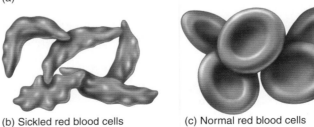

(b) Sickled red blood cells   (c) Normal red blood cells

Figure 15.16   Why the sickle-cell mutation causes hemoglobin to clump.

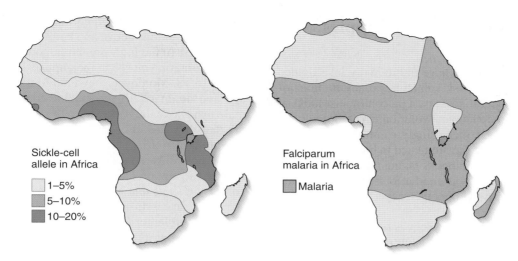

Figure 15.17 How stabilizing selection maintains sickle-cell anemia.
The diagrams show the frequency of the sickle-cell allele (*left*) and the distribution of falciparum malaria (*right*). Falciparum malaria is one of the most devastating forms of the often fatal disease. As you can see, its distribution in Africa is closely correlated with that of the allele of the sickle-cell characteristic.

If Darwin is right, and natural selection drives evolution, then why has natural selection not acted against the defective allele in Africa and eliminated it from the human population there? Why is this potentially fatal allele instead very common there?

The Answer: Stabilizing Selection

The defective allele has not been eliminated from Central Africa because people who are heterozygous for the sickle-cell allele are much less susceptible to malaria, one of the leading causes of death in Central Africa. Examine the maps in figure 15.17, and you will see the relationship between sickle-cell anemia and malaria clearly. On the left map you have the frequency of the sickle-cell allele, the darker green areas indicating a 10% to 20% frequency of the allele. The map on the right indicates the distribution of malaria in dark orange. Clearly, the areas that are colored in darker green on the left map overlap many of the dark orange areas in the map on the right. Even though the population pays a high price—the many individuals in each generation who are homozygous for the sickle-cell allele die—the deaths are far fewer than would occur due to malaria if the heterozygous individuals were not malaria resistant. One in 5 individuals (20%) are heterozygous and survive malaria, while only 1 in 100 (1%) are homozygous and die of sickle-cell anemia. Similar inheritance patterns of the sickle-cell allele are found in other countries frequently exposed to malaria, such as areas around the Mediterranean, India, and Indonesia. Natural selection has favored the sickle-cell allele in Central Africa and other areas hit by malaria because the payoff in survival of heterozygotes more than makes up for the price in death of homozygotes. This phenomenon is an example of **heterozygote advantage.**

Stabilizing selection (also called *balancing selection*) is thus acting on the sickle-cell allele: (1) selection tends to elimi-

nate the sickle-cell allele because of its lethal effects on homozygous individuals, and (2) selection tends to favor the sickle-cell allele because it protects heterozygotes from malaria. Like a manager balancing a store's inventory, natural selection increases the frequency of an allele in a species as long as there is something to be gained by it, until the cost balances the benefit.

Stabilizing selection occurs because malarial resistance counterbalances lethal anemia. Malaria is a tropical disease that has essentially been eradicated in the United States since the early 1950s, and stabilizing selection has not favored the sickle-cell allele here. Africans brought to America several centuries ago have not gained any evolutionary advantage in all that time from being heterozygous for the sickle-cell allele. There is no benefit to being resistant to malaria if there is no danger of getting malaria anyway. As a result, the selection against the sickle-cell allele in America is not counterbalanced by any advantage, and the allele has become far less common among African Americans than among native Africans in Central Africa.

Stabilizing selection is thought to have influenced many other human genes in a similar fashion. The recessive *cf* allele causing cystic fibrosis is unusually common in northwestern Europeans. Apparently, the bacterium causing typhoid fever uses the healthy version of the CFTR protein (see page 84) to enter the cells it infects, but it cannot use the cystic fibrosis version of the protein. As with sickle-cell anemia, heterozygotes are protected.

> **15.6 The prevalence of sickle-cell anemia in African populations is thought to reflect the action of natural selection. Natural selection favors individuals carrying one copy of the sickle-cell allele, because they are resistant to malaria, common in Africa.**

15.7 Selection on Color in Guppies

To study evolution, biologists have traditionally investigated what has happened in the past, sometimes many millions of years ago. To learn about dinosaurs, a paleontologist looks at dinosaur fossils. To study human evolution, an anthropologist looks at human fossils and, increasingly, examines the "family tree" of mutations that have accumulated in human DNA over millions of years. For the biologists taking this traditional approach, evolutionary biology is similar to astronomy and history, relying on observation rather than experiment to examine ideas about past events.

Nonetheless, evolutionary biology is not entirely an observational science. Darwin was right about many things, but one area in which he was mistaken concerns the pace at which evolution occurs. Darwin thought that evolution occurred at a very slow, almost imperceptible, pace. However, in recent years many case studies have demonstrated that in some circumstances evolutionary change can occur rapidly. Consequently, it is possible to establish experimental studies to test evolutionary hypotheses. Although laboratory studies on fruit flies and other organisms have been common for more than 50 years, it has only been in recent years that scientists have started conducting experimental studies of evolution in nature. One excellent example of how observations of the natural world can be combined with rigorous experiments in the lab and in the field concerns research on the guppy, *Poecilia reticulata.*

Guppies Live in Different Environments

The guppy is a popular aquarium fish because of its bright coloration and prolific reproduction. In nature, guppies are found in small streams in northeastern South America and the nearby island of Trinidad. In Trinidad, guppies are found in many mountain streams. One interesting feature of several streams is that they have waterfalls. Amazingly, guppies and some other fish are capable of colonizing portions of the stream above the waterfall. The killifish, *Rivulus hartii,* is a particularly good colonizer; apparently on rainy nights, it will wriggle out of the stream and move through the damp leaf litter. Guppies are not so proficient, but they are good at swimming upstream. During flood seasons, rivers sometimes overflow their banks, creating secondary channels that move through the forest. During these occasions, guppies may be able to move upstream and invade pools above waterfalls. By contrast, not all species are capable of such dispersal and thus are only found in these streams below the first waterfall. One species whose distribution is restricted by waterfalls is the pike cichlid, *Crenicichla alta,* a voracious predator that feeds on other fish, including guppies.

Because of these barriers to dispersal, guppies can be found in two very different environments. The guppies you see living in pools just below the waterfalls in figure 15.18 are faced with predation by the pike cichlid. This substantial risk keeps rates of survival relatively low. By contrast, in similar pools just above the waterfall, the only predator present is the killifish, which only rarely preys on guppies. Guppy populations above and below waterfalls exhibit many differences. In the high-predation pools, guppies exhibit the drab coloration you see in the guppies below the waterfall in figure 15.18. Moreover, they tend to reproduce at a younger age and attain relatively smaller adult sizes. By contrast, male fish above the waterfall in the figure display gaudy colors that they use to court females. Adults mature later and grow to larger sizes.

These differences suggest the function of natural selection. In the low-predation environment, males display gaudy colors and spots that help in mating. Moreover, larger males are most successful at holding territories and mating with females, and larger females lay more eggs. Thus, in the absence of predators, larger and more colorful fish may have produced more offspring, leading to the evolution of those traits. In pools below the waterfall, however, natural selection would favor different traits. Colorful males are likely to attract the attention of the pike cichlid, and high predation rates mean that most fish live short lives; thus, individuals that are more drab and shunt energy into early reproduction, rather than into growth to a larger size, are likely to be favored by natural selection.

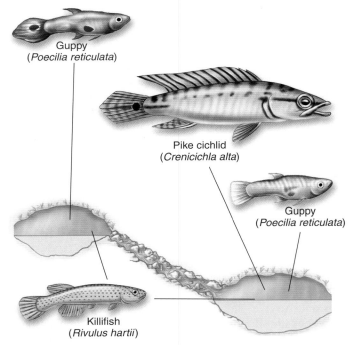

Guppy
(*Poecilia reticulata*)

Pike cichlid
(*Crenicichla alta*)

Guppy
(*Poecilia reticulata*)

Killifish
(*Rivulus hartii*)

Figure 15.18 The evolution of protective coloration in guppies.

In pools below waterfalls where predation is high, male guppies (*Poecilia reticulata*) are drab colored. In the absence of the highly predatory pike cichlid (*Crenicichla alta*), male guppies in pools above waterfalls are much more colorful and attractive to females. The killifish (*Rivulus hartii*) is also a predator but only rarely eats guppies. The evolution of these differences in guppies can be experimentally tested.

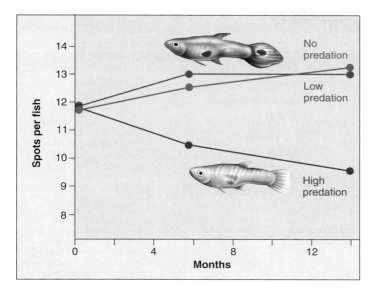

Figure 15.19 Evolutionary change in spot number.
Guppies raised in low-predation or no predation environments in laboratory greenhouses had a greater number of spots, whereas selection in more dangerous environments, like the pools with the highly predatory pike cichlid, led to less conspicuous fish. The same results are seen in field experiments conducted in pools above and below waterfalls (photo).

The Experiments

Although the differences between guppies living above and below the waterfalls suggest that they represent evolutionary responses to differences in the strength of predation, alternative explanations are possible. Perhaps, for example, only very large fish are capable of swimming upstream past the waterfall to colonize pools. If this were the case, then a founder effect would occur in which the new population was established solely by individuals with genes for large size.

Laboratory Experiment. The only way to rule out such alternative possibilities is to conduct a controlled experiment. John Endler, now of the University of California, Santa Barbara, conducted the first experiments in large pools in laboratory greenhouses. At the start of the experiment, a group of 2,000 guppies was divided equally among 10 large pools. Six months later, pike cichlids were added to four of the pools and killifish to another four, with the remaining two pools left to serve as "no predator" controls. Fourteen months later (which corresponds to 10 guppy generations), the scientists compared the populations. You can see their results in figure 15.19. The guppies in the killifish pool (the blue line) and control pools (the green line) were notably large, brightly colored fish with about 13 colorful spots per individual. In contrast, the guppies in the pike cichlid pools (the red line) were smaller and drab in coloration, with a reduced number of spots (about 9 per fish). These results clearly suggest that predation can lead to rapid evolutionary change, but do these laboratory experiments reflect what occurs in nature?

Field Experiment. To find out, Endler and colleagues—including David Reznick, now at the University of California, Riverside—located two streams that had guppies in pools below a waterfall, but not above it (you can see a photograph of the waterfalls in one of these streams on the right side of figure 15.19). As in other Trinidadian streams, the pike cichlid was present in the lower pools, but only the killifish was found above the waterfalls. The scientists then transplanted guppies to the upper pools and returned at several-year intervals to monitor the populations. Despite originating from populations in which predation levels were high, the transplanted populations rapidly evolved the traits characteristic of low-predation guppies: they matured late, attained greater size and had brighter colors. Control populations in the lower pools, by contrast, continued to be drab and matured early and at smaller sizes. Laboratory studies confirmed that the differences between the populations were the result of genetic differences. These results demonstrate that substantial evolutionary change can occur in less than 12 years. More generally, these studies indicate how scientists can formulate hypotheses about how evolution occurs and then test these hypotheses in natural conditions. The results give strong support to the theory of evolution by natural selection.

> **15.7 Experiments can be conducted in nature to test hypotheses about how evolution occurs. Such studies reveal that natural selection can lead to rapid evolutionary change.**

15.8 The Biological Species Concept

A key aspect of Darwin's theory of evolution is his proposal that adaptation (microevolution) leads ultimately to large-scale changes leading to species formation and higher taxonomic groups (macroevolution). The way natural selection leads to the formation of new species has been thoroughly documented by biologists, who have observed the stages of the species-forming process, or **speciation**, in many different plants, animals, and microorganisms. Speciation usually involves successive change: first, local populations become increasingly specialized; then, if they become different enough, natural selection may act to keep them that way.

Before we can discuss how one species gives rise to another, we need to understand exactly what a species is. The evolutionary biologist Ernst Mayr coined the **biological species concept,** which defines species as "groups of actually or potentially interbreeding natural populations which are reproductively isolated from other such groups."

In other words, the biological species concept says that a species is composed of populations whose members mate with each other and produce fertile offspring—or would do so if they came into contact. Conversely, populations whose members do not mate with each other or who cannot produce fertile offspring are said to be **reproductively isolated** and, thus, members of different species.

What causes reproductive isolation? If organisms cannot interbreed or cannot produce fertile offspring, they clearly belong to different species. However, some populations that are considered to be separate species can interbreed and produce fertile offspring, but they ordinarily do not do so under natural conditions. They are still considered to be reproductively isolated in that genes from one species generally will not be able to enter the gene pool of the other species. Table 15.2 summarizes the steps at which barriers to successful reproduction may occur. Examine this table carefully. We will return to it throughout our discussion of species formation. Such barriers are termed **reproductive isolating mechanisms** because they prevent genetic exchange between species. We will first discuss **prezygotic isolating mechanisms,** those that prevent the formation of zygotes. Then we will examine **postzygotic isolating mechanisms,** those that prevent the proper functioning of zygotes after they have formed.

Even though the definition of what constitutes a species is of fundamental importance to evolutionary biology, this issue has still not been completely settled and is currently the subject of considerable research and debate.

> **15.8** A species is generally defined as a group of similar organisms that does not exchange genes extensively with other groups in nature.

TABLE 15.2 ISOLATING MECHANISMS

Mechanism		Description
Prezygotic Isolating Mechanisms		
Geographic isolation		Species occur in different areas, which are often separated by a physical barrier such as a river or mountain range.
Ecological isolation		Species occur in the same area, but they occupy different habitats. Survival of hybrids is low because they are not adapted to either environment of their parents.
Temporal isolation		Species reproduce in different seasons or at different times of the day.
Behavioral isolation		Species differ in their mating rituals.
Mechanical isolation		Structural differences between species prevent mating.
Prevention of gamete fusion		Gametes of one species function poorly with the gametes of another species or within the reproductive tract of another species.
Postzygotic Isolating Mechanisms		
Hybrid inviability or infertility		Hybrid embryos do not develop properly, hybrid adults do not survive in nature, or hybrid adults are sterile or have reduced fertility.

Author's Corner

Are Bird-Killing Cats Nature's Way of Making Better Birds?

Death is not pretty, early in the morning on the doorstep. A small dead bird was left at our front door one morning, lying by the newspaper as if it might at any moment fly away. I knew it would not. Like other birds before it, it was a gift to our household by Feisty, a cat who lives with us. Feisty is a killer of birds, and every so often he leaves one for us, like rent.

We have four cats, and the other three, true housecats, would not know what to do with a bird. Feisty is different, a long-haired gray Persian with the soul of a hunter. While the other three cats sleep safely in the house with us, Feisty spends most nights outside, prowling.

Feisty's nocturnal donations are not well received by my family. More than once it has been suggested, as we donate the bird to the trashman, that perhaps Feisty would be happier living in the country.

As a biologist I try to take a more scientific view. I tell my girls that getting rid of Feisty is unwarranted, because hunting cats like Feisty actually help birds, in a Darwinian sort of way. Like an evolutionary quality control check, I explain, predators ensure that only those individuals of a population that are better-suited to their environment contribute to the next generation, by the simple expedient of removing the lesser-suited. By taking the birds who are least able to escape predation—the sick and the old—Feisty culls the local bird population, leaving it on average a little better off.

That's what I tell my girls. It all makes sense, from a biological point of view, and it is a story they have heard before, in movies like *Never Cry Wolf*, and *The Lion King*. So Feisty is given a reprieve, and survives to hunt another night.

What I haven't told my girls is how little evidence actually backs up this pretty defense of Feisty's behavior. My explanation may be couched in scientific language, but without proof this "predator-as-purifier" tale is no more than a hypothesis. It might be true, and then again it might not. By such thin string has Feisty's future with our family hung.

Recently the string became a strong cable. Two French biologists put the hypothesis I had been using to defend Feisty to the test. To my great relief, it was supported.

Drs. Anders Møller and Johannes Erritzoe of the Université Pierre et Marie Curie in Paris devised a simple way to test the hypothesis. They compared the health of birds killed by domestic cats like Feisty with that of birds killed in accidents such as flying into glass windows or moving cars. Glass windows do not select for the weak or infirm—a sickly bird flies into a glass window and breaks its neck just as easily as a healthy bird. If cats are actually selecting the less-healthy birds, then their prey should include a larger proportion of sickly individuals than those felled by flying into glass windows.

How can we know what birds are sickly? Drs. Møller and Erritzoe examined the size of the dead bird's spleens. The size of its spleen is a good indicator of how healthy a bird is. Birds experiencing a lot of infections, or harboring a lot of parasites, have smaller spleens than healthy birds.

They examined 18 species of birds, more than 500 individuals. In all but two species (robins and goldcrests) they found that the spleens of birds killed by cats were significantly smaller than those killed accidently. We're not splitting hairs here, talking about some minor statistical difference. Spleens were on average a third smaller in cat-killed birds. In five bird species (blackcaps, house sparrows, lesser whitethroats, skylarks, and spotted flycatchers), the spleens of birds pounced on by cats were less than half the size of those killed by flying at speed into glass windows or moving cars.

As a control to be sure that additional factors were not operating, the Paris biologists checked for other differences between birds killed by cats and birds killed accidentally. Weight, sex, and wing length, all of which you could imagine might be important, were not significant. Cat-killed birds had, on average, the same weight, proportion of females, and wing length as accident-killed birds.

One other factor did make a difference: age. About 50% of the birds killed accidently were young, while fully 70% of the birds killed by cats were. Apparently it's not quite so easy to catch an experienced old codger as it is a callow youth.

So Feisty was just doing Darwin's duty, I pleaded, informing my girls that the birds he catches would soon have died anyway. But a dead bird on a doorstep argues louder than any science, and they remained unconvinced.

They are my daughters, and thus not ones to give in without a fight. Scouring the Internet, they assembled this counter-argument: Predatory house cats not unlike Feisty, as well as feral cats (domesticated cats that have been abandoned to the wild), are causing major problems for native bird populations of England, New Zealand, and Australia, as well as here in the United States. Although house cats like Feisty have the predatory instincts of their ancestors, they seem to lack the restraint that their wild relatives have. Most wild cats hunt only when hungry, but pet and feral cats seem to "love the kill," not killing for food but for sport.

So Darwin and I lost this argument. It seems I must restrict Feisty's hunting expeditions after all. While a little pruning may benefit a bird population, wholesale slaughter only devastates it. I will always see a lion whenever I look at Feisty on the prowl, but it will be a lion restricted to indoor hunting.

Prezygotic Isolating Mechanisms

Geographical Isolation. This mechanism is perhaps the easiest to understand; species that exist in different areas are not able to interbreed. The two populations of flowers in the first panel of table 15.2 are separated by a mountain range and so would not be capable of interbreeding.

Ecological Isolation. Even if two species occur in the same area, they may utilize different portions of the environment and thus not hybridize because they do not encounter each other, like the lizards in the second panel of table 15.2. One lives on the ground and the other in the trees. Another example in nature is the ranges of lions and tigers in India. Their ranges overlapped until about 150 years ago. Even when they did overlap, however, there were no records of natural hybrids. Lions stayed mainly in the open grassland and hunted in groups called prides; tigers tended to be solitary creatures of the forest. Because of their ecological and behavioral differences, lions and tigers rarely came into direct contact with each other, even though their ranges overlapped thousands of square kilometers. Figure 15.20 shows that hybrids are possible; the tigon shown in figure 15.20c is a hybrid of a lion and tiger. These matings do not occur in the wild but can happen in artificial environments such as zoos.

Temporal Isolation. *Lactuca graminifolia* and *L. canadensis*, two species of wild lettuce, grow together along roadsides throughout the southeastern United States. Hybrids between these two species are easily made experimentally and are completely fertile. But such hybrids are rare in nature because *L. graminifolia* flowers in early spring and *L. canadensis* flowers in summer. This is called temporal isolation and is shown in the third panel in table 15.2. When the blooming periods of these two species overlap, as they do occasionally, the two species do form hybrids, which may become locally abundant.

Behavioral Isolation. In chapter 22, we will consider the often elaborate courtship and mating rituals of some groups of animals, which tend to keep these species distinct in nature even if they inhabit the same places. This behavioral isolation is discussed in the fourth panel of table 15.2. For example, mallard and pintail ducks are perhaps the two most common freshwater ducks in North America. In captivity, they produce completely fertile offspring, but in nature they nest side-by-side and rarely hybridize.

Mechanical Isolation. Structural differences that prevent mating between related species of animals and plants is called mechanical isolation and is shown in panel five of table 15.2. Flowers of related species of plants often differ significantly in their proportions and structures. Some of these differences

(a)

(b)

(c)

Figure 15.20 Lions and tigers are ecologically isolated.

The ranges of lions and tigers used to overlap in India. However, lions and tigers do not hybridize in the wild because they utilize different portions of the habitat. (a) Tigers are solitary animals that live in the forest, whereas (b) lions live in open grassland. (c) Hybrids, such as this tigon, have been successfully produced in captivity, but hybridization does not occur in the wild.

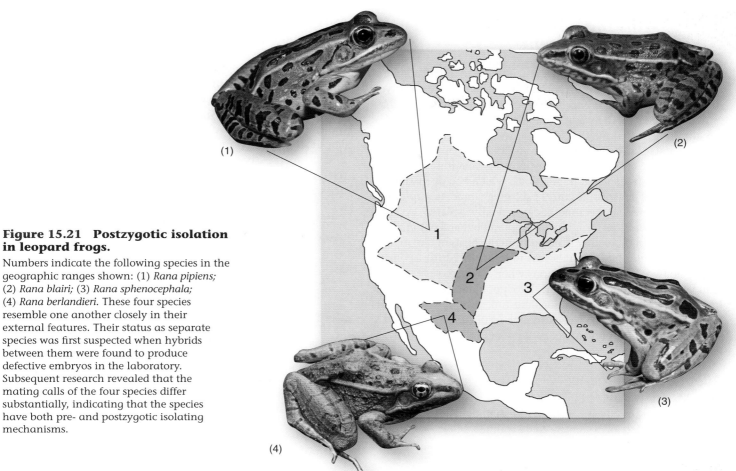

Figure 15.21   Postzygotic isolation in leopard frogs.
Numbers indicate the following species in the geographic ranges shown: (1) *Rana pipiens*; (2) *Rana blairi*; (3) *Rana sphenocephala*; (4) *Rana berlandieri*. These four species resemble one another closely in their external features. Their status as separate species was first suspected when hybrids between them were found to produce defective embryos in the laboratory. Subsequent research revealed that the mating calls of the four species differ substantially, indicating that the species have both pre- and postzygotic isolating mechanisms.

limit the transfer of pollen from one plant species to another. For example, bees may pick up the pollen of one species on a certain place on their bodies; if this area does not come into contact with the receptive structures of the flowers of another plant species, the pollen is not transferred.

Prevention of Gamete Fusion.   In animals that shed their gametes directly into water, eggs and sperm derived from different species may not attract one another. Many land animals may not hybridize successfully because the sperm of one species may function so poorly within the reproductive tract of another that fertilization never takes place. In plants, the growth of pollen tubes may be impeded in hybrids between different species. In both plants and animals, the operation of such isolating mechanisms prevents the union of gametes even following successful mating. The sixth panel in table 15.2 discusses this isolating mechanism.

Postzygotic Isolating Mechanisms

All of the factors we have discussed up to this point tend to prevent hybridization. If hybrid matings do occur, and zygotes are produced, many factors may still prevent those zygotes from developing into normally functioning, fertile individuals. Development in any species is a complex process. In hybrids, the genetic complements of two species may be so different that they cannot function together normally in embryonic development. For example, hybridization between sheep and goats usually produces embryos that die in the earliest developmental stages.

Figure 15.21 shows four species of leopard frogs (*Rana pipiens* complex) and their ranges throughout North America. It was assumed for a long time that they constituted a single species. However, careful examination revealed that although the frogs appear similar, successful mating between them is rare because of problems that occur as the fertilized eggs develop. Many of the hybrid combinations cannot be produced even in the laboratory. Examples of this kind, in which similar species have been recognized only as a result of hybridization experiments, are common in plants.

Even if hybrids survive the embryo stage, however, they may not develop normally. If the hybrids are weaker than their parents, they will almost certainly be eliminated in nature. Even if they are vigorous and strong, as in the case of the mule, a hybrid between a female horse and a male donkey, they may still be sterile and thus incapable of contributing to succeeding generations. Sterility may result in hybrids because the development of sex organs may be abnormal, because the chromosomes derived from the respective parents may not pair properly, or from a variety of other causes.

15.9   Prezygotic isolating mechanisms lead to reproductive isolation by preventing the formation of hybrid zygotes. Postzygotic mechanisms lead to the failure of hybrid zygotes to develop normally, or prevent hybrids from becoming established in nature.

Working with the Biological Species Concept

Speciation is a two-part process. First, initially identical populations must diverge. Second, reproductive isolation must evolve to maintain these differences, overcoming the homogenizing effect of gene flow between populations, which acts to erase any differences that may arise, either by genetic drift or natural selection. However, gene flow occurs only between populations that are in contact, so speciation is much more likely in geographically isolated populations, which become isolated for a variety of reasons. Figure 15.22 gives three examples of how populations can become geographically isolated. A bird or population of birds flying from the mainland and colonizing an island is geographically isolated from the birds on the mainland. The population of wild horses in figure15.22*b* splits into two different populations when a volcano reduces movement between the two populations. In figure 15.22*c*, two populations are linked by an intermediate population. The extinction of the intermediate population prevents gene flow and isolates the two more distant populations.

Allopatric Divergence

Ernst Mayr was the first biologist to strongly make the case for **allopatric speciation.** Marshalling data from a wide variety of organisms and localities, Mayr was clearly able to demonstrate that geographically separated, or *allopatric,* populations appear much more likely to have evolved substantial differences leading to speciation. For example, the Papuan kingfisher, *Tanysiptera hydrocharis,* varies little throughout its wide range in New Guinea despite the great variation in the island's topography and climate. By contrast, isolated populations on nearby islands are strikingly different from each other and from the mainland population.

Sympatric Speciation

Sympatric speciation is one species splitting into two at a single locality, without the new species ever having been geographically separated. Instantaneous sympatric speciation occurs when an individual is reproductively isolated from all other members of its species through the process of **polyploidy,** which occurs commonly in plants. A polyploid individual has more than two sets of chromosomes.

Polyploids can arise in two ways. In **autopolyploidy,** all sets of the chromosomes are from the same species. Typically, the chromosomes in a gamete of a diploid species fail to separate in meiosis. The gamete thus possesses two sets of chromosomes rather than one. This gamete cannot produce fertile offspring with a normal gamete because such offspring are triploid (three sets of chromosomes) and so are sterile. Occasionally, however, the chromosomes of such a gamete spontaneously double, forming a viable zygote. Such individuals, termed tetraploids, can fertilize themselves or mate with other tetraploids. A new species has formed.

A more common type of polyploid speciation is **allopolyploidy,** which occurs sometimes when two species hybridize. The resulting offspring, with one copy of the chromosomes of each species, is usually infertile because the chromosomes have no pairing partners in meiosis. However, such individuals are often otherwise healthy and can reproduce asexually, and if the chromosomes of such an individual spontaneously double, as just described, the resulting tetraploid would have two copies of each set of chromosomes. Consequently, pair-

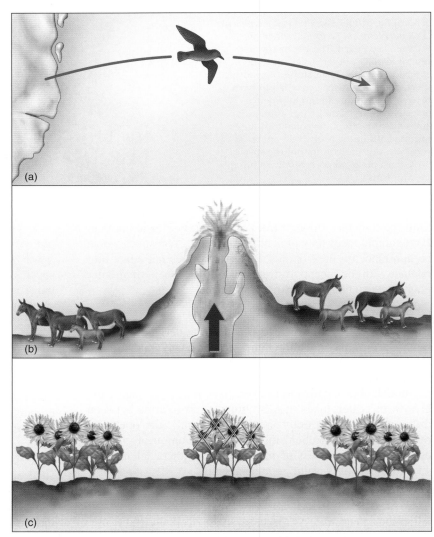

Figure 15.22 Populations can become geographically isolated for a variety of reasons.

(a) Colonization of distant areas by one or a few individuals can establish populations in a distant place. (b) Barriers to movement can split an ancestral population into two isolated populations. (c) Extinction of intermediate populations can leave the remaining populations isolated from each other.

ing would no longer be a problem in meiosis: each chromosome could pair with its double. As a result, such tetraploids would be able to intermate, and a new species would have been created.

It is estimated that about half of the approximately 260,000 species of plants have a polyploid episode in their history, including many of great commercial importance, such as bread wheat, cotton, tobacco, sugarcane, bananas, and potatoes. Although much rarer than in plants, speciation by polyploidy is also known from a variety of animals, including insects, fish, and salamanders.

Problems with the Biological Species Concept

The biological species concept has proven to be an effective way of understanding the existence of species in nature. Nonetheless, it has a number of problems that have led some scientists to propose alternative species concepts.

Reproductive Isolation in Plants. One criticism concerns the extent to which all species truly are reproductively isolated. By definition, under the biological species concept, species should not interbreed and produce fertile offspring. Nonetheless, in recent years, biologists have detected much greater amounts of hybridization than previously realized between populations that seem to coexist as distinct biological entities. Botanists have always been aware that species can often experience substantial amounts of hybridization. For example, more than 50% of California plant species included in one study were not well defined by genetic isolation. Such coexistence without genetic isolation can be long-lasting: fossil data show that balsam poplars and cottonwoods have been phenotypically distinct for 12 million years but have routinely produced hybrids throughout this time. Consequently, many botanists have long felt that the biological species concept only applies to animals, mainly vertebrates.

Reproductive Isolation in Animals. What is becoming increasingly evident, however, is that hybridization is not all that uncommon in animals, either. One recent survey indicated that almost 10% of the world's 9,500 bird species are known to have hybridized in nature. Recent years have seen the documentation of many cases of substantial hybridization between animal species. Galápagos finches provide a particularly well-studied example. Three species on the island of Daphne Major—the medium ground finch, the cactus finch, and the small ground finch—are clearly distinct morphologically and occupy different ecological niches. Studies over the past 20 years by Peter and Rosemary Grant found that, on average, 2% of the medium ground finches and 1% of the cactus finches mated with other species every year. Furthermore, hybrid offspring appeared to be at no disadvantage in terms of survival or subsequent reproduction. This is not a trivial amount of genetic exchange, and one might expect to see the species coalesce into one genetically variable population, but the species are maintaining their distinctiveness.

This is not to say hybridization is rampant throughout the animal world. Most bird species do not hybridize, and even fewer probably experience significant amounts of hybridization. Still, it is common enough to cast doubt about whether reproductive isolation is the only force maintaining the integrity of species.

Other Problems with the Biological Species Concept. The biological species concept has been criticized for other reasons as well. For example, it can be difficult to apply the concept to populations that do not occur together in nature or together in time, such as extinct species. Because individuals of these populations do not encounter each other, it is not possible to observe whether they would interbreed naturally. Although experiments can determine whether fertile hybrids can be produced, this information is not enough because many species that will coexist without interbreeding in nature will readily hybridize in the artificial settings of the laboratory or zoo. Consequently, evaluating whether such populations constitute different species is ultimately a judgment call. In addition, the concept is more limited than its name would imply. The vast majority of organisms on this earth are asexual and reproduce without mating; reproductive isolation has no meaning for such organisms.

Natural Selection and the Ecological Species Concept

An alternative hypothesis is that the distinctions among species are maintained by natural selection. The idea is that each species has adapted to its own specific part of the environment. Stabilizing selection then maintains the species' adaptations; hybridization has little effect because alleles introduced into the gene pool from other species are quickly eliminated by natural selection.

This and a variety of other ideas put forward to establish criteria for defining species are specific to a particular type of organism, and none has universal applicability. In truth, it may be that there is no single explanation for what maintains the identity of species. Given the incredible variation evident in plants, animals, and microorganisms in all aspects of their biology, it is perhaps not surprising that different processes are operating in different organisms. In addition, some scientists have turned from emphasizing the processes that maintain species distinctions to examining the history of populations. These genealogical species concepts are currently a topic of great debate. The study of species concepts is thus an area of active research that demonstrates the dynamic nature of the field of evolutionary biology.

> **15.10** Speciation occurs much more readily in the absence of gene flow among populations. However, speciation can occur in sympatry by means of polyploidy. Because of the diversity of living organisms, no single definition of what constitutes a species may be universally applicable.

Does Natural Selection Act on Enzyme Polymorphism?

The essence of Darwin's theory of evolution is that, in nature, selection favors some gene alternatives over others. Many studies of natural selection have focused on genes encoding enzymes because populations in nature tend to possess many alternative alleles of their enzymes (a phenomenon called *enzyme polymorphism*). Often investigators have looked to see if weather influences which alleles are more common in natural populations. A particularly nice example of such a study was carried out on a fish, the mummichog (*Fundulus heteroclitus*), which ranges along the East Coast of North America. Researchers studied allele frequencies of the gene encoding the enzyme lactate dehydrogenase, which catalyzes the conversion of pyruvate to lactate. As you learned in chapter 8, this reaction is a key step in energy metabolism, particularly when oxygen is in short supply. There are two common alleles of lactate dehydrogenase in these fish populations, with allele *a* being a better catalyst at lower temperatures than allele *b*.

In an experiment, investigators sampled the frequency of allele *a* in 41 fish populations located over 14 degrees of latitude, from Jacksonville, Florida (31° North), to Bar Harbor, Maine (44° North). Annual mean water temperatures change 1° C per degree change in latitude. The survey is designed to test a prediction of the hypothesis that natural selection acts on this enzyme polymorphism. If it does, then you would expect that allele *a*, producing a better "low-temperature" enzyme, would be more common in the colder waters of the more northern latitudes. The graph on the right presents the results of this survey. The points on the graph are derived from pie chart data such as shown for 20 populations in the map (a **pie chart diagram** assigns a slice of the pie to each variable; the size of the slice is proportional to the contribution made by that variable to the total). The blue line on the graph is the line that best fits the data (a **best-fit" line,** also called a **regression line,** is determined statistically by a process called *regression analysis*).

1. **Applying Concepts**
   a. **Variable.** In the graph, what is the dependent variable?
   b. **Reading pie charts.** In the fish population located at 35° N latitude, what is the frequency of the *a* allele? Locate this point on the graph.
   c. **Analyzing a continuous variable.** Compare the frequency of allele *a* among fish captured in waters at 44° N latitude with the frequency among fish captured at 31° N latitude. Is there a pattern? Describe it.
2. **Interpreting Data** At what latitude do fish populations exhibit the greatest variability in allele *a* frequency?
3. **Making Inferences**
   a. Are fish populations in cold waters at 44° N latitude

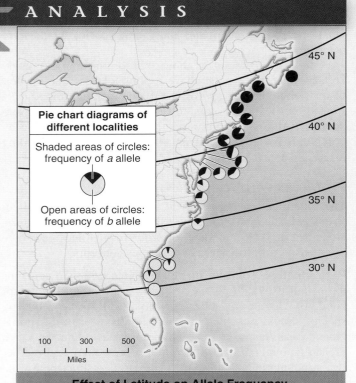

Pie chart diagrams of different localities

Shaded areas of circles: frequency of *a* allele

Open areas of circles: frequency of *b* allele

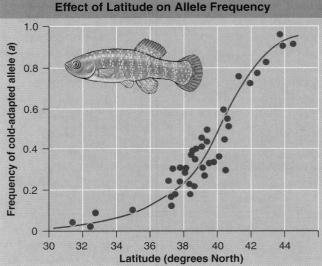

Effect of Latitude on Allele Frequency

more or less likely to contain heterozygous individuals than fish populations in warm waters at 31° N latitude? Why this difference, or lack of it?
   b. Where along this latitudinal gradient in the frequency of allele *a* would you expect to find the highest frequency of heterozygous individuals? Why?
4. **Drawing Conclusions** Are the differences in population frequencies of allele *a* consistent with the hypothesis that natural selection is acting on the alleles encoding this enzyme? Explain.
5. **Further Analysis** If you were to release fish captured at 32° N into populations located at 44° N, so that the local population now had equal frequencies of the two alleles, what would you expect to happen in future generations? How might you test this prediction?

The Theory of Evolution

15.1 Evolution: Getting from There to Here

- Darwin's proposal of evolution by natural selection focuses primarily on microevolution. Microevolution occurs as changes in allele frequencies in a population. Over time microevolution can lead to macroevolution, the formation of new species (**figure 15.1**).

- Microevolution can only occur when there is variation in the population, different alleles in the population on which natural selection can act. Without variation in alleles there can be no natural selection and no microevolution.

- The rate of evolution isn't the same in all populations. Some evolve at a uniform rate, gradualism, while others evolve in spurts, punctuated equilibrium (**figure 15.2**).

15.2 The Evidence for Evolution

- The evidence for evolution includes the fossil record (**figure 15.3**). The evolution of a species from an ancestral organism can be traced through the fossil record. The fossil record reveals organisms that are intermediate in form between the ancestral species and the present-day species.

- The evidence for evolution includes the anatomical record, which reveals similarities in structures between species (**figure 15.4**). Homologous structures are similar in structure but differ in their functions (**figure 15.5**). Analogous structures are similar in function but differ in their underlying structure. Analogous structures arise from convergent evolution (**figure 15.6**).

- The evidence for evolution includes the molecular record. The molecular record traces changes in the genomes of organisms, with organisms that are more distantly related showing more differences in their genomes (**figures 15.7** and **15.8**).

15.3 Evolution's Critics

- Darwin's theory of evolution through natural selection has always had its critics. More recently, the teaching of evolution in schools in the United States is coming under attack from groups who disagree with it and want equal treatment of other proposals. Their criticisms of evolution through natural selection are without scientific merit (**figures 15.9** and **15.10**).

How Populations Evolve

15.4 Genetic Change Within Populations: The Hardy-Weinberg Rule

- If a population follows the five assumptions of Hardy-Weinberg, the frequencies of alleles within the population will not change (**figure 15.11**). However, if a population is small, has selective mating, mutations, immigrations or emigrations, or is under the influence of natural selection, allele frequencies will be different in future populations.

15.5 Agents of Evolution

- There are five factors that act on populations to change their allele and genotype frequencies (**table 15.1**): mutations, migrations, genetic drift, nonrandom mating, and selection.

- Mutations are changes in DNA. Migrations are the movement of individuals or alleles into or out of the population. Genetic drift is the random loss of alleles in a population due to chance occurrences and not due to fitness. Nonrandom mating occurs when individuals seek out mates based on certain traits. Selection occurs when individuals with certain traits leave more offspring because their traits allow them to better respond to the challenges of their environment.

- Selection can change a population such that certain traits are more common than others. When a trait is controlled by more than one gene, selection can act on the genes in that population in several different ways (**figure 15.13**).

- Stabilizing selection tends to reduce extreme phenotypes, making the intermediate phenotype more common (**figure 15.14a**). Disruptive selection tends to reduce intermediate phenotypes, leaving extreme phenotypes in the population (**figure 15.14b**). Directional selection tends to reduce one extreme phenotype from the population (**figure 15.14c**).

Adaptation Within Populations

15.6 Sickle-Cell Anemia

- Sickle-cell anemia is an example of heterozygous advantage, where individuals who are heterozygous for a trait tend to survive better in areas with malaria than individuals with either of the two homozygous phenotypes (**figures 15.16** and **15.17**).

15.7 Selection on Color in Guppies

- Experimentation has been able to show evolutionary change due to natural selection on certain traits in a population of guppies. Although this isn't macroevolution, it is microevolution in action (**figures 15.18** and **15.19**).

How Species Form

15.8 The Biological Species Concept

- The biological species concept states that a species is a group of organisms that mate with each other and produce fertile offspring, or would do so if in contact with each other. If they cannot mate, or mate but cannot produce fertile offspring, they are said to be reproductively isolated. Many barriers to successful reproduction lead to reproductive isolation (**table 15.2**).

15.9 Isolating Mechanisms

- There are two types of isolating mechanisms, prezygotic and postzygotic. Prezygotic isolating mechanisms prevent the formation of a hybrid zygote. Postzygotic isolating mechanisms prevent normal development of a hybrid zygote or result in sterile offspring (**figure 15.21**).

15.10 Working with the Biological Species Concept

- Speciation is more likely to occur between populations that are geographically isolated from each other, called allopatric speciation (**figure 15.22**). However, sympatric speciation, where one species splits into two at a single location, is more common in plants than in animals.

- The biological species concept doesn't explain all situations of speciation, and so other concepts have been proposed, but no single concept is adequate.

1. Changes in organisms over a long period of time, including the formation of new species, is known as
   a. natural selection.
   b. microevolution.
   c. macroevolution.
   d. punctuated equilibrium.

2. One of the major sources of evidence for evolution is in the comparative anatomy of organisms. Features that have a similar look but different structural origin are called
   a. homologous structures.
   b. analogous structures.
   c. vestigial structures.
   d. equivalent structures.

3. Homologous structures in organisms are the result of
   a. divergence.
   b convergent evolution.
   c. stasis.
   d. polyploidy.

4. The Hardy-Weinberg equilibrium allows scientists to study
   a. natural selection.
   b. microevolution.
   c. macroevolution.
   d. punctuated equilibrium

5. A large group of organisms lives in a large, stable ecosystem. There is no competition for resources. Individuals mate at random. All organisms appear to be identical except for a few individuals in the most recent generation of offspring that exhibit a different fur coat color and pattern. The ecosystem and population are geographically isolated from other populations of the same organism. Which Hardy-Weinberg assumption has been violated?
   a. large population size
   b. random mating within the population
   c. no mutation within the population
   d. no input of new alleles from outside or loss of alleles

6. A population of 1,000 individuals has 200 individuals who show a homozygous recessive phenotype and 800 individuals who express the dominant phenotype. What is the frequency of homozygous dominant individuals (p^2) in this population?
   a. $p^2 = 0.20$
   b. $p^2 = 0.30$
   c. $p^2 = 0.45$
   d. $p^2 = 0.55$

7. A chance event occurs that causes a population to lose some individuals (they died)—hence, a loss of alleles in the population results from
   a. mutation.
   b. migration.
   c. selection.
   d. genetic drift.

8. Selection that causes an extreme phenotype to be more frequent in a population is an example of
   a. disruptive selection.
   b. stabilizing selection.
   c. directional selection.
   d. equivalent selection.

9. A key element of Ernst Mayr's biological species concept is
   a. homologous isolation.
   b. divergent isolation.
   c. convergent isolation.
   d. reproductive isolation.

10. Sympatric species are populations of species that live in the same habitat. Sympatric species of deer mice (*Peromyscus* sp.) are externally identical. However, the males of different species have a differently shaped baculum, or bone found in the penis. This is an example of which type of reproductive isolation?
    a. temporal
    b. mechanical
    c. behavioral
    d. ecological

Visual Understanding

1. **Figure 15.13** Because of prolonged drought, the trees on an island are producing nuts that are much smaller with thicker and harder shells. What will happen to the birds that depend on the nuts for food?

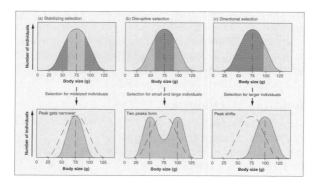

2. **Table 15.2** A very heavy rainstorm floods a mountain river, changing its course and digging a deep canyon through the soft soils of the meadow in the valley below. How could mice populations on either side of the valley be affected?

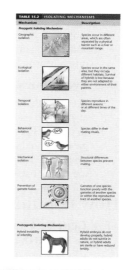

Challenge Questions

1. How does microevolution differ from macroevolution?
2. How likely is it that any naturally occurring population on a continent (not on an island) will meet all the assumptions of the Hardy-Weinberg equations? Explain your reasoning.
3. Use Hardy-Weinberg and the tenets of natural selection to explain why there are colorful, large guppies in pools without pike cichlids, and drab, small guppies when pike cichlids are present.
4. In a courtroom in 2005, biologist Ken Miller criticized the claims of intelligent design. After noting that 99.9% of the organisms that have ever lived on earth are now extinct, he said that "an intelligent designer who designed things, 99.9% of which didn't last, certainly wouldn't be very intelligent." Evaluate Miller's criticism.

16

Exploring Biological Diversity

In 1799, the skin of a most unusual animal was sent to England by Captain John Hunter, governor of the British penal colony in New South Wales (Australia). Covered in soft fur, it was less than 2 feet long. As it had mammary glands with which to suckle its young, it was clearly a mammal, but in other ways it seemed very like a reptile. Males have internal testes, and females have a shared urinary and reproductive tract opening called a cloaca, lay eggs as reptiles do, and like reptilian eggs, the yolk of the fertilized egg does not divide. It thus seemed a confusing mixture of mammalian and reptilian traits. Adding to this impression was its appearance: it has a tail not unlike that of a beaver, a bill not unlike that of a duck, and webbed feet! It was as if a child had mixed together body parts at random—a most unusual animal. Individuals like the one pictured here are abundant in freshwater streams of eastern Australia today. What does one call such a beast? In its original 1799 description, it was named *Platypus anatinus* (flatfooted ducklike animal), which was later changed to *Ornithorhynchus anatinus* (ducklike animal with a bird's snout)—informally, the duckbill platypus. How biologists assign names to the organisms they discover is the subject of this chapter. You will be surprised at how much information is crammed into the two words of a scientific name.

16.1 The Invention of the Linnaean System

It is estimated that our world is populated by some 10 to 100 million different kinds of organisms. To talk about them and study them, it is necessary to give them names, just as it is necessary that people have names. Of course, no one can remember the name of every kind of organism, so biologists use a kind of multilevel grouping of individuals called **classification.**

Organisms were first classified more than 2,000 years ago by the Greek philosopher Aristotle, who categorized living things as either plants or animals. He classified animals as either land, water, or air dwellers, and he divided plants into three kinds based on stem differences. This simple classification system was expanded by the Greeks and Romans, who grouped animals and plants into basic units such as cats, horses, and oaks. Eventually, these units began to be called **genera** (singular, **genus**), the Latin word for "group." Starting in the Middle Ages, these names began to be systematically written down, using Latin, the language used by scholars at that time. Thus, cats were assigned to the genus *Felis*, horses to *Equus*, and oaks to *Quercus*—names that the Romans had applied to these groups. For genera that were not known to the Romans, new names were invented.

The classification system of the Middle Ages, called the *polynomial system*, was used virtually unchanged for hundreds of years, until it was replaced about 250 years ago by the **binomial system** introduced by Linnaeus.

The Polynomial System

Until the mid-1700s, biologists usually added a series of descriptive terms to the name of the genus when they wanted to refer to a particular kind of organism, which they called a **species.** These phrases, starting with the name of the genus, came to be known as **polynomials** (*poly,* many, and *nomial,* name), strings of Latin words and phrases consisting of up to 12 or more words. For example, the common wild briar rose was called *Rosa sylvestris inodora seu canina* by some and *Rosa sylvestris alba cum rubore, folio glabro* by others. This would be like the mayor of New York referring to a particular citizen as "Brooklyn resident: Democrat, male, Caucasian, middle income, Protestant, elderly, likely voter, short, bald, heavyset, wears glasses, works in the Bronx selling shoes." As you can imagine, these polynomial names were cumbersome. Even more worrisome, the names were altered at will by later authors, so that a given organism really did not have a single name that was its alone, as was the case with the briar rose.

The Binomial System

A much simpler system of naming animals, plants, and other organisms stems from the work of the Swedish biologist Carolus Linnaeus (1707–78). Linnaeus devoted his life to a challenge that had defeated many biologists before him—cataloging all the different kinds of organisms. Linnaeus, a

(a) *Quercus phellos*
(Willow oak)

(b) *Quercus rubra*
(Red oak)

Figure 16.1 How Linnaeus named two species of oaks.

(a) Willow oak, *Quercus phellos.* (b) Red oak, *Quercus rubra.* Although they are clearly oaks (members of the genus *Quercus*), these two species differ sharply in the shapes and sizes of their leaves and in many other features, including their overall geographical distributions.

botanist studying the plants of Sweden and from around the world, developed a plant classification system based on grouping plants based on their reproductive structures. This system resulted in some seemingly unnatural groupings and therefore was never universally accepted. However, in the 1750s he produced several major works that, like his earlier books, employed the polynomial system. But as a kind of shorthand, Linnaeus also included in these books a two-part name for each species (others had also occasionally done this, but Linnaeus used these shorthand names consistently). These two-part names, or **binomials** (*bi* is the Latin prefix for "two"), have become our standard way of designating species. For example, he designated the willow oak (shown in figure 16.1*a* with its smaller, unlobed leaves) *Quercus phellos* and the red oak (with the larger deeply lobed leaves in figure 16.1*b*) *Quercus rubra,* even though he also included the polynomial name for these species. We also use binomial names for ourselves, our so-called given and family names. So, this naming system is like the mayor of New York calling the Brooklyn resident Sylvester Kingston.

Linnaeus took the naming of organisms a step further, grouping similar organisms into higher-level categories based on similar characteristics (discussed later). Although not intended to show evolutionary connections between different organisms, this hierarchical system acknowledged that there were broad similarities shared by groups of species that distinguished them from other groups.

> **16.1** Two-part (binomial) Latin names, first used by Linnaeus, are now universally employed by biologists to name particular organisms.

16.2 Species Names

A group of organisms at a particular level in a classification system is called a **taxon** (plural, **taxa**), and the branch of biology that identifies and names such groups of organisms is called **taxonomy.** Taxonomists are in a real sense detectives, biologists who must use clues of appearance and behavior to identify and assign names to organisms.

By formal agreement among taxonomists throughout the world, no two organisms can have the same name. So that no one country is favored, a language spoken by no country—Latin—is used for the names. Because the scientific name of an organism is the same anywhere in the world, this system provides a standard and precise way of communicating, whether the language of a particular biologist is Chinese, Arabic, Spanish, or English. This is a great improvement over the use of common names, which often vary from one place to the next. As you can see in figure 16.2, in America corn refers to the plant in the upper photo on the left, but in Europe it refers to the plant Americans call wheat, the lower photo on the left. A bear is a large placental omnivore in the United States (the upper middle photo), but in Australia it is a koala, a vegetarian marsupial (the lower middle photo). A robin in North America (the upper right photo) is a very different bird in Europe (the lower right photo).

By convention, the first word of the binomial name is the genus to which the organism belongs. This word is always capitalized. The second word, called the *specific epithet*, refers to the particular species and is not capitalized. The two words together are called the **scientific name,** or species name, and are written in italics. The system of naming animals, plants, and other organisms established by Linnaeus has served the science of biology well for nearly 250 years.

> **16.2** By convention, the first part of a binomial species name identifies the genus to which the species belongs, and the second part distinguishes that particular species from other species in the genus.

(a)   (b)   (c)

Figure 16.2 Common names make poor labels.
The common names corn (a), bear (b), and robin (c) bring clear images to our minds (photos on *top*), but the images would be very different to someone living in Europe or Australia (photos on *bottom*). There, the same common names are used to label very different species.

16.3 Higher Categories

Like the mayor of New York, a biologist needs more than two categories to classify all the world's living things. Taxonomists group the genera with similar properties into a cluster called a **family.** For example, the Eastern gray squirrel at the center in figure 16.3 is placed in a family with other squirrel-like animals including prairie dogs, marmots, and chipmunks. Similarly, families that share major characteristics are placed into the same **order** (for example, squirrels placed in with other rodents). Orders with common properties are placed into the same **class** (squirrels in the class Mammalia), and classes with similar characteristics into the same **phylum** (plural, **phyla**) such as the Chordata. Botanists (that is, those who study plants) also call plant phyla "divisions." Finally, the phyla are assigned to one of several gigantic groups, the **kingdoms.** Biologists currently recognize six kingdoms: two kinds of prokaryotes (Archaea and Bacteria), a largely unicellular group of eukaryotes (Protista), and three multicellular groups (Fungi, Plantae, and Animalia). To remember the seven categories in their proper order, it may prove useful to memorize a phrase such as "**k**indly **p**ay **c**ash **o**r **f**urnish **g**ood **s**ecurity" or "**K**ing **P**hilip **c**ame **o**ver **f**or **g**reen **s**paghetti" (**k**ingdom–**p**hylum–**c**lass–**o**rder–**f**amily–**g**enus–**s**pecies).

In addition, an eighth level of classification, called *domains,* is sometimes used. Domains are the broadest and most inclusive taxa, and biologists recognize three of them, Bacteria, Archaea, and Eukarya—which are discussed later in this chapter.

Each of the categories in this **Linnaean system of classification** is loaded with information. For example, consider a honeybee:

Level 1: Its species name, *Apis mellifera,* identifies the particular species of honey bee.

Level 2: Its genus name, *Apis,* tells you it is a honey bee.

Level 3: Its family, Apidae, are all bees, some solitary, others living in hives as *A. mellifera* does.

Level 4: Its order, Hymenoptera, tells you that it is likely able to sting and may live in colonies.

Level 5: Its class, Insecta, says that *A. mellifera* has three major body segments, with wings and three pairs of legs attached to the middle segment.

Level 6: Its phylum, Arthropoda, tells us that it has a hard cuticle of chitin and jointed appendages.

Level 7: Its kingdom, Animalia, says that it is a multicellular heterotroph whose cells lack cell walls.

Level 8: An addition to the Linnaean system, its domain, Eukarya, says that its cells contain membrane-bounded organelles.

> **16.3** A hierarchical system is used to classify organisms, in which higher categories convey more general information about the group.

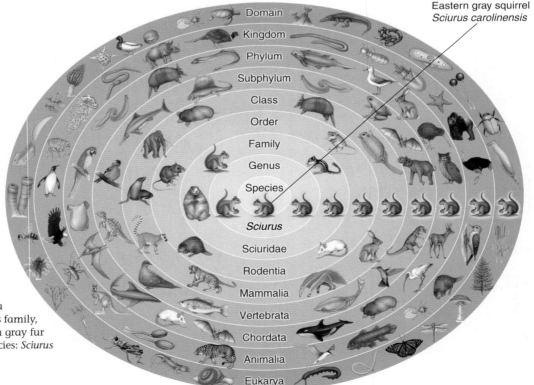

Eastern gray squirrel
Sciurus carolinensis

Figure 16.3 The hierarchical system used to classify an organism.

In this example, the organism is first recognized as a eukaryote (domain: Eukarya). Second, within this domain, it is an animal (kingdom: Animalia). Among the different phyla of animals, it is a vertebrate (phylum: Chordata, subphylum: Vertebrata). The organism's fur characterizes it as a mammal (class: Mammalia). Within this class, it is distinguished by its gnawing teeth (order: Rodentia). Next, because it has four front toes and five back toes, it is a squirrel (family: Sciuridae). Within this family, it is a tree squirrel (genus: *Sciurus*), with gray fur and white-tipped hairs on the tail (species: *Sciurus carolinensis,* the eastern gray squirrel).

Biodiversity Behind Bars

There is something about a child that doesn't like bars. When I was seven, I knew with a searing certainty that no person, no animal, should have to live caged, peering out behind bars at a free world it cannot reach. And I acted on that certainty. I lived at 60 Hopkins Street, Hilton Village, a workers' suburb of the shipbuilding town of Newport News, Virginia. Across the street from me resided a public health nurse who maintained in her backyard a large colony of guinea pigs used in medical testing and research. Nowadays such an informal arrangement would surely violate some rule or regulation, but 1949 was a simpler time. I used to visit her guinea pigs in their outdoor cage, seeing in their liquid eyes all the despair that only a seven-year-old can feel. And one day I picked up a rock, broke the lock, and set the guinea pigs free. Out into the neighborhood they shot, 97 furry little lightning bolts, and I was proud to have struck this blow for freedom. I am amazed, looking back, that my parents kept me.

What they did was buy me a dog. That, and explain to me that some bars have a purpose. These guinea pigs were part of an effort to help sick people, they patiently explained, their freedom the price of a much greater good. It was only as I grew to adulthood that I came to understand that bitter truth. And now I find I must explain it to my children. At 13 years of age, my daughter Caitlin came to me, incandescent with anger, demanding to know why the apes at the Saint Louis Zoo weren't transported back to Africa and released to freedom. I looked into her eyes and saw myself 50 years ago (can it be so long?) and knew her question demanded an answer. So here it is.

Ape captivity, Caitlin, is no more justifiable than human captivity, for apes are our very close cousins, with loving families and complex cultures. But bars sometimes have a necessary purpose. The apes in the Saint Louis Zoo are captive for two reasons: to preserve them as a species, and to educate people about them.

Zoos as Conservators of a Disappearing World

There are only four kinds of living apes: gibbons, orangutans, gorillas, and chimpanzees. All are rare and highly endangered, living in relatively small areas, and their natural habitat is rapidly disappearing as human activities destroy the mountain forests of Central Africa and Asia. In your lifetime, I told Caitlin, most of these wild ape populations will be gone, extinct. No amount of regret can change that sad fact. All that will be left of humanity's last link to its evolutionary past will be the captive ape populations living in zoos like ours.

That is the primary role of zoos today, not entertainment to distract children on hot summer days but preservation of an invaluable biological heritage that otherwise would be lost. Zoos are the conservators

of a rapidly disappearing natural world, a living library in which we store biodiversity. Zoos take this role seriously. The breeding of endangered species is overseen collectively by the nation's zoos. Species preservation programs move animals from one zoo to another in carefully managed breeding programs that prevent unnecessary inbreeding.

Sometimes a captive species can be introduced back into the wild. Serious efforts are being made to establish large preserves, where some of the natural habitat will be saved for the animals. For most endangered animals, however, there are no such rosy hopes, no wild habitat reserved for their return. Captivity, for better or worse, is their only future, surely a better choice than extinction.

Zoos as Educators of You and Me

The second role of zoos today, more subtle but equally important, is to make biodiversity immediate and real for the general public, for you and me. It was once wisely said that we will not preserve what we do not understand. Seeing with your own eyes a chimpanzee return your gesture, or watching with your own eyes a baby giraffe take its first awkward steps—no words or film have that impact, that immediate concrete contact with our animal relations. Like many zoos around the country, our Saint Louis Zoo has a very active education department, letting busloads of schoolchildren experience animals firsthand. But the bigger educational effort is simply our seeing the animals ourselves. Every zoo is a teaching institution; we the public are the students, and our response to the animals is the lesson we are being taught. The test we must pass, as the citizens upon whose support the world's zoos depend, will be graded severely. It is no less than the survival of this rich animal biodiversity for our children's children to see and share in turn.

16.6 The Kingdoms of Life

Classification systems have gone through their own evolution of sorts, as illustrated in figure 16.8. The earliest classification systems recognized only two kingdoms of living things: animals, shown in blue in part *a*, and plants, shown in green. But as biologists discovered microorganisms (the yellow-colored boxes in part *b*) and learned more about other organisms like the protists (in teal) and the fungi (in light brown), they added kingdoms in recognition of fundamental differences. Most biologists now use a six-kingdom system (indicated by the six different-colored boxes in part *c*) first proposed by Carl Woese of the University of Illinois.

In this system, four kingdoms consist of eukaryotic organisms. The two most familiar kingdoms, **Animalia** and **Plantae,** contain only organisms that are multicellular during most of their life cycle. These groups of animals and plants are no doubt familiar to you. The kingdom **Fungi** contains multicellular forms, such as mushrooms and molds, and single-celled yeasts, which are thought to have multicellular ancestors. Fundamental differences divide these three kingdoms. Plants are mainly stationary, but some have motile sperm; fungi have no motile cells; animals are mainly motile. Animals ingest their food, plants manufacture it, and fungi digest it by means of secreted extracellular enzymes. Each of these kingdoms probably evolved from a different single-celled ancestor.

The large number of unicellular eukaryotes are arbitrarily grouped into a single kingdom called **Protista** (see chapter 17). They include the algae and many kinds of microscopic aquatic organisms. This kingdom is an artificial group in that many of these organisms are only distantly related, and the classification of the protists is in flux.

The remaining two kingdoms, **Archaea** and **Bacteria,** consist of prokaryotic organisms, which are vastly different from all other living things (see chapter 17). The prokaryotes with which you are most familiar, those that cause disease or are used in industry, are members of the kingdom Bacteria. Archaea are a diverse group including the methanogens and extreme thermophiles, and they differ greatly from bacteria in many ways. The characteristics of these six kingdoms are presented in table 16.1.

Domains

As biologists have learned more about the archaea, it has become increasingly clear that this ancient group is very different from all other organisms. When the full genomic DNA sequences of an archaean and a bacterium were first compared in 1996, the differences proved striking. Archaea are as different from bacteria as bacteria are from eukaryotes. Recognizing this, biologists have in recent years adopted a taxonomic level higher than kingdom that recognizes three **domains** (figure 16.8*d*). Archaea (red-colored box) are in one domain, bacteria (yellow-colored box) in a second, and eukaryotes (the four purple boxes representing the four eukaryotic kingdoms) in the third. While the domain Eukarya contains four kingdoms of organisms, the domains Bacteria and Archaea contain only one kingdom in each. Because of this, the kingdom level of classification for Bacteria and Archaea is now often omitted, biologists using just their domain and phyla names.

> **16.6** Living organisms are grouped into three categories called domains. One of the domains, Eukarya, is divided into four kingdoms: Protista, Fungi, Plantae, and Animalia.

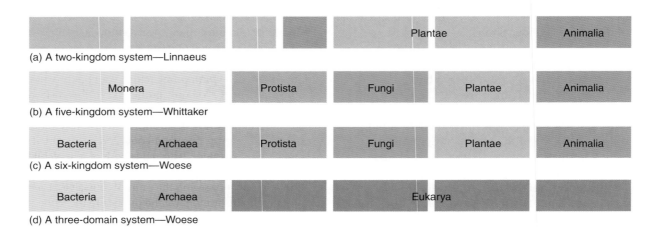

Figure 16.8 Different approaches to classifying living organisms.

(a) Linnaeus popularized a two-kingdom approach, in which the fungi and the photosynthetic protists were classified as plants and the nonphotosynthetic protists as animals; when prokaryotes were described, they too were considered plants. (b) Whittaker in 1969 proposed a five-kingdom system that soon became widely accepted. (c) Woese has championed splitting the prokaryotes into two kingdoms for a total of six kingdoms or even assigning them separate domains, with a third domain containing the four eukaryotic kingdoms (d).

16.7 Domain Bacteria

The domain Bacteria contains one kingdom of the same name, Bacteria. The bacteria are the most abundant organisms on earth. There are more living bacteria in your mouth than there are mammals living on earth. Although too tiny to see with the unaided eye, bacteria play critical roles throughout the biosphere. They extract from the air all the nitrogen used by organisms, and they play key roles in cycling carbon and sulfur.

There are many different kinds of bacteria, and the evolutionary links between them are not well understood. Al-though there is considerable disagreement among taxonomists about the details of bacterial classification, most recognize 12 to 15 major groups of bacteria. Comparisons of the nucleotide sequences of rRNA molecules are beginning to reveal how these groups are related to each other and to the other two domains. The archaea and eukaryotes are more closely related to each other than to bacteria and are on a separate evolutionary branch of the tree (as seen in figure 16.9), even though archaea and bacteria are both prokaryotes.

16.7 Bacteria are as different from archaea as they are from eukaryotes.

TABLE 16.1	CHARACTERISTICS OF THE SIX KINGDOMS					
	Bacteria	**Archaea**	**Protista**	**Plantae**	**Fungi**	**Animalia**
Cell type	Prokaryotic	Prokaryotic	Eukaryotic	Eukaryotic	Eukaryotic	Eukaryotic
Nuclear envelope	Absent	Absent	Present	Present	Present	Present
Mitochondria	Absent	Absent	Present or absent	Present	Present or absent	Present
Chloroplasts	None (photosynthetic membranes in some types)	None (bacteriorhodopsin in one species)	Present in some forms	Present	Absent	Absent
Cell wall	Present in most; peptidoglycan	Present in most; polysaccharide, glycoprotein, or protein	Present in some forms; various types	Cellulose and other polysaccharides	Chitin and other noncellulose polysaccharides	Absent
Means of genetic recombination, if present	Conjugation, transduction, transformation	Conjugation, transduction, transformation	Fertilization and meiosis	Fertilization and meiosis	Fertilization and meiosis	Fertilization and meiosis
Mode of nutrition	Autotrophic (chemosynthetic, photosynthetic) or heterotrophic	Autotrophic (photosynthesis in one species) or heterotrophic	Photosynthetic or heterotrophic or combination of both	Photosynthetic, chlorophylls a and b	Absorption	Digestion
Motility	Bacterial flagella, gliding, or nonmotile	Unique flagella in some	9 + 2 cilia and flagella; amoeboid, contractile fibrils	None in most forms, 9 + 2 cilia and flagella in gametes of some forms	Nonmotile	9 + 2 cilia and flagella, contractile fibrils
Multicellularity	Absent	Absent	Absent in most forms	Present in all forms	Present in most forms	Present in all forms

16.8 Domain Archaea

The domain Archaea contains one kingdom by the same name, the Archaea. The term *archaea* (Greek, *archaio*, ancient) refers to the ancient origin of this group of prokaryotes, which most likely diverged very early from the bacteria. Notice in figure 16.9 that the Archaea, in red, branched off from a line of prokaryotic ancestors that lead to the evolution of eukaryotes. Today, archaea inhabit some of the most extreme environments on earth. Though a diverse group, all archaea share certain key characteristics. Their cell walls lack the peptidoglycan characteristic of the cell walls of bacteria. They possess very unusual lipids and characteristic ribosomal RNA (rRNA) sequences. Some of their genes possess introns, unlike those of bacteria.

Archaea are grouped into three general categories: methanogens, extremophiles, and nonextreme archaea.

Methanogens obtain their energy by using hydrogen gas (H_2) to reduce carbon dioxide (CO_2) to methane gas (CH_4). They are strict anaerobes, poisoned by even traces of oxygen. They live in swamps, marshes, and the intestines of mammals. Methanogens release about 2 billion tons of methane gas into the atmosphere each year.

Extremophiles are able to grow under conditions that seem extreme to us.

Thermophiles ("heat lovers") live in very hot places, typically from 60° to 80°C. Many thermophiles have metabolisms based on sulfur. Thus, the *Sulfolobus* inhabiting the hot sulfur springs of Yellowstone National Park at 70° to 75°C obtain their energy by oxidizing elemental sulfur to sulfuric acid. The recently described *Pyrolobus fumarii* holds the current record for heat stability, temperature optimum (106°C) and temperature maximum (113°C). These extreme temperatures are characteristic of the deep-sea hydrothermal vents where this organism was discovered. *P. fumarii* is so heat-tolerant that it is not killed by a one-hour treatment in an autoclave (121°C)!

Halophiles ("salt lovers") live in very salty places like the Great Salt Lake in Utah, Mono Lake in California, and the Dead Sea in Israel. Whereas the salinity of seawater is around 3%, these prokaryotes thrive in, and indeed require, water with a salinity of 15% to 20%.

pH-tolerant archaea grow in highly acidic (pH = 0.7) and very basic (pH = 11) environments.

Pressure-tolerant archaea have been isolated from ocean depths that require at least 300 atmospheres of pressure to survive, and tolerate up to 800 atmospheres!

Nonextreme archaea grow in the same environments bacteria do. As the genomes of archaea have become better known, microbiologists have been able to identify **signature sequences** of DNA present in all archaea and in no other organisms. When samples from soil or seawater are tested for genes matching these signature sequences, many of the prokaryotes living there prove to be archaea. Clearly, archaea are not restricted to extreme habitats, as microbiologists used to think.

> **16.8** Archaea are unique prokaryotes that inhabit diverse environments, some of them extreme.

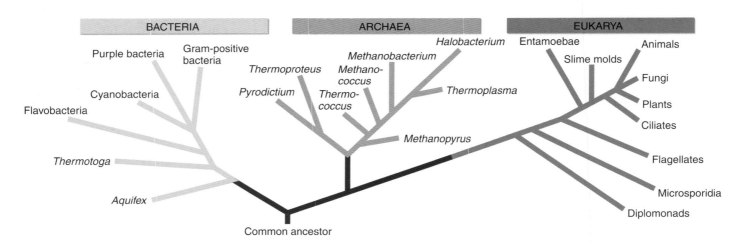

Figure 16.9 A tree of life.
This phylogeny, prepared from rRNA analyses, shows the evolutionary relationships among the three domains. The base of the tree was determined by examining genes that are duplicated in all three domains, the duplication presumably having occurred in the common ancestor. When one of the duplicates is used to construct the tree, the other can be used to root it. This approach clearly indicates that the root of the tree is within the bacterial domain. Archaea and eukaryotes diverged later and are more closely related to each other than either is to bacteria.

16.9 Domain Eukarya

For at least 1 billion years, prokaryotes ruled the earth. No other organisms existed to eat them or compete with them, and their tiny cells formed the world's oldest fossils. The third great domain of life, the eukaryotes, appear in the fossil record much later, only about 1.5 billion years ago. Metabolically, eukaryotes are more uniform than prokaryotes. Each of the two domains of prokaryotic organisms has far more metabolic diversity than all eukaryotic organisms taken together.

Three Largely Multicellular Kingdoms

Fungi, plants, and animals are well-defined evolutionary groups, each of them clearly stemming from a different single-celled eukaryotic ancestor. They are largely multicellular, each a distinct evolutionary line from an ancestor that would be classified in the kingdom Protista.

The amount of diversity among the protists, however, is much greater than that within or between the three largely multicellular kingdoms derived from the protists. Because of the size and ecological dominance of plants, animals, and fungi, and because they are predominantly multicellular, we recognize them as kingdoms distinct from Protista.

A Fourth Very Diverse Kingdom

When multicellularity evolved, the diverse kinds of single-celled organisms that existed at that time did not simply become extinct. A wide variety of unicellular eukaryotes and their relatives exists today, grouped together in the kingdom Protista solely because they are not fungi, plants, or animals. Protists are a fascinating group containing many organisms of intense interest and great biological significance.

Symbiosis and the Origin of Eukaryotes

The hallmark of eukaryotes is complex cellular organization, highlighted by an extensive endomembrane system that subdivides the eukaryotic cell into functional compartments called organelles (see chapter 5). Not all of these organelles, however, are derived from the endomembrane system. Mitochondria and chloroplasts are both believed to have entered early eukaryotic cells by a process called endosymbiosis (*endo,* inside) where an organism such as a bacterium is taken into the cell and remains functional inside the cell. Figure 16.10 shows how mitochondria (the brown arrows) and chloroplasts (the green arrows) evolved in ancestral eukaryotic cells through endosymbiosis.

With few exceptions, all modern eukaryotic cells possess energy-producing organelles, the mitochondria—notice that the brown branching arrows in figure 16.10 extend to all four eukaryotic kingdoms. Mitochondria are about the size of bacteria and contain DNA. Comparison of the nucleotide sequence of this DNA with that of a variety of organisms indicates clearly that mitochondria are the descendants of purple bacteria that were incorporated into eukaryotic cells early in the history of the group. Some protist phyla have in addition acquired chloroplasts during the course of their evolution and thus are photosynthetic as indicated by the green arrow in figure 16.10 that branches into the different kinds of algae. These chloroplasts are derived from

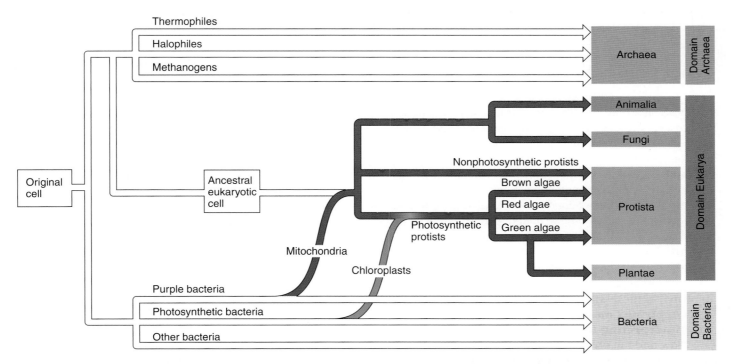

Figure 16.10 Diagram of the evolutionary relationship among the six kingdoms of organisms.
The colored lines indicate symbiotic events. Mitochondria are present in essentially all eukaryotic organisms. Chloroplasts are present in a subset of eukaryotic organisms, those that are photosynthetic.

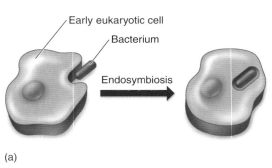

(a)

(b)

Figure 16.11   Endosymbiosis.

(a) This figure shows how an organelle could have arisen in early eukaryotic cells through a process called endosymbiosis. An organism, such as a bacterium, is taken into the cell through a process similar to endocytosis but remains functional inside the host cell. (b) Many corals contain endosymbionts, algae called zooxanthellae that carry out photosynthesis and provide the coral with nutrients. In this photograph, the zooxanthellae are the golden-brown spheres packed into the tentacles of a coral animal.

cyanobacteria that became symbiotic in several groups of protists early in their history. Figure 16.11a shows how this could have happened, with the green cyanobacterium being engulfed by an early protist. Some of these photosynthetic protists gave rise to land plants. Endosymbiosis is not strictly an ancient process but still happens today. Some photosynthetic protists are endosymbionts of some eukaryotic organisms, such as certain species of sponges, jellyfish, corals (the green structures inside the coral in figure 16.11b are endosymbiotic protists), octopuses, and others.

We discussed the theory of the endosymbiotic origin of mitochondria and chloroplasts in chapter 5, and we will revisit it in chapter 17.

16.9   Eukaryotic cells acquired mitochondria and chloroplasts by endosymbiosis, mitochondria being derived from purple bacteria and chloroplasts from cyanobacteria.

INQUIRY & ANALYSIS

What Causes New Forms to Arise?

Biologists once presumed that new forms—genera, families, and orders—arose most often during times of massive geological disturbance, stimulated by the resulting environmental changes. But no such relationship exists. An alternative hypothesis was proposed by evolutionist George Simpson in 1953. He proposed that diversification followed new evolutionary innovations, "inventions" that permitted an organism to occupy a new "adaptive zone." After a burst of new orders that define the major groups, subsequent specialization would lead to new genera.

The graph shows the evolutionary history of the class Osteichthyes, the bony fishes, since they first appeared in the Silurian some 420 million years ago.

1. **Applying Concepts** What is the dependent variable?
2. **Interpreting Data**   Three great innovations in jaw and tail occur during the history of the bony fishes, producing the superorders represented by sturgeons, then gars, and then teleost fishes. In what period did each innovation occur?
3. **Making Inferences** Do bursts of new genera appear at these same three times, or later?

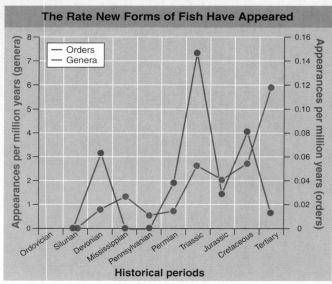

4. **Drawing Conclusions** Does the data presented in the graph support Simpson's hypothesis? Explain.
5. **Further Analysis** If you were to plot on the graph the rate at which new families of fishes appeared, what general pattern would you expect to see, relative to new orders, if Simpson is right? Explain.

The Classification of Organisms

16.1 The Invention of the Linnaean System

- Scientists use a system of grouping similar organisms together, called classification. Latin is used because it was the language used by earlier philosophers and scientists.

- The polynomial system of classification named an organism by using a list of adjectives that described the organism. The binomial system, using a two-part name, was originally developed as a "shorthand" reference to the polynomial name. Linnaeus used this two-part naming system consistently and its use became widespread (**figure 16.1**).

16.2 Species Names

- Taxonomy is the area of biology involved in identifying, naming, and grouping organisms. Scientific names that consist of two parts, the genus and species, are standardized, universal names that are less confusing than common names (**figure 16.2**).

16.3 Higher Categories

- In addition to the genus and species names, an organism is also assigned to higher levels of classification. The higher categories convey more general information about the organisms in a particular group. The most general category, domain, is the largest grouping followed by ever-increasingly specific information that is used to group organisms into a kingdom, phylum, class, order, family, genus, and species (**figure 16.3**).

16.4 What Is a Species?

- The biological species concept states that a species is a group of organisms that is reproductively isolated, meaning that the individuals mate and produce fertile offspring with each other but not with those of other species.

- This concept works well to define animal species because animals regularly outcross (mate with other individuals—**figure 16.4**). However, the concept does not apply to other organisms (fungi, protists, plants, and prokaryotes) that regularly reproduce without mating through asexual reproduction. The classification of these organisms relies more on physical, behavioral, and genetic characteristics.

Inferring Phylogeny

16.5 How to Build a Family Tree

- In addition to organizing a great number of organisms, the study of taxonomy also gives us a glimpse of the evolutionary history of life on earth. Organisms with similar characteristics are more likely to be related to each other. The evolutionary history of an organism and its relationship to other species is called phylogeny, and relationships are often mapped out using phylogenetic trees.

- Phylogenetic trees can be created using key characteristics that are shared by some organisms, presumably having been inherited from a common ancestor. A group of organisms that have shared characteristics is called a clade, and a phylogenetic tree organized in this manner is called a cladogram (**figure 16.5**).

A cladogram suggests the order in which evolutionary changes occurred.

- Cladograms can sometimes be misleading when the characteristics are weighted, placing more importance on a characteristic that seems to have a more significant impact on evolution. The problem with this system is that some characteristics may turn out to be less important than first thought. For this reason, cladograms work best when all characteristics are weighted equally.

- Traditional taxonomy focuses more on the significance or evolutionary impact of a characteristic and not just on the commonality of the characteristic (**figures 16.6** and **16.7**).

- Each approach has its merits. Traditional taxonomy is best used when there is a great deal of information available to correctly weight different characteristics. Cladograms are best used when there is little information about the importance of the characteristic to the life of the organism.

Kingdoms and Domains

16.6 The Kingdoms of Life

- The designation of kingdoms, the second-highest category used in classification, has changed over the years as more and more information about organisms has been uncovered. Originally there were two kingdoms, Plantae and Animalia, but as more information was obtained, biologists began to classify organisms into other kingdoms as well: Fungi, Protista, Archaea, and Bacteria (**figure 16.8**).

- The domain level of classification was added in the mid-1990s, recognizing three fundamentally different types of cells: Eukarya (eukaryotic cells), Archaea (prokaryotic archaea), and Bacteria (prokaryotic bacteria) (**figure 16.9**).

16.7 Domain Bacteria

- The domain Bacteria contains prokaryotic organisms in the kingdom Bacteria. These single-celled organisms play key roles in ecology (**table 16.1**).

16.8 Domain Archaea

- The domain Archaea contains prokaryotic organisms in the kingdom Archaea. These single-celled organisms are found in diverse environments but most interestingly, in very extreme environments.

16.9 Domain Eukarya

- The domain Eukarya contains very diverse organisms from four kingdoms but are similar in that they are all eukaryotes. They contain cellular organelles that were most likely acquired through endosymbiosis (**figures 16.10** and **16.11**).

1. The wolf, domestic dog, and red fox are all in the same family, Canidae. The scientific name for the wolf is *Canis lupus,* the domestic dog is *Canis familiaris,* and the red fox is *Vulpes vulpes.* This means that
   a. the red fox is in the same family, but different genus than dogs and wolves.
   b. the dog is in the same family, but different genus than red fox and wolves.
   c. the wolf is in the same family, but different genus than dogs and red foxes.
   d. all three organisms are in different genera.
2. The evolutionary relationship of an organism, and its relationship to other species, are its
   a. taxonomy.
   b. phylogeny.
   c. ontogeny.
   d. systematics.
3. All classification systems for organisms are based on
   a. physical and chemical characteristics.
   b where the organism lives.
   c. what the organism eats.
   d. the size of the organism.
4. The six kingdoms of organisms can be organized into three domains based on
   a. where the organism lives.
   b. what the organism eats.
   c. cell structure.
   d. cell structure and DNA sequence.
5. All of the extremophiles belong to the domain of
   a. Bacteria.
   b. Archaea.
   c. Prokarya.
   d. Eukarya.

6. Bacteria are also known as prokaryotic cells because they
   a. are chemosynthetic.
   b. are unicellular.
   c. cause disease.
   d. do not have an internal membrane system.
7. Organisms in the domain Bacteria are different than organisms in the domain Archaea because they
   a. are prokaryotes.
   b. have a nucleus.
   c. have cell walls that are made of different materials.
   d. have mitochondria.
8. It is theorized that the ancestral organism for the plants, animals, and fungi originated in the kingdom
   a. Bacteria.
   b. Archaea.
   c. Protista.
   d. All of the above, each giving rise to one of the three kingdoms listed.
9. One difference between the kingdom Protista and the other three kingdoms in the domain Eukarya is that the other kingdoms are mostly
   a. chemosynthetic.
   b. multicellular.
   c. eukaryotic.
   d. unicellular.
10. It is thought that two important organelles of eukaryotic cells came from
    a. development of the internal membrane system.
    b. ingestion of endosymbiotic protists.
    c. mutation.
    d. ingestion of endosymbiotic bacteria.

1. **Figure 16.5** Which of the organisms shown have amniotic membranes that surround the fetus?

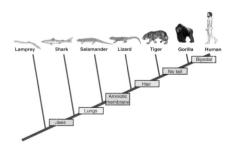

2. **Figure 16.9** Compared to the bacteria and archaea, how similar are animals and plants?

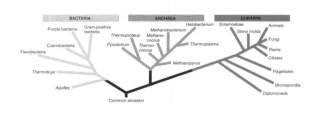

1. Your friend, Julio, wants to know what the big deal is—everyone knows that a rose is a rose, why bother with all the fancy Latin stuff, like *Rosa odorata?* What do you tell him?

2. Why are birds classified so differently in traditional phylogeny and in cladistics?
3. If we already have things divided into kingdoms, why do we also need domains?

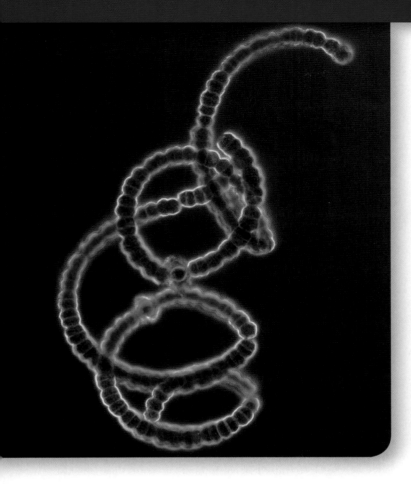

17

Evolution of Microbial Life

Much of the living world is composed of organisms too small for you to see without a microscope to magnify them. The beautiful jewel-like creature you see above is a prokaryote, the cyanobacterium *Anabaena*, in which cells adhere in filaments. The larger cells (areas on the filament that seem to be bulging) are heterocysts, specialized cells in which nitrogen fixation occurs. These organisms exhibit one of the closest approaches to multicellularity among the prokaryotes, and in the nineteenth century some biologists considered *Anabaena* to be a very simple sort of plant. Today, however, most biologists consider *Anabaena* to be a colonial prokaryote. In this chapter we will explore microbes—the sorts of creatures too small to see with the unaided eye. Most living organisms are microbes, and they have an enormous impact on human life. Some of the other creatures you will encounter in this chapter are near relatives of microbes that are multicellular and large. Others will not be organisms at all, but viruses, which are renegade segments of genomes that infect living cells. All of these creatures—microbes, multicellular relatives, and viruses—are important components of the living world.

17.1 How Cells Arose

All living organisms are constructed of the same four kinds of macromolecules discussed in chapter 4, the bricks and mortar of cells. Where the first macromolecules came from and how they came to be assembled together into cells are among the least understood questions in biology—questions that address the very origin of life itself.

No one knows for sure where the first organisms (thought to be like today's bacteria) came from. It is not possible to go back in time and watch how life originated, nor are there any witnesses. Nevertheless, it is difficult to avoid being curious about the origin of life, about what, or who, is responsible for the appearance of the first living organisms on earth. There are, in principle, at least three possibilities:

1. **Extraterrestrial origin.** Life may not have originated on earth at all but instead may have been carried to it, perhaps as an extraterrestrial infection of spores originating on a planet of a distant star. How life came to exist on that planet is a question we cannot hope to answer soon.

2. **Special creation.** Life-forms may have been put on earth by supernatural or divine forces. This viewpoint, called *creationism,* is common to most Western religions and is the oldest hypothesis. However, almost all scientists reject creationism, because to accept its supernatural explanation requires abandoning the scientific approach.

3. **Evolution.** Life may have evolved from inanimate matter, with associations among molecules becoming more and more complex. In this view, the force leading to life was selection; changes in molecules that increased their stability caused the molecules to persist longer.

In this text, we focus on the third possibility and attempt to understand whether the forces of evolution could have led to the origin of life and, if so, how the process might have occurred. This is not to say that the third possibility, evolution, is definitely the correct one. Any one of the three possibilities might be true. Nor does the third possibility preclude religion: a divine agency might have acted via evolution. Rather, we are limiting the scope of our inquiry to scientific matters. Of the three possibilities, only the third permits testable hypotheses to be constructed and so provides the only scientific explanation—that is, one that could potentially be disproved by experiment.

Forming Life's Building Blocks

If we look at the development of living organisms as a 24-hour clock of biological time shown in figure 17.1, with the formation of the earth 4.5 billion years ago being midnight, humans do not appear until the day is almost all over, only minutes before its end. How can we learn about the origin of the first cells? One way is to try to reconstruct what the earth was like when life originated 2.5 billion years ago. We know

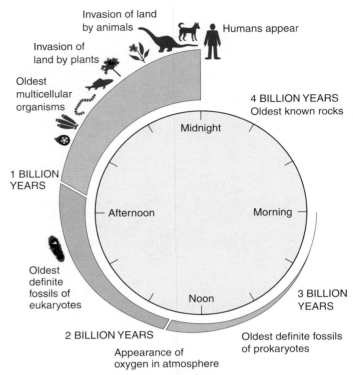

Figure 17.1 A clock of biological time.

A billion seconds ago, most students using this text had not yet been born. A billion minutes ago, Jesus was alive and walking in Galilee. A billion hours ago, the first human had not been born. A billion days ago, no biped walked on earth. A billion months ago, the last dinosaurs had not yet been hatched. A billion years ago, no creature had ever walked on the surface of the earth.

from rocks that there was little or no oxygen in the earth's atmosphere then and more of the hydrogen-rich gases hydrogen sulfide (SH_2), ammonia (NH_3), and methane (CH_4). Electrons in these gases would have been frequently pushed to higher energy levels by photons crashing into them from the sun or by electrical energy in lightning. Today, high-energy electrons are quickly soaked up by the oxygen in earth's atmosphere (air is 21% oxygen, all of it contributed by photosynthesis) because oxygen atoms have a great "thirst" for such electrons. But in the absence of oxygen, high-energy electrons would have been free to help form biological molecules.

When the scientists Stanley Miller and Harold Urey reconstructed the oxygen-free atmosphere of the early earth in their laboratory and subjected it to the lightning and UV radiation it would have experienced then, they found that many of the building blocks of organisms, such as amino acids and nucleotides, formed spontaneously. They concluded that life may have evolved in a "primordial soup" of biological molecules formed in the ancient earth's oceans.

Recently, concerns have been raised regarding the "primordial soup" hypothesis as the origin of life on earth. If the earth's atmosphere had no oxygen soon after it was formed, as Miller and Urey assumed (and most evidence supports this

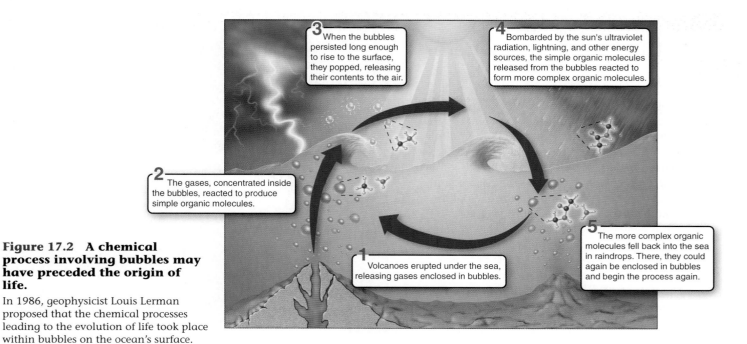

Figure 17.2 A chemical process involving bubbles may have preceded the origin of life.
In 1986, geophysicist Louis Lerman proposed that the chemical processes leading to the evolution of life took place within bubbles on the ocean's surface.

3 When the bubbles persisted long enough to rise to the surface, they popped, releasing their contents to the air.

4 Bombarded by the sun's ultraviolet radiation, lightning, and other energy sources, the simple organic molecules released from the bubbles reacted to form more complex organic molecules.

2 The gases, concentrated inside the bubbles, reacted to produce simple organic molecules.

5 The more complex organic molecules fell back into the sea in raindrops. There, they could again be enclosed in bubbles and begin the process again.

1 Volcanoes erupted under the sea, releasing gases enclosed in bubbles.

assumption), then there would have been no protective layer of ozone to shield the earth's surface from the sun's damaging UV radiation. Without an ozone layer, scientists think UV radiation would have destroyed any ammonia and methane present in the atmosphere. When these gases are missing, the **Miller-Urey experiment** does not produce key biological molecules such as amino acids. If the necessary ammonia and methane were not in the atmosphere, where were they?

In the last two decades, support has grown among scientists for what has been called the **bubble model.** The bubble model, shown in figure 17.2, proposes that the key chemical processes generating the building blocks of life took place not in a primordial soup but rather within bubbles on the ocean's surface. Bubbles produced by wind, wave action, the impact of raindrops, and the eruption of volcanoes cover about 5% of the ocean's surface at any given time. Because water molecules are polar, water bubbles tend to attract other polar molecules, in effect concentrating them within the bubbles. Chemical reactions would proceed much faster in bubbles, where polar reactants would be concentrated. The bubble model solves a key problem with the primordial soup hypothesis. Inside the bubbles, the methane and ammonia required to produce amino acids would have been protected from destruction by UV radiation.

The First Cells

We don't know how the first cells formed, but most scientists suspect they aggregated spontaneously. When complex carbon-containing macromolecules are present in water, they tend to gather together, much as the people from the same foreign country tend to aggregate within a large city. Sometimes the aggregations form a cluster big enough to see. Try vigorously shaking a bottle of oil-and-vinegar salad dressing—tiny

bubbles called **microspheres** form spontaneously, suspended in the vinegar. Similar microspheres might have represented the first step in the evolution of cellular organization. Such microspheres have many cell-like properties—their outer boundary resembles the membranes of a cell in that it has two layers (see figure 4.16), and the microspheres can increase in size and divide. Over millions of years, those microspheres better able to incorporate molecules and energy would have tended to persist longer than others.

Scientists suspect that the first macromolecules to form were RNA molecules, and with the recent discovery that RNA molecules can behave as enzymes, catalyzing their own assembly, this provides a possible early mechanism of inheritance. Eventually DNA may have taken the place of RNA as the storage molecule for genetic information because the double-stranded DNA would have been more stable than single-stranded RNA.

As you can see, the scientific vision of life's origin is at best a hazy outline. Many different scenarios seem possible, and some have solid support from experiments. Deep-sea hydrothermal vents are an interesting possibility; the prokaryotes populating these vents are among the most primitive of living organisms. Other researchers have proposed that life originated deep in the earth's crust. How life might have originated naturally and spontaneously remains a subject of intense interest, research, and discussion among scientists.

17.1 Life appeared on earth 2.5 billion years ago. It may have arisen spontaneously, although the nature of the process is not clearly understood. Little is know about how the first cells originated.

Today's *Biology*

Has Life Evolved Elsewhere?

We should not overlook the possibility that life processes might have evolved in different ways on other planets. A functional genetic system, capable of accumulating and replicating changes and thus of adaptation and evolution, could theoretically evolve from molecules other than carbon, hydrogen, nitrogen, and oxygen in a different environment. Silicon, like carbon, needs four electrons to fill its outer energy level, and ammonia is even more polar than water. Perhaps under radically different temperatures and pressures, these elements might form molecules as diverse and flexible as those carbon has formed on earth.

The universe has 10^{20} (100,000,000,000,000, 000,000) stars similar to our sun. We don't know how many of these stars have planets, but it seems increasingly likely that many do. Since 1996, astronomers have been detecting planets orbiting distant stars. At least 10% of stars are thought to have planetary systems. If only 1 in 10,000 of these planets is the right size and at the right distance from its star to duplicate the conditions in which life originated on earth, the "life experiment" will have been repeated 10^{15} times (that is, a million billion times). It does not seem likely that we are alone.

A dull gray chunk of rock collected in 1984 in Antarctica ignited an uproar about ancient life on Mars with the report that the rock contains evidence of possible life. Analysis of gases trapped within small pockets of the rock indicate it is a meteorite from Mars. It is, in fact, the oldest rock known to science—fully 4.5 billion years old. Evidence collected by the 2004 NASA Mars mission (the photo below of the Martian surface was taken by the rover Spirit) suggests that the surface, now cold and arid, was much warmer when the Antarctic meteorite formed 4.5 billion years ago, that water flowed over its surface, and that it had a carbon dioxide atmosphere—conditions not too different from those that spawned life on earth.

When examined with powerful electron microscopes, carbonate patches within the meteorite exhibit what look like microfossils, some 20 to 100 nanometers in length. One hundred times smaller than any known bacteria, it is not clear they actually are fossils, but the resemblance to bacteria is striking.

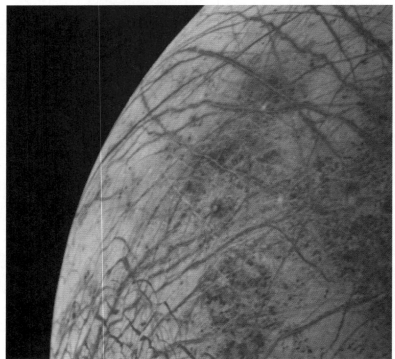

Viewed as a whole, the evidence of bacterial life associated with the Mars meteorite is not compelling. Clearly, more painstaking research remains to be done before the discovery can claim a scientific consensus. However, while there is no conclusive evidence of bacterial life associated with this meteorite, it seems very possible that life has evolved on other worlds in addition to our own.

There are planets other than ancient Mars with conditions not unlike those on earth. Europa, a large moon of Jupiter, is a promising candidate (photo above). Europa is covered with ice, and photos taken in close orbit in the winter of 1998 reveal seas of liquid water beneath a thin skin of ice. Additional satellite photos taken in 1999 suggest that a few miles under the ice lies a liquid ocean of water larger than earth's, warmed by the push and pull of the gravitational attraction of Jupiter's many large satellite moons. The conditions on Europa now are far less hostile to life than the conditions that existed in the oceans of the primitive earth. In coming decades, satellite missions are scheduled to explore this ocean for life.

The Origin of Viral Diseases

Sometimes viruses that originate in one organism pass to another, causing a disease in the new host. Thus, influenza is fundamentally a bird virus, and smallpox is thought to have passed from cattle to humans when cows were first domesticated.

New pathogens arising in this way, called *emerging viruses,* represent a greater threat today than in the past, as air travel and world trade in animals allows infected individuals and animals to move about the world quickly, spreading an infection. The widespread conversion of tropical forests into agricultural land has greatly increased the contact between people and wild animals, amplifying the opportunity for the introduction into people of novel viruses.

Influenza. Perhaps the most lethal virus in human history has been the influenza virus. Over 20 million worldwide died of flu within 19 months in 1918 and 1919—an astonishing number. The natural reservoir of influenza virus is in ducks (and pigs) in Central Asia. Major flu pandemics (that is, worldwide epidemics) arise in Asian ducks through recombination within multiple infected individuals, putting together novel combinations of virus surface proteins unrecognizable by human immune defenses. The Asian flu of 1957 killed over 100,000 Americans. The Hong Kong flu of 1968 infected 50 million people in the United States alone, of which 70,000 died.

AIDS (HIV virus). The virus that causes AIDS first entered humans from chimpanzees somewhere in Central Africa, probably between 1910 and 1950. The chimpanzee virus, called simian immunodeficiency virus, or SIV, mutates rapidly, at a rate of 1% a year, and in humans it continued to do so, soon becoming what we now know as HIV and spreading widely, mostly through sexual contact with an infected person. Where did chimpanzees acquire SIV? SIV viruses are rampant in African monkeys, and chimpanzees eat monkeys. Study of the nucleotide sequences of monkey SIVs revealed in 2001 that one end of the chimp virus RNA closely resembles the SIV found in red-capped mangabey monkeys, while the other end resembles the virus from the greater spot-nosed monkey. It thus seems certain that chimpanzees acquired SIV from monkeys they ate.

Ebola virus. Among the most lethal of emerging viruses are a collection of filamentous viruses arising in Central Africa that attack human connective tissue. With lethality rates in excess of 50%, these so-called filoviruses cause some of the most lethal infectious diseases known. One, Ebola virus, has exhibited lethality rates in excess of 90% in isolated outbreaks in Central Africa. Luckily, victims die too fast to spread the disease very far. Extensive searches among wild and domestic animals have failed to reveal for certain the identity of the natural host of Ebola virus. Researchers in 2005 reported evidence implicating fruit bats, eaten for food everywhere in Central Africa that outbreaks have occurred.

Hantavirus. A sudden outbreak of a highly fatal hemorrhagic infection in the southwestern United States in 1993 was soon attributed to a species of hantavirus, an RNA virus associated with rodents. This species was eventually traced to deer mice. The deer mouse hantavirus is transmitted to humans through fecal contamination in areas of human habitation. Control of deer mouse populations has limited the disease.

SARS. A recently emerged species of coronavirus was responsible for a worldwide outbreak in 2003 of *severe acute respiratory syndrome* (SARS), a respiratory infection with pneumonia-like symptoms that in over 8% of cases is fatal. When the 29,751-nucleotide RNA genome of the SARS virus was sequenced, it proved to be a completely new form of coronavirus, not closely related to any of the three previously described forms. Virologists in 2005 identified the Chinese horseshoe bat as the natural host of the SARS virus. Because these bats are healthy carriers not sickened by the virus, and occur commonly throughout Asia, it will be difficult to prevent future outbreaks.

West Nile Virus. A mosquito-borne virus, West Nile virus, first infected people in North America in 1999, killing four people in Queens, New York. Carried by infected crows and other birds, the virus proceeded to spread across the country, with 4,156 cases at its peak in 2002, 284 of whom died. Fewer cases were reported in 2004, and by 2005 the wave of infection had greatly lessened. First detected in humans in Uganda, Africa, in 1937, the virus is common among birds, and is thought to have been transmitted to humans by mosquitos that had previously bitten infected birds, much as it is being transmitted now. Earlier spread of the virus through Europe also abated after several years.

> **17.4** Viruses are genomes of DNA or RNA, encased in a protein shell, that can infect cells and replicate within them. They are chemical assemblies, not cells, and are not alive. Viruses are responsible for some of the most lethal diseases of humans.

17.5 The Origin of Eukaryotic Cells

The First Eukaryotic Cells

All fossils more than 1.7 billion years old are small, simple cells, similar to the bacteria of today. In rocks about 1.7 billion years old, we begin to see the first microfossils, which are noticeably larger than bacteria and have internal membranes and thicker walls. A new kind of organism had appeared, called a **eukaryote,** from the Greek words for "true" and "nucleus." Eukaryotic cells possess an internal structure called a nucleus.

Many bacteria have infoldings of their outer membranes extending into the interior that serve as passageways to the surface. The network of internal membranes in eukaryotes called the endoplasmic reticulum (ER) is thought to have evolved from such infoldings, as did the nuclear envelope.

Endosymbiosis

The **endosymbiotic theory,** now widely accepted, suggests that at a critical stage in the evolution of eukaryotic cells, energy-producing bacteria came to reside symbiotically (that is, cooperatively) within larger early eukaryotic cells, eventually evolving into the cell organelles we now know as mitochondria. You can see this step illustrated in the upper part of figure 17.7. Similarly, photosynthetic bacteria came to live within some of these early eukaryotic cells, leading to the evolution of chloroplasts, the photosynthetic organelles of plants and algae. This step is illustrated in the lower part of the figure. Present-day mitochondria and chloroplasts still contain their own DNA, remarkably similar to the DNA of bacteria in size and character.

Mitochondria. Mitochondria, the energy-generating organelle in eukaryotic cells, are sausage-shaped organelles about 1 to 3 micrometers long, about the same size as most bacteria. Mitochondria are bounded by *two* membranes. The outer membrane is smooth and was apparently derived from the host cell as it wrapped around the bacterium, shown in the upper portion of figure 17.7. The inner membrane is folded into numerous layers, embedded within which are the proteins of oxidative metabolism.

During the billion-and-a-half years in which mitochondria have existed as endosymbionts within eukaryotic cells, most of their genes have been transferred to the chromosomes of the host cells—but not all. Each mitochondrion still has its own genome, a circular, closed molecule of DNA similar to that found in bacteria, on which is located genes encoding some of the essential proteins of oxidative metabolism. Mitochondria divide by simple fission, just as bacteria do, and can divide on their own without the cell nucleus dividing. However, the cell's nuclear genes direct the process, and mitochondria cannot be grown outside of the eukaryotic cell, in cell-free culture.

Chloroplasts. Many eukaryotic cells contain other endosymbiotic bacteria in addition to mitochondria. Plants and algae contain chloroplasts, bacteria-like organelles that were apparently derived from symbiotic photosynthetic bacteria, shown in the lower portion of figure 17.7. Chloroplasts have a complex system of inner membranes and a circle of DNA.

While all mitochondria are thought to have arisen from a single symbiotic event, it is difficult to be sure with chloroplasts. Three biochemically distinct classes of chloroplasts exist, but all appear to have their origin in the cyanobacteria.

> **17.5** The theory of endosymbiosis proposes that mitochondria originated as symbiotic aerobic bacteria and chloroplasts originated from a second endosymbiotic event.

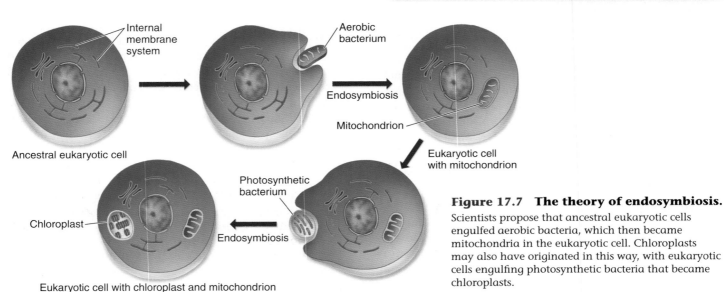

Figure 17.7 The theory of endosymbiosis.
Scientists propose that ancestral eukaryotic cells engulfed aerobic bacteria, which then became mitochondria in the eukaryotic cell. Chloroplasts may also have originated in this way, with eukaryotic cells engulfing photosynthetic bacteria that became chloroplasts.

17.6 General Biology of Protists

Protists are eukaryotes united on the basis of a single negative characteristic: they are not fungi, plants, or animals. In all other respects, they are highly variable with no uniting features. Many are unicellular, like the *Vorticella* you see in figure 17.8 with its contractible stalk, but there are numerous colonial and multicellular groups. Most are microscopic, but some are as large as trees.

The Cell Surface

Protists possess varied types of cell surfaces. All protists have plasma membranes. But some protists, like algae and molds, are additionally encased within strong cell walls. Still others, like diatoms and radiolarians, secrete glassy shells of silica.

Locomotor Organelles

Movement in protists is also accomplished by diverse mechanisms. Protists move by cilia, flagella, pseudopods, or gliding mechanisms. Many protists wave one or more flagella to propel themselves through the water, whereas others use banks of short, flagella-like structures called cilia to create

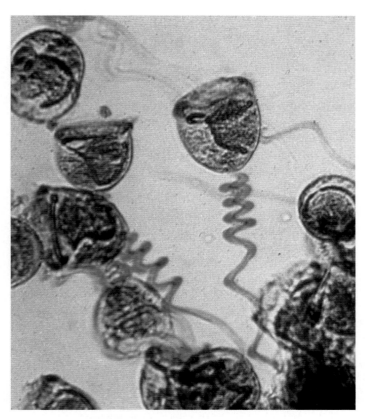

Figure 17.8 A unicellular protist.
The protist kingdom is a catch-all kingdom for many different groups of unicellular organisms, such as this *Vorticella* (phylum Ciliophora), which is heterotrophic, feeding on bacteria, and has a retractable stalk.

water currents for their feeding or propulsion. Pseudopodia are the chief means of locomotion among amoeba, whose pseudopods are large, blunt extensions of the cell body called lobopodia. Other related protists extend thin, branching protrusions called filopodia. Still other protists extend long, thin pseudopodia called axopodia supported by axial rods of microtubules. Axopodia can be extended or retracted. Because the tips can adhere to adjacent surfaces, the cell can move by a rolling motion, shortening the axopodia in front and extending those in the rear.

Cyst Formation

Many protists with delicate surfaces are successful in quite harsh habitats. How do they manage to survive so well? They survive inhospitable conditions by forming **cysts.** A cyst is a dormant form of a cell with a resistant outer covering in which cell metabolism is more or less completely shut down. Amoebic parasites in vertebrates, for example, form cysts that are quite resistant to gastric acidity (although they will not tolerate desiccation or high temperature).

Nutrition

Protists employ every form of nutritional acquisition except chemoautotrophy, which has so far been observed only in prokaryotes. Some protists are photosynthetic autotrophs and are called **phototrophs.** Others are heterotrophs that obtain energy from organic molecules synthesized by other organisms. Among heterotrophic protists, those that ingest visible particles of food are called **phagotrophs,** or **holozoic feeders.** Those ingesting food in soluble form are called **osmotrophs,** or **saprozoic feeders.**

Phagotrophs ingest food particles into intracellular vesicles called **food vacuoles,** or **phagosomes.** Lysosomes fuse with the food vacuoles, introducing enzymes that digest the food particles within. As the digested molecules are absorbed across the vacuolar membrane, the food vacuole becomes progressively smaller.

Reproduction

Protists typically reproduce asexually, most reproducing sexually only in times of stress. Asexual reproduction involves mitosis, but the process is often somewhat different from the mitosis that occurs in multicellular animals. The nuclear membrane, for example, often persists throughout mitosis, with the microtubular spindle forming within it. In some groups, asexual reproduction involves spore formation, in others fission. The most common type of fission is **binary,** in which a cell simply splits into nearly equal halves. When the progeny cell is considerably smaller than its parent, and then grows to adult size, the fission is called **budding.** In multiple fission, common among some protists, fission is preceded by several nuclear divisions, so that fission produces several individuals almost simultaneously.

Sexual reproduction also takes place in many forms among the protists. In ciliates and some flagellates, **gametic meiosis** occurs just before gamete formation, as it does in most animals. In the sporozoans, **zygotic meiosis** occurs directly *after* fertilization, and all the individuals that are produced are haploid until the next zygote is formed. In algae, there is **sporic meiosis,** producing an alternation of generations similar to that seen in plants, with significant portions of the life cycle spent as haploid as well as diploid.

Multicellularity

A single cell has limits. It can only be so big without encountering serious surface-to-volume problems. Said simply, as a cell becomes larger, there is too little surface area for so much volume. The evolution of multicellular individuals composed of many cells solved this problem. **Multicellularity** is a condition in which an organism is composed of many cells, permanently associated with one another, that integrate their activities. The key advantage of multicellularity is that it allows specialization—distinct types of cells, tissues, and organs can be differentiated within an individual's body, each with a different function. With such functional "division of labor" within its body, a multicellular organism can possess cells devoted specifically to protecting the body, others to moving it about, still others to seeking mates and prey, and yet others to carry on a host of other activities. This allows the organism to function on a scale and with a complexity that would have been impossible for its unicellular ancestors. In just this way, a small city of 50,000 inhabitants is vastly more complex and capable than a crowd of 50,000 people in a football stadium—each city dweller is specialized in a particular activity that is interrelated to everyone else's, rather than just being another body in a crowd.

Colonies. A **colonial organism** is a collection of cells that are permanently associated but in which little or no integration of cell activities occurs. Many protists form colonial assemblies, consisting of many cells with little differentiation or integration. In some protists, the distinction between colonial and multicellular is blurred. For example, in the green algae *Volvox* shown in figure 17.9, individual motile cells aggregate into a hollow ball of cells that moves by a coordinated beating of the flagella of the individual cells—like scores of rowers all pulling their oars in concert. A few cells near the rear of the moving colony are reproductive cells, but most are relatively undifferentiated.

Aggregates. An **aggregation** is a more transient collection of cells that come together for a period of time and then separate. Cellular slime molds, for example, are unicellular organisms that spend most of their lives moving about and feeding as single-celled amoebas. They are common in damp soil and on rotting logs, where they move about, ingesting bacteria and other small organisms. When the individual amoebas exhaust the supply of bacteria in a given area and are near starvation, all of the individual organisms in that immediate area aggregate into a large moving mass of cells called a slug. By

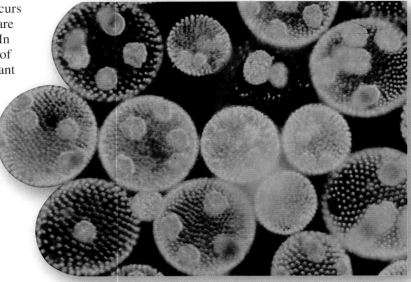

Figure 17.9   A colonial protist.
Individual, motile, unicellular green algae are united in the protist *Volvox* as a hollow colony of cells that moves by the beating of the flagella of its individual cells. Some species of *Volvox* have cytoplasmic connections between the cells that help coordinate colony activities. *Volvox* is a highly complex form of colony that has many of the properties of multicellular life.

moving to a different location, the aggregation increases the chance that food will be found.

Multicellular Individuals. True multicellularity, in which the activities of the individual cells are coordinated and the cells themselves are in contact, occurs only in eukaryotes and is one of their major characteristics. Three groups of protists have independently attained true but simple multicellularity—the brown algae (phylum Phaeophyta), green algae (phylum Chlorophyta), and red algae (phylum Rhodophyta). In **multicellular organisms,** individuals are composed of many cells that interact with one another and coordinate their activities.

Simple multicellularity does not imply small size or limited adaptability. Some marine algae grow to be enormous. An individual kelp, one of the brown algae, may grow to tens of meters in length—some taller than a redwood! Red algae grow at great depths in the sea, far below where kelp or other algae are found. Not all algae are multicellular. Green algae, for example, include many kinds of multicellular organisms but an even larger number of unicellular ones.

17.6   Protists exhibit a wide range of forms, locomotion, nutrition, and reproduction. Their cells form clusters with varying degrees of specialization, from transient aggregations to more persistent colonies to permanently multicellular organisms.

17.7 Kinds of Protists

Protists are the most diverse of the four kingdoms in the domain Eukarya. The kingdom Protista contains many unicellular, colonial, and multicellular groups. Probably the most important statement we can make about the kingdom Protista is that it is an artificial group; as a matter of convenience, single-celled eukaryotic organisms have typically been grouped together into this kingdom. This lumps many very different and only distantly related forms together. The "single-kingdom" classification of the Protista is not representative of any evolutionary relationships. The diagram shown in figure 17.10 merely identifies the five major groups and does not suggest that one group arose from or is related to another. The phyla of protists are, with very few exceptions, only distantly related to one another.

New applications of a wide variety of molecular methods are providing important insights into the relationships among the protists. Of all the groups of organisms biologists study, protists are probably in the greatest state of flux when it comes to classification. There is little consensus, even among experts, as to how the different kinds of protists should be classified. Are they a single, very diverse kingdom, or are they better considered as several different kingdoms, each of equal rank with animals, plants, and fungi?

Because the Protista are still predominantly considered part of one diverse, ununified group, that is how we will treat them in this chapter, bearing in mind that biologists are rapidly gaining a better understanding of the evolutionary relationships among members of the kingdom Protista. It seems likely that within a few years, the traditional kingdom Protista will be replaced by another more illuminating arrangement.

Five Groups of Protists

There are some 15 distinct phyla of protists. It is difficult to encompass their great diversity with any simple scheme. Traditionally, texts have grouped them artificially (as was done in the nineteenth century) into photosynthesizers (algae), heterotrophs (protozoa), and absorbers (funguslike protists).

In this text, we group the protists into five general groups according to some of the major shared characteristics (see figure 17.10). These are characteristics that taxonomists are using today in broad attempts to classify the kingdom Protista. These include (1) the presence or absence and type of cilia or flagella, (2) the pres-

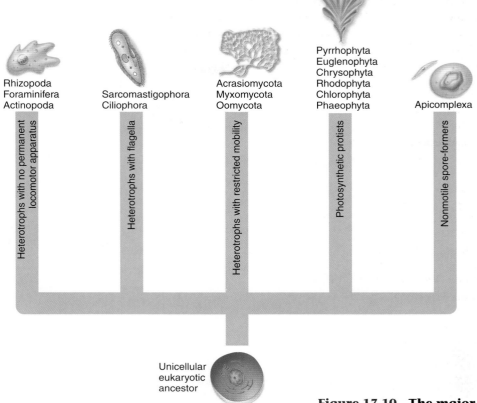

Figure 17.10 The major protist groups.

ence and kinds of pigments, (3) the type of mitosis, (4) the kinds of cristae present in the mitochondria, (5) the molecular genetics of the ribosomal "S" subunit, (6) the kind of inclusions the protist may have, (7) overall body form (amoeboid, coccoid, and so forth), (8) whether the protist has any kind of shell or other body "armor," and (9) modes of nutrition and movement. These represent only some of the characters used to define phylogenetic relationships.

The criteria we have chosen to define the five general groups are not the only ones that might be chosen, and there is no broad agreement among biologists as to which set of criteria is preferable. As molecular analysis gives us a clearer picture of the phylogenetic relationships among the protists, more evolutionarily suitable groupings will without a doubt replace the one represented here. Table 17.2 on the next page summarizes some of the general characteristics and groupings of the 15 major phyla of protists. It is important to remember that while the phyla of protists discussed here are generally accepted taxa, the collections of phyla you see gathered together into categories such as "heterotrophs with no permanent locomotor apparatus" are functional groupings, not phylogenetic ones.

17.7 The 15 protist phyla can be grouped into five categories according to major shared characteristics.

TABLE 17.2 KINDS OF PROTISTS

Group	Phylum	Typical Examples		Key Characteristics
Heterotrophs with no Permanent Locomotor Apparatus				
Amoebas	Rhizopoda	_Amoeba_		Move by pseudopodia
Forams	Foraminifera	Forams		Rigid shells; move by protoplasmic streaming
Radiolarians	Actinopoda	Radiolarians		Glassy skeletons; needlelike pseudopods
Heterotrophs with Flagella				
Zoomastigotes	Sarcomastigophora	Trypanosomes		Heterotrophic; unicellular
Ciliates	Ciliophora	_Paramecium_		Heterotrophic unicellular protists with cells of fixed shape possessing two nuclei and many cilia; many cells also contain highly complex and specialized organelles
Heterotrophs with Restricted Mobility				
Cellular slime molds	Acrasiomycota	_Dictyostelium_		Colonial aggregations of individual cells; most closely related to amoebas
Plasmodial slime molds	Myxomycota	_Fuligo_		Stream along as a multinucleate mass of cytoplasm
Water molds	Oomycota	Water molds and downy mildew		Terrestrial and freshwater
Photosynthetic Protists				
Dinoflagellates	Pyrrhophyta	Red tides		Unicellular; two flagella; contain chlorophylls _a_ and _b_
Euglenoids	Euglenophyta	_Euglena_		Unicellular; some photosynthetic; others heterotrophic; contain chlorophylls _a_ and _b_ or none
Diatoms	Chrysophyta	_Diatoma_		Unicellular; manufacture the carbohydrate chrysolaminarin; unique double shells of silica; contain chlorophylls _a_ and _c_
Golden algae	Chrysophyta	_Dinobryon_		Unicellular, but often colonial; manufacture the carbohydrate chrysolaminarin; contain chlorophylls _a_ and _c_
Red algae	Rhodophyta	Coralline algae		Most multicellular; contain chlorophyll _a_ and a red pigment
Green algae	Chlorophyta	_Chlamydomonas_		Unicellular or multicellular; contain chlorophylls _a_ and _b_
Brown algae	Phaeophyta	Kelp		Multicellular; contain chlorophylls _a_ and _c_
Nonmotile Spore Formers				
Sporozoans	Apicomplexa	_Plasmodium_		Nonmotile; unicellular; the apical end of the spores contains a complex mass of organelles

17.8 A Fungus Is Not a Plant

The fungi are a distinct kingdom of organisms, comprising about 74,000 named species. **Mycologists,** scientists who study fungi, believe there may be many more species in existence. Although fungi were at one time included in the plant kingdom, they lack chlorophyll and resemble plants only in their general appearance and lack of mobility. Significant differences between fungi and plants include the following:

Fungi are heterotrophs. Perhaps most obviously, a mushroom is not green because it does not contain chlorophyll. Virtually all plants are photosynthesizers, whereas no fungi carry out photosynthesis. Instead, fungi obtain their food by secreting digestive enzymes onto whatever they are attached to and then absorbing into their bodies the organic molecules that are released by the enzymes.

Fungi have filamentous bodies. A plant is built of groups of functionally different cells called tissues, with different parts typically made of several different tissues. Fungi by contrast are basically filamentous in their growth form (that is, their body consists entirely of cells organized into long, slender filaments called *hyphae*), even though these filaments may be packed together to form a mass, called a *mycelium.*

Fungi have nonmotile sperm. Some plants have motile sperm with flagella. The majority of fungi do not.

Fungi have cell walls made of chitin. The cell walls of fungi contain chitin, the same tough material that a crab shell is made of. The cell walls of plants are made of cellulose, also a strong building material. Chitin, however, is far more resistant to microbial degradation than is cellulose.

Fungi have nuclear mitosis. Mitosis in fungi is different from plants and most other eukaryotes in one key respect: The nuclear envelope does not break down and re-form. Instead, all of mitosis takes place *within* the nucleus. A spindle apparatus forms there, dragging chromosomes to opposite poles of the *nucleus* (not the cell, as in all other eukaryotes).

You could build a much longer list, but already the take-home lesson is clear: Fungi are not like plants at all! Their many unique features are strong evidence that fungi are not closely related to any other group of organisms.

The Body of a Fungus

Fungi exist mainly in the form of slender filaments, barely visible with the naked eye, called **hyphae** (singular, **hypha**). A hypha is basically a long string of cells. Different hyphae then associate with each other to form much larger structures, like the two mushrooms you see in figure 17.11.

The main body of a fungus is not the mushroom, which is a temporary reproductive structure, but rather the hyphae. A mass of hyphae is called a **mycelium** (plural, **mycelia**) and may contain many meters of individual hyphae. Keep in mind the differences in scale between the hyphae and mycelia; the hypha is about 3.4 μm across whereas the mycelia, like these mushrooms, are visible with the naked eye.

Fungal cells are able to exhibit a high degree of communication within such structures, because although most cells of fungal hyphae are separated by cross-walls called *septa* (singular, *septum*), these septa rarely form a complete barrier. Openings in the septa allow cytoplasm to flow throughout the hyphae from one cell to another.

Because of such cytoplasmic streaming, proteins synthesized throughout the hyphae can be carried to the hyphal tips. This novel body plan is perhaps the most important innovation of the fungal kingdom. As a result of it, fungi can respond quickly to environmental changes, growing very rapidly when food and water are plentiful and the temperature is optimal. This body organization creates a unique relationship between the fungus and its environment. All parts of the fungal body are metabolically active, secreting digestive enzymes and actively attempting to digest and absorb any organic material with which the fungus comes in contact.

Also due to cytoplasmic streaming, many nuclei may be connected by the shared cytoplasm of a fungal mycelium. None of them (except for reproductive cells) are isolated in any one cell; all of them are linked cytoplasmically with every cell of the mycelium. Indeed, the entire concept of multicellularity takes on a new meaning among the fungi, the ultimate communal sharers among the multicellular organisms.

(a)   (b)

Figure 17.11   Mushrooms.
Most of the body of a fungus is belowground, a network of fine threads that penetrate the ground, often over great distances. Mushrooms are aboveground reproductive structures that release spores into the air, allowing the fungus to invade new habitats. (a) Yellow morel. (b) *Amanita* mushroom.

How Fungi Reproduce

Fungi reproduce both asexually and sexually. All fungal nuclei except for the zygote are haploid. Often in the sexual reproduction of fungi, individuals of different "mating types" must participate, much as two sexes are required for human reproduction. Sexual reproduction is initiated when two hyphae of genetically different mating types come in contact, and the hyphae fuse. What happens next? In animals and plants, when the two haploid gametes fuse, the two haploid nuclei immediately fuse to form the diploid nucleus of the zygote. As you might by now expect, fungi handle things differently. In most fungi, the two nuclei do not fuse immediately. Instead, they remain unmarried inhabitants of the same house, coexisting in a common cytoplasm for most of the life of the fungus! A fungal hypha that has two nuclei is called **dikaryotic.** If the nuclei are derived from two genetically different individuals, it is called a **heterokaryon** (Greek, *heteros,* other, and *karyon,* kernel or nucleus). A fungal hypha in which all the nuclei are genetically similar is said to be a **homokaryon** (Greek, *homo,* one).

When reproductive structures are formed in fungi, complete septa form between cells, the only exception to the free flow of cytoplasm between cells of the fungal body. There are three kinds of reproductive structures: (1) **gametangia** form haploid gametes, which fuse to give rise to a zygote that undergoes meiosis; (2) **sporangia** produce haploid spores that can be dispersed; and (3) **conidiophores** produce asexual spores called **conidia** that can be produced quickly and allow for the rapid colonization of a new food source.

Spores are a common means of reproduction among the fungi. The puffball fungus in figure 17.12 is releasing spores in a somewhat explosive manner. Spores are well suited to the needs of an organism anchored to one place. They are so

Figure 17.13 The oyster mushroom.
This species, *Pleurotus ostreatus,* immobilizes nematodes, which the fungus uses as a source of food.

small and light that they may remain suspended in the air for long periods of time and may be carried great distances. When a spore lands in a suitable place, it germinates and begins to divide, soon giving rise to a new fungal hypha.

How Fungi Obtain Nutrients

All fungi obtain their food by secreting digestive enzymes into their surroundings and then absorbing back into the fungus the organic molecules produced by this **external digestion.** Many fungi are able to break down the cellulose in wood, cleaving the linkages between glucose subunits and then absorbing the glucose molecules as food. That is why fungi are so often seen growing on trees.

Just as some plants like the Venus's-flytrap are active carnivores, so some fungi are active predators. For example, the edible oyster fungus *Pleurotus ostreatus,* shown growing on a tree in figure 17.13, attracts tiny roundworms known as nematodes that feed on it—and secretes a substance that anesthetizes the nematodes. When the worms become sluggish and inactive, the fungal hyphae envelop and penetrate their bodies and absorb their contents, a rich source of nitrogen (always in short supply in natural ecosystems). Other fungi are even more active predators, snaring or trapping prey or firing projectiles into nematodes, rotifers, and other small animals that come near.

Figure 17.12 Many fungi produce spores.
Spores explode from the surface of a puffball fungus.

> **17.8 Fungi are not at all like plants. The fungal body is basically long strings of cells, often interconnected. Fungi reproduce both asexually and sexually. They obtain their nutrients by secreting digestive enzymes into their surroundings and then absorbing the digested molecules back into the fungal body.**

17.9 Kinds of Fungi

Fungi are an ancient group of organisms at least 400 million years old. There are nearly 74,000 described species, in five groups (described in table 17.3), and many more awaiting discovery. Many fungi are harmful because they decay, rot, and spoil many different materials as they obtain food and because they cause serious diseases in animals and particularly in plants. Other fungi, however, are extremely useful. The manufacture of both bread and beer depends on the biochemical activities of yeasts, single-celled fungi that produce abundant quantities of carbon dioxide and ethanol. Fungi are used on a major scale in industry to convert one complex organic molecule into another; many commercially important steroids are synthesized in this way.

The four fungal phyla, distinguished from one another primarily by their mode of sexual reproduction, are the zygomycetes, the ascomycetes, the basidiomycetes, and the chytridiomycetes. A fifth group, the imperfect fungi, is an artificial grouping of fungi in which sexual reproduction has not been observed; these organisms are assigned to an appropriate group once their mode of sexual reproduction is identified. Molecular data is contributing to our understanding of the fungal phylogeny and as additional molecular evidence is acquired, a new fungal phylogeny is likely to appear. However, it already seems clear that fungi are more closely related to animals than to plants.

Ecological Roles as Decomposers

Fungi, together with bacteria, are the principal decomposers in the biosphere. They break down organic materials and return the substances that had been locked in those molecules to circulation in the ecosystem. Fungi are virtually the only organisms capable of breaking down lignin, one of the major constituents of wood. By breaking down such substances, fungi release critical building blocks, such as carbon, nitrogen, and phosphorus, from the bodies of dead organisms and make them available to other organisms.

In breaking down organic matter, some fungi attack living plants and animals as a source of organic molecules, whereas others attack dead ones. Fungi often act as disease-causing organisms for both animals and plants and are responsible for billions of dollars in agricultural losses every year. Not only are fungi

TABLE 17.3	FUNGI		
Phylum	Typical Examples	Key Characteristics	Approximate Number of Living Species
Zygomycota	*Rhizopus* (black bread mold)	Reproduce sexually and asexually; multinucleate hyphae lack septa, except for reproductive structures; fusion of hyphae leads directly to formation of a zygote, in which meiosis occurs just before it germinates	1,050
Ascomycota	Yeasts, truffles, morels	Reproduce by sexual means; ascospores are formed inside a sac called an ascus; asexual reproduction is also common	32,000
Basidiomycota	Mushrooms, toadstools, rusts	Reproduce by sexual means; basidiospores are borne on club-shaped structures called basidia; the terminal hyphal cell that produces spores is called a basidium; asexual reproduction occurs occasionally	22,000
Chytridiomycota	*Allomyces*	Produce flagellated gametes (zoospores); predominately aquatic, some freshwater and some marine; oldest group of fungi	1,500
Imperfect fungi (not a phylum)	*Aspergillus*, *Penicillium*	Sexual reproduction has not been observed; most are thought to be ascomycetes that have lost the ability to reproduce sexually	17,000

the most harmful pests of living plants, but they also attack food products once they have been harvested and stored. In addition, fungi often secrete substances into the foods they are attacking that make these foods unpalatable or poisonous.

Commercial Uses

The same aggressive metabolism that makes fungi ecologically important has been put to commercial use in many ways. The manufacture of both bread and beer depends on the biochemical activities of yeasts, single-celled fungi that produce abundant quantities of ethanol and carbon dioxide. Cheese and wine achieve their delicate flavors because of the metabolic processes of certain fungi, and others make possible the manufacture of soy sauce. Vast industries depend on the biochemical manufacture of organic substances such as citric acid by fungi in culture, and yeasts are now used on a large scale to produce protein for the enrichment of animal food. Many antibiotics, including the first one that was used on a wide scale, penicillin, are derived from fungi.

Some fungi are used to convert one complex organic molecule into another, cleaning up toxic substances in the environment. For example, at least three species of fungi have been isolated that combine selenium, accumulated at the San Luis National Wildlife Refuge in California's San Joaquin Valley, with harmless volatile chemicals—thus removing it from the soil.

Fungal Associations

Two kinds of mutualistic associations between fungi and autotrophic organisms are ecologically important: mycorrhizae and lichens. In each case, a photosynthetic organism fixes atmospheric carbon dioxide and thus makes organic material

Figure 17.14   Lichens growing on a rock.

available to the fungi. The metabolic activities of the fungi, in turn, enhance the overall ability of the symbiotic association to exist in a particular habitat. Mycorrhizae are symbiotic associations between fungi and the roots of plants where the fungal partner expedites the plant's absorption of essential nutrients such as phosphorus. Lichens are symbiotic associations between fungi and either green algae or cyanobacteria. They are prominent nearly everywhere in the world, especially in unusually harsh habitats such as bare rock (figure 17.14).

> **17.9   The fungal phyla are distinguished primarily by their modes of sexual reproduction. Fungi are key decomposers within almost all terrestrial ecosystems. In mycorrhizae and lichens, fungi form ecologically important associations with autotrophic organisms.**

INQUIRY & ANALYSIS

Are Chytrids Killing the Frogs?

Fungi in the phylum Chytridiomycota, referred to as chytrids, are thought to be playing a major role in a worldwide wave of amphibian extinctions. Discussed in much more detail in chapter 23 (page 467), this global extinction has many contributing causes. Initial attempts to assess the potential role of chytrids involved experiments in which some frogs of the genus *Dendrobates* were exposed to chytrids and others were not. After three weeks, all frogs were examined for shed skin, a clinical sign of the frog-killing disease. The results are seen in the pie charts.

1. **Applying Concepts** In this study, is there a dependent variable? If so, what is it?
2. **Making Inferences** What is the incidence of disease in non-exposed frogs? in exposed frogs?
3. **Drawing Conclusions** What is the impact of exposure to chytrids upon development of the frog-killing disease?

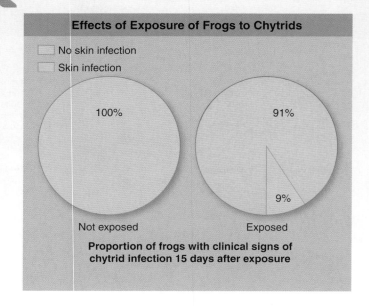

Effects of Exposure of Frogs to Chytrids

☐ No skin infection
☐ Skin infection

100%   Not exposed

91%
9%   Exposed

Proportion of frogs with clinical signs of chytrid infection 15 days after exposure

Four key evolutionary innovations serve to trace the evolution of the plant kingdom:

1. **Alternation of generations.** Although algae exhibit a haploid and diploid phase, the diploid phase is not a significant portion of their life cycle. By contrast, even in early plants (the nonvascular plants indicated by the first vertical bar in figure 18.5) the diploid sporophyte is a larger structure and offers protection for the egg and developing embryo. The dominance of the sporophyte, both in size and the proportion of time devoted to it in the life cycle, becomes greater throughout the evolutionary history of plants.

2. **Vascular tissue.** A second key innovation was the emergence of vascular tissue. Vascular tissue transports water and nutrients throughout the plant body and provides structural support. With the evolution of vascular tissue, plants were able to supply the upper portions of their bodies with water absorbed from the soil and had some rigidity, allowing the plants to grow larger and in drier conditions. The first vascular plants were the seedless vascular plants, the second vertical bar in figure 18.5.

3. **Seeds.** The evolution of seeds was a key innovation that allowed plants to dominate their terrestrial environments. Seeds provide nutrients and a tough, durable cover that protects the embryo until it encounters favorable growing conditions. The first plants with seeds were the gymnosperms, the third vertical bar in figure 18.5.

4. **Flowers and fruits.** The evolution of flowers and fruits were key innovations that improved the chances of successful mating in sedentary organisms and facilitated the dispersal of their seeds. Flowers both protected the egg and improved the odds of its fertilization, allowing plants that were located at considerable distances to mate successfully. Fruit, which surrounds the seed and aids in its dispersal, allows plant species to better invade new and possibly more favorable environments. The angiosperms, the fourth vertical bar in figure 18.5, are the only plants to produce flowers and fruits.

18.2 Plants evolved from freshwater green algae and eventually developed more dominant diploid phases of the life cycle, conducting systems of vascular tissue, seeds that protected the embryo, and flowers and fruits that aided in fertilization and distribution of the seeds.

TABLE 18.1	(continued)		
Phylum	**Typical Examples**	**Key Characteristics**	**Approximate Number of Living Species**
Seed Plants			
Coniferophyta (conifers)	Pines, spruce, fir, redwood, cedar	Gymnosperms; wind pollinated; ovules partially exposed at time of pollination; flowerless; seeds are dispersed by the wind; sperm lack flagella; sporophyte is dominant structure in life cycle; leaves are needlelike or scalelike; most species are evergreens and live in dense stands; among the most common trees on earth	550
Cycadophyta (cycads)	Cycads, sago palms	Gymnosperms; wind pollination or possibly insect-pollination; very slow growing, palmlike trees; sperm have flagella: trees are either male or female; sporophyte dominant in the life cycle	140
Gnetophyta (shrub teas)	Mormon tea, *Welwitschia*	Gymnosperms; nonmotile sperm; shrubs and vines; wind pollination and possibly insect pollination; plants are either male or female; sporophyte is dominant in the life cycle	70
Ginkgophyta (ginkgo)	Ginkgo trees	Gymnosperms; fanlike leaves that are dropped in winter (deciduous); seeds fleshy and ill-scented; motile sperm; trees are either male or female; sporophyte is dominant in the life cycle	1
Anthophyta (flowering plants, also called angiosperms)	Oak trees, corn, wheat, roses	Flowering; pollination by wind, animal, and water; characterized by ovules that are fully enclosed by the carpel; fertilization involves two sperm nuclei; one forms the embryo, the other fuses with the polar body to form endosperm for the seed; after fertilization, carpels and the fertilized ovules (now seeds) mature to become fruit; sporophyte is dominant in life cycle	235,000

18.3 Nonvascular Plants

Liverworts and Hornworts

The first successful land plants had no vascular system—no tubes or pipes to transport water and nutrients throughout the plant. This greatly limited the maximum size of the plant body because all materials had to be transported by osmosis and diffusion. Only two phyla of living plants, the **liverworts** (phylum Hepaticophyta) and the **hornworts** (phylum Anthocerophyta), completely lack a vascular system. The word *wort* meant *herb* in medieval Anglo-Saxon when these plants were named. Liverworts are the simplest of all living plants. About 6,000 species of liverworts and 100 species of hornworts survive today, usually growing in moist and shady places.

Primitive Conducting Systems: Mosses

Another phylum of plants, the **mosses** (phylum Bryophyta), were the first plants to evolve strands of specialized cells that conduct water and carbohydrates up the stem of the gametophyte. The conducting cells do not have specialized wall thickenings; instead they are like nonrigid pipes and cannot carry water very high. Because these conducting cells could at the most be considered a primitive vascular system, mosses are usually grouped by botanists with the liverworts and hornworts as "nonvascular" plants. Today about 9,500 species of mosses grow in moist places all over the world. The life cycle of a moss is illustrated in figure 18.6. You can see that the majority of the life cycle consists of the haploid gametophyte generation (the green part of the plant), which exists as male or female plants. The diploid sporophyte (the brown stalk with the swollen head, shown in light blue area of the cycle) grows out of the gametophyte of the female plant ❸, after the egg cell has been fertilized. The photo shows this stage of the life cycle, with the sporophytes growing out of the tops of the gametophytes. Cells within the sporophyte undergo meiosis to produce haploid spores ❹ that grow into gametophytes ❺.

> **18.3** While liverworts and hornworts totally lack a vascular system, mosses have simple soft strands of conducting cells.

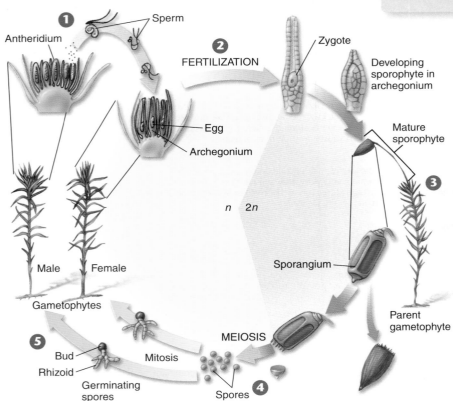

A hair-cup moss, *Polytrichum*

Figure 18.6 The life cycle of a moss.

On the gametophytes, which are haploid, sperm are released from each antheridium (sperm-producing structure) ❶. They then swim through free water to an archegonium (egg-producing structure) and down its neck to the egg. Fertilization takes place there ❷; the resulting zygote develops into a sporophyte, which is diploid. The sporophyte grows out of the archegonium and differentiates into a slender, basal stalk with a swollen capsule, the sporangium, at its apex ❸. The capsule is covered, at least at first, with a cap formed from the swollen archegonium. The sporophyte grows on the gametophyte and eventually produces spores as a result of meiosis. The spores are shed from the capsule ❹. The spores germinate, giving rise to gametophytes ❺. The gametophytes initially are threadlike; they grow along the ground. Ultimately, buds form on them, from which leafy gametophytes arise. The photo shows sporophytes growing from the tops of the leafy green gametophytes.

18.4 The Evolution of Vascular Tissue

The remaining seven phyla of plants, which have efficient vascular systems made of highly specialized cells, are called **vascular plants.** The first vascular plant appeared approximately 430 million years ago, but only incomplete fossils have been found. The first vascular plants for which we have relatively complete fossils, the extinct phylum Rhyniophyta, lived 410 million years ago. Among them is the oldest known vascular plant, *Cooksonia.* The fossil in figure 18.7 clearly shows that the plant had branched, leafless shoots that formed spores at the tips in structures called *sporangia.*

Cooksonia and the other early plants that followed became successful colonizers of the land through the development of efficient water- and food-conducting systems known as **vascular tissues** (Latin, *vasculum,* vessel or duct). These tissues, discussed in detail in chapter 33, consist of strands of specialized cylindrical or elongated cells that form a network throughout a plant, extending from near the tips of the roots (when present), through the stems, and into the leaves (when present). Figure 18.8 shows the network of vascular tissue in a leaf. The surrounding leaf tissue has been cleared away in this preparation, leaving only the vascular tissue. The presence of a cuticle and stomata are also characteristic of vascular plants.

Most early vascular plants seem to have grown by cell division at the tips of the stem and roots. Imagine stacking dishes—the stack can get taller but not wider! This sort of growth is called **primary growth** and was quite successful. During the so-called Coal Age (between 350 and 290 million

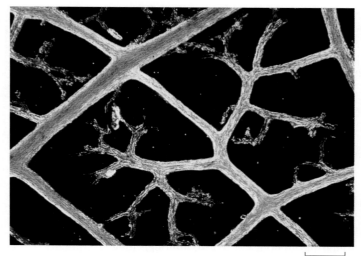

260 µm

Figure 18.8   The vascular system of a leaf.
The veins of a vascular plant contain strands of specialized cells for conducting food and water. The veins from a leaf are shown in this magnified view.

years ago), when much of the world's fossil fuel was formed, the lowland swamps that covered Europe and North America were dominated by an early form of seedless tree called a lycophyte. Lycophyte trees grew to heights of 10 to 35 meters (33 to 115 ft), and their trunks did not branch until they attained most of their total height. The pace of evolution was rapid during this period, for the world's climate was changing, growing dryer and colder. As the world's swamplands began to dry up, the lycophyte trees vanished, disappearing abruptly from the fossil record. They were replaced by tree-sized ferns, a form of vascular plant that will be described in detail in section 18.5. Tree ferns grew to heights of more than 20 meters (66 ft) with trunks 30 centimeters (12 in) thick. Like the lycophytes, the trunks of tree ferns were formed entirely by primary growth.

About 380 million years ago, vascular plants developed a new pattern of growth, in which a cylinder of cells beneath the bark divides, producing new cells in regions around the plant's periphery. This growth is called **secondary growth.** Secondary growth makes it possible for a plant stem to increase in diameter. Only after the evolution of secondary growth could vascular plants become thick-trunked and therefore tall. Redwood trees today reach heights of up to 117 meters (384 ft) and trunk diameters in excess of 11 meters (36 ft). This evolutionary advance made possible the dominance of the tall forests that today cover northern North America. You are familiar with the product of plant secondary growth as **wood.** The growth rings that are so visible in cross sections of trees are zones of secondary growth (spring–summer) spaced by zones of little growth (fall–winter).

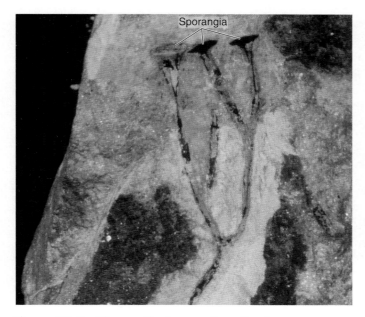

Sporangia

Figure 18.7   The earliest vascular plant.
The earliest vascular plant of which we have complete fossils is *Cooksonia.* This fossil shows a plant that lived some 410 million years ago; its upright branched stems terminated in spore-producing sporangia at the tips.

> **18.4**   Vascular plants have specialized vascular tissue composed of hollow tubes that conduct water to the leaves and cells that form cylinders that conduct food from them.

18.5 Seedless Vascular Plants

The earliest vascular plants lacked seeds, and two of the seven phyla of modern-day vascular plants do not have them. The two phyla of living seedless vascular plants include the ferns, phylum Pterophyta. This phylum includes the typical ferns seen growing on forests floors, like those shown in figure 18.9a, b. It also includes the whisk ferns (figure 18.9c) and the horsetails (figure 18.9d). The other phylum, Lycophyta, contains the club mosses (figure 18.9e). These phyla have free-swimming sperm that require the presence of free water for fertilization.

By far the most abundant of seedless vascular plants are the **ferns,** with about 11,000 living species. Ferns are found throughout the world, although they are much more abundant in the tropics than elsewhere. Many are small, only a few centimeters in diameter, but some of the largest plants that live today are also ferns. Descendants of ancient tree ferns, they can have trunks more than 24 meters (79 ft) tall and leaves up to 5 meters (16 ft) long!

The Life of a Fern

In ferns, the life cycle of plants begins a revolutionary change that culminates later with seed plants. Nonvascular plants like mosses are made largely of gametophyte (haploid) tissue. Vascular seedless plants like ferns have both gametophyte and sporophyte individuals, each independent and self-sufficient. The gametophyte (the heart-shaped plant at the top of the life cycle in figure 18.10) produces eggs and sperm; after sperm swim through water and fertilize the egg,

(a)

(b)

(c)

(d)

(e)

Figure 18.9 Seedless vascular plants.
(a) A tree fern in the forests of Malaysia (phylum Pterophyta). The ferns are by far the largest group of spore-producing vascular plants. (b) Ferns on the floor of a redwood forest. (c) A whisk fern. Whisk ferns have no roots or leaves. (d) A horsetail, *Equisetum telmateia.* This species forms two kinds of erect stems; one is green and photosynthetic, and the other, which terminates in a spore-producing "cone," is mostly light brown. (e) The club moss *Lycopodium lucidulum,* recently renamed *Huperzia lucidula* (phylum Lycophyta). Although superficially similar to the gametophytes of mosses, the conspicuous club moss plants shown here are sporophytes.

the zygote grows into a sporophyte (the longer lobed leaves at the bottom of the life cycle). The sporophyte bears haploid spores on the underside of their leaves (in the brown clusters called sori (singular, sorus)). The spores are released from the sorus and float to the ground where they germinate, growing into haploid gametophytes. The fern gametophytes are small, thin, heart-shaped photosynthetic plants, usually no more than a centimeter in length, that live in moist places. The fern spo-

rophytes are much larger and more complex, with long vertical leaves called **fronds.** When you see a fern, you are almost always looking at the sporophyte.

> **18.5   Ferns are among the vascular plants that lack seeds, reproducing with spores as nonvascular plants do.**

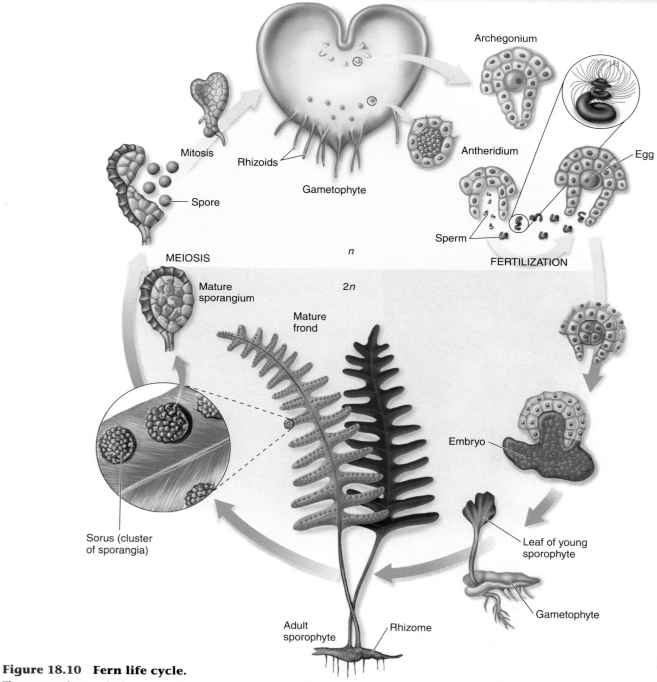

Figure 18.10   Fern life cycle.
The gametophytes, which are haploid, grow in moist places. Rhizoids (anchoring structures) project from their lower surface. Eggs develop in an archegonium, and sperm develop in an antheridium, both located on the gametophyte's lower surface. The sperm, when released, swim through free water to the mouth of the archegonium, entering and fertilizing the single egg. Following the fusion of egg and sperm to form a zygote—the first cell of the diploid sporophyte generation—the zygote starts to grow within the archegonium. Eventually, the sporophyte, the fern plant, becomes much larger than the gametophyte. Most ferns have more or less horizontal stems, called rhizomes, that creep along below ground. On the sporophyte's leaves, called fronds, occur clusters (called sori; singular, sorus) of sporangia, within which meiosis occurs and spores are formed. The release of these spores, which is explosive in many ferns, and their germination lead to the development of new gametophytes.

18.6   Evolution of Seed Plants

A key evolutionary advance among the vascular plants was the development of a protective cover for the embryo called a seed. The seed is a crucial adaptation to life on land because it protects the embryonic plant when it is at its most vulnerable stage. The plant in figure 18.11 is a cycad and its seeds (the green balls) develop on the edges of the scales of the cone. The embryonic plants are inside the seeds where they are protected. The evolution of the seed was a critical step in the domination of the land by plants.

The change in the life cycle of vascular plants in favor of the sporophyte (diploid) generation reaches its full force with the advent of the seed plants. Seed plants produce two kinds of gametophytes—male and female, each of which consists of just a few cells. Both kinds of gametophytes develop separately within the sporophyte and are completely dependent on it for their nutrition. Male gametophytes, commonly referred to as **pollen grains,** arise from **micro-**

spores. The pollen grains become mature when sperm are produced. The sperm are carried to the egg in the female gametophyte without using free water in the environment. A female gametophyte contains the egg and develops from a **megaspore** produced within an **ovule.** The transfer of pollen to an ovule by insects, wind, or other agents is referred to as **pollination.** The pollen grain then cracks open and sprouts, or germinates, and the pollen tube, containing the sperm cells, grows out, transporting the sperm directly to the egg. Thus there is no need for free water in the pollination and fertilization process.

Botanists generally agree that all seed plants are derived from a single common ancestor. There are five living phyla. In four of them, collectively called the **gymnosperms** (Greek, *gymnos,* naked, and *sperma,* seed), the ovules are not completely enclosed by sporophyte tissue at the time of pollination. Gymnosperms were the first seed plants. From gymnosperms evolved the fifth group of seed plants, called **angiosperms** (Greek, *angion,* vessel, and *sperma,* seed), phylum Anthophyta. Angiosperms, or flowering plants, are the most recently evolved of all the plant phyla. Angiosperms differ from all gymnosperms in that their ovules are completely enclosed by a vessel of sporophyte tissue in the flower called the **carpel** at the time they are pollinated. We will discuss gymnosperms and angiosperms later in this chapter.

The Structure of a Seed

A seed has three parts that are visible in the corn and bean seeds shown in figure 18.12: (1) a sporophyte plant embryo, (2) a source of food for the developing embryo called **endosperm**

Figure 18.11   A seed plant.
The seeds of this cycad, like all seeds, consist of a plant embryo and a protective covering. A cycad is a gymnosperm (naked-seeded plant), and its seeds develop out in the open on the edges of the cone scales.

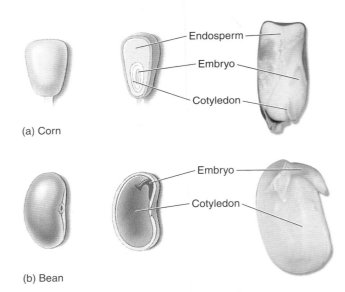

(a) Corn

(b) Bean

Figure 18.12   Basic structure of seeds.
A seed contains a sporophyte (diploid) embryo and a source of food, either endosperm (a) or food stored in the cotyledons (b). A seed coat, formed of sporophytic tissue from the parent, surrounds the seed and protects the embryo.

Figure 18.13 Seeds allow plants to bypass the dry season.
When it does rain, seeds can germinate, and plants can grow rapidly to take advantage of the relatively short periods when water is available. This palo verde desert tree (*Cercidium floridum*) has tough seeds (*inset*) that germinate only after they are cracked. Rains leach out the chemicals in the seed coats that inhibit germination, and the hard coats of the seeds may be cracked when they are washed down gullies in temporary floods.

in flowering plants (the endosperm makes up most of the seed in corn and is the white part of popcorn), and (3) a drought-resistant protective cover. In some seeds, the endosperm is used up during the development of the embryo and is stored as food by the embryo in thick "leaflike" structures called **cotyledons.** The endosperm is replaced by the cotyledon, as in the bean seed. Seeds are one way in which plants, anchored by their roots to one place in the ground, solve the problem of dispersing their progeny to new locations. The hard cover of the seed (formed from the tissue of the parent plant) protects the seed while it travels to a new location. The seed travels by many means such as by air, water, and animals. Many airborne seeds have devices to aid in carrying them farther. Most species of pine, for example, have seeds with thin flat wings attached. These wings help catch air currents, which carry the seeds to new areas. Seed dispersal will be discussed further later in this chapter and in chapter 34.

Once a seed has fallen to the ground, it may lie there, dormant, for many years. When conditions are favorable, however, and particularly when moisture is present, the seed germinates and begins to grow into a young plant (figure 18.13). Most seeds have abundant food stored in them to provide a ready source of energy for the new plant as it starts its growth.

The advent of seeds had an enormous influence on the evolution of plants. Seeds have greatly improved the adaptation of plants to living on land in at least four respects:

1. **Dispersal.** Most important, seeds facilitate the migration and dispersal of plant offspring into new habitats.
2. **Dormancy.** Seeds permit plants to postpone development when conditions are unfavorable, as during a drought, and to remain dormant until conditions improve.
3. **Germination.** By making the reinitiation of development dependent upon environmental factors such as temperature, seeds permit the course of embryonic development to be synchronized with critical aspects of the plant's habitat, such as the season of the year.
4. **Nourishment.** The seed offers the young plant nourishment during the critical period just after germination, when the seedling must establish itself.

18.6 A seed is a dormant diploid embryo encased with food reserves in a hard protective coat. Seeds play critical roles in improving a plant's chances of successfully reproducing in a varied environment.

18.7 Gymnosperms

Four phyla constitute the gymnosperms: the conifers (Coniferophyta), the cycads (Cycadophyta), the gnetophytes (Gnetophyta), and the ginkgo (Ginkgophyta). The conifers are the most familiar of the four phyla of gymnosperms and include pine, spruce, hemlock, cedar, redwood, yew, cypress, and fir trees, such as the Douglas firs in figure 18.14. Conifers are trees that produce their seeds in cones. The seeds (ovules) of conifers develop on scales within the cones and are exposed at the time of pollination. Most of the conifers have needlelike leaves, an evolutionary adaptation for retarding water loss. Conifers are often found growing in moderately dry regions of the world, including the vast taiga forests of the northern latitudes. Many are very important as sources of timber and pulp.

There are about 550 living species of conifers. The tallest living vascular plant, the coastal sequoia (*Sequoia sempervirens*), found in coastal California and Oregon, is a conifer and reaches over 100 meters (328 ft). The biggest redwood, however, is the mountain sequoia redwood species (*Sequoiadendron gigantea*) of the Sierra Nevadas. The largest individual tree is nicknamed after General Sherman of the Civil War, and it stands more than 83 meters (274 ft) tall while measuring 31 meters (102 ft) around its base. Another much smaller type of conifer, the bristlecone pines in Nevada, may be the oldest trees in the world—about 5,000 years old.

The other three gymnosperm phyla are much less widespread. Cycads (figure 18.15*a*), the predominant land plant in the golden age of dinosaurs, the Jurassic period (213–144 million years ago), have short stems and

Figure 18.14 A familiar group of gymnosperms: conifers (phylum Coniferophyta).

These Douglas fir trees, a type of conifer, often occur in vast forests.

(a)

(b)

(c)

Figure 18.15 Three more phyla of gymnosperms.

(a) An African cycad, *Encephalartos transvenosus*, phylum Cycadophyta. The cycads have fernlike leaves and seed-forming cones, like the ones shown here. (b) *Welwitschia mirabilis*, phylum Gnetophyta, is found in the extremely dry deserts of southwestern Africa. In *Welwitschia*, two enormous, beltlike leaves grow from a circular zone of cell division that surrounds the apex of the carrot-shaped root. (c) Maidenhair tree, *Ginkgo biloba*, the only living representative of the phylum Ginkgophyta, a group of plants that was abundant 200 million years ago. Among living seed plants, only the cycads and ginkgo have swimming sperm.

palmlike leaves. They are still widespread throughout the tropics. The gnetophytes, phylum Gnetophyta, contains only three kinds of plants, all unusual. One of them is perhaps the most bizarre of all plants, *Welwitschia,* shown in figure 18.15b, which grows on the exposed sands of the harsh Namibian Desert of southwestern Africa. *Welwitschia* acts like a plant standing on its head! Its two beltlike, leathery leaves are generated continuously from their base, splitting as they grow out over the desert sands. There is only one living species of ginkgo, the maidenhair tree, which has fan-shaped leaves (shown in figure 18.15c) shed in the autumn. Because ginkgos are resistant to air pollution, they are commonly planted along city streets.

The fossil record indicates that members of the ginkgo phylum were once widely distributed, particularly in the Northern Hemisphere; today, only one living species, the maidenhair tree (*Ginkgo biloba*), remains. The reproductive structures of ginkgos are produced on separate trees. The fleshy outer coverings of the seeds of female ginkgo plants exude the foul smell of rancid butter caused by butyric and isobutyric acids. In many Asian countries, however, the seeds are considered a delicacy. In Western countries, because of the seed odor, male plants vegetatively propagated are preferred for cultivation.

The Life of a Gymnosperm

We will examine conifers as typical gymnosperms. The conifer life cycle is illustrated in figure 18.16. Conifer trees form two kinds of cones. Seed cones ❸contain the female gametophytes, with their egg cells; pollen cones ❶contain pollen grains. Conifer pollen grains ❷ are small and light and are carried by the wind to seed cones. Because it is very unlikely that any particular pollen grain will succeed in being carried to a seed cone (the wind can take it anywhere), a great many pollen grains are needed to be sure that at least a few succeed in pollinating seed cones. For this reason, pollen grains are shed from their cones in huge quantities, often appearing as a sticky yellow layer on the surfaces of ponds and lakes—and even on windshields.

When a grain of pollen settles down on a scale of a female cone, a slender tube grows out of the pollen cell up into the scale, delivering the male gamete (sperm cell in ❺) to the female gametophyte containing the egg, or ovum. Fertilization occurs when the sperm cell fuses with the egg, forming a zygote that develops into an embryo. This zygote is the beginning of the sporophyte generation. What happens next is the essential improvement in reproduction achieved by seed plants. Instead of

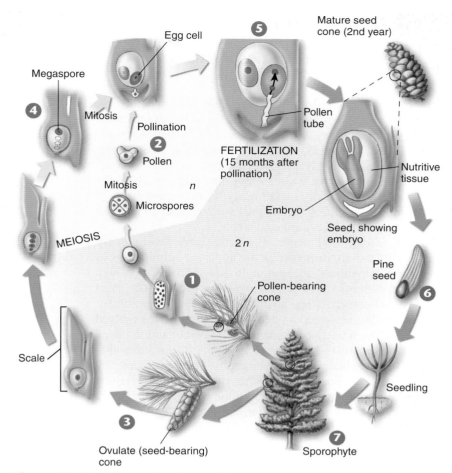

Figure 18.16 Life cycle of a conifer.

In all seed plants, the gametophyte generation is greatly reduced. In conifers such as pines, the relatively delicate pollen-bearing cones ❶ contain microspores, which give rise to pollen grains ❷, the male gametophytes. The familiar seed-bearing cones of pines ❸ are much heavier and more substantial structures than the pollen-bearing cones. Two ovules, and ultimately two seeds, are borne on the upper surface of each scale, which contains the megaspores that give rise to the female gametophytes ❹. After a pollen grain has reached a scale, it germinates, and a slender pollen tube grows toward the egg. When the pollen tube grows to the vicinity of the female gametophyte, sperm are released ❺, fertilizing the egg and producing a zygote there. The development of the zygote into an embryo takes place within the ovule, which matures into a seed ❻. Eventually, the seed falls from the cone and germinates, the embryo resuming growth and becoming a new pine tree ❼.

the zygote growing immediately into an adult sporophyte—just as you grow directly into an adult from a fertilized zygote—the fertilized ovule forms a seed ❻ . The pine seed contains a saillike structure that helps the seed be carried by the wind. The seed can then be dispersed into new habitats. If conditions are favorable where the seed lands, it will germinate and begin to grow, forming a new sporophyte plant ❼ .

> **18.7 Gymnosperms are seed plants in which the ovules are not completely enclosed by diploid tissue at pollination. Gymnosperms do not have flowers.**

18.8 **Rise of the Angiosperms**

Angiosperms, plants in which the ovule is completely enclosed by sporophyte tissue when it is fertilized, are the most successful of all plants. Ninety percent of all living plants are angiosperms, over 235,000 species, including many trees, shrubs, herbs, grasses, vegetables, and grains—in short, nearly all of the plants that we see every day. Virtually all of our food is derived, directly or indirectly, from angiosperms. In fact, more than half of the calories we consume come from just three species: rice, corn, and wheat.

In a very real sense, the remarkable evolutionary success of the angiosperms is the apparent culmination of the plant kingdom's adaptation to life on land, although plants continue to evolve. Angiosperms successfully meet the last difficult challenge posed by terrestrial living: the inherent conflict between the need to obtain nutrients (solved by roots, which anchor the plant to one place) and the need to find mates (solved by accessing plants of the same species). This challenge has never really been overcome by gymnosperms, whose pollen grains are carried passively by the wind on the chance that they might by luck encounter a female cone. Think about how inefficient this is! Angiosperms are also able to deliver their pollen directly, as if in an addressed envelope, from one individual of a species to another. How? *By inducing insects and other animals to carry it for them!* The tool that makes this animal-dictated pollination possible, the great advance of the angiosperms, is the flower. While some very successful later-evolving angiosperms like grasses have reverted to wind pollination, the directed pollination of flowering plants has led to phenomenal evolutionary success.

The Flower

Flowers are the reproductive organs of angiosperm plants. A flower is a sophisticated pollination machine. It employs bright colors to attract the attention of insects (or birds or small mammals), nectar to induce the insect to enter the flower, and structures that coat the insect with pollen grains while it is visiting. Then, when the insect visits another flower, it carries the pollen with it into that flower.

The basic structure of a flower consists of four concentric circles, or **whorls,** connected to a base called the **receptacle:**

1. The outermost whorl, called the **sepals** of the flower, typically serves to protect the flower from physical damage. These are the green leaflike structures in figure 18.17*a* and are in effect modified leaves that protect the flower while it is a bud. The sepals can be seen surrounding the flower bud to the lower left of the flower in figure 18.17*b*.

2. The second whorl, called the **petals** of the flower, serves to attract particular pollinators. Petals have

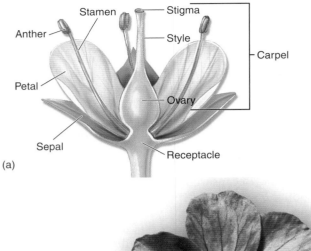

Figure 18.17   An angiosperm flower.

(a) The basic structure of a flower is a series of four concentric circles, or whorls: the sepals, petals, stamens, and the carpel (ovary, style, and stigma). (b) This flower of the wild woodland plant *Geranium* shows the five free petals, 10 stamens, and a fused carpel typical of angiosperm flowers.

particular pigments, often vividly colored like the light purple color in figure 18.17.

3. The third whorl, called the **stamens** of the flower, contains the "male" parts that produce the pollen grains. Stamens are the slender, threadlike filaments in figure 18.17 with a swollen **anther** at the tip containing pollen.

4. The fourth and innermost whorl, called the **carpel** of the flower, contains the "female" parts that produce eggs. The carpel is the vase-shaped structure in figure 18.17. The carpel is sporophyte tissue that completely encases the ovules within which the egg cell develops. The ovules occur in the bulging lower portion of the carpel, called the **ovary;** usually there is a slender stalk rising from the ovary called the **style,** with a sticky tip called a **stigma,** which receives pollen. When the flower is pollinated, a pollen tube grows down from the pollen grain on the stigma through the style to the ovary to fertilize the egg.

18.8   **Angiosperms are seed plants in which the ovule is completely enclosed by diploid tissue at pollination.**

18.9 Why Are There Different Kinds of Flowers?

If you were to watch insects visiting flowers, you would quickly discover that the visits are not random. Instead, certain insects are attracted by particular flowers. Insects recognize a particular color pattern and odor and search for flowers that look similar. Insects and plants have coevolved (see chapters 21 and 34) so that certain insects specialize in visiting particular kinds of flowers. As a result, a particular insect carries pollen from one individual flower to another *of the same species*. It is this keying to particular species that makes insect pollination so effective.

Of all insect pollinators, the most numerous are bees. Bees evolved soon after flowering plants, some 125 million years ago. Today there are over 20,000 species. Bees locate sources of nectar largely by odor at first (that is why flowers smell sweet) and then focus in on the flower's color and shape. Bee-pollinated flowers are usually yellow or blue, like the yellow flower in figure 18.18a. They frequently have guiding stripes or lines of dots to indicate the position in the flower of the nectar, usually in the throat of the flower, but these markings may not always be visible to the human eye. For example, the yellow flower in figure 18.18a looks very different through an ultraviolet filter in 18.18b. The UV rays show a dark area in the middle of the flower, where the nectar is located. Why have hidden signals? Because they are not hidden from bees that can detect UV rays. While inside the flower, the bee becomes coated with pollen, as in figure 18.18c. When the bee leaves this flower and visits another, it takes the pollen along for the ride, pollinating a neighboring flower.

Many other insects pollinate flowers. Butterflies tend to visit flowers of plants like phlox that have "landing platforms" on which they can perch. These flowers typically have long, slender floral tubes filled with nectar that a butterfly can reach by uncoiling its long proboscis (a hoselike tube extending out from the mouth). Moths, which visit flowers at night, are attracted to white or very pale-colored flowers, often heavily scented, that are easy to locate in dim light. Flowers pollinated by flies, such as members of the milkweed family, are usually brownish in color and foul smelling.

Red flowers, interestingly, are not typically visited by insects, most of which cannot "see" red as a distinct color. Who pollinates these flowers? Hummingbirds and sunbirds (figure 18.19)! To these birds, red is a very conspicuous color, just as it is to us. Birds do not have a well-developed sense of smell, and do not orient to odor, which is why red flowers often do not have a strong smell.

Some angiosperms have reverted to the wind pollination practiced by their ancestors, notably oaks, birches, and, most important, the grasses. The flowers of these plants are small, greenish, and odorless. Other angiosperm species are aquatic, and while some have developed specialized pollination systems where the pollen is transported underwater or floats from one plant to another, most aquatic angiosperms are either wind pollinated or insect pollinated like their terrestrial ancestors. Their flowers extend up above the surface of the water.

Figure 18.18 How a bee sees a flower.
(a) The yellow flower of *Ludwigia peruviana* (Onagraceae) photographed in normal light and (b) with a filter that selectively transmits ultraviolet light. The outer sections of the petals reflect both yellow and ultraviolet, a mixture of colors called "bee's purple"; the inner portions of the petals reflect yellow only and therefore appear dark in the photograph that emphasizes ultraviolet reflection. To a bee, this flower appears as if it has a conspicuous central bull's-eye. (c) When inside the flower, the bee becomes covered in pollen, which it takes to a neighboring flower.

Figure 18.19 Red flowers are pollinated by hummingbirds.
This long-tailed hermit hummingbird is extracting nectar from the red flowers of *Heliconia imbricata* in the forests of Costa Rica. Note the pollen on the bird's beak.

> **18.9** Flowers can be viewed as pollinator-attracting devices, with different kinds of pollinators attracted to different kinds of flowers.

18.10 Improving Seeds: Double Fertilization

The seeds of gymnosperms often contain food to nourish the developing plant in the critical time immediately after germination, but the seeds of angiosperms have greatly improved on this aspect of seed function. Angiosperms produce a special, highly nutritious tissue called **endosperm** within their seeds. Here is how it happens. The angiosperm life cycle is presented in figure 18.20, but there are actually two parts to the cycle, a male and a female part, indicated by the two sets of arrows at the top of the cycle. We'll begin with the flower of the sporophyte on the left side of the cycle ❶. The development of the male gametophyte (the pollen grain) occurs in the anthers and is indicated in the upper set of arrows. The anthers are shown in cross section in ❷ so you can see the microspores mother cells that develop into pollen grains. The pollen grain contains two haploid sperm. Upon adhering to the stigma at the top of the carpel (the female organ in where the egg cell is produced), the pollen begins to form a pollen tube ❹. The yellow pollen tube grows down into the carpel until it reaches the ovule in the ovary ❺. The two sperm (the small purplish cells)

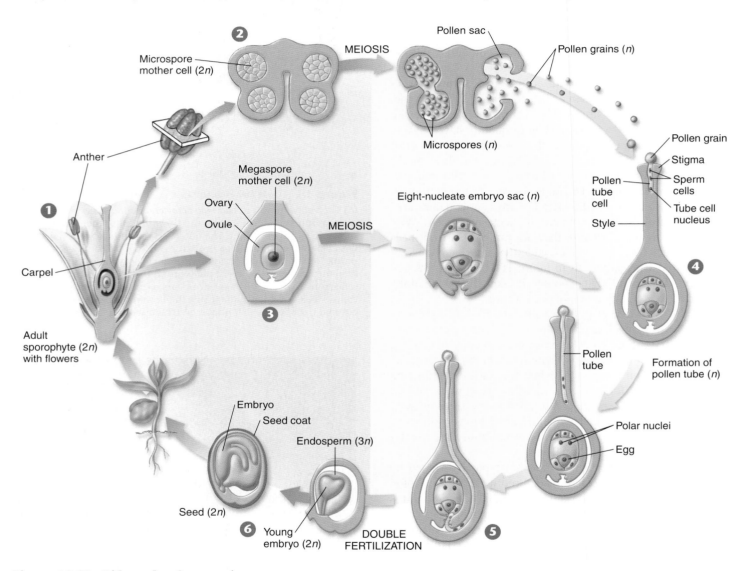

Figure 18.20   Life cycle of an angiosperm.

In angiosperms, as in gymnosperms, the sporophyte is the dominant generation. Eggs form within the embryo sac, inside the ovules ❸, which, in turn, are enclosed in the carpels. The carpel is differentiated in most angiosperms into a slender portion, or style, ending in a stigma, the surface on which the pollen grains germinate ❹. The pollen grains, meanwhile, are formed within the anthers ❷ and complete their differentiation to the mature, three-celled stage either before or after grains are shed. Fertilization is distinctive in angiosperms, being a double process ❺. A sperm and an egg come together, producing a zygote; at the same time, another sperm fuses with the two polar nuclei, producing the primary endosperm nucleus, which is triploid. The zygote and the primary endosperm nucleus divide mitotically, giving rise, respectively, to the embryo and the endosperm ❻. The endosperm is the tissue, almost exclusive to angiosperms, that nourishes the embryo and young plant.

travel down the pollen tube and into the ovary. The first sperm fuses with the egg (the green cell at the base of the ovary), as in all sexually reproducing organisms, forming the zygote that develops into the embryo. The other sperm cell fuses with two other products of meiosis, called *polar nuclei,* to form a triploid (three copies of the chromosomes, $3n$) endosperm cell. This cell divides much more rapidly than the zygote, giving rise to the nutritive endosperm tissue within the seed (the tan material surrounding the embryo ❻). This process of fertilization with two sperm to produce both a zygote and endosperm is called **double fertilization.** Double fertilization forming endosperm is exclusive to angiosperms.

In some angiosperms, such as the common pea or bean, the endosperm is fully used up by the time the seed is mature. Food reserves are stored by the embryo in swollen, fleshy leaves called *cotyledons,* or seed leaves. In other angiosperms, such as corn, the mature seed contains abundant endosperm, which is used after germination. It also contains a cotyledon, but its seed leaf is used to protect the plant during germination and not as a food source.

Some angiosperm embryos have two cotyledons, and are called *dicotyledons,* or **dicots.** The first angiosperms were like this. Dicots typically have leaves with netlike branching of veins and flowers with four to five parts per whorl (figure 18.21, *top*). Oak and maple trees are dicots, as are many shrubs.

The embryos of other angiosperms, which evolved somewhat later, have a single cotyledon and are called *monocotyledons,* or **monocots.** Monocots typically have leaves with parallel veins and flowers with three parts per whorl (figure 18.21, *bottom*). Grasses, one of the most abundant of all plants, are wind-pollinated monocots. There are also differences in the organization of vascular tissue in the stems of monocots and dicots, which will be compared in more detail in chapter 33.

18.10 Two sperm fertilize each angiosperm ovule. One fuses with the egg to form the zygote, the other with two polar nuclei to form triploid ($3n$) nutritious endosperm.

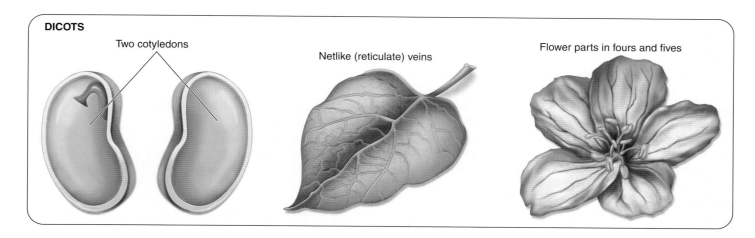

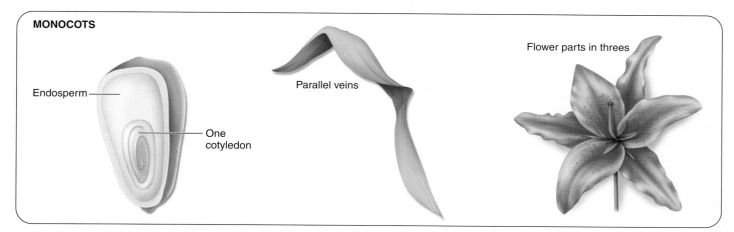

Figure 18.21 Dicots and monocots.
Dicots have two cotyledons and netlike (reticulate) veins. Their flower parts occur in fours and fives. Monocots are characterized by one cotyledon, parallel veins, and the occurrence of flower parts in threes (or multiples of three).

(a)

(b)

(c)

Figure 18.22 Different ways of dispersing fruits.
(a) Berries are fruits that are dispersed by animals. (b) Coconuts are fruits dispersed by water, where they are carried off to new island habitats.
(c) The fruits of maples are dry and winged, carried by the wind like small helicopters floating through the air.

18.11 Improving Seed Dispersal: Fruits

Just as a mature ovule becomes a seed, so a mature ovary that surrounds the ovule becomes all or a part of the **fruit.** This is why fruit forms in angiosperm and not in gymnosperm. Compare the life cycles of the gymnosperm in figure 18.16 and that of the angiosperm in figure 18.20. Both types of plants have tissue that surrounds the egg, called the ovule. The ovule becomes the seed in both. But, in the angiosperm life cycle, you will notice that the ovule is surrounded by another layer of tissue, the ovary, which develops into the fruit. A fruit is a mature ripened ovary containing fertilized seeds. Fruits provide angiosperms with a second way of dispersing their progeny than simply sending their seeds off on the wind. Instead, just as in pollination, they employ animals. By making fruits fleshy and tasty to animals, like the berries in figure 18.22a, angiosperms encourage animals to eat them. The seeds within the fruit are resistant to chewing and digestion. They pass out of the animal with the feces, undamaged and ready to germinate at a new location far from the parent plant.

Although many fruits are dispersed by animals, some fruits are dispersed by water, like the coconut in figure 18.22b, and many plant fruits are specialized for wind dispersal. The small, nonfleshy fruits of the dandelion, for example, have a plumelike structure that allows them to be carried long distances on wind currents. The fruits of many grasses are small particles, so light wind bears them easily. Maples have long wings attached to the fruit, as in figure 18.22c, that allows them to be carried by the wind before reaching the ground. In tumbleweeds, the whole plant breaks off and is blown across open country by the wind, scattering seeds as it moves. Fruits will be discussed in more detail in chapter 34.

> **18.11** A fruit is a mature ovary containing fertilized seeds, often specialized to aid in seed dispersal.

INQUIRY & ANALYSIS

How Does Arrowgrass Tolerate Salt?

Seaside arrowgrass (*Triglochin maritima*) plants are able to grow in very salty soils, where few other plants survive. How do they manage? Researchers found that their roots do not take up salt. Because of this, you would expect root cells to lose water to the surrounding soil. How then do the roots achieve osmotic balance with their surrounding soil? In an attempt to find out, researchers grew arrowgrass plants in nonsalty soil for two weeks, then transferred them to one of several soils, which differed in salt level. After ten days, shoots were harvested and analyzed for amino acids, because accumulating amino acids could be one way that the cells maintain osmotic balance. Results are presented in the graph.

1. **Applying Concepts** What is the dependent variable?
2. **Making Inferences** What is the effect of soil salt concentration on arrowgrass plants' accumulation of

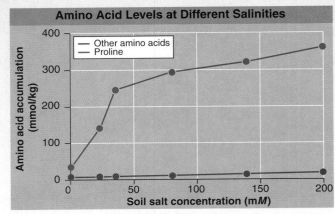

Amino Acid Levels at Different Salinities

- Other amino acids
- Proline

x-axis: Soil salt concentration (m*M*)
y-axis: Amino acid accumulation (mmol/kg)

the amino acid proline? of other amino acids?

3. **Drawing Conclusions** Do these results support the hypothesis that arrowgrass accumulates proline to achieve osmotic balance with salty soils?

Plants

18.1 Adapting to Terrestrial Living

- Plants are complex multicellular autotrophs, producing their own food through photosynthesis. Plants acquire water and minerals from the soil through their roots. They control water loss through a watertight layer called the cuticle, with openings called stomata, which allow for gas exchange with the air (**figure 18.2**).

- Plant life cycles involve an alternation of generations where a haploid gametophyte alternates with a diploid sporophyte, with the sporophyte dominating the life cycle more and more as plants evolved (**figure 18.3**).

18.2 Plant Evolution

- Plants evolved from green algae. Four innovations illustrate the challenges that were overcome to invade the land. These innovations include an alternation of generation life cycle, the evolution of vascular tissue, seeds, and flower and fruits (**table 18.1** and **figure 18.5**).

- Plants exhibit a life cycle with an alternation of generations where the plant spends significant portions of its life in a multicellular haploid phase and in a multicellular diploid phase.

- The evolution of vascular tissue allowed plants to transport water and minerals from the ground up through their bodies and to grow taller.

- Seeds provide protection and nutrients for the developing embryo.

- Flowers and fruits improved the chances of successful mating and the distribution of seeds.

Seedless Plants

18.3 Nonvascular Plants

- Liverworts and hornworts don't have vascular tissue, which limits how large they can grow. Mosses have specialized cells for conducting water and carbohydrates in the plant, but not rigid vascular tissue. These plants are grouped together as nonvascular plants. Their life cycles are dominated by the haploid gametophyte (**figure 18.6**).

18.4 The Evolution of Vascular Tissue

- Vascular systems consist of specialized cylindrical or elongated cells that form a network throughout the plant. These cells conduct water from the roots throughout the plant and carbohydrates manufactured in the leaves throughout the plant (**figure 18.8**).

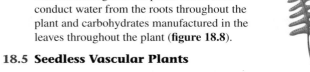

18.5 Seedless Vascular Plants

- Among the most primitive vascular plants are the Pterophyta (ferns) and Lycophyta (**figure 18.9**). These plants can grow very tall due to

vascularization but lack seeds and require water for fertilization. The gametophyte and sporophyte both make up significant portions of the life cycle (**figure 18.10**).

The Advent of Seeds

18.6 Evolution of Seed Plants

- The seed was an evolutionary innovation that provided protection for the plant embryo. The seed contains the embryo and food for the developing embryo, inside a drought-resistant cover (**figure 18.12**). In seed plants, the sporophyte also becomes the dominant structure in the life cycle.

- Seeds are a key adaptation to land-dwelling plants by improving the dispersal of embryos, providing dormancy when needed, allowing for germination when conditions are favorable, and nourishment during germination.

18.7 Gymnosperms

- Gymnosperms are nonflowering seed plants and include conifers, cycads, gnetophytes, and the ginkgo. The seeds are produced in cones where pollination occurs. Pollen, produced in smaller, pollen-bearing cones, travels by wind to the seed-bearing cones containing the egg cells and fertilizes the egg. The ovules are not completely enclosed by diploid tissue as is the case with flowering plants. The seed forms and is released from the cone for dispersal (**figure 18.16**).

The Evolution of Flowers

18.8 Rise of the Angiosperms

- Angiosperms, flowering plants, are plants in which the ovule is completely enclosed by sporophyte tissue. The flowers are the reproductive structures, containing the pollen on the anthers and the egg within the ovary (**figure 18.17**). They improved the efficiency of mating.

18.9 Why Are There Different Kinds of Flowers?

- Flowers vary greatly in size, shape, and color so that they are identifiable by a particular pollinator. This improves the likelihood that the pollen will be carried to the appropriate mate (**figures 18.18** and **18.19**).

18.10 Improving Seeds: Double Fertilization

- Double fertilization provides food for the germinating plant. The pollen grain produces two sperm cells, one fuses with the egg and the other fuses with two polar nuclei, producing endosperm (**figure 18.20**). The angiosperms are divided into two groups: monocots and dicots (**figure 18.21**).

18.11 Improving Seed Dispersal: Fruits

- A further evolutionary advancement in angiosperms was the development of the ovary into fruit tissue. Fruits aid in the dispersal of seeds to new habitats (**figure 18.22**).

1. A major consideration for the evolution of terrestrial plants is the problem of
   a. lack of nutrients.
   b. predators.
   c. dehydration.
   d. not enough carbon.

2. Which of the following structures or systems does *not* give the plants that have them an evolutionary advantage?
   a. chloroplasts
   b. vascular tissue
   c. seeds
   d. flowers and fruits

3. Mosses, liverworts, and hornworts do not reach a large size because
   a. they lack chlorophyll.
   b. they do not have specialized vascular tissue to transport water very high.
   c. photosynthesis does not take place at a very fast rate.
   d. alternation of generations does not allow the plants to grow very tall before reproduction.

4. One characteristic that separates ferns from complex vascular plants is that ferns do not have
   a. a vascular system.
   b. chloroplasts.
   c. alternation of generations in their life cycle.
   d. seeds.

5. As seed plants evolved, the _____ form became more visible.
   a. gametophyte
   b. gymnosperm
   c. sporophyte
   d. angiosperm

6. In seeds, the seed coat aids in
   a. germination.
   b. photosynthesis.
   c. nourishment.
   d. dispersal.

7. What separates the gymnosperms from the rest of the vascular plants is/are
   a. a vascular system.
   b. ovules completely covered by the gametophyte.
   c. ovules not completely covered by the sporophyte.
   d. fruits and flowers.

8. What separates the angiosperms from the rest of the vascular plants is/are
   a. a vascular system.
   b. ovules completely covered by the gametophyte.
   c. fruits and flowers.
   d. ovules not completely covered by the sporophyte.

9. Flower shape and color can be linked to the process of
   a. pollination.
   b. photosynthesis.
   c. senescence.
   d. secondary growth.

10. If the seeds of a plant are encased in a fleshy fruit, then the most likely form of dispersal is
    a. to attach to an animal's fur or skin.
    b. wind.
    c. an animal's digestive system and processes.
    d. water.

1. **Figure 18.5** Explain which key innovation lead to the evolution of each group and why the innovation was significant.

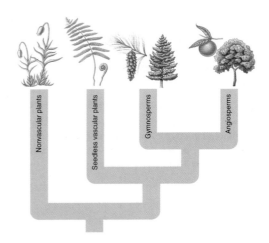

2. **Figure 18.12** Why do you think these two seeds were the most important foods of the Native Americans of this country? *Hint:* Think about the purpose of a seed.

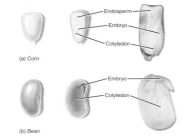

(a) Corn

(b) Bean

3. **Figure 18.13** These seeds fall from the tree and are scattered about on the ground nearby. What are the benefits or drawbacks to having (a) chemicals in the seed to inhibit sprouting, and (b) hard seed coats?

1. Why do most plant pollinators have wings?
2. Why are mosses limited to moist places?
3. Why does a pine tree, which might live more than 100 years, produce hundreds of pine cones each year, each with 30 to 50 seeds?

4. Your friend, Jessica, is admiring the wildflowers one spring day, and wonders aloud why in the world plants evolved such beautiful colors and shapes and produce such wonderful scents. What do you tell her?

19
Evolution of Animals

Animals are the most diverse in appearance of all eukaryotes. *Polistes,* the common paper wasp you see here, is a member of the most diverse of all animal groups, the insects. It has been a major challenge for biologists to sort out the millions of kinds of animals. This wasp has a segmented external skeleton and jointed appendages, and so, based on these characteristics, is classified as an arthropod. But how are arthropods related to mollusks such as snails, and to segmented worms like earthworms? Until recently, biologists grouped all three kinds of animals together, as they all share a true body cavity called a coelom, a character assumed to be so fundamental it could have evolved only once. Now, molecular analyses suggest this assumption may be wrong. Instead, mollusks and segmented worms are grouped together with other animals that grow the same way you do, by adding additional mass to an existing body, while arthropods are grouped with other molting animals. These animals increase in size by molting their external skeletons, an ability that seems to have evolved only once. Thus we learn that even in a long-established field like taxonomy, biology is constantly changing. In this chapter we will explore the most diverse of all kingdoms, the animals.

19.1 General Features of Animals

Animals are the eaters or consumers of the earth. A very diverse group, no one criterion fits all, but several characteristics are of major importance (table 19.1): (1) Animals are heterotrophs and must ingest plants, algae, or other animals for nourishment. (2) All animals are multicellular, and unlike plants and protists animal cells lack cell walls. (3) Animals are able to move from place to place. (4) Animals are very diverse in form and habitat. (5) Most animals reproduce sexually. (6) Animals have characteristic pattern of embryonic development, and possess unique tissues.

> **19.1** Animals are complex multicellular organisms typically characterized by high mobility and heterotrophy. Most animals also possess internal tissues, and reproduce sexually.

TABLE 19.1	GENERAL FEATURES OF ANIMALS
Heterotrophs. Unlike autotrophic plants and algae, animals cannot construct organic molecules from inorganic chemicals. All animals are heterotrophs—that is, they obtain energy and organic molecules by ingesting other organisms. Some animals (herbivores) consume autotrophs, other animals (carnivores) consume heterotrophs, and still others (detritivores) consume decomposing organisms.	
Multicellular. All animals are multicellular, often with complex bodies like that of this brittlestar (*right*). The unicellular heterotrophic organisms called protozoa, which were at one time regarded as simple animals, are now considered members of the large and diverse kingdom Protista, discussed in chapter 17.	
No Cell Walls. Animal cells are distinct among those of multicellular organisms because they lack rigid cell walls and are usually quite flexible, as are these cancer cells. The many cells of animal bodies are held together by extracellular lattices of structural proteins such as collagen. Other proteins form a collection of unique intercellular junctions between animal cells.	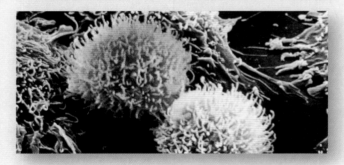
Active Movement. The ability of animals to move more rapidly and in more complex ways than members of other kingdoms is perhaps their most striking characteristic, one that is directly related to the flexibility of their cells and the evolution of nerve and muscle tissues. A remarkable form of movement unique to animals is flying, an ability that is well developed among vertebrates and insects like this butterfly. The only terrestrial vertebrate group never to have evolved flight is amphibians.	

TABLE 19.1 *(continued)*

Diverse in Form. Almost all animals (99%) are **invertebrates,** which, like this millipede, lack a backbone. Of the estimated 10 million living animal species, only 42,500 have a backbone and are referred to as **vertebrates.** Animals are very diverse in form, ranging in size from organisms too small to see with the unaided eye to enormous whales and giant squids.

Diverse in Habitat. The animal kingdom includes about 35 phyla, most of which, like these jellyfish (phylum Cnidaria), occur in the sea. Far fewer phyla occur in freshwater and fewer still occur on land. Members of three successful marine phyla, Arthropoda (insects), Mollusca (snails), and Chordata (vertebrates), dominate animal life on land.

Sexual Reproduction. Most animals reproduce sexually, as these tortoises are doing. Animal eggs, which are nonmotile, are much larger than the small, usually flagellated sperm. In animals, cells formed in meiosis function directly as gametes. The haploid cells do not divide by mitosis first, as they do in plants and fungi, but rather fuse directly with each other to form the zygote. Consequently, with a few exceptions, there is no counterpart among animals to the alternation of haploid (gametophyte) and diploid (sporophyte) generations characteristic of plants.

Embryonic Development. Most animals have a similar pattern of embryonic development. The zygote first undergoes a series of mitotic divisions, called *cleavage,* and, like this dividing frog egg, becomes a solid ball of cells, the **morula,** then a hollow ball of cells, the **blastula.** In most animals, the blastula folds inward at one point to form a hollow sac with an opening at one end called the **blastopore.** An embryo at this stage is called a **gastrula.** The subsequent growth and movement of the cells of the gastrula differ widely from one phylum of animals to another.

Unique Tissues. The cells of all animals except sponges are organized into structural and functional units called **tissues,** collections of cells that have joined together and are specialized to perform a specific function. Animals are unique in having two tissues associated with movement: (1) muscle tissue, which powers animal movement, and (2) nervous tissue, which conducts signals among cells. Neuromuscular junctions, where nerves connect with muscle tissue, are shown here.

Five Key Transitions in Body Plan

The evolution of animals is marked by five key transitions: the evolution of tissues, bilateral symmetry, a body cavity, deuterostome development, and segmentation. These five body transitions are indicated at the branchpoints of the animal evolutionary tree in figure 19.1.

1. Evolution of Tissues

The simplest animals, the Parazoa, lack both defined tissues and organs. Characterized by the sponges, these animals exist as aggregates of cells with minimal intercellular coordination. All other animals, the Eumetazoa, have distinct tissues with highly specialized cells. The evolution of tissues is the first key transition in the animal body plan.

2. Evolution of Bilateral Symmetry

Sponges also lack any definite symmetry, growing asymmetrically as irregular masses. Virtually all other animals have a definite shape and symmetry that can be defined along an imaginary axis drawn through the animal's body.

Radial Symmetry. Symmetrical bodies first evolved in marine animals exhibiting **radial symmetry.** The parts of their bodies are arranged around a central axis in such a way that any plane passing through the central axis divides the organism into halves that are approximate mirror images.

Bilateral Symmetry. The bodies of all other animals are marked by a fundamental **bilateral symmetry,** a body design in which the body has a right and a left half that are mirror images of each other. Bilateral symmetry constitutes the second major evolutionary advance in the animal body plan. This unique form of organization allows parts of the body to evolve in different ways, permitting different organs to be located in different parts of the body. Also, bilaterally symmetrical animals move from place to place more efficiently than radially symmetrical ones, which, in general, lead a sessile or passively floating existence. Due to their increased mobility, bilaterally symmetrical animals are efficient in seeking food, locating mates, and avoiding predators.

3. Evolution of a Body Cavity

A third key transition in the evolution of the animal body plan was the evolution of the body cavity. The evolution of efficient organ systems within the animal body was not possible until a body cavity evolved for supporting organs, distributing materials, and fostering complex developmental interactions.

The presence of a body cavity allows the digestive tract to be larger and longer. This longer passage allows for storage of undigested food, longer exposure to enzymes for more complete digestion. Such an arrangement allows an animal to eat a great deal when it is safe to do so and then to hide during the digestive process, thus limiting the animal's exposure to predators. The body cavity architecture is also more flexible, thus allowing the animal greater freedom to move.

An internal body cavity also provides space within which the gonads (ovaries and testes) can expand, allowing the accumulation of large numbers of eggs and sperm. Such storage capacity allows the diverse modifications of breeding strategy that characterize the more advanced phyla of animals. Furthermore, large numbers of gametes can be stored and released when the conditions are as favorable as possible for the survival of the young animals.

4. The Evolution of Deuterostome Development

Bilateral animals can be divided into two groups based on differences in the basic pattern of development. One group is called the **protostomes** (from the Greek words *protos,* first, and *stoma,* "mouth") and includes the flatworms, nematodes, mollusks, annelids, and arthropods. Two outwardly dissimilar groups, the echinoderms and the chordates, together with a few other smaller related phyla, comprise the second group, the **deuterostomes** (Greek, *deuteros,* second, and *stoma,* mouth). Protostomes and deuterostomes differ in several aspects of embryo growth and will be discussed later in the chapter.

Deuterostomes evolved from protostomes more than 630 million years ago, and the consistency of deuterostome development, and its distinctiveness from that of the protostomes suggests that it evolved once, in a common ancestor to all of the phyla that exhibit it.

5. The Evolution of Segmentation

The fifth key transition in the animal body plan involved the subdivision of the body into **segments.** Just as it is efficient for workers to construct a tunnel from a series of identical prefabricated parts, so segmented animals are assembled from a succession of identical segments. During the animal's early development, these segments become most obvious in the mesoderm but later are reflected in the ectoderm and endoderm as well. Two advantages result from early embryonic segmentation:

1. In annelids and other highly segmented animals, each segment may go on to develop a more or less complete set of adult organ systems. Damage to any one segment need not be fatal to the individual, since the other segments duplicate that segment's functions.
2. Locomotion is far more effective when individual segments can move independently because the animal as a whole has more flexibility of movement. Because the separations isolate each segment into an individual skeletal unit, each is able to contract or expand autonomously in response to changes in hydrostatic pressure. Therefore, a long body can move in ways that are often quite complex.

These key evolutionary changes lead to the large diversity of animals presented in table 19.2.

> **19.2 Five key transitions in body design are responsible for most of the differences we see among the major animal phyla.**

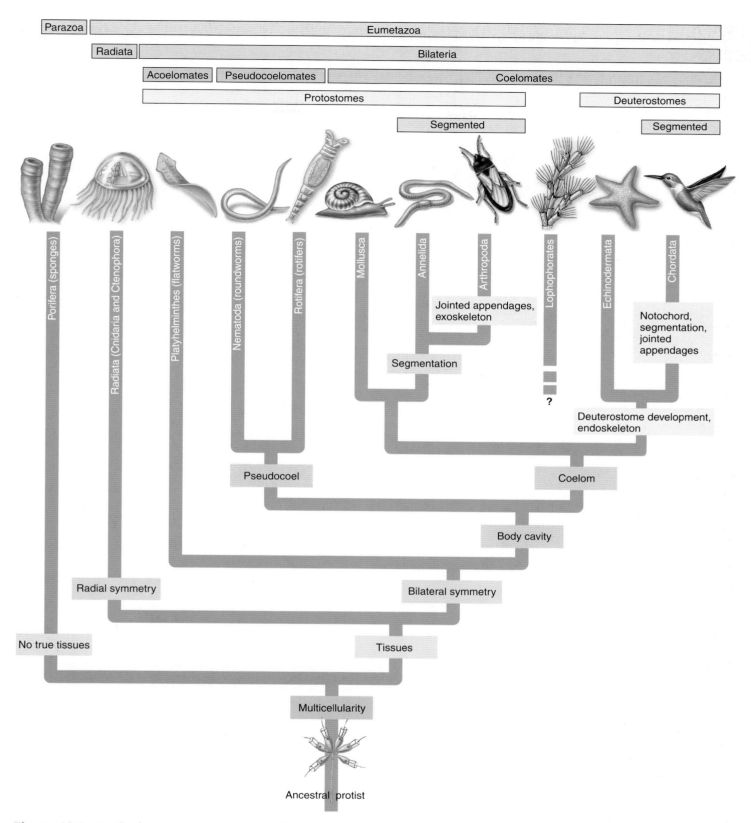

Figure 19.1 Evolutionary trends among the animals.

In this chapter, we examine the diversity of animals and a series of key evolutionary innovations in the animal body plan, shown here along the branches.

TABLE 19.2 THE MAJOR ANIMAL PHYLA

Phylum	Typical Examples		Key Characteristics	Approximate Number of Named Species
Arthropoda (arthropods)	Insects, crabs, spiders, millipedes		Most successful of all animal phyla; chitinous exoskeleton covering segmented bodies with paired, jointed appendages; most insect groups have wings; nearly all are freshwater or terrestrial	1,000,000
Mollusca (mollusks)	Snails, clams, octopuses, nudibranchs		Soft-bodied coelomates whose bodies are divided into three parts: head-foot, visceral mass, and mantle; many have shells; almost all possess a unique rasping tongue called a radula; most are marine or freshwater but 35,000 species are terrestrial	110,000
Chordata (chordates)	Mammals, fish, reptiles, birds, amphibians		Segmented coelomates with a notochord; possess a dorsal nerve cord, pharyngeal pouches, and a tail at some stage of life; in vertebrates, the notochord is replaced during development by the spinal column; most are marine, many are freshwater, and 20,000 species are terrestrial	56,000
Platyhelminthes (flatworms)	*Planaria*, tapeworms, flukes		Solid, unsegmented, bilaterally symmetrical worms; no body cavity; digestive cavity, if present, has only one opening; marine, freshwater, or parasitic	20,000
Nematoda (roundworms)	*Ascaris*, pinworms, hookworms, *Filaria*		Pseudocoelomate, unsegmented, bilaterally symmetrical worms; tubular digestive tract passing from mouth to anus; tiny; without cilia; live in great numbers in soil and aquatic sediments; some are important animal parasites	20,000
Annelida (segmented worms)	Earthworms, marine worms, leeches		Coelomate, serially segmented, bilaterally symmetrical worms; complete digestive tract; most have bristles called setae on each segment that anchor them during crawling; marine, freshwater, and terrestrial	12,000

TABLE 19.2 *(continued)*

Phylum	Typical Examples		Key Characteristics	Approximate Number of Named Species
Cnidaria (cnidarians)	Jellyfish, hydra, corals, sea anemones		Soft, gelatinous, radially symmetrical bodies whose digestive cavity has a single opening; possess tentacles armed with stinging cells called cnidocytes that shoot sharp harpoons called nematocysts; almost entirely marine	10,000
Echinodermata (echinoderms)	Sea stars, sea urchins, sand dollars, sea cucumbers		Deuterostomes with radially symmetrical adult bodies; endoskeleton of calcium plates; pentamerous (five-part) body plan and unique water vascular system with tube feet; able to regenerate lost body parts; all are marine	6,000
Porifera (sponges)	Barrel sponges, boring sponges, basket sponges, vase sponges		Asymmetrical bodies without distinct tissues or organs; saclike body consists of two layers breached by many pores; internal cavity lined with food-filtering cells called choanocytes; most marine (150 species live in freshwater)	5,150
Lophophorates (moss animals, also called Bryozoa)	*Bowerbankia, Plumatella*, sea mats, sea moss		Microscopic, aquatic deuterostomes that form branching colonies, possess circular or U-shaped row of ciliated tentacles for feeding called a lophophore that usually protrudes through pores in a hard exoskeleton; also called Ectoprocta because the anus, or proct, is external to the lophophore; marine or freshwater	4,000
Rotifera (wheel animals)	Rotifers		Small, aquatic pseudocoelomates with a crown of cilia around the mouth resembling a wheel; almost all live in freshwater	2,000

19.3 Sponges and Cnidarians: The Simplest Animals

The kingdom Animalia consists of two subkingdoms: (1) *Parazoa,* animals that lack a definite symmetry and possess neither tissues nor organs, and (2) *Eumetazoa,* animals that have a definite shape and symmetry, and in most cases tissues organized into organs. The subkingdom Parazoa consists primarily of the sponges, phylum Porifera. The other animals, comprising about 35 phyla, belong to the subkingdom Eumetazoa.

Sponges

Sponges, members of the phylum Porifera, are the simplest animals. Most sponges completely lack symmetry, and although some of their cells are highly specialized, they are not organized into tissues. The bodies of sponges consist of little more than masses of specialized cells embedded in a gel-like matrix, like chopped fruit in Jell-O. However, sponge cells do possess a key property of animal cells: cell recognition. For example, when a sponge is passed through a fine silk mesh, individual cells separate and then reaggregate on the other side to re-form the sponge. Clumps of cells disassociated from a sponge can give rise to entirely new sponges.

About 5,000 species exist, almost all in the sea. Some are tiny, and others are more than 2 meters in diameter (the diver in figure 19.2*a* could almost crawl inside the sponge shown). The body of an adult sponge is anchored in place on the seafloor and is shaped like a vase (as you can see in figure 19.2*b*). The outside of the sponge is covered with a skin of flattened cells called epithelial cells that protect the sponge.

The body of the sponge is perforated by tiny holes. The name of the phylum, Porifera, refers to this system of pores. Unique flagellated cells called **choanocytes,** or collar cells, line the body cavity of the sponge. The beating of the flagella of the many choanocytes draws water in through the pores and through the cavity. One cubic centimeter of sponge tissue can propel more than 20 liters of water a day in and out of the sponge body! Why all this moving of water? The sponge is a "filter-feeder." The beating of each choanocyte's flagellum draws water down through its collar, made of small hairlike projections resembling a picket fence. Any food particles in the water, such as protists and tiny animals, are trapped in the fence and later ingested by the choanocyte or other cells of the sponge.

The choanocytes of sponges very closely resemble a kind of protist called choanoflagellates, which seem almost certain to have been the ancestors of sponges. Indeed, they may be the ancestors of *all* animals, although it is difficult to be certain that sponges are the direct ancestors of the other more complex phyla of animals.

All animals other than sponges have both symmetry and tissues and thus are eumetazoans.

(a)

(b)

Figure 19.2 Sponges.
These two marine sponges are barrel sponges. They are among the largest of sponges, with well-organized forms. Many are more than 2 meters in diameter (a), while others are smaller (b).

Cnidarians

All eumetazoans form distinct embryonic layers. The **radially symmetrical** (that is, with body parts arranged around a central axis) eumetazoans have two embryonic layers, an outer **ectoderm,** which gives rise to the epidermis, and an inner **endoderm,** which gives rise to the gastrodermis. A jellylike layer called the *mesoglea* forms between the epidermis and gastrodermis. These layers give rise to the basic body plan, differentiating into the many tissues of the body. No such layers are present in sponges.

The most primitive eumetazoans to exhibit symmetry and tissues are two radially symmetrical phyla whose bodies are organized around an oral-aboral axis, like the petals of a daisy. The oral side of the animal contains the "mouth." Radial symmetry offers advantages to animals that either remain attached or closely associated to the surface or to animals that are free-floating. These animals don't pass through their environment, but rather they interact with their environment on all sides. These two phyla are Cnidaria (pronounced ni-DAH-ree-ah), which includes hydra (figure 19.3), jellyfish, corals, and sea anemones, and Ctenophora (pronounced tea-NO-fo-rah), a minor phylum that includes the comb jellies. These two phyla together are called the Radiata. The bodies of all other eumetazoans, the Bilateria, are marked by a fundamental bilateral symmetry (discussed in section 19.4). Even sea stars, which exhibit radial symmetry as adults, are bilaterally symmetrical when young.

A major evolutionary innovation among the radiates is the **extracellular digestion** of food. In sponges, food trapped by a choanocyte is taken directly into that cell, or into a circulating amoeboid cell, by endocytosis. In radiates, digestion begins *outside of cells,* in a gut cavity, called the *gastrovascular cavity* (the yellow areas in figure 19.4). After the food is broken down into smaller pieces, cells lining the gut cavity will complete digestion intracellularly. Extracellular digestion is the same heterotrophic strategy pursued by fungi, except that fungi digest food outside their bodies, while animals digest it within their bodies, in a cavity. This evolutionary advance has been retained by all of the more advanced groups of animals. For the first time it became possible to digest an animal larger than oneself.

Cnidarians (phylum Cnidaria) are carnivores that capture their prey, which include fishes and shellfish, with tentacles that ring their mouths. These tentacles, and sometimes the body surface, bear stinging cells called **cnidocytes,** which are unique to this group and give the phylum its name. Within each cnidocyte is a small but powerful harpoon called a **nematocyst,** which cnidarians use to spear their prey and then draw the harpooned prey back to the tentacle containing the cnidocyte. The cnidocyte builds up a very high internal osmotic pressure and uses it to push the nematocyst outward so explosively that the barb can penetrate the hard shell of a crab.

Cnidarians have two basic body forms, **medusae,** the floating form in figure 19.4, and **polyps,** the sessile form. Some cnidarians exist only as medusae, others only as polyps, and still others alternate between these two phases during

Figure 19.3 A cnidarian.
Hydroids are a group of cnidarians that are mostly marine and colonial. However, *Hydra,* shown here, is a freshwater genus whose members exist as solitary polyps.

Figure 19.4   Two basic body forms of cnidarians.

The medusa (*top*) and polyp (*bottom*) are the two phases that alternate in the life cycles of many cnidarians, but several species (corals and sea anemones, for example) exist only as polyps.

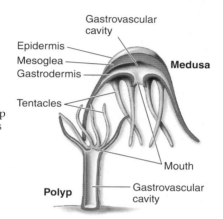

the course of their life cycles. Medusae are free-floating, gelatinous, and often umbrella-shaped forms that produce gametes. Their mouths point downward, with a ring of tentacles hanging down around the edges (hence the radial symmetry). Medusae are commonly called "jellyfish" because of their gelatinous interior or "stinging nettles" because of their nematocysts. Polyps are cylindrical, pipe-shaped animals that usually attach to a rock. They also exhibit radial symmetry. *Hydra* (shown in figure 19.3), sea anemones, and corals are examples of polyps. In polyps, the mouth faces away from the rock and therefore is often directed upward. Polyps are primarily sessile, and although *Hydra* are able to move around using a "cartwheel-type" motion, many polyps remain fixed and exposed to predators. For shelter and protection, corals deposit an external "skeleton" of calcium carbonate within which they live. This is the structure usually identified as coral.

> **19.3** Sponges have a multicellular body with specialized cells but lack definite symmetry and organized tissues. Cnidarians possess radial symmetry and specialized tissues and carry out extracellular digestion.

19.4 The Advent of Bilateral Symmetry

All eumetazoans other than cnidarians and ctenophores are **bilaterally symmetrical**—that is, they have a right half and a left half that are mirror images of each other. This is apparent when you compare the radially symmetrical sea anemone in figure 19.5 with the bilaterally symmetrical squirrel. Any of the three planes that cut the sea anemone in half produce mirror images, but only one plane, the green sagittal plane, produces mirror images of the squirrel. In looking at a bilaterally symmetrical animal, you refer to the top half of the animal as **dorsal** and the bottom half as **ventral.** The front is called **anterior** and the back **posterior.** Bilateral symmetry was a major evolutionary advance among the animals because it allows different parts of the body to become specialized in different ways. For example, most bilaterally symmetrical animals have evolved a definite head end, a process called **cephalization.** Animals that have heads are often active and mobile, moving through their environment headfirst, with sensory organs concentrated in front so the animal can test for food, danger, and mates, as it enters new surroundings.

The bilaterally symmetrical eumetazoans produce three embryonic layers that develop into the tissues of the body: an outer ectoderm (colored blue in the drawing of a flatworm in figure 19.6), an inner endoderm (colored yellow), and a third layer, the **mesoderm** (colored red), that lies between the ectoderm and endoderm. In general, the outer coverings of the body and the nervous system develop from the ectoderm, the digestive organs and intestines develop from the endoderm, and the skeleton and muscles develop from the mesoderm.

The simplest of all bilaterally symmetrical animals are the **solid worms.** By far the largest phylum of these, with about 20,000 species, is Platyhelminthes (pronounced plat-ee-hel-MIN-theeze), which includes the flatworms. Flatworms are the simplest animals in which organs occur. An organ is a collection of different tissues that function as a unit. The testes and uterus of flatworms are reproductive organs, for example. The dark spots on the head are eyespots that can detect light, although they cannot focus an image like your eyes can.

Solid worms lack any internal cavity other than the digestive tract. Flatworms are soft-bodied animals flattened from top to bottom, like a piece of tape or ribbon. If you were to cut a flatworm in half across its body, you would see that the gut is completely surrounded by tissues and organs. This solid body construction is called **acoelomate,** meaning without a body cavity.

(a)

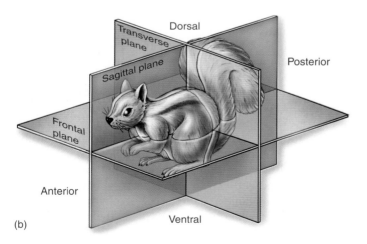

(b)

Figure 19.5   How radial and bilateral symmetry differ.

(a) Radial symmetry is the regular arrangement of parts around a central axis. (b) Bilateral symmetry is reflected in a body form that has a left and right half.

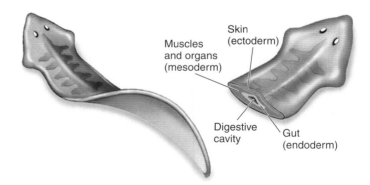

Figure 19.6   Body plan of a solid worm.

All bilaterally symmetrical eumetazoans produce three layers during embryonic development: an outer ectoderm, a middle mesoderm, and an inner endoderm. These layers differentiate to form the skin, muscles and organs, and gut, respectively, in the adult animal.

Flatworms

Although flatworms have a simple body design, they do have a definite head at the anterior end and they do possess organs. Flatworms range in size from a millimeter or less to many meters long, as in some tapeworms. Most species of flatworms are parasitic, occurring within the bodies of many other kinds of animals. Other flatworms are free-living, occurring in a wide variety of marine and freshwater habitats, as well as moist places on land (figure 19.7). Free-living flatworms are carnivores and scavengers; they eat various small animals and bits of organic debris. They move from place to place by means of ciliated epithelial cells concentrated on their ventral surfaces.

There are two classes of parasitic flatworms, which live within the bodies of other animals: flukes and tapeworms. Both groups of worms have epithelial layers resistant to the digestive enzymes and immune defenses produced by their hosts—an important feature in their parasitic way of life. Some parasitic flatworms require only one host, but many flukes require two or more hosts to complete their life cycles. The parasitic lifestyle has resulted in the eventual loss of features not used or needed by the parasite. Parasitic flatworms lack certain features of the free-living flatworms, such as cilia in the adult stage, eyespots, and other sensory organs that lack adaptive significance for an organism that lives within the body of another animal, a loss sometimes dubbed "degenerative evolution."

Characteristics of Flatworms

Those flatworms that have a digestive cavity have an incomplete gut, one with only one opening. As a result, they cannot eat, digest, and eliminate undigested particles of food simultaneously. Thus, flatworms cannot feed continuously, as more advanced animals can. The gut is branched and extends throughout the body (the gut is the green structure in the *Planaria* in figure 19.6), functioning in both digestion and transport of food. Cells that line the gut engulf most of the food particles by phagocytosis and digest them; but, as in the cnidarians, some of these particles are partly digested extracellularly. Tapeworms, which are parasitic flatworms, lack digestive systems. They absorb their food directly through their body walls.

Unlike cnidarians, flatworms have an excretory system, which consists of a network of fine tubules (little tubes) that runs throughout the body. Cilia line the hollow centers of bulblike **flame cells,** which are located on the side branches of the tubules. Cilia in the flame cells move water and excretory substances into the tubules and then to exit pores located between the epidermal cells. Flame cells were named because of the flickering movements of the tuft of cilia within them. They primarily regulate the water balance of the organism. The excretory function of flame cells appears to be a secondary one. A large proportion of the metabolic wastes excreted by flatworms probably diffuses directly into the gut and is eliminated through the mouth.

Like sponges, cnidarians, and ctenophorans, flatworms lack a **circulatory system,** a network of vessels that carries fluids, oxygen, and food molecules to parts of the body. Consequently, all

(a)   (b)

Figure 19.7   Flatworms.
(a) A common flatworm, *Planaria.* (b) A marine free-living flatworm.

flatworm cells must be within diffusion distance of oxygen and food. Flatworms have thin bodies and highly branched digestive cavities that make such a relationship possible.

The nervous system of flatworms is very simple. Some primitive flatworms have only a loosely organized nerve net. However, most members of this phylum have longitudinal nerve cords that constitute a simple central nervous system. Between the longitudinal cords are cross connections, so that the flatworm nervous system resembles a ladder.

Free-living flatworms use sensory pits or tentacles along the sides of their heads to detect food, chemicals, or movements of the fluid in which they are living. Free-living members of this phylum also have eyespots on their heads. These are inverted, pigmented cups containing light-sensitive cells connected to the nervous system. These eyespots enable the worms to distinguish light from dark. Flatworms are far more active than cnidarians or ctenophores. Such activity is characteristic of bilaterally symmetrical animals. In flatworms, this activity seems to be related to the greater concentration of sensory organs and, to some degree, the nervous system elements in the heads of these animals.

The reproductive systems of flatworms are complex. Most flatworms are **hermaphroditic,** with each individual containing both male and female sexual structures. Some genera of flatworms are also capable of asexual regeneration; when a single individual is divided into two or more parts, each part can regenerate an entirely new flatworm.

> **19.4   Flatworms have internal organs, bilateral symmetry, and a distinct head. They do not have a body cavity.**

The Advent of a Body Cavity

A key transition in the evolution of the animal body plan was the evolution of the body cavity. All bilaterally symmetrical animals other than solid worms have a cavity within their body. The evolution of an internal body cavity was an important improvement in animal body design for three reasons:

1. **Circulation.** Fluids that move within the body cavity can serve as a circulatory system, permitting the rapid passage of materials from one part of the body to another and opening the way to larger bodies.
2. **Movement.** Fluid in the cavity makes the animal's body rigid, permitting resistance to muscle contraction and thus opening the way to muscle-driven body movement.
3. **Organ function.** In a fluid-filled enclosure, body organs can function without being deformed by surrounding muscles. For example, food can pass freely through a gut suspended within a cavity, at a rate not controlled by when the animal moves.

Kinds of Body Cavities

There are three basic kinds of body plans found in bilaterally symmetrical animals. Acoelomates, such as solid worms that we discussed in the previous section and that are shown at the top of figure 19.8, have no body cavity. **Pseudocoelomates,** shown in the middle of the figure, have a body cavity called the **pseudocoel** located between the mesoderm (red layer) and endoderm (yellow layer). A third way of organizing the body is one in which the fluid-filled body cavity develops not between endoderm and mesoderm but rather entirely within the mesoderm. Such a body cavity is called a **coelom** (the two arch-shaped cavities in the worm at the bottom of the figure), and animals that possess such a cavity are called **coelomates.** In coelomates, the gut is suspended, along with other organ systems of the animal, within the coelom; the coelom, in turn, is surrounded by a layer of epithelial cells entirely derived from the mesoderm.

The development of a body cavity poses a problem—circulation—solved in pseudocoelomates by churning the fluid within the body cavity. In coelomates, the gut is again surrounded by tissue that presents a barrier to diffusion, just as it was in solid worms. This problem is solved among coelomates by the development of a circulatory system. The circulating fluid, or blood, carries nutrients and oxygen to the tissues and removes wastes and carbon dioxide. Blood is usually pushed through the circulatory system by contraction of one or more muscular hearts. In an open circulatory system, the blood passes from vessels into sinuses, mixes with body fluid, and then reenters the vessels later in another location. In a closed circulatory system, the blood is separate from the body fluid and can be separately controlled. Also, blood moves through a closed circulatory system faster and more efficiently than it does through an open system.

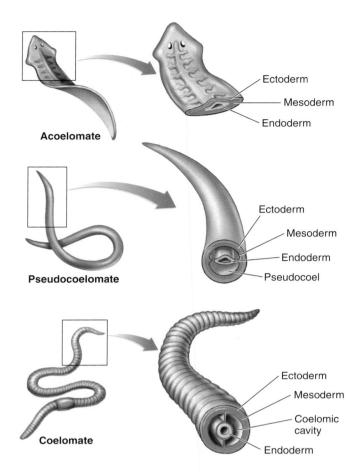

Figure 19.8   Three body plans for bilaterally symmetrical animals.

Acoelomates, such as flatworms, have no body cavity between the digestive tract (endoderm) and the outer body layer (ectoderm). Pseudocoelomates have a body cavity, the pseudocoel, between the endoderm and the mesoderm. Coelomates have a body cavity, the coelom, that develops entirely within the mesoderm, and so is lined on both sides by mesoderm tissue.

The evolutionary relationship among coelomates, pseudocoelomates, and acoelomates is not clear. Acoelomates, for example, could have given rise to coelomates, but scientists also cannot rule out the possibility that acoelomates were derived from coelomates. The different phyla of pseudocoelomates form two groups that do not appear to be closely related.

Pseudocoelomates

As we have noted, all bilaterally symmetrical animals except solid worms possess an internal body cavity. Among them, seven phyla are characterized by their possession of a pseudocoel. In all pseudocoelomates, the pseudocoel serves as a hydrostatic skeleton—one that gains its rigidity from being filled with fluid under pressure. The animal's muscles can work against this "skeleton," thus making the movements of pseudocoelomates far more efficient than those of the acoelomates.

Only one of the seven pseudocoelomate phyla includes a large number of species. This phylum, Nematoda, includes some 20,000 recognized species of **nematodes,** eelworms, and other roundworms. Scientists estimate that the actual number might approach 100 times that many. Members of this phylum are found everywhere. Nematodes are abundant and diverse in marine and freshwater habitats, and many members of this phylum are parasites of animals and plants, like the intestinal roundworm in figure 19.9a. Many nematodes are microscopic and live in soil. It has been estimated that a spadeful of fertile soil may contain, on the average, a million nematodes.

A second phylum consisting of animals with a pseudocoelomate body plan is Rotifera, the rotifers. **Rotifers** are common, small, aquatic animals that have a crown of cilia at their heads, which can just barely be seen in figure 19.9b; they range from 0.04 to 2 millimeters long. About 2,000 species exist throughout the world. Bilaterally symmetrical and covered with chitin, rotifers depend on their cilia for both locomotion and feeding, ingesting bacteria, protists, and small animals.

All pseudocoelomates lack a defined circulatory system; this role is performed by the fluids that move within the pseudocoel. Most pseudocoelomates have a complete, one-way digestive tract that acts like an assembly line. Food is broken down, then absorbed, and then treated and stored.

Phylum Nematoda: The Roundworms

Nematodes are bilaterally symmetrical, cylindrical, unsegmented worms. They are covered by a flexible, thick cuticle, which is molted as they grow. Their muscles constitute a layer beneath the epidermis and extend along the length of the worm, rather than encircling its body. These longitudinal muscles pull both against the cuticle and the pseudocoel, which forms a hydrostatic skeleton. When nematodes move, their bodies whip about from side to side.

Near the mouth of a nematode, at its anterior end, are usually 16 raised, hairlike sensory organs. The mouth is often equipped with piercing organs called **stylets.** Food passes through the mouth as a result of the sucking action of a muscular chamber called the **pharynx.** After passing through a short corridor into the pharynx, food continues through the other portions of the digestive tract, where it is broken down and then digested. Some of the water with which the food has been mixed is reabsorbed near the end of the digestive tract, and waste material is eliminated through the anus.

Nematodes completely lack flagella or cilia, even on sperm cells. Reproduction in nematodes is sexual, with sexes usually separate. Their development is simple, and the adults consist of very few cells. For this reason, nematodes have become extremely important subjects for genetic and developmental studies. The 1-millimeter-long *Caenorhabditis elegans* matures in only three days, its body is transparent, and it has only 959 cells. It is the only animal whose complete developmental cellular anatomy is known, and the first animal whose genome (97 million DNA bases encoding over 21,000 different genes) was fully sequenced.

About 50 species of nematodes, including several that are rather common in the United States, regularly parasitize

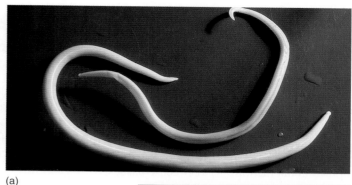

(a)

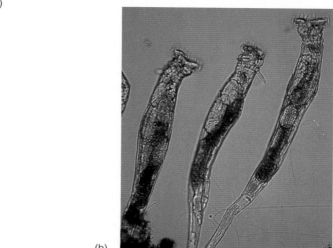

(b)

Figure 19.9 Pseudocoelomates.

(a) These nematodes (phylum Nematoda) are intestinal roundworms that infect humans and some other animals. Their fertilized eggs pass out with feces and can remain viable in soil for years. (b) Rotifers (phylum Rotifera) are common aquatic animals that depend on their crown of cilia for feeding and locomotion.

human beings. Trichinosis, a nematode-caused disease in temperate regions, is caused by worms of the genus *Trichinella*. These worms live in the small intestine of pigs, where fertilized female worms burrow into the intestinal wall. Once it has penetrated these tissues, each female produces about 1,500 live young. The young enter the lymph channels and travel to muscle tissue throughout the body, where they mature and form cysts. Infection in human beings or other animals arises from eating undercooked or raw pork in which the cysts of *Trichinella* are present. If the worms are abundant, a fatal infection can result, but such infections are rare.

A more prevalent human parasitic nematode is *Ascaris lumbricoides*. This intestinal worm infects approximately one of six people worldwide but is rare in areas with modern plumbing. These worms live in the intestines and spread their fertilized eggs in feces, which can remain viable in the soil for years. Adult females, which are up to 30 centimeters long, contain up to 30 million eggs, and can lay up to 20,000 of them each day.

Coelomates

Even though acoelomates and pseudocoelomates have proven very successful, the bulk of the animal kingdom consists of coelomates. A major advantage of the coelomate body plan is that it allows contact between mesoderm and endoderm during development, permitting localized portions of the digestive tract to develop into complex, highly specialized regions like the stomach. In pseudocoelomates, mesoderm and endoderm are separated by the body cavity, limiting developmental interactions between these tissues.

Mollusks

The only major phylum of coelomates without segmented bodies are the Mollusca. The **mollusks** are the second largest animal phylum, behind the arthropods, with over 110,000 species. Mollusks are mostly marine, but occur almost everywhere.

Mollusks include three general groups with outwardly different body plans. However, the seeming differences hide a basically similar body design. The body of mollusks is composed of three distinct parts: a head-foot, a central section called the visceral mass that contains the body's organs, and a mantle. The foot of a mollusk is muscular and may be adapted for locomotion, attachment, food capture (in squids and octopuses), or various combinations of these functions. The **mantle** is a heavy fold of tissue wrapped around the visceral mass like a cape, with the gills positioned on its inner surface like the lining of a coat. The **gills** are filamentous projections of tissue, rich in blood vessels, that capture oxygen from the water circulating between the mantle and visceral mass and release carbon dioxide.

The three major groups of mollusks, all different variations upon this same basic design, are gastropods, bivalves, and cephalopods.

1. **Gastropods** (snails, like the one shown in figure 19.10*a,* and slugs) use the muscular foot to crawl, and their mantle often secretes a single, hard protective shell. All terrestrial mollusks are gastropods.
2. **Bivalves** (clams, oysters, and scallops, like the one in figure 19.10*b*) secrete a two-part shell with a hinge, as their name implies. They filter-feed by drawing water into their shell.
3. **Cephalopods** (octopuses, like the one shown in figure 19.10*c,* and squids) have modified the mantle cavity to create a jet propulsion system that can propel them rapidly through the water. In most groups, the shell is greatly reduced to an internal structure or is absent.

One of the most characteristic features of mollusks is the **radula,** a rasping, tonguelike organ. With rows of pointed, backward-curving teeth, the radula is used by some snails to scrape algae off rocks. The small holes often seen in oyster shells are produced by gastropods that have bored holes to kill the oyster and extract its body.

(a)

(b)

(c)

Figure 19.10   Three major groups of mollusks.
(a) A gastropod. (b) A bivalve. (c) A cephalopod.

In most mollusks, as stated earlier, the outer surface of the mantle also secretes a protective shell. The shell consists of a horny outer layer, rich in protein, which protects the two underlying calcium-rich layers from erosion. The inner layer is pearly and is used as mother-of-pearl. Pearls themselves are formed when a foreign object, such as a grain of sand, becomes lodged between the mantle and the inner shell layer of a bivalve, including clams and oysters. The mantle coats the foreign object with layer upon layer of shell material to reduce irritation. The shell serves primarily for protection with some mollusks withdrawing into their shell when threatened.

Annelids

One of the early key innovations in body plan to arise among the coelomates was **segmentation,** the building of a body from a series of similar segments. The first segmented animals to evolve were the **annelid worms,** phylum Annelida. These advanced coelomates are assembled as a chain of nearly identical segments, like the boxcars of a train. The great advantage of such segmentation is the evolutionary flexibility it offers—a small change in an existing segment can produce a new kind of segment with a different function. Thus, some segments are modified for reproduction, some for feeding, and others for eliminating wastes.

Two-thirds of all annelids live in the sea (about 8,000 species including the colorful red shiny bristle worm in figure 19.11*b*) but some live in freshwater, and most of the rest—some 3,100 species—are earthworms (shown emerging from underground in figure 19.11*a*). The basic body plan of an annelid is a tube within a tube: the digestive tract is suspended within the coelom, which is itself a tube running from mouth to anus. There are three characteristics of this organization:

1. **Repeated segments.** The body segments of an annelid are visible as a series of ringlike structures running the length of the body, looking like a stack of doughnuts (which can be seen externally in figure 19.11). The segments are divided internally from one another by partitions, just as walls separate the rooms of a building. In each of the cylindrical segments, the excretory and locomotor organs are repeated. The body fluid within the coelom of each segment creates a hydrostatic (liquid-supported) skeleton that gives the segment rigidity, like an inflated balloon. Muscles within each segment pull against the fluid in the coelom. Because each segment is separate, each is able to expand or contract independently. This lets the worm's body move in ways that are quite complex. When an earthworm crawls on a flat surface, for example, it lengthens some parts of its body while shortening others.

2. **Specialized segments.** The anterior (front) segments of annelids contain the sensory organs of the worm. Elaborate eyes with lenses and retinas have evolved in some annelids. One anterior segment contains a well-developed cerebral ganglion, or brain.

3. **Connections.** Because partitions separate the segments, it is necessary to provide ways for materials and information to pass between segments. A circulatory system carries blood from one segment to another, while nerve cords connect the nerve centers located in each segment with each other and the brain. The brain can then coordinate the worm's activities.

Segmentation underlies the body organization of all complex coelomate animals, not only annelids but also arthropods (crustaceans, spiders, and insects) and chordates (mostly vertebrates). For example, vertebrate muscles develop from repeated blocks of tissue called somites that occur in the embryo. Vertebrate segmentation is also seen in the vertebral column, which is a stack of very similar vertebrae.

(a) (b)

Figure 19.11 Representative annelids.

(a) Earthworms are the terrestrial annelids. This night crawler, *Lumbricus terrestris,* is in its burrow. (b) Shiny bristle worm, *Oenone fulgida,* is an aquatic annelid, a polychaete.

Arthropods

The evolution of segmentation among the annelids marked a major innovation in body structure among the coelomates. An even more profound innovation marks the origin of the body plan characteristic of the most successful of all animal groups, the **arthropods,** phylum Arthropoda. This innovation was the development of jointed appendages. Jointed appendages lead to the evolution of insects, which make up nearly 80% of all arthropod species (represented by five of the eight areas in the pie chart in figure 19.12). Jointed appendages are also present in crustaceans (represented by the orange area), spiders (represented by the white area), and other types of arthropods (represented by the darker blue area).

Jointed Appendages. The name *arthropod* comes from two Greek words, *arthros,* jointed, and *podes,* feet. All arthropods have jointed appendages (figure 19.13). Some are legs, and others may be modified for other uses. To gain some idea of the importance of jointed appendages, imagine yourself without them—no hips, knees, ankles, shoulders, elbows, wrists, or knuckles. Without jointed appendages, you could not walk or grasp an object. Arthropods use jointed appendages as legs and wings for moving, as antennae to sense their environment, and as mouthparts for sucking, ripping, and chewing prey. A scorpion, for example, seizes and tears apart its prey with mouthpart appendages modified as large pincers.

Rigid Exoskeleton. The arthropod body plan has a second great innovation: arthropods have a rigid external skeleton, or **exoskeleton,** made of chitin. In any animal, a key function of the skeleton is to provide places for muscle attachment, and in arthropods the muscles attach to the interior surface of the hard chitin shell, which also protects the animal from predators and impedes water loss.

However, while chitin is hard and tough, it is also brittle and cannot support great weight. As a result, the exoskeleton must be much thicker to bear the pull of the muscles in large insects than in small ones, so there is a limit to how big an arthropod body can be. That is why you don't see beetles as big as birds or crabs the size of a cow—the exoskeleton would be so thick the animal couldn't move its great weight. In fact, the great majority of arthropod species consist of small animals—mostly about a millimeter in length—but members of the phylum range in adult size from about 80 micrometers long (some parasitic mites) to 3.6 meters across (a gigantic crab found in the sea off Japan). Some lobsters are nearly a meter in length. The largest living insects are about 33 centimeters long, but the giant dragonflies that lived 300 million years ago had wingspans of as much as 60 centimeters (2 feet)!

Because this size limitation is inherent in the body design of arthropods, no arthropods have ever grown to great size. As we will see, to overcome this limitation, a strong, flexible endoskeleton is required.

Arthropods have proven very successful. About two-thirds of all named species on earth are arthropods. Among the major kinds are arachnids (including spiders, ticks, mites, scorpions, and daddy longlegs), crustaceans (including lobsters, crabs, shrimp, and barnacles), centipedes and millipedes,

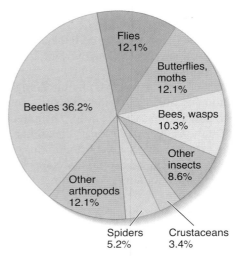

Figure 19.12 Arthropods are a successful group.
About two-thirds of all named species are arthropods. About 80% of all arthropods are insects, and about half of the named species of insects are beetles.

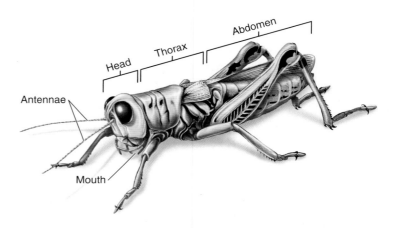

Figure 19.13 The body plan of an insect.
This grasshopper illustrates the body segmentation found in adult insects. The many segments present in most larval stages of insects become fused in the adult, giving rise to three adult body segments: the head, thorax, and abdomen. The appendages—legs, wings, mouthparts, antennae—are jointed.

and insects (figure 19.14). Scientists estimate that a quintillion (a billion billion) insects are alive at any one time—200 million insects for each living human!

> **19.5** Roundworms have a pseudocoel body cavity, while mollusks have a coelom body cavity; neither is segmented. Annelids are mostly marine segmented worms. Arthropods, the most successful animal phylum, have jointed appendages, a rigid exoskeleton, and, in the case of insects, wings.

Figure 19.14 Insect diversity.

(a) Some insects have a tough exoskeleton, like this South American scarab beetle, *Dilobderus abderus* (order Coleoptera). (b) Human flea, *Pulex irritans* (order Siphonaptera). Fleas are flattened laterally, slipping easily through hair. (c) The honeybee, *Apis mellifera* (order Hymenoptera), is a widely domesticated and efficient pollinator of flowering plants. (d) This pot dragonfly (order Odonata) has a fragile exoskeleton. (e) A true bug, *Edessa rufomarginata* (order Hemiptera), in Panama. (f) Copulating grasshoppers (order Orthoptera). (g) Luna moth, *Actias luna,* in Virginia. Luna moths and their relatives are among the most spectacular insects (order Lepidoptera).

19.6 Redesigning the Embryo

There are two major kinds of coelomate animals representing two distinct evolutionary lines. All the coelomates we have met so far have essentially the same kind of embryonic development. Cell divisions of the fertilized egg produce a hollow ball of cells, a blastula, which indents to form a two-layer-thick ball with a blastopore opening to the outside. In mollusks, annelids, and arthropods, the mouth (stoma) develops from or near the blastopore. This same pattern of development, in a general sense, is seen in all noncoelomate animals. An animal whose mouth develops in this way is called a **protostome** (figure 19.15, *top*). If such an animal has a distinct anus or anal pore, it develops later in another region of the embryo.

A second distinct pattern of embryological development occurs in the echinoderm, and chordates. In these animals, the anus forms from or near the blastopore, and the mouth forms subsequently on another part of the blastula. This group of phyla consists of animals that are called the **deuterostomes** (figure 19.15, *bottom*).

Deuterostomes represent a revolution in embryonic development. In addition to the pattern of blastopore formation, deuterostomes differ from protostomes in three other fundamental embryological features:

1. The progressive division of cells during embryonic growth is called *cleavage*. The cleavage pattern relative to the embryo's polar axis determines how the cells array. In nearly all protostomes, each new cell buds off at an angle oblique to the polar axis. As a result, a new cell nestles into the space between the older ones in a closely packed array (see the 16-cell stage in the upper row of cells). This pattern is called **spiral cleavage** because the orientation of the dividing cells spirals around the polar axis (indicated by the curving blue arrow in the 32-cell stage).

In deuterostomes, the cells divide parallel to and at right angles to the polar axis. As a result, the pairs of cells from each division are positioned directly above and below one another (see the 16-cell stage in the lower row of cells); this process gives rise to a loosely packed array of cells. This pattern is called **radial cleavage** because the orientation of the dividing cells is in a radius outward from the polar axis (indicated by the straight blue arrow in the 32-cell stage).

2. In protostomes, the developmental fate of each cell in the embryo is fixed when that cell first appears. Even at the four-celled stage, each cell is different, and no one cell, if separated from the others, can develop into a complete animal because the chemicals that act as developmental signals have already been localized in different parts of the egg. In deuterostomes, on the other hand, the first cleavage divisions of the fertilized embryo produce identical daughter cells, and any single cell, if separated, can develop into a complete organism.

3. In all coelomates, the coelom originates from mesoderm. In protostomes, this occurs simply and directly: the red mesoderm cells in the upper row simply move away from one another as the coelomic cavity expands within the mesoderm. However, in deuterostomes, whole groups of cells usually move around to form new tissue associations. The coelom is normally produced by an evagination of the **archenteron**—the main cavity within the gastrula, also called the primitive gut. This cavity, lined with endoderm, opens to the outside via the blastopore and eventually becomes the gut cavity. The evaginating cells give rise to the red mesodermal cells in the lower row of cells. The mesoderm expands to form the coelom.

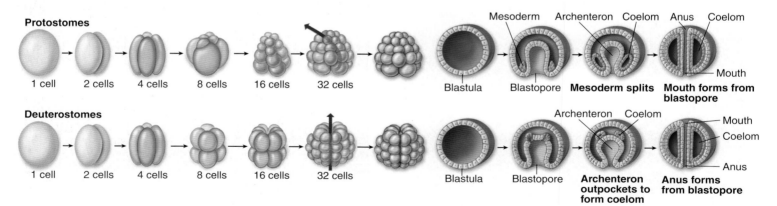

Figure 19.15 Embryonic development in protostomes and deuterostomes.
Cleavage of the egg produces a hollow ball of cells called the blastula. Invagination, or infolding, of the blastula produces the blastopore. In protostomes, embryonic cells cleave in a spiral pattern and become tightly packed. The blastopore becomes the animal's mouth, and the coelom originates from a mesodermal split. In deuterostomes, embryonic cells cleave radially and form a loosely packed array. The blastopore becomes the animal's anus, and the mouth develops at the other end. The coelom originates from an evagination, or outpouching, of the archenteron in deuterostomes.

Echinoderms

The first deuterostomes, marine animals called **echinoderms** in the phylum Echinodermata, appeared more than 650 million years ago. The term *echinoderm* means "spiny skin" and refers to an **endoskeleton** composed of hard, calcium-rich plates called ossicles just beneath the delicate skin. When they are first formed, the plates are enclosed in living tissue, and so are truly an endoskeleton, although in adults they may fuse, forming a hard shell. About 6,000 species of echinoderms are living today, almost all of them on the ocean bottom. Many of the most familiar animals seen along the seashore are echinoderms (figure 19.16), including sea cucumbers, feather stars and brittle stars, sand dollars, sea urchins, and sea stars (starfish).

The body plan of echinoderms undergoes a fundamental shift during development: all echinoderms are bilaterally symmetrical as larvae but become radially symmetrical as adults. Many biologists believe that early echinoderms were sessile and evolved adult radiality as an adaptation to the sessile existence. Bilaterality is of adaptive value to an animal that travels through its environment, whereas radiality is of value to an animal whose environment meets it on all sides. Adult echinoderms have a five-part body plan, easily seen in the five arms of a sea star. Its nervous system consists of a central ring of nerves from which five branches arise—while the animal is capable of complex response patterns, there is no centralization of function, no "brain." Some echinoderms like feather stars have 10 or 15 arms, but always multiples of five.

A key evolutionary innovation of echinoderms is the development of a hydraulic system to aid movement. Called a **water vascular system,** this fluid-filled system is composed of a central ring canal from which five radial canals extend out into the arms. From each radial canal, tiny vessels extend through short side branches into thousands of tiny, hollow **tube feet.** At the base of each tube foot is a fluid-filled muscular sac that acts as a valve. When a sac contracts, its fluid is prevented from reentering the radial canal and instead is forced into the tube foot, thus extending it. When extended, the tube foot attaches itself to the ocean bottom, often aided by suckers. The sea star can then pull against these tube feet and so haul itself over the seafloor.

Most echinoderms reproduce sexually, but they have the ability to regenerate lost parts, which can lead to asexual reproduction. In a few sea stars, asexual reproduction takes place by splitting, and the broken parts of the sea star can sometimes regenerate whole animals.

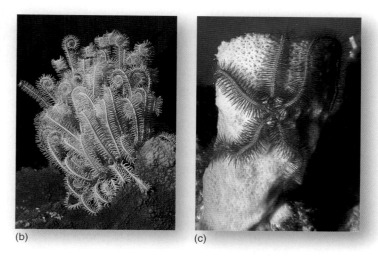

Figure 19.16 Diversity in echinoderms.

(a) Warty sea cucumber, *Parastichopus parvimensis.* (b) Feather star (class Crinoidea) on the Great Barrier Reef in Australia. (c) Brittle star, *Ophiothrix* (class Ophiuroidea). (d) Sand dollar, *Echinarachnius parma.* (e) Giant red sea urchin, *Strongylocentrotus franciscanus.* (f) Sea star, *Oreaster occidentalis* (class Asteroidea), in the Gulf of California, Mexico.

Chordates

Chordates (phylum Chordata) are deuterostome coelomates whose nearest relations in the animal kingdom are the echinoderms, also deuterostomes. Chordates exhibit great improvements in the endoskeleton over what is seen in echinoderms. The endoskeleton of echinoderms is functionally similar to the exoskeleton of arthropods, in that it is a hard shell that encases the body, with muscles attached to its inner surface. Chordates employ a very different kind of endoskeleton, one that is truly internal. Members of the phylum Chordata are characterized by a flexible rod called a **notochord** that develops along the back of the embryo. Muscles attached to this rod allowed early chordates to swing their bodies back and forth, swimming through the water. This key evolutionary innovation, attaching muscles to an internal element, started chordates along an evolutionary path that leads to the vertebrates and for the first time to truly large animals.

The approximately 56,000 species of chordates are distinguished by four principal features:

1. **Notochord.** A stiff, but flexible, rod that forms beneath the nerve cord in the early embryo.
2. **Nerve cord.** A single dorsal (along the back) hollow nerve cord, to which the nerves that reach the different parts of the body are attached.
3. **Pharyngeal pouches.** A series of pouches behind the mouth that develop into slits in some animals. The slits open into the pharynx, which is a muscular tube that connects the mouth to the digestive tract and windpipe.
4. **Postanal tail.** Chordates have a postanal tail, a tail that extends beyond the anus. A postanal tail is present at least during their embryonic development if not in the adult. Nearly all other animals have a terminal anus.

All chordates have all four of these characteristics at some time in their lives. For example, the tunicate in figure 19.17 looks more like a sponge than a chordate, but its larval stage, which resembles a tadpole, has all four features listed above. Human embryos have pharyngeal pouches, a nerve cord, a notochord, and a postanal tail as embryos. The nerve cord remains in the adult, differentiating into the brain and spinal cord. The pharyngeal pouches and postanal tail disappear during human development, and the notochord is replaced with the vertebral column.

Vertebrates

In their body plan, chordates are segmented, and distinct blocks of muscles can be seen clearly in many forms (figures 19.18 and 19.19). With the exception of tunicates (figure 19.17) and lancelets (figure 19.18), all chordates are **vertebrates.** Vertebrates differ from tunicates and lancelets in two important respects:

1. **Backbone.** The notochord becomes surrounded and then replaced during the course of the embryo's

Figure 19.17   A tunicate.
This beautiful blue and gold tunicate is one type of nonvertebrate chordate.

Figure 19.18   Lancelets.
Two lancelets, *Branchiostoma lanceolatum*, partly buried in shell gravel, with their anterior ends protruding. The muscle segments are clearly visible in this photograph. Lancelets are a second type of nonvertebrate chordate.

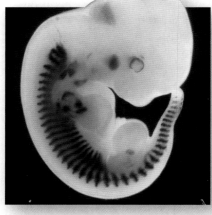

Figure 19.19 A mouse embryo.
At 11.5 days of development, the muscle is already divided into segments called somites (stained *dark* in this photo), reflecting the fundamentally segmented nature of all chordates.

development by a bony vertebral column, a tube of hollow bones called *vertebrae* that encloses the dorsal nerve cord like a sleeve and protects it.
2. **Head.** All vertebrates except the earliest fishes have a distinct and well-differentiated head, with a skull and brain.

All vertebrates have an internal skeleton made of bone or cartilage against which the muscles work. This endoskeleton makes possible the great size and extraordinary powers of movement that characterize the vertebrates.

> **19.6   Echinoderms and chordates are deuterostomes. The deuterostome development differs from protosome development in that deuterostome eggs cleave radially, and the blastopore of their embryo becomes the animal's anus. Echinoderms have an endoskeleton of hard plates, often fused together. Adults are radially symmetrical. Chordates have a notochord at some stage of their development. In adult vertebrates, the notochord is replaced by a backbone.**

Overview of Vertebrate Evolution

When scientists first began to study and date fossils, they had to find some way to organize the different time periods from which the fossils came. They divided the earth's past into large blocks of time called **eras.** Eras are further subdivided into smaller blocks of time called **periods,** and some periods, in turn, are subdivided into **epochs,** which can be divided into **ages.**

Virtually all of the major groups of animals that survive at the present time originated in the sea at the beginning of the **Paleozoic era,** during or soon after the Cambrian period (545–490 M.Y.A.). Thus, the diversification of animal life on earth is basically a marine record, and the fossils from the Paleozoic era all originated in the sea. Many of the animal phyla that appeared in the Cambrian period have no surviving close relatives.

The first vertebrates evolved about 500 million years ago in the oceans—fishes without jaws. They didn't have paired fins either—many of them looked something like a flat hot dog with a hole at one end and a fin at the other. For over 100 million years, a parade of different kinds of fishes were the only vertebrates on earth. They became the dominant creatures in the sea, some as large as 10 meters, larger than most cars.

Invasion of the Land

Only a few of the animal phyla that evolved in the Cambrian seas have invaded the land successfully; most others have remained exclusively marine. The first organisms to colonize the land were fungi and plants, over 500 million years ago. The ancestors of plants were specialized members of a group of photosynthetic protists known as the green algae. It seems probable that plants first occupied the land in symbiotic association with fungi, as discussed in chapter 18.

The first invasion of the land by animals, and perhaps the most successful invasion of the land, was accomplished by the arthropods, a phylum of hard-shelled animals with jointed legs and a segmented body. This invasion occurred about 410 million years ago.

Vertebrates invaded the land during the Carboniferous period (360–280 M.Y.A.). The first vertebrates to live on land were the amphibians, represented today by frogs, toads, salamanders, and caecilians (legless amphibians). The earliest amphibians known are from the Devonian period, and among their descendants are the reptiles, which became the ancestors of the dinosaurs, birds, and mammals.

Mass Extinctions

The history of life on earth has been marked by periodic episodes of extinction, where the loss of species outpaces the formation of new species. Particularly sharp declines in species diversity are called **mass extinctions.** Five mass extinctions have occurred, the first of them near the end of the Ordovician period about 438 million years ago. At that time, most of the existing families of trilobites, a very common type of marine arthropod, became extinct. Another mass extinction occurred about 360 million years ago at the end of the Devonian period.

The third and most drastic mass extinction in the history of life on earth happened during the last 10 million years of the Permian period, marking the end of the Paleozoic era. It is estimated that 96% of all species of marine animals that were living at that time became extinct! All of the trilobites disappeared forever. Brachiopods, marine animals resembling mollusks but with a different filter-feeding system, were extremely diverse and widespread during the Permian; only a few species survived. Bryozoans, marine filter-feeders that formed coral-like colonies in oceans throughout the world in the Permian, became rare afterward.

Mass extinctions left vacant many ecological opportunities, and for this reason they were followed by rapid evolution among the relatively few plants, animals, and other organisms that survived the extinction. Little is known about the causes of major extinctions. In the case of the Permian mass extinction, some scientists argue that the extinction was brought on by a gradual accumulation of carbon dioxide in ocean waters, the result of large-scale volcanism due to the collision of the earth's landmasses during formation of the single large "super-continent" of Pangaea. Such an increase in carbon dioxide would have severely disrupted the ability of animals to carry out metabolism and form their shells.

The most famous and well-studied extinction, though not as drastic, occurred at the end of the Cretaceous period (65 million years ago), at which time the dinosaurs and a variety of other organisms went extinct. Recent findings have supported the hypothesis that this fifth mass extinction event was triggered when a large asteroid slammed into the earth, perhaps causing global forest fires and obscuring the sun for months by throwing particles into the air.

We are living during a new sixth mass extinction event. The number of species in the world is greater today than it has ever been. Unfortunately, that number is decreasing at an alarming rate due to human activity. Some estimate that as many as one-fourth of all species will become extinct in the near future, a rate of extinction not seen on earth since the Cretaceous mass extinction.

19.7 The diversification of the animal phyla occurred in the sea. Only two animal phyla have invaded the land successfully: arthropods and chordates (the vertebrates).

Fishes Dominate the Sea

A series of key evolutionary advances allowed vertebrates to first conquer the sea and then the land. Figure 19.20 shows a phylogeny of the vertebrates. Branch points in the family tree indicate key adaptations that lead to great diversity. About half of all vertebrates are **fishes.** The most diverse and successful vertebrate group, they provided the evolutionary base for invasion of land by amphibians.

Characteristics of Fishes

From whale sharks that are 12 meters long to tiny cichlids no larger than your fingernail, fishes vary considerably in size, shape, color, and appearance. However varied, all fishes have four important characteristics in common:

1. **Gills.** Fish are water-dwelling creatures, and they use gills to extract dissolved oxygen gas from the water around them. Gills are fine filaments of tissue rich in blood vessels. When water passes over the gills in the back of the mouth, oxygen gas diffuses from the water into the fish's blood.
2. **Vertebral column.** All fishes have an internal skeleton with a spine surrounding the dorsal nerve cord, although it may not necessarily be made of bone. The brain is fully encased within a protective box, the skull or cranium, made of bone or cartilage.
3. **Single-loop blood circulation.** Blood is pumped from the heart to the gills. From the gills, the oxygenated blood passes to the rest of the body and then returns to the heart.
4. **Nutritional deficiencies.** Fishes are unable to synthesize the aromatic amino acids and must consume them in their diet. This inability has been inherited by all their vertebrate descendants.

The earliest fishes are now extinct. Only their head-shields were made of bone; their elaborate internal skeletons were constructed of cartilage. Wriggling through the water, jawless and toothless, these fishes sucked up small food particles from the ocean floor. One group of jawless fishes, the agnathans, survive today as hagfish and parasitic lampreys. These first fishes were eventually replaced by larger, heavier jawed fishes that were predators. Jaws seem to have evolved from the frontmost of a series of arch supports made of cartilage that were used to reinforce the tissue between gill slits, holding the slits open. These armored fish were eventually replaced by fishes that moved through the water faster—the sharks and the bony fishes.

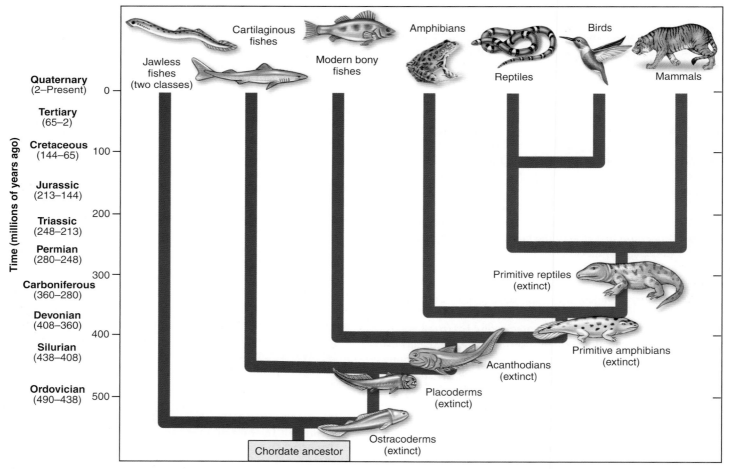

Figure 19.20 Vertebrate family tree.
Primitive amphibians arose from fishes. Primitive reptiles arose from amphibians and gave rise to mammals and to dinosaurs, which survive today as birds.

Figure 19.21 Chondrichthyes.
The Galápagos shark is a member of the class Chondrichthyes, which are mainly predators or scavengers and spend most of their time in graceful motion. As they move, they create a flow of water past their gills, from which they extract oxygen.

Sharks

Sharks are fast and maneuverable swimmers due to their light and flexible cartilaginous skeleton. Members of this group, the class Chondrichthyes, consist of sharks, skates, and rays. Sharks are very powerful swimmers, with a back fin, a tail fin, and two sets of paired side fins for controlled thrusting through the water (figure 19.21). Skates and rays are flattened sharks that are bottom-dwellers. Today there are about 750 species of sharks, skates, and rays.

Some of the largest sharks filter their food from the water like jawless fishes, but most are predators, their mouths armed with rows of hard, sharp teeth. Reproduction among the Chondrichthyes is the most advanced of any fish. Shark eggs are fertilized internally. During mating, the male grasps the female with modified fins called claspers. Sperm pass from the male into the female through grooves in the claspers. About 40% of sharks, skates, and rays lay fertilized eggs. The eggs of other species develop within the female's body, and the pups are born alive.

Bony Fishes

Bony fish are also fast and maneuverable swimmers but instead of gaining speed through lightness, as sharks did, bony fishes adopted a heavy internal skeleton made completely of bone. Such an internal skeleton is very strong, providing a base against which very strong muscles could pull. Bony fishes are still buoyant though because they possess a **swim bladder**. The swim bladder is a gas-filled sac that allows fish to regulate their buoyant density and so remain effortlessly suspended at any depth in the water. You can explore how a swim bladder works by examining the enlarged drawing in figure 19.22. The amount of air in the swim bladder is adjusted by extracting gases from the blood passing through blood vessels near the swim bladder, or by releasing gases back into the blood. Using swim bladders, a bony fish can rise up and down in the water the same way a submarine does. Sharks, by contrast, increase buoyancy with oil in their liver, but still must move

through the water or sink, because their bodies are denser than water. The swim bladder solution to the challenge of swimming has proven to be a great success in bony fish.

Bony fishes are the most successful of all fishes, indeed of all vertebrates. Of the nearly 30,800 living species of fishes in the world today, about 30,000 species are bony fishes with swim bladders. That's more species than all other kinds of vertebrates combined!

The remarkable success of the bony fishes has resulted from a series of significant adaptations. In addition to the swim bladder, they have a highly developed **lateral line system,** a sensory system that enables them to detect changes in water pressure and thus the movement of predators and prey in the water. Also, most bony fishes have a hard plate called the **operculum** that covers the gills on each side of the head. Flexing the operculum permits bony fishes to pump water over their gills. Using the operculum as very efficient bellows, bony fishes can pass water over the gills while stationary in the water. That is what a goldfish in a fish tank is doing when it seems to be gulping.

> **19.8 Fishes are characterized by gills, a vertebral column, and a simple, single-loop circulatory system. Sharks are fast swimmers, whereas the very successful bony fishes have unique characteristics such as swim bladders and lateral line systems.**

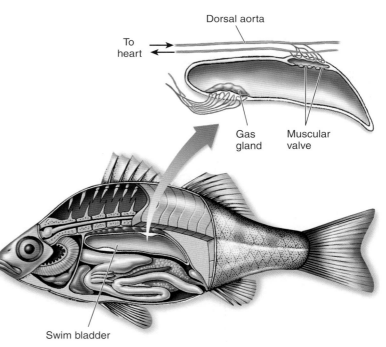

Figure 19.22 Diagram of a swim bladder.
The bony fishes use this structure, which evolved as a dorsal outpocketing of the pharynx, to control their buoyancy in water. The swim bladder can be filled with or drained of gas to allow the fish to control buoyancy. Gases are taken from the blood, and the gas gland secretes the gases into the swim bladder; gas is released from the bladder by a muscular valve.

19.9 Amphibians and Reptiles Invade the Land

Frogs, salamanders, and caecilians, the damp-skinned vertebrates, are direct descendants of fishes. They are the sole survivors of a very successful group, the **amphibians,** the first vertebrates to walk on land. The red-eyed tree frog staring at you in figure 19.23 is an example of a very successful group of amphibians, the order Anura that includes frogs and toads. Amphibians almost certainly evolved from the lobe-finned fishes, fish with paired fins that consist of a long fleshy muscular lobe supported by a central core of bones that form fully articulated joints with one another. The bones of the fins can be seen in the enlargements in the upper panel of figure 19.24.

Characteristics of Amphibians

Amphibians have five key characteristics that allowed them to successfully invade the land:

1. **Legs.** Frogs and salamanders have four legs and can move about on land quite well. The way in which legs are thought to have evolved from fins is illustrated in figure 19.24. Compare the legs and bones of an early amphibian's legs with the fins and bones of a lobe-finned fish in figure 19.24a. Notice that the arrangement of the bones in the early amphibian limbs is similar to the arrangement found in the lobe-finned fish.

Figure 19.23   A representative amphibian.

This red-eyed tree frog, *Agalychnis callidryas,* is a member of the group of amphibians that includes frogs and toads (order Anura).

2. **Lungs.** Most amphibians possess a pair of lungs, although the internal surfaces are poorly developed. Lungs were necessary because the delicate structure of fish gills requires the buoyancy of water to support it. Also, there is far more oxygen in air than water and so lungs provide a more efficient means of respiration.

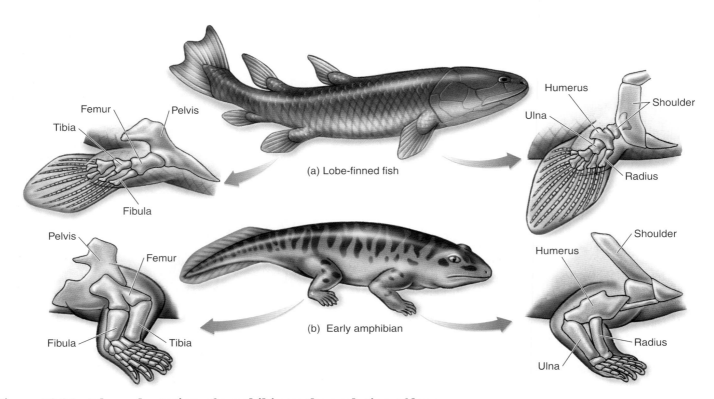

Figure 19.24   A key adaptation of amphibians: the evolution of legs.

(a) The limbs of a lobe-finned fish. Some lobe-finned fishes could move out onto land. (b) The limbs of an early amphibian. As illustrated by their skeletal structure, the legs of primitive amphibians could clearly function on land better than could the fins of lobe-finned fishes.

3. **Cutaneous respiration.** Frogs, salamanders, and caecilians all supplement the use of lungs by respiring directly across their skin, which is kept moist and provides an extensive surface area. This mode of respiration limits the body size of amphibians, because it is only efficient for a high surface-to-volume ratio.

4. **Pulmonary veins.** After blood is pumped through the lungs, two large veins called pulmonary veins return the aerated blood to the heart for repumping. This allows the aerated blood to be pumped to the tissues at a much higher pressure than when it leaves the lungs.

5. **Partially divided heart.** The heart evolved to deliver greater amounts of oxygen to the amphibian tissues, because greater amounts of oxygen are required by muscles for walking. The direction of blood flow through the amphibian heart helps prevent aerated blood from the lungs from mixing with nonaerated blood being returned to the heart from the rest of the body. This separates the blood circulation into two separate paths, pulmonary and systemic. The separation is imperfect; the third chamber has no dividing wall.

History of Amphibians

Amphibians were the dominant land vertebrates for 100 million years. They first became common in the late Paleozoic era, when much of North America was covered by lowland tropical swamps. Amphibians reached their greatest diversity during the mid-Permian period, when 40 families existed. Sixty percent of them were fully terrestrial, with bony plates and armor covering their bodies and many grew to be very large—some as big as a pony! After the great Permian extinction, the terrestrial forms began to decline, and by the time dinosaurs evolved, only 15 families remained, all aquatic. Only two of these families survived the Age of Dinosaurs, both aquatic: the anurans (frogs and toads) and the urodeles (salamanders).

Approximately 4,850 species of amphibians exist today, in 37 different families, all aquatic and all descended from the two aquatic families that survived the Age of Dinosaurs. Three orders constitute the class Amphibia: Anura, frogs and toads; Urodela, salamanders and newts; and Apoda, caecilians. Most of today's amphibians must reproduce in water and live the early part of their lives there, so amphibians are not completely terrestrial. However, in most habitats, particularly in the tropics, they are often today the most abundant and successful vertebrates to be found.

Reptiles

If we think of amphibians as the "first draft" of a manuscript about survival on land, then **reptiles** were the published book. All 7,000 living species of reptiles share certain fundamental characteristics, features they retained from the time when they replaced amphibians as the dominant terrestrial vertebrates. Among the most important are:

1. **Amniotic egg.** Amphibians never succeeded in becoming fully terrestrial because amphibian eggs must be laid in

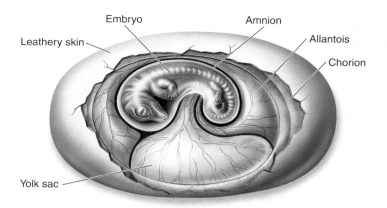

Figure 19.25   A key adaptation of reptiles: watertight eggs.
The watertight amniotic egg allows reptiles to live in a wide variety of terrestrial habitats.

water to avoid drying out. Most reptiles lay watertight eggs that offer various layers of protection from drying out, as shown in the cutaway portion of the amniotic egg in figure 19.25. The reptilian egg contains a food source (the yellow yolk) and a series of four membranes—the chorion (the outermost layer), the amnion (the membrane surrounding the embryo), the yolk sac (containing the yolk), and the allantois (the red structure). Each membrane plays a role in making the egg an independent life-support system. The **chorion** allows oxygen to enter the porous shell but retains water within the egg. The **amnion** encases the developing embryo within a fluid-filled cavity. The **yolk sac** provides food from the yolk for the embryo via blood vessels connecting to the embryo's gut. The **allantois** surrounds a cavity into which waste products from the embryo are excreted.

2. **Dry skin.** Amphibians have a moist skin and must remain in moist places to avoid drying out. Like their ancestors, reptiles have dry skin. A layer of scales or armor covers their bodies, preventing water loss.

3. **Thoracic breathing.** Amphibians breathe by squeezing their throat to pump air into their lungs; this limits their breathing capacity to the volume of their mouth. Reptiles developed pulmonary breathing, expanding and contracting the rib cage to move air in and out of the lungs.

In addition, reptiles improved on the innovations first attempted by amphibians. Legs were arranged to more effectively support the body's weight, allowing reptile bodies to be bigger and to run. Also, the lungs and heart were altered to make them far more efficient.

> **19.9   Amphibians were the first vertebrates to successfully invade land, helped by legs, lungs, and two pathways of blood circulation, pulmonary and systemic. Reptiles have three characteristics that suit them well for life on land: a watertight (amniotic) egg, dry skin, and thoracic breathing.**

19.10 Birds Master the Air

Birds evolved from small bipedal dinosaurs about 150 million years ago, but they were not common until the flying reptiles called pterosaurs became extinct along with the dinosaurs. Unlike pterosaurs, birds are insulated with feathers. Birds are so structurally similar to dinosaurs in all other respects that many scientists consider birds to be simply feathered dinosaurs.

Characteristics of Birds

Modern birds lack teeth and have only vestigial tails, but they still retain many reptilian characteristics. For instance, birds lay amniotic eggs, although the shells of bird eggs are hard rather than leathery. Also, reptilian scales are present on the feet and lower legs of birds. What makes birds unique? What distinguishes them from living reptiles?

1. **Feathers.** Derived from reptilian scales, feathers are the ideal adaptation for flight, lightweight and easily replaced if damaged. Feathers consist of a center shaft with barbs extending out, as pictured in figure 19.26. The barbs are held together with barbules that hook over each other, which reinforces that structure of the feather without added much weight to it.

2. **Flight skeleton.** The bones of birds are thin and hollow. Many of the bones are fused, making the bird skeleton more rigid than a reptilian skeleton, forming a sturdy frame that anchors muscles during flight. The power for active flight comes from large breast muscles that can make up 30% of a bird's total body weight. They stretch down from the wing and attach to the breastbone, which is greatly enlarged and bears a prominent keel for muscle attachment. They also attach to the fused collarbones that form the so-called wishbone. No other living vertebrates have a fused collarbone or a keeled breastbone.

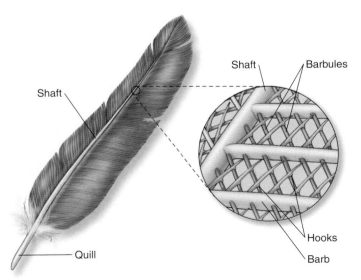

Figure 19.26 A key adaptation of birds: feathers.
The barbs off the main shaft of a feather have secondary branches called barbules that attach to one another by microscopic hooks.

Labels: Shaft, Shaft, Barbules, Hooks, Barb, Quill

Figure 19.27 *Archaeopteryx.*
An artist's reconstruction of *Archaeopteryx* based on fossil records.

Birds, like mammals, are endothermic. They generate enough heat through metabolism to maintain a high body temperature. Birds maintain body temperatures significantly higher than most mammals. The high body temperature permits a faster metabolism, necessary to satisfy the large energy requirements of flight.

History of Birds

The oldest bird of which there is a clear fossil is ***Archaeopteryx*** (meaning "ancient wing" and shown in figure 19.27), which was about the size of a crow and shared many features with small, bipedal, carnivorous dinosaurs. For example, it had teeth and a long reptilian tail. And unlike the hollow bones of today's birds, its bones were solid. Because of its many dinosaur features, several *Archaeopteryx* fossils were originally classified as *Compsognathus,* a small theropod dinosaur of similar size—until feathers were discovered on the fossils. What made *Archaeopteryx* distinctly avian was the presence of feathers on its wings and tail. It also had other birdlike features, notably the presence of a wishbone. Dinosaurs lacked a wishbone, although thecodonts had them.

By the early Cretaceous, only a few million years after *Archaeopteryx,* a diverse array of birds had evolved, with many of the features of modern birds. Fossils in Mongolia, Spain, and China discovered within the last few years reveal a diverse collection of toothed birds with the hollow bones and breastbones necessary for sustained flight. The diverse birds of the Cretaceous shared the skies with pterosaurs for 70 million years.

Today about 8,600 species of birds (class Aves) occupy a variety of habitats all over the world. You can tell a great deal about birds by examining their beaks. For example, carnivorous birds such as hawks have a sharp beak for tearing apart meat, the beaks of ducks are flat for shoveling through mud, and the beaks of finches are short and thick for crushing seeds.

> **19.10 Birds are essentially dinosaurs with feathers. Feathers and a strong, light skeleton make flight possible.**

19.11 Mammals Adapt to Colder Times

Characteristics of Mammals

The **mammals** (class Mammalia) that evolved about 220 million years ago side by side with the dinosaurs would look strange to you, not at all like modern-day lions and tigers and bears. They share three key characteristics with mammals today:

1. **Mammary glands.** Female mammals have mammary glands, which produce milk to nurse the newborns. Even baby whales are nursed by their mother's milk. Milk is a very-high-calorie food (human milk has 750 kcal per liter), important because of the high energy needs of a rapidly growing newborn mammal.
2. **Hair.** Among living vertebrates, only mammals have hair, and all mammals do (even whales and dolphins have a few sensitive bristles on their snout). A hair is a filament composed of dead cells filled with the protein keratin. The primary function of hair is insulation. The insulation provided by fur may have ensured the survival of mammals when the dinosaurs perished.
3. **Middle ear.** All mammals have three middle-ear bones, which evolved from bones in the reptile jaw. These bones play a key role in hearing by amplifying vibrations created by sound waves that beat upon the eardrum.

History of the Mammals

Mammals have been around since the time of the dinosaurs, although they were always small until dinosaurs disappeared. We have learned a lot about the evolutionary history of mammals from their fossils. The first mammals were tiny, shrew-like creatures that lived in trees chasing insects. For 155 million years, all the time the dinosaurs flourished, mammals were a minor group that changed little.

Today, over 4,500 species of mammals occupy all the large-body niches that dinosaurs once claimed, among many others. They range in size from 1.5-gram shrews to 100-ton whales. Almost half of all mammals are rodents—mice and their relatives. Almost one-quarter of all mammals are bats! Mammals have even invaded the seas, as plesiosaur and ichthyosaur reptiles did so successfully millions of years earlier—79 species of whales and dolphins live in today's oceans. The placental mammals that walked the earth during the ice ages were even larger than today's versions; the world's climate has warmed again in recent times, favoring smaller bodies, which are easier to cool.

Primates, the order to which we belong, are not a particularly large group; there are only 233 known species. Human beings evolved only recently, less than 2 million years ago. There have been several species of humans, but ours, *Homo sapiens,* is the only one that survives today. We are notable among primates for having less hair, walking upright, making complex tools, and having complicated language.

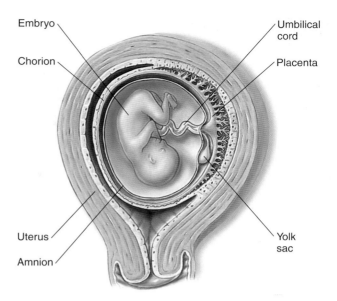

Figure 19.28 The placenta.
The placenta is characteristic of the largest group of mammals, the placental mammals. It evolved from membranes in the amniotic egg. The umbilical cord evolved from the allantois. The chorion, or outermost part of the amniotic egg, forms most of the placenta itself. The placenta serves as the provisional lungs, intestine, and kidneys of the embryo, without ever mixing maternal and fetal blood.

Other Characteristics of Modern Mammals

Endothermy. Mammals are endothermic, a crucial adaptation that allows them to be active at any time of the day or night and to colonize severe environments, from deserts to ice fields. Many characteristics, such as hair or feathers to provide insulation, play important roles in making endothermy possible. Also, the more efficient blood circulation and breathing found in mammals make possible the higher metabolic rate upon which endothermy depends.

Placenta. In most mammal species, females carry their young in the uterus during development, nourishing them by a placenta, and give birth to live young. The placenta is a specialized organ within the womb of the mother that brings the bloodstream of the fetus into close contact with the bloodstream of the mother. Figure 19.28 shows a drawing of a fetus within the uterus. The placenta is to the right side, attached to the umbilical cord. Food, water, and oxygen can pass across from mother to child, and wastes can pass over to the mother's blood and be carried away.

Teeth. Reptiles have homodont dentition: their teeth are all the same. However, mammals have heterodont dentition, with different types of teeth that are highly specialized to match particular eating habits.

Hooves and Horns. Keratin, the protein of hair, is also the structural building material in claws, fingernails, and hooves. Also, the horns of cattle, sheep, and antelope are composed of a core of bone surrounded by a sheath of keratin.

Today's Mammals

Monotremes: Egg-Laying Mammals. The duck-billed platypus and two species of echidna, or spiny anteater (figure 19.29a), are the only living monotremes. The monotremes have many reptilian features, including laying shelled eggs, but they also have both of the defining mammalian features: hair and functioning mammary glands. Females lack well-developed nipples, so the newly hatched babies cannot suckle. Instead, the milk oozes onto the mother's fur, and the babies lap it off with their tongues. The platypus, found only in Australia, lives much of its life in the water and is a good swimmer. It uses its bill much as a duck does, rooting in the mud for worms and other small animals.

Marsupials: Pouched Mammals. The major difference between marsupials (figure 19.29b) and other mammals is their pattern of embryonic development. In marsupials, a fertilized egg is surrounded by chorion and amniotic membranes, but no shell forms around the egg as it does in monotremes. The marsupial embryo is nourished by an abundant yolk within the shell-less egg. Shortly before birth, a short-lived placenta forms from the chorion membrane. After the embryo is born, tiny and hairless, it crawls into the marsupial pouch where it latches onto a nipple and continues its development.

Placental Mammals. As stated earlier, mammals that produce a true placenta, which nourishes the embryo throughout its entire development, are called placental mammals (figure 19.29c). Most species of mammals living today, including humans, are in this group. Early in the course of embryonic development, the placenta forms from both fetal and maternal tissues. Unlike marsupials, the young undergo a considerable period of development before they are born.

(a)

(b)

(c)

Figure 19.29 Today's mammals.
(a) This echidna, *Tachyglossus aculeatus,* is a monotreme.
(b) Marsupials include kangaroos, like this adult with young in its pouch. (c) This female African lion, *Panthera leo* (order Carnivora), is a placental mammal.

> **19.11 Mammals are endotherms that nurse their young with milk and exhibit a variety of different kinds of teeth. All mammals have at least some hair.**

INQUIRY & ANALYSIS

Are Extinction Rates Constant?

Evolutionist Lee Van Valen put forth the hypothesis in 1973 that extinction is usually due to random events unrelated to a species' particular adaptations. If this were the case, then the likelihood that a species will go extinct would be expected to be virtually constant, when viewed over long periods of time. This hypothesis has been tested for a variety of groups. In the graph to the right you see an examination of the 200-million-year fossil record of echinoids (sea urchins and sand dollars). Data are presented as the number of echinoid families that have survived for over a period of 200 million years. The red dashed line shows a theoretical constant extinction rate, as postulated by Van Valen. The blue line is a curve determined by statistical regression analysis.

1. **Applying Concepts** What is the dependent variable?

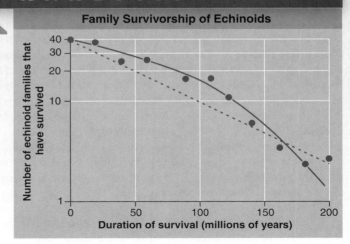

Family Survivorship of Echinoids

y-axis: Number of echinoid families that have survived
x-axis: Duration of survival (millions of years)

2. **Making Inferences** Which of the two lines best represents the data?
3. **Drawing Conclusions** Is Van Valen's hypothesis supported by this analysis?

20.3 The Water Cycle

Unlike energy, which flows through the earth's ecosystems in one direction (from the sun to producers to consumers), the physical components of ecosystems are passed around and reused within ecosystems. Ecologists speak of such constant reuse as recycling or, more commonly, **cycling.** Materials that are constantly recycled include all the chemicals that make up the soil, water, and air. While many are important and will be considered later, the proper cycling of four materials is particularly critical to the health of any ecosystem: water, carbon, and the soil nutrients nitrogen and phosphorus.

The paths of water, carbon, and soil nutrients as they pass from the environment to living organisms and back form closed circles, or cycles. In each cycle, the chemical resides for a time in an organism and then returns to the nonliving environment, often referred to as a *biogeochemical cycle.*

Of all the nonliving components of an ecosystem, water has the greatest influence on the living portion. The availability of water and the way in which it cycles in an ecosystem in large measure determines the biological richness of that eco-system—how many different kinds of creatures live there and how many of each.

Water cycles within an ecosystem in two ways: the environmental water cycle and the organismic water cycle. Both cycles are shown in figure 20.7.

The Environmental Water Cycle

In the environmental water cycle, water vapor in the atmosphere condenses and falls to the earth's surface as rain or snow (called precipitation in figure 20.7). Heated there by the sun, it reenters the atmosphere by **evaporation** from lakes, rivers, and oceans, where it condenses and falls to the earth again.

The Organismic Water Cycle

In the organismic water cycle, surface water does not return directly to the atmosphere. Instead, it is taken up by the roots of plants. After passing through the plant, the water reenters the atmosphere through tiny openings (stomata) in the leaves, evaporating from their surface. This evaporation from leaf surfaces is called **transpiration.** Transpiration is also driven by the sun: The sun's heat creates wind currents that draw moisture from the plant by passing air over the leaves.

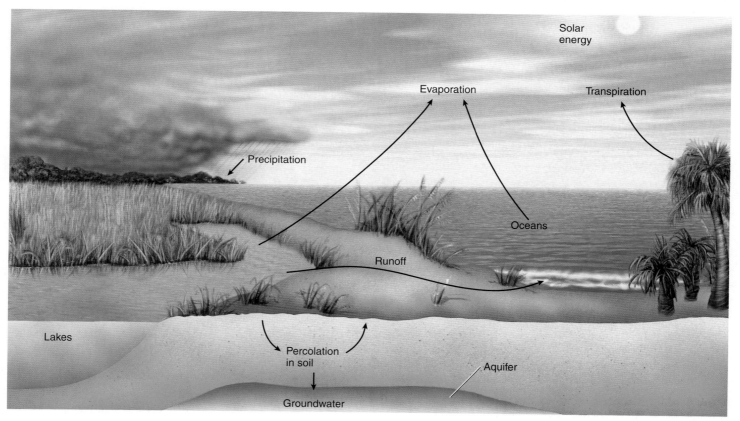

Figure 20.7 The water cycle.

Precipitation on land eventually makes its way to the ocean via groundwater, lakes, and finally, rivers. Solar energy causes evaporation, adding water to the sky. Plants give off excess water through transpiration, also adding water to the atmosphere. Atmospheric water falls as rain or snow over land and oceans, completing the water cycle.

Breaking the Cycle

In very dense forest ecosystems, such as tropical rain forests, more than 90% of the moisture in the ecosystem is taken up by plants and then transpired back into the air. Because so many plants in a rain forest are doing this, the vegetation is the primary source of local rainfall. In a very real sense, these plants create their own rain: the moisture that travels up from the plants into the atmosphere falls back to earth as rain.

Where forests are cut down, the organismic water cycle is broken, and moisture is not returned to the atmosphere. Water drains off to the sea instead of rising to the clouds and falling again on the forest. During his expeditions from 1799–1805, the great German explorer Alexander von Humboldt reported that stripping the trees from a tropical rain forest in Colombia prevented water from returning to the atmosphere and created a semiarid desert. It is a tragedy of our time that just such a transformation is occurring in many tropical areas, as tropical and temperate rain forests are being clear-cut or burned in the name of "development" (figure 20.8).

Groundwater

Much less obvious than the surface waters seen in streams, lakes, and ponds is the groundwater, which occurs in permeable, saturated, underground layers of rock, sand, and gravel called *aquifers*. In many areas, groundwater is the most important water reservoir; for example, in the United States, more than 96% of all freshwater is groundwater. Groundwater flows much more slowly than surface water, anywhere from a few millimeters to as much as a meter or so per day. In the United States, groundwater provides about 25% of the water used for all purposes and provides about 50% of the population with drinking water. Rural areas tend to depend on groundwater almost exclusively, and its use is growing at about twice the rate of surface water use.

Because of the greater rate at which groundwater is being used, the increasing chemical pollution of groundwater is a very serious problem. Pesticides, herbicides, and fertilizers are key sources of groundwater pollution. Because of the large volume of water, its slow rate of turnover, and its inaccessibility, removing pollutants from aquifers is virtually impossible.

> **20.3** Water cycles through ecosystems in the atmosphere via precipitation and evaporation, some of it passing through plants on the way.

Figure 20.8 Burning or clear-cutting forests breaks the water cycle.
The high density and large size of plants in a forest translate into great quantities of water being transpired to the atmosphere, creating rain over the forests. In this way rain forests perpetuate the wet climate that supports them. Tropical deforestation permanently alters the climate in these areas, creating arid zones.

20.4 The Carbon Cycle

The earth's atmosphere contains plentiful carbon, present as carbon dioxide (CO_2) gas. This carbon cycles between the atmosphere and living organisms, often being locked up for long periods of time in organisms or deep underground. The cycle is begun by plants that use CO_2 in photosynthesis to build organic molecules—in effect, they trap the carbon atoms of CO_2 within the living world. The carbon atoms are returned to the atmosphere's pool of CO_2 through respiration, combustion, and erosion. The carbon cycle, seemingly more complex than the water cycle, is shown in figure 20.9.

Respiration

Most of the organisms in ecosystems respire—that is, they extract energy from organic food molecules by stripping away the carbon atoms and combining them with oxygen to form CO_2. At times, plants respire, as do the herbivores, which eat the plants, and the carnivores, which eat the herbivores. All of these organisms use oxygen to extract energy from food, and CO_2 is what is left when they are done. This by-product of respiration is released into the atmosphere.

Combustion

A lot of carbon is tied up in wood, and it may stay trapped there for many years, only returning to the atmosphere when the wood is burned. Sometimes the duration of the carbon's visit to the organic world is long indeed. Plants that become buried in sediment, for example, may be gradually transformed by pressure into coal or oil. The carbon originally trapped by these plants is only released back into the atmosphere when the coal or oil (called **fossil fuels**) is burned.

Erosion

Very large amounts of carbon are present in seawater as dissolved CO_2. Substantial amounts of this carbon are extracted from the water by marine organisms, which use it to build their calcium carbonate shells. When these marine organisms die, their shells sink to the ocean floor, become covered with sediments, and form limestone. Eventually, the ocean recedes and the limestone becomes exposed to weather and erodes; as a result the carbon washes back and is dissolved in oceans where it is returned to the cycle through diffusion.

> **20.4** Carbon captured from the atmosphere by photosynthesis is returned to it through respiration, combustion, and erosion.

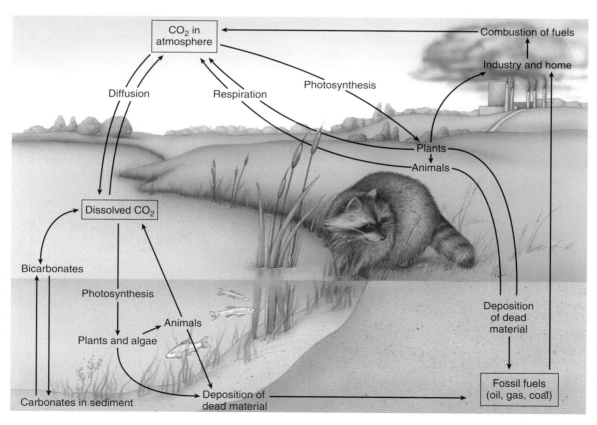

Figure 20.9 The carbon cycle.
Carbon from the atmosphere and from water is fixed by photosynthetic organisms and returned through respiration, combustion, and erosion.

20.5 | Soil Nutrients and Other Chemical Cycles

The Nitrogen Cycle

Organisms contain a lot of nitrogen (a principal component of protein) and so does the atmosphere, which is 78.08% nitrogen gas (N_2). However, the chemical connection between these two reservoirs is very delicate, because most living organisms are unable to use the N_2 so plentifully available in the air surrounding them. The two nitrogen atoms of N_2 are bound together by a particularly strong "triple" covalent bond that is very difficult to break. Luckily, a few kinds of bacteria can break the nitrogen triple bond and bind its nitrogen atoms to hydrogen (forming "fixed" nitrogen, ammonia [NH_3], which becomes ammonium ion [NH_4^+]) in a process called **nitrogen fixation.**

Bacteria evolved the ability to fix nitrogen early in the history of life, before photosynthesis had introduced oxygen gas into the earth's atmosphere, and that is still the only way the bacteria are able to do it—even a trace of oxygen poisons the process. In today's world, awash with oxygen, these bacteria live encased within bubbles called cysts that admit no oxygen or within special airtight cells in nodules of tissue on the roots of beans, aspen trees, and a few other plants. Figure 20.10 shows the workings of the nitrogen cycle. Bacteria make needed nitrogen available to other organisms. The nitrogen moves up the food chain as one organism eats another and eventually returns following their deaths or through their excretions. Decomposing bacteria and then ammonifying bacteria return the nitrogen to ammonia and ammonium ion forms. Continuing the cycle, nitrifying bacteria can convert ammonium ion into nitrate (NO_3^-), and denitrifying bacteria are able to convert nitrate back into atmospheric nitrogen (N_2).

The growth of plants in ecosystems is often severely limited by the availability of "fixed" nitrogen in the soil, which is why farmers fertilize fields. This agricultural practice is a very old one, known even to primitive societies—the American Indians instructed the pilgrims to bury fish, a rich source of fixed nitrogen, with their corn seeds. Today most fixed nitrogen added to soils by farmers is not organic but instead is produced in factories by industrial rather than bacterial nitrogen fixation, a process that accounts for a prodigious 30% of the entire nitrogen cycle.

The Phosphorus Cycle

Phosphorus is an essential element in all living organisms, a key part of both ATP and DNA. Phosphorus is often in very limited supply in the soil of particular ecosystems, and because phosphorus does not form a gas, none is available in the atmosphere. Most phosphorus exists in soil and rock as the mineral calcium phosphate, which, as shown in figure 20.11, dissolves in water to form phosphate ions (Coca-Cola is a sweetened solution of phosphate ions). These phosphate ions are absorbed by the roots of plants and used by them to build organic molecules like ATP and DNA. When the plants and animals die and decay, bacteria in the soil convert the organic phosphorus back into phosphorus ions, completing the cycle.

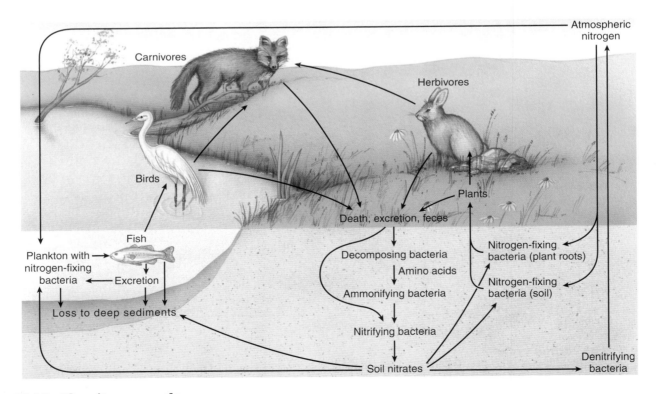

Figure 20.10 The nitrogen cycle.

Relatively few kinds of organisms—all of them bacteria—can convert atmospheric nitrogen into forms that can be used for biological processes.

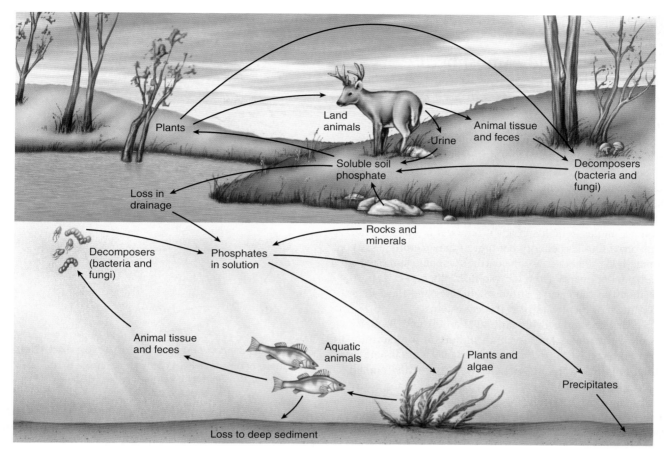

Figure 20.11   The phosphorus cycle.
Phosphorus plays a critical role in plant nutrition; next to nitrogen, phosphorus is the element most likely to be so scarce that it limits plant growth.

The phosphorus level in freshwater lake ecosystems is often quite low, preventing much growth of photosynthetic algae in these systems. Such ecosystems are particularly vulnerable to the inadvertent addition of phosphorus by human activity. For example, agricultural fertilizers and many commercial detergents are rich in phosphorus. Pollution of a lake by the addition of phosphorus to its waters first produces a green scum of algal growth on the surface of the lake, which then, if the pollution continues, proceeds to "kill" the lake. After the initial bloom of rapid algal growth, aging algae die, and bacteria feeding on the dead algae cells use up so much of the lake's dissolved oxygen that fish and invertebrate animals suffocate. Such rapid, uncontrolled growth caused by excessive nutrients in an aquatic ecosystem is called **eutrophication.**

The Cycling of Other Chemicals

Many other chemicals cycle through an ecosystem and must be maintained in a balanced state for the ecosystem to be healthy. Proper balance is important. Some chemicals can become harmful when their concentrations exceed normal levels for cycling, as we saw with phosphorus. Other chemicals, when in excess of normal cycling levels, can have similar devastating effects on an ecosystem.

Sulfur, a chemical that cycles through the atmosphere, can harm an ecosystem when large amounts of it are pumped into the atmosphere through coal-burning power plants. The excess sulfur combines with water vapor and oxygen, producing sulfuric acid. This acid then reenters the ecosystem as precipitation. This "acid rain" is discussed further in chapter 23.

Heavy metals, which include mercury, cadmium, and lead, are particularly damaging as they cycle through biological food chains, as they tend to progressively accumulate in organisms of higher trophic levels. This process, called *biological magnification,* is discussed further in chapter 23.

20.5 Most of the earth's atmosphere is diatomic nitrogen gas that cannot be used by most organisms. Certain bacteria are able to convert this nitrogen gas into ammonia through nitrogen fixation. These nitrogen atoms then cycle through the earth's ecosystem. Phosphorus, critical to organisms, is available in soil and dissolved in water. It cycles between organisms and the environment and is often the limiting factor in determining what organisms are able to live in an ecosystem.

20.6 The Sun and Atmospheric Circulation

The world contains a great diversity of ecosystems because its climate varies a great deal from place to place. On a given day, Miami and Boston often have very different weather. There is no mystery about this. The tropics are warmer than the temperate regions because the sun's rays arrive almost perpendicular (that is, dead on) at regions near the equator. As you move from the equator into temperate latitudes, sunlight strikes the earth at more oblique angles, which spreads it out over a much greater area, thus providing less energy per unit of area (figure 20.12). This simple fact—that because the earth is a sphere some parts of it receive more energy from the sun than others—is responsible for much of the earth's different climates and thus, indirectly, for much of the diversity of its ecosystems.

The earth's annual orbit around the sun and its daily rotation on its own axis are also both important in determining world climate. Because of the daily cycle, the climate at a given latitude is relatively constant. Because of the annual cycle and the inclination of the earth's axis, all parts away from the equator experience a progression of seasons. In summer in the Southern Hemisphere, the earth is tilted toward the sun as shown in figure 20.12, and rays hit more directly lead-

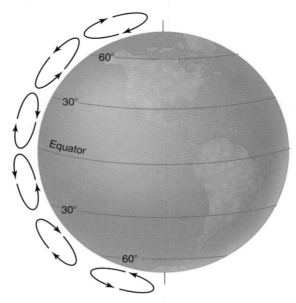

Figure 20.13 Air rises at the equator and then falls.
The pattern of air movement out from and back to the earth's surface forms three pairs of great cycles.

ing to higher temperatures; in the winter, the tilt of the earth is opposite, with the Northern Hemisphere nearer to the sun, experiencing summer.

The major atmospheric circulation patterns result from the interactions between six large air masses. These great air masses (shown as circulating arrows in figure 20.13) occur in pairs, with one air mass of the pair occurring in the northern latitudes and the other occurring in the southern latitudes. These air masses affect climate because the rising and falling of an air mass influence its temperature, which, in turn, influences its moisture-holding capacity.

Near the equator, warm air rises and flows toward the poles (indicated by arrows at the equator that rise and circle toward the poles). As it rises and cools, this air loses most of its moisture because cool air holds less water vapor than warm air. (This explains why it rains so much in the tropics where the air is warm.) When this air has traveled to about 30 degrees north and south latitudes, the cool, dry air sinks and becomes reheated, soaking up water like a sponge as it warms, producing a broad zone of low rainfall. It is no accident that all of the great deserts of the world lie near 30 degrees north or 30 degrees south latitude. Air at these latitudes is still warmer than it is in the polar regions, and thus it continues to flow toward the poles. At about 60 degrees north and south latitudes, air rises and cools and sheds its moisture, and such are the locations of the great temperate forests of the world. Finally, this rising air descends near the poles, producing zones of very low precipitation.

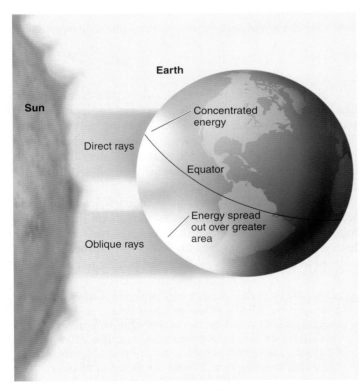

Figure 20.12 Latitude affects climate.
The relationship between the earth and sun is critical in determining the nature and distribution of life on earth. The tropics are warmer than the temperate regions because the sun's rays strike at a direct angle, providing more energy per unit of area.

20.6 The sun drives circulation of the atmosphere, causing rain in the tropics and a band of deserts at 30 degrees latitude.

Temperatures are higher in tropical ecosystems for a simple reason: more sunlight per unit area falls on tropical latitudes (see figure 20.12). Solar radiation is most intense when the sun is directly overhead, and this occurs only in the tropics, where sunlight strikes the equator perpendicularly. Temperature also varies with elevation, with higher altitudes becoming progressively colder. At any given latitude, air temperature falls about 6°C for every 1,000-meter increase in elevation. The ecological consequences of temperature varying with elevation are the same as temperature varying with latitude. Figure 20.14 illustrates this principle comparing changes in ecosystems that occur with increasing latitudes in North America with the ecosystem changes that occur with increasing elevation at the tropics. A 1,000-meter increase in elevation on a mountain in southern Mexico (figure 20.14b) results in a temperature drop equal to that of an 880-kilometer increase in latitude on the North American continent (figure 20.14a). This is why "timberline" (the elevation above which trees do not grow) occurs at progressively lower elevations as one moves farther from the equator.

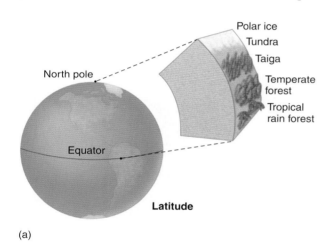

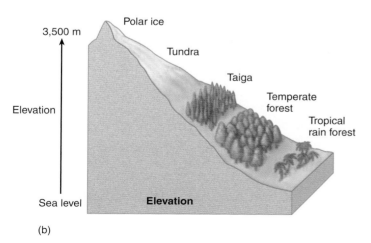

Figure 20.14 How elevation affects ecosystems.
The same land ecosystems that normally occur as latitude increases north and south of the equator at sea level (a) can occur in the tropics as elevation increases (b).

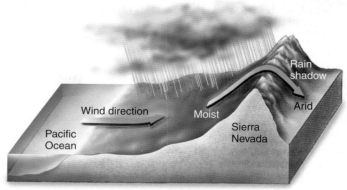

Figure 20.15 The rain shadow effect.
Moisture-laden winds from the Pacific Ocean rise and are cooled when they encounter the Sierra Nevada. As they cool, their moisture-holding capacity decreases and precipitation occurs. As the air descends on the east side of the range, it warms, its moisture-holding capacity increases, and the air picks up moisture from its surroundings. As a result, arid conditions prevail on the east side of these mountains.

Rain Shadows

When a moving body of air encounters a mountain (figure 20.15), it is forced upward, and as it is cooled at higher elevations the air's moisture-holding capacity decreases, producing the rain you see on the windward side of the mountains—the side from which the wind is blowing. The effect on the other side of the mountain—the leeward side—is quite different. As the air passes the peak and descends on the far side of the mountains, it is warmed, so its moisture-holding capacity increases. Sucking up all available moisture, the air dries the surrounding landscape, often producing a desert. This effect, called a **rain shadow,** is responsible for deserts such as Death Valley, which is in the rain shadow of Mount Whitney, the tallest mountain in the Sierra Nevada.

Similar effects can occur on a larger scale. Regional climates are areas that are located on different parts of the globe but share similar climates because of similar geography. A so-called Mediterranean climate results when winds blow from a cool ocean onto warm land during the summer. As a result, the air's moisture-holding capacity is increased and precipitation is blocked, similar to what occurs on the leeward side of mountains. This effect accounts for dry, hot summers and cool, moist winters in areas with a Mediterranean climate such as portions of southern California or Oregon, central Chile, southwestern Australia, and the Cape region of South Africa. Such a climate is unusual on a world scale. In the regions where it occurs, many unusual kinds of endemic (local in distribution) plants and animals have evolved.

> **20.7 Temperatures fall with increasing latitude and also with increasing elevation. Rainfall is higher on the windward side of mountains, with air losing its moisture as it rises up the mountain; descending on the far side, the dry air warms and sucks up moisture, creating deserts.**

20.8 Patterns of Circulation in the Ocean

Patterns of ocean circulation are determined by the patterns of atmospheric circulation, but they are modified by the locations of landmasses. Oceanic circulation is dominated by the movement of surface waters in huge spiral patterns called gyres, which move around the subtropical zones of high pressure between approximately 30 degrees north and south latitudes. These gyres, indicated by the red and blue arrows in figure 20.16, move clockwise in the Northern Hemisphere and counterclockwise in the Southern Hemisphere. The ways they redistribute heat profoundly affects life not only in the oceans but also on coastal lands. For example, the Gulf Stream, in the North Atlantic (the red-colored arrow, meaning it carries warm waters), swings away from North America near Cape Hatteras, North Carolina, and reaches Europe near the southern British Isles. Because of the Gulf Stream, western Europe is much warmer and more temperate than eastern North America at similar latitudes. As a general principle, western sides of continents in temperate zones of the Northern Hemisphere are warmer than their eastern sides; the opposite is true of the Southern Hemisphere.

Off the western coast of South America, the Humboldt Current carries phosphorus-rich cold water northward up the west coast. Phosphorus is brought up from the ocean depths by the upwelling of cool water that occurs as offshore winds blow from the mountainous slopes that border the Pacific Ocean. This nutrient-rich current helps make possible the abundance of marine life that supports the fisheries of Peru and northern Chile. Marine birds, which feed on these organisms, are responsible for the commercially important, phosphorus-rich guano deposits on the seacoasts of these countries.

El Niño Southern Oscillations and Ocean Ecology

Every Christmas a warm current sweeps down the coast of Peru and Ecuador from the tropics, reducing the fish population slightly and giving local fishers some time off. The local fishers named this Christmas current *El Niño* ("The Christ Child"). Now, though, the term is reserved for a catastrophic version of the same phenomenon, one that occurs every two to seven years and is not only felt locally but on a global scale.

Scientists now have a pretty good idea of what goes on in an El Niño. Normally the Pacific Ocean is fanned by constantly blowing east-to-west trade winds. These winds push warm surface water away from the ocean's eastern

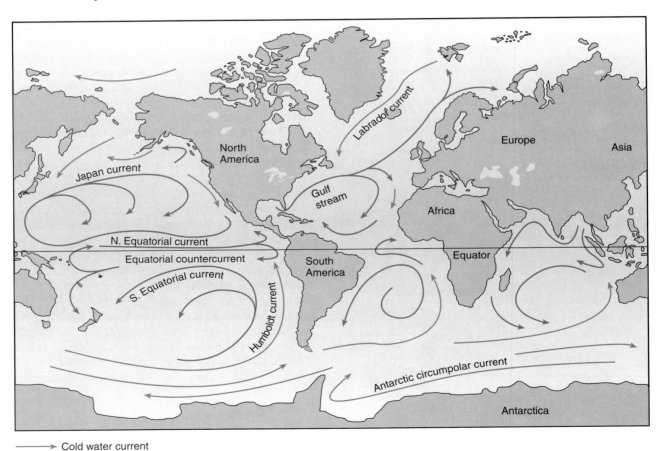

Figure 20.16 Oceanic circulation.
The circulation in the oceans moves in great surface spiral patterns called gyres; oceanic circulation affects the climate on adjacent lands.

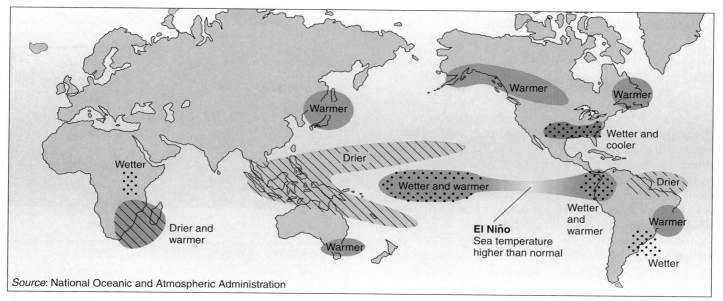

Source: National Oceanic and Atmospheric Administration

Figure 20.17 An El Niño winter.
El Niño currents produce unusual weather patterns all over the world as warm waters from the western Pacific move eastward.

side (Peru, Ecuador, and Chile) and allow cold water to well up from the depths in its place, carrying nutrients that feed plankton and hence fish. This warm surface water piles up in the west, around Australia and the Philippines, making it several degrees warmer and a meter or so higher than the eastern side of the ocean. But if the winds slacken briefly, warm water begins to slosh back across the ocean (indicated by the darker red band stretching between Australia and the northern coast of South America, as shown in figure 20.17), causing an El Niño.

Once this happens, ocean and atmosphere conspire to ensure it keeps happening. The warmer the eastern ocean gets, the warmer and lighter the air above it becomes, and hence more similar to the air on the western side. This reduces the difference in pressure across the ocean. Because a pressure difference is what makes winds blow, the easterly trades weaken further, letting the warm water continue its eastward advance.

The end result is to shift the weather systems of the western Pacific Ocean 6,000 kilometers eastward. The tropical rainstorms that usually drench Indonesia and the Philippines are caused when warm seawater abutting these islands causes the air above it to rise, cool, and condense its moisture into clouds. When the warm water moves east, so do the clouds, leaving the previously rainy areas in drought (indicated by the light pink, hatched areas in figure 20.17). Conversely, the western edge of South America, its coastal waters usually too cold to trigger much rain, gets a soaking (the dotted red areas), while the upwelling slows down because of the warm water. During an El Niño, commercial fish stocks virtually disappear from the waters of Peru and northern Chile, and plankton drops to a twentieth of its normal abundance. The commercially valuable anchovy fisheries of Peru were essentially destroyed by the 1972 and 1997 El Niños.

That is just the beginning. El Niño's effects are propagated across the world's weather systems. Violent winter storms lash the coast of California, accompanied by flooding, and El Niño produces colder and wetter winters than normal in Florida and along the Gulf Coast. The U.S. Midwest experiences heavier than normal rains (the blue dotted area).

Though the effects of an El Niño are now fairly clear, what triggers them still remains a mystery. Models of these weather disturbances suggest that the climatic change that triggers an El Niño is "chaotic." Wind and ocean currents return again and again to the same condition, but never in a regular pattern, and small nudges can send them off in many different directions—including an El Niño.

La Niña. El Niño is an extreme phase of a naturally occurring climatic cycle, but as in all cycles, there is an opposite side to it. While El Niño is characterized by unusually warm ocean temperatures in the eastern Pacific, *La Niña* is characterized by unusually cold ocean temperatures in the eastern Pacific. The strengthening of the east-to-west trade winds intensifies the cold upwelling along the eastern Pacific, with coastal water temperatures along the South American coast falling as much as 7°F below normal. Although not as well known as El Niño, La Niña causes equally extreme effects that are nearly opposite to those of El Niño. In the United States, the effects of La Niña are most apparent during the winter months.

20.8 The world's oceans circulate in huge gyres deflected by continental landmasses. Disturbances in ocean currents like an El Niño and La Niña can have profound influences on world climate.

20.9 Ocean Ecosystems

Most of the earth's surface—nearly three-quarters—is covered by water. The seas have an average depth of more than 3 kilometers, and they are, for the most part, cold and dark. Photosynthetic organisms are confined to the upper few hundred meters (the light blue area in figure 20.18) because light does not penetrate any deeper. Almost all organisms that live below this level feed on organic debris that rains downward. The three main kinds of marine ecosystems are shallow waters, open-sea surface, and deep-sea waters (figure 20.18).

Shallow Waters

Very little of the earth's ocean surface is shallow—mostly that along the shoreline—but this small area contains many more species than other parts of the ocean (figure 20.19a). The world's great commercial fisheries occur on banks in the coastal zones, where nutrients derived from the land are more abundant than in the open ocean. Part of this zone consists of the **intertidal region,** which is exposed to the air whenever the tides recede. Partly enclosed bodies of water, such as those that often form at river mouths and in coastal bays, where the salinity is intermediate between that of seawater and freshwater, are called **estuaries.** Estuaries are among the most naturally fertile areas in the world, often containing rich stands of submerged and emergent plants, algae, and microscopic organisms. They provide the breeding grounds for most of the coastal fish and shellfish that are harvested both in the estuaries and in open water.

(a)

(b)

Figure 20.19   Shallow waters and open sea surface.
(a) Fishes and many other kinds of animals find food and shelter among the coral in the coastal waters of some regions. (b) The upper layers of the open ocean contain plankton and large schools of fish, like these yellow tail grunts.

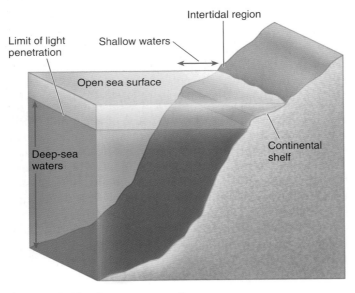

Figure 20.18   Ocean ecosystems.
There are three primary ecosystems found in the earth's oceans. Shallow water ecosystems occur along the shoreline and at areas of coral reefs. Open sea surface ecosystems occur in the upper 100–200 meters where light can penetrate. Finally, deep-sea water ecosystems are areas below 300 meters.

Open-Sea Surface

Drifting freely in the upper, better-illuminated waters of the ocean is a diverse biological community of microscopic organisms. Most of the plankton occurs in the top 100 meters of the sea. Many fishes swim in these waters as well, feeding on the plankton and one another (figure 20.19b). Some members of the plankton, including algae and some bacteria, are photosynthetic and are called phytoplankton. Collectively, these organisms are responsible for about 40% of all photosynthesis that takes place on earth. Over half of this is carried out by organisms less than 10 micrometers in diameter—at the lower limits of size for organisms—and almost all of it near the surface of the sea, in the zone into which light from the surface penetrates freely.

Deep-Sea Waters

In the deep waters of the sea, below the top 300 meters, little light penetrates. Very few organisms live there, compared to the rest of the ocean, but those that do include some of the most bizarre organisms found anywhere on earth. Many deep-sea inhabitants have bioluminescent (light-producing) body parts that they use to communicate or to attract prey (figure 20.20a).

The supply of oxygen can often be critical in the deep ocean, and as water temperatures become warmer, the water holds less oxygen. For this reason, the amount of available oxygen becomes an important limiting factor for deep-sea organisms in warmer marine regions of the globe. Carbon dioxide, in contrast, is almost never limited in the deep ocean. The distribution of minerals is much more uniform in the ocean than it is on land, where individual soils reflect the composition of the parent rocks from which they have weathered.

Frigid and bare, the floors of the deep sea have long been considered a biological desert. Recent close-up looks taken by marine biologists, however, paint a different picture (figure 20.20c). The ocean floor is teeming with life. Often kilometers deep, thriving in pitch darkness under enormous pressure, crowds of marine invertebrates have been found in hundreds of deep samples from the Atlantic and Pacific. Rough estimates of deep-sea diversity have soared to hundreds of thousands of species. Many appear endemic (local). The diversity of species is so high it may rival that of tropical rain forests! This profusion is unexpected. New species usually require some kind of barrier to diverge (see chapter 15), and the ocean floor seems boringly uniform. However, little migration occurs among deep populations, and this lack of movement may encourage local specialization and species formation. A patchy environment may also contribute to species formation there; deep-sea ecologists find evidence that fine but nonetheless formidable resource barriers arise in the deep sea.

No light falls in the deep ocean. From where do deep-sea organisms obtain their energy? While some utilize energy falling to the ocean floor as debris from above, other deep-sea organisms are autotrophic, gaining their energy from **hydrothermal vent systems,** areas in which seawater circulates through porous rock surrounding fissures where molten material from beneath the earth's crust comes close to the surface. Hydrothermal vent systems, also called deep-sea vents, support a broad array of heterotrophic life (figure 20.20b). Water in the area of these hydrothermal vents is heated to temperatures in excess of 350°C, and contains high concentrations of hydrogen sulfide. Prokaryotes that live by these deep-sea vents obtain energy and produce carbohydrates through chemosynthesis instead of photosynthesis. Like plants, they are autotrophs; they extract energy from hydrogen sulfide to manufacture food, much as a plant extracts energy from the sun to manufacture its food. These prokaryotes live symbiotically within the tissues of heterotrophs that live around the deep-sea vents. The animals provide a place for the prokaryotes to live and obtain nutrients, and in turn the prokaryotes supply the animal with organic compounds to use as food.

Despite the many new forms of small invertebrates now being discovered on the seafloor, and the huge biomass that occurs in the sea, more than 90% of all *described* species of organisms occur on land. Each of the largest groups of organisms, including insects, mites, nematodes, fungi, and plants, has marine representatives, but they constitute only a very small fraction of the total number of described species.

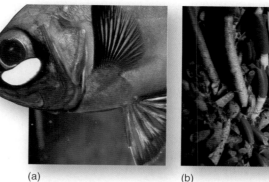

(a)

(b)

(c)

Figure 20.20   Deep-sea waters.

(a) The luminous spot below the eye of this deep-sea fish results from the presence of a symbiotic colony of luminous bacteria. (b) These giant beardworms live along vents where water jets from fissures at 350°C and then cools to the 2°C of the surrounding water. (c) Looking for all the world like some undersea sunflower, these two sea anemones (actually animals) use a glass-sponge stalk to catch "marine snow," food particles raining down on the ocean floor from the ocean surface several kilometers above.

20.9   The three principal ocean ecosystems occur in shallow water, in the open-sea surface, and along the deep-sea bottom. Both intertidal shallows and deep-sea communities are very diverse.

Freshwater Ecosystems

Freshwater ecosystems (lakes, ponds, rivers, and wetlands) are distinct from both ocean and land ecosystems, and they are very limited in area. Inland lakes cover about 1.8% of the earth's surface and rivers, streams, and wetlands about 0.4%. All freshwater habitats are strongly connected to land habitats, with marshes and swamps (wetlands) constituting intermediate habitats. In addition, a large amount of organic and inorganic material continually enters bodies of freshwater from communities growing on the land nearby (figure 20.21). Many kinds of organisms are restricted to freshwater habitats (figure 20.22). When they occur in rivers and streams, they must be able to attach themselves in such a way as to resist or avoid the effects of current or risk being swept away.

Like the ocean, ponds and lakes have three zones in which organisms live: a shallow "edge" zone (the littoral zone in figure 20.23), an open-water surface zone (the limnetic zone), and a deep-water zone where light does not penetrate (the profundal zone). **Thermal stratification,** characteristic of the larger lakes in temperate regions, is the process whereby water at a temperature of 4°C (which is when water is most dense) sinks beneath water that is either warmer or cooler. Follow through the changes in a large lake in figure 20.24 beginning in winter ❶, where water at 4°C sinks beneath cooler water that freezes at the surface at 0°C. Below the ice, the water remains between 0° and 4°C, and plants and animals survive there. In spring ❷, as the ice melts, the surface water is warmed to 4°C and sinks below the cooler water, bringing the cooler water to the top with nutrients from the lake's lower regions. This process is known as the *spring overturn.*

In summer ❸, warmer water forms a layer over the cooler water that lies below. In the area between these two layers, called the *thermocline,* temperature changes abruptly. You may have experienced the existence of these layers if you have dived into a pond in temperate regions in the summer. Depending on the climate of the particular area, the warm upper layer may become as much as 20 meters thick during the summer. In

(a)

Figure 20.21   A nutrient-rich stream.
In this stream in the northern coastal mountains of California, as in all streams, much organic material falls or seeps into the water from communities along the edges. This input is responsible for much of the stream's biological productivity.

(b)

Figure 20.22   Freshwater organisms.
(a) This speckled darter and (b) this giant waterbug with eggs on its back can only live in freshwater habitats.

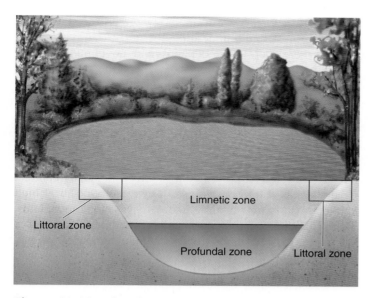

Figure 20.23　The three zones in ponds and lakes.
A shallow "edge" (littoral) zone lines the periphery of the lake where attached algae and their insect herbivores live. An open-water surface (limnetic) zone lies across the entire lake and is inhabited by floating algae, zooplankton, and fish. A dark, deep-water (profundal) zone overlies the sediments at the bottom of the lake. The profundal zone contains numerous bacteria and wormlike organisms that consume dead debris settling at the bottom of the lake.

autumn ❹, its surface temperature drops until it reaches that of the cooler layer underneath—4°C. When this occurs, the upper and lower layers mix—a process called the *fall overturn.* Therefore, colder waters reach the surfaces of lakes in the spring and fall, bringing up fresh supplies of dissolved nutrients.

Lakes can be divided into two categories, based on their production of organic material. **Eutrophic lakes** have an abundant supply of minerals and organic matter. Oxygen is depleted below the thermocline in the summer because of the abundant organic material and high rate at which aerobic decomposers in the lower layer use oxygen. These stagnant waters again reach the surface after the fall overturn and are then infused with more oxygen. In **oligotrophic lakes,** on the other hand, organic matter and nutrients are relatively scarce. Such lakes are often deeper than eutrophic ones, and their deep waters are always rich in oxygen. Oligotrophic lakes are highly susceptible to pollution from excess phosphorus from such sources as fertilizer runoff, sewage, and detergents.

> **20.10**　Freshwater ecosystems cover only about 2% of the earth's surface; all are strongly tied to adjacent terrestrial ecosystems. In some, organic materials are common, and in others, scarce. The temperature zones in lakes overturn twice a year, in spring and fall.

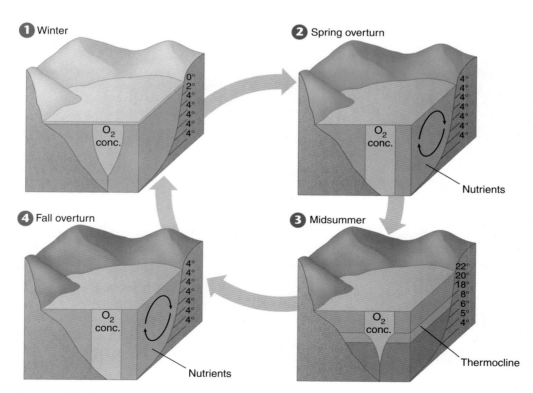

Figure 20.24　Spring and fall overturns in freshwater ponds or lakes.
The pattern of stratification in a large pond or lake in temperate regions is upset in the spring and fall overturns. Of the three layers of water shown in midsummer (*lower right*), the densest water occurs at 4°C. The warmer water at the surface is less dense. The thermocline is the zone of abrupt change in temperature that lies between them.

Living on land ourselves, we humans tend to focus much of our attention on terrestrial ecosystems. A **biome** is a terrestrial ecosystem that occurs over a broad area. Each biome is characterized by a particular climate and a defined group of organisms.

While biomes can be classified in a number of ways, the seven most widely occurring biomes (color-coded in figure 20.25) are (1) tropical rain forest (dark green), (2) savanna (pink), (3) desert (pale yellow), (4) temperate grassland (tan), (5) temperate deciduous forest (brown), (6) taiga (purple), and (7) tundra (light blue). The reason that there are seven primary biomes, and not one or 80, is that they have evolved to suit the climate of the region, and the earth has seven principal climates. The seven biomes differ remarkably from one another but show many consistencies within; a particular biome often looks similar, with many of the same types of creatures living there, wherever it occurs on earth.

There are seven other less widespread biomes also shown in figure 20.25: chaparral; polar ice; mountain zone; temperate evergreen forest; warm, moist evergreen forest; tropical monsoon forest; and semidesert.

If there were no mountains and no climatic effects caused by the irregular outlines of the continents and by different sea temperatures, each biome would form an even belt around the globe. In fact, their distribution is greatly affected by these factors, especially by elevation. Thus, the summits of the Rocky Mountains are covered with a vegetation type that resembles tundra, whereas other forest types that resemble taiga occur farther down. It is for reasons such as these that the distributions of the biomes are so irregular. One trend that is apparent is that those biomes that normally occur at high latitudes also follow an altitudinal gradient along mountains. That is, biomes found far north and far south of the equator at sea level also occur in the tropics but at high mountain elevations (see figure 20.14).

Distinctive features of the seven major biomes—tropical rain forest, savanna, desert, temperate grassland, temperate deciduous forest, taiga, and tundra—along with several of the less widespread biomes are now discussed in more detail.

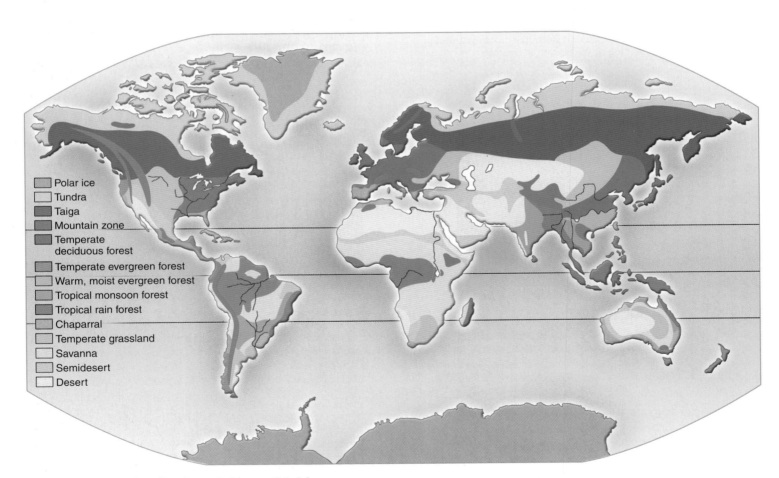

Figure 20.25   Distribution of the earth's biomes.
The seven primary types of biomes are tropical rain forest, savanna, desert, temperate grassland, temperate deciduous forest, taiga, and tundra. In addition, seven less widespread biomes are shown.

Lush Tropical Rain Forests

Rain forests, which experience over 250 centimeters of rain a year, are the richest ecosystems on earth (figure 20.26). They contain at least half of the earth's species of terrestrial plants and animals—more than 2 million species! In a single square mile of tropical forest in Rondonia, Brazil, there are 1,200 species of butterflies—twice the total number found in the United States and Canada combined. The communities that make up tropical rain forests are diverse in that each kind of animal, plant, or microorganism is often represented in a given area by very few individuals. There are extensive tropical rain forests in South America, Africa, and Southeast Asia. But the world's tropical rain forests are being destroyed, and with them, countless species, many of them never seen by humans. Perhaps a quarter of the world's species will disappear with the rain forests during the lifetime of many of us.

Figure 20.26   Tropical rain forest.

Savannas: Dry Tropical Grasslands

In the dry climates that border the tropics are found the world's great grasslands, called **savannas.** Landscapes are open, often with widely spaced trees, and rainfall (75 to 125 cm annually) is seasonal. Many of the animals and plants are active only during the rainy season. The huge herds of grazing animals that inhabit the African savanna are familiar to all of us (figure 20.27). Such animal communities occurred in the temperate grasslands of North America during the Pleistocene epoch but have persisted mainly in Africa. On a global scale, the savanna biome is transitional between tropical rain forest and desert. As these savannas are increasingly converted to agricultural use to feed rapidly expanding human populations in subtropical areas, their inhabitants are finding it difficult to survive. The elephant and rhino are now endangered species; the lion, giraffe, and cheetah will soon follow them.

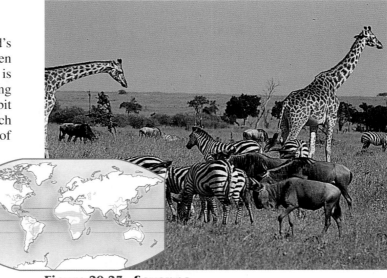

Figure 20.27   Savanna.

Deserts: Burning Hot Sands

In the interior of continents are found the world's great deserts, especially in Africa (the Sahara), Asia (the Gobi), and Australia (the Great Sandy Desert). **Deserts** are dry places where less than 25 centimeters of rain falls in a year—an amount so low that vegetation is sparse and survival depends on water conservation (figure 20.28). Plants and animals may restrict their activity to favorable times of the year, when water is present. To avoid high temperatures, most desert vertebrates live in deep, cool, and sometimes even somewhat moist burrows. Those that are active over a greater portion of the year emerge only at night, when temperatures are relatively cool. Some, such as camels, can drink large quantities of water when it is available and then survive long, dry periods. Many animals simply migrate to or through the desert, where they exploit food that may be abundant seasonally.

Figure 20.28   Desert.

Grasslands: Seas of Grass

Halfway between the equator and the poles are temperate regions where rich **grasslands** grow. These grasslands once covered much of the interior of North America, and they were widespread in Eurasia and South America as well. Such grasslands are often highly productive when converted to agriculture. Many of the rich agricultural lands in the United States and southern Canada were originally occupied by **prairies,** another name for temperate grasslands. The roots of perennial grasses characteristically penetrate far into the soil, and grassland soils tend to be deep and fertile. Temperate grasslands are often populated by herds of grazing mammals. In North America, the prairies were once inhabited by huge herds of bison and pronghorns (figure 20.29). The herds are almost all gone now, with most of the prairies having been converted to the richest agricultural region on earth.

Figure 20.29   Temperate grassland.

Deciduous Forests: Rich Hardwood Forests

Mild climates (warm summers and cool winters) and plentiful rains promote the growth of **deciduous** ("hardwood") **forests** in Eurasia, the northeastern United States, and eastern Canada (figure 20.30). A deciduous tree is one that drops its leaves in the winter. Deer, bears, beavers, and raccoons are the familiar animals of the temperate regions. Because the temperate deciduous forests represent the remnants of more extensive forests that stretched across North America and Eurasia several million years ago, these remaining areas—especially those in eastern Asia and eastern North America—share animals and plants that were once more widespread. Alligators, for example, are found only in China and in the southeastern United States. The deciduous forest in eastern Asia is rich in species because climatic conditions have remained constant.

Figure 20.30   Temperate deciduous forest.

Taiga: Trackless Conifer Forests

A great ring of northern forests of coniferous trees (spruce, hemlock, larch, and fir) extends across vast areas of Asia and North America. Coniferous trees are ones with leaves like needles that are kept all year long. This ecosystem, called **taiga,** is one of the largest on earth (figure 20.31). Here, the winters are long and cold. Rain, often as little as in hot deserts, falls in the summer. Because it has too short a growing season for farming, few people live there. Many large mammals, including elk, moose, deer, and such carnivores as wolves, bears, lynx, and wolverines, live in the taiga. Traditionally, fur trapping has been extensive in this region. Lumber production is also important. Marshes, lakes, and ponds are common and are often fringed by willows or birches. Most of the trees occur in dense stands of one or a few species.

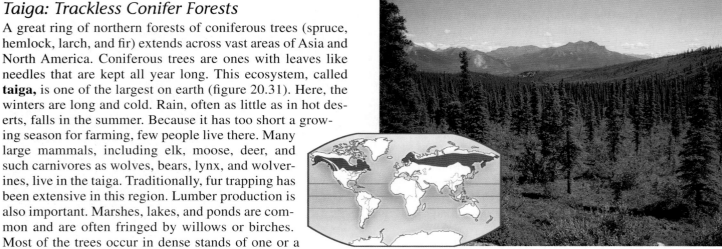

Figure 20.31   Taiga.

Tundra: Cold Boggy Plains

In the far north, above the great coniferous forests and below the polar ice, there are few trees. There the grassland, called **tundra,** is open, windswept, and often boggy (figure 20.32). Enormous in extent, this ecosystem covers one-fifth of the earth's land surface. Very little rain or snow falls. When rain does fall during the brief arctic summer, it sits on frozen ground, creating a sea of boggy ground. **Permafrost,** or permanent ice, usually exists within a meter of the surface. Trees are small and are mostly confined to the margins of streams and lakes. Large grazing mammals, including musk-oxen, caribou, reindeer, and carnivores such as wolves, foxes, and lynx, live in the tundra. Lemming populations rise and fall on a long-term cycle, with important effects on the animals that prey on them.

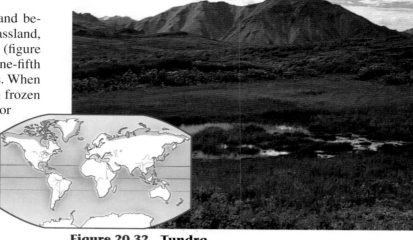

Figure 20.32   Tundra.

Chaparral

Chaparral consists of evergreen, often spiny shrubs and low trees that form communities in regions with what is called a "Mediterranean," dry summer climate: the Mediterranean area itself, California, central Chile, the Cape region of South Africa, and southwestern Australia (figure 20.33). Many plant species found in chaparral can germinate only when they have been exposed to the hot temperatures generated during a fire. The chaparral of California and adjacent regions is historically derived from deciduous forests.

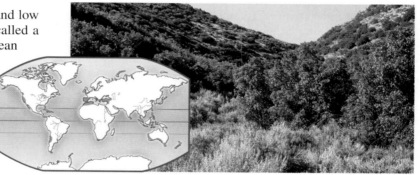

Figure 20.33   Chaparral.

Polar Ice Caps

Polar ice caps lie over the Arctic Ocean in the north and Antarctica in the south (figure 20.34). The poles receive almost no precipitation, so although ice is abundant, freshwater is scarce. The sun barely rises in the winter months. Life in Antarctica is largely limited to the coasts. Because the Antarctic ice cap lies over a landmass, it is not warmed by the latent heat of circulating ocean water and becomes very cold. As a result, only prokaryotes, algae, and some small insects inhabit the vast Antarctic interior.

Figure 20.34   Polar ice.

Tropical Upland Forest

Tropical upland forests (figure 20.35) occur in the tropics and semitropics at slightly higher latitudes than rain forests or where local climates are drier. Most trees in these forests are deciduous, losing many of their leaves during the dry season. This loss of leaves allows sunlight to penetrate to the understory and ground levels of the forest, where a dense layer of shrubs and small trees grow rapidly. Leaves are smaller, and photosynthesis is depressed because of lower temperatures. Rainfall is typically very seasonal, measuring several inches daily in the monsoon season and approaching drought conditions in the dry season, particularly in locations far from oceans, such as in central India. Tropical upland forests are found in Bangladesh, southeastern Asia, China, the West Indies, Central and South America, and Australia.

Figure 20.35   Tropical monsoon forest.

Semidesert

Semidesert areas occur in regions with less rain than monsoon forests but more rain than savannas (figure 20.36). Vegetation is dominated by bushes and trees with thorns and spikes, which is why these regions are also known as thornwood forests. Little or no rain falls for eight or nine months during the winter. Plants survive on one or a few short heavy rains received during the summer wet season, growing intensively in response to the moisture. The brief rain is followed by the long dry season, when leaves fall from plants and little or no growth occurs. Semideserts are found along the edges of desert biomes such as in Africa, Asia, and Australia.

Other Biomes

Other biomes include the mountain (alpine) zone, temperate evergreen forests, and warm moist evergreen forests. Mountain zone areas are similar to tundras because the increasing altitude produces many of the same changes in temperature and moisture as seen with increasing latitudes (see figure 20.14). The tops of mountains have a typical windswept vegetation similar in many respects to tundra. Few if any trees are able to grow in this alpine zone, which, like the polar tundra, is alive with life in the warm summer months. During the harsh winters, little grows. Temperate evergreen forests occur in regions where winters are cold and there is a strong, seasonal dry period. The pine forests of the western United States, the California oak woodlands, and the Australian eucalyptus forests are typical temperate evergreen forests. Many of these forests are endangered by overlogging, particularly in the western United

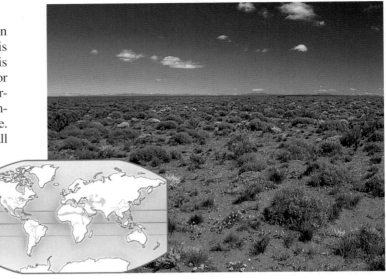

Figure 20.36 Semidesert.

States. Warm, moist evergreen forests occur in temperate regions where winters are mild and moisture is plentiful. These can be seen in central China, in the pine forests covering much of the southeastern United States, and in the coastal redwood forest of northern California.

> **20.11 Biomes are major terrestrial communities defined largely by temperature and rainfall patterns.**

INQUIRY & ANALYSIS

How Do Ecosystems Cycle Nitrogen?

The way nutrients cycle in ecosystems has been studied in a classic experiment at the Hubbard Brook Experimental Forest in New Hampshire. Concrete dams were constructed across the six streams that drain the forest, and the runoff examined. The forest proved very efficient at retaining nitrogen and other nutrients. In the winter of 1965 the investigators felled all the trees and shrubs in 48 acres drained by one stream (as shown in the photo), and examined the water running off. The red line in the graph shows nitrogen minerals leaving the ecosystem in the runoff water from this stream; the blue line shows the nitrogen runoff in a neighboring stream draining an uncut portion of the forest.

1. **Applying Concepts** What is the dependent variable?
2. **Making Inferences**
   a. Is there any yearly pattern to the nitrogen runoff in the uncut forest? Can you explain it?
   b. How does the loss of nitrogen from the ecosystem in the cut forest compare with nitrogen loss from the uncut forest?

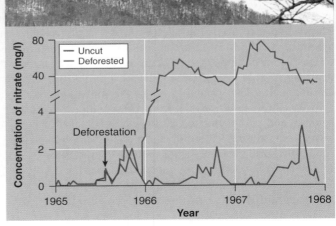

The Hubbard Brook Deforestation Experiment

3. **Drawing Conclusions** What is the impact of this forest's trees upon its ability to retain nitrogen?

The Energy in Ecosystems

20.1 Energy Flows Through Ecosystems

- Energy flows through an ecosystem by way of food chains. Energy from the sun is captured by photosynthetic producers, which are eaten by herbivores, which are in turn eaten by carnivores. Organisms at all trophic levels die and are consumed by detritivores and decomposers (**figures 20.1** and **20.2**). Energy is lost at every level, such that only about 10% of available energy in an organism is passed on to organisms at the next trophic level (**figures 20.4** and **20.5**).

20.2 Ecological Pyramids

- Because energy is lost as it passes up through the trophic levels of the food chain, there tends to be more individuals at the lower trophic levels (that is, there are more producers than herbivores and more herbivores than carnivores). Ecological pyramids illustrate this distribution of energy and biomass (**figure 20.6**).

Materials Cycle Within Ecosystems

20.3 The Water Cycle

- Physical components of the ecosystem cycle through the ecosystem, being used then recycled then reused. This cycling of materials often involves living organisms and is referred to as a biogeochemical cycle.

- Water availability determines how many and what kinds of organisms can be supported in the ecosystem. Water cycles from the atmosphere as precipitation, where it falls to the earth and is heated. Upon heating, water reenters the atmosphere through evaporation. Water also cycles through plants, entering through the roots and leaving as water vapor in transpiration (**figure 20.7**).

20.4 The Carbon Cycle

- Carbon cycles through plants—via carbon fixation in photosynthesis—which are then eaten by animals. It returns to the atmosphere as CO_2 from cellular respiration. Carbon also cycles through the ecosystem by diffusion and by the burning of fossil fuels (**figure 20.9**).

20.5 Soil Nutrients and Other Chemical Cycles

- Nitrogen gas in the atmosphere cannot be readily used by organisms and needs to be fixed by certain types of bacteria into ammonia and nitrate. Animals eat plants that have taken up the fixed nitrogen. Nitrogen reenters the ecosystem through animal excretion and decomposition through detritivores and decomposers (**figure 20.10**).

- Phosphorus also cycles through the ecosystem and may limit growth when not available (**figure 20.11**).

How Weather Shapes Ecosystems

20.6 The Sun and Atmospheric Circulation

- The heating power of the sun affects evaporation and air currents, causing certain parts of the globe, such as the tropics, to have larger amounts of precipitation (**figures 20.12** and **20.13**).

20.7 Latitude and Elevation

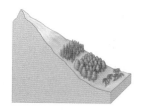

- Temperature and precipitation are similarly affected by elevation and latitude. Changes in ecosystems seen going from the equator to the poles are similarly reflected in the changes in ecosystems seen going from sea level to mountaintops (**figure 20.14**). The changes in temperature also cause the rain shadow effect, where precipitation is deposited on the windward side of mountains (**figure 20.15**).

20.8 Patterns of Circulation in the Ocean

- The earth's oceans circulate in patterns that distribute warmer and cooler waters to different areas of the world. These ocean patterns affect climates across the globe (**figures 20.16** and **20.17**).

Major Kinds of Ecosystems

20.9 Ocean Ecosystems

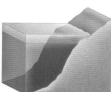

- There are three primary ocean ecosystems: shallow waters, open-sea surfaces, and deep-sea bottoms (**figure 20.18**). Each is affected by light and temperature.

20.10 Freshwater Ecosystems

- Freshwater ecosystems are closely tied to and affected by the terrestrial environments that surround them. Freshwater ecosystems are affected by light, temperature, and nutrients (**figures 20.23** and **20.24**).

20.11 Land Ecosystems

- Biomes are terrestrial communities found throughout the world. Each biome contains its own characteristic representation of plants and animals based on temperature and rainfall patterns (**figure 20.25**).

- There are 11 primary biomes: tropical rain forest, savanna, desert, grassland, deciduous forest, taiga, tundra, chaparral, polar ice cap, tropical monsoon forest, and semidesert.

1. Energy from the sun is converted into chemical energy for organisms by
   a. herbivores.
   b. carnivores.
   c. producers.
   d. detritivores.

2. As energy is transferred from one trophic level to the next, substantial amounts of energy are lost to/as
   a. undigestible biomass.
   b. heat.
   c. metabolism.
   d. All answers are correct.

3. The number of carnivores found at the top of an ecological pyramid is limited by the
   a. number of organisms below the top carnivores.
   b. number of trophic levels below the top carnivores.
   c. amount of biomass below the top carnivores.
   d. amount of energy transferred to the top carnivores.

4. Hydrologists, scientists who study the movements and cycles of water, refer to the return of water from the ground to the air as evapotranspiration. The first part of the word refers to evaporation. The second part of the word refers to transpiration, which is evaporation of water
   a. from plants.
   b. through animal perspiration.
   c. off the ground shaded by plants.
   d. from the surface of rivers.

5. The carbon cycle includes the gathering of
   a. materials to make proteins.
   b. light energy through photosynthesis.
   c. phosphorus to make DNA.
   d. No answer is correct.

6. The element phosphorus is needed in organisms to build
   a. proteins.
   b. carbohydrates.
   c. ATP.
   d. steroids.

7. A rain shadow results in
   a. extremely wet conditions due to loss of moisture from winds rising over a mountain range.
   b. dry air moving toward the poles that cools and sinks in regions 15 to 30 degrees north/south latitude.
   c. global polar regions that rarely receive moisture from the warmer, tropical regions, and are therefore dryer.
   d. desert conditions on the downwind side of a mountain due to increased moisture-holding capacity of the winds as the air heats up.

8. As one travels from northern Canada south to the United States, the timberline increases in elevation. This is because as latitude
   a. increases, temperature decreases.
   b. decreases, temperature decreases.
   c. increases, humidity decreases.
   d. decreases, humidity increases.

9. In freshwater lakes during the summer, layers of sudden temperature change called _____ form.
   a. eutrophy
   b. thermal restratification
   c. oligotrophy
   d. thermocline

10. A temperate evergreen forest can exist in the southwest deserts of the United States if the forest exists
    a. in desert riparian (stream or river) areas.
    b. near natural springs.
    c. near the top of a desert mountain range.
    d. in the desert flatlands.

1. **Figure 20.5** A. One thousand calories of energy is produced by a certain amount of algae. Explain the probable efficiency of having humans eat, respectively, (1) the algae itself, (2) the heterotrophs, (3) the smelt, or (4) the trout. B. How many more people could be fed if we all ate algae? C. Why can't we survive just eating algae?

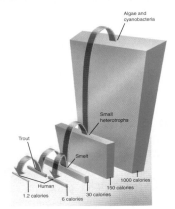

2. **Figure 20.7** Imagine that you are a molecule of water. Create a journey for yourself, starting with falling as part of a drop of rain onto the earth, and ending in a cloud ready to fall once again. Take a trip through a plant along the way.

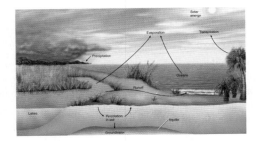

1. Given the amount of sunlight that hits the plants on our planet, and the ability of plants for rapid growth and reproduction, how come we aren't all hip deep in dead plants?

2. Plants need both carbon (C) and nitrogen (N) in large amounts. Compare and contrast how plants obtain these two important elements.

3. Why do increasing latitude and increasing elevation affect, in the same way, which plant species grow at a place?

4. Pick two very different land ecosystems. Compare and contrast the sunshine, rainfall, major temperature features, and something about the plants and animals.

21

Populations and Communities

Often the most significant events that occur in an ecosystem involve the organisms that inhabit it. The swarming insects you see here are migratory locusts, *Locusta migratoria,* moving across farmland in North Africa in 1988. In most years, the locusts are not plentiful and do not swarm. In particularly favorable years, however, when food is plentiful and the weather mild, the abundance of resources leads to greater-than-usual growth of locust populations. When high population densities are reached, the locusts exhibit different hormonal and physical characteristics and take off as a swarm. Moving over the landscape, the swarm eats every available plant, denuding the landscape. Swarming locusts, although not common in North America, are a legendary plague of large areas of Africa and Eurasia. In this chapter, we examine how natural populations grow, and what factors limit this growth. The organisms of the living world have evolved many accommodations to facilitate living together, creating complex evolutionary arrangements. When these arrangements are disturbed by unusual weather—or human intervention—the consequences can be catastrophic.

Successfully reproducing is the very essence of evolutionary success, and few organisms do it alone. Rather, organisms live as members of **populations,** groups of individuals of a species that live together and influence each other's survival. In this chapter, we will explore the properties of populations, focusing on the factors that influence whether a population will grow or shrink, and at what rate. Although we humans picture ourselves as different from populations of animals living in the wild, factors that affect wild populations, such as population densities, dispersion, growth, competition, and sharing resources, affect human populations in similar ways.

21.1 Population Growth

One of the critical properties of any population is its **population size**—the number of individuals in the population. For example, if an entire species consists of only one or a few small populations, that species is likely to become extinct, especially if it occurs in areas that have been or are being radically changed. In addition to population size, **population density**—the number of individuals that occur in a unit area, such as per square kilometer—is often an important characteristic. The density of a population, how closely individuals associate with each other, is an indication of how they live. Animals that live in large groups, such as herds of wildebeests or zebras, may find safety in numbers. A third significant property is **population dispersion,** the scatter of individual organisms within the population's range. As you can see in figure 21.1, individuals may be spaced in clumps (due to uneven distribution of resources or in response to social interaction), uniformly (as a result of competition for resources), or randomly, which is less common in nature. In addition to size, density, and dispersion, another key characteristic of any population is its capacity to grow. To understand populations we must consider this, and what factors in nature limit **population growth.**

The Exponential Growth Model

The simplest model of population growth assumes a population growing without limits at its maximal rate. This rate, symbolized r and called the **biotic potential,** is the rate at which a population of a given species will increase when no limits are placed on its rate of growth. In mathematical terms, this is defined by the following formula:

$$growth\ rate = dN/dt = r_i N$$

where N is the number of individuals in the population, dN/dt is the rate of change in its numbers over time, and r_i is the *intrinsic* rate of natural increase for that population—its innate capacity for growth.

The *actual* rate of population increase r is defined as the difference between the birthrate b and the death rate d corrected for any movement of individuals in or out of the popu-

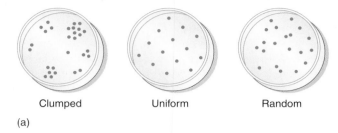

Clumped   Uniform   Random

(a)

(b)

Figure 21.1 Population dispersion.
(a) Different arrangements of bacterial colonies, and (b) starlings distributed uniformly along telephone wires.

lation, whether net emigration (movement out of the area e) or net immigration (movement into the area i). Thus,

$$r = (b - d) + (i - e)$$

Movements of individuals can have a major impact on population growth rates. For example, the increase in human population in the United States during the closing decades of the twentieth century was mostly due to immigrants. Less than half of the increase came from the reproduction of the people already living there.

The innate capacity for growth of any population is exponential, and is called *exponential growth.* Even when the *rate* of increase remains constant, the actual increase in the *number* of individuals accelerates rapidly as the size of the population grows. Rapid exponential growth is indicated by the red line in figure 21.2. This sort of growth pattern is similar to that obtained by compounding interest on an investment. In practice, such patterns prevail only for short periods, usually when an organism reaches a new habitat with abundant resources. Natural examples include dandelions reaching the fields, lawns, and meadows of North America from Europe for the first time; algae colonizing a newly formed pond; or the first plants arriving on an island recently thrust up from the sea.

Carrying Capacity

No matter how rapidly populations grow, they eventually reach a limit imposed by shortages of important environmental factors such as space, light, water, or nutrients. A population ultimately stabilizes at a certain size, called the **carrying capacity** of the particular place where it lives, and the size of the population levels off, like the blue line in figure 21.2. The carrying capacity, symbolized by K, is the maximum number of individuals that an area can support.

The Logistic Growth Model

As a population approaches its carrying capacity, its rate of growth slows greatly, because fewer resources remain for each new individual to use. The growth curve of such a population, which is always limited by one or more factors in the environment, can be approximated by the following **logistic growth equation** that adjusts the growth rate to account for the lessening availability of limiting factors:

$$dN/dt = rN \left(\frac{K - N}{K} \right)$$

In this logistic model of population growth, the growth rate of the population (dN/dt) equals its rate of increase (r multiplied by N, the number of individuals present at any one time), adjusted for the amount of resources available. The adjustment is made by multiplying rN by the fraction of K still unused (K minus N, divided by K). As N increases (the population grows in size), the fraction by which r is multiplied (the remaining resources) becomes smaller and smaller, and the rate of increase of the population declines.

In mathematical terms, as N approaches K, the rate of population growth (dN/dt) begins to slow, until it reaches 0 when $N = K$ (the blue line in figure 21.2). In practical terms, factors such as increasing competition among more individuals for a given set of resources, the buildup of waste, or an increased rate of predation causes the decline in the rate of population growth.

Graphically, if you plot N versus t (time), you obtain an S-shaped **sigmoid growth curve** characteristic of most biological populations. The curve is called "sigmoid" because its shape has a double curve like the letter S. As the size of a population stabilizes at the carrying capacity, its rate of growth slows down, eventually coming to a halt. The fur seal population in figure 21.3 has a carrying capacity of about 10,000 seals.

Processes such as competition for resources, emigration, and the accumulation of toxic wastes all tend to increase as a population approaches its carrying capacity for a particular habitat. The resources for which the members of the population are competing may be food, shelter, light, mating sites, mates, or any other factor needed to survive and reproduce.

> **21.1** The size at which a population stabilizes in a particular place is defined as the carrying capacity of that place for that species.
> Populations grow to the carrying capacity of their environment.

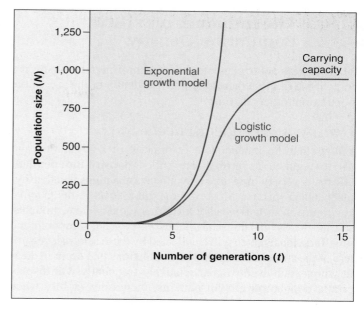

Figure 21.2 Two models of population growth.
The *red line* illustrates the exponential growth model for a population with an r of 1.0. The *blue line* illustrates the logistic growth model in a population with $r = 1.0$ and $K = 1,000$ individuals. At first, logistic growth accelerates exponentially, and then, as resources become limiting, the death rate increases and growth slows. Growth ceases when the death rate equals the birthrate. The carrying capacity (K) ultimately depends on the resources available in the environment.

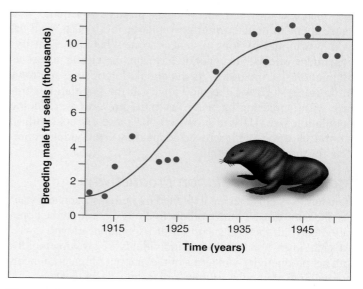

Figure 21.3 Most natural populations exhibit logistic growth.
These data present the history of a fur seal (*Callorhinus ursinus*) population on St. Paul Island, Alaska. Driven almost to extinction by hunting in the late 1800s, the fur seal made a comeback after hunting was banned in 1911. Today the number of breeding males with "harems" oscillates around 10,000 individuals, presumably the carrying capacity of the island for fur seals.

The Influence of Population Density

Many factors act to regulate the growth of populations in nature. Some of these factors act independently of the size of the population; others do not.

Density-Independent Effects

Effects that are independent of the size of a population and act to regulate its growth are called **density-independent effects.** A variety of factors may affect populations in a density-independent manner. Most of these are aspects of the external environment, such as weather (extremely cold winters, droughts, storms, floods) and physical disruptions (volcanic eruptions and fire). Individuals often will be affected by these activities regardless of the size of the population. Populations that occur in areas in which such events occur relatively frequently will display erratic population growth patterns, increasing rapidly when conditions are relatively good, but suffering extreme reductions whenever the environment turns hostile.

Density-Dependent Effects

Effects that are dependent on the size of the population and act to regulate its growth are called **density-dependent effects.** Among animals, these effects may be accompanied by hormonal changes that can alter behavior that will directly affect the ultimate size of the population. One striking example occurs in migratory locusts ("short-horned" grasshoppers, which you encountered at the beginning of this chapter). When they become crowded, the locusts produce hormones that cause them to enter a migratory phase; the locusts take off as a swarm and fly long distances to new habitats. Density-dependent effects, in general, have an increasing effect as population size increases. As the population of song sparrows in figure 21.4 grows, the individuals in the population compete with increasing intensity for limited resources, reducing population size. Darwin proposed that these effects result in natural selection and improved adaptation as individuals compete for the limiting factors.

Maximizing Population Productivity

In natural systems that are exploited by humans, such as fisheries, the aim is to maximize productivity by exploiting the population early in the rising portion of its sigmoid growth curve. At such times, populations and individuals are growing rapidly, and net productivity—in terms of the amount of material incorporated into the bodies of these organisms—is highest.

Commercial fisheries attempt to operate so that they are always harvesting populations in the steep, rapidly growing parts of the curve. The point of *maximal sustainable yield* (the red line in figure 21.5) lies partway up the sigmoid curve. Harvesting the population of an economically desirable species near this point will result in the best sustained yields. Overharvesting a population that is smaller than this critical size can destroy its productivity for many years or even drive it to extinction. This evidently happened in the Peruvian anchovy fishery after the populations had been depressed by the 1972

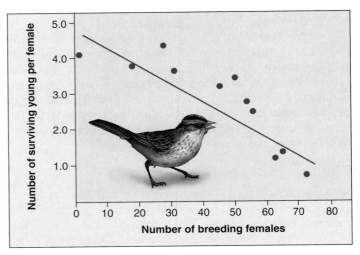

Figure 21.4   Density-dependent effects.
Reproductive success of the song sparrow (*Melospiza melodia*) decreases as population size increases.

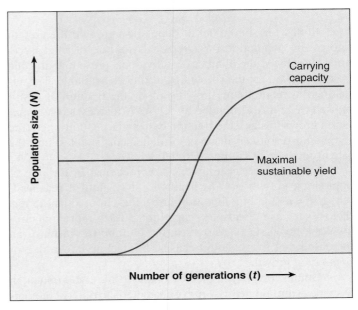

Figure 21.5   Maximal sustainable yield.
The goal of harvesting organisms for commercial purposes is to harvest just enough organisms to maximize current yields but also to sustain the population for future yields. Harvesting the organisms when the population is in the rapid growth phase of the sigmoidal curve, but not overharvesting, will result in sustained yields.

El Niño. It is often difficult to determine population levels of commercially valuable species, and without this information it is equally difficult to determine the yield most suitable for long-term, productive harvesting.

> **21.2   Density-independent effects are controlled by factors that operate regardless of population size; density-dependent effects are caused by factors that come into play particularly when the population size is larger.**

21.6 The Niche and Competition

Each organism in an ecosystem confronts the challenge of survival in a different way. As we discussed in chapter 2, the **niche** an organism occupies is the sum total of all the ways it utilizes the resources of its environment. A niche may be described in terms of space utilization, food consumption, temperature range, appropriate conditions for mating, requirements for moisture, and other factors.

Sometimes species are not able to occupy their entire niche because of the presence or absence of other species. Species can interact with each other in a number of ways, and these interactions can either have positive or negative effects. **Competition** describes the interaction when two organisms attempt to use the same resource when there is not enough of the resource to satisfy both.

Competition between individuals of different species is called **interspecific competition.** As mentioned in chapter 2, interspecific competition is often greatest between organisms that obtain their food in similar ways and between organisms that are more similar. Another type of competition can occur, called **intraspecific competition.** Intraspecific competition is competition between individuals of the same species, whereas interspecific competition occurs between individuals of different species.

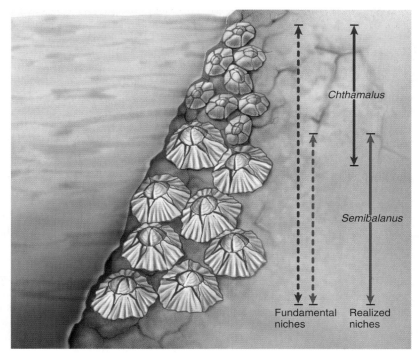

Figure 21.9   Competition among two species of barnacles limits niche use.

Chthamalus can live in both deep and shallow zones (its fundamental niche), but *Semibalanus* forces *Chthamalus* out of the part of its fundamental niche that overlaps the realized niche of *Semibalanus*.

The Realized Niche

Because of competition, organisms may not be able to occupy the entire niche they are theoretically capable of using, called the **fundamental niche** (or theoretical niche). The actual niche the organism is able to occupy in the presence of competitors is called its **realized niche.**

In a classic study, J. H. Connell of the University of California, Santa Barbara, investigated competitive interactions between two species of barnacles that grow together on rocks along the coast of Scotland. Barnacles are marine animals (crustaceans) that have free-swimming larvae. The larvae eventually settle down, cementing themselves to rocks and remaining attached for the rest of their lives. Of the two species Connell studied, *Chthamalus stellatus* (the smaller barnacle in figure 21.9) lives in shallower water, where tidal action often exposes it to air, and *Semibalanus balanoides* (the larger barnacle) lives lower down, where it is rarely exposed to the atmosphere. In the deeper zone, *Semibalanus* could always outcompete *Chthamalus* by crowding it off the rocks, undercutting it, and replacing it even where it had begun to grow. When Connell removed *Semibalanus* from the area, however, *Chthamalus* was easily able to occupy the deeper zone, indicating that no physiological or other general obstacles prevented it from becoming established there. In contrast, *Semibalanus* could not survive in the shallow-water habitats where *Chthamalus* normally occurs; it evidently does not have the special physiological and morphological adaptations that allow *Chthamalus* to occupy this zone. Thus, the fundamental niche of the barnacle *Chthamalus* in Connell's experiments in Scotland included that of *Semibalanus* (the red dashed arrow), but its realized niche was much narrower (the red solid arrow) because *Chthamalus* was outcompeted by *Semibalanus* in its fundamental niche.

Predators, as well as competitors, can limit the realized niche of a species. In the previous example, *Chthamalus* was able to fully occupy its fundamental niche when there was no competition. This is often the case when a species first enters a new, very favorable habitat that presents it with adequate resources, little or no competition, and no predators. However, once resources begin to become limiting as the population approaches carrying capacity and other species begin to compete for them, and as predators begin to more frequently recognize them, the population will be forced into its realized niche. For example, a plant called the St. John's-wort was introduced and became widespread in open rangeland habitats in California. It occupied all of its fundamental niche until a species of beetle that feeds on the plant was introduced into the habitat. Populations of the plant then quickly decreased, and it is now only found in shady sites where the beetle cannot thrive.

Competitive Exclusion

In classic experiments carried out between 1934 and 1935, Russian ecologist G. F. Gause studied competition among three species of *Paramecium,* a tiny protist. All three species grew well alone in culture tubes (each represented by a graph in figure 21.10*a*), preying on bacteria and yeasts that fed on oatmeal suspended in the culture fluid. However, when Gause grew *P. aurelia* together with *P. caudatum* in the same culture tube (graph in figure 21.10*b*), the numbers of *P. caudatum* (the green line) always declined to extinction, leaving *P. aurelia* the only survivor. Why? Gause found *P. aurelia* was able to grow six times faster than its competitor, *P. caudatum,* because it was able to better use the limited available resources.

From experiments such as this, Gause formulated what is now called the *principle of competitive exclusion.* This principle states that if two species are competing for a resource, the species that uses the resource more efficiently will eventually eliminate the other locally—no two species with the same niche can coexist. This overlapping of niches leads to competition, but is one competitor always eliminated? No, as we shall soon see.

Niche Overlap

In a revealing experiment, Gause challenged *P. caudatum*—the defeated species in his earlier experiments—with a third species, *P. bursaria.* Because he expected these two species to also compete for the limited bacterial food supply, Gause thought one would win out, as had happened in his previous experiments. But that's not what happened. Instead, both species survived in the culture tubes (graph in figure 21.10*c*); the paramecia found a way to divide the food resources. How did they do it? In the upper part of the culture tubes, where

the oxygen concentration and bacterial density were high, *P. caudatum* dominated because it was better able to feed on bacteria. However, in the lower part of the tubes, the lower oxygen concentration favored the growth of a different potential food, yeast, and *P. bursaria* was better able to eat this food. The fundamental niche of each species was the whole culture tube, but the realized niche of each species was only a portion of the tube. This graph also demonstrates the negative effect competition had on the participants: competition was always detrimental to both species involved. Both species more than doubled their densities when grown without a competitor as when grown together.

Gause's principle of competitive exclusion can be restated to say that no two species can occupy the same niche indefinitely when resources are limiting. Certainly species can and do coexist while competing for the same resources; we have seen many examples of such relationships. Nevertheless, Gause's theory predicts that when two species are able to coexist on a long-term basis, either resources must not be limited or their niches will always differ in one or more features; otherwise, one species outcompetes the other and the extinction of the second species inevitably results, a process referred to as **competitive exclusion.**

Niche is a complex concept, involving all facets of the environment that are important to individual species. In recent years, a vigorous debate has arisen concerning the role of competitive exclusion, not only in determining the structure of communities but also in setting the course of evolution. When one or more resources suddenly become sharply limiting, as in periods of drought, the role of competition becomes much more obvious. However, in less immediately stressful situations, species act to avoid competition whenever possible. When their niches overlap, two outcomes are possible: competitive exclusion (winner takes

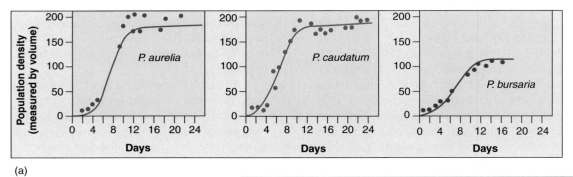

(a)

Figure 21.10 Competitive exclusion among three species of *Paramecium*.

In the microscopic world, *Paramecium* is a ferocious predator. *Paramecia* eat by ingesting their prey; their plasma membranes surround bacterial or yeast cells, forming a food vacuole containing the prey cell. In his experiments, (a) Gause found that three species of *Paramecium* grew well alone in culture tubes. (b) However, *P. caudatum* declined to extinction when grown with *P. aurelia* because they shared the same realized niche, and *P. aurelia* outcompeted *P. caudatum* for food resources. (c) However, *P. caudatum* and *P. bursaria* were able to coexist, although in smaller populations, because the two have different realized niches and thus avoid competition.

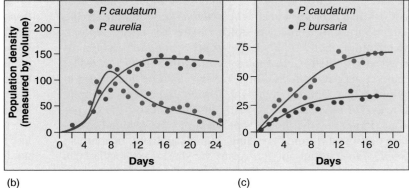

(b)   (c)

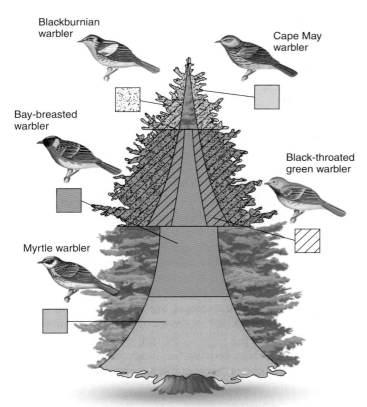

Figure 21.11 Resource partitioning in warblers.
Five different species of warblers appear to be competing for the
same resources, but they actually feed in different parts of the tree.

all), or **resource partitioning** (dividing up resources to create
two realized niches). It is only through resource partitioning, the
topic of the next section, that the two species can continue to
coexist over long periods.

Resource Partitioning

Gause's exclusion principle has a very important consequence:
persistent competition between two species is rare in natural
communities. Either one species drives the other to extinction,
or natural selection reduces the competition between them.
When the late Princeton ecologist Robert MacArthur studied
five species of warblers, small insect-eating forest songbirds,
he found that they all appeared to be competing for the same
resources. However, when he studied them more carefully, he
found that each species actually fed in a different part of
spruce trees and so ate different subsets of insects. His observations shown in figure 21.11 reveal some overlapping but, in
general: one species fed on insects near the tips of branches
(speckled areas), a second within the dense foliage (red areas),
a third on the lower branches (orange areas), a fourth high on
the trees (hatched areas), and a fifth at the very apex of the
trees (yellow areas). Thus, the species of warblers had *subdivided the niche*, partitioning the available resource so as to
avoid direct competition with one another.

Resource partitioning can often be seen in similar species that occupy the same geographical area. Called **sympatric**

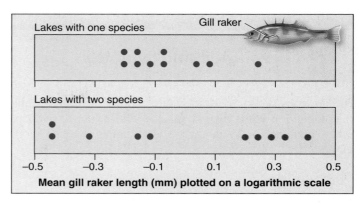

**Figure 21.12 Character displacement in stickleback
fish.**
In lakes that have two species of three-spined stickleback (*lower*
graph), one species (*blue* dots) feeds on plankton in open water and
has long gill rakers (structures on the gills of fish) that act like a
sieve to filter the plankton out. The other species (*red* dots) is more
of a bottom-dweller, has shorter gill rakers, and feeds on larger prey.
However, in lakes with only one species of stickleback (*upper* graph),
the fish take advantage of both food resources, and the gill rakers
are intermediate in length.

species (Greek, *syn*, same, and *patria*, country), these species
avoid competition by living in different portions of the habitat
or by using different food or other resources. For example, as
we saw in chapter 2, sympatric *Anolis* lizards species are able
to partition the tree habitat (see figure 2.20). Species that do
not live in the same geographical area, called **allopatric species** (Greek, *allos,* other, and *patria,* country), often use the
same habitat locations and food resources—because they are
not in competition, natural selection does not favor evolutionary changes that subdivide their niche.

When a pair of species occupy the same habitat (that
is, when they are sympatric), they tend to exhibit greater differences in morphology and behavior than the same two species do when living in different habitats (that is, when they
are allopatric). Called **character displacement,** the differences evident between sympatric species are thought to have
been favored by natural selection as a mechanism to facilitate
habitat partitioning and thus reduce competition. Thus, the
stickleback fish in figure 21.12 have different-sized gill rakers (structures of the fish gill) where the fish are sympatric, in
the lower graph where two species sharing the same lake but
partitioning the food resources. But the gill rakers are intermediate in size in lakes with only one species of stickleback
(the upper graph), the fish taking advantage of all the available
food resources.

> **21.6 A niche may be defined as the way in which
> an organism uses its environment. No two species
> can occupy the same niche indefinitely without
> competition driving one to extinction if resources
> are limiting. Sympatric species partition available
> resources, reducing competition between them.**

21.7 Coevolution and Symbiosis

The previous section described the "winner take all" results of competition between two species whose niches overlap. Other relationships in nature are less competitive and more cooperative.

Coevolution

The plants, animals, protists, fungi, and prokaryotes that live together in communities have changed and adjusted to one another continually over millions of years. For example, many features of flowering plants have evolved in relation to the dispersal of the plant's gametes by animals (figure 21.13). These animals, in turn, have evolved a number of special traits that enable them to obtain food or other resources efficiently from the plants they visit, often from their flowers. In addition, the seeds of many flowering plants have features that make them more likely to be dispersed to new areas of favorable habitat.

Figure 21.13 Pollination by bat.
Many flowers have coevolved with other species to facilitate pollen transfer. Insects are widely known as pollinators, but they're not the only ones. Notice the cargo of pollen on the bat's snout.

Such interactions, which involve the long-term, mutual evolutionary adjustment of the characteristics of the members of biological communities, are examples of **coevolution.** Coevolution is the adaptation of a species not only to its physical environment but also to the other organisms that share it. In this section, we consider some examples of coevolution, including symbiotic relationships and predator-prey interactions.

Symbiosis Is Widespread

In symbiotic relationships, two or more kinds of organisms live together in often elaborate and more or less permanent relationships. All symbiotic relationships carry the potential for coevolution between the organisms involved, and in many instances the results of this coevolution are fascinating. Examples of symbiosis include lichens, which are associations of certain fungi with green algae or cyanobacteria (see chapter 17). Another important example are mycorrhizae, the association between fungi and the roots of most kinds of plants. The fungi expedite the plant's absorption of certain nutrients, and the plants in turn provide the fungi with carbohydrates. Similarly, root nodules that occur in legumes and certain other kinds of plants contain bacteria that fix atmospheric nitrogen and make it available to their host plants.

In the tropics, leaf-cutter ants are often so abundant that they can remove a quarter or more of the total leaf surface of the plants in a given area. They do not eat these leaves directly; rather, they take them to underground nests, where they chew them up and inoculate them with the spores of particular fungi. These fungi are cultivated by the ants and brought from one specially prepared bed to another, where they grow and reproduce. In turn, the fungi constitute the primary food of the ants and their larvae. The relationship between leaf-cutter ants and these fungi is an excellent example of symbiosis.

The major kinds of symbiotic relationships include (1) mutualism, in which both participating species benefit; (2) parasitism, in which one species benefits but the other is harmed; and (3) commensalism, in which one species benefits while the other neither benefits nor is harmed. Parasitism can also be viewed as a form of predation, although the organism that is preyed upon does not necessarily die.

Mutualism

Mutualism is a symbiotic relationship among organisms in which both species benefit. Examples of mutualism are of fundamental importance in determining the structure of biological communities. Some of the most spectacular examples of mutualism occur among flowering plants and their animal visitors, including insects, birds, and bats. As we discussed in chapter 18, during the course of their evolution, the characteristics of flowers have evolved in large part in relation to the characteristics of the animals that visit them for food and, in doing so, spread their pollen from individual to individual. At

Figure 21.14 Mutualism: ants and aphids.
These ants are tending to aphids (small oval organisms), feeding on the "honeydew" that the aphids excrete continuously, moving the aphids from place to place, and protecting them from potential predators.

Figure 21.15 Mutualism: ants and acacias.
Ants of the genus *Pseudomyrmex* live within the hollow thorns of certain species of acacia trees. The nectaries at the bases of the leaves and the Beltian bodies at the ends of the leaflets (yellow structures) provide food for the ants. The ants, in turn, supply the acacias with organic nutrients, protect the trees from herbivores, and reduce shading by other plants.

the same time, characteristics of the animals have changed, increasing their specialization for obtaining food or other substances from particular kinds of flowers.

Another example of mutualism involves ants and aphids. Aphids, also called greenflies, are small insects that suck fluids with their piercing mouthparts from the phloem of living plants. They extract a certain amount of the sucrose and other nutrients from this fluid, but they excrete much of it in an altered form through their anus. Certain ants have taken advantage of this—in effect, domesticating the aphids (figure 21.14). The ants carry the aphids to new plants, where they come into contact with new sources of food, and then consume as food the "honeydew" that the aphids excrete.

Ants and Acacias. A particularly striking example of mutualism involves ants and certain Latin American species of the plant genus *Acacia*. In these species, certain leaf parts, called stipules, are modified as paired, hollow thorns; these particular species are called "bull's horn acacias." The thorns are inhabited by stinging ants of the genus *Pseudomyrmex*, which do not nest anywhere else. Like all thorns that occur on plants, the acacia horns serve to deter herbivores.

At the tip of the leaflets of these acacias are unique, protein-rich bodies called Beltian bodies, named after Thomas Belt, a nineteenth-century British naturalist who first wrote about them after seeing them in Nicaragua. Beltian bodies do not occur in species of *Acacia* that are not inhabited by ants, and their role is clear: they serve as a primary food for the ants (figure 21.15). In addition, the plants secrete nectar from glands near the bases of their leaves. The ants consume this nectar as well, feeding it and the Beltian bodies to their larvae.

Obviously, this association is beneficial to the ants, and one can readily see why they inhabit acacias. The ants and their larvae are protected within the swollen thorns, and the trees provide a

balanced diet, including the sugar-rich nectar and the protein-rich Beltian bodies. What, if anything, do the ants do for the plants? This question had fascinated observers for nearly a century until it was answered by Daniel Janzen, then a graduate student at the University of California, Berkeley, in a beautifully conceived and executed series of field experiments.

Whenever any herbivore lands on the branches or leaves of an acacia inhabited by ants, the ants immediately attack and devour the herbivore. Thus, the ants protect the acacias from being eaten, and the herbivore also provides additional food for the ants, which continually patrol the acacia's branches. Related species of acacias that do not have the special features of the bull's horn acacias and are not protected by ants have bitter-tasting substances in their leaves that the bull's horn acacias lack. Evidently, these bitter-tasting substances protect the acacias in which they occur from herbivores in a different way.

The ants that live in the bull's horn acacias also help their hosts to compete with other plants. The ants cut away any branches of other plants that touch the bull's horn acacia in which they are living. They create, in effect, a tunnel of light through which the acacia can grow, even in the lush deciduous forests of lowland Central America. Without the ants, as Janzen showed experimentally by poisoning the ant colonies that inhabited individual plants, the acacia is unable to compete successfully in this habitat. Finally, the ants bring organic material into their nests. The parts they do not consume, together with their excretions, provide the acacias with an abundant source of nitrogen.

Parasitism

Parasitism is a symbiotic relationship that may be regarded as a special form of predator/prey relationship, discussed later. In this symbiotic relationship the predator, or parasite, is much smaller than the prey, or host, and remains closely associated with it. Parasitism is harmful to the host organism and beneficial to the parasite, but unlike a predator/prey relationship, a parasite often does not kill its host. The concept of parasitism seems obvious, but individual instances are often surprisingly difficult to distinguish from predation and from other kinds of symbiosis.

External Parasites. Parasites that feed on the exterior surface of an organism are external parasites, or **ectoparasites.** Many instances of external parasitism are known (figure 21.16*a*). Lice, which live their entire lives on the bodies of vertebrates—mainly birds and mammals—are normally considered parasites. Mosquitoes are not considered parasites, even though they draw food from birds and mammals in a similar manner to lice, because their interaction with their host is so brief.

Parasitoids are insects that lay eggs on living hosts. This behavior is common among wasps, whose larvae feed on the body of the unfortunate host, often killing it.

Internal Parasites. Vertebrates are parasitized internally by **endoparasites,** members of many different phyla of animals and protists (figure 21.16*b*). Invertebrates also have many kinds of parasites that live within their bodies. Bacteria and viruses are not usually considered parasites, even though they fit our definition precisely.

Internal parasitism is generally marked by much more extreme specialization than external parasitism, as shown by the many protist and invertebrate parasites that infect humans. The more closely the life of the parasite is linked with that of its host, the more its morphology and behavior are likely to have been modified during the course of its evolution. The same is true of symbiotic relationships of all sorts. Conditions within the body of an organism are different from those encountered outside and are apt to be much more constant. Consequently, the structure of an internal parasite is often simplified, and unnecessary armaments and structures are lost as it evolves.

Many parasites have complex life cycles that require several different hosts for growth to adulthood and reproduction.

Brood Parasitism. Not all parasites consume the body of their host. In brood parasitism, birds like cowbirds and European cuckoos lay their eggs in the nests of other species (figure 21.16*c*). The host parents raise the brood parasite as if it were one of their own clutch, in many cases investing more in feeding the imposter than in feeding their own offspring. The brood parasite reduces the reproductive success of the foster parent hosts, so it is not surprising that natural selection has increased the hosts' ability to detect parasite eggs and reject them.

(a)

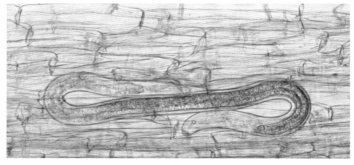

(b)

(c)

Figure 21.16 Parasitism.

(a) Dodder (*Cuscuta*) is a parasitic plant that has lost its chlorophyll and its leaves in the course of its evolution and is heterotrophic—unable to manufacture its own food. Instead, it obtains its food from the host plants it grows on. (b) This lesion nematode (*Pratylenchus penetrans*) is an endoparasite that infects a number of different kinds of plants and damages their roots. (c) Brood parasitism. Cuckoos lay their eggs in the nests of other species of birds. Oftentimes the young cuckoo, which can be larger than the adult that feeds it, ejects the other nestlings from the nest to lessen competition for food.

Commensalism

Commensalism is a symbiotic relationship that benefits one species and neither hurts nor helps the other. In nature, individuals of one species are often physically attached to members of another. For example, epiphytes are plants that grow on the branches of other plants. In general, the host plant is unharmed, and the epiphyte that grows on it benefits. Similarly, various marine animals, such as barnacles, grow on other, often actively moving, sea animals like whales and thus are carried passively from place to place without harming their hosts. These "passengers" presumably gain more protection from predation than they would if they were fixed in one place, and they also reach new sources of food. The increased water circulation that such animals receive as their host moves around may be of great importance, particularly if the passengers are filter-feeders.

Examples of Commensalism. The best-known examples of commensalism involve the relationships between certain small tropical fishes and sea anemones, marine animals that have stinging tentacles (see chapter 19). These fish have evolved the ability to live among the tentacles of sea anemones, even though these tentacles would quickly paralyze other fishes that touched them (figure 21.17). The anemone fishes feed on the detritus left from the meals of the host anemone, remaining uninjured under remarkable circumstances.

Figure 21.18   Commensalism between oxpeckers and African cape buffalo.

Oxpeckers eat insects off the buffalo.

On land, an analogous relationship exists between birds called oxpeckers and grazing animals such as cattle or rhinoceroses. The birds, like the oxpecker in figure 21.18, spend most of their time clinging to the animals, picking off parasites and other insects, carrying out their entire life cycles in close association with the host animals.

When Is Commensalism Commensalism? In each of these instances, it is difficult to be certain whether the second partner receives a benefit or not; there is no clear-cut boundary between commensalism and mutualism. For instance, it may be advantageous to the sea anemone to have particles of food removed from its tentacles; it may then be better able to catch other prey. Similarly, the grazing animals may benefit from the relationship with the oxpeckers or cattle egrets if the bugs that are picked off of them are harmful, such as ticks or fleas. If so, then the relationship is mutualism. However, if the birds also pick at scabs, causing bleeding and possibly infections, the relationship may be parasitic. In true commensalism, only one of the partners benefits and the other neither benefits nor is harmed. If the grazing animals are not harmed by either the ticks that are eaten off of their bodies or by the oxpeckers that feed off of them, then it is an example of commensalism.

Figure 21.17   Commensalism in the sea.

Clownfishes, such as this *Amphiprion perideraion* in Guam, often form symbiotic associations with sea anemones, gaining protection by remaining among their tentacles and gleaning scraps from their food. Different species of anemones secrete different chemical mediators; these attract particular species of fishes and may be toxic to the fish species that occur symbiotically with other species of anemones in the same habitat. There are 26 species of clownfishes, all found only in association with sea anemones; 10 species of anemones are involved in such associations, so that some of the anemone species are host to more than one species of clownfish.

21.7   Coevolution is a term that describes the long-term evolutionary adjustments of species to one another. In symbiosis, two or more species live together. Mutualism involves cooperation between species, to the mutual benefit of both. In parasitism, one organism serves as a host to another organism, usually to the host's disadvantage. Commensalism is the benign use of one organism by another.

Predator-Prey Interactions

In the previous section, we considered parasitism, a symbiotic relationship, a specialized form of **predator/prey interaction** in which the predator is much smaller than its prey and does not generally kill it. **Predation** is the consuming of one organism by another, usually of a similar or larger size. In this sense, predation includes everything from a leopard capturing and eating an antelope to a whale grazing on microscopic ocean plankton.

In nature, predators often have large effects on prey populations. Some of the most dramatic examples involve situations in which humans have either added or eliminated predators from an area. For example, the elimination of large carnivores from much of the eastern United States has led to populations explosions of white-tailed deer, which strip the habitat of all edible plant life within their reach. Similarly, when sea otters were hunted to near extinction on the western coast of the United States, populations of sea urchins, a principal prey item of the otters exploded. Appearances, however, sometimes can be deceiving. On Isle Royale in Lake Superior, moose reached the island by crossing over ice in an unusually cold winter and multiplied freely there in isolation. When wolves later reached the island by crossing over the ice, naturalists widely assumed that the wolves were playing a key role in controlling the moose population. More careful studies have demonstrated that this is not in fact the case. The moose that the wolves eat are, for the most part, old or diseased animals that would not survive long anyway. In general, the moose are controlled by food availability, disease, and other factors rather than by the wolves (figure 21.19).

Figure 21.19 Wolves chasing a moose—what will the outcome be?

On Isle Royale, Michigan, a large pack of wolves pursue a moose. They chased this moose for almost 2 kilometers; it then turned and faced the wolves, who by that time were exhausted from running through chest-deep snow. The wolves lay down, and the moose walked away.

Predator-Prey Cycles

When experimental populations are set up under simple laboratory conditions, the predator (the red *Didinium* in figure 21.20) often exterminates its prey (the blue *Paramecium*) and then becomes extinct itself, having nothing left to eat. However, if refuges are provided for the prey, its population drops to low levels but not to extinction. Low prey population levels then provide inadequate food for the predators, causing the predator population to decrease. When this occurs, the prey population can recover. In this situation the predator and prey populations may continue in this cyclical pattern for some time.

Cycles in Hare Populations: A Case Study

Population cycles are characteristic of some species of small mammals, such as lemmings, and they appear to be stimulated, at least in some situations, by their predators. Ecologists have studied cycles in hare populations since the 1920s. They have

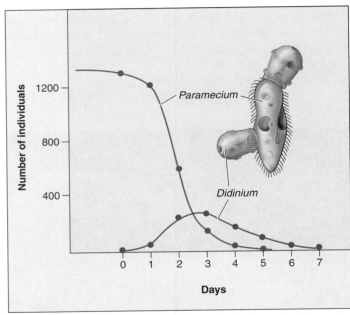

Figure 21.20 Predator-prey in the microscopic world.

When the predatory *Didinium* is added to a *Paramecium* population, the numbers of *Didinium* initially rise, while the numbers of *Paramecium* steadily fall. As the *Paramecium* population is depleted, however, the *Didinium* individuals also die.

found that the North American snowshoe hare, *Lepus americanus*, follows a "10-year cycle" (in reality, it varies from 8 to 11 years). Its numbers fall 10-fold to 30-fold in a typical cycle, and 100-fold changes can occur. Two factors appear to be generating the cycle: food plants and predators.

1. **Food plants.** The preferred foods of snowshoe hares are willow and birch twigs. As hare density increases, the quantity of these twigs decreases,

leading to a precipitous decline in willow and birch twig abundance. The result is that hares are forced to feed on high-fiber (low-quality) food, causing lower birthrates, low juvenile survivorship, low growth rates, and a corresponding fall in hare abundance. It takes two to three years for the quantity of mature twigs to recover.

2. **Predators.** A key predator of the snowshoe hare is the Canada lynx, *Lynx canadensis.* The Canada lynx shows a "10-year cycle" of abundance that seems remarkably entrained to the hare abundance cycle (compare the hare population cycle, the blue line in figure 21.21 with the lynx population, the red line). As hare numbers increase, lynx numbers do, too, rising in response to the increased availability of lynx food. When hare numbers fall, so do lynx numbers, their food supply depleted.

Which factor is responsible for the predator-prey oscillations? Do increasing numbers of hares lead to overharvesting of plants (a hare-plant cycle), or do increasing numbers of lynx lead to overharvesting of hares (a hare-lynx cycle)? Field experiments carried out by C. Krebs and coworkers in 1992 provide an answer. Krebs set up experimental plots in Canada's Yukon containing hare populations. If food is added and predators excluded from an experimental area, hare numbers increase 10-fold and stay there—the cycle is lost. However, the cycle is retained if either of the factors is allowed to operate alone: exclude predators but don't add food, or add food in presence of predators. Thus, both factors can affect the cycle, which, in practice, seems to be generated by the interaction between the two factors.

Predation Reduces Competition

Predator-prey interactions are an essential factor in the maintenance of communities that are rich and diverse in species. The predators prevent or greatly reduce competitive exclusion by reducing the numbers of individuals of competing species. For example, in preying selectively on bivalves in marine intertidal habitats, sea stars prevent bivalves from monopolizing such habitats, opening up space for many other organisms. When sea stars are removed, species diversity falls precipitously, the sea-floor community coming to be dominated by a few species of bivalves (figure 21.22). Because predation tends to reduce competition in natural communities, it is usually a mistake to attempt to eliminate a major predator such as sea stars, wolves, or mountain lions from a community. The result is to decrease rather than increase the biological diversity of the community, the opposite of what is intended. Predators such as these are examples of **keystone species,** species that play key roles in their communities.

> **21.8** Predators and their prey often show similar cyclic oscillations, often promoted by refuges that prevent prey populations from being driven to extinction.

(a)

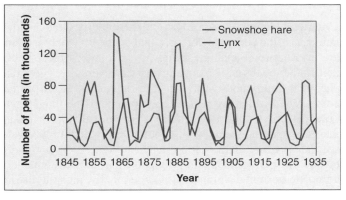

(b)

Figure 21.21   A predator-prey cycle.

(a) A snowshoe hare being chased by a lynx. (b) The numbers of lynxes and snowshoe hares oscillate in tune with each other in northern Canada. The data are based on numbers of animal pelts from 1845 to 1935. As the number of hares grows, so does the number of lynxes, with the cycle repeating about every nine years. Both predators (lynxes) and available food resources control the number of hares. The number of lynxes is controlled by the availability of prey (snowshoe hares).

Figure 21.22   Predation reduces competition.

When a key predator, starfish (*Pisaster*), is removed from a coastal ecosystem, fiercely competitive mussels explode in growth, effectively crowding out seven other indigenous species.

Plant Defenses

As we have noted, predator-prey interactions are interactions between organisms in which one organism uses the other for food. Plants have evolved many mechanisms to defend themselves from herbivores. The most obvious are morphological (structural) defenses: thorns, spines, and prickles play an important role in discouraging browsers, and plant hairs, especially those that have a glandular, sticky tip, deter insect herbivores. Some plants, such as grasses, deposit silica in their leaves, both strengthening and protecting themselves. If enough silica is present in their cells, these plants are simply too tough to eat.

Chemical Defenses. Significant as these morphological adaptations are, the chemical defenses that occur so widely in plants are even more crucial. Best known and perhaps most important in the defenses of plants against herbivores are secondary chemical compounds. These are distinguished from primary compounds, which are regular components of the major metabolic pathways, such as respiration. Virtually all plants, and apparently many algae as well, contain very structurally diverse secondary compounds that are either toxic to most herbivores or disturb their metabolism so greatly that they are unable to complete normal development. Consequently, most herbivores tend to avoid the plants that possess these compounds.

The mustard family (Brassicaceae) is characterized by a group of chemicals known as mustard oils. These are the substances that give the pungent aromas and tastes to such plants as mustard, cabbage, watercress, radish, and horseradish. The same tastes we enjoy signal the presence of toxic chemicals to many groups of insects. Similarly, plants of the milkweed family (Asclepiadaceae) and the related dogbane family (Apocynaceae) produce a milky sap that deters herbivores from eating them. In addition, these plants usually contain cardiac glycosides, molecules named for their drastic effect on heart function in vertebrates.

The Evolutionary Response of Herbivores. Certain groups of herbivores are associated with each family or group of plants protected by a particular kind of secondary compound. These herbivores are able to feed on these plants without harm, often as their exclusive food source. For example, cabbage butterfly caterpillars (subfamily Pierinae) feed almost exclusively on plants of the mustard and caper families, as well as on a few other small families of plants that also contain mustard oils (figure 21.23). Similarly, caterpillars of monarch butterflies and their relatives (subfamily Danainae) feed on plants of the milkweed and dogbane families. How do these animals manage to avoid the chemical defenses of the plants, and what are the evolutionary precursors and ecological consequences of such patterns of specialization?

We can offer a potential explanation for the evolution of these particular patterns. Once the ability to manufacture

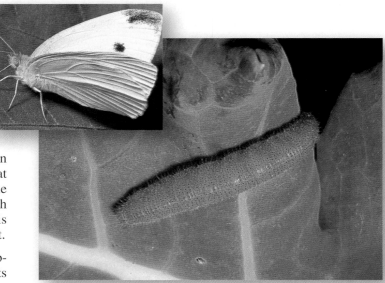

Figure 21.23   Insect herbivores are well suited to their hosts.
The green caterpillars of the cabbage butterfly, Pieris rapae, are camouflaged on the leaves of cabbage and other plants on which they feed. Although mustard oils protect these plants against most herbivores, the cabbage butterfly caterpillars are able to break down the mustard oil compounds. *Inset:* An adult cabbage butterfly.

mustard oils evolved in the ancestors of the caper and mustard families, the plants were protected for a time against most or all herbivores that were feeding on other plants in their area. At some point, certain groups of insects—for example, the caterpillar of the cabbage butterfly—developed the ability to break down mustard oils and thus feed on these plants without harming themselves. Having evolved this ability, the butterflies were able to use a new resource without competing with other herbivores for it.

Animal Defenses

Some animals that feed on plants rich in secondary chemical compounds receive an extra benefit. When the caterpillars of monarch butterflies feed on plants of the milkweed family, they do not break down the cardiac glycosides that protect these plants from herbivores. Instead, the caterpillars concentrate and store the cardiac glycosides in fat bodies; they then pass them through the chrysalis stage to the adult and even to the eggs of the next generation. The incorporation of cardiac glycosides thus protects all stages of the monarch life cycle from predators. A bird that eats a monarch butterfly quickly regurgitates it and in the future avoids the conspicuous orange-and-black pattern that characterizes the adult monarch (figure 21.24). Some birds, however, appear to have acquired the ability to tolerate the protective chemicals. These birds eat the monarchs.

Defensive Coloration. Many insects that feed on milkweed plants are brightly colored; they advertise their poisonous nature using an ecological strategy known as **warning coloration** or **aposematic coloration.** Showy coloration is characteristic of animals that use poisons (see figure 21.26) and stings to repel predators, whereas organisms that lack specific chemical defenses are seldom brightly colored. In fact, many have **cryptic coloration**—color that blends with the surroundings and thus hides the individual from predators (figure 21.25). Camouflaged animals usually do not live together in groups because a predator that discovers one individual gains a valuable clue to the presence of others.

Chemical Defenses. Animals also manufacture and use a startling array of substances to perform a variety of defensive functions. Bees, wasps, predatory bugs, scorpions, spiders, and many other arthropods use chemicals to defend themselves and to kill their prey. In addition, various chemical defenses have evolved among marine animals and the vertebrates, including jellyfish, venomous snakes, lizards, fishes, and some birds. The poison-dart frogs of the family Dendrobatidae produce toxic

(a)   (b)

Figure 21.24 A blue jay learns that monarch butterflies taste bad.

(a) This cage-reared jay, which had never seen a monarch butterfly before, tried eating one. (b) The same jay regurgitated the butterfly a few minutes later. This bird is not likely to attempt to eat an orange-and-black insect again.

Figure 21.25 Cryptic coloration.

An inchworm caterpillar (*Necophora quernaria*) closely resembles a twig.

Figure 21.26 Vertebrate chemical defenses.

Dendrobatid frogs advertise their toxicity with aposematic coloration.

alkaloids in the mucus that covers their brightly colored skin (figure 21.26). Some of these toxins are so powerful that a few micrograms will kill a person if injected into the bloodstream. More than 200 different alkaloids have been isolated from these frogs, and some are playing important roles in neuromuscular research. There is an intensive investigation of marine animals, algae, and flowering plants for new drugs to fight cancer and other diseases, and as sources of antibiotics.

Coevolution of Predator and Prey

Predation can exert strong selective pressures on prey populations. Any feature that acts to decrease the probability of capture should thus be strongly favored by natural selection. In turn, the evolution of such features by prey would be expected to encourage natural selection to then favor counteradaptations in their predator populations. In this way, a coevolutionary arms race may ensue, in which predators and prey are continually evolving better defenses and better means of circumventing these defenses. Herbivores that are able to feed on mustard, milkweed, and dogbane are predators that have acquired counteradaptations to the chemicals produced by the plants. And, birds that are able to eat monarch butterflies seem to be counteradapting to the defense of their animal prey.

21.9 The members of many groups of plants are protected from most herbivores by their secondary compounds. Once the members of a particular herbivore group evolve the ability to feed on them, these herbivores gain access to a new resource. Animals defend themselves against predators with warning coloration, camouflage, and chemical defenses such as poisons and stings.

Mimicry

During the course of their evolution, many unprotected (nonpoisonous) species have come to resemble distasteful ones that exhibit aposematic coloration. The unprotected mimic gains an advantage by looking like the distasteful model. Two types of mimicry have been identified: Batesian mimicry, as described here, and Müllerian mimicry, which is a type of "group" protection.

Batesian Mimicry

Batesian mimicry is named for Henry Bates, the nineteenth-century British naturalist who first brought this type of mimicry to general attention in 1857. In his journeys to the Amazon region of South America, Bates discovered many instances of

(a) Model

(b) Batesian mimic

Figure 21.27 A Batesian mimic.

(a) The model. Monarch butterflies (*Danaus plexippus*) are protected from birds and other predators by the cardiac glycosides they incorporate from the milkweeds and dogbanes they feed on as larvae. Adult monarch butterflies advertise their poisonous nature with warning coloration. (b) The mimic. Viceroy butterflies, *Limenitis archippus,* are Batesian mimics of the poisonous monarch. Although the viceroy is not related to the monarch, it looks a lot like it, so predators that have learned not to eat distasteful monarchs avoid viceroys, too.

palatable insects that resembled brightly colored, distasteful species. He reasoned that the mimics are avoided by predators, who are fooled by the disguise into thinking the mimic actually is the distasteful model.

Many of the best-known examples of Batesian mimicry occur among butterflies and moths. Obviously, predators in systems of this kind must use visual cues to hunt for their prey; otherwise, similar color patterns would not matter to potential predators. There is also increasing evidence indicating that Batesian mimicry can also involve nonvisual cues, such as olfaction, although such examples are less obvious to humans.

The kinds of butterflies that provide the models in Batesian mimicry are, not surprisingly, members of groups whose caterpillars feed only on one or a few closely related plant families. The plant families on which they feed are strongly protected by toxic chemicals. The model butterflies incorporate the poisonous molecules from these plants into their bodies. The mimic butterflies, in contrast, belong to groups in which the feeding habits of the caterpillars are not so restricted. As caterpillars, these butterflies feed on a number of different plant families unprotected by toxic chemicals.

One often-studied mimic among North American butterflies is the viceroy, *Limenitis archippus* (figure 21.27*b*). This butterfly, which resembles the poisonous monarch (in figure 21.27*a*), ranges from central Canada through much of the United States and into Mexico. The caterpillars feed on willows and cottonwoods, and neither caterpillars nor adults were thought to be distasteful to birds, although recent findings may dispute this. Interestingly, the Batesian mimicry seen in the adult viceroy butterfly does not extend to the caterpillars: viceroy caterpillars are camouflaged on leaves, resembling bird droppings, whereas the monarch's distasteful caterpillars are very conspicuous.

Batesian mimicry also occurs in vertebrates. Probably the most famous case is the scarlet king snake, whose red, black, and yellow bands mimic those of the venomous coral snake.

Müllerian Mimicry

Another kind of mimicry, **Müllerian mimicry,** was named for German biologist Fritz Müller, who first described it in 1878. In Müllerian mimicry, several unrelated but protected animal species come to resemble one another (figure 21.28). Thus, different kinds of stinging wasps have yellow-and-black-striped abdomens, but they may not all be descended from a common yellow-and-black-striped ancestor. In general, yellow-and-black and bright red tend to be common color patterns that warn predators relying on vision. If animals that resemble one another are all poisonous or dangerous, they gain an advantage because a predator learns more quickly to avoid them.

In both Batesian and Müllerian mimicry, mimic and model must not only look alike but also act alike if predators are to be deceived. For example, the members of several families of insects that resemble wasps behave surprisingly like the wasps they mimic, flying often and actively from place to place.

(a)

(b)

(c)

(d)

**Figure 21.28
Müllerian mimics.**
Because the color patterns of these insects are very similar, and because they all sting, they are Müllerian mimics. The yellow jacket (a), the masarid wasp (b), the sand wasp (c), and the anthidiine bee (d) all act as models for each other, strongly reinforcing the color pattern that they all share.

Self Mimicry

Another type of mimicry, called **self mimicry,** involves adaptations in which one animal body part comes to resemble another body part. This type of mimicry is used by both prey and predators. In prey, it is used to increase survival during an attack. For example, many moths, butterflies, and fish appear to have "eyespots," which are large dark circular markings, like those shown in figure 21.29. In some cases, these eyespots startle a predator, allowing the prey time to escape, or they present a false target for the predator's attack, attacking the tail where the eyespots appear instead of the head. Predators may use mimicry to simulate bait to lure prey in. For example, some predators have accessory body parts that look like food, such as the tongue of the alligator snapping turtle that looks like a wriggly worm. The wormlike tongue attracts prey and brings them in close enough for a successful attack.

Figure 21.29   Self mimicry
The eyespots on the wings of this moth may startle predators or may be a pattern that predators are instinctively afraid of, as the eyespots resemble the eyes of predators' predators.

> **21.10   In Batesian mimicry, unprotected species resemble others that are distasteful. Both species exhibit aposematic coloration. In Müllerian mimicry, two or more unrelated but protected species resemble one another, thus achieving a kind of group defense. In self mimicry, one body part resembles another, helping either prey or predators.**

21.11 | Ecological Succession

Competition, predation, and cooperation often produce dramatic changes in communities. This results in changes in ecosystems, by the orderly replacement of one community with another, from simple to complex in a process known as **succession.** This process is familiar to anyone who has seen a vacant lot or cleared woods slowly become occupied by an increasing number of plants, or a pond become dry land as it is filled with vegetation encroaching from the sides.

Secondary Succession

If a wooded area is cleared and left alone, plants slowly reclaim the area. Eventually, traces of the clearing disappear and the area is again woods. Similarly, a stream that experiences intense flooding may clear the stream bed of all organisms, leaving sand and rock; the bed is progressively reinhabited by protists, invertebrates, and other aquatic organisms. This kind of succession, which occurs in areas where an existing community has been disturbed, is called **secondary succession.** Humans are often responsible for initiating secondary succession, which may also take place when a fire has burned off an area or in abandoned agricultural fields.

Primary Succession

In contrast, **primary succession** occurs on bare, lifeless substrate, such as rocks. Primary succession occurs in lakes left behind after the retreat of glaciers, on volcanic islands that rise above the sea, and on land exposed by retreating glaciers. Primary succession on glacial moraines provides an example. The graph in figure 21.30 shows how the soil changes in its concentration of nitrogen as primary succession occurs. The letters on the graph correspond to the photos in figure 21.31. On bare, mineral-poor soil, lichens grow first (figure 21.31b), forming small pockets of soil. Acidic secretions from the lichens help to break down the substrate and add to the accumulation of soil. Mosses then colonize these pockets of soil (figure 21.31c), eventually building up enough nutrients in the soil for alder shrubs to take hold (figure 21.31d). These first plants to appear form a **pioneering community.** Over 100 years, the alders (figure 21.31e) build up the soil nitrogen levels until spruce are able to thrive, eventually crowding out the alder and forming a dense spruce forest (figure 21.31f).

Primary successions end with a community called a **climax community,** whose populations remain relatively stable, and that is characteristic of the region as a whole. However, because local climate keeps changing, the process of succession is often very slow, and human activities have a major impact, so many successions do not reach climax.

Why Succession Happens

Succession happens because species alter the habitat and the resources available in it, often in ways that favor other species. Three dynamic concepts are of critical importance in the process:

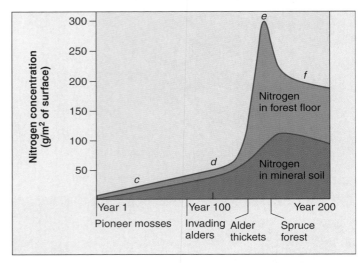

Figure 21.30 Plant succession produces progressive changes in the soil.

Initially the glacial moraine at Glacier Bay, Alaska, portrayed in figure 21.31, had little soil nitrogen, but nitrogen-fixing alders led to a buildup of nitrogen in the soil, encouraging the subsequent growth of the conifer forest. Letters in the graph above correspond to photographs in figure 21.31 c–f.

1. **Tolerance.** Early successional stages are characterized by weedy r-selected species that do not compete well in established communities but are tolerant of the harsh, abiotic conditions in barren areas.
2. **Facilitation.** The weedy early successional stages introduce local changes in the habitat that favor other, less weedy species. Thus, the mosses in the Glacier Bay succession of figure 21.30 fix nitrogen, which allows alders to invade. The alders in turn lower soil pH as their fallen leaves decompose, allowing spruce and hemlock, which require acidic soil, to be able to invade.
3. **Inhibition.** Sometimes the changes in the habitat caused by one species, while favoring other species, inhibit the growth of the species that caused them. Alders, for example, do not grow as well in acidic soil as the spruce and hemlock that replace them.

As ecosystems mature, and more K-selected species replace r-selected ones, species richness and total biomass increase but net productivity decreases. Because earlier successional stages are more productive than later ones, agricultural systems are intentionally maintained in early successional stages to keep net productivity high.

> **21.11 In succession, communities change through time, often in a predictable sequence.**

(a)   Retreating Glacier   (b)   Barren Morraine

(c)   Pioneering Mosses   (d)   Invading Alders

(e)   Alder Thickets   (f)   Spruce Forest

Figure 21.31   Primary succession at Alaska's Glacier Bay.
The sides of the glacier (a) have been retreating at a rate of some 8 meters a year, leaving behind exposed soil (b) from which nitrogen and other minerals have been leached out. The first invaders of these exposed sites are pioneer moss species (c) with nitrogen-fixing mutualistic microbes. Within 20 years, young alder shrubs take hold (d). Rapidly fixing nitrogen, they soon form dense thickets (e). As soil nitrogen levels rise, spruce crowd out the mature alders, forming a forest (f).

Are Island Populations of Song Sparrows Density Dependent?

When island populations are isolated, receiving no visitors from other populations, they provide an attractive opportunity to test the degree to which a population's growth rate is affected by its size. A population's size can influence the rate at which it grows because increased numbers of individuals within a population tend to deplete available resources, leading to an increased risk of death by deprivation. Also, predators tend to focus their attention on common prey, resulting in increasing rates of mortality as populations grow. However, simply knowing that a population is decreasing in numbers does not tell you that the decrease has been caused by the size of the population. Many factors such as severe weather, volcanic eruption, and human disturbance can influence island population sizes too.

The graph to the right displays data collected from 13 song sparrow populations on Mandarte Island (see map below). In an attempt to gauge the impact of population size on the evolutionary success of these populations, each population was censused, and its juvenile mortality rate estimated. On the graph, these juvenile mortality rates have been plotted against the number of breeding adults in each population. Although the data appear scattered, the "best-fit" regression line is statistically significant (**statistically significant** means that there is a less than 5% chance that there is in fact no correlation between dependent and independent variables).

1. **Applying Concepts**
   a. Variable. In the graph, what is the dependent variable?
   b. Analyzing Scattered Data. What is the size of the song sparrow population (based on breeding adults) with the least juvenile mortality? with the most?

2. **Interpreting Data**
   a. What is the average juvenile mortality of all 13 populations, estimated from the 13 points on the graph?

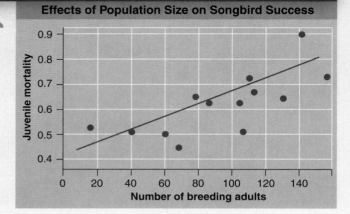

Effects of Population Size on Songbird Success

(graph: y-axis "Juvenile mortality" 0.4 to 0.9; x-axis "Number of breeding adults" 0 to 140)

b. How many populations were observed to have juvenile mortality rates below this average value? What is the average size of these populations?
c. How many populations were observed to have juvenile mortality rates above this average value? What is the average size of these populations?

3. **Making Inferences** Are the populations with lower juvenile mortality bigger or smaller than the populations with higher juvenile mortality?

4. **Drawing Conclusions** Do the population sizes of these song sparrows appear to exhibit density dependence?

5. **Further Analysis**
   a. The fact that the song sparrow populations with lower juvenile mortality are a different size than those with higher juvenile mortality does not in itself establish that the difference is statistically significant. How would you go about testing these data to see if the relationship between juvenile mortality and population size is real?
   b. What would you expect to happen if the researchers supplemented the food available to the birds? Explain.
   c. What would you expect to happen if the researchers removed individuals from populations with more than 100 breeding adults, reducing each to 100?

Population Dynamics

21.1 Population Growth

- Several factors affect population growth including population size, density, and dispersion (**figure 21.1**). A population often grows exponentially when no factors are limiting its growth. As resources are used up, its growth slows and stabilizes at a size called the carrying capacity (**figures 21.2** and **21.3**).

21.2 The Influence of Population Density

- Factors, such as weather and physical disruptions, are density-independent effects and act on population growth, regardless of population size. Density-dependent effects are factors, such as resources, that are affected by increases in population size (**figure 21.4**).

21.3 Life History Adaptations

- Populations whose resources are limitless experience little competition and reproduce rapidly; these organisms exhibit *r*-selected adaptations. Populations that experience competition over limited resources tend to be more reproductively efficient and exhibit *K*-selected adaptations (**table 21.1**).

21.4 Population Demography

- Survivorship curves illustrate the impact of mortality rates among different age groups in a population (**figure 21.7**).

How Competition Shapes Communities

21.5 Communities

- The array of organisms that live together in an area is called a community. Different species in a community compete and cooperate with each other to make the community stable. (**figure 21.8**).

21.6 The Niche and Competition

- A niche is the way an organism uses all available resources in its environment. Competition limits an organism from using its entire niche (**figure 21.9**). Two species cannot use the same niche; one will either outcompete the other, driving it to extinction, called competitive exclusion (**figure 21.10***a,b*), or they will divide the niche into two smaller niches, called resource partitioning (**figures 21.10***c* and **21.11**).

- Resource partitioning can affect morphological characteristics, as each species adapts to its portion of the niche, called character displacement (**figure 21.12**).

How Coevolution Shapes Communities

21.7 Coevolution and Symbiosis

- Coevolution is the process whereby one species changes as a consequence of a relationship with another, and the other species changes in response. Symbiotic relationships such as mutualism (**figures 21.14** and **21.15**), parasitism (**figure 21.16**), and commensalism (**figures 21.17** and **21.18**) can lead to coevolution.

21.8 Predator-Prey Interactions

- In predatory-prey relationships, the predator kills and consumes the prey.

- Predator and prey populations often exhibit cyclic oscillations, the prey population being hunted to a low number, which begins to negatively affect the predator population. When the predator population decreases in numbers, the prey population rebounds (**figure 21.21**).

21.9 Plant and Animal Defenses

- Plants have evolved defense mechanisms, such as toxic or distasteful chemical compounds, but this in turn led to counteradaptations in certain herbivores that allow them to consume the plants (**figure 21.23**).

- Animals have also evolved chemical defense mechanisms and warning coloration in order to let potential predators know that they should "stay away" (**figures 21.24** and **21.26**). Cryptic coloration is also a defense mechanism in animals to camouflage them to avoid predation (**figure 21.25**).

21.10 Mimicry

- Mimicry is where one organism takes advantage of the warning coloration of another organism.

- Batesian mimicry is where a harmless species has come to resemble a harmful species (**figure 21.27**).

- Müllerian mimicry is where a group of harmful species has a similar warning coloration pattern (**figure 21.28**).

- Self mimicry is where one body part mimics another (**figure 21.29**).

Community Stability

21.11 Ecological Succession

- Succession is the ordered replacement of one community with another as in secondary succession. Primary succession is the emergence of a pioneering community where no life existed before (**figures 21.30** and **21.31**).

1. When the number of organisms remains more or less the same over time in the specific place where these organisms live, it is said that this population of organisms has reached the _____ of that place.
   a. dispersion
   b. biotic potential
   c. carrying capacity
   d. population density

2. Which of the following is a density-dependent effect on a population?
   a. earthquake
   b. increased competition for food
   c. habitat destruction by humans
   d. seasonal flooding

3. Which of the following traits is *not* a characteristic of an organism that has *K*-selected adaptations?
   a. short life span
   b. few offspring per breeding season
   c. extensive parental care of offspring
   d. low mortality rate

4. If the age structure of a population shows more older organisms than younger organisms, then the fecundity
   a. will increase, and the mortality will decrease.
   b. will decrease, and the mortality will increase.
   c. and the mortality will be equal.
   d. and the mortality will not change.

5. All the animals and plants that live in the same location make up a(n)
   a. biome.
   b. population.
   c. ecosystem.
   d. community.

6. For similar species to occupy the same space, their niches must be different in some way. One way for these species to both survive is
   a. competitive exclusion.
   b. interspecific competition.
   c. resource partitioning.
   d. intraspecific competition.

7. A relationship between two species where one species benefits and the other is neither hurt nor helped is known as
   a. parasitism.
   b. commensalism.
   c. mutualism.
   d. competition.

8. Predators can assist in maintaining the species diversity of an area by
   a. increasing competitive exclusion between prey species.
   b. decreasing competitive exclusion between prey species.
   c. not affecting competitive exclusion between prey species.
   d. decreasing resource partitioning between prey species.

9. The bright colors of poison-dart frogs and Gila monsters are examples of
   a. aposematic coloration.
   b. Müllerian mimicry.
   c. cryptic coloration.
   d. Batesian mimicry.

10. Succession that occurs on abandoned agricultural fields is best described as
    a. coevolution.
    b. primary succession.
    c. secondary succession.
    d. prairie succession.

1. **Figure 21.2** Speculate on what happens with populations that follow the exponential growth model.

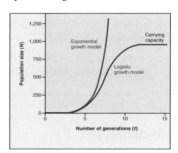

2. **Figure 21.9** How would the niche change for *Chthamalus* if *Semibalanus* was removed from the community? How would the niche change for *Semibalanus* if *Chthamalus* was removed from the community?

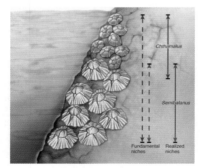

1. Give at least two examples, from your own area, of limiting factors on a population that are density-dependent factors, and two that are density-independent factors on the same population.

2. Use the terms from this section: interspecific competition, fundamental niche, realized niche, niche overlap, competitive exclusion, and resource partitioning to discuss Robert MacArthur's warbler research.

3. Briefly discuss and give examples of the three types of symbiosis.

4. Many U.S. communities struggle with issues of deer overpopulation. Explain how human activities have created this situation.

22

Behavior and the Environment

This dog's bow is its way of saying to another dog, "Let's play." Its rear shoved high in the air, its forelegs flat on the ground, the pup looks up, hopeful that its partner will agree. Every dog invites play in exactly this way, a terrier like this, a retriever—even a wolf. The bow is an innate behavior that they all share, part of what it means to be a canine. In other ways, this dog may behave like no other dog. It may learn to "sit" or "roll over"—or to catch Frisbees, herd sheep, or bring in the morning paper. It can solve surprisingly complex problems. Wolves and other wild canines are social animals, living in packs that cooperate in hunting and child rearing. Of course, there are some things that a dog cannot learn, no matter how much effort is expended in trying. Although a dog can bark, howl, or whine, it will never be able to speak. Behavioral biologists investigate how animals behave, and why they behave in one way and not another. They have learned that there are surprisingly few differences other than language between humans and apes, for example. Even you and this puppy share more behavior than you might expect.

22.1 Approaches to the Study of Behavior

Animals respond to their environment in many ways. Beavers build dams in the fall that create lakes, and birds sing in the spring. Bees search for honey, and when they find it, they fly back to the hive and spread the good news. To understand these behaviors, we need to appreciate both the internal factors that shape the way an animal behaves, as well as the aspects of the external environment that trigger an individual behavior.

Behavior can be defined as the way an animal responds to stimuli in its environment. These stimuli might be as simple as the odor of food. In this sense, a bacterial cell "behaves" by moving toward higher concentrations of sugar. In animals, behaviors are far more complex than this simple bacterial one. This is particularly true for animals with nervous systems. Using eyes, ears, and a variety of other sense organs, they are able to perceive environmental stimuli, process the information, and dictate appropriate body responses, which can be both complex and subtle.

Explaining Behavior

When we observe animal behavior, we can examine it in two different ways. First, we might ask *how* it all works. How do the animal's senses, nerve networks, and internal state act together physiologically to produce the behavior. Like a mechanic studying the behavior of a car, we are asking how the machine works. A psychologist would say we would be asking a question of **proximate causation.** To analyze the proximate cause of behavior, we might measure hormone levels, or record the impulse activity of particular neurons in the brain. The field of psychology often focuses on proximate causes.

We might also ask *why* it all works the way it does. Why did the behavior evolve in this way? What is the adaptive value of this particular response to the environment? This is a question of **ultimate causation.** To study the ultimate cause of a behavior, we would attempt to find how it influenced the animal's survival or reproductive success. The field of ethology (evolutionary behavior) typically focuses on ultimate causes.

Any animal behavior can be looked at either way. For example, a male songbird may sing during the breeding season (figure 22.1). Why? One explanation is that longer spring days have induced in his body elevated levels of the steroid sex hormone testosterone, which at these high levels bind to hormone receptors in the songbird's brain and trigger the production of song. Elevated testosterone would be the proximate cause of the male songbird's song.

Another explanation is that the male is exhibiting a pattern of behavior produced by natural selection to better adapt it to its environment. Seen in this light, the male songbird sings to defend a territory from other males, and to attract a female to reproduce. These reproductive motives are the ultimate, or evolutionary, explanation of the male songbird's behavior.

Figure 22.1   Two ways to look at a behavior.
This male songbird can be said to be singing because his levels of testosterone are elevated, triggering innate "song" programs in his brain. Viewed in a different light, he is singing to defend his territory and attract a mate, behaviors that have evolved to increase his reproductive fitness.

A Controversial Field of Biology

The study of behavior has had a long history of controversy. One source of controversy has been the question of whether an animal's behavior is determined more by an individual's genes, or by its learning and experience. In other words, is behavior the result of nature (instinct) or nurture (learning)? In the past, this question has been considered an "either-or" proposition, but we now know that instinct and learning both play significant roles, often interacting in complex ways to produce the final behavior. We will begin our study of animal behavior by examining more closely the scientific study of instinct and learning, and the ways in which they interact to determine behavior, both proximately and ultimately.

22.1   Animal behavior is the way an animal responds to stimuli in its environment. Some biologists study the physiological mechanisms producing the behavior, others the evolutionary forces responsible for its development.

Instinctive Behavioral Patterns

Early research in the field of animal behavior focused on behavioral patterns that appeared to be instinctive or innate. Because behavior in animals is often stereotyped (that is, appearing in the same way in different individuals of a species), these behavioral scientists argued that the behaviors must be based on preset paths in the nervous system. In their view, these neural paths are structured from genetic blueprints and cause animals to show essentially the same behavior from the first time it is produced throughout their lives.

This study of the instinctive nature of animal behavior was typically carried out in the field rather than on laboratory animals. The study of animal behavior in natural conditions is called **ethology.** The three scientists most responsible for founding the field of ethology were Karl von Frisch, Konrad Lorenz, and Niko Tinbergen. These scientists were awarded the Nobel Prize in Physiology or Medicine in 1973 for their path-making contributions.

An Example of Innate Behavior

The study by Konrad Lorenz of the egg retrieval behavior of geese provides a clear example of what ethologists mean by innate behavior. The drawings in figure 22.2*a* illustrate how when a goose is incubating its eggs in a nest and it notices that an egg has been knocked out of the nest, it will extend its neck toward the egg, get up, and roll the egg back into the nest with a side-to-side motion of its neck while the egg is tucked beneath its bill. Even if the egg is removed during retrieval, the goose completes the behavior, as if driven by a program released by the initial sight of the egg outside the nest.

According to ethologists like Lorenz, egg retrieval behavior is triggered by a **sign stimulus** (also called a key stimulus), which in this case is the appearance of an egg outside of the nest. The way in which nerves are connected in the goose's brain, the **innate releasing mechanism,** responds to the sign stimulus by providing the neural instructions for the motor program, or **fixed action pattern,** which causes the goose to carry out the intricate egg retrieval behavior.

More generally, the sign stimulus is a "signal" in the environment that triggers a behavior. The innate releasing mechanism is the hard-wired element of the brain, and the fixed action pattern is the stereotyped act.

Studying fixed action patterns in birds and other animals, ethologists have discovered that in some situations, a wide variety of objects will trigger a fixed action pattern. For example, geese will attempt to roll baseballs, and even beer cans, back into their nests!

A clear example of the general nature of some sign stimuli is seen in the mating behavior of male stickleback fish studied by Niko Tinbergen. During the breeding season, males develop bright red coloration on their undersides. The males are very territorial, reacting aggressively to the approach of other males. They first perform an aggressive display (shown on the right in figure 22.2*b*), and if the invading male is not deterred, they attack it.

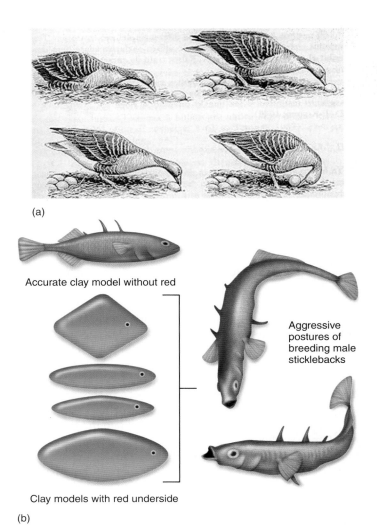

Figure 22.2 Sign stimulus and fixed action pattern.
(a) The series of movements used by a goose to retrieve an egg is a fixed action pattern. Once it detects the sign stimulus (in this case, an egg outside the nest), the goose goes through the entire set of movements: it will extend its neck toward the egg, get up, and roll the egg back into the nest with a side-to-side motion of its neck while the egg is tucked underneath its bill. (b) In stickleback fish, a red color acts as a sign stimulus to trigger a fixed action pattern in males: aggressive threat displays or postures. When the models above are presented to a male stickleback, he will display more often to the last four models than to the first model, which looks like a male stickleback but lacks the red belly characteristic of males.

However, when Niko Tinbergen observed a male stickleback in a laboratory aquarium display aggressively when a red firetruck passed by the window, he realized that the red color was the sign stimulus. In experiments using models shown on the left in figure 22.2*b*, he was able to produce the aggressive display in males by challenging them with the many unfishlike models, so long as the models had a red strip.

22.2 The ethological approach to studying animal behavior has emphasized innate, instinctive behaviors that are the result of preset pathways in the nervous system.

Genetic Effects on Behavior

Although most animal behaviors are not "hard-wired" instincts, such as those studied by the early ethologists, animal behaviorists who followed these pioneer researchers have clearly demonstrated that many animal behaviors are strongly influenced by genes passed from parent to offspring. In other words, "nature" plays a key role in determining patterns of behavior.

If genes determine behavior, then it should be possible to study their inheritance, much as Mendel studied the inheritance of flower color in garden peas. This sort of investigation is called **behavioral genetics.**

Studies of Genetic Hybrids

Behavioral genetic studies have revealed many examples of behaviors that seem to be inherited in a Mendelian manner. William Dilger of Cornell University examined two species of lovebirds that differ in the way they carry twigs, paper, and other materials used to build a nest. One species, fisher's lovebird, holds nest materials in its beak, while another, the peachfaced lovebird, carries material tucked beneath its flank (tail) feathers. When Dilger crossed the two species to produce hybrids, he found that the hybrids carry nest material in a way that is intermediate between that of the parents: they repeatedly shift material between the bill and the flank feathers. Other studies conducted on courtship songs in crickets and tree frogs also demonstrate the intermediate nature of hybrid behavior; hybrids, possessing alleles from both parental species produce songs that are a combination of the songs or their parents.

Studies of Twins

The influence of genes on behavior can also be seen in humans by comparing the behavior of identical twins. Identical twins are, as their name implies, genetically identical. Because most sets of identical twins are raised together, any similarities in their behavior might result either from identical genes, or from shared experiences as they grow up. However, in some instances twins have been separated at birth, and raised apart in different families. A recent study of 50 such sets of twins revealed many similarities in personality, temperament, and even leisure-time activities, even though the twins were often raised in very different circumstances. These results show that genes play a key role in determining human behavior, although the relative importance of genes versus environment is still hotly debated.

A Detailed Look at How One Gene Affects a Behavior

One well-studied gene mutation in mice provides a clear look at how a particular gene influences a behavior. In 1996 behavioral geneticists discovered a new gene, *fosB*, that seems to determine whether or not female mice will nurture their young. Females with both *fosB* alleles knocked out (experimentally removed) will initially investigate their newborn babies, but

(a)

(b)

Figure 22.3 A gene alters maternal care.
In mice, normal mothers (a) take very good care of their offspring, retrieving them if they move away and crouching over them. Mothers with the mutant *fosB* allele (b) perform neither of these behaviors, leaving their pups exposed.

then ignore them, in stark contrast to the caring and protective maternal care provided by normal females (figure 22.3).

This inattentiveness appears to result from a chain reaction. When mothers of new babies initially inspect them, information from their auditory, olfactory, and tactile senses is transmitted to the hypothalamus. There, *fosB* alleles are activated, producing a particular protein, which in turn activates both enzymes and other genes that affect the neural circuitry within the hypothalamus. These modifications within the brain cause the female to react maternally toward her offspring. In a general way, the information gained from inspecting the newborn babies can be viewed as acting like a sign stimulus, the *fosB* gene as an innate releasing mechanism, and the maternal behavior as the resulting action pattern.

In mothers lacking the *fosB* allele, this innate behavioral pattern is stopped midway. No protein is activated, the brain's neural circuitry is not rewired, and maternal behavior does not result.

> **22.3** The conclusion that genes play a key role in many behaviors is supported by a broad range of studies in many animals, including humans.

22.4 How Animals Learn

Many of the behavioral patterns displayed by animals are not the result solely of instinct. In many cases, animals alter their behavior as a result of previous experiences, a process termed **learning.** The simplest type of learning, **nonassociative learning,** does not require an animal to form an association between two stimuli, or between a stimulus and a response. One form of nonassociative learning is *sensitization,* in which repeating a stimulus produces a greater response. Another form of nonassociative learning is *habituation,* a decrease in response to a repeated stimulus. In many cases, the stimulus evokes a strong response when it is first encountered, but the magnitude of the response gradually declines with repeated exposure. As an everyday example, are you still conscious of the chair you are sitting in? Habituation can be thought of as learning not to respond to a stimulus. Being able to ignore unimportant stimuli is critical when facing a barrage of stimuli in a complex environment.

A change in behavior that involves an association between two stimuli, or between a stimulus and a response, is called **associative learning.** The behavior is modified, or *conditioned,* through the association. This form of learning is more complex than habituation. The two major types of associative learning are called classical conditioning and operant conditioning. They differ in the way the association is established.

Classical Conditioning

In **classical conditioning,** the paired presentation of two kinds of stimuli causes the animal to form an association between the stimuli. When the Russian psychologist Ivan Pavlov presented meat powder, an *unconditioned stimulus,* to a dog, the dog responded by salivating. If an unrelated stimulus, such as the ringing of a bell, was present at the same time as the meat powder, then over repeated trials the dog would come to salivate in response to the sound of the bell alone. The dog had learned to associate the unrelated sound stimulus with the meat powder stimulus. Its response to the sound stimulus had become conditioned, the sound of the bell now a *conditioned stimulus.*

Operant Conditioning

In **operant conditioning,** an animal learns to associate its behavioral response with a reward or punishment. Psychologist B. F. Skinner studied operant conditioning in rats by placing them in an experimental cage nicknamed a "Skinner box." As the rat explored the box, it would occasionally press a lever by accident, causing a pellet of food to appear. At first, the rat would ignore the lever, eat the food pellet, and continue to move about. Soon, however, it would learn to associate pressing the lever (the behavioral response) with food (the reward). When a conditioned rat was hungry, it would spend all its time pressing the lever. This sort of trial-and-error learning is of major importance to most vertebrates.

Figure 22.4   An unlikely parent.
The eager goslings follow ethologist Konrad Lorenz as if he were their mother. He is the first object they saw when they hatched, and they have used him as a model for imprinting.

Imprinting

As an animal matures, it may form preferences or social attachments to other individuals that will profoundly influence behavior later in life. This process, called **imprinting,** is sometimes considered a type of learning. In *filial imprinting,* social attachments form between parents and offspring. For example, young birds of some species begin to follow their mother within a few hours after hatching, forming a strong bond between mother and young. This is a form of associative learning—it is the association the young bird forms during a critical window of time (roughly 13 to 16 hours in geese, for example) that determines how the imprint will be established. Birds will follow the first object they see after hatching, and direct their social behavior toward that object as their mother. Ethologist Konrad Lorenz raised geese from eggs, and when he offered himself as a model for imprinting, the goslings treated him as if he were their parent, following him dutifully (figure 22.4).

> **22.4   Habituation and sensitization are simple forms of learning in which there is no association between stimulus and response. In contrast, associative learning (conditioning and imprinting) involves the formation of an association between two stimuli or between a stimulus and a response.**

22.5 Instinct and Learning Interact to Determine Behavior

Some animals have innate predispositions toward forming certain associations. Certain pairs of stimuli can be linked by operant conditioning, others not. For example, pigeons can learn to associate food with colors but not with sounds; on the other hand, they can associate danger with sounds but not with colors. This sort of *learning preparedness* demonstrates that what an animal can learn is biologically influenced—that is, learning is possible only within the boundaries set by instinct.

The innate programs that make up an animal's instincts have evolved because each of them reinforces an adaptive response. The seed a pigeon eats may have a distinctive color that the pigeon can see, but it makes no sound the pigeon can hear. The approach of a predator a pigeon fears may generate noise but involve no distinctive color.

Behavior Often Reflects Ecological Factors

Knowledge of an animal's ecology is key to understanding its behavior, as the genetic component of behaviors has evolved to match animals to their habitats. For example, some species of birds, like Clark's nutcracker, feed on seeds. These birds store seeds in caches they bury when seeds are abundant so they will have food during the winter. Thousands of seed caches may be buried by a bird and then later recovered, sometimes as much as nine months later. One would expect these birds to have an extraordinary spatial memory, and this is indeed what researchers have found. One Clark's nutcracker can remember the locations of up to 2,000 seed caches, using features of the landscape and other surrounding objects as spatial references to memorize the locations of the caches. When examined, Clark's nutcrackers turn out to have an unusually large hippocampus, the center for memory storage in the brain.

The Interaction Between Instinct and Learning

The way in which white-crowned sparrows first acquire their courtship songs provides an excellent example of the interaction between instinct and learning in the development of behavior. Courtship songs are sung by mature birds and are species specific. By rearing male birds in soundproof incubators provided with speakers and microphones, animal behaviorist Peter Marler could control what a bird heard as it matured, and then record the song it produced as an adult, a recording called a sonogram. When compared to a normal sonogram, in figure 22.5*a*, he found that white-crowned sparrows that heard no song at all during development had a poorly developed song as adults, shown in the sonogram in figure 22.5*b*. The same thing happened if they heard only the song of a different species, the song sparrow. But birds that heard the song of their own species sang a fully developed white-crowned sparrow song as adults. This was true even if the young birds also heard the song sparrow song along with their own.

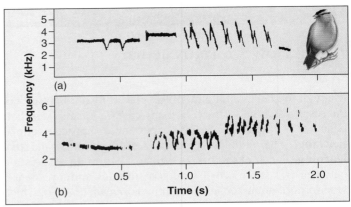

Figure 22.5 Song development in birds involves both instinct and learning.
The sonograms of songs produced by male white-crowned sparrows that had been exposed to their own species' song during development (a) are different from those of male sparrows that heard no song during rearing (b). This difference indicates that the genetic program itself is insufficient to produce a normal song.

Marler's results suggest that these birds have a genetic template, or instinctive program, that guides them to learn the appropriate song. During a critical period in development, the template will accept the correct song as a model. Thus, song acquisition depends on learning, but only the song of the correct species can be learned.

Although the song template is genetically determined, Marler found that learning also plays a prominent role in song development. If a young white-crowned sparrow becomes deaf *after* it hears its species' song during the critical period, it will sing a poorly developed song as an adult. The bird must "practice" listening to himself sing, matching what he hears to the model his template has accepted.

The males of some bird species have no opportunity to hear the song of their own species. In such cases, it appears that the males instinctively "know" their own species song. For example, cuckoos are brood parasites; females lay their eggs in the nest of another species of bird, and the young that hatch are reared by the foster parents. When the cuckoos become adults, they sing the song of their own species rather than that of their foster parents. Because male brood parasites would hear the song of their host species during development, it is adaptive for them to ignore such "incorrect" stimuli. They hear no adult males of their own species singing, so no correct song models are available. In this species, natural selection has provided the male with a genetically programmed song totally guided by instinct.

> **22.5** Behavior is both instinct (influenced by genes) and learned through experience. Genes are thought to limit the extent to which behavior can be modified and the types of associations that can be made.

22.6 Animal Cognition

For many decades, students of animal behavior flatly rejected the notion that nonhuman animals can think. Instead, the prevailing approach was to treat animals as though they responded to the environment through instinct and simple innately programmed learning.

In recent years, serious attention has been given by researchers to the topic of animal awareness. The central question is whether animals other than humans show **cognitive behavior**—that is, do they process information and respond in a manner that suggests thinking?

Evidence of Conscious Planning

What kinds of behavior would demonstrate cognition? Some birds in urban areas remove the foil caps from nonhomogenized milk bottles to get at the cream beneath. Japanese macaques (a kind of monkey) learn to float grain on water to separate it from sand, and teach other macaques to do it. A chimpanzee pulls the leaves off of a tree branch and uses the stick to probe the entrance to a termite nest and gather termites, suggesting that the ape is consciously planning ahead, with full knowledge of what it intends to do. A sea otter will use a rock as an "anvil," against which it bashes a clam to break it open, often keeping a favorite rock for a long time, as though it had a clear idea of its future use of the rock.

Problem Solving

Some instances of problem solving by animals are hard to explain in any other way than as a result of some sort of cognitive process—of what, if we were doing it, we would call reasoning. For example, in a series of classic experiments conducted in the 1920s, a chimpanzee was left in a room with a banana hanging from the ceiling out of reach. Also in the room were several boxes, each lying on the floor. After unsuccessful attempts to jump up and grab the banana, the chimpanzee suddenly looked at the boxes and immediately proceeded to move them underneath the banana, placing one on top of another, and climbed up the boxes to claim its prize. Many humans would not have solved the problem so quickly.

It is not surprising to find obvious intelligence in animals so closely related to us as chimpanzees. Perhaps more surprising, however, are recent studies finding that other animals also show evidence of cognition. Ravens have always been considered among the most intelligent of birds. Bernd Heinrich of the University of Vermont conducted an experiment using a group of hand-reared ravens that lived in an outdoor aviary. Heinrich placed a piece of meat on the end of a string and hung it from a branch in the aviary. The birds like to eat meat, but had never seen string before and were unable to get at the meat. After several hours, during which time the birds periodically looked at the meat but did nothing else, one bird flew to the branch, reached down, grabbed the string with

Figure 22.6   Problem solving by a raven.
Confronted with a problem it had never previously encountered, the raven figures out how to get the meat at the end of the string by repeatedly pulling up a bit of string and stepping on it.

its beak, pulled it up, and placed it under his foot. He then reached down and pulled up another length of the string, repeating this action over and over, each time bringing the meat closer (figure 22.6). Eventually the meat was within reach, and was grasped and eaten by the bird. The raven, presented with a completely novel problem, had devised a solution. Eventually, three of the other five ravens also figured out how to get the meat. This result can leave little doubt that ravens have advanced cognitive abilities.

> **22.6** Research on the cognitive abilities of animals is in its infancy, but some examples argue compellingly that animals can reason.

22.7 Behavioral Ecology

The investigation of animal behavior can be conveniently divided into three sorts of questions: (1) *The study of its development.* Lorenz's study of imprinting in geese was a study of this sort. (2) *The study of its physiological basis.* Analysis of the impact of the *fosB* gene on maternal behavior in mice was a study of this sort. (3) *The study of its function* (that is, its evolutionary significance). This third sort of question is addressed by biologists working in the field of **behavioral ecology.** Behavioral ecology is the study of how natural selection shapes behavior.

Behavioral ecology examines the survival value of behavior. How does an animal's behavior allow it to stay alive and reproduce, or keep its offspring alive to reproduce? Research in behavioral ecology thus focuses on a behavior's adaptive significance—that is, on the contribution a behavior makes to an animal's reproductive success, or fitness.

It is important to remember that all genetic differences in behavior need not have survival value. Many genetic differences in natural populations are the result of random mutations that accidentally become common, a process called genetic drift. It is only by experiment that we can learn if a particular behavior has been favored by natural selection.

Nobel laureate Niko Tinbergen's pioneering study of seagull nesting provides an excellent example of how a behavioral ecologist investigates the potential evolutionary significance of a behavior. Tinbergen observed that after gull nestlings hatched from their eggs, the parent birds quickly remove the eggshells from the nest. Why? What possible evolutionary advantage would this behavior confer on the birds?

To investigate this, Tinbergen camouflaged chicken eggs by painting them to resemble gulls' eggs that blend in with the natural background where the gull nests were located (figure 22.7), and distributed them on the ground throughout the nesting area. He placed broken eggshells next to some of the eggs, and, as a control, he left other camouflaged eggs alone without eggshells. He then watched to see which eggs were found more easily by crows. Because the crows could use the white interior of a broken eggshell as a cue, they repeatedly ate the camouflaged eggs that were near broken eggshells, and tended to ignore solitary camouflaged eggs that sat on the ground in plain sight. Tinbergen concluded that eggshell removal behavior is adaptive, that it does confer an evolutionary advantage on birds. Removing broken eggshells from the nest reduces predation of unhatched eggs (and probably of newborn chicks) and thus increases the chance that offspring will survive.

Figure 22.7 The adaptive value of eggshell removal.
Niko Tinbergen painted chicken eggs to resemble the mottled brown camouflage of gull eggs. Mottled eggs look like the rocky ground around a gull nest. The eggs were used to test the hypothesis that camouflaged eggs are more difficult for a predator to find and thus increase the young's chance of survival. He then placed broken eggshells near the camouflaged ones to test the hypothesis that the white interior of broken eggshells attracts predators.

It is not always so easy to learn how an adaptive trait confers its evolutionary advantage. Some behaviors, like eggshell removal, reduce predation. Other behaviors enhance energy intake, allowing an increased number of offspring to be supported. Still others reduce exposure or increase resistance to disease, enhance the ability to acquire a mate, or in some other way increase an individual's fitness, its ability to contribute offspring to the next generation.

> **22.7** Behavioral ecology is the study of how natural selection shapes behavior.

22.8 A Cost-Benefit Analysis of Behavior

One important way in which behavioral ecologists examine the evolutionary advantage of a behavior is to ask if it provides an evolutionary benefit greater than its cost. Thus, for example, a behavior may be favored by natural selection if it increases the intake of food by parents. This is a clear adaptive benefit when it increases survival of offspring, but it comes at a cost. Searching for food or defending food supplies can expose a parent to predation, decreasing the probability that the parent will survive to raise its offspring. To understand these sorts of behaviors, it is necessary to carefully evaluate their costs and benefits.

Foraging Behavior

For many animals, the food that they eat can be found in many sizes, and in many places. An animal must choose what food to select, and how far to go seeking it. These choices are called the animal's **foraging behavior.** Each choice involves benefits and associated costs. Thus, although a larger food may contain more energy, it may be harder to capture and less abundant. In addition, more desirable foods may be farther away than other types. Hence, an animal's foraging behavior involves a trade-off between a food's energy content and the cost of obtaining it.

The net energy (in calories) gained by feeding on each kind of food available to a foraging animal is simply the energy content of the food minus the energy costs of pursuing and handling it. At first glance, one might expect that evolution would favor foraging behaviors that are as energetically efficient as possible. This sort of reasoning has led to what is known as the **optimal foraging theory,** which predicts that animals will select food items that maximize their net energy intake per unit of foraging time.

Is the optimal foraging theory correct? Many foragers do preferentially use food items that maximize the energy return per unit time. Shore crabs, for example, tend to feed primarily on intermediate-sized mussels, which provide the greatest energy return. Larger mussels provide more energy, but also take considerably more energy to crack open. Many other animals also behave to maximize energy acquisition.

The key question, however, is whether increased energy resources acquired by optimal foraging leads to increased reproductive success. In many cases, it does. In a diverse group of animals that includes ground squirrels, zebra finches, and orb-weaving spiders, the number of offspring raised successfully increases when parents have access to more food energy.

In other cases, however, the costs of foraging seem to outweigh the benefits. An animal in danger of being eaten itself is often better off to minimize the amount of time it spends foraging. Many animals alter their foraging behavior when predators are present, reflecting this trade-off between food and danger.

Figure 22.8 The benefit of territoriality.
Sunbirds, found in Africa and ecologically similar to hummingbirds, increase nectar availability by defending flowers. A sunbird will expend 3,000 calories per hour chasing intruders away.

Territorial Behavior

Animals often move over a large area, or *home range*. In many species, the home range of several individuals overlap, but each individual defends only a portion of its home range and uses it exclusively. This behavior is called **territoriality.**

Territories are defended by displays that advertise that the territories are occupied, and by overt aggression. A bird sings from its perch within its territory to prevent invasion of its territory by a neighboring bird. If the intruding bird is not deterred by the song, the territory owner may attack and attempt to drive the invader away.

Why aren't all animals territorial? The answer involves a cost-benefit analysis. The actual adaptive value of an animal's territorial behavior depends on the trade-off between the behavior's benefits and its costs. Territoriality offers clear benefits, including increased food intake from nearby resources (figure 22.8), access to refuges from predators, and exclusive access to mates. The costs of territorial behavior, however, may also be significant. The singing of a bird, for example, is energetically expensive, and attacks of competitors can lead to injury. In addition, advertisement through song or visual display can reveal one's location to a predator. In many instances, particularly when food sources are abundant, defending easily obtained resources is simply not worth the cost.

22.8 Natural selection tends to favor the evolution of foraging and territorial behaviors that maximize energy gain, although other considerations such as avoiding predators are also important.

22.9 Migratory Behavior

Many animals breed in one part of the world, and spend the rest of the year in another. Long-range two-way annual movements like this are called **migrations.** Migratory behavior is particularly common in birds. Ducks and geese migrate southward along flyways from northern Canada across the United States each fall, overwinter, and then return northward each spring to nest. Warblers and many other insect-eating songbirds winter in the tropics and breed in the United States and Canada in spring and summer, when insects are plentiful. Monarch butterflies migrate each fall from central and eastern North America to overwinter in several small, geographically isolated areas of coniferous forests in the mountains of central Mexico. Gray whales feed in summer in the Arctic Ocean, then swim 10,000 kilometers to the warm waters off Baja, California, where they breed during winter months.

Biologists have studied migration with great interest. In attempting to understand how animals are able to navigate accurately over such long distances, it is important to understand the difference between *compass sense* (an innate ability to move in a particular direction, called following a bearing) and *map sense* (a learned ability to adjust a bearing depending on the animal's location). Experiments on starlings shown in figure 22.9 indicate that inexperienced birds migrate using a compass sense, and older birds that have migrated previously also employ a map sense to help them navigate—in essence, they learn the route. Migrating birds were captured in Holland, the halfway point of their migration, and were taken to Switzerland where they were released. Inexperienced birds (the red arrows) kept flying in their original direction, while experienced birds (the blue arrow) were able to adjust course and reached their normal wintering grounds.

The Compass Sense

In birds we now have a good understanding of how the compass sense is achieved. Many migrating birds have the ability to detect the earth's magnetic field and to orient themselves with respect to it. In a closed indoor cage, they will attempt to move in the correct geographical direction, even though there are no visible external clues. However, the placement of a powerful magnet near the cage can alter the direction in which the birds attempt to move.

The first migration of young birds appears to be innately guided by the earth's magnetic field. Inexperienced birds also use the sun and particularly the stars to orient themselves (migrating birds fly mainly at night).

The indigo bunting, which flies during the day and uses the sun to set its bearing, compensates for the movement of the sun in the sky as the day progresses by reference to the North Star, which does not move in the sky. Starlings compensate for the sun's apparent movement in the sky by using an internal clock. If captive starlings are shown an experimental sun in a fixed position, they will change their orientation to it at a constant rate of about 15 degrees per hour—the same rate the sun moves across the sky.

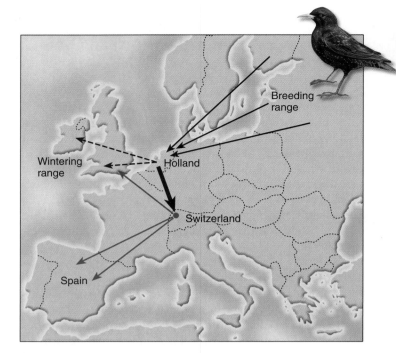

Figure 22.9 Starlings learn how to navigate.
The navigational abilities of inexperienced birds differ from those of adults that have made the migratory journey before. Starlings were captured in Holland, halfway along their full migratory route from Baltic breeding grounds to wintering grounds in the British Isles. These captured birds were transported to Switzerland and released. Experienced older birds compensated for the displacement and flew toward the normal wintering grounds (*blue* arrow). Inexperienced young birds kept flying in the same direction as before, on a course that took them toward Spain (*red* arrows). These observations indicate that inexperienced birds migrate with an innate compass sense, while experienced birds utilize a learned map sense to aid their navigation.

The Map Sense

Much less is known about how migrating birds and other animals acquire their map sense. During their first migration, young birds move with a flock of experienced older birds that know the route, and during the course of the journey they appear to learn to recognize certain cues, such as the position of mountains and coastline.

Animals that migrate through featureless terrain present more of a puzzle. Consider the green sea turtle. Every year great numbers of these large 400-pound turtles migrate with incredible precision from Brazil halfway across the Atlantic to Ascension Island, over 1,400 miles of open ocean, where females lay their eggs. Plowing head down through the waves, how do they find this tiny rocky island, over the horizon, more than a thousand miles away? No one knows for sure, although recent studies suggest that the direction of wave action provides an important navigational clue.

> **22.9** Many animals migrate in predictable ways, navigating by looking at the sun and stars, and in some cases by detecting magnetic fields. In many instances, young individuals learn the route by following experienced ones.

The Great Pigeon Race Disaster of 1997 Suggests an Answer to an Enduring Mystery

Homing pigeons (*Columba livia*) are small gray birds with a remarkable ability to find their way home from distant, unfamiliar places. A homing pigeon released hundreds of miles from where it lives, at a location it has never seen, will unerringly set out on a beeline homeward, flying through darkness and storm at speeds that often average 50 miles per hour, without stopping, until it arrives at its home loft hours or even days later. How can an unassuming little bird carry out a feat a normal person could only match with a satellite GPS navigation system? Navigating accurately over unfamiliar terrain as homing pigeons do demands of the birds both a sense of direction (compass sense) and a sense of location (map sense).

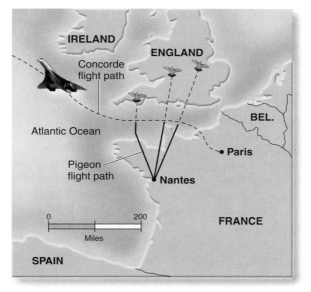

The compass sense of migrating birds is pretty well understood by biologists. Their usual daytime compass is the sun, but they can also navigate on cloudy days because they possess as well a magnetic sense that they can use to locate "north."

However, how does a homing pigeon determine its location, so it can select the correct homeward bearing? This has until now remained a mystery. They don't do it by memorizing landmarks—birds fitted with frosted contact lenses find their way home just fine. Nor do they sense local variations in the earth's magnetic field, or smell particular odors.

How then do homing pigeons figure out where they are? A researcher at the U.S. Geological Survey, Jonathan Hagstrum, has come up with a novel suggestion. It involves, of all things, pigeon races.

In Europe, and to a lesser extent in the United States, pigeon racing has become an international sport for which birds are carefully bred and trained. Birds from many lofts are taken to a common distant location, released together, and their return speeds timed. The most rapid return wins. Some birds are a little tardy, but typically over 90% of the birds return within a few days, and eventually almost all do.

On Sunday, June 29, 1997, a great race was held to celebrate the centenary of the Royal Pigeon Racing Association. More than 60,000 homing pigeons were released at 6:30 a.m. from a field in Nantes (southern France), flying to lofts all over southern England 400 to 500 miles away. By 11:00 a.m., the majority of the racing birds had made it out of France and were over the English Channel. They should have arrived at their lofts by early afternoon. They didn't.

A few thousand of the birds straggled in over the next few days. Most were never seen again. In pigeon racing terms, the loss of so many birds was practically unheard of, a disaster. Any one bird could get lost, but tens of thousands?

Hagstrum, in studying this event, noticed an odd fact. At the very same time the racing pigeons were crossing the Channel, 11:00 a.m., the Concorde supersonic transport (SST) airliner was flying along the Channel on its morning flight from Paris to New York (a flight discontinued in 2004). In flight, the SST generates a shock wave that pounds down toward the earth, a carpet of sound almost a hundred miles wide. The racing pigeons flying below the Concorde could not have escaped the intense wave of sound. The birds that did eventually arrive at their lofts were lucky enough to be very slow racers—they were still south of the Channel when the SST passed over, ahead of them.

Perhaps, Hagstrum suggests, racing pigeons locate where they are using atmospheric infrasounds that the SST obliterated. These very-low-frequency sounds travel thousands of miles from their sources. That's why you can hear distant thunder. Pigeons can hear infrasounds very well, because a pigeon ear is particularly good at detecting very-low-frequency sounds.

What sort of infrasounds are available to guide pigeons? All over the world, there is one infrasound pigeons should all be able to hear—the very-low-frequency acoustic shock waves generated by ocean waves banging against one another! Like an acoustic beacon, a constant stream of these tiny seismic waves would always say where the ocean is. Even more valuable to a racing pigeon looking for home, infrasounds reflect from cliffs, mountains, and other steep-sided features of the earth's surface. Ocean wave infrasounds reflecting off of local terrain could provide a pigeon with a detailed sound picture of its surroundings, near and far.

Hagstrum's suggestion explains neatly what happened to the lost pigeons in the great race of 1997. The enormous wave of infrasound generated by the SST's sonic boom would have blotted out all of the normal oceanic infrasound information. Any bird flying in its path would lose all orientation.

22.10 Reproductive Behaviors

Animals exhibit many different types of behaviors (table 22.1) but reproductive behavior is complex encompassing a variety of animal behaviors, including courtship and parental care. The reproductive success of an animal is influenced by a number of factors that in turn are directly affected by its behavior: how long the individual lives, how frequently it mates, and how many offspring it produces per mating. The second of these factors, competition for mating opportunities, has been termed **sexual selection.**

Sexual selection involves both *intrasexual selection,* or interactions between members of one sex ("the power to conquer other males in battle," as Darwin put it), and *intersexual selection* ("the power to charm").

Intrasexual selection leads to the evolution of structures used in combat with other males (such as a deer's antlers or a ram's horns). Combat for mates, whether ritual or real, is a form of *agonistic behavior,* a confrontation waged by threats, displays, or actual combat.

Intersexual selection, also called **mate choice,** leads to the evolution of complex courtship behaviors, and of ornaments used to "persuade" members of the opposite sex to mate, such as long tail feathers or bright plumage. The male peacock will parade in front of females with tail feathers displayed. Figure 22.10 shows the more eyespots on a male's tail feathers, the more mates he will attract.

The Benefits of Mate Choice

Why did mating preferences evolve? What is their adaptive value? Biologists have proposed several reasons:

1. In many species of birds and mammals, males help raise the offspring. In these cases, females would benefit by choosing the male that can provide the best care—the better the male parent, the more offspring she is likely to rear successfully.
2. In other species, males provide no care, but maintain territories that provide food, nesting sites, and predator refuges. In such species, females that choose males with the best territories will maximize reproductive success.
3. In some species, males provide no direct benefits of any kind to the female. If a female selects a more vigorous male, probably at least to some degree the result of a good genetic makeup, the female will be ensuring that her offspring receive good genes from their father.

Mating Systems

Reproductive behavior in animals varies a lot from one species to the next. Some animals mate with many partners during the breeding season, others with only one. The typical number of mates an animal has during its breeding season is called the **mating system.** Among animals, there are three principal mating systems: *monogamy* (one male mates with one female), *polygyny* (one male mates with more than one female), and *polyandry* (one female mates with more than one male).

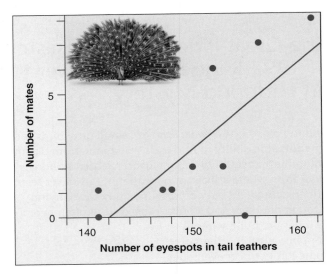

Figure 22.10 The male peacock's feather is a product of sexual selection.
While female tail feathers are drab, the male tail feathers are decorated with colorful "eyespots." Experiments show that female peahens prefer to mate with males with greater numbers of eyespots in their tail feathers.

Like mate choice, mating systems have evolved to maximize reproductive fitness. For instance, a male may defend a territory with resources sufficient for more than one female.

A male holding such a high-quality territory may already have a mate, but it is still more advantageous for a female to breed with that male than with an unmated male that defends a low-quality territory. In this way, natural selection would favor the evolution of polygyny.

Although polygyny is much more common in animals, polyandrous systems—in which one female mates with several males—are known in a variety of animals. For example, in the spotted sandpiper, a seashore bird, males take care of all incubation and parenting, and females mate and leave eggs with two or more males.

Reproductive Strategies

During the breeding season, animals make several important "decisions" concerning their choice of mates (mate choice), how many mates to have (mating system), and how much time and energy to devote to rearing offspring (parenting). These decisions are all aspects of an animal's reproductive strategy, a set of behaviors that has evolved to maximize that species' reproductive success. The two sexes of a species often have different reproductive strategies, reflecting the different parental investment each sex makes in producing and rearing offspring. In most animal species, females exercise far more mate choice, and in these species female parental investment is much greater than that of males. In species with biparental care, male and female parental investments are roughly equal, and both sexes tend to exercise mate choice.

> **22.10** Natural selection has favored the evolution of mate choice, mating system, and parenting behaviors that maximize reproductive success.

TABLE 22.1 ANIMAL BEHAVIORS

Foraging Behavior

Select, obtain, and consume food

Oystercatchers forage for food by stabbing the ground or rocks in search of arthropods or mollusks.

Territorial Behavior

Defend portion of home range and use it exclusively

Male elephant seals fight with each other for possession of territories. Only the largest males can hold territories, which contain many females.

Migratory Behavior

Move to a new location for part of the year

Wildebeests undergo an annual migration in search of new pastures and sources of water. Migratory herds can contain up to a million individuals and can stretch thousands of miles.

Courtship

Attract and communicate with potential mates

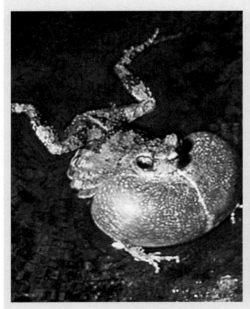

This male frog is vocalizing, producing a call that attracts females.

Parental Care

Produce and rear offspring

Female lions share the responsibility of raising the pride's young, increasing the probability that the young will survive into adulthood.

Social Behavior

Communicate information and interact with members of a social group

These leaf-cutter ants are members of different castes (or worker classes) in an insect society. The large ant is a worker, carrying leaves to the nest, and the smaller ants are protecting the worker from attack.

22.11 | Communication Within Social Groups

Many insects, fish, birds, and mammals live in social groups in which information is communicated between group members. For example, some individuals in mammalian societies serve as "guards." When a predator appears, the guards give an **alarm call,** and group members respond by seeking shelter. Social insects, such as ants and honeybees, secrete chemicals called **alarm pheromones** that trigger attack behavior. Ants also deposit **trail pheromones** between the nest and a food source to lead other colony members to food (figure 22.11). Honeybees have an extremely complex **dance language** that directs nestmates to rich nectar sources.

The Dance Language of the Honeybee

The European honeybee, *Apis mellifera,* lives in hives consisting of 30,000 to 40,000 individuals whose behaviors are integrated into a complex colony. Worker bees may forage miles from the hive, collecting nectar and pollen from a variety of plants on the basis of how energetically rewarding their food is. The food sources used by bees tend to occur in patches, and each patch offers much more food than a single bee can transport to the hive. A colony is able to exploit the resources of a patch because of the behavior of scout bees, which locate patches and communicate their location to hivemates through a *dance language.* Over many years, Nobel laureate Karl von Frisch was able to unravel the details of this communication system.

After a successful scout bee returns to the hive, she performs a remarkable behavior pattern called a *waggle dance* on a vertical comb (figure 22.12). The path of the bee during the dance resembles a figure-eight. On the straight part of the path, the bee vibrates or waggles her abdomen while producing bursts of sound. She may stop periodically to give her hivemates a sam-

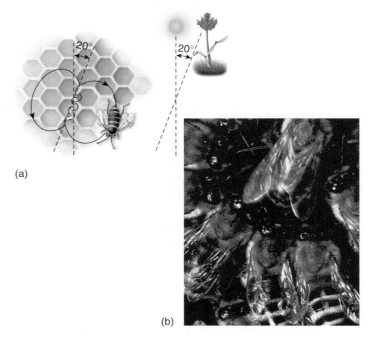

(a)

(b)

Figure 22.12   The waggle dance of honeybees.
(a) The angle between the food source, the nest, and the sun is represented by a dancing bee as the angle between the straight part of the dance and vertical. Here the food is 20 degrees to the right of the sun, and the straight part of the bee's dance on the hive is 20 degrees to the right of vertical. (b) A scout bee dancing in the hive.

ple of the nectar she has carried back to the hive in her crop. As she dances, she is followed closely by other bees, which soon appear as foragers at the new food source.

Von Frisch and his colleagues claimed that the other bees use information in the waggle dance to locate the food source. According to their explanation, the scout bee indicates the *direction* of the food source by representing the angle between the food source, the hive, and the sun as the deviation from vertical of the straight part of the dance performed on the hive wall (that is, if the bee moved straight, then the food source would be in the direction of the sun, but if the food were at a 30 degree angle relative to the sun's position, then the bee would move upward at a 30 degree angle). The *distance* to the food source is indicated by the tempo, or degree of vigor, of the dance.

Adrian Wenner, a scientist at the University of California, did not believe that the dance language communicated anything about the location of food, and he challenged von Frisch's explanation. Wenner maintained that flower odor was the most important cue allowing recruited bees to arrive at a new food source. A heated controversy ensued as the two groups of researchers published articles supporting their positions.

Such controversies can be very beneficial, because they often generate innovative experiments. In this case, the "dance language controversy" was resolved (in the minds of most scientists) in the mid-1970s by the creative research of James L.

Figure 22.11   Ants following a pheromone trail.
Trail pheromones organize cooperative foraging. The trails taken by the first ants to travel to a food source are soon followed by most of the other ants due to the release of pheromones by the first ants.

(a)

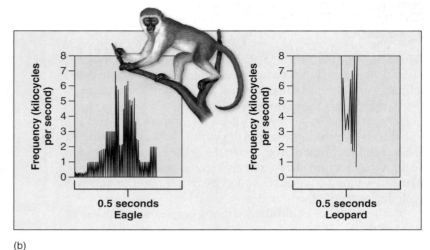

(b)

Figure 22.13 Primate semantics.

(a) This leopard, an efficient predator of primates, has attacked and will eat a vervet monkey. (b) Escaping a leopard attack presents a very different challenge to a vervet monkey than avoiding an eagle, another key predator of monkeys. Vervet monkeys give different alarm calls when leopards are sighted by troup members than they do when a member sees an eagle. Each distinctive call elicits a different and adaptive escape behavior.

Gould. Gould devised an experiment in which hive members were tricked into misinterpreting the directions given by the scout bee's dance. As a result, Gould was able to manipulate where the hive members would go if they were using visual signals. If odor were the cue they were using, hive members would have appeared at the food source, but instead they appeared exactly where Gould predicted. This confirmed von Frisch's ideas.

Recently, researchers have extended the study of the honeybee dance language by building robot bees whose dances can be completely controlled. Their dances are programmed by a computer and perfectly reproduce the natural honeybee dance; the robots even stop to give food samples! The use of robot bees has allowed scientists to determine precisely which cues direct hivemates to food sources.

Primate Language

Some primates have a "vocabulary" that allows individuals to communicate the identity of specific predators. The vocalizations of African vervet monkeys, for example, distinguish eagles, leopards, and snakes. The two sonograms in figure 22.13*b* show the alarm calls for an eagle and a leopard, each eliciting different responses in other members of the troup. Chimpanzees and gorillas can learn to recognize a large number of symbols and use them to communicate abstract concepts. The complexity of human language would at first appear to defy biological explanation, but closer examination suggests that the differences are in fact superficial—all languages share many basic structural similarities. All of the roughly 3,000 human languages draw from the same set of 40 consonant sounds (English uses two dozen of them), and any human can learn them. Researchers believe these similarities

reflect the way our brains handle abstract information, a genetically determined characteristic of all humans.

Language develops at an early age in humans. Human infants are capable of recognizing the 40 consonant sounds characteristic of speech, including those not present in the particular language they will learn, while they ignore other sounds. In contrast, individuals who have not heard certain consonant sounds as infants can only rarely distinguish or produce them as adults. That is why English speakers have difficulty mastering the throaty French "r," French speakers typically replace the English "th" with "z," and native Japanese often substitute "r" for the unfamiliar English "l." Children go through a "babbling" phase, in which they learn by trial and error how to make the sounds of language. Even deaf children go through a babbling phase using sign language. Next, children quickly and easily learn a vocabulary of thousands of words. Like babbling, this phase of rapid learning seems to be genetically programmed. It is followed by a stage in which children form simple sentences that, though they may be grammatically incorrect, can convey information. Learning the rules of grammar constitutes the final step in language acquisition.

While language is the primary channel of human communication, odor and other nonverbal signals (such as "body language") may also convey information. However, it is difficult to determine the relative importance of these other communication channels in humans.

22.11 The study of animal communication involves analysis of the specificity of signals, their information content, and the methods used to produce and receive them.

Altruism—the performance of an action that benefits another individual at a cost to the actor—occurs in many guises in the animal world. In many bird species, for example, parents are assisted in raising their young by other birds, which are called *helpers at the nest.* In species of both mammals and birds, individuals that spy a predator will give an alarm call, alerting other members of their group, even though such an act would seem to call the predator's attention to the caller. Finally, lionesses with cubs will allow all cubs in the pride to nurse, including cubs of other females.

The existence of altruism has long perplexed evolutionary biologists. If altruism imposes a cost to an individual, how could an allele for altruism be favored by natural selection? One would expect such alleles to be at a disadvantage, and thus their frequency in the gene pool should decrease through time.

A number of explanations have been put forward to explain the evolution of altruism. One suggestion often heard on television documentaries is that such traits evolve for the good of the species. The problem with such explanations is that natural selection operates on individuals within species, not on species themselves. Thus, natural selection will not favor alleles that lead an individual to act in ways that benefit others at a cost to itself; it is even possible for traits to evolve that are detrimental to the species as a whole, as long as they benefit the individual.

In some cases, selection can operate on groups of individuals, but this is rare. For example, if an allele for supercannibalism evolved within a population, individuals with that allele would be favored, as they would have more to eat; however, the group might eventually eat itself to extinction, and the allele would be removed from the species. In certain circumstances, such **group selection** can occur, but the conditions for it to occur are rarely met in nature. In most cases, consequently, the "good of the species" cannot explain the evolution of altruistic traits.

Another possibility is that seemingly altruistic acts aren't altruistic after all. For example, helpers at the nest are often young and gain valuable parenting experience by assisting established breeders. Moreover, by hanging around an area, such individuals may inherit the territory when the established breeders die. Similarly, alarm callers (figure 22.14) may actually benefit by causing other animals to panic. In the ensuing confusion, the caller may be able to slip off undetected. Detailed field studies in recent years have demonstrated that some acts truly are altruistic, but others are not as they seemed.

Reciprocity

Robert Trivers, now of Rutgers University, proposed that individuals may form "partnerships" in which mutual exchanges of altruistic acts occur because it benefits both participants to do so. In the evolution of such reciprocal altruism, "cheaters" (nonreciprocators) are discriminated against and are cut off

Figure 22.14 An altruistic act—or is it?
A meerkat sentinel on duty. Meerkats, *Suricata suricata,* are a species of highly social mongoose living in the semiarid sands of the Kalahari Desert in southern Africa. This meerkat is taking its turn to act as a lookout for predators. Under the security of its vigilance, the other members of the group can focus their attention on foraging. The sentinel puts his own life at risk when he gives an alarm, an apparent example of altruistic behavior.

from receiving future aid. According to Trivers, if the altruistic act is relatively inexpensive, the small benefit a cheater receives by not reciprocating is far outweighed by the potential cost of not receiving future aid. Under these conditions, cheating should not occur.

For example, vampire bats roost in hollow trees in groups of 8 to 12 individuals. Because these bats have a high metabolic rate, individuals that have not fed recently may die. Bats that have found a host imbibe a great deal of blood; giving up a small amount presents no great energy cost to the donor, and it can keep a roostmate from starvation. Vampire bats tend to share blood with past reciprocators. If an individual fails to give blood to a bat from which it had received blood in the past, it will be excluded from future bloodsharing.

Kin Selection

The most influential explanation for the origin of altruism was presented by William D. Hamilton in 1964. It is perhaps best introduced by quoting a passing remark made in a pub in 1932 by the great population geneticist J. B. S. Haldane. Haldane said that he would willingly lay down his life for two brothers or eight first cousins. Evolutionarily speaking, Haldane's statement makes sense, because for each allele Haldane received from his parents, his brothers each had a 50% chance of receiving the same allele. Consequently, it is statistically expected that two of his brothers would pass on as many of Haldane's particular combination of alleles to the next generation as Haldane himself would. Similarly, Haldane and a first cousin would share an eighth of their alleles. Their parents, which are siblings, would each share half their alleles, and each of their children would receive half of these, of which half on the average would be in common: $1/2 \times 1/2 \times 1/2 = 1/8$. Eight first cousins would therefore pass on as many of those alleles to the next generation as Haldane himself would. Hamilton saw Haldane's point clearly: Natural selection will favor any strategy that increases the net flow of an individual's alleles to the next generation.

Hamilton showed that by directing aid toward kin, or close genetic relatives, an altruist may increase the reproductive success of its relatives enough to compensate for the reduction in its own fitness. Because the altruist's behavior increases the propagation of its own alleles in relatives, it will be favored by natural selection. Selection that favors altruism directed toward relatives is called **kin selection.** Although the behaviors being favored are cooperative, the genes are actually "behaving selfishly," because they encourage the organisms to support copies of themselves in other individuals. In other words, if an individual has a dominant allele that causes altruism, any action that increases the frequency of this allele in future generations will be favored, even if that action is detrimental to the particular individual taking the action.

Examples of Kin Selection

Hamilton's kin selection model predicts that altruism is likely to be directed toward close relatives. The more closely related two individuals are, the greater the potential genetic payoff.

Many examples of kin selection are known from the animal world. Belding's ground squirrels give alarm calls when they spot a predator such as a coyote or a badger. Such predators may attack a calling squirrel, so giving a signal places the caller at risk. The social unit of a ground squirrel colony consists of a female and her daughters, sisters, aunts, and nieces. When they mature, males disperse long distances from where they are born; thus, adult males in the colony are not genetically related to the females. By marking all squirrels in a colony with an individual dye pattern on their fur and by recording which individuals gave calls and the social circumstances of their calling, researchers found that females who have relatives living nearby are more likely to give alarm calls

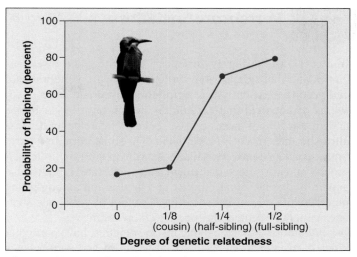

Figure 22.15 Kin selection is common among vertebrates.

In the white-fronted bee-eaters (*Merops bullockoides*), nonbreeding individuals will help raise the offspring of others. Most helpers are close relatives, and the probability that a bird will help another increases with genetic relatedness.

than females with no kin nearby. Males tend to call much less frequently, as would be expected because they are not related to most colony members.

Another example of kin selection comes from a bird called the white-fronted bee-eater, which lives along rivers in Africa in colonies of 100 to 200 birds. In contrast to the ground squirrels, it is the males that usually remain in the colony in which they were born, and the females that disperse to join new colonies. Many bee-eaters do not raise their own offspring, but rather help others. Many of these birds are relatively young, but helpers also include older birds whose nesting attempts have failed. The presence of a single helper, on average, doubles the number of offspring that survive. Two lines of evidence support the idea that kin selection is important in determining helping behavior in this species. First, helpers are normally males, which are usually related to other birds in the colony, and not females, which are not related. Second, when birds have the choice of helping different parents, they almost invariably choose the parents to which they are most closely related. The graph in figure 22.15 compares the probability of a bird helping in a nest (on the *y* axis) and the relationship of the helper (on the *x* axis). The more closely related the helper (toward the right side of the graph), the higher the probability that it will be a helper in the nest.

22.12 Many factors could be responsible for the evolution of altruistic behaviors. Individuals may benefit directly if altruistic acts are reciprocated; kin selection explains how alleles for altruism can increase in frequency if altruistic acts are directed toward relatives.

22.13 Vertebrate Societies

In contrast to the highly structured and integrated insect societies and their remarkable forms of altruism, vertebrate social groups are usually less rigidly organized and cohesive. It seems paradoxical that vertebrates, which have larger brains and are capable of more complex behaviors, are generally less altruistic than insects. Nevertheless, in some complex vertebrate social systems, individuals may be exhibiting both reciprocity and kin-selected altruism. But vertebrate societies also generally display more conflict and aggression among group members than do insect societies. Conflict in vertebrate societies generally centers on access to food and mates.

Vertebrate societies, like insect societies, have particular types of organization. Each social group of vertebrates has a certain size, stability of members, number of breeding males and females, and type of mating system. Behavioral ecologists have learned that the way a group is organized is influenced most often by ecological factors such as food type and predation.

African weaver birds, which construct nests from vegetation, provide an excellent example to illustrate the relationship between ecology and social organization. Their roughly 90 species can be divided according to the type of social group they form. One set of species lives in the forest and builds camouflaged, solitary nests. Males and females are monogamous; they forage for insects to feed their young. The second group of species nests in colonies in trees on the savanna. They are polygynous and feed in flocks on seeds. The feeding and nesting habits of these two sets of species are correlated with their mating systems. In the forest, insects are hard to find, and both parents must cooperate in feeding the young. The camouflaged nests do not call the attention of predators to their brood. On the open savanna, building a hidden nest is not an option. Rather, savanna-dwelling weaver birds protect their young from predators by nesting in trees, which are not very abundant. This shortage of safe nest sites means that birds must nest together in colonies. Because seeds occur abundantly, a female can acquire all the food needed to rear young without a male's help. The male, free from the duties of parenting, spends his time courting many females—a polygynous mating system (figure 22.16).

One exception to the general rule that vertebrate societies are not organized like those of insects is the naked mole rat, a small, hairless rodent that lives in and near East Africa. Unlike other kinds of mole rats, which live alone or in small family groups, naked mole rats form large underground colonies with a far-ranging system of tunnels and a central nesting area. It is not unusual for a colony to contain 80 individuals.

Naked mole rats feed on bulbs, roots, and tubers, which they locate by constant tunneling. As in insect societies, there is a division of labor among the colony members, with some mole rats working as tunnelers while others perform different tasks, depending upon the size of their body. Large mole rats defend the colony and dig tunnels.

Figure 22.16 Savanna-dwelling African weaver birds form colonial nests.

Some species of African weaver birds live in the savanna and build colonial nests, as safe nest sites are hard to find. Food resources are abundant on the savanna, and a polygynous mating system forms. Other species of weaver birds live in the forest and build solitary, camouflaged nests. Monogamous pairs form and cooperate in obtaining hard-to-find food sources.

Figure 22.17 Naked mole rats live in structured colonies.

Naked mole rats live in East Africa and are rare among vertebrates in that they form large, underground colonies with a division of labor. A "queen" does all of the breeding, and workers, consisting of both sexes, divide up tasks, such as tunneling, foraging, and defense.

Naked mole rat colonies have a *reproductive* division of labor similar to the one normally associated with the social insects. All of the breeding is done by a single female or "queen," who has one or two male consorts (figure 22.17). The workers, consisting of both sexes, keep the tunnels clear and forage for food.

> **22.13** Vertebrate societies exhibit both altruism and conflict. The extent of vertebrate sociality is affected by environmental conditions.

22.14 Human Social Behavior

One of the most profound lessons of biology is that we human beings are animals, quite close relatives of the chimpanzee (figure 22.18), and not some special form of life set apart from the earth's other creatures. This biological view of human life raises an important issue as we conclude this chapter. To what degree are the animal behaviors we have described in this chapter characteristic of the social behavior of humans?

Genes and Human Behavior

As we have seen repeatedly, genes play a key role in determining many aspects of animal behavior. From maternal behavior in mice to migratory behavior in songbirds, changes in genes have been shown to have a profound impact. This leads to an important prediction. If behaviors have a genetic basis, and if behaviors affect an animal's ability to survive and raise young, then surely behaviors must be subject to natural selection, as are any other gene-mediated traits that impact survival. Ethology, the study of animal behavior in natural conditions, has provided ample evidence that behaviors do indeed evolve.

To the degree that social behaviors are determined by genes (and both theory and evidence strongly suggest they are), the complex social behaviors of animals, including humans, should also evolve. The study of this facet of animal behavior, often called **sociobiology,** was pioneered by E. O. Wilson of Harvard University. It has proven highly controversial.

Genes certainly affect human behavior in important ways. Human facial expressions are similar the world over, no matter the culture or language, arguing that they have a deep genetic basis. Infants born blind still smile and frown, even though they have never seen these expressions on another face. Similarly, studies of identical twins demonstrate that personality and intelligence are highly heritable—well over 50% of the variance (the "scatter" among individuals) in these traits is due to the contribution of genes.

It is also true, however, that learning has a huge impact on how humans behave. Of the 40 different consonant sounds that humans can make, babies in the United States learn some 24, and then most quickly lose the ability to make the others. Just as birds learn songs and migration patterns from experienced adults, so human babies learn to speak from listening to the adults around them.

Figure 22.18 A chimpanzee family group.
Chimpanzees, like humans, have close family ties. Chimps usually stay with their mothers for about 7 years and help rear younger sisters and brothers. Many human family behaviors, from mother's love to sibling rivalry, are seen strongly displayed in these chimpanzee families.

Diversity Is the Hallmark of Human Culture

When a group of biologists that studied social behavior in chimpanzees among seven widely separated areas compared notes, they identified 39 behaviors involving such things as social behavior, courtship, and tool use that were common in some groups but absent in others. Each population seemed to have a distinct repertoire of behaviors. It seems that each population has developed its own collection of customary behaviors, which each generation teaches its children. In a word, chimpanzees have culture.

No animal, not even our close relative the chimpanzee, exhibits cultural differences to the degree seen in human populations (figure 22.19). There are polyandrous, polygynous, and monogamous societies. There are human groups in which warfare is common, others in which it never occurs. In some cultures, marriage to first cousins is forbidden; in others, it is encouraged. The variation in social behavior of other species is vanishingly small compared to the enormous diversity of human cultures.

To what degree human culture is the product of evolution acting on behavior-determining genes is a question that is hotly debated by behavioral biologists. However one chooses to answer this question, it is perfectly clear that much of our behavior is molded by experience. Human cultures, and the behaviors that produce them, change very rapidly, far too rapidly to reflect genetic evolution. Human cultural diversity, a hallmark of our species, is certainly strongly influenced, if not largely determined, by learning and experience.

Figure 22.19   A New York City street scene.

22.14   **In humans, just as in other animals, behavior is molded by experience within limits set by our genes. Both heredity and learning play key roles in determining how we behave.**

INQUIRY & ANALYSIS

Do Crabs Eat Sensibly?

Many behavioral ecologists claim that animals exhibit so-called optimal foraging behavior. The idea is that because an animal's choice in seeking food involves a trade-off between the food's energy content and the cost of obtaining it, evolution should favor foraging behaviors that optimize the trade-off. In the experiment to the right, an investigator looked to see if shore crabs in fact feed on those mussels which provide the most energy, as the theory predicts. The blue curve describes the net energy which the crab gets from feeding on different-sized mussels. The red histogram shows the numbers of mussels of each size actually eaten.

1. **Applying Concepts** What is the dependent variable in the curve? in the histogram?
2. **Making Inferences**
   a. What is the most energetically optimal mussel size for the crabs to eat, in mm?
   b. What size mussel is most frequently eaten by crabs, in mm?
3. **Drawing Conclusions** Do shore crabs tend to feed on those mussels that provide the most energy?

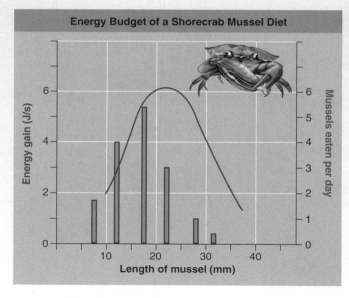

4. **Further Analysis** What factors might be responsible for the slight difference in peak prey length relative to the length optimal for maximal energy gain?

Some Behavior Is Genetically Determined

22.1 Approaches to the Study of Behavior

- The study of animal behavior, how animals respond to stimuli in their environment, includes examining how and why the behavior occurs. Instinct and learning play significant roles in behavior.

22.2 Instinctive Behavioral Patterns

- Instinctive, or innate, behaviors are those that are the same in all individuals of a species and appear to be caused by preset pathways in the nervous system. A sign stimulus triggers the behavior, called a fixed action pattern, such as egg retrieval in geese (**figure 22.2**).

22.3 Genetic Effects on Behavior

- Many behaviors are strongly influenced by genes and as such, can be studied as inherited traits. Hybrids, twins, and "knockout" mice (**figure 22.3**) have been used to study genetically influenced behaviors.

Behavior Can Also Be Influenced by Learning

22.4 How Animals Learn

- Many behaviors are learned, meaning that they have been formed or altered based on previous experiences. Classical conditioning results when two stimuli are paired such that the animal learns to associate the two stimuli. Operant conditioning results when an animal associates a behavior with a reward or punishment. Imprinting is when an animal forms social attachments, usually during a critical window of time (**figure 22.4**).

22.5 Instinct and Learning Interact to Determine Behavior

- Behavior is often both genetically determined and modified by learning. Genes often limit the extent to which a behavior can be modified through learning. Ecology has a lot to do with behavior, and knowing an animal's ecological niche can reveal much about its behavior (**figure 22.5**).

22.6 Animal Cognition

- While humans have evolved a great capacity for cognitive thought, studies show that other animals possess varying degrees of cognitive abilities. Some animals presented with a novel situation show problem-solving abilities to respond to the situation (**figure 22.6**).

Evolutionary Forces Shape Behavior

22.7 Behavioral Ecology

- Behavioral ecology is the study of how natural selection shapes behavior. Only behaviors that have a genetic basis and offer some advantage for survival or reproduction can be acted upon through natural selection (**figure 22.7**).

22.8 A Cost-Benefit Analysis of Behavior

- For every behavior that offers an individual an advantage for survival, there is usually an associated cost. For example, foraging and territorial behaviors offer a benefit by providing food and shelter for offspring but may endanger the parents through predation or expenditure of energy (**figure 22.8**). The benefits have to outweigh the costs for the behavior to be favored by natural selection.

22.9 Migratory Behavior

- Migration is a behavior that changes throughout the life of an animal. Inexperienced animals seem to rely on compass sense (following a direction), while experienced animals may rely more on map sense (learned ability to alter the path based on location) (**figure 22.9**).

22.10 Reproductive Behaviors

- Behaviors that maximize reproduction, such as mate choice, mating systems, and parenting behaviors, are favored by natural selection (**figure 22.10**).

Social Behavior

22.11 Communication Within Social Groups

- Communication is a behavior found in animals that live in groups or societies. Some animals secrete chemical pheromones to communicate information to others (**figure 22.11**). Others use movements (**figure 22.12**), or auditory signals (**figure 22.13**).

22.12 Altruism and Group Living

- Altruistic behaviors evolved in societies. The reason why may involve the reciprocation of altruistic acts or to benefit relatives, called kin selection (**figure 22.15**).

22.13 Vertebrate Societies

- Vertebrates are less altruistic than societal insects, but they do engage in cooperative breeding and in alarm calling.

22.14 Human Social Behavior

- Both genetics and learning play key roles in human behaviors, but the extent of each is hotly debated.

1. Innate behavior patterns
   a. can be modified if the stimulus changes.
   b. cannot be modified, as these behaviors seem built into the brain and nervous system.
   c. can be modified if environmental conditions begin to vary over a long period, a year or more.
   d. cannot be modified, as these behaviors are learned while very young.
2. The study of mouse behavior shows a clearer genetic effect on behavior than lovebird or human twin studies because
   a. there is a clear link between presence or absence of a specific gene, a specific metabolic pathway, and a specific behavior.
   b. the behavior of mice was less complex and easier to study than lovebirds or human twins.
   c. parental care has a larger influence on mice than on lovebirds and humans.
   d. No answer is correct.
3. Training a dog to perform tricks using verbal commands and treats is an example of
   a. nonassociative learning.
   b. operant conditioning.
   c. classical conditioning.
   d. imprinting.
4. Behavior can be tied to ecology and evolution by considering what the behavior does to increase
   a. body size.
   b. number of breeding sites.
   c. reproductive fitness.
   d. territory size.
5. The selection of foods and the journey to seek those foods is called
   a. territoriality.
   b. optimal foraging theory.
   c. migratory behavior.
   d. foraging behavior.
6. Courtship rituals are thought to have come about through
   a. intrasexual selection.
   b. competition.
   c. intersexual selection.
   d. kin selection.
7. A mating system where the female mates with more than one male is
   a. protandry.
   b. polyandry.
   c. polygyny.
   d. monogamy.
8. Sharing of blood meals between fed and hungry vampire bats and the shunning of a bat who takes but doesn't share are examples of
   a. helpers.
   b. kin selection.
   c. reciprocity.
   d. group selection.
9. Bird offspring who help their parents care for younger offspring are showing
   a. brood parasitism.
   b. kin selection.
   c. reciprocity.
   d. group selection.
10. While vertebrate animal societies are more loosely structured than insect societies, the organization of these societies is most influenced by
    a. which females are receptive to reproduction.
    b. migration patterns.
    c. how large the territory is compared to neighboring societies.
    d. ecological factors such as food type and predation.

Visual Understanding

1. **Figure 22.2b** What does this figure imply about certain types of fixed or innate behaviors?

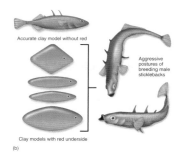

Accurate clay model without red

Aggressive postures of breeding male sticklebacks

Clay models with red underside

(b)

2. **Figure 22.13b** What does this graph imply about communication in nonhuman primates?

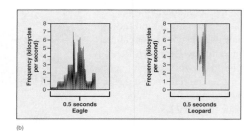

(b)

Challenge Questions

1. Nature versus nurture is an old argument regarding which one predominates in behavior. How have studies of hybrids and of human twins helped us learn more?
2. Some behavior requires both genetic (nature) and environmental (nurture) input to be carried out. Explain how these two forces interact to affect the song of the white-crowned sparrow.
3. Optimal foraging theory balances costs and benefits of the various methods of finding and hoarding food items, and the danger of being exposed to predators. Discuss the probable differences in foraging behavior of a skunk, a porcupine, and a field mouse.
4. What are some of the reasons advanced for altruism among groups of animals? Give some examples.

23
Planet Under Stress

The girl gazing from this now-classic *National Geographic* photo faces an uncertain future. An Afghani refugee, the whims of war destroyed her home, her family, and all that was familiar to her. Her expression carries a message about our own future: The problems humanity faces on an increasingly unstable, overcrowded, and polluted earth are no longer hypothetical. They are with us today and demand solutions. This chapter provides an overview of the problems and then focuses on solutions—on what can be done to address very real problems. As a concerned citizen, your first task must be to clearly understand the nature of the problem. You cannot hope to preserve what you do not understand. The world's environmental problems are acute, and a knowledge of biology is an essential tool you will need to contribute to the effort to solve them. It has been said that we do not inherit the earth from our parents—we borrow it from our children. We must preserve for them a world in which they can live. That is our challenge for the future, and it is a challenge that must be met soon. In many parts of the world, the future is happening right now.

23.1 Pollution

Our world is one ecological continent, one highly interactive biosphere, and damage done to any one ecosystem can have ill effects on many others. Burning high-sulfur coal in Illinois kills trees in Vermont, while dumping refrigerator coolants in New York destroys atmospheric ozone over Antarctica and leads to increased skin cancer in Madrid. Biologists call such widespread effects on the worldwide ecosystem **global change.** The pattern of global change that has become evident within recent years, including chemical pollution, acid precipitation, the ozone hole, the greenhouse effect, and the loss of biodiversity, is one of the most serious problems facing humanity's future.

Chemical Pollution

The problem posed by chemical pollution has grown very serious in recent years, both because of the growth of heavy industry and because of an overly casual attitude in industrialized countries. In one example, a poorly piloted oil tanker named the *Exxon Valdez* ran aground in Alaska in 1989 and heavily polluted with oil many kilometers of North American coastline, and the organisms that live there. If the tanker had been loaded no higher than the waterline, little oil would have been lost, but it was loaded far higher than that, and the weight of the above-waterline oil forced thousands of tons of oil out the hole in the ship's hull. Why do policies permit overloading like this?

Chemicals are released into both the air and into water; therefore, their effects are far reaching.

Air Pollution. Air pollution is a major problem in the world's cities. In Mexico City, oxygen is sold routinely on corners for patrons to inhale. Cities such as New York, Boston, and Philadelphia are known as gray-air cities because the pollutants in the air are usually sulfur oxides emitted by industry. Cities such as Los Angeles, however, are called brown-air cities because the pollutants in the air undergo chemical reactions in the sunlight to form smog.

Water Pollution. Water pollution is a very serious consequence of our casual attitude about pollution. "Flushing it down the sink" doesn't work in today's crowded world. There is simply not enough water available to dilute the many substances that the enormous human population produces continuously. Despite improved methods of sewage treatment, lakes and rivers throughout the world are becoming increasingly polluted with sewage. In addition, fertilizers and insecticides also get washed from the land to the water in great quantities.

Agricultural Chemicals

The spread of "modern" agriculture, and particularly the Green Revolution, which brought high-intensity farming to developing countries, has caused very large amounts of many kinds of new chemicals to be introduced into the global ecosystem,

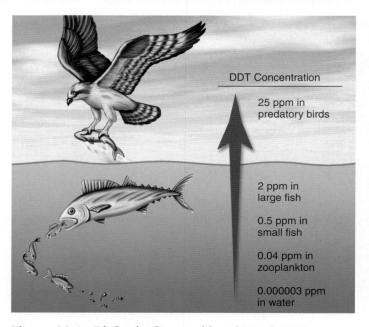

Figure 23.1 Biological magnification of DDT.
Because DDT accumulates in animal fat, the compound becomes increasingly concentrated in higher levels of the food chain.

particularly pesticides, herbicides, and fertilizers. Industrialized countries like the United States now attempt to carefully monitor side effects of these chemicals. Unfortunately, large quantities of many toxic chemicals, although no longer manufactured, still circulate in the ecosystem.

For example, the chlorinated hydrocarbons, a class of compounds that includes DDT, chlordane, lindane, and dieldrin, have all been banned for normal use in the United States, where they were once widely used. They are still manufactured in the United States and exported to other countries, where their use continues. Chlorinated hydrocarbon molecules break down slowly and accumulate in animal fat tissue. Furthermore, as they pass through a food chain, they become increasingly concentrated in a process called **biological magnification.** Figure 23.1 shows how a minute concentration of DDT in plankton increases to significant levels as it is passed up through this aquatic food chain. DDT caused serious ecological problems by leading to the production of thin, fragile eggshells in many predatory bird species, such as peregrine falcons, bald eagles, osprey, and brown pelicans, in the United States and elsewhere until the late 1960s, when it was banned in time to save the birds from extinction. Chlorinated compounds have other undesirable side effects and exhibit hormonelike activities in the bodies of animals.

> **23.1 All over the globe, increasing industrialization is leading to higher levels of pollution.**

23.2 Acid Precipitation

The smokestacks you see in figure 23.2 are those of the Four Corners power plant in New Mexico. This facility burns coal, sending the smoke high into the atmosphere through these tall stacks. The smoke contains high concentrations of sulfur dioxide and other sulfates, which produce acid when they combine with water vapor in the air. The first tall stacks were introduced in Britain in the mid-1950s, and the design rapidly spread through Europe and the United States. The intent of having tall smokestacks was to release the sulfur-rich smoke high in the atmosphere, where winds would disperse and dilute it, carrying the acids far away.

However, in the 1970s, scientists began noticing that the acids from the sulfur-rich smoke were having devastating effects. Throughout northern Europe, lakes were reported to have suffered drastic drops in biodiversity, some even becoming devoid of life. The trees of the great Black Forest of Germany seemed to be dying—and the damage was not limited to Europe. In the eastern United States and Canada, many of the forests and lakes in the eastern United States and Canada have been seriously damaged.

It turns out that when the sulfur introduced into the upper atmosphere combined with water vapor to produce sulfuric acid, the acid was taken far from its source, but it later fell along with water as acidic rain and snow. This pollution-acidified precipitation is called **acid rain** (but the term acid precipitation is actually more correct). Natural rainwater rarely has a pH lower than 5.6; however, rain and snow in many areas of the United States have pH values less than 5.3, and in the northeastern U.S., pHs of 4.2 or below have been recorded, with occasional storms as low as 3.0.

Acid precipitation destroys life. Many of the forests of the northeastern United States and Canada have been seriously damaged. In fact, it is now estimated that at least 1.4

Figure 23.2 Tall stacks export pollution.
Tall stacks like those of the Four Corners coal-burning power plant in New Mexico send pollution far up into the atmosphere.

Figure 23.3 Acid precipitation.
Acid precipitation is killing many of the trees in North American and European forests. Much of the damage is done to the mycorrhizae, fungi growing within the cells of the tree roots. Trees need mycorrhizae in order to extract nutrients from the soil.

million acres of forests in the Northern Hemisphere have been adversely affected by acid precipitation (figure 23.3). In addition, thousands of lakes in Sweden and Norway no longer support fish—these lakes are now eerily clear. In the northeastern United States and Canada, tens of thousands of lakes are dying biologically as their pH levels fall to below 5.0. At pH levels below 5.0, many fish species and other aquatic animals die, unable to reproduce.

The solution seems like it would be easy—clean up the sulfur emissions. But there have been serious problems with implementing this solution. First, it is expensive. Estimates of the cost of installing and maintaining the necessary "scrubbers" in the United States are on the order of $5 billion a year. An additional difficulty is that the polluter and the recipient of the pollution are far from one another, and neither wants to pay so much for what they view as someone else's problem. The Clean Air Act revisions of 1990 have begun to address this problem by mandating some cleaning of emissions in the United States, although much still remains to be done worldwide.

> **23.2** Pollution-acidified precipitation—loosely called acid rain—is destroying forest and lake ecosystems in Europe and North America. The solution is to clean up the emissions.

23.3   The Ozone Hole

For 2 billion years, life was trapped in the oceans because radiation from the sun seared the earth's surface unchecked. Nothing could survive that bath of destructive energy. Living things were able to leave the oceans and colonize the surface of the earth only after a protective shield of ozone had been added to the atmosphere by photosynthesis. Imagine if that shield were taken away. Alarmingly, it appears that we are destroying it ourselves. Starting in 1975, the earth's ozone shield began to disintegrate. Over the South Pole in September of that year, satellite photos revealed that the ozone concentration was unexpectedly less than elsewhere in the earth's atmosphere. It was as if some "ozone eater" were chewing it up in the Antarctic sky, leaving a mysterious zone of lower-than-normal ozone concentration, an **ozone hole.** Every year after that, more of the ozone has been depleted, and the hole grows bigger and deeper. The satellite image in figure 23.4 shows lower levels of ozone as light purple (Antarctica is also colored purple, indicating that the ozone hole completely covers it). The graph indicates the size of the ozone hole over a 10-year period, with the largest hole appearing in September of 2000 (the blue line).

What is eating the ozone? Scientists soon discovered that the culprit was a class of chemicals that everyone had thought to be harmless: **chlorofluorocarbons (CFCs).** CFCs were invented in the 1920s, a miracle chemical that was stable, harmless, and a near-ideal heat exchanger. Throughout the world, CFCs are used in large amounts as coolants in refrigerators and air conditioners, as the gas in aerosol dispensers, and as the foaming agent in Styrofoam containers. All of these CFCs eventually escape into the atmosphere, but no one worried about this until recently, both because CFCs were thought to be chemically inert and because everyone tends to think of the atmosphere as limitless. But CFCs are very stable chemicals, and have continually accumulated in the atmosphere.

It turned out that the CFCs were causing mischief the chemists had not imagined. High over the South and North Poles, nearly 50 kilometers up, where it was very, very cold, the CFCs stuck to frozen water vapor and began to act as catalysts of a chemical reaction. Just as an enzyme carries out a reaction in your cells without being changed itself, so the CFCs began to catalyze the conversion of ozone (O_3) into oxygen (O_2) without being used up themselves. Very stable, the CFCs in the atmosphere just kept at it—little machines that never stop. They are still there, still doing it, today. The drop in ozone worldwide is now over 3%.

Ultraviolet radiation is a serious human health concern. Every 1% drop in the atmospheric ozone content is estimated to lead to a 6% increase in the incidence of skin cancers. At middle latitudes, the drop of approximately 3% that has occurred worldwide is estimated to have led to an increase of perhaps as much as 20% in lethal melanoma skin cancers.

Experts generally agree that levels of ozone-killing chemicals in the upper atmosphere are leveling off since more than 180 countries in the 1980s signed an international agreement, which phases out the manufacture of most CFCs. The 2005 ozone hole peaked at about 25 million square kilometers (the size of North America), below the 2000 record size of about 28.4 million square kilometers. Current computer models suggest the Antarctic ozone hole should recover by 2065, and the lesser-damaged ozone layer over the Arctic by about 2023.

> **23.3** CFCs and other chemicals are catalytically destroying the ozone in the upper atmosphere, exposing the earth's surface to dangerous radiation. International attempts to solve the problem appear to be succeeding.

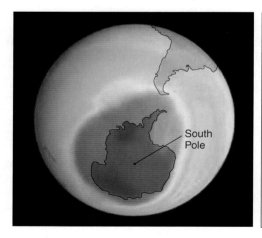

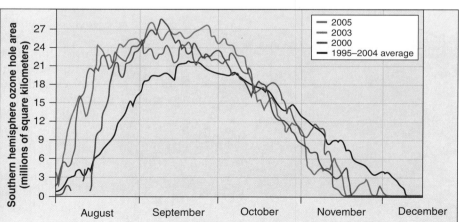

Figure 23.4   The ozone hole over Antarctica.
For decades NASA satellites have tracked the extent of ozone depletion over Antarctica. Every year since 1975 an ozone "hole" has appeared in August when sunlight triggers chemical reactions in cold air trapped over the South Pole during Antarctic winter. The hole intensifies during September before tailing off as temperatures rise in November–December. In 2000, the 28.4-million-square-kilometer hole (*purple* in the satellite image) covered an area larger than the United States, Canada, and Mexico combined, the largest hole ever recorded. In September 2000, the hole extended over Punta Arenas, a city of about 120,000 people in southern Chile, exposing residents to very high levels of UV radiation.

Global Warming

For over 150 years, the growth of our industrial society has been fueled by cheap energy, much of it obtained by burning fossil fuels—coal, oil, and gas. Coal, oil, and gas are the remains of ancient plants, transformed by pressure and time into carbon-rich "fossil fuels." When such fossil fuels are burned, this carbon is combined with oxygen atoms, producing carbon dioxide (CO_2). Industrial society's burning of fossil fuels has released huge amounts of carbon dioxide into the atmosphere. As with CFCs, no one paid any attention to this because the carbon dioxide was thought to be harmless and because the atmosphere was thought to be a limitless reservoir, able to absorb and disperse any amount. It turns out neither assumption was true, and in recent decades, the levels of carbon dioxide in the atmosphere have risen sharply and continue to rise.

What is alarming is that the carbon dioxide doesn't just sit in the air doing nothing. The chemical bonds in carbon dioxide molecules transmit radiant energy from the sun but trap the longer wavelengths of infrared light, or heat, and prevent them from radiating into space. This creates what is known as the **greenhouse effect.** Planets that lack this type of "trapping" atmosphere are much colder than those that possess one. If the earth did not have a "trapping" atmosphere, the average earth temperature would be about –20°C, instead of the actual +15°C.

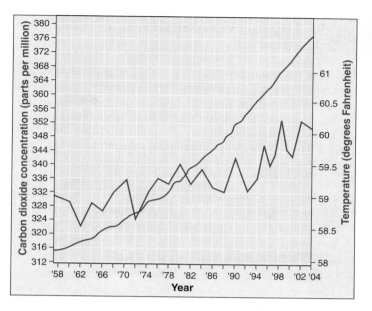

Figure 23.5   The greenhouse effect.
The concentration of carbon dioxide in the atmosphere has shown a steady increase for many years *(blue line)*. The *red line* shows the average global temperature for the same period of time. Note the general increase in temperature since the 1950s and, specifically, the sharp rise beginning in the 1980s. Data from Geophysical Monograph, American Geophysical Union, National Academy of Sciences, and National Center for Atmospheric Research.

Global Warming

The earth's greenhouse effect is intensifying with increased fossil fuel combustion and certain types of waste disposal. These activities are increasing the amounts of carbon dioxide, CFCs, nitrogen oxides, and methane—all "greenhouse gases"—in the atmosphere. The rise in average global temperatures during recent decades (shown as the red line in figure 23.5) is consistent with increased carbon dioxide concentrations in the atmosphere (the blue line). The idea of **global warming** due to accumulation of greenhouse gases in the earth's atmosphere has been controversial, because correlations do not prove a cause-and-effect relationship. However, as more data become available, a growing consensus of scientists accept global warming as an unwelcome reality.

Increases in the amounts of greenhouse gases could increase average global temperatures from 1° to 4°C, which could have serious impact on rain patterns in prime agricultural lands, and in changes in sea levels.

Effects on Rain Patterns.   Global warming is predicted to have a major effect on rainfall patterns. Areas that have already been experiencing droughts may see even less rain, contributing to even greater water shortages. Recent increases in the frequency of El Niño events (see chapter 20) and catastrophic hurricanes may indicate that global warming climatic changes are already beginning to occur.

Effects on Agriculture.   Both positive and negative effects of global warming on agriculture are predicted. Warmer temperatures and increased levels of carbon dioxide in the atmosphere would be expected to increase the yields of some crops, while having a negative impact on others. Droughts that may result from global warming will also negatively affect crops. Plants in the tropics are growing at maximal temperature limits; any further increases in temperature will probably begin to have a negative impact on agricultural yields of tropical farms.

Rising Sea Levels.   Much of the water on earth is locked into ice in glaciers and polar ice caps. As global temperatures increase, these large stores of ice have begun to melt. Most of the water from the melted ice ends up in the oceans, causing water levels to rise (but because the Arctic ice cap floats, its melting will not raise sea levels, any more than melting ice raises the level of water in a glass). Higher water levels can be expected to cause increased flooding of low-lying lands.

There is considerable disagreement among governments about what ought to be done about global warming. The Clean Air Act of 1990 and the Kyoto Treaty have established goals for reducing the emission of greenhouse gases. Countries across the globe are making progress toward reducing emissions, but much more needs to be done.

> **23.4   Humanity's burning of fossil fuels has greatly increased atmospheric levels of CO_2, leading to global warming.**

23.5 Loss of Biodiversity

Just as death is as necessary to a normal life cycle as reproduction, so extinction is as normal and necessary to a stable world ecosystem as species formation. Most species, probably all, go extinct eventually. More than 99% of species known to science (most from the fossil record) are now extinct. However, current rates of extinctions are alarmingly high. The extinction rate for birds and mammals was about one species every decade from 1600 to 1700, but it rose to one species every year during the period from 1850 to 1950, and four species per year between 1986 and 1990. It is this increase in the rate of extinction that is the heart of the **biodiversity** crisis.

Factors Responsible for Extinction

What factors are responsible for extinction? Studying a wide array of recorded extinctions, and many species currently threatened with extinction, biologists have identified three factors that seem to play a key role in many extinctions: habitat loss, species overexploitation, and introduced species (figure 23.6).

Habitat Loss. Habitat loss is the single most important cause of extinction. Given the tremendous amounts of ongoing destruction of all types of habitat, from rain forest to ocean floor, this should come as no surprise. Natural habitats may be adversely affected by human influences in four ways: (1) destruction, (2) pollution, (3) human disruption, and (4) habitat fragmentation (dividing up the habitat into small isolated areas). Habitat destruction is rapidly endangering species on Madagascar. You can see the loss of rain forest habitat, colored in green on the overlain maps in figure 23.7.

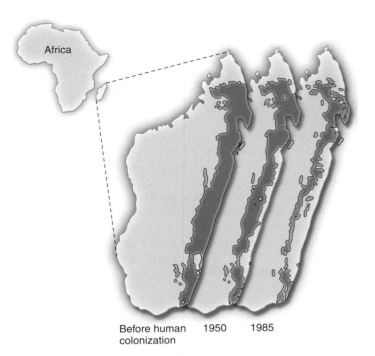

Figure 23.7   Extinction and habitat destruction.

The rain forest covering the eastern coast of Madagascar, an island off the coast of East Africa, has been progressively destroyed as the island's human population has grown. Ninety percent of the original forest cover is now gone. Many species have become extinct, and many others are threatened, including 16 of Madagascar's 31 primate species.

Species Overexploitation. Species that are hunted or harvested by humans have historically been at grave risk of extinction, even when the species populations are initially very abundant. There are many examples in our recent history of overexploitation: passenger pigeons, bison, many species of whales, commercial fish such as Atlantic bluefin tuna, and mahogany trees in the West Indies are but a few.

Introduced Species. Occasionally, a new species will enter a habitat and colonize it, usually at the expense of native species. Colonization occurs in nature, but it is rare; however, humans have made this process more common with devastating ecological consequence. The introduction of exotic species has wiped out or threatened many native populations. African bees (page 419) are an obvious example. Species introductions occur in many ways, usually unintentionally. Plants and animals can be transported in nursery plants, in the ballast of large ocean vessels, as stowaways in boats, cars, and planes, and as beetle larvae within wood products. These species enter new environments where they have no native predators to keep their population sizes in check. Free to populate the habitat, they crowd out native species.

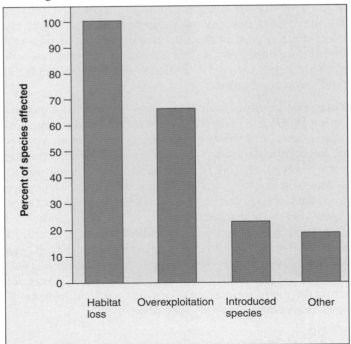

Figure 23.6   Factors responsible for animal extinction.

These data represent known extinctions of mammals in Australia, Asia, and the Americas.

> **23.5   The loss of biodiversity can usually be attributed to one of a few main causes, including habitat loss, overexploitation, and introduced species.**

Today's *Biology*

The Global Decline in Amphibians

Sometimes important things happen, right under our eyes, without anyone noticing. That thought occurred to David Bradford as he stood looking at a quiet lake high in the Sierra Nevada Mountains of California in the summer of 1988. Bradford, a biologist, had hiked all day to get to the lake, and when he got there his worst fears were confirmed. The lake was on a list of mountain lakes that Bradford had been visiting that summer in Sequoia-Kings Canyon National Parks while looking for a little frog with yellow legs. The frog's scientific name was *Rana muscosa,* and it had lived in the lakes of the parks for as long as anyone had kept records. But this silent summer evening, the little frog was gone. The last major census of the frog's populations within the parks had been taken in the mid-1970s, and *R. muscosa* had been everywhere, a common inhabitant of the many freshwater ponds and lakes within the parks. Now, for some reason Bradford did not understand, the frogs had disappeared from 98% of the ponds that had been their homes.

After Bradford reported this puzzling disappearance to other biologists, an alarming pattern soon became evident. Throughout the world, local populations of amphibians (frogs, toads, and salamanders) were becoming extinct. Waves of extinction have swept through high-elevation amphibian populations in the western United States, and have also cut through the frogs of Central America and coastal Australia.

Amphibians have been around for 350 million years, since long before the dinosaurs. Their sudden disappearance from so many of their natural homes sounded an alarm among biologists. What are we doing to our world? If amphibians cannot survive the world we are making, can we?

In 1998 the U.S. National Research Council brought scientists together from many disciplines in a serious attempt to address the problem. After years of intensive investigation, they have begun to sort out the reasons for the global decline in amphibians. Like many important questions in science, this one does not have a simple answer.

Four factors seem to be contributing in a major way to the worldwide amphibian decline: (1) habitat deterioration and destruction, particularly clear-cutting of forests, which drastically lowers the humidity (water in the air) that amphibians require; (2) the introduction of exotic species that outcompete local amphibian populations; (3) chemical pollutants that are toxic to amphibians; and (4) fatal infections by pathogens.

Infection by parasites appears to have played a particularly important role in the western United States and coastal Australia. Amphibian ecology expert James Collins of Arizona State University has reported one clear instance of infection leading to amphibian decline. When Collins examined populations of salamanders living on the Kaibab Plateau along the Grand Canyon rim, he found many sick salamanders. Their skin was covered with white pustules, and most infected ones died, their hearts and spleens collapsed. The infectious agent proved to be a virus common in fish called a ranavirus. Ranavirus isolated by Collins from one sick salamander would cause the disease in a healthy salamander, so there was no doubt that ranavirus was the culprit responsible for the salamander decline on the Kaibab Plateau.

Ranavirus outbreaks eliminate small populations, but in larger ones a few individuals survive infection, sloughing off their pustule-laden skin. These populations slowly recover.

A second kind of infection, very common in Australia but also seen in the United States, is the real species killer. Populations infected with this microbe, a kind of fungus called a chytrid (pronounced "kit-rid," see chapter 17), do not recover. Usually a harmless soil fungus that decomposes plant material, this particular chytrid (with the Latin name of *Batrachochytrium dendrobatidis*) is far from harmless to amphibians. It dissolves and absorbs the chitinous mouthparts of amphibian larvae, killing them.

This killer chytrid was introduced to Australia near Melbourne in the early 1980s. Now almost all Australia is affected. How did the disease spread so rapidly? Apparently it traveled by truck. Infected frogs moved all across Australia in wooden boxes with bunches of bananas. In one year, 5,000 frogs were collected from banana crates in one Melbourne market alone.

In other parts of the world, infection does not seem to play as important a role as acid precipitation, habitat loss, and introduction of exotic species. This complex pattern of cause and effect only serves to emphasize the take-home lesson: worldwide amphibian decline has no one culprit. Instead, all four factors play important roles. It is their total impact that has shifted the worldwide balance toward extinction.

To reverse the trend toward extinction, we must work to lessen the impact of all four factors. It is important that we not get discouraged at the size of the job, however. Any progress we make on any one factor will help shift the balance back toward survival. Extinction is only inevitable if we let it be.

23.6 Reducing Pollution

The pattern of global change that is overtaking our world is very disturbing. Human activities are placing a severe stress on the biosphere, and we must quickly find ways to reduce the harmful impact. There are four key areas in which it will be particularly important to meet the challenge successfully: reducing pollution, finding other sources of energy, preserving nonreplaceable resources, and curbing population growth.

To solve the problem of industrial pollution, it is first necessary to understand the cause of the problem. In essence, it is a failure of our economy to set a proper price on environmental health. To understand how this happens, we must think for a moment about money. The economy of the United States (and much of the rest of the industrial world) is based on a simple feedback system of supply and demand. As a commodity gets scarce, its price goes up, and this added profit acts as an incentive for more of the item to be produced; if too much is produced, the price falls and less of it is made because it is no longer so profitable to produce it.

This system works very well and is responsible for the economic strength of our nation, but it has one great weakness. If demand is set by price, then it is very important that all the costs be included in the price. Imagine that the person selling the item were able to pass off part of the production cost to a third person. The seller would then be able to set a lower price and sell more of the item! Driven by the lower price, the buyer would purchase more than if all the costs had been added into the price.

Unfortunately, that sort of pricing error is what has driven the pollution of the environment by industry. The true costs of energy and of the many products of industry are composed of direct production costs, such as materials and wages, and of indirect costs, such as pollution of the ecosystem. Economists have identified an "optimum" amount of pollution based on how much it costs to reduce pollution versus the social and environmental cost of allowing pollution. The economically optimum amount of pollution is indicated by the blue dot in figure 23.8. If more pollution than the optimum is allowed, the social cost is too high, but if less than the optimum is allowed, the economic cost is too high.

The indirect costs of pollution are usually not taken into account. However, the indirect costs do not disappear because we ignore them. They are simply passed on to future generations, which must pay the bill in terms of damage to the ecosystems on which we all depend. Increasingly, the future is now. Our world, unable to support more damage, is demanding that something be done—that we finally pay up.

Antipollution Laws

Two effective approaches have been devised to curb pollution in this country. The first is to pass laws forbidding it. In the last 20 years, laws have begun to significantly curb

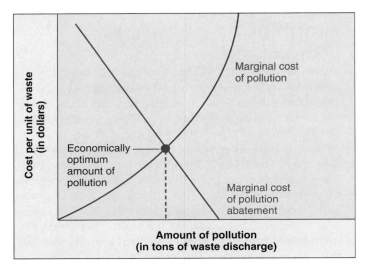

Figure 23.8 Is there an optimum amount of pollution?
Economists identify the "optimum" amount of pollution as the point at which eliminating the next unit of pollution (the marginal cost of pollution abatement) equals the cost in damages caused by that unit of pollution (the marginal cost of pollution).

the spread of pollution by setting stiff standards for what can be released into the environment. For example, all cars are required to have effective catalytic converters to eliminate automobile smog. Similarly, the Clean Air Act of 1990 requires that power plants eliminate sulfur emissions. They can accomplish this by either installing scrubbers on their smokestacks or by burning low-sulfur coal (clean-coal technology), which is more expensive. The effect is that the consumer pays to avoid polluting the environment. The cost of the converters makes cars more expensive, and the cost of the scrubbers increases the price of the energy. The new, higher costs are closer to the true costs, lowering consumption to more appropriate levels.

Pollution Taxes

A second approach to curbing pollution has been to increase the consumer costs directly by placing a tax on the pollution, in effect an artificial price hike imposed by the government as a tax added to the price of production. This added cost lowers consumption too, but by adjusting the tax, the government can attempt to balance the conflicting demands of environmental safety and economic growth. Such taxes, often imposed as "pollution permits," are becoming an increasingly important part of antipollution laws.

> **23.6 Free market economies often foster pollution when prices do not include environmental costs. Laws and taxes are being designed in an attempt to compensate.**

23.7 Finding Other Sources of Energy

The pollution generated by burning coal and oil, the increasing scarcity of oil, and the potential contributions of carbon dioxide to global warming all make it desirable to find alternative energy sources. Many countries are turning to nuclear power for their growing energy needs. In less than 50 years, nuclear power has become a leading source of energy. In 1995, more than 500 nuclear reactors were producing power worldwide. Over 70% of France's electricity is now produced by nuclear power plants.

Nuclear power plants have not been as popular in this country as in the rest of the world, because we have ample access to cheap coal and because the public fears the consequences of an accident. A reactor partial meltdown at the Three Mile Island nuclear plant in Pennsylvania in 1979 released little radiation into the environment but galvanized these fears. There has been little nuclear power development in this country since then (figure 23.9).

In theory, nuclear power can provide plentiful, cheap energy, but the reality is less encouraging. Nuclear power presents several problems—safety, waste disposal, security— that must be overcome if it is to provide a significant portion of the energy that will fuel our future world.

Alternative Energy Sources

A variety of other sources of energy can help reduce our use of fossil fuels, chief among these is solar power. A variety of technologies has been developed that are increasingly ef-

Figure 23.9 Three Mile Island nuclear power plant.
Since a nuclear accident here in 1979, the building of nuclear power stations in the United States has slowed dramatically.

(a)

(b)

Figure 23.10 Alternate energy sources.
(a) Solar energy uses large mirrors to collect energy from the sun. These solar panels absorb heat from the sun that is used to boil water (or other fluids), which creates steam. The steam turns large turbines (not pictured), generating electricity. (b) Wind-powered energy is an old technology modernized for large-scale use. Large wind fields harness the kinetic energy in wind, converting it into electricity.

ficient at capturing the energy in the sun's rays. The large solar panels in figure 23.10a capture this energy to heat water or other fluids to make steam that turns a turbine, generating electricity. Smaller applications use solar panels connected to photovoltaic cells, which convert solar energy directly into electricity.

Other sources of energy include capturing the energy in the wind, accomplished by wind farms like the one shown in figure 23.10b, and tapping the earth's heat in places, such as near hot springs, where it rises to the earth's surface. In the longer run, nuclear fusion reactors and automobiles running on hydrogen gas may provide great amounts of energy at lower cost with little pollution. Currently, these sources of energy contribute relatively little to our energy budget. However, as technologies improve and the cost of fossil fuels rises, these sources will become increasingly important.

> **23.7 Safety, security, and particularly waste disposal remain serious obstacles to the widespread use of nuclear power. Alternate sources of energy are becoming increasingly important.**

Preserving Nonreplaceable Resources

Among the many ways ecosystems are being damaged, one class of problem stands out as more serious than all the rest: consuming or destroying resources that we all share in common (figure 23.11) but cannot replace in the future. Although a polluted stream can be cleaned, no one can restore an extinct species. In the United States, three sorts of nonreplaceable resources are being reduced at alarming rates: topsoil, groundwater, and biodiversity.

Topsoil

Soil is composed of a mixture of rocks and minerals with partially decayed organic matter called humus. Plant growth is strongly affected by soil composition. Minerals like nitrogen and phosphorus are critical to plant growth, and are abundant in humus-rich soils.

The United States is one of the most productive agricultural countries on earth, largely because much of it is covered with particularly fertile soils. Our midwestern farm belt sits astride what was once a great prairie. The soil of that ecosystem accumulated bit by bit from countless generations of animals and plants until, by the time humans came to plow, the humus-rich soil extended down several feet.

We cannot replace this rich **topsoil,** the capital upon which our country's greatness is built, yet we are allowing it to be lost at a rate of centimeters every decade. Our country has lost one-quarter of its topsoil since 1950! By repeatedly tilling (turning the soil over) to eliminate weeds, we permit rain to wash more and more of the topsoil away, into rivers and eventually out to sea. New approaches are desperately needed to lessen the reliance on intensive cultivation. Some possible solutions include using genetic engineering to make crops resistant to weed-killing herbicides and terracing to recapture lost topsoil.

Groundwater

A second resource that we cannot replace is **groundwater,** water trapped beneath the soil within porous rock reservoirs called aquifers. This water seeped into its underground reservoir very slowly during the last ice age over 12,000 years ago. We should not waste this treasure, for we cannot replace it.

In most areas of the United States, local governments exert relatively little control over the use of groundwater. As a result, a very large portion is wasted watering lawns, washing cars, and running fountains. A great deal more is inadvertently being polluted by poor disposal of chemical wastes—and once pollution enters the groundwater, there is no effective means of removing it. Some cities, like Phoenix and Las Vegas, may completely deplete their groundwater within several decades.

"Freedom in a Commons Brings Ruin to All"

The essence of Hardin's original essay:

Picture a pasture open to all. It is expected that each herdsman will try to keep as many cattle as possible on [this] commons....What is the utility...of adding one more animal?...Since the herdsman receives all the proceeds from the sale of the additional animal, the positive utility [to the herdsman] is nearly +1.... Since, however, the effects of overgrazing are shared by all the herdsmen, the negative utility for any particular decision-making herdsman is only a fraction of -1. Adding together the...partial utilities, the rational herdsman concludes that the only sensible course for him to pursue is to add another animal to [the] herd. And another; and another.... Therein is the tragedy. Each man is locked into a system that [causes] him to increase his herd without limit—in a world that is limited....Freedom in a commons brings ruin to all.

—G. Hardin, "The Tragedy of the Commons,"
Science **162,** 1243 (1968), p. 1244

Figure 23.11 The tragedy of the commons.
In a now-famous essay, ecologist Garrett Hardin argues that destruction of the environment is driven by freedom without responsibility.
Reprinted with permission from "The Tragedy of the Commons," by G. Hardin, Science, 162, p. 1244. Copyright 1968 AAAS.

(a)

(b)

(c)

Figure 23.12 Tropical rain forest destruction.
(a) These fires are destroying the rain forest in Brazil, which is being cleared for cattle pasture. (b) The flames are so widespread and so high that their smoke can be viewed from space. (c) The consequences of deforestation can be seen on these middle-elevation slopes in Ecuador. The slopes now support only low-grade pastures where they used to support highly productive forest.

Biodiversity

The number of species in danger of extinction during your lifetime is far greater than the number that became extinct with the dinosaurs. This disastrous loss of biodiversity is important to every one of us, because as these species disappear, so does our chance to learn about them and their possible benefits for ourselves. The fact that our entire supply of food is based on 20 kinds of plants, out of the 250,000 available, should give us pause. Like burning a library without reading the books, we don't know what it is we waste. All we can be sure of is that we cannot retrieve it. Extinct is forever.

Over the last 20 years, about half of the world's tropical rain forests have been either burned to make pasture land or cut for timber (figure 23.12). Over 6 million square kilometers have been destroyed. Every year the rate of loss increases as the human population of the tropics grows. About 160,000 square kilometers were cut each year in the 1990s, a rate greater than 0.6 hectares (1.5 acres) per second! At this rate, all the rain forests of the world will be gone in your lifetime. In the process, it is estimated that one-fifth or more of the world's species of animals and plants will become extinct—more than a million species. This would be an extinction event unparalleled for at least 65 million years, since the Age of Dinosaurs.

You should not be lulled into thinking that loss of biodiversity is a problem limited to the tropics. The ancient forests of the Pacific Northwest are being cut at a ferocious rate today, largely to supply jobs (the lumber is exported), with much of the cost of cutting it down subsidized by our government (the Forest Service builds the necessary access roads, for example). At the current rate, very little will remain in a decade. Nor is the problem restricted to one area. Throughout our country, natural forests are being "clear-cut," replaced by pure stands of lumber trees planted in rows like so many lines of corn. It is difficult to scold those living in the tropics when we ourselves do such a poor job of preserving our own country's biodiversity.

But what is so bad about losing species? What is the value of biodiversity? Loss of a species entails three costs: (1) the direct economic value of the products we might have obtained from species; (2) the indirect economic value of benefits produced by species without our consuming them, such as nutrient recycling in ecosystems; and (3) their ethical and aesthetic value. It is not difficult to see the value in protecting species that we use to obtain food, medicine, clothing, energy, and shelter, but other species are vitally important to maintaining healthy ecosystems; by destroying biodiversity, we are creating conditions of instability and lessened productivity. Other species add beauty to the living world, no less crucial because it is hard to set a price upon.

> **23.8 Nonreplaceable resources are being consumed at an alarming rate all over the world, key among them topsoil, groundwater, and biodiversity.**

23.9 Curbing Population Growth

If we were to solve all the problems mentioned in this chapter, we would merely buy time to address the fundamental problem: there are getting to be too many of us.

Humans first reached North America at least 12,000 to 13,000 years ago, crossing the narrow straits between Siberia and Alaska and moving swiftly to the southern tip of South America. By 10,000 years ago, when the continental ice sheets withdrew and agriculture first developed, about 5 million people lived on earth, distributed over all the continents except Antarctica. With the new and much more dependable sources of food that became available through agriculture, the human population began to grow more rapidly. By the time of Christ, 2,000 years ago, an estimated 130 million people lived on earth. By the year 1650, the world's population had doubled, and doubled again, reaching 500 million. Starting in the early 1700s, changes in technology have given humans more control over their food supply, enabled them to develop superior weapons to ward off predators, and led to the development of cures for many diseases. At the same time, improvements in shelter and storage capabilities have made humans less vulnerable to climatic uncertainties. Recall from chapter 21, that populations grow exponentially until they reach the limits of their environment, called the carrying capacity. These changes since the 1700s allowed humans to expand the carrying capacity of the habitats in which they lived and thus to escape the confines of logistic growth and reenter the exponential phase of the sigmoidal growth curve, shown by the explosive growth in figure 23.13.

Although the human population has grown explosively for the last 300 years, the average human birthrate has stabilized at about 21 births per year per 1,000 people worldwide. However, with the spread of better sanitation and improved medical techniques, the death rate has fallen steadily, to its present level of 9 per 1,000 per year. The difference between birth and death rates amounts to a population growth rate of 1.2% per year, which seems like a small number, but it is not, given the large population size.

The world population reached 6.5 billion people in 2004, and the annual increase now amounts to about 78 million people, which leads to a doubling of the world population in about 58 years. Put another way, more than 214,000 people are added to the world population each day, or almost 150 every minute. At this rate, the world's population will continue to grow and perhaps stabilize at a figure between 7.3 billion and 10.7 billion. Such growth cannot continue, because our world cannot support it. Just as a cancer cannot grow unabated in your body without eventually killing you, so humanity cannot continue to grow unchecked in the biosphere without killing it.

One of the most alarming trends taking place in developing countries is the massive movement to urban centers. For example, Mexico City, one of the largest cities in the world, is plagued by smog, traffic, inadequate waste disposal, and other problems; it has a population of about 26 million people (fig-

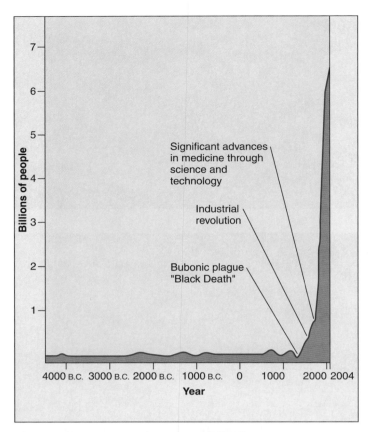

Figure 23.13 Growth curve of the human population.
Over the past 300 years, the world population has been growing steadily. Currently, there are over 6 billion people on the earth.

ure 23.14). The prospects of supplying adequate food, water, and sanitation to this city's people are almost unimaginable. The lot of the rural poor, mainly farmers, in Mexico is even worse.

In view of the limited resources available to the human population, and the need to learn how to manage those resources well, the first and most necessary step toward global prosperity is to stabilize the human population. One of the surest signs of the pressure being placed on the environment is human use of about 40% of the total net global photosynthetic productivity on land. Given that statistic, a doubling of the human population in 58 years poses extraordinarily severe problems. The facts virtually demand restraint in population growth. If and when technology is developed that would allow greater numbers of people to inhabit the earth in a stable condition, the human population can be increased to whatever level might be appropriate.

A key element in the world's population growth is its uneven distribution among countries. Figure 23.15 gives a general breakdown of the distribution of the human population across the globe. At the beginning of the last century, the population was fairly evenly distributed between the developing countries (peach-colored area) and developed countries (red area). Of the billion people added to the world's population in the 1990s, 80% to 90% live in developing countries and of that number, about 60% of the people in the world live in countries that are at least partly tropical or subtropical. An additional 20% live in China. The remaining 20% live in the so-called developed, or industrialized, countries: Europe, Russia, Japan, the United States, Canada, Australia, and New Zealand. Whereas the populations of the developed countries are growing at an annual rate of only about 0.1%, those of the less developed, mostly tropical countries (excluding China) are growing at an annual rate estimated to be about 1.9%.

Most countries are devoting considerable attention to slowing the growth rate of their populations, and there are genuine signs of progress but the world population may still gain another 1 to 4 billion people before it stabilizes. No one knows whether the world can support so many people indefinitely. Finding a way to do so is the greatest task facing humanity. The quality of life that will be available for your children in this new century will depend to a large extent on our success.

Population Growth Rate Has Been Declining

The world population growth rate has been declining, from a high of 2.0% in the period 1965–70 to 1.2% in 2004. Nonetheless, because of the larger population, this amounts to an increase of 78 million people per year to the world population, compared to 53 million per year in the 1960s.

The United Nations attributes the decline to increased family planning efforts and the increased economic power and social status of women. Although the United Nations applauds the United States for leading the world in funding family planning programs abroad, some oppose spending money on international family planning. The opposition states that money is better spent on improving education and the economy in other countries, leading to an increased awareness and lowered fertility rates. The United Nations certainly supports the improvement of education programs in developing countries, but, interestingly, it has reported increased education levels *following* a decrease in family size as a result of family planning.

Slowing population growth will help sustain the world's resources, but per capita consumption is also important. Surprisingly, the majority of resource consumption occurs in developed countries, even though the majority of the world's population is in developing countries. It is necessary that those in the developed world do a better job of lessening the impact each of us makes.

No one knows whether the world can sustain today's population of over 6.5 billion people, much less the far greater numbers expected in the future. We cannot reasonably expect to expand the world's carrying capacity indefinitely. The population will begin scaling back in size, as predicted by logistic growth models; indeed it is already happening. In the sub-Saharan area of Africa, population projections for the year 2025 have been scaled back from 1.33 billion to 1.05 billion because of the impact of AIDS. If we are to avoid catastrophic increases in death rates, such as the tragedy we are seeing in the sub-Saharan, the birthrates must continue to fall dramatically.

Figure 23.14 The world's population is centered in mega-cities.
Mexico City, one of the world's largest cities, has about 26 million inhabitants.

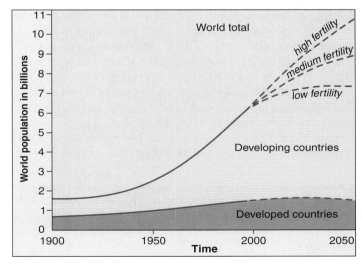

Figure 23.15 Distribution of population growth.
Most of the worldwide increase in population since 1950 has occurred in developing countries. This trend will likely increase in the near future. World population in 2050 is estimated to be around 9 billion, according to recent projections. Depending on fertility rates, the population at that time will either be increasing rapidly or slightly, or in the best case, declining slightly.

Population Pyramids

While the human population as a whole continues to grow rapidly, this growth is not occurring uniformly over the planet. Some countries, like Mexico, are currently growing rapidly. Figure 23.16 shows how Mexico's birthrate, while declining (the blue line), still greatly exceeds its death rate (the red line), which is also declining. There is often a correlation in how developed a country is and how rapidly its population grows. Table 23.1 compares three countries that differ in how well developed they are. Ethiopia, a developing country, has a higher fertility rate, which results in a higher birthrate than either Brazil or the United States but it also has a much higher infant mortality rate and a lower life expectancy. But overall, it will double its population much more quickly than Brazil or the United States. The rate at which a population can be expected to grow in the future can be assessed graphically by means of a population pyramid—a bar graph displaying the numbers of people in each age category (some examples are shown in figure 23.17). Males are conventionally shown to the left of the vertical age axis (colored blue here) and females to the right (colored red). In most human population pyramids, the number of older females is disproportionately large compared with the number of older males, because females in most regions have a longer life expectancy than males. This is apparent in the upper portion of the 2005 U.S. pyramid.

Viewing such a pyramid, one can predict demographic trends in births and deaths. In general, rectangular pyramids are characteristic of countries whose populations are stable; their numbers are neither growing nor shrinking. A triangular pyramid, like the 2005 Kenya pyramid, is characteristic of a country that will exhibit rapid future growth, as most of its population has not yet entered the child-bearing years. Inverted triangles are characteristic of populations that are shrinking.

Compare the differences in the population pyramids for the United States and Kenya in figure 23.17. In the somewhat more rectangular population pyramid for the United States in 2005, the cohort (group of individuals) 40 to 59 years old represents the "baby boom," the large number of babies born following World War II. When the media refers to the "graying of America," they are referring to the aging of this disproportionately large cohort that will impact the health-care system and other age-related systems in the future. The very triangular pyramid of Kenya, by contrast, predicts explosive future growth. The population of Kenya is predicted to double in less than 20 years. However, it is impor-tant to note that these estimates do not take into account the huge impact that natural disasters such as the AIDS epidemic will have on population sizes. In sub-Sahara Africa, the AIDS epidemic has reduced the life expectancy at birth by 20 years. Figure 23.18 shows two population pyramid projections for Botswana, Africa, where over 36% of the population is living with HIV or AIDS. The uncolored portions of the bars indicate what the population would be like in 2025 without the effect of the AIDS epidemic, and the colored bars reflect actual projections.

The Level of Consumption in the Developed World Is Also a Problem

The world population is expected to stabilize sometime in this century at about 10 billion. We in the developed countries of the world need to pay more attention to lessening the impact of our resource consumption. Indeed, the wealthiest 20% of the world's population accounts for 86% of the world's consumption of resources and produces 53% of the world's carbon dioxide emissions, whereas the poorest 20% of the world is responsible for only 1.3% of consumption and 3% of CO_2 emissions.

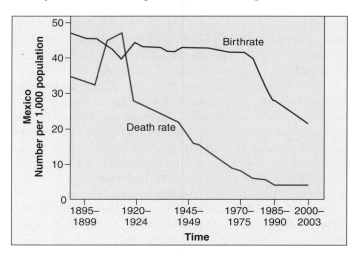

Figure 23.16 Why Mexico's population is growing.
The death rate (*red line*) in Mexico has been falling, while the birthrate (*blue line*) remained fairly steady until 1970. The difference between birth and death rates has fueled a high growth rate. Efforts begun in 1970 to reduce the birthrate have been quite successful. Although the growth rate remains rapid, it is expected to begin leveling off in the near future as the birthrate continues to drop.

TABLE 23.1	A COMPARISON OF 2003 POPULATION DATA IN DEVELOPED AND DEVELOPING COUNTRIES		
	United States (highly developed)	**Brazil (moderately developed)**	**Ethiopia (developing)**
Fertility rate	2.1	2.0	5.5
Doubling time at current rate (yr)	115	57.5	35
Infant mortality rate (infant deaths/1,000 births)	6.8	32	103
Life expectancy (yr)	77	71	41
Per capita income (U.S. dollar equivalent)	$36,300	$7,600	$700

Source: Population Reference Bureau

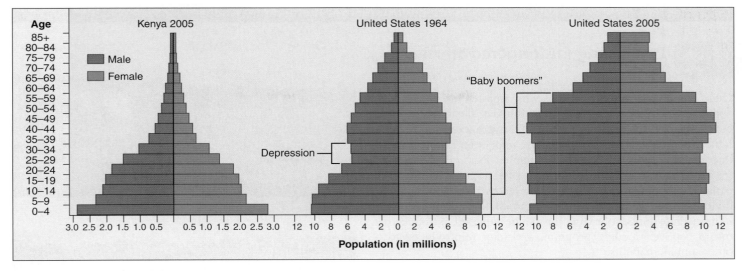

Figure 23.17 Population pyramids.
Population pyramids are graphed according to a population's age distribution. Kenya's pyramid has a broad base because of the great number of individuals below child-bearing age. When all of the young people begin to bear children, the population will experience rapid growth. The 2005 U.S. pyramid demonstrates a larger number of individuals in the "baby boom" cohort—the pyramid bulges because of an increase in births between 1945 and 1964, as shown at the base of the 1964 pyramid. The 25 to 34 cohort in the 1964 pyramid represents people born during the Depression and is smaller in size than the cohorts in the preceding and following years.

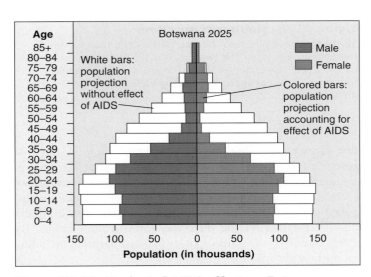

Figure 23.18 Projected AIDS effect on Botswana population (year 2025).

One way of quantifying this disparity is by calculating what has been termed the **ecological footprint,** which is the amount of productive land required to support an individual at the standard of living of a particular population through the course of his or her life. As figure 23.19 illustrates, the ecological footprint of an individual in the United States is more than 10 times greater than that of someone in India. Based on these measurements, researchers have calculated that resource use by humans is now 1/3 greater than the amount that nature can sustainably replace; if all humans lived at the standard of living in the developed world, two additional planet earths would be needed.

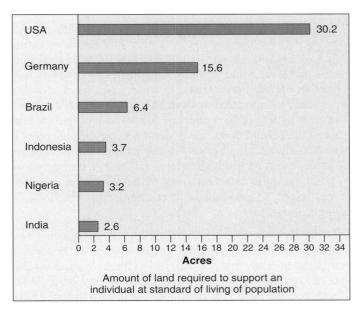

Figure 23.19 Ecological footprint of individuals in different countries.
The ecological footprint calculates how much land is required to support a person through his or her life, including the acreage used for production of food, forest products, and housing, in addition to the forest required to absorb the carbon dioxide produced by the combustion of fossil fuels.

23.9 The problem at the core of all other environmental concerns is the rapid growth of the world's human population. Serious efforts are being made to slow its growth.

23.10 | Preserving Endangered Species

Once you understand the reasons why a particular species is endangered, it becomes possible to think of designing a recovery plan. If the cause is commercial overharvesting, regulations can be designed to lessen the impact and protect the threatened species. If the cause is habitat loss, plans can be instituted to restore lost habitat. Loss of genetic variability in isolated subpopulations can be countered by transplanting individuals from genetically different populations. Populations in immediate danger of extinction can be captured, introduced into a captive breeding program, and later reintroduced to other suitable habitats.

Of course, all of these solutions are extremely expensive. As Bruce Babbitt, Interior Secretary in the Clinton administration, noted, it is much more economical to prevent such "environmental trainwrecks" from occurring than it is to clean them up afterward. Preserving ecosystems and monitoring species before they are threatened is the most effective means of protecting the environment and preventing extinctions.

Habitat Restoration

Conservation biology typically concerns itself with preserving populations and species in danger of decline or extinction. Conservation, however, requires that there be something left to preserve, while in many situations, conservation is no longer an option. Species, and in some cases whole communities, have disappeared or have been irretrievably modified. The clear-cutting of the temperate forests of Washington State leaves little behind to conserve; nor does converting a piece of land into a wheat field or an asphalt parking lot. Redeeming these situations requires restoration rather than conservation.

Three quite different sorts of habitat restoration programs might be undertaken, depending very much on the cause of the habitat loss.

Pristine Restoration. In situations where all species have been effectively removed, one might attempt to restore the plants and animals that are believed to be the natural inhabitants of the area, when such information is available. When abandoned farmland is to be restored to prairie (figure 23.20), how do you know what to plant? Although it is in principle possible to reestablish each of the original species in their original proportions, rebuilding a community requires that you know the identity of all of the original inhabitants, and the ecologies of each of the species. We rarely ever have this much information, so no restoration is truly pristine.

Removing Introduced Species. Sometimes the habitat of a species has been destroyed by a single introduced species. In such a case, habitat restoration involves removal of the introduced species. For example, Lake Victoria, Africa, was home to over 300 species of cichlid fishes, small perchlike fishes

(a)

(b)

Figure 23.20 Habitat restoration.
The University of Wisconsin-Madison Arboretum has pioneered restoration ecology. (a) The restoration of the prairie was at an early stage in November, 1935. (b) The prairie as it looks today. This picture was taken at approximately the same location as the 1935 photograph.

that display incredible diversity. However, in 1954, the Nile perch, a commercial fish with a voracious appetite, was introduced into Lake Victoria. For decades, these perch did not seem to have a significant impact, and then something happened to cause the Nile perch to explode and spread rapidly through the lake, eating their way through the cichlids. By 1986, over 70% of cichlid species had disappeared, including all open-water species.

So what happened to kick-start the mass extinction of the cichlids? The trigger seems to have been eutrophication. High inputs of nutrients from agricultural runoff and sewage from towns and villages led to algal blooms that severely depleted oxygen levels in deeper parts of the lake. This is thought to have led to an increase in cichlids that feed on algae, and a subsequent explosion of Nile perch numbers. The situation has been compounded by a second factor, the introduction into Lake Victoria of a floating water weed from South America, the water hyacinth *Eichornia crassipes*. Extremely prolific under eutrophic conditions, thick mats of water hyacinth soon covered entire bays and inlets, choking off the coastal habitats of non-open-water cichlids.

Restoration of the once-diverse cichlid fishes to Lake Victoria will require more than breeding and restocking the endangered species. Eutrophication will have to be reversed, and the introduced water hyacinth and Nile perch populations brought under control or removed.

Cleanup and Rehabilitation. Habitats seriously degraded by chemical pollution cannot be restored until the pollution is cleaned up. The successful restoration of the Nashua River in New England, discussed later in this chapter, is one example of how a concerted effort can succeed in restoring a heavily polluted habitat to a relatively pristine condition.

Captive Propagation

Recovery programs, particularly those focused on one or a few species, often must involve direct intervention in natural populations to avoid an immediate threat of extinction. Introducing wild-caught individuals into captive breeding programs is being used in an attempt to save the black-footed ferret and California condor populations in immediate danger of disappearing. Several other such captive propagation programs have had success.

Case History: The Peregrine Falcon. U.S. populations of birds of prey such as the Peregrine falcon (*Falco peregrinus*) began an abrupt decline shortly after World War II. Of the approximately 350 breeding pairs east of the Mississippi River in 1942, all had disappeared by 1960. The culprit proved to be the chemical pesticide DDT and related organochlorine pesticides. Birds of prey are particularly vulnerable to DDT because they feed at the top of the food chain, where DDT becomes concentrated. DDT interferes with the deposition of calcium in the bird's eggshells, causing most of the eggs to break before they hatch.

The use of DDT was banned by federal law in 1972, causing levels in the eastern United States to fall quickly. There were no peregrine falcons left in the eastern United States to reestablish a natural population, however. Falcons from other parts of the country were used to establish a captive breeding program at Cornell University in 1970, with the intent of reestablishing the peregrine falcon in the eastern United States by releasing offspring of these birds. By the end of 1986, over 850 birds had been released in 13 eastern states, producing an astonishingly strong recovery.

Sustaining Genetic Diversity

One of the chief obstacles to a successful species recovery program is that a species is generally in serious trouble by the time a recovery program is instituted. When populations become very small, much of their genetic diversity is lost. If a program is to have any chance of success, every effort must be made to sustain as much genetic diversity as possible.

Case History: The Black Rhino. All five species of rhinoceros are critically endangered. The three Asian species live in a forest habitat that is rapidly being destroyed, while the two African species are illegally killed for their horns. Fewer than 11,000 individuals of all five species survive today. The problem is intensified by the fact that many of the remaining animals live in very small, isolated populations. The 2,400 wild-living individuals of the black rhino, *Diceros bicornis* (figure 23.21), live in approximately 75 small, widely separated groups consisting of six subspecies adapted to local conditions throughout the species' range. All of these subspecies appear to have low genetic variability; in three of the subspecies, only a few dozen animals remain. Analysis of mitochondrial DNA suggests that in these populations most individuals are genetically very similar.

This lack of genetic variability represents one of the greatest challenges to the future of the species. Much of the range of the black rhino is still open and not yet subject to human encroachment. To have any significant chance of success, a species recovery program will have to find a way to sustain the genetic diversity that remains in this species. Heterozygosity could be best maintained by bringing all black rhinos together in a single breeding population, but this is not a practical possibility. A more feasible solution would be to move individuals between populations. Managing the black rhino populations for genetic diversity could prevent the loss of genetic variation, which might prove fatal to this species.

Placing black rhinos from a number of different locations together in a sanctuary to increase genetic diversity raises a potential problem: local subspecies may be adapted in different ways to their immediate habitats—what if these

Figure 23.21 Sustaining genetic diversity.
The black rhino is highly endangered, living in 75 small, widely separated populations. Only about 2,400 individuals survive in the wild. Conservation biologists have the difficult job of finding ways to preserve genetic diversity in small, isolated populations.

local adaptations are crucial to their survival? Homogenizing the black rhino populations by pooling their genes risks destroying such local adaptations, if they exist, perhaps at great cost to survival.

Preserving Keystone Species

Keystone species are species that exert a particularly strong influence on the structure and functioning of their ecosystem. Their removal can have disastrous consequences.

Case History: Flying Foxes. The severe decline of many species of pteropodid bats, or "flying foxes," in the Old World tropics is an example of how the loss of a keystone species can have dramatic effects on the other species living within an ecosystem, sometimes even leading to a cascade of further extinctions (figure 23.22). These bats have very close relationships with important plant species on the islands of the Pacific and Indian Oceans. The family Pteropodidae contains nearly 200 species, approximately a quarter of them in the genus *Pteropus,* and is widespread on the islands of the South Pacific, where they are the most important—and often the only—pollinators and seed dispersers. A study in Samoa found that 80% to 100% of the seeds landing on the ground during the dry season were deposited by flying foxes. Many species are entirely dependent on these bats for pollination.

Figure 23.22   Preserving keystone species.
The flying fox is a keystone species in many Old World tropical islands. It pollinates many of the plants, and is a key disperser of seeds. Its elimination by hunting and habitat loss is having a devastating effect on the ecosystems of many South Pacific islands.

In Guam, where the two local species of flying fox have recently been driven extinct or nearly so, the impact on the ecosystem appears to be substantial. Many plant species are not fruiting, or are doing so only marginally, with fewer fruits than normal. Fruits are not being dispersed away from parent plants, so offspring shoots are being crowded out by the adults.

Flying foxes are being driven to extinction by human hunting. They are hunted for food, for sport, and by orchard farmers, who consider them pests. Flying foxes are particularly vulnerable because they live in large, easily seen groups of up to a million individuals. Because they move in regular and predictable patterns and can be easily tracked to their home roost, hunters can easily bag thousands at a time.

Species preservation programs aimed at preserving particular species of flying foxes are only just beginning. One particularly successful example is the program to save the Rodrigues fruit bat, *Pteropus rodricensis,* which occurs only on Rodrigues Island in the Indian Ocean near Madagascar. The population dropped from about 1,000 individuals in 1955 to fewer than 100 by 1974, the drop reflecting largely the loss of the fruit bat's forest habitat to farming. Since 1974 the species has been legally protected, and the forest area of the island is being increased through a tree-planting program. Eleven captive breeding colonies have been established, and the bat population is now increasing rapidly. The combination of legal protection, habitat restoration, and captive breeding has in this instance produced a very effective preservation program.

Conservation of Ecosystems

Habitat fragmentation is one of the most pervasive enemies of biodiversity conservation efforts. Some species simply require large patches of habitat to thrive, and conservation efforts that cannot provide suitable habitat of such a size are doomed to failure. As it has become clear that isolated patches of habitat lose species far more rapidly than large preserves do, conservation biologists have promoted the creation, particularly in the tropics, of so-called megareserves, large areas of land containing a core of one or more undisturbed habitats.

In addition to this focus on maintaining large enough reserves, in recent years, conservation biologists also have recognized that the best way to preserve biodiversity is to focus on preserving intact ecosystems, rather than focusing on particular species. For this reason, attention in many cases is turning to identifying those ecosystems most in need of preservation and devising the means to protect not only the species within the ecosystem, but the functioning of the ecosystem itself.

23.10   Recovery programs at the species level must deal with habitat loss and fragmentation, and often with a marked reduction in genetic diversity. Captive breeding programs that stabilize genetic diversity and pay careful attention to habitat preservation and restoration are typically involved in successful recoveries.

Individuals Can Make the Difference

The development of appropriate solutions to the world's environmental problems must rest partly on the shoulders of politicians, economists, bankers, engineers—many kinds of public and commercial activity will be required. However, it is important not to lose sight of the key role often played by informed individuals in solving environmental problems. Often one person has made the difference; two examples serve to illustrate the point.

The Nashua River

Running through the heart of New England, the Nashua River was severely polluted by mills established in Massachusetts in the early 1900s. By the 1960s, the river was clogged with pollution and declared ecologically dead. When Marion Stoddart moved to a town along the river in 1962, she was appalled. She approached the state about setting aside a "greenway" (trees running the length of the river on both sides), but the state wasn't interested in buying land along a filthy river. So Stoddart organized the Nashua River Cleanup Committee and began a campaign to ban the dumping of chemicals and wastes into the river. The committee presented bottles of dirty river water to politicians, spoke at town meetings, recruited businesspeople to help finance a waste treatment plant, and began to clean garbage from the Nashua's banks. This citizen's campaign, coordinated by Stoddart, greatly aided passage of the Massachusetts Clean Water Act of 1966. Industrial dumping into the river is now banned, and the river has largely recovered (figure 23.23).

Lake Washington

A large, 86-square-kilometer freshwater lake east of Seattle, Lake Washington became surrounded by Seattle suburbs in the building boom following the Second World War. Be-tween 1940 and 1953, a ring of 10 municipal sewage plants discharged their treated effluent into the lake. Safe enough to drink, the effluent was believed "harmless." By the mid-1950s a great deal of effluent had been dumped into the lake (try multiplying 80 million liters/day × 365 days/year × 10 years). In 1954, an ecology professor at the University of Washington in Seattle, W. T. Edmondson, noted that his research students were reporting filamentous blue-green algae growing in the lake. Such algae require plentiful nutrients, which deep freshwater lakes usually lack—the sewage had been fertilizing the lake! Edmondson, alarmed, began a campaign in 1956 to educate public officials to the danger: bacteria decomposing dead algae would soon so deplete the lake's oxygen that the lake would die. After five years, joint municipal taxes financed the building of a sewer to carry the effluent out to sea. The lake is now clean (figure 23.24).

Solving Environmental Problems

It is easy to become discouraged when considering the world's many environmental problems, but do not lose track of the single most important conclusion that emerges from our examination of these problems—the fact that each is solvable. A polluted lake can be cleaned; a dirty smokestack can be altered to remove noxious gas; waste of key resources can be stopped. What is required is a clear understanding of the problem and a commitment to doing something about it. The extent to which U.S. families **recycle** aluminum cans and newspapers is evidence of the degree to which people want to become part of the solution, rather than part of the problem.

> **23.11** In solving environmental problems, the commitment of one person can make a critical difference. Biological literacy is no longer a luxury for scientists—it has become a necessity for all of us.

Figure 23.23 Cleaning up the Nashua River.
The Nashua River, seen on the left in the 1960s, was severely polluted because factories set up along its banks dumped their wastes directly into the river. Seen on the right today, the river is mostly clean.

Figure 23.24 Lake Washington, Seattle.
Lake Washington in Seattle is surrounded by residences, businesses, and industries. By the 1950s, the dumping of sewage and the runoff of fertilizers had caused an algal bloom in the lake, which would eventually deplete the lake's oxygen. Efforts to reverse this effect and clean up the lake were started by W. T. Edmondson of the University of Washington in 1956. The lake is now clean.

How Real Is Global Warming?

The controversy over global warming has two aspects. The first contentious issue is the claim that global temperatures are rising significantly, a profound change in the earth's atmosphere and oceans referred to as "global warming." The second contentious issue is the assertion that global warming is the consequence of elevated concentrations of carbon dioxide in the atmosphere as a consequence of the widespread burning of fossil fuels.

Resolution of the second issue requires detailed science and is only now reaching consensus acceptance. Resolution of the first issue is a simpler proposition, because it is, in essence, a data statement. The graph to the right displays the data in question, global air temperatures for the last century and a half. Temperature data is collected from measuring stations across the globe and averaged, as shown in the image below. The bars of the histogram represent mean yearly global air temperatures for each year since 1850. In order to dampen the effects of random year-to-year variations and so better reveal accumulating influences, the data are presented as an anomaly histogram (in an **anomoly histogram**, each bar presents the deviation of the value during that period from the average value determined for some standard period). In this instance, the anomaly histogram shows the deviation of each year's global mean air temperature from the mean of these values observed over a standard 30-year period between 1961 and 1990.

1. **Applying Concepts**
   a. Variable. In the plot, is there a dependent variable? If so, what is it?
   b. Anomaly Histograms. What fraction of the 155 years do not deviate from the 1961–90 mean value? What fraction deviates more than +0.2°C? more than –0.2°C? more than +0.4°C? more than –4°C?
2. **Interpreting Data**
   a. Of the years that deviate more than +0.2°C, how many are before 1940? between 1940 and 1980? after 1980? What fraction occur after 1980?
   b. Of the years that deviate more than +0.4°C, how many are before 1980? after 2000? What fraction occur after 2000?
   c. Of the years that deviate more than –0.2°C, how many are before 1940? between 1940 and 1980? after 1980? What fraction occur before 1940?

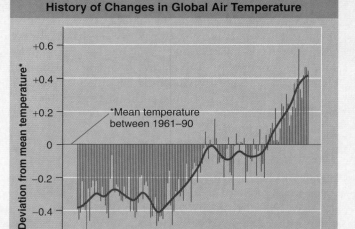

History of Changes in Global Air Temperature

*Mean temperature between 1961–90

   d. Of the years that deviate more than –0.4°C, how many are before 1940? 1900? What fraction occur before 1900?
3. **Making Inferences** If you were to pick a year at random between 1850 and 1900, would it be most likely to deviate +0.2, +0.4, 0, –0.2, or –0.4? a year between 1900 and 1940? a year between 1940 and 1980? a year after 1980? a year after 2000?
4. **Drawing Conclusions** Has the global air temperature been warming progressively over the last century and a half?
5. **Further Analysis** Correlation does not establish causation. How would you go about investigating the validity of the second assertion, that global warming is the direct consequence of elevated levels of carbon dioxide in the earth's atmosphere?

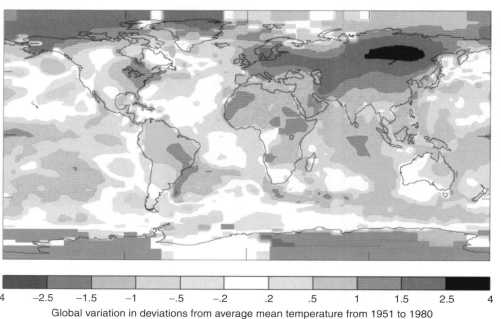

| –4 | –2.5 | –1.5 | –1 | –.5 | –.2 | .2 | .5 | 1 | 1.5 | 2.5 | 4 |

Global variation in deviations from average mean temperature from 1951 to 1980

in which the embryo grows. The blastopore of a protostome becomes the animal's mouth (on the left), and the anus develops at the other end. In a deuterostome, by contrast, the blastopore becomes the animal's anus (on the right), and the mouth develops at the other end.

3. **How the developmental fate of the embryo is fixed.** Most protostomes undergo determinate cleavage, which rigidly fixes the developmental fate of each cell very early. No one cell isolated at

even the four-cell stage can go on to form a normal individual. In marked contrast, deuterostomes undergo indeterminate cleavage, with each cell retaining the capacity to develop into a complete individual.

24.1 Several key innovations in animal body design set the stage for a great diversity in the phylum.

TABLE 24.1	INNOVATIONS IN BODY DESIGN		
Innovation		**Difference in Design**	**Organisms**
❶ Radial versus bilateral symmetry		In radial symmetry, the parts of the body are arranged around a central axis, usually a mouth. In bilateral symmetry, the body is divided into right and left halves that are mirror images of each other. Bilateral symmetry results in specialization in areas of the body such as a head region.	Radial symmetry: cnidarians and adult echinoderms; bilateral symmetry: all other eumetazoan animals
❷ No body cavity versus body cavity Ectoderm / Mesoderm / Endoderm Ectoderm / Mesoderm / Endoderm / Pseudocoel Ectoderm / Mesoderm / Coelom / Endoderm		The evolution of a body cavity resulted in the expansion of organ systems in the body, supported within the body cavity. Animals that lack a body cavity are acoelomates; the body cavity of pseudocoelomates forms between germ layers; the body cavity of coelomates forms within the mesoderm.	Acoelomates: flatworms; pseudocoelomates: roundworms; coelomates: mollusks, arthropods, echinoderms, and chordates
❸ Nonsegmented versus segmented body Segments		In segmented animals, the body is subdivided into compartments, called segments. Segmentation resulted in redundant structures that serve as "backups" if some segments are damaged. Segmented animals also move around more efficiently.	Nonsegmented coelomate animals: mollusks and echinoderms; segmented animals: annelids, arthropods, and chordates
❹ Protostome versus deuterostome development Spiral cleavage Radial cleavage Blastopore becomes mouth Blastopore becomes anus Protostomes Deuterostomes		In protostome development, the cells undergo spiral cleavage, the blastopore becomes the mouth, and the developmental fate of each embryonic cell is determined early. In deuterostome development, cells undergo radial cleavage, the blastopore becomes the anus, and the developmental fate of each embryonic cell is flexible (each cell retains the capacity to develop into a complete individual).	Deuterostomes: echinoderms and chordates; protostomes: all other bilaterally symmetrical animals

Organization of the Vertebrate Body

All vertebrates have the same general architecture: a long internal tube that extends from mouth to anus, which is suspended within an internal body cavity called the *coelom*. The coelom of many terrestrial vertebrates is divided into two parts: the *thoracic cavity,* which contains the heart and lungs, and the *abdominal cavity,* which contains the stomach, intestines, and liver. The vertebrate body is supported by an internal scaffold, or skeleton, made up of jointed bones. A bony skull surrounds and protects the brain, while a column of bones, the vertebrae, surrounds the spinal cord.

Like all animals, the vertebrate body is composed of cells—over 100 trillion of them in your body. It's difficult to picture how many 100 trillion actually is. A line with 100 trillion cars in it would stretch from the earth to the sun and back 50 million times! Not all of these 100 trillion cells are the same, of course. If they were, we would not be bodies but amorphous blobs. Vertebrate bodies contain over 100 different kinds of cells.

Tissues

Groups of cells of the same type are organized within the body into **tissues,** which are the structural and functional units of the vertebrate body. A tissue is a group of cells of the same type that performs a particular function in the body.

Tissues form as the vertebrate body develops. Early in development, the growing mass of cells that will become a mature animal differentiates into three fundamental layers of cells: endoderm, mesoderm, and ectoderm. These three kinds of embryonic cell layers, in turn, differentiate into the more than 100 different kinds of cells in the adult body.

It is possible to assemble many different kinds of tissue from 100 cell types, but biologists have traditionally grouped adult tissues into four general classes: *epithelial, connective, muscle,* and *nerve tissue.* The bird pictured in figure 24.1 contains all four classes of tissues, and you can see by the circled enlargements, each class of tissue contains different types of cells. Of these, connective tissues (indicated by the light green arrows) are particularly diverse.

Organs

Organs are body structures composed of several different tissues grouped together into a larger structural and functional unit, just as a factory is a group of people with different jobs who work together to make something. The heart is an organ. It contains cardiac muscle tissue wrapped in connective tissue and joined to many nerves. All of these tissues work together to pump blood through the body: the cardiac muscles contract, which squeezes the heart to push the blood; the connective tissues act as a bag to hold the heart in the proper shape and ensure that the different chambers of the heart squeeze in the proper order; and the nerves control the rate at which the heart beats. No single tissue can do the job of the heart, any more than one piston can do the job of an automobile engine.

You are probably familiar with many of the major organs of a vertebrate body. Lungs are organs that terrestrial vertebrates use to extract oxygen from the air. Fish use gills to accomplish the same task from water. The stomach is an organ that digests food, and the liver an organ that controls the level of sugar and other chemicals in the blood. Organs are the machines of the vertebrate body, each built from several different tissues and each doing a particular job. How many others can you name?

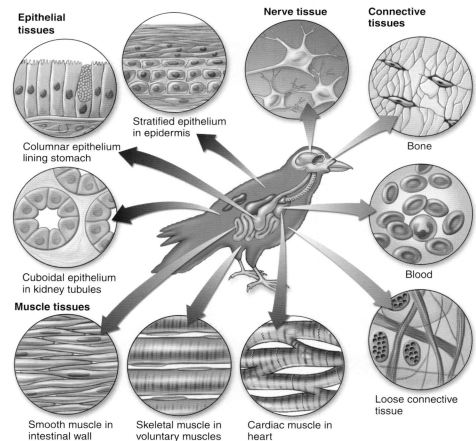

Epithelial tissues

Columnar epithelium lining stomach

Cuboidal epithelium in kidney tubules

Stratified epithelium in epidermis

Nerve tissue

Connective tissues

Bone

Blood

Loose connective tissue

Muscle tissues

Smooth muscle in intestinal wall

Skeletal muscle in voluntary muscles

Cardiac muscle in heart

Figure 24.1
Vertebrate tissue types.

The four basic classes of tissue are epithelial, nerve, connective, and muscle.

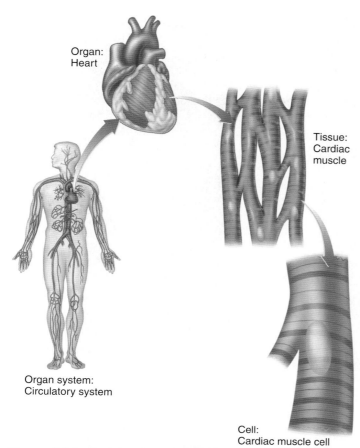

Organ:
Heart

Tissue:
Cardiac
muscle

Organ system:
Circulatory system

Cell:
Cardiac muscle cell

Figure 24.2 Levels of organization within the vertebrate body.
Similar cell types operate together and form tissues. Tissues functioning together form organs. Several organs working together to carry out a function for the body are called an organ system. The circulatory system is an example of an organ system.

Organ Systems

An **organ system** is a group of organs that work together to carry out an important function. For example, the vertebrate digestive system is an organ system composed of individual organs that break up the food (beaks or teeth), pass the food to the stomach (esophagus), break down the food (stomach and intestine), absorb the food (intestine), and expel the solid residue (rectum). If all of these organs do their job right, the body obtains energy and necessary building materials from food. The digestive system is a particularly complex organ system with many different organs consisting of many different types of cells, all working together to carry out a complex function. The circulatory system illustrated in figure 24.2 involves fewer different types of organs, but the level of organization is the same, organ systems are made up of organs, that are made up of tissues, that are made up of cells.

The vertebrate body contains 11 principal organ systems (figure 24.3):

1. **Skeletal.** Perhaps the most distinguishing feature of the vertebrate body is its bony internal skeleton.

The skeletal system protects the body and provides support for locomotion and movement. Its principal components are bones, skull, cartilage, and ligaments. Like arthropods, vertebrates have jointed appendages—the arms, hands, legs, and feet.

2. **Circulatory.** The circulatory system transports oxygen, nutrients, and chemical signals to the cells of the body and removes carbon dioxide, chemical wastes, and water. Its principal components are the heart, blood vessels, and blood.

3. **Endocrine.** The endocrine system releases hormones to coordinates and integrates the activities of the body. Its principal components are the pituitary, adrenal, thyroid, and other ductless glands.

4. **Nervous.** The activities of the body are coordinated by the nervous system. Its principal components are the nerves, sense organs, brain, and spinal cord.

5. **Respiratory.** The respiratory system captures oxygen and exchanges gases and is composed of the lungs, trachea, and other air passageways.

6. **Immune and lymphatic.** The immune system removes foreign bodies from the bloodstream using special cells, such as lymphocytes, macrophages, and antibodies. The lymphatic system provides vessels that transport extracellular fluid and fats to the circulatory system but also provides sites (lymph nodes and thymus, tonsils, and spleen) for the storage of immune cells.

7. **Digestive.** The digestive system captures soluble nutrients from ingested food. Its principal components are the mouth, esophagus, stomach, intestines, liver, and pancreas.

8. **Urinary.** The urinary system removes metabolic wastes from the bloodstream. Its principal components are the kidneys, bladder, and associated ducts.

9. **Muscular.** The muscular system produces movement, both within the body and of its limbs. Its principal components are skeletal muscle, cardiac muscle, and smooth muscle.

10. **Reproductive.** The reproductive system carries out reproduction. Its principal components are the testes in males, ovaries in females, and associated reproductive structures.

11. **Integumentary.** The integumentary system covers and protects the body. Its principal components are the skin, hair, nails, and sweat glands.

24.2 Groups of cells of the same type are organized in the vertebrate body into tissues. Organs are body structures composed of several different tissues. An organ system is a group of organs that work together to carry out an important function.

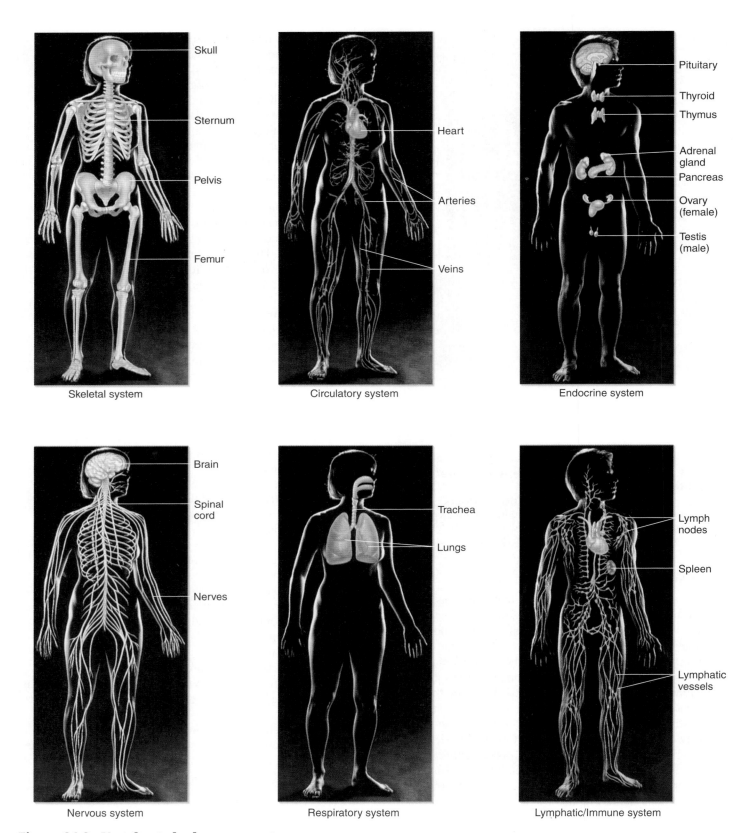

Figure 24.3 Vertebrate body organ systems.

The 11 principal organ systems of the human body are shown, including both male and female reproductive systems.

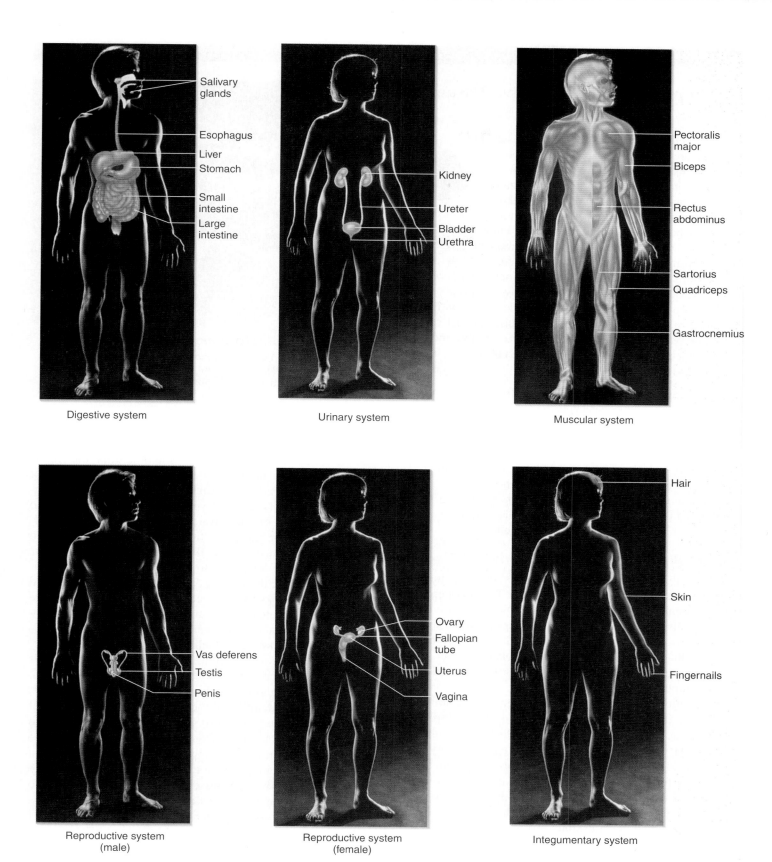

Salivary glands
Esophagus
Liver
Stomach
Small intestine
Large intestine

Digestive system

Kidney
Ureter
Bladder
Urethra

Urinary system

Pectoralis major
Biceps
Rectus abdominus
Sartorius
Quadriceps
Gastrocnemius

Muscular system

Vas deferens
Testis
Penis

Reproductive system (male)

Ovary
Fallopian tube
Uterus
Vagina

Reproductive system (female)

Hair
Skin
Fingernails

Integumentary system

Figure 24.3   (continued)

24.3 Epithelium Is Protective Tissue

We begin our discussion of tissues on the outside. Epithelial cells are the guards and protectors of the body. They cover its surface and determine which substances enter it and which do not. The organization of the vertebrate body is fundamentally tubular, with one tube (the digestive tract) suspended inside another (the body cavity or coelom) like an inner tube inside a tire. The outside of the body is covered with cells (skin) that develop from embryonic *ectoderm* tissue; the body cavity is lined with cells that develop from embryonic *mesoderm* tissue; and the hollow inner core of the digestive tract (the gut) is lined with cells that develop from embryonic *endoderm* tissue. All three germ layers give rise to epithelial cells. Although different in embryonic origin, all epithelial cells are broadly similar in form and function and together are called the **epithelium.**

The body's epithelial layers function in three ways:

1. They *protect the tissues beneath them* from dehydration (water loss) and mechanical damage. Because epithelium encases all the body's surfaces, every substance that enters or leaves the body must cross an epithelial layer, even one as thick as the gila monster's in figure 24.4.
2. They *provide sensory surfaces.* Many of a vertebrate's sense organs are in fact modified epithelial cells.
3. They *secrete materials.* Most secretory glands are derived from pockets of epithelial cells that pinch together during embryonic development.

Types of Epithelial Cells and Epithelial Tissues

Epithelial cells are classified into three types according to their shapes: squamous, cuboidal, and columnar. Layers of epithelial tissue are usually only one or a few cells thick. Individual epithelial cells possess only a small amount of cytoplasm and have a relatively low metabolic rate. A characteristic of all epithelia is that sheets of cells are tightly bound together, with very little space between them. This forms the barrier that is key to the functioning of the epithelium.

Epithelium possesses remarkable regenerative abilities. The cells of epithelial layers are constantly being replaced throughout the life of the organism. The cells lining the digestive tract, for example, are continuously replaced every few days. The epidermis, the epithelium that forms the skin, is renewed every two weeks. The liver, which is a football-sized gland formed of epithelial tissue, can readily regenerate substantial portions of itself removed during surgery.

There are two general kinds of epithelial tissue. First, the membranes that line the lungs and the major cavities of the body are a **simple epithelium** only a single cell layer thick. The first three entries in table 24.2 (**❶**, **❷**, **❸**) are simple epithelium. When you read the functions of each, you can see why these layers are only one cell thick—these are surfaces across which many materials must pass, entering and leaving

Figure 24.4   The epithelium prevents dehydration.
The tough, scaly skin of this gila monster provides a layer of protection against dehydration and injury. For all land-dwelling vertebrates, the relative impermeability of the surface epithelium (the epidermis) to water offers essential protection from dehydration and from airborne pathogens (disease-causing organisms).

the body's compartments, and it is important that the "road" into and out of the body not be too long. Second, the skin, or epidermis, is a **stratified epithelium** composed of more complex epithelial cells several layers thick, as seen in the photo in **❹** in the table. Several layers are necessary to provide adequate cushioning and protection and to enable the skin to continuously replace its cells. The epithelium that lines parts of the respiratory tract **❺** also looks like stratified epithelium, but it is actually a single layer and of **pseudostratified epithelium:** it looks like several layers because the nuclei are positioned in different places in the cells and so gives the appearance of several layers of cells.

A type of simple epithelial tissue that has a secretory function is cuboidal epithelium, which is found in the **glands** of the body. Endocrine glands secrete hormones into the blood. Exocrine glands (those with ducts that open to the body's outside) secrete sweat, milk, saliva, and digestive enzymes out of the body. Exocrine glands also secrete digestive enzymes into the stomach. If you think about it, the stomach and digestive tract are *outside* the body, because they are the inner canal that passes right through the body. It is possible for a substance to pass all the way through this digestive tract, from mouth to anus, and never enter the body at all. A substance must cross an epithelial layer to truly enter the body.

> **24.3** Epithelial tissue is the protective tissue of the vertebrate body. In addition to providing protection and support, vertebrate epithelial tissues provide sensory surfaces and secrete key materials.

TABLE 24.2	EPITHELIAL TISSUE		
Tissue		**Typical Location**	**Tissue Function**

Simple Epithelium

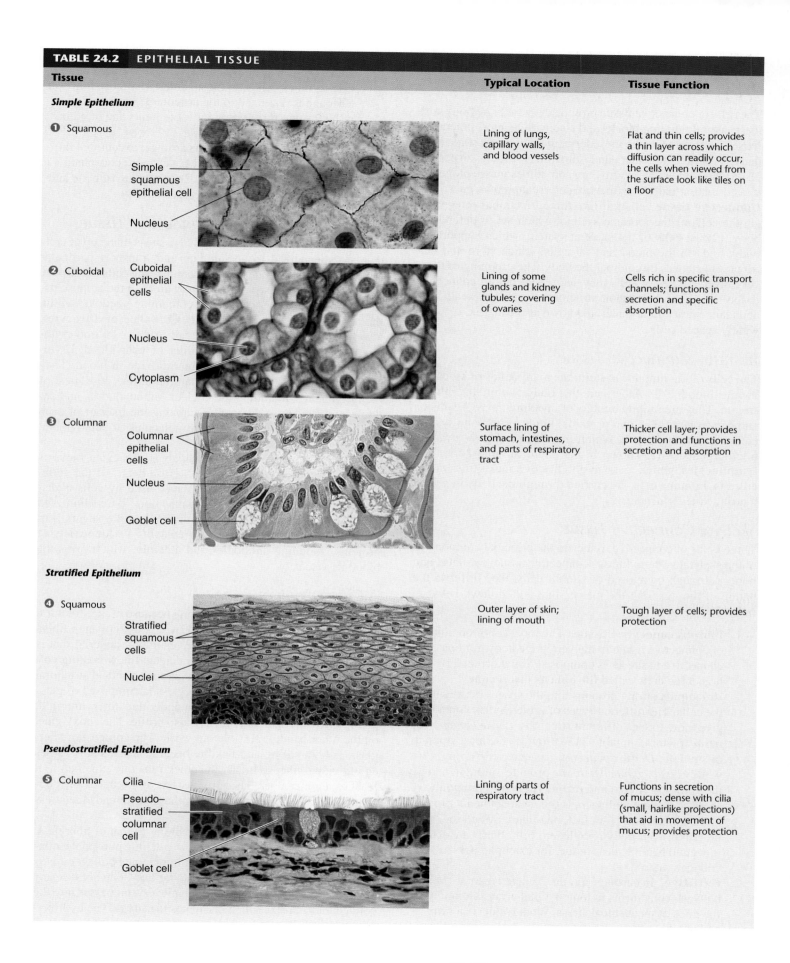

❶ Squamous

Simple squamous epithelial cell
Nucleus

Lining of lungs, capillary walls, and blood vessels

Flat and thin cells; provides a thin layer across which diffusion can readily occur; the cells when viewed from the surface look like tiles on a floor

❷ Cuboidal

Cuboidal epithelial cells
Nucleus
Cytoplasm

Lining of some glands and kidney tubules; covering of ovaries

Cells rich in specific transport channels; functions in secretion and specific absorption

❸ Columnar

Columnar epithelial cells
Nucleus
Goblet cell

Surface lining of stomach, intestines, and parts of respiratory tract

Thicker cell layer; provides protection and functions in secretion and absorption

Stratified Epithelium

❹ Squamous

Stratified squamous cells
Nuclei

Outer layer of skin; lining of mouth

Tough layer of cells; provides protection

Pseudostratified Epithelium

❺ Columnar

Cilia
Pseudo–stratified columnar cell
Goblet cell

Lining of parts of respiratory tract

Functions in secretion of mucus; dense with cilia (small, hairlike projections) that aid in movement of mucus; provides protection

24.4 | Connective Tissue Supports the Body

The cells of connective tissue provide the vertebrate body with its structural building blocks and also with its most potent defenses. Derived from the mesoderm, these cells are sometimes densely packed together, and sometimes widely dispersed, just as the soldiers of an army are sometimes massed together in a formation and sometimes widely scattered as guerrillas. **Connective tissue** cells fall into three functional categories: (1) the cells of the immune system, which act to defend the body; (2) the cells of the skeletal system, which support the body; and (3) the blood and fat cells, which store and distribute substances throughout the body. The grouping of these very diverse types of cells may seem odd, but all connective tissues do share a common structural feature: they all have abundant extracellular material, known as the matrix, between widely spaced cells.

Immune Connective Tissue

The cells of the immune system, the many kinds of so-called "white blood cells ❶," roam the body within the bloodstream. They are mobile hunters of invading microorganisms and cancer cells. The two principal kinds of immune system cells are **macrophages,** which engulf and digest invading microorganisms (as shown in the photo in table 24.3), and **lymphocytes**, which make antibodies or attack virus-infected cells. Immune cells are carried through the body in a fluid matrix, called *plasma.*

Skeletal Connective Tissue

Three kinds of connective tissue are the principal components of the skeletal system: fibrous connective tissue, cartilage, and bone. Although composed of similar cells, they differ in the nature of the material, the matrix, that is laid down between individual cells.

1. **Fibrous connective tissue.** The most common kind of connective tissue in the vertebrate body, fibrous connective tissue ❷ is composed of flat, irregularly branching cells called **fibroblasts** that secrete structurally strong proteins into the spaces between the cells. The different types of proteins they contain give these tissues different strengths. *Loose connective tissue*, pictured in table 24.3, provides the least amount of strength. *Dense connective tissue* is very strong, while *elastic connective tissue* provides strength and the ability to stretch and recoil. The most commonly secreted protein, collagen, is the most abundant protein in the human body—in fact, one-quarter of all the protein in your body is collagen! Fibroblasts are active in wound healing; scar tissue, for example, possesses a collagen matrix.

2. **Cartilage.** In cartilage ❸, the collagen matrix between cells forms in long parallel arrays along the lines of mechanical stress. What results is a firm and flexible tissue of great strength, just as strands of nylon molecules laid down in long, parallel arrays produce strong, flexible ropes. Cartilage makes up the entire skeletal system of the modern agnathans and cartilaginous fishes. In most adult vertebrates, however, cartilage is restricted to the articular (joint) surfaces of bones that form freely movable joints and to other specific locations.

3. **Bone.** Bone ❹ is similar to cartilage, except that the collagen fibers are coated with a calcium phosphate salt, making the tissue rigid. The structure of bone and the way it is formed are discussed shortly.

Storage and Transport Connective Tissue

The third general class of connective tissue is made up of cells that are specialized to accumulate and transport particular molecules. They include the fat-accumulating cells of **adipose tissue ❺**. The large seemingly empty areas in the adipose tissue in table 24.3 are actually fat-containing vacuoles within the adipose cells. Red blood cells ❻, called **erythrocytes,** also function in transport and storage. About 5 billion erythrocytes are present in every milliliter of your blood. Erythrocytes transport oxygen and carbon dioxide in blood. They are unusual in that during their maturation they lose most of their organelles, including the nucleus, mitochondria, and endoplasmic reticulum. Instead, occupying the interior of each erythrocyte are about 300 million molecules of hemoglobin, the protein that carries oxygen.

The fluid, or **plasma,** in which erythrocytes move is both the "banquet table" and the "refuse heap" of the vertebrate body. Practically every substance used by cells is dissolved in plasma, including inorganic salts like sodium and calcium, body wastes, and food molecules like sugars, lipids, and amino acids. Plasma also contains a wide variety of proteins, including antibodies and albumin, which gives the blood its viscosity.

A Closer Look at Bone

The vertebrate endoskeleton is strong because of the structural nature of bone. Bone is produced by coating collagen fibers with a calcium phosphate salt; the result is a material that is strong without being brittle. To understand how coating collagen fibers with calcium salts makes such an ideal structural material, consider fiberglass. Fiberglass is composed of glass fibers embedded in epoxy glue. The individual fibers are rigid, giving great strength, but they are also brittle. The epoxy glue, on the other hand, is flexible but weak. The composite, fiberglass, is both strong and flexible because when stress causes an individual fiber to break, the crack runs into glue before it reaches another fiber. The glue distorts and reduces the concentration of the stress—in effect, the glue spreads the stress over many fibers.

The construction of bone is similar to that of fiberglass: Small, needle-shaped crystals of a calcium phosphate mineral, hydroxyapatite, surround and impregnate collagen fibers within bone. No crack can penetrate far into bone because any stress that breaks a hard hydroxyapatite crystal passes into the collagenous matrix, which dissipates the stress. The hydroxyapatite mineral provides rigidity, whereas the collagen "glue" provides flexibility.

TABLE 24.3 CONNECTIVE TISSUE

Tissue		Typical Location	Tissue Function	Characteristic Cell Types
Immune				
1 White blood cells		Circulatory system	Attack invading microorganisms and virus-infected cells	Macrophages; lymphocytes; mast cells
Skeletal				
2 Fibrous connective tissue	Loose	Beneath skin and other epithelial tissues	Support; provide a fluid reservoir for epithelium	Fibroblasts
	Dense	Tendons; sheath around muscles; kidney; liver; dermis of skin	Provide flexible, strong connections	Fibroblasts
	Elastic	Ligaments; large arteries; lung tissue; skin	Enable tissues to expand and then return to	Fibroblasts
3 Cartilage		Spinal disks; knees and other joints; ear; nose; tracheal rings	Provides flexible support; functions in shock absorption and reduction of friction on load-bearing surfaces	Chondrocytes (specialized fibroblast-like cells)
4 Bone		Most of skeleton	Protects internal organs; provides rigid support for muscle attachment	Osteocytes (specialized fibroblast-like cells)
Storage and Transport				
5 Adipose tissue		Beneath skin	Stores fat	Specialized fibroblasts (adipocytes)
6 Red blood cells		In plasma	Transport oxygen	Red blood cells (erythrocytes)

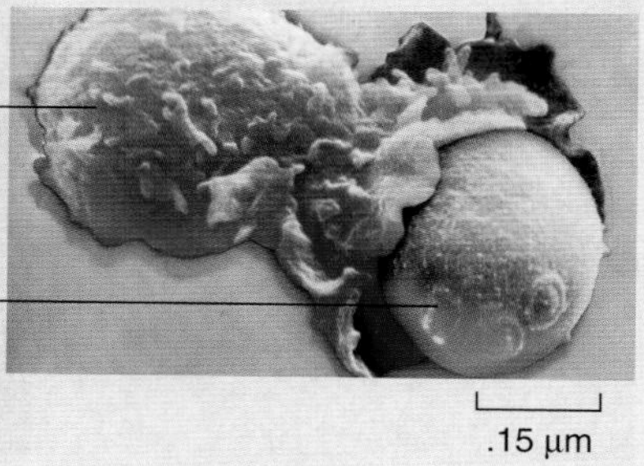

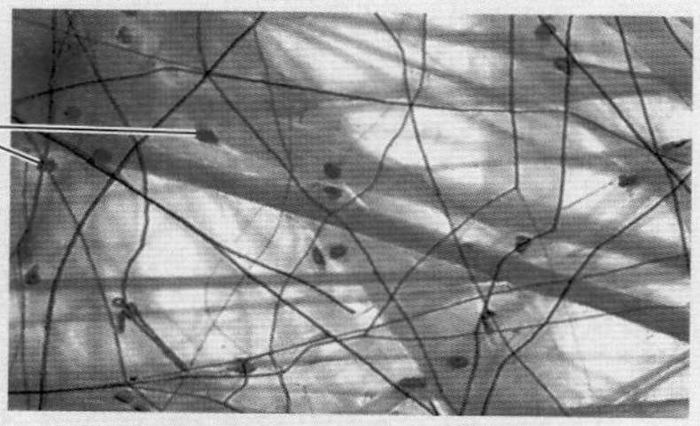

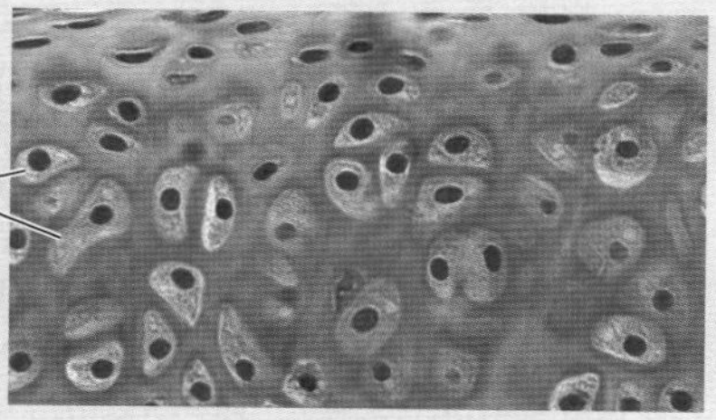

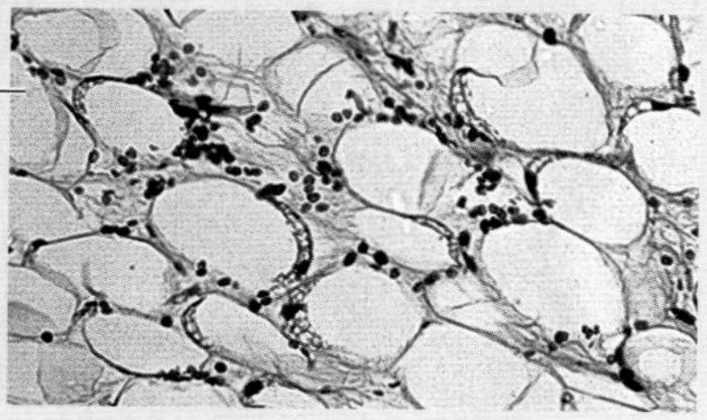

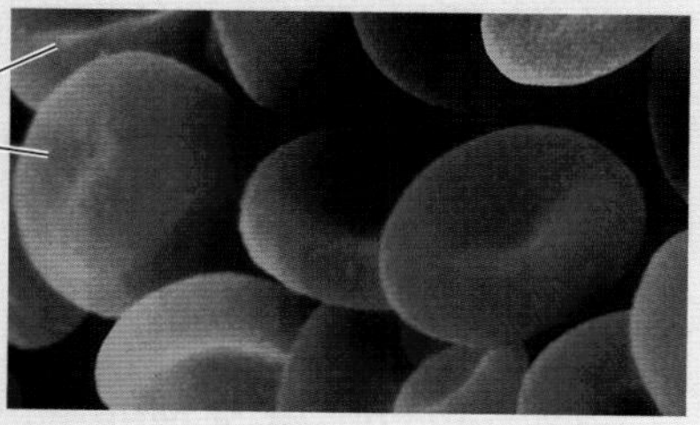

White blood cell

Invading micro-organism

.15 µm

Loose connective tissue (fibroblasts)

Chondrocytes (cartilage cells)

Adipose tissue

Red blood cells

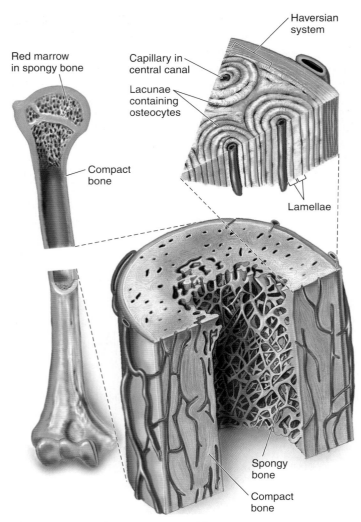

Figure 24.5   The structure of bone.
Some parts of bones are dense and compact, giving the bone strength. Other parts are spongy, with a more open lattice; red blood cells form in the bone marrow.

Most of us think of bones as solid and rocklike. But actually, bone is a dynamic tissue that is constantly being reconstructed. The cross section through a bone in figure 24.5 shows the outer layer of bone is very dense and compact and so is called **compact bone.** The interior is less compact, with a more open lattice structure, and is called **spongy bone.** Red blood cells form in the red marrow of spongy bone. New bone is formed in two stages: first, collagen is secreted by cells called **osteoblasts,** which lay down a matrix of fibers along lines of stress. Then calcium minerals impregnate the fibers. Bone is laid down in thin, concentric layers, like layers of paint on an old pipe. The layers form as a series of tubes around a narrow central channel called a **central canal,** also called a *Haversian canal,* which runs parallel to the length of the bone. You can see the central canals and the concentric rings surrounding the canals in the enlargement in figure 24.5. The many central canals within a bone, all interconnected, contain blood vessels and nerves that provide a lifeline to its living, bone-forming cells.

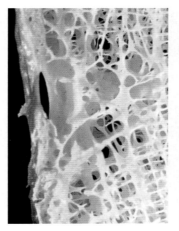
(a) Normal bone tissue

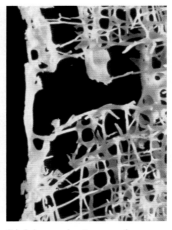
(b) Advanced osteoporosis

Figure 24.6   Osteoporosis.
Common in older women, osteoporosis is a bone disorder in which bones progressively lose minerals.

When bone is first formed in the embryo, osteoblasts use a cartilage skeleton as a template for bone formation. During childhood, bones grow actively. The total bone mass in a healthy young adult, by contrast, does not change much from one year to the next. This does not mean change is not occurring. Large amounts of calcium and thousands of *osteocytes* (former osteoblasts) are constantly being removed and replaced, but total bone mass does not change because deposit and removal take place at about the same rate.

Two cell types are responsible for this dynamic bone "remodeling": *osteoblasts* deposit bone, and *osteoclasts* secrete enzymes that digest the organic matrix of bone, liberating calcium for reabsorption by the bloodstream. The dynamic remodeling of bone adjusts bone strength to workload, new bone being formed along lines of stress. When a bone is subjected to compression, mineral deposition by osteoblasts exceeds withdrawals by osteoclasts. That is why long-distance runners must slowly increase the distances they attempt, to allow their bones to strengthen along lines of stress; otherwise, stress fractures can cripple them.

As a person ages, the backbone and other bones tend to decline in mass. Excessive bone loss is a condition called **osteoporosis.** After the onset of osteoporosis, the replacement of calcium and other minerals lags behind withdrawal, causing the bone tissue to gradually erode. Compare the normal bone in figure 24.6*a* with bone from a person with osteoporosis in figure 24.6*b.* Eventually the bones become brittle and easily broken.

24.4   Connective tissues support the vertebrate body and consist of cells embedded in an extracellular matrix. They include cells of the immune system, cells of the skeletal system, and cells found throughout the body like blood and fat cells. Bone is a type of connective tissue.

24.5 Muscle Tissue Lets the Body Move

Muscle cells are the motors of the vertebrate body. The distinguishing characteristic of muscle cells, the thing that makes them unique, is the abundance of contractible protein fibers within them. These fibers, called **myofilaments,** are made of the proteins actin and myosin. Vertebrate cells have a fine network of these myofilaments, but muscle cells have many more than other cells. Crammed in like the fibers of a rope, they take up practically the entire volume of the muscle cell. When actin and myosin slide past each other, the muscle contracts. Like slamming a spring-loaded door, the shortening of all of these fibers together within a muscle cell can produce considerable force. The process of muscle contraction will be discussed later in this chapter.

The vertebrate body possesses three different kinds of muscle cells: **smooth muscle, skeletal muscle,** and **cardiac muscle** (table 24.4). In smooth muscle, the myofilaments are only loosely organized (seen under ❶ in the table). In skeletal and cardiac muscle, the myofilaments are bunched together into fibers called *myofibrils.* Each myofibril contains many thousands of myofilaments, all aligned to provide maximum force when they simultaneously slide passed each other. Skeletal and cardiac muscle are often called striated muscles because, as you can see under ❷ and ❸ in the table, their cells appear to have transverse stripes when viewed in longitudinal section under the microscope.

Smooth Muscle

Smooth muscle cells are long and spindle-shaped, each containing a single nucleus. However, the individual myofilaments are not aligned into orderly assemblies as they are in

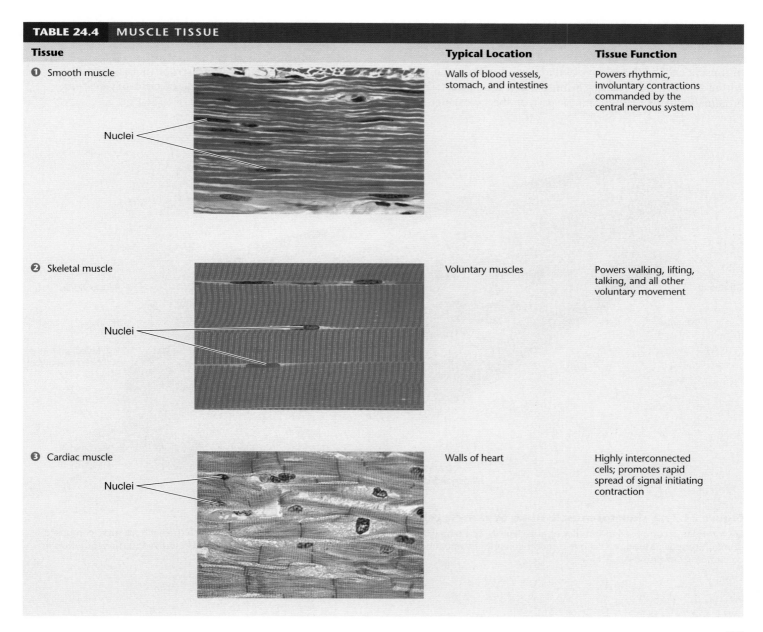

TABLE 24.4 MUSCLE TISSUE		
Tissue	**Typical Location**	**Tissue Function**
❶ Smooth muscle	Walls of blood vessels, stomach, and intestines	Powers rhythmic, involuntary contractions commanded by the central nervous system
Nuclei		
❷ Skeletal muscle	Voluntary muscles	Powers walking, lifting, talking, and all other voluntary movement
Nuclei		
❸ Cardiac muscle	Walls of heart	Highly interconnected cells; promotes rapid spread of signal initiating contraction
Nuclei		

skeletal and cardiac muscles. Smooth muscle tissue is organized into sheets of cells. In some tissues, smooth muscle cells contract only when they are stimulated by a nerve or hormone. Examples are the muscles that line the walls of many blood vessels and those that make up the iris of the vertebrate eye. In other smooth muscle tissue, such as that found in the wall of the gut, the individual cells contract spontaneously, leading to a slow, steady contraction of the tissue.

Skeletal Muscle

Skeletal muscles move the bones of the skeleton. Skeletal muscle cells are produced during development by the fusion of several cells at their ends to form a very long fiber. Each of these muscle cell fibers still contains all the original nuclei, pushed out to the periphery of the cytoplasm. Looking at figure 24.7, you can see how the nuclei are positioned to the outside of the muscle fiber. Each **muscle fiber** consists of many elongated **myofibrils,** and each myofibril is, in turn, composed of many myofilaments, the protein filaments actin and myosin. Myofibrils and myofilaments have been pulled out from the muscle cells in figure 24.7 so you can see the levels of organization in the muscle fiber. The key property of muscle cells is the relative abundance of actin and myosin within them, which enables a muscle cell to contract. These protein filaments are present as part of the cytoskeleton of all eukaryotic cells, but they are far more abundant and more highly organized in muscle cells.

Cardiac Muscle

The vertebrate heart is composed of striated muscle fibers arranged very differently from the fibers of skeletal muscle. Instead of very long, multinucleate cells running the length of the muscle, heart muscle is composed of chains of single cells, each with its own nucleus (refer back to the photo under ❸ in table 24.4). Chains of cells are organized into fibers that branch and interconnect, forming a latticework. This lattice structure is critical to the way heart muscle functions. Each heart muscle cell is coupled to its neighbors electrically by tiny holes called *gap junctions* that pierce the plasma membranes in regions where the cells touch each other. Heart contraction is initiated at one location by the opening of transmembrane channels that conduct ions across the membrane. This changes the electrical properties of the membrane. An electrical impulse then passes from cell to cell across the gap junctions, causing the heart to contract in an orderly pulsation.

> **24.5** Muscle tissue is the tool the vertebrate body uses to move its limbs, contract its organs, and pump the blood through its circulatory system.

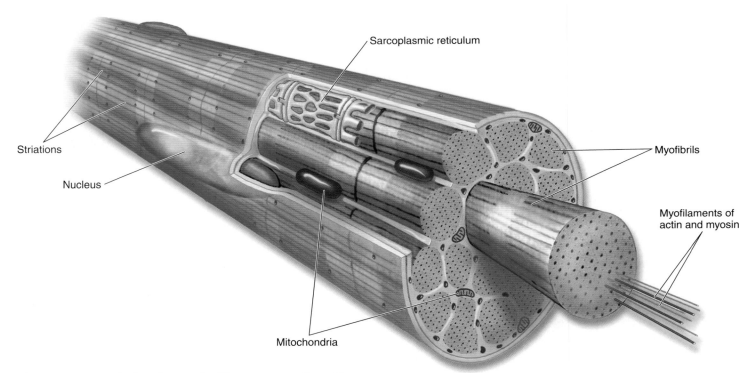

Figure 24.7 A skeletal muscle fiber, or muscle cell.
Each muscle is composed of bundles of muscle cells, or fibers. Each fiber is composed of many myofibrils, which are each, in turn, composed of myofilaments. Muscle cells have a modified endoplasmic reticulum called the sarcoplasmic reticulum that is involved in the regulation of calcium ions in muscles.

24.6 Nerve Tissue Conducts Signals Rapidly

Nerve cells carry information rapidly from one vertebrate organ to another. Nerve tissue, the fourth major class of vertebrate tissue, is composed of two kinds of cells: (1) **neurons,** which are specialized for the transmission of nerve impulses, and (2) supporting **glial cells,** which supply the neurons with nutrients, support, and insulation.

Neurons have a highly specialized cell architecture that enables them to conduct signals rapidly throughout the body. Their plasma membranes are rich in ion-selective channels that maintain a voltage difference between the interior and the exterior of the cell, the equivalent of a battery. When ion channels in a local area of the membrane open, ions flood in from the exterior, temporarily wiping out the charge difference. This process, called depolarization, tends to open nearby voltage-sensitive channels in the neuron membrane, resulting in a wave of electrical activity that travels down the entire length of the neuron as a nerve impulse.

Each neuron is composed of three parts, as illustrated in figure 24.8: (1) a **cell body,** which contains the nucleus; (2) threadlike extensions called **dendrites** extending from the cell body, which act as antennae, bringing nerve impulses to the cell body from other cells or sensory systems; and (3) a single, long extension called an **axon,** which carries nerve impulses away from the cell body. Axons often carry nerve impulses for considerable distances: The axons that extend from the skull to the pelvis in a giraffe are about 3 meters long!

The body contains neurons of various sizes and shapes. Some are tiny and have only a few projections, others are bushy and have more projections, and still others have extensions that are meters long. However, all fit into one of three general categories of neurons as shown in table 24.5. *Sensory neurons* ❶ generally carry electrical impulses from the body to the central nervous system (CNS), the brain and spinal cord. *Motor neurons* ❷ generally carry electrical impulses from the central nervous system to the muscles. *Association neurons* ❸ occur within the central nervous system and act as a "connector" between sensory and motor neurons. These will be discussed in more detail in chapter 30. Neurons are not normally in direct contact with one another. Instead, a tiny gap called a **synapse** separates them. Neurons communicate with other neurons by passing chemical signals called **neurotransmitters** across the gap.

Vertebrate nerves appear as fine white threads when viewed with the naked eye, but they are actually composed of bundles of axons. Like a telephone trunk cable, nerves include large numbers of independent communication channels— bundles composed of hundreds of axons, each connecting a nerve cell to a muscle fiber or other type of cell. It is important not to confuse a nerve with a neuron. A nerve is made up of the axons of many neurons, just as a cable is made of many wires.

> **24.6** Nerve tissue provides the vertebrate body with a means of communication and coordination.

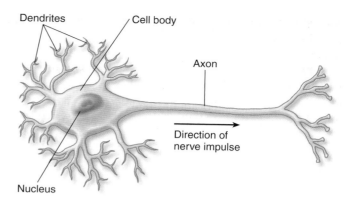

Figure 24.8 Neurons carry nerve impulses.
Neurons carry nerve impulses, which are electrical signals, from their initiation in dendrites, to the cell body, and down the length of the axon, where they may pass the signal to a neighboring cell.

TABLE 24.5	NERVE TISSUE	
Tissue	**Typical Location**	**Tissue Function**
❶ Sensory neurons	Eyes; ears; surface of skin	Receive information about body's condition and external environment; send impulses from sensory receptors to CNS
❷ Motor neurons	Brain and spinal cord	Stimulate muscles and glands; conduct impulses out of CNS toward muscles and glands
❸ Association neurons	Brain and spinal cord	Integrate information; conduct impulses between neurons within CNS

24.7   Types of Skeletons

With muscles alone, the animal body could not move—it would simply pulsate as its muscles contracted and relaxed in futile cycles. For a muscle to produce movement, it must direct its force against another object. Animals are able to move because the opposite ends of their muscles are attached to a rigid scaffold, or **skeleton,** so that the muscles have something to pull against. There are three types of skeletal systems in the animal kingdom: hydraulic skeletons, exoskeletons, and endoskeletons.

Hydraulic skeletons are found in soft-bodied invertebrates such as earthworms and jellyfish. In this case, a fluid-filled cavity is encircled by muscle fibers that raise the pressure of the fluid when they contract. The earthworm in figure 24.9 moves forward by a wave of contractions of circular muscles that begins anteriorly and compresses the body, so that the fluid pressure pushes it forward. Contractions of longitudinal muscles then pull the rest of the body.

Figure 24.10   Crustaceans have an exoskeleton.

The exoskeleton of this rock crab is bright orange.

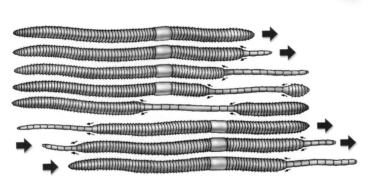

Figure 24.9   Earthworms have a hydraulic skeleton.

When an earthworm's circular muscles contract, the internal fluid presses on the longitudinal muscles, which then stretch to elongate segments of the earthworm. A wave of contractions down the body of the earthworm produces forward movement.

Exoskeletons surround the body as a rigid hard case to which muscles attach internally. When a muscle contracts, it moves the section of exoskeleton to which it is attached. Arthropods, such as crustaceans (figure 24.10) and insects, have exoskeletons made of the polysaccharide *chitin*. An animal with an exoskeleton cannot get too large because its exoskeleton would have to become thicker and heavier to prevent collapse. If an insect were the size of an elephant, its exoskeleton would have to be so thick and heavy it would hardly be able to move.

Endoskeletons, found in vertebrates and echinoderms, are rigid internal skeletons to which muscles are attached. Vertebrates have a soft, flexible exterior that stretches to accommodate the movements of their skeleton. The endoskeleton of vertebrates is composed of bone (figure 24.11). Unlike chitin, bone is a cellular, living tissue capable of growth, self-repair, and remodeling in response to physical stresses.

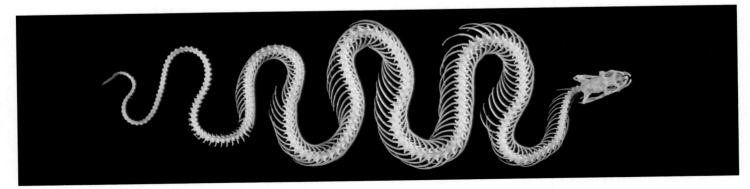

Figure 24.11   Snakes have an endoskeleton.

The endoskeleton of most vertebrates is made of bone. A snake's skeleton is specialized for quick lateral movement.

A Vertebrate Endoskeleton: The Human Skeleton

The human skeleton is made up of 206 individual bones. If you saw them as a pile of bones jumbled together, it would be hard to make any sense of them. To understand the skeleton, it is necessary to group the 206 bones according to their function and position in the body. The 80 bones of the **axial skeleton,** the purple-colored bones in figure 24.12, support the main body axis, while the remaining 126 bones of the **appendicular skeleton,** the tan-colored bones, support the arms and legs. These two skeletons function more or less independently—that is, the muscles controlling the axial skeleton (postural muscles) are managed by the brain separately from those controlling the appendages (manipulatory muscles).

The Axial Skeleton

The axial skeleton is made up of the skull, backbone, and rib cage. Of the skull's 28 bones, only 8 form the cranium, which encases the brain; the rest are facial bones and middle ear bones.

The skull is attached to the upper end of the backbone, which is also called the *spine,* or **vertebral column.** The spine is made up of 26 vertebrae, stacked one on top of the other to provide a flexible column surrounding and protecting the spinal cord. Curving forward from the vertebrae are 12 pairs of ribs, attached at the front to the breastbone, or sternum, and forming a protective cage around the heart and lungs.

The Appendicular Skeleton

The 126 bones of the appendicular skeleton are attached to the axial skeleton at the shoulders and hips. The shoulder, or **pectoral girdle,** is composed of two large, flat shoulder blades, each connected to the top of the sternum by a slender, curved collarbone (clavicle). The arms are attached to the pectoral girdle; each arm and hand contains 30 bones. The clavicle is the most frequently broken bone of the body. Can you guess why? Because if you fall on an outstretched arm, a large component of the force is transmitted to the clavicle.

The **pelvic girdle** forms a bowl that provides strong connections for the legs, which must bear the weight of the body. Each leg and foot contains a total of 30 bones.

Joints, points where two bones come together, colored green in the figure, confer flexibility to the rigid endoskeleton, allowing a range of motion determined by the type of joint. Slightly movable joints bridged by cartilage allow the bones of the axial skeleton's spine some movement, whereas the freely movable joints of the appendicular skeleton's limbs are encased within fibrous capsules filled with a lubricating fluid.

> **24.7 The animal skeletal system provides a framework against which the body's muscles can pull. Many soft-bodied invertebrates employ a hydraulic skeleton, whereas arthropods have a rigid, hard exoskeleton surrounding their body. Echinoderms and vertebrates have an internal endoskeleton to which muscles attach.**

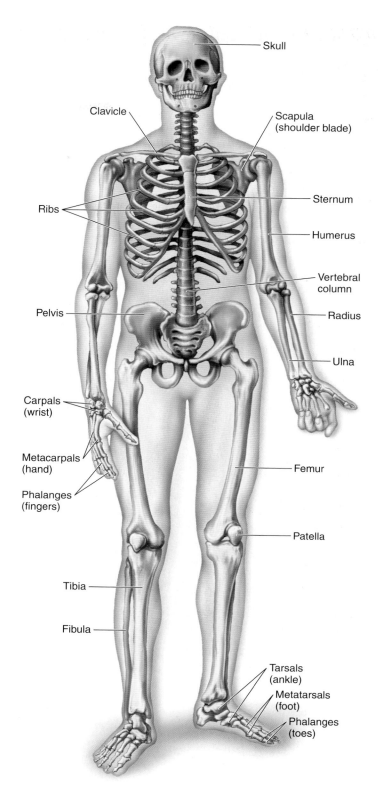

Figure 24.12 Axial and appendicular skeletons.
The axial skeleton is shown in *purple,* and the appendicular skeleton is shown in *tan.*

Muscles and How They Work

Three kinds of muscle together form the vertebrate muscular system. As we have discussed, the vertebrate body is able to move because *skeletal muscles* pull the bones with considerable force. The heart pumps because of the contraction of *cardiac muscle.* Food moves through the intestines because of the rhythmic contractions of *smooth muscle.*

Actions of Skeletal Muscle

Skeletal muscles move the bones of the skeleton. Some of the major human muscles are labeled on the right in figure 24.13. Muscles are attached to bones by straps of dense connective

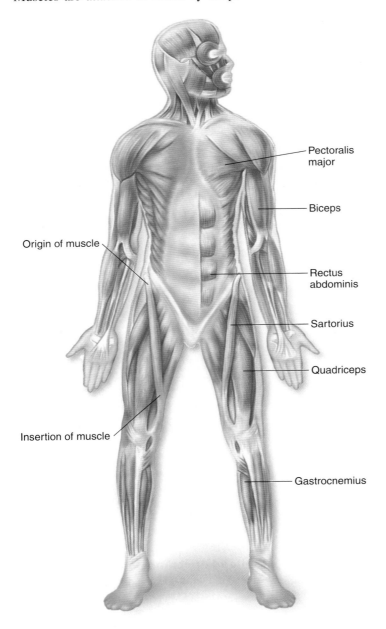

Pectoralis major

Biceps

Origin of muscle

Rectus abdominis

Sartorius

Insertion of muscle

Quadriceps

Gastrocnemius

Figure 24.13 The muscular system.
Some of the major muscles in the human body are labeled.

tissue called **tendons.** Bones pivot about flexible connections called *joints,* pulled back and forth by the muscles attached to them. Each muscle pulls on a specific bone. One end of the muscle, the *origin,* is attached by a tendon to a bone that remains stationary during a contraction. This provides an object against which the muscle can pull. The other end of the muscle, the *insertion,* is attached to a bone that moves if the muscle contracts. For example, origin and insertion for the sartorius muscle is labeled on the left in figure 24.13. This muscle helps bend the leg at the hip, bringing the knee to the chest. The origin of the muscle is at the hip and stays stationary. The insertion is at the knee, such that when the muscle contracts (gets shorter) the knee is pulled up toward the chest.

Muscles can only pull, not push, because myofibrils contract rather than expand. For this reason, the muscles in the movable joints of vertebrate are attached in opposing pairs, called flexors and extensors, which when contracted, move the bones in different directions. As you can see in figure 24.14, when the **flexor** muscle at the back of your upper leg contracts, the lower leg is moved closer to the thigh. When the **extensor** muscle at the front of your upper leg contracts, the lower leg is moved in the opposite direction, away from the thigh.

All muscles contract, but there are two types of muscle contractions, isotonic and isometric contractions. In **isotonic contractions,** the muscle shortens, moving the bones as just described. In **isometric contractions,** a force is exerted by the muscle, but the muscle does not shorten. This occurs when you try to lift something very heavy. Eventually, if your muscles generate enough force and you are able to lift the object, the isometric contraction becomes isotonic.

Muscle Contraction

Recall from figure 24.7 that myofibrils are composed of bundles of myofilaments. Far too fine to see with the naked eye, the individual myofilaments of vertebrate muscles are only eight to 12 nanometers thick. Each is a long, threadlike filaments of the proteins actin or myosin. An **actin filament** consists of two strings of actin molecules wrapped around one another, like two strands

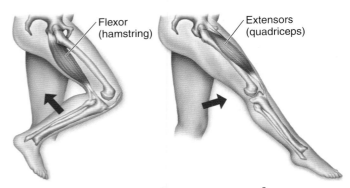

Flexor (hamstring)

Extensors (quadriceps)

Figure 24.14 Flexor and extensor muscles.
Limb movement is always the result of muscle contraction, never muscle extension. Muscles that retract limbs are called flexors; those that extend limbs are called extensors.

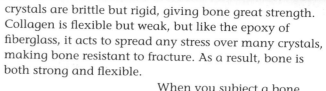

Running Improperly Provides a Painful Lesson in the Biology of Bones and Muscles

No one seeing the ring of fat decorating my middle would take me for a runner. Only in my memory do I get up with the robins, lace on my running shoes, bounce out the front door, and run the streets around Washington University before going to work. Now my 5-K runs are 25-year-old memories. Any mention I make of my running in a race only evokes screams of laughter from my daughters, and an arch look from my wife. Memory is cruelest when it is accurate.

I remember clearly the day I stopped running. It was a cool fall morning in 1978, and I was part of a mob running a 5-K (that's 5 kilometers for the uninitiated) race, winding around the hills near the university. I started to get flashes of pain in my legs below the knees— like shin splints, but much worse. Imagine fire pouring on your bones. Did I stop running? No. Like a bonehead I kept going, "working through the pain," and finished the race. I have never run a race since.

I had pulled a muscle in my thigh, which caused part of the pain. But that wasn't all. The pain in my lower legs wasn't shin splints, and didn't go away. A trip to the doctor revealed multiple stress fractures in both legs. The X rays of my legs looked like tiny threads had been wrapped around the shaft of each bone, like the red stripe on a barber's pole. It was summer before I could walk without pain.

What went wrong? Isn't running supposed to be GOOD for you? Not if you run improperly. In my enthusiasm to be healthy, I ignored some simple rules and paid the price. The biology lesson I ignored had to do with how bones grow. The long bones of your legs are not made of stone, solid and permanent. They are dynamic structures, constantly being re-formed and strengthened in response to the stresses to which you subject them.

To understand how bone grows, we first need to know a bit about what bone is like. Bone is made of fibers of a flexible protein called collagen stuck together to form cartilage. While an embryo, all your bones are made of cartilage. As your adult body develops, the collagen fibers become impregnated with tiny needle-shaped crystals of calcium phosphate, turning the cartilage into bone. The crystals are brittle but rigid, giving bone great strength. Collagen is flexible but weak, but like the epoxy of fiberglass, it acts to spread any stress over many crystals, making bone resistant to fracture. As a result, bone is both strong and flexible.

When you subject a bone in your body to stress—say, by running—the bone grows so as to withstand the greater workload. How does the bone "know" just where to add more material? When stress deforms the collagen fibers of a leg bone, the interior of the collagen fibers becomes exposed, like opening your jacket and exposing your shirt. The fiber interior has a minute electrical charge. Cells called fibroblasts are attracted to the electricity like bugs to night lights, and secrete more collagen there. As a result, new collagen fibers are laid down on a bone along the lines of stress. Slowly, over months, calcium phosphate crystals convert the new collagen to new bone. In your legs, the new bone forms along the long stress lines that curve down along the shank of the bone.

Now go back 25 years, and visualize me pounding happily down the concrete pavement each morning. I had only recently begun to run on the sidewalk, and for an hour or more at a stretch. Every stride I took those mornings was a blow to my shinbones, a stress to which my bones no doubt began to respond by forming collagen along the spiral lines of stress. Had I run on a softer surface, the daily stress would have been far less severe. Had I gradually increased my running, new bone would have had time to form properly in response to the added stress. I gave my leg bones a lot of stress, and no time to respond to it. I pushed them too hard, too fast, and they gave way.

Nor was my improper running limited to overstressed leg bones. Remember that pulled thigh muscle? In my excessive enthusiasm, I never warmed up before I ran. I was having too much fun to worry about such details. Wiser now, I am sure the pulled thigh muscle was a direct result of failing to properly stretch before running.

I was reminded of that pulled muscle recently, listening to a good friend of my wife's describe how she sets out early each morning for a long run without stretching or warming up. I can see her in my mind's eye, bundled up warmly on the cooler mornings, an enthusiastic gazelle pounding down the pavement in search of health. Unless she uses more sense than I did, she may fail to find it.

of pearls loosely wound together. A **myosin filament** is also composed of two strings of protein wound about each other, but a myosin filament is about twice as long as an actin filament, and the myosin strings have a very unusual shape. One end of a myosin filament consists of a very long rod, while the other end consists of a double-headed globular region, or "head." Overall, a myosin filament looks a bit like a two-headed snake. This odd structure is the key to how muscles work.

How Myofilaments Contract

The **sliding filament model** of muscle contraction, illustrated in figure 24.15, describes how actin and myosin cause muscles to contract. Focus on the knob-shaped myosin head in panel 1. When a muscle contraction begins, the heads of the myosin filaments move first. Like flexing your hand downward at the wrist, the heads bend backward and inward as in panel 2. This moves them closer to their rodlike backbones and several nanometers in the direction of the flex. In itself, this myosin head-flex accomplishes nothing—but the myosin head is attached to the actin filament! As a result, the actin filament is pulled along with the myosin head as it flexes, causing the actin filament to slide by the myosin filament in the direction of the flex (the dotted circles in panel 2 indicate the movement of the actin filament). As one after another myosin head flexes, the myosin in effect "walks" step by step along the actin. Each step uses a molecule of ATP to recock the myosin head (in panel 3) before it attaches to the actin again in panel 4, ready for the next flex.

How does this sliding of actin past myosin lead to myofibril contraction and muscle cell movement? The actin filament is anchored at one end, at a position in striated muscle called the Z line, indicated by the lavender-colored bars toward the edges in figure 24.16. Two Z lines with the actin and myosin filaments in between make up a contractile unit called a **sarcomere.** Because it is tethered like this, the actin cannot simply move off. Instead, the actin pulls the anchor with it! As actin moves past myosin, it drags the Z line toward the myosin. The secret of muscle contraction is that each myosin is interposed between two pairs of actin filaments, which are anchored at both ends to Z lines, as shown in panel 1 of figure 24.16. One moving to the left and the other to the right, the two pairs of actin molecules drag the Z lines toward each other as they slide past the myosin core, shown progressively in panel 2 and panel 3. As the Z lines are pulled closer together, the plasma membranes to which they are attached move toward one another, and the cell contracts.

How Nerves Signal Muscles to Contract

In vertebrate skeletal muscle, contraction is initiated by a nerve impulse. When a signal reaches the end of a neuron, the neuron releases a chemical, called a neurotransmitter, into the tiny gap separating neuron from muscle. The neurotransmitter, in this case acetylcholine, passes across the gap to the muscle plasma membrane. Acetylcholine binds to receptors that open ion channels and allow ions to pass into the cell. The ions make that area of the cell membrane more positively charged inside. This electrical change causes calcium ion channels to open, and Ca^{++} floods into the cytoplasm.

Myofilament Contraction

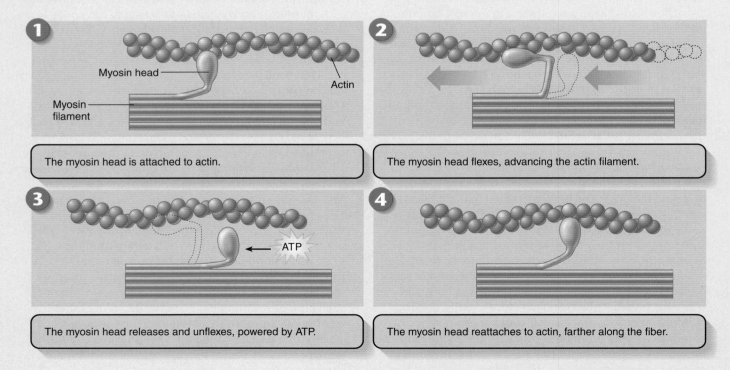

Figure 24.15   How myofilament contraction works.

The Sliding Filament Model of Muscle Contraction

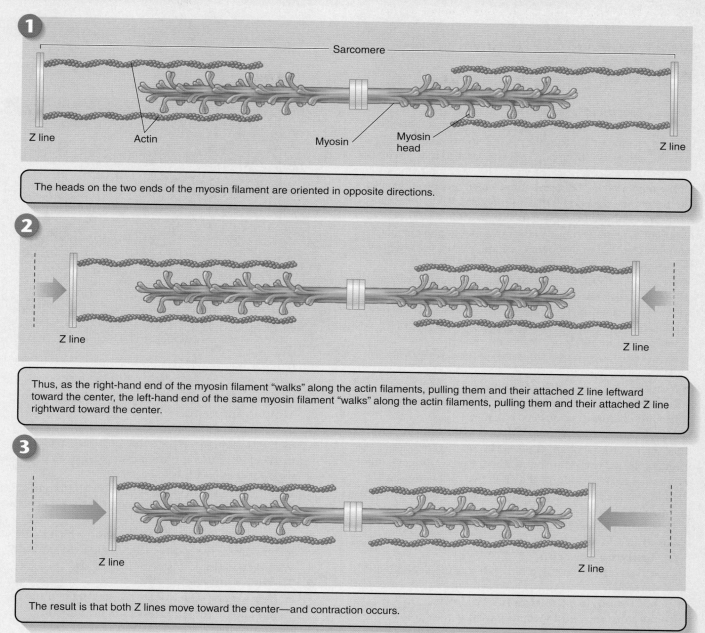

1

Sarcomere

Z line   Actin   Myosin   Myosin head   Z line

The heads on the two ends of the myosin filament are oriented in opposite directions.

2

Z line   Z line

Thus, as the right-hand end of the myosin filament "walks" along the actin filaments, pulling them and their attached Z line leftward toward the center, the left-hand end of the same myosin filament "walks" along the actin filaments, pulling them and their attached Z line rightward toward the center.

3

Z line   Z line

The result is that both Z lines move toward the center—and contraction occurs.

Figure 24.16 How actin and myosin filaments interact.

The Role of Calcium Ions in Contraction

When a muscle is relaxed, its myosin heads are "cocked" and ready, but are unable to bind to actin because the attachment sites for the myosin heads are physically blocked by the protein **tropomyosin.** For a muscle to contract, this tropomyosin must first be moved out of the way.

When the Ca^{++} concentration of the muscle cell cytoplasm is raised, this causes the tropomyosin to be shifted away from the attachment sites for the myosin heads on the actin. When this repositioning has occurred, myosin heads attach to actin and, using ATP energy, move along the actin in a stepwise fashion to shorten the myofibril.

Where does the Ca^{++} come from? Muscle fibers store Ca^{++} in a modified endoplasmic reticulum called the **sarcoplasmic reticulum.** When a muscle fiber is stimulated to contract, Ca^{++} is released from the sarcoplasmic reticulum and diffuses into the myofibrils, where it causes contraction.

> **24.8 Muscles are made of many tiny threadlike filaments of actin and myosin called myofilaments. Muscles work by using ATP to power the sliding of myosin along actin, causing the myofilaments to contract.**

Which Mode of Locomotion Is the Most Efficient?

Running, flying, and swimming require more energy than sitting still, but how do they compare? The greatest differences between moving on land, in the air, and in water result from the differences in support and resistance to movement provided by water and air. The weight of swimming animals is fully supported by the surrounding water, and no effort goes into supporting the body, while running and flying animals must support the full weight of their bodies. On the other hand, water presents considerable resistance to movement, air much less, so that flying and running require less energy to push the medium out of the way.

A simple way to compare the costs of moving for different animals is to determine how much energy it takes to move. The energy cost to run, fly, or swim is in each case the energy required to move one unit of body mass over one unit of distance with that mode of locomotion. (Energy is measured in the **metric system** as a **kilocalorie** (**kcal**) or technically 4.184 kilojoules [note that the Calorie measured in food diets and written with a capital C is equivalent to 1 kcal]; body mass is measured in kilograms, where one kilogram [**kg**] is 2.2 pounds; distance is measured in kilometers, where one kilometer [**km**] is 0.62 miles). The graph to the right displays three such "cost-of-motion" studies. The blue squares are running, the red circles are flying, and the green triangles are swimming. In each study, the line is drawn as the statistical "best-fit" for the points. Some animals like humans have data in two lines, as they both run (well) and swim (poorly). Ducks have data in all three lines, as they not only fly (very well), but also run and swim (poorly).

1. **Applying Concepts**
   a. Variables. In the graph, what is the dependent variable?
   b. Comparing Continuous Variables. Do the three modes of locomotion have the same or different costs?

2. **Interpreting Data**
   a. For any given mode of locomotion, what is the impact of body size on cost of moving?

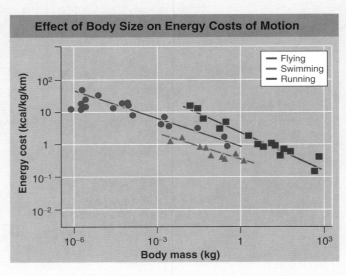

b. Is the impact of body mass the same for all three modes of locomotion? If not, which mode's cost is least affected by body mass? Why do you think this is so?

3. **Making Inferences**
   a. Comparing the energy costs of running versus flying for animals of the same body size, which mode of locomotion is the most expensive? Why would you expect this to be so?
   b. Comparing the energy costs of swimming to flying, which use the least energy? Why would you expect this to be so?

4. **Drawing Conclusions** In general, which mode of locomotion is the most efficient? the least efficient? Why do you think this is so?

5. **Further Analysis**
   a. How would you expect the slithering of a snake to compare to the three modes of locomotion examined here? Why?
   b. Do you think the costs of running by an athlete decrease with training? Why? How might you go about testing this?

The Animal Body Plan

24.1 Innovations in Body Design

- Several innovations have been key to the diversity seen in the animal kingdom: no tissues versus tissues, radial versus bilateral symmetry, no body cavity versus a body cavity, nonsegmentation versus segmentation, and protostome versus deuterostome development (**table 24.1**).

24.2 Organization of the Vertebrate Body

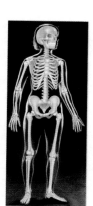

- The animal body is composed of cells that exhibit specialization of function and intercellular communication. Animal bodies have increasing levels of structural and functional complexity.

- Cells with similar functions group together into tissues, which act as functional units (**figure 24.1**).

- Organs of the body are composed of several different tissues that act together to perform a higher level of function.

- Organs work together in an organ system to perform larger-scale body functions (**figures 24.2** and **24.3**).

Tissues of the Vertebrate Body

24.3 Epithelium Is Protective Tissue

- Epithelial tissue is composed of different types of epithelial cells (**table 24.2**). It covers surfaces of the body, both internal and external surfaces, providing protection. Many sensory surfaces and glands are lined with a layer of epithelium and, in these cases, the epithelial tissue has sensory or secretory functions.

- Epithelium can be simple, a single layer of cells through which substances can pass, or it can be stratified, with multiple layers, providing protection.

24.4 Connective Tissue Supports the Body

- The connective tissues of the body are very diverse in structure and function, but all are composed of cells embedded in an extracellular matrix (**table 24.3**). The matrix may be hard, as in bone (**figure 24.5**), or may be fluid, as in blood. Connective tissues function in providing the body structural support, storage, and transportation of nutrients, gases, and immune cells.

24.5 Muscle Tissue Lets the Body Move

- Muscle tissue is dynamic tissue; it contracts, causing the body to move. There are three types of muscle tissue: smooth, skeletal, and cardiac muscle (**table 24.4**). All three types of muscle contain actin and myosin myofilaments but differ in the organization of the myofilaments.

- Smooth muscle cells are organized into sheets of cells with little alignment of the cells. This results in muscle contractions that are less coordinated. Smooth muscle is found in the walls of blood vessels, the digestive system, and so forth.

- Skeletal muscle cells are aligned but independent of each other (**figure 24.7**). They contract as small units when stimulated by nerves. Skeletal muscle is attached to the skeleton, so when the muscle contracts, the skeleton moves.

- Cells of cardiac muscle, found in the heart, are aligned and interconnected so that they contract together as one unit.

24.6 Nerve Tissue Conducts Signals Rapidly

- Nerve tissue is composed of neurons and supporting glial cells (**table 24.5**). Neurons carry electrical impulses from one area of the body to another (**figure 24.8**), providing the body a means of communication and coordination.

The Skeletal and Muscular Systems

24.7 Types of Skeletons

- The skeletal system provides a framework on which muscles act to move the body. Soft-bodied invertebrates have hydraulic skeletons, where muscles act on a fluid-filled cavity (**figure 24.9**).

- Arthropods have exoskeletons, where muscles attach from within to the hard outer covering of the body (**figure 24.10**).

- Vertebrates and echinoderms have endoskeletons, where muscles attach to bones or cartilage inside the body (**figures 24.11** and **24.12**).

24.8 Muscles and How They Work

- Skeletal muscles attach to bone at two points and cause the skeleton to move at joints. The end of the muscle that attaches to the stationary bone is called the origin. The muscle passes over a joint and attaches to another bone that moves as the muscle contracts. That anchoring point of the muscle is called the insertion. As the muscle contracts and shortens, the insertion is brought closer to the origin and the joint flexes. Muscles act to flex or extend a joint (**figure 24.14**).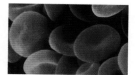

- The contraction of a muscle is due to molecular forces between actin and myosin myofilaments within the muscle cells.

- Actin and myosin filaments stack on top of each other, creating an orderly array. Myosin attaches to actin and pulls the actin along its length, causing the myofilaments to slide past each other. Energy from ATP causes the myofilaments to dissociate and reattach, triggering the sliding motion again. This is called the sliding filament model (**figures 24.15** and **24.16**).

- This process is controlled by Ca^{++} concentrations in the muscle cells. Muscle contraction is inhibited by tropomyosin, a protein that blocks myosin binding sites on actin. When Ca^{++} is present, it causes the tropomyosin to reposition so that the myosin binding sites are exposed and myosin can bind actin, triggering muscle contraction. Ca^{++} is stored in modified endoplasmic reticulum called sarcoplasmic reticulum.

1. One of the innovations in animal body design, segmentation, allowed for
   a. development of efficient internal organ systems.
   b. more flexible movement as individual segments can move independently of each other.
   c. locating organs in different areas of the body.
   d. early determination of embryonic cells.

2. Which of the following is the correct organization sequence from smallest to largest in animals?
   a. cells, tissues, organs, organ systems, organism
   b. organism, organ systems, organs, tissues, cells
   c. tissues, organs, cells, organ systems, organism
   d. organs, tissues, cells, organism, organ systems

3. Which of the following is *not* a function of the epithelial tissue?
   a. secrete materials
   b. provide sensory surfaces
   c. move the body
   d. protect underlying tissue from damage and dehydration

4. An example of connective tissue is
   a. nerve cells in your fingers.
   b. skin cells.
   c. brain cells.
   d. red blood cells.

5. When a person has osteoporosis, the work of _____ falls behind the work of _____.
   a. osteoclasts; osteoblasts
   b. osteoclasts; collagen
   c. osteoblasts; osteoclasts
   d. osteoblasts; collagen

6. Nerve impulses jump between nerve cells through the use of
   a. hormones.
   b. neurotransmitters.
   c. pheromones.
   d. calcium ions.

7. The type of muscle used for the movement of the skeleton is
   a. skeletal.
   b. cardiac.
   c. smooth.
   d. squamous.

8. The vertebral column is part of the
   a. appendicular skeleton.
   b. axial skeleton.
   c. hydrostatic skeleton.
   d. exoskeleton.

9. Movement requires a pair of muscles because
   a. a single muscle can only pull and not push.
   b. a single muscle can only push and not pull.
   c. both muscles are required to move a limb in any single direction.
   d. No answer is correct.

10. The role of calcium in the process of muscle contraction is to
   a. gather ATP for the myosin to use.
   b. cause the myosin head to shift position, contracting the myofibril.
   c. cause the myosin head to detach from the actin, causing the muscle to relax.
   d. expose myosin attachment sites on actin.

Visual Understanding

1. **Figure 24.6a** Normal bone is composed of an open lattice framework of minerals, including calcium. Why wouldn't it be more sensible to have bones be almost solid mineral, and be more resistant to breakage?

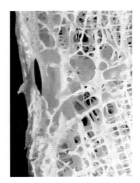

2. **Figure 24.12** The bones of your head and trunk form three cagelike structures; these are the skull, the rib cage, and the pelvic girdle. What is the importance of these structures?

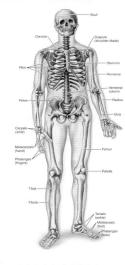

Challenge Questions

1. Your friends are having a friendly debate about which organ systems are the most important to them. Lucas says it's his muscles and bones so that he can play soccer. Brigit insists that her nervous system and respiratory system are the most important so that she can think, feel, and talk. Joseph is sure that it's his digestive system; otherwise he couldn't enjoy foods like pizza, tacos, hamburgers, and curry. Pick one or two systems and explain why you think they are the most important to your life.

2. Imagine that you are designing a living organism, some type of vertebrate. Explain briefly how the four types of tissue are all necessary to your design.

3. When you are exercising rapidly, such as playing tennis, dancing to fast music, or doing aerobics, you begin to breathe rapidly and your heart rate increases. If you continue, you might even say that you are "out of breath." Why is rapid breathing and heart rate important when you are giving your muscles a lot of work to do?

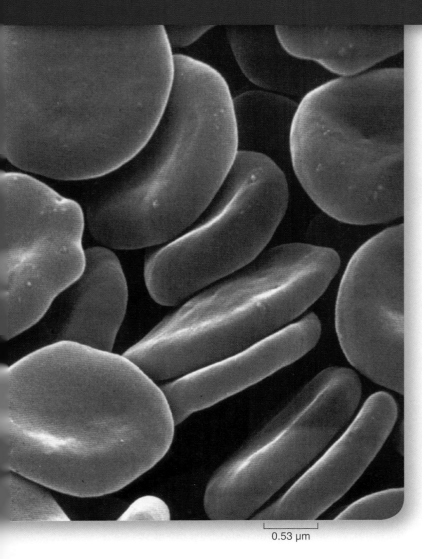

0.53 μm

25
Circulation

Blood has been called the river of life. Among all the vertebrate body's tissues, blood is the only liquid tissue, a fluid highway transporting gases, nutrients, hormones, antibodies, and wastes throughout the body. The material that makes blood a liquid is a protein-rich fluid called plasma that makes up approximately 55% of blood. The other 45% is made up mostly of red and white blood cells. Red blood cells like those seen above are the oxygen transporters of the vertebrate circulatory system. There are approximately 5 million of them in each microliter (1 μl) of blood! Each red blood cell is shaped like a rounded cushion, squashed in the center, and is crammed full of a protein called hemoglobin, an iron-containing molecule that gives blood its red color. Oxygen binds easily to the iron in hemoglobin, making red blood cells efficient oxygen carriers. A single red blood cell contains about 250 million hemoglobin molecules, and each red blood cell can carry about 1 billion molecules of oxygen at one time. The average life span of a red blood cell is only 120 days—about 2 million new ones are produced in the bone marrow every second to replace those that die or are worn out. In this chapter we will examine the ways in which vertebrates use cells like these, and the fluid surrounding them, to transport oxygen, food, and information throughout the body.

Every cell in the vertebrate body must acquire the energy it needs for living from organic molecules outside the body. Like residents of a city whose food is imported from farms in the countryside, the cells of the body need trucks to carry the food, highways for the trucks to travel on, and a means to cook the food when it arrives. In the vertebrate body, the trucks are blood, the highways are blood vessels, and oxygen molecules are used to cook the food. Remember from chapter 8 that cells obtain energy by "burning" sugars like glucose, using up oxygen and generating carbon dioxide. In animals, the organ system that provides the trucks and highways is called the *circulatory system,* while the organ system that acquires the oxygen fuel and disposes of the carbon dioxide waste is called the *respiratory system.* We discuss the functions of the circulatory system in this chapter and of the respiratory system in chapter 26.

25.1 Open and Closed Circulatory Systems

Among the unicellular protists, oxygen and nutrients are obtained directly by simple diffusion from the aqueous external environment. Cnidarians, such as *Hydra,* and flatworms, such as *Planaria,* have cells that are directly exposed to either the external environment or to a body cavity called the **gastrovascular cavity** that functions in both digestion and circulation. The gastrovascular cavity of *Hydra* extends even into the tentacles, and that of *Planaria,* colored green in figure 25.1*a*, branches extensively to supply every cell with oxygen and the nourishment obtained by digestion. Larger animals, however, have tissues that are several cell layers thick, so that many cells are too far away from the body surface or digestive cavity to exchange materials directly with the environment. Instead, oxygen and nutrients are transported from the environment and digestive cavity to the body cells by an internal fluid within a **circulatory system.**

There are two main types of circulatory systems: *open* or *closed.* In an **open circulatory system,** such as that found in arthropods and many mollusks, there is no distinction between the circulating fluid (blood) and the extracellular fluid of the body tissues (interstitial fluid or lymph). This fluid is thus called **hemolymph.** Insects, like the fly in figure 25.1*b*, have a muscular tube that serves as a heart to pump the hemolymph through a network of open-ended channels that empty into the cavities in the body (downward-pointing arrows). There, the hemolymph delivers nutrients to the cells of the body. It then reenters the circulatory system through pores in the heart (upward-pointing arrows). The pores close when the heart pumps to keep the hemolymph from flowing back out into the body cavity.

In a **closed circulatory system,** the circulating fluid, or *blood,* is always enclosed within blood vessels that transport blood away from and back to a pump, the *heart.* Annelids and all vertebrates have a closed circulatory system. In annelids such as the earthworm in figure 25.1*c*, a dorsal blood vessel contracts rhythmically to function as a pump. Blood is pumped through five small connecting vessels called lateral hearts, which also function as pumps, to a ventral blood vessel, which transports the blood posteriorly (lower arrows) until it eventually reenters the dorsal blood vessel (upper arrows). Smaller vessels branch between the ventral and dorsal blood vessels to supply the tissues of the earthworm with oxygen and nutrients and to transport waste products.

In vertebrates, blood vessels form a tubular network that permits blood to flow from the heart to all the cells of the body and then back to the heart. *Arteries* carry blood away from the heart, whereas *veins* return blood to the heart. Blood passes from the arterial to the venous system in *capillaries,* which are the thinnest and most numerous of the blood vessels.

As blood plasma passes through capillaries, the pressure of the blood forces some of this fluid out of the capillary walls. Fluid derived this way is called **interstitial fluid.** Some

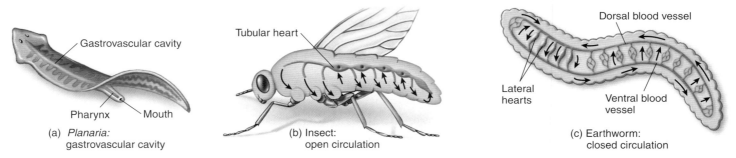

Figure 25.1 Three types of circulatory systems found in the animal kingdom.

(a) The gastrovascular cavity of *Planaria* serves as both a digestive and circulatory system, delivering nutrients directly to the tissue cells by diffusion from the digestive cavity. (b) In the open circulation of an insect, hemolymph is pumped from a tubular heart into cavities in the insect's body; the hemolymph then returns to the blood vessels so that it can be recirculated. (c) In the closed circulation of the earthworm, blood pumped from the hearts remains within a system of vessels that returns it to the hearts. All vertebrates also have closed circulatory systems.

of this fluid returns directly to capillaries, and some enters into **lymph vessels,** located in the connective tissues around the blood vessels. This fluid, now called *lymph,* is returned to the venous blood at specific sites. The lymphatic system is considered a part of the circulatory system and is discussed later in this chapter.

The Functions of Vertebrate Circulatory Systems

The functions of the circulatory system can be divided into three areas: transportation, regulation, and protection.

1. **Transportation.** Substances essential for cellular functions are transported by the circulatory system. These substances can be categorized as follows:

   Respiratory. Red blood cells, or erythrocytes, transport oxygen to the tissue cells. In the capillaries of the lungs or gills, oxygen attaches to hemoglobin molecules within the erythrocytes and is transported to the cells for aerobic respiration. Carbon dioxide produced by cell respiration is carried by the blood to the lungs or gills for elimination.

   Nutritive. The digestive system is responsible for the breakdown of food so that nutrients can be absorbed through the intestinal wall and into the blood vessels of the circulatory system. The blood then carries these absorbed products of digestion through the liver and to the cells of the body.

   Excretory. Metabolic wastes, excessive water and ions, and other molecules in plasma (the fluid portion of blood) are filtered through the capillaries of the kidneys and excreted in urine.

   Endocrine. The blood carries hormones from the endocrine glands, where they are secreted, to the distant target organs they regulate.

2. **Regulation.** The cardiovascular system participates in temperature regulation.

   Temperature regulation. In warm-blooded vertebrates, or homeotherms, a constant body temperature is maintained, regardless of the surrounding temperature. This is accomplished in part by blood vessels located just under the epidermis. When the ambient temperature is cold, the superficial vessels constrict to divert the warm blood to deeper vessels. When the ambient temperature is warm, the superficial vessels dilate so that the warmth of the blood can be lost by radiation.

   Some vertebrates also retain heat in a cold environment by using a **countercurrent heat exchange.** Figure 25.2 shows how a countercurrent heat exchange system works in the flipper of a killer whale. In this process, a vessel carrying warm blood from deep within the body (colored red) passes next to a vessel carrying cold blood from the surface of the body (colored blue). The warm blood going out heats

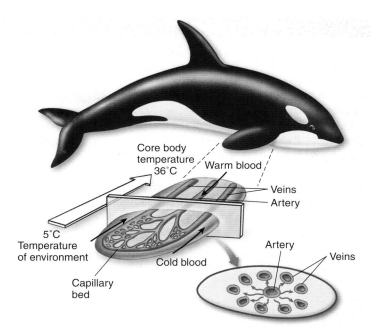

Figure 25.2 Countercurrent heat exchange.
Many marine mammals, such as this killer whale, limit heat loss in cold water by countercurrent flow that allows heat exchange between arteries and veins. The warm blood pumped from within the body in arteries warms the cold blood returning from the skin in veins, so that the core body temperature can remain constant in cold water. The cut-away portion in the figure shows how the veins surround the artery, maximizing the heat exchange between the artery and the veins.

the cold blood returning from the body surface (heat indicated by the red arrows), so that this blood is no longer cold when it reaches the interior of the body, helping to maintain a stable core body temperature.

3. **Protection.** The circulatory system protects against injury and foreign microbes or toxins introduced into the body.

   Blood clotting. The clotting mechanism protects against blood loss when vessels are damaged. This clotting mechanism involves both proteins from the blood plasma and cell structures called platelets (discussed in section 25.4).

   Immune defense. The blood contains white blood cells, or leukocytes, that provide immunity against many disease-causing agents. Some white blood cells are phagocytic, some produce antibodies, and some act by other mechanisms to protect the body.

> **25.1 Circulatory systems may be open or closed.** All vertebrates have a closed circulatory system, in which blood circulates away from the heart in arteries and back to the heart in veins. The circulatory system serves a variety of functions, including transportation, regulation, and protection.

25.2 Architecture of the Vertebrate Circulatory System

The vertebrate circulatory system, also called the **cardiovascular system,** is made up of three elements: (1) the **heart,** a muscular pump that pushes blood through the body; (2) the **blood vessels,** a network of tubes through which the blood moves; and (3) the **blood,** which circulates within these vessels.

Blood moves through the body in a cycle, from the heart, through a system of vessels: from the arteries and arterioles, into the capillaries, and then back to the heart through the venules and veins as shown here:

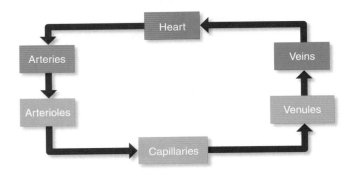

Blood leaves the heart through vessels known as **arteries.** From the arteries, the blood passes into a network of smaller arteries called **arterioles,** shown in figure 25.3*a* ❶. From these, it is eventually forced through a capillary bed ❷, a fine latticework of very narrow tubes called **capillaries** (from the Latin, *capillus,* "a hair"). While passing through the capillaries, the blood exchanges gases and metabolites (glucose, vitamins, hor-mones) with the cells of the body. Capillary beds can be opened or closed, based on the physiological needs of the tissues, by the relaxation or contraction of small circular muscles called *precapillary sphincters.* The closing of the precapillary sphincters by contraction is shown in figure 25.3*b* with the blood being diverted from the capillary bed. After traversing the capillaries, the blood passes into a fourth kind of vessel, the **venules,** or small veins, ❸. A network of venules empties into larger **veins** that collect the circulating blood and carry it back to the heart.

The capillaries have a much smaller diameter than the other blood vessels of the body. Blood leaves the mammalian heart through a large artery, the aorta, a tube that in your body has a diameter of about 2 centimeters (about the same as your thumb), but when it reaches the capillaries it passes through vessels with an average diameter of only 8 micrometers, a reduction in radius of some 1,250 times! This decrease in size of blood vessels has a very important consequence.

Although each capillary is very narrow, there are so many of them that the capillaries have the greatest *total* cross-sectional area of any other type of vessel. Consequently, this allows more time for blood to exchange materials with the surrounding extracellular fluid. By the time the blood reaches the end of a capillary, it has released some of its oxygen and nutrients and picked up carbon dioxide and other waste products. Blood loses most of its pressure and velocity in passing through the vast capillary networks, and so is under very low pressure when it enters the veins. The blood flow through the capillaries is like water flowing out of the sprinkler head of a watering can, the stream of blood spreads out to many small streams. These smaller streams don't flow with as much force, nor as quickly, as the larger stream that entered the capillary bed.

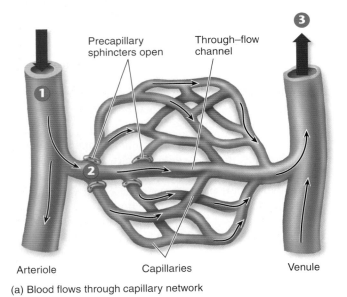

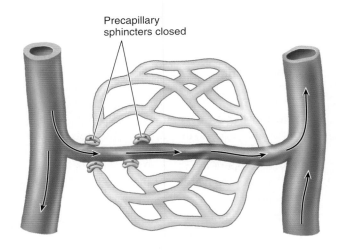

(a) Blood flows through capillary network

(b) Blood flow in capillary network is limited

Figure 25.3 The capillary network connects arteries with veins.

Through-flow channels connect arterioles directly to venules. Branching from these through-flow channels is a network of finer channels, the capillaries. Most of the exchange between the body tissues and the red blood cells occurs while they are in this capillary network. The flow of blood into the capillaries is controlled by bands of muscle called precapillary sphincters located at the entrance to each capillary. (a) When a sphincter is open, blood flows through that capillary. (b) When a sphincter contracts, it closes off the capillary.

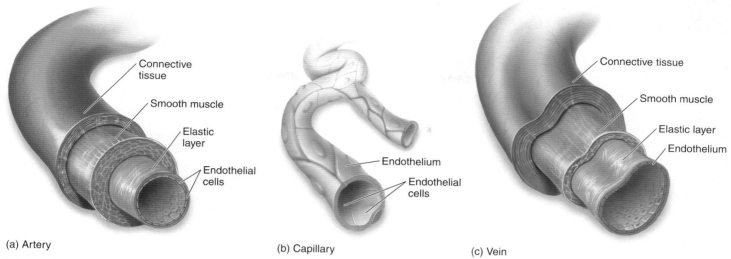

Connective tissue
Smooth muscle
Elastic layer
Endothelial cells

(a) Artery

Endothelium
Endothelial cells

(b) Capillary

Connective tissue
Smooth muscle
Elastic layer
Endothelium

(c) Vein

Figure 25.4   The structure of blood vessels.
(a) Arteries, which carry blood away from the heart, are expandable and are composed of layers of tissue. (b) Capillaries are simple tubes whose thin walls facilitate the exchange of materials between the blood and the cells of the body. (c) Veins, which transport blood back to the heart, do not need to be as sturdy as arteries. The walls of veins have thinner muscle layers than arteries, and they collapse when empty. Note, this drawing is not to scale; as stated in the text, arteries can be up to two centimeters in diameter, the capillaries are only about eight micrometers in diameter, and the largest veins can be up to three centimeters in diameter.

Arteries: Highways from the Heart

The arterial system, composed of arteries and arterioles, carries blood away from the heart. An artery is more than simply a pipe. Blood comes from the heart not in a smooth flow but rather in pulses, slammed into the artery in great big slugs as the heart forcefully ejects its contents with each contraction. The artery has to be able to *expand* to withstand the pressure caused by each contraction of the heart. An artery, then, is designed as an expandable tube, with its walls made up of three layers of tissue. Figure 25.4a shows these layers pulled out from the artery, like a telescope so they are more easily seen. The innermost thin layer is composed of endothelial cells. Surrounding them is a layer of elastic fibers and then a thick layer of smooth muscle, which in turn is encased within an envelope of protective connective tissue. Because this sheath and envelope are elastic, the artery is able to expand its volume considerably when the heart contracts, shoving a new volume of blood into the artery—just as a tubular balloon expands when you blow more air into it. The steady contraction of the smooth muscle layer strengthens the wall of the vessel against overexpansion.

Arterioles differ from arteries in two ways. They are smaller in diameter, and the muscle layer that surrounds an arteriole can be relaxed under the influence of hormones to enlarge the diameter. When the diameter increases, the blood flow also increases, an advantage during times of high body activity. Most arterioles are also in contact with nerve fibers. When stimulated by these nerves, the muscle lining of the arteriole contracts, constricting the diameter of the vessel. Such contraction limits the flow of blood to the extremities during periods of low temperature or stress. You turn pale when you are scared or cold because the arterioles

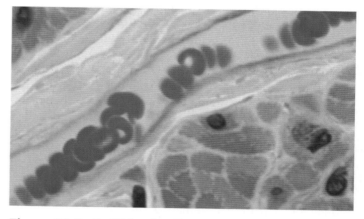

Figure 25.5   Red blood cells within a capillary.
The red blood cells in this capillary pass along in single file. Red blood cells can pass through capillaries even narrower than their own diameter, pushed along by the pressure of the pumping heart.

in your skin are constricting. You blush for just the opposite reason. When you overheat or are embarrassed, the nerve fibers connected to muscles surrounding the arterioles are inhibited, which relaxes the smooth muscle and causes the arterioles in the skin to expand, bringing heat to the surface for escape.

Capillaries: Where Exchange Takes Place

Capillaries are where oxygen and food molecules are transferred from the blood to the body's cells and where waste carbon dioxide is picked up. To facilitate this back-and-forth traffic, capillaries are narrow, as shown in figure 25.5, and have thin walls across which gases and metabolites pass

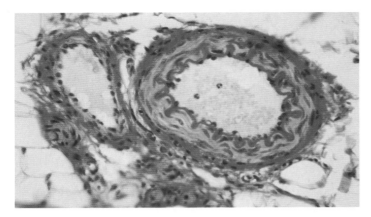

Figure 25.6 Veins and arteries.
The vein (*left*) has the same general structure as the artery (*right*), but an artery retains its shape when empty, while a vein collapses.

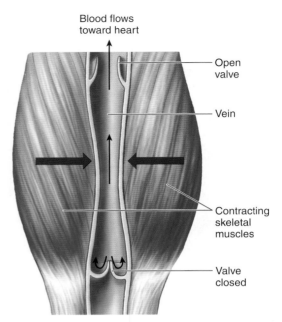

Figure 25.7 Flow of blood through veins.
Venous valves ensure that blood moves through the veins in only one direction back to the heart. This movement of blood is aided by the contraction of skeletal muscles surrounding the veins.

easily. Capillaries have the simplest structure of any element in the cardiovascular system. They are built like a soft-drink straw, simple tubes with walls only one cell thick (see figure 25.4*b*). The average capillary is about 1 millimeter long and connects an arteriole with a venule. All capillaries are very narrow, with an internal diameter of about 8 micrometers, just bigger than the diameter of a red blood cell (5 to 7 micrometers). This design is critical to the function of capillaries. By bumping against the sides of the vessel as they pass through (like the cells in figure 25.5), the red blood cells are forced into close contact with the capillary walls, making exchange easier. Examining this photo, you can better understand the complications that result from sickle-cell anemia, discussed in chapters 11 and 15. The red blood cells of people with sickle-cell anemia take on an elongated shape, and do not easily flow through the narrow passageways of the capillaries, causing the capillaries to become blocked.

Almost all cells of the vertebrate body are no more than 100 micrometers from a capillary. At any one moment, about 5% of the circulating blood is in capillaries, a network that amounts to several thousand miles in overall length. If all the capillaries in your body were laid end to end, they would extend across the United States! Individual capillaries have high resistance to flow because of their small diameters. However, the total cross-sectional area of the extensive capillary network (that is, the sum of all the diameters of all the capillaries, expressed as area) is greater than that of the arteries leading to it. As a result, the blood pressure is actually far lower in the capillaries than in the arteries. This is important, because the walls of capillaries are not strong, and they would burst if exposed to the pressures that arteries routinely withstand.

Veins: Returning Blood to the Heart

Veins are vessels that return blood to the heart. Veins do not have to accommodate the pulsing pressures that arteries do, because much of the force of the heartbeat is weakened by the high resistance and great cross-sectional area of the capillary network. For this reason, the walls of veins have much thinner

layers of muscle and elastic fiber as seen in figure 25.4*c*. An empty artery will stay open, like a pipe, but when a vein is empty, its walls collapse like an empty balloon. In figure 25.6 you can see photographs of a vein and an artery side-by-side. The vein on the left is partially collapsed, while the artery on the right still holds its shape.

Because the pressure of the blood flowing within veins is low, it becomes important to avoid any further resistance to flow, lest there not be enough pressure to get the blood back to the heart. Because a wide tube presents much less resistance to flow than a narrow one, the internal passageway of veins is often quite large, requiring only a small pressure difference to return blood to the heart. The diameters of the largest veins in the human body, the venae cavae, which lead into the heart, are fully 3 centimeters; this is wider than your thumb! Pressure alone cannot force the blood in the veins back to the heart but several features provide help. Most significantly, when skeletal muscles surrounding the veins contract, they move blood by squeezing the veins. Veins also have unidirectional valves (the small flaps within the vein in figure 25.7) that ensure the return of this blood by preventing it from flowing backward. These structural features keep the blood flowing in a cycle through the circulatory system.

> **25.2 The vertebrate circulatory system is composed of arteries and arterioles, which are elastic and carry blood away from the heart; a fine network of capillaries across whose thin walls the exchange of gases and food molecules takes place; and venules and veins, which return blood from the capillaries to the heart.**

25.6 Amphibian and Reptile Circulation

The advent of lungs involved a major change in the pattern of circulation. After blood is pumped by the heart to the lungs, it does not go directly to the tissues of the body but instead returns to the heart. This results in two circulations: one that goes between the heart and the lungs, called the **pulmonary circulation,** and one that goes between the heart and the rest of the body, called the **systemic circulation.**

If no changes had occurred in the structure of the heart, the oxygenated blood from the lungs would be mixed in the heart with the deoxygenated blood returning from the rest of the body. Consequently, the heart would pump a mixture of oxygenated and deoxygenated blood rather than fully oxygenated blood. The amphibian heart has several structural features that help reduce this mixing. First, the atrium is divided by a *septum,* or dividing wall, into two chambers: the right atrium (the blue-colored area in figure 25.13*a*) receives deoxygenated blood from the systemic circulation, and the left atrium (colored red) receives oxygenated blood from the lungs. The septum prevents the two stores of blood from mixing in the atria, but some mixing might be expected when the contents of each atrium enter the single, common ventricle (the purple-colored area). Surprisingly, however, little mixing actually occurs. Two factors contribute to the separation of the two blood supplies. First, the ventricle in some amphibians is lined with folds that help direct the flow of blood from the atria. Second, the conus arteriosus (the branched structure in the foreground) is partially separated by another septum that directs deoxygenated blood into the *pulmonary arteries* toward the lungs, and oxygenated blood into the *aorta,* the major artery of the systemic circulation to the body. To trace the blood flow through the heart, follow the deoxygenated blood, the blue arrows, from the body to the lungs and then the oxygenated blood, the red arrows, from the lungs out to the body.

Amphibians in water supplement the oxygenation of the blood by obtaining additional oxygen by diffusion through their skin. This process is called **cutaneous respiration.**

Among reptiles, additional modifications have reduced the mixing of blood in the heart still further. In addition to having two separate atria, reptiles have a septum that partially subdivides the ventricle, with red and blue halves of the ventricle in figure 25.13*b*. This results in an even greater separation of oxygenated and deoxygenated blood within the heart. The separation is complete in one order of reptiles, the crocodiles, which have two separate ventricles divided by a complete septum. Crocodiles therefore have a completely divided pulmonary and systemic circulation. Another change in the circulation of reptiles is that the conus arteriosus has become incorporated into the trunks of the large arteries leaving the heart.

> **25.6** Amphibians and reptiles have two circulations, pulmonary and systemic, that deliver blood to the lungs and to the rest of the body, respectively.

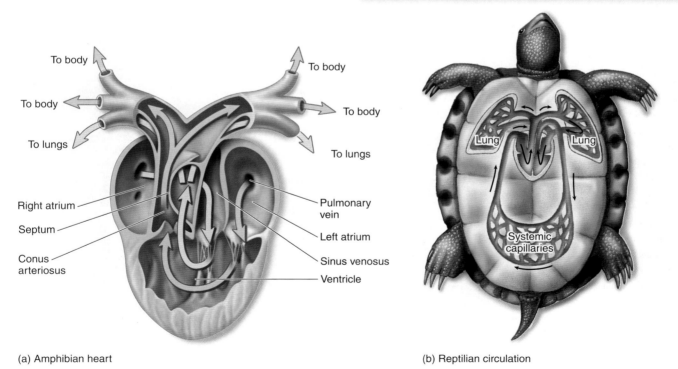

(a) Amphibian heart

(b) Reptilian circulation

Figure 25.13 The amphibian heart and circulation in reptiles.
(a) The frog heart has two atria but only one ventricle, which pumps blood both to the lungs and to the body. Despite the potential for mixing, the oxygenated and deoxygenated bloods (*red* and *blue,* respectively) mix very little as they are pumped to the body and lungs. (b) In reptiles, not only are there two separate atria, but the ventricle is also partially divided.

Mammalian and Bird Circulation

Mammals, birds, and crocodiles have a four-chambered heart that is really two separate pumping systems operating together within a single unit. One of these pumps blood to the lungs, while the other pumps blood to the rest of the body. The left side has two connected chambers, and so does the right, but the two sides are not connected with one another. The increased efficiency of the double circulatory system in mammals and birds is thought to have been important in the evolution of endothermy (warm-bloodedness), because a more efficient circulation is necessary to support the high metabolic rate required.

Circulation Through the Heart

Let's follow the journey of blood through the mammalian heart in figure 25.14*a*, starting with the entry of oxygen-rich blood into the heart from the lungs. Oxygenated blood (the red arrows) from the lungs enters the left side of the heart (which is on the right as you look at the figure), emptying directly into the **left atrium** through large vessels called the **pulmonary veins.** From the atrium, blood flows through an opening into the adjoining chamber, the **left ventricle.** Most of this flow, roughly 80%, occurs while the heart is relaxed. When the heart starts to contract, the atrium contracts first, pushing the remaining 20% of its blood into the ventricle.

After a slight delay, the ventricle contracts. The walls of the ventricle are far more muscular than those of the atrium (as seen in this cross section), and thus this contraction is much stronger. It forces most of the blood out of the ventricle in a single strong pulse. The blood is prevented from going back into the atrium by a large, one-way valve, the **bicuspid (mitral) valve,** or left atrioventricular valve, whose two flaps are pushed shut as the ventricle contracts.

Prevented from reentering the atrium, the blood within the contracting left ventricle takes the only other passage out (partially covered by the large blue vessel). It moves through a second opening that leads into a large blood vessel called the **aorta.** The aorta is separated from the left ventricle by a one-way valve, the **aortic semilunar valve.** The aortic valve is oriented to permit the flow of the blood *out* of the ventricle. Once this outward flow has occurred, the aortic valve closes, preventing the reentry of blood from the aorta into the heart.

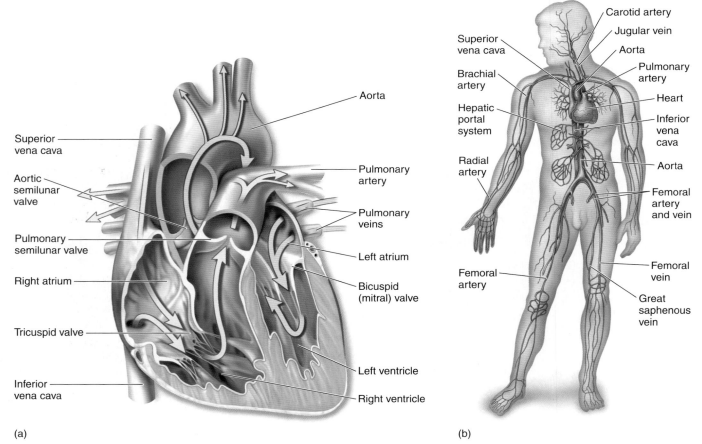

(a)   (b)

Figure 25.14 The heart and circulation of mammals and birds.

(a) Unlike the amphibian heart, this heart has a septum dividing the ventricle into left and right ventricles. Oxygenated blood from the lungs enters the left atrium of the heart by way of the pulmonary veins. This blood then enters the left ventricle, from which it passes into the aorta to circulate throughout the body and deliver oxygen to the tissues. When gas exchange has taken place at the tissues, veins return blood to the heart. After entering the right atrium by way of the superior and inferior venae cavae, deoxygenated blood passes into the right ventricle and then through the pulmonary valve to the lungs by way of the pulmonary artery. (b) Some of the major arteries and veins in the human circulatory system are shown.

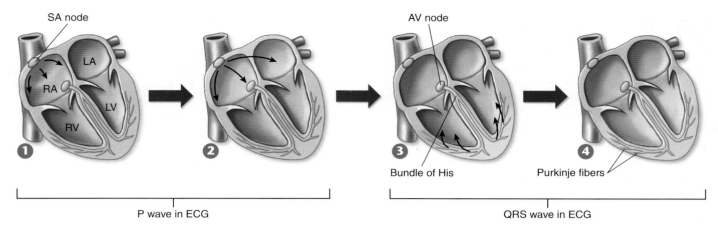

P wave in ECG

QRS wave in ECG

Figure 25.15   How the mammalian heart contracts.

❶ Contraction of the mammalian heart is initiated by a wave of depolarization that begins at the SA node. ❷ After passing over the right and left atria and causing their contraction (forming the P wave on the ECG), ❸ the wave of depolarization reaches the AV node, from which it passes to the ventricles by the bundle of His. ❹ The depolarization is then conducted rapidly over the surface of the ventricles by a set of finer fibers called Purkinje fibers, causing the ventricles to contract (forming the QRS wave on the ECG). The T wave on the ECG corresponds to the repolarization of the ventricles. This ECG is showing a heart rate of 60 beats per minute. The time between peaks varies with changes in heart rates.

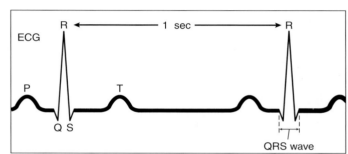

The aorta and its many branches are systemic arteries and carry oxygen-rich blood to all parts of the body.

Eventually, this blood returns to the heart after delivering its cargo of oxygen to the cells of the body. In returning, it passes through a series of progressively larger veins, ending in two large veins that empty into the right atrium of the heart. The **superior vena cava** drains the upper body (extending from the heart to the top of the figure), and the **inferior vena cava** drains the lower body (extending from the heart downward).

The right side of the heart is similar in organization to the left side. Blood (the blue-colored arrows) passes from the **right atrium** into the **right ventricle** through a one-way valve, the **tricuspid valve** or right atrioventricular valve. It passes out of the contracting right ventricle (the large branched blue vessel) through a second valve, the **pulmonary semilunar valve,** into the **pulmonary arteries,** which carry the deoxygenated blood to the lungs. The blood then returns from the lungs to the left side of the heart with a new cargo of oxygen, which is pumped to the rest of the body. Figure 25.14*b* shows the major veins and arteries in the human body. Veins carry blood to the heart and arteries carry blood away from the heart.

How the Heart Contracts

Throughout the evolutionary history of the vertebrate heart, the sinus venosus has served as a pacemaker, the site where the impulses that produce the heartbeat originate. Although it constitutes a major chamber in the fish heart, it is reduced in size in amphibians and further reduced in reptiles. In mammals and birds, the sinus venosus is no longer a separate chamber, but some of its tissue remains in the wall of the right

atrium, near the point where the systemic veins empty into the atrium. This tissue, which is called the **sinoatrial (SA) node** (indicated in panel ❶ of figure 25.15), is still the site where each heartbeat originates.

The contraction of the heart consists of a carefully orchestrated series of muscle contractions. First, the atria contract together, followed by the ventricles. Contraction is initiated by the sinoatrial (SA) node. Its membranes spontaneously depolarize (that is, admit ions that cause it to become more positively charged) with a regular rhythm that determines the rhythm of the heart's beating. Each depolarization initiated within this pacemaker region passes quickly from one heart muscle cell to another in a wave that envelops the left and the right atria almost simultaneously (indicated by the yellow coloring in panels ❶ and ❷).

But the wave of depolarization does not immediately spread to the ventricles. There is a pause before the lower half of the heart starts to contract. The reason for the delay is that the atria of the heart are separated from the ventricles by connective tissue that does not propagate a depolarization wave. The depolarization would not pass to the ventricles at all except for a slender connection of cardiac muscle cells known as the **atrioventricular (AV) node** (labeled in panel ❸), which connects across the gap to a strand of specialized muscle known as the atrioventricular bundle, or **bundle of His.** Bundle branches divide into fast-conducting **Purkinje fibers,** which initiate the almost simultaneous contraction of all the cells of the right and left ventricles about 0.1 seconds after the atria contract. This delay permits the atria to finish emptying their contents into the corresponding ventricles before those ventricles start to contract. The contraction of the ventricles begins at the apex (the bottom of the heart) where the depolarization of Purkinje fibers begins. The contraction then spreads up toward the atria (indicated by the yellow coloring in panels ❸ and ❹). This results in a "wringing out" of the ventricle, forcing the blood up and out of the heart.

Because the vertebrate body basically consists of water, it conducts electrical currents rather well. A wave of membrane depolarization passing over the surface of the heart generates an electrical current that passes in a wave throughout the body. The magnitude of this electrical pulse is tiny, but it can be detected by sensors placed on the skin. The recording, called an **electrocardiogram** (**ECG** or **EKG**), shows how the cells of the heart depolarize and repolarize during the cardiac cycle (figure 25.15). Depolarization causes contraction of the heart, while repolarization causes relaxation. The first peak in the recording, P, is produced by the depolarization of the atria. The second, larger peak, QRS, is produced by ventricular depolarization; during this time, the ventricles contract and eject blood into the arteries. The last peak, T, reflects ventricular repolarization.

Because the overall circulatory system is closed, the same volume of blood must move through the pulmonary circulation as through the much larger systemic circulation with each heartbeat. Therefore, the right and left ventricles must pump the same amount of blood each time they contract. If the output of one ventricle did not match that of the other, fluid would accumulate and pressure would increase in one of the circuits. The result would be increased filtration out of the capillaries and edema (as occurs in congestive heart failure, for example). Although the volume of blood pumped by the two ventricles is the same, the pressure they generate is not. The left ventricle, which pumps blood through the systemic pathway, is more muscular and generates more pressure than the right ventricle.

Monitoring the Heart's Performance

As you can see, the heartbeat is not simply a squeeze-release, squeeze-release cycle but rather a series of events that occur in a predictable order. The simplest way to monitor heartbeat is to listen to the heart at work, using a stethoscope. The first sound you hear, a low-pitched *lub,* is the closing of the bicuspid and tricuspid valves at the start of ventricular contraction. A little later, you hear a higher-pitched *dub,* the closing of the pulmonary and aortic valves at the end of ventricular contraction. If the valves are not closing fully, or if they open incompletely, a turbulence is created within the heart. This turbulence can be heard as a **heart murmur,** a liquid sloshing sound.

A second way to examine the events of the heartbeat is to monitor the blood pressure. This is done using a device called a sphygmomanometer, which measures the blood pressure in the brachial artery found on the inside part of the arm, at the elbow (figure 25.16). A cuff wrapped around the upper arm is tightened enough to stop the flow of blood to the lower part of the arm ❶. As the cuff is loosened, blood begins pulsating through the artery and can be detected using

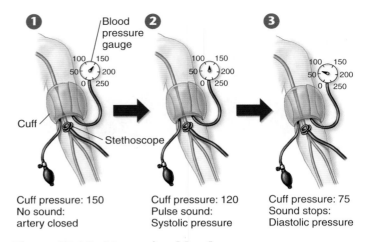

Cuff pressure: 150
No sound:
artery closed

Cuff pressure: 120
Pulse sound:
Systolic pressure

Cuff pressure: 75
Sound stops:
Diastolic pressure

Figure 25.16 Measuring blood pressure.
The blood pressure cuff is tightened to stop the blood flow through the brachial artery ❶. As the cuff is loosened, the systolic pressure is recorded as the pressure at which a pulse is heard through a stethoscope ❷. The diastolic pressure is recorded as the pressure at which a sound is no longer heard ❸.

a stethoscope. Two measurements are recorded: the systolic pressure ❷ is recorded when a pulse is heard, and the diastolic pressure ❸ is recorded when the pressure in the cuff is so low that the sound stops.

To understand these measurements it is important to remember what is happening in the heart. During the first part of the heartbeat, the atria are filling. At this time the pressure in the arteries leading from the left side of the heart out to the tissues of the body decreases slightly. This low pressure is referred to as the **diastolic** pressure. During the contraction of the left ventricle, a pulse of blood is forced into the systemic arterial system, immediately raising the blood pressure within these vessels. The high blood pressure produced in this pushing period, which ends with the closing of the aortic valve, is referred to as the **systolic** pressure. Normal blood pressure values are 70 to 90 diastolic and 110 to 130 systolic. When the inner walls of the arteries accumulate fats, as they do in the condition known as *atherosclerosis,* the diameters of the passageways are narrowed. If this occurs, the systolic blood pressure is elevated.

The evolution of multicellular organisms has depended critically on the ability to circulate materials throughout the body efficiently. Vertebrates carefully regulate the operation of their circulatory systems and are able to integrate their body activities. Indeed, the metabolic demands of different vertebrates have shaped the evolution of circulatory systems.

25.7 The mammalian heart is a two-cycle pump. The left side pumps oxygenated blood to the body's tissues, while the right side pumps O_2-depleted blood to the lung. The performance of the heart can be monitored in many ways.

25.8 Cardiovascular Diseases

Cardiovascular diseases (heart disease and other diseases of the circulatory system) are the leading cause of death in the United States, accounting for nearly 40% of all deaths in 2004; more than 71 million people have some form of cardiovascular disease. Heart attacks are the main cause of cardiovascular deaths in the United States. They result from an insufficient supply of blood reaching one or more parts of the heart muscle, which causes myocardial cells in those parts to die. Heart attacks, also called myocardial infarctions, may be caused by a blood clot forming somewhere in the coronary arteries (the arteries that supply the heart muscle with blood) and blocking the passage of blood through those vessels. They may also result if an artery is blocked by atherosclerosis (discussed later). Recovery from a heart attack is possible if the portion of the heart that was damaged is small. **Angina pectoris,** which literally means "chest pain," can occur when the supply of blood to the heart is reduced. The pain may occur in the heart and often also in the left arm and shoulder. Angina pectoris is a warning sign that the blood supply to the heart is inadequate but still sufficient to avoid myocardial cell death.

Strokes are caused by an interference with the blood supply to the brain. They may occur when a blood vessel bursts in the brain or when blood flow in a cerebral artery is blocked by a thrombus (blood clot) or by atherosclerosis. The effects of a stroke depend on how severe the damage is and where in the brain the stroke occurs.

Atherosclerosis is an accumulation within the arteries of fatty materials, abnormal amounts of smooth muscle, deposits of cholesterol or fibrin, or various kinds of cellular debris. These accumulations cause blood flow to be reduced. The lumen (interior) of the artery may be further reduced in size by a clot that forms as a result of the atherosclerosis. Three different arteries are shown in figure 25.17 with different degrees of blockage. In the severest cases (as in figure 25.17c), the artery may be blocked completely. Atherosclerosis is promoted by genetic factors, smoking, hypertension (high blood pressure), and high blood cholesterol levels. Diets low in cholesterol and saturated fats (from which cholesterol can be made) can help lower the level of blood cholesterol, and therapy for hypertension can reduce that risk factor. Stopping smoking, however, is the single most effective action a smoker can take to reduce the risk of atherosclerosis.

Arteriosclerosis, or hardening of the arteries, occurs when calcium is deposited in arterial walls. It tends to occur when atherosclerosis is severe. Not only do such arteries have restricted blood flow, but they also lack the ability to expand as normal arteries do to accommodate the volume of blood pumped out by the heart. This inflexibility forces the heart to work harder.

Treatments of Blocked Coronary Arteries

Atherosclerosis is treated both with medication and with invasive procedures. Medications include enzymes, which help dissolve clots; anticoagulants, which prevent clots from forming (aspirin works as a weak anticoagulant); and nitroglycerin, which dilates blood vessels.

Invasive treatments include reducing the blockage with *angioplasty.* Angioplasty is a procedure where a small balloon is threaded into a partially blocked coronary artery. Once in the blocked artery, the balloon is inflated, flattening the atherosclerosis deposit against the side of the artery. In some cases, a small metal mesh sleeve called a *stent* may also be inserted to prop the artery open. More aggressive treatments include *coronary bypass surgery,* where healthy segments of blood vessels are patched into a coronary artery, which diverts the flow of blood around a blocked section of artery; and *heart transplants,* where the damaged heart is replaced by a donor heart.

> **25.8** Humans are subject to a variety of cardiovascular diseases, many of them associated with the accumulation of fatty materials on the inner surfaces of arteries.

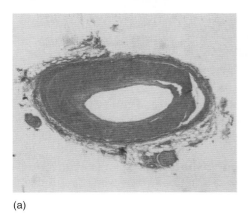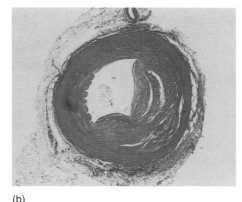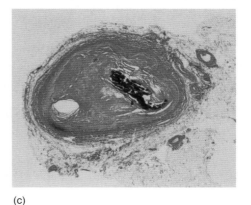

(a) (b) (c)

Figure 25.17 The path to a heart attack.
(a) The coronary artery shows only minor blockage. (b) The artery exhibits severe atherosclerosis—much of the passage is blocked by buildup on the interior walls of the artery. (c) The coronary artery is essentially completely blocked.

Do Big Hearts Beat Faster?

Small animals live at a much faster pace than large animals. They reproduce more quickly, and live shorter lives. As a rule, they tend to move about more quickly, and so to consume more oxygen per unit body weight. Interestingly, small and large mammals have about the same size heart, relative to body size. (about 0.6% of body mass). It is interesting to ask whether all mammalian hearts beat at the same rate. The heart of a 7,000-kilogram (a kilogram is 1,000 grams) African bull elephant must push a far greater volume of blood through its body than the heart of a 3-gram mouse, but the elephant is able to do it through much-larger-diameter arteries, which impose far less resistance to the blood's flow. Does the elephant's heart beat faster? Or does the mouse's, in order to deliver more oxygen to its muscles? Or perhaps the mouse's heart beats more slowly, because of increased resistance to flow through narrower blood vessels?

The graph to the right displays the pulse rate of a number of mammals of different body sizes (the **pulse rate** is the number of heartbeats counted per minute, a measure of how rapidly the heart is beating). For comparison, the pulse rate of an adult human at rest is about 70 beats per minute. The largest mammal is the blue whale, as big as a super-sized moving van with a body mass as great as 136,000 kilograms; the smallest is the pygmy shrew, smaller than a cockroach with a body mass of a few hundredths of a gram.

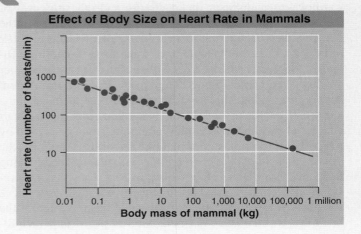

1. **Applying Concepts**
   a. Variable. In the graph, what is the dependent variable?
   b. Reading a Line Graph. All mammals have the same size hearts relative to their body size. Do their hearts beat at the same rate?
2. **Interpreting Data**
   a. What is the resting pulse rate of a 7,000-kilogram African bull elephant?
   b. What is the resting pulse rate of a 3-gram shrew? How many complete heartbeats is that per second?
   c. What general statement can be made regarding the effect of body size on heart rate in mammals?
3. **Making Inferences**
   a. The data in the graph, plotted on logarithmic coordinates (that is, the scale rises in powers of 10), fall nicely upon a straight line. How would you expect them to look, plotted on linear coordinates?

b. As you walk through the graph from left to right, the line slopes down, this is called a negative slope. What does the negative slope of the line signify?
4. **Drawing Conclusions** If you plotted data for an experiment measuring body mass versus resting oxygen consumption, you would get exactly the same slope of the line as shown in this graph. What does this tell us about why body size affects heart rate in mammals as it does?
5. **Further Analysis** What would you expect to be the relationship between lung size and body mass in mammals? Explain your reasoning.

Circulation

25.1 Open and Closed Circulatory Systems

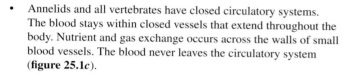

- Cnidarians and flatworms have a gastrovascular cavity that functions in digestion and circulation (**figure 25.1a**).

- Mollusks and arthropods have open circulatory systems. Blood vessels open into a body cavity and hemolymph is pumped from a tubular heart into the body cavity where cells acquire nutrients from the hemolymph. The hemolymph then reenters the circulatory system through pores in the tubular heart (**figure 25.1b**).

- Annelids and all vertebrates have closed circulatory systems. The blood stays within closed vessels that extend throughout the body. Nutrient and gas exchange occurs across the walls of small blood vessels. The blood never leaves the circulatory system (**figure 25.1c**).

- The circulatory system in vertebrates functions in the transportation of substances throughout the body, in the regulation of the body through hormones and temperature control (**figure 25.2**), and in the protection of the body through blood clotting and immunity.

25.2 Architecture of the Vertebrate Circulatory System

- In the vertebrate circulatory system, blood circulates from the heart through arteries and arterioles to capillaries. The capillary is the site of gas, nutrient, and waste exchange. Blood flows from the capillaries back to the heart through venules and larger veins (**figures 25.3** and **25.4**).

25.3 The Lymphatic System: Recovering Lost Fluid

- Needed fluid is lost to the body tissues through leaky capillary walls. The fluid is recovered by lymphatic capillaries that drain into lymphatic vessels. Lymphatic vessels return the fluid back to the circulatory system (**figure 25.9**).

25.4 Blood

- Blood is a salty, protein-rich fluid that circulates through the body in the circulatory system. Blood contains red blood cells that are involved in gas exchange, and various types of white blood cells, called leukocytes, that defend the body against infection. Platelets are fragments of cells involved in blood clotting (**figure 25.11**).

Evolution of Vertebrate Circulatory Systems

25.5 Fish Circulation

- The fish heart consists of a series of chambers. Blood enters the heart and is collected in the first two chambers, the sinus venosus and the atrium. Blood is then pumped from the third and fourth chambers, the ventricle and conus arteriosus, to the body. It passes to the gills, where gas exchange occurs, and then oxygenated blood travels throughout the body (**figure 25.12**). Deoxygenated blood returns to the heart, where it is again pumped to the gills, where gas exchange occurs.

25.6 Amphibian and Reptile Circulation

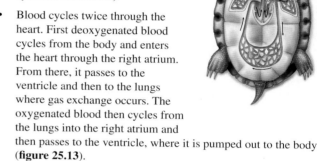

- Amphibians and reptiles have lungs and so the blood pumps in two cycles: between the heart and lungs (the pulmonary circulation) and between the heart and the body (the systemic circulation).

- Blood cycles twice through the heart. First deoxygenated blood cycles from the body and enters the heart through the right atrium. From there, it passes to the ventricle and then to the lungs where gas exchange occurs. The oxygenated blood then cycles from the lungs into the right atrium and then passes to the ventricle, where it is pumped out to the body (**figure 25.13**).

25.7 Mammalian and Bird Circulation

- Mammalian and bird hearts are two-cycle pumps—but, unlike amphibians and reptiles, the ventricle is separated so that the heart has four chambers (**figure 25.14**).

- Deoxygenated blood enters the right side of the heart through the right atrium and then to the right ventricle, where it is pumped to the lungs. Oxygenated blood from the lungs enters the left side of the heart through the left atrium and then to the left ventricle, where it is pumped throughout the body.

- A pacemaker, called the sinoatrial (SA) node, controls heartbeat rate. An electrical impulse that begins in the SA node triggers muscles in the atrium to contract. The electrical impulse passes through the atria and stimulates the AV node. An electrical impulse then travels from the AV node down to the apex of the heart, where it initiates another wave of muscle contraction of the ventricles. These ventricle contractions pump the blood out from the heart. The electrical impulses of the heart can be detected and are recorded as an electrocardiogram (ECG) (**figure 25.15**).

- The activity of the heart is also monitored by measuring blood pressure. Measurements of the systolic and diastolic pressures indicate how hard the heart is having to work to pump blood through the body (**figure 25.16**). Heart disease is often associated with high blood pressure.

25.8 Cardiovascular Diseases

- Cardiovascular diseases such as heart attacks, angina pectoris, strokes, atherosclerosis, and arteriosclerosis are primarily caused by fatty deposits in arteries that interfere with blood flow (**figure 25.17**).

1. Which of the following is *not* a function of the circulatory system?
   a. regulation of some body processes and characteristics
   b. protection against injury, foreign toxins, and microbes
   c. transportation of materials in the body
   d. All of the above are functions of the circulatory system.

2. How do some vertebrates maintain body temperature in cold environments?
   a. by mixing cold blood with warm blood in the heart
   b. by pumping more warm blood to the extremities
   c. by passing warm blood near cold blood in the extremities to warm the blood
   d. All of the above are used by vertebrates in cold environments.

3. Exchange of waste material, oxygen, carbon dioxide, and metabolites such as salts and food molecules, takes place in the
   a. capillaries.
   b. venules.
   c. arterioles.
   d. arteries.

4. The lymphatic system is like the circulatory system in that they both
   a. have nodes that filter out pathogens.
   b. are made up of arteries.
   c. deliver blood to the heart.
   d. carry fluids.

5. The most numerous blood cell is the
   a. macrophage.
   b. leukocyte.
   c. platelet.
   d. erythrocyte.

6. What advancement in fish led to a more efficient circulatory system?
   a. A muscular pump is first seen in fishes.
   b. A closed circulatory system is first seen in fishes.
   c. A heart with separate chambers is first seen in fishes.
   d. A double loop system is first seen in fishes.

7. Additional septa in the amphibian and reptile heart allows for
   a. higher blood pressure to move blood faster.
   b. better separation of oxygenated and deoxygenated blood.
   c. better body temperature regulation.
   d. better transport of food to needy tissues.

8. Which of the following statements is *false?*
   a. Only arteries carry oxygenated blood.
   b. Both arteries and veins have a layer of smooth muscle.
   c. Capillary beds lie between arteries and veins.
   d. Sphincters regulate the flow of blood through capillaries.

9. The four-chambered heart and double-loop vessel system is thought to be important in the evolution of
   a. locomotion.
   b. ectothermy.
   c. exothermy.
   d. endothermy.

10. Which of the following statements is *false?*
   a. Cardiovascular disease is the number one killer in the United States.
   b. Cardiovascular disease is limited to problems with the heart.
   c. Atherosclerosis leads to blocked arteries.
   d. Changes in lifestyle can reduce your risk for cardiovascular disease.

1. **Figure 25.7** Does being a dedicated "couch potato" or "video game addict" ever cause circulatory problems? Explain.

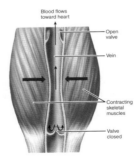

Blood flows toward heart
Open valve
Vein
Contracting skeletal muscles
Valve closed

2. **Figure 25.12** By comparing the fish's circulatory system to that of an amphibian, reptile, or mammal, explain how the fish's circulatory system could be more efficient without a fish having to develop lungs.

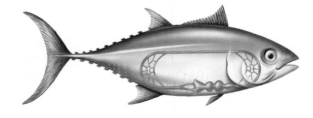

1. Your friend, James, has a painful paper cut on his finger and it's bleeding. Disgusted, because it already stained his shirt, he asks you why the blood is so important, anyway. What do you tell him?

2. Explain how the mnemonic "VAVA lung, VAVA body" describes the path of the blood in humans.

26

Respiration

All animals obtain the energy that fuels their lives by consuming other organisms, harvesting energy-rich electrons from the organic molecules of these creatures, and then using these electrons to drive the synthesis of ATP and other molecules. Afterward, the spent electrons are donated to oxygen gas (O_2) to form water (H_2O), while the carbon atoms left over after the electrons were stripped from them combine with oxygen to form carbon dioxide (CO_2). Capturing energy by animals is in effect a process that utilizes oxygen and produces carbon dioxide. The uptake of oxygen and the release of carbon dioxide together are called respiration, neatly defining one of the principle evolutionary challenges facing all animals—how to obtain oxygen and dispose of carbon dioxide. The evolution of respiratory mechanisms among the vertebrates has favored changes that maximize the exchange of these two gases. The most efficient respiratory mechanism to evolve in water is the gill, used by bony fishes and also by the shark you see swimming toward you (*upper left*). The earliest land vertebrates, amphibians like the frog appearing at the *lower left,* utilized simple lungs, and also respired through their moist skin. Reptiles like the crocodile invented expandable rib cages to draw air into the lungs, while mammals like the squirrel seen at the *upper right* greatly increased the interior surface area of the lung, making it a more powerful respiratory machine. Birds improved the respiratory design of the lung even more by rearranging its plumbing.

26.1   Types of Respiratory Systems

As mentioned previously, animals obtain the energy they need by oxidizing molecules rich in energy-laden carbon–hydrogen bonds. This oxidative metabolism requires a ready supply of oxygen. The uptake of oxygen and the simultaneous release of carbon dioxide together constitute a form of gas exchange called **respiration.**

Most of the primitive phyla of organisms obtain oxygen by direct diffusion from their aquatic environments, which contains about 10 milliliters of dissolved oxygen per liter. Sponges, cnidarians, many flatworms and roundworms, and some annelid worms all obtain their oxygen by diffusion from surrounding water. Oxygen and carbon dioxide diffuse across the surface of the body as shown in the flatworm in figure 26.1a. Similarly, some members of the vertebrate class Amphibia conduct gas exchange by direct diffusion through their moist skin.

The more advanced marine invertebrates (mollusks, arthropods, and echinoderms) possess special respiratory organs called gills that increase the surface area available for diffusion of oxygen. A **gill** is basically a thin sheet of tissue that waves through the water. Gills can be simple, as in the papulae of echinoderms, or complex, as in the highly convoluted gills of fish. In fish, the gills are protected by a covering called an operculum (removed in figure 26.1b). Because of this, their gills do not wave in the water, instead water is pumped over the gills and gas exchange occurs across the walls of capillaries contained in the gills. Terrestrial arthropods do not have a single major respiratory organ like a gill. Instead, a network of air ducts called **tracheae** (the purple tubes in figure 26.1c), branching into smaller and smaller tubes, carries air to every part of the body. The openings of tracheae to the outside are through special structures called **spiracles,** which can be closed and opened. Terrestrial vertebrates, except for some amphibians, do have a single respiratory organ, called the **lung,** with gas exchange occurring across the walls of capillaries and air sacs called *alveoli,* shown in the enlargement in figure 26.1d.

> **26.1   Aquatic animals extract oxygen dissolved in water, some by direct diffusion, others with gills. Terrestrial animals use tracheae or lungs.**

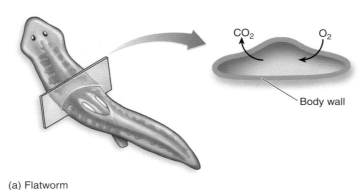

(a) Flatworm

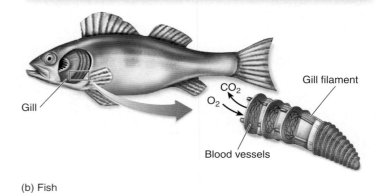

(b) Fish

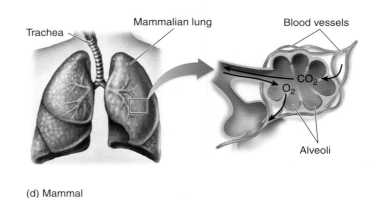

(c) Terrestrial arthropod

(d) Mammal

Figure 26.1   Gas exchange in animals.
(a) Gases diffuse directly across the body wall in many invertebrates, including flatworms, and in some species of amphibians. (b) Fish gills provide a very large respiratory surface and employ countercurrent flow, which will be discussed later. (c) Terrestrial arthropods respire through tracheae, which open to the outside through spiracles. (d) Mammalian lungs provide a large respiratory surface but do not permit countercurrent flow.

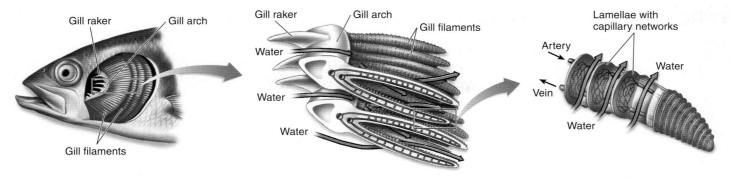

Figure 26.2 Structure of a fish gill.

Water passes from the gill arch over the filaments (from *left* to *right* in the diagram). Water always passes the lamellae in the same direction, which is opposite to the direction the blood circulates across the lamellae.

26.2 Respiration in Aquatic Vertebrates

Have you ever seen the face of a swimming fish up close? A fish swimming in water continuously opens and closes its mouth, pushing water through the mouth cavity and out a slit at the rear of the mouth—and (this is the whole point) past the gills on its one-way journey.

This swallowing process, which seems so awkward, is at the heart of a great advance in gill design achieved by the fishes. What is important about the swallowing is that it causes the water to always move past the fish's gills *in one and always the same direction.* Moving the water past the gills in the same direction permits **countercurrent flow,** which is a supremely efficient way of extracting oxygen. Here is how it works:

Each gill is composed of two rows of gill filaments (two sections of gills are shown in the middle panel in figure 26.2, each with two rows of gill filaments). The gill filaments are made of thin membranous plates stacked one on top of the other and projecting out into the flow of water. As water flows past the filaments from front to back (indicated by the blue arrows in the enlargements), oxygen diffuses from the water into blood circulating within the gill filament. Within each filament the blood circulation is arranged so that the blood is carried in the direction opposite the movement of the water, from the back of the filament to the front. The advantage of the countercurrent flow system is that blood in the blood vessels of the gill filaments always encounters water that has a higher oxygen concentration, resulting in the diffusion of oxygen into the blood vessels. To understand this, compare the countercurrent exchange system in figure 26.3*a* to a concurrent exchange system in figure 26.3*b*. In the countercurrent system, when blood and water flow in opposite directions, the initial oxygen concentration difference at the bottom is not large (10% in the blood and 15% in water), but it is sufficient for oxygen diffusion. As the blood oxygen concentration increases as it travels upward, the blood continually encounters water with a higher oxygen concentration. Even at 85% oxygen concentration in the blood, it is still encountering oxygen concentrations in water of 100%. In the concurrent exchange system, oxygen diffusion is rapid at first, because of the large difference in oxygen concentrations

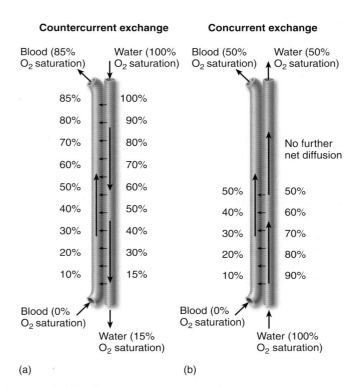

Figure 26.3 Countercurrent flow.

between the blood and water (0% versus 100%), but quickly slows as the difference becomes less until it reaches equilibrium at 50% saturation. Thus, countercurrent flow ensures that an oxygen concentration gradient remains between blood and water throughout the flow, permitting oxygen diffusion along the entire length of the filament.

Because of the countercurrent flow, the blood in the fish's gills can build up oxygen concentrations as high as those of the water entering the gills. The gills of bony fishes are the most efficient respiratory machines that have ever evolved among organisms.

> **26.2 Fish gills achieve countercurrent flow, making them very efficient at extracting oxygen.**

26.3 Respiration in Terrestrial Vertebrates

Amphibians Get Oxygen from Air with Lungs

One of the major challenges facing the first land vertebrates was obtaining oxygen from air. Fish gills, which are superb oxygen-gathering machines in water, don't work in air. The gill's system of delicate membranes has no means of support in air, and the membranes collapse on top of one another—that's why a fish dies when kept out of water, literally suffocating in air for lack of oxygen.

Unlike a fish, if you lift a frog out of water and place it on dry ground, it doesn't suffocate. Partly this is because the frog is able to respire through its moist skin, but mainly it is because the frog has lungs. A **lung** is a respiratory organ designed like a bag. The amphibian lung is hardly more than a sac with a convoluted internal membrane that opens up to a central cavity (the convoluted membrane is shown in figure 26.4a). The air moves into the sac through a tubular passage from the head and then back out again through the same passage. Lungs are not as efficient as gills because new air that is inhaled mixes with old air already in the lung. But, air contains about 210 milliliters of oxygen per liter, over 20 times as much as seawater. So, because there is so much more oxygen *in* air, the lung doesn't have to be as efficient as the gill.

Reptiles and Mammals Increase the Lung Surface

Reptiles are far more active than amphibians, so they need more oxygen. But reptiles cannot rely on their skin for respiration the way amphibians can; their dry scaly skin is "watertight" to avoid water loss. Instead, the lungs of reptiles contain a larger surface area. The internal membrane is also convoluted but the central cavity has many small air chambers, shown as partitions in figure 26.4b, which greatly increase the surface area of the lung available for diffusion of oxygen.

Because mammals maintain a constant body temperature by heating their bodies metabolically, they have even greater metabolic demands for oxygen than do reptiles. The problem of harvesting more oxygen is solved by increasing the diffusion surface area within the lung even more. The lungs of mammals possess on their inner surface many small chambers called **alveoli** that look like clusters of grapes in figure 26.4c. Each cluster is connected to the main air sac in the lung by a short passageway called a **bronchiole.** Air within the lung passes through the bronchioles to the alveoli, where all oxygen uptake and carbon dioxide disposal takes place. In more active mammals, the individual alveoli are smaller and more numerous, increasing the diffusion surface area even more. Humans have about 300 million alveoli in each of their lungs, for a total surface area devoted to diffusion of about 80 square meters (about 42 times the surface area of the body)!

Birds Perfect the Lung

There is a limit to how much efficiency can be improved by increasing the surface area of the lung, a limit that has already been reached by the more active mammals. This efficiency is not enough for the metabolic needs of birds. Flying creates a respiratory demand for oxygen that exceeds the capacity of the saclike lungs of even the most active mammal. Unlike bats, whose flight involves considerable gliding, most birds beat their wings rapidly as they fly, often for quite a long time. This intensive wing beating uses up a lot of energy quickly, because the wing muscles must contract very frequently. Flying birds thus must carry out very active oxidative respiration within their cells to replenish the ATP expended by their flight muscles, and this requires a great deal of oxygen.

A novel way to improve the efficiency of the lung, one that does not involve further increases in its surface area, evolved in birds' lungs. This higher-efficiency lung copes with

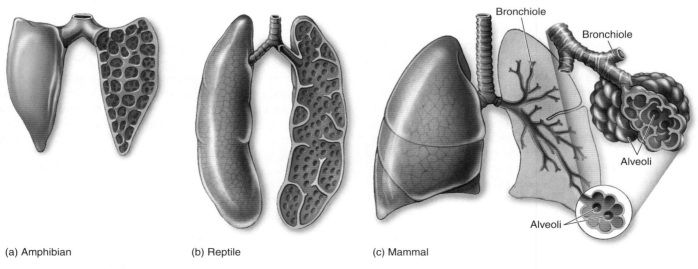

(a) Amphibian   (b) Reptile   (c) Mammal

Figure 26.4 Evolution of the vertebrate lung.

Gas Exchange During Respiration

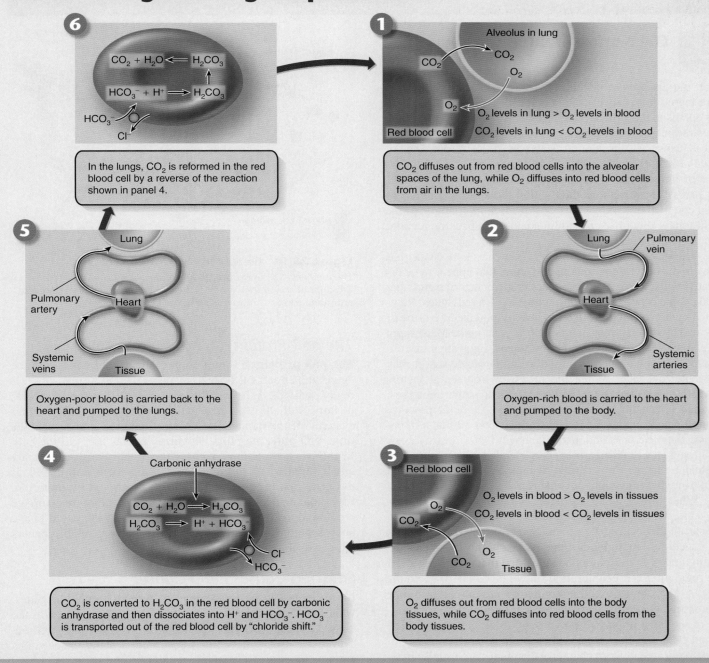

6 In the lungs, CO_2 is reformed in the red blood cell by a reverse of the reaction shown in panel 4.

$$CO_2 + H_2O \longleftarrow H_2CO_3$$
$$HCO_3^- + H^+ \longrightarrow H_2CO_3$$
HCO_3^- Cl^-

1 Alveolus in lung
Red blood cell
O_2 levels in lung > O_2 levels in blood
CO_2 levels in lung < CO_2 levels in blood

CO_2 diffuses out from red blood cells into the alveolar spaces of the lung, while O_2 diffuses into red blood cells from air in the lungs.

5 Lung, Pulmonary artery, Heart, Systemic veins, Tissue

Oxygen-poor blood is carried back to the heart and pumped to the lungs.

2 Lung, Pulmonary vein, Heart, Systemic arteries, Tissue

Oxygen-rich blood is carried to the heart and pumped to the body.

4 Carbonic anhydrase
$$CO_2 + H_2O \longrightarrow H_2CO_3$$
$$H_2CO_3 \longrightarrow H^+ + HCO_3^-$$
Cl^- HCO_3^-

CO_2 is converted to H_2CO_3 in the red blood cell by carbonic anhydrase and then dissociates into H^+ and HCO_3^-. HCO_3^- is transported out of the red blood cell by "chloride shift."

3 Red blood cell
O_2 levels in blood > O_2 levels in tissues
CO_2 levels in blood < CO_2 levels in tissues
CO_2 O_2 Tissue

O_2 diffuses out from red blood cells into the body tissues, while CO_2 diffuses into red blood cells from the body tissues.

Figure 26.9   How respiratory gas exchange works.

blood flow and blood pressure are regulated by the amount of NO released into the bloodstream.

Hemoglobin carries NO in a special form called super nitric oxide. In this form, NO has acquired an extra electron and is able to bind to an amino acid, called cysteine, present in hemoglobin. In the lungs, hemoglobin that is dumping CO_2 and picking up O_2 also picks up NO as super nitric oxide. In tissues, hemoglobin that is releasing its O_2 and picking up CO_2 may also release super nitric oxide as NO into the blood, making blood vessels expand, thereby increasing blood flow to the tissue. Alternatively, hemoglobin may trap any excesses of NO on its iron atoms left va-

cant by the release of oxygen, causing blood vessels to constrict, which reduces blood flow to the tissue. When the red blood cells return to the lungs, hemoglobin dumps its CO_2 and the NO bound to the iron atoms. It is then ready to pick up O_2 and super nitric oxide and continue the cycle.

26.5 Oxygen and NO move through the circulatory system carried by the protein hemoglobin within red blood cells. Most CO_2 is transported in the plasma as bicarbonate.

26.6 The Nature of Lung Cancer

Of all the diseases to which humans are susceptible, none is more feared than cancer (see chapter 9). Nearly one in every four deaths in the United States was caused by cancer in 2004. The American Cancer Society estimates that 563,700 people died of cancer in the United States in 2004. About 28% of these—160,440 people—died of **lung cancer.** About 140,000 cases of lung cancer were diagnosed each year in the 1980s, and 90% of these persons died within three years. Lung cancer is one of the leading causes of death among adults in the world today. What has caused lung cancer to become a major killer of Americans?

The search for a cause of cancers such as lung cancer has uncovered a host of environmental factors that appear to be associated with cancer. For example, the incidence of cancer per 1,000 people is not uniform throughout the United States. Rather, it is centered in cities, like the heavily populated Northeast indicated by the red areas in figure 26.10, and in the Mississippi Delta, indicated by the red and brown areas, suggesting that environmental factors such as pollution and pesticide runoff may contribute to cancer. When the many environmental factors associated with cancer are analyzed, a clear pattern emerges: most cancer-causing agents, or carcinogens, share the property of being potent mutagens. Recall from chapter 12 that a mutagen is a chemical or radiation that damages DNA, destroying or changing genes (a change in a gene is called a mutation). The conclusion that cancer is caused by mutation is now supported by an overwhelming body of evidence.

What sort of genes are being mutated? In the last several years, researchers have found that mutation of only a few genes is all that is needed to transform normally dividing cells into cancerous ones. Identifying and isolating these cancer-causing genes, investigators have learned that all are involved with regulating cell proliferation (how fast cells grow and divide). A key element in this regulation are so-called tumor suppressors, genes that actively prevent tumors from forming. Two of the most important tumor-suppressor genes are called *Rb* and *p53,* and the proteins they produce are Rb and p53, respectively (recall from chapter 9 that genes are usually indicated in italics and proteins in regular type).

The Rb Protein

The **Rb protein** (named after retinoblastoma, the rare eye cancer in which it was first discovered) acts as a brake on cell division, attaching itself to the machinery the cell uses to replicate its DNA, and preventing it from doing so. When the cell wants to divide, a growth factor molecule ties up Rb so that it is not available to act as a brake on the division process. If the gene that produces Rb is disabled, there are no brakes to prevent the cell from replicating its DNA and dividing. The control switch is locked in the "ON" position.

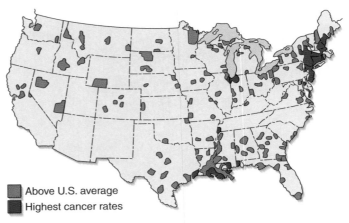

Above U.S. average

Highest cancer rates

Figure 26.10 Cancer in the United States.
The incidence of cancer per 1,000 people is not uniform throughout the United States. It is centered in cities where chemical manufacturing is common and in the Mississippi Delta.

The p53 Protein

The **p53 protein,** a tumor suppressor sometimes called the "guardian angel" of the cell, inspects the DNA to ensure it is ready to divide. When p53 detects damaged or foreign DNA, it stops cell division and activates the cell's DNA repair systems. If the damage doesn't get repaired in a reasonable time, p53 pulls the plug, triggering events that kill the cell. In this way, mutations such as those that cause cancer are either repaired or the cells containing them eliminated. If the gene that produces p53 is itself destroyed by mutation, future damage accumulates unrepaired. Among this damage are mutations that lead to cancer, mutations that would have been repaired by healthy p53. Fifty percent of all cancers have a disabled *p53* gene.

Smoking Causes Lung Cancer

If cancer is caused by damage to growth-regulating genes, what then has led to the rapid increase in lung cancer in the United States? Two lines of evidence are particularly telling. The first consists of detailed information about cancer rates among smokers. The annual incidence of lung cancer among nonsmokers is only a few per 100,000 but increases with the number of cigarettes smoked per day to a staggering 300 per 100,000 for those smoking 30 cigarettes a day.

A second line of evidence consists of changes in the incidence of lung cancer that mirror changes in smoking habits. Look carefully at the data presented in figure 26.11. The upper graph is compiled from data on men and shows the incidence of smoking (blue line) and of lung cancer (red line) in the United States since 1900. As late as 1920, lung cancer was a rare disease. About 30 years after the incidence of smoking began to increase among men, lung cancer also started to become more common. Now look at the lower graph, which presents data on women. Because of social mores, significant numbers of women in the United States did not smoke until after World War II (see blue line), when many social conventions changed. As late as 1963, only 6,588 women had died of lung cancer. But as women's frequency of smoking has increased, so has their incidence of lung cancer (red line), again with a lag of about 30 years. Women today have achieved equality with men in the number of cigarettes they smoke, and their lung cancer death rates are now rapidly approaching those for men. In 2004, an estimated 68,500 American women died of lung cancer.

How does smoking cause cancer? Cigarette smoke contains many powerful mutagens, among them benzo[a]pyrene, and smoking introduces these mutagens to the lung tissues. Benzo[a]pyrene, for example, binds to three sites on the p53 gene and causes mutations at these sites that inactivate the gene. In 1997, scientists studying this tumor-suppressor gene demonstrated a direct link between cigarettes and lung cancer. They found that the p53 gene is inactivated in 70% of all lung cancers. When these inactivated p53 genes are examined, they prove to have mutations at just the three sites where benzo[a]pyrene binds! Clearly, the chemical in cigarette smoke is responsible for the lung cancer.

In the face of these facts, why do so many people continue to smoke? Because the nicotine in cigarette smoke is an addictive drug. Researchers have identified the receptor on central nervous system neurons that it binds to, and they have demonstrated that smoking leads to a decrease in the brain's population of these receptors, leading to a craving for more cigarettes. This mechanism of nicotine addiction is very similar to that of cocaine addiction; once a person becomes addicted, it is difficult to quit. About half of those who try eventually succeed. Because your life is at stake, it is well worth the effort.

Clearly, an effective way to avoid lung cancer is not to smoke. Life insurance companies have computed that, on a statistical basis, smoking a single cigarette lowers your life expectancy 10.7 minutes (more than the time it takes to smoke the cigarette!). Every pack of 20 cigarettes bears an unwritten label: *The price of smoking this pack of cigarettes is 3 1/2 hours of your life.* Smoking a cigarette is very much like going into a totally dark room with a person who has a gun and standing still. The person with the gun cannot see you, does not know where you are, and shoots once in a random direction. A hit is unlikely, and most shots miss. As the person keeps shooting, however, the chance of eventually scoring a hit becomes more likely. Every time an individual smokes a cigarette, mutagens are being shot at his or her genes. Nor do statistics protect any one individual: nothing says the first shot will not hit. Older people are not the only ones who die of lung cancer.

> **26.6** Cancer results from the destruction of genes by mutations that, when healthy, enable the cell to regulate cell division. To avoid lung cancer, don't smoke.

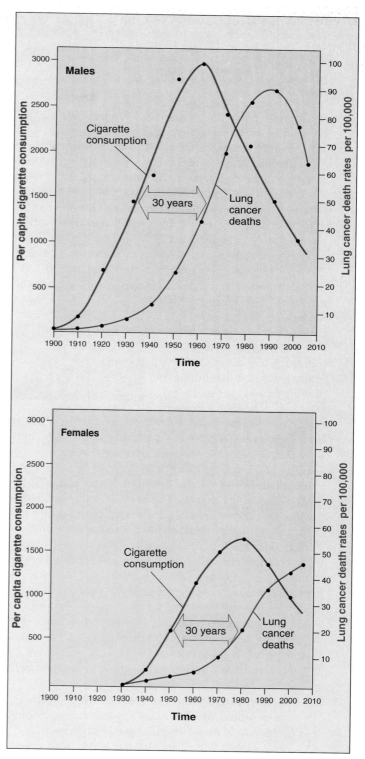

Figure 26.11 Incidence of lung cancer in men and women.

Lung cancer was a rare disease a century ago. As men in the United States increased smoking in the early 1900s, the incidence of lung cancer also increased. Women followed suit years later, and in 2004, more than 68,500 women died of lung cancer, a rate of 46 per 100,000. In that year, about 20% of women smoked and about 25% of men.

How Do Llamas Live So High Up?

Because of mixing, the air animals breathe is 21% oxygen everywhere, even way up into the sky 100 km above earth's surface. However, the amount of air (the number of molecules in a unit volume) decreases sharply with altitude, as shown in the upper graph. Air pressure at 5,000 meters is half that at sea level. This lack of air presents a serious problem to humans, as mountain climbers know. The amount of oxygen in the air (measured as oxygen partial pressure) is lower, so there is simply too little oxygen to fuel a climber's muscles. To combat this problem, high-altitude climbers typically spend months acclimating to high altitude, a period in which their bodies greatly increase the amount of hemoglobin in their red blood cells and so increase the amount of oxygen the red blood cells can capture. Many mammals live their entire lives at high altitudes. The llama and the vicuna (pictured here) both live in the high Andes of South America, often above 5,000 meters. Do they stuff extra hemoglobin into their red blood cells too, or are they able to solve the problem of low oxygen in another way?

The graph on the lower right displays three "oxygen loading curves" that reveal the effectiveness with which hemoglobin binds oxygen. The more effective the binding, the less oxygen required before hemoglobin becomes fully loaded. In the graph, the percent hemoglobin saturation (that is, how much of the hemoglobin is bound to oxygen) is presented on the y axis, and the oxygen partial pressure (a measure of the amount of oxygen available to the hemoglobin molecules) is presented on the x axis. Oxygen-loading curves are presented for three mammalian species: humans living at sea level, and llamas and vicunas, each living in the Andes above 5,000 meters.

1. **Applying Concepts**
   a. Variable. In the graph on the lower right, what is the dependent variable(s)?
   b. Comparing Curves. Extrapolating on the lower graph, which species possesses hemoglobin able to load oxygen well at sea level partial pressures (160 mm Hg)? Which of the three species possesses hemoglobin better able to load oxygen on Mount Everest?

2. **Interpreting Data**
   a. The partial pressure of oxygen in human muscle tissue at sea level is about 40 mm Hg. What is the percent hemoglobin bound to O_2 for each of the three species at this partial pressure? What percent of the human hemoglobin has released its oxygen? of the llama? of the vicuna?

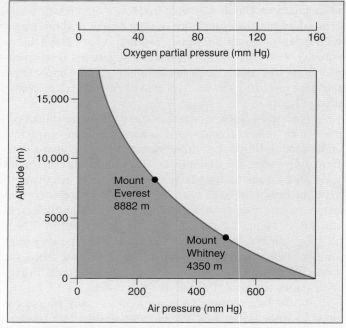

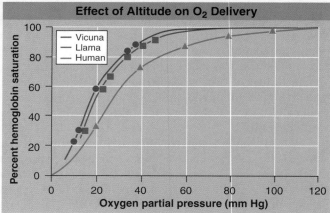

b. Are there any significant differences in the hemoglobin saturation values for the two high-altitude species?

3. **Making Inferences** At an elevation of 5,000 meters, the partial pressure of oxygen is 80 mm Hg (half of what it is at sea level). At this elevation, how much of human hemoglobin has succeeded in binding oxygen? How much of llama hemoglobin? of vicuna?

4. **Drawing Conclusions** What is the effect of shifting the oxygen loading curve to the left? What general statement can be made regarding the affinity of hemoglobin for oxygen in the three species?

5. **Further Analysis** What saturation values would you expect in llamas raised from birth in the National Zoo at Washington, D.C.? Why would you expect this? How might you test your prediction?

Respiration

26.1 Types of Respiratory Systems

- Some aquatic animals extract oxygen across the skin and others use gills. Terrestrial animals extract oxygen from the air with tracheae or lungs (**figure 26.1**).

26.2 Respiration in Aquatic Vertebrates

- Fishes' gills (**figure 26.2**) are very efficient by using a countercurrent flow system. Oxygenated water flows through the gills in a direction opposite of blood flow through the gills such that the water always has a higher concentration of oxygen driving the diffusion of oxygen into the blood (**figure 26.3**).

26.3 Respiration in Terrestrial Vertebrates

- Terrestrial vertebrate lungs (**figure 26.4**) are less efficient than a fish's gills but work well in terrestrial environments. Lungs evolved to increase efficiency from amphibian lungs to reptilian lungs to mammalian lungs by increasing surface area. Mammals require more oxygen and have larger surface areas.

- Bird lungs use a crosscurrent flow system that is more efficient than other terrestrial animals' lungs. Air flows through the bird lung in one direction, after first entering posterior air sacs. Air passes through the lungs in a direction perpendicular to the flow of blood, a crosscurrent flow, which allows the lung to extract more oxygen than the uniform pool lung found in mammals, but not as efficiently as the countercurrent system found in fish (**figure 26.5**).

26.4 The Mammalian Respiratory System

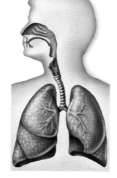

- Mammalian lungs are positioned within an internal cavity, called the thoracic cavity. A layer of fluid lies between the membrane lining the inner layer of the thoracic cavity and the lungs. This causes the lungs to stick to the walls of the thoracic cavity.

- The airways of the mammalian lung end in clusters of structures called alveoli, which are surrounded by capillaries. Gas exchange occurs across the single cell layers of the alveoli and capillaries (**figure 26.6**).

- Mammals breathe by contracting muscles that line the thoracic cavity that holds the lungs. Contraction of these muscles, including the diaphragm, expands the space in the lungs, causing air to rush into the lungs. Relaxation of the muscles causes air to be exhaled (**figure 26.7**).

26.5 How Respiration Works: Gas Exchange

- Hemoglobin carries oxygen from the lungs to the cells of the body. Oxygen binds to iron atoms contained within the heme groups of hemoglobin in the lungs where oxygen concentrations are high. The oxygen is then released to metabolizing cells at distant areas of the body where oxygen concentrations are lower (**figure 26.8**).

- Carbon dioxide is primarily carried as bicarbonate and hydrogen ions. Carbon dioxide enters the red blood cells and is converted into carbonic acid by the actions of carbonic anhydrase. Carbonic acid dissociates into bicarbonate and hydrogen ions. The hydrogen ions bind to hemoglobin, but the bicarbonate is transported out of the cell into the plasma with the counter transport of a chloride ion. This is called the chloride shift (**figure 26.9**).

- The conversion of carbon dioxide to bicarbonate increases the rate of diffusion of carbon dioxide from the cells of tissues into the blood.

- In the lungs, the carbonic anhydrase reaction is reversed, converting bicarbonate and hydrogen ions back to carbon dioxide where it passes into the lungs and is exhaled (**figure 26.9**).

Lung Cancer and Smoking

26.6 The Nature of Lung Cancer

- Cancer results from mutations of the DNA, often caused by chemicals in the environment (**figure 26.10**).

- Many cancers have been linked to mutations of two key tumor suppressor genes, *Rb* and *p53*. The protein products of these two genes control cell division, keeping the cell from dividing when it is not supposed to. If mutations damage either of these genes, the cells are able to divide uncontrollably.

- Lung cancer results from harmful mutations caused by chemicals in cigarette smoke. The incidence of cigarette smoking and lung cancer mirror each other with a separation of about 30 years, time for the accumulation of mutations that lead to cancer (**figure 26.11**).

1. In the gills of fish, countercurrent flow allows
   a. the animal's blood to be continually exposed to water of lower oxygen concentration.
   b. the animal's blood to be continually exposed to water of equal oxygen concentration.
   c. the animal's blood to be continually exposed to water of higher oxygen concentration.
   d. the animal to be endothermic.

2. The purpose of gill filaments and the lung alveoli is to
   a. decrease the surface area available for gas exchange.
   b. increase the surface area available for gas exchange.
   c. decrease the volume available for gas exchange.
   d. increase the volume available for gas exchange.

3. In general, the need for alveoli increases as the
   a. energy need for different vertebrate classes increases.
   b. energy need for different vertebrate classes decreases.
   c. habitat need for different vertebrate classes increases.
   d. nutrition need for different vertebrate classes increases.

4. Which group of animals is the most efficient at extracting oxygen?
   a. reptiles
   c. fish
   b. birds
   d. mammals

5. Which of the following statements about the bird's respiratory system is *false*?
   a. Birds breathe using a crosscurrent flow of air and blood.
   b. Birds have air sacs in addition to a lung.
   c. It takes three cycles of breathing for air to pass through the bird's respiratory system.
   d. The bird's respiratory system has two locations where air is held during inhalation.

6. When you take a deep breath, your stomach moves out because
   a. swallowing air increases the volume of the thoracic cavity.
   b. your stomach shouldn't move out when you take a deep breath because you want the volume of your chest cavity to increase, not your abdominal cavity.
   c. contracting your abdominal muscles pushes your stomach out, generating negative pressure in your lungs.
   d. when your diaphragm contracts, it moves down, pressing your abdominal cavity out.

7. Oxygen is transported by
   a. hemoglobin in red blood cells.
   b. dissolving it in the blood plasma.
   c. proteins in the blood plasma.
   d. platelets in the blood plasma.

8. Most of the carbon dioxide is transported by
   a. hemoglobin in red blood cells.
   b. the blood plasma as biocarbonate.
   c. proteins in the blood plasma.
   d. dissolving it in red blood cells.

9. Which of the following is not carried by hemoglobin?
   a. bicarbonate
   b. hydrogen ions
   c. oxygen
   d. super nitric oxide

10. Which of the following can lead to cancer?
    a. smoking
    b. pollution
    c. mutations of *Rb* and *p53*
    d. All of the above.

1. **Figure 26.9** Explain what would happen to carbon dioxide transport if a person was poisoned with a chemical that blocked the actions of carbonic anhydrase?

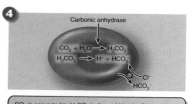

CO₂ is converted to H₂CO₃ in the red blood cell by carbonic anhydrase and then dissociates into H⁺ and HCO₃⁻. HCO₃⁻ is transported out of the red blood cell by "chloride shift."

2. **Figure 26.11** What conclusions can you draw about cigarette smoking and women's health? Explain.

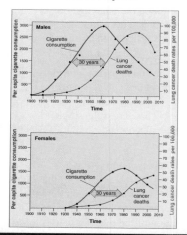

1. Sometimes when people are eating, they take a bite that is too big, or is not completely chewed, and when they swallow it becomes stuck partway down the esophagus near the epiglottis (a flap that covers the trachea when swallowing). When food is stuck in this location, a person usually can't breathe. In this case, people have been trained to do the Heimlich maneuver, which is a method of pushing up rapidly on the diaphragm, compressing the lungs. Why might this help?

2. How is it that cigarette smoking can be linked to an increased incidence of many kinds of cancer?

27

The Path of Food Through the Animal Body

There are no photosynthetic animals. Animals are heterotrophs, gaining the energy to power their lives by consuming and oxidizing organic molecules present in other organisms. All animals must continuously consume plant material or other animals in order to live. The grass in this prairie dog's mouth will be consumed and converted within its cells to body tissue, energy, and refuse. Most of the molecules in the grass are far too large to be conveniently absorbed by the prairie dog's cells, and so they are first broken down into smaller pieces: carbohydrates are broken down into simple sugars, proteins into amino acids, fats into fatty acids. This process, called digestion, is the focus of this chapter. The prairie dog's digestive system is a long tube passing from mouth to anus, with specialized segments for digestion and for the subsequent absorption into its body of the resulting sugars, amino acids, and fatty acids. Whatever is left is excreted from the body as feces. In this chapter you will follow the path of food as it moves through the vertebrate body. As you will see, it is a surprisingly interesting journey.

27.1 Food for Energy and Growth

The food animals eat provides both a source of energy and essential molecules such as certain amino acids and fats that the animal body is not able to manufacture for itself. An optimal diet contains more carbohydrates than fats and also a significant amount of protein, as recommended by the federal government's "pyramid of nutrition" in figure 27.1. The pyramid is intended as a general guideline of what a person should eat. A healthy diet should include more of the foods indicated by the larger sections—for example, the orange section indicates grains and cereals with an emphasis on whole grains. Fats (the yellow section) are recommended in much smaller amounts because they have a far greater number of energy-rich carbon–hydrogen bonds and thus a much higher energy content per gram than carbohydrates or proteins, which contain more carbon–oxygen bonds that are already oxidized. For this reason, fats are a very efficient way to store energy. When food is consumed, it is either metabolized by muscles and other cells of the body, or it is converted into fat and stored in fat cells.

Carbohydrates are obtained primarily from cereals, grains, and breads (the orange section), fruits (the red section), and vegetables (the green section). On the average, carbohydrates contain 4.1 calories per gram; fats, by comparison, contain 9.3 calories per gram, over twice as much. Dietary fats are obtained from oils, margarine, and butter and are abundant in fried foods, meats, and processed snack foods, such as po-

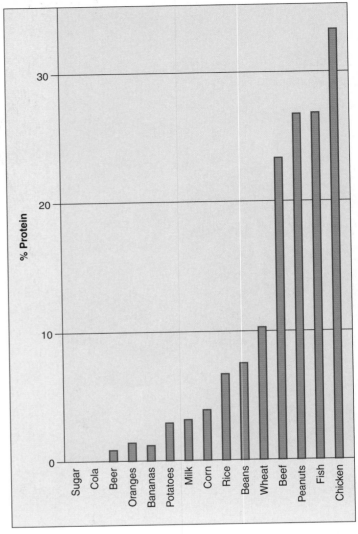

Figure 27.2   The protein content of a variety of common foods.

Humans and many other vertebrates must eat protein to obtain the amino acids they are unable to synthesize.

tato chips and crackers. Like carbohydrates, proteins (in the aqua and purple sections) have 4.1 calories per gram and can be obtained from many foods, as figure 27.2 indicates, including dairy products, poultry, fish, meat, and grains.

The body uses carbohydrates for energy and fats to construct cell membranes and other cell structures, to insulate nervous tissue, and to provide energy. Fat-soluble vitamins that are essential for proper health are also absorbed with fats. Proteins are used for energy and as building materials for cell structures, enzymes, hemoglobin, hormones, and muscle and bone tissue.

In wealthy countries, such as those of North America and Europe, being significantly overweight is common, the result of habitual overeating and high-fat diets, in which fats constitute over 35% of the total caloric intake. The international standard

Figure 27.1   The pyramid of nutrition.

The width of each section indicates how much you should consume of that food group. The orange section is grains, green is vegetables, red is fruits, yellow is fats and sweets, aqua is dairy, and purple is meats and beans. However, one size does not fit all; go to www.mypyramid.gov to customize the food pyramid that is right for you. *Source: U.S. Department of Agriculture, U.S. Department of Health and Human Services.*

| | 100 | 105 | 110 | 115 | 120 | 125 | 130 | 135 | 140 | 145 | 150 | 155 | 160 | 165 | 170 | 175 | 180 | 185 | 190 | 195 | 200 | 205 |

25 OVERWEIGHT LIMIT   OVERWEIGHT

WEIGHT	100	105	110	115	120	125	130	135	140	145	150	155	160	165	170	175	180	185	190	195	200	205
HEIGHT																						
5' 0"	20	21	21	22	23	24	25	26	27	28	29	30	31	32	33	34	35	36	37	38	39	40
5' 1"	19	20	21	22	23	24	25	26	26	27	28	29	30	31	32	33	34	35	36	37	38	39
5' 2"	18	19	20	21	22	23	24	25	26	27	27	28	29	30	31	32	33	34	35	36	37	37
5' 3"	18	19	19	20	21	22	23	24	25	26	27	27	28	29	30	31	32	33	34	35	35	36
5' 4"	17	18	19	20	21	21	22	23	24	25	26	27	27	28	29	30	31	32	33	33	34	35
5' 5"	17	17	18	19	20	21	22	22	23	24	25	26	27	27	28	29	30	31	32	32	33	34
5' 6"	16	17	18	19	19	20	21	22	23	23	24	25	26	27	27	28	29	30	31	31	32	33
5' 7"	16	16	17	18	19	20	20	21	22	23	23	24	25	26	27	27	28	29	30	31	31	32
5' 8"	15	16	17	17	18	19	20	21	21	22	23	23	24	25	26	27	27	28	29	30	30	31
5' 9"	15	16	16	17	18	18	19	20	21	21	22	23	24	24	25	26	27	27	28	29	30	30
5' 10"	14	15	16	17	17	18	19	19	20	21	22	22	23	24	24	25	26	27	27	28	29	29
5' 11"	14	15	15	16	17	17	18	19	20	20	21	22	22	23	24	24	25	26	26	27	28	29
6' 0"	14	14	15	16	16	17	18	18	19	20	20	21	22	22	23	24	24	25	26	26	27	28
6' 1"	13	14	15	15	16	16	17	18	18	19	20	20	21	22	22	23	24	24	25	26	26	27
6' 2"	13	13	14	15	16	16	17	17	18	19	19	20	21	21	22	22	23	24	24	25	26	26
6' 3"	12	13	14	14	15	16	16	17	17	18	19	19	20	21	21	22	22	23	24	24	25	26
6' 4"	12	13	13	14	15	16	16	16	17	18	18	19	19	20	21	21	22	23	23	24	24	25

Figure 27.3   Are you overweight?

This chart presents the body mass index (BMI) values used by federal health authorities to determine who is overweight. Your body mass index is at the intersection of your height and weight.

Source: "Shape Up America" National Institutes of Health.

measure of appropriate body weight is the body mass index (BMI), estimated as your body weight in kilograms, divided by your height in meters squared. A BMI chart is presented in figure 27.3. To determine your BMI, find your height in the left hand column (in feet and inches) and trace it across to the column with your weight in pounds. A BMI value of 25 (dark blue boxes) and above is considered overweight and 30 or over is considered obese. In the United States, the National Institutes of Health estimated in 2002 that 64.5% of adults, 134.8 million Americans, were overweight, with a body mass index of 25 or more. Of those individuals, 63.1 million were considered obese with a body mass index of 30 or greater. Being overweight is highly correlated with coronary heart disease, diabetes, and many other disorders. A BMI of less than 18.5 is also unhealthy, often resulting from eating disorders including anorexia nervosa.

One essential characteristic of food is its fiber content. Fiber is the part of plant food that cannot be digested by humans and is found in fruits, vegetables, and grains such as in breads and cereals. Other animals, however, have evolved many different ways to process food that has a relatively high fiber content. Diets that are low in fiber, now common in the United States, result in a slower passage of food through the colon. This low dietary fiber content is thought to be associated with incidences of colon cancer in the United States, which are among the highest levels in the world.

Essential Substances for Growth

Over the course of their evolution, many animals have lost the ability to manufacture certain substances they need, substances that often play critical roles in their metabolism. Mosquitoes and many other blood-sucking insects, for example, cannot manufacture cholesterol, but they obtain it in their diet because human blood is rich in cholesterol. Many vertebrates are unable to manufacture one or more of the 20 amino acids used to make proteins. Humans are unable to synthesize eight amino acids: lysine, tryptophan, threonine, methionine, phenylalanine, leucine, isoleucine, and valine. These amino acids, called **essential amino acids,** must therefore be obtained from proteins in the food we eat. For this reason, it is important to eat so-called complete proteins—that is, ones containing all the essential amino acids. In addition, all vertebrates have also lost the ability to synthesize certain polyunsaturated fats that provide backbones for the many kinds of fats their bodies manufacture.

Trace Elements. In addition to supplying energy, food that is consumed must also supply the body with essential minerals such as calcium and phosphorus, as well as a wide variety of **trace elements,** which are minerals required in very small amounts. Among the trace elements are iodine (a component of thyroid hormone), cobalt (a component of vitamin B_{12}), zinc and molybdenum (components of enzymes), manganese, and selenium. All of these, with the possible exception of selenium, are also essential for plant growth; animals obtain them directly from plants that they eat or indirectly from animals that have eaten plants, or plant eaters.

Vitamins. Essential organic substances that are used in trace amounts are called **vitamins.** Humans require at least 13 different vitamins listed in table 27.1. Many vitamins are required cofactors for cellular enzymes. Humans, monkeys, and guinea pigs, for example, have lost the ability to synthesize ascorbic acid (vitamin C) and will develop the potentially fatal disease called scurvy—characterized by weakness, spongy gums, and bleeding of the skin and mucous membranes—if vitamin C is not supplied in their diets. All other mammals are able to synthesize ascorbic acid.

> **27.1** Food is an essential source of calories. It is important to maintain a proper balance of carbohydrate, protein, and fat. Individuals with a body mass index of 25 or more are considered overweight. Food also provides key amino acids that the body cannot manufacture for itself, as well as necessary trace elements and vitamins.

TABLE 27.1	MAJOR VITAMINS			
Vitamin	**Function**	**Dietary Source**	**Recommended Daily Allowance (milligrams)**	**Deficiency Symptoms**
Vitamin A	Used in making visual pigments, maintenance of epithelial tissues	Green vegetables, carrots, milk products, liver	1	Night blindness, flaky skin
B-complex vitamins				
B₁	Coenzyme in CO_2 removal during cellular respiration	Meat, grains, legumes	1.5	Beriberi, weakening of heart, edema
B₂ (riboflavin)	Part of coenzymes FAD and FMN, which play metabolic roles	In many different kinds of foods	1.8	Inflammation and breakdown of skin, eye irritation
B₃ (niacin)	Part of coenzymes NAD⁺ and NADP⁺	Liver, lean meats, grains	20	Pellagra, inflammation of nerves, mental disorders
B₅ (pantothenic acid)	Part of coenzyme A, a key connection between carbohydrate and fat metabolism	In many different kinds of foods	5 to 10	Rare: fatigue, loss of coordination
B₆ (pyridoxine)	Coenzyme in many phases of amino acid metabolism	Cereals, vegetables, meats	2	Anemia, convulsions, irritability
B₁₂ (cyanocobalamin)	Coenzyme in the production of nucleic acids	Red meat, dairy products	0.003	Pernicious anemia
Biotin	Coenzyme in fat synthesis and amino acid metabolism	Meat, vegetables	Minute	Rare: depression, nausea
Folic acid	Coenzyme in amino acid and nucleic acid metabolism	Green leafy vegetables, whole-grain products	0.4	Anemia, diarrhea (also spina bifida, if deficient in mother)
Vitamin C	Important in forming collagen, cement of bone, teeth, connective tissue of blood vessels; may help maintain resistance to infection	Fruit, green leafy vegetables	45	Scurvy; breakdown of skin, blood vessels
Vitamin D (calciferol)	Increases absorption of calcium and promotes bone formation	Dairy products, cod liver oil	0.01	Rickets, bone deformities
Vitamin E (tocopherol)	Protects fatty acids and cell membranes from oxidation	Margarine, seeds, green leafy vegetables	15	Rare
Vitamin K	Essential to blood clotting	Green leafy vegetables	0.03	Severe bleeding

Today's *Biology*

Closing in on the Long-Sought Link Between Diabetes and Obesity

We Americans love to eat, but in 2004 the Centers for Disease Control and Prevention released a report warning we are eating ourselves into a diabetes epidemic. Diabetes affected 7 million Americans in 1991. At the end of 2002, the number was 18.2 million, over 6% of all Americans. This represents an increase of 62% in just over 10 years! Over that same period, the obesity rate increased from 12% of the population to 30%.

Diabetes is a disorder in which the body's cells fail to take up glucose from the blood. Tissues waste away as glucose-starved cells are forced to consume their own proteins. Diabetes is the leading cause of kidney failure, blindness, and amputation in adults. Almost all the increase in diabetes in the last decade is in the 85% of diabetics who suffer from type II, or "adult-onset," diabetes. These individuals lack the ability to use the hormone insulin.

Your body manufactures insulin after a meal as a way to alert cells that higher levels of glucose are coming soon. The insulin signal attaches to special receptors on the cell surfaces, which respond by causing the cell to turn on its glucose-transporting machinery.

Individuals who suffer from type II diabetes have normal or even elevated levels of insulin in their blood, and normal insulin receptors, but for some reason the binding of insulin to their cell receptors does not turn on the glucose-transporting machinery like it is supposed to do. For 30 years researchers have been trying to figure out why not.

How does insulin act to turn on a normal cell's glucose transporting machinery? Proteins called IRS proteins (the names refer not to tax collectors, but to *insulin receptor substrate*) snuggle up against the insulin receptor inside the cell. When insulin attaches to the receptor protein, the receptor responds by adding a phosphate group onto the IRS molecules. Like being touched by a red-hot poker, this galvanizes the IRS molecules into action. Dashing about, they activate a variety of processes, including an enzyme that turns on the glucose-transporting machinery.

When the IRS genes are deliberately taken out of action in so-called "knockout" mice, type II diabetes results. Are defects in the genes for IRS proteins responsible for type II diabetes? Probably not. When researchers look for IRS gene mutations in inherited type II diabetes, they don't find them. The IRS genes are normal.

This suggests that in type II diabetes something is interfering with the action of the IRS proteins. What might it be? An estimated 80% of those who develop type II diabetes are obese, a tantalizing clue.

What is the link between diabetes and obesity? Recent research suggests an answer to this key question. A team

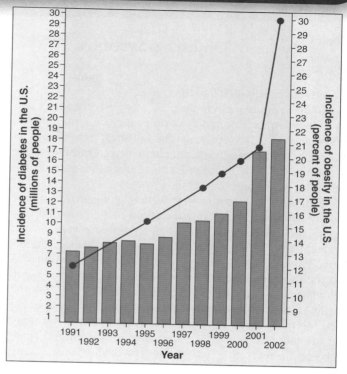

of scientists at the University of Pennsylvania School of Medicine had been investigating why a class of drugs called thiazolidinediones (TZDs) helped combat diabetes. They found that TZDs cause the body's cells to use insulin more effectively, and this suggested to them that the TZD drug might be targeting a hormone.

The researchers then set out to see if they could find such a hormone in mice. In search of a clue, they started by looking to see which mouse genes were activated or deactivated by TZD. Several were. Examining them, they were able to zero in on the hormone they sought. Dubbed *resistin,* the hormone is produced by fat cells and prompts tissues to resist insulin. The same resistin gene is present in humans too. The researchers speculate that resistin may have evolved to help the body deal with periods of famine.

Mice given resistin by the researchers lost much of their ability to take up blood sugar. When given a drug that lowers resistin levels, these mice recovered the lost glucose-transporting ability.

Researchers don't yet know how resistin acts to lower insulin sensitivity, although blocking the action of IRS proteins seems a likely possibility.

Importantly, dramatically high levels of the hormone were found in mice obese from overeating. Finding this sort of result is like ringing a dinner bell to diabetes researchers. If obesity is causing high resistin levels in humans, leading to type II diabetes, then resistin-lowering drugs might offer a diabetes cure!

On the scent of something important, resistin researchers are now shifting their efforts from mice to humans. Much needs to be checked, as there are no guarantees that what works in a mouse will do so in the same way in a human. Still, the excitement is tangible.

Types of Digestive Systems

Heterotrophs are divided into three groups on the basis of their food sources. Animals that eat plants exclusively are classified as **herbivores;** common examples include cows, horses, rabbits, and sparrows. Animals that are meat eaters, such as cats, eagles, trout, and frogs, are **carnivores. Omnivores** are animals that eat both plants and other animals. We humans are omnivores, as are pigs, bears, and crows.

Single-celled organisms (as well as sponges) digest their food intracellularly, breaking down food particles with digestive enzymes inside their cells. Other animals digest their food extracellularly, within a digestive cavity. In this case, the digestive enzymes are released into a cavity that is continuous with the animal's external environment. In flatworms (such as *Planaria*) and cnidarians, like the hydra in figure 27.4, the digestive cavity in the center of the body has only one opening at the top that serves as both mouth (the red arrow bringing food in) and anus (the blue arrow passing waste out). There can be no specialization within this type of digestive system, called a *gastrovascular cavity,* because every cell is exposed to all stages of food digestion.

Specialization occurs when the digestive tract, or alimentary canal, has a separate mouth and anus, so that transport of food is one way. Three examples are shown in figure

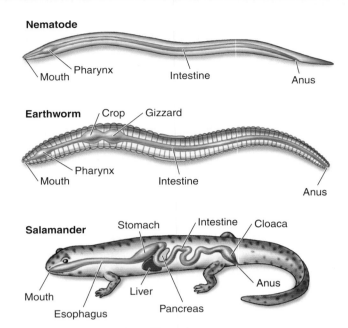

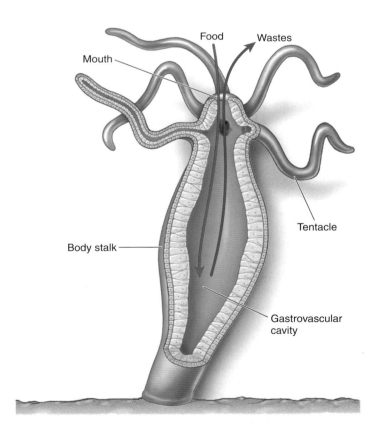

Figure 27.4 The gastrovascular cavity of *Hydra*.

Figure 27.5 One-way digestive tracts.

One-way movement through the digestive tract allows different regions of the digestive system to become specialized for different functions.

27.5. The most primitive digestive tract is seen in nematodes (phylum Nematoda), where it is simply a tubular *gut* lined by an epithelial membrane. Earthworms (phylum Annelida) have a digestive tract specialized in different regions for the ingestion, storage (crop), fragmentation (gizzard), digestion and absorption of food (intestine). All higher animals, like this salamander, show similar specializations.

The ingested food may be stored in a specialized region of the digestive tract or may first be subjected to physical fragmentation through the chewing action of teeth (in the mouth of many vertebrates) or the grinding action of pebbles (in the gizzard of earthworms and birds). Chemical **digestion** then occurs primarily in the intestine, breaking down the larger food molecules of polysaccharides, fats, and proteins into smaller subunits. Carbohydrate digestion begins in the mouth of some animals, and protein digestion begins in the stomach in some animals. Chemical digestion involves hydrolysis reactions that liberate the subunits—primarily monosaccharides, amino acids, and fatty acids—from the food. These products of chemical digestion pass through the epithelial lining of the gut and ultimately into the blood, in a process known as absorption. Any molecules in the food that are not absorbed cannot be used by the animal. These wastes are excreted from the anus.

> **27.2 Most animals digest their food extracellularly. A digestive tract with a one-way transport of food allows specialization of regions for different functions.**

Vertebrate Digestive Systems

In humans and other vertebrates, the digestive system consists of a tubular gastrointestinal tract and accessory digestive organs (figure 27.6). Working through the figure from the top down, the initial components of the gastrointestinal tract are the mouth and the pharynx, which is the common passage of the oral and nasal cavities. The pharynx leads to the esophagus, a muscular tube that delivers food to the stomach, where some preliminary digestion occurs. From the stomach, food passes to the first part of the small intestine, where a battery of digestive enzymes continues the digestive process. The products of digestion then pass across the wall of the small intestine into the bloodstream. The small intestine empties what remains into the large intestine, where water and minerals are absorbed. In most vertebrates other than mammals, the waste products emerge from the large intestine into a cavity called the cloaca (see the salamander in figure 27.5), which also receives the products of the urinary and reproductive systems. In mammals, the urogenital products are separated from the fecal material in the large intestine, also called the colon; the fecal material enters the rectum and is expelled through the anus.

In general, carnivores have shorter intestines for their size than do herbivores. A short intestine is adequate for a carnivore, but herbivores ingest a large amount of plant cellulose, which resists digestion. These animals have a long, convoluted small intestine. In addition, mammals called *ruminants* (such as cows) that consume grass and other vegetation have stomachs with multiple chambers, where bacteria aid in the digestion of cellulose. Other herbivores, including rabbits and horses, digest cellulose (with the aid of bacteria) in a blind pouch called the **cecum** located at the beginning of the large intestine (these will be discussed in more detail in section 27.7). Accessory digestive organs described later in this chapter include the liver, the gallbladder, and the pancreas.

The tubular gastrointestinal tract of a vertebrate, as you can see in figure 27.7, has a characteristic layered structure. Working from the inside (the lumen) outward, the innermost layer (light pink) is the mucosa, an epithelium that lines the lumen. The next major tissue layer, composed of connective tissue, is called the submucosa (darker pink). Just outside the submucosa is the muscularis, which consists of a double layer of smooth muscles. The muscles in the inner layer have a circular orientation, and those in the outer layer are arranged longitudinally. An outer connective tissue layer, the serosa, covers the external surface of the tract. Nerves, intertwined in regions called *plexuses,* are located in the submucosa and help regulate the gastrointestinal activities.

> **27.3** The vertebrate digestive system consists of a tubular gastrointestinal tract, which is modified in different animals, composed of a series of tissue layers.

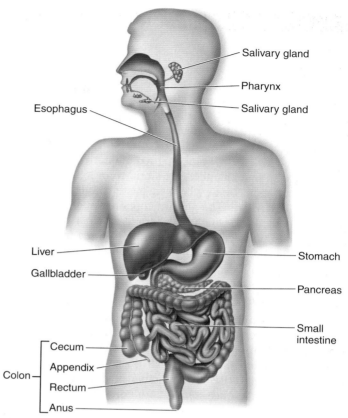

Figure 27.6   The human digestive system.
The tubular gastrointestinal tract and accessory digestive organs are shown. The colon extends from the cecum to the anus.

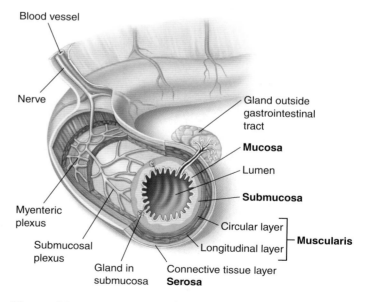

Figure 27.7   The layers of the gastrointestinal tract.
The mucosa contains a lining epithelium, the submucosa is composed of connective tissue (as is the outer serosa layer), and the muscularis consists of smooth muscles.

27.4 The Mouth and Teeth

Specializations of the digestive systems in different kinds of vertebrates reflect differences in the way these animals live. Fishes have a large pharynx with gill slits, whereas air-breathing vertebrates have a greatly reduced pharynx. Many vertebrates have teeth (figure 27.8), and chewing (*mastication*) breaks up food into small particles and mixes it with fluid secretions. Birds, which lack teeth, break up food in their two-chambered stomachs. The first chamber, the stomach in figure 27.9, produces digestive enzymes, which are passed along with the food into the gizzard. The gizzard contains small pebbles ingested by the bird, which are churned together with the food by muscular action. This churning grinds up the seeds and other hard plant material into smaller chunks that can be digested more easily in the intestine.

Vertebrate Teeth

Carnivorous mammals have pointed teeth that lack flat grinding surfaces. Such teeth are adapted for cutting and shearing. Carnivores often tear off pieces of their prey but have little need to chew them, because digestive enzymes can act directly on animal cells. (Recall how a cat or dog gulps down its food.) By contrast, grass-eating herbivores, such as cows and horses, must pulverize the cellulose cell walls of plant tissue before digesting it. These animals have large, flat teeth with complex ridges well suited for grinding.

The four front teeth in the upper and lower jaws of vertebrates are chisel-shaped incisors used for biting (figure 27.8). On each side of the incisors are sharp, pointed teeth (each point called a cuspid), the first, referred to as "canine" teeth, are used for tearing food. Behind the canines are premolars

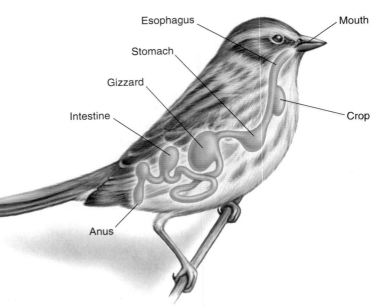

Figure 27.9 The digestive tract of birds.
In birds, food enters the mouth and is stored in the crop. Because birds lack teeth, they swallow gritty objects or pebbles, which lodge in the gizzard, to help pulverize food. Digestive enzymes produced in the stomach are churned up with the food and gritty objects in the gizzard before passing into the intestine.

(bicuspids) and then molars, which have flattened, ridged surfaces for grinding and crushing food.

Humans are omnivores, and human teeth are specialized for eating both plant and animal food. Viewed simply, humans are carnivores in the front of the mouth and herbivores in the back. Children have only 20 teeth, but these deciduous teeth are lost during childhood and are replaced by 32 adult teeth. The third molars are the wisdom teeth, which usually grow in during the late teens or early twenties, when a person is assumed to have gained a little "wisdom."

As you can see in figure 27.10, the tooth is a living organ, composed of connective tissue, nerves, and blood vessels, held in place by cementum, a bonelike substance that anchors the tooth in the jaw. The interior of the tooth contains connective tissue called pulp that extends into the root canals and contains nerves and blood vessels. A layer of calcified tissue called dentin surrounds the pulp cavity. The portion of the tooth that projects above the gums is called the crown and is covered with an extremely hard, nonliving substance called enamel. Enamel protects the tooth against abrasion and acids that are produced by bacteria living in the mouth. Cavities form when bacterial acids break down the enamel, allowing bacteria to infect the inner tissues of the tooth.

Processing Food in the Mouth

Inside the mouth, the tongue mixes food with a mucous solution, **saliva.** In humans, three pairs of salivary glands secrete saliva into the mouth through ducts in the mouth's mucosal

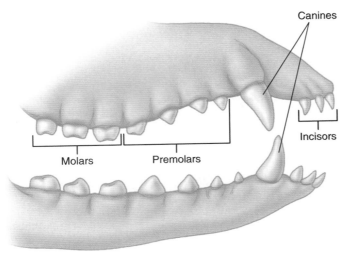

Figure 27.8 Diagram of generalized vertebrate dentition.
Different vertebrates have specific variations from this generalized pattern, depending on whether the vertebrate is an herbivore, carnivore, or omnivore.

lining. Saliva moistens and lubricates the food so that it is easier to swallow and does not abrade the tissue it passes on its way through the esophagus. Saliva also contains the hydrolytic enzyme salivary **amylase,** which initiates the breakdown of the polysaccharide starch into the disaccharide maltose. This digestion is usually minimal in humans, however, because most people don't chew their food very long.

The secretions of the salivary glands are controlled by the nervous system, which in humans maintains a constant flow of about half a milliliter per minute when the mouth is empty of food. This continuous secretion keeps the mouth moist. The presence of food in the mouth triggers an increased rate of secretion, as taste-sensitive neurons in the mouth send impulses to the brain, which responds by stimulating the salivary glands. The most potent stimuli are acidic solutions; lemon juice, for example, can increase the rate of salivation eightfold. The sight, sound, or smell of food can stimulate salivation markedly in dogs, but in humans, these stimuli are much less effective than thinking or talking about food.

Swallowing

When food is ready to be swallowed, the tongue moves it to the back of the mouth. In mammals, the process of swallowing begins when the soft palate elevates, pushing against the back wall of the pharynx (figure 27.11). Elevation of the soft palate seals off the nasal cavity and prevents food from entering it ❶. Pressure against the pharynx stimulates neurons within its walls, which send impulses to the swallowing center in the brain. In response, muscles are stimulated to contract and raise the *larynx* (voice box). This pushes the *glottis,* the opening from the larynx into the trachea (windpipe), against a flap of tissue called the *epiglottis* ❷. These actions keep food out of the respiratory tract, directing it instead into the esophagus ❸.

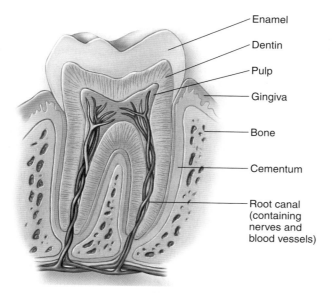

Figure 27.10   Human teeth.

Each vertebrate tooth is alive, with a central pulp containing nerves and blood vessels. The actual chewing surface is a hard enamel layered over the softer dentin, which forms the body of the tooth.

Enamel
Dentin
Pulp
Gingiva
Bone
Cementum
Root canal (containing nerves and blood vessels)

27.4   In many vertebrates, ingested food is fragmented through the tearing or grinding action of specialized teeth. In birds, this is accomplished through the grinding action of pebbles in the gizzard. Food mixed with saliva is swallowed and enters the esophagus.

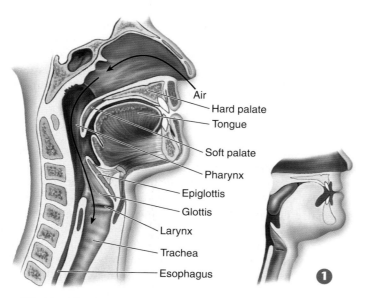
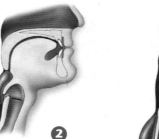
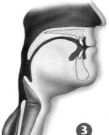

Air
Hard palate
Tongue
Soft palate
Pharynx
Epiglottis
Glottis
Larynx
Trachea
Esophagus

Figure 27.11   The human pharynx, palate, and larynx.

27.5 The Esophagus and Stomach

Structure and Function of the Esophagus

Swallowed food enters a muscular tube called the **esophagus,** which connects the pharynx to the stomach. In adult humans, the esophagus is about 25 centimeters long; the upper third is enveloped in skeletal muscle, for voluntary control of swallowing, while the lower two-thirds is surrounded by involuntary smooth muscle. The swallowing center stimulates successive waves of contraction in these muscles that move food along the esophagus to the stomach. The muscles relax ahead of the food, allowing it to pass freely, and contract behind the food to push it along, as shown in figure 27.12. These rhythmic waves of muscular contraction are called **peristalsis;** they enable humans and other vertebrates to swallow even if they are upside down.

In many vertebrates, the movement of food from the esophagus into the stomach is controlled by a ring of circular smooth muscle, the **sphincter,** that opens in response to the pressure exerted by the food. Contraction of this sphincter prevents food in the stomach from moving back into the esophagus. Rodents and horses have a true sphincter at this site, and thus stomach contents cannot move back out. Humans lack a true sphincter, and stomach contents can be brought back out during vomiting, when the sphincter between the stomach and esophagus is relaxed and the contents of the stomach are forcefully expelled through the mouth. The relaxing of this sphincter can also result in the movement of stomach acid into the esophagus, causing an irritation called *heartburn.* Chronic and severe heartburn is a condition known as *acid reflux.*

Structure and Function of the Stomach

The **stomach** is a saclike portion of the digestive tract. Its inner surface is highly convoluted, enabling it to fold up when empty and open out like an expanding balloon as it fills with food. Thus, while the human stomach has a volume of only about 50 milliliters when empty, it may expand to contain 2 to 4 liters of food when full.

The stomach contains an extra layer of smooth muscle for churning food and mixing it with *gastric juice,* an acidic secretion of the tubular gastric glands of the mucosa. The gastric glands lie at the bottom of deep depressions, the gastric pits shown in the enlargement in figure 27.13. These exocrine glands contain two kinds of secretory cells: *parietal cells,* which secrete hydrochloric acid (HCl); and *chief cells,* which secrete pepsinogen, a weak protease (protein-digesting enzyme) that requires a very low pH to be active. This low pH is provided by the HCl. Activated pepsinogen molecules then cleave each other at specific sites, producing a much more active protease, pepsin. This process of secreting a relatively inactive enzyme that is then converted into a more active enzyme outside the cell prevents the chief cells from digesting themselves. It should be noted that only proteins are partially digested in the stomach—there is no significant digestion of carbohydrates or fats.

Action of Acid

The human stomach produces about 2 liters of HCl and other gastric secretions every day, creating a very acidic solution inside the stomach. The concentration of HCl in this solution is about 10 millimolar, corresponding to a pH of 2. Thus, gastric juice is about 250,000 times more acidic than blood, whose normal pH is 7.4. The low pH in the stomach helps denature food proteins, making them easier to digest, and keeps pepsin maximally active. Active pepsin hydrolyzes food proteins into shorter chains of polypeptides that are not fully digested until the mixture enters the small intestine. The mixture of partially digested food and gastric juice is called **chyme.**

The acidic solution within the stomach also kills most of the bacteria that are ingested with the food. The few bacteria

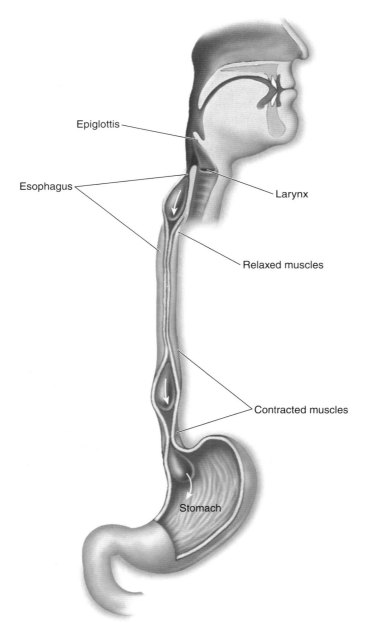

Figure 27.12 The esophagus and peristalsis.

Epiglottis

Esophagus

Larynx

Relaxed muscles

Contracted muscles

Stomach

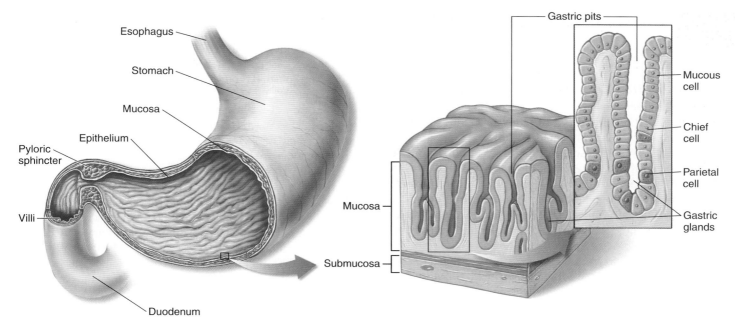

Figure 27.13 The stomach and gastric glands.
Food enters the stomach from the esophagus. The epithelial walls of the stomach are dotted with gastric pits, which contain glands that secrete hydrochloric acid (HCl) and the enzyme pepsinogen. The gastric glands consist of mucous cells, chief cells that secrete pepsinogen, and parietal cells that secrete HCl. Gastric pits are the openings of the gastric glands.

that survive the stomach and enter the intestine intact are able to grow and multiply there, particularly in the large intestine. In fact, most vertebrates harbor thriving colonies of bacteria within their intestines, and bacteria are a major component of feces. As we discuss later, bacteria that live within the digestive tract of cows and other ruminants play a key role in the ability of these mammals to digest cellulose.

Ulcers

It is important that the stomach not produce too much acid. If it did, the body could not neutralize the acid later in the small intestine, a step essential for the final stage of digestion. Production of acid is controlled by hormones. These hormones are produced by endocrine cells scattered within the walls of the stomach. The hormone gastrin regulates the synthesis of HCl by the parietal cells of the gastric pits, permitting HCl to be made only when the pH of the stomach is higher than about 1.5.

Overproduction of gastric acid can occasionally eat a hole through the wall of the stomach. Such gastric **ulcers** are rare, however, because epithelial cells in the mucosa of the stomach are protected somewhat by a layer of alkaline mucus, and because those cells are rapidly replaced by cell division if they become damaged (gastric epithelial cells are replaced every two to three days). Over 90% of gastrointestinal ulcers are duodenal ulcers, which are ulcers of the small intestine. These may be produced when excessive amounts of acidic chyme are delivered into the duodenum, so that the acid cannot be properly neutralized through the action of alkaline pancreatic juice

(described later). Susceptibility to ulcers is increased when the mucosal barriers to self-digestion are weakened by an infection of the bacterium *Helicobacter pylori*. Modern antibiotic treatments can reduce symptoms and often cure the ulcer.

In addition to producing HCl, the parietal cells of the stomach also secrete intrinsic factor, a polypeptide needed for the intestinal absorption of vitamin B_{12}. Because this vitamin is required for the production of red blood cells, persons who lack sufficient intrinsic factor develop a type of anemia (low red blood cell count) called *pernicious anemia.*

Leaving the Stomach

Chyme leaves the stomach through the *pyloric sphincter,* shown at the base of the stomach in figure 27.13, to enter the small intestine. This is where all terminal digestion of carbohydrates, fats, and proteins occurs, and where the products of digestion—amino acids, glucose, and fatty acids—are absorbed into the blood. Only water from chyme and a few substances such as aspirin and alcohol are absorbed through the wall of the stomach.

27.5 Peristaltic waves of contraction propel food along the esophagus to the stomach. Gastric juice contains strong hydrochloric acid and the protein-digesting enzyme pepsin, which begins the digestion of proteins into shorter polypeptides. The acidic chyme is then transferred through the pyloric sphincter to the small intestine.

27.6   The Small and Large Intestines

Digestion and Absorption: The Small Intestine

The digestive tract exits from the stomach into the **small intestine,** where the breaking down of large molecules into small ones occurs. Only relatively small portions of food are introduced into the small intestine at one time, to allow time for acid to be neutralized and enzymes to act. The small intestine is the true digestive vat of the body. Within it, carbohydrates are broken down into simple sugars, proteins into amino acids, and fats into fatty acids. Once these small molecules have been produced, they pass across the epithelial wall of the small intestine into the bloodstream.

Some of the enzymes necessary for these digestive processes are secreted by the cells of the intestinal wall. Most, however, are made in a large gland called the *pancreas* (discussed in section 27.8), situated near the junction of the stomach and the small intestine. It is one of the body's major exocrine (secreting through ducts) glands. The pancreas sends its secretions into the small intestine through a duct that empties into its initial segment, the **duodenum.** Your small intestine is approximately 6 meters long—unwound and stood on its end, it would be far taller than you are! Only the first 25 centimeters, about 4% of the total length, is the duodenum. It is within this initial segment, where the pancreatic enzymes enter the small intestine, that digestion occurs.

Much of the food energy the vertebrate body harvests is obtained from fats. The digestion of fats is carried out by a collection of molecules known as *bile salts* secreted into the duodenum from the *liver* (discussed in section 27.8). Because fats are insoluble in water, they enter the intestine as drops within the watery chyme. The bile salts, which are partly lipid-soluble and partly water-soluble, work like detergents. They combine with fats to form microscopic droplets in a process called emulsification. These tiny droplets have greater surface areas upon which the enzyme that breaks down fats, called lipase, can work. This allows the digestion of fats to proceed more rapidly.

Two areas make up the rest of the small intestine (96% of its length), the **jejunum** and the **ileum.** Digestion continues into the jejunum, but the ileum is devoted to absorbing water and the products of digestion into the bloodstream. The lining of the small intestine is folded into ridges, seen in the cutaway portion of the intestine in figure 27.14*a.* The ridges are covered with fine fingerlike projections called **villi** (singular, **villus**), shown in the first enlarged view, but each too small to see with the naked eye. In turn, each of the cells covering a villus is covered on its outer surface by a field of cytoplasmic projections called **microvilli.** The enlargement of the villus shows epithelial cells lining the villus, and the further enlargement of these cells shows the microvilli on the surface side of the cells. Scanning and transmission electron micrographs in figure 27.14*b, c* give you different perspectives of the microvilli. Both villi and microvilli greatly increase the absorptive surface of the lining of the small intestine. The average surface area of the small intestine of an adult human is about 300 square meters, more than the surface of many swimming pools!

The amount of material passing through the small intestine is startlingly large. Per day, an average human consumes about 800 grams of solid food, and 1,200 milliliters of water, for a total volume of about 2 liters. To this amount is added about 1.5 liters of fluid from the salivary glands, 2 liters from the gastric secretions of the stomach, 1.5 liters from the pancreas, 0.5 liters from the liver, and 1.5 liters of intestinal secretions. The total adds up to a remarkable 9 liters—more than 10% of the total volume of your body! However, although the flux is great, the *net* passage is small. Almost all these fluids and solids are reabsorbed during their passage through the small intestine—about 8.5 liters across the walls of the small intestine and 0.35 liters across the wall of the large intestine. Of the 800 grams of solids and 9 liters of liquids that enter the digestive tract each day, only about 50 grams of solids and 100 milliliters of liquids leave the body as feces. The fluid absorption efficiency of the digestive tract thus approaches 99%, very high indeed.

Concentration of Solids: The Large Intestine

The **large intestine,** or **colon,** is much shorter than the small intestine, approximately 1 meter long, but it is called the large intestine because of its larger diameter. The small intestine empties directly into the large intestine at a junction where the cecum and the appendix are located, which are two structures no longer actively used in humans (see figure 27.6). No digestion takes place within the large intestine, and only about 6% to 7% of fluid absorption occurs there. The large intestine is not convoluted, lying instead in three relatively straight segments, and its inner surface does not possess villi. As a consequence, the large intestine has only one-thirtieth the absorptive surface area of the small intestine. Although some water, sodium, and vitamin K are absorbed across its walls, the primary function of the large intestine is to act as a refuse dump. Within it, undigested material, including large amounts of plant fiber and cellulose, is compacted and stored. Many bacteria live and actively divide within the large intestine, where they play a role in the processing of undigested material into the final excretory product, **feces.** Bacterial fermentation produces gas within the colon at a rate of about 500 milliliters per day. This rate increases greatly after the consumption of beans or other vegetable matter because the passage of undigested plant material (fiber) into the large intestine provides substrates for fermentation.

The final segment of the digestive tract is a short extension of the large intestine called the **rectum.** Compact solids within the colon pass through the rectum as a result of the peristaltic contractions of the muscles encasing the large intestine, and then out of the body through the **anus.**

> **27.6**   Most digestion occurs in the initial upper portion of the small intestine, called the duodenum. The rest of the small intestine is devoted to absorption of water and the products of digestion. The large intestine compacts residual solid wastes.

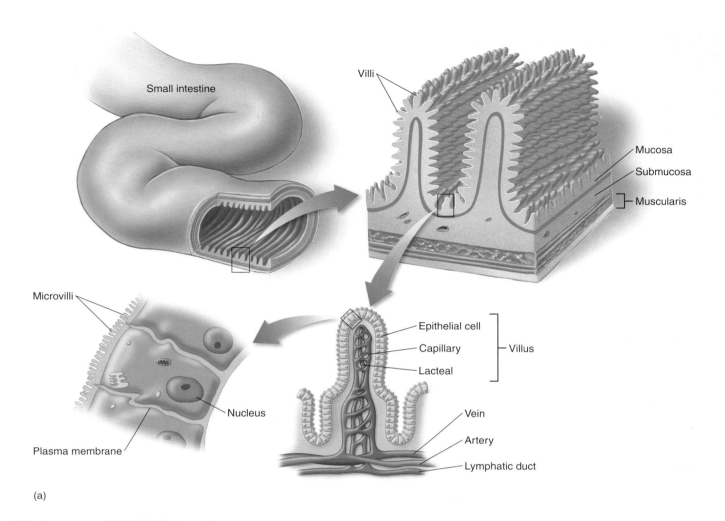

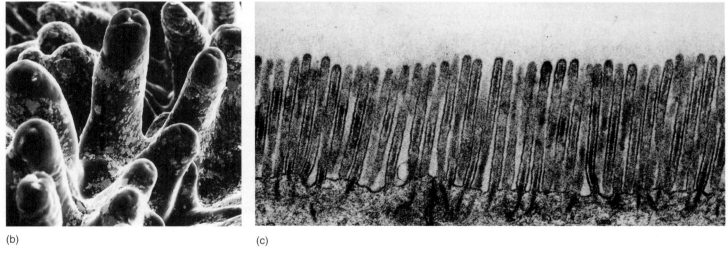

(a)

(b)

(c)

Figure 27.14 The small intestine.

(a) Cross section of the small intestine with details showing villi structure. (b) Microvilli, shown in a scanning electron micrograph, are very densely clustered, giving the small intestine an enormous surface area, which is very important for efficient absorption. (c) Intestinal microvilli as shown in a transmission electron micrograph.

27.7 Variations in Vertebrate Digestive Systems

Most animals lack the enzymes necessary to digest cellulose, the carbohydrate that functions as the chief structural component of plants. The digestive tracts of some animals, however, contain prokaryotes and protists that convert cellulose into substances the host can digest. Although digestion by gastrointestinal microorganisms plays a relatively small role in human nutrition, it is an essential element in the nutrition of many other kinds of animals, including insects like termites and cockroaches and a few groups of herbivorous mammals. The relationships between these microorganisms and their animal hosts are mutually beneficial and provide an excellent example of symbiosis.

Cows, deer, and other herbivores called *ruminants* have large, divided stomachs. By following the path food takes in figure 27.15, we can explore the areas of the stomach. Food enters the stomach by way of the rumen ❶. The rumen, which may hold up to 50 gallons, serves as a fermentation vat in which prokaryotes and protists convert cellulose and other molecules into a variety of simpler compounds. The location of the rumen at the front of the four chambers is important because it allows the animal to regurgitate and rechew the contents of the rumen (see how the arrow leaves the stomach after looping through the rumen and reenters), an activity called *rumination,* or "chewing the cud." The cud is then swallowed and enters the reticulum ❷, from which it passes to the omasum ❸ and then the abomasum ❹, where it is finally mixed with gastric juice. Hence, only the abomasum is equivalent to the human stomach in its function. This process leads to a far more efficient digestion of cellulose in ruminants than in mammals that lack a rumen, such as horses.

In some animals such as rodents, horses, and lagomorphs (rabbits and hares), the digestion of cellulose by microorganisms takes place in the cecum, which is greatly enlarged in these animals (see the rabbit, a nonruminant herbivore, in figure 27.16). Because the cecum is located beyond the stomach, regurgitation of its contents is impossible. However, rodents and lagomorphs have evolved another way to digest cellulose that achieves a degree of efficiency similar to that of ruminant digestion. They do this by eating their feces, thus passing the food through their digestive tract a second time. The second passage makes it possible for the animal to absorb the nutrients produced by the microorganisms in its cecum. Animals that engage in this practice of **coprophagy** (from the Greek words *copros,* excrement, and *phagein,* eat) cannot remain healthy if they are prevented from eating their feces. The organization of the digestive system reflects the diet of the animal. Thus the large cecum of the rabbit reflects a diet of plants. In contrast, the insectivore and carnivore in figure 27.16 digest primarily protein from animal bodies; therefore, they have a reduced or absent cecum. Ruminant herbivores, as described earlier, have

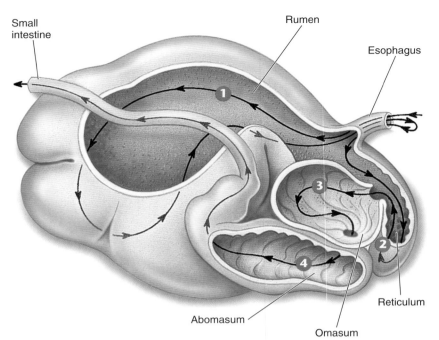

Figure 27.15 Four-chambered stomach of a ruminant.
The grass and other plants that a ruminant, such as a cow, eats enter the rumen, where they are partially digested. From there, the food may be regurgitated and rechewed. The food is then transferred through the last three chambers. Only the abomasum secretes gastric juice.

a large four-chambered stomach and also a cecum, although most digestion of vegetation occurs in the stomach.

Cellulose is not the only plant product that vertebrates can use as a food source because of the digestive activities of intestinal microorganisms. Wax, a substance indigestible by most terrestrial animals, is digested by symbiotic bacteria living in the gut of honeyguides, African birds that eat the wax in bees' nests. In the marine food chain, wax is a major constituent of copepods (crustaceans in the plankton), and many marine fish and birds appear to be able to digest wax with the aid of symbiotic microorganisms.

Another example of the way intestinal microorganisms function in the metabolism of their animal hosts is provided by the synthesis of vitamin K. All mammals rely on intestinal bacteria to synthesize this vitamin, which is necessary for the clotting of blood. Birds, which lack these bacteria, must consume the required quantities of vitamin K in their food. In humans, prolonged treatment with antibiotics greatly reduces the populations of bacteria in the intestine; under such circumstances, it may be necessary to provide supplementary vitamin K.

27.7 Much of the food value of plants is tied up in cellulose, and the digestive tract of many animals harbors colonies of cellulose-digestive microorganisms. Intestinal microorganisms also produce molecules such as vitamin K that are important to the well-being of their vertebrate hosts.

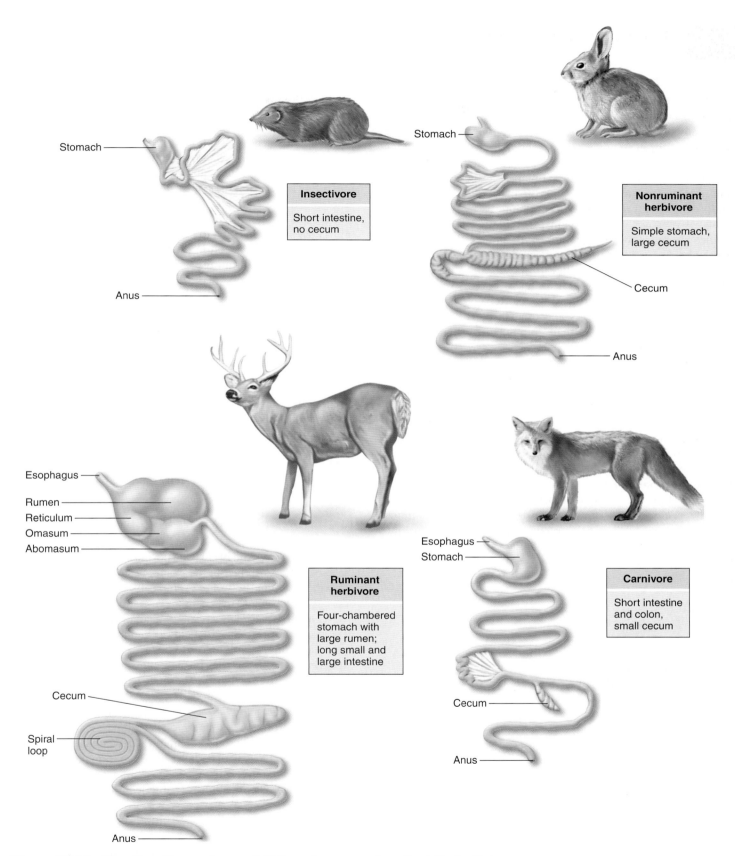

Figure 27.16 The digestive systems of different mammals reflect their diets.
Herbivores require long digestive tracts with specialized compartments for the breakdown of plant matter. Protein diets are more easily digested; thus, insectivorous and carnivorous mammals have short digestive tracts with few specialized pouches.

The Pancreas

The **pancreas,** a large gland situated near the junction of the stomach and the small intestine (see figure 27.6), is one of the accessory organs that contribute secretions to the digestive tract. Fluid from the pancreas is secreted into the duodenum through the *pancreatic duct* shown in figure 27.17. Note that the pancreatic duct joins with another duct, the common bile duct (discussed later), before entering the small intestine. This fluid contains a host of enzymes, including trypsin and chymotrypsin, which digest proteins. Inactive forms of these enzymes are released into the duodenum and are then activated by the enzymes of the intestine. Pancreatic fluid also contains pancreatic amylase, which digests starch; and lipase, which digests fats. Pancreatic enzymes digest proteins into smaller polypeptides, polysaccharides into shorter chains of sugars, and fat into free fatty acids and other products. The digestion of these molecules is then completed by the intestinal enzymes.

Pancreatic fluid also contains bicarbonate, which neutralizes the HCl from the stomach and gives the chyme in the duodenum a slightly alkaline pH. The digestive enzymes and bicarbonate are produced by clusters of secretory cells known as *acini.*

In addition to its exocrine role in digestion, the pancreas also functions as an endocrine gland, secreting several hormones into the blood that control the blood levels of glucose and other nutrients. These hormones are produced in the **islets of Langerhans,** clusters of endocrine cells scattered throughout the pancreas and shown in the enlarged view in figure 27.17. The two most important pancreatic hormones, insulin and glucagon, are discussed in chapters 28 and 31.

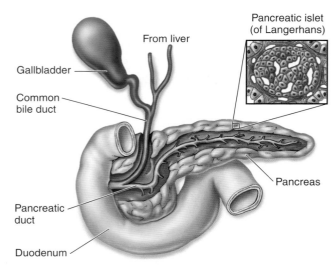

Figure 27.17   The pancreatic and bile ducts empty into the duodenum.

The Liver and Gallbladder

The **liver** is the largest internal organ of the body. In an adult human, the liver weighs about 1.5 kilograms and is the size of a football. The main exocrine secretion of the liver is **bile,** a fluid mixture consisting of *bile pigments* and *bile salts* that is delivered into the duodenum during the digestion of a meal.

The bile salts play a very important role in the digestion of fats. As explained earlier, fats are insoluble in water, and so they enter the intestine as drops within the watery chyme. The bile salts work like detergents, dispersing the large drops of fat into a fine suspension of smaller droplets. This breaking up, or emulsification, of the fat into droplets produces a greater surface area of fat upon which the lipase enzymes can act, and thus allows the digestion of fat to proceed more rapidly.

After it is produced in the liver, bile is stored and concentrated in the **gallbladder** (the green organ in figure 27.17). The arrival of fatty food in the duodenum triggers a neural and endocrine reflex that stimulates the gallbladder to contract, causing bile to be transported through the common bile duct and injected into the duodenum. If the bile duct is blocked by a *gallstone* (formed from a hardened precipitate of cholesterol), contraction of the gallbladder causes pain that is generally felt under the right scapula (shoulder blade).

The digestive system is highly specialized and involves the interactions of many different organs. Figure 27.18 overviews the different functional areas in the digestive system and the different organs involved. The colored circles indicate the primary areas of digestion and enzyme production: red for protein digestion, orange for carbohydrate digestion, green for fat digestion, and blue for nucleic acid digestion (not really discussed in this chapter as nucleic acids are not a major source of calories in the diet).

Regulatory Functions of the Liver

Because a large vein carries blood from the stomach and intestine directly to the liver, the liver is in a position to chemically modify the substances absorbed in the gastrointestinal tract before they reach the rest of the body. For example, ingested alcohol and other drugs are taken into liver cells and metabolized; this is why the liver is often damaged as a result of alcohol and drug abuse. The liver also removes toxins, pesticides, carcinogens, and other poisons, converting them into less toxic forms (figure 27.18). Also, excess amino acids that may be present in the blood are converted to glucose by liver enzymes. The first step in this conversion is the removal of the amino group ($-NH_2$) from the amino acid, a process called *deamination.* Unlike plants, animals cannot reuse the nitrogen from these amino groups and must excrete it as nitrogenous waste. The product of amino acid deamination, ammonia (NH_3), combines with carbon dioxide to form urea. The urea is released by the liver into the bloodstream, where—as you will learn in chapter 28—the kidneys subsequently remove it.

> **27.8**   The pancreas secretes digestive enzymes and bicarbonate into the pancreatic duct. The liver produces bile, which is stored and concentrated in the gallbladder. The liver and the pancreatic hormones regulate blood glucose concentration.

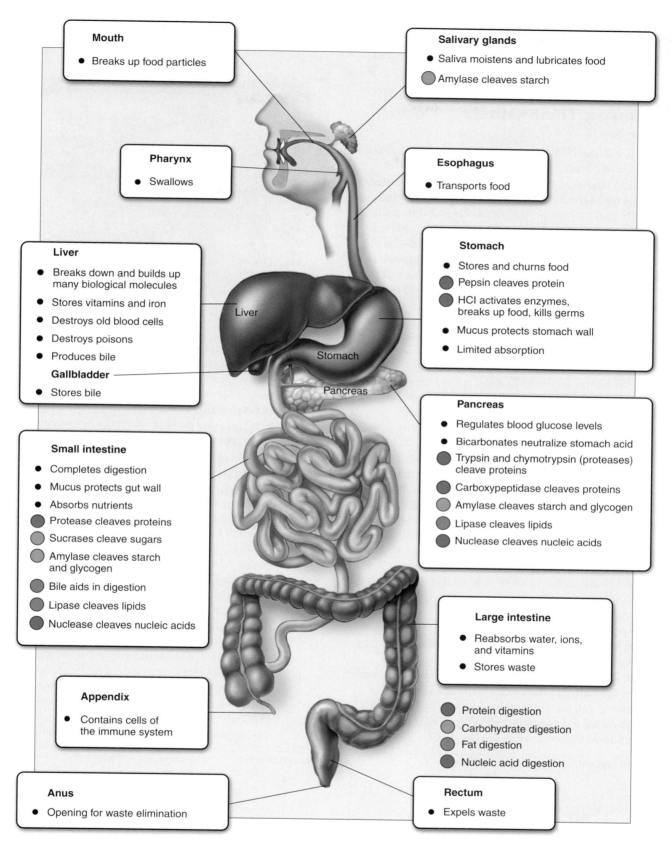

Mouth
- Breaks up food particles

Salivary glands
- Saliva moistens and lubricates food
- Amylase cleaves starch

Pharynx
- Swallows

Esophagus
- Transports food

Liver
- Breaks down and builds up many biological molecules
- Stores vitamins and iron
- Destroys old blood cells
- Destroys poisons
- Produces bile

Gallbladder
- Stores bile

Stomach
- Stores and churns food
- Pepsin cleaves protein
- HCl activates enzymes, breaks up food, kills germs
- Mucus protects stomach wall
- Limited absorption

Pancreas
- Regulates blood glucose levels
- Bicarbonates neutralize stomach acid
- Trypsin and chymotrypsin (proteases) cleave proteins
- Carboxypeptidase cleaves proteins
- Amylase cleaves starch and glycogen
- Lipase cleaves lipids
- Nuclease cleaves nucleic acids

Small intestine
- Completes digestion
- Mucus protects gut wall
- Absorbs nutrients
- Protease cleaves proteins
- Sucrases cleave sugars
- Amylase cleaves starch and glycogen
- Bile aids in digestion
- Lipase cleaves lipids
- Nuclease cleaves nucleic acids

Large intestine
- Reabsorbs water, ions, and vitamins
- Stores waste

- Protein digestion
- Carbohydrate digestion
- Fat digestion
- Nucleic acid digestion

Appendix
- Contains cells of the immune system

Anus
- Opening for waste elimination

Rectum
- Expels waste

Liver

Stomach

Pancreas

Figure 27.18 The organs of the digestive system and their functions.
The digestive system contains some dozen different organs that act on the food that is consumed, starting with the mouth and ending with the anus. All of these organs must work properly for the body to effectively obtain nutrients.

Why Do Diabetics Excrete Glucose in Their Urine?

Late-onset diabetes is a serious and increasingly common disorder in which the body's cells lose their ability to respond to insulin, a hormone which is needed to trigger their uptake of glucose. As illustrated below, the binding of insulin to a receptor in the plasma membrane causes the rapid insertion of glucose transporter channels into the plasma membrane, allowing the cell to take up glucose. In diabetics, however, glucose molecules accumulate in the blood while the body's cells starve for the lack of them. In mild cases, blood glucose levels rise to several times the normal value of 4 m*M;* in severe untreated cases, blood glucose levels may become enormously elevated, up to 25 times the normal value. A characteristic symptom of even mild diabetes is the excretion of large amounts of glucose in the urine. The name of the disorder, *diabetes mellitus*, means "excessive secretion of sweet urine." In normal individuals, by contrast, only trace amounts of glucose are excreted. The kidney very efficiently reabsorbs glucose molecules from the fluid passing through it. Why doesn't it do so in diabetic individuals?

The graph on the upper right displays so-called glucose tolerance curves for a normal person (*blue line*) and a diabetic (*red line*). After a night without food, each individual drank a test dose of 100 grams of glucose dissolved in water. Blood glucose levels were then monitored at 30-minute and one-hour intervals. The dotted line indicates the kidney threshold, the maximum concentration of blood glucose molecules (about 10 m*M*) that the kidney is able to retrieve from the fluid passing through it when all of its glucose-transporting channels are being utilized full-bore.

1. **Applying Concepts**
    a. Variable. In the graph, what is the dependent variable?
    b. Reading a Curve. What is the immediate impact on the normal individual's blood glucose levels of consuming the test dose of glucose? How long does it take for the normal person's blood glucose level to return to the level before the test dose?

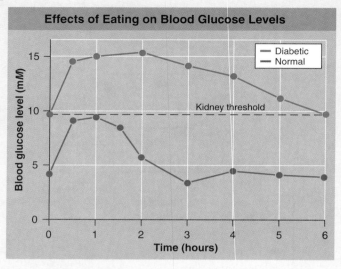

Effects of Eating on Blood Glucose Levels

c. **Comparing Curves.** Is the impact any different for the diabetic person? How long does it take for the diabetic person's blood glucose levels to return to the level before the test dose?
2. **Interpreting Data**
    a. Is there any point at which the normal individual's blood glucose levels exceed the kidney threshold?
    b. Is there any point at which the diabetic individual's blood glucose levels do *not* exceed the kidney threshold?
3. **Making Inferences**
    a. Why do you suppose the diabetic individual took so much longer to recover from the test dose?
    b. Would you expect the normal individual to excrete glucose? Explain. The diabetic individual? Explain.
4. **Drawing Conclusions** Why do diabetic individuals secrete sweet urine?
5. **Further Analysis**
    a. If glucose molecules are being excreted in the urine, then they are not being converted to fatty acids for storage as fat. This would imply that severe diabetics would lose weight, even if on a high-calorie diet. How would you go about testing this prediction?
    b. Denied glucose, the cells of a diabetic might be expected to turn in desperation to the cell's proteins as a food source. How might you test this hypothesis?

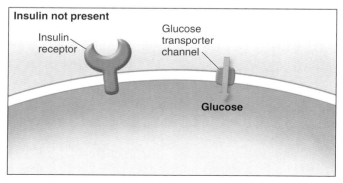

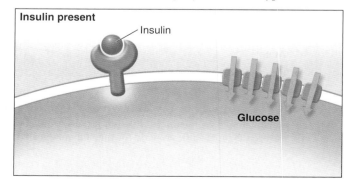

Food Energy and Essential Nutrients

27.1 Food for Energy and Growth

- Animals consume food as a source of energy and as a source of essential molecules and minerals.

- When food is consumed, it is either used up in metabolic processes or stored as fat. A balanced diet that is higher in complex carbohydrates, fruits, and vegetables and low in fats and sweets is recommended (**figure 27.1**).

- Consuming excess calories that the body doesn't use results in gaining weight and obesity, a major health problem (**figure 27.3**).

- Many animals must consume proteins, fruits, and vegetables to obtain essential amino acids and minerals that the body needs but cannot produce itself (**table 27.1**).

Digestion

27.2 Types of Digestive Systems

- Single-celled organisms and lower-level animals, such as sponges, digest food intracellularly. Food is taken up by individual cells and is digested in those cells.

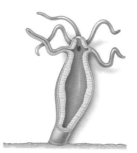

- All other animals digest food extracellularly, in a cavity or tract that contains digestive enzymes (**figure 27.4**). The products of digestion are then absorbed by the body.

- Specialization of the digestive tract, where different regions of the tract are involved in different digestive functions, was possible with the evolution of one-way digestive tracts (**figure 27.5**).

27.3 Vertebrate Digestive Systems

- Vertebrate digestion occurs in a gastrointestinal tract (**figure 27.6**). Areas of the tract are specialized for different digestive functions, and the areas vary in different animal groups depending on the animal's diet.

27.4 The Mouth and Teeth

- Digestion begins in the mouth where teeth are used to chew up food, breaking it into smaller pieces (**figures 27.8** and **27.10**). The chewed food mixes with saliva in the mouth, which contains the enzyme salivary amylase that begins the digestion of starches.

- Birds don't have teeth but break up food in the gizzard, where the food is churned and ground up with pebbles that the bird has swallowed (**figure 27.9**).

- The moistened food is then swallowed, passing from the mouth into the esophagus (**figure 27.11**).

27.5 The Esophagus and Stomach

- Food entering the esophagus is moved along by peristaltic waves of muscle contractions, moving food down the esophagus to the stomach (**figure 27.12**).

- In the stomach, muscle contractions churn up the food with gastric juice, which contains hydrochloric acid and pepsin, a protein-digesting enzyme activated by HCl (**figure 27.13**). Proteins are partially digested in the stomach.

27.6 The Small and Large Intestines

- The acidic chyme from the stomach passes into the upper portion of the small intestine, where it is neutralized and mixed with other digestive enzymes (**figure 27.14**). Some enzymes are secreted by the cells that line the walls of the intestine, but most enzymes and other digestive substances are produced in the pancreas or other accessory organs. The rest of the small intestine is involved in absorption of food molecules and water.

- The large intestine collects and compacts solid waste and releases the waste from the body through the rectum and anus.

27.7 Variations in Vertebrate Digestive Systems

- In ruminants, cellulose-digesting microorganisms live in a chamber of the stomach called the rumen (**figure 27.15**). Food enters the rumen where microorganisms begin digesting cellulose, and is then regurgitated, chewed again and is reswallowed, entering the reticulum. The products of cellulose digestion pass through the other areas of the stomach and then to the small intestine where they are absorbed.

- In other animals, cellulose digestion occurs in the cecum. To gain the nutritional value of cellulose digestion, these animals eat their feces. The digestive systems of animals differ, based on their diets (**figure 27.16**).

27.8 Accessory Digestive Organs

- The pancreas produces the protein-digesting enzymes trypsin and chymotrypsin, the starch-digesting enzyme pancreatic amylase, and the fat-digesting enzyme lipase, which are released into the small intestine.

- The liver produces bile (a mixture of bile pigments and bile salts), which breaks down fats. Bile is stored in the gallbladder and released into the small intestine (**figure 27.17**). All of the organs of digestion work together (**figure 27.18**).

1. One-way passage of food through the digestive system of many animal groups allows
   a. intracellular digestion.
   b. specialization of different regions of the digestive system.
   c. release of digestive enzymes into the gut.
   d. extracellular digestion.
2. Organisms with longer digestive systems, which help break down difficult to digest food, are usually
   a. herbivores.
   b. carnivores.
   c. omnivores.
   d. detritivores.
3. The purpose of a gizzard, like teeth, is to
   a. hold on to prey.
   b. begin the chemical digestion of food.
   c. release enzymes.
   d. begin the physical digestion of food.
4. When a mammal swallows food, it is prevented from going up into the nasal cavity by the
   a. esophagus.
   b. tongue.
   c. soft palate.
   d. epiglottis.
5. The first site of protein digestion in the digestive system occurs in the
   a. mouth.
   b. esophagus.
   c. stomach.
   d. small intestine.
6. Which of the following statements is *false*?
   a. Carnivores have a reduced or absent cecum.
   b. Only ruminants are able to digest cellulose.
   c. The human digestive system contains bacteria but is not able to gain nutritional value from the digestion of cellulose.
   d. Ruminants are able to regurgitate food.
7. Most of the absorption of food molecules takes place in the
   a. stomach.
   b. liver.
   c. small intestine.
   d. large intestine.
8. The purpose of the villi and microvilli in the small intestine is to
   a. neutralize stomach acid.
   b. produce bile.
   c. produce digestive enzymes.
   d. increase the surface area of the small intestine for absorption of nutrients.
9. The primary function of the large intestine is
   a. the breakdown and absorption of fats.
   b. the absorption of water.
   c. the concentration of solid wastes.
   d. the absorption of vitamin C.
10. The _____ secretes digestive enzymes and bicarbonate solution into the small intestine to aid digestion.
    a. pancreas
    b. liver
    c. gallbladder
    d. All of these are correct.

1. **Page 543** Discuss the relationship of obesity and diabetes in the United States, and our ability to make headway against the disease.

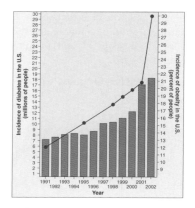

2. **Figure 27.11** "Don't talk with your mouth full" is parental advice that is important for not only social reasons (manners) but for medical reasons. Talking while eating can result in choking. Can you explain how this could occur?

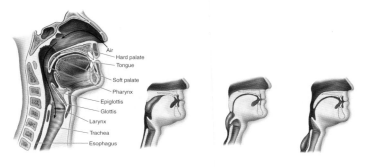

1. Why is it so important for Popeye, and you, to eat your green leafy vegetables (spinach, chard, turnip and mustard greens, bok choi, broccoli, cauliflower, cabbage)?
2. You're going on a backpacking trip with three friends. Lisa wants to pack trail snacks of chocolate cupcakes; Chris wants to take beef jerky. Andre insists that you should all pack GORP (*g*ood *o*ld *r*aisins and *p*eanuts—sometimes known as trail mix). They turn to you to decide. Explain which one is best for a quick snack to keep up your energy on a long hike, and why.

28

Maintaining the Internal Environment

T he man above is sweating, something each of us has done when our bodies have become overheated from too much exercise or sun. The evaporation of the sweat cools our skin, a clever mechanism to remove heat. Your body, like that of all birds and mammals, tries to maintain a constant body temperature, no matter how hot or cold the surrounding air might be, and sweating is one of the ways it does it. If your body begins to heat above 37°C, you start to sweat and release heat; if you instead begin to cool below 37°C, you shiver and generate heat. Keeping your body temperature constant is only one example of a much broader physiological strategy: vertebrates maintain relatively constant physiological conditions within their bodies. The pH of your blood, the rate at which you breathe, your blood pressure, the concentrations of water, salt, and glucose in your blood—all are monitored carefully by the brain, which acts continuously to keep each within narrow bounds. Your body is constantly making dynamic adjustments in these parameters to counter changes caused by outside factors that would alter the body's internal environment. This steady-state balance of internal conditions, known as homeostasis, is the subject of this chapter. A major objective of your study of biology will be to learn how animals maintain homeostasis.

How the Animal Body Maintains Homeostasis

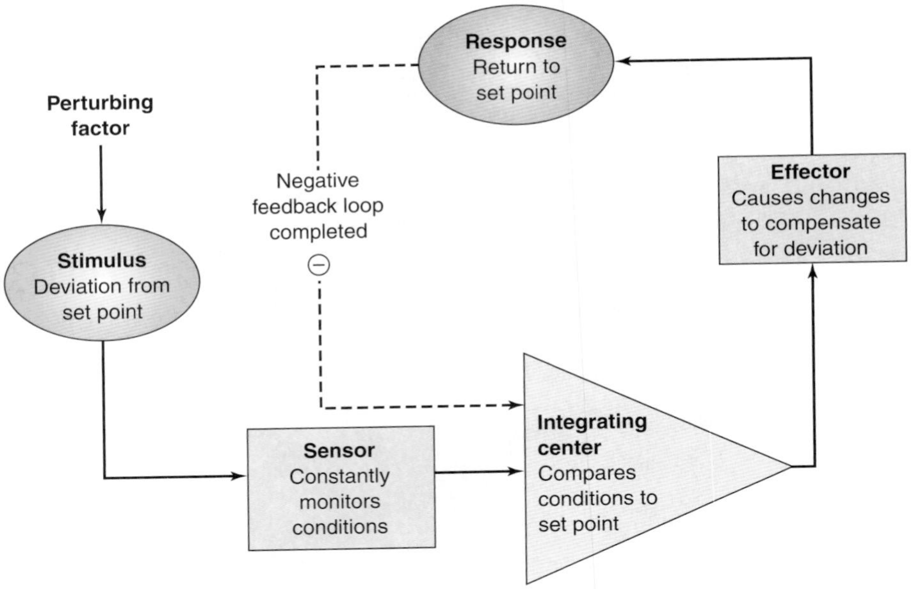

As the animal body has evolved, specialization has increased. Each cell is a sophisticated machine, finely tuned to carry out a precise role within the body. Such specialization of cell function is possible only when extracellular conditions are kept within narrow limits. Temperature, pH, the concentration of glucose and oxygen, and many other factors must be held fairly constant for cells to function efficiently and interact properly with one another.

Homeostasis may be defined as the dynamic constancy of the internal environment. The term *dynamic* is used because conditions are never absolutely constant, but fluctuate continuously within narrow limits. Homeostasis is essential for life, and most of the regulatory mechanisms of the vertebrate body that are not devoted to reproduction are concerned with maintaining homeostasis.

Figure 28.1   A generalized diagram of a negative feedback loop.
Negative feedback loops maintain a state of homeostasis, or dynamic constancy of the internal environment, by correcting deviations from a set point.

Negative Feedback Loops

To maintain internal constancy, the vertebrate body must have **sensors** that are able to measure each condition of the internal environment (the green box in figure 28.1). These constantly monitor the extracellular conditions and relay this information (usually via nerve signals) to an **integrating center** (the yellow triangle), which contains the *set point* (the proper value for that condition). This set point is analogous to the temperature setting on a house thermostat. In a similar manner, there are set points for body temperature, blood glucose concentration, the tension on a tendon, and so on. The integrating center is often a particular region of the brain or spinal cord, but in some cases it can also be cells of endocrine glands. It receives messages from several sensors, weighs the relative strengths of each sensor input, and then determines whether the value of the condition is deviating from the set point. When a deviation in a condition occurs (the "stimulus" indicated by the red oval), sensors detect this condition, and the integrating center sends a message to increase or decrease the activity of particular effectors. **Effectors** (the blue box) are generally muscle or glands, and can change the value of the condition in question back toward the set point value, which is "the response" (the purple oval).

To return to the idea of a home thermostat, suppose you set the thermostat at a set point of 70°F. If the temperature of the house rises sufficiently above the set point, the thermostat

(equivalent to an integrating center) receives this input from a temperature sensor, like a thermometer within the wall unit. It compares the actual temperature to its set point. When these are different, it sends a signal to an effector. The effector in this case may be an air conditioner, which acts to reverse the deviation from the set point.

In a human, if the body temperature exceeds the set point of 37°C, sensors in a part of the brain detect this deviation. Acting via an integrating center (also in the brain), these sensors stimulate effectors (including sweat glands) that lower the temperature. One can think of the effectors as "defending" the set points of the body against deviations. Because the activity of the effectors is influenced by the effects they produce, and because this regulation is in a negative, or reverse, direction, this type of control system is known as a **negative feedback loop** (figure 28.1).

The nature of the negative feedback loop becomes clear when we again refer to the analogy of the thermostat and air conditioner. After the air conditioner has been on for some time, the room temperature may fall significantly below the set point of the thermostat. When this occurs, the air conditioner will be turned off. The effector (air conditioner) is turned on by high temperature; and when activated, it produces a negative change (lowering of the temperature) that ultimately causes the effector to be turned off. In this way, constancy is maintained.

Regulating Body Temperature

Humans, together with other mammals and with birds, are endothermic; they can maintain relatively constant body temperatures independent of the environmental temperature. When the temperature of your blood exceeds 37°C (98.6°F), neurons in a part of the brain called the hypothalamus (discussed in chapters 30 and 31) detect the temperature change. Acting through the control of neurons, the hypothalamus responds by promoting the dissipation of heat through sweating, dilation of blood vessels in the skin, and other mechanisms. These responses tend to counteract the rise in body temperature. When body temperature falls, the hypothalamus coordinates a different set of responses, such as shivering and the constriction of blood vessels in the skin, which help to raise body temperature and correct the initial challenge to homeostasis.

Vertebrates other than mammals and birds are ectothermic; their body temperatures are more or less dependent on the environmental temperature. However, to the extent that it is possible, many ectothermic vertebrates attempt to maintain some degree of temperature homeostasis. Certain large fish, including tuna, swordfish, and some sharks, for example, can maintain parts of their body at a significantly higher temperature than that of the water. Reptiles attempt to maintain a constant body temperature through behavioral means—by placing themselves in varying locations of sun and shade. That's why you frequently see lizards basking in the sun. Sick lizards even give themselves a "fever" by seeking warmer locations!

Most invertebrates, like reptiles, modify behaviors to adjust their body temperature. Many butterflies, for example, must reach a certain body temperature before they can fly. In the cool of the morning, they orient their bodies to maximize their absorption of sunlight. Moths and other insects use a shivering reflex to warm their flight muscles.

Regulating Blood Glucose

When you digest a carbohydrate-containing meal, you absorb glucose into your blood. This causes a temporary rise in the blood glucose concentration, which is brought back down in a few hours. What counteracts the rise in blood glucose following a meal?

Glucose levels within the blood are constantly monitored by a sensor, cells called the islets of Langerhans in the pancreas (discussed in chapter 27). When glucose levels increase (the condition of "high blood sugar" in figure 28.2a), the islets secrete the hormone *insulin*, which stimulates the uptake of blood glucose into muscles, liver, and adipose tissue (not shown). The muscles and liver can convert the glucose into the polysaccharide glycogen. Adipose cells can convert glucose into fat. These actions lower the blood glucose and help to store energy in forms that the body can use later. When enough blood glucose is absorbed, reaching the set point indicated by the green box, the release of insulin stops. When blood glucose levels decrease below the set point, as they do between meals, during periods of fasting, and during exercise, the liver secretes glucose into the blood (the center arrow in figure 28.2b). This glucose is obtained in part from the breakdown of liver glycogen. This breakdown of liver glycogen is stimulated in two ways: by another hormone, *glucagon*, which is also secreted by the islets of Langerhans, and by adrenaline secreted from the adrenal gland (discussed in much more detail in chapter 31).

> **28.1 Negative feedback mechanisms correct deviations from a set point for different internal variables. In this way, body temperature and blood glucose, for example, are kept within a normal range.**

Figure 28.2 Control of blood glucose levels.

(a) When blood glucose levels are high, cells within the pancreas produce the hormone insulin, which stimulates the liver and muscles to convert blood glucose into glycogen. (b) When blood glucose levels are low, other cells within the pancreas release the hormone glucagon into the bloodstream; in addition, cells within the adrenal gland release the hormone adrenaline into the bloodstream. When they reach the liver, these two hormones act to increase the liver's breakdown of glycogen to glucose.

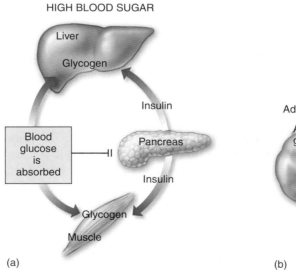

(a)

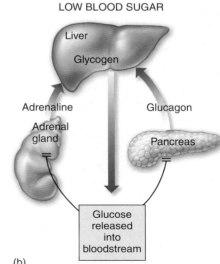

(b)

28.2 Regulating the Body's Water Content

Animals must also carefully monitor the water content of their bodies. The first animals evolved in seawater, and the physiology of all animals reflects this origin. Approximately two-thirds of every vertebrate's body is water. If the amount of water in the body of a vertebrate falls much lower than this, the animal dies. Animals use various mechanisms for **osmoregulation,** the regulation of the body's osmotic composition, or how much water and salt it contains. The proper operation of many vertebrate organ systems of the body requires that the osmotic concentration of the blood—the concentration of solutes dissolved within it—be kept within narrow bounds.

Animals have evolved a variety of mechanisms to cope with problems of water balance. In many animals and single-celled organisms, the removal of water or salts from the body is coupled with the removal of metabolic wastes through the excretory system. Protists, like the *Paramecium* in figure 28.3, employ contractile vacuoles for this purpose, as do sponges. Water and metabolic wastes are collected by the endoplasmic reticulum and pass through feeder canals to the contractile vacuole. The water and wastes are expelled when the vacuole contracts and releases its contents out a pore. Multicellular animals have a system of excretory tubules (little tubes) that expel fluid and wastes from the body.

In flatworms, these tubules are called *protonephridia* (the green structures you see in figure 28.4). They branch throughout the body into bulblike **flame cells,** shown in the enlargement. While these simple excretory structures open to the outside of the body, they do not open to the inside of the body. Rather, the beating action of cilia within the flame cells draw in fluid from the body, which is passed into a collecting tube. Water and metabolites are then reabsorbed, and the substances to be excreted are expelled through excretory pores.

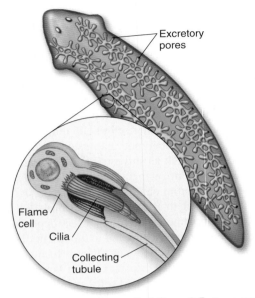

Figure 28.4   The protonephridia of flatworms.

A branching system of tubules, bulblike flame cells, and excretory pores make up the protonephridia of flatworms.

Other invertebrates have a system of tubules that open both to the inside and to the outside of the body. In the earthworm, these tubules are known as *nephridia,* the blue structures you see in figure 28.5. The nephridia obtain fluid from the body cavity through a process of filtration into funnel-shaped structures called *nephrostomes.* The term *filtration* is used because the fluid is formed under pressure and passes through small openings, so that molecules larger than a certain size are excluded. This filtered fluid is isotonic (having the same osmotic concentration) to the fluid in the coelom, but as it passes through the tubules of the nephridia, NaCl is removed by active transport processes. A general term for transport out of the tubule and into the surrounding body fluids is *reabsorption.* Because salt is reabsorbed from the filtrate, the urine excreted is more dilute than the body fluids (meaning it is hypotonic).

The excretory organs in insects are called **Malpighian tubules,** the green structures you see in figure 28.6. Malpighian tubules are extensions of the digestive tract that branch off anterior to the hindgut. Urine is not formed by filtration in these tubules, because there is no pressure difference between the blood in the body cavity and the tubule. Instead, waste molecules and potassium ions (K^+) are secreted into the tubules by active transport. *Secretion* is the opposite of reabsorption—ions or molecules are transported from the body fluid into the tubule. The secretion of K^+ creates an osmotic gradient that causes water to enter the tubules by osmosis from the body's open circulatory system. Most of the water and K^+ is then reabsorbed into the circulatory system through the epithelium of the hindgut, leaving only small molecules and waste products to be excreted from the rectum along with feces. Malpighian tubules thus provide a very efficient means of water conservation.

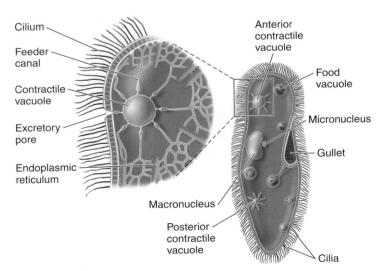

Figure 28.3   Contractile vacuoles in *Paramecium.*

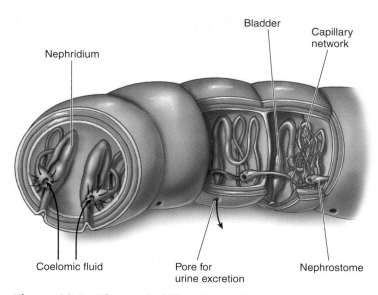

Figure 28.5 The nephridia of annelids.

Most invertebrates, such as the annelid shown here, have nephridia. These consist of tubules that receive a filtrate of coelomic fluid, which enters the funnel-like nephrostomes. Salt can be reabsorbed from these tubules, and the fluid that remains, urine, is released from pores into the external environment.

Kidneys are the excretory organs in vertebrates and is the topic for the rest of the chapter. Unlike the Malpighian tubules of insects, kidneys create a tubular fluid by filtration of the blood under pressure. In addition to containing waste products and water, the filtrate contains many small molecules that are of value to the animal, including glucose, amino acids, and vitamins. These molecules and most of the water are reabsorbed from the tubules into the blood, while wastes remain in the filtrate. Additional wastes may be secreted by the tubules and added to the filtrate, and the final waste product, urine, is eliminated from the body.

It may seem odd that the vertebrate kidney should filter out almost everything from blood plasma (except proteins, which are too large to be filtered) and then spend energy to take back or reabsorb what the body needs. But selective reabsorption provides great flexibility; various vertebrate groups have evolved the ability to reabsorb different molecules that are especially valuable in particular habitats. This flexibility is a key factor underlying the successful colonization of many diverse environments by the vertebrates.

28.2 Many invertebrates filter fluid into a system of tubules and then reabsorb ions and water, leaving waste products for excretion. Insects create an excretory fluid by secreting K^+ into tubules, which draws water osmotically. The vertebrate kidney produces a filtrate that enters tubules from which water is reabsorbed.

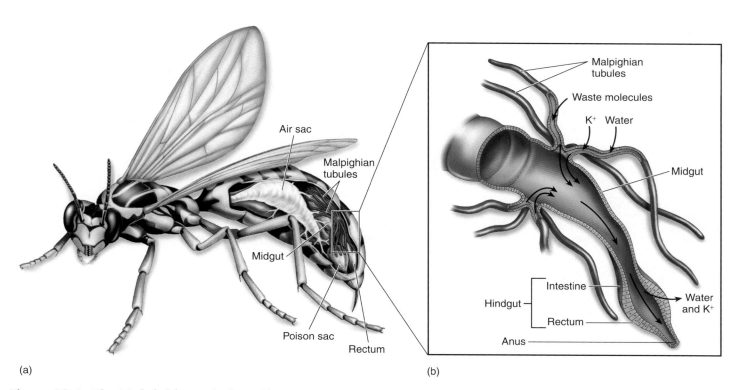

(a) (b)

Figure 28.6 The Malpighian tubules of insects.

(a) The Malpighian tubules of insects are extensions of the digestive tract that collect water and wastes from the body's circulatory system.
(b) K^+ is secreted into these tubules, drawing water with it osmotically. Much of this water is reabsorbed across the wall of the hindgut.

28.3 Evolution of the Vertebrate Kidney

The kidney, first evolving in freshwater fish, is a complex organ consisting of up to a million repeating disposal units called **nephrons.** The nephron pictured in figure 28.7 is representative of those found in mammals and birds; the nephron of other vertebrates lacks a looped portion. Blood pressure forces the fluid in blood through a capillary bed, called the *glomerulus,* at the top of each nephron. The glomerulus retains blood cells, proteins, and other useful large molecules in the blood but allows the water, and the small molecules and wastes dissolved in it, to pass into a cuplike structure surrounding the glomerulus and then into the nephron tube. As the filtered fluid passes through the first part of the nephron tube (labeled as the proximal arm in figure 28.7), useful sugars, amino acids, and ions (such as Ca^{++}) are recovered from it by active transport, leaving the water and metabolic wastes dissolved in a fluid, called urine. Water and salts are reabsorbed later in the nephron.

Although the same basic design has been retained in all vertebrate kidneys, there have been a few modifications. Because the original glomerular filtrate is isotonic to blood, all vertebrates can produce a urine that is isotonic to (by reabsorbing ions) or hypotonic to (more dilute than) blood. Only birds and mammals can reabsorb water from their glomerular filtrate to produce a urine that is hypertonic to (more concentrated than) blood.

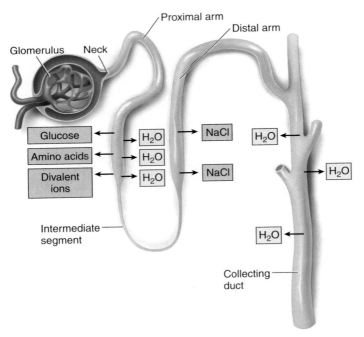

Figure 28.7 Basic organization of the vertebrate nephron.

Freshwater Fish

Kidneys are thought to have evolved first among the freshwater teleosts, or bony fish. Because the body fluids of a freshwater fish have a greater osmotic concentration than the surrounding water, these animals face two serious problems because of osmosis and diffusion: (1) water tends to enter the body from the environment; and (2) solutes tend to leave the body and enter the environment. Freshwater fish, as shown at the top of figure 28.8, address the first problem by *not* drinking water (water enters the mouth but passes out through the gills—it is not swallowed) and by excreting a large volume of dilute urine, which is hypotonic to their body fluids. They address the second problem by reabsorbing ions (NaCl) across the nephron tubules, from the glomerular filtrate back into the blood. In addition, they actively transport ions (NaCl) across their gills from the surrounding water into the blood.

Marine Bony Fish

Although most groups of animals seem to have evolved first in the sea, marine bony fish (teleosts) probably evolved from freshwater ancestors. They faced significant new problems in making the transition to the sea because their body fluids are hypotonic to the surrounding seawater. Consequently, water tends to leave their bodies by osmosis across their gills, and they also lose water in their urine. To compensate for this continuous water loss, marine fish, the reddish-colored fish in figure 28.8, drink large amounts of seawater.

Many of the divalent cations in the seawater that a marine fish drinks (principally Ca^{++} and Mg^{++} in the form of $MgSO_4$) remain in the digestive tract and are eliminated through the anus. Some, however, are absorbed into the blood, as are the monovalent ions K^+, Na^+, and Cl^-. Most of the monovalent ions are actively transported out of the blood across the gills, while the divalent ions that enter the blood (represented by $MgSO_4$ in the figure) are secreted into the nephron tubules and excreted in the urine. In these two ways, marine bony fish eliminate the ions they get from the seawater they drink. The urine they excrete is isotonic to their body fluids. It is more concentrated than the urine of freshwater fish but not as concentrated as that of birds and mammals.

Cartilaginous Fish

The elasmobranchs—sharks, rays, and skates—are by far the most common subclass in the class Chondrichthyes (cartilaginous fish). Elasmobranchs have solved the osmotic problem posed by their seawater environment in a different way than have the bony fish. Instead of having body fluids that are hypotonic to seawater, so that they have to continuously drink seawater and actively pump out ions, the elasmobranchs reabsorb urea from the nephron tubules (as shown in the enlarged view that is circled in figure 28.9) and maintain a blood urea concentration that is 100 times higher than that of mammals.

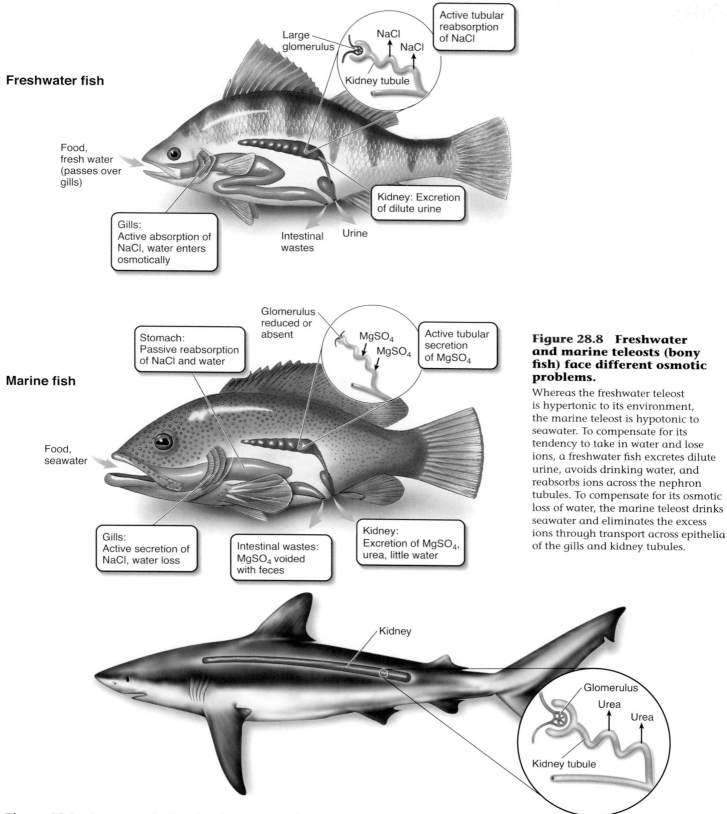

Freshwater fish

Large glomerulus

NaCl

NaCl

Active tubular reabsorption of NaCl

Kidney tubule

Food, fresh water (passes over gills)

Gills: Active absorption of NaCl, water enters osmotically

Intestinal wastes

Urine

Kidney: Excretion of dilute urine

Marine fish

Stomach: Passive reabsorption of NaCl and water

Glomerulus reduced or absent

$MgSO_4$

$MgSO_4$

Active tubular secretion of $MgSO_4$

Food, seawater

Gills: Active secretion of NaCl, water loss

Intestinal wastes: $MgSO_4$ voided with feces

Kidney: Excretion of $MgSO_4$, urea, little water

Figure 28.8 Freshwater and marine teleosts (bony fish) face different osmotic problems.

Whereas the freshwater teleost is hypertonic to its environment, the marine teleost is hypotonic to seawater. To compensate for its tendency to take in water and lose ions, a freshwater fish excretes dilute urine, avoids drinking water, and reabsorbs ions across the nephron tubules. To compensate for its osmotic loss of water, the marine teleost drinks seawater and eliminates the excess ions through transport across epithelia of the gills and kidney tubules.

Kidney

Glomerulus

Urea

Urea

Kidney tubule

Figure 28.9 Osmoregulation in elasmobranchs.

The elasmobranchs control osmotic balance in seawater by reabsorbing urea from the nephron, thereby maintaining an internal osmotic concentration that is the same as the surrounding seawater.

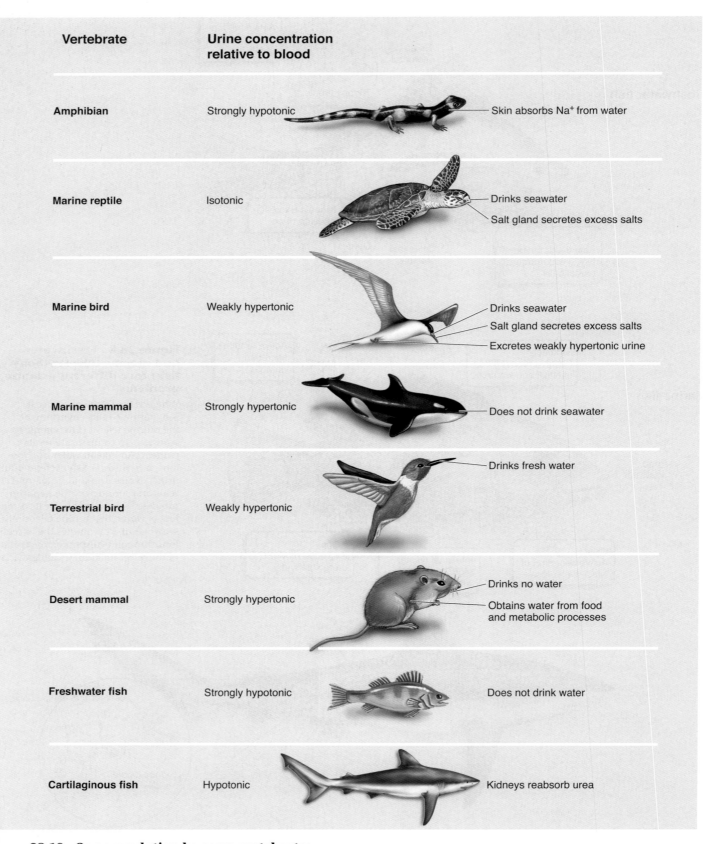

Vertebrate	Urine concentration relative to blood	
Amphibian	Strongly hypotonic	Skin absorbs Na⁺ from water
Marine reptile	Isotonic	Drinks seawater / Salt gland secretes excess salts
Marine bird	Weakly hypertonic	Drinks seawater / Salt gland secretes excess salts / Excretes weakly hypertonic urine
Marine mammal	Strongly hypertonic	Does not drink seawater
Terrestrial bird	Weakly hypertonic	Drinks fresh water
Desert mammal	Strongly hypertonic	Drinks no water / Obtains water from food and metabolic processes
Freshwater fish	Strongly hypotonic	Does not drink water
Cartilaginous fish	Hypotonic	Kidneys reabsorb urea

Figure 28.10 Osmoregulation by some vertebrates.
Only birds and mammals can produce a hypertonic urine and thereby retain water efficiently, but marine reptiles and birds can drink seawater and excrete the excess salt through salt glands.

This added urea makes their blood approximately isotonic to the surrounding sea. Because there is no net water movement between isotonic solutions, water loss is prevented. Hence, these fish do not need to drink seawater for osmotic balance, and their kidneys and gills do not have to remove large amounts of ions from their bodies. The enzymes and tissues of the cartilaginous fish have evolved to tolerate the high urea concentrations.

Amphibians and Reptiles

The first terrestrial vertebrates were the amphibians (pictured at the top of figure 28.10), and the amphibian kidney is identical to that of freshwater fish. This is not surprising because amphibians spend a significant portion of their time in freshwater, and when on land, they generally stay in wet places. Like their freshwater ancestors, amphibians produce a very dilute urine and they compensate for their loss of Na^+ by actively transporting Na^+ across their skin from the surrounding water.

Reptiles, on the other hand, live in diverse habitats. Those living mainly in freshwater, like some of the crocodilians, occupy a habitat in many ways similar to that of the freshwater fish and amphibians, and thus have similar kidneys. Marine reptiles, which consist of other crocodilians, turtles (the second entry in figure 28.10), sea snakes, and one lizard, possess kidneys similar to those of their freshwater relatives but face opposite problems; they tend to lose water and take in salts. Like marine teleosts (bony fish), they drink the seawater and excrete an isotonic urine. Marine teleosts eliminate the excess salt by transport across their gills, while marine reptiles eliminate excess salt through salt glands near the nose or eye.

The kidneys of terrestrial reptiles also reabsorb much of the salt and water in the nephron tubules, helping somewhat to conserve blood volume in dry environments. Like fish and amphibians, they cannot produce urine that is more concentrated than the blood plasma. However, when their urine enters their cloaca (the common exit of the digestive and urinary tracts), additional water can be reabsorbed.

Mammals and Birds

Mammals and birds are the only vertebrates able to produce urine with a higher osmotic concentration than their body fluids. This allows these vertebrates to excrete their waste products in a small volume of water, so that more water can be retained in the body. Human kidneys can produce urine that is as much as 4.2 times as concentrated as blood plasma, but the kidneys of some other mammals are even more efficient at conserving water. For example, camels, gerbils, and pocket mice, *Perognathus*, can excrete urine 8, 14, and 22 times as concentrated as their blood plasma, respectively. The kidneys of the kangaroo rat shown in figure 28.11 are so efficient it never has to drink water; it can obtain all the water it needs from its food and from water produced in aerobic cell respiration!

The production of hypertonic urine is accomplished by the looped portion of the nephron (look ahead to figure 28.13), found only in mammals and birds. A nephron with a long loop, called the *loop of Henle*, extends deeper into

Figure 28.11 A desert mammal.
The kangaroo rat *(Dipodomys panamintensis)* has very efficient kidneys that can concentrate urine to a high degree by reabsorbing water, thereby minimizing water loss from the body. This feature is extremely important to the kangaroo rat's survival in dry or desert habitats.

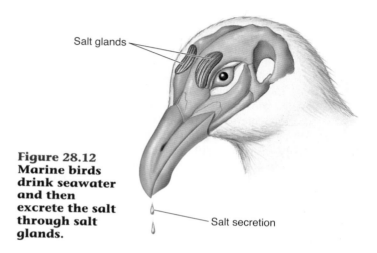

Figure 28.12 Marine birds drink seawater and then excrete the salt through salt glands.

Salt glands

Salt secretion

the tissue of the kidney and can produce more concentrated urine. Most mammals have some nephrons with short loops and other nephrons with loops that are much longer. Birds, however, have relatively few or no nephrons with long loops, so they cannot produce urine that is as concentrated as that of mammals. At most, they can only reabsorb enough water to produce a urine that is about twice the concentration of their blood. Marine birds solve the problem of water loss by drinking seawater and then excreting the excess salt from salt glands near the eyes, which dribbles down the beak as shown in figure 28.12.

The moderately hypertonic urine of a bird is delivered to its cloaca, along with the fecal material from its digestive tract. If needed, additional water can be absorbed across the wall of the cloaca to produce a semisolid white paste or pellet, which is excreted.

> 28.3 The kidneys of freshwater fish must excrete copious amounts of very dilute urine, whereas marine teleosts drink seawater and excrete an isotonic urine. The basic design and function of the nephron of freshwater fish have been retained in the terrestrial vertebrates. Modifications, particularly the loop of Henle, allow mammals and birds to reabsorb water and produce a hypertonic urine.

The Mammalian Kidney

In humans, the kidneys are fist-sized organs located in the region of the lower back (figure 28.13a). Each kidney receives blood from a renal artery, and it is from this blood that urine is produced. Urine drains from each kidney through a **ureter,** which carries the urine to a **urinary bladder.** Urine passes out of the body through the **urethra.** Within the kidney, the mouth of the ureter flares open to form a funnel-like structure, the *renal pelvis.* The renal pelvis, in turn, has cup-shaped extensions that receive urine from the renal tissue. This tissue is divided into an outer **renal cortex** (containing blood vessels in figure 28.13b) and an inner **renal medulla** (containing the cup-shaped structures). Together, these structures perform filtration, reabsorption, secretion, and excretion.

The mammalian kidney is composed of roughly 1 million nephrons (figure 28.13c), each of which is composed of three regions:

1. **Filter.** The filtration device at the top of each nephron is called a **Bowman's capsule.** Within each capsule an arteriole enters and splits into a fine network of vessels called a **glomerulus** (labeled ❶ in the figure). The walls of these capillaries act as a filtration device. Blood pressure forces fluid through the capillary walls. These walls withhold proteins and other large molecules in the blood, while passing water, small molecules, ions, and urea, the primary waste product of metabolism.

2. **Tube.** The Bowman's capsule is connected to a long narrow tube called a renal tubule (labeled ❷ through ❹), which is bent back on itself in its center, called the **loop of Henle.** This long hairpin loop is a reabsorption device. Like the mammalian small intestine, it extracts from the filtrate passing through the tube molecules useful to the body, such as glucose and a variety of ions.

3. **Duct.** The tube empties into a large collection tube called a **collecting duct** (❺). The collecting duct operates as a water conservation device, reclaiming

water from the urine so that it is not lost from the body. Human urine is four times as concentrated as blood plasma—that is, the collecting ducts remove much of the water from the filtrate passing through the kidney. Your kidneys achieve this remarkable degree of water conservation by a simple but superbly designed mechanism: they bend the duct back alongside the nephron tube and make the duct permeable to urea. Urea passes out of the collecting duct by diffusion. This greatly increases the local salt (urea) concentration in the tissue surrounding the tube, causing water in urine to pass out of the tube by osmosis. The salty tissue sucks up water from the urine like blotting paper, passing it on to blood vessels that carry it out of the kidneys and back to the bloodstream.

The Kidney at Work

The formation of urine within the mammalian kidney involves the movement of several kinds of molecules between nephrons and the capillaries that surround them. Five steps are involved and indicated in figure 28.13c: pressure filtration ❶, reabsorption of water ❷, selective reabsorption of ions and nutrients ❸, tubular secretion ❹, and further reabsorption of water ❺.

Pressure Filtration. Driven by the blood pressure, small molecules are pushed across the thin walls of the glomerulus

Figure 28.13 The mammalian urinary system contains two kidneys, each of which contain about a million nephrons that lie in the renal cortex and renal medulla.

(a) The urinary system consists of the kidneys, the ureter, which transports urine from the kidneys to the urinary bladder, and the urethra. (b) The kidney is a bean-shaped reddish brown organ and contains about 1 million nephrons. (c) The glomerulus is enclosed within a filtration device called a Bowman's capsule. Blood pressure forces liquid from blood through the glomerulus and into the proximal tubule of the nephron, where glucose and small proteins are reabsorbed from the filtrate. The filtrate then passes through a double-loop arrangement consisting of the loop of Henle and the collecting duct, which act to remove water from the filtrate. The water is then collected by blood vessels and transported out of the kidney to the systemic (body) circulation.

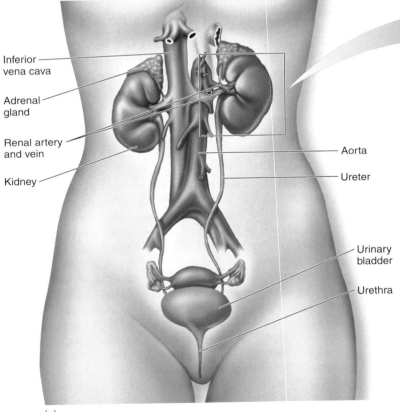

Inferior vena cava

Adrenal gland

Renal artery and vein

Kidney

Aorta

Ureter

Urinary bladder

Urethra

(a)

to the inside of the Bowman's capsule **1**. Blood cells and large molecules like proteins cannot pass through, and as a result the blood that enters the glomerulus is divided into two paths: nonfilterable blood components that are retained and leave the glomerulus in the bloodstream and filterable components that pass across and leave the glomerulus in the urine. This filterable stream is called the **glomerular filtrate.** It contains water, nitrogenous wastes (principally urea), nutrients (principally glucose and amino acids), and a variety of ions.

Reabsorption of Water. Filtrate from the glomerulus passes down the proximal tube into the descending arm of the loop of Henle. The walls of this portion of the tube are impermeable to either salts or urea but are freely permeable to water. Because (for reasons we discuss later) the surrounding tissue has a high concentration of urea, water passes out of the descending arm by osmosis **2**, leaving behind a more concentrated filtrate.

Selective Reabsorption. At the turn in the loop, the walls of the tubule become permeable to salts but much less permeable to water. As the concentrated filtrate passes up this ascending arm, these nutrients pass out into the surrounding tissue **3**, where they are carried away by blood vessels. In the upper region of the ascending arm are active transport channels that pump out salt (NaCl). Left behind in the filtrate is the urea that initially passed through the glomerulus as nitrogenous waste.

At this point in the tubule, the urea concentration is becoming very high.

Tubular Secretion. In the ascending loop, substances are also added to the urine by a process called tubular secretion **4**. This active transport process secretes into the urine other nitrogenous wastes such as uric acid and ammonia, as well as excess hydrogen ions.

Further Reabsorption of Water. The tubule then empties into a collecting duct that passes back through the tissue of the kidney. Unlike the tubule, the lower portions of the collecting duct are permeable to urea, some of which diffuses out into the surrounding tissue (that is why the tissue surrounding the descending arm of the loop of Henle has a high urea concentration indicated by the darker pink color in the figure). The high urea concentration in the tissue causes even more water to pass outward from the filtrate by osmosis **5**. The filtrate that is left in the collecting duct after salts, nutrients, and water have been removed is urine.

> **28.4** The mammalian kidney pushes waste molecules through a filter and then reclaims water and useful metabolites and ions from the filtrate before eliminating the residual urine.

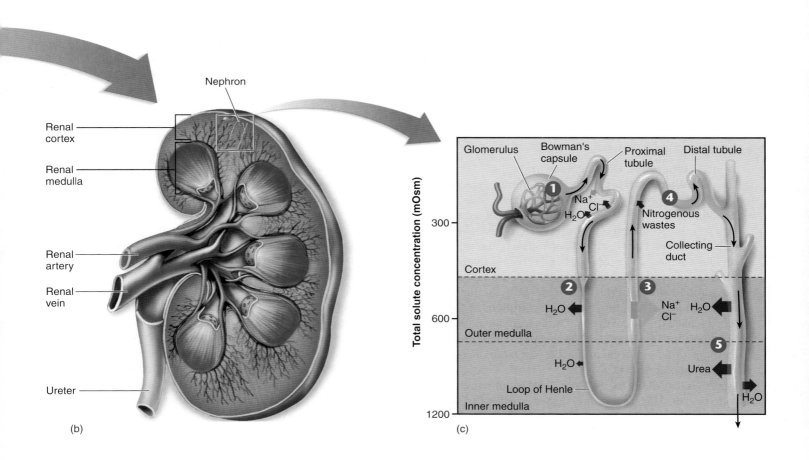

(b)

(c)

How Hormones Control Your Kidney's Functions

In humans, like all mammals and birds, the amount of water excreted in the urine varies according to the changing needs of the body. Acting through the mechanisms described in this chapter, your kidneys excrete a hypertonic urine when your body needs to conserve water. If you drink a lot of water, your kidneys will excrete a hypotonic urine. As a result, the volume of your blood, your blood pressure, and the salt levels of your blood plasma are all maintained relatively constant by the kidneys, no matter how much water you drink. The kidneys also regulate the plasma Na⁺ and K⁺ concentrations and blood pH within narrow limits. These homeostatic functions of the kidneys are coordinated primarily by hormones, which are chemical signals produced in one part of the body but influence events in other parts. Hormones are the subject of chapter 31.

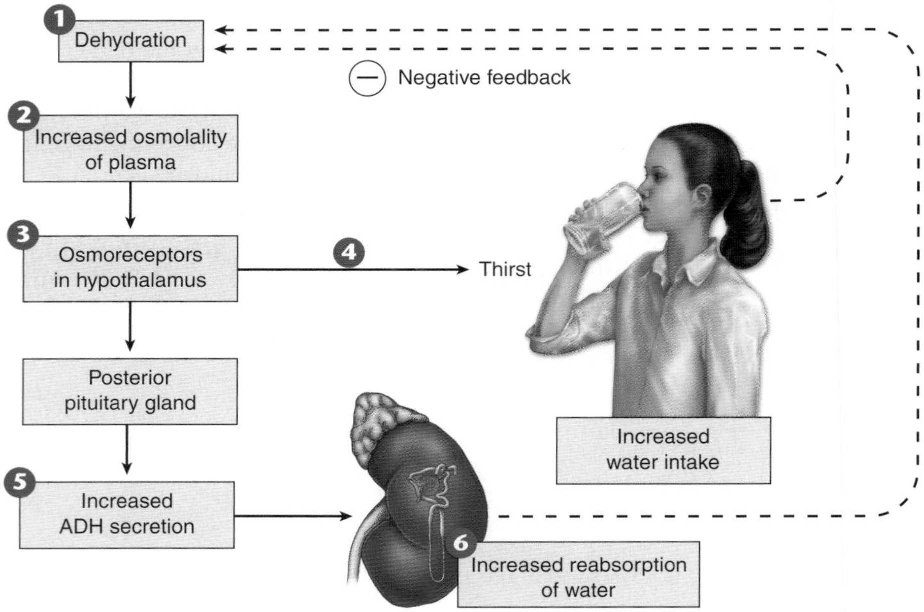

Antidiuretic Hormone. Antidiuretic hormone (ADH) is produced by a portion of your brain called the hypothalamus. The primary stimulus for ADH secretion into your blood stream is an increase in the osmolality (concentration of salt) of the blood plasma. The salt in each milliliter of plasma increases when you are dehydrated or when you eat salty food. Consider the stimulus of dehydration shown in the figure here. Dehydration ❶ causes an increase in the solute concentrations in the blood (called the *osmolality*) ❷. Osmoreceptors in the hypothalamus ❸ respond to the elevated blood osmolality by triggering a sensation of thirst ❹ and by an increase in the secretion of ADH ❺.

ADH causes the walls of the distal tubules and collecting ducts in your kidneys (see figure 28.13c) to become more permeable to water, thus increasing the amount of water they absorb from the urine as it passes through your kidneys ❻. The increased absorption of water from the kidneys feeds back to the osmoreceptors (the dashed line), causing a reduction in the secretion of ADH. When the secretion of ADH is reduced, the walls become less permeable, and you excrete more water in your urine.

Under conditions of maximal ADH secretion, you excrete only 600 milliliters of highly concentrated urine per day. A person who lacks ADH has the disorder known as diabetes insipidus and constantly excretes a large volume of dilute urine. Such a person may become severely dehydrated and succumb to dangerously low blood pressure.

Aldosterone. Sodium ion is the major solute in your blood plasma. When the blood concentration of Na⁺ falls, the blood osmolality also falls. This drop in osmolality inhibits ADH secretion, causing more water to remain in the collecting duct for excretion in your urine. As a result, your blood volume and blood pressure decrease. A decrease in extracellular Na⁺ also causes more water to be drawn into your cells by osmosis, partially offsetting the drop in plasma osmolality but further decreasing your blood volume and blood pressure. If Na⁺ deprivation is severe, the blood volume may fall so low that there is insufficient blood pressure to sustain your life. For this reason, salt is necessary for life. Many animals have a "salt hunger" and actively seek salt. This is why a "salt lick" will attract deer.

A drop in blood Na⁺ concentration is normally compensated by the kidneys under the influence of another hormone, aldosterone, which is also secreted by your brain. Indeed, under conditions of maximal aldosterone secretion, Na⁺ may be completely absent from the urine. The reabsorption of Na⁺ is followed by Cl⁻ and by water, so aldosterone has the net effect of promoting the retention of both salt and water. It thereby helps to maintain blood volume, osmolality, and pressure.

Antidiuretic hormone stimulates the reabsorption of water by the kidneys.

This action completes a negative feedback loop and helps to maintain homeostasis of blood volume and osmolality.

Figure labels:
- ❶ Dehydration
- ⊖ Negative feedback
- ❷ Increased osmolality of plasma
- ❸ Osmoreceptors in hypothalamus
- ❹ Thirst
- Posterior pituitary gland
- ❺ Increased ADH secretion
- Increased water intake
- ❻ Increased reabsorption of water

28.5 Eliminating Nitrogenous Wastes

Amino acids and nucleic acids are nitrogen-containing molecules. When animals catabolize these molecules for energy or convert them into carbohydrates or lipids, they produce nitrogen-containing by-products called **nitrogenous wastes** that must be eliminated from the body.

The first step in the metabolism of amino acids and nucleic acids is the removal of the amino ($-NH_2$) group and its combination with H^+ to form **ammonia** (NH_3) in the liver, ❶ in figure 28.14. Ammonia is quite toxic to cells and therefore is safe only in very dilute concentrations. The excretion of ammonia is not a problem for the bony fish and tadpoles, which eliminate most of it by diffusion through the gills and the rest by excretion in very dilute urine ❷. In sharks, adult amphibians, and mammals, the nitrogenous wastes are eliminated in the far less toxic form of **urea** ❸. Urea is water-soluble and so can be excreted in large amounts in the urine. It is carried in the bloodstream from its place of synthesis in the liver to the kidneys, where it is excreted in the urine.

Reptiles, birds, and insects excrete nitrogenous wastes in the form of **uric acid** ❹, which is only slightly soluble in water. As a result of its low solubility, uric acid precipitates and thus can be excreted using very little water. Uric acid forms the pasty white material in bird droppings called *guano*. The ability to synthesize uric acid in these groups of animals is also important because their eggs are encased within shells, and nitrogenous wastes build up as the embryo grows within the egg. The formation of uric acid, while a lengthy process that requires considerable energy, produces a compound that crystallizes and precipitates. As a precipitate, it is unable to affect the embryo's development even though it is still inside the egg.

Mammals also produce some uric acid, but it is a waste product of the degradation of purine nucleotides (see chapter 4), not of amino acids. Most mammals have an enzyme called *uricase*, which converts uric acid into a more soluble derivative, **allantoin.** Only humans, apes, and the dalmatian dog lack this enzyme and so must excrete the uric acid. In humans, excessive accumulation of uric acid in the joints produces a condition known as *gout*.

> **28.5** The metabolic breakdown of amino acids and nucleic acids produces ammonia as a by-product. Ammonia is excreted by bony fish, but other vertebrates convert nitrogenous wastes into urea and uric acid, which are less toxic nitrogenous wastes.

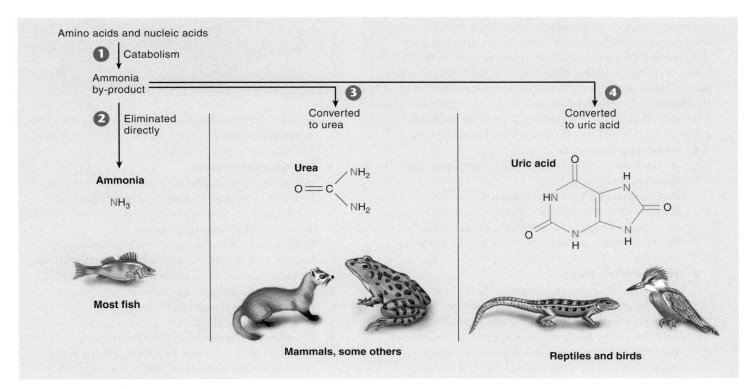

Figure 28.14 Nitrogenous wastes.
When amino acids and nucleic acids are metabolized, the immediate nitrogen by-product is ammonia ❶, which is quite toxic but can be eliminated through the gills of bony fish ❷. Mammals convert ammonia into urea ❸, which is less toxic. Birds and terrestrial reptiles convert it instead into uric acid ❹, which is insoluble in water.

How Do Sleeping Birds Stay Warm?

Mammals and birds are homeotherms—they maintain body temperatures regardless of the temperature of their surroundings. This lets them reliably run their metabolism even when external temperatures fall—the rates of most enzyme-catalyzed reactions slow two- to three-fold for every 10°C temperature drop. Your body keeps its temperature within narrow bounds at 37°C (98.6°F), and birds maintain even higher temperatures. To stay warm like this, mammals and birds continuously carry out oxidative metabolism, which generates heat. This requires a several-fold increase in metabolic rate, which is expensive, particularly when the animal is not active. The logical solution is to give up the struggle to keep warm and let the body temperature drop during sleep, a condition known as *torpor*. While humans don't adopt this approach, many other mammals and birds do. This raises an interesting question: What prevents a sleeping bird in torpor from freezing? Does its body simply adopt the temperature of its surroundings, or is there a body temperature below which metabolic heating kicks in to avoid freezing?

The graph to the right displays an experiment examining this issue in the tropical hummingbird *Eulampis*. The study examines the effect on metabolic rate (measured as oxygen consumption) of decreasing air temperature. Oxygen consumption was assessed over a range of air temperatures from 3° to 37°C, for two contrasting physiological states: the blue data were collected from birds that were awake, the red data from sleeping torpid birds. The blue and red lines, called regression lines, were plotted using curve-fitting statistics that provide the best fit to the data.

1. **Applying Concepts**
   a. Variable. In the graph, what is the dependent variable?
   b. Comparing Two Data Sets. Do awake hummingbirds maintain the same metabolic rate at all body temperatures? sleeping torpid ones? At a given temperature, which has the higher metabolic rate, an awake bird or a sleeping one?
2. **Interpreting Data**
   a. How does the oxygen consumption of awake hummingbirds change as air temperature falls? Why do you think this is so? Is the change consistent over the entire range of air temperatures examined?
   b. How does the oxygen consumption of sleeping torpid birds change as air temperature falls? Is this change consistent over the entire range of air temperatures examined? Explain any difference you detect.

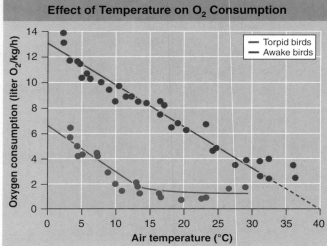

c. Are there any significant differences in the slope of the two regression lines below 15°C? What does this suggest to you?
3. **Making Inferences**
   a. For each five-degree air temperature interval, estimate the average oxygen consumption for awake and for sleeping birds, and plot the difference as a function of air temperature.
   b. Based on this curve, what would you expect to happen to a sleeping bird's body temperature as air temperatures fall from 30° to 20°C? from 15° to 5°C?
4. **Drawing Conclusions** How do *Eulampis* hummingbirds avoid becoming chilled while sleeping on cold nights?
5. **Further Analysis** Flying hummingbirds would be expected to use more metabolic energy than perched awake ones. How would you expect the level of activity to influence the birds' regulation of body temperature? Explain.

Homeostasis

28.1 How the Animal Body Maintains Homeostasis

- Animals maintain relatively constant internal conditions, a process called homeostasis. Homeostasis refers to a dynamic constancy of the internal environment within narrow ranges around set points.

- A negative feedback loop occurs as the activity of effectors, called the response, feeds back to control the effector, usually inhibiting the effector, which is why it is called negative feedback (**figure 28.1**).

- Examples of homeostasis include body temperature and blood glucose levels (**figure 28.2**).

Osmoregulation

28.2 Regulating the Body's Water Content

- Osmotic balance in the body is important, and organisms have evolved various mechanisms to control water balance. *Paramecium* utilize contractile vacuoles, where waste and excess water are collected in vacuoles that contract and expel their contents outside the cell (**figure 28.3**).

- Many invertebrates use systems of tubules that collect fluid and reabsorb ions and water. Flatworms use a network of tubules called protonephridia that collect fluid and ions, directed into the tubules by flame cells and expel waste through openings called excretory pores (**figure 28.4**).

- Nephridia, such as those in earthworms, open to the interior of the body through which fluids are filtered. NaCl is pumped out of the nephridia and reabsorbed by the body. Waste and fluids are excreted as urine through pores (**figure 28.5**).

- Insects create an osmotic gradient by secreting K⁺ into Malpighian tubules, causing water to enter the tubules by osmosis. Water and K⁺ are reabsorbed through the epithelium of the hindgut (**figure 28.6**). The kidneys serve this excretory function in vertebrates.

Osmoregulation in Vertebrates

28.3 Evolution of the Vertebrate Kidney

- The kidney has evolved in different animals, adapting to different environmental conditions. The nephron is the basic unit of the kidney (**figure 28.7**).

- The kidneys of freshwater fish excrete dilute urine, whereas marine bony fish drink seawater and excrete an isotonic urine (**figure 28.8**).

- Cartilaginous fish reabsorb urea so that their blood is isotonic to seawater so they don't lose water. Their bodies have evolved the ability to function with high urea concentrations (**figure 28.9**).

- The kidneys of amphibians and reptiles are similar to those found in fish. Kidneys of mammals and birds differ in that they have a loop of Henle that acts in concentrating the urine to be hypertonic to the blood. Marine birds drink seawater to obtain water, and excrete excess salt from salt glands (**figure 28.12**).

28.4 The Mammalian Kidney

- Each kidney receives blood from a renal artery and produces urine, which then drains from each kidney through a ureter and is carried to the urinary bladder (**figure 28.13a**).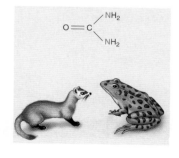

- The blood is first filtered from the glomerulus, a network of blood vessels encapsulated by a structure called Bowman's capsule. Fluids are filtered across the wall of the capillaries and collected by the Bowman's capsule. Water, small molecules, ions, and urea pass into the capsule (**figure 28.13c**).

- The Bowman's capsule is connected to a long tubule where selected molecules and ions are reabsorbed by the body. The tubule forms a hairpin loop called the loop of Henle. The loop of Henle is longer in mammals, allowing the formation of a hypertonic urine. The tubule empties into a collecting duct, which recovers water from the filtrate.

- Different areas of the loop of Henle are permeable to different substances. In the descending arm of the loop of Henle, water passes back into the surrounding tissue by osmosis, but the loop at this point is impermeable to either salts or urea.

- At the turn of the loop, the walls become permeable to salts and other molecules, which pass out of the tubules and into the surrounding tissue. This results in a higher concentration of solutes in the tissue surrounding the nephron tubule.

- As the filtrate passes into the collecting duct, whose walls are permeable to water, water is absorbed back into the tissues through osmosis. Osmosis occurs because the surrounding tissue is highly concentrated with solutes such as ions and other molecules that were absorbed from the ascending loop.

- The fluid that remains in the collecting duct is a highly hypertonic urine that passes to the ureter and is excreted from the body.

28.5 Eliminating Nitrogenous Wastes

- Metabolic breakdown of amino acids and nucleic acids produces ammonia in the liver. Ammonia is toxic and must be eliminated from the body. Different animals use different means of excretion (**figure 28.14**).

- Ammonia is excreted directly in bony fish and tadpoles through gills and in dilute urine.

- In sharks, adult amphibians, and mammals the ammonia is converted to the less toxic urea, which is carried in the bloodstream to the kidneys where it is excreted in the urine.

- Reptiles, birds, and insects excrete nitrogenous wastes as uric acid, which has low toxicity and is only slightly soluble in water and so can be excreted using only small amounts of water.

1. The monitoring and adjusting of the body's condition, such as temperature and pH, is known as
   a. exothermy.
   b. homeostasis.
   c. osmoregulation.
   d. ectothermy.
2. Which of the following describes the method of regulation whereby the response of an effector inhibits the actions of the effector?
   a. inhibitory regulation
   b. homeostasis
   c. positive feedback loop
   d. negative feedback loop
3. Which of the following is *not* involved in osmoregulation?
   a. pancreas
   b. nephridia
   c. Malpighian tubules
   d. flame cells
4. Which of the following animals use Malpighian tubules for excretion?
   a. birds
   b. ants
   c. kangaroo rats
   d. earthworms
5. If your blood sugar is too low, the hormone glucagon is released by the pancreas. This hormone will cause
   a. the release of insulin.
   b. glycogen to break down.
   c. glycogen to be formed.
   d. fat to be formed.

6. To keep the proper concentrations of water and solutes in their blood, freshwater bony fish must drink
   a. lots of water and excrete large volumes of urine that are hypotonic to body fluids.
   b. no water and excrete large volumes of urine that are hypotonic to body fluids.
   c. lots of water and excrete large volumes of urine that are isotonic to body fluids.
   d. no water and excrete large volumes of urine that are isotonic to body fluids.
7. Which of the following animals has the least concentrated urine relative to its blood plasma?
   a. bird
   b. freshwater fish
   c. camel
   d. shark
8. Selective reabsorption of components of the glomerular filtrate occurs where?
   a. Bowman's capsule
   b. glomerulus
   c. loop of Henle
   d. collecting duct
9. Water is removed from kidney filtrate by the process of
   a. diffusion.   c. facilitated diffusion.
   b. active transport. d. osmosis.
10. Humans excrete their excess nitrogenous wastes as
   a. uric acid crystals.
   b. compounds containing protein.
   c. ammonia.
   d. urea.

1. **Figure 28.1** Explain the regulation of blood glucose levels using figure 28.1. First indicate what the various components are, assuming the stimulus is high blood glucose after eating a meal. Then describe what happens to each of the components after the stimulus is detected.

2. **Figure 28.13c** When the body becomes dehydrated it produces a hormone called ADH (see "A Closer Look" on page 570). ADH affects the body by making the collecting duct of the kidney more permeable to water. How would this help a dehydrated person? How would the person's urine be affected?

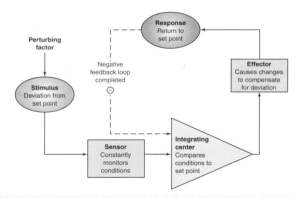

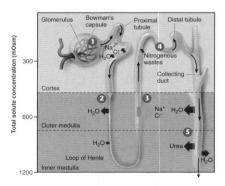

1. Hummingbirds, pocket mice, and young Chiricahua leopard frogs are all about the same size, and all live in the southwestern deserts. Why is there such a difference in their urine?

2. The glomerulus filters out a tremendous amount of water and molecules needed by the body, which must be reabsorbed by the body in the kidney. The Malpighian tubules of insects might seem to function more logically, secreting molecules and ions that need to be excreted. What advantages might a filtration-reabsorption process provide over a strictly secretion process of elimination?

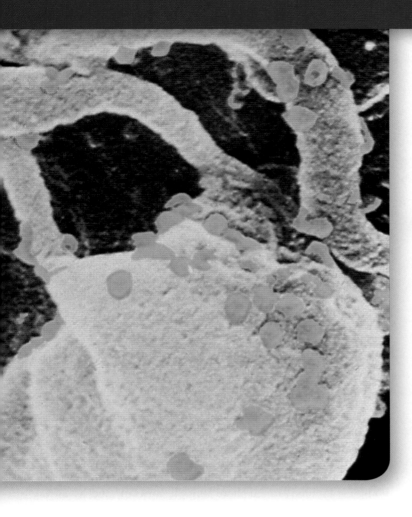

29

How the Animal Body Defends Itself

A ll animals are constantly at war with bacteria and viruses that attempt to use the rich resources of the cellular environment to fuel their own reproduction. Another war is fought on a very different front by animals against their own cells when cells become cancerous and begin to grow without restraint. Both of these wars, against invading microbes and against cancer, are fought with the same defensive weapons, the immune system. This chapter focuses on the vertebrate immune system and how it defends the body in the face of these onslaughts. Sometimes the immune system itself is the target of infection, leading to a loss of the body's ability to defend itself. AIDS results from just this sort of infection by a virus called HIV (human immunodeficiency virus). The cell you see here is a human immune system cell called a lymphocyte that has been infected by HIV. Progeny HIV viruses are being released from the infected cell, budding out from its surface. These viruses soon spread to neighboring lymphocytes, infecting them in turn. Eventually a large majority of lymphocytes become infected, and the immune defense is destroyed. The individual viruses, in buds colored blue in this scanning electron micrograph, are extremely small; over 200 million would fit on the period at the end of this sentence.

29.1 Skin: The First Line of Defense

Multicellular bodies offer a feast of nutrients for tiny, single-celled creatures, as well as a warm, sheltered environment in which they can grow and reproduce. We live in a world awash with microbes, and no animal can long withstand their onslaught unprotected. Animals survive because they have a variety of very effective defenses against this constant attack.

Overview of the Three Lines of Defense

The vertebrate is defended from infection the same way knights defended medieval cities. "Walls and moats" make entry difficult; "roaming patrols" attack strangers; and "sentries" challenge anyone wandering about and call patrols if a proper "ID" is not presented.

1. **Walls and moats.** The outermost layer of the vertebrate body, the skin, is the first barrier to penetration by microbes. Mucous membranes in the respiratory and digestive tracts are also important barriers that protect the body from invasion.
2. **Roaming patrols.** If the first line of defense is penetrated, the response of the body is to mount a **cellular counterattack,** using a battery of cells and chemicals that kill microbes. These defenses act very rapidly after the onset of infection.
3. **Sentries.** Lastly, the body is also guarded by mobile cells that patrol the bloodstream, scanning the surfaces of every cell they encounter. They are part of the **specific immune response.** One kind of immune cell aggressively attacks and kills any cell identified as foreign, whereas the other type marks the foreign cell or virus for elimination by the roaming patrols.

The Skin

Skin, like the thick, tough skin of the elephants in figure 29.1, is the outermost layer of the vertebrate body and provides the first defense against invasion by microbes. Skin is our largest organ, comprising some 15% of our total weight. One square centimeter of skin from your forearm (about the size of a dime) contains 200 nerve endings, 10 hairs and muscles, 100 sweat glands, 15 oil glands, 3 blood vessels, 12 heat-sensing organs, 2 cold-sensing organs, and 25 pressure-sensing organs. The section of skin you see in figure 29.2 has two distinct layers: an outer **epidermis** and a lower **dermis.** A **subcutaneous layer** lies underneath the dermis. Cells of the outer epidermis are continually being worn away and replaced by cells moving up from below—in one hour your body loses and replaces approximately 1.5 million skin cells!

The epidermis of skin is from 10 to 30 cells thick, about as thick as this page. The outer layer, called the **stratum corneum,** is the one you see when you look at your arm or face. Cells from this layer are continuously subjected to damage. They are abraded, injured, and worn by friction and stress dur-

Figure 29.1 Skin is the vertebrate body's first line of defense.

This young elephant has tough, leathery skin thicker than a belt, allowing it to follow the herd through dense thickets without injury.

ing the body's many activities. They also lose moisture and dry out. The body deals with this damage not by repairing cells but by replacing them. Cells from the stratum corneum are shed continuously, replaced by new cells produced deep within the epidermis (the dark layer of cells at the border of the epidermis and dermis). The cells of this inner **basal layer** are among the most actively dividing cells of the vertebrate body. New cells formed there migrate upward, and as they move they manufacture keratin protein, which makes them tough. Each cell eventually arrives at the outer surface and takes its turn in the stratum corneum, residing there for about a month before it is shed and replaced by a newer cell. Persistent dandruff (psoriasis) is a chronic skin disorder in which new cells reach the epidermal surface every three or four days, about eight times faster than normal.

The dermis of the skin is from 15 to 40 times thicker than the epidermis. It provides structural support for the epidermis, as well as a matrix for many specialized cells residing within the skin. The wrinkling that occurs as we grow older occurs here. The leather used to manufacture belts and shoes is derived from thick animal dermis. The layer of subcutaneous tissue below the dermis is composed of fat-rich cells that act as shock absorbers and provide insulation, which conserves body heat.

The skin not only defends the body by providing a nearly impermeable barrier, but it also reinforces this defense with chemical weapons. For example, the oil glands that occur along the shaft of the hair and sweat glands, which look like yellow coiled spaghetti, make the skin's surface very acidic (pH of between 4 and 5.5), which inhibits the growth of many microbes. Sweat also contains the enzyme lysozyme, which attacks and digests the cell walls of many bacteria.

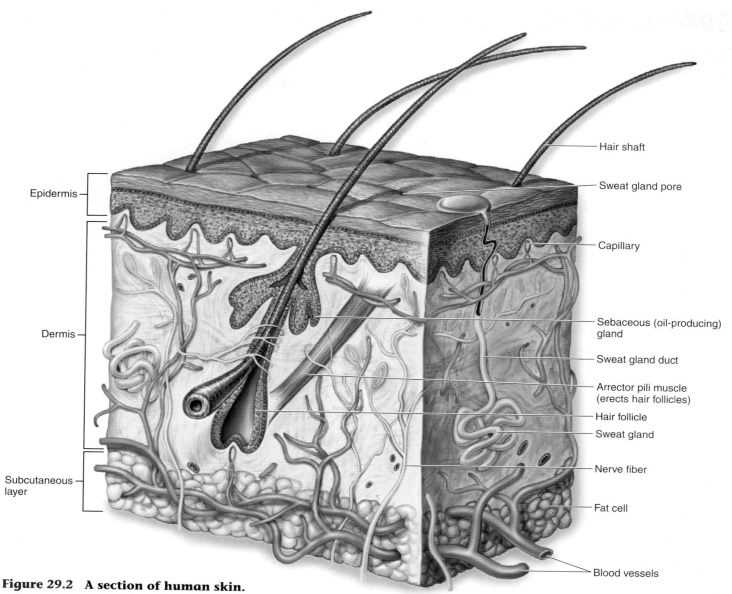

Epidermis

Dermis

Subcutaneous layer

Hair shaft

Sweat gland pore

Capillary

Sebaceous (oil-producing) gland

Sweat gland duct

Arrector pili muscle (erects hair follicles)

Hair follicle

Sweat gland

Nerve fiber

Fat cell

Blood vessels

Figure 29.2 A section of human skin.
The skin defends the body by providing a barrier and sweat and oil glands whose secretions make the skin's surface acidic enough to inhibit the growth of microorganisms.

Other External Surfaces

In addition to the skin, other surfaces, such as the eyes, are exposed to the outside. Like sweat, the tears that bathe the eyes contain lysozyme to fight bacterial infections. Two other potential routes of entry by viruses and microorganisms must be guarded: the *digestive tract* and the *respiratory tract.* Microbes are present in food, but many are killed by saliva (which also contains lysozyme), by the very acidic environment of the stomach (pH of 2), and by digestive enzymes in the intestine. Microorganisms are also present in inhaled air. The cells lining the smaller bronchi and bronchioles secrete a layer of sticky mucus that traps most microorganisms before they can reach the warm, moist lungs, which would provide ideal breeding grounds for them. Other cells lining these passages have cilia that continually sweep the mucus up toward the glottis in the throat, like an escalator. There it can be swallowed, carrying potential invaders out of the lungs and into the digestive tract.

The surface defenses are very effective, but they are occasionally breached. Through breathing, eating, or cuts and nicks, bacteria and viruses now and then enter our bodies. When these invaders reach deeper tissue, a second line of defense comes into play, a cellular counterattack.

> **29.1 Skin and the mucous membranes lining the digestive and respiratory tracts are the body's first defenses.**

Cellular Counterattack: The Second Line of Defense

When the body's interior is invaded, a host of cellular and chemical defenses swing into action. Four are of particular importance: (1) cells that kill invading microbes; (2) proteins that kill invading microbes; (3) the inflammatory response, which speeds defending cells to the point of infection; and (4) the temperature response, which elevates body temperature to slow the growth of invading bacteria.

Although these cells and proteins roam through the body, there is a central location for their storage and distribution; it is called the **lymphatic system.** The lymphatic system consists of the structures shown in figure 29.3: lymph nodes, lymphatic organs, and a network of lymphatic capillaries that drain into lymphatic vessels. Although it has other functions involved with circulation (see chapter 25), it is also involved in the immune response.

Cells That Kill Invading Microbes

The most important counterattack to infection is mounted by white blood cells, which attack invading microbes. These cells

Figure 29.4   A macrophage in action.

In this scanning electron micrograph, a macrophage is "fishing" for rod-shaped bacteria.

patrol the bloodstream and await invaders within the tissues. The three basic kinds of killing cells are macrophages and neutrophils, which are phagocytes, and natural killer cells. Each uses a different tactic to kill invading microbes.

Macrophages. White blood cells called **macrophages** (Greek, "big eaters") kill bacteria by ingesting them, much as an amoeba ingests a food particle. The macrophage in figure 29.4 is sending out long, sticky cytoplasmic extensions that catch the sausage-shaped bacteria and draw them back to the cell where they are engulfed. Although some macrophages are anchored within particular organs, particularly the spleen, most patrol the byways of the body, circulating as precursor cells called **monocytes** in the blood, lymph, and fluid between cells. Macrophages are among the most actively mobile cells of the body.

Neutrophils. Other white blood cells called **neutrophils** act like kamikazes. In addition to ingesting microbes, they release chemicals (identical to household bleach) to "neutralize" the entire area, killing any bacteria in the neighborhood—and themselves in the process. A neutrophil is like a grenade tossed into an infection. It kills everything in the vicinity. Macrophages, by contrast, kill only one invading cell at a time but live to keep on doing it.

Natural Killer Cells. A third kind of white blood cell, called a **natural killer cell,** does not attack invading microbes but rather the body cells that are infected by them. Natural killer cells puncture the membrane of the infected target cell with a molecule called *perforin.* The natural killer cell in figure 29.5 is releasing several perforin molecules that insert in the membrane of the infected cell, like boards on a picket fence, forming a pore that allows water to rush in, causing the cell to swell and burst. Natural killer cells are particularly effective at detecting and attacking body cells that have been infected with viruses. They are also one of the body's most potent defenses against cancer. The cancer cell in figure 29.6 was killed before it had a chance to develop into a tumor.

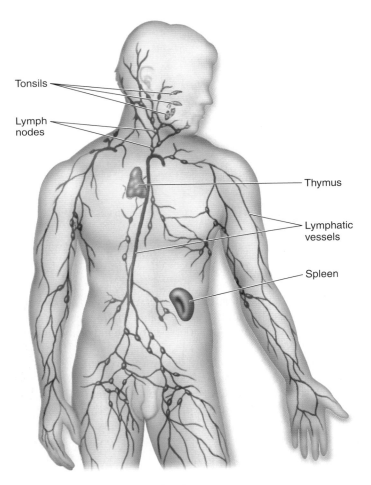

Tonsils

Lymph nodes

Thymus

Lymphatic vessels

Spleen

Figure 29.3   The lymphatic system.

The major lymphatic vessels, organs, and nodes are shown.

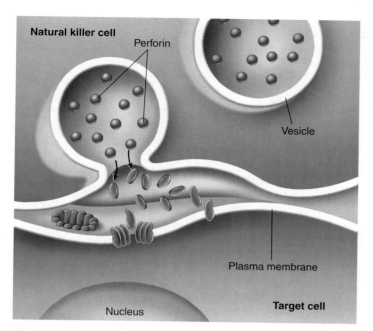

Figure 29.5 Natural killer cells attack target cells.
The initial event is the tight binding of the natural killer cell to the target cell. Binding initiates a chain of events within the killer cell in which vesicles loaded with perforin molecules move to the outer plasma membrane and expel their contents into the intercellular space over the target. The perforin molecules insert into the membrane forming a pore that admits water and ruptures the cell.

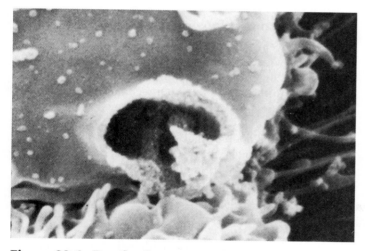

Figure 29.6 Death of a tumor cell.
A natural killer cell has attacked this cancer cell, punching a hole in its plasma membrane. Water has rushed in, making it balloon out. Soon it will burst.

These three kinds of cells can distinguish the body's cells (self) from foreign cells (nonself) because the body's cells contain self-identifying MHC proteins (discussed later in this chapter). When the body's defensive cells fail to make the self versus nonself distinction correctly, they may attack the body's own tissues. Diseases resulting from this failure are known as *autoimmune diseases.*

Proteins That Kill Invading Microbes

The cellular defenses of vertebrates are complemented by a very effective chemical defense called the **complement system.** This system consists of approximately 20 different proteins that circulate freely in the blood plasma in an inactive state. Their defensive activity is triggered when they encounter the cell walls of bacteria or fungi. The complement proteins then aggregate to form a *membrane attack complex* that inserts itself into the foreign cell's plasma membrane, forming a pore like that produced by natural killer cells. Like the perforin pore, the membrane attack complex in figure 29.7 allows water to enter the foreign cell, causing the cell to swell and burst. The difference between the two is that perforin attacks host cells that have become infected, while complement attacks the foreign cell directly. Aggregation of the complement proteins is also triggered by the binding of antibodies to invading microbes, as we'll see in a later section.

The proteins of the complement system can augment the effects of other body defenses. Some amplify the inflammatory response (discussed next) by stimulating histamine release; others attract phagocytes (monocytes and neutrophils) to the area of infection; and still others coat invading microbes, roughening the microbes' surfaces so that phagocytes may attach to them more readily.

Another class of proteins that play a key role in body defense are interferons. There are three major categories of interferons: alpha, beta, and gamma. These polypeptides act as messengers that protect other cells in the vicinity from viral infection. The viruses are still able to penetrate the other cells, but the ability of the viruses to replicate and assemble new virus particles is inhibited.

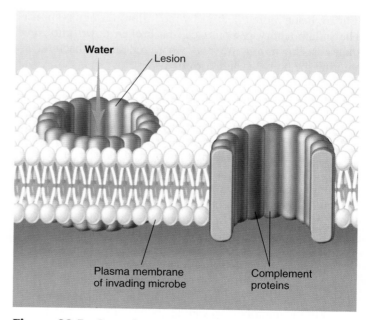

Figure 29.7 Complement proteins attack invaders.
Complement proteins form a transmembrane channel resembling the perforin-lined lesion made by natural killer cells, but complement proteins are free-floating, and they attach to the invading microbe directly, while perforin molecules insert into infected body cells.

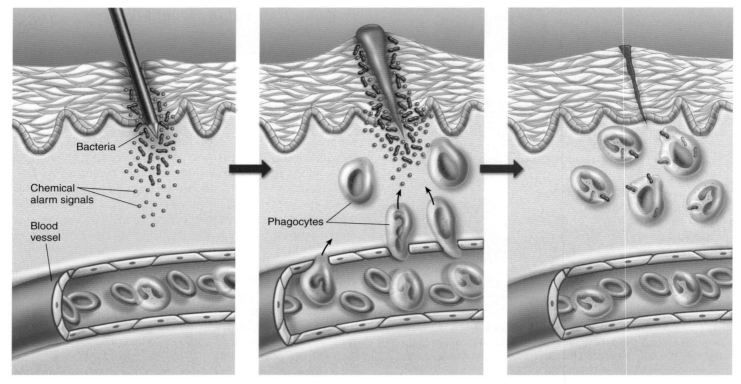

Figure 29.8 The events in a local inflammation.
When an invading microbe has penetrated the skin, chemicals, such as histamine and prostaglandins, act as alarm signals that cause nearby blood vessels to dilate. Increased blood flow brings a wave of phagocytic cells that attack and engulf invading bacteria.

The Inflammatory Response

The aggressive cellular and chemical counterattacks to infection are made more effective by the **inflammatory response.** The inflammatory response can be broken down into three stages as shown in figure 29.8, where rod-shaped bacteria are entering the body through a wound:

1. In the first panel, infected or injured cells release chemical alarm signals, most notably histamine and prostaglandins.
2. These chemical alarm signals cause blood vessels to expand, both increasing the flow of blood to the site of infection or injury and, by stretching their thin walls, making the capillaries more permeable. This produces the redness and swelling so often associated with infection.
3. In the second panel, the increased blood flow through larger, leakier capillaries promotes the migration of phagocytes from the blood to the site of infection, squeezing between cells in the capillary walls. Neutrophils arrive first, spilling out chemicals that kill the microbes (as well as tissue cells in the vicinity and themselves). Monocytes follow and become macrophages, which engulf the pathogens and remains of all the dead cells, as shown in the third panel. This counterattack takes a considerable toll; the pus associated with infections is a mixture of dead or dying neutrophils, tissue cells, and pathogens.

The Temperature Response

Human pathogenic bacteria do not grow well at high temperatures. Thus, when macrophages initiate their counterattack, they increase the odds in their favor by sending a message to the brain to raise the body's temperature. The cluster of brain cells that serves as the body's thermostat responds to the chemical signal by boosting the body temperature several degrees above the normal value of 37°C (98.6°F). The higher-than-normal temperature that results is called a **fever.** Although fever is quite effective at inhibiting microbial growth, very high fevers are dangerous because excessive heat can inactivate critical cellular enzymes. In general, temperatures greater than 39.4°C (103°F) are considered dangerous; those greater than 40.6°C (105°F) are often fatal.

The second line of defense, with both chemical and cellular weapons, provides a sophisticated defense against microbial infection. Only occasionally do bacteria or viruses overwhelm this defense. When this happens, they face yet a third line of defense, more difficult to evade than any they have encountered. It is the specific immune response, the most elaborate of the body's defenses.

> **29.2** Vertebrates respond to infection with a battery of cellular and chemical weapons, including cells and proteins that kill invading microbes, and inflammatory and temperature responses.

29.3 Specific Immunity: The Third Line of Defense

Specific immune defense mechanisms of the body involve the actions of white blood cells, or leukocytes. They are very numerous—of the 100 trillion cells of your body, two in every 100 are white blood cells! Macrophages are white blood cells, as are neutrophils and natural killer cells. In addition, there are T cells, B cells, plasma cells, mast cells, and monocytes (table 29.1). T cells and B cells are called **lymphocytes** and are critical to the specific immune response.

After their origin in the bone marrow, **T cells** migrate to the thymus (hence the designation "T"), a gland just above the heart (see figure 29.3). There they develop the ability to identify microorganisms and viruses by the antigens exposed on their surfaces. An **antigen** is a molecule that provokes a specific immune response. Antigens are large, complex molecules, such as proteins, and they are generally foreign to the body, usually belonging to bacteria and viruses. Tens of millions of different T cells are made, each specializing in the recognition of one particular antigen. No invader can escape being recognized by at least a few T cells. There are four principal kinds of T cells: helper T cells (often symbolized T_H) initiate the immune response; memory T cells (T_M) provide a quick response to a previously encountered antigen; cytotoxic ("cell-poisoning") T cells (T_C) lyse cells that have been infected by viruses; and suppressor T cells (T_S), also called regulatory T cells, terminate the immune response.

Unlike T cells, **B cells** do not travel to the thymus; they complete their maturation in the bone marrow. (B cells are so named because they were originally characterized in a region of chickens called the bursa.) From the bone marrow, B cells are released to circulate in the blood and lymph. Individual B cells, like T cells, are specialized to recognize particular foreign antigens. When a B cell encounters the antigen to which it is targeted, it begins to divide rapidly, and its progeny differentiate into plasma cells and memory cells. Each plasma cell is a miniature factory producing markers called *antibodies*. These antibodies stick like flags to that antigen wherever it occurs in the body, marking any cell bearing the antigen for destruction. So, B cells don't kill foreign invaders directly, but rather they mark these cells so that they are more easily recognized by the other white blood cells that do the dirty work.

B cells and T cells also produce memory cells that provide the body with the ability to recall a previous exposure to an antigen and mount an attack against that antigen very quickly. As described later in this chapter, the initial specific immune response to an antigen encountered for the first time is delayed, which allows the pathogen time to infect the body. A second infection is halted much earlier due to the presence of memory cells, which respond to the pathogen more quickly.

> **29.3** T cells develop in the thymus, whereas B cells develop in the bone marrow. T cells can attack cells that carry antigens. When a B cell encounters a specific antigen, it gives rise to plasma cells that produce antibodies.

TABLE 29.1 CELLS OF THE IMMUNE SYSTEM

Cell Type	Function
Helper T cell	Commander of the immune response; detects infection and sounds the alarm, initiating both T cell and B cell responses
Memory T cell	Provides a quick and effective response to an antigen previously encountered by the body
Cytotoxic T cell	Detects and kills infected body cells; recruited by helper T cells
Suppressor T cell	Dampens the activity of T and B cells, scaling back the defense after the infection has been checked
B cell	Precursor of plasma and memory cells; specialized to recognize specific foreign antigens
Memory B cell	Provides a quick and effective response to an antigen previously encountered by the body
Neutrophil	Engulfs invading bacteria and releases chemicals that kill neighboring bacteria
Plasma cell	Biochemical factory devoted to the production of antibodies directed against specific foreign antigens
Mast cell	Initiator of the inflammatory response, which aids the arrival of leukocytes at a site of infection; secretes histamine and is important in allergic responses
Monocyte	Precursor of macrophage
Macrophage	The body's first cellular line of defense; also serves as antigen-presenting cell to B and T cells and engulfs antibody-covered cells
Natural killer cell	Recognizes and kills infected body cells; natural killer cell detects and kills cells infected by a broad range of invaders

29.4 Initiating the Immune Response

To understand how this third line of defense works, imagine you have just come down with the flu. Influenza viruses enter your body in small water droplets inhaled into your respiratory system. If they avoid becoming ensnared in the mucus lining the respiratory membranes (first line of defense), and avoid consumption by macrophages (second line of defense), the viruses infect and kill mucous membrane cells.

At this point, macrophages initiate the immune response. Macrophages inspect the surfaces of all cells they encounter. Every cell in the body carries special marker proteins on its surface, called major histocompatibility proteins, or **MHC proteins.** The MHC proteins are different for each individual, much as fingerprints are. The MHC protein on the cell in figure 29.9*a* is exactly the same on all cells in that person's body. As a result, the MHC proteins on the tissue cells serve as "self" markers that enable the individual's immune system to distinguish its cells from foreign cells. For example, the foreign microbe in figure 29.9*b* has different surface proteins that are recognized as antigens.

When a foreign particle infects the body, it is taken in by cells and partially digested. Within the cells, the viral antigens are processed and moved to the surface of the plasma membrane, as shown in figure 29.9*c*. The cells that perform this function are called **antigen-presenting cells** and are usually macrophages. At the membrane, the processed antigens are complexed with the MHC proteins. This process is critical for the function of T cells because T cell receptors can be called into action only when antigens are presented in this way. B cells can interact with free antigens directly.

Macrophages that encounter pathogens—either a foreign cell such as a bacterial cell, which lacks proper MHC proteins, or a virus-infected body cell with telltale viral proteins stuck to its surface—respond by secreting a chemical alarm signal. The alarm signal is a protein called **interleukin-1** (Latin for "between white blood cells"). This protein stimulates **helper T cells.** The helper T cells respond to the interleukin-1 alarm by simultaneously initiating two different parallel lines of immune system defense: the cellular immune response carried out by T cells and the antibody or humoral response carried out by B cells. The immune response carried out by T cells is called the *cellular response* because the T lymphocytes attack the cells that carry antigens. The B cell response is called the *humoral response* because antibodies are secreted into the blood and body fluids (*humor* refers to a body fluid).

> **29.4** When macrophages encounter cells without the proper MHC proteins, they secrete a chemical alarm that initiates the immune defense.

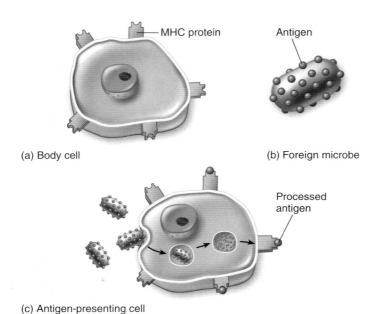

(a) Body cell

(b) Foreign microbe

(c) Antigen-presenting cell

MHC protein

Antigen

Processed antigen

Macrophage

Lymphocyte

(d)

Figure 29.9 How antigens are presented.
(a) Cells of the body have MHC proteins on their surfaces that identify them as "self" cells. Immune system cells do not attack these cells.
(b) Foreign cells or microbes have antigens on their surfaces. (c) T cells can bind to the antigens to initiate an attack only after the antigens are processed and complexed with MHC proteins on the surface of an antigen-presenting cell. B cells recognize the antigens directly, not requiring an antigen-presenting cell. (d) In this electron micrograph, a lymphocyte (*right*) contacts a macrophage (*left*), an antigen-presenting cell.

T Cells: The Cellular Response

When macrophages process the foreign antigens, they trigger the **cellular immune response,** illustrated in figure 29.10. As shown in step ❶, macrophages secrete interleukin-1, which stimulates cell division and proliferation of T cells. Helper T cells become activated when they bind to the complex of MHC proteins and antigens presented to them by the macrophages. The helper T cells then secrete **interleukin-2 ❷**, which stimulates the proliferation of **cytotoxic T cells ❸**, which recognize and destroy infected body cells. Cytotoxic T cells destroy infected cells only if those cells display the foreign antigen together with their MHC proteins ❹.

The body makes millions of different versions of T cells. Each version bears a single, unique kind of receptor protein on its membrane, a receptor able to bind to a particular antigen-MHC protein complex on the surface of an antigen-presenting cell. Any cytotoxic T cell whose receptor fits the particular antigen-MHC protein complex present in the body begins to multiply rapidly, soon forming large numbers of T cells ❸ capable of recognizing the complex containing the particular foreign antigen. Large numbers of infected cells can be quickly eliminated, because the single T cell able to recognize the invading virus is amplified in number to form a large clone of identical T cells, all able to carry out the attack. Any of the body's cells that bear traces of viral infection are destroyed. The method used by cytotoxic T cells to kill infected body cells is similar to that used by natural killer cells and complement—they puncture the plasma membrane of the infected cell. Following an infection, some activated T cells give rise to memory T cells ❺ that remain in the body, ready to mount an attack quickly if the antigen is encountered again.

Cytotoxic T cells will also attack any foreign version of the MHC proteins. Thus even though vertebrates did not evolve the immune system as a defense against tissue transplants, their immune systems will attack transplanted tissue and cause graft rejection. It is for this reason that relatives are often sought for kidney transplants—their MHC proteins are genetically closer to the recipient. The drug cyclosporine is often given to transplant patients because it inactivates cytotoxic T cells.

There is some evidence that cancer cells alter their "self" markers in a way that can be detected by immune cells, creating so-called "cancer-specific antigens," but the potential role of such modified cell surface markers in immunological surveillance against cancer is not well understood.

> **29.5** The cellular immune response is carried out by T cells, which mount an immediate attack on infecting and infected cells, killing any that present unusual surface antigens.

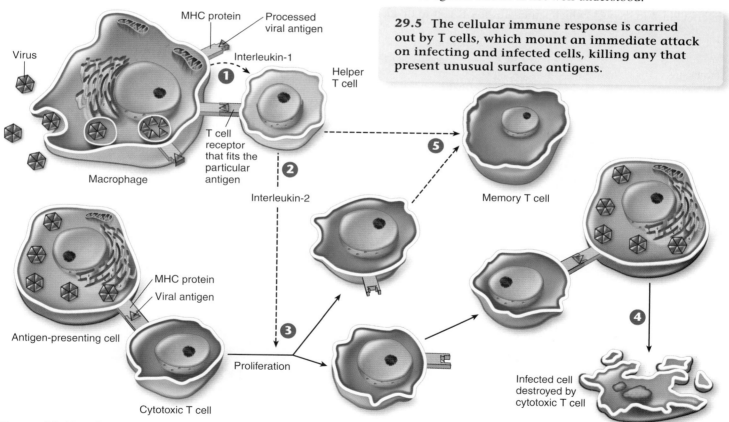

Figure 29.10 The T cell immune defense.

After a macrophage has processed an antigen, it releases interleukin-1, signaling helper T cells to bind to the antigen-MHC protein complex. This triggers the helper T cell to release interleukin-2, which stimulates the multiplication of cytotoxic T cells. In addition, proliferation of cytotoxic T cells is stimulated when a T cell with a receptor that fits the antigen displayed by an antigen-presenting cell binds to the antigen-MHC protein complex. Body cells that have been infected by the antigen are destroyed by the cytotoxic T cells. As the infection subsides, suppressor T cells "turn off" the immune response.

B Cells: The Humoral Response

B cells also respond to helper T cells activated by interleukin-1. Like cytotoxic T cells, B cells have receptor proteins on their surfaces, a different receptor for each version of B cell. B cells recognize invading microbes much as cytotoxic T cells recognize infected cells, but unlike cytotoxic T cells, they do not go on the attack themselves. Rather, they mark the pathogen for destruction by mechanisms that have no "ID check" system of their own. Early in the immune response known as the **humoral immune response,** the markers placed by B cells alert complement proteins to attack the cells carrying them. Later in the response, the markers placed by B cells activate macrophages and natural killer cells.

The way B cells do their marking is simple and foolproof. Unlike the receptors on T cells, which bind only to antigen-MHC protein complexes on antigen-presenting cells, B cell receptors can bind to free, unprocessed antigens, shown in step ❶ of figure 29.11. When a B cell encounters an antigen, antigen particles enter the B cell by endocytosis and get processed and placed on the surface complexed with MHC proteins. Helper T cells that are able to recognize the specific antigen bind to the antigen-MHC protein complex on the B cell and release interleukin-2 in step ❷, which stimulates the B cell to divide.

In addition, free, unprocessed antigens stick to antibodies (the green Y-shaped structures) on the B cell surface. This antigen exposure triggers even more B cell proliferation. B cells divide to produce plasma cells that serve as short-lived antibody factories ❸ and long-lived memory B cells ❹ that remain in the body after the initial infection and are available to mount a quick attack if the antigen enters the body again.

Antibodies are proteins in a class called **immunoglobulins** (abbreviated *Ig*), which is divided into subclasses based on the structures and functions of the antibodies. The five different immunoglobulin subclasses are as follows:

1. **IgM.** This is the first type of antibody to be secreted into the blood during the primary response and to serve as a receptor on the B cell surface. These antibodies also promote agglutination reactions (causing antigen-containing particles to stick together, or agglutinate).
2. **IgG.** This is the major form of antibody secreted in a secondary response and the major one in the blood plasma.
3. **IgD.** These antibodies serve as receptors for antigens on the B cell surface. Their other functions are unknown.
4. **IgA.** This is the form of antibody in external secretions, such as saliva, mucus, and mother's milk.
5. **IgE.** This form of antibody promotes the release of histamine and other agents that produce allergic symptoms, such as those of hay fever.

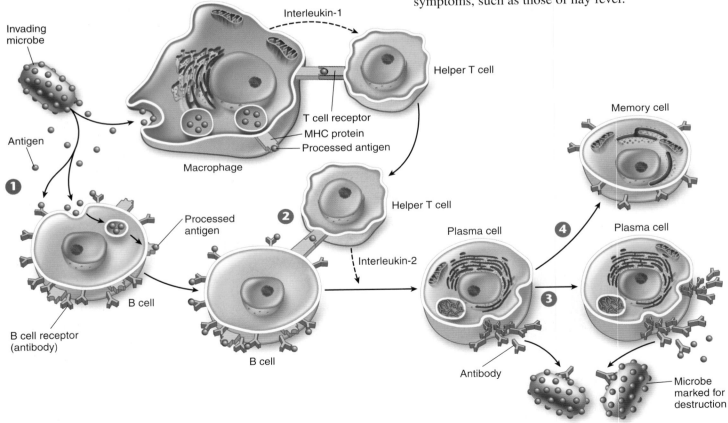

Figure 29.11 The B cell immune defense.
Invading particles are bound by B cells, which interact with helper T cells and are activated to divide. The multiplying B cells produce either memory B cells or plasma cells that secrete antibodies that bind to invading microbes and tag them for destruction by macrophages.

The **plasma cells** that are derived from B cells produce lots of the same particular antibody that was able to bind the antigen in the initial immune response. Flooding through the bloodstream, these antibody proteins (figure 29.12) are able to stick to antigens on any cells or microbes that presents them, flagging those cells and microbes for destruction. Complement proteins, macrophages, or natural killer cells then destroy the antibody-displaying cells and microbes.

The B cell defense is very powerful because it amplifies the reaction to an initial pathogen encounter a millionfold. It is also a very long-lived defense in that a few of the multiplying B cells do not become antibody producers. Instead they become a line of **memory B cells** that continue to patrol your body's tissues, circulating through your blood and lymph for a long time—sometimes for the rest of your life.

Antibody Diversity

The vertebrate immune system is capable of recognizing as foreign practically any nonself molecule presented to it—literally millions of different antigens. Although vertebrate chromosomes contain only a few hundred receptor-encoding genes, it is estimated that human B cells can make between 10^6 and 10^9 different antibody molecules. How do vertebrates generate millions of different antigen receptors when their chromosomes contain only a few hundred versions of the genes encoding those receptors?

The answer to this question is that the millions of immune receptor genes do not exist as single sequences of nucleotides. Rather, they are assembled by stitching together three or four DNA segments that code for different parts of the receptor molecule. When an antibody is assembled, these different sequences of DNA are brought together to form a composite gene. The antibody molecule in figure 29.13 was produced by a composite gene, different sections of DNA were used to produce the constant regions (green), the joining regions (red), the diversity regions (blue), and the variable regions (yellow). This process is called **somatic rearrangement.**

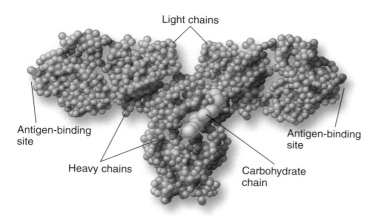

Figure 29.12 An antibody molecule.
In this molecular model of an antibody molecule, each amino acid is represented by a small sphere. Each molecule consists of four protein chains, two "light" (*red*) and two "heavy" (*blue*). The four protein chains wind around one another to form a Y shape. Foreign molecules, called antigens, bind to the arms of the Y.

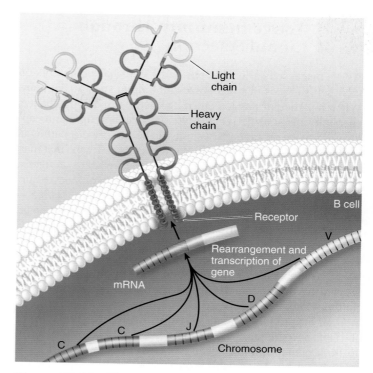

Figure 29.13 The antibody molecule is produced by a composite gene.
Different regions of the DNA code for different regions of the antibody (C, constant regions; J, joining regions; D, diversity regions; and V, variable regions) and are brought together to make a composite gene that codes for the antibody. By combining segments, an enormous number of different antibodies can be produced.

Two other processes generate even more sequences. First, the DNA segments are often joined together with one or two nucleotides off-register, shifting the reading frame during gene transcription and so generating a totally different sequence of amino acids in the protein. Second, random mistakes occur during successive DNA replications as the lymphocytes divide during clonal expansion. Both mutational processes produce changes in amino acid sequences, a phenomenon known as **somatic mutation** because it takes place in a somatic cell rather than in a gamete.

Because a cell may end up with any heavy chain gene and any light chain gene during its maturation, the total number of different antibodies possible is staggering: 16,000 heavy chain combinations × 1,200 light chain combinations = 19 million different possible antibodies. If one also takes into account the changes induced by somatic mutation, the total can exceed 200 million! It should be understood that although this discussion has centered on B cells and their receptors, the receptors on T cells are as diverse as those on B cells because they also are subject to similar somatic rearrangements and mutations.

29.6 In the humoral immune response, B cells label infecting and infected cells with antibodies for destruction by complement proteins, natural killer cells, and macrophages.

Active Immunity Through Clonal Selection

As we discussed earlier, B cells and T cells have receptors on their cell surfaces that recognize and bind to specific antigens. When a particular antigen enters the body, it must, by chance, encounter the specific lymphocyte with the appropriate receptor to provoke an immune response. The first time a pathogen invades the body, there are only a few B cells or T cells that may have the receptors that can recognize the invader's antigens. Binding of the antigen to its receptor on the lymphocyte surface, however, stimulates cell division and produces a *clone* (a population of genetically identical cells). This process is known as **clonal selection.** For example, in the first encounter with a chicken pox virus in figure 29.14 there are only a few cells that can mount an immune response, and the response is relatively weak. This is called a **primary immune response** and is indicated by the first curve, which shows the initial amount of antibody produced upon exposure to the virus.

If the primary immune response involves B cells, some of the cells become plasma cells that secrete antibodies (taking 10 to 14 days to clear the chicken pox virus from the system), and some become memory cells. Some of the T cells involved in the primary response also become memory cells. Because a clone of memory cells specific for that antigen develops after the primary response, the immune response to a second infection by the same pathogen is swifter and stronger, as shown by the second curve. Many memory cells can be produced following the primary response, providing a jump start for the production of antibodies should a second exposure occur. The next time the body is invaded by the same pathogen, the immune system is ready. As a result of the first infection, there is now a large clone of lymphocytes that can recognize that pathogen. This more effective response, elicited by subsequent exposures to an antigen, is called a **secondary immune response.** The "Inquiry and Analysis" feature at the end of this chapter further explores the nature of the secondary immune response.

Memory cells can survive for several decades, which is why people rarely contract chicken pox a second time after they have had it once. Memory cells are also the reason that vaccinations are effective. The viruses causing childhood diseases have surface antigens that change little from year to year, so the same antibody is effective for decades. Other diseases, such as influenza, are caused by viruses whose genes that encode surface proteins mutate rapidly. This rapid genetic change causes new strains to appear every year or so that are not recognized by memory cells from previous infections.

Although the cellular and humoral immune responses were discussed separately, they occur simultaneously in the body. Figure 29.15 follows the steps of a viral infection and shows how the cellular and humoral lines of defense work together to produce the body's specific immune response.

When a virus invades the body, viral proteins are displayed on the surfaces of infected cells as shown in step 1. Viruses and infected cells are taken up by macrophages through phagocytosis (2), and viral proteins are displayed on the surface of the macrophage attached to MHC proteins. Stimulated in this way, macrophages release interleukin-1 (3). Interleukin-1 is an alarm signal that stimulates helper T cells (4). Activated helper T cells release interleukin-2, which triggers both the cellular (T cell) and humoral (B cell) responses (5). The cellular response follows the green arrows, and the humoral response follows the red arrows.

While some activated T cells become memory T cells (5a) that remain in the body and are able to more quickly fight future infections by the same virus, interleukin-2 also activates cytotoxic T cells. The cytotoxic T cells bind to infected cells that carry the viral antigen and kill them (6).

Interleukin-2 also activates B cells (7) that multiply in the cell. Some B cells become memory cells (8) that remain in the body for future infections by the same virus. Other activated B cells become plasma cells (9) that produce antibodies directed against the viral surface proteins. The antibodies released into the body will bind to viral proteins that are displayed on the surface of infected body cells (10) or that are present on the surface of the viruses. Cells or viruses that are tagged with antibodies are destroyed by macrophages that are circulating in the body (11). As you can see, both arms of the immune response work together very effectively to rid the body of invaders.

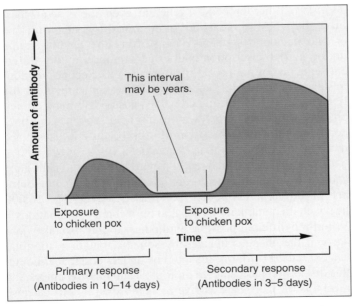

Figure 29.14 The development of active immunity.
Immunity to chicken pox occurs because the first exposure stimulated the development of lymphocyte clones with receptors for the chicken pox virus. As a result of clonal selection, a second exposure stimulates the immune system to produce large amounts of the antibody more rapidly than before, keeping the person from getting sick again.

> **29.7 A strong immune response is possible because infecting cells stimulate the few responding B cells and T cells to divide repeatedly, forming clones of responding cells.**

The Immune Response

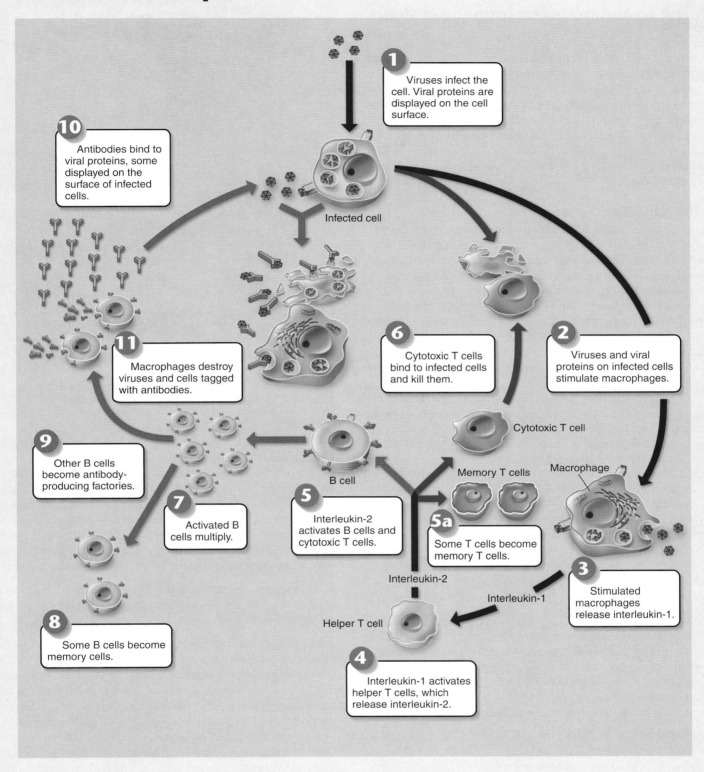

1 Viruses infect the cell. Viral proteins are displayed on the cell surface.

10 Antibodies bind to viral proteins, some displayed on the surface of infected cells.

Infected cell

11 Macrophages destroy viruses and cells tagged with antibodies.

6 Cytotoxic T cells bind to infected cells and kill them.

2 Viruses and viral proteins on infected cells stimulate macrophages.

Cytotoxic T cell

9 Other B cells become antibody-producing factories.

Memory T cells

Macrophage

7 Activated B cells multiply.

B cell

5 Interleukin-2 activates B cells and cytotoxic T cells.

5a Some T cells become memory T cells.

Interleukin-1

3 Stimulated macrophages release interleukin-1.

Interleukin-2

8 Some B cells become memory cells.

Helper T cell

4 Interleukin-1 activates helper T cells, which release interleukin-2.

Figure 29.15 How the immune response works.

Evolution of the Immune System

All organisms possess mechanisms to protect themselves from the onslaught of smaller organisms and viruses. Bacteria defend against viral invasion by means of *restriction endonucleases*. Recall from chapter 14 that these are enzymes that degrade any foreign DNA lacking the specific pattern of DNA methylation characteristic of that bacterium. Multicellular organisms face a more difficult problem in defense because their bodies often take up whole viruses, bacteria, or fungi instead of naked DNA.

Invertebrates

Invertebrate animals solve this problem by marking the surfaces of their cells with proteins that serve as self labels. Special amoeboid cells in the invertebrate attack and engulf any invading cells that lack such labels. By looking for the absence of specific markers, invertebrates employ a *negative* test to recognize foreign cells and viruses. This method provides invertebrates with a very effective surveillance system, although it has one great weakness: any microorganism or virus with a surface protein resembling the invertebrate self marker will not be recognized as foreign. An invertebrate has no defense against such a "copycat" invader.

In 1882, Russian zoologist Elie Metchnikoff became the first to recognize that invertebrate animals possess immune defenses. On a beach in Sicily, he collected the tiny transparent larva of a common sea star. Carefully he pierced it with a rose thorn. When he looked at the larva the next morning, he saw a host of tiny cells covering the surface of the thorn as if trying to engulf it (figure 29.16). The cells were attempting to defend the larva by ingesting the invader by phagocytosis (see figure 5.28). For this discovery of what came to be known as the cellular immune response, Metchnikoff was awarded the 1908 Nobel Prize in Physiology or Medicine, along with Paul Ehrlich for his work on the other major part of the immune defense, the antibody or humoral immune response. The invertebrate immune response shares several elements with the vertebrate immune response.

Phagocytes. All animals possess phagocytic cells that attack invading microbes. These phagocytic cells travel through the animal's circulatory system or circulate within the fluid-filled body cavity. In simple animals like sponges, which lack either a circulatory system or a body cavity, the phagocytic cells circulate among the spaces between cells.

Distinguishing Self from Nonself. The ability to recognize the difference between cells of one's own body and those of another individual appears to have evolved early in the history of life. Sponges, thought to be the oldest animals, attack grafts from other sponges, as do insects and sea stars. None of these invertebrates, however, exhibit any evidence of immunological memory; apparently, the antibody-based humoral immune defense did not evolve until the vertebrates.

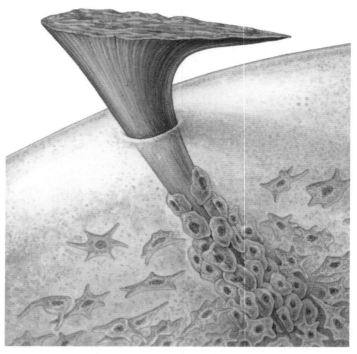

Figure 29.16   Discovering the cellular immune response in invertebrates.
In a Nobel Prize–winning experiment, the Russian zoologist Elie Metchnikoff pierced the larva of a sea star with a rose thorn and the next day found tiny phagocytic cells covering the thorn.

Complement. While invertebrates lack complement, many arthropods (including crabs and a variety of insects) possess an analogous nonspecific defense called the prophenyloxidase (proPO) system. Like the vertebrate complement defense, the proPO defense is activated by a cascade of enzyme reactions, the last of which converts the inactive protein prophenyloxidase into the active enzyme phenyloxidase. Phenyloxidase both kills microbes and aids in encapsulating foreign objects.

Lymphocytes. Invertebrates also lack lymphocytes, but annelid earthworms and other invertebrates do possess lymphocyte-like cells that may be evolutionary precursors of lymphocytes.

Antibodies. All invertebrates possess proteins called lectins that may be the evolutionary forerunners of antibodies. Lectins bind to sugar molecules on cells, making the cells stick to one another. Lectins isolated from sea urchins, mollusks, annelids, and insects appear to tag invading microorganisms, enhancing phagocytosis. The genes encoding vertebrate antibodies are part of a very ancient gene family, the immunoglobulin superfamily. Proteins in this group all have a characteristic recognition structure called the *immunoglobulin fold*. The fold probably evolved as a self-recognition molecule in early metazoans. Insect immunoglobulins that bind to microbial surfaces and promote their destruction by phagocytes have been described in moths, grasshoppers, and flies. The antibody immune response appears to have evolved from these earlier, less complex systems.

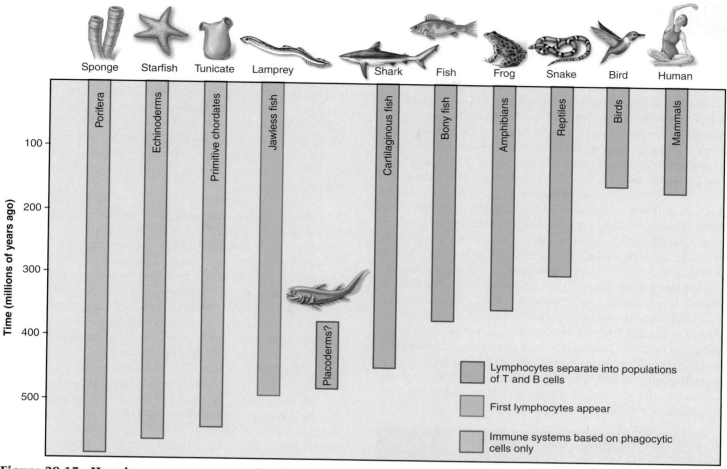

Figure 29.17 How immune systems evolved in vertebrates.

Lampreys were the first vertebrates to possess an immune system based on lymphocytes, although distinct B cells and T cells did not appear until the jawed fishes evolved. By the time sharks and other cartilaginous fish appeared, the vertebrate immune response was fully formed.

Vertebrates

Ancestors of the vertebrates (the tan bars in figure 29.17) contain phagocytic cells, which are the only means of immune defense they have. The earliest vertebrates of which we have any clear information, the jawless lampreys that first evolved some 500 million years ago, possess an immune system based on lymphocytes (the green bar). At this early stage of vertebrate evolution, however, lampreys lack distinct populations of B cells and T cells such as those found in all higher vertebrates (the pink bars).

The early jawed fishes, including placoderms and spiny fishes, are now extinct and so we can only speculate on their immune systems. The oldest surviving group of jawed fishes are the sharks, which evolved some 450 million years ago. By then, the vertebrate immune defense had fully evolved. So we can assume that two separate populations of lymphocytes emerge in the early jawed fishes. Sharks have an immune response much like that seen in mammals, with a cellular response carried out

by T cell lymphocytes and an antibody-mediated humoral response carried out by B cells. The similarities of the cellular and humoral immune defenses are far more striking than the differences. Both sharks and mammals possess a thymus that produces T cells and a spleen that is a rich source of B cells. Evolution spanning 450 million years did little to change the antibody molecule—the amino acid sequences of shark and human antibody molecules are very similar. The most notable difference between sharks and mammals is that their antibody-encoding genes are arrayed somewhat differently.

29.8 The sophisticated two-part immune defense of mammals evolved about the time jawed fishes appeared. Before then, animals used a simpler immune defense based on mobile phagocytic cells. Many of the elements of today's mammalian immune response can be recognized in analogous systems in invertebrates.

29.9 Vaccination

In 1796, an English country doctor named Edward Jenner carried out an experiment that marks the beginning of the study of immunology. Smallpox was a common and deadly disease in those days. Jenner observed, however, that milkmaids who had caught a much milder form of "the pox" called cowpox (presumably from cows) rarely caught smallpox. Jenner set out to test the idea that cowpox conferred protection against smallpox. He infected people with mild cowpox (figure 29.18), and as he had predicted, many of them became immune to smallpox.

We now know that smallpox and cowpox are caused by two different but similar viruses. Jenner's patients who were injected with the cowpox virus mounted a defense that was also effective against a later infection of the smallpox virus. Jenner's procedure of injecting a harmless microbe to confer resistance to a dangerous one is called vaccination. **Vaccination** is the introduction into your body of a dead or disabled pathogen or, more commonly these days, of a harmless microbe with pathogen proteins displayed on its surface. The vaccination triggers an immune response against the pathogen, without an infection ever occurring. Afterward the bloodstream of the vaccinated person contains circulating memory cells (B and T cells) directed against that specific pathogen. The vaccinated person is said to have been "immunized" against the disease.

Figure 29.18 The birth of immunology.
This famous painting shows Edward Jenner inoculating patients with cowpox in the 1790s and thus protecting them from smallpox.

Through genetic engineering, scientists are now routinely able to produce "piggyback," or subunit, vaccines. These vaccines are made of harmless viruses that contain in their DNA a single gene cut out of a pathogen, a gene encoding a protein normally exposed on the pathogen's surface. By splicing the pathogen gene into the DNA of the harmless host, that host is induced to display the protein on its surface. The harmless virus displaying the pathogen protein is like a sheep in wolf's clothing, unable to hurt you but raising alarm as if it could. Your body responds to its presence by making an antibody directed against the pathogen protein that acts like an alarm to the immune system, should that pathogen ever visit your body.

If the activities of memory cells provide such an effective defense against future infection, why can you catch some diseases like flu more than once? The reason you don't stay immune to flu is that the flu virus has evolved a way to evade the immune system—it changes. The genes encoding the surface proteins of the flu virus mutate very rapidly. Thus, the shapes of these surface proteins alter swiftly. Your memory cells do not recognize viruses with altered surface proteins as being the same viruses they have already successfully defeated or been vaccinated against, because the memory cells' receptors no longer "fit" the new shape of the flu surface proteins. When the new version of flu virus invades your body, you need to mount an entirely new immune defense.

Sometimes the flu virus surface proteins possess shapes that the immune system does not readily recognize. When mutations arose in a bird flu in 1918 that allowed this flu virus to pass easily from one infected human to another, over 20 million Americans and Europeans died in 18 months (figure 29.19). Less profound changes in flu virus surface proteins occur periodically, resulting in new strains of flu for which we are not immune. The annual flu shots are vaccines against these new strains. Researchers are able to predict the current year's strain of the flu by examining pre-season flu reports from across the globe and determining which strain seems to be dominant. Researchers then prepare a vaccine against this year's strain. Sometimes, as in the case of the 1918 influenza pandemic, entirely new flu virus strains infect humans from birds. The possibility that the deadly H5N1 bird flu will mutate so that it is able to pass between infected individuals is worrying health officials.

One of the most intensive efforts in the history of medicine is currently under way to develop an effective vaccine against HIV, the virus responsible for AIDS. Researchers are using the "piggyback" method shown in figure 29.20. Steps ❶ through ❸ show how the gene that encodes an HIV surface protein is isolated, and then it is inserted into the DNA of a harmless vaccinia virus (steps ❹ and ❺). The genetically engineered vaccinia virus is injected into the body ❻, which triggers the body to begin producing antibodies and memory cells against the HIV surface protein antigen ❼. The HIV virus has nine genes, encoding a variety of proteins. Initial efforts focused on producing a subunit vaccine containing the HIV *env* (envelope) gene, which encodes the protein on the outside of the virus.

Figure 29.19 The flu epidemic of 1918 killed over 20 million in 18 months.

With 25 million Americans alone infected during the influenza epidemic, it was hard to provide care for everyone. The Red Cross often worked around the clock.

Unfortunately for these initial attempts to develop an AIDS vaccine, the HIV virus mutates even more rapidly than the flu virus. Even vaccines that work in the laboratory are not effective outside the laboratory, where new strains of HIV are encountered. This high mutation rate has been the single biggest obstacle to developing a successful AIDS vaccine.

New vaccine approaches look more promising. Although mutations of the *env* gene are frequent, they are random events. Because no two virus particles create the same new mutations at the same instant, a vaccine simultaneously directed against three different HIV proteins has a good chance of working—it is very unlikely that a single HIV particle will mutate all three genes simultaneously. To further strengthen new vaccines, multiprotein subunit vaccines are being supplemented with HIV DNA to activate T cell immune defenses. These new AIDS vaccines work well in monkeys, and over two dozen vaccines have entered human clinical trials.

As you can see, the immune system, often aided by vaccination, can respond in a variety of different ways to different kinds of pathogens. However, as discussed later in this chapter, its ability to function normally and efficiently is often disrupted.

> **29.9 Vaccines introduce antigens similar or identical to those of pathogens, eliciting an immune response that defends against the pathogen too.**

Figure 29.20 Researchers are attempting to construct an AIDS vaccine.

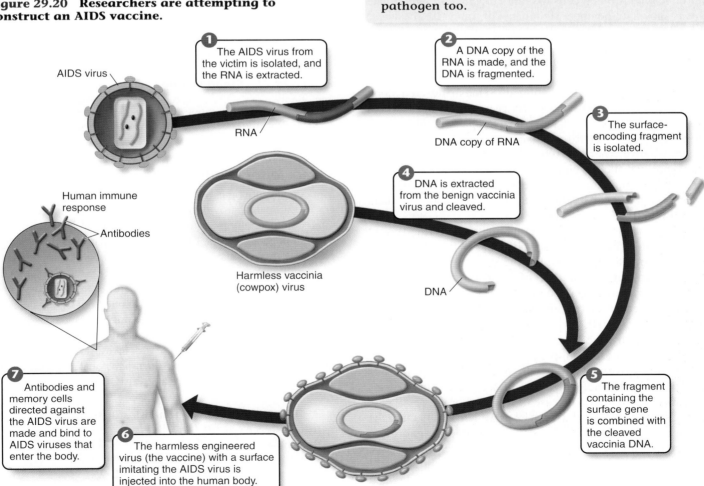

AIDS virus

1 The AIDS virus from the victim is isolated, and the RNA is extracted.

RNA

2 A DNA copy of the RNA is made, and the DNA is fragmented.

DNA copy of RNA

3 The surface-encoding fragment is isolated.

Human immune response

Antibodies

4 DNA is extracted from the benign vaccinia virus and cleaved.

Harmless vaccinia (cowpox) virus

DNA

5 The fragment containing the surface gene is combined with the cleaved vaccinia DNA.

7 Antibodies and memory cells directed against the AIDS virus are made and bind to AIDS viruses that enter the body.

6 The harmless engineered virus (the vaccine) with a surface imitating the AIDS virus is injected into the human body.

Blood Typing

A person's blood type indicates the class of antigens found on the red blood cell surface. Red blood cell antigens are clinically important because their types must be matched between donors and recipients for blood transfusions. There are several groups of red blood cell antigens, but the major group is known as the **ABO system.** In terms of the antigens present on the red blood cell surface, a person may be *type A* (with only A antigens), *type B* (with only B antigens), *type AB* (with both A and B antigens), or *type O* (with neither A nor B antigens).

The immune system is tolerant to its own red blood cell antigens. A person who is type A, for example, does not produce anti-A antibodies. However, people with type A blood do make antibodies against the B antigen, and people with blood type B make antibodies against the A antigen. People who are type AB develop tolerance to both antigens and thus do not produce either anti-A or anti-B antibodies. Those who are type O make both anti-A and anti-B antibodies.

If type A blood is mixed on a glass slide with serum from a person with type B blood, the anti-A antibodies in the serum cause the type A blood cells to clump together, or agglutinate (this is shown in the upper right panel of figure 29.21). These tests allow the blood types to be matched prior to transfusions, so that agglutination will not occur in the blood vessels, where it could lead to inflammation and organ damage.

Rh Factor. Another group of antigens found in most red blood cells is the *Rh factor* (Rh stands for rhesus monkey, in which these antigens were first discovered). People who have these antigens are said to be Rh-positive, whereas those who do not are Rh-negative. There are fewer Rh-negative people because the Rh-positive allele is clinically dominant to the Rh-negative allele and is more common in the human population. The Rh factor is of particular significance when Rh-negative mothers give birth to Rh-positive babies.

Because the fetal and maternal blood are normally kept separate across the placenta (see chapter 32), the Rh-negative mother is not usually exposed to the Rh antigen of the fetus during pregnancy. At the time of birth, however, a varying degree of exposure may occur, and the Rh-negative mother's immune system may become sensitized and produce antibodies against the Rh antigen. If the woman does produce antibodies against the Rh factor, these antibodies can cross the placenta in subsequent pregnancies and cause hemolysis of the Rh-positive red blood cells of the fetus. The baby is therefore born anemic, with a condition called erythroblastosis fetalis, or hemolytic disease of the newborn.

Erythroblastosis fetalis can be prevented by injecting the Rh-negative mother with an antibody preparation against the Rh factor within 72 hours after the birth of each Rh-positive baby. The injected antibodies inactivate the Rh antigens and thus prevent the mother from becoming actively immunized to them.

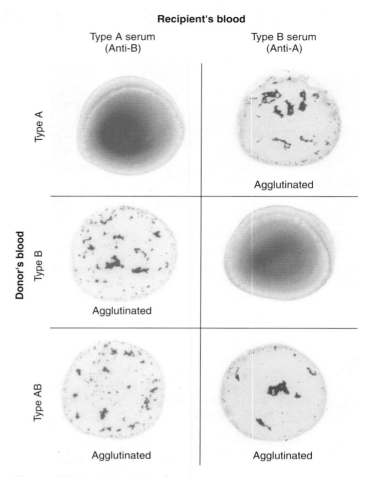

Recipient's blood

Type A serum (Anti-B)   Type B serum (Anti-A)

Type A

Agglutinated

Donor's blood

Type B

Agglutinated

Type AB

Agglutinated   Agglutinated

Figure 29.21   Blood typing.
Agglutination of the red blood cells is seen when blood types are mixed with sera containing antibodies against the A and B antigens. Note that no agglutination would be seen if type O blood (not shown) was used.

Monoclonal Antibodies

Monoclonal antibodies are antibodies that are specific to one antigen. Because they provide a very sensitive assay, monoclonals are often commercially prepared for use in clinical laboratory tests. Modern pregnancy tests, for example, use particles that are covered with monoclonal antibodies produced against a pregnancy hormone (abbreviated hCG—see chapter 32) as the antigen. In the blood pregnancy test, these particles are mixed with a sample from a pregnant woman. If the sample contains a significant level of the hCG hormone, it reacts with the antibody and causes a visible agglutination of the particles, indicating a positive test result. Over-the-counter pregnancy tests work in a similar way. hCG in a pregnant woman's urine binds to the monoclonal antibodies within the testing strip and indicates a positive result.

> **29.10   Agglutination occurs because different antibodies exist for the ABO and Rh factor antigens on the surface of red blood cells. Monoclonal antibodies are commercially produced antibodies that react against one specific antigen.**

29.11 Overactive Immune System

Figure 29.22 The house dust mite.

This tiny animal, *Dermatophagoides*, causes an allergic reaction in many people.

Although the immune system is one of the most sophisticated systems of the vertebrate body, it is still not perfect. Many of the major diseases we face, and some minor irritations as well, reflect an overactive immune system.

Autoimmune Diseases

The ability of T cells and B cells to distinguish cells of your own body—"self" cells—from nonself cells is the key ability of the immune system that makes your body's third line of defense so effective. In certain diseases, this ability breaks down, and the body attacks its own tissues. Such diseases are called **autoimmune diseases.**

Multiple sclerosis is an autoimmune disease that usually strikes people between the ages of 20 and 40. In multiple sclerosis, the immune system attacks and destroys a sheath of fatty material, called myelin (see chapter 30), that insulates motor nerves (like the rubber covering electrical wires). Recall from section 24.6 that nerve impulses travel along the length of the nerve cell, and so degeneration of the myelin sheath interferes with transmission of nerve impulses, until eventually they cannot travel at all. Voluntary functions, such as movement of limbs, and involuntary functions, such as bladder control, are lost, leading finally to paralysis and death. Scientists do not know what stimulates the immune system to attack myelin.

Another autoimmune disease is type I diabetes, in which cells are unable to take in glucose because the pancreas fails to produce insulin (recall from chapter 28 that insulin plays a key role in the liver's regulation of levels of glucose in the blood). Type I diabetes is thought to result from an immune attack on the insulin-manufacturing cells of the pancreas. No one knows why the attack occurs. Other autoimmune diseases are rheumatoid arthritis (an immune system attack on the tissues of the joints), lupus (in which the connective tissue and kidneys are attacked), and Graves' disease (in which the thyroid is attacked).

Allergies

Although your immune system provides very effective protection against fungi, parasites, bacteria, and viruses, sometimes it does its job too well, mounting a major defense against a harmless substance. Such an immune response is called an **allergy.** Hay fever, sensitivity to even tiny amounts of plant pollen, is a familiar example of an allergy. Many people are allergic to proteins released from the feces of a minute mite that lives on grains of house dust (figure 29.22). The dust that the mite calls home is present in mattresses and pillows, and the mite goes out on foraging expeditions and consumes the dead skin cells that we all shed in large quantities daily. Many people sensitive to feather pillows are in reality allergic to the mites that are residents of the feathers.

What makes an allergic reaction uncomfortable, and sometimes dangerous, is the involvement of antibodies attached to a kind of white blood cell called a **mast cell.** It is the job of the mast cells in an immune response to initiate an inflammatory response. Figure 29.23 shows what happens when a mast cell encounters something that matches its antibody. Mast cells release histamines and other chemicals that cause capillaries to swell. **Histamines** also increase mucus production by cells of the mucous membranes, resulting in runny noses and nasal congestion (all the symptoms of hay fever). Most allergy medicines relieve these symptoms with antihistamines, chemicals that block the action of histamines.

Asthma is a form of allergic response in which histamines cause the narrowing of air passages in the lungs. People who have asthma have trouble breathing when exposed to substances to which they are allergic.

> **29.11** Autoimmune diseases are inappropriate responses to "self" cells, whereas allergies are inappropriate immune responses to harmless antigens.

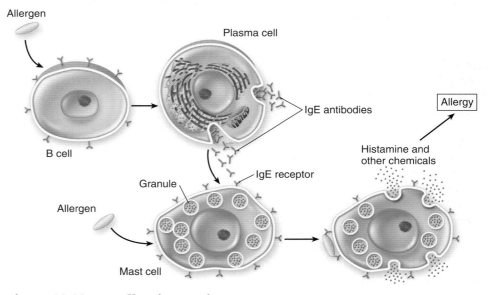

Figure 29.23 An allergic reaction.

In an allergic response, B cells secrete IgE antibodies (see page 584) that attach to the plasma membranes of mast cells, which secrete histamine in response to antigen-antibody binding.

29.12 AIDS: Immune System Collapse

AIDS (acquired immunodeficiency syndrome) was first recognized as a disease in 1981. By the end of 2004 in the United States, more than 529,113 individuals had died of AIDS, and more than 2 million others were thought to be infected with **HIV** (human immunodeficiency virus), the virus that causes the disease. Worldwide, 42 million are infected, and 25 million have died. HIV apparently evolved from a very similar virus that infects chimpanzees in Africa when a mutation arose that allowed the virus to recognize a human cell surface receptor called **CD4**. This receptor is present in the human body on certain immune system cells, notably macrophages and helper T cells. It is the identity of these immune system cells that leads to the devastating nature of the disease.

How HIV Attacks the Immune System

HIV attacks and cripples the immune system by inactivating cells that have CD4 receptors (CD4$^+$ cells), such as helper T cells. Review figures 29.10, 29.11, and 29.15 to see the role of helper T cells in the immune responses and you will understand the significance of this inactivation of helper T cells. This leaves the immune system unable to mount a response to *any* foreign antigen. AIDS is a deadly disease for just this reason. The AIDS-causing HIV virus mounts a direct attack on CD4$^+$ T cells because it recognizes their CD4 receptors.

HIV's attack on CD4$^+$ T cells progressively cripples the immune system, because HIV-infected cells die after releasing replicated viruses that proceed to infect other CD4$^+$ T cells. Over time, the body's entire population of CD4$^+$ T cells is destroyed (figure 29.24). In a normal individual, CD4$^+$ T cells make up 60% to 80% of circulating T cells; in AIDS patients, CD4$^+$ T cells often become too rare to detect, wiping out the human immune defense. With no defense against infection, any of a variety of otherwise commonplace infections proves fatal. With no ability to recognize and destroy cancer cells when they arise, death by cancer becomes far more likely. Indeed, AIDS was first recognized because of a cluster of cases of a rare cancer, Karposi's sarcoma. More AIDS victims die of cancer than from any other cause.

The fatality rate of AIDS is 100%; no patient exhibiting the symptoms of AIDS has recovered. The disease is *not* highly contagious, because it is only transmitted from one individual to another through the transfer of internal body fluids, typically in semen and vaginal fluid during sexual intercourse and in blood transmitted by needles during drug use. However, symptoms of AIDS do not usually show up for several years after infection with HIV, which means that infected individuals can unknowingly spread the virus to others. Education and awareness campaigns in the United States have helped reduce the numbers of new AIDS cases (figure 29.25).

A variety of drugs inhibit HIV in the test tube. These include AZT and its analogs (which inhibit virus nucleic acid replication) and protease inhibitors (which inhibit the cleavage of the large virus proteins into functional segments). A combination of a protease inhibitor and two AZT analog drugs entirely eliminates the HIV virus from many patients' bloodstreams. Widespread use of this *combination therapy* has cut the U.S. AIDS death rate by

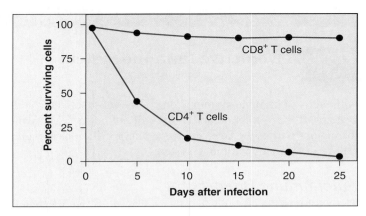

Figure 29.24 Survival of T cells after exposure to HIV.

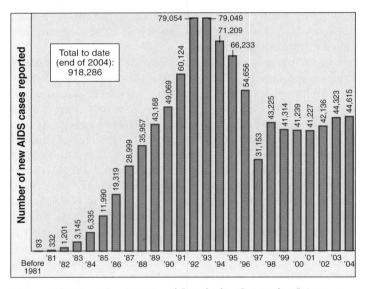

Figure 29.25 The AIDS epidemic in the United States.
The U.S. Centers for Disease Control (CDC) reports that 44,323 new AIDS cases were reported in 2003 and 44,615 new cases in 2004 in the United States, with a total of 918,286 cases and 529,113 deaths. Over 2 million other individuals are thought to be infected with the HIV virus in the United States and 42 million worldwide. The 100,000th AIDS case was reported in August 1989, eight years into the epidemic; the next 100,000 cases took just 26 months; the third 100,000 cases took barely 19 months (May 1993), and the fourth 100,000 took only 13 months (June 1994).
Source: Data from U.S. Centers for Disease Control and Prevention, Atlanta, GA.

almost two-thirds since its introduction in the mid-1990s, from 43,000 AIDS deaths in 1995 to 31,000 in 1996, and eight years later in 2004, deaths remained low, at approximately 15,700.

Unfortunately, this sort of combination therapy does not appear to actually succeed in eliminating HIV from the body. While the virus disappears from the bloodstream, traces of it can still be detected in lymph tissue of the patients. When combination therapy is discontinued, virus levels in the bloodstream once again rise. Because of demanding therapy schedules and many side effects, long-term combination therapy does not seem a promising approach.

> **29.12 HIV cripples the vertebrate immune defense by infecting and killing key lymphocytes.**

The Search for an Effective AIDS Vaccine Looks More Promising

Since the AIDS epidemic burst upon us in 1981, scientists have feverishly sought a vaccine to protect people from this deadly and incurable disease. But the path to a vaccine has not been easy. Over twenty years and nearly one million American AIDS cases later, an effective AIDS vaccine still eludes the best efforts of researchers.

One problem has thwarted all efforts: the HIV virus that causes AIDS has an unusually mistake-prone DNA copying enzyme. Making mistakes here and there as it produces offspring virus particles, HIV generates mutations at a prodigious rate. That is why few of those infected with HIV have exactly the same virus. A vaccine targeted against one version is ineffective against others.

Like a thief of many disguises, at least a few of the HIV particles in an infection are able to dodge any antibody a vaccine throws against it.

For twenty years this problem has seemed insurmountable, but recently researchers have devised a solution. They seized on a property of mutation so obvious it had been largely overlooked— mutations are accidents. A strain of HIV may produce numerous mutations as it proliferates, but each of these mutations occurs as a random mistake in a different virus particle. No two particles undergo exactly the same mutation.

The key to an effective vaccine, then, is to use more than one HIV protein. In initial experiments, researchers used three. Any one virus might have mutated to be different from one of the three proteins, but the probability of all three proteins being mutated in the same virus particle is about the same as lightning striking the same person three times—real, but very very small.

Not content with this innovation, researchers set out to boost the power of the immune defenses. The human body has another line of immune defense, the cell-mediated response. Why not bring this to bear as well?

In a double-barreled attack on HIV, the researchers first infected 24 monkeys with a circular piece of naked DNA containing three HIV genes. Such naked DNA

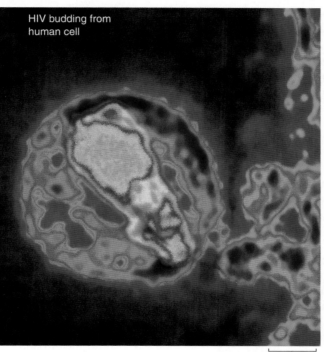

HIV budding from human cell

.02 μm

readily infects human cells. Once inside, it directs the production of the three HIV proteins. Passing to the cell surface, these proteins create a mock-HIV infection that kicks cell-mediated immunity into action. To the immune system, the cells look HIV-infected. Rushing to the body's defense, the immune system produces killer cells that wait in ambush to attack and destroy any body cells displaying one of the three proteins.

A few weeks later, the researchers administered the same three HIV genes to the monkeys, but this time the HIV genes are stitched into a harmless virus. Introduced this way, they activate a monkey's antibody defenses instead of its cell-mediated immunity. Within a few days, a burst of antibodies is produced directed against the three HIV proteins.

Seven months later, the 24 vaccinated animals were "challenged" by inoculating them with a laboratory version of HIV. Called SHIV, it is a genetically engineered version that is part HIV (to make it lethal) and part simian AIDS virus (to enable it to infect monkeys).

The vaccine worked! For more than a year, after the SHIV challenge, 23 of the infected monkeys continued to control the infection, and still had healthy immune systems. Four unvaccinated control animals, by contrast, all developed simian AIDS and died.

It was too soon to consider the vaccine a complete success, however. The monkeys remained HIV-infected, and after a year the HIV in the monkeys had accumulated mutations that allowed the infection to circumvent the immune defenses provided by the vaccine, and the monkeys began to die of simian AIDS. The vaccine had bought time, but not solved the basic problem of ridding the body of the virus.

Learning from this partial success, researchers have tweaked the approach in numerous ways. Although there were only two (unpromising) vaccines in clinical trials in the year 2000, now more than 2 dozen different AIDS vaccines have started clinical trials, employing this double-barreled, multigene approach in different ways.

It is difficult to avoid hope for these exciting developments. However, keep in mind that the initial studies were of an SHIV vaccine developed for monkeys. There is no guarantee that it will work with humans. Many exciting discoveries fail to clear this last crucial hurdle. But in more than 20 years, no candidate AIDS vaccines have looked so promising.

Is Immunity Antigen-Specific?

The immune response provides a valuable protection against infection, because it can remember prior experiences. We develop lifelong immunity to many infectious diseases after one childhood exposure. This long-term immunity is why vaccines work. A key question about immune protection is whether or not it is specific. Does exposure to one pathogen confer immunity to only that one, or is the immunity you acquire a more general response, protecting you from a range of infections?

The graph to the right displays the results of an experiment designed to answer this question. A colony of rabbits is immunized once with antigen A, and the level of antibody directed against this antigen monitored in each individual. After 40 days, each rabbit is reinjected, some with antigen A and others with antigen B, and the level of antibody directed against the reinjected antigens is monitored. The red line is typical of results for antigen A, the blue line for antigen B.

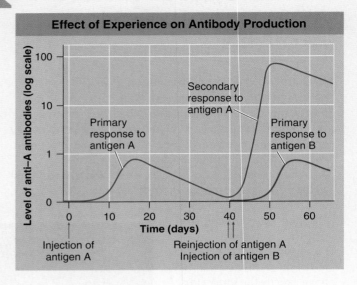

Effect of Experience on Antibody Production

1. **Applying Concepts**
   a. Variable. In the graph, what is the dependent variable?
   b. Reading a Continuous Curve. Does each injection of antigen A result in detectable antibody production? antigen B?
2. **Interpreting Data**
   a. The initial response to antigen A is called the "primary" response, and the second response to antigen A administered 40 days later is called the "secondary" response. Compare the speed of the primary and secondary responses—which reaches maximal antibody response quicker?
   b. Compare the magnitude of the primary and secondary immune responses to antigen A—are they similar, or is one response of greater magnitude?

3. **Making Inferences**
   a. Why is the secondary response induced by a second exposure to antigen A different from the primary response?
   b. Is the response to antigen B more similar to the primary or secondary response of antigen A?
4. **Drawing Conclusions**
   a. Does the prior exposure to antigen A have any impact on the speed or magnitude of the primary response to antigen B?
   b. Is the immune response to these antigens antigen-specific?
5. **Further Analysis** If you were to inject both sets of rabbits with antigen B on day 80, what would you expect the results to be? Explain the difference you would expect in the immune responses of the two groups of rabbits to this injection, if any.

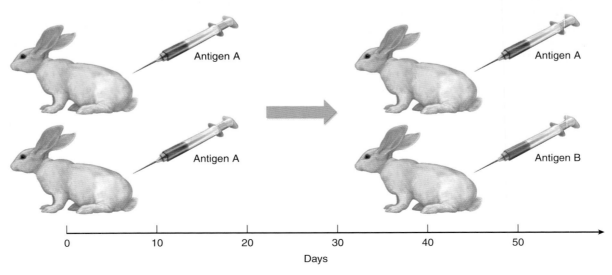

Three Lines of Defense

29.1 Skin: The First Line of Defense

- The body has three lines of defense against infection, the first being skin (**figure 29.2**) and the mucous membranes that line the digestive and respiratory tracts. Skin and mucous membranes form barriers to pathogens.

29.2 Cellular Counterattack: The Second Line of Defense

- The second line of defense is a nonspecific cellular attack. The cells and chemicals used in this line of defense attack all foreign agents that they encounter. Macrophages and neutrophils attack the invading pathogen, while natural killer cells attack infected cells, killing them before the pathogen can spread to other cells (**figures 29.4** and **29.5**). Free-floating proteins in the blood, called complement, insert into the membranes of foreign cells, killing them (**figure 29.7**). The inflammatory and temperature responses are also part of the second line of defense (**figure 29.8**).

29.3 Specific Immunity: The Third Line of Defense

- The third line of defense is a specific immune response (**table 29.1**). T cells and B cells are "programmed" by exposure to specific antigens and once programmed, will seek out the antigens or cells carrying those antigens and destroy them.

The Immune Response

29.4 Initiating the Immune Response

- Macrophages survey cells for "self" MHC proteins. A cell that exhibits "nonself" MHC proteins is engulfed by a macrophage, which secretes interleukin-1. Interleukin-1 stimulates helper T cells that trigger the cellular and humoral immune responses. These macrophages also insert foreign antigens in their membranes and present these antigens to T cells, activating the T cell response. These macrophages are called antigen-presenting cells (**figure 29.9**).

29.5 T Cells: The Cellular Response

- The cellular response involves T cells. Antigen-presenting cells activate helper T cells that release interleukin-2. Interleukin-2 stimulates the cloning of cytotoxic T cells that recognize and kill cells with the specific antigen found on the antigen-presenting cells (**figure 29.10**). Following the infection, memory T cells form and remain in the body to fight subsequent reinfections.

29.6 B Cells: The Humoral Response

- The humoral response involves B cells. B cells are able to recognize foreign antigens on the surfaces of pathogens and when activated by interleukin-2, released from helper T cells,

they "mark" the foreign invaders with specific antibody proteins. The "marked" cells are then attacked by the nonspecific immune response. Activated B cells divide to form plasma cells, which produce and release the antibodies, and memory cells, that circulate in the blood. Memory cells quickly become plasma cells in the event of a second infection by the same antigen (**figure 29.11**).

29.7 Active Immunity Through Clonal Selection

- The initial immune response triggered by infection is called the primary response. It is a delayed and somewhat weak response. But through clonal selection, a large population of memory cells is present in the body, such that a second infection by the same antigen will trigger a quicker and larger response, called the secondary response (**figures 29.14** and **29.15**).

29.8 Evolution of the Immune System

- The immune system evolved from a nonspecific immune response, involving phagocytic cells only, present in invertebrates, to a two-part immune response involving B cells and T cells, present in vertebrates (**figure 29.17**).

29.9 Vaccination

- Vaccination takes advantage of the mechanism of the primary and secondary immune responses. Vaccines introduce harmless antigens present on the pathogenic cell into the body. The body produces memory B cells toward the antigen, such that with an actual infection, the body elicits a swift and large immune response (**figure 29.20**).

29.10 Antibodies in Medical Diagnosis

- Antibodies are keen detectors of antigens, and so are used in various medical diagnostic applications such as blood typing and monoclonal antibody assays (**figure 29.21**).

Defeat of the Immune System

29.11 Overactive Immune System

- Sometimes the immune system attacks antigens that are not foreign or pathogenic. In autoimmune responses, the body attacks its own cells. In allergic reactions, the body mounts an attack against a harmless substance (**figure 29.23**).

29.12 AIDS: Immune System Collapse

- AIDS is a fatal disease caused by infection with the HIV virus. HIV attacks macrophages and helper T cells, eventually destroying the cells that protect the body from other infections (**figures 29.24** and **29.25**).

1. Skin is a physical barrier but also provides chemical protection generated from
   a. sweat and oil glands.
   b. mucus.
   c. the skin's basal layer.
   d. the stratum corneum of the skin.

2. The immune system can identify foreign cells in the bloodstream because these foreign cells
   a. are observed destroying other cells by the immune system.
   b. do not have the proper cell surface proteins that identify the cell as the body's own.
   c. are observed causing inflammatory response.
   d. are observed causing temperature response.

3. The purpose of the inflammatory response is to
   a. increase the temperature of an infected area.
   b. reduce pain of an infected area.
   c. increase the number of immune system cells in an infected area.
   d. increase physical protection of an infected area.

4. Increasing human body temperature—that is, causing a fever—assists the immune system because
   a. increased temperature speeds up the chemical reactions used by the immune system.
   b. pathogenic bacteria do not grow well at high temperatures.
   c. increased temperature causes foreign proteins to denature.
   d. All of these answers are correct.

5. Immune responses tailored to specific pathogens involve
   a. T cells.
   b. macrophages.
   c. monocytes.
   d. neutrophils.

6. Antibody production takes place in
   a. T cells.
   b. natural killer cells.
   c. B cells.
   d. mast cells.

7. Cytotoxic T cells
   a. produce antibodies.
   b. destroy pathogens directly.
   c. destroy foreign antigens floating freely in the bloodstream.
   d. destroy cells infected by pathogens.

8. Immunity to future invasion of a specific pathogen is accomplished by production of
   a. plasma cells.
   b. memory T and B cells.
   c. helper T cells.
   d. monocytes.

9. When a body's immune system attacks the body's own cells, this is called
   a. an inflammatory response.
   b. a temperature response.
   c. an autoimmune response.
   d. an allergic response.

10. HIV-infected people with advanced AIDS usually die of an infectious disease or cancer. This is because HIV attacks
    a. helper T cells.
    b. neutrophils.
    c. memory T and B cells.
    d. mast cells.

Visual Understanding

1. **Figure 29.14** Certain diseases are considered to be primarily "childhood" diseases—measles, mumps, chicken pox, for instance—and once someone has had these illnesses as a child, they don't catch them again when, as parents, they take care of their own children who are sick. Use the graph to help explain the reason.

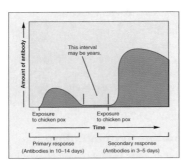

2. **Figure 29.25** What might account for the large first curve in the AIDS epidemic, from 1981 to 1992, then the decline until 1997, and now the essentially steady state or flat line, neither growing nor diminishing by very much, for the past seven years?

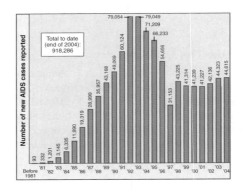

Challenge Questions

1. Your friend, Sadako, fell yesterday while she was skateboarding and has a deep cut on her forearm from a piece of wire. She shows it to you and complains about how sore it is. You can see the skin is red and swollen around the small puncture wound. Explain to her what is happening.

2. A doctor will give you an antibiotic to treat a bacterial infection, but you are given a vaccine to keep a virus (and some bacteria) from making you ill. What is the difference between these approaches?

3. What are two major reasons why we have been unable to develop a good vaccine for AIDS/HIV?

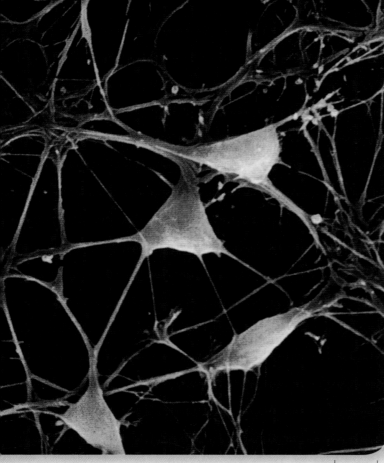

19.94 μm

30

The Nervous System

In vertebrates, the central nervous system coordinates and regulates the diverse activities of the body, using a network of specialized cells called neurons to direct the voluntary muscles, and a second network not under voluntary control to direct cardiac and smooth muscles. All sensory information is acquired through depolarization of sensory nerve endings. From a knowledge of which neurons are sending signals, and how often they are doing so, the brain builds a picture of the body's internal condition and of the external environment. The network of neurons (nerve cells) seen here, magnified over a thousand times, is transmitting signals within the portion of the brain called the cerebral cortex. The vertebrate brain contains a staggering number of neurons—the human brain contains an estimated 100 billion. The cerebral cortex is a layer of gray matter only a few millimeters thick on the brain's outer surface. Densely packed with neurons and highly convoluted, it is the site of higher mental activities.

30.1 Evolution of the Animal Nervous System

An animal must be able to respond to environmental stimuli. To do this, it must have sensory receptors that can detect the stimulus and motor *effectors* that can respond to it. In most invertebrate phyla and in all vertebrate classes, sensory receptors and motor effectors are linked by way of the **nervous system.** As described in chapter 24, the nervous system consists of neurons and supporting cells. One type of neuron, called **association neurons** (or **interneurons**), is present in the nervous systems of most invertebrates and all vertebrates. These neurons are located in the brain and spinal cord of vertebrates, together called the **central nervous system (CNS),** the yellow circle in figure 30.1. They help provide more complex reflexes and in the case of the brain, higher associative functions, including learning and memory, which require integration of many sensory inputs.

There are two other types of neurons. **Sensory** (or **afferent**) **neurons** (❶ in figure 30.2) carry impulses from sensory receptors to the CNS. **Motor** (or **efferent**) **neurons** ❸ carry impulses away from the CNS to effectors—muscles and glands. The association neurons ❷ link these two types of neurons together in the CNS. Together, motor and sensory neurons constitute the **peripheral nervous system (PNS)** of vertebrates (the bracketed tan circles in figure 30.1).

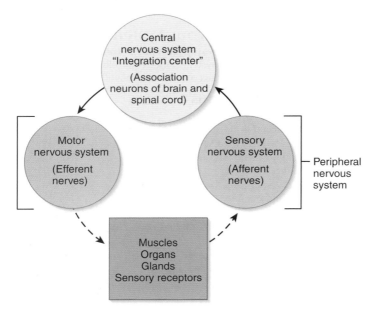

Figure 30.1   Organization of the vertebrate nervous system.

The central nervous system, consisting of the brain and spinal cord, issues commands via the motor nervous system and receives information from the sensory nervous system. The motor and sensory nervous systems together make up the peripheral nervous system.

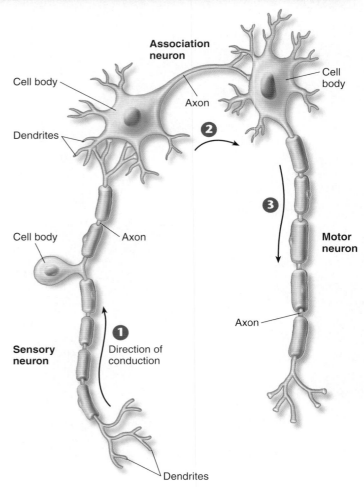

Figure 30.2   Three types of neurons.

Sensory neurons carry information about the environment to the brain and spinal cord. *Association neurons* are found in the brain and spinal cord and often provide links between sensory and motor neurons. *Motor neurons* carry impulses to muscles and glands (effectors).

Invertebrate Nervous Systems

Sponges are the only major phylum of multicellular animals that lack nerves. If you prick a sponge, the nearby surface contracts slowly. The cytoplasm of each individual cell conducts an impulse that fades within a few millimeters. No messages dart from one part of the sponge body to another, as they do in all other multicellular animals.

The Simplest Nervous Systems: Reflexes. The simplest nervous systems occur among cnidarians, like the *Hydra* ❶ in figure 30.3: all neurons are similar, each having fibers of approximately equal length. Cnidarian neurons are linked to one another in a web, or *nerve net,* dispersed through the body. Although conduction is slow, a stimulus anywhere can eventually spread through the whole net. There is no associative activity, no control of complex actions, and little coordination.

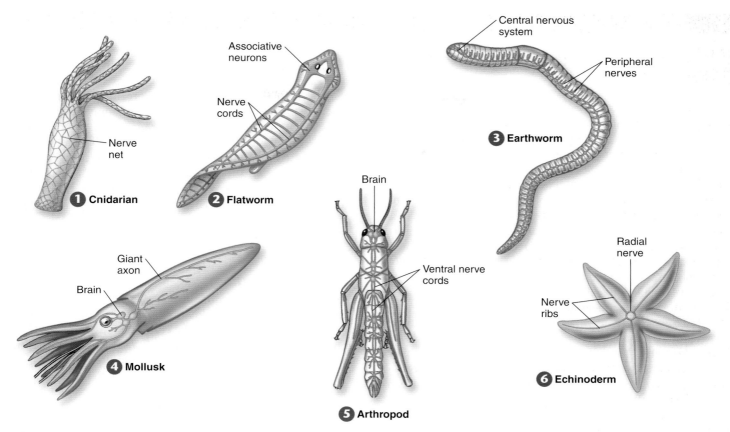

Figure 30.3 Evolution of the nervous system.
Invertebrates exhibit a progressive elaboration of organized nerve cords and the centralization of complex responses in the front end of the nerve cord. The adult echinoderm nervous system is the exception, seeming to become less complex and decentralized. The larval stage of echinoderms exhibits more of the higher-order animal characteristics, including the characteristics of the nervous system.

Any motion that results is called a **reflex** because it is an automatic consequence of the nerve stimulation.

More Complex Nervous Systems: Associative Activities. The first associative activity in nervous systems is seen in the free-living flatworms ❷, phylum Platyhelminthes. Running down the bodies of these flatworms are two nerve cords, looking like the uprights of a ladder; peripheral nerves extend outward to the muscles of the body. The two nerve cords converge at the front end of the body, forming an enlarged mass of nervous tissue that also contains associative neurons that connect neurons to one another. This primitive "brain" is a rudimentary central nervous system and permits a far more complex control of muscular responses than is possible in cnidarians.

The Evolutionary Path to the Vertebrates. All of the subsequent evolutionary changes in nervous systems can be viewed as a series of elaborations on the characteristics already present in flatworms. Five trends can be identified, each becoming progressively more pronounced as nervous systems evolved greater complexity.

1. *More sophisticated sensory mechanisms.* Particularly among the vertebrates, sensory systems become highly complex.

2. *Differentiation into central and peripheral nervous systems.* For example, earthworms ❸ exhibit a central nervous system that is connected to all other parts of the body by peripheral nerves.

3. *Differentiation of sensory and motor nerves.* Neurons operating in particular directions (sensory signals traveling to the brain, or motor signals traveling from the brain) become increasingly specialized.

4. *Increased complexity of association.* Central nervous systems with more numerous interneurons evolved, increasing association capabilities dramatically.

5. *Elaboration of the brain.* Coordination of body activities became increasingly localized in mollusks ❹, arthropods ❺, and vertebrates in the front end of the nerve cord, which evolved into the vertebrate brain discussed later in the chapter.

> **30.1 As nervous systems became more complex, there was a progressive increase in associative activity, increasingly localized in a brain.**

Neurons

The basic structural unit of the nervous system, whether central, motor, or sensory, is the nerve cell, or **neuron.** All neurons have the same basic structure as you can see by comparing the three cell types in figure 30.2 and the generalized cell in figure 30.4a. The **cell body** in figure 30.4a is the flat region of the neuron containing the nucleus. Short, slender branches called **dendrites** extend from one end of a neuron's cell body. Dendrites are input channels. Nerve impulses travel inward along them, toward the cell body. Motor and association neurons possess a profusion of highly branched dendrites, enabling those cells to receive information from many different sources simultaneously. Projecting out from the other end of the cell body is a single, long, tubelike extension called an **axon.** Axons are output channels. Nerve impulses travel outward along them, away from the cell body, toward other neurons or to muscles or glands.

Most neurons are unable to survive alone for long; they require the nutritional support provided by companion **neuroglial cells.** More than half the volume of the human nervous system is composed of supporting neuroglial cells. Two of the most important kinds of supporting neurons are the **Schwann cells** and **oligodendrocytes,** which envelop the axons of many neurons with a sheath of fatty material called myelin, which acts as an electrical insulator. Schwann cells

produce myelin in the PNS, while oligodendrocytes produce myelin in the CNS. During development, these cells associate with the axon, as shown at the top in figure 30.4b, and begin to wrap themselves around the axon several times to form a **myelin sheath,** an insulating covering consisting of multiple layers of membrane. Axons that have myelin sheaths are said to be myelinated, and those that don't are unmyelinated. The myelin sheath is interrupted at intervals, leaving uninsulated gaps called **nodes of Ranvier** (the nodes are where the yellow underlying axon can be seen). At the node regions, the axon is in direct contact with the surrounding fluid. The nerve impulse jumps from node to node, speeding its travel down the axon. Multiple sclerosis (see chapter 29) and Tay-Sachs (see chapter 11) are debilitating clinical disorders, and in the case of Tay-Sachs is fatal. They result from the degeneration of the myelin sheath.

The Nerve Impulse

When a neuron is "at rest," not carrying an impulse, active transport channels in the neuron's plasma membrane transport sodium ions (Na^+) out of the cell and potassium ions (K^+) in. This sodium-potassium pump was described in chapter 5. Sodium ions cannot easily move back into the cell once they are pumped out, so the concentration of sodium ions builds up outside the cell. Similarly, potassium ions accumulate inside the cell, although they are not as highly concentrated because many potassium ions are able to diffuse out through open channels. This resting phase is indicated by the yellow coloring in panel 1 of figure 30.5. The result is to make the outside of the neuron more positive than the inside, a condition called the *resting membrane potential.* The resting plasma membrane is said to be "polarized."

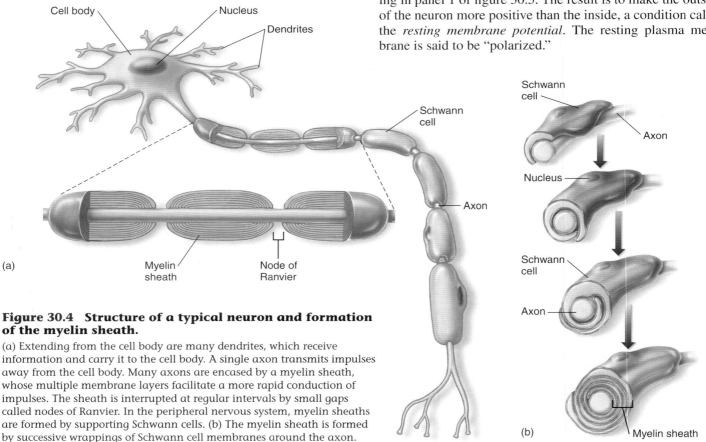

(a)

Cell body · Nucleus · Dendrites · Schwann cell · Axon · Myelin sheath · Node of Ranvier

Schwann cell · Axon · Nucleus · Schwann cell · Axon · Myelin sheath

(b)

Figure 30.4 Structure of a typical neuron and formation of the myelin sheath.

(a) Extending from the cell body are many dendrites, which receive information and carry it to the cell body. A single axon transmits impulses away from the cell body. Many axons are encased by a myelin sheath, whose multiple membrane layers facilitate a more rapid conduction of impulses. The sheath is interrupted at regular intervals by small gaps called nodes of Ranvier. In the peripheral nervous system, myelin sheaths are formed by supporting Schwann cells. (b) The myelin sheath is formed by successive wrappings of Schwann cell membranes around the axon.

Neurons are constantly expending energy to pump sodium ions out of the cell, in order to maintain the resting membrane potential. The net negative charge of most proteins within the cell also adds to this charge difference. Using sophisticated instruments, scientists have been able to measure the voltage difference between the neuron interior and exterior as –70 millivolts (thousandth of a volt). The resting membrane potential is the starting point for a nerve impulse.

A nerve impulse travels along the axon and dendrites as an electrical current caused by ions moving in and out of the neuron through **voltage-gated channels** (that is, protein channels in the neuron membrane that open and close in response to an electrical voltage). The impulse starts when pressure or other sensory inputs disturb a neuron's plasma membrane, causing sodium channels on a dendrite to open (the purple channels in panel 2). As a result, sodium ions flood into the neuron from outside, down its concentration gradient, and for a brief moment a localized area inside of the membrane is "depolarized," becoming more positive than the outside in that immediate area of the axon (indicated by the pink coloring in panel 2).

The sodium channels in the small patch of depolarized membrane remain open for only about a half a millisecond. However, if the change in voltage is large enough, it causes nearby voltage-gated sodium and potassium channels to open (panel 3). The sodium channels open first, which starts a wave of depolarization moving down the neuron. The opening of the gated channels causes nearby voltage-gated channels to open, like a chain of falling dominoes. This local reversal of voltage moving along the axon is called an **action potential.** An action potential follows an all-or-none law: a large enough depolarization produces either a full action potential or none at all, because the voltage-gated Na^+ channels open completely or not at all. Once they open, an action potential occurs. After a slight delay, potassium voltage-gated channels open and K^+ flows out of the cells down its concentration gradient, making the inside of the cell more negative. The increasingly negative membrane potential (colored green in panel 4) causes the voltage-gated sodium channels to snap closed again. This period of time after the action potential has passed and before the resting membrane potential is restored, is called the *refractory period.* A second action potential cannot fire during the refractory period, not until the resting potential is restored by the actions of the sodium-potassium pump.

The depolarization and restoration of the resting membrane potential takes only about 5 milliseconds. Fully 100 such cycles could occur, one after another, in the time it takes to say the word *nerve.*

30.2 **Neurons are cells specialized to conduct impulses. Signals typically arrive along any of numerous dendrites, pass over the cell body's surface, and travel outward on a single long axon. Nerve impulses result from ion movements across the neuron plasma membrane through special protein channels that open and close in response to chemical or electrical stimulation.**

The Nerve Impulse

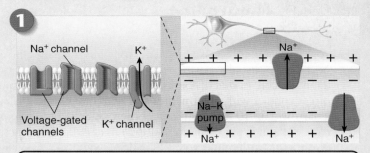

At the resting membrane potential, the inside of the axon is negatively charged because the sodium-potassium pump keeps a higher concentration of Na^+ outside. Voltage-gated ion channels are closed, but there is some leakage of K^+.

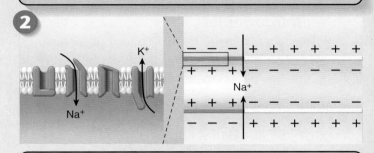

In response to a stimulus, the membrane depolarizes: voltage-gated Na^+ channels open, Na^+ flows into the cell, and the inside becomes more positive.

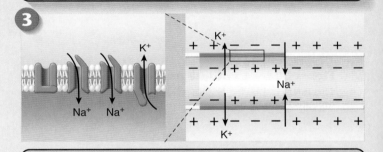

The local change in voltage opens adjacent voltage-gated Na^+ channels, and an action potential is produced.

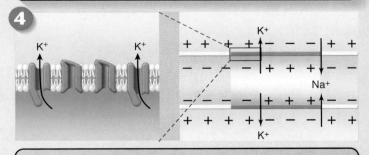

As the action potential travels farther down the axon, voltage-gated Na^+ channels close and K^+ channels open, allowing K^+ to flow out of the cell and restoring the negative charge inside the cell. Ultimately, the sodium-potassium pump restores the resting membrane potential.

Figure 30.5 How an action potential works.

30.3　The Synapse

A nerve impulse travels along a neuron until it reaches the end of the axon, usually positioned very close to another neuron, a muscle cell, or gland. Axons, however, do not actually make direct contact with other neurons or with other cells. Instead, a narrow gap, 10 to 20 nanometers across, called the *synaptic cleft,* separates the axon tip and the target cell. This junction of an axon with another cell is called a **synapse.** A synapse is shown in figure 30.6. The membrane on the axon side of the synapse (on the left here) is called the **presynaptic membrane;** the membrane on the receiving side of the synapse (on the right) is called the **postsynaptic membrane.**

Neurotransmitters

When a nerve impulse reaches the end of an axon, its message must cross the synapse if it is to continue. Messages do not "jump" across synapses. Instead, they are carried across by chemical messengers called **neurotransmitters.** These chemicals are packaged in tiny sacs, or vesicles, at the tip of the axon. When a nerve impulse arrives at the tip, it causes the sacs to release their contents into the synapse, as shown in figure 30.7a. The neurotransmitters diffuse across the synaptic cleft and bind to receptors (the purple structures) in the postsynaptic membrane. The signal passes to the postsynaptic cell when the binding of the neurotransmitter opens special ion channels, allowing ions to enter the postsynaptic cell and so causing a change in electrical charge across its membrane. The enlarged view of figure 30.7a shows how the channel opens and the ion (the yellow ball) enters the cell. Because these channels open when stimulated by a chemical, they are said to be *chemically gated.*

Why go to all this trouble? Why not just wire the neurons directly together? For the same reason that the wires of your house are not all connected but instead are separated by a host of switches. When you turn on one light switch, you don't want every light in the house to go on, the toaster to start heating, and the television to come on! If every neuron in your body were connected to every other neuron, it would be impossible to move your hand without moving every other part of your body at the same time. Synapses are the control switches of the nervous system. However, the control switch must be turned off at some point by getting rid of the neurotransmitter, or the postsynaptic cell would keep firing action potentials. In some cases, the neurotransmitter molecules diffuse away from the synapse. In other cases, the neurotransmitter molecules are either reabsorbed by the presynaptic cell, or are degraded in the synaptic cleft.

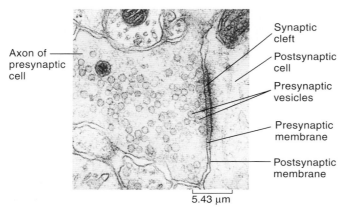

Figure 30.6　A synapse between two neurons.
This micrograph clearly shows the space between the presynaptic and postsynaptic membranes, which is called the synaptic cleft.

Kinds of Synapses

The vertebrate nervous system uses dozens of different kinds of neurotransmitters, each recognized by specific receptors

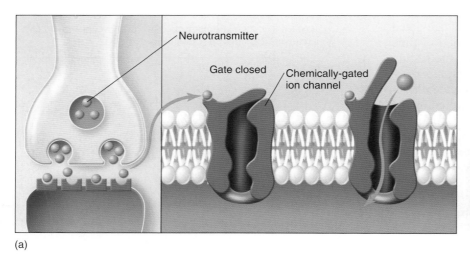

(a)

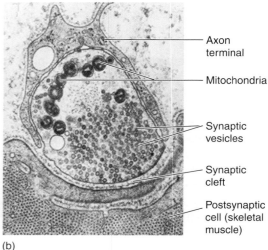

(b)

Figure 30.7　Events at the synapse.
(a) When a nerve impulse reaches the end of an axon, it releases a neurotransmitter into the synaptic cleft. The neurotransmitter molecules diffuse across the synapse and bind to receptors on the postsynaptic cell, opening ion channels. (b) A transmission electron micrograph of the tip of an axon filled with synaptic vesicles.

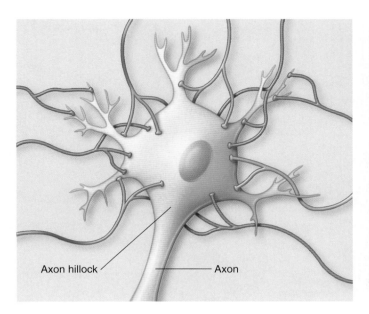

Figure 30.8 Integration.
Many different axons synapse with the cell body and dendrites of the postsynaptic neuron illustrated here. Excitatory synapses are shown in *red* and inhibitory synapses are shown in *blue*. The summed influence of their input at the axon hillock determines whether or not a nerve impulse will be sent down the axon extending below. The scanning electron micrograph shows a neuronal cell body with numerous synapses.

on receiving cells. They fall into two classes, depending on whether they excite or inhibit the postsynaptic cell.

In an excitatory synapse, the receptor protein is usually a chemically gated sodium channel, meaning that a sodium channel through the membrane is opened by the neurotransmitter. On binding with a neurotransmitter whose shape fits it, the sodium channel opens, allowing sodium ions to flood inward. If enough sodium ion channels are opened by neurotransmitters, an action potential begins.

In an *inhibitory synapse,* the receptor protein is a chemically gated potassium or chloride channel. Binding with its neurotransmitter opens these channels, leading to the exit of positively charged potassium ions or the influx of negatively charged chloride ions, resulting in a more negative interior in the receiving cell. This inhibits the start of an action potential, because the negative voltage change inside means that even more sodium ion channels must be opened to get a domino effect started among voltage-gated sodium channels, and so start an action potential.

An individual nerve cell, like the neuron in figure 30.8, can possess both kinds of synaptic connections to other nerve cells. In the drawing, the excitatory synapses are colored red and the inhibitory synapses are colored blue. When signals from both excitatory and inhibitory synapses reach the cell body of a neuron, the excitatory effects (which cause less internal negative charge) and the inhibitory effects (which cause more internal negative charge) interact with one another. The result is a process of **integration** in which the various excitatory and inhibitory electrical effects tend to cancel or reinforce one another. An area at the base of the axon, called the **axon hillock,** is the site of this integration process. If the result of the integration is a large enough depolarization (that is, the inside of the cell becomes more positive), an action potential will

fire. Neurons often receive many inputs. A single motor neuron in the spinal cord may have as many as 50,000 synapses on it!

Neurotransmitters and Their Functions

Acetylcholine (ACh) is the neurotransmitter released at the neuromuscular junction, the synapse that forms between a neuron and a muscle fiber. ACh forms an excitatory synapse with skeletal muscle but has the opposite effect on cardiac muscle, causing an inhibitory synapse.

Glycine and *GABA* are inhibitory neurotransmitters. This inhibitory effect is very important for neural control of body movements and other brain functions. Interestingly, the drug diazepam (Valium) causes its sedative and other effects by enhancing the binding of GABA to its receptors.

Biogenic amines are a group of neurotransmitters that include *dopamine, norepinephrine, serotonin,* and the hormone *epinephrine.* These neurotransmitters have various effects on the body: dopamine is important in controlling body movements, norepinephrine and epinephrine are involved in the autonomic nervous system, which will be discussed later, and, serotonin is involved in sleep regulation and other emotional states. The drug PCP (angel dust) elicits its actions by blocking the elimination of biogenic amines from the synapse. The symptoms vary depending on the dosage.

30.3 A synapse is the junction of an axon with another cell, a gap across which neurotransmitters carry a signal either facilitating or inhibiting transmission of a signal, depending on which ion channels they open.

Addictive Drugs Act on Chemical Synapses

Neuromodulators

The body sometimes deliberately prolongs the transmission of a signal across a synapse by slowing the destruction of neurotransmitters. It does this by releasing into the synapse special long-lasting chemicals called **neuromodulators.** Some neuromodulators aid the release of neurotransmitters into the synapse; others inhibit the reabsorption of neurotransmitters so that they remain in the synapse; still others delay the breakdown of neurotransmitters after their reabsorption, leaving them in the tip to be released back into the synapse when the next signal arrives.

Mood, pleasure, pain, and other mental states are determined by particular groups of neurons in the brain that use special sets of neurotransmitters and neuromodulators. Mood, for example, is strongly influenced by the neurotransmitter serotonin. Many researchers think that depression results from a shortage of serotonin. Prozac, the world's bestselling antidepressant, inhibits the reabsorption of serotonin, thus increasing the amount in the synapse. The synapse in figure 30.9 illustrates the effects of Prozac. The red serotonin molecules released into the synapse are usually reabsorbed by the presynaptic cell. As the circled enlargement shows, Prozac inhibits this reabsorption, leaving serotonin in the synapse.

Drug Addiction

When a cell of the body is exposed to a chemical signal for a prolonged period, it tends to lose its ability to respond to the stimulus with its original intensity. (You are familiar with this loss of sensitivity—when you sit in a chair, how long are you aware of the chair?) Nerve cells are particularly prone to this loss of sensitivity. If receptor proteins within synapses are exposed to high levels of neurotransmitter molecules for prolonged periods, that nerve cell often responds by inserting fewer receptor proteins into the membrane. This feedback is a normal part of the functioning of all neurons, a simple mechanism that has evolved to make the cell more efficient by adjusting the number of "tools" (receptor proteins) in the membrane "workshop" to suit the workload.

Cocaine. The drug cocaine is a neuromodulator that causes abnormally large amounts of neurotransmitters to remain in the synapses for long periods of time. Cocaine affects nerve cells in the brain's pleasure pathways (the so-called limbic system, an area of the brain discussed in section 30.6). These cells transmit pleasure messages using the neurotransmitter dopamine. Panel 1 of figure 30.10 shows normal activity at the synapse, where dopamine molecules, the red balls, are reabsorbed by transporters in the presynaptic cell. Using radioactively labeled cocaine molecules, investigators found that cocaine binds tightly to the transporter proteins on presynaptic membranes (panel 2). These proteins normally remove the neurotransmitter dopamine after it has acted. Like a game of musical chairs in which all the chairs become occupied, eventually there are no unoccupied transporter proteins avail-

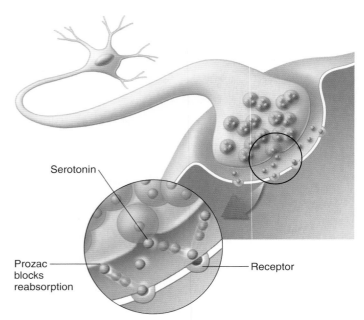

Figure 30.9 Drugs alter transmission of impulses across the synapse.

Depression can result from a shortage of the neurotransmitter serotonin. The antidepressant drug Prozac works by blocking reabsorption of serotonin, keeping serotonin in the synapse longer.

able to the dopamine molecules, so the dopamine stays in the synapse, firing the receptors again and again. As new signals arrive, more and more dopamine is added, firing the pleasure pathway more and more often.

When receptor proteins on limbic system nerve cells are exposed to high levels of dopamine neurotransmitter molecules for prolonged periods of time, the nerve cells "turn down the volume" of the signal by lowering the number of receptor proteins on their surfaces (panel 3). They respond to the greater number of neurotransmitter molecules by simply reducing the number of targets available for these molecules to hit. The cocaine user is now addicted. **Addiction** occurs when chronic exposure to a drug induces the nervous system to adapt physiologically. With so few receptors, as in panel 4, normal levels of dopamine are not able to trigger an action potential in the postsynaptic cell and so the user needs the drug to maintain even normal levels of limbic activity.

Is Addiction to Smoking Cigarettes Drug Addiction?

Investigators attempting to explore the habit-forming nature of smoking cigarettes used what had been learned about cocaine to carry out what seems a reasonable experiment—they introduced radioactively labeled nicotine from tobacco into the brain and looked to see what sort of transporter protein it attached itself to. To their great surprise, the nicotine ignored proteins on the presynaptic membrane and instead bound directly to a specific receptor on the postsynaptic cell! This was totally unexpected, as nicotine does not normally occur in the brain—why should it have a receptor there?

Drug Addiction

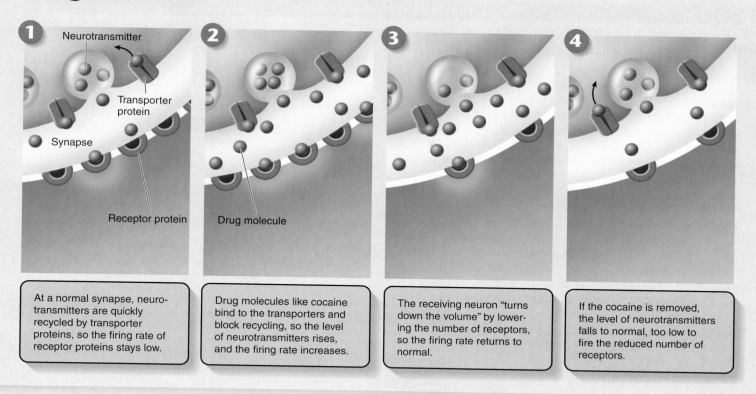

1 At a normal synapse, neuro-transmitters are quickly recycled by transporter proteins, so the firing rate of receptor proteins stays low.

2 Drug molecules like cocaine bind to the transporters and block recycling, so the level of neurotransmitters rises, and the firing rate increases.

3 The receiving neuron "turns down the volume" by lowering the number of receptors, so the firing rate returns to normal.

4 If the cocaine is removed, the level of neurotransmitters falls to normal, too low to fire the reduced number of receptors.

Figure 30.10   How drug addiction works.

Intensive research followed, and researchers soon learned that the "nicotine receptors" normally served to bind the neurotransmitter acetylcholine. It was just an accident of nature that nicotine, an obscure chemical from a tobacco plant, was also able to bind to them. What, then, is the normal function of these receptors? The target of considerable research, these receptors turned out to be one of the brain's most important tools. The brain uses them to coordinate the activities of many other kinds of receptors, acting to "fine-tune" the sensitivity of a wide variety of behaviors.

When neurobiologists compare the limbic system nerve cells of smokers to those of nonsmokers, they find changes in both the number of nicotine receptors and in the levels of RNA used to make the receptors. They have found that the brain adjusts to prolonged exposure to nicotine by "turning down the volume" in two ways: (1) by making fewer receptor proteins to which nicotine can bind; and (2) by altering the pattern of activation of the nicotine receptors (that is, their sensitivity to neurotransmitters).

It is this second adjustment that is responsible for the profound effect smoking has on the brain's activities. By overriding the normal system used by the brain to coordinate its many activities, nicotine alters the pattern of release of many neurotransmitters into synaptic clefts, including acetylcholine, dopamine, serotonin, and many others. As a result, changes in level of activity occur in a wide variety of nerve pathways within the brain.

Addiction to nicotine occurs because the brain compensates for the many changes nicotine induces by making other

changes. Adjustments are made to the numbers and sensitivities of many kinds of receptors within the brain, restoring an appropriate balance of activity.

Now what happens if you stop smoking? Everything is out of whack! The newly coordinated system *requires* nicotine to achieve an appropriate balance of nerve pathway activities. This is addiction in any sensible use of the term. The body's physiological response is profound and unavoidable. There is no way to prevent addiction to nicotine with willpower, any more than willpower can stop a bullet when playing Russian roulette with a loaded gun. If you smoke cigarettes for a prolonged period, you will become addicted.

What do you do if you are addicted to smoking cigarettes and you want to stop? When use of an addictive drug like nicotine is stopped, the level of signaling changes to levels far from normal. If the drug is not reintroduced, the altered level of signaling eventually induces the nerve cells to once again make compensatory changes that restore an appropriate balance of activities within the brain. Over time, receptor numbers, their sensitivity, and patterns of release of neurotransmitters all revert to normal, once again producing normal levels of signaling along the pathways.

30.4   Cigarette smokers find it difficult to quit because they have become addicted to nicotine, a powerful neuromodulator.

30.5 Evolution of the Vertebrate Brain

The structure and function of the vertebrate brain have long been the subject of scientific inquiry. Despite ongoing research, scientists are still not sure how the brain performs many of its functions. For instance, scientists continue to look for the mechanism the brain employs to store memories, and they do not understand how some memories can be "locked away," only to surface in times of stress. The brain is the most complex vertebrate organ ever to evolve, and it can perform a bewildering variety of complex functions (figure 30.11).

Casts of the interior braincases of fossil agnathans, fishes that swam 500 million years ago, have revealed much about the early evolutionary stages of the vertebrate brain. Although small, these brains already had the three divisions, shown in figure 30.12, that characterize the brains of all contemporary vertebrates: (1) the hindbrain, or rhombencephalon (colored yellow); (2) the midbrain, or mesencephalon (colored green); and (3) the forebrain, or prosencephalon (colored blue and purple).

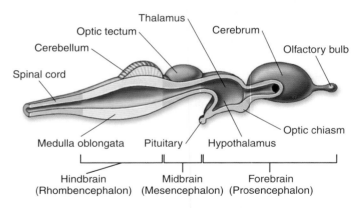

Figure 30.12   The brain of a primitive fish.

The basic organization of the vertebrate brain can be seen in the brains of primitive fishes. The brain is divided into three regions that are found in differing proportions in all vertebrates: the hindbrain, which is the largest portion of the brain in fishes; the midbrain, which in fishes is devoted primarily to processing visual information; and the forebrain, which is concerned mainly with olfaction (the sense of smell) in fishes. In terrestrial vertebrates, the forebrain plays a far more dominant role in neural processing than it does in fishes.

Figure 30.11   Singing well takes practice.

This baby coyote is greeting the approaching evening. His howling is not as impressive as his dad's—a good performance takes practice. His brain is learning by repetition how to control the vocal cords properly.

The hindbrain was the major component of these early brains, as it still is in fishes today. Composed of the *cerebellum* and *medulla oblongata*, the fish hindbrain may be considered an extension of the spinal cord devoted primarily to coordinating motor reflexes. Tracts containing large numbers of axons run like cables up and down the spinal cord to the hindbrain. The hindbrain integrates the many sensory signals coming from the muscles and coordinates the pattern of motor responses.

Much of this coordination is carried on within a small extension of the hindbrain called the cerebellum ("little cerebrum"). In more advanced vertebrates, the cerebellum plays an increasingly important role as a coordinating center and is correspondingly larger than it is in the fishes. In all vertebrates, the cerebellum processes data on the current position and movement of each limb, the state of relaxation or contraction of the muscles involved, and the general position of the body and its relation to the outside world. These data are gathered in the cerebellum and synthesized, and the resulting commands are issued to efferent pathways.

In fishes, the remainder of the brain is devoted to the reception and processing of sensory information. The midbrain is composed primarily of the **optic lobes** (also called the optic tectum), which receive and process visual information, while the forebrain is devoted to the processing of *olfactory* (smell) information. The brains of fishes continue growing throughout their lives. This continued growth is in marked contrast to the brains of other classes of vertebrates, which generally complete their development by infancy. The human brain continues to develop through early childhood, but no new neurons are produced once development has ceased, except in the hippocampus, involved in long-term memory.

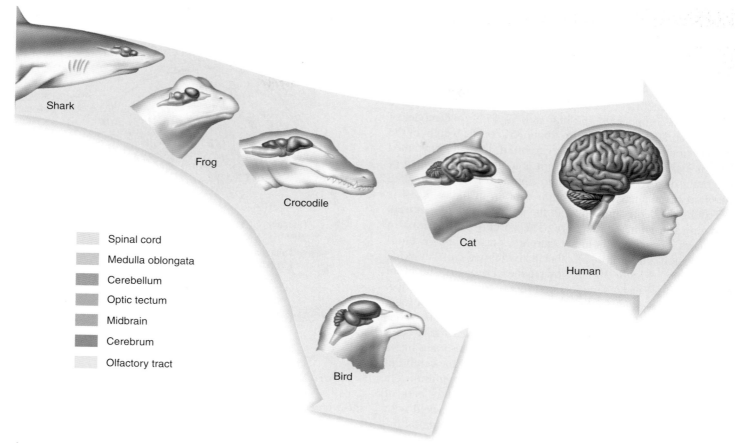

Figure 30.13 The evolution of the vertebrate brain.
In sharks and other fishes, the hindbrain is predominant, and the rest of the brain serves primarily to process sensory information. In amphibians and reptiles, the forebrain is far larger, and it contains a larger cerebrum devoted to associative activity. In birds, which evolved from reptiles, the cerebrum is even more pronounced. In mammals, the cerebrum covers the optic tectum and is the largest portion of the brain. The dominance of the cerebrum is greatest in humans, where it envelops much of the rest of the brain.

Legend:
- Spinal cord
- Medulla oblongata
- Cerebellum
- Optic tectum
- Midbrain
- Cerebrum
- Olfactory tract

The Dominant Forebrain

Starting with the amphibians and continuing more prominently in the reptiles, sensory information is increasingly centered in the forebrain. This pattern was the dominant evolutionary trend in the further development of the vertebrate brain. The areas of the brain are color-coded in figure 30.13 so that you can follow this trend in brain development, with the cerebrum of the forebrain (colored purple) becoming larger in mammals.

The forebrain in reptiles, amphibians, birds, and mammals is composed of two elements that have distinct functions. The *diencephalon* (Greek, *dia*, between) consists of the thalamus and hypothalamus (colored blue in figure 30.12). The **thalamus** is an integrating and relay center between incoming sensory information and the cerebrum. The **hypothalamus** participates in basic drives and emotions and controls the secretions of the pituitary gland, which in turn regulates many of the other endocrine glands of the body (see chapter 31). Through its connections with the nervous system and endocrine system, the hypothalamus helps coordinate the neural and hormonal responses to many internal stimuli and emotions. The *telencephalon*, or "end brain" (Greek, *telos*, end), is located at the front of the forebrain and is devoted largely to as-sociative activity (colored purple in figures 30.12 and 30.13). In mammals, the telencephalon is called the **cerebrum.**

The Expansion of the Cerebrum

If you examine the relationship between brain size and body size in the animals pictured in figure 30.13, you can see a remarkable difference between fishes, amphibians, and reptiles (to the left of the branch point in the figure), and birds and mammals (to the right of the branch point in the figure). Mammals in particular have brains that are particularly large relative to their body size. This is especially true of porpoises and humans. The increase in brain size in the mammals largely reflects the great enlargement of the cerebrum, the dominant part of the mammalian brain. The cerebrum is the center for correlation, association, and learning in the mammalian brain. It receives sensory data from throughout the body and issues motor commands to the body.

30.5 In fishes, the hindbrain forms much of the brain; as terrestrial vertebrates evolved, the forebrain became increasingly more prominent.

The Cerebrum Is the Control Center of the Brain

Although vertebrate brains differ in the relative importance of different components, the human brain is a good model of how vertebrate brains function. About 85% of the weight of the human brain is made up of the cerebrum, the tan convoluted area in figure 30.14. The cerebrum is a large rounded area of the brain divided by a groove into right and left halves called cerebral hemispheres. The sectioned brain in figure 30.14 is cut along the center groove, with the left hemisphere removed, showing the right hemisphere. The cerebrum functions in language, conscious thought, memory, personality development, vision, and a host of other activities we call "thinking and feeling." Figure 30.15 shows general areas of the brain color-coded for easy identification (yellow for the frontal lobe, orange for the parietal lobe, light green for the occipital lobe, and light purple for temporal lobe) and the functions they control. The cerebrum, which looks like a wrinkled mushroom, is positioned over and surrounding the rest of the brain, like a hand holding a fist. Much of the neural activity of the cerebrum occurs within a thin, gray outer layer only a few millimeters thick called the **cerebral cortex** (*cortex* is Latin for "bark of a tree"). This layer is gray because it is densely packed with neuron cell bodies. The human cerebral cortex contains the cell bodies of more than 10 billion nerve cells, roughly 10% of all the neurons in the brain. The wrinkles in the surface of the cerebral cortex increase its surface area (and number of cell bodies) threefold. Underneath the cortex is a solid white region of myelinated nerve fibers that shuttle information between the cortex and the rest of the brain.

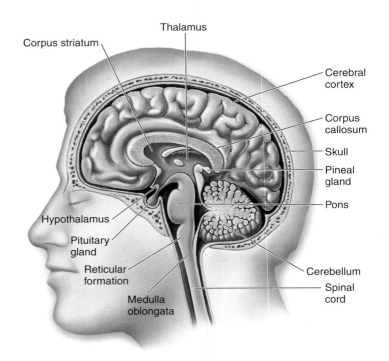

Figure 30.14   A section through the human brain.
The cerebrum occupies most of the brain. Only its outer layer, the cerebral cortex, is visible on the surface.

The right and left cerebral hemispheres are linked by bundles of neurons called **tracts.** These tracts serve as information highways, telling each half of the brain what the other half is doing. Because these tracts cross over, in the area of

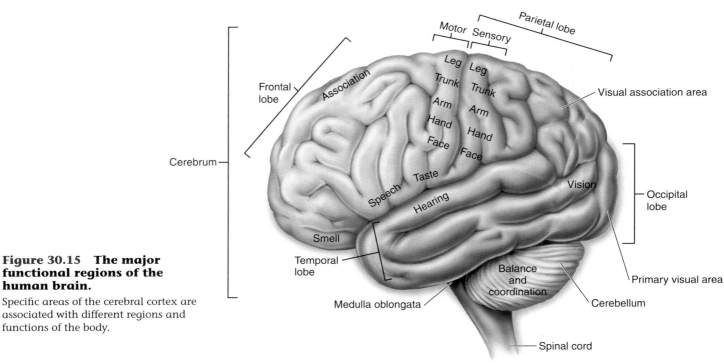

Figure 30.15   The major functional regions of the human brain.

Specific areas of the cerebral cortex are associated with different regions and functions of the body.

the brain called the *corpus callosum* (the blue-colored band in figure 30.14), each half of the brain controls muscles and glands on the opposite side of the body. In general, the left brain is associated with language, speech, and mathematical abilities, whereas the right brain is associated with intuitive, musical, and artistic abilities.

Researchers have found that the two sides of the cerebrum can operate as two different brains. For instance, in some people the tract between the two hemispheres has been cut by accident or surgery. In laboratory experiments, one eye of an individual with such a "split brain" is covered and a stranger is introduced. If the other eye is then covered instead, the person does not recognize the stranger who was just introduced!

Sometimes blood vessels in the brain are blocked by blood clots, causing a disorder called a **stroke.** During a stroke, circulation to an area in the brain is blocked and the brain tissue dies. A severe stroke in one side of the cerebrum may cause paralysis of the other side of the body.

The Thalamus and Hypothalamus Process Information

Beneath the cerebrum are the thalamus and hypothalamus, important centers for information processing. The thalamus is the major site of sensory processing in the brain. Auditory (sound), visual, and other information from sensory receptors enter the thalamus and then are passed to the sensory areas of the cerebral cortex. The thalamus also controls balance. Information about posture, derived from the muscles, and information about orientation, derived from sensors within the ear, combine with information from the cerebellum and pass to the thalamus. The thalamus processes the information and channels it to the appropriate motor center on the cerebral cortex.

The hypothalamus integrates all the internal activities. It controls centers in the brain stem that in turn regulate body temperature, blood pressure, respiration, and heartbeat. It also directs the secretions of the brain's major hormone-producing gland, the pituitary gland. The hypothalamus is linked by an extensive network of neurons to some areas of the cerebral cortex. This network, along with parts of the hypothalamus and areas of the brain called the *hippocampus* and *amygdala,* make up the **limbic system.** The areas highlighted in green in figure 30.16 indicate the components of the limbic system. The operations of the limbic system are responsible for many of the most deep-seated drives and emotions of vertebrates, including pain, anger, sex, hunger, thirst, and pleasure, centered in the amygdala. You'll recall on page 606 that the limbic system is the area of the brain affected by cocaine. It is also involved in memory, centered in the hippocampus.

The Cerebellum Coordinates Muscle Movements

Extending back from the base of the brain is a structure known as the **cerebellum.** The cerebellum controls balance, posture, and muscular coordination. This small, cauliflower-

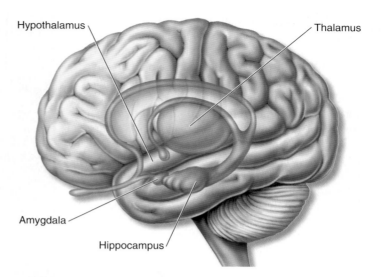

Figure 30.16 The limbic system.
The hippocampus and the amygdala are the major components of the limbic system, which controls our most deep-seated drives and emotions.

shaped structure, while well developed in humans and other mammals, is even better developed in birds. Birds perform more complicated feats of balance than we do, because they move through the air in three dimensions. Imagine the kind of balance and coordination needed for a bird to land on a branch, stopping at precisely the right moment without crashing into it.

The Brain Stem Controls Vital Body Processes

The **brain stem,** a term used to collectively refer to the midbrain, pons, and medulla oblongata, connects the rest of the brain to the spinal cord. This stalklike structure contains nerves that control your breathing, swallowing, and digestive processes, as well as the beating of your heart and the diameter of your blood vessels. A network of nerves called the **reticular formation** runs through the brain stem and connects to other parts of the brain. Their widespread connections make these nerves essential to consciousness, awareness, and sleep. One part of the reticular formation filters sensory input, enabling you to sleep through repetitive noises such as traffic yet awaken instantly when a telephone rings.

Language and Other Higher Functions

Although the two cerebral hemispheres seem structurally similar, they are responsible for different activities. The most thoroughly investigated example of this lateralization of function is language. The left hemisphere is the "dominant" hemisphere for language—the hemisphere in which most neural processing related to language is performed—in 90% of right-handed people and nearly two-thirds of left-handed people. There are two language areas in the dominant hemisphere: One is important for language comprehension and the formulation of thoughts into speech, and the other is responsible for the generation of motor output needed for language communication. Different language activities in figure 30.17 confirm that different parts of the brain are involved.

While the dominant hemisphere for language is adept at sequential reasoning, like that needed to formulate a sentence, the nondominant hemisphere (the right hemisphere in most people) is adept at spatial reasoning, the type of reasoning needed to assemble a puzzle or draw a picture. It is also the hemisphere primarily involved in musical ability—a person with damage to the speech area in the left hemisphere may not be able to speak but may retain the ability to sing! Damage to the nondominant hemisphere may lead to an inability to appreciate spatial relationships and may impair musical activities such as singing. Reading, writing, and oral comprehension remain normal. The nondominant hemisphere is also important for the consolidation of memories of nonverbal experiences.

One of the great mysteries of the brain is the basis of memory and learning. There is no one part of the brain in which all aspects of a memory appear to reside. Although memory is impaired if portions of the brain, particularly the temporal lobes, are removed, it is not lost entirely. Many memories persist in spite of the damage, and the ability to access them gradually recovers with time. Therefore, investigators who have tried to probe the physical mechanisms underlying memory often have felt that they were grasping at a shadow. Although we still do not have complete understanding, we have learned a good deal about the basic processes in which memories are formed.

There appear to be fundamental differences between short-term and long-term memory. Short-term memory is transient, lasting only a few moments. Such memories can readily be erased by the application of an electrical shock, leaving previously stored long-term memories intact. This result suggests that short-term memories are stored electrically in the form of a transient neural excitation. Long-term memory, in contrast, appears to involve structural changes in certain neural connections within the brain. Two parts of the temporal lobes, the hippocampus and the amygdala, are involved in both short-term memory and its consolidation into long-term memory. Damage to these structures impairs the ability to process recent events into long-term memories.

The Mechanism of Alzheimer Disease Still a Mystery

In the past, little was known about Alzheimer disease, a condition in which the memory and thought processes of the brain become dysfunctional. Drug companies are eager to develop new products for the treatment of Alzheimer disease, but they have little concrete evidence to go on. Scientists disagree about the biological nature of the disease and its cause. Two hypotheses have been proposed: one suggests that nerve cells

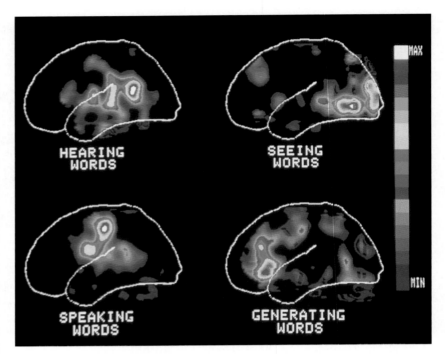

Figure 30.17   Different brain regions control various activities.
This illustration shows how the brain reacts in human subjects asked to listen to a spoken word, to read that same word silently, to repeat the word out loud, and then to speak a word related to the first. Regions of white, red, and yellow show the greatest activity. Compare this to figure 30.15 to see how regions of the brain are mapped.

in the brain are killed from the outside in, the other from the inside out.

In the first hypothesis, external proteins called β-amyloid peptides kill nerve cells. A mistake in protein processing produces an abnormal form of the peptide, which then forms aggregates, or plaques. The plaques begin to fill-in the brain and then damage and kill nerve cells. However, these amyloid plaques have been found in autopsies of people who did not have Alzheimer disease.

The second hypothesis maintains that the nerve cells are killed by an abnormal form of an internal protein. This protein, called tau (τ), normally functions to maintain protein transport microtubules. Abnormal forms of τ assemble into helical segments that form tangles, which interfere with the normal functioning of the nerve cells. Researchers continue to study whether tangles and plaques are causes or effects of Alzheimer disease.

Progress has been made in identifying genes that increase the likelihood of developing Alzheimer disease and genes that, when mutated, can cause the disorder. Most Alzheimer patients do not have these mutated genes, but for those that do, the symptoms of Alzheimer are expressed much earlier in life.

30.6 The associative activity of the brain is centered in the wrinkled cerebral cortex, which lies over the cerebrum. Beneath, the thalamus and hypothalamus process information and integrate body activities. At the base of the brain, the cerebellum coordinates muscle movements.

The **spinal cord** is a cable of neurons extending from the brain down through the backbone, which is the view in figure 30.18. The cross section through the spinal cord in figure 30.19 shows a darker gray area in the center formed by neuron cell bodies, which form a column down the length of the cord, surrounded by a sheath of axons and dendrites, which make the outer edges of the cord white because they are coated with myelin. The spinal cord is surrounded and protected by a series of bones called the vertebrae. Spinal nerves pass out to the body from between the vertebrae. Messages from the body and the brain run up and down the spinal cord, an information highway.

In each segment of the spine, motor nerves extend out of the spinal cord to the muscles. Motor nerves from the spine control most of the muscles below the head. This is why injuries to the spinal cord often paralyze the lower part of the body. A muscle is paralyzed and cannot move if its motor neurons are damaged.

Spinal Cord Regeneration

In the past, scientists have tried to repair severed spinal cords by installing nerves from another part of the body to bridge the gap and act as guides for the spinal cord to regenerate. But most of these experiments have failed because the nerve bridges did not go from white matter to gray matter. Also, there is a factor that inhibits nerve growth in the spinal cord. After discovering that fibroblast growth factor stimulates nerve growth, neurobiologists tried gluing on the nerves, from white to gray matter, with fibrin that had been mixed with the fibroblast growth factor.

Three months later, rats with the nerve bridges began to show movement in their lower bodies. In further analy-

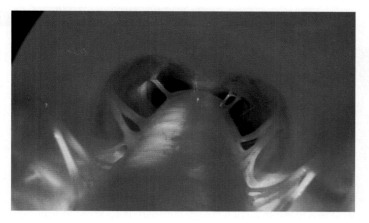

Figure 30.18   A view down the human spinal cord.
Pairs of spinal nerves can be seen extending out from the spinal cord. Along these nerves, the brain and spinal cord communicate with the body.

ses of the experimental animals, dye tests indicated that the spinal cord nerves had regrown from both sides of the gap. Many scientists are encouraged by the potential to use a similar treatment in human medicine. However, most spinal cord injuries in humans do not involve a completely severed spinal cord; often, nerves are crushed. Also, although the rats with nerve bridges did regain some locomotory ability, tests indicated that they were barely able to walk or stand.

> **30.7 The spinal cord, protected in vertebrates by a backbone, extends motor nerves to the muscles below the head.**

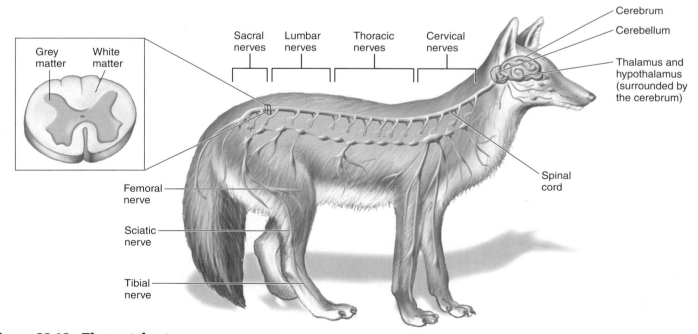

Figure 30.19   The vertebrate nervous system.
The brain is colored *tan* and the spinal cord and nerves are colored *yellow.*

30.8 Voluntary and Autonomic Nervous Systems

As you learned in the opening discussion in section 30.1, the nervous system is divided into two main parts: the central nervous system (the pink boxes in figure 30.20, which include the brain and spinal cord) and the peripheral nervous system (the blue boxes, which include the motor and sensory pathways). The motor pathways of the peripheral nervous system of a vertebrate can be further subdivided into the **somatic (voluntary) nervous system,** which relays commands to skeletal muscles, and the **autonomic (involuntary) nervous system,** which stimulates glands and relays commands to the smooth muscles of the body and to cardiac muscle. The voluntary nervous system can be controlled by conscious thought. You can, for example, command your hand to move. The autonomic nervous system, by contrast, cannot be controlled by conscious thought. You cannot, for example, tell the smooth muscles in your digestive tract to speed up their action. The central nervous system issues commands over both voluntary and autonomic systems, but you are conscious of only the voluntary commands.

Voluntary Nervous System

Motor neurons of the voluntary nervous system stimulate skeletal muscles to contract in two ways. First, motor neurons may stimulate the skeletal muscles of the body to contract in response to conscious commands. For example, if you want to bounce a basketball, your CNS sends messages through motor neurons to the muscles in your arms and hands. However,

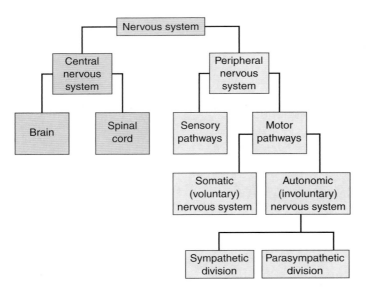

Figure 30.20   The divisions of the vertebrate nervous system.
The motor pathways of the peripheral nervous system are the somatic (voluntary) and autonomic nervous systems.

skeletal muscle can also be stimulated as part of reflexes that do not require conscious control.

Reflexes Enable Quick Action.   The motor neurons of the body have been wired to enable the body to act particularly quickly in time of danger—even before the animal is consciously aware of the threat. These sudden, involuntary movements are called reflexes. A **reflex** produces a rapid motor response to a stimulus because the sensory neuron bringing information about the threat passes the information directly to a motor neuron. The escape reaction of a fly about to be swatted is a reflex. One of the most frequently used reflexes in your body is blinking, a reflex that protects your eyes. If anything, such as an insect or a cloud of dust, approaches your eye, the eyelid blinks closed even before you realize what has happened. The reflex occurs before the cerebrum is aware the eye is in danger.

Because they involve passing information between few neurons, reflexes are very fast. Many reflexes never reach the brain. The "danger" nerve impulse travels only as far as the

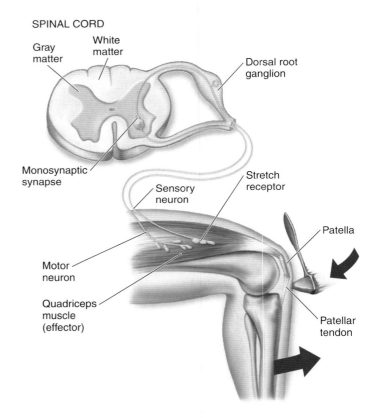

Figure 30.21   The knee-jerk reflex.
The most famous involuntary response, the knee jerk, is produced by activating stretch receptors in the quadriceps muscle. When a rubber mallet taps the patellar tendon, the muscle and stretch receptors in the muscle are stretched. A signal travels up a sensory neuron to the spinal cord, where the sensory neuron stimulates a motor neuron, which sends a signal to the quadriceps muscle to contract.

spinal cord and then comes right back as a motor response. Most reflexes involve a single connecting interneuron between the sensory neuron and the motor neuron. A few, like the knee-jerk reflex in figure 30.21, are monosynaptic reflex arcs. You see in the figure that the sensory neuron, a stretch receptor embedded in a muscle, "senses" the stretching of the muscle when the tendon is tapped. This stretching could harm the muscle and so a nerve impulse is sent to the spinal cord where it synapses directly with a motor neuron—there is no interneuron intermediary between them. If you step on something sharp, your leg jerks away from the danger: in the same way, the prick causes nerve impulses in sensory neurons, which pass to the spinal cord directly to motor neurons, which cause your leg muscles to contract, jerking your leg up.

Autonomic Nervous System

Some motor neurons are active all the time, even during sleep. These neurons carry messages from the CNS that keep the body going even when it is not active. These neurons are called the **autonomic nervous system.** The word *autonomic* means involuntary. The autonomic nervous system carries messages to muscles and glands that work without the animal noticing.

The autonomic nervous system is the command network the CNS uses to maintain the body's homeostasis. Using it, the CNS regulates heartbeat and controls muscle contractions in the walls of the blood vessels. It directs the muscles that control blood pressure, breathing, and the movement of food through the digestive system. It also carries messages that help stimulate glands to secrete tears, mucus, and digestive enzymes.

The autonomic nervous system is composed of two elements that act in opposition to one another. One division, the **sympathetic nervous system,** dominates in times of stress. It controls the "fight-or-flight" reaction, increasing blood pressure, heart rate, breathing rate, and blood flow to the muscles. The sympathetic nervous system is colored pink in figure 30.22 and consists of a network of short motor axons extending out from the spinal cord to

clusters of neuron cell bodies, called **ganglia,** indicated by the darker-colored band, located just to the right of the spinal cord in the figure. You can also see this chain of ganglia in figure 30.19. Long motor neurons extend from the ganglia directly to each target organ. Another division, the **parasympathetic nervous system,** has the opposite effect. It conserves energy by slowing the heartbeat and breathing rate and by promoting digestion and elimination. The parasympathetic nervous system is colored in blue and consists of a network of long axons extending out from motor neurons within the spinal cord; these axons extend to ganglia in the immediate vicinity of an organ. It also consists of short motor neurons extending from the ganglia to the nearby organ.

Most glands, smooth muscles, and cardiac muscles get constant input from *both* the sympathetic and parasympathetic systems. The CNS controls activity by varying the ratio of the two signals to either stimulate or inhibit the organ.

> **30.8** The voluntary nervous system relays commands to skeletal muscles and can be controlled by conscious thought. The autonomic nervous system relays commands to muscles and glands that cannot be controlled by conscious thought.

Figure 30.22 How the sympathetic and parasympathetic nervous systems interact.

A nerve path runs from both of the systems to every organ indicated except the adrenal gland, which is only innervated by the sympathetic nervous system.

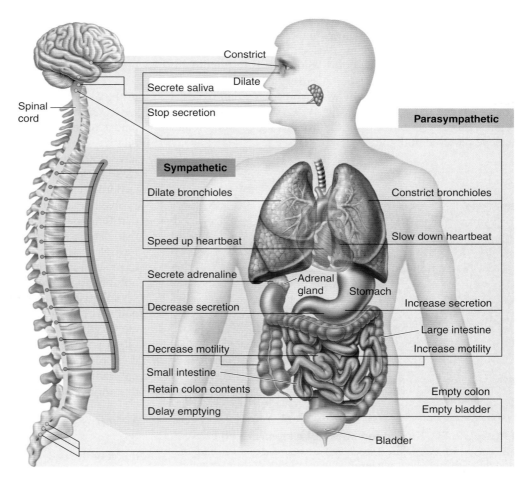

Constrict
Dilate
Secrete saliva
Stop secretion
Spinal cord

Parasympathetic

Sympathetic

Dilate bronchioles — Constrict bronchioles

Speed up heartbeat — Slow down heartbeat

Secrete adrenaline
Adrenal gland
Stomach
Decrease secretion — Increase secretion

Large intestine
Decrease motility — Increase motility

Small intestine
Retain colon contents — Empty colon
Delay emptying — Empty bladder
Bladder

30.9 Sensory Perception

Did you ever wonder what it would be like not to know anything about what is going on around you? Imagine if you couldn't hear, or see, or feel, or smell. After a while, a human goes mad if completely deprived of sensory input. The senses are the bridge to experience and perceive the way the body relates to everything around it.

Sensory Receptors

The **sensory nervous system** tells the central nervous system what is happening. Sensory neurons carry impulses to the CNS from more than a dozen different types of sensory cells that detect changes outside and inside the body. Called **sensory receptors,** these specialized sensory cells detect many different things, including changes in blood pressure, strain on ligaments, and smells in the air. Particularly complex sensory receptors, made up of many cell and tissue types, are called **sensory organs.** The eyes and ears (figure 30.23) are sensory organs, and so are the taste buds in your mouth.

How does the brain know whether an incoming nerve impulse is light, sound, or pain? This information is built into the "wiring"—into which neurons interact while passing the information to the CNS and into the location in the brain where the information is sent. The brain "knows" it is responding to light because the message from a sensory neuron is wired to light receptor cells. That is why when you press your fingertips gently against the corners of your eyes, you "see stars"—the brain treats any impulse from the eyes as light, even though the eye received no light.

Figure 30.23 Kangaroo rats have specialized ears.
The ears of kangaroo rats (*Dipodomys*) are adapted to nocturnal life and allow them to hear the low-frequency sounds of their predators, such as an owl's wingbeats or a sidewinder rattlesnake's scales rubbing against the ground. Also, the ears seem to be adapted to the poor sound-carrying quality of dry, desert air.

The Path of Sensory Information

The path of sensory information to the CNS is a simple one, composed of three stages:

1. **Stimulation.** A physical stimulus impinges on a sensory receptor.
2. **Transduction.** Sensory receptors change the stimulus into an electrical potential by initiating the opening or closing of ion channels in a sensory neuron.
3. **Transmission.** The sensory neuron conducts a nerve impulse along an afferent pathway to the CNS.

All sensory receptors are able to initiate nerve impulses by opening or closing **stimulus-gated channels** within sensory neuron membranes. Except for visual photoreceptors, these channels are sodium ion channels that depolarize the membrane and so start an electrical signal. If the stimulus is large enough, the depolarization will trigger an action potential. The greater the sensory stimulus, the greater the depolarization of the sensory receptor and the higher the frequency of action potentials. The channels are opened by chemical or mechanical stimulation, often a disturbance such as touch, heat, or cold. The receptors differ from one another in the nature of the environmental input that triggers the opening of the channel. The body contains many sorts of receptors, each sensitive to a different aspect of the body's condition or to a different quality of the external environment.

Exteroceptors are receptors that sense stimuli that arise in the external environment. Almost all of a vertebrate's exterior senses evolved in water before vertebrates invaded the land. Consequently, many senses of terrestrial vertebrates emphasize stimuli that travel well in water, using receptors that have been retained in the transition from the sea to the land. Hearing, for example, converts an airborne stimulus into a waterborne one, using receptors similar to those that originally evolved in aquatic animals. A few vertebrate sensory systems that function well in the water, such as the electric organs of fish, cannot function in the air and are not found among terrestrial vertebrates. On the other hand, some land dwellers have sensory systems, such as infrared receptors, that could not function in the sea.

Interoceptors sense stimuli that arise from within the body. These internal receptors detect stimuli related to muscle length and tension, limb position, pain, blood chemistry, blood volume and pressure, and body temperature. Many of these receptors are simpler than those that monitor the external environment and are believed to bear a closer resemblance to primitive sensory receptors.

Sensing the Internal Environment

Sensory receptors inside the body inform the CNS about the condition of the body. Much of this information passes to a coordinating center in the brain, the hypothalamus, the part of the brain responsible for maintaining the body's homeostasis—that is, keeping the body's internal environment constant. The vertebrate body uses a variety of different sensory receptors to respond to different aspects of its internal environment.

Temperature change. Two kinds of nerve endings in the skin are sensitive to changes in temperature, one stimulated by cold, the other by warmth. By comparing information from the two, the CNS can learn what the temperature is and if it is changing.

Blood chemistry. Receptors in the walls of arteries sense CO_2 levels in the blood. The brain uses this information to regulate the body's respiration rate, increasing it when CO_2 levels rise above normal.

Pain. Damage to tissue is detected by special nerve endings within tissues, usually near the surface, where damage is most likely to occur. When these nerve endings are physically damaged or deformed, the CNS responds by reflexively withdrawing the body segment and often by changing heartbeat and blood pressure as well.

Muscle contraction. Buried deep within muscles are sensory receptors called stretch receptors. You heard about these receptors in the discussion on reflexes. In each, the end of a sensory neuron is wrapped around a

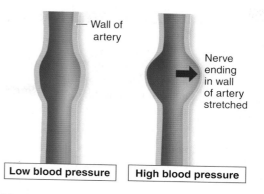

Low blood pressure | High blood pressure

Figure 30.25 How a baroreceptor works.
A network of nerve endings covers a region where the wall of the artery is thin. High blood pressure causes the wall to balloon out there, stretching the nerve endings and causing them to fire impulses.

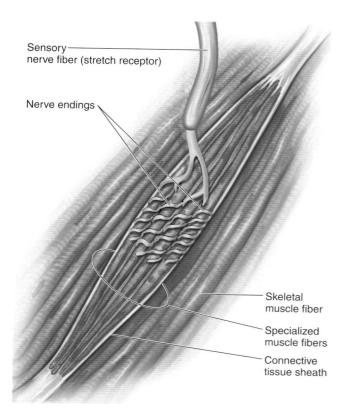

Sensory nerve fiber (stretch receptor)

Nerve endings

Skeletal muscle fiber

Specialized muscle fibers

Connective tissue sheath

Figure 30.24 A stretch receptor embedded within skeletal muscle.
Stretching the muscle elongates the specialized muscle fibers, which deforms the nerve endings, causing them to send a nerve impulse out along the nerve fiber.

muscle fiber, like the receptor shown in figure 30.24. When the muscle is stretched, the fiber elongates, stretching the spiral nerve ending (like stretching a spring) and causing repeated nerve impulses to be sent to the brain. From these signals the brain can determine the rate of change of muscle length at any given moment. The CNS uses this information to control movements that require the combined action of several muscles, such as those that carry out breathing or locomotion.

Blood pressure. Blood pressure is sensed by neurons called baroreceptors with highly branched nerve endings within the walls of major arteries. When blood pressure increases, the stretching of the arterial wall, like the expansion of the artery in figure 30.25, causes the sensory neuron to increase the rate at which it sends nerve impulses to the CNS. When the wall of the artery is not stretched, the rate of firing of the sensory neuron goes down. Thus, the frequency of impulses provides the CNS with a continuous measure of blood pressure.

Touch. Touch is sensed by pressure receptors buried below the surface of the skin. There are a variety of different types, some specialized to detect rapid changes in pressure, others to measure the duration and extent to which pressure is applied, and still others sensitive to vibrations.

30.9 Sensory receptors initiate nerve impulses in response to stimulation. All sensory nerve impulses are the same, differing only in the stimulus that fires them and their destination in the brain. A variety of different sensory receptors inform the hypothalamus about different aspects of the body's internal environment, enabling it to maintain the body's homeostasis.

Sensing Gravity and Motion

Two types of receptors in the ear inform the brain where the body is in three dimensions. This knowledge enables an animal to move freely and maintain its balance. Figure 30.26 shows the anatomy of the inner ear and the locations of these receptors.

Balance. To keep the body's balance the brain needs a frame of reference, and the reference point it uses is gravity. The sensory receptors that detect gravity are hair cells that extend from the floor of the utricle and the saccule (see the enlarged view of the inner ear). The tips of the hair cells project into a gelatinous matrix with embedded particles called **otoliths** (figure 30.26b). To illustrate how these receptors work, imagine a pencil standing in a glass. No matter which way you tip the glass, the pencil rolls along the rim, applying pressure to the lip of the glass. If you want to know the direction the glass is tipped, you need only ask where on the rim pressure is being applied. Similarly, the otoliths will shift in the matrix in response to the pull of gravity stimulating hair cells, which the brain uses to determine vertical positioning.

Motion. The brain senses motion by employing a receptor in which fluid deflects cilia in a direction opposite that of the motion. Within the inner ear are three fluid-filled **semicircular canals,** each oriented in a different plane at right angles to the other two (see the enlarged view of the inner ear) so that motion in any direction can be detected. Protruding into the canal are groups of cilia from sensory cells. The cilia from each cell are arranged in a tentlike assembly called a *cupula,* shown in figure 30.26c, which is pushed when fluid in the semicircular canals moves in a direction opposite that of the head's movement. Because the three canals are oriented in all three planes, movement in any plane is sensed by at least one of them, and the brain is able to analyze complex movements by comparing the sensory inputs from each canal.

The semicircular canals do not react if the body moves in a straight line because the fluid in the canals does not move. That is why traveling in a car or airplane at a constant speed in one direction gives no sense of motion.

> **30.10** The body senses gravity and acceleration by the deflection of cilia by moving objects or fluid. The body cannot sense motion at a constant velocity and direction.

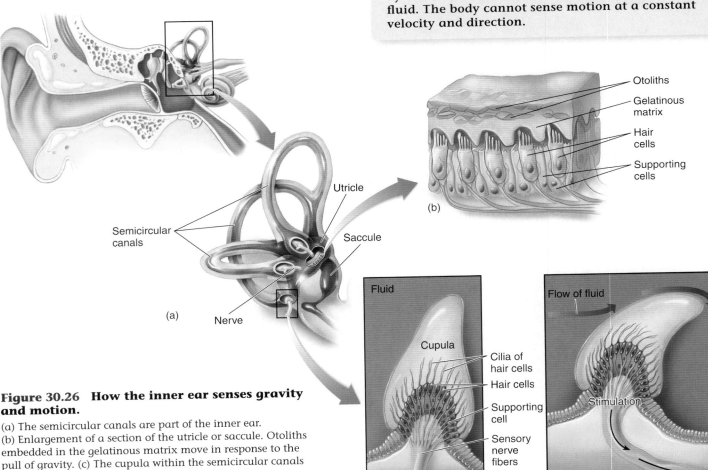

Figure 30.26 How the inner ear senses gravity and motion.
(a) The semicircular canals are part of the inner ear.
(b) Enlargement of a section of the utricle or saccule. Otoliths embedded in the gelatinous matrix move in response to the pull of gravity. (c) The cupula within the semicircular canals are surrounded by fluid and contain hair cells. Movement in a particular direction causes fluid in the semicircular canal of that plane to move; the cupula is displaced, thereby stimulating the hair cells.

30.11 Sensing Chemicals: Taste and Smell

Vertebrates are able to detect many of the chemicals in air and in food.

Taste. Embedded within the surface of the tongue are *taste buds* located within *papillae*, which are the raised areas on the tongue in figure 30.27. Taste buds are onion-shaped structures in the enlarged view of the papillae that contain many taste receptor cells, each of which has fingerlike microvilli that project into an opening called the taste pore. Chemicals from food dissolve in saliva and contact the taste cells through the taste pore. Salty, sour, sweet, bitter, and umami (a "meaty" taste) are perceived because chemicals in food are detected in different ways by taste buds. When the tongue encounters a chemical, information from the taste cells passes to sensory neurons, which transmit the signals to the brain.

Smell. In the nose are chemically sensitive neurons whose cell bodies are embedded within the epithelium of the nasal passage, shown in cross section in figure 30.28.

When they detect chemicals, these sensory neurons (the red cells in the enlarged view) transmit information to a location in the brain where smell information is processed and analyzed. For discovering how this works, American scientists Richard Axel and Linda Buck were awarded the Noble Prize in Physiology or Medicine in 2004. In many vertebrates (dogs are a familiar example), these neurons are far more sensitive than in humans.

Smell as well as taste is very important in telling an animal about its food. That is why when you have a bad cold and your nose is stuffed up, your food has little taste. Other receptors also play a role. For example, the "hot" sensation of foods such as chili peppers is detected by pain receptors, not chemical receptors.

> **30.11 Taste and smell are chemical senses. In many vertebrates, the sense of smell is very well developed.**

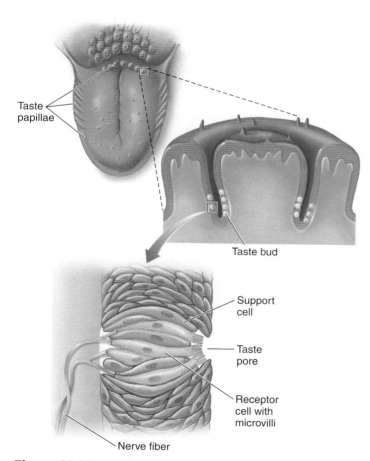

Figure 30.27 Taste.

Taste buds on the human tongue are typically grouped into projections called papillae. Individual taste buds are bulb-shaped collections of taste receptor cells that open out into the mouth through a taste pore.

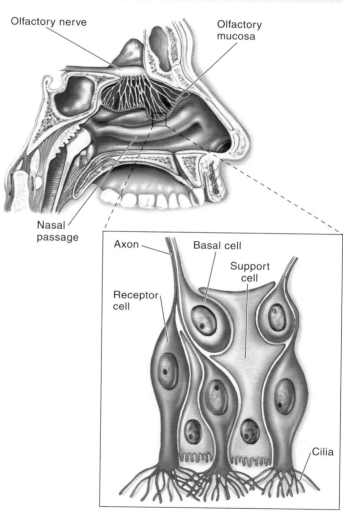

Figure 30.28 Smell.

Humans smell by using receptor cells located in the lining of the nasal passage. The receptor cells are neurons. Axons from these sensory neurons project back through the olfactory nerve directly to the brain.

30.12 Sensing Sounds: Hearing

When you hear a sound, you are detecting the air vibrating—waves of pressure in the air beating against your ear, pushing a membrane called the eardrum in and out. As you can see in figure 30.29, on the inner side of the eardrum are three small bones, called ossicles, that act as a lever system to increase the force of the vibration. They transfer the amplified vibration across a second membrane to fluid within the inner ear. The fluid-filled chamber of the inner ear is shaped like a tightly coiled snail shell and is called the cochlea, from the Latin name for "snail." The middle ear, where the ossicles are located, is connected to the throat by the eustachian tube in such a way that there is no difference in air pressure between the middle ear and the outside. That is why your ears sometimes "pop" when landing in an airplane—the pressure is equalizing between the two sides of the eardrum. This equalized pressure is necessary for the eardrum to work.

The sound receptors within the cochlea are hair cells that rest on a membrane that runs up and down the middle of the spiraling chamber, separating it into two halves like a wall, into the upper and lower fluid-filled canals in the enlarged view. The hair cells do not project into the fluid-filled canals of the cochlea; instead, they are covered by a second membrane (the dark blue membrane in the figure). When a sound wave enters the cochlea, it causes the fluid in the chambers to move. The moving fluid causes this membrane "sandwich" to vibrate, bending the hairs pressed against the upper membrane and causing them to send nerve impulses to sensory neurons that travel to the brain.

Sounds of different frequencies cause different parts of the membrane to vibrate, and thus fire different sensory neurons—the identity of the sensory neuron being fired tells the CNS what the frequency of the sound is. Sound waves of higher frequencies, about 20,000 vibrations (or cycles) per second, also called hertz (Hz), move the membrane in the area closest to the middle ear. Medium-length frequencies, about 2,000 Hz, move the membrane in the area about midway down the length of the cochlea. The lowest-frequency sound waves, about 500 Hz, move the membrane near the tip of the cochlea.

The intensity of the sound is determined by how *often* the neurons fire. Our ability to hear depends upon the flexibility of the membranes within the cochlea. Humans cannot hear low-pitched sounds, below 20 vibrations (or cycles) per second, although some vertebrates can. As children, we can hear high-pitched sounds, up to 20,000 cycles per second, but this ability decreases as we get older. Other vertebrates can hear sounds at far higher frequencies. Dogs readily hear sounds of 40,000 cycles per second and so respond to a high-pitched dog whistle that seems silent to a human.

Frequent or prolonged exposure to loud noises can result in damage of the hair cells and membrane, especially in the high-frequency area of the cochlea. The loss of the ability to detect high-frequency sounds affects a person's ability to hear certain sounds, especially in a noisy setting.

The Lateral Line System

A lateral line system supplements the fish's sense of hearing. It is performed by a different sensory structure, and provides a sense of "distant touch." A fish is able to sense objects that reflect pressure waves and low-frequency vibrations and thus can detect prey, for example, and swim in synchrony with the rest of its school. The lateral line system also enables a blind cave fish to sense its environment by monitoring changes in the patterns of water flow past the lateral line receptors. The same system is found in amphibian larvae, but it is lost at metamorphosis and is not present in any terrestrial vertebrate.

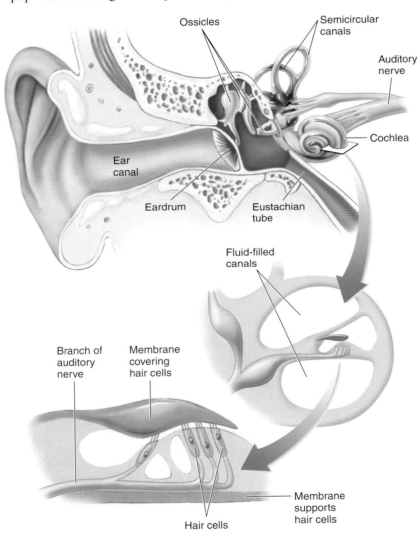

Figure 30.29 Structure and function of the human ear.
Sound waves passing through the ear canal beat on the eardrum, pushing a set of three small bones, or ossicles, against an inner membrane. This sets up a wave motion in the fluid filling the canals within the cochlea. The wave causes the membrane covering the hair cells to move back and forth against the hair cells, which causes associated neurons to fire impulses.

The lateral line system consists of sensory structures within a longitudinal canal in the fish's skin, shown in figure 30.30, that extends along each side of the body and within several canals in the head. As you can see in the enlarged view, openings lead into the canal, which is lined with sensory structures known as hair cells because they have hairlike processes at their surface. The processes of the hair cells project into a gelatinous membrane called a *cupula* (Latin, "little cup"). The hair cells are innervated by sensory neurons that transmit impulses to the brain. Vibrations carried through the fish's environment and down into the canal produce movements of the cupula, which cause the hairs to bend. When the hair cells bend, the associated sensory neurons are stimulated and generate a nerve impulse that is sent to the brain.

Figure 30.31   Using ultrasound to locate a moth.
This bat is emitting high-frequency "chirps" as it flies. It then listens for the sound's reflection against the moth. By timing how long it takes for a sound to return, the bat can "see" the moth even in total darkness.

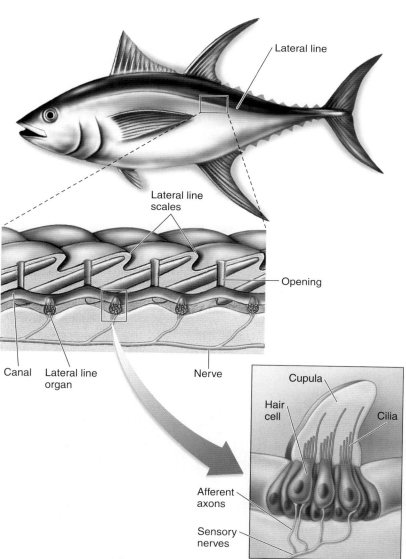

Figure 30.30   The lateral line system.
This system consists of canals running the length of the fish's body beneath the surface of the skin. Within these canals are sensory structures containing hair cells with cilia that project into a gelatinous cupula. Pressure waves traveling through the water in the canals deflect the cilia and depolarize the sensory neurons associated with the hair cells.

Sonar

A few groups of mammals that live and obtain their food in dark environments have circumvented the limitations of darkness. A bat flying in a completely dark room easily avoids objects that are placed in its path—even a wire less than a millimeter in diameter. Shrews use a similar form of "lightless vision" beneath the ground, as do whales and dolphins beneath the sea. All of these mammals perceive distance by means of sonar. They emit sounds and then determine the time it takes these sounds to reach an object and return to the animal. This process is called **echolocation.** The bat in figure 30.31, for example, produces clicks that last 2 to 3 milliseconds and are repeated several hundred times per second. The three-dimensional imaging achieved with such an auditory sonar system can be quite sophisticated, and will help this bat find the moth.

30.12   Sound receptors detect the air vibrating as waves of pressure push against the membrane covering the ear. Inside, these waves are amplified and press down hair cells that send signals to the brain. Fish sense pressure waves in water much as an ear senses sound. Many vertebrates sense distant objects by bouncing sounds off of them.

30.13 Sensing Light: Vision

No other stimulus provides as much detailed information about the environment as light. Vision, the perception of light, is carried out by a special sensory apparatus called an eye. All the sensory receptors described to this point have been chemical or mechanical ones. Eyes contain sensory receptors called rods and cones that respond to photons of light. The light energy is absorbed by pigments in the rods and cones, which respond by triggering nerve impulses in sensory neurons.

Evolution of the Eye

Vision begins with the capture of light energy by photoreceptors. Because light travels in a straight line and arrives virtually instantaneously, visual information can be used to determine both the direction and the distance of an object. No other stimulus provides as much detailed information.

Many invertebrates have simple visual systems with photoreceptors clustered in an eyespot. The flatworm in figure 30.32 has an eyespot, consisting of pigment molecules that are stimulated by light, triggering a nerve impulse in photoreceptor

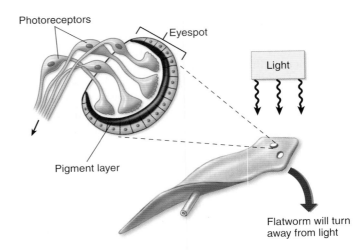

Figure 30.32   Simple eyespots in the flatworm.
Eyespots will detect the direction of light because a pigmented layer on one side of the eyespot screens out light coming from the back of the animal. Light is thus detected more readily coming from the front of the animal; flatworms will respond by turning away from the light.

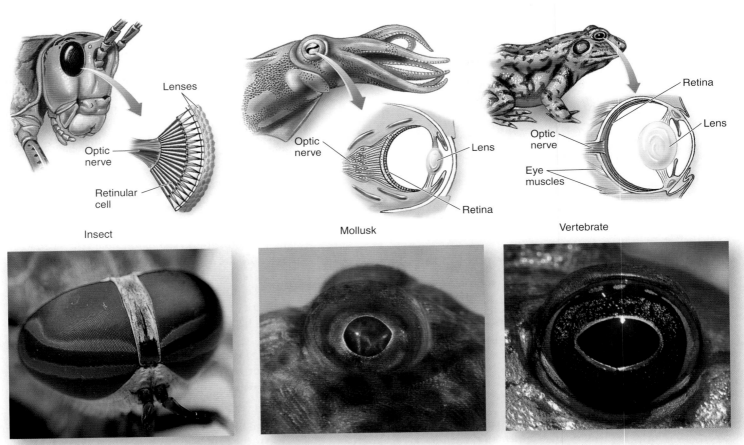

Figure 30.33   Eyes in three phyla of animals.
Although they are superficially similar, these eyes differ greatly in structure and are not homologous. Each has evolved separately and, despite the apparent structural complexity, has done so from simpler structures.

cells. Although an eyespot can perceive the direction of light, it cannot be used to construct a visual image. The members of four phyla—annelids, mollusks, arthropods, and vertebrates—have evolved well-developed, image-forming eyes. True image-forming eyes in these phyla, though they at first seem similar (compare the eyes of an arthropod, mollusk, and vertebrate in figure 30.33), are believed to have evolved independently. Interestingly, the photoreceptors in all of them use the same light-capturing molecule, suggesting that not many alternative molecules are able to play this role.

Structure of the Vertebrate Eye

The vertebrate eye works like a lens-focused camera. Light first passes through a transparent protective covering, the light blue **cornea** in figure 30.34, which begins to focus the light onto the rear of the eye. The beam of light then passes through the **lens,** which completes the focusing. The lens is the fat disc-shaped structure, somewhat resembling a flattened balloon. It is attached by stringlike suspending ligaments to **ciliary muscles.** When these muscles contract, they change the shape of the lens and thus the point of focus on the rear of the eye. The amount of light entering the eye is controlled by a shutter, called the **iris** (the colored part of your eye), between the cornea and the lens. The transparent zone in the middle of the iris, the **pupil,** gets larger in dim light and smaller in bright light.

The light that passes through the pupil is focused by the lens onto the back of the eye. An array of light-sensitive receptor cells line the back surface of the eye, called the **retina.** The retina is the light-sensing portion of the eye. The vertebrate retina contains two kinds of photoreceptors, called **rods** and **cones,** which, when stimulated by light, generate nerve im-

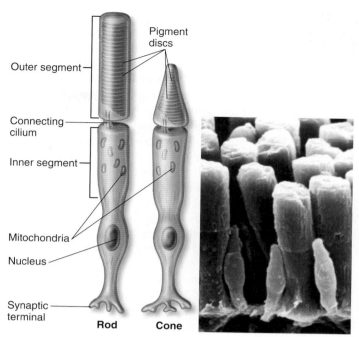

Figure 30.35 Rods and cones.
The broad tubular cell on the *left* is a rod. The shorter, tapered cell next to it is a cone. Electron micrograph of rods and cones.

pulses that travel to the brain along a short, thick nerve pathway called the optic nerve. Rods, the taller, flat-topped cell in figure 30.35, are receptor cells that are extremely sensitive to light, and they can detect various shades of gray even in dim light. However, they cannot distinguish colors, and because they do not detect edges well, they produce poorly defined images. Cones, the pointed-topped cells, are receptor cells that detect color and are sensitive to edges so that they produce sharp images. The center of the vertebrate retina contains a tiny pit, called the **fovea,** densely packed with some 3 million cones. This area produces the sharpest image, which is why we tend to move our eyes so that the image of an object we want to see clearly falls on this area.

The lens of the vertebrate eye is constructed to filter out short-wavelength light. This solves a difficult optical problem: any uniform lens bends short wavelengths more than it does longer ones, a phenomenon known as chromatic aberration. Consequently, these short wavelengths cannot be brought into focus simultaneously with longer wavelengths. Unable to focus the short wavelengths, the vertebrate eye eliminates them. Insects, whose eyes do not focus light, are able to see these lower, ultraviolet wavelengths quite well and often use them to locate food or mates.

How Rods and Cones Work

A rod or cone cell in the eye is able to detect a single photon of light. How can it be so sensitive? The primary sensing event of vision is the absorption of a photon of light by a pigment. The pigments in rods and cones are made from plant pigments called carotenoids. That is why eating carrots is said to be

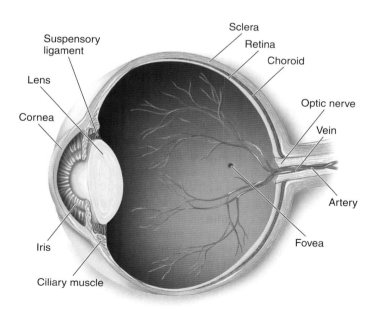

Figure 30.34 The structure of the human eye.
Light passes through the transparent cornea and is focused by the lens on the rear surface of the eye, the retina, at a particular location called the fovea. The retina is rich in photoreceptors.

good for night vision—the orange color of carrots is due to the presence of carotenoids called carotenes. The visual pigment in the human eye is a fragment of carotene called ***cis*-retinal.** The pigment is attached to a protein called **opsin** to form a light-detecting complex called **rhodopsin.**

When it receives a photon of light, the pigment undergoes a change in shape. This change in shape must be large enough to alter the shape of the opsin protein attached to it. When light is absorbed by the *cis*-retinal pigment (the upper molecule in figure 30.36), the linear end of the molecule rotates sharply upward, straightening out that end of the molecule. The new form of the pigment is referred to as ***trans*-retinal** and the dashed outline in the figure shows the shape before it was stimulated by light. This radical change in the pigment's shape induces a change in the shape of the protein opsin to which the pigment is bound, initiating a chain of events that leads to the generation of a nerve impulse.

Each rhodopsin activates several hundred molecules of a protein called transducin. Each of these activates several hundred molecules of an enzyme whose product stimulates sodium channels in the photoreceptor membrane at a rate of about 1,000 per second. This cascade of events allows a single photon to have a large effect on the receptor.

Color Vision

Three kinds of **cone cells** provide us with color vision. Each possesses a different version of the opsin protein (that is, one with a distinctive amino acid sequence and thus a different shape). These differences in shape affect the flexibility of the attached retinal pigment, shifting the wavelength at which it absorbs light. The absorption spectrum in figure 30.37 shows the wavelength of light that is absorbed by each cone and rod cell. In rods, light is absorbed at 500 nanometers. In cones, the three versions of opsin absorb light at 420 nanometers (blue-absorbing), 530 nanometers (green-absorbing), or 560

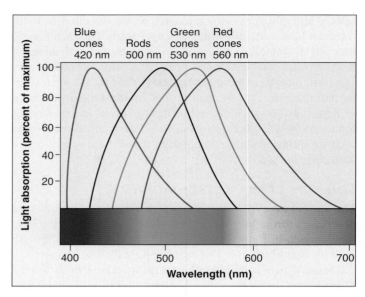

Figure 30.37 Color vision.
The absorption spectrum of *cis*-retinal is shifted in cone cells from the 500 nanometers characteristic of rod cells. The amount of the shift determines what color the cone absorbs: 420 nanometers yields blue absorption; 530 nanometers yields green absorption; and 560 nanometers yields red absorption. Red cones do not peak in the red part of the spectrum, but they are the only cones that absorb red light.

nanometers (red-absorbing). By comparing the relative intensities of the signals from the three cones, the brain can calculate the intensity of other colors.

Some people are not able to see all three colors, a condition referred to as *color blindness.* Color blindness is typically due to an inherited lack of one or more types of cones. People with normal vision have all three types of cones. People with only two types of cones lack the ability to detect the third color. For example, people with red color blindness lack red cones and have difficulty distinguishing red from green (figure 30.38). Color blindness is a sex-linked trait (see chapter 11), and so men are far more likely to be color blind than women.

Most vertebrates, particularly those that are diurnal (active during the day), have color vision, as do many insects. Indeed, honeybees can see light in the near-ultraviolet range, which is invisible to the human eye. Color vision requires the presence of more than one photopigment in different receptor cells, but not all animals with color vision have the three-cone system characteristic of humans and other primates. Fish, turtles, and birds, for example, have four or five kinds of cones; the "extra" cones enable these animals to see near-ultraviolet light. Many mammals (such as squirrels), on the other hand, have only two types of cones.

Conveying the Light Information to the Brain

The path of light through each eye is the reverse of what you might expect. The rods and cones are at the rear of the retina, not the front. If you track the path that light would take in figure 30.39, you will see that light passes through several layers of ganglion and bipolar cells before it reaches the rods and

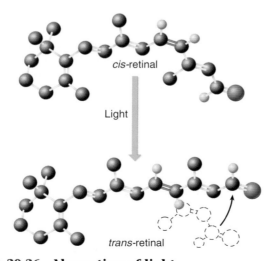

Figure 30.36 Absorption of light.
When light is absorbed by *cis*-retinal, the pigment undergoes a change in shape and becomes *trans*-retinal.

cones. Once the photoreceptors are activated, they stimulate bipolar cells, which in turn stimulate ganglion cells. The direction of nerve impulses in the retina is thus opposite to the direction of light.

Action potentials propagated along the axons of ganglion cells are relayed through structures called the *lateral geniculate nuclei* of the thalamus and projected to the occipital lobe of the cerebral cortex. There the brain interprets this information as light in a specific region of the eye's receptive field. The pattern of activity among the ganglion cells across the retina encodes a point-to-point map of the receptive field, allowing the retina and brain to image objects in visual space. In addition, the frequency of impulses in each ganglion cell provides information about the light intensity at each point, while the relative activity of ganglion cells connected (through bipolar cells) with the three types of cones provides color information.

Binocular Vision

Primates (including humans) and most predators have two eyes, one located on each side of the face. When both eyes are trained on the same object, the image that each sees is slightly different because each eye views the object from a different angle. This slight displacement of the images permits **binocular vision,** the ability to perceive three-dimensional images and to sense depth or the distance to an object. Having their eyes facing forward maximizes the field of overlap in which this stereoscopic vision occurs, as seen by the overlapping blue triangles in the human in figure 30.40. The triangles are the field of view for each eye.

In contrast, prey animals generally have eyes located to the sides of the head, preventing binocular vision but enlarging the overall receptive field. Depth perception is less important to prey than detection of potential enemies from any quarter. The eyes of the American woodcock, for example, are located at exactly opposite sides of its skull so that it has a 360-degree field of view without turning its head! Most birds have laterally placed eyes and, as an adaptation, have two foveas in each retina. One fovea provides sharp frontal vision, like the single fovea in the retina of mammals, and the other fovea provides sharper lateral vision.

30.13 Vision receptors detect reflected light; binocular vision allows the brain to form three-dimensional images of objects.

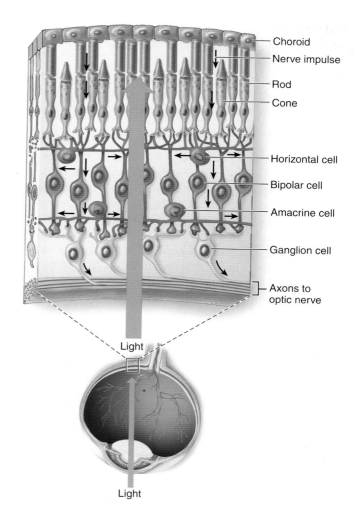

Figure 30.39 Structure of the retina.
The rods and cones are at the rear of the retina, not the front. Light passes over four other types of cells in the retina before it reaches the rods and cones. Nerve impulses then travel through the bipolar cells to the ganglion cells and on to the optic nerve.

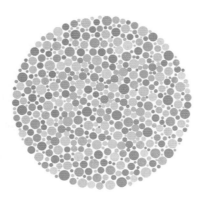

Figure 30.38 Test for color blindness.
People with normal color vision see the number 16, but people that are red-green color blind see just spots and no discernible number.

Figure 30.40 Binocular vision.
When the eyes are located on the sides of the head (as on the *left*), the two vision fields do not overlap and binocular vision does not occur. When both eyes are located toward the front of the head (as on the *right*) so that the two fields of vision overlap, depth can be perceived.

30.14 Other Types of Sensory Reception

Although vision is the primary sense used by all vertebrates that live in a light-filled environment, other parts of the electromagnetic spectrum are also used by some vertebrates.

Heat

Infrared ("below red") radiation, what we normally think of as radiant heat, is used by certain snakes called pit vipers to sense their environment. Pit vipers possess a pair of heat-detecting pit organs located on either side of the head between the eye and the nostril (figure 30.41). The pit organs permit a blindfolded rattlesnake to accurately strike at a warm, dead rat. Each pit organ is composed of two chambers (colored red in the drawing) separated by a membrane. The infrared radiation falls on the membrane and warms it. The organ apparently operates by comparing the temperatures of the two chambers.

Electricity

While air does not readily conduct an electrical current, water is a good conductor. All aquatic animals generate electrical currents from contractions of their muscles. A number of different groups of fishes can detect these electrical currents. The *electric fish* even have the ability to produce electrical discharges from specialized electric organs. Electric fish use these weak discharges to locate their prey and mates and to construct a three-dimensional image of their environment even in murky water.

Magnetism

Eels, sharks, and many birds appear to navigate along the magnetic field lines of the earth. Even some bacteria use such forces to orient themselves. Birds kept in blind cages, with no visual cues

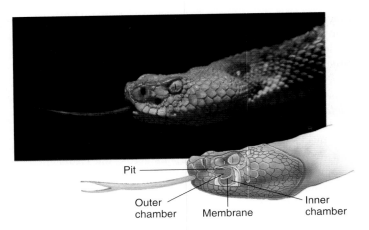

Figure 30.41 "Seeing" heat.
The depression between the nostril and the eye of this rattlesnake opens into the pit organ. In the cutaway portion of the diagram, you can see that the organ is composed of two chambers separated by a membrane.

to guide them, will peck and attempt to move in the direction in which they would normally migrate at the appropriate time of the year. They will not do so, however, if the cage is shielded from magnetic fields by steel. There has been much speculation about the nature of the magnetic receptors in these vertebrates, but the mechanism is still very poorly understood (see chapter 22).

> **30.14** Pit vipers can locate warm prey by infrared radiation (heat), and many aquatic vertebrates can locate prey and ascertain the contours of their environment by means of electroreceptors. Some vertebrates can even orient themselves with respect to the earth's magnetic field.

INQUIRY & ANALYSIS

Are Bigger Nerves Faster?

Physics tells us that there should be a relationship between the conduction velocity of a nerve and its diameter. To be precise, there should be a 10-fold increase in conduction speed for a 100-fold increase in fiber diameter. Squids have giant axons 100 times thicker than most animal axons, so they can contract their mantles fast enough to power their jet propulsion. Is there a similar relationship for the very thin myelinated nerve fibers of vertebrates? The graph shows the speed of conduction in myelinated fibers of a cat, plotted against fiber diameter.

1. **Applying Concepts** What is the dependent variable?
2. **Making Inferences** Is conduction velocity faster for larger-diameter fibers?
3. **Drawing Conclusions** Do these data support the conclusion that conduction velocity is directly proportional to fiber diameter?

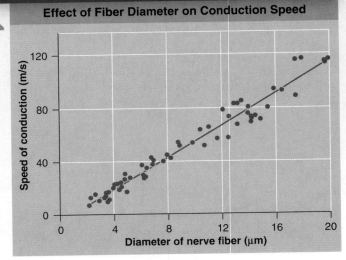

Effect of Fiber Diameter on Conduction Speed

4. **Further Analysis** Form an explanation of why conduction velocity might be greater in larger-diameter myelinated axons. How would you test it?

Neurons and How They Work

30.1 Evolution of the Animal Nervous System

- The nervous system is the communication network in the body (**figure 30.1**), evolving from nerve nets to more and more complex systems, with specialized cell types and localization of integration centers in the brain (**figure 30.3**).

30.2 Neurons Generate Nerve Impulses

- Neurons are cells that conduct electrical impulses (**figure 30.4**). Electrical signals begin in dendrites and travel down an axon. Nerve impulses result from the movement of ions across the plasma membranes through specialized channels. The movement of ions in one area of the membrane causes a change in electrical properties, called depolarization, which causes the opening of adjacent ion channels, spreading the electrical impulse down the length of the axon (**figure 30.5**).

30.3 The Synapse

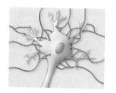

- When a nerve impulse reaches the end of the axon, it triggers the release of neurotransmitters that pass across a small gap, called a synaptic cleft, between the neuron and another cell. Neurotransmitter molecules bind to receptors on the postsynaptic cell, causing ion channels to open, creating electrical impulses in the postsynaptic cell (**figure 30.7**). All neural inputs are integrated in the postsynaptic cell, producing an overall positive or negative change in membrane potential (**figure 30.8**).

30.4 Addictive Drugs Act on Chemical Synapses

- Molecules called neuromodulators increase or decrease the effects of neurotransmitters at a synapse (**figure 30.9**). Many addictive drugs act as neuromodulators (**figure 30.10**).

The Central Nervous System

30.5 Evolution of the Vertebrate Brain

- The evolution of the brain resulted in a larger forebrain, leading to more complex functions (**figure 30.13**).

30.6 How the Brain Works

- The cerebral cortex lies over the cerebrum and is the site of neural activities such as language, conscious thought, memory, personality development, vision, and many other higher-level activities (**figure 30.15**). The thalamus and hypothalamus integrate bodily functions and lie underneath the cerebrum (**figure 30.16**). The brain stem controls vital functions, and the cerebellum controls muscular coordination.

30.7 The Spinal Cord

- The spinal cord is a cable of neurons that extends down the back and is encased in the bony vertebrae of the backbone (**figure**

30.19). Motor nerves carry impulses from the spinal cord and brain out to the body, and sensory nerves carry impulses from the body to the brain and spinal cord.

The Peripheral Nervous System

30.8 Voluntary and Autonomic Nervous Systems

- The voluntary nervous system relays commands between the CNS and skeletal muscles and can be consciously controlled; however, reflexes work without conscious control (**figure 30.21**). The autonomic nervous system consists of opposing sympathetic and parasympathetic divisions that unconsciously relay commands between the CNS and muscles and glands (**figure 30.22**).

The Sensory Nervous System

30.9 Sensory Perception

- Neurons called sensory receptors initiate and carry nerve impulses to the CNS. Different sensory cells are stimulated by different stimuli. Exteroceptors sense stimuli from the external environment, and interoceptors sense stimuli within the body (**figures 30.24** and **30.25**).

30.10 Sensing Gravity and Motion

- Sensory receptors in the ear sense gravity and acceleration. The otolith sensory receptors detect gravity by the deflection of hair cells caused by the movement of otoliths in a gelatin-like matrix (**figure 30.26b**). Motion is detected by the deflection of hair cells in the cupula of the semicircular canals (**figure 30.26c**).

30.11 Sensing Chemicals: Taste and Smell

- Chemicals are detected through the senses of taste, using taste buds on the tongue and smell, using olfactory receptors that line the nasal passages (**figures 30.27** and **30.28**).

30.12 Sensing Sound: Hearing

- Sound receptors detect vibrations of air and water through the deflection of hair cells. These senses include hearing in terrestrial vertebrates (**figure 30.29**) and the lateral line system in fishes (**figure 30.30**).

30.13 Sensing Light: Vision

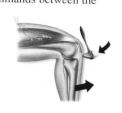

- Sensory receptors in the eye detect light. Light receptors evolved in several different animal phyla (**figures 30.32–30.34**). Rod cells detect the intensity of light, while cone cells detect different colors of light (**figures 30.35–30.39**).

30.14 Other Types of Sensory Reception

- Other sensory receptors, for specialized functions, detect heat, electricity, and magnetic fields.

1. Complexity in animal nervous systems evolved with an increase in
   a. animal body size.
   b. animal body nutritive requirements.
   c. the amount of associative neurons that eventually formed the "brain."
   d. types of animal behavior.
2. A nerve impulse is caused by a quick reversal of polarity in a nerve cell resulting from the
   a. exchange of sodium and potassium ions.
   b. outflow of sodium ions.
   c. inflow of potassium ions.
   d. exchange of sodium and chloride ions.
3. Excitatory neurotransmitters initiate an action potential in a postsynaptic neuron by opening
   a. sodium ion gates in the postsynaptic cell.
   b. potassium ion gates in the postsynaptic cell.
   c. chloride ion gates in the postsynaptic cell.
   d. calcium ion gates in the postsynaptic cell.
4. The purpose of the hindbrain is to
   a. coordinate olfactory information from the nose with the eyes.
   b. process critical thought and learning.
   c. coordinate optic information with the muscles.
   d. coordinate sensory information from muscles and motor responses.
5 Involuntary body activities such as breathing are controlled by the
   a. cerebrum.
   b. cerebellum.
   c. hypothalamus.
   d. brain stem.

6. The purpose of the autonomic nervous system is to do all of the following *except*
   a. stimulate glands.
   b. relay messages to skeletal muscles.
   c. relay messages to cardiac and smooth muscles.
   d. regulate the body's homeostasis.
7. When arm muscles hurt after heavy exercise, the pain is detected by
   a. neurotransmitters.
   b. interoceptors.
   c. associative neurons.
   d. exteroceptors.
8. Motion and orientation to gravity are sensed in the ear, specifically in the
   a. pinnae.
   b. ear bones (hammer, anvil, stirrup).
   c. semicircular canals.
   d. eustachian tubes.
9. Which of the following statements is incorrect?
   a. Vertebrates focus the eye by changing the shape of the lens.
   b. The eyes of arthropods and vertebrates use the same light-capturing molecule.
   c. The vertebrate eye adjusts the amount of light entering the eye by contracting the ciliary muscles.
   d. Light changes *cis*-retinal into *trans*-retinal.
10. The lateral line system in fish detects
   a. pressure waves.
   b. light.
   c. taste.
   d. magnetism.

1. **Figure 30.10** Describe the qualitative feelings that might be experienced by someone who goes through the four stages related to drug addiction shown in the drawings.

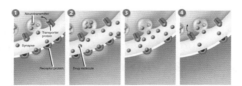

2. **Figure 30.13** Look at the comparative size of the olfactory tract and the cerebrum in the various animals. What does the figure tell you about the possible importance of olfaction in humans?

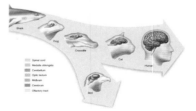

1. There is a lot of concern today about high sodium (Na) levels in our diet, and the resulting high blood pressure for many people. Your friend, Henry, suggests that if a low-sodium diet is healthy, then a no-sodium diet should be even better. What do you tell him?
2. When people become very angry, they are said to be operating from their "dinosaur brain"—not thinking clearly. Explain what this means.

3. Why is it important that the neurons in the autonomic nervous system are always active, not shutting down at night while you're asleep?
4. If you only taste a few things (salt, bitter, sweet, sour, umami), why do foods all seem so distinct?

31
Chemical Signaling Within the Animal Body

In vertebrates and most other animals, the central nervous system coordinates and regulates the diverse activities of the body by using chemical signals called hormones to effect changes in physiological activities. Many hormones—adrenaline (epinephrine), estrogen, testosterone, insulin, thyroid hormone—are probably familiar to you. Some of these hormones have very different roles in other animals, however. For example, thyroid hormone is needed in amphibians for the metamorphosis of larvae into adults. If the thyroid gland is removed from a tadpole, it will not change into a frog. Conversely, if an immature tadpole is fed pieces of a thyroid gland, it will undergo premature metamorphosis and become a miniature frog! Similarly, melanocyte-stimulating hormone, a peptide hormone, is present in mammals, but we don't know what it does. In reptiles and amphibians, this hormone stimulates color changes. The green anole (*Anolis carolinensis*) shown here in the upper photo has changed its color to tan in the lower photo, in response to an environmental or physiological cue. This color change is the result of dispersal of pigment-containing granules from the center of the anole's skin cells into extensions of the cells, darkening the skin. Triggered by melanocyte-stimulating hormone, the reversible color change takes 5 to 10 minutes. In this chapter you will encounter many other hormones, some familiar, others less so, but all used by vertebrates to regulate their body condition.

31.1 Hormones

A **hormone** is a chemical signal produced in one part of the body that is stable enough to be transported in active form far from where it is produced and that typically acts at a distant site. There are three big advantages to using chemical hormones as messengers rather than speedy electrical signals (like those used in nerves) to control body organs. First, chemical molecules can spread to all tissues via the blood (imagine trying to wire every cell with its own nerve!) and are usually required in only small amounts. Second, chemical signals can persist much longer than electrical ones, a great advantage for hormones controlling slow processes like growth and development. Third, many different kinds of chemicals can act as hormones, so different hormone molecules can be targeted at different tissues. For all these reasons, hormones are excellent messengers for signaling widespread, slow-onset, long duration responses.

Hormones, in general, are produced by glands, most of which are controlled by the central nervous system. Because these glands are completely enclosed in tissue rather than having ducts that empty to the outside, they are called **endocrine glands** (from the Greek, *endon,* within). Hormones are secreted from them directly into the bloodstream (this is in contrast to **exocrine glands,** which, like sweat glands, have ducts). Your body has a dozen principal endocrine glands, shown in figure 31.1, that together make up the endocrine system.

The *endocrine system* and the *motor nervous system* are the two main routes the central nervous system (CNS) uses to issue commands to the organs of the body. The two are so closely linked that they are often considered a single system—the **neuroendocrine system.** The **hypothalamus** can be considered the main switchboard of the neuroendocrine system. The hypothalamus is continually checking conditions inside the body to maintain a constant internal environment, a condition known as homeostasis. Is the body too hot or too cold? Is it running out of fuel? Is the blood pressure too high? If homeostasis is no longer maintained, the hypothalamus has several ways to set things right again. For example, if the hypothalamus needs to speed up the heart rate, it can send a nerve signal to the medulla oblongata, or it can use a chemical command, causing the adrenal gland to produce the hormone adrenaline, which also speeds up the heart rate. Which command the hypothalamus uses depends on the desired duration of the effect. A chemical message is typically far longer lasting than a nerve signal.

The Chain of Command

The hypothalamus issues commands to a nearby gland, the pituitary, which in turn sends chemical signals to the various hormone-producing glands of the body. The pituitary is suspended from the hypothalamus by a short stalk, across which chemical messages pass from the hypothalamus to the pituitary. The first of these chemical messages to be discovered

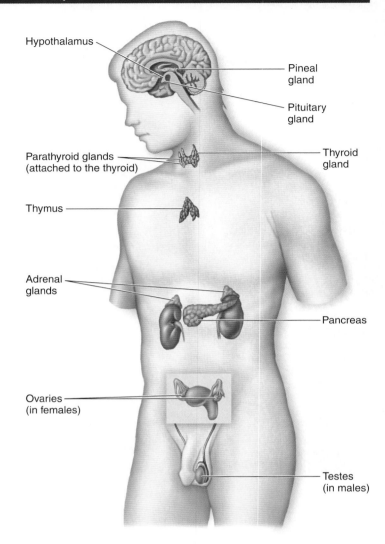

Figure 31.1 Major glands of the human endocrine system.
Hormone-secreting cells are clustered in endocrine glands. The pituitary and adrenal glands are each composed of two glands.

was a short peptide called thyrotropin-releasing hormone (TRH), which was isolated in 1969. The release of TRH from the hypothalamus triggers the pituitary to release a hormone called thyrotropin, or thyroid-stimulating hormone (TSH), which travels to the thyroid and causes the thyroid gland to release thyroid hormones.

Several other hypothalamic hormones have since been isolated, which together govern the pituitary. Thus, the CNS regulates the body's hormones through a chain of command. The "releasing" hormones made by the hypothalamus cause the pituitary to synthesize a corresponding pituitary hormone, which travels to a distant endocrine gland and causes that gland to begin producing its particular endocrine hormone. The hypothalamus also secretes inhibiting hormones that keep the pituitary from secreting specific pituitary hormones.

Hormonal Communication

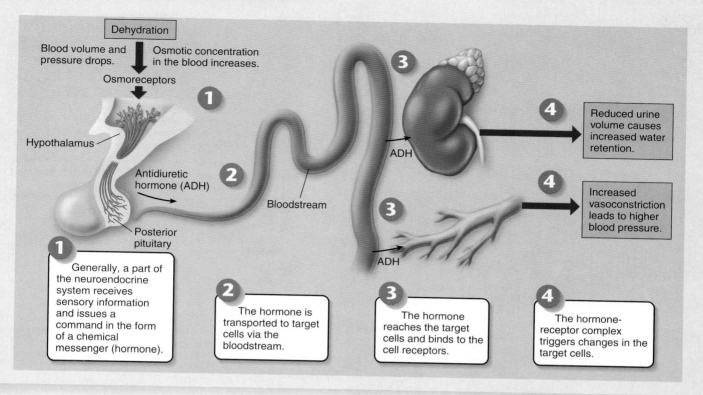

Figure 31.2 How hormonal communication works.

How Hormones Work

The key reason why hormones are effective messengers within the body is because a particular hormone can influence a specific target cell. How does the target cell recognize that hormone, ignoring all others? Embedded in the plasma membrane or within the target cell are receptor proteins that match the shape of the potential signal hormone like a hand fits a glove. As you recall from chapter 30, nerve cells have highly specific receptors within their synapses, each receptor shaped to "respond" to a different neurotransmitter molecule. Cells that the body has targeted to respond to a particular hormone have receptor proteins shaped to fit that hormone and no other. Thus, chemical communication within the body involves *two* elements: a molecular signal (the hormone) and a protein receptor on or in target cells. The system is highly specific because each protein receptor has a shape that only a particular hormone fits.

Hormones secreted by endocrine glands belong to four different chemical categories:

1. **Polypeptides** are composed of chains of amino acids that are shorter than about 100 amino acids. Some important examples include insulin and antidiuretic hormone (ADH).
2. **Glycoproteins** are composed of polypeptides significantly longer than 100 amino acids to which are attached carbohydrates. Examples include follicle-stimulating hormone (FSH) and luteinizing hormone (LH).

3. **Amines,** derived from the amino acids tyrosine and tryptophan, include hormones secreted by the adrenal medulla, thyroid, and pineal glands.
4. **Steroids** are lipids derived from cholesterol, and include the hormones testosterone, estrogen, progesterone, aldosterone, and cortisol.

The path of communication taken by a hormonal signal can be visualized as the series of simple steps shown in figure 31.2:

1. **Issuing the command.** The hypothalamus of the CNS controls the release of many hormones. Hormones produced in cells in the hypothalamus are stored in the posterior pituitary and are released into the bloodstream in response to a signal from the brain.
2. **Transporting the signal.** While hormones can act on an adjacent cell, most are transported throughout the body by the bloodstream.
3. **Hitting the target.** When a hormone encounters a cell with a matching receptor, called a target cell, the hormone binds to that receptor.
4. **Having an effect.** When the hormone binds to the receptor protein, the protein responds by changing shape, which triggers a change in cell activity.

> **31.1 Hormones are effective because they are recognized by specific receptors. Thus, only cells possessing the appropriate receptor will respond to a particular hormone.**

Steroid Hormones

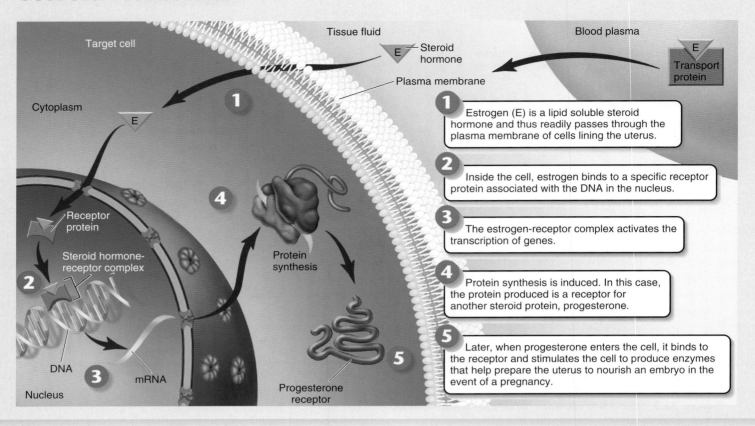

Figure 31.3 How steroid hormones work.

The figure contains the following labels:

- Target cell
- Cytoplasm
- Receptor protein
- Steroid hormone-receptor complex
- DNA
- mRNA
- Nucleus
- Tissue fluid
- Steroid hormone
- Plasma membrane
- Protein synthesis
- Progesterone receptor
- Blood plasma
- Transport protein

1. Estrogen (E) is a lipid soluble steroid hormone and thus readily passes through the plasma membrane of cells lining the uterus.

2. Inside the cell, estrogen binds to a specific receptor protein associated with the DNA in the nucleus.

3. The estrogen-receptor complex activates the transcription of genes.

4. Protein synthesis is induced. In this case, the protein produced is a receptor for another steroid protein, progesterone.

5. Later, when progesterone enters the cell, it binds to the receptor and stimulates the cell to produce enzymes that help prepare the uterus to nourish an embryo in the event of a pregnancy.

31.2 How Hormones Target Cells

Steroid Hormones Enter Cells

Some protein receptors designed to recognize hormones are located in the cytoplasm or nucleus of the target cell. The hormones in these cases are lipid-soluble molecules, typically **steroid hormones.** The chemical shapes of these molecules are ring-structures resembling chicken wire. All steroid hormones are manufactured from cholesterol, a complex molecule composed of four rings. The hormones that promote the development of secondary sexual characteristics are steroids. They include testosterone, as well as estrogen and progesterone, which control the female reproductive system and are discussed in chapter 32. Cortisol is also a steroid hormone.

Steroid hormones, like estrogen, "E" in figure 31.3, can pass across the lipid bilayer of the cell plasma membrane (1), and bind to receptors within the cell and often, as in the case with estrogen, within the nucleus. This complex of receptor and hormone then binds to the DNA in the nucleus (2), and activates the gene for a progesterone receptor protein, which is transcribed (3). The protein is synthesized (4), and the receptor is available to bind progesterone when it enters the cell (5), which itself activates another set of genes.

Anabolic steroids, which are used by some weight lifters and other athletes, are synthetic compounds that re-semble the male sex hormone testosterone. Their injection into muscles activates growth genes and causes the muscle cells to produce more protein, resulting in bigger muscles and increased strength. However, anabolic steroids have many dangerous side effects in both men and women, including liver damage, heart disease, high blood pressure, acne, balding, and psychological disorders. Men can also experience the suppression of testicular function and feminization, and women can undergo masculinization. Use in adolescents can also result in stunted growth and accelerated puberty changes. Anabolic steroids are illegal, and for many sports, athletes are tested for their presence.

Peptide Hormones Act at the Cell Surface

Other hormone receptors are embedded within the plasma membrane, with their recognition region directed outward from the cell surface. Peptide hormones like the one binding to the receptor in figure 31.4 (1), are typically short peptide chains (although some are full-sized proteins). The binding of the **peptide hormone** to the receptor triggers a change in the cytoplasmic end of the receptor protein. This change then triggers events within the cell cytoplasm, usually through intermediate within-cell signals called **second messengers** (2), which greatly amplify the original signal and results in changes in the cell (3).

How does a second messenger amplify a hormone's signal? Second messengers activate enzymes. One of the most common second messengers, cyclic AMP, is shown in figure

31.5. Cyclic AMP is made from ATP by an enzyme that removes two phosphate units, forming AMP; the ends of the AMP join, forming a circle. A single hormone molecule binding to a receptor in the plasma membrane can result in the formation of many second messengers in the cytoplasm. Each second messenger can activate many molecules of a certain enzyme, and sometimes each of these enzymes can in turn activate many other enzymes. Thus, second messengers enable each hormone molecule to have a tremendous effect inside the cell, far greater than if the hormone had simply entered the cell and sought out a single target.

The hormone insulin is one of many hormones that act through second messenger systems, and provides a well-studied example of how peptide hormones achieve their effects within target cells. Most human cells have insulin receptors in their membranes—typically only a few hundred but far more in tissues involved in glucose metabolism. A single liver cell, for example, may have 100,000 of them. When insulin binds to one of these insulin receptors, the receptor protein changes its shape, prodding an adjacent signal-modulating protein on the cell interior to activate the release of Ca^{++} ions. The Ca^{++} acts as a second messenger, activating a variety of cellular enzymes in a cascading series of events that greatly amplifies the strength of the original signal.

31.2 Steroid hormones pass through the cell's plasma membrane and bind to receptors in the cell, forming a complex that alters the transcription of specific genes. Peptide hormones do not enter cells. Instead, they bind to receptors on the target cell surface, triggering a cascade of enzymic activations within the cell.

Peptide Hormones

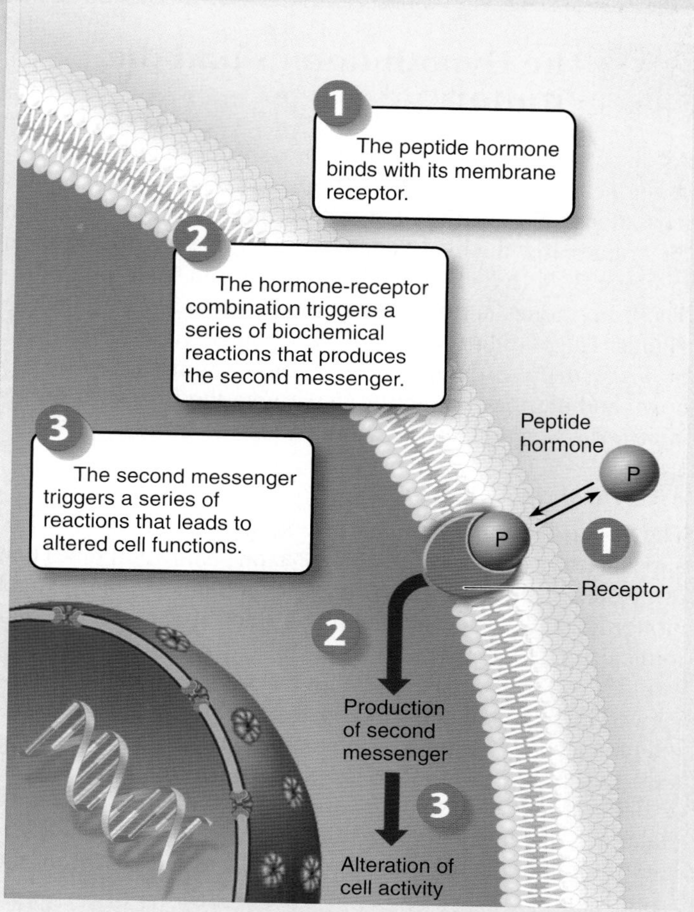

1 The peptide hormone binds with its membrane receptor.

2 The hormone-receptor combination triggers a series of biochemical reactions that produces the second messenger.

3 The second messenger triggers a series of reactions that leads to altered cell functions.

Peptide hormone

Receptor

1

2 Production of second messenger

3 Alteration of cell activity

Figure 31.4 How peptide hormones work.

Second Messengers

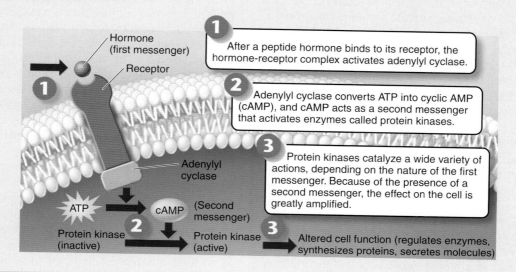

Hormone (first messenger)

Receptor

1 After a peptide hormone binds to its receptor, the hormone-receptor complex activates adenylyl cyclase.

2 Adenylyl cyclase converts ATP into cyclic AMP (cAMP), and cAMP acts as a second messenger that activates enzymes called protein kinases.

3 Protein kinases catalyze a wide variety of actions, depending on the nature of the first messenger. Because of the presence of a second messenger, the effect on the cell is greatly amplified.

Adenylyl cyclase

ATP

cAMP (Second messenger)

Protein kinase (inactive)

Protein kinase (active)

Altered cell function (regulates enzymes, synthesizes proteins, secretes molecules)

Figure 31.5 How second messengers work.

31.3 The Hypothalamus and the Pituitary

As discussed earlier, the hypothalamus is the "control center" of the neuroendocrine system. It exerts control over the pituitary gland. The **pituitary gland,** located in a bony recess in the brain below the hypothalamus (see the brain cross section in figure 31.6), is where nine major hormones are produced. These hormones act principally to influence other endocrine glands. The pituitary is actually two glands. The back portion, or *posterior lobe,* regulates water conservation, also milk letdown and uterine contraction in women; the front portion, or *anterior lobe,* regulates the other endocrine glands.

The Posterior Pituitary

The posterior pituitary, seen in an enlarged view in figure 31.6, appears fibrous because it contains axons (shown in blue) that originate in cell bodies within the hypothalamus and extend along the stalk of the pituitary as a tract of fibers. This anatomical relationship results from the way that the posterior pituitary is formed from the brain during embryonic development. The hypothalamus and posterior pituitary thus become interconnected by a tract of axons. In figure 31.6 you can see that the hormones released from the posterior pituitary are actually produced by neuron cell bodies located in the hypothalamus. The hormones are transported to the posterior pituitary through axon tracts and are stored and released from the posterior pituitary.

    The role of the posterior pituitary first became evident in 1912, when a remarkable medical case was reported: A man who had been shot in the head developed a surprising disorder—he began to urinate every 30 minutes, unceasingly. The bullet had lodged in his pituitary gland, and subsequent research demonstrated that surgical removal of the pituitary also produces these unusual symptoms. Pituitary extracts were shown to contain a substance that makes the kidneys conserve water, and in the early 1950s the peptide hormone **vasopressin** (also called antidiuretic hormone, **ADH**) was isolated. As you learned in chapter 28, ADH regulates the kidney's retention of water. When ADH is missing, the kidneys cannot retain water, which is why the bullet caused excessive urination. Excessive alcohol and caffeine consumption, which inhibit ADH secretion, have a similar effect.

    The posterior pituitary also releases a second hormone, oxytocin, of very similar structure—both are short peptides composed of nine amino acids—but very different function. Oxytocin initiates uterine contraction during childbirth and milk release in mothers. Here is how milk release works: Sensory receptors in the mother's nipples, when stimulated by sucking, send messages to the hypothalamus, causing the hypothalamus to stimulate the release of oxytocin from the posterior pituitary. The oxytocin travels in the bloodstream to the breasts, where it stimulates contraction of the muscles around the ducts into which the mammary glands secrete milk. Both oxytocin and ADH are produced in the cell bod-

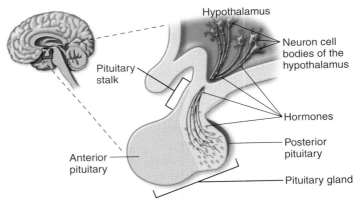

Figure 31.6 The posterior pituitary contains cells that originate in the hypothalamus.

A tract of nerve cells originates in the hypothalamus and extends down along the pituitary stalk and ends in the posterior pituitary. The cell bodies of the neurons produce hormones, which travel down the axons and are stored in the posterior pituitary. Thus, the hormones released from the posterior pituitary are not produced by the gland, but are actually synthesized in neurons in the hypothalamus.

ies of the hypothalamus but stored and released from the posterior pituitary.

The Anterior Pituitary

The anterior pituitary, unlike the posterior pituitary, does not develop as an extension of the brain; instead, it develops from a pouch of epithelial tissue that pinches off from the roof of the embryo's mouth. Because it is epithelial tissue, the anterior pituitary is a complete gland—it produces the hormones it secretes. The key role of the anterior pituitary first became understood in 1909, when a 38-year-old South Dakota farmer was cured of the growth disorder called acromegaly by the surgical removal of a pituitary tumor. Acromegaly is a form of giantism in which the jaw begins to protrude and the features thicken. It turned out that giantism is almost always associated with pituitary tumors. Robert Wadlow, born in Alton, Illinois, in 1928, grew to a height of 8 feet, 11 inches, and weighed 475 pounds before he died from infection at age 22—the tallest human being ever recorded. Skull X rays showed he had a pituitary tumor. Pituitary hormones have also proven to be the cause of several other well-known cases of giantism, such as the 8-foot, 2-inch Irish giant Charles Byrne, born in 1761; his skeleton, preserved in the Royal College of Surgeons, London, shows the effects of a pituitary tumor.

    Why did removal of the pituitary tumor cure the South Dakota farmer? Pituitary tumors produce giants because the tumor cells, which were pituitary cells that started to divide uncontrollably and so still act like pituitary cells, produce enormous amounts of a growth-promoting hormone. This **growth hormone (GH),** a long peptide of 191 amino acids, is normally produced in only minute amounts by the anterior pituitary gland and during periods of body growth, such as infancy and puberty.

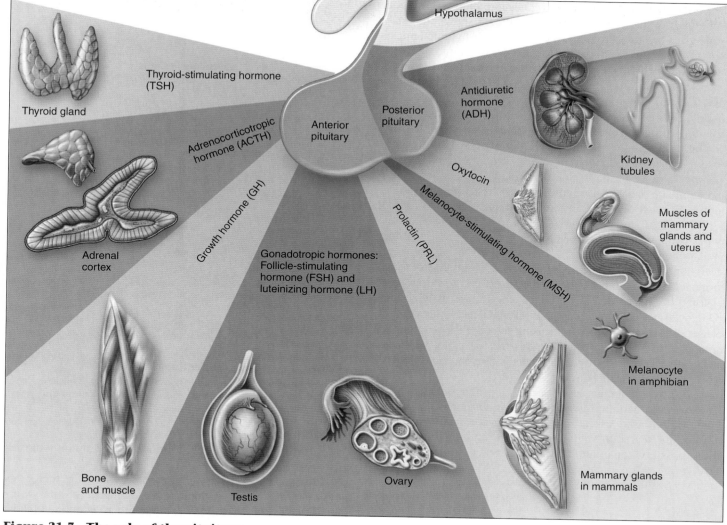

Figure 31.7   The role of the pituitary.

The anterior pituitary gland (highlighted in blue in figure 31.7, with the posterior pituitary colored tan) produces seven major peptide hormones, each controlled by a particular releasing signal secreted from cells in the hypothalamus. All seven of these hormones have major endocrine roles, as well as poorly understood roles in the central nervous system. Beginning in the upper left of the figure:

1. **Thyroid-stimulating hormone (TSH).** TSH stimulates the thyroid gland to produce the thyroid hormone thyroxine, which in turn stimulates oxidative respiration.
2. **Adrenocorticotropic hormone (ACTH).** ACTH stimulates the adrenal gland to produce a variety of steroid hormones. Some regulate the production of glucose from fat; others regulate the balance of sodium and potassium ions in the blood.
3. **Growth hormone (GH).** GH stimulates the growth of muscle and bone throughout the body.
4. **Follicle-stimulating hormone (FSH).** FSH is significant in the female menstrual cycle by triggering the maturation of egg cells and stimulating the release

of estrogen. In males, it stimulates cells in the testes, regulating development of the sperm.
5. **Luteinizing hormone (LH).** LH plays an important role in the female menstrual cycle by triggering ovulation, the release of a mature egg. It also stimulates the male gonads to produce testosterone, which initiates and maintains the development of male secondary sexual characteristics not involved directly in reproduction.
6. **Prolactin (PRL).** Prolactin stimulates the breasts to produce milk, which is released in response to oxytocin.
7. **Melanocyte-stimulating hormone (MSH).** In reptiles and amphibians, melatonin stimulates color changes in the epidermis. The function of this hormone in humans is still poorly understood.

How the Hypothalamus Controls the Anterior Pituitary

As noted earlier, the hypothalamus controls production and secretion of the anterior pituitary hormones by means of a family of special hormones. Neurons in the hypothalamus secrete these releasing and inhibiting hormones into blood capillaries at the base of the hypothalamus. Figure 31.8 shows the relationship of the two groups of neurons in the hypothalamus. As you've already seen, the blue neurons extend into the posterior pituitary where axons deliver the hormones for storage and release. The yellow neurons of the hypothalamus produce releasing and inhibiting hormones and release them into capillaries, colored red. These capillaries drain into small veins that run within the stalk of the pituitary to a second bed of capillaries in the anterior pituitary, colored blue. This unusual system of vessels is known as the *hypothalamohypophyseal portal system*. It is called a portal system because it has a second capillary bed downstream from the first; the only other body location with a similar system is the liver.

Each releasing hormone delivered to the anterior pituitary by this portal system regulates the secretion of a specific anterior pituitary hormone. For example, thyrotropin-releasing hormone (TRH) stimulates the release of TSH, corticotropin-releasing hormone (CRH) stimulates the release of ACTH, gonadotropin-releasing hormone (GnRH) stimulates the release of FSH and LH, growth-hormone-releasing hormone (GHRH) stimulates the release of GH, and prolactin-releasing factor (PRF) stimulates the release of prolactin—however, this factor has not yet been chemically characterized and may actually be a chemical similar to thyrotropin-releasing hormone.

The hypothalamus also secretes hormones that inhibit the release of certain anterior pituitary hormones. To date, three such hormones have been discovered: somatostatin inhibits the secretion of GH; prolactin-inhibiting hormone (PIH), possibly dopamine, inhibits the secretion of prolactin;

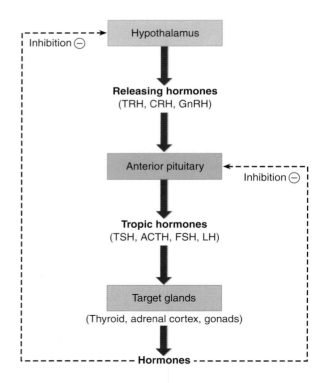

Figure 31.9   Negative feedback.
The hormones secreted by some endocrine glands feed back to inhibit the secretion of hypothalamic-releasing hormones and anterior pituitary tropic hormones.

and melanotropin-inhibiting hormone (MIH) inhibits the secretion of MSH.

Because hypothalamic hormones control the secretions of the anterior pituitary gland, and the anterior pituitary hormones control the secretions of some other endocrine glands, it may seem that the hypothalamus functions as a "master gland," in charge of hormonal secretion in the body. This idea is not generally valid, however, for two reasons. First, a number of endocrine organs, such as the adrenal medulla and the pancreas, are not directly regulated by this control system. Second, the hypothalamus and the anterior pituitary gland are themselves controlled by the very hormones whose secretion they stimulate! In most cases this is an inhibitory control. Figure 31.9 shows how the target gland hormones inhibit the hypothalamus and anterior pituitary. When enough of the target gland hormone has been produced, the hormone then feeds back and inhibits the release of the stimulating hormones from the hypothalamus and anterior pituitary, indicated by the dashed lines. This type of control system is an example of **negative feedback (or feedback inhibition),** which was also discussed in chapter 28.

> **31.3   The posterior pituitary gland contains axons originating from neurons in the hypothalamus. The anterior pituitary responds to hormonal signals from the hypothalamus by itself producing a family of pituitary hormones that are carried to distant glands and that induce them to produce specific hormones.**

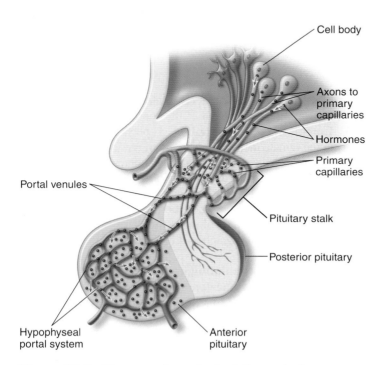

Figure 31.8   Hormonal control of the anterior pituitary gland by the hypothalamus.
Neurons in the hypothalamus secrete hormones that are carried by short blood vessels to the anterior pituitary gland, where they either stimulate or inhibit the secretion of anterior pituitary hormones.

31.4 The Pancreas

The **pancreas** gland is located behind the stomach and is connected to the front end of the small intestine by a narrow tube. It secretes a variety of digestive enzymes into the digestive tract through this tube, and for a long time it was thought to be solely an exocrine gland. In 1869, however, a German medical student named Paul Langerhans described some unusual clusters of cells scattered throughout the pancreas. In 1893, doctors concluded that these clusters of cells, which came to be called islets of Langerhans, produced a substance that prevented diabetes mellitus. **Diabetes mellitus** is a serious disorder in which affected individuals' cells are unable to take up glucose from the blood, even though their levels of blood glucose become very high. Some individuals lose weight and literally starve; others develop poor circulation, sometime resulting in amputation of limbs with restricted circulation. Diabetes is the leading cause of blindness among adults, and it accounts for one-third of all kidney failures. It is the seventh leading cause of death in the United States.

The substance produced by the islets of Langerhans, which we now know to be the peptide hormone *insulin,* was not isolated until 1922. Two young doctors working in a Toronto hospital injected an extract purified from beef pancreas glands into a 13-year-old boy, a diabetic whose weight had fallen to 29 kilograms (65 pounds) and who was not expected to survive. The hospital record gives no indication of the historic importance of the trial, only stating, "15 cc of MacLeod's serum. 7-1/2 cc into each buttock." With this single injection, the glucose level in the boy's blood fell 25%—his cells were taking up glucose. A more potent extract soon brought levels down to near normal.

This was the first instance of successful insulin therapy. The islets of Langerhans in the pancreas produce two hormones that interact to govern the levels of glucose in the blood. These hormones are *insulin* and *glucagon.* Insulin is a storage hormone, designed to put away nutrients for leaner times. It promotes the accumulation of glycogen in the liver and triglycerides in fat cells. When food is consumed (left side of figure 31.10), beta cells in the islets of Langerhans secrete insulin, causing the cells of the body to take up and store glucose as glycogen and triglycerides to be used later. When body activity causes the level of glucose in the blood to fall as it is used as fuel (right side of figure 31.10), other cells in the islets of Langerhans, called alpha cells, secrete glucagon, which causes liver cells to release stored glucose and fat cells to break down triglycerides for energy use. The two hormones work together to keep glucose levels in the blood within a narrow range.

Over 18 million people in the United States, and over 194 million people worldwide, have **diabetes.** There are *two* kinds of diabetes mellitus. About 5% to 10% of affected individuals suffer from type I diabetes, a hereditary autoimmune disease in which the immune system attacks the islets of Langerhans, resulting in abnormally low insulin secretion. Called juvenile-onset diabetes, this type usually develops before age 20. Affected individuals can be treated by daily injections of

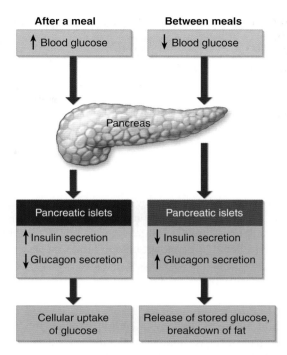

Figure 31.10 Insulin and glucagon secreted by the pancreas regulate blood glucose levels.
After a meal, an increased secretion of insulin by the beta cells of the pancreatic islets of Langerhans promotes the movement of glucose from blood into tissue cells. Between meals, an increased secretion of glucagon by the alpha cells of the pancreatic islets and decreased secretion of insulin cause the release of stored glucose and the breakdown of fat.

insulin. Active research on the possibility of transplanting islets of Langerhans holds much promise as a lasting treatment for this form of diabetes.

In type II diabetes, the level of insulin in the blood is often higher than normal but the number of insulin receptors on the target tissue is abnormally low. This form of diabetes usually develops in people over 40 years of age. It is almost always a consequence of excessive weight; in the United States, 90% of those who develop type II diabetes are obese. Cells of these individuals, overwhelmed with food, adjust their appetite for glucose downward by reducing their sensitivity to insulin. As a drug addict's neurons reduce their number of neurotransmitter receptors after continued exposure to a drug, the obese individual's cells reduce their number of insulin receptors. To compensate, the pancreas pumps out ever-more insulin, and, in some people, the insulin-producing cells are unable to keep up with the ever-heavier workload and stop functioning. Type II diabetes is usually treatable by diet and exercise, and most affected individuals do not need daily injections of insulin.

> **31.4** Clusters of cells within the pancreas secrete the hormones insulin and glucagon. Insulin stimulates the storage of glucose as glycogen, while glucagon stimulates glycogen breakdown to glucose. Working together, these hormones keep glucose levels within a narrow range.

31.5 The Thyroid, Parathyroid, and Adrenal Glands

The Thyroid: A Metabolic Thermostat

The **thyroid gland** is shaped like a shield (its name comes from *thyros,* the Greek word for "shield") and lies just below the Adam's apple in the front of the neck. The thyroid makes several hormones, the two most important of which are **thyroxine,** which increases metabolic rate and promotes growth, and **calcitonin,** which inhibits the release of calcium from bones.

Thyroxine regulates the level of metabolism in the body in several important ways. Without adequate thyroxine, growth is retarded. For example, children with underactive thyroid glands are not able to carry out carbohydrate breakdown and protein synthesis at normal rates, a condition called cretinism, which results in stunted growth. Mental retardation can also result, because thyroxine is needed for normal development of the central nervous system. The thyroid is stimulated to produce thyroxine by the hypothalamus, which is inhibited by thyroxine via negative feedback. Recall from page 636 and from chapter 28 that in negative feedback, the target gland's hormone inhibits the stimulation of the gland. The dashed lines in figure 31.11a indicate that thyroxine inhibits the release of TRH and TSH from the hypothalamus and anterior pituitary, respectively. Thyroxine contains iodine, and if the amount of iodine in the diet is too low, the thyroid cannot make adequate amounts of thyroxine to keep the hypothalamus inhibited. The hypothalamus will then continue to stimulate the thyroid, which will grow larger in a futile attempt to manufacture more thyroxine. The greatly enlarged thyroid gland that results is called a goiter (figure 31.11b). This need for iodine in the diet is why iodine is added to table salt. You probably have had your thyroid examined. A doctor will probe the area around your throat, feeling for lumps or enlarged areas. An overactive thyroid, which produces too much thyroxine, can cause hyperthyroidism. The overproduction of thyroxine results in elevated metabolism, which causes increased heart rate, weight loss, elevated body temperature—all symptoms of elevated metabolism.

Calcitonin, which will be discussed later, plays a key role in maintaining proper calcium levels in the body.

The Parathyroids: Regulating Calcium

The **parathyroid glands** are four small glands attached to the thyroid. Small and unobtrusive, they were ignored by researchers until well into the last century. The first suggestion that the parathyroids produce a hormone came from experiments in which they were removed from dogs: The concentration of calcium in the dogs' blood plummeted to less than half the normal level. However, if an extract of the parathyroid gland was administered, calcium levels returned to normal. If an excess was administered, calcium levels in the blood became *too* high, and the bones of the dogs were literally dismantled by the extract. It was clear that the parathyroid glands were producing a hormone that acted on calcium uptake into and out of the bones.

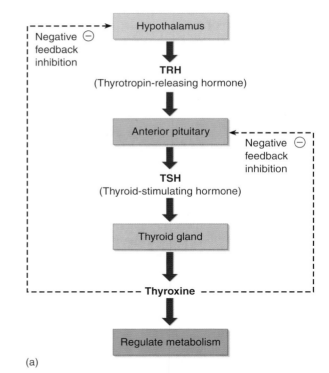

(a)

(b)

Figure 31.11 The thyroid gland secretes thyroxine.
(a) Thyroxine exerts negative feedback control of the hypothalamus and anterior pituitary. (b) A goiter is caused by a lack of iodine in the diet: thyroxine secretion is low, there is less negative feedback, and TSH is not inhibited from stimulating the thyroid gland.

The hormone produced by the parathyroids is **parathyroid hormone (PTH).** It is one of only two hormones in the body that is absolutely essential for survival (the other is aldosterone, a hormone produced by the adrenal glands, discussed on the facing page). PTH regulates the level of calcium in blood. Recall that calcium ions are the key actors in muscle contraction—by initiating calcium release, nerve impulses cause muscles to contract. A vertebrate cannot live without the muscles that pump the heart and drive the body, and these muscles cannot function if calcium levels are not kept within narrow limits.

PTH acts as a fail-safe to make sure calcium levels never fall too low. If they do (figure 31.12a), PTH is released into the bloodstream, travels to the bones, and acts on the osteoclast cells (the blue cells) within bones, stimulating them to dismantle bone tissue and release calcium into the bloodstream. PTH also acts on the kidneys to reabsorb calcium ions from the filtrate and leads to activation of vitamin D, necessary for calcium absorption by the intestine. A diet deficient in vitamin D leads to poor bone formation, a condition called rickets. When PTH is synthesized by the parathyroids in response to falling levels of calcium in the blood, the body is essentially sacrificing bone to keep calcium levels within the narrow limits necessary for proper functioning of muscle and nerve tissues. Calcitonin, a hormone referred to earlier, is released from the thyroid gland and acts in reverse of PTH. When calcium levels in the blood rise (figure 31.12b), calcitonin activates osteoblast cells (the orange cells) to take up calcium, and rebuild bone.

The Adrenals: Two Glands in One

Mammals have two **adrenal glands,** one located just above each kidney (see table 31.1 on the next page). Each adrenal gland is composed of two parts: (1) an inner core, the **medulla,** which produces the hormones adrenaline (also called epinephrine) and norepinephrine; and (2) an outer shell, the **cortex,** which produces the steroid hormones cortisol and aldosterone.

The Adrenal Medulla: Emergency Warning Siren

The medulla releases **epinephrine** (adrenaline) and **norepinephrine** in times of stress. These hormones act as emergency signals that stimulate rapid deployment of body fuel. The "alarm" response these hormones produce throughout the body is identical to the individual effects achieved by the sympathetic nervous system, but it is much longer lasting. Among the effects of these hormones are an accelerated heartbeat, increased blood pressure, higher levels of blood sugar, and increased blood flow to the heart and lungs.

The Adrenal Cortex: Maintaining the Proper Amount of Salt

The adrenal cortex produces the steroid hormone **cortisol.** Cortisol (also called hydrocortisone) acts on many different cells in the body to maintain nutritional well-being. It stimulates carbohydrate metabolism and reduces inflammation. Synthetic derivatives of this hormone, such as prednisone, have widespread medical use as anti-inflammatory agents. Cortisol is also called the *stress hormone,* released in times of stress to help the body deal with acute stress. Problems arise when the body experiences chronic stress and cortisol levels remain high in the body. This can lead to problems with maintaining blood sugar, high blood pressure, reduced immune function, fat accumulation, among others. These chronic effects of cortisol are unhealthy.

The adrenal cortex also produces **aldosterone.** Aldosterone acts primarily in the kidney to promote the uptake of sodium and other salts from the urine, which also increases the reabsorption of water. Sodium ions play crucial roles in nerve conduction and many other body functions. Water is needed to maintain blood volume and blood pressure. Aldosterone is, with PTH, one of the two endocrine hormones essential for survival. Removal of the adrenal glands is invariably fatal.

Table 31.1 on pages 640 and 641 summarizes the actions of the principal endocrine glands.

> **31.5** The thyroid acts as a metabolic thermostat, secreting hormones that adjust metabolic rate. Parathyroid hormone regulates calcium levels in the blood. The adrenal hormone aldosterone promotes the uptake of sodium and other salts in the kidney.

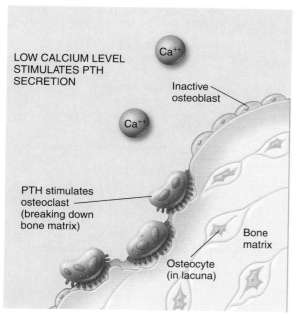

(a)

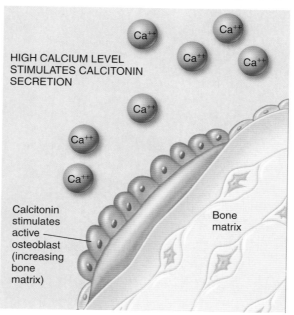

(b)

Figure 31.12 Maintenance of proper calcium levels in the blood.

(a) When calcium levels in the blood become too low, the parathyroid gland produces additional amounts of PTH, which stimulates the breakdown of bone, releasing calcium. (b) Conversely, abnormally high levels of calcium in the blood trigger the thyroid gland to secrete calcitonin, which inhibits the release of calcium from bone, and promotes the activity of osteoblasts to remove calcium from the blood and deposit it in bone.

TABLE 31.1   THE PRINCIPAL ENDOCRINE GLANDS

Endocrine Gland and Hormone	Target	Principal Actions
Adrenal Cortex		
Aldosterone	Kidney tubules	Maintains proper balance of sodium and potassium ions
Cortisol	General	Adaptation to long-term stress; raises blood glucose level; mobilizes fat
Adrenal Medulla		
Epinephrine (adrenaline) and norepinephrine (noradrenaline)	Smooth muscle, cardiac muscle, blood vessels, skeletal muscle	Initiate stress responses; increase heart rate, blood pressure, metabolic rate; dilate blood vessels; mobilize fat; raise blood glucose level
Hypothalamus		
Thyrotropin-releasing hormone (TRH)	Anterior pituitary	Stimulates TSH release from anterior pituitary
Corticotropin-releasing hormone (CRH)	Anterior pituitary	Stimulates ACTH release from anterior pituitary
Gonadotropin-releasing hormone (GnRH)	Anterior pituitary	Stimulates FSH and LH release from anterior pituitary
Prolactin-releasing factor (PRF)	Anterior pituitary	Stimulates PRL release from anterior pituitary
Growth-hormone-releasing hormone (GHRH)	Anterior pituitary	Stimulates GH release from anterior pituitary
Prolactin-inhibiting hormone (PIH)	Anterior pituitary	Inhibits PRL release from anterior pituitary
Growth-hormone-inhibiting hormone (somatostatin)	Anterior pituitary	Inhibits GH release from anterior pituitary
Melanotropin-inhibiting hormone (MIH)	Anterior pituitary	Inhibits MSH release from anterior pituitary
Ovary		
Estrogen	General; female reproductive structures	Stimulates development of secondary sex characteristics in females and growth of sex organs at puberty; prompts monthly preparation of uterus for pregnancy
Progesterone	Uterus, breasts	Completes preparation of uterus for pregnancy; stimulates development of breasts
Pancreas		
Insulin	General	Lowers blood glucose level; increases storage of glycogen in liver
Glucagon	Liver, adipose tissue	Raises blood glucose level; stimulates breakdown of glycogen in liver
Parathyroid Glands		
Parathyroid hormone (PTH)	Bone, kidneys, digestive tract	Increases blood calcium level by stimulating bone breakdown; stimulates calcium reabsorption in kidneys; activates vitamin D

TABLE 31.1 (continued)

Endocrine Gland and Hormone	Target	Principal Actions
Pineal Gland		
Melatonin	Hypothalamus	Function not well understood; may help control onset of puberty in humans and help regulate sleep cycle
Posterior Lobe of Pituitary		
Oxytocin (OT)	Uterus	Stimulates contraction of uterus
	Mammary glands	Stimulates ejection of milk
Vasopressin (antidiuretic hormone, ADH)	Kidneys	Conserves water; increases blood pressure
Anterior Lobe of Pituitary		
Growth hormone (GH)	General	Stimulates growth by promoting protein synthesis and breakdown of fatty acids
Prolactin (PRL)	Mammary glands	Sustains milk production after birth
Thyroid-stimulating hormone (TSH)	Thyroid gland	Stimulates secretion of thyroid hormones
Adrenocorticotropic hormone (ACTH)	Adrenal cortex	Stimulates secretion of adrenal cortical hormones
Follicle-stimulating hormone (FSH)	Gonads	Stimulates ovarian follicle growth and secretion of estrogen in females; stimulates production of sperm cells in males
Luteinizing hormone (LH)	Ovaries and testes	Stimulates ovulation and corpus luteum formation in females; stimulates secretion of testosterone in males
Melanocyte-stimulating hormone (MSH)	Skin	Stimulates color change in reptiles and amphibians; unknown function in mammals
Testes		
Testosterone	General; male reproductive structures	Stimulates development of secondary sex characteristics in males and growth spurt at puberty; stimulates development of sex organs; stimulates sperm production
Thyroid Gland		
Thyroid hormone (thyroxine, T4, and others)	General	Stimulates metabolic rate; essential to normal growth and development
Calcitonin	Bone	Lowers blood calcium level by inhibiting release of calcium from bone
Thymus		
Thymosin	White blood cells	Promotes production and maturation of white blood cells

31.6 A Host of Other Hormones

Sexual Development

The ovaries and testes are important endocrine glands, producing the steroid sex hormones (including estrogen, progesterone, and testosterone), to be described in detail in chapter 32.

Other Hormones

There are a variety of hormones secreted by organs whose primary functions are not endocrine in nature. The thymus is the site of production of particular lymphocytes called T cells, and it secretes a number of hormones that function in the regulation of the immune system. The right atrium of the heart secretes atrial natriuretic hormone, which stimulates the kidneys to excrete salt and water in the urine. The kidneys secrete erythropoietin, a hormone that stimulates the bone marrow to produce red blood cells. Even the skin has an endocrine function, because it secretes vitamin D.

Molting and Metamorphosis in Insects

In insects, hormonal secretions influence both metamorphosis and molting. You can see in figure 31.13 that prior to molting, neurosecretory cells on the surface of the brain secrete brain hormone ❶, which in turn stimulates a gland in the thorax called the prothoracic gland to produce **molting hormone,** or *ecdysone* ❷. Another hormone, called juvenile hormone, is produced by structures near the brain called the corpora allata ❸. For insects to molt, both ecdysone and juvenile hormone must be present, but the level of juvenile hormone determines the result of a particular molt. When juvenile hormone levels are high ❹, the molt produces another larva. At the late stages of metamorphosis, juvenile hormone levels decrease, indicated by dashed lines ❺, and the molt produces a pupa and eventually an adult insect.

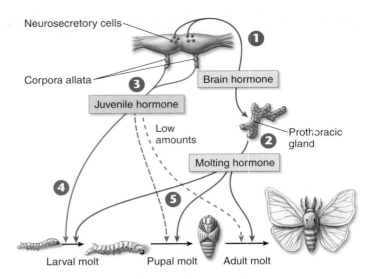

Figure 31.13 The hormonal control of metamorphosis in the silkworm moth, *Bombyx mori*.
While molting hormone (ecdysone), produced by the prothoracic gland, triggers when molting occurs, juvenile hormone, produced by structures near the brain called the corpora allata, determines the result of a particular molt. High levels of juvenile hormone inhibit the formation of the pupa and adult forms. At the late stages of metamorphosis, therefore, it is important that the corpora allata not produce large amounts of juvenile hormone.

> **31.6** Sex steroid hormones from the gonads regulate reproduction. Other hormones are released from nontraditional endocrine glands. Molting hormone, or ecdysone, and juvenile hormone regulate metamorphosis and molting in insects.

INQUIRY & ANALYSIS

Are You Pregnant?

Early in pregnancy, cells from the new embryo secrete a hormone called human chorionic gonadotropin (hCG), which helps maintain the uterus and so prevents spontaneous abortion. hCG is soon present in all of a pregnant woman's body fluids. Levels of hCG reach a peak at 50 to 60 days, then fall to a much lower level for the remainder of the pregnancy.

1. **Applying Concepts** In the graph, what is the dependent variable?
2. **Making Inferences** Is hCG detectable on the graph when a woman is not pregnant?
3. **Drawing Conclusions** Would a test for the presence of hCG in a woman's urine, if sensitive, provide a reliable test of pregnancy?

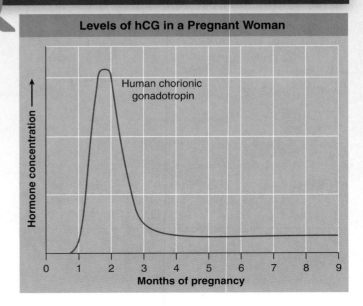

Levels of hCG in a Pregnant Woman

The Neuroendocrine System

31.1 Hormones

- Hormones are chemical signals produced in glands or other endocrine tissues (**figure 31.1**) and transported to distant sites in the body. Endocrine glands produce hormones and release the hormones into the bloodstream.

- Endocrine glands and tissues are under the control of the central nervous system, primarily the hypothalamus. A command from the hypothalamus causes the release of a hormone from an endocrine gland. Only cells that have receptors for the hormone respond and are called "target cells." The hormone binds to the receptor and elicits a response in the cell, often a change in cellular activity or genetic expression (**figure 31.2**).

31.2 How Hormones Target Cells

- Steroid hormones are lipid-soluble molecules. They pass through the plasma membrane of the target cell and bind to receptors in the cytoplasm or nucleus. The hormone-receptor complex binds to DNA, causing a change in gene expression that alters cell function (**figure 31.3**).

- Peptide hormones are not lipid-soluble and therefore travel through the bloodstream unaided but are not able to pass through the plasma membrane. Instead, they bind to transmembrane protein receptors. The binding of the hormone causes a change in the internal side of the receptor, triggering a change in cellular activity (**figure 31.4**).

- The change in cellular activity caused by peptide hormones is facilitated by a second messenger system, which is a cascade of reactions that amplifies the signal (**figure 31.5**).

The Major Endocrine Glands

31.3 The Hypothalamus and the Pituitary

- The pituitary is actually two glands: the anterior and posterior pituitary glands. The posterior pituitary develops as an extension of the hypothalamus and contains axons that extend from cell bodies in the hypothalamus (**figure 31.6**).

- The hormones released from the posterior pituitary are actually produced in the hypothalamus and transported by the axons to the posterior pituitary for storage and release. The hormones of the posterior pituitary include antidiuretic hormone (ADH), which regulates water retention in the kidneys, and oxytocin, which initiates uterine contractions during childbirth and milk release in the mother (**figure 31.7**).

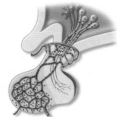

- The anterior pituitary originates from epithelial tissue and produces the hormones it releases, although it is controlled by the hypothalamus. Hormones released from the hypothalamus control the release of anterior pituitary hormones (**figure 31.8**).

- Seven hormones are released from the anterior pituitary; some have direct effects on target cells, such as LH, FSH, prolactin, and MSH, while others, such as TSH and ACTH, stimulate target cells to release hormones (**figure 31.7**).

- Many hormones are controlled by negative feedback, where the hormone itself feeds back to inhibit the process (**figure 31.9**).

31.4 The Pancreas

- The pancreas secretes two hormones—insulin and glucagon—that interact to maintain stable blood glucose levels. Insulin stimulates cell uptake of glucose from the blood. Glucagon stimulates the breakdown of glycogen to glucose.

- These hormones work opposite to each other (**figure 31.10**). When insulin is not available or cells fail to respond to insulin, diabetes can result.

31.5 The Thyroid, Parathyroid, and Adrenal Glands

- The two most important hormones produced by the thyroid are thyroxine, which increases metabolism and growth, and calcitonin, which stimulates calcium uptake by bones. Thyroxine is controlled by negative feedback, and the over- or underproduction of thyroxine can lead to serious health problems (**figure 31.11**).

- The parathyroid glands produce parathyroid hormone, a hormone necessary for survival. PTH regulates the levels of calcium in the blood (**figure 31.12**).

- The adrenal gland is actually two glands: the adrenal medulla is the inner core, and the adrenal cortex is the outer shell. The adrenal medulla secretes epinephrine and norepinephrine, and the adrenal cortex secretes aldosterone, which, like PTH, is necessary for survival. It promotes the uptake of sodium and water from urine.

31.6 A Host of Other Hormones

- The sex steroid hormones, including estrogen, progesterone, and testosterone, are released from the gonads and regulate sexual development and reproduction. Hormones are also released from other nontraditional endocrine tissues.

- In insects, molting hormone, ecdysone, and juvenile hormone regulate metamorphosis (**figure 31.13**).

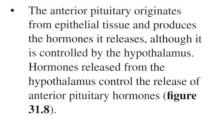

1 One advantage chemical signaling has over nervous signaling is that
  a. reaction to stimuli can happen very quickly.
  b. although it takes large amounts of chemicals, the chemical signals are efficient.
  c. chemical signals stick around longer than nervous signals and can be used for slow processes.
  d. chemical signals are used in reaction to external and internal stimuli.

2. A coordination center for some of the endocrine system is the
  a. hypothalamus.   c. thyroid gland.
  b. adrenal gland.   d. pancreas.

3. Hormones are very similar to neurotransmitters and antibodies because they
  a. all fit into receptors specifically shaped for them.
  b. are all proteins.
  c. appear in response to a signal from the brain.
  d. All of the above.

4. The action of steroid hormones is different from peptide hormones because
  a. peptide hormones must enter the cell to begin action, whereas steroid hormones must begin action on the external surface of the cell membrane.
  b. steroid hormones must enter the cell to begin action, whereas peptide hormones must begin action on the external surface of the cell membrane.
  c. peptide hormones produce a hormone receptor complex that works directly on the DNA, whereas steroid hormones cause the release of a secondary messenger that triggers enzymes.
  d. No answer is correct.

5. Regulation of kidney function is done by the
  a. thyroid gland.
  b. thymus.
  c. anterior pituitary gland.
  d. posterior pituitary gland.

6. _____ is the hormone that stimulates the adrenal gland to produce a number of steroid hormones.
  a. ACTH
  b. LH
  c. TSH
  d. MSH

7. Type I diabetes is caused by an abnormality in endocrine cells of the
  a. pancreas.   c. adrenal glands.
  b. thymus.   d. hypothalamus.

8. The release of the thyroid hormone calcitonin is triggered by
  a. too much glucose in the blood.
  b. too much sodium in the blood.
  c. too much calcium in the blood.
  d. too little calcium in the blood.

9. Epinephrine mimics the effects of the
  a. autonomic nervous system.
  b. central nervous system.
  c. parasympathetic nervous system.
  d. sympathetic nervous system.

10. A person afflicted with acromegaly, a form of giantism, has too much
  a. GH.   c. oxytocin.
  b. GnRH.   d. thyroxine.

Visual Understanding

1. **Figure 31.1** Some of the body systems are located primarily in one area of the body, or are obviously connected. The respiratory system, for instance, is located in the head and upper portion of the body. The skeletal system is articulated, every bone connected to others. The endocrine system, however, is spread out, a batch of glands that do not appear connected with one another. Speculate on why this is so.

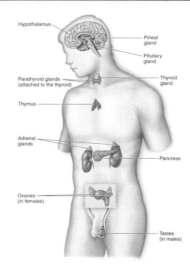

2. **Figure 31.7** A hypothetical patient has a disorder of the hypothalamus in which it can no longer secrete its "inhibiting" hormones. Which of the anterior pituitary hormones will be affected?

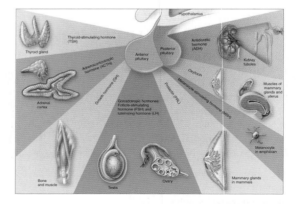

Challenge Questions

1. Tad, the younger brother of your friend, Sofia, wants to be a sports star in high school. Only in the eighth grade, he brags that he is taking steroids he gets from a friend in order to "bulk up" for next year. He asks if you have ever heard of any problems for kids his age—he only wants to take them "a couple of years" to get a football scholarship so his family can afford for him to go to college. How would you advise him?

2. Since your bones are so very important, why would you ever need a system that includes osteoclasts, which literally break down your bone tissue?

32

Reproduction and Development

Few subjects pervade our everyday thinking more than sex; few urges are more insistent. They are no accident, these strong feelings. They are a natural part of being human. All animals share them. The cry of a cat in heat, insects chirping outside the windows, frogs croaking in swamps, wolves howling in a frozen northern scene—all these are the sounds of the living world's essential act, an urgent desire to reproduce that has been patterned by a long history of evolution. It is a pattern that each of us shares. The reproduction of our families spontaneously elicits in us a sense of rightness and fulfillment. It is difficult not to return the smile of a new infant, not to feel warmed by it and by the look of wonder and delight to be seen on the faces of parents like this nursing mother. This chapter deals with sex and reproduction among the vertebrates, of which we human beings are one kind. Few subjects are of more direct concern to students than sex. Because many students must make important decisions about sex, the subject is of far more than academic interest, and is one about which all students need to be well informed.

32.1 Asexual and Sexual Reproduction

Not all reproduction involves two parents. Asexual reproduction, in which the offspring are genetically identical to one parent, is the primary means of reproduction among protists, cnidarians, and tunicates, and also occurs in some of the more complex animals.

Through mitosis, genetically identical cells are produced from a single parent cell. This permits asexual reproduction to occur in the *Euglena* in figure 32.1 by division of the organism, or **fission**. The DNA replicates and organs, such as the flagellum, duplicate. The nucleus divides with identical nuclei going to each daughter cell. Cnidaria commonly reproduce by **budding**, where a part of the parent's body becomes separated from the rest and differentiates into a new individual. The new individual may become an independent animal or may remain attached to the parent, forming a colony.

Unlike asexual reproduction, sexual reproduction occurs when a new individual is formed by the union of *two* cells. These cells are called **gametes,** and the two kinds that combine are generally called *sperm* and *eggs* (or *ova*). The union of a sperm and an egg produces a fertilized egg, or **zygote,** that develops by mitotic division into a new multicellular organism. The zygote and the cells it forms by mitosis are diploid; they contain both members of each homologous pair of chromosomes. The gametes, formed by meiosis in the sex organs, or **gonads**—the *testes* and *ovaries*—are haploid (see chapter 10). The processes of spermatogenesis (sperm formation) and oogenesis (egg formation) are described in later sections.

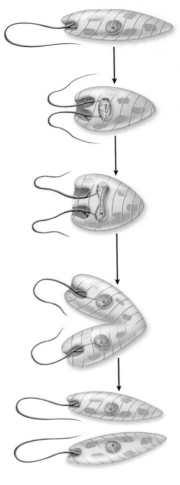

Figure 32.1 Asexual reproduction in protists.

This protist *Euglena* reproduces asexually: a mature individual divides by fission, and two complete individuals result.

vertebrates. He observed that some populations of small lizards of the genus *Lacerta* were exclusively female, and he suggested that these lizards could lay eggs that were viable even if they were not fertilized. In other words, they were capable of asexual reproduction in the absence of sperm, a type of parthenogenesis. Further work has shown that parthenogenesis occurs among populations of other lizard genera.

Another variation in reproductive strategies is **hermaphroditism,** when one individual has both testes and ovaries and so can produce both sperm and eggs. The hamlet bass in figure 32.2*a* are hermaphrodite, producing both eggs and sperm. During mating each fish switches from producing eggs that are fertilized by its partner, to producing sperm that fertilizes its partner's eggs. A tapeworm is hermaphroditic and can fertilize itself as well as cross fertilize, a useful strategy because it is unlikely to encounter another tapeworm living inside its host. Most hermaphroditic animals, however, require another individual to reproduce. Two earthworms, for example, are required for reproduction—like the hamlet bass, each functions as both male and female. Each leaves the encounter with fertilized eggs.

Numerous fish genera include species in which individuals can change their sex, a process called *sequential hermaphroditism.* Among coral reef fish, for example, both **protogyny** ("first female," a change from female to male) and **protandry** ("first male," a change from male to female) occur. In the protogynous bluehead wrasse in figure 32.2*b*, the sex change appears to be under social control. These fish commonly live in large groups, or schools, where successful reproduction is typically limited to one or a few large, dominant males. If those males are removed, the largest female rapidly changes sex and becomes a dominant male (the blue-headed fish in the photo).

Different Approaches to Sex

Parthenogenesis, a type of reproduction in which offspring are produced from unfertilized eggs, is common in many species of arthropods; some species are exclusively parthenogenic, whereas others switch between sexual reproduction and parthenogenesis in different generations. In honeybees, for example, a queen bee mates only once and stores the sperm. She then can control the release of sperm. If no sperm are released, the eggs develop parthenogenetically into drones, which are males; if sperm are allowed to fertilize the eggs, the fertilized eggs develop into other queens or worker bees, which are female.

The Russian biologist Ilya Darevsky reported in 1958 one of the first cases of unusual modes of reproduction among

Sex Determination

Among the fish just described, and in some species of reptiles, environmental changes can cause changes in the sex of the animal. In mammals, the sex is determined early in embryonic development. The reproductive systems of human males and females appear similar for the first 40 days after conception. During this time, the cells that will give rise to ova or sperm migrate to the embryonic gonads, which have the potential to become either ovaries in females or testes in males. If the embryo is XY, it is a male and will carry a gene on the Y chromosome whose product converts the gonads into testes

(a)

(b)

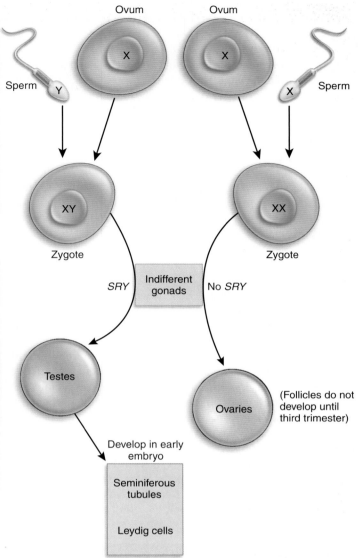

Figure 32.3 Sex determination.

Sex determination in mammals is made by a gene of the Y chromosome designated *SRY*. Testes are formed when the Y chromosome and *SRY* are present; ovaries are formed when they are absent.

Figure 32.2 Hermaphroditism and protogyny.

(a) The hamlet bass (genus *Hypoplectrus*) is a deep-sea fish that is a hermaphrodite. In the course of a single pair-mating, one fish may switch sexual roles as many as four times. Here, the fish acting as a male curves around its motionless partner, fertilizing the upward-floating eggs. (b) The bluehead wrasse *Thalassoma bifasciatium* is protogynous. Here a large male, or sex-changed female, is seen among females, typically much smaller.

(as on the left in figure 32.3). In females, who are XX, this Y chromosome gene and the protein it encodes are absent, and the gonads become ovaries (as on the right). Recent evidence suggests that the sex-determining gene may be one known as **SRY** (for "*s*ex-determining *r*egion of the *Y* chromosome"). The *SRY* gene appears to have been highly conserved during the evolution of different vertebrate groups.

Once testes form in the embryo, they secrete testosterone and other hormones that promote the development of the male external genitalia and accessory reproductive organs (indicated in the blue box). If testes do not form, the embryo develops female external genitalia and accessory reproductive organs. The ovaries do not promote this development of female organs because the ovaries are nonfunctional at this stage. In other words, all mammalian embryos will develop female sex accessory organs and external genitalia by default unless they are masculinized by the secretions of the testes.

> **32.1 Sexual reproduction is most common among animals, but many reproduce asexually by fission, budding, or parthenogenesis. Sexual reproduction generally involves the fusion of gametes derived from different individuals of a species, but some species are hermaphroditic.**

32.2 Evolution of Reproduction Among the Vertebrates

Vertebrate sexual reproduction evolved in the ocean before vertebrates colonized the land. The females of most species of marine bony fish produce eggs or ova in batches and release them into the water. The males generally release their sperm into the water containing the eggs, where the union of the free gametes occurs. This process is known as **external fertilization.**

Although seawater is not a hostile environment for gametes, it does cause the gametes to disperse rapidly, so their release by females and males must be almost simultaneous. Thus, most marine fish restrict the release of their eggs and sperm to a few brief and well-defined periods. Some reproduce just once a year, while others do so more frequently. There are few seasonal cues in the ocean that organisms can use as signals for synchronizing reproduction, but one all-pervasive signal is the cycle of the moon. Once each month, the moon approaches closer to the earth than usual, and when it does, its increased gravitational attraction causes somewhat higher tides. Many marine organisms sense the tidal changes and entrain the production and release of their gametes to the lunar cycle.

Fertilization is external in most fish but internal in most other vertebrates. The invasion of land posed the new danger of desiccation (drying out), a problem that was especially severe for the small and vulnerable gametes. On land, the gametes could not simply be released near each other, because they would soon dry up and perish. Consequently, there was intense selective pressure for terrestrial vertebrates (as well as some groups of fish) to evolve **internal fertilization,** that is, the introduction of male gametes into the female reproductive tract. By this means, fertilization still occurs in a nondesiccating environment, even when the adult animals are fully terrestrial. The vertebrates that practice internal fertilization have three strategies for embryonic and fetal development. Depending upon the relationship of the developing embryo to the mother and egg, those vertebrates with internal fertilization may be classified as oviparous, ovoviviparous, or viviparous.

Oviparity. In oviparity, the eggs, after being fertilized internally, are deposited outside the mother's body to complete their development. This is found in some bony fish, most reptiles, some cartilaginous fish, some amphibians, a few mammals, and all birds.

Ovoviviparity. In ovoviviparity, the fertilized eggs are retained within the mother to complete their development, but the embryos still obtain all of their nourishment from the egg yolk. The young are fully developed when they are hatched and released from the mother. This is found in some bony fish (including mollies, guppies, and mosquito fish), some cartilaginous fish, and many reptiles.

Viviparity. In viviparity, the young develop within the mother and obtain nourishment directly from their mother's blood, rather than from the egg yolk (figure 32.4). This is found in most cartilaginous fish, some amphibians, a few reptiles, and almost all mammals.

Fish and Amphibians

Most fish and amphibians, unlike other vertebrates, reproduce by means of external fertilization.

Fish. Fertilization in most species of bony fish (teleosts) is external, and the eggs contain only enough yolk to sustain the developing embryo for a short time. After the initial supply of yolk has been exhausted, the young fish must seek its

Figure 32.4 Viviparous vertebrates carry live, mobile young within their bodies.
The young complete their development within the body of the mother and are then released. Here a lemon shark has just given birth to a young shark, which is still attached by the umbilical cord.

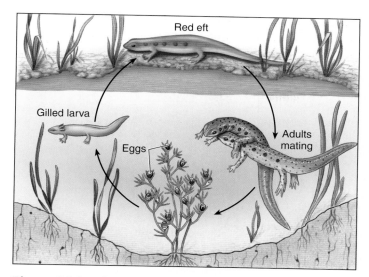

Figure 32.5 Life cycle of the red-spotted newt.
Many salamanders have both aquatic and terrestrial stages in their life cycle. In the red-spotted newt (*Notophthalmus viridescens*), eggs are laid in water and hatch into aquatic larvae with external gills and a finlike tail. After a period of growth, the larvae can metamorphose into a terrestrial "red eft" stage that later metamorphose again to produce aquatic, breeding adults. In some environments, the "red eft" stage is skipped, and the populations remain entirely aquatic.

Figure 32.6 The eggs of frogs are fertilized externally.
When frogs mate, as these two are doing, the clasp of the male induces the female to release a large mass of mature eggs, over which the male discharges his sperm.

food from the waters around it. Development is speedy, and the young that survive mature rapidly. Although thousands of eggs are fertilized in a single mating, many of the resulting individuals succumb to microbial infection or predation, and few grow to maturity.

In marked contrast to the bony fish, fertilization in most cartilaginous fish is internal. The male introduces sperm into the female through a modified pelvic fin. Development of the young in these vertebrates is generally viviparous.

Amphibians. The amphibians invaded the land without fully adapting to the terrestrial environment, and their life cycle is still tied to the water. Amphibians, like the red-spotted newt in figure 32.5, reproduce in the water and have aquatic larval stages before moving to the land. Fertilization is external in most amphibians, just as it is in most species of bony fish. Gametes from both males and females are released through the cloaca, a common opening used by the digestive, reproductive, and urinary systems. Among the frogs and toads, the male grasps the female and discharges fluid containing the sperm onto the eggs as they are released into the water (figure 32.6). Although the eggs of most amphibians develop in the water, there are some interesting exceptions shown in figure 32.7. In two species of frogs (one being the Darwin's frog in figure 32.7d) the eggs develop in the vocal sacs and the stomach, and the young frogs leave through their parent's mouth!

The time required for development of most amphibians is much longer than that for fish, but amphibian eggs do not include a significantly greater amount of yolk. Instead, the process of development in most amphibians is divided

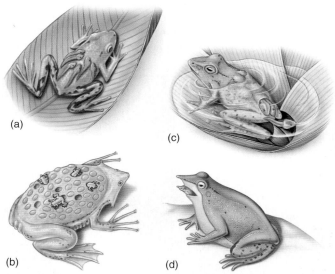

Figure 32.7 Different ways young develop in frogs.
(a) In the poison arrow frog, the male carries the tadpoles on his back. (b) In the female Surinam frog, froglets develop from eggs in special brooding pouches on the back. (c) In the South American pygmy marsupial frog, the female carries the developing larvae in a pouch on her back. (d) Tadpoles of the Darwin's frog develop into froglets in the vocal pouch of the male and emerge from the mouth.

into embryonic, larval, and adult stages (see figure 32.5), in a way reminiscent of the life cycles found in some insects. The embryo develops within the egg, obtaining nutrients from the yolk. After hatching from the egg, the aquatic larva then functions as a free-swimming, food-gathering machine, often for a considerable period of time. The larvae may increase in size rapidly; some tadpoles, which are the larvae of frogs and toads, grow in a matter of weeks from creatures no bigger than the tip of a pencil into individuals as big as a goldfish. When the larva has grown large enough, it undergoes a developmental transition, or metamorphosis, into the terrestrial adult form.

Reptiles and Birds

Most reptiles and all birds are oviparous, laying amniotic eggs that are protected by watertight membranes from desiccation. After the eggs are fertilized internally, they are deposited outside of the mother's body to complete their development. Like most vertebrates that fertilize internally, most male reptiles use a cylindrical organ, the penis, to inject sperm into the female, a process called copulation (figure 32.8). The penis, containing erectile tissue, can become quite rigid and penetrate far into the female reproductive tract. Reptiles exhibit all three types of internal fertilization. Most are oviparous, laying eggs and then abandoning them. These eggs are surrounded by a leathery shell that is deposited as the egg passes through the oviduct, the part of the female reproductive tract leading from the ovary. Other species of reptiles are ovoviviparous (forming eggs that develop into embryos and hatch within the body of the mother) or viviparous (developing inside the mother while being nourished by her rather than by the yolk of an egg).

All birds practice internal fertilization, though most male birds lack a penis. In some of the larger birds (including swans, geese, and ostriches), however, the male cloaca extends to form a false penis. Figure 32.9*a* shows the formation of a bird's egg. As the fertilized egg (ovum) passes along the

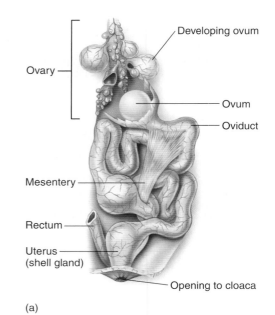

(a)

(b)

Figure 32.9   Egg formation and incubation in birds.
(a) In birds, fertilization of the egg (ovum) takes place within the female, in the upper portion of the oviduct. As the fertilized egg passes down the oviduct, albumin (egg white), shell membranes, and the shell is secreted around the egg. (b) This nesting pair of crested penguins is changing the parental guard in a stylized ritual.

oviduct (from top to bottom here), glands secrete albumin proteins (the egg white) and the hard, calcareous shell that distinguishes bird eggs from reptilian eggs. While modern reptiles are poikilotherms (animals whose body temperature varies with the temperature of their environment), birds are homeotherms (animals that maintain a relatively constant body temperature independent of environmental temperatures). Hence, most birds incubate their eggs after laying them to keep them warm (figure 32.9*b*).

The shelled eggs of reptiles and birds constitute one of the most important adaptations of these vertebrates to life on land, because shelled eggs can be laid in dry places. Such eggs are known as amniotic eggs because the embryo develops

Figure 32.8   The introduction of sperm by the male into the female's body is called copulation.

(a) Monotremes (b) Marsupials (c) Placentals

Figure 32.10 Reproduction in mammals.

(a) Monotremes, like the duck-billed platypus shown here, lay eggs in a nest. (b) Marsupials, such as this kangaroo, give birth to small fetuses that complete their development in a pouch. (c) In placental mammals, such as this doe nursing her fawn, the young remain inside the mother's uterus for a longer period of time and are born relatively more developed.

within a fluid-filled cavity surrounded by a membrane called the amnion. The amnion is an extraembryonic membrane—that is, a membrane formed from embryonic cells but located outside the body of the embryo. Amniotic eggs contain three other extraembryonic membranes, one of which is a yolk sac. In contrast, the eggs of fish and amphibians contain only one extraembryonic membrane, the yolk sac.

Mammals

Some mammals are seasonal breeders, reproducing only once a year, such as dogs, foxes, and bears, whereas others, such as horses and sheep, have multiple short reproductive cycles throughout a given time of the year. Among the latter, the females generally undergo the reproductive cycles, whereas the males are more constant in their reproductive activity. Cycling in females involves the periodic release of a mature ovum from the ovary in a process known as ovulation. Most female mammals are "in heat," or sexually receptive to males, only around the time of ovulation. This period of sexual receptivity is called **estrus,** and the reproductive cycle is therefore called an **estrous cycle.** Females continue to cycle until they become pregnant.

In the estrous cycle of most mammals, changes in the secretion of follicle-stimulating hormone (FSH) and luteinizing hormone (LH) by the anterior pituitary gland cause changes in egg cell development and hormone secretion in the ovaries. Humans and apes have menstrual cycles that are similar to the estrous cycles of other mammals in their cyclic pattern of hormone secretion and ovulation. Unlike mammals with estrous cycles, however, human and some ape females bleed when they shed the inner lining of their uterus, a process called menstruation, and may engage in copulation at any time during the cycle.

Rabbits and cats differ from most other mammals in that they are **induced ovulators.** Instead of ovulating in a cyclic fashion regardless of sexual activity, the females ovulate only after copulation as a result of a reflex stimulation of LH secretion (described later). This makes them extremely fertile.

The most primitive mammals, the **monotremes** (consisting solely of the duck-billed platypus and the echidna), are oviparous, like the reptiles from which they evolved. They incubate their eggs in a nest (figure 32.10a) or specialized pouch, and the young hatchlings obtain milk from their mother's mammary glands by licking her skin, because monotremes lack nipples. All other mammals are viviparous and are divided into two subcategories based on how they nourish their young. The **marsupials,** a group that includes opossums and kangaroos, give birth to fetuses that are incompletely developed. They complete their development in a pouch of their mother's skin, where they obtain nourishment from nipples of her mammary glands (figure 32.10b). The **placental mammals** (figure 32.10c) retain their young for a much longer period of development within the mother's uterus. The fetuses are nourished by a structure known as the placenta, which is derived from both an extraembryonic membrane and from the mother's uterine lining. Because the fetal and maternal blood vessels are in very close proximity in the placenta, the fetus can obtain nutrients by diffusion from the mother's blood. The placenta is discussed in more detail later in this chapter.

32.2 Fertilization is external in frogs and most bony fish and internal in other vertebrates. Birds and most reptiles lay watertight eggs, as do monotreme mammals. All other mammals are viviparous, giving birth to live young.

32.3 Males

The human male gamete, or **sperm,** is highly specialized for its role as a carrier of genetic information. Produced after meiosis, sperm cells have 23 chromosomes instead of the 46 found in other cells of the male body. Sperm do not successfully complete their development at 37°C (98.6°F), the normal human body temperature. The sperm-producing organs, the **testes** (singular, **testis**), move during the course of fetal development into a sac called the **scrotum** (figure 32.11), which hangs between the legs of the male, maintaining the two testes at a temperature about 3°C cooler than the rest of the body. The testes contain cells that secrete the male sex hormone **testosterone.**

Male Gametes Are Formed in the Testes

An internal view of the testes in figure 32.12 ❶ shows that they are composed of several hundred compartments, each packed with large numbers of tightly coiled tubes called **seminiferous tubules** (seen in cross section in ❷). Sperm production, *spermatogenesis,* takes place inside the tubules. The process

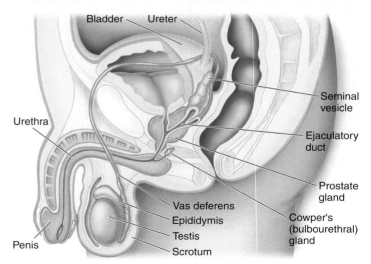

Figure 32.11   The male reproductive organs.
The testis is where sperm are formed. Cupped above the testis is the epididymis, a highly coiled passageway within which sperm complete their maturation. Extending away from the epididymis is a long tube, the vas deferens.

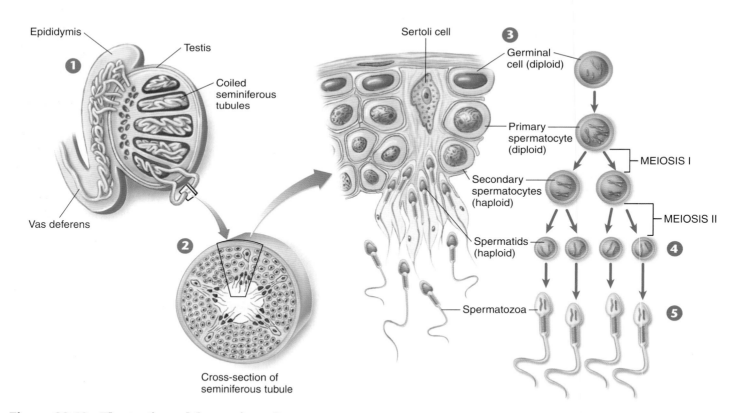

Figure 32.12   The testis and formation of sperm.
Inside the testis ❶, the seminiferous tubules ❷ are the sites of sperm formation. Germinal cells in the seminiferous tubules ❸ give rise to primary spermatocytes (diploid), which undergo meiosis to form haploid spermatids ❹. Spermatids develop into mobile spermatozoa, or sperm ❺. Sertoli cells are nongerminal cells within the walls of the seminiferous tubules. They assist spermatogenesis in several ways, such as helping to convert spermatids into spermatozoa.

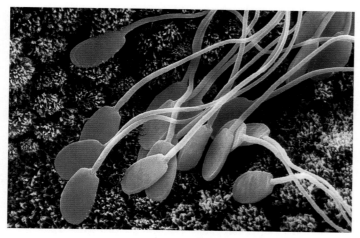

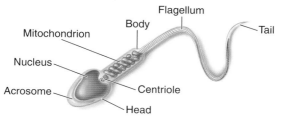

Figure 32.13 Human sperm cells.

Each sperm possesses a long tail that propels the sperm and a head that contains the nucleus. The tip, or acrosome, contains enzymes to help the sperm cell digest a passageway into the egg for fertilization.

of spermatogenesis begins in germinal cells toward the outside of the tubule (shown in the enlarged view in ❸). As the cells undergo meiosis, they move toward the lumen of the tubule, with spermatozoa being released into the tubule. The number of sperm produced is truly incredible. A typical adult male produces several hundred million sperm each day of his life! Those that are not ejaculated from the body are broken down, and their materials are resorbed and recycled.

After a sperm cell is manufactured within the testes through intermediate stages of meiosis, it is delivered to a long, coiled tube called the **epididymis** (see figure 32.11), where it matures. The sperm cell, shown in figure 32.13, is not motile when it arrives in the epididymis, and it must remain there for at least 18 hours before its motility develops. From there, the sperm is delivered to another long tube, the **vas deferens.** When sperm are released during intercourse, they travel through a tube from the vas deferens to the **urethra,** where the reproductive and urinary tracts join, emptying through the penis. Sperm is released in a fluid called semen that contains fluids produced in glands that empty into the urethra.

Male Gametes Are Delivered by the Penis

In the case of humans and some other mammals, the **penis** is an external tube containing two long cylinders of spongy tissue side by side (seen in cross section in figure 32.14). Below and between them runs a third cylinder of spongy tissue that contains

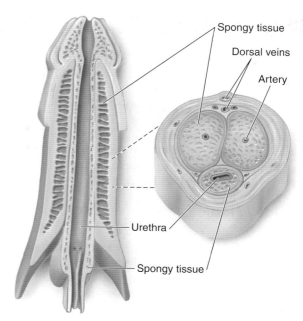

Figure 32.14 Structure of the penis.

(*Left*) Longitudinal section; (*right*) cross section.

in its center a tube called the urethra, through which both semen (during ejaculation) and urine (during urination) pass. Why this unusual design? The penis is designed to inflate. The spongy tissues that make up the three cylinders are riddled with small spaces between the cells, and when nerve impulses from the CNS cause the arterioles leading into this tissue to expand, blood collects within these spaces. Like blowing up a balloon, this causes the penis to become erect and rigid. Continued stimulation by the CNS is required for erection to continue.

Erection can be achieved without any physical stimulation of the penis, but physical stimulation is required for semen to be delivered. Stimulation of the penis, as by repeated thrusts into the vagina of a female, leads first to the mobilization of the sperm. In this process, muscles encircling the vas deferens contract, moving the sperm along the vas deferens into the urethra. The stimulation then leads to the strong contraction of the muscles at the base of the penis. The result is **ejaculation,** the forceful ejection of 2 to 5 milliliters of semen. Semen contains sperm and a collection of secretions from the prostate and other glands that provides metabolic energy sources for the sperm. Within this small 5-milliliter volume are several hundred million sperm. Because the odds against any one individual sperm cell successfully completing the long journey to the egg and fertilizing it are extraordinarily high, successful fertilization requires a high sperm count. Males with fewer than 20 million sperm per milliliter are generally considered sterile.

32.3 Male testes continuously produce large numbers of male gametes, sperm, which mature in the epididymis, are stored in the vas deferens, and are delivered through the penis into the female.

32.4 Females

In females, eggs develop from cells called **oocytes,** located in the outer layer of compact masses of cells called **ovaries** within the abdominal cavity (figure 32.15). Recall that in males the gamete-producing cells are constantly dividing. In females all of the oocytes needed for a lifetime are already present at birth. During each reproductive cycle, one or a few of these oocytes are initiated to continue their development in a process called **ovulation;** the others remain in developmental holding patterns.

Only One Female Gamete Matures Each Month

At birth, a female's ovaries contain some 2 million oocytes, all of which have begun the first meiotic division. At this stage they are called primary oocytes (❶ in figure 32.16). Each primary oocyte waits to receive the proper developmental "go" signal before continuing on with meiosis. Until then, its meiosis remains arrested in prophase of the first meiotic division. Very few ever receive the awaited signal, which turns out to be the hormone FSH, which you recall was discussed in chapter 31.

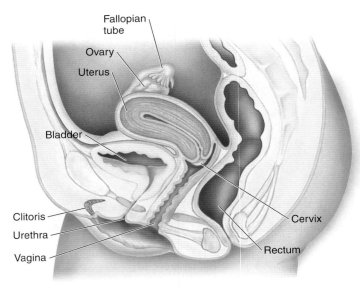

Figure 32.15 The female reproductive system.
The organs of the female reproductive system are specialized to produce gametes and to provide a site for embryonic development if the gamete is fertilized.

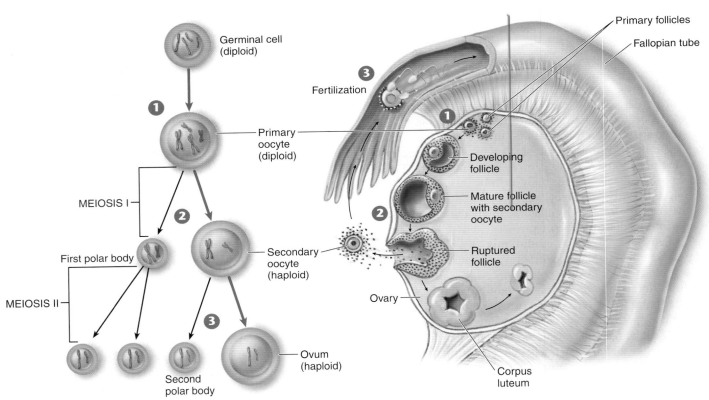

Figure 32.16 The ovary and formation of an ovum.
In this figure, the maturation of the ovum through meiosis is shown on the left, and the developmental journey of the ovum is on the right, with corresponding stages numbered on each. At birth, a human female's ovaries contain about 2 million egg-forming cells called oocytes, which have begun the first meiotic division and stopped. At this stage, they are called primary oocytes ❶, and their further development is halted until they receive the proper developmental signal, which is the hormone FSH. When FSH is released at puberty, meiosis resumes in a few oocytes, but only one oocyte usually continues to mature while the others regress. The primary oocyte (diploid) completes the first meiotic division, and one division product is eliminated as a nonfunctional polar body. The other product, the secondary oocyte, is released during ovulation ❷. The secondary oocyte does not complete the second meiotic division until fertilization ❸; that division yields a second nonfunctional polar body and a single haploid egg, or ovum. Fusion of the haploid egg with a haploid sperm during fertilization produces a diploid zygote.

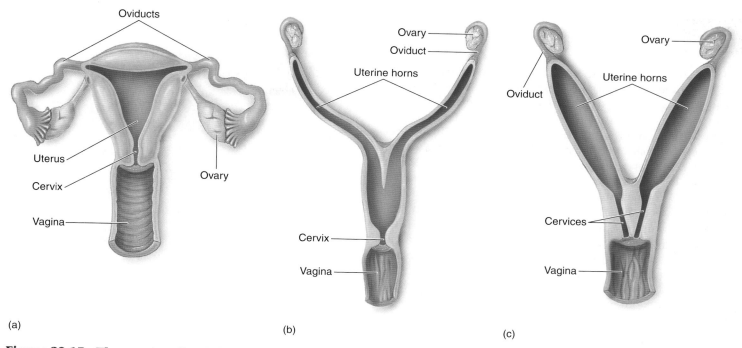

Figure 32.17　The mammalian uterus—several examples.
(a) Humans and other primates; (b) cats, dogs, and cows; and (c) rats, mice, and rabbits.

With the onset of puberty, females mature sexually. At this time, the release of FSH initiates the resumption of the first meiotic division in a few oocytes, but in humans, usually only a single oocyte continues to mature and becomes an ovum (❸), and the others regress. In some instances more than one oocyte develops; if both are fertilized, they become fraternal twins. Approximately every 28 days after that, another oocyte matures and is ovulated, although the exact timing may vary from month to month. Only about 400 of the approximately 2 million oocytes a woman is born with mature and are ovulated during her lifetime. When they mature, the egg cells are called **ova** (singular, **ovum**), the Latin word for "egg."

Fertilization Occurs in the Oviducts

The **fallopian tubes** (also called uterine tubes, or **oviducts**) transport ova from the ovaries to the **uterus.** In humans, the uterus is a muscular, pear-shaped organ about the size of a fist that narrows to a muscular ring called the **cervix,** which leads to the vagina (figure 32.17*a*). The uterus is lined with a stratified epithelial membrane, the **endometrium.** The surface of the endometrium is shed during menstruation, while the underlying portion remains to generate a new surface during the next cycle.

Mammals other than primates have more complex female reproductive tracts, where part of the uterus divides to form uterine "horns," each of which leads to an oviduct (figure 32.17*b, c*). In cats, dogs, and cows, for example, there is one cervix but two uterine horns separated by a septum, or wall. Marsupials, such as opossums, carry the split even further, with two unconnected uterine horns, two cervices, and two vaginas. A male marsupial has a forked penis that can enter both vaginas simultaneously.

Smooth muscles lining the fallopian tubes contract rhythmically, moving the egg down the tube to the uterus in much the same way that food is moved down through your intestines, pushing it along by squeezing the tube behind it. The journey of the egg through the fallopian tube is a slow one, taking from five to seven days to complete. If the egg is unfertilized, it loses its capacity to develop within a few days. Any egg that arrives at the uterus unfertilized can never become so. For this reason the sperm cannot simply lie in wait within the uterus. To fertilize an egg successfully, a sperm must make its way far up the fallopian tube, a long passage that few survive.

Sperm are deposited within the vagina, a thin-walled muscular tube about 7 centimeters long that leads to the mouth of the uterus. Sperm entering the uterus swim up to and enter the fallopian tube. They swim upward against the current generated by the tube's contractions, which are carrying the ovum downward toward the uterus.

When a sperm succeeds in fertilizing an egg high in the fallopian tube, the fertilized egg—called a zygote—travels down the fallopian tube. When it reaches the uterus, it attaches itself to the endometrial lining and continues the long developmental journey that eventually leads to the birth of a child.

> **32.4**　In human females, hormones trigger the maturation of one or a few oocytes each 28 days. When mature, the egg cells travel down the fallopian tubes and, if fertilized during their journey, implant in the wall of the uterus and initiate embryonic development.

32.5 Hormones Coordinate the Reproductive Cycle

The female reproductive cycle, called a **menstrual cycle,** is composed of two distinct phases, the follicular phase, in which an egg reaches maturation and is ovulated, and the luteal phase, where the body continues to prepare for pregnancy. These phases are coordinated by a family of hormones. Hormones play many roles in human reproduction. Sexual development is initiated by hormones, released from the anterior pituitary and ovary, that coordinate simultaneous sexual development in many kinds of tissues. The production of gametes is another closely orchestrated process, involving a series of carefully timed developmental events. Successful fertilization initiates yet another developmental "program," in which the female body continues its preparation for the many changes of pregnancy.

Production of the sex hormones that coordinate all these processes is coordinated by the hypothalamus, which sends releasing hormones to the pituitary, directing it to produce particular sex hormones. Negative feedback, discussed in chapter 31, plays a key role in regulating these activities of the hypothalamus. When target organs receive a pituitary hormone, they begin to produce a hormone of their own, which circulates back to the hypothalamus, shutting down production of the pituitary hormone.

Triggering the Maturation of an Egg

The first, or **follicular, phase** of the reproductive cycle is when the egg develops within the ovary, and corresponds to days 0 through 14 in figure 32.18. This development is carefully regulated by hormones. The anterior pituitary, after receiving a chemical signal (GnRH) from the hypothalamus, starts the cycle by secreting **follicle-stimulating hormone (FSH**—the light blue curve in panel ❶), which binds to receptors on the surface of cells surrounding the egg (the oocyte and its surrounding mass of tissue is called a **follicle**) and triggers resumption of meiosis. Several follicles are stimulated to grow under FSH stimulation, but only one achieves full maturity ❷. FSH levels then fall, so just a few eggs are stimulated to grow.

The fall of FSH levels is achieved by a feedback command to the pituitary. FSH does not itself carry out this feedback—instead, the ovary sends another hormone as a messenger. FSH not only triggers final egg development, it also causes the ovary to start producing the female sex hormone **estrogen,** more technically known as *estradiol* (the dark green curve in panel ❸). Rising levels of estrogen in the bloodstream feed back to the hypothalamus, which responds to the rising estrogen by commanding the anterior pituitary to cut off the further production of FSH. This "shuts the door" on further FSH-induced egg development. The rise in estrogen levels signals the completion of the follicular phase of the reproductive cycle.

Preparing the Body for Fertilization

The second, or **luteal, phase** of the cycle (days 14 through 28) follows smoothly from the first. The hypothalamus responds

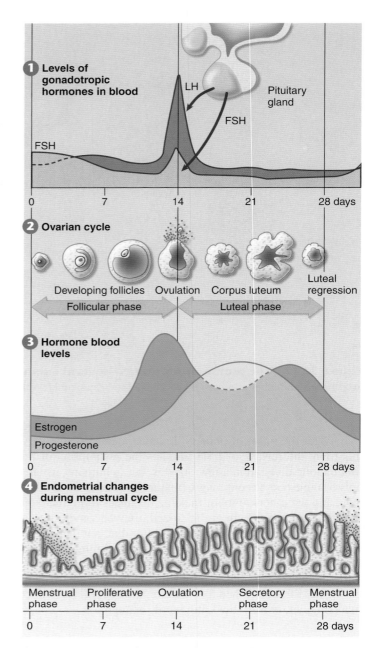

Figure 32.18 The human menstrual cycle.
Ovulation and the preparation of the uterine lining for implantation is controlled by a group of four hormones during the menstrual cycle.

to estrogen not only by shutting down the pituitary's FSH production but also by causing the anterior pituitary to begin secreting a second hormone, called **luteinizing hormone (LH).** LH (the dark blue curve in panel ❶) is the hormone that causes ovulation (seen in ❷). The peak in its secretion sends the egg on its journey toward fertilization. LH is carried in the bloodstream to the developing follicle, where it inhibits further estrogen production and causes the wall of the follicle to burst. The egg within the follicle is released into one of the fallopian tubes extending from the ovary to the uterus (see ❶ in figure 32.19).

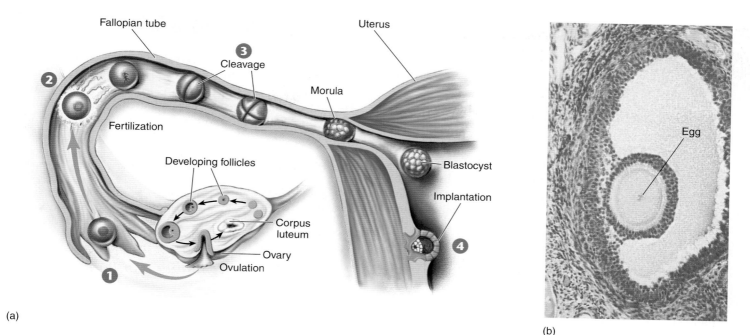

(a)

(b)

Figure 32.19 The journey of an ovum.

(a) Produced within a follicle and released at ovulation, an ovum is swept up into a fallopian tube ❶ and carried down by waves of contraction of the tube walls. Fertilization occurs within the tube ❷, by sperm journeying upward. Several mitotic divisions occur while the fertilized ovum undergoes cleavage and continues its journey down the fallopian tube ❸, becoming first a morula then a blastocyst. The blastocyst implants itself within the wall of the uterus ❹, where it continues its development. (b) A mature egg within an ovarian follicle. In each menstrual cycle, a few follicles are stimulated to grow under the influence of FSH, but usually only one achieves full maturity and ovulation.

After the egg's release and departure, LH directs the repair of the ruptured follicle that fills in and becomes yellowish. In this condition, it is called the **corpus luteum,** which is simply the Latin phrase for "yellow body." The corpus luteum soon begins to secrete a hormone, **progesterone,** which inhibits FSH (a backup for estrogen in preventing further ovulations). Progesterone (the light green curve in figure 32.18 ❸) completes the body's preparation of the uterus for fertilization including the thickening of the endometrium (in panel ❹). If fertilization does *not* occur soon after ovulation, however, production of progesterone slows and eventually ceases, marking the end of the luteal phase. The decreasing levels of progesterone cause the thickened layer of blood-rich tissue to be sloughed off, a process that results in the bleeding associated with menstruation. **Menstruation,** or "having a period," usually occurs about midway between successive ovulations (shown in figure 32.18 at 28 days), although its timing varies widely for individual females.

At the end of the luteal phase, neither estrogen nor progesterone is being produced. In their absence, the anterior pituitary can again initiate production of FSH, thus starting another reproductive cycle. Each cycle begins immediately after the preceding one ends. A cycle usually occurs every 28 days, or a little more frequently than once a month, although this varies in individual cases. The Latin word for "month" is *mens,* which is why the reproductive cycle is called the menstrual cycle, or monthly cycle.

If fertilization does occur high in the fallopian tube (❷ in figure 32.19), the zygote undergoes a series of cell divi-

sions called cleavage ❸, while traveling toward the uterus. At the blastocyst stage, it implants in the lining of the uterus ❹. The tiny embryo secretes human chorionic gonadotropin (hCG), an LH-like hormone, that maintains the corpus luteum and prevents subsequent menstruation. By maintaining the corpus luteum, hCG keeps the levels of estrogen and progesterone high, thereby preventing menstruation, which would terminate the pregnancy. Because hCG comes from the embryo and not from the mother, it is hCG that is tested in all pregnancy tests.

Two other hormones, both secreted by the pituitary, are important in the female reproductive system. For the first couple of days after childbirth, the mammary glands produce a fluid called colostrum, which contains protein and lactose but little fat. Then milk production is stimulated by the anterior pituitary hormone **prolactin,** usually by the third day after delivery. When the infant suckles at the breast, the posterior pituitary hormone **oxytocin** is released, initiating milk release. Earlier, in combination with chemicals released from the uterus, oxytocin initiates labor and delivery.

32.5 Humans and apes have menstrual cycles driven by cyclic patterns of hormone secretion and ovulation. The cycle is composed of two distinct phases, follicular and luteal, coordinated by a family of four sex hormones.

32.6 Embryonic Development

Cleavage: Setting the Stage for Development

Fertilization begins a carefully orchestrated series of developmental events. Table 32.1 traces the major stages of mammalian development, beginning with fertilization. You should follow down the table as the stages of development are discussed here.

The first major event in human embryonic development is the rapid division of the zygote into a larger and larger number of smaller and smaller cells, becoming first 2 cells, then 4, then 8, and so on. The first of these divisions occurs about 30 hours after union of the egg and the sperm, and the second, 30 hours later. During this period of division, called **cleavage,** the overall size does not increase from that of the zygote. The resulting tightly packed mass of about 32 cells is called a **morula,** and each individual cell in the morula is referred to as a **blastomere.** The cells of the morula continue to divide, each cell secreting a fluid into the center of the cell mass. Eventually, a hollow ball of 500 to 2,000 cells is formed. This is the **blastocyst,** which contains a fluid-filled cavity called the **blastocoel.** Within the ball is an *inner cell mass* concentrated at one pole that goes on to form the developing embryo. The outer sphere of cells, called the *trophoblast,* releases the hCG hormone, discussed earlier.

During cleavage, the morula journeys down the mother's fallopian tube. On about the sixth day, the blastocyst has formed and reaches the uterus; it attaches to the uterine lining, and penetrates into the tissue of the lining. The blastocyst now begins to grow rapidly, initiating the formation of the membranes that will later surround, protect, and nourish it. One of these membranes, the **amnion,** will enclose the developing embryo, whereas another, the **chorion,** which forms from the trophoblast, will interact with uterine tissue to form the **placenta,** which will nourish the growing embryo. The placenta connects the developing embryo to the blood supply of the mother. Fully 61 of the cells at the 64-celled stage develop into the trophoblast and only 3 into the embryo proper.

Gastrulation: The Onset of Developmental Change

Ten to 11 days after fertilization, certain groups of cells move inward from the surface of the cell mass in a carefully orchestrated migration called **gastrulation.** First, the lower cell layer of the blastocyst cell mass differentiates into **endoderm,** one of the three primary embryonic tissues, and the upper layer into **ectoderm.** Just after this differentiation, much of the **mesoderm** arises by the invagination of cells that move from the upper layer of the cell mass *inward,* along the edges of a furrow that appears at the embryo midline, the **primitive streak.**

During gastrulation, about half of the cells of the blastocyst cell mass move into the interior of the human embryo. This movement largely determines the future development of the embryo. By the end of gastrulation, distribution of cells into the three primary germ layers has been completed. The ectoderm is destined to form the epidermis and neural tissue. The mesoderm is destined to form the connective tissue, muscle, and vascular elements. The endoderm forms the lining of the gut and its derivative organs. The fate of all three primary germ layers are:

Ectoderm	Epidermis, central nervous system, sense organs, neural crest
Mesoderm	Skeleton, muscles, blood vessels, heart, gonads
Endoderm	Lining of digestive and respiratory tracts; liver, pancreas

Neurulation: Determination of Body Architecture

In the third week of embryonic development, the three primary cell types begin their development into the tissues and organs of the body. This stage in development is called **neurulation.**

The first characteristic vertebrate feature to form is the **notochord,** a flexible rod. Soon after gastrulation is complete, it forms from mesoderm tissue along the midline of the embryo, below its dorsal surface. After the notochord has been formed, the second characteristic vertebrate feature, the **neural tube,** forms from the region of the ectoderm that is located above the notochord and later differentiates into the spinal cord and brain. Just before the neural tube closes, two strips of cells break away and form the **neural crest.** These neural crest cells give rise to neural structures found in the vertebrate body.

While the neural tube is forming from ectoderm, the rest of the basic architecture of the human body is being rapidly determined by changes in the mesoderm. On either side of the developing notochord, segmented blocks of tissue form. Ultimately, these blocks, or **somites,** give rise to the muscles, vertebrae, and connective tissues. As development continues, more and more somites are formed. Within another strip of mesoderm that runs alongside the somites, many of the significant glands of the body, including the kidneys, adrenal glands, and gonads, develop. The remainder of the mesoderm layer moves out and around the inner endoderm layer of cells and eventually surrounds it entirely. As a result, the mesoderm forms two layers. The outer layer is associated with the body wall and the inner layer is associated with the gut. Between these two layers of mesoderm is the **coelom,** which becomes the body cavity of the adult.

By the end of the third week, over a dozen somites are evident, and the blood vessels and gut have begun to develop. At this point the embryo is about 2 millimeters (less than a tenth of an inch) long.

32.6 The vertebrate embryo develops in three stages: cleavage, a hollow ball of cells forms; gastrulation, cells move into the interior, forming the primary tissues; neurulation, organs begin to form.

TABLE 32.1   STAGES OF MAMMALIAN DEVELOPMENT

	Stage (age)	Description
Sperm Ovum	Fertilization (day 1)	The haploid male and female gametes fuse to form a diploid zygote.
Blastocyst Inner cell mass Trophoblast Blastocoel	Cleavage (days 2–10)	The zygote rapidly divides into many cells, with no overall increase in size. These divisions affect future development, because different cells receive different portions of the egg cytoplasm and, hence, different regulatory signals.
Amniotic cavity Ectoderm Endoderm	Gastrulation (days 11–15)	The cells of the embryo move, forming three primary germ layers: ectoderm and endoderm form first, followed by the formation of mesoderm.
Primitive streak Ectoderm Mesoderm Endoderm Formation of extraembryonic membranes		
Neural groove Notochord	Neurulation (days 16–25)	In all chordates, the first organ to form is the notochord; the second is the neural tube.
Neural crest Neural tube Notochord		During neurulation, the neural crest is produced as the neural tube is formed. The neural crest gives rise to several uniquely vertebrate structures such as sensory neurons, sympathetic neurons, Schwann cells, and other cell types.
	Organogenesis (days 26+)	Cells from the three primary cell layers combine in various ways to produce the organs of the body.

Fetal Development

The Fourth Week: Organ Formation

In the fourth week of pregnancy, the body organs begin to form, a process called **organogenesis** (figure 32.20a, the drawing helps identify the structures in the photo). The eyes form, and the heart begins a rhythmic beating and develops four chambers. At 70 beats per minute, the little heart is des-

tined to beat more than 2.5 billion times during a lifetime of about 70 years. More than 30 pairs of somites are visible by the end of the fourth week, and the arm and leg buds have begun to form. The embryo more than doubles in length during this week, reaching about 5 millimeters.

By the end of the fourth week, the developmental scenario is far advanced, although most women are not yet aware that they are pregnant. This is a crucial time in development because the proper course of events can be interrupted easily.

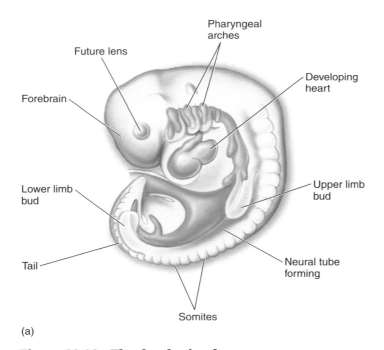

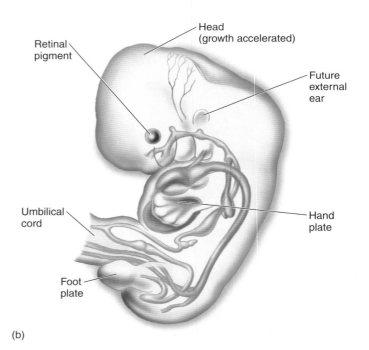

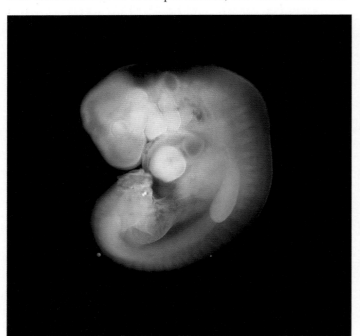

(a)

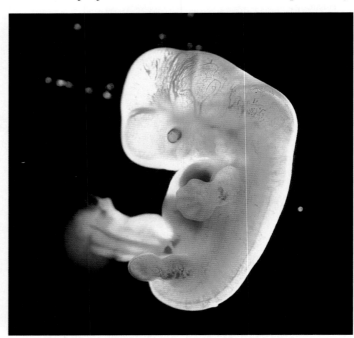

(b)

Figure 32.20 The developing human.

(a) Four weeks; (b) seven weeks; (c) three months; and (d) four months.

For example, alcohol use by pregnant women during the first months of pregnancy is one of the leading causes of birth defects, producing **fetal alcohol syndrome,** in which the baby is born with a deformed face and often severe mental retardation. One in 250 newborns in the United States is affected with fetal alcohol syndrome. Also, most spontaneous abortions (miscarriages) occur during this period.

The Second Month: The Embryo Takes Shape

During the second month of pregnancy, great changes in morphology occur as the embryo takes shape (figure 32.20b). The miniature limbs of the embryo assume their adult shapes. The arms, legs, knees, elbows, fingers, and toes can all be seen as well as a short, bony tail. The bones of the embryonic tail, an evolutionary reminder of our past, later fuse to form the coccyx,

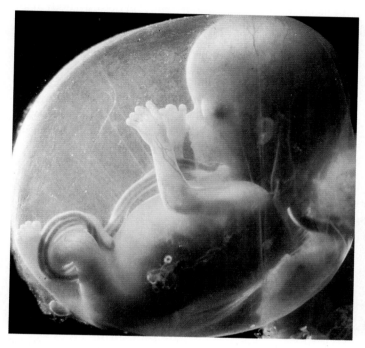

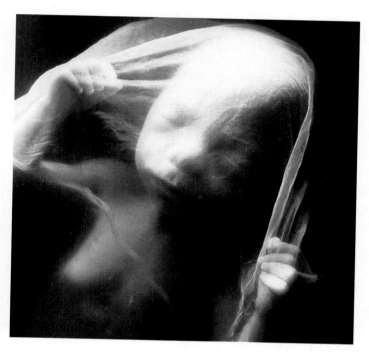

Development is essentially complete.

The developing human is now referred to as a fetus.

Facial expressions and primitive reflexes are carried out.

All of the major body organs have been established.

Arms and legs begin to move.

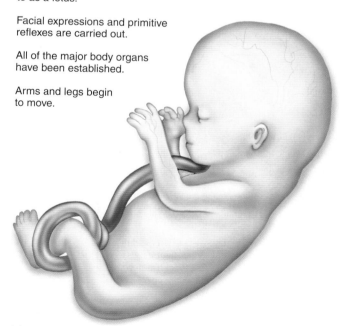

Bones actively enlarge.

Mother can feel baby kicking.

Following a period of rapid growth, the fetus is born.

Neurological growth continues after birth.

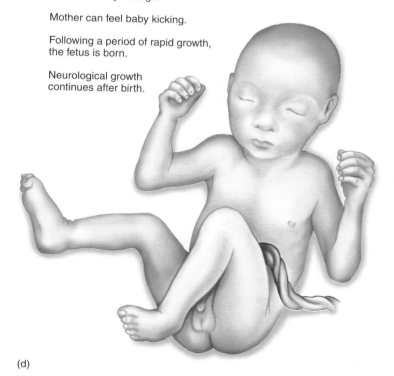

(c)

Figure 32.20 (continued)

(d)

or tailbone. Within the body cavity, the major internal organs are evident, including the liver and pancreas. By the end of the second month, the embryo has grown to about 25 millimeters in length—it is 1 inch long. It weighs perhaps a gram and is beginning to look distinctly human.

The Third Month: Completion of Development

Development of the embryo is essentially complete except for the lungs and brain. The lungs don't complete development until the third trimester, and the brain continues to develop even after birth. From this point on, the developing human is referred to as a **fetus** rather than an embryo. What remains is essentially growth. The nervous system and sense organs develop during the third month. The fetus begins to show facial expressions and carries out primitive reflexes such as the startle reflex and sucking. By the end of the third month, all of the major organs of the body have been established and the arms and legs begin to move (figure 32.20c).

The Second Trimester: The Fetus Grows in Earnest

The second trimester is a time of growth. In the fourth (figure 32.20d) and fifth months of pregnancy, the fetus grows to about 175 millimeters in length (almost 7 in long), with a body weight of about 225 grams. Bone formation occurs actively during the fourth month. During the fifth month, the head and body become covered with fine hair. This downy body hair, called **lanugo**, is another evolutionary relic and is lost later in development. By the end of the fourth month, the mother can feel the baby kicking; by the end of the fifth month, she can hear its rapid heartbeat with a stethoscope.

In the sixth month, growth accelerates. By the end of the sixth month, the baby is over 0.3 meters (1 ft) long and weighs 0.6 kilograms (about 1.5 lb)—and most of its prebirth growth is still to come. At this stage, the fetus cannot yet survive outside the uterus without special medical intervention.

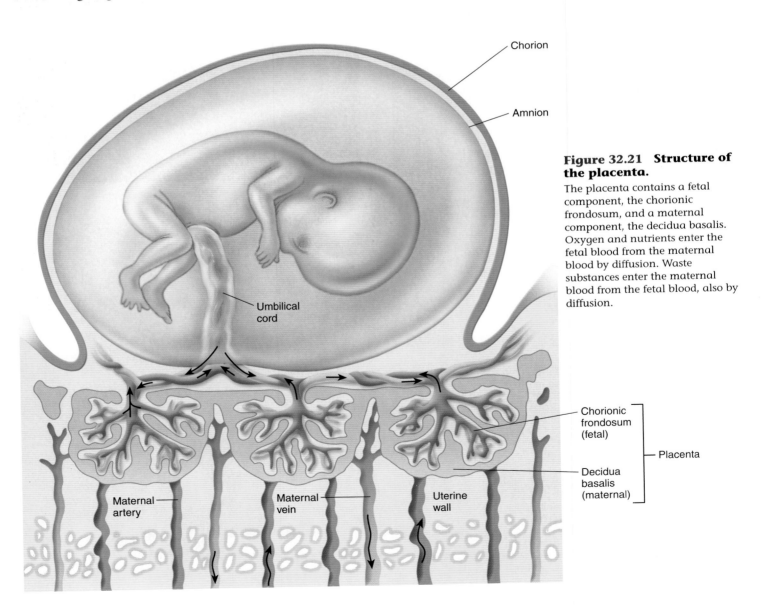

Figure 32.21   Structure of the placenta.

The placenta contains a fetal component, the chorionic frondosum, and a maternal component, the decidua basalis. Oxygen and nutrients enter the fetal blood from the maternal blood by diffusion. Waste substances enter the maternal blood from the fetal blood, also by diffusion.

Chorion

Amnion

Umbilical cord

Chorionic frondosum (fetal)

Decidua basalis (maternal)

Placenta

Maternal artery

Maternal vein

Uterine wall

The Third Trimester: The Pace of Growth Accelerates

The third trimester is a period of rapid growth. In the seventh, eighth, and ninth months of pregnancy, the weight of the fetus more than doubles. This increase in bulk is not the only kind of growth that occurs. Most of the major nerve tracts are formed within the brain during this period, as are new brain cells.

All of this growth is fueled by nutrients provided by the mother's bloodstream, passing into the fetal blood supply within the placenta. The placenta in figure 32.21 contains blood vessels that extend from the umbilical cord into tissues that line the uterus (the *decidua basalis* in the figure). The mother's blood bathes this tissue so that nutrients can pass from the mother's blood into the blood vessels that carry it back to the fetus without the two blood systems ever mixing blood. The undernourishment of the fetus by a malnourished mother can adversely affect this growth and result in severe retardation of the infant. Retardation resulting from fetal malnourishment is a severe problem in many underdeveloped countries, where poverty and hunger are common.

By the end of the third trimester, the neurological growth of the fetus is far from complete and, in fact, continues long after birth. But by this time the fetus is able to exist on its own. Why doesn't the fetus continue to develop within the uterus until its neurological development is complete? What's the rush to get out and be born? Because physical growth is continuing as well, and the fetus is about as large as it can get and still be delivered through the pelvis without damage to mother or child. As any woman who has had a baby can testify, it is a tight fit. Birth takes place as soon as the probability of survival is high.

Postnatal Development

Growth continues rapidly after birth. Babies typically double their birth weight within a few months. Different organs grow at different rates, however. The reason that adult body proportions are different from infant ones is that different parts of the body grow or cease growing at different times, a growth pattern called **allometric growth**. In figure 32.22 you can see that the lower part of the human jaw (on the right) grows at a faster rate than the rest of the skull, which results in the jaw extending further below the base of the skull. By comparison, the allometric growth of the chimpanzee jaw (on the left) is much more pronounced. At birth, the developing nervous system of humans is generating new nerve cells at an average rate of more than 250,000 per minute. Then, about six months after birth, this astonishing production of new neurons essentially ceases (recent research has shown new neurons continue to be produced into adulthood in a few small regions of the brain). Because both jaw and skull continue to grow at the same rate, the proportions of the head do not change after birth. That is why a young human fetus seems so incredibly adultlike. The fact that the human brain continues to grow significantly for the first few years of postnatal life means that adequate nutrition and a safe environment are particularly crucial during this period for the full development of a person's intellectual potential.

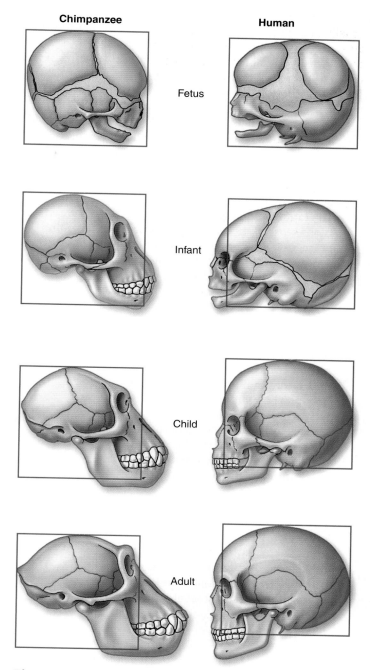

Figure 32.22 Allometric growth.
In young chimpanzees, the jaw grows at a faster rate than the rest of the head. As a result, the adult chimpanzee head shape differs greatly from its head shape as a newborn. In humans, the difference in growth between the jaw and the rest of the head is much smaller, and the adult head shape is similar to that of the newborn.

32.7 Most of the key events in fetal development occur early. Organs begin to form in the fourth week, and by the end of the second month the developing body looks distinctly human. The development of the embryo is essentially complete before the woman may even know she is pregnant. What remains in the second and third trimester is primarily growth.

The Scientific Process

Why You Age and Cancer Cells Don't

Death is an integral part of the life cycle, its inevitable end. All organisms die. However, while each of us knows we shall die someday, few of us can escape wishing we could delay the process. Some succeed. The oldest documented living person, Marie-Louise Febronie Meilleur of Ontario, Canada, reached the age of 117 years in 1997. The tantalizing possibility of long life that she represents is one reason why there is such interest in the aging process—if we knew enough about it, perhaps we could slow it. A wide variety of theories have been advanced to explain why we age. In recent years, scientists have come a long way toward unraveling one aspect of the puzzle.

The first clue was the discovery that cells appear to die on schedule, as if following a script. In a famous experiment carried out in 1961, geneticist Leonard Hayflick demonstrated that fibroblast cells growing in tissue culture will divide only a certain number of times. After about 50 population doublings, cell division stops, the cell cycle blocked just before DNA replication. If a cell sample is taken after 20 doublings and frozen, when thawed it resumes growth for 30 more doublings, then stops.

An explanation of the "Hayflick limit" was suggested in 1986 when cell biologist Howard Cooke first glimpsed an extra length of DNA at the ends of chromosomes. These so-called telomeric regions, about 5,000 nucleotides long, are each composed of several thousand repeats of the sequence TTAGGG. Cooke found the telomere region to be substantially shorter in body tissue chromosomes than in those of germ-line cells, eggs and sperm. He speculated that in body cells a portion of the telomere cap was lost by a chromosome during each cycle of DNA replication.

Cooke was right. The cell machinery that replicates the DNA of each chromosome sits on the last 100 units of DNA at the chromosome's tip, and so cannot copy that bit. As a consequence, each time the cell divides, its chromosomes get a little shorter. Eventually, after some 50 replication cycles, the telomeric cap is used up, and the cell line then enters senescence (old age), no longer able to proliferate.

How do our germ-line cells that produce sperm and egg avoid this trap, dividing continuously for decades? Scientists have recently learned that all human cells contain an enzyme, telomerase, which lengthens telomeres. This enzyme is active in germ-line cells, maintaining their chromosome telomeres at a constant length of 5,000 nucleotides. In body cells, by contrast, the telomerase gene is silent.

Research published in the last few years has confirmed Cooke's hypothesis, providing direct evidence of a causal relationship between telomeric shortening and cell aging. Using genetic engineering, teams of researchers from California and Texas transferred into human body cell cultures a DNA fragment that unleashes each cell's telomerase gene. The result was unequivocal. New telomeric caps were added to the chromosomes of the cells, and the cells with the artificially elongated telomeres did not senesce at the Hayflick limit, continuing to divide in a healthy and vigorous manner for more than 20 additional generations (see the graph).

This research confirms the hypothesis that loss of telomere DNA eventually restrains the ability of human cells to proliferate—and yet every human cell possesses a copy of the telomerase gene that, if expressed, would rebuild the telomere. Why do our cells accept aging, if they need not? The answer, it seems, is to avoid cancer. By limiting the number of cell divisions allotted to human cell lines, the body

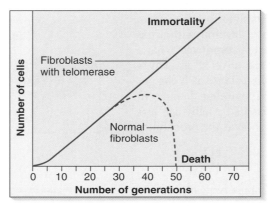

ensures that no cell can continue to divide indefinitely. Most adult cells are but a few cell generations from the Hayflick limit. Uncontrolled proliferation of a normal cell would soon grind to a halt. When scientists examine cancer cells, they commonly find that the telomerase genes in the cancer cells have been activated, and are maintaining telomeres at full length. Thus, the telomere shortening that occurs in normal cells is a tumor-suppressing mechanism and one of your body's key safeguards against cancer.

Aging, then, is at least partly a strategy to avoid the inevitable consequence of wear and tear to our genes, producing mutations that sooner or later lead to cancer. A powerful source of such mutations are free radicals, highly reactive fragments of atoms with unpaired electrons that shear through DNA like a shotgun blast. Free radicals are an unavoidable by-product of how our bodies use oxygen to metabolize food. Because free radicals are so destructive, every cell has numerous mechanisms to control and eliminate them. If all else fails to eliminate free radicals, the cell activates a second key safeguard against cancer. A special fail-safe gene called *p66* pulls the plug, causing the cell to commit suicide rather than permit free radicals to damage the DNA and produce cancer.

Interestingly, Italian researchers have found that mice with disabled *p66* genes lived longer than normal mice. Their cells were no longer being zapped by the hair-trigger *p66* self-destruct mechanism. Inhibitory drugs of the p66 class of proteins are known. If you are willing to accept the added risk of cancer, perhaps taking them would, like the Italian mice, let you live longer.

32.8 Contraception and Sexually Transmitted Diseases

Contraception

Not all couples want to initiate a pregnancy every time they have sex, yet sexual intercourse may be a necessary and important part of their emotional lives together. The solution to this dilemma is to find a way to avoid reproduction without avoiding sexual intercourse, an approach that is commonly called **birth control,** or contraception. Several different birth control methods are currently available (figure 32.23). These methods differ from one another in their effectiveness and in their acceptability to different couples.

Abstinence. The simplest and most reliable way to avoid pregnancy is not to have sex at all. Of all methods of birth control, this is the most certain—and the most limiting, because it denies a couple the emotional support of a sexual relationship. A variant of this approach is to avoid sex only on the two days preceding and following ovulation, because this is the period during which successful fertilization is likely to occur. The rest of the sexual cycle is considered relatively "safe" for intercourse. This approach, called the rhythm method, or **natural family planning** when other indicators are also monitored, is satisfactory in principle but difficult in application because ovulation is not easy to predict and may occur unexpectedly. Failure rate can be as high as 20% to 30%.

Prevention of Egg Maturation. Since about 1960, a widespread form of birth control in the United States has been the daily ingestion of hormones, or **birth control pills.** These pills contain estrogen and progesterone. These hormones shut down production of the pituitary hormones FSH and LH, fooling the body into acting as if ovulation had already occurred, when in fact it has not. The ovarian follicles do not ripen in the absence of FSH, and ovulation does not occur in the absence of LH. Other methods of delivering hormones that prevent egg maturation include medroxy progesterone (Depo-Provera), which is injected every one to three months, the seven-day birth control patch, which releases the hormones through the skin, and surgically implanted capsules that release hormones. Failure rates of these methods are less than 2%.

Prevention of Embryo Implantation. The insertion of a coil or other irregularly shaped object into the uterus is an effective means of birth control because the irritation in the uterus prevents the implantation of the descending embryo within the uterine wall. Such **intrauterine devices (IUDs)** are very effective, with a failure rate of less than 2%. Their high degree of effectiveness, like surgically implanted hormones, undoubtedly reflects their being "no-brainers"—once they are inserted, they can be forgotten.

A chemical means of preventing embryo implantation or ending a early pregnancy is RU486. This pill, taken

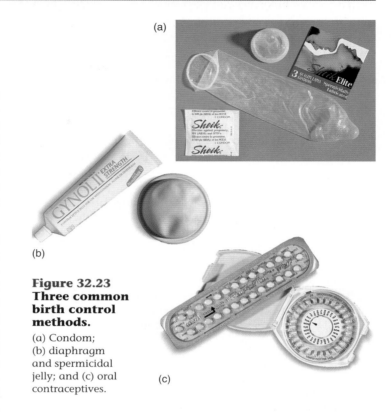

Figure 32.23 Three common birth control methods.
(a) Condom; (b) diaphragm and spermicidal jelly; and (c) oral contraceptives.

after sexual intercourse, blocks the action of progesterone, causing the endometrium to slough off. RU486 must be administered under a doctor's care because of potentially serious side effects.

Sperm Blockage. If sperm is not delivered to the uterus, fertilization cannot occur. One way to prevent the delivery of sperm is to encase the penis within a thin rubber bag, or **condom.** In principle, this method is easy to apply and foolproof, but in practice, it proves to be less effective than you might expect, with a failure rate of up to 15%. A second way to prevent the entry of sperm is to cover the cervix with a rubber dome called a **diaphragm,** inserted immediately before intercourse. Because the dimensions of individual cervices vary, diaphragms must be fitted by a physician. Failure rates average 20%.

Sperm Destruction. A third general approach to birth control is to remove or destroy the sperm after ejaculation. Sperm can be destroyed within the vagina with **spermicidal jellies, suppositories,** and **foams.** These require application immediately before intercourse. The failure rate varies widely, from 3% to 22%.

Sterilization. Sterilization is the surgical removal of portions of the tubes that transport the gametes from the testes or ovaries. A vasectomy in males involves removing a segment of the vas deferens from each testis (see figure 32.11). A tubal ligation in females involves removal of a section of each fallopian tube (see figure 32.16).

Sexually Transmitted Diseases

Sexually transmitted diseases (STDs) are diseases that spread from one person to another through sexual contact. AIDS, discussed in chapter 29, is a deadly viral STD. Other significant sexually transmitted diseases include:

Gonorrhea. Caused by the bacterium *Neisseria gonorrhoeae,* this disease is on the increase worldwide, but decreasing in the United States. The primary symptom is discharge from the penis or vagina. It can be treated with antibiotics. If left untreated in women, gonorrhea can cause pelvic inflammatory disease (PID), a condition in which the fallopian tubes become scarred and blocked. PID can eventually lead to sterility.

Chlamydia. Caused by the bacterium *Chlamydia trachomatis,* this disease is sometimes called the "silent STD" because women usually experience no symptoms until after the infection has become established. Like gonorrhea, chlamydia can cause PID in women if left untreated with antibiotics.

Syphilis. Caused by the bacterium *Treponema pallidum,* this disease is one of the most potentially devastating STDs. The disease progresses in four stages. In the first stage, a small, painless lesion called a chancre appears on the site of infection, usually on the penis or hidden in the vagina. In the second stage, a pink rash appears over the body. In the third stage, also called latent stage, symptoms often disappear for several years but the person is still contagious and can still infect a partner. The fourth stage is the most debilitating, with heart disease, mental deficiency, and nerve damage that may include loss of motor function or blindness.

Genital herpes. Caused by the herpes simplex virus type 2 (HSV-2), this disease is the most common STD in the United States. The virus causes red blisters on the penis or on the labia, vagina, or cervix that rupture and scab over. The lesions heal, but the virus travels to the dorsal root ganglion by way of the sensory neurons. They become dormant in the dorsal root ganglion, from which they can spread out to other parts of the body forming similar lesions. An infected person is always contagious when the lesions are present, but can also be contagious even when not exhibiting lesions. Medication can reduce the symptoms but cannot prevent the spread or recurrences of outbreaks.

This summary of STDs may give the impression that sexual activity is fraught with danger, and when practiced with unknown partners it is. In light of this, it is folly not to take precautions to avoid STDs. The surest way to avoid STDs is to avoid sex. If sexually active, it is important to know one's sexual partner well enough to discuss the possible presence of an STD. Condom use can also reduce transmission of most STDs. Anyone responsible enough to have sex should be responsible enough to protect themselves.

> **32.8** A variety of birth control methods are available, many of them quite effective. Sexually transmitted diseases are spread through sexual contact.

INQUIRY & ANALYSIS

Why Do STDs Vary in Frequency?

As a general rule, the incidence of a sexually transmitted disease is expected to increase with increasing frequencies of unprotected sexual contact. Intense publicity and education can lessen such dangerous behavior. However, the level of one STD sometimes rises while another falls. The simplest explanation of such a difference is that the two STDs are occurring in different populations, and one population has rising levels of sexual activity, while the other has falling levels. Another possible explanation is a change in the infectivity of one of the STDs. A third possible explanation can be seen by examining trends in the incidence in the United States of the STDs gonorrhea, chlamydia, and syphilis, reported in the graph to the right. As you can see, the trends for these three STDs have been quite different over the past two decades.

1. **Applying Concepts** What is the dependent variable?
2. **Making Inferences**
   a. *Gonorrhea:* What is the incidence in 1985? in 1995? Has the frequency declined or increased?

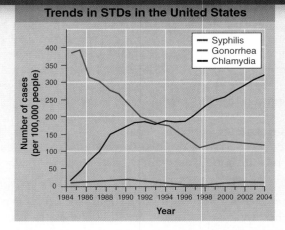

Trends in STDs in the United States

Are individuals aware they are infected when they transmit the STD?

b. *Chlamydia:* What is the incidence in 1985? in 1995? Has the frequency declined or increased? Are individuals aware they are infected when they transmit the STD?

3. **Drawing Conclusions** How might heightened public awareness explain why the trend in levels of gonorrhea differs from that of chlamydia?

Vertebrate Reproduction

32.1 Asexual and Sexual Reproduction

- Asexual reproduction through fission (**figure 32.1**) or budding is the primary means of reproduction among protists and some animals, but most animals reproduce sexually.

- Although sexual reproduction is most common in animals, parthenogenesis and hermaphroditism are variations seen in the animal kingdom. Parthenogenesis is reproduction where offspring are produced from unfertilized eggs. Hermaphroditism is a reproductive strategy where an individual has both testes and ovaries, producing both sperm and eggs. Some hermaphrodites can self-fertilize.

32.2 Evolution of Reproduction Among the Vertebrates

- Fertilization in most fish and amphibians occurs externally (**figures 32.5** and **32.6**), but internal fertilization occurs in most other vertebrates. Even with internal fertilization, development may occur inside or outside of the mother.

- Birds and reptiles lay watertight eggs (**figure 32.9**), whereas almost all mammals are viviparous, giving birth to live young.

The Human Reproductive System

32.3 Males

- Sperm is produced in the testes (**figure 32.11**). The testes contain a large number of tightly coiled tubes called seminiferous tubules, where sperm develop in a process called spermatogenesis (**figure 32.12**). As sperm develop, they move to the center of the tubules, where they pass into the epididymis. Once matured, they are stored in the vas deferens. During sexual intercourse, sperm are delivered through the penis (**figure 32.14**) into the female.

32.4 Females

- Female gametes, eggs or ova, develop in the ovary from oocytes (**figure 32.15**). A female is born with some 2 million oocytes, all arrested in meiosis. The hormone FSH initiates the resumption of meiosis in a few oocytes, but usually only one completes meiosis and becomes a mature egg each month.

- The egg ruptures from the ovary, called ovulation, and enters the fallopian tube (**figure 32.16**). Sperm deposited in the vagina travel up into the fallopian tube, where fertilization occurs.

32.5 Hormones Coordinate the Reproductive Cycle

- The follicular phase begins with the secretion of FSH, which triggers oocyte maturation and secretion of estrogen. The luteal phase begins with the secretion of LH, which causes ovulation and the formation of the corpus luteum (**figure 32.18**).

- The corpus luteum begins secreting progesterone, which acts in preparing the uterus for implantation of the zygote (**figures 32.18** and **32.19**).

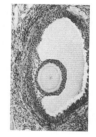

- If a zygote implants in the lining of the uterus (**figure 32.19**), estrogen and progesterone levels remain high due to the release of human chorionic gonadotropin (hCG) from the embryo. The uterus is maintained and no further egg maturation occurs.

- If fertilization does not occur, estrogen and progesterone levels drop and the endometrial lining of the uterus sloughs off (menstruation).

The Course of Development

32.6 Embryonic Development

- The vertebrate embryo develops in three stages. The first stage, called cleavage, involves hundreds of cell divisions, eventually producing a hollow ball of cells called a blastocyst (**table 32.1**).

- The second stage, called gastrulation, involves the orchestrated movement of cells, forming the three germ layers.

- The third stage is neurulation, where the notochord and neural tube form.

32.7 Fetal Development

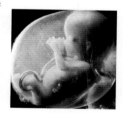

- Organs begin forming by the fourth week, and by the end of the second month the embryo looks distinctly human (**figure 32.20**). By the end of the third month, all major organs except the brain and lungs are developed.

- The second and third trimesters are periods of considerable growth.

Birth Control and Sexually Transmitted Diseases

32.8 Contraception and Sexually Transmitted Diseases

- Various birth control methods are available and work by preventing egg maturation, preventing embryo implantation, blocking or killing sperm, or sterilization.

- Sexually transmitted diseases are spread through sexual contact. AIDS is a deadly STD, but there are other STDs that may not be as fatal as AIDS but are quite destructive, especially if left untreated.

1. If the offspring are genetically similar but not identical to each other or the parent, then the organism reproduces through
   a. fission.
   b. sexual reproduction.
   c. budding.
   d. All of the above.

2. For terrestrial vertebrates, dehydration was a selection pressure that caused the evolution of
   a. parthenogenesis.
   b. external fertilization.
   c. budding.
   d. internal fertilization.

3. Dogs show
   a. viviparity.
   b. ovoviviparity.
   c. oviparity.
   d. parthenogenesis.

4. Temperature regulation of spermatogenesis in human males is controlled by the position of the
   a. seminiferous tubules.
   b. epididymis.
   c. vas deferens.
   d. scrotum.

5. Gametogenesis in human females requires the hormones
   a. estrogen and testosterone.
   b. FSH and LH.
   c. progesterone and testosterone.
   d. oxytocin and prolactin.

6. When pregnancy occurs, the endometrium is maintained by the
   a. embryo releasing hCG.
   b. decrease in levels of progesterone.
   c. hypothalamus releasing GnRH.
   d. increasing levels of FSH.

7. Initiation of labor and childbirth is controlled by the hormone
   a. estrogen.
   b. prolactin.
   c. oxytocin.
   d. progesterone.

8. A human embryo has formed the three germ layers from which all tissues form by the time
   a. the blastula forms.
   b. neurulation is complete.
   c. the blastocyst forms.
   d. gastrulation is complete.

9. In a developing human, the first tissues to begin forming are the
   a. skeletal.
   b. muscular.
   c. nervous.
   d. digestive.

10. Childbirth before the third trimester is dangerous for a human baby because the
    a. respiratory system is not fully formed.
    b. circulatory system is not fully formed.
    c. digestive system is not fully formed.
    d. excretory system is not fully formed.

1. **Figures 32.7** and **32.10b** These figures represent some of the more unusual methods that some organisms use to ensure the survival of their offspring. Briefly describe them.

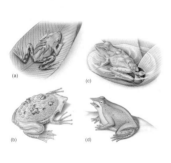

2. **Figure 32.17** Observe the shapes of the uteri shown in the figure. Speculate on the number of offspring the individuals are able to gestate at one time.

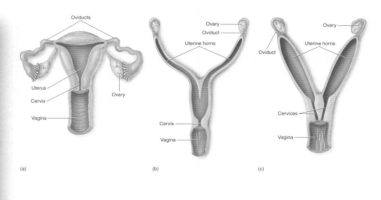

1. Compare the advantages and disadvantages of external and internal fertilization.

2. Speculate on why males produce so many sperm and females have, comparatively, so few eggs.

3. Explain some of the ways that a woman's use of drugs—alcohol, cocaine, prescription drugs, over-the-counter drugs—can affect the embryo or fetus that she is carrying.

4. Your friend, Paloma, tells you that the most reliable method of contraception and prevention of sexually transmitted disease is communication. Explain what she means.

33
Plant Form and Function

Of all the many kinds of plants, trees are the largest, rising high above the surrounding vegetation to capture the sun's rays. This hardwood forest, green with the leaves of spring, can be thought of as an enormous photosynthesis machine, each of its trees competing with its neighbors for light and soil nutrients, and putting the raw materials it captures to work producing the organic molecules necessary for growth and reproduction. Typical of plants, a tree captures light with the green pigment chlorophyll, which gives its leaves their characteristic color. A tree captures soil nutrients with its roots, which spread out in a fine network through the surrounding soil. Connecting the leaves of a tree with its roots is the stem, the massive, tall woody cylinder that makes up most of the mass of the tree. The stem of a tall tree is an engineering marvel, piping water and dissolved soil nutrients to the leaves many meters higher up in the air, and sending back down to the roots the carbohydrates produced by photosynthesis in the leaves. In this chapter, we will journey through the plant body, one of nature's most interesting creations.

Organization of a Vascular Plant

The similarities between a cactus, an orchid, and a pine tree may not at first be obvious. However, most plants possess the same fundamental architecture and the same three major groups of organs—roots, stems, and leaves. The organs of vascular plants have an outer covering of protective tissue and an inner matrix of tissue. Within the inner matrix is embedded vascular tissue, which conducts water, minerals, and food throughout the plant. The cells and tissues of vascular plants, and how the plant body carries out the functions of living, are the focus of this chapter. We discuss the fundamental differ-

ences between the roots, which are usually belowground, and the shoot, which is typically aboveground. We also examine the structural and functional relationships between them.

A vascular plant is organized along a vertical axis. The part belowground is called the **root,** and the part aboveground is called the **shoot** (although in some instances roots may extend above the ground, and some shoots can extend below it) (figure 33.1). Although roots and shoots differ in their basic structure, growth at the tips throughout the life of the individual is characteristic of both. The root penetrates the soil and absorbs water and various ions, which are crucial for plant nutrition. It also anchors the plant. The shoot consists of stem and leaves. The **stem** serves as a framework for the positioning of the **leaves,** where most photosynthesis takes place. The arrangement, size, and other characteristics of the leaves are critically important in the plant's production of food. Flowers, and ultimately fruits and seeds, are also formed on the shoot.

Meristems

Animals grow all over, but plants do not. As children grow into adults, their torsos grow at the same time their legs do. If, instead, children grew in only one place, with only their legs getting longer and longer, they would be growing in a way similar to the way plants grow.

Plants contain growth zones of unspecialized cells called **meristems.** Meristems are areas with actively dividing cells that result in plant growth but also continually replenish themselves. That is, one cell divides to give rise to two cells. One remains meristematic, while the other is free to differentiate and contribute to the plant body, resulting in plant growth. In this way, meristem cells function much like "stem cells" in animals (see chapter 14). Molecular genetic evidence supports the hypothesis that animal stem cells and plant meristem cells may also share some common pathways of gene expression.

In plants, **primary growth** is initiated at the tips by the **apical meristems,** regions of active cell division that occur at the tips of roots and shoots, colored lime green in figure 33.1. The growth of these meristems results primarily in the extension of the plant body. As it elongates, it forms what is known as the primary plant body, which is made up of the primary tissues.

Growth in thickness, **secondary growth,** involves the activity of the **lateral meristems,** which are cylinders of meristematic tissue, labeled in figure 33.1. The continued division of their cells results primarily in the thickening of the plant body. There are two kinds of lateral meristems: the **vascular cambium,** which gives rise to ultimately thick accumulations of secondary xylem and phloem, and the **cork cambium,** from which arise the outer layers of bark on both roots and shoots.

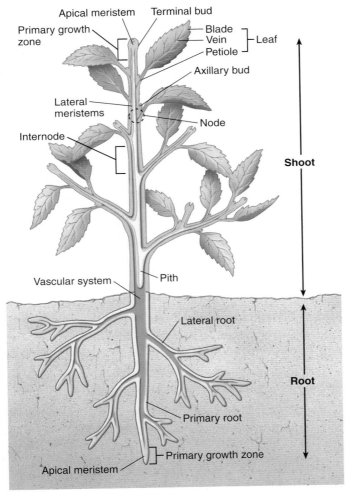

Figure 33.1   The body of a plant.
The body of this dicot plant consists of an aboveground portion called the shoot (stems and leaves) and a belowground portion called the root. Elongation of the plant, so-called primary growth, takes place when clusters of cells called the apical meristems *(lime green areas)* divide at the ends of the roots and the stems. Thickening of the plant, so-called secondary growth, takes place in the lateral meristems of the stem, allowing the plant to enlarge in girth like letting out a belt.

> **33.1   The body of a vascular plant is a continuous structure, a grouping of tubes connecting roots to leaves, with growth zones called meristems.**

Plant Tissue Types

The organs of a plant—the roots, stem, leaves, and in some cases, flowers and fruits—are composed of different combinations of tissues, just as your legs are composed of bone, muscle, and connective tissue. A tissue is a group of similar cells—cells that are specialized in the same way—organized into a structural and functional unit. Most plants have three major tissue types: (1) *ground tissue,* in which the vascular tissue is embedded; (2) *dermal tissue,* the outer protective covering of the plant; and (3) *vascular tissue,* which conducts water and dissolved minerals up the plant and conducts the products of photosynthesis throughout.

Each major tissue type is composed of distinctive kinds of cells, whose structures are related to the functions of the tissues in which they occur. For example, vascular tissue is composed of *xylem,* which conducts water and dissolved minerals, and *phloem,* which conducts carbohydrates (mostly sugars), which the plant uses as food.

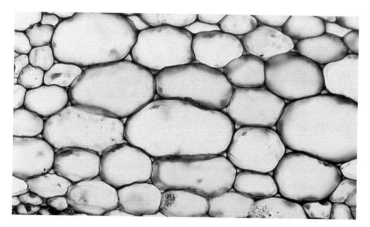

Figure 33.2   Parenchyma cells.
Cross section of parenchyma cells from grass. Only thin primary cell walls are seen in this living tissue.

Ground Tissue

Parenchyma cells are the least specialized and the most common of all plant cell types; they form masses in leaves, stems, and roots. Parenchyma cells, unlike some other cell types, are characteristically alive at maturity, with fully functional cytoplasm and a nucleus. They are the cells that carry out the basic functions of living including photosynthesis, cellular respiration, and food and water storage. The edible parts of most fruits and vegetables are composed of parenchyma cells. They are capable of cell division and are important in cell regeneration and wound healing. Most parenchyma cells have only thin cell walls, as seen in figure 33.2, called **primary cell walls,** which are mostly cellulose that is laid down while the cells are still growing.

Collenchyma cells, which are also living at maturity, form strands or continuous cylinders beneath the epidermis of stems or leaf stalks and along veins in leaves. They are usually elongated, with unevenly thickened primary walls, which are clearly visible in figure 33.3, and are their distinguishing feature. Strands of collenchyma provide much of the support for plant organs in which secondary growth has not occurred.

In contrast to parenchyma and collenchyma cells, **sclerenchyma cells** have tough, thick cell walls called **secondary cell walls;** they usually do not contain living cytoplasm when mature. The secondary cell wall is laid down inside of the primary cell wall after the cell has stopped growing and expanding in size. The secondary cell wall provides cells with strength and rigidity. There are two types of sclerenchyma: **fibers,** which are long, slender cells that usually form strands, and **sclereids,** which are variable in shape but often branched. Sclereids, the reddish-colored cells in figure 33.4, are sometimes called "stone cells." Clusters of sclereids form the gritty texture you feel in the flesh of pears. Both fibers and sclereids are thick-walled and strengthen the tissues in which they occur. Compare the thickness of the cell walls in the sclereid cells in figure 33.4 with the green-stained parenchyma cells that surround them.

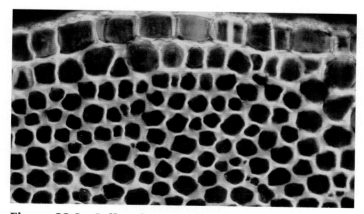

Figure 33.3   Collenchyma cells.
Cross section of collenchyma cells, with thickened side walls, from a young branch of elderberry (*Sambucus*). In some collenchyma cells, the thickened areas may occur at the corners of the cells or in others, as strands.

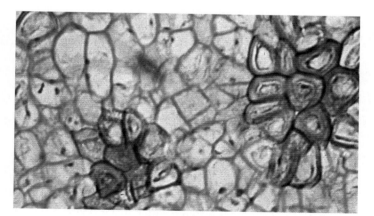

Figure 33.4   Sclerenchyma cells in sclereids.
Clusters of sclereids ("stone cells"), stained *red* in this preparation, in the pulp of a pear. These sclereid clusters give pears their gritty texture. The surrounding thin-walled cells, stained *green,* are parenchyma cells.

Dermal Tissue

All parts of the outer layer of a primary plant body are covered by flattened epidermal cells, which are often covered with a thick, waxy layer called the **cuticle.** These are the most abundant cells in the plant epidermis, or outer covering. They protect the plant and provide an effective barrier against water loss. One type of specialized cell that occurs among the epidermal cells is the guard cell.

Guard cells are the paired cells in figure 33.5a with an opening that lies between them called **stoma** (plural, **stomata**). Guard cells and stomata occur frequently in the epidermis of leaves and occasionally on other parts of the shoot, such as on stems or fruits. Oxygen, carbon dioxide, and water vapor pass across the epidermis almost exclusively through the stomata, which open and shut in response to such external factors as supply of moisture and light.

Trichomes are outgrowths of the epidermis that occur on the shoot, on the surfaces of stems and leaves. Trichomes vary greatly in form in different kinds of plants, from the rounded tip trichome in figure 33.5a to the pointed and globular-tipped ones in figure 33.5b. A "fuzzy" or "woolly" leaf is covered with trichomes, which when viewed under the microscope look like a thicket of fibers. Trichomes play an important role in regulating the heat and water balance of the leaf, much as the hairs of a fur coat provide insulation. Other trichomes are glandular, secreting sticky or toxic substances that may deter potential herbivores.

Other outgrowths of the epidermis occur belowground, on the surface of roots near their tips. Called **root hairs,** these tubular extensions of single epidermal cells keep the root in intimate contact with the particles of soil (see figure 33.21). Root hairs play an important role in the absorption of water and minerals from the soil, by increasing the surface area of the root.

Vascular Tissue

Vascular plants contain two kinds of conducting, or vascular, tissue: the xylem and the phloem. **Xylem** is the plant's principal water-conducting tissue, forming a continuous system that runs throughout the plant body. Within this system, water (and minerals dissolved in it) passes from the roots up through the shoot in an unbroken stream. When water reaches the leaves, much of it passes into the air as water vapor, through the stomata.

The two principal types of conducting cells in the xylem are **tracheids** and **vessel elements,** both of which have thick secondary walls that are laid down inside the primary cell wall, are elongated, and have no living cytoplasm (they are dead) at maturity. Tracheids are elongated cells that overlap at the ends as shown in figure 33.6a. In conducting elements composed of tracheids, water flows from tracheid to tracheid through openings called *pits* in the secondary walls. In contrast, vessel elements are elongated cells that line up end-on-end as shown in figure 33.6b, c. The end walls of vessel elements may be almost completely open or may have bars or strips of wall material perforated by pores through which water flows. A linked row of vessel elements forms a vessel. Primitive angiosperms and other vascular plants have only tracheids, but the majority of angiosperms have vessels. Vessels conduct water much more efficiently than do strands of tracheids.

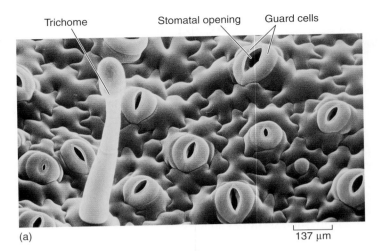

Trichome   Stomatal opening   Guard cells

(a)   137 μm

Figure 33.5 Guard cells and trichomes.
(a) Numerous stomata occur among the leaf epidermal cells of this tobacco (*Nicotiana tabacum*) leaf. A trichome is also visible. (b) Trichomes on a tomato plant (*Lycopersicon lycopersicum*).

(b)   280 μm

Phloem is the principal food-conducting tissue in vascular plants. Food conduction in phloem is carried out through two kinds of elongated cells: **sieve cells** and **sieve-tube members.** The cells differ in the extent of the perforations between the cells, with the sieve cells having smaller perforations between cells. Seedless vascular plants and gymnosperms have only sieve cells; most angiosperms have sieve-tube members. Clusters of pores known as *sieve areas* occur on both kinds of cells and connect the cytoplasms of adjoining sieve cells and sieve-tube members. Both cell types are living, but their nuclei are lost during maturation.

In sieve-tube members, some sieve areas have larger pores and are called *sieve plates*. Sieve-tube members occur end to end, as shown in the drawing and photo of figure 33.7, forming longitudinal series called **sieve tubes.** Specialized parenchyma cells known as **companion cells** occur regularly in association with sieve-tube members. The companion cells can be seen in figure 33.7 associated with the left side of the sieve-tube members. Companion cells apparently carry out some of the metabolic functions that are needed to maintain the associated sieve-tube member; their cytoplasms are connected to the sieve-tube members through openings called *plasmodesmata*.

33.2 Plants contain a variety of ground, dermal, and vascular tissues.

areas in figure 33.13) and the phloem (the light green area). The cylindrical form of the vascular cambium is completed by the differentiation of some of the parenchyma cells that lie between the bundles. Once established, the vascular cambium consists of elongated, somewhat flattened cells with large vacuoles. The cells that divide from the vascular cambium outwardly, toward the bark, become secondary phloem; those that divide from it inwardly become secondary xylem.

While the vascular cambium is becoming established, a second kind of lateral cambium, the cork cambium, develops in the stem's outer layers. The cork cambium usually consists of plates of dividing cells that move deeper and deeper into the stem as they divide. Outwardly, the cork cambium splits off densely packed **cork cells;** they contain a fatty substance and are nearly impermeable to water. Cork cells are dead at maturity. Inwardly, the cork cambium divides to produce a layer of parenchyma cells. The cork, the cork cambium that produces it, and this layer of parenchyma cells make up a layer called the **periderm** (shown by a bracket in figure 33.13), which is the plant's outer protective covering.

Cork covers the surfaces of mature stems or roots. The term **bark** refers to all of the tissues of a mature stem or root outside of the vascular cambium. Because the vascular cambium has the thinnest-walled cells that occur anywhere in a secondary plant body, it is the layer at which bark breaks away from the accumulated secondary xylem.

Wood is one of the most useful, economically important, and beautiful products obtained from plants. Anatomically, wood is accumulated secondary xylem (the light purple pie-shaped areas in figure 33.13). As the secondary xylem ages, its cells become infiltrated with gums and resins, and the wood may become darker. For this reason, the wood located nearer the central regions of a given trunk, called heartwood, can be darker and denser than the wood nearer the vascular cambium, called sapwood, which is still actively involved in water transport within the plant.

Because of the way it is accumulated, wood often displays rings. The rings that you see in the section of pine in figure 33.14 reflect the fact that the vascular cambium of trees divides more actively in the spring and summer, when water is plentiful and temperatures are suitable for growth, than in the fall and winter, when water is scarce and the weather is cold. As a result, layers of larger, thinner-walled cells formed during the growing season (the lighter rings) alternate with the smaller, darker layers of thick-walled cells formed during the rest of the year. New rings are laid down each year toward the outer edge of the stem. A count of such annual rings in a tree trunk can be used to calculate

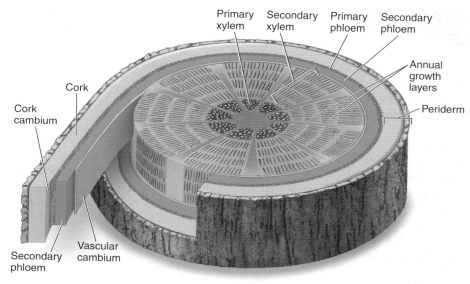

Figure 33.13 Vascular cambium and secondary growth.
The vascular cambium and cork cambium (lateral meristems) produce secondary tissues, causing the stem's girth to increase. Each year, a new layer of secondary tissue is laid down, forming rings in the wood.

Figure 33.14 Annual rings in a section of pine.
To test your understanding of how they form, answer this question: Are the broad inner rings older or younger than the narrow outer rings?

the tree's age, and the width of rings can reveal information about environmental factors. For example, the region of thinner rings could indicate a period of prolonged drought conditions that was followed by wetter years. Can you estimate the age of the tree shown in figure 33.14?

33.4 Stems, the aboveground framework of the plant body, grow both at their tips and in circumference.

Leaves are usually the most prominent shoot organs and are structurally diverse (figure 33.15). As outgrowths of the shoot apex, leaves are the major light-capturing organs of most plants. Most of the chloroplast-containing cells of a plant are within its leaves, and it is there where the bulk of photosynthesis occurs (see chapter 7). Exceptions to this are found in some plants, such as cacti, whose green stems have largely taken over the function of photosynthesis for the plant. Photosynthesis is conducted mainly by the "greener" parts of plants because they contain more chlorophyll, the most efficient photosynthetic pigment. In some plants, other pigments may also be present, giving the leaves a color other than green. Recall in chapter 7, we described accessory pigments that absorb light of other wavelengths. Thus, although coleus plants and red maple trees have leaves that are reddish in color, these leaves still contain chlorophyll and are the primary sites of photosynthetic activity in the plant.

The apical meristems of stems and roots are capable of growing indefinitely under appropriate conditions. Leaves, in contrast, grow by means of **marginal meristems,** which flank their thick central portions. These marginal meristems grow outward and ultimately form the **blade** (flattened portion) of the leaf, while the central portion becomes the midrib. Once a leaf is fully expanded, its marginal meristems cease to grow.

In addition to the flattened blade, most leaves have a slender stalk, the **petiole.** Two leaflike organs, the **stipules,** may flank the base of the petiole where it joins the stem (see figure 33.11). Veins, consisting of both xylem and phloem, run through the leaves. As mentioned in chapter 18, in most dicots the pattern is net, or reticulate, venation—as you can see in the red-stained African violet pictured in figure 33.16a; in many monocots, the veins are parallel, like the parallel veins that pass vertically up through the monocot leaf of the cabbage palmetto in figure 33.16b.

Leaf blades come in a variety of forms from oval to deeply lobed to having separate leaflets

(a)

(b)

(c)

(d)

(e)

(f)

Figure 33.15 Leaves.
Leaves are stunningly variable. (a) *Simple leaves* from a gray birch, in which there is a single blade. (b) A simple leaf, its margin lobed, from the vine maple. (c) A *pinnately compound* leaf of a black walnut tree, where leaflets occur in pairs along the central axis of the main vein. (d) *Palmately compound* leaves of a California buckeye, in which the leaflets radiate out from a single point. (e) The leaves of pine trees are tough and needlelike. (f) Many unusual types of modified leaves occur in different kinds of plants. For example, some plants produce floral leaves or bracts; the most conspicuous parts of this poinsettia flower are the red bracts, which are modified leaves that surround the small yellowish true flowers in the center.

(the blade being divided but attached to a single petiole like the black walnut leaf in figure 33.15c). In **simple leaves** (see figure 33.15a, b), such as those of birch or maple trees, there is a single blade, undivided, but some simple leaves may have teeth, indentations, or lobes, such as the leaves of maples and oaks. In **compound leaves,** such as those of ashes, box elders, and walnuts, the blade is divided into leaflets. If the leaflets are arranged in pairs along a common axis—the equivalent of the main central vein, or *midrib,* in simple leaves—the leaf is **pinnately compound,** such as in the black walnut (see figure 33.15c). If, however, the leaflets radiate out from a common point at the blade end of the petiole, the leaf is **palmately compound,** such as in buckeyes (figure 33.15d) and Virginia creepers. Leaves may be **alternately** arranged (alternate leaves usually spiral around a shoot, like the ivy shown to the right) or they may be in **op-** **posite** pairs (like the periwinkle). Less often, three or more leaves may be in a **whorl,** a circle of leaves at the same level at a node (like the sweet woodruff).

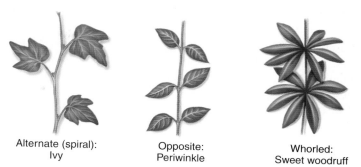

Alternate (spiral): Ivy

Opposite: Periwinkle

Whorled: Sweet woodruff

A typical leaf contains masses of parenchyma, called **mesophyll** ("middle leaf"), through which the vascular bundles, or veins, run. Beneath the upper epidermis of a leaf are one or more layers of closely packed, columnlike parenchyma cells called **palisade mesophyll** (the red-stained cells in the photo in figure 33.17). These cells contain more chloroplasts than other cells in the leaf and so are more capable of carrying out photosynthesis. This makes sense when you consider that the cells on the surface receive more sun. The rest of the leaf interior, except for the veins, consists of a tissue called **spongy mesophyll.** Between the spongy mesophyll cells are large intercellular spaces that function in gas exchange and particularly in the passage of carbon dioxide from the atmosphere to the mesophyll cells. You can see the spongy mesophyll in the photo of figure 33.17, but the air spaces that are the basis of this tissue's function might be easier to see in the drawing. These intercellular spaces are connected, directly or indirectly, with the stomata in the lower epidermis.

33.5 Leaves, the photosynthetic organs of the plant body, are varied in shape and arrangement.

Figure 33.16 Dicot and monocot leaves.

The leaves of dicots, such as the *red*-stained leaf in (a), have netted, or reticulate, veins; (b) those of monocots have parallel veins.

(a)   (b)

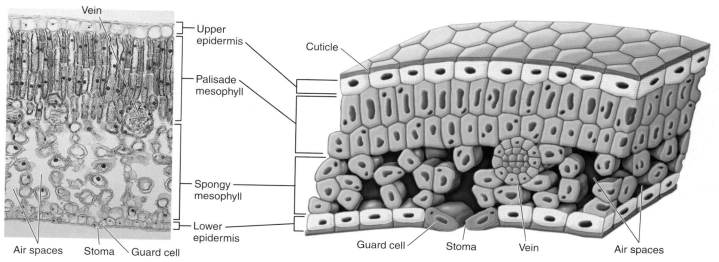

Vein

Upper epidermis

Palisade mesophyll

Spongy mesophyll

Lower epidermis

Air spaces   Stoma   Guard cell

Cuticle

Guard cell   Stoma   Vein   Air spaces

Figure 33.17 A leaf in cross section.

Cross section of a leaf, showing the arrangement of palisade and spongy mesophyll, a vascular bundle or vein, and the epidermis, with paired guard cells flanking the stoma.

33.6 Water Movement

Vascular plants have a conducting system, as humans do, for transporting fluids and nutrients from one part to another. Functionally, a plant is essentially a bundle of tubes with its base embedded in the ground. At the base of the tubes are roots, and at their tops are leaves. For a plant to function, two kinds of transport processes must occur: first, the carbohydrate molecules produced in the leaves by photosynthesis must be carried to all of the other living plant cells. To accomplish this, liquid, with these carbohydrate molecules dissolved in it, must move both up and down the tubes. Second, minerals and water in the ground must be taken up by the roots and ferried to the leaves and other plant cells. In this process, liquid moves up the tubes. Plants accomplish these two processes by using chains of specialized cells: those of the phloem transport photosynthetically produced carbohydrates up and down the plant, and those of the xylem carry water and minerals upward.

Cohesion-Adhesion-Tension Theory

Many of the leaves of a large tree may be more than 10 stories off the ground. How does a tree manage to raise water so high? Several factors are at work to move water up the height of a plant. The initial movement of water into the roots of a plant involves osmosis. Water moves into the cells of the root because the fluid in the xylem contains more solutes than the surroundings—recall from chapter 5 that water will move across a membrane from an area of lower solute concentration to an area of higher solute concentration. However, this force, called *root pressure,* is not by itself strong enough to "push" water up a plant's stem.

Capillary action adds a "pull." *Capillary action* results from the tiny electrical attractions of polar water molecules to surfaces that carry an electrical charge, a process called *adhesion.* In the laboratory, a column of water rises up a tube of glass because the attraction of the water molecules to the charged molecules on the interior surface of the glass tube "pull" the water up in the tube. In figure 33.18, which illustrates this process, why does the water travel higher up in the narrower tube? The water molecules are attracted to the glass molecules, and the water travels up farther in the narrower tube because the amount of surface area available for adhesion is greater than in the larger-diameter tube.

However, although capillary action can produce enough force to raise water a meter or two, it cannot account for the movement of water to the tops of tall trees. A second very strong "pull" accomplishes this, provided by transpiration. Opening up the tube and blowing air across its upper end demonstrates how transpiration draws water up a plant stem. The stream of relatively dry air causes water molecules at the water column's exposed top surface to evaporate from the tube. The water level in the tube does not fall, because as water molecules are drawn from the top, they are replenished by

Figure 33.18 Capillary action.
Capillary action causes the water within a narrow tube to rise above the surrounding water; the attraction of the water molecules to the glass surface, which draws water upwards, is stronger than the force of gravity, which tends to draw it down. The narrower the tube, the greater the surface area available for adhesion for a given volume of water, and the higher the water rises in the tube.

new water molecules pulled up from the bottom. This, in essence, is what happens in plants. The passage of air across leaf surfaces results in the loss of water by evaporation, creating a "pull" at the open upper end of the plant. New water molecules that enter the roots are pulled up the plant. Adhesion of water molecules to the walls of the narrow vessels in plants also helps to maintain water flow to the tops of plants.

A column of water in a tall tree does not collapse simply because of its weight because water molecules have an inherent strength that arises from their tendency to form hydrogen bonds with one another. These hydrogen bonds cause *cohesion* of the water molecules; in other words, a column of water resists separation. The beading of water droplets illustrates the property of cohesion. This resistance, called *tensile strength,* varies inversely with the diameter of the column; that is, the smaller the diameter of the column, the greater the tensile strength. Therefore, plants must have very narrow transporting vessels to take advantage of tensile strength.

How the combination of gravity, tensile strength, and cohesion affects water movement in plants is called the **cohesion-adhesion-tension theory.** It is important to note that the movement of water up through a plant is a passive process and requires no expenditure of energy on the part of the plant.

Transpiration

The process by which water leaves a plant is called **transpiration.** More than 90% of the water taken in by plant roots is ultimately lost to the atmosphere, almost all of it from the leaves. It passes out primarily through the stomata in the evaporation of water vapor, as you can see in panel 1 of figure 33.19. On its journey from the plant's interior to the outside, a molecule of water first passes into the pockets of air within the leaf by evaporating from the walls of the spongy mesophyll that lines the intercellular spaces. These intercellular spaces open to the outside of the leaf by way of the stomata (see figure 33.17). The water that evaporates from these surfaces of the spongy mesophyll cells is continuously replenished from the tips of the veinlets in the leaves. Because the strands of xylem conduct water within the plant in an unbroken stream all the way from the roots to the leaves, when a portion of the water vapor in the intercellular spaces passes out through the stomata, the

Transpiration

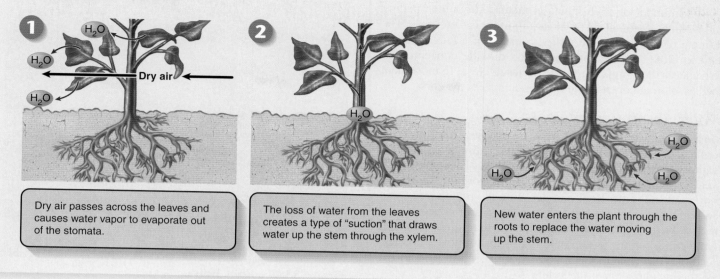

Figure 33.19 How transpiration works.
Water evaporating from the leaves through the stomata causes the movement of water upward in the xylem and the entrance of water through the roots.

supply of water vapor in these spaces is continually renewed from lower down in the column (panel 2) and ultimately from the roots (panel 3). Because the process of transpiration is dependent upon evaporation, factors that affect evaporation also affect transpiration. In addition to the movement of air across the stomata, mentioned earlier, humidity levels in the air will affect the rate of evaporation—high humidity reducing it and low humidity increasing it. Temperature will also affect the rate of evaporation—high temperatures increasing it and lower temperatures reducing it. This temperature effect is especially important because evaporation also acts to cool plant tissues.

Structural features such as the stomata, the cuticle, and the intercellular spaces in leaves have evolved in response to one or both of two contradictory requirements: minimizing the loss of water to the atmosphere, on the one hand, and admitting carbon dioxide, which is essential for photosynthesis, on the other. How plants resolve this problem is discussed next.

Regulation of Transpiration: Open and Closed Stomata

The only way plants can control water loss on a short-term basis is to close their stomata. Many plants can do this when subjected to water stress. But the stomata must be open at least part of the time so that carbon dioxide, which is necessary for photosynthesis, can enter the plant. In its pattern of opening or closing its stomata, a plant must respond to both the need to conserve water and the need to admit carbon dioxide.

The stomata open and close because of changes in the water pressure of their guard cells. Stomatal guard cells are long, sausage-shaped cells attached at their ends. These are the green cells in figure 33.20. The cellulose microfibrils of

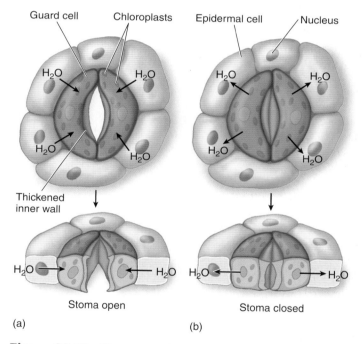

Figure 33.20 How guard cells regulate the opening and closing of stomata.
(a) When guard cells contain a high level of solutes, water enters the guard cells, causing them to swell and bow outward. This bowing opens the stoma. (b) When guard cells contain a low level of solutes, water leaves the guard cells, causing them to become flaccid. This flaccidity closes the stoma.

their cell wall wrap around the cell such that when the guard cells are **turgid** (plump and swollen with water), they expand in length, causing the cells to bow, thus opening the stomata as wide as possible, shown on the left side of the figure.

A number of environmental factors affect the opening and closing of stomata. The most important is water loss. The stomata of plants that are wilted because of a lack of water tend to close. An increase in carbon dioxide concentration also causes the stomata of most species to close. In most plant species, stomata open in the light and close in the dark.

Water Absorption by Roots

Most of the water absorbed by plants comes in through the root hairs, extensions of epidermal cells. These give a root the feathery appearance shown in figure 33.21. These root hairs greatly increase the surface area and therefore the absorptive powers of the roots. Root hairs are turgid—plump and swollen with water—because they contain a higher concentration of dissolved minerals and other solutes than does the water in the soil solution; water, therefore, tends to move into them steadily. Once inside the roots, water passes inward to the conducting elements of the xylem.

Water is not the only substance that enters the roots by passing into the cells of root hairs. Minerals also enter the root. Membranes of root hair cells contain a variety of ion transport channels that actively pump specific ions into the plant, even against large concentration gradients. These ions, many of which are plant nutrients, are then transported throughout the plant as a component of the water flowing through the xylem. In figure 33.22 the transport of water and minerals is indicated by the blue arrows. Carbohydrate transport is indicated by the red arrows and is discussed in section 33.7.

Figure 33.21 Root hairs.
Abundant fine root hairs can be seen in the back of the root apex of this germinating seedling of radish, *Raphanus sativus*.

33.6 Water is drawn up the plant stem from the roots by transpiration from the leaves.

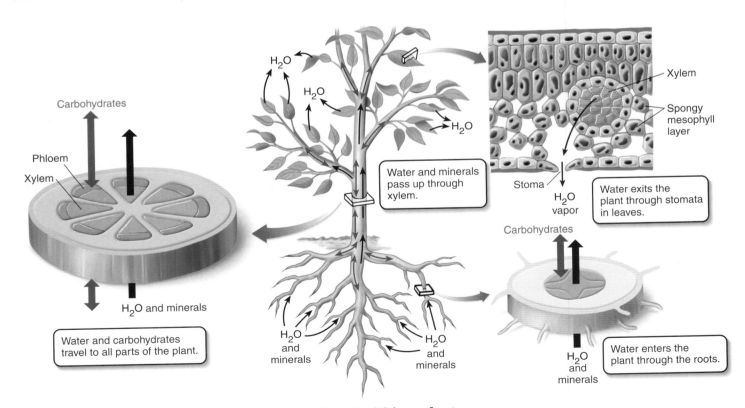

Figure 33.22 The flow of materials into, out of, and within a plant.
Water and minerals enter through the roots of a plant and are transported through the xylem to all parts of the plant body (*blue arrows*). Water leaves the plant through the stomata in the leaves. Carbohydrates synthesized in the leaves are circulated throughout the plant by the phloem (*red arrows*).

Most of the carbohydrates manufactured in plant leaves and other green parts are moved through the phloem to other parts of the plant. This process, known as **translocation,** makes suitable carbohydrate building blocks available at the plant's actively growing regions. The carbohydrates are concentrated in storage organs such as underground stems (potatoes), roots (carrots), and leaves (onions and cabbage), often in the form of starch. The starch is converted into transportable molecules, such as sucrose, and moved through the plant.

The pathway that sugars and other substances travel within the plant has been demonstrated precisely by using radioactive isotopes and aphids, a group of insects that suck the sap of plants. Aphids thrust their piercing mouthparts into the phloem cells of leaves and stems to obtain the abundant sugars there. When the aphids are cut off of the leaf, the liquid continues to flow from the detached mouthparts protruding from the plant tissue and is thus available in pure form for analysis. The liquid in the phloem contains 10% to 25% dissolved solid matter, almost all of which is usually sucrose. The harvesting of sap from maple trees uses a similar process. The starch, stored over the winter is converted into sap that is carried up throughout the plant in the xylem. A hole is drilled in the tree and the sugar-rich xylem fluid is drained from the tree using tubing and collected in buckets. The sap is then processed into maple syrup.

Using aphids to obtain the critical samples and radioactive tracers to mark them, researchers have learned that movement of substances in the phloem can be remarkably fast—rates of 50 to 100 centimeters per hour have been measured. This translocation movement is a passive process that does not require the expenditure of energy by the plant. The **mass flow** of materials transported in the phloem occurs because of water pressure, which develops as a result of osmosis. Figure 33.23 walks you through the process of translocation. Sucrose produced as a result of photosynthesis is actively "loaded" into the sieve tubes (or sieve cells) of the vascular bundles (panel 1). This loading increases the solute concentration of the sieve tubes, so water passes into them by osmosis (panel 2). An area where the sucrose is made is called a *source;* an area where sucrose is delivered from the sieve tubes is called a *sink.* Sinks include the roots and other regions of the plant that are not photosynthetic, such as young leaves and fruits. Water flowing into the phloem forces the sugary substance in the phloem to flow down the plant (panel 3). The sucrose is unloaded and stored in sink areas (panel 4). There the solute concentration of the sieve tubes is decreased as the sucrose is removed. As a result of these processes, water moves through the sieve tubes from the areas where sucrose is being added into those areas where it is being withdrawn, and the sucrose moves passively with the water. This is called the *pressure-flow hypothesis.*

33.7 Carbohydrates move through the plant by the passive osmotic process of translocation.

Translocation

1 Leaf cells · Sugar · Phloem · Xylem · Root cells

Sugar created in the leaves by photosynthesis ("source") enters the phloem by active transport.

2 H_2O

When the sugar concentration in the phloem increases, water is drawn into phloem cells from the xylem by osmosis.

3 Sugar

The addition of water from the xylem causes pressure to build up inside the phloem and pushes the sugar down.

4 Sugar

Sugar from the phloem enters the root cells ("sink") by active transport.

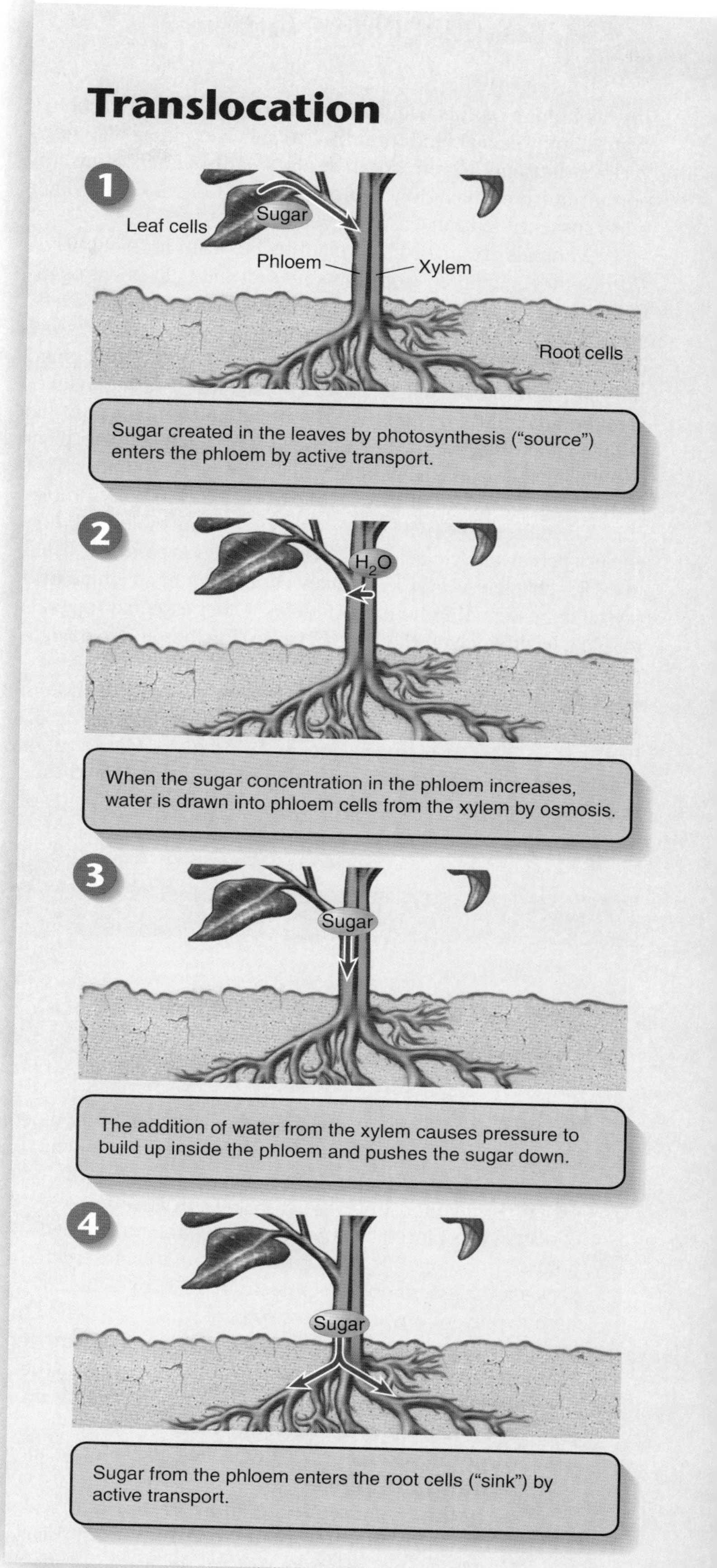

Figure 33.23   How translocation works.

33.8 Essential Plant Nutrients

Just as human beings need certain nutrients, such as carbohydrates, amino acids, and vitamins, to survive, plants also need various nutrients to remain alive and healthy. Lack of an important nutrient may slow a plant's growth or make the plant more susceptible to disease or even death.

Minerals are involved in plant metabolism in many ways. *Nitrogen* (N), acquired from the soil with the help of nitrogen-fixing bacteria, is an essential part of proteins and nucleic acids. *Potassium* (K) ions regulate the **turgor pressure** (the pressure within a cell that results from water moving into the cell) of guard cells and therefore the rate at which the plant loses water and takes in carbon dioxide. *Calcium* (Ca) is an essential component of the middle lamellae, the structural elements laid down between plant cell walls, and it also helps to maintain the physical integrity of membranes. *Magnesium* (Mg) is a part of the chlorophyll molecule. The presence of *phosphorus* (P) in many key biological molecules such as nucleic acids and ATP has been explored in detail in earlier chapters. *Sulfur* (S) is a key component of an amino acid (cysteine), essential in building proteins. Other essential minerals for plant health include chlorine (Cl), iron (Fe), boron (B), manganese (Mn), zinc (Zn), copper (Cu), and molybdenum (Mo).

Most plants acquire minerals from the soil, although some carnivorous plants are able to use other organisms directly as sources of nitrogen, just as animals do. Carnivorous plants lure and trap insects and other small animals and then digest their prey with enzymes secreted from various kinds of

Figure 33.24 A carnivorous plant.
Venus's-flytrap, *Dionaea muscipula*, inhabits low boggy ground in North and South Carolina. The Venus's-flytrap leaf on the right has snapped together, imprisoning a fly.

glands. The Venus's-flytrap (*Dionaea muscipula*) has sensitive hairs on each side of each leaf, which, when touched by the fly in figure 33.24, triggers the two halves of the leaf to snap together. The unfortunate trapped fly is digested and its minerals are absorbed by the plant.

> **33.8** Plants require ample supplies of nitrogen, phosphorus, and potassium, and smaller amounts of many other nutrients.

INQUIRY & ANALYSIS

Does Water Move Up a Tree Through Phloem or Xylem?

The stem of a tree has two vessel systems, xylem and phloem, either of which could provide the plumbing through which water moves up a tree trunk or other stem. An elegant experiment demonstrates which vessel carries water. A section of a stem was placed in water containing the radioactive potassium isotope ^{42}K. Wax paper was carefully inserted between the xylem and the phloem in a 23-cm section of stem to prevent any lateral transport of water. After enough time had elapsed to allow water movement up the stem, the 23-cm section was removed, cut into six segments, and the amounts of ^{42}K measured in each segment. The results are presented in the graph on the right.

1. **Applying Concepts** What is the dependent variable?
2. **Making Inferences**
   a. In the 23-cm section, is more ^{42}K found in xylem or phloem? What might you conclude from this?
   b. Above and below the 23-cm section, is more ^{42}K found in xylem or phloem? How would you account for this?

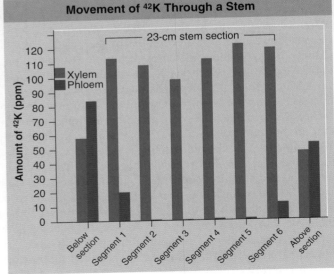

Movement of ^{42}K Through a Stem

c. Within the 23-cm section, the phloem segments at the ends exhibits far more ^{42}K than interior segments. What best accounts for this?
3. **Drawing Conclusions** Does water move up a stem through phloem or xylem? Explain.

Structure and Function of Plant Tissues

33.1 Organization of a Vascular Plant

- Most plants possess the same three major groups of organs: roots, stems, and leaves, although they may appear very different in different types of plants (**figure 33.1**). Vascular tissue extends throughout the plant, connecting roots, stems, and leaves.

- Growth occurs in regions called meristems. The tips of the roots and shoots contain apical meristem, which is the site of primary growth, extending the plant body in vertical directions. Extending the thickness or girth of the plant, called secondary growth, occurs at the lateral meristem, which are cylinders of meristematic tissue.

33.2 Plant Tissue Types

- Plants contain different types of tissue. A tissue is a group of similar cells that functions as a unit. There are three main types of tissues in plants: ground tissue, dermal tissue, and vascular tissue.

- Ground tissue makes up the main body of the plant and contains several different cell types. Parenchyma cells form the masses of tissue in leaves, stems, and roots. They carry out functions such as photosynthesis, food, and water storage. 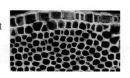 The edible parts of fruits and vegetables are primarily parenchyma cells (**figure 33.2**). Collenchyma cells form strands or cylinders that provide support (**figure 33.3**). Sclerenchyma cells have thick, secondary cell walls. They provide strength and rigidity to the plant body. They are organized into long fibers or branched structures called sclereids or "stone cells" (**figure 33.4**). Mature sclerenchyma cells are dead, while parenchyma and collenchyma cells are living when mature.

- Dermal tissue makes up the outer layer of the plant body, which consists of epidermal cells covered with a waxy layer called the cuticle (roots have epidermal cells but no cuticle). This outer layer protects the plant and the cuticle provides a barrier to water loss. Guard cells are specialized epidermal cells that provide openings for gas exchange. Trichomes and root hairs are extensions of epidermal cells (**figure 33.5**).

- Vascular tissue is composed of xylem and phloem. Xylem contains water-conducting cells, tracheids and vessel elements, which are dead at maturity (**figure 33.6**). Phloem contains food-conduction cells, sieve cells and sieve-tube members, both of which are living but lack nuclei when mature. Companion cells associated with the phloem cells carry out metabolic functions needed to maintain the sieve-tube members (**figure 33.7**).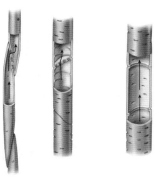

The Plant Body

33.3 Roots

- Roots are organs adapted to absorb water and minerals from the soil (**figure 33.8**). A waxy Casparian strip keeps water from passing between endodermal cells and forces the water to pass through the cells into the xylem cells (**figure 33.9**).

33.4 Stems

- The stem serves as a framework for positioning the leaves. Primary growth occurs at the apical meristem, laying down the stem tissue behind the meristem. Leaves grow out of the stems at node areas (**figure 33.11**). Secondary growth occurs at the lateral meristems with the differentiation of vascular cambium, which gives rise to xylem and phloem and cork cambium, which gives rise to the layers of cork inside the bark (**figures 33.12–33.14**).

33.5 Leaves

- Leaves are the primary site for photosynthesis. They vary in size, shape, and arrangement (**figures 33.15–33.17**).

Plant Transport and Nutrition

33.6 Water Movement

- Water enters the plant through the roots by osmosis. Root pressure and capillary action cause the water to pass up farther into the tissues (**figure 33.18**). However, for water to travel up the length of the stem, it requires a stronger force, the combination of cohesion and adhesion, known as the cohesion-adhesion-tension theory. Transpiration, the  evaporation of water vapor from the leaves, creates the "pull" that raises the water through the plant (**figures 33.19** and **33.20**). Minerals follow the flow of water into the plant (**figure 33.22**).

33.7 Carbohydrate Transport

- Carbohydrates produced in the leaves through photosynthesis travel throughout the plant in phloem tissue. The process, called translocation, involves osmotic movement of water into the phloem cells, forcing the sugars to "sinks" for carbohydrate storage (**figure 33.23**).

33.8 Essential Plant Nutrients

- Plants require minerals for metabolic functions that they obtain from the soil. However, some carnivorous plants obtain nitrogen directly from animals that are digested by plant enzymes (**figure 33.24**).

1. Growth in vascular plants is regulated and coordinated by
   a. photosynthetic tissue.
   c. meristematic tissue.
   b. root tissue.
   d. leaf epidermal tissue.
2. In vascular plants, phloem tissue
   a. transports water.
   b. transports carbohydrates.
   c. transports minerals.
   d. supports the plant.
3. In vascular plant leaves, gases enter and leave the plant through pores called
   a. stomata.
   c. chloroplasts.
   b. guard cells.
   d. trichomes.
4. In roots, growth of lateral branches begins
   a. on the root epidermis.
   c. at the ground meristem.
   b. on the root hairs.
   d. at the pericycle.
5. In stems, the tissue responsible for secondary growth is the
   a. collenchyma.
   b. pith.
   c. cambium.
   d. cortex.

6. One difference between monocot and dicot plant stems is the
   a. absence of buds in monocots.
   b. organization of vascular tissue.
   c. presence of guard cells.
   d. absence of stomata.
7. In leaves, gas exchange takes place in the
   a. cuticle.
   c. spongy mesophyll.
   b. palisade mesophyll.
   d. guard cells.
8. Which of the following is *not* a process that directly assists in water movement from the roots to the leaves?
   a. photosynthesis
   c. capillary action
   b. root pressure
   d. transpiration
9. The passive process of moving carbohydrates throughout a plant is called
   a. transpiration.
   c. translation.
   b. translocation.
   d. evaporation.
10. What important nutrient does the Venus's-flytrap acquire from its insect dinner?
    a. potassium
    c. magnesium
    b. calcium
    d. nitrogen

1. **Figure 33.8** What is the purpose of the root cap covering the apical meristem of the root?

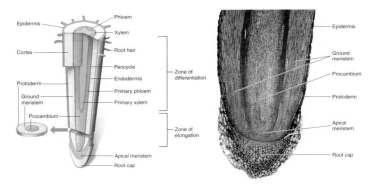

2. **Figure 33.14** Besides wet and dry years, and the age of the tree, what else might the study of tree rings tell us?

3. **Figure 33.20** If you went on vacation for several days and left your house plants in a warm, stuffy apartment, would the stomata look like the one on the left or the one on the right? Explain.

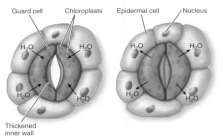

1. Why do land plants need sclerenchyma cells?
2. Some plants in cold climates lose their leaves in the winter. In desert climates some plants, such as palo verde and ocotillo, may lose some or all of their leaves in the hot, dry summer to minimize water loss. These plants have green stems. Why?

3. Your friend, Alex, just returned from a family trip to northern Michigan, where he visited a maple tree farm where they made maple syrup. On the maple trees, they made just one small hole (or two small holes on larger trees) and hung a bucket beneath to catch the sap. Why, he asks you, wouldn't they just make a cut completely around the tree on the diagonal and collect much more sap, much faster? Why isn't this feasible?

34
Plant Reproduction and Growth

S eeds are one of the cleverest adaptations of plants. Carried by wind or animals, they are capable of transporting the next generation to distant locations and so ensuring the plant an opportunity to occupy any available habitats. A seed is a protected package of genetic information, an embryonic individual kept in a dormant state by a variety of mechanisms such as a watertight covering that keeps the seed's interior free of water. When the seed is deposited in suitable soil, the watertight covering splits open and the embryo begins to grow, a process called germination. These seeds of a soybean plant are germinating, with leaves thrusting upward and roots downward toward the soil. For many seeds, moisture and moderate temperatures are sufficient to trigger germination. For some seeds, however, more extreme cues are required. The seeds of many species of pine tree, for example, will not germinate unless exposed to extreme temperatures of the sort experienced in forest fires; periodic forest fires provide openings for sunlight to reach the seedlings, and abundant nutrients enter the soil from the tissues of fire-killed trees.

34.1 Angiosperm Reproduction

Although reproduction varies greatly among the members of the plant kingdom, we focus in this chapter on reproduction among flowering plants. While the evolution of their unique sexual reproductive features, flowers and fruits, have contributed to their success, angiosperms also reproduce asexually.

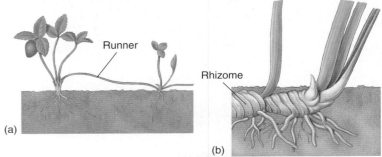

Figure 34.1 Vegetative reproduction.
(a) Runners are slender stems that grow along the ground, sending out roots and shoots at the nodes. (b) Rhizomes are underground horizontal stems that give rise to new shoots. (c) Small plants arise from notches along the leaves of the house plant *Kalanchoë daigremontiana.*

Asexual Reproduction

In **asexual reproduction,** an individual inherits all of its chromosomes from a single parent and is, therefore, genetically identical to that parent. Asexual reproduction produces a "clone" of the parent.

In a stable environment, asexual reproduction may prove more advantageous than sexual reproduction because it allows individuals to reproduce with a lower investment of energy than sexual reproduction. A common type of asexual reproduction called *vegetative reproduction* results when new individuals are simply cloned from parts of the parent.

The forms of vegetative reproduction in plants are many and varied:

Runners. Some plants reproduce by means of runners—long, slender stems that grow along the surface of the soil. The strawberry plant shown in figure 34.1*a* reproduces by runners. Notice that at node regions on the stem, adventitious roots form, extending into the soil. Leaves and flowers form, and a new shoot is sent out, continuing the runner.

Rhizomes. Rhizomes are underground horizontal stems that create a network underground. You'll see in figure 34.1*b* that, like in runners, nodes give rise to new flowering shoots. The noxious character of many weeds results from this type of growth pattern, but so do grasses and many garden plants such as irises. Other specialized stems, called tubers, function in food storage and reproduction. White potatoes are specialized underground stems that store food, and the "eyes" give rise to new plants.

Suckers. The roots of some plants produce "suckers," or sprouts, which give rise to new plants, such as found in cherry, apple, raspberry, and blackberry plants. When the root of a dandelion is broken, which may occur if one tries to pull it from the ground, each root fragment may give rise to a new plant.

Adventitious Plantlets. In a few species, even the leaves are reproductive. One example is the house plant *Kalanchoë daigremontiana,* familiar to many people as the "maternity plant," or "mother of thousands." The common names of this plant are based on the fact that numerous plantlets arise from meristematic tissue located in notches along the leaves. You can see the numerous little plantlets in figure 34.1*c*. The maternity plant is ordinarily propagated by means of these small

plants, which, when they mature, drop to the soil and take root.

Sexual Reproduction

Plant sexual life cycles are characterized by an alternation of generations, in which a diploid sporophyte generation gives rise to a haploid gametophyte generation. In angiosperms, the developing gametophyte generation is completely enclosed within the tissues of the parent sporophyte (see figure 18.20). The male gametophytes are **pollen grains,** and they develop from *microspores.* The female gametophyte is the **embryo sac,** which develops from a *megaspore.* Pollen grains and the embryo sac both are produced in separate, specialized structures of the angiosperm flower.

Like animals, angiosperms have separate structures for producing male and female gametes, but the reproductive organs of angiosperms are different from those of animals in two ways. First, in angiosperms, both male and female structures usually, but not always, occur together in the same individual flower. Second, angiosperm reproductive structures are not permanent parts of the adult individual. Angiosperm flowers and reproductive organs develop seasonally; these flowering seasons correspond to times of the year most favorable for pollination.

Structure of the Flower. Flowers contain male and female parts. The male parts, called *stamens,* are the long filament structures you see in the cutaway flower in figure 34.2. The female part, called the *carpel,* is the vase-shaped structure in the figure. Often flowers contain both stamens and carpels, but there are some exceptions. In various species of flowering plants, for example, willows and some mulberries, flowers containing only male or only female parts, called *imperfect*

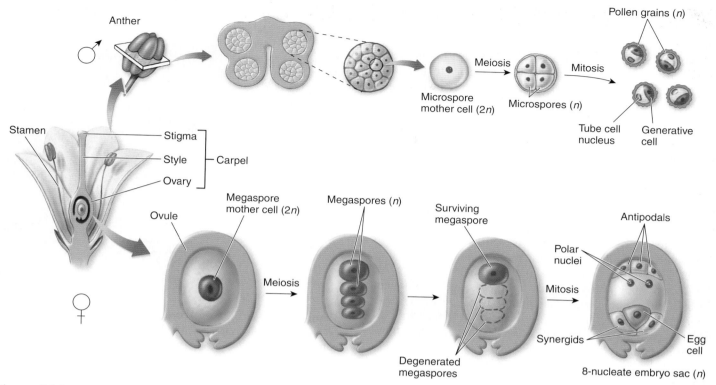

Figure 34.2 Formation of pollen and egg.
Diploid microspore mother cells are housed in the anther and divide by meiosis to form four haploid (*n*) microspores. Each microspore develops by mitosis into a pollen grain with a generative cell and a tube cell nucleus. The generative cell will later divide to form two sperm cells. Within the ovule, one diploid megaspore mother cell divides by meiosis to produce four haploid megaspores. Usually only one megaspore survives, and the other three degenerate. The surviving megaspore divides by mitosis to produce an embryo sac with eight nuclei. Upon fertilization, the egg cell becomes the embryo and the polar nuclei become the endosperm.

flowers, occur on separate plants. Plants that contain imperfect flowers that produce only ovules or only pollen are called *dioecious,* from the Greek words for "two houses." In other plants, there are separate male and female flowers, but they occur on the same plant. These plants are called *monoecious,* meaning "one house." In monoecious plants, the male and female flowers may mature at different times, which keeps the plant from pollinating itself. Even if, as is usually the case, functional stamens and carpels are both present in each flower of a particular plant, these organs may reach maturity at different times, which keeps the plant from pollinating itself.

Pollen Formation. If you were to cut an anther in half, as shown in the upper portion of figure 34.2, you would see four pollen sacs, each sac containing a collection of microspores. Pollen grains develop from these microspores. Each pollen sac contains specialized chambers in which the microspore mother cells are enclosed and protected. Each microspore mother cell undergoes meiosis to form four haploid microspores. Subsequently, mitotic divisions form pollen grains that contain a generative cell (the purple cells in the pollen grain) and a tube cell nucleus. The tube cell nucleus forms the pollen tube; the generative cell will later divide to form two sperm cells.

Pollination. Pollination is the process by which pollen is transferred from the anther to the stigma (the top of the carpel). The pollen may be carried to the flower by wind or by animals, or it may originate within the individual flower itself. When pollen from a flower's anther pollinates the same flower's stigma, the process is called *self-pollination,* which can lead to *self-fertilization.* For some plants, self-pollination and self-fertilization occur because self-pollination eliminates the need for animal pollinators and maintains beneficial phenotypes in stable environments. However, other plants are adapted to *outcrossing,* the crossing of two different plants. As described earlier, the presence of only male or female flowers on a plant requires outcrossing, as does the different timing of appearance of the male and female parts on a flower. Even when a flower's stamen and stigma mature at the same time, some plants exhibit **self-incompatibility.** Self-incompatibility results when the pollen and stigma recognize each other as being genetically related, and respond by blocking the fertilization of the flower.

In many angiosperms, the pollen grains are carried from flower to flower by insects and other animals that visit the flowers for food or other rewards or are deceived into doing so because the flower's characteristics suggest such rewards.

A liquid called **nectar,** which is rich in sugar as well as amino acids and other substances, is often the reward sought by animals. Successful pollination depends on the plants attracting insects and other animals regularly enough that the pollen is carried from one flower of that particular species to another.

The relationship between such animals, known as *pollinators,* and flowering plants has been important to the evolution of both groups, a process called **coevolution.** By using insects to transfer pollen, the flowering plants can disperse their gametes on a regular and more or less controlled basis, despite their being anchored to the ground. The more attractive the plant is to the pollinator, the more frequently the plant will be visited. Therefore, any changes in the phenotype of the plant that result in more visits by pollinators offer a selective advantage. This has resulted in the evolution of a wide variety of angiosperm species. Furthermore, animals that could obtain food from the flowers became more abundant and diverse as their food supply increased and diversified.

For pollination by animals to be effective, a particular insect or other animal must visit plant individuals of the same species. A flower's color and form have been shaped by evolution to promote such specialization. Yellow and blue flowers are particularly attractive to bees (figure 34.3*a*), whereas red flowers attract birds but are not particularly noticed by most insects. Some flowers have very long floral tubes with the nectar produced deep within them; only the long, slender beaks of hummingbirds or the long, coiled proboscis of moths or butterflies can reach such nectar supplies. You can see the long proboscis in the butterfly in figure 34.3*b* reaching down inside a flower.

In certain angiosperms and all gymnosperms, pollen is blown about by the wind and reaches the stigmas passively. For such a system to operate efficiently, the individuals of a given plant species must grow where there is ample wind and grow relatively close together because wind does not carry pollen very far or very precisely, compared to transport by insects or other animals. Because gymnosperms, such as spruces or pines, grow in dense stands, wind pollination is very effective. Wind-pollinated angiosperms, such as birches, grasses, and ragweed, also tend to grow in dense stands. The flowers of wind-pollinated angiosperms are usually small, greenish, and odorless, and their petals are either reduced in size or absent altogether. They typically produce large quantities of pollen.

Egg Formation. Eggs develop in the **ovules** of the angiosperm flower, which form the base of the carpel (the bulge in the vase-shaped structure in the bottom portion of figure 34.2). Within each ovule is a megaspore mother cell (the large brown cell). Each megaspore mother cell undergoes meiosis to produce four haploid megaspores (the column of four cells). In most plants, only one of these megaspores, however, survives; the rest are absorbed by the ovule. The lone remaining megaspore undergoes repeated mitotic divisions to produce eight haploid nuclei, which are enclosed within a structure called an *embryo sac.* Within the embryo sac, the eight nuclei are arranged in precise positions. One nucleus (the green egg cell) is located near the opening of the embryo sac. Two nuclei are

(a)

(b)

Figure 34.3   Insect pollination.
(a) Bees are usually attracted to yellow flowers. (b) This alfalfa butterfly (*Colias eurytheme*) has a long proboscis that allows it to feed on nectar deep in the flower.

located in a single cell above the egg cell and are called polar nuclei. A nucleus resides in each of the two cells that flank the egg cell, which are called synergids; and the other three nuclei are located in cells at the top of the embryo sac, opposite the egg cell, which are called antipodals.

Fertilization. Once a pollen grain has been spread by wind, an animal, or self-pollination, it adheres to the sticky, sugary substance that covers the stigma and begins to grow a **pollen tube,** which pierces the style. The pollen tube, nourished by the sugary substance, grows until it reaches the ovule in the ovary.

Figure 34.4 traces the steps from fertilization through seed formation. When the pollen tube reaches the entry to the embryo sac in the ovule ❶, the tip of the tube bursts and releases the two sperm cells that form from the generative cell. In ❷ you can see that one of the sperm cells fertilizes the egg cell, forming a zygote. The other sperm cell fuses with the two polar nuclei located at the center of the embryo sac, forming the triploid ($3n$) primary endosperm nucleus. This process of fertilization in angiosperms in which two sperm cells are used is called **double fertilization.** The primary endosperm nucleus eventually develops into the endosperm, shown beginning to form in ❹, which nourishes the embryo.

> **34.1 Reproduction in angiosperms involves asexual and sexual reproduction. In sexual reproduction, pollen is transferred to the female stigma. Double fertilization leads to the development of an embryo and endosperm.**

34.2 Seeds

The entire series of events that occurs between fertilization and maturity is called *development*. During development, cells become progressively more specialized, or differentiated. Continuing with figure 34.4, look at the lower series of panels and you can see that the first stage in the development of a plant is active cell division to form an organized mass of cells, the embryo ❻. In angiosperms, the differentiation of cell types within the embryo begins almost immediately after fertilization. By the fifth day, the principal tissue systems can be detected within the embryo mass, and within another day, the root and shoot apical meristems can be detected, as shown in ❽.

Early in the development of an angiosperm embryo, a profoundly significant event occurs: the embryo simply stops developing and becomes dormant as a result of drying. In many plants, embryo development is arrested at the point shown in ❽, soon after apical meristems and the seed leaves,

or **cotyledons,** are differentiated. The integuments—the outermost covering of the ovule—develop into a relatively impermeable seed coat, which encloses the dormant embryo within the seed, together with a source of stored food.

Once the seed coat fully develops around the embryo, most of the embryo's metabolic activities cease; a mature seed contains only about 10% water. Under these conditions, the seed and the young plant within it are very stable.

Germination, or the resumption of metabolic activities that leads to the growth of a mature plant, cannot take place until water and oxygen reach the embryo, a process that sometimes involves cracking the seed. Seeds of some plants have been known to remain viable for hundreds, and in some cases thousands, of years. The seed will germinate when conditions are favorable for the plant's survival.

> **34.2 A seed contains a dormant embryo and substantial food reserves, encased within a tough drought-resistant coat.**

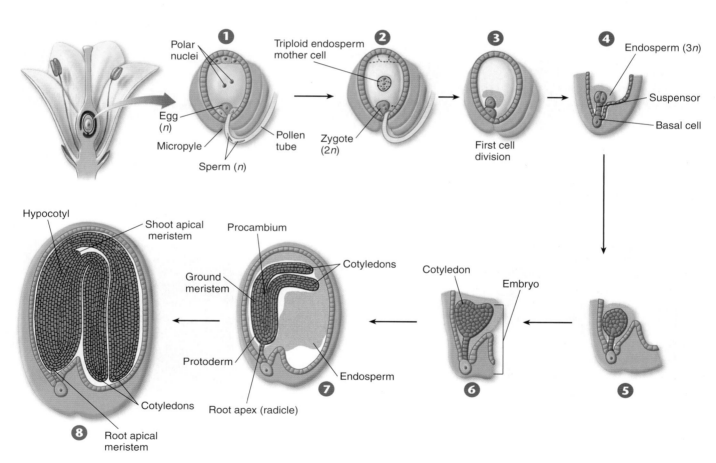

Figure 34.4   Development in an angiosperm embryo.
After the zygote forms, the first cell division is asymmetric ❸. After another division, the basal cell, the one nearest the opening through which the pollen tube entered, undergoes a series of divisions and forms a narrow column of cells called the suspensor ❹. The other three cells continue to divide and form a mass of cells arranged in layers ❺. By about the fifth day of cell division, the principal tissue systems of the developing plant can be detected within this mass ❼.

34.3 Fruit

During seed formation, the flower ovary begins to develop into fruit. The evolution of flowers was key to the success and diversification of the angiosperms. But of equal importance to angiosperm success has been the evolution of these fruits in response to their modes of dispersal. Fruits form in many ways and exhibit a wide array of modes of specialization.

Three layers of ovary wall can have distinct fates and account for the diversity of fruit types, from fleshy to dry and hard. There are three main kinds of fleshy fruits: berries, drupes, and pomes. In *berries*—such as grapes, tomatoes (figure 34.5*a*), and peppers—which are typically many-seeded, the inner layers of the ovary wall are fleshy. In *drupes*—such as peaches (figure 34.5*b*), olives, plums, and cherries—the inner layer of the fruit is stony and adheres tightly to a single seed. In *pomes*—such as apples (figure 34.5*c*) and pears—the fleshy portion of the fruit forms from the portion of the flower that is embedded in the receptacle (the swollen end of the flower stem that holds the petals and sepals). The inner layer of the ovary is a tough, leathery membrane that encloses the seeds.

Fruits that have fleshy coverings, often black, bright blue, or red (as in figure 34.5*d*), are normally dispersed by birds and other vertebrates. By feeding on these fruits, the animals carry seeds from place to place before excreting the seeds as solid waste. The seeds, not harmed by the animal digestive system, thus are transferred from one suitable habitat to another. Other fruits that are dispersed by wind, or by attaching themselves to the fur of mammals or the feathers of birds, are called dry fruits because they lack the fleshy tissue of edible fruits, with their ovaries forming hard layers rather than fleshy tissue. Dry fruits can have structures that aid in their dispersal, as seen in the plumed dandelion in figure 34.5*e* or the spiny cocklebur in figure 34.5*f*, which catches onto fur (or socks or pants!) and is carried to new habitats. Still other fruits, such as those of mangroves, coconuts, and certain other plants that characteristically occur on or near beaches or swamps, are spread from place to place by water.

> **34.3** Fruits are specialized to achieve widespread dispersal by wind, by water, by attachment to animals, or, in the case of fleshy fruits, by being eaten.

(a) Berries (b) Drupes (c) Pomes

(d) Eaten by animals (e) Dispersed by wind (f) Dispersed by attaching to animals

Figure 34.5 Types of fruits and common modes of dispersion.
(a) Tomatoes are a type of fleshy fruit called berries that have multiple seeds. (b) Peaches are a type of fleshy fruit called drupes that contain a single large seed. (c) Apples are a type of fleshy fruit that contains multiple seeds. (d) The bright red berries of this honeysuckle, *Lonicera*, are highly attractive to birds. Birds may carry the berry seeds either internally or stuck to their feet for great distances. (e) The seeds of this dandelion, *Taraxacum officinale*, are enclosed in a dry fruit with a "parachute" structure that aids its dispersal by wind. (f) The spiny fruits of this cocklebur, *Xanthium strumarium*, adhere readily to any passing animal.

34.4 Germination

What happens to a seed when it encounters conditions suitable for its germination? First, it absorbs water. Seed tissues are so dry at the start of germination that the seed takes up water with great force, after which metabolism resumes. Initially, the metabolism may be anaerobic, but when the seed coat ruptures, aerobic metabolism takes over. At this point, oxygen must be available to the developing embryo because plants, which drown for the same reason people do, require oxygen for active growth (see chapter 8). Few plants produce seeds that germinate successfully underwater, although some, such as rice, have evolved a tolerance of anaerobic conditions and can initially respire anaerobically. Fig-ure 34.6 shows the development of a dicot (on the left) and monocot (on the right) from germination through early stages. The first stage in both cases is the emergence of the roots. Following that, in dicots, the cotyledons emerge from underground along with the stem. The cotyledons eventually wither and the first leaves begin the process of photosynthesis. In monocots, the cotyledon doesn't emerge from underground; instead a structure called the *coleoptile* (a sheath wrapped around the emerging shoot) pushes through to the surface where the first leaves emerge and begin photosynthesis.

34.4 Germination is the resumption of a seed's growth and reproduction, triggered by water.

Figure 34.6 Development of angiosperms.

Dicot development in a soybean. The first structure to emerge is the embryonic root followed by the two cotyledons of the dicot. The cotyledons are pulled up through the soil along with the hypocotyl (the stem below the cotyledons). The cotyledons are the seed leaves that provide nutrients to the growing plant. As other leaves develop, they provide nutrients through photosynthesis, and the cotyledons shrivel and fall off the stem. Flowers develop in buds at the nodes.
Monocot development in corn. The first structure to emerge is the radicle or primary root. Monocots have one cotyledon, which does not emerge from underground. The coleoptile is a tubular sheath; it encloses and protects the shoot and leaves as they push their way up through the soil.

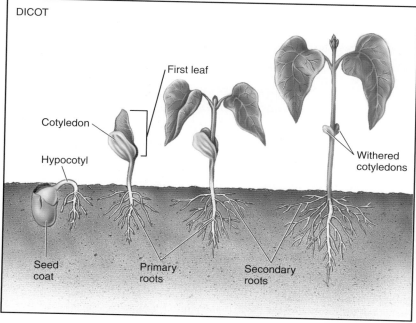

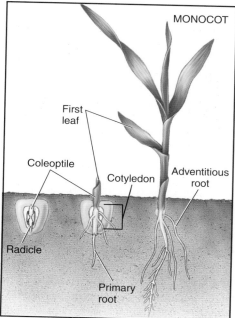

34.5 Plant Hormones

After a seed germinates, the pattern of growth and differentiation that was established in the embryo is repeated indefinitely until the plant dies. But differentiation in plants, unlike that in animals, is largely reversible. Botanists first demonstrated in the 1950s that individual differentiated cells isolated from mature individuals could give rise to entire individuals. F. C. Steward's experiment is shown in figure 34.7 where he was able to induce isolated bits of phloem tissue taken from carrots to form new plants, plants that were normal in appearance and fully fertile. Regeneration of entire plants from differentiated tissue has since been carried out in many plants, including cotton, tomatoes, and cherries. These experiments clearly demonstrate that the original differentiated phloem tissue still contains cells that retain all of the genetic potential needed for the differentiation of entire plants. No information is lost during plant tissue differentiation in these cells, and no irreversible steps are taken.

Once a seed has germinated, the plant's further development depends on the activities of the meristematic tissues, which interact with the environment through hormones, which we will discuss next. The shoot and root apical meristems give rise to all of the other cells of the adult plant. Differentiation, or the formation of specialized tissues, occurs in five stages in plants and shown in figure 34.8. The establishment of the shoot and root apical meristems occurs at stage 2; after that point, the tissues become more and more differentiated.

The tissue regeneration experiments of Steward and many others have led to the general conclusion that some nucleated cells in differentiated plant tissue are capable of

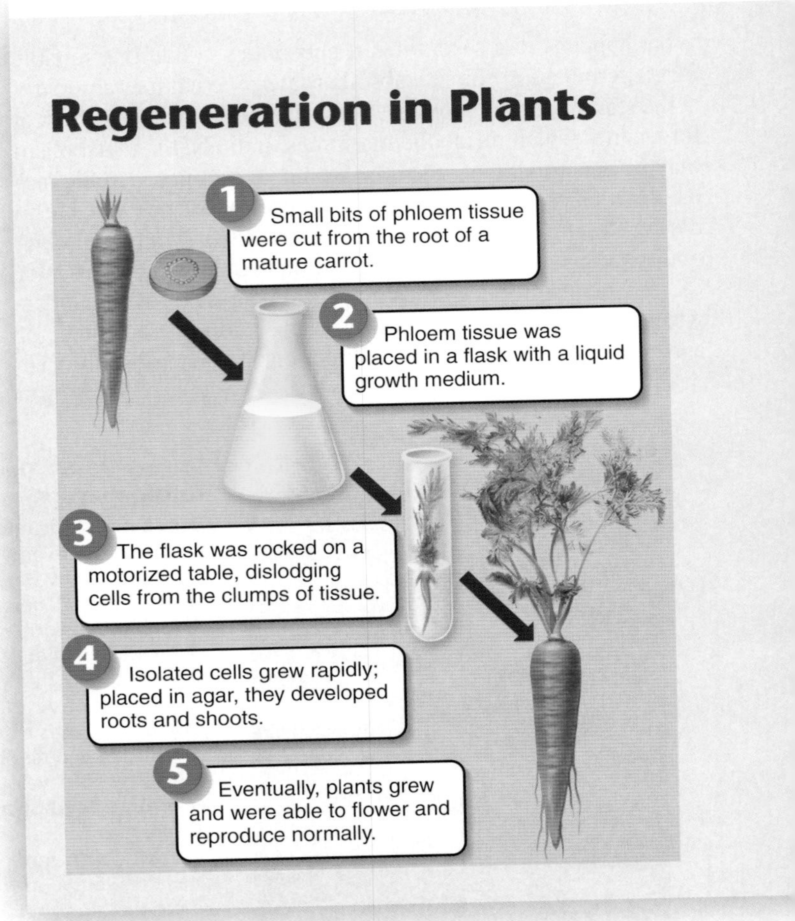

Regeneration in Plants

1. Small bits of phloem tissue were cut from the root of a mature carrot.

2. Phloem tissue was placed in a flask with a liquid growth medium.

3. The flask was rocked on a motorized table, dislodging cells from the clumps of tissue.

4. Isolated cells grew rapidly; placed in agar, they developed roots and shoots.

5. Eventually, plants grew and were able to flower and reproduce normally.

Figure 34.7   How Steward regenerated a plant from differentiated tissue.

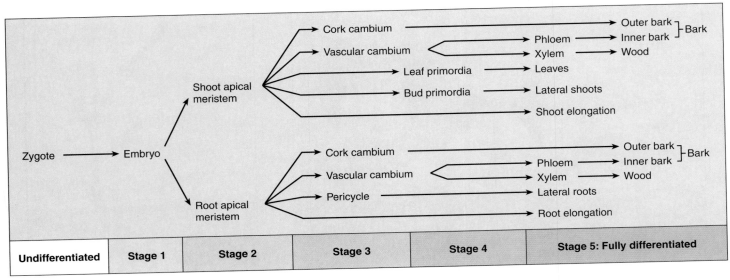

Figure 34.8   Stages of plant differentiation.

As this diagram shows, the different cells and tissues in a plant all originate from the shoot and root apical meristems. It is important to remember, however, that this is showing the origin of the tissue, not the location of the tissue in the plant. For example, the vascular tissues of xylem and phloem arise from the vascular cambium, but these tissues are present throughout the plant, in the leaves, shoots, and roots.

expressing their hidden genetic information when provided with suitable environmental signals. What halts the expression of genetic potential when the same kinds of cells are incorporated into normal, growing plants? As we will see, the expression of some of these genes is controlled by plant hormones.

Hormones are chemical substances produced in small (often minute) quantities in one part of an organism and then transported to another part of the organism, where they stimulate certain physiological processes and inhibit others. How they act in a particular instance is influenced both by which hormones are present and by how they affect the particular tissue that receives their message.

In animals, there are several organs, called endocrine glands, that are solely involved with hormone production (hormones are produced in other organs as well). In plants, on the other hand, all hormones are produced in tissues that are not specialized for that purpose and carry out many other functions.

At least five major kinds of hormones are found in plants: auxin, gibberellins, cytokinins, ethylene, and abscisic acid. Their chemical structures and descriptions are provided in table 34.1. Other kinds of plant hormones certainly exist but are less well understood. Hormones have multiple functions in the plant; the same hormone may work differently in different parts of the plant, at different times, and interact with other hormones in different ways. The study of plant hormones, especially how hormones produce their effects, is today an active and important field of research.

> **34.5** The development of plant tissues is controlled by the actions of hormones. They act on the plant by regulating the expression of key genes.

TABLE 34.1 FUNCTIONS OF THE MAJOR PLANT HORMONES

Hormone	Major Functions	Where Produced or Found in Plant	Practical Applications
Auxin (IAA)	Promotes stem elongation and growth; forms adventitious roots; inhibits leaf abscission; promotes cell division (with cytokinins); induces ethylene production; promotes lateral bud dormancy	Apical meristems; other immature parts of plants	Seedless fruit production; synthetic auxins act as herbicides
Gibberellins (GA₁, GA₂, GA₃, etc.)	Promotes stem elongation; stimulates enzyme production in germinating seeds	Root and shoot tips; young leaves; seeds	Uniform seed germination for production of barley malt used in brewing; early seed production of biennial plants; increasing size of grapes by allowing more space for growth
Cytokinins	Stimulates cell division, but only in the presence of auxin; promotes chloroplast development; delays leaf aging; promotes bud formation	Root apical meristems; immature fruits	Tissue culture and biotechnology; pruning trees and shrubs, which cause them to "fill out"
Ethylene	Controls leaf, flower, and fruit abscission; promotes fruit ripening	Roots, shoot apical meristems; leaf nodes; aging flowers; ripening fruits	Fruit ripening of agricultural products that are picked early to retain freshness
Abscisic acid (ABA)	Controls stomatal closure; some control of seed dormancy; inhibits effects of other hormones	Leaves, fruits, root caps, seeds	Research on stress tolerance in plants, specifically drought tolerance

34.6 Auxin

In his later years, the great evolutionist Charles Darwin became increasingly devoted to the study of plants. In 1881, he and his son Francis published a book called *The Power of Movement in Plants,* in which they reported their systematic experiments concerning the way in which growing plants bend toward light, a phenomenon known as **phototropism.**

After conducting a series of experiments shown in figure 34.9, they observed that plants grew toward light (panel 1). If the tip of the seedling was covered, the plant didn't bend toward the light (panel 2). A control experiment showed that the cap was not interfering with the directional growth pattern (panel 3). Another control showed that covering the lower portions of the plant did not block the directional growth (panel 4). The Darwins hypothesized that when plant shoots were illuminated from one side, an "influence" that arose in the uppermost part of the shoot was then transmitted downward, causing the shoot to bend. Later, several botanists conducted a series of experiments that demonstrated that the substance causing the shoots to bend was a chemical we call **auxin.**

How auxin controls plant growth was discovered in 1926 by Frits Went, a Dutch plant physiologist, in the course of studies for his doctoral dissertation. From his experiments, described in figure 34.10, Went was able to show that the substance that flowed into the agar from the tips of the light-grown grass seedlings (steps 1 and 2) enhanced cell elongation (shown in step 3). This chemical messenger caused the tissues on the side of the seedling into which it flowed to grow more than those on the opposite side (step 4). Control experiments indicated that the effects were not due to properties of the agar (steps 1*a* and 2*a*). He named the substance that he had discovered auxin, from the Greek word *auxin,* meaning "to increase."

Went's experiments provided a basis for understanding the responses the Darwins had obtained some 45 years earlier: grass seedlings bend toward the light because the side of the shoot that is in the shade has more auxin; therefore, its cells elongate more than those on the lighted side, bending the plant toward the light as in the enlargement in figure 34.11. Later experiments showed that auxin in normal plants migrates away from the illuminated side toward the dark side in response to light and thus causes the plant to bend toward the light.

The Discovery of Auxin

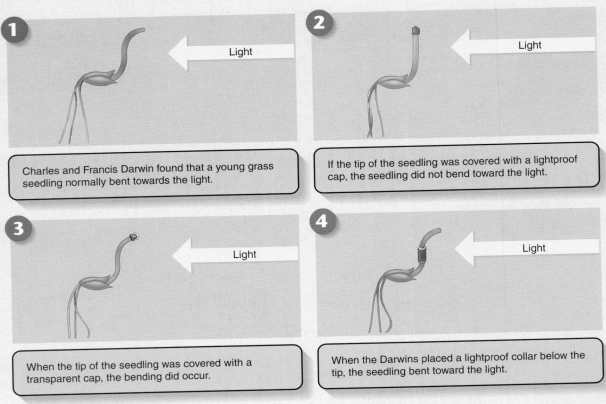

1 Charles and Francis Darwin found that a young grass seedling normally bent towards the light.

2 If the tip of the seedling was covered with a lightproof cap, the seedling did not bend toward the light.

3 When the tip of the seedling was covered with a transparent cap, the bending did occur.

4 When the Darwins placed a lightproof collar below the tip, the seedling bent toward the light.

Figure 34.9 The Darwins' experiment with phototropism.
From these experiments, the Darwins concluded that, in response to light, an "influence" that causes bending was transmitted from the tip of the seedling to the area below the tip, where bending usually occurs.

Auxin Promotes Plant Growth

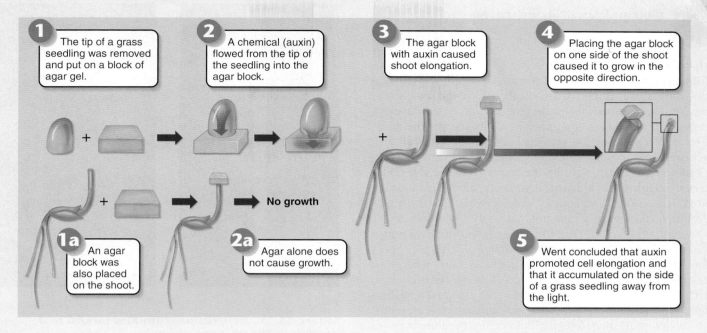

1. The tip of a grass seedling was removed and put on a block of agar gel.

2. A chemical (auxin) flowed from the tip of the seedling into the agar block.

3. The agar block with auxin caused shoot elongation.

4. Placing the agar block on one side of the shoot caused it to grow in the opposite direction.

1a. An agar block was also placed on the shoot.

No growth

2a. Agar alone does not cause growth.

5. Went concluded that auxin promoted cell elongation and that it accumulated on the side of a grass seedling away from the light.

Figure 34.10   How Went demonstrated the effects of auxin on plant growth.
The experiment showed how a chemical at the tip of the seedling caused the shoot to elongate and to bend. Steps 1a and 2a show the control experiment.

Auxin appears to act by increasing the stretchability of the plant cell wall within minutes of its application. Researchers speculate that the covalent bonds linking the polysaccharides of the cell wall to one another change extensively in response to auxin, allowing the cells to take up water and thus enlarge.

Synthetic auxins are routinely used to control weeds. When applied as herbicides, they are used in higher concentrations than those at which auxin normally occurs in plants. One of the most important of the synthetic auxins used in this way is 2,4-dichlorophenoxyacetic acid, usually known as 2,4-D. It kills weeds in lawns without harming the grass because 2,4-D affects only broad-leaved dicots. When treated, the weeds literally "grow to death," rapidly reducing ATP production so that no energy remains for transport or other essential functions.

Closely related to 2,4-D is the herbicide 2,4,5-trichlorophenoxyacetic acid (2,4,5-T), which is widely used to kill woody seedlings and weeds. Notorious as the Agent Orange of the Vietnam War, 2,4,5-T is easily contaminated with a by-product of its manufacture, dioxin. Dioxin is harmful to people because it is an **endocrine disrupter,** a chemical that interferes with the course of human development. The growing release of endocrine disrupters as by-products of modern chemical manufacturing is a subject of great environmental concern.

> **34.6   The primary growth-promoting hormone of plants is auxin, which increases the plasticity of plant cell walls, allowing growth in specific directions.**

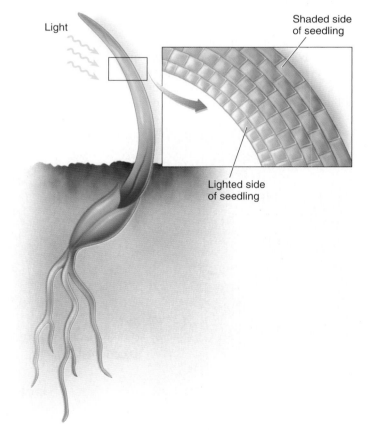

Figure 34.11   Auxin causes cells to elongate.
Plant cells that are on the shaded side have more auxin and grow faster, elongating more, than cells on the lighted side, causing the plant to bend toward light.

34.7 Other Plant Hormones

Gibberellins

Gibberellins are a large class of over 100 naturally occurring hormones, abbreviated GA and numbered. Synthesized in the apical portions of both shoots and roots, gibberellins have important effects on stem elongation in plants and play the leading role in controlling this process in mature trees and shrubs. In these plants, the application of gibberellins characteristically promotes elongation within the spaces between leaf nodes on stems. This is very apparent if you compare the plant on the left in figure 34.12, a mutant that is defective in producing gibberellin, and a similarly mutant plant on the right, which was exposed to gibberellin. This effect is enhanced if auxin is also present. Gibberellins also affect a number of other aspects of plant growth and development. The application of gibberellins can often induce biennial plants (plants that live for two years) to flower early during their first year of growth. These hormones also hasten seed germination, apparently because they can substitute for the effects of cold or light requirements in this process. Gibberellins are used commercially to space out the flowers of grape vines by extending the internode lengths so the fruits have more room to grow and become larger.

Cytokinins

A **cytokinin** is a plant hormone that, in combination with auxin, stimulates cell division in plants and determines the course of differentiation. Substances with these properties are widespread, both in bacteria and in eukaryotes. In vascular plants, most cytokinins seem to be produced in the roots, from which they are then transported throughout the rest of the plant. Cytokinins apparently stimulate cell division by influencing the

Figure 34.12   The effect of a gibberellin.
This rosette mutant (*left*) of the mustard family plant (*Brassica rapa*) is defective in producing gibberellins, but it can be rescued by applying gibberellin (*right*).

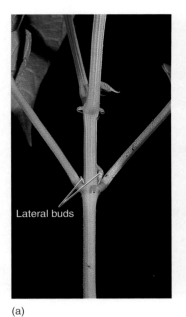 

(a)   (b)

Figure 34.13   Cytokinins stimulate lateral bud growth in the absence of auxin.
(a) When the apical meristem of a plant is intact, auxin from the apical bud will inhibit the growth of lateral buds. (b) When the apical bud is removed, cytokinins are able to promote the growth of lateral buds.

synthesis or activation of proteins specifically required for mitosis and are therefore used in tissue culture.

Cytokinins promote growth of lateral buds into branches but only when the influence of auxin is removed. Auxin released from the apical meristem inhibits the formation of lateral branches as in figure 34.13*a*, but when the apical meristem is removed as in figure 34.13*b*, cytokinins are able to promote lateral branch growth. This is why pruning bushes and trees makes the plants fuller. Thus, along with auxin and ethylene (discussed next), cytokinins play a role in the control of apical dominance and lateral bud growth. Cytokinins inhibit formation of lateral roots, while auxins promote their formation. As a consequence of these relationships, the balance between cytokinins and auxin, along with other factors, determines the appearance of a mature plant.

Ethylene

Ethylene is a gas that is produced in relatively large quantities during a certain phase of fruit ripening, when the fruit's respiration is proceeding at its most rapid rate. At this phase, complex carbohydrates are broken down into simple sugars, cell walls become soft, and the volatile compounds associated with flavor and scent in the ripe fruits are produced. When applied to fruits, ethylene hastens their ripening.

One of the first lines of evidence that led to the recognition of ethylene as a plant hormone was the observation that gases from oranges caused premature ripening in bananas. Such relationships have led to major commercial uses. Tomatoes are often picked green and then artificially ripened as desired by the

Figure 34.14 The effects of ethylene.
Ethylene released from the ripe apple on the *right* causes the holly twig to drop its leaves, while the twig on the *left* keeps its leaves.

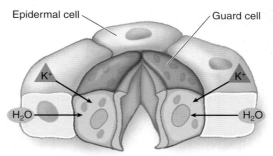

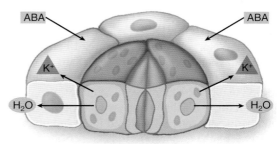

In the presence of ABA—stoma is closed

Figure 34.15 Effects of abscisic acid.
Abscisic acid (ABA) affects the closing of stomata by influencing the movement of potassium ions out of guard cells. As potassium ions are transported out of the guard cells, water also passes out of the cells due to osmosis, causing the stoma to close.

application of ethylene. Ethylene is widely used to speed the color formation of lemons and oranges as well.

Genetic engineers, using techniques described in chapter 14, have placed genes that interfere with the synthesis of ethylene into tomatoes to slow the ripening process. Until now, commercial tomatoes had to be picked very early in order to get them to market before they become overripe. The ripening of genetically engineered tomatoes is delayed, so the tomatoes can be left on the vine longer, improving their taste.

Ethylene also plays an important ecological role. Ethylene production increases rapidly when a plant is exposed to ozone and other toxic chemicals, temperature extremes, drought, attack by pathogens or herbivores, and other stresses. The increased production of ethylene that occurs can accelerate the abscission (the dropping) of leaves or fruits that have been damaged by these stresses. The experiment in figure 34.14 demonstrates this property of ethylene. The holly twig on the left hasn't lost its leaves, even after a week. The holly twig on the right, exposed to ethylene from the ripe apple, lost its leaves. It now appears that some of the damage associated with exposure to ozone is due to the ethylene produced by the plants. Some studies suggest that the production of ethylene by plants subjected to attack by herbivores or infected with diseases may be a signal to activate the defense mechanisms of the plants. Such mechanisms may include the production of molecules toxic to the animals or pests attacking them. A full understanding of these relationships is obviously important for agriculture and forestry.

Abscisic Acid

Abscisic acid (ABA) is a naturally occurring plant hormone that is synthesized mainly in mature green leaves, fruits, and root caps. The hormone was given its name because it was thought that it stimulated leaves to age rapidly and fall off (the process of abscission), but evidence that abscisic acid plays an important natural role in this process is scant. In fact, it is believed that abscisic acid may cause ethylene synthesis, and that it is actually the

ethylene that promotes senescence and abscission. When abscisic acid is applied to a green leaf, the areas of contact turn yellow. Thus, abscisic acid has the exact opposite effect on a leaf from that of the cytokinins; a yellowing leaf remains green in an area where cytokinins are applied.

Abscisic acid was also initially thought to induce the formation of winter buds—dormant buds that remain through the winter—by suppressing growth, but recent evidence does not support this. Abscisic acid levels increase during seed development and decrease during germination, and so it is likely that ABA plays a role in causing the dormancy of many seeds. ABA may also function in transpiration. During drought conditions, leaves produce large amounts of ABA, which induces the closing of stomata. ABA's effects on stomata occur on the order of minutes, suggesting that this action does not involve turning genes on and off, but rather occurs by influencing the membrane permeability of guard cells. Stomata open when water flows into guard cells, causing them to swell. Water flows into the guard cells osmotically, driven by the influx of potassium ions (the green triangles in figure 34.15) into the guard cells. The lower panel shows how ABA most likely stimulates the transport of potassium ions out of the guard cells, causing water to pass out of the guard cells by osmosis. This loss of water causes the stomata to close.

34.7 Other plant hormones work with auxin and each other to control growth. These include gibberellins, cytokinins, ethylene, and abscisic acid.

34.8 Photoperiodism and Dormancy

Plants respond to different environmental stimuli in a variety of ways. As discussed earlier in the chapter, plants bend toward light as they grow in response to this environmental stimulus. A host of other plant responses, including flowering, dropping of leaves, and yellowing of leaves due to loss of chlorophyll, are also prompted by various environmental stimuli.

Photoperiodism

Essentially all eukaryotic organisms are affected by the cycle of night and day, and many features of plant growth and development are keyed to changes in the proportions of light and dark in the daily 24-hour cycle. Such responses constitute **photoperiodism,** a mechanism by which organisms measure seasonal changes in relative day and night length. One of the most obvious of these photoperiodic reactions concerns angiosperm flower production.

Day length changes with the seasons; the farther from the equator you are, the greater the variation. Plants' flowering responses fall into three basic categories in relation to day length: long-day plants, short-day plants, and day-neutral plants. Long-day plants like the iris in figure 34.16, panel 1, initiate flowers in the summer, when nights become shorter than a certain length (and days become longer). Short-day plants, on the other hand, begin to form flowers when nights become longer than a critical length (and days become shorter); the goldenrod in panel 2 doesn't flower in summer, but instead flowers in fall. Thus, many spring and early summer flowers are long-day plants, and many fall flowers are short-day plants. The "interrupted night" experiment in panel 3 makes it clear that it is the length of uninterrupted dark that is the flowering trigger. The flash of light during a long night triggers flowering in the iris and inhibits flowering in the goldenrod, even though the day is shorter.

In addition to long-day and short-day plants, a number of plants are described as day-neutral. Day-neutral plants produce flowers without regard to day length.

Photoperiodism

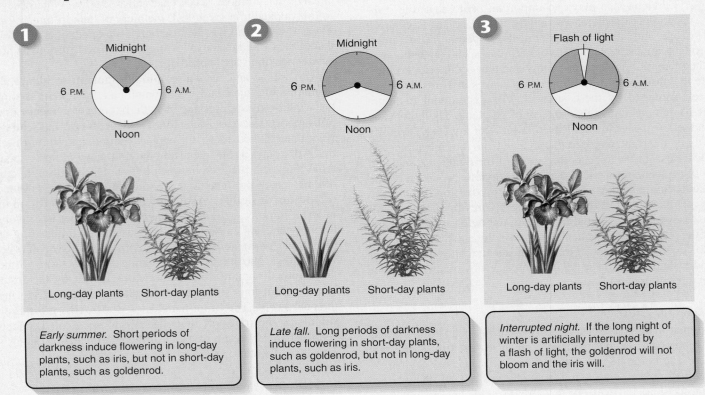

Figure 34.16 How photoperiodism works in plants.

The Chemical Basis of Photoperiodism

Flowering responses to daylight and darkness are controlled by several chemicals that interact in complex ways. Although the nature of some of these chemicals has been deduced, how the various chemicals work together to promote or inhibit flowering responses is still being debated.

Plants contain a pigment, **phytochrome,** that exists in two interconvertible forms, P_r (inactive) and P_{fr} (active). When P_{fr} is present, biological reactions like flowering are influenced by phytochrome. When P_r absorbs red light (660 nm—orangish red), it is instantly converted into P_{fr}. Conversely, when P_{fr} is left in the dark or absorbs far-red light (730 nm—deep red) it is instantly converted to P_r and the biological response ceases.

In short-day plants, the presence of P_{fr} leads to a biological reaction that suppresses flowering. The amount of P_{fr} steadily declines in darkness, the molecules converting to P_r. When the period of darkness is long enough, the suppression reaction ceases and the flowering response is triggered. However, a single flash of red light at a wavelength of about 660 nanometers converts most of the molecules of P_r to P_{fr}, and the flowering reaction is blocked.

Dormancy

Plants respond to their external environment largely by changes in growth rate. Plants' ability to stop growing altogether when conditions are not favorable—to become dormant—is critical to their survival.

In temperate regions, dormancy is generally associated with winter, when low temperatures and the unavailability of water because of freezing make it impossible for plants to grow. During this season, the buds of deciduous trees and shrubs remain dormant, and the apical meristems remain well protected inside enfolding scales. Perennial herbs spend the winter underground as stout stems or roots packed with stored food. Many other kinds of plants, including most annuals, pass the winter as seeds.

> **34.8 Plant growth and reproduction are sensitive to photoperiod, using chemicals to link flowering to season.**

34.9 Tropisms

Tropisms are directional and irreversible growth responses of plants to external stimuli. They control patterns of plant growth and thus plant appearance. Three major classes of plant tropisms include phototropism (figure 34.17a, and discussed earlier), gravitropism, and thigmotropism.

Gravitropism

Gravitropism causes stems to grow upward and roots downward. Both of these responses clearly have adaptive significance: stems, like the one growing from a test tube in figure 34.17b, grow upward and are apt to receive more light than those that do not; roots that grow downward are more apt to encounter a more favorable environment than those that do not. The phenomenon is called gravitropism because it is clearly a response to gravity.

Thigmotropism

Still another commonly observed response of plants is **thigmotropism,** a name derived from the Greek root *thigma,* meaning "touch." Thigmotropism is defined as the response of plants to touch. Examples include plant tendrils, which rapidly curl around and cling to stems or other objects, and twining plants, such as bindweed, which also coil around objects (figure 34.17c). These behaviors result from rapid growth responses to touch. Specialized groups of cells in the plant epidermis appear to be concerned with thigmotropic reactions, but again, their exact mode of action is not well understood.

> **34.9 Growth of the plant body is often sensitive to light, gravity, or touch.**

(a)

(b)

(c)

Figure 34.17 Tropism guides plant growth.

(a) Phototropism is exhibited by this *Impatiens* plant growing toward light. (b) Gravitropism is exhibited by this plant, *Zebrina pendula*. Note the negative gravitational response of the shoot. (c) The thigmotropic response of these twining stems causes them to coil around the object with which they have come in contact.

Does Auxin Use Ethylene to Inhibit Root Growth?

The plant hormone auxin (the name is from the Greek word *auxein*, "to increase") promotes plant growth. Released at low concentrations from cells at the growing tip of a plant, auxin diffuses downward, causing the stem to elongate. At similarly low concentrations, auxin also promotes elongation of roots, and formation of new roots at cut surfaces. In the photo, the stalk of the leaf on the left sits in a solution of auxin, the one on the right in water. You can see auxin has promoted root growth. However, at high concentrations auxin has quite the opposite effect, inhibiting root growth (**inhibition** is the stopping or restraining of a process). It has long been assumed that auxin achieves this inhibition by triggering the production of the plant hormone ethylene, a well-known inhibitor of growth. Supporting this hypothesis is the observation that auxin does stimulate many kinds of plant cells to produce ethylene. But just because the "suspect is present at the scene of a crime" does not prove the suspect is guilty of the crime. Perhaps auxin is directly inhibiting root growth by some unknown mechanism.

   The graph to the right displays the results of an experiment designed to test the hypothesis that high concentrations of auxin inhibit root growth by stimulating ethylene production. Seedlings with roots were grown in varying concentrations of auxin, with ethylene assays taken at the end of the experiment. The rate of pea root elongation (blue points) and the production of ethylene in the pea roots (red points) are determined for a range of auxin concentrations. Auxin concentrations are plotted on a log scale (a **log** or logarithmic **scale** is a series of numbers plotted as powers of ten; because the scale is exponential [1, 10, 100, 1000...] rather than linear [1, 2, 3, 4...], a broad range of values can be visualized on a single graph).

1. **Applying Concepts**
   a. Variable. In the graph, which variable or variables are dependent variables?
   b. Inhibition. What is the rate of pea root elongation when there is no auxin present? at the highest auxin concentration tested? Is there any concentration of auxin in this study that promotes root growth? Why?
   c. Log scale. What is the effect of auxin on the rate of pea root elongation over the range of concentrations studied? What would the elongation rate curve look like plotted on a linear scale? What does this say about the sensitivity of the inhibition to auxin concentration?
2. **Interpreting Data** Assess the effect of these high auxin concentrations on the production of ethylene by pea root cells. Do the cells produce more ethylene when exposed to higher concentrations of auxin?
3. **Making Inferences**
   a. Construct a plot of ethylene concentration (*y* axis) versus elongation rate (*x* axis). Does this plot

Effects of Auxin on Root Elongation

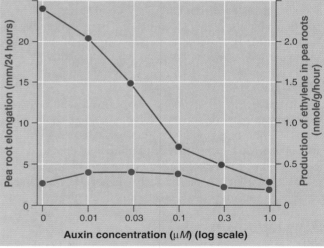

provide any evidence that these two variables are being influenced in the same way by auxin? that they are *not* being influenced in the same way?
   b. Is the inhibition of root elongation by high concentrations of auxin accompanied by corresponding increases in the concentration of ethylene?
4. **Drawing Conclusions** Does auxin exert its inhibition of root elongation by stimulating ethylene production?
5. **Further Analysis** Roots of intact plants exhibit significant elongation when no auxin is supplied. Does this mean that elongation requires no auxin? How would you test the hypothesis that pea roots have the ability to synthesize enough auxins for their growth?

Flowering Plant Reproduction

34.1 Angiosperm Reproduction

- Angiosperms reproduce sexually and asexually. In asexual reproduction, offspring are genetically identical to the parent; this often involves cloning of the parent, a process called vegetative reproduction. Examples of vegetative reproduction include runners, rhizomes, suckers, and adventitious plants (**figure 34.1**).

- Sexual reproduction involves the pollination and fertilization of flowers. Flowers have male and female parts. The male parts include the stamen and the pollen-producing anthers.

- The female parts include the stigma, style, and ovary, which make up the carpel. The egg cell and two polar nuclei are produced in the ovule, within the ovary (**figure 34.2**).

- Pollen grains are carried by animals or wind to a flower of the same species (**figure 34.3**). The pollen, after landing on the stigma, extends a pollen tube through the style to the base of the ovule. Two sperm cells travel down the pollen tube, into the ovule. One sperm fertilizes the egg, giving rise to the zygote, and the other fuses with the two polar nuclei, giving rise to the endosperm. This process is called double fertilization.

34.2 Seeds

- Development begins after fertilization. The fertilized egg begins dividing, forming the embryo. After a few days, the embryo stops growing and becomes dormant in a structure called a seed.

- The endosperm may remain intact or may be consumed by the embryo, forming cotyledons (seed leaves) that becomes the food source. The outer layer of the ovule becomes the seed coat, enclosing the embryo and food source (**figure 34.4**).

- When conditions are favorable for the plant's survival, when water and oxygen are available to the seed, development resumes through the process of germination.

34.3 Fruit

- The evolution of fruit was another key advancement in the angiosperms. During seed formation, the flower's ovary begins to develop into fruit that surrounds the seed.

- Fruits vary from fleshy to dry, depending on their means of dispersal. Fleshy fruits are usually eaten by animals and then dispersed. Dry fruits are usually dispersed by wind, water, or carried on animals (**figure 34.5**).

34.4 Germination

- When a seed encounters favorable conditions, it will resume growth, a process called germination. The seed absorbs water and uses the endosperm or cotyledons as a food source. Eventually, the seed coat cracks open and the plant begins to grow, sending out roots and shoots. The overall process is similar in monocots and dicots but there are differences in the structures involved (**figure 34.6**). Dicots use their fleshy cotyledons as a food source, and monocots use their stored endosperm and their single cotyledon doesn't emerge from underground.

Regulating Plant Growth

34.5 Plant Hormones

- As a plant grows, its cells differentiate, but differentiation in plants is largely reversible. New plants can be grown from parts of adult plants through regeneration (**figure 34.7**). Differentiation results from the activation or suppression of particular genes, controlled by hormones (**table 34.1**).

34.6 Auxin

- Early researchers, including Darwin, described a process, now called phototropism, where plants grow toward light (**figure 34.9**). Went identified a chemical he called auxin as the hormone involved in phototropism (**figure 34.10**).

- When exposed to light on one side, auxin is released from the tip of the shoot and causes cells on the shady side of the plant to elongate, causing the plant to grow toward the light (**figure 34.11**).

34.7 Other Plant Hormones

- Other plant hormones include gibberellins, cytokinins, ethylene, and abscisic acid.

- Gibberellins affect stem elongation, promoting elongation between the node regions (**figure 34.12**).

- Cytokinin stimulates cell division and acts in combination with auxin. Cytokinin also stimulates lateral bud growth in the absence of auxin (**figure 34.13**).

- Ethylene is released as a gas during fruit ripening. It also accelerates abscission (falling off) of leaves (**figure 34.14**).

- Abscisic acid (ABA) works in combination with other hormones but seems to play a role in seed dormancy and the closing of stomata (**figure 34.15**).

Plant Responses to Environmental Stimuli

34.8 Photoperiodism and Dormancy

- The length of daylight affects flowering, a process called photoperiodism (**figure 34.16**) and dormancy. Some plants flower in response to short days, some to long days, and some are day-neutral. A plant pigment, phytochrome, exists in two forms converted by darkness. P_{fr} inhibits flowering, while P_r allows flowering to occur.

34.9 Tropisms

- Tropisms are directed growth responses in plants that are irreversible. Phototropism is a growth response toward light; gravitropism is growth in response to the pull of gravity; thigmotropism is growth in response to touch (**figure 34.17**).

1. Sexual reproduction in angiosperms requires
   - a. pollen.
   - b. runners.
   - c. rhizomes.
   - d. suckers.
2. Angiosperm plants that contain both male and female flowers are called
   - a. dioecious.
   - b. monoecious.
   - c. imperfect.
   - d. incomplete.
3. The flower shape, scent, color, and presence of nectar in the flowers of some angiosperms are related to the plant's
   - a. predators.
   - b. animal pollinators.
   - c. insect pests.
   - d. symbionts.
4. For a seed to germinate, the dormant plant embryo must get
   - a. carbon dioxide and water.
   - b. nitrogen and water.
   - c. oxygen and nitrogen.
   - d. oxygen and water.
5. Fruit forms from a flower's
   - a. ovary.
   - b. sepals.
   - c. carpels.
   - d. stigma.

6. Meristematic tissue regulates plant growth and development through the use of
   - a. carbohydrates.
   - b. amount of available water.
   - c. hormones.
   - d. phototropism.
7. The hormone auxin causes
   - a. cells in plant stems to shorten by releasing water.
   - b. fruit to ripen.
   - c. cells in plant stems to elongate through absorption of water.
   - d. growth of more lateral branches.
8. The hormone ethylene causes
   - a. cells in plant stems to shorten by releasing water.
   - b. fruit to ripen.
   - c. increase in fruit size.
   - d. growth of more lateral branches.
9. Angiosperm flower production is controlled by
   - a. temperature.
   - b. gibberellins.
   - c. photoperiod.
   - d. magnesium.
10. Sensitivity of plants to touch is known as
    - a. thigmotropism.
    - b. photoperiodism.
    - c. phototropism.
    - d. gravitropism.

Visual Understanding

1. **Figure 34.3** Bees tend to pollinate yellow flowers; hummingbirds cue in on red flowers. Why are there so many white flowers?

2. **Figure 34.5** If fruits are produced by plants to encourage organisms, especially birds and mammals, to eat them and carry the seeds elsewhere, then speculate on the purpose of fruits such as peaches, shown here, mangoes, and avocados.

(a) Berries   (b) Drupes   (c) Pomes

(d) Eaten by animals   (e) Dispersed by wind   (f) Dispersed by attaching to animals

Challenge Questions

1. Your friend, Shonille, says her teacher made the comment that "nectar is to pollination as fruit is to dispersal." She asks you to explain.
2. An old saying is "one bad apple spoils the whole barrel." We even refer to troublemakers as "a bad apple." Can you explain the scientific basis for these statements?
3. If it is out in the open, trailing ivy grows along the ground and can form dense mats of ground cover. If planted in a hanging pot, the stems dangle down, forming a graceful flow. If ivy is planted next to a building, however, its stems attach to the side of the building and, in time, it will cover the building. What causes this?

Appendix A

Answers to Self-Test Questions

Chapter 1: The Science of Biology

1. a. kingdoms 2. c. cellular organization 3. b. atom, molecule, organelle, cell, tissue, organ, organ system, organism, population, species, community, ecosystem 4. d. emergent properties 5. b. evolution, energy flow, cooperation, structure determines function, and homeostasis 6. a. inductive reasoning 7. b. test each hypothesis, using appropriate controls, to rule out as many as possible 8. c. After sufficient testing, you can accept it as probable, being aware that it may be revised or rejected in the future 9. d. all living organisms consist of cells, and all cells come from other cells 10. c. is contained in a long molecule called DNA

Chapter 2: Evolution and Ecology

1. c. populations can change over time, sometimes forming new species 2. b. the characteristics of a species varied in different places; there were geographic patterns 3. c. populations are capable of geometric increase, yet remain at constant levels 4. a. natural selection 5. b. seems to agree with Darwin's original ideas 6. d. niche 7. a. a population 8. d. energy flows through once and is lost, while materials cycle and recycle 9. d. competitive exclusion 10. b. predation

Chapter 3: The Chemistry of Life

1. b. an atom 2. c. an ion 3. d. ionic, covalent, and hydrogen 4. b. it can form four single covalent bonds 5. d. All of these are correct 6. a. hydrogen bonds between the individual water molecules 7. c. heat storage and heat of vaporization 8. a. cohesion 9. a. (1) acids and (2) bases 10. c. to keep pH from ever changing

Chapter 4: Molecules of Life

1. b. proteins, carbohydrates, lipids, and nucleic acids 2. c. polypeptides 3. c. structure, function 4. d. All of these 5. d. are information storage devices found in body cells 6. a. Adenine forms hydrogen bonds with thymine 7. a. structure and for energy 8. a. glycogen 9. b. that they are insoluble in water 10. c. energy storage and for some hormones

Chapter 5: Cells

1. a. cells are the smallest living things. Nothing smaller than a cell is considered alive 2. b. a double lipid layer with proteins inserted in it, which surrounds every cell individually 3. d. prokaryotes, eukaryotes 4. a. a nucleolus 5. c. endoplasmic reticulum and the Golgi complex 6. d. mitochondria and the chloroplasts 7. b. Eukaryotic cells in plants and fungi, and all prokaryotes, have a cell wall 8. d. slowly disperse throughout the water; this is because of diffusion 9. b. endocytosis and phagocytosis 10. c. energy and specialized pumps or channels

Chapter 6: Energy and Life

1. c. energy 2. d. says that energy can change forms, but cannot be made or destroyed 3. b. says that entropy, or disorder, continually increases in a closed system 4. a. exergonic and release energy 5. b. enzymes 6. c. temperature and pH 7. d. All of these are true 8. d. produces repressor molecules that alter the enzyme shape 9. c. an inhibitor molecule competes with the substrate for the same binding site on the enzyme 10. a. ATP molecules

Chapter 7: Photosynthesis: Acquiring Energy from the Sun

1. c. photosynthesis 2. b. with molecules called pigments that absorb photons and use their energy 3. c. a small portion in the middle of the spectrum 4. a. red, blue, and orange 5. d. All of these are true 6. c. only go through the system once; they are obtained by splitting a water molecule 7. b. chemiosmosis 8. a. electron transport system of photosystem I/Calvin cycle 9. c. build sugar molecules 10. b. use C_4 photosynthesis or CAM

Chapter 8: How Cells Harvest Energy from Food

1. a. breaking down the organic molecules that were consumed 2. b. fermentation 3. c. substrate-level phosphorylation 4. b. glycolysis 5. b. makes ATP by splitting glucose and capturing the energy 6. b. NAD^+, electron transport chain 7. a. pyruvate 8. c. mitochondria of the cell and are broken down in the presence of O_2 to make more ATP 9. d. during the electron transport chain 10. c. each type of macromolecule is broken down into its subunits, which enter the oxidative respiration pathway

Chapter 9: Mitosis

1. a. copying DNA then undergoing binary fission 2. c. in the production of daughter cells 3. c. and most eukaryotes have between 10 and 50 pairs of chromosomes 4. c. carry information about the same traits located in the same places on the chromosome 5. b. metaphase 6. c. cytokinesis 7. a. a series of checkpoints 8. b. cancer 9. c. an oncogene 10. d. All of the above

Chapter 10: Meiosis

1. d. the germ cells went through meiosis; the egg and sperm only have half the parental chromosomes 2. d. 23 3. a. $1n$ gametes (haploid), then next there are $2n$ zygotes (diploid) 4. b. randomly separate the homologous pairs, called independent assortment 5. c. separate the duplicated sister chromatids 6. a. prophase I 7. c. homologous chromosomes become closely associated 8. a. make diploid cells/make haploid cells 9. c. has a lot of genetic reassortment due to processes in meiosis I 10. a. sexual reproduction

Chapter 11: Foundations of Genetics

1. d. All of these 2. a. all purple flowers 3. c. $^3/_4$ purple and $^1/_4$ white flowers 4. b. some factor, or information, about traits to their offspring and it may or may not be expressed 5. a. dihybrid cross

6. d. multiple genes 7. c. codominant traits 8. b. sex-linked eye color in fruit flies 9. d. All of the above. 10. d. genetic screening and prenatal diagnosis is now available

Chapter 12: DNA: The Genetic Material

1. b. hereditary information can be added to cells from other cells 2. d. DNA 3. a. structure of DNA 4. c. the type of nitrogen base 5. d. adenine and guanine 6. b. TAACGTA 7. b. splits down the middle into two single helices, and each one then acts as a template to build its complement 8. a. a primer 9. c. by mutation or by recombination 10. a. germ-line tissues and be passed on to future generations

Chapter 13: How Genes Work

1. a. nRNA (nuclear RNA) 2. d. a codon 3. c. 64 4. c. transcription 5. a. promoter 6. d. translation 7. c. AUG 8. b. a ribosome 9. d. some genes are always off unless a repressor is not bound 10. c. repressors block transcription by binding to the DNA

Chapter 14: The New Biology

1. c. genome 2. b. the exons used to make a specific mRNA can be rearranged to form different proteins 3. c. restriction enzyme 4. b. cleaving DNA, producing recombinant DNA, cloning, and screening 5. a. the drug to be produced in far larger amounts than in the past 6. a. increased yield 7. d. harm to the crop itself from mutations 8. d. methylation or demethylation of DNA 9. a. immunological rejection of the tissue by the patient 10. c. viruses

Chapter 15: Evolution and Natural Selection

1. c. macroevolution 2. b. analogous structures 3. a. divergence 4. b. microevolution 5. c. no mutation within the population 6. b. $p^2 = 0.30$ 7. d. genetic drift 8. c. directional selection 9. d. reproductive isolation 10. b. mechanical

Chapter 16: Exploring Biological Diversity

1. a. the red fox is in the same family, but different genus than dogs and wolves 2. b. phylogeny 3. a. physical and chemical characteristics 4. d. cell structure and DNA sequence 5. b. Archaea 6. d. they do not have an internal membrane system 7. c. have cell walls that are made of different materials 8. c. Protista 9. b. multicellular 10. d. ingestion of endosymbiotic bacteria

Chapter 17: Evolution of Microbial Life

1. c. RNA 2. d. All answers are correct 3. b. chemoautotrophs 4. a. protein coats that contain DNA or RNA 5. b. mitochondria and chloroplasts have their own DNA 6. c. cysts 7. d. any organism that is not plant, animal, or fungi is a protist 8. d. mycelium 9. a. both sexually and asexually 10. b. algae and fungi

Chapter 18: Evolution of Plants

1. c. dehydration 2. a. chloroplasts 3. b. they do not have specialized vascular tissue to transport water very high 4. d. seeds 5. c. sporophyte 6. d. dispersal 7. c. ovules not completely covered by the sporophyte 8. c. fruits and flowers 9. a. pollination 10. c. an animal's digestive system and processes

Chapter 19: Evolution of Animals

1. b. choanocytes 2. c. specialization of digestive tract 3. b. endoderm; mesoderm 4. c. weight of the thick exoskeleton needed to support very large insects. 5. b. an animal with a sessile lifestyle, rather than one that moves through the environment 6. b. jaws 7. c. an internal skeleton made of cartilage 8. c. middle ear bones 9. d. thin hollow bones in the skeleton 10. c. a placenta

Chapter 20: Ecosystems

1. c. producers 2. d. All answers are correct 3. d. amount of energy transferred to the top carnivores 4. a. from plants 5. b. light energy through photosynthesis 6. c. ATP 7. d. desert conditions on the downwind side of a mountain due to increased moisture-holding capacity of the winds as the air heats up 8. a. increases, temperature decreases 9. d. thermocline 10. c. near the top of a desert mountain range

Chapter 21: Populations and Communities

1. c. carrying capacity 2. b. increased competition for food 3. a. short life span 4. b. will decrease, and the mortality will increase 5. d. community 6. c. resource partitioning 7. b. commensalism 8. b. decreasing competitive exclusion between prey species 9. a. aposematic coloration 10. c. secondary succession

Chapter 22: Behavior and the Environment

1. b. cannot be modified, as these behaviors seem built into the brain and nervous system 2. a. there is a clear link between presence or absence of a specific gene, a specific metabolic pathway, and a specific behavior 3. b. operant conditioning 4. c. reproductive fitness 5. d. foraging behavior 6. c. intersexual selection 7. b. polyandry 8. c. reciprocity 9. b. kin selection 10. d. ecological factors such as food type and predation

Chapter 23: Planet Under Stress

1. d. sulfur oxides as a major air pollutant 2. b. coal-powered industry 3. c. chlorofluorocarbons 4. a. ammonia 5. d. All answers are correct 6. a. environmental costs are hardly ever recognized as part of the economy 7. a. needed to preserve possible direct value from species, such as new medicines 8. c. increasing amounts of open space as countries develop 9. d. captive propagation 10. a. keystone species

Chapter 24: The Animal Body and How It Moves

1. b. more flexible movement as individual segments can move independently of each other 2. a. cells, tissues, organs, organ systems, organism 3. c. move the body 4. d. red blood cells 5. c. osteoblasts; osteoclasts 6. b. neurotransmitters 7. a. skeletal 8. b. axial skeleton 9. a. a single muscle can only pull and not push 10. d. expose myosin attachment sites on actin

Chapter 25: Circulation

1. d. All of the above are functions of the circulatory system 2. c. by passing warm blood near cold blood in the extremities to warm the blood 3. a. capillaries 4. d. carry fluids 5. d. erythrocyte 6. c. A heart with separate chambers is first seen in fishes 7. b. better separation of oxygenated and deoxygenated blood 8. a. Only arteries carry oxygenated blood 9. d. endothermy

endocrine gland (Gr. *endon*, within + *krinein*, to separate) A ductless gland producing hormonal secretions that pass directly into the bloodstream or lymph.

endocrine system The dozen or so major endocrine glands of a vertebrate.

endocytosis (Gr. *endon*, within + *kytos*, cell) The process by which the edges of plasma membranes fuse together and form an enclosed chamber called a vesicle. It involves the incorporation of a portion of an exterior medium into the cytoplasm of the cell by capturing it within the vesicle.

endoderm (Gr. *endon*, outside + *derma*, skin) One of three embryonic germ layers that forms in the gastrula; giving rise to the epithelium that lines internal organs and most of the digestive and respiratory tracts.

endoskeleton (Gr. *endon*, within + *skeletos*, hard) In vertebrates, an internal scaffold of bone or cartilage to which muscles are attached.

endosperm (Gr. *endon*, within + *sperma*, seed) A nutritive tissue characteristic of the seeds of angiosperms that develops from the union of a male nucleus and the polar nuclei of the embryo sac. The endosperm is either digested by the growing embryo or retained in the mature seed to nourish the germinating seedling.

endosymbiotic (Gr. *endon*, within + *bios*, life) theory A theory that proposes how eukaryotic cells arose from large prokaryotic cells that engulfed smaller ones of a different species. The smaller cells were not consumed but continued to live and function within the larger host cell. Organelles that are believed to have entered larger cells in this way are mitochondria and chloroplasts.

endothermic The ability of animals to maintain a constant body temperature.

energy The capacity to bring about change, to do work.

enhancer A site of regulatory protein binding on the DNA molecule distant from the promoter and start site for a gene's transcription.

entropy (Gr. *en*, in + *tropos*, change in manner) A measure of the disorder of a system. A measure of energy that has become so randomized and uniform in a system that the energy is no longer available to do work.

enzyme (Gr. *enzymos*, leavened; from *en*, in + *zyme*, leaven) A protein capable of speeding up specific chemical reactions by lowering the energy required to activate or start the reaction but that remains unaltered in the process.

epidermis (Gr. *epi*, on or over + *derma*, skin) The outermost layer of cells. In vertebrates, the nonvascular external layer of skin of ectodermal origin; in invertebrates, a single layer of ectodermal epithelium; in plants, the flattened, skinlike outer layer of cells.

epistasis (Gr. *epistasis*, a standing still) An interaction between the products of two genes in which one modifies the phenotypic expression produced by the other.

epithelium (Gr. *epi*, on + *thele*, nipple) A thin layer of cells forming a tissue that covers the internal and external surfaces of the body. Simple epithelium consists of the membranes that line the lungs and major body cavities and that are a single cell layer thick. Stratified epithelium (the skin or epidermis) is composed of more complex epithelial cells that are several cell layers thick.

erythrocyte (Gr. *erythros*, red + *kytos*, hollow vessel) A red blood cell, the carrier of hemoglobin. Erythrocytes act as the transporters of oxygen in the vertebrate body. During the process of their maturation in mammals, they lose their nuclei and mitochondria, and their endoplasmic reticulum is reabsorbed.

estrus (L. *oestrus*, frenzy) The period of maximum female sexual receptivity. Associated with ovulation of the egg. Being "in heat."

estuary (L. *aestus*, tide) A partly enclosed body of water, such as those that often form at river mouths and in coastal bays, where the salinity is intermediate between that of saltwater and freshwater.

ethology (Gr. *ethos*, habit or custom + *logos*, discourse) The study of patterns of animal behavior in nature.

euchromatin (Gr. *eu*, true + *chroma*, color) Chromatin that is extended except during cell division, from which RNA is transcribed.

eukaryote (Gr. *eu*, true + *karyon*, kernel) A cell that possesses membrane-bounded organelles, most notably a cell nucleus, and chromosomes whose DNA is associated with proteins; an organism composed of such cells. The appearance of eukaryotes marks a major event in the evolution of life, as all organisms on earth other than bacteria and archaea are eukaryotes.

eumetazoan (Gr. *eu*, true + *meta*, with + *zoion*, animal) A "true animal." An animal with a definite shape and symmetry and nearly always distinct tissues.

eutrophic (Gr. *eutrophos*, thriving) Refers to a lake in which an abundant supply of minerals and organic matter exists.

evaporation The escape of water molecules from the liquid to the gas phase at the surface of a body of water.

evolution (L. *evolvere*, to unfold) Genetic change in a population of organisms over time (generations). Darwin proposed that natural selection was the mechanism of evolution.

exergonic (L. *ex*, out + Gr. *ergon*, work) Any reaction that produces products that contain less free energy than that possessed by the original reactants and that tends to proceed spontaneously.

exocytosis (Gr. *ex*, out of + *kytos*, cell) The extrusion of material from a cell by discharging it from vesicles at the cell surface. The reverse of endocytosis.

exoskeleton (Gr. *exo*, outside + *skeletos*, hard) An external hard shell that encases a body. In arthropods, comprised mainly of chitin.

experiment The test of a hypothesis. An experiment that tests one or more alternative hypotheses and those that are demonstrated to be inconsistent with experimental observation and are rejected.

F

facilitated diffusion The transport of molecules across a membrane by a carrier protein in the direction of lowest concentration.

family A taxonomic group ranking below an order and above a genus.

feedback inhibition A regulatory mechanism in which a biochemical pathway is regulated by the amount of the product that the pathway produces.

fermentation (L. *fermentum*, ferment) A catabolic process in which the final electron acceptor is an organic molecule.

fertilization (L. *ferre*, to bear) The union of male and female gametes to form a zygote.

fitness The genetic contribution of an individual to succeeding generations, relative to the contributions of other individuals in the population.

flagellum, *pl.* flagella (L. *flagellum*, whip) A fine, long, threadlike organelle protruding from the surface of a cell. In bacteria, a single protein fiber capable of rotary motion that propels the cell through the water. In eukaryotes, an array of microtubules with a characteristic internal $9 + 2$ microtubule structure that is capable of vibratory but not rotary motion. Used in locomotion and feeding. Common in protists and motile gametes. A cilium is a short flagellum.

food web The food relationships within a community. A diagram of who eats whom.

founder effect The effect by which rare alleles and combinations of alleles may be enhanced in new populations.

frequency In statistics, defined as the proportion of individuals in a certain category, relative to the total number of individuals being considered.

fruit In angiosperms, a mature, ripened ovary (or group of ovaries) containing the seeds.

G

gamete (Gr. wife) A haploid reproductive cell. Upon fertilization, its nucleus fuses with that of another gamete of the opposite sex. The resulting diploid cell (zygote) may develop into a new diploid individual, or in some protists and fungi, may undergo meiosis to form haploid somatic cells.

gametophyte (Gr. *gamete*, wife + *phyton*, plant) In plants, the haploid (n), gamete-producing generation, which alternates with the diploid ($2n$) sporophyte.

ganglion, *pl.* ganglia (Gr. a swelling) A group of nerve cells forming a nerve center in the peripheral nervous system.

gastrulation The inward movement of certain cell groups from the surface of the blastula.

gene (Gr. *genos*, birth, race) The basic unit of heredity. A sequence of DNA nucleotides on a chromosome that encodes a polypeptide or RNA molecule and so determines the nature of an individual's inherited traits.

gene expression The process in which an RNA copy of each active gene is made, and the RNA copy directs the sequential assembly of a chain of amino acids at a ribosome.

gene frequency The frequency with which individuals in a population possess a particular gene. Often confused with allele frequency.

genetic code The "language" of the genes. The mRNA codons specific for the 20 common amino acids constitute the genetic code.

genetic drift Random fluctuations in allele frequencies in a small population over time.

genetic map A diagram showing the relative positions of genes.

genetics (Gr. *genos*, birth, race) The study of the way in which an individual's traits are transmitted from one generation to the next.

genome (Gr. *genos*, offspring + L. *oma*, abstract group) The genetic information of an organism.

genomics The study of genomes as opposed to individual genes.

genotype (Gr. *genos*, offspring + *typos*, form) The total set of genes present in the cells of an organism. Also used to refer to the set of alleles at a single gene locus.

genus, *pl.* genera (L. race) A taxonomic group that ranks below a family and above a species.

germination (L. *germinare*, to sprout) The resumption of growth and development by a spore or seed.

gland (L. *glandis*, acorn) Any of several organs in the body, such as exocrine or endocrine, that secrete substances for use in the body. Glands are composed of epithelial tissue.

glomerulus (L. a little ball) A network of capillaries in a vertebrate kidney, whose walls act as a filtration device.

glycolysis (Gr. *glykys*, sweet + *lyein*, to loosen) The anaerobic breakdown of glucose; this enzyme-catalyzed process yields two molecules of pyruvate with a net of two molecules of ATP.

gravitropism (L. *gravis*, heavy + *tropes*, turning) The response of a plant to gravity, which generally causes shoots to grow up and roots to grow down.

greenhouse effect The process in which carbon dioxide and certain other gases, such as methane, that occur in the earth's atmosphere transmit radiant energy from the sun but trap the longer wavelengths of infrared light, or heat, and prevent them from radiating into space.

guard cells Pairs of specialized epidermal cells that surround a stoma. When the guard cells are turgid, the stoma is open; when they are flaccid, it is closed.

gymnosperm (Gr. *gymnos*, naked + *sperma*, seed) A seed plant with seeds not enclosed in an ovary. The conifers are the most familiar group.

H

habitat (L. *habitare*, to inhabit) The place where individuals of a species live.

half-life The length of time it takes for half of a radioactive substance to decay.

haploid (Gr. *haploos*, single + *eidos*, form) The gametes of a cell or an individual with only one set of chromosomes.

Hardy-Weinberg equilibrium After G. H. Hardy, English mathematician, and G. Weinberg, German physician. A mathematical description of the fact that the relative frequencies of two or more alleles in a population do not change because of Mendelian segregation. Allele and genotype frequencies remain constant in a random-mating population in the absence of inbreeding, selection, or other evolutionary forces. Usually stated as: If the frequency of allele *A* is *p* and the frequency of allele *a* is *q*, then the genotype frequencies after one generation of random mating will always be $(p + q)^2 = p^2 + 2pq + q^2$.

Haversian canal After Clopton Havers, English anatomist. Narrow channels that run parallel to the length of a bone and contain blood vessels and nerve cells.

helper T cell A class of white blood cells that initiates both the cell-mediated immune response and the humoral immune response; helper T cells are the targets of the AIDS virus (HIV).

hemoglobin (Gr. *haima*, blood + L. *globus*, a ball) A globular protein in vertebrate red blood cells and in the plasma of many invertebrates that carries oxygen and carbon dioxide.

herbivore (L. *herba*, grass + *vorare*, to devour) Any organism that eats only plants.

heredity (L. *heredis*, heir) The transmission of characteristics from parent to offspring.

heterochromatin (Gr. *heteros*, different + *chroma*, color) That portion of a eukaryotic chromosome that remains permanently condensed and therefore is not transcribed into RNA. Most centromere regions are heterochromatic.

heterokaryon (Gr. *heteros*, other + *karyon*, kernel) A fungal hypha that has two or more genetically distinct types of nuclei.

heterotroph (Gr. *heteros*, other + *trophos*, feeder) An organism that does not have the ability to produce its own food. *See also* autotroph.

heterozygote (Gr. *heteros*, other + *zygotos*, a pair) A diploid individual carrying two different alleles of a gene on its two homologous chromosomes.

hierarchical (Gr. *hieros*, sacred + *archos*, leader) Refers to a system of classification in which successively smaller units of classification are included within one another.

histone (Gr. *histos*, tissue) A complex of small, very basic polypeptides rich in the amino acids arginine and lysine. A basic part of chromosomes, histones form the core around which DNA is wrapped.

homeostasis (Gr. *homeos*, similar + *stasis*, standing) The maintaining of a relatively stable internal physiological environment in an organism or steady-state equilibrium in a population or ecosystem.

homeotherm (Gr. *homeo*, similar + *therme*, heat) An organism, such as a bird or mammal, capable of maintaining a stable body temperature independent of the environmental temperature. "Warm-blooded."

hominid (L. *homo*, man) Human beings and their direct ancestors. A member of the family Hominidae. *Homo sapiens* is the only living member.

homologous chromosome (Gr. *homologia*, agreement) One of the two nearly identical versions of each chromosome. Chromosomes that associate in pairs in the first stage of meiosis. In diploid cells, one chromosome of a pair that carries equivalent genes.

homology (Gr. *homologia*, agreement) A condition in which the similarity between two structures or functions is indicative of a common evolutionary origin.

homozygote (Gr. *homos*, same or similar + *zygotos*, a pair) A diploid individual whose two copies of a gene are the same. An individual carrying identical alleles on both homologous chromosomes is said to be homozygous for that gene.

hormone (Gr. *hormaein*, to excite) A chemical messenger, often a steroid or peptide, produced in a small quantity in one part of an organism and then transported to another part of the organism, where it brings about a physiological response.

hybrid (L. *hybrida*, the offspring of a tame sow and a wild boar) A plant or animal that results from the crossing of dissimilar parents.

hybridization The mating of unlike parents of different taxa.

hydrogen bond A molecular force formed by the attraction of the partial positive charge of one hydrogen atom of a water molecule with the partial negative charge of the oxygen atom of another.

hydrolysis reaction (Gr. *hydro*, water + *lyse*, break) The process of tearing down a polymer by adding a molecule of water. A hydrogen is attached to one subunit and a hydroxyl to the other, which breaks the covalent bond. Essentially the reverse of a dehydration reaction.

hydrophobic (Gr. *hydro*, water + *phobos*, hating) Nonpolar molecules, which do not form hydrogen bonds with water and therefore are not soluble in water.

hydroskeleton (Gr. *hydro*, water + *skeletos*, hard) The skeleton of most soft-bodied invertebrates that have neither an internal nor an external skeleton. They use the relative incompressibility of the water within their bodies as a kind of skeleton.

hypertonic (Gr. *hyper*, above + *tonos*, tension) A cell that contains a higher concentration of solutes than its surrounding solution.

hypha, *pl.* hyphae (Gr. *hyphe*, web) A filament of a fungus. A mass of hyphae comprises a mycelium.

hypothalamus (Gr. *hypo*, under + *thalamos*, inner room) The region of the brain under the thalamus that controls temperature, hunger, and thirst and that produces hormones that influence the pituitary gland.

hypothesis (Gr. *hypo*, under + *tithenai*, to put) A proposal that might be true. No hypothesis is ever proven correct. All hypotheses are provisional—proposals that are retained for the time being as useful but that may be rejected in the future if found to be inconsistent with new information. A hypothesis that stands the test of time—often tested and never rejected—is called a theory.

hypotonic (Gr. *hypo*, under + *tonos*, tension) A solution surrounding a cell that has a lower concentration of solutes than does the cell.

I

inbreeding The breeding of genetically related plants or animals. In plants, inbreeding results from self-pollination. In animals, inbreeding results from matings between relatives. Inbreeding tends to increase homozygosity.

incomplete dominance The ability of two alleles to produce a heterozygous phenotype that is different from either homozygous phenotype.

independent assortment Mendel's second law: The principle that segregation of alternative alleles at one locus into gametes is independent of the segregation of alleles at other loci. Only true for gene loci located on different chromosomes or those so far apart on one chromosome that crossing over is very frequent between the loci.

industrial melanism (Gr. *melas*, black) The evolutionary process in which a population of initially light-colored organisms becomes a population of dark organisms as a result of natural selection.

inflammatory response (L. *inflammare*, to flame) A generalized nonspecific response to infection that acts to clear an infected area of infecting microbes and dead tissue cells so that tissue repair can begin.

integument (L. *integumentum*, covering) The natural outer covering layers of an animal. Develops from the ectoderm.

interneuron A nerve cell found only in the CNS that acts as a functional link between sensory neurons and motor neurons. Also called association neuron.

internode The region of a plant stem between nodes where stems and leaves attach.

interoception (L. *interus*, inner + Eng. [re]ceptive) The sensing of information that relates to the body itself, its internal condition, and its position.

interphase That portion of the cell cycle preceding mitosis. It includes the G_1 phase, when cells grow, the S phase, when a replica of the genome is synthesized, and a G_2 phase, when preparations are made for genomic separation.

intron (L. *intra*, within) A segment of DNA transcribed into mRNA but removed before translation. These untranslated regions make up the bulk of most eukaryotic genes.

ion An atom in which the number of electrons does not equal the number of protons. An ion carries an electrical charge.

ionic bond A chemical bond formed between ions as a result of the attraction of opposite electrical charges.

ionizing radiation High-energy radiation, such as X rays and gamma rays.

isolating mechanisms Mechanisms that prevent genetic exchange between individuals of different populations or species.

isotonic (Gr. *isos*, equal + *tonos*, tension) A cell with the same concentration of solutes as its environment.

isotope (Gr. *isos*, equal + *topos*, place) An atom that has the same number of protons but different numbers of neutrons.

J

joint The part of a vertebrate where one bone meets and moves on another.

K

karyotype (Gr. *karyon*, kernel + *typos*, stamp or print) The particular array of chromosomes that an individual possesses.

kinetic energy The energy of motion.

kinetochore (Gr. *kinetikos*, putting in motion + *choros*, chorus) A disk of protein bound to the centromere to which microtubules attach during cell division, linking chromatids to the spindle.

kingdom The chief taxonomic category. This book recognizes six kingdoms: Archaea, Bacteria, Protista, Fungi, Animalia, and Plantae.

L

lamella, *pl.* lamellae (L. a little plate) A thin, platelike structure. In chloroplasts, a layer of chlorophyll-containing membranes. In bivalve mollusks, one of the two plates forming a gill. In vertebrates, one of the thin layers of bone laid concentrically around the Haversian canals.

ligament (L. *ligare*, to bind) A band or sheet of connective tissue that links bone to bone.

linkage The patterns of assortment of genes that are located on the same chromosome. Important because if the genes are located relatively far apart, crossing over is more likely to occur between them than if they are close together.

lipid (Gr. *lipos*, fat) A loosely defined group of molecules that are insoluble in water but soluble in oil. Oils such as olive, corn, and coconut are lipids, as well as waxes, such as beeswax and earwax.

lipid bilayer The basic foundation of all biological membranes. In such a layer, the nonpolar tails of phospholipid molecules point inward, forming a nonpolar zone in the interior of the bilayers. Lipid bilayers are selectively permeable and do not permit the diffusion of water-soluble molecules into the cell.

littoral (L. *litus*, shore) Referring to the shoreline zone of a lake or pond or the ocean

that is exposed to the air whenever water recedes.

locus, *pl.* loci (L. place) The position on a chromosome where a gene is located.

loop of Henle After F. G. J. Henle, German anatomist. A hairpin loop formed by a urine-conveying tubule when it enters the inner layer of the kidney and then turns around to pass up again into the outer layer of the kidney.

lymph (L. *lympha*, clear water) In animals, a colorless fluid derived from blood by filtration through capillary walls in the tissues.

lymphatic system An open circulatory system composed of a network of vessels that function to collect the water within blood plasma forced out during passage through the capillaries and to return it to the bloodstream. The lymphatic system also returns proteins to the circulation, transports fats absorbed from the intestine, and carries bacteria and dead blood cells to the lymph nodes and spleen for destruction.

lymphocyte (Gr. *lympha*, water + Gr. *kytos*, hollow vessel) A white blood cell. A cell of the immune system that either synthesizes antibodies (B cells) or attacks virus-infected cells (T cells).

lyse (Gr. *lysis*, loosening) To disintegrate a cell by rupturing its plasma membrane.

M

macromolecule (Gr. *makros*, large + L. *moliculus*, a little mass) An extremely large molecule. Refers specifically to carbohydrates, lipids, proteins, and nucleic acids.

macrophage (Gr. *makros*, large + -*phage*, eat) A phagocytic cell of the immune system able to engulf and digest invading bacteria, fungi, and other microorganisms, as well as cellular debris.

marrow The soft tissue that fills the cavities of most bones and is the source of red blood cells.

mass flow The overall process by which materials move in the phloem of plants.

mass number The mass number of an atom consists of the combined mass of all of its protons and neutrons.

meiosis (Gr. *meioun*, to make smaller) A special form of nuclear division that precedes gamete formation in sexually reproducing eukaryotes. It results in four haploid daughter cells.

Mendelian ratio After Gregor Mendel, Austrian monk. Refers to the characteristic 3:1 segregation ratio that Mendel observed, in which pairs of alternative traits are expressed in the F_2 generation in the ratio of three-fourths dominant to one-fourth recessive.

menstruation (L. *mens*, month) Periodic sloughing off of the blood-enriched lining of the uterus when pregnancy does not occur.

meristem (Gr. *merizein*, to divide) In plants, a zone of unspecialized cells whose only function is to divide.

mesoderm (Gr. *mesos*, middle + *derma*, skin) One of the three embryonic germ layers that form in the gastrula. Gives rise to muscle, bone, and other connective tissue; the peritoneum; the circulatory system; and most of the excretory and reproductive systems.

mesophyll (Gr. *mesos*, middle + *phyllon*, leaf) The photosynthetic parenchyma of a leaf, located within the epidermis. The vascular strands (veins) run through the mesophyll.

metabolism (Gr. *metabole*, change) The process by which all living things assimilate energy and use it to grow.

metamorphosis (Gr. *meta*, after + *morphe*, form + *osis*, state of) Process in which form changes markedly during postembryonic development—for example, tadpole to frog or larval insect to adult.

metaphase (Gr. *meta*, middle + *phasis*, form) The stage of mitosis characterized by the alignment of the chromosomes on a plane in the center of the cell.

metastasis, *pl.* metastases (Gr. to place in another way) The spread of cancerous cells to other parts of the body, forming new tumors at distant sites.

microevolution (Gr. *mikros*, small + L. *evolvere*, to unfold) Refers to the evolutionary process itself. Evolution within a species. Also called adaptation.

microtubule (Gr. *mikros*, small + L. *tubulus*, little pipe) In eukaryotic cells, a long, hollow cylinder about 25 nanometers in diameter and composed of the protein tubulin. Microtubules influence cell shape, move the chromosomes in cell division, and provide the functional internal structure of cilia and flagella.

mimicry (Gr. *mimos*, mime) The resemblance in form, color, or behavior of certain organisms (mimics) to other more powerful or more protected ones (models), which results in the mimics being protected in some way.

mitochondrion, *pl.* mitochondria (Gr. *mitos*, thread + *chondrion*, small grain) A tubular or sausage-shaped organelle 1 to 3 micrometers long. Bounded by two membranes, mitochondria closely resemble the aerobic bacteria from which they were originally derived. As chemical furnaces of the cell, they carry out its oxidative metabolism.

mitosis (Gr. *mitos*, thread) The M phase of cell division in which the microtubular apparatus is assembled, binds to the chromosomes, and moves them apart. This phase is the essential step in the separation of the two daughter cell genomes.

mole (L. *moles*, mass) The atomic weight of a substance, expressed in grams. One mole is defined as the mass of 6.0222×10^{23} atoms.

molecule (L. *moliculus*, a small mass) The smallest unit of a compound that displays the properties of that compound.

monocot Short for monocotyledon; flowering plant in which the embryos have only one cotyledon, the flower parts are often in threes, and the leaves typically are parallel-veined.

monosaccharide (Gr. *monos*, one + *sakcharon*, sugar) A simple sugar.

morphogenesis (Gr. *morphe*, form + *genesis*, origin) The formation of shape. The growth and differentiation of cells and tissues during development.

motor endplate The point where a neuron attaches to a muscle. A neuromuscular synapse.

multicellularity A condition in which the activities of the individual cells are coordinated and the cells themselves are in contact. A property of eukaryotes alone and one of their major characteristics.

muscle (L. *musculus*, mouse) The tissue in the body of humans and animals that can be contracted and relaxed to make the body move.

muscle cell A long, cylindrical, multinucleated cell that contains numerous myofibrils and is capable of contraction when stimulated.

muscle spindle A sensory organ that is attached to a muscle and sensitive to stretching.

mutagen (L. *mutare*, to change) A chemical capable of damaging DNA.

mutation (L. *mutare*, to change) A change in a cell's genetic message.

mutualism (L. *mutuus*, lent, borrowed) A symbiotic relationship in which both participating species benefit.

mycelium, *pl.* mycelia (Gr. *mykes*, fungus) In fungi, a mass of hyphae.

mycology (Gr. *mykes*, fungus) The study of fungi. A person who studies fungi is called a mycologist.

mycorrhiza, *pl.* mycorrhizae (Gr. *mykes*, fungus + *rhiza*, root) A symbiotic association between fungi and plant roots.

myofibril (Gr. *myos*, muscle + L. *fibrilla*, little fiber) An elongated structure in a muscle fiber, composed of myosin and actin.

myosin (Gr. *myos*, muscle + *in*, belonging to) One of two protein components of myofilaments. (The other is actin.)

N

natural selection The differential reproduction of genotypes caused by factors in the environment. Leads to evolutionary change.

nematocyst (Gr. *nema*, thread + *kystos*, bladder) A coiled, threadlike stinging structure of cnidarians that is discharged to capture prey and for defense.

nephron (Gr. *nephros*, kidney) The functional unit of the vertebrate kidney. A human kidney has more than 1 million nephrons that filter waste matter from the blood. Each nephron consists of a Bowman's capsule, glomerulus, and tubule.

nerve A bundle of axons with accompanying supportive cells, held together by connective tissue.

nerve impulse A rapid, transient, self-propagating reversal in electrical potential that travels along the membrane of a neuron.

neuromodulator A chemical transmitter that mediates effects that are slow and longer lasting and that typically involve second messengers within the cell.

neuromuscular junction The structure formed when the tips of axons contact (innervate) a muscle fiber.

neuron (Gr. nerve) A nerve cell specialized for signal transmission.

neurotransmitter (Gr. *neuron*, nerve + L. *trans*, across + *mitere*, to send) A chemical released at an axon tip that travels across the synapse and binds a specific receptor protein in the membrane on the far side.

neurulation (Gr. *neuron*, nerve) The elaboration of a notochord and a dorsal nerve cord that marks the evolution of the chordates.

neutron (L. *neuter*, neither) A subatomic particle located within the nucleus of an atom. Similar to a proton in mass, but as its name implies, a neutron is neutral and possesses no charge.

neutrophil An abundant type of white blood cell capable of engulfing microorganisms and other foreign particles.

niche (L. *nidus*, nest) The role an organism plays in the environment; realized niche is the niche that an organism occupies under natural circumstances; fundamental niche is the niche an organism would occupy if competitors were not present.

nitrogen fixation The incorporation of atmospheric nitrogen into nitrogen compounds, a process that can be carried out only by certain microorganisms.

nocturnal (L. *nocturnus*, night) Active primarily at night.

node (L. *nodus*, knot) The place on the stem where a leaf is formed.

node of Ranvier After L. A. Ranvier, French histologist. A gap formed at the point where two Schwann cells meet and where the axon is in direct contact with the surrounding intercellular fluid.

nondisjunction The failure of homologous chromosomes to separate in meiosis I. The cause of Down syndrome.

nonrandom mating A phenomenon in which individuals with certain genotypes sometimes mate with one another more commonly than would be expected on a random basis.

notochord (Gr. *noto*, back + L. *chorda*, cord) In chordates, a dorsal rod of cartilage that forms between the nerve cord and the developing gut in the early embryo.

nucleic acid A nucleotide polymer. A long chain of nucleotides. Chief types are deoxyribonucleic acid (DNA), which is double-stranded, and ribonucleic acid (RNA), which is typically single-stranded.

nucleosome (L. *nucleus*, kernel + *soma*, body) The basic packaging unit of eukaryotic chromosomes, in which the DNA molecule is wound around a ball of histone proteins. Chromatin is composed of long strings of nucleosomes, like beads on a string.

nucleotide A single unit of nucleic acid, composed of a phosphate, a five-carbon sugar (either ribose or deoxyribose), and a purine or a pyrimidine.

nucleus (L. *a kernel*, dim. Fr. *nux*, nut) A spherical organelle (structure) characteristic of eukaryotic cells. The repository of the genetic information that directs all activities of a living cell. In atoms, the central core, containing positively charged protons and (in all but hydrogen) electrically neutral neutrons.

O

oocyte (Gr. oion, egg + kytos, vessel) A cell in the outer layer of the ovary that gives rise to an ovum. A primary oocyte is any of the 2 million oocytes a female is born with, all of which have begun the first meiotic division.

operon (L. operis, work) A cluster of functionally related genes transcribed onto a single mRNA molecule. A common mode of gene regulation in prokaryotes; it is rare in eukaryotes other than fungi.

order A taxonomic category ranking below a class and above a family.

organ (L. organon, tool) A complex body structure composed of several different kinds of tissue grouped together in a structural and functional unit.

organelle (Gr. organella, little tool) A specialized compartment of a cell. Mitochondria are organelles.

organism Any individual living creature, either unicellular or multicellular.

organ system A group of organs that function together to carry out the principal activities of the body.

osmoconformer An animal that maintains the osmotic concentration of its body fluids at about the same level as that of the medium in which it is living.

osmoregulation The maintenance of a constant internal solute concentration by an organism, regardless of the environment in which it lives.

osmosis (Gr. osmos, act of pushing, thrust) The diffusion of water across a membrane that permits the free passage of water but not that of one or more solutes. Water moves from an area of low solute concentration to an area with higher solute concentration.

osmotic pressure The increase of hydrostatic water pressure within a cell as a result of water molecules that continue to diffuse inward toward the area of lower water concentration (the water concentration is lower inside than outside the cell because of the dissolved solutes in the cell).

osteoblast (Gr. osteon, bone + blastos, bud) A bone-forming cell.

osteocyte (Gr. osteon, bone + kytos, hollow vessel) A mature osteoblast.

outcross A term used to describe species that interbreed with individuals other than those like themselves.

oviparous (L. ovum, egg + parere, to bring forth) Refers to reproduction in which the eggs are developed after leaving the body of the mother, as in reptiles.

ovulation The successful development and release of an egg by the ovary.

ovule (L. ovulum, a little egg) A structure in a seed plant that becomes a seed when mature.

ovum, pl. ova (L. egg) A mature egg cell. A female gamete.

oxidation (Fr. oxider, to oxidize) The loss of an electron during a chemical reaction from one atom to another. Occurs simultaneously with reduction. Is the second stage of the 10 reactions of glycolysis.

oxidative metabolism A collective term for metabolic reactions requiring oxygen.

oxidative respiration Respiration in which the final electron acceptor is molecular oxygen.

P

parasitism (Gr. para, beside + sitos, food) A symbiotic relationship in which one organism benefits and the other is harmed.

parthenogenesis (Gr. parthenos, virgin + Eng. genesis, beginning) The development of an adult from an unfertilized egg. A common form of reproduction in insects.

partial pressures (P) The components of each individual gas—such as nitrogen, oxygen, and carbon dioxide—that together constitute the total air pressure.

pathogen (Gr. pathos, suffering + Eng. genesis, beginning) A disease-causing organism.

pedigree (L. pes, foot + grus, crane) A family tree. The patterns of inheritance observed in family histories. Used to determine the mode of inheritance of a particular trait.

peptide (Gr. peptein, to soften, digest) Two or more amino acids linked by peptide bonds.

peptide bond A covalent bond linking two amino acids. Formed when the positive (amino, or NH_2) group at one end and a negative (carboxyl, or COOH) group at the other end undergo a chemical reaction and lose a molecule of water.

peristalsis (Gr. peri, around + stellein, to wrap) The rhythmic sequences of waves of muscular contraction in the walls of a tube.

pH Refers to the concentration of H^+ ions in a solution. The numerical value of the pH is the negative of the exponent of the molar concentration. Low pH values indicate high concentrations of H^+ ions (acids), and high pH values indicate low concentrations (bases).

phagocyte (Gr. phagein, to eat + kytos, hollow vessel) A cell that kills invading cells by engulfing them. Includes neutrophils and macrophages.

phagocytosis (Gr. phagein, to eat + kytos, hollow vessel) A form of endocytosis in which cells engulf organisms or fragments of organisms.

phenotype (Gr. phainein, to show + typos, stamp or print) The realized expression of the genotype. The observable expression of a trait (affecting an individual's structure, physiology, or behavior) that results from the biological activity of proteins or RNA molecules transcribed from the DNA.

pheromone (Gr. pherein, to carry + [hor]mone) A chemical signal emitted by certain animals as a means of communication.

phloem (Gr. phloos, bark) In vascular plants, a food-conducting tissue basically composed of sieve elements, various kinds of parenchyma cells, fibers, and sclereids.

phosphodiester bond The bond that results from the formation of a nucleic acid chain in which individual sugars are linked together in a line by the phosphate groups. The phosphate group of one sugar binds to the hydroxyl group of another, forming an—O—P—O bond.

photon (Gr. photos, light) The unit of light energy.

photoperiodism (Gr. photos, light + periodos, a period) A mechanism that organisms use to measure seasonal changes in relative day and night length.

photorespiration A process in which carbon dioxide is released without the production of ATP or NADPH. Because it produces neither ATP nor NADPH, photorespiration acts to undo the work of photosynthesis.

photosynthesis (Gr. photos, light + -syn, together + tithenai, to place) The process by which plants, algae, and some bacteria use the energy of sunlight to create from carbon dioxide (CO_2) and water (H_2O) the more complicated molecules that make up living organisms.

phototropism (Gr. photos, light + trope, turning to light) A plant's growth response to a unidirectional light source.

phylogeny (Gr. phylon, race, tribe) The evolutionary relationships among any group of organisms.

phylum, pl. phyla (Gr. phylon, race, tribe) A major taxonomic category, ranking above a class.

physiology (Gr. physis, nature + logos, a discourse) The study of the function of cells, tissues, and organs.

pigment (L. pigmentum, paint) A molecule that absorbs light.

pinocytosis (Gr. pinein, to drink + kytos, cell) A form of endocytosis in which the material brought into the cell is a liquid containing dissolved molecules.

pistil (L. pistillum, pestle) Central organ of flowers, typically consisting of ovary, style, and stigma; a pistil may consist of one or more fused carpels and is more technically and better known as the gynoecium.

plankton (Gr. planktos, wandering) The small organisms that float or drift in water, especially at or near the surface.

plasma (Gr. form) The fluid of vertebrate blood. Contains dissolved salts, metabolic wastes, hormones, and a variety of proteins, including antibodies and albumin. Blood minus the blood cells.

plasma membrane A lipid bilayer with embedded proteins that control the cell's permeability to water and dissolved substances.

plasmid (Gr. plasma, a form or something molded) A small fragment of DNA that replicates independently of the bacterial chromosome.

platelet (Gr. dim of plattus, flat) In mammals, a fragment of a white blood cell that circulates in the blood and functions in the formation of blood clots at sites of injury.

pleiotropy (Gr. pleros, more + trope, a turning) A gene that produces more than one phenotypic effect.

polarization The charge difference of a neuron so that the interior of the cell is negative with respect to the exterior.

polar molecule A molecule with positively and negatively charged ends. One portion of a polar molecule attracts electrons more strongly than another portion, with the result that the molecule has electron-rich (−) and electron-poor (+) regions, giving it magnetlike positive and negative poles. Water is one of the most polar molecules known.

pollen (L. fine dust) A fine, yellowish powder consisting of grains or microspores, each of which contains a mature or immature male gametophyte. In flowering plants, pollen is released from the anthers of flowers and fertilizes the pistils.

pollen tube A tube that grows from a pollen grain. Male reproductive cells move through the pollen tube into the ovule.

pollination The transfer of pollen from the anthers to the stigmas of flowers for fertilization, as by insects or the wind.

polygyny (Gr. poly, many + gyne, woman, wife) A mating choice in which a male mates with more than one female.

polymer (Gr. polus, many + meris, part) A large molecule formed of long chains of similar molecules called subunits.

polymerase chain reaction (PCR) A process by which DNA polymerase is used to copy a sequence of DNA repeatedly, making millions of copies of the same DNA.

polymorphism (Gr. polys, many + morphe, form) The presence in a population of more than one allele of a gene at a frequency greater than that of newly arising mutations.

polynomial system (Gr. polys, many + [bi]nomial) Before Linnaeus, naming a genus by use of a cumbersome string of Latin words and phrases.

polyp A cylindrical, pipe-shaped cnidarian usually attached to a rock with the mouth facing away from the rock on which it is growing. Coral is made up of polyps.

polypeptide (Gr. polys, many + peptein, to digest) A general term for a long chain of amino acids linked end to end by peptide bonds. A protein is a long, complex polypeptide.

polysaccharide (Gr. polys, many + sakcharon, sugar) A sugar polymer. A carbohydrate composed of many monosaccharide sugar subunits linked together in a long chain.

population (L. populus, the people) Any group of individuals of a single species, occupying a given area at the same time.

posterior (L. post, after) Situated behind or farther back.

potential difference A difference in electrical charge on two sides of a membrane caused by an unequal distribution of ions.

potential energy Energy with the potential to do work. Stored energy.

predation (L. praeda, prey) The eating of other organisms. The one doing the eating is called a predator, and the one being consumed is called the prey.

primary growth In vascular plants, growth originating in the apical meristems of shoots and roots, as contrasted with secondary growth; results in an increase in length.

primary plant body The part of a plant that arises from the apical meristems.

primary producers Photosynthetic organisms, including plants, algae, and photosynthetic bacteria.

primary structure of a protein The sequence of amino acids that makes up a particular polypeptide chain.

primordium, pl. primordia (L. primus, first + ordiri, begin) The first cells in the earliest stages of the development of an organ or structure.

productivity The total amount of energy of an ecosystem fixed by photosynthesis per unit of time. Net productivity is productivity minus that which is expended by the metabolic activity of the organisms in the community.

prokaryote (Gr. pro, before + karyon, kernel) A simple organism that is small, single-celled, and has little evidence of internal structure.

promoter An RNA polymerase binding site. The nucleotide sequence at the end of a gene to which RNA polymerase attaches to initiate transcription of mRNA.

prophase (Gr. pro, before + phasis, form) The first stage of mitosis during which the chromosomes become more condensed, the nuclear envelope is reabsorbed, and a network of microtubules (called the spindle) forms between opposite poles of the cell.

protein (Gr. proteios, primary) A long chain of amino acids linked end to end by peptide bonds. Because the 20 amino acids that occur in proteins have side groups with very different chemical properties, the function and shape of a protein is critically affected by its particular sequence of amino acids.

protist (Gr. protos, first) A member of the kingdom Protista, which includes unicellular eukaryotic organisms and some multicellular lines derived from them.

proton A subatomic particle in the nucleus of an atom that carries a positive charge. The number of protons determines the chemical character of the atom because it dictates the number of electrons orbiting the nucleus and available for chemical activity.

protostome (Gr. protos, first + stoma, mouth) An animal in whose embryonic development the mouth forms at or near the blastopore. Also characterized by spiral cleavage.

protozoa (Gr. protos, first + zoion, animal) The traditional name given to heterotrophic protists.

pseudocoel (Gr. pseudos, false + koiloma, cavity) A body cavity similar to the coelom except that it forms between the mesoderm and endoderm.

punctuated equilibrium A hypothesis of the mechanism of evolutionary change that proposes that long periods of little or no change are punctuated by periods of rapid evolution.

Q

quaternary structure of a protein A term to describe the way multiple protein subunits are assembled into a whole.

R

radial symmetry (L. radius, a spoke of a wheel + Gr. summetros, symmetry) The regular arrangement of parts around a central axis so that any plane passing through the central axis divides the organism into halves that are approximate mirror images.

radioactivity The emission of nuclear particles and rays by unstable atoms as they decay into more stable forms. Measured in curies, with 1 curie equal to 37 billion disintegrations a second.

radula (L. scraper) A rasping, tonguelike organ characteristic of most mollusks.

recessive allele An allele whose phenotype effects are masked in heterozygotes by the presence of a dominant allele.

recombination The formation of new gene combinations. In bacteria, it is accomplished by the transfer of genes into cells, often in association with viruses. In eukaryotes, it is accomplished by reassortment of chromosomes during meiosis and by crossing over.

reducing power The use of light energy to extract hydrogen atoms from water.

reduction (L. reductio, a bringing back; originally, "bringing back" a metal from its oxide) The gain of an electron during a chemical reaction from one atom to another. Occurs simultaneously with oxidation.

reflex (L. reflectere, to bend back) An automatic consequence of a nerve stimulation. The motion that results from a nerve impulse passing through the system of neurons, eventually reaching the body muscles and causing them to contract.

refractory period The recovery period after membrane depolarization during which the membrane is unable to respond to additional stimulation.

renal (L. renes, kidneys) Pertaining to the kidney.

repression (L. reprimere, to press back, keep back) The process of blocking transcription by the placement of the regulatory protein between the polymerase and the gene, thus blocking movement of the polymerase to the gene.

repressor (L. reprimere, to press back, keep back) A protein that regulates transcription of mRNA from DNA by binding to the operator and so preventing RNA polymerase from attaching to the promoter.

resolving power The ability of a microscope to distinguish two points as separate.

respiration (L. respirare, to breathe) The utilization of oxygen. In terrestrial vertebrates, the inhalation of oxygen and the exhalation of carbon dioxide.

resting membrane potential The charge difference that exists across a neuron's membrane at rest (about 70 millivolts).

restriction endonuclease A special kind of enzyme that can recognize and cleave DNA molecules into fragments. One of the basic tools of genetic engineering.

restriction fragment-length polymorphism (RFLP) An associated genetic mutation marker detected because the mutation alters the length of DNA segments.

retrovirus (L. retro, turning back) A virus whose genetic material is RNA rather than DNA. When a retrovirus infects a cell, it makes a DNA copy of itself, which it can then insert into the cellular DNA as if it were a cellular gene.

ribose A five-carbon sugar.

ribosome A cell structure composed of protein and RNA that translates RNA copies of genes into protein.

RNA polymerase The enzyme that transcribes RNA from DNA.

S

saltatory conduction A very fast form of nerve impulse conduction in which the impulses leap from node to node over insulated portions.

sarcoma (Gr. sarx, flesh) A cancerous tumor that involves connective or hard tissue, such as muscle.

sarcomere (Gr. sarx, flesh + meris, part of) The fundamental unit of contraction in skeletal muscle. The repeating bands of actin and myosin that appear between two Z lines.

sarcoplasmic reticulum (Gr. sarx, flesh + plassein, to form, mold; L. reticulum, network) The endoplasmic reticulum of a muscle cell. A sleeve of membrane that wraps around each myofilament.

scientific creationism A view that the biblical account of the origin of the earth is literally true, that the earth is much younger than most scientists believe, and that all species of organisms were individually created just as they are today.

secondary growth In vascular plants, growth that results from the division of a cylinder of cells around the plant's periphery. Secondary growth causes a plant to grow in diameter.

second messenger An intermediary compound that couples extracellular signals to intracellular processes and also amplifies a hormonal signal.

seed A structure that develops from the mature ovule of a seed plant. Contains an embryo and a food source surrounded by a protective coat.

selection The process by which some organisms leave more offspring than competing ones and their genetic traits tend to appear in greater proportions among members of succeeding generations than the traits of those individuals that leave fewer offspring.

self-fertilization The transfer of pollen from an anther to a stigma in the same flower or to another flower of the same plant.

sepal (L. sepalum, a covering) A member of the outermost whorl of a flowering plant. Collectively, the sepals constitute the calyx.

septum, pl. septa (L. saeptum, a fence) A partition or cross-wall, such as those that divide fungal hyphae into cells.

sex chromosomes In humans, the X and Y chromosomes, which are different in the two sexes and are involved in sex determination.

sex-linked characteristic A genetic characteristic that is determined by genes located on the sex chromosomes.

sexual reproduction Reproduction that involves the regular alternation between syngamy and meiosis. Its outstanding characteristic is that an individual offspring inherits genes from two parent individuals.

shoot In vascular plants, the aboveground parts, such as the stem and leaves.

sieve cell In the phloem (food-conducting tissue) of vascular plants, a long, slender sieve element with relatively unspecialized sieve areas and with tapering end walls that lack sieve plates. Found in all vascular plants except angiosperms, which have sieve-tube members.

soluble Refers to polar molecules that dissolve in water and are surrounded by a hydration shell.

solute The molecules dissolved in a solution. See also solution, solvent.

solution A mixture of molecules, such as sugars, amino acids, and ions, dissolved in water.

solvent The most common of the molecules in a solution. Usually a liquid, commonly water.

somatic cells (Gr. soma, body) All the diploid body cells of an animal that are not involved in gamete formation.

somite A segmented block of tissue on either side of a developing notochord.

species, pl. species (L. kind, sort) A level of taxonomic hierarchy; a species ranks next below a genus.

sperm (Gr. sperma, sperm, seed) A sperm cell. The male gamete.

spindle The mitotic assembly that carries out the separation of chromosomes during cell division. Composed of microtubules and assembled during prophase at the centrioles of the dividing cell.

spore (Gr. spora, seed) A haploid reproductive cell, usually unicellular, that is capable of developing into an adult without fusion with another cell. Spores result from meiosis, as do gametes, but gametes fuse immediately to produce a new diploid cell.

sporophyte (Gr. spora, seed + phyton, plant) The spore-producing, diploid ($2n$) phase in the life cycle of a plant having alternation of generations.

stabilizing selection A form of selection in which selection acts to eliminate both extremes from a range of phenotypes.

stamen (L. thread) The part of the flower that contains the pollen. Consists of a slender filament that supports the anther. A flower that produces only pollen is called staminate and is functionally male.

steroid (Gr. stereos, solid + L. ol, from oleum, oil) A kind of lipid. Many of the molecules that function as messengers and pass across cell membranes are steroids, such as the male and female sex hormones and cholesterol.

steroid hormone A hormone derived from cholesterol. Those that promote the development of the secondary sexual characteristics are steroids.

stigma (Gr. mark) A specialized area of the carpel of a flowering plant that receives the pollen.

stoma, pl. stomata (Gr. mouth) A specialized opening in the leaves of some plants that allows carbon dioxide to pass into the plant body and allows water vapor and oxygen to pass out of them.

stratum corneum The outer layer of the epidermis of the skin of the vertebrate body.

substrate (L. substratus, strewn under) A molecule on which an enzyme acts.

substrate-level phosphorylation The generation of ATP by coupling its synthesis to a strongly exergonic (energy-yielding) reaction.

succession In ecology, the slow, orderly progression of changes in community composition that takes place through time. Primary succession occurs in nature on bare substrates, over long periods of time. Secondary succession occurs when a climax community has been disturbed.

sugar Any monosaccharide or disaccharide.

surface tension A tautness of the surface of a liquid, caused by the cohesion of the liquid molecules. Water has an extremely high surface tension.

surface-to-volume ratio Describes cell size increases. Cell volume grows much more rapidly than surface area.

symbiosis (Gr. syn, together with + bios, life) The condition in which two or more dissimilar organisms live together in close association; includes parasitism, commensalism, and mutualism.

synapse (Gr. synapsis, a union) A junction between a neuron and another neuron or muscle cell. The two cells do not touch. Instead, neurotransmitters cross the narrow space between them.

synapsis (Gr. synapsis, contact, union) The close pairing of homologous chromosomes that occurs early in prophase I of meiosis. With the genes of the chromosomes thus aligned, a DNA strand of one homologue can pair with the complementary DNA strand of the other.

syngamy (Gr. syn, together with + gamos, marriage) Fertilization. The union of male and female gametes.

T

taxonomy (Gr. taxis, arrangement + nomos, law) The science of the classification of organisms.

T cell A type of lymphocyte involved in cell-mediated immune responses and interactions with B cells. Also called a T lymphocyte.

tendon (Gr. tenon, stretch) A strap of connective tissue that attaches muscle to bone.

tertiary structure of a protein The three-dimensional shape of a protein. Primarily the result of hydrophobic interactions of amino acid side groups and, to a lesser extent, of hydrogen bonds between them. Forms spontaneously.

test cross A cross between a heterozygote and a recessive homozygote. A procedure Mendel used to further test his hypotheses.

theory (Gr. *theorein*, to look at) A well-tested hypothesis supported by a great deal of evidence.

thigmotropism (Gr. *thigma*, touch + *trope*, a turning) The growth response of a plant to touch.

thorax (Gr. a breastplate) The part of the body between the head and the abdomen.

thylakoid (Gr. *thylakos*, sac + *-oides*, like) A flattened, saclike membrane in the chloroplast of a eukaryote. Thylakoids are stacked on top of one another in arrangements called grana and are the sites of photosystem reactions.

tissue (L. *texere*, to weave) A group of similar cells organized into a structural and functional unit.

trachea, *pl.* tracheae (L. windpipe) In vertebrates, the windpipe.

tracheid (Gr. *tracheia*, rough) An elongated cell with thick, perforated walls that carries water and dissolved minerals through a plant and provides support. Tracheids form an essential element of the xylem of vascular plants.

transcription (L. *trans*, across + *scribere*, to write) The first stage of gene expression in which the RNA polymerase enzyme synthesizes an mRNA molecule whose sequence is complementary to the DNA.

translation (L. *trans*, across + *latus*, that which is carried) The second stage of gene expression in which a ribosome assembles a polypeptide, using the mRNA to specify the amino acids.

translocation (L. *trans*, across + *locare*, to put or place) In plants, the process in which most of the carbohydrates manufactured in the leaves and other green parts of the plant are moved through the phloem to other parts of the plant.

transpiration (L. *trans*, across + *spirare*, to breathe) The loss of water vapor by plant parts, primarily through the stomata.

transposon (L. *transponere*, to change the position of) A DNA sequence carrying one or more genes and flanked by insertion sequences that confer the ability to move from one DNA molecule to another. An element capable of transposition (the changing of chromosomal location).

trophic level (Gr. *trophos*, feeder) A step in the movement of energy through an ecosystem.

tropism (Gr. *trop*, turning) A plant's response to external stimuli. A positive tropism is one in which the movement or reaction is in the direction of the source of the stimulus. A negative tropism is one in which the movement or growth is in the opposite direction.

turgor pressure (L. *turgor*, a swelling) The pressure within a cell that results from the movement of water into the cell. A cell with high turgor pressure is said to be turgid.

U

unicellular Composed of a single cell.

urea (Gr. *ouron*, urine) An organic molecule formed in the vertebrate liver. The principal form of disposal of nitrogenous wastes by mammals.

urine (Gr. *ouron*, urine) The liquid waste filtered from the blood by the kidneys.

V

vaccination The injection of a harmless microbe into a person or animal to confer resistance to a dangerous microbe.

vacuole (L. *vacuus*, empty) A cavity in the cytoplasm of a cell that is bound by a single membrane and contains water and waste products of cell metabolism. Typically found in plant cells.

variable Any factor that influences a process. In evaluating alternative hypotheses about one variable, all other variables are held constant so that the investigator is not misled or confused by other influences.

vascular bundle In vascular plants, a strand of tissue containing primary xylem and primary phloem. These bundles of elongated cells conduct water with dissolved minerals and carbohydrates throughout the plant body.

vascular cambium In vascular plants, the meristematic layer of cells that gives rise to secondary phloem and secondary xylem. The activity of the vascular cambium increases stem or root diameter.

ventral (L. *venter*, belly) Refers to the bottom portion of an animal. Opposite of dorsal.

vertebrate An animal having a backbone made of bony segments called vertebrae.

vesicle (L. *vesicula*, a little (ladder) Membrane-enclosed sacs within eukaryotic cells.

vessel element In vascular plants, a typically elongated cell, dead at maturity, that conducts water and solutes in the xylem.

villus, *pl.* villi (L. a tuft of hair) In vertebrates, fine, microscopic, fingerlike projections on epithelial cells lining the small intestine that serve to increase the absorptive surface area of the intestine.

vitamin (L. *vita*, life + *amine*, of chemical origin) An organic substance that the organism cannot synthesize, but is required in minute quantities by an organism for growth and activity.

viviparous (L. *vivus*, alive + *parere*, to bring forth) Refers to reproduction in which eggs develop within the mother's body and young are born free-living.

voltage-gated channel A transmembrane pathway for an ion that is opened or closed by a change in the voltage, or charge difference, across the cell membrane.

W

water vascular system The system of water-filled canals connecting the tube feet of echinoderms.

whorl A circle of leaves or of flower parts present at a single level along an axis.

wood Accumulated secondary xylem. Heartwood is the central, nonliving wood in the trunk of a tree. Hardwood is the wood of dicots, regardless of how hard or soft it actually is. Softwood is the wood of conifers.

X

xylem (Gr. *xylon*, wood) In vascular plants, a specialized tissue, composed primarily of elongate, thick-walled conducting cells, that transports water and solutes through the plant body.

Y

yolk (O.E. *geolu*, yellow) The stored substance in egg cells that provides the embryo's primary food supply.

Z

zygote (Gr. *zygotos*, paired together) The diploid (2n) cell resulting from the fusion of male and female gametes (fertilization).

Credits

Photographs

Chapter 1

Opener: © Frans Lanting; **1.1(top left):** © R. Robinson/Visuals Unlimited; **1.1(top middle):** © Alfred Pasieka /Science Photo Library/Photo Researchers; **1.1(top right):** © Corbis/Volume 64; **1.1(bottom left):** © PhotoDisc/BS Volume 15; **1.1(bottom middle):** © Corbis/Volume 46; **1.1(bottom right):** © PhotoDisc/Volume 44; **1.2:** © T.E. Adams/Visuals Unlimited; **1.3:** © Corbis/Volume 6; **p. 5(crane):** © Jim Bailey; **p. 5(goose):** © Jim Bailey; **p. 5(top & middle left):** © Raymond Gehman/Corbis; **p. 5(middle right):** © Royalty-Free/Corbis; **p. 5(bottom both):** © Winfried Wisniewski/zefa/Corbis; **p. 7(top left):** © Joe McDonald/Animals Animals /Earth Scenes; **p.7(middle left):** © Tom McHugh/Photo Researchers; **p. 7(top center):** © Kenneth Fink/Photo Researchers; **p. 7(top right):** © Michael Fogden/DRK Photo; **p. 7(middle):** © Royalty-Free/Corbis; **p. 7(bottom left):** © Royalty-free/Corbis; **p. 7(bottom right):** © Runk/Schoenberger/Grant Heilman Photography; **p. 8:** Courtesy of Bill Ober; **1.8:** Ozone Processing Team, Goddard Space Flight Center, NASA; **1.9:** © Laurel Hungerford; **1.10:** © Dennis Kunkel / Phototake; **1.13:** © Leonard Lessin/Peter Arnold; **1.14:** © Photo Disc/Vol. 6; **1.15(top left):** © Nature Picture Library/Alamy Images; **1.15(top middle):** © James Gregory/ Alamy Images; **1.15(top right):** © Royalty-Free/Corbis; **1.15(second row left):** © Royalty-Free/Corbis; **1.15(second row middle):** © Royalty-Free/Corbis; **1.15(second row right):** © Corbis/Vol. 6; **1.15(third row left & middle):** © Royalty-Free/Corbis; **1.15(third row right):** © Corbis/Vol.6; **1.15(fourth row left):** © Royalty-Free/Corbis; **1.15(fourth row right):** © PhotoDisc/Getty Images; **1.15(bottom left):** © Dr. T.J. Beveridge/Visuals Unlimited; **1.15(bottom middle):** © John D. Cunningham/Visuals Unlimited; **1.15(bottom right):** © Dennis Kunkel Microscopy, Inc.

Chapter 2

Opener(top left): Courtesy of Dr. Robert H. Rothman, Department of Biological Sciences, Rochester Institute of Technology; **(top right):** © Tui De Roy/ Minden Pictures; **(bottom left):** © D. Parera & E. Parera-Cook/Auscape/ Minden Pictures; **(bottom right):** Courtesy of Dr. Robert H. Rothman, Department of Biological Sciences, Rochester Institute of Technology; **2.1:** © Huntington Library/Superstock; **2.2:** From Darwin, the Life of a Tormented Evolutionist, by Adrian Desmond; **2.8:** © Mary Evans Picture Library/Photo Researchers; **p. 29:** Chas. McRae/Visuals Unlimited; **2.10(both):** Courtesy of Dr. Arkhat Abzhanov, Harvard School of Dental Medicine, Photographer: Dr. Peter Grant of Princeton University; **2.13:** Cleveland Hickman; **2.16a:** © Vanessa Vick/Photo Researchers; **2.16b:** © Edward S. Ross; **2.16c:** © Ken Lucas/Visuals Unlimited; **2.16d:** © Edward S. Ross; **2.17:** © BrandX/Getty Images; **2.18:** © H. Richard Johnston/Getty Images; **2.19:** © Royalty-Free/Corbis; **2.20(all):** Courtesy of J.B. Losos; **2.22:** © Parks Canada; photographer: Lynch, W.–1996; **2.23:** © Michael & Patricia Fogden/Corbis; **2.24:** © Paulo De Oliveira/Getty Images; **2.25:** © Jeremy Woodhouse/PhotoDisc Blue/Getty Images; **2.26:** © Fred Bavendam/Minden Pictures; **2.27:** © B. Runk/S. Schoenberger/Grant Heilman Photography; **p. 40:** © Michael & Patricia Fogden/Corbis

Chapter 3

Opener: © Ernest H. Robl; **p. 50:** Courtesy of Who Zoo (whozoo.org); **3.6(both):** Courtesy of PETNET Solutions, A Siemens Company; **3.12a:** © Royalty-Free/Corbis; **3.12b:** © Hermann Eisenbeiss/National Audubon Society Collection/Photo Researchers; **3.14(top):** © The McGraw-Hill Companies, Inc./Bob Coyle, photographer; **3.14(middle):** © The McGraw-Hill Companies, Inc./Jacques Cornell photographer; **3.14(bottom):** © The McGraw-Hill Companies, Inc./Jill Braaten photographer; **p. 54:** © Gilbert S. Grant/National Audubon Society Collection/Photo Researchers; **p. 56(both):** © Photo Archives South Tyrol Museum of Archaeology–www.iceman.it

Chapter 4

Opener: © Michael Viard/ Peter Arnold, Inc.; **4.7a:** © Manfred Kage/Peter Arnold, Inc.; **4.7b:** © PhotoDisc/Volume 6; **4.7c:** © Scott Blackman/Tom Stack & Associates; **4.7d:** © PhotoDisc/Volume 6; **p. 69:** © A.C. Barrington Brown/Photo Researchers, Inc.; **4.15:** © J.D. Litvay/Visuals Unlimited; **p. 71(first, second):** © Royalty-Free/Corbis; **p. 71(third):** © PhotoDisc/Getty; **p. 71(fourth):** © Royalty-Free/Corbis; **p. 71(fifth):** © Scott Johnson/Animals Animals/Earth Scenes; **4.17b:** © Getty Images; **4.17c:** © C Squared Studios/Getty Images; **4.18c,d:** © Brand X Pictures/PunchStock; **p. 74:** © Dr. David M. Phillips / Visuals Unlimited

Chapter 5

Opener: © Manfred Kage/ Peter Arnold, Inc.; **p. 81(bottom left):** © Microworks/Phototake; **p.81(top left):** © David M. Phillips/Visuals Unlimited; **p. 81(second row left):** © M. Abbey/Visuals Unlimited; **p. 81(third row left):** © David M. Phillips/Visuals Unlimited; **p. 81(top right):** © Mike Abbey/Visuals Unlimited; **p. 81(second row right):** © K.G. Murti/Visuals Unlimited; **p. 81(third row right):** © David Becker/Science Photo Library/Photo Researchers; **p. 81(bottom right):** © Stanley Flegler/Visuals Unlimited; **p. 84:** © SIU/Visuals Unlimited; **5.8a:** © Andrew Syred / Photo Researchers, Inc; **5.8b:** © Alfred Pasieka/Photo Researchers, Inc; **5.8c:** © Microfield Scientific Ltd / Photo Researchers, Inc; **5.13:** © R. Bolender & D. Fawcett/Visuals Unlimited; **5.14b:** Courtesy of Dr. Charles Flickinger, Medical Cellular Biology, W.B. Saunders, 1979; **5.16b:** © Don W. Fawcett/Visuals Unlimited; **5.17:** Courtesy of Dr. Kenneth Miller, Brown University; **5.21b:** © Stanley Flegler/Visuals Unlimited; **5.23:** © Biophoto Associates/Photo Researchers; **5.24:** © Biophoto Associates/Photo Researchers; **5.29b:** Courtesy of Dr. Birgit Satir, Albert Einstein College of Medicine; **5.30(all):** Courtesy of M.M. Perry & A.B. Gilbert, Cell Science, 39-257, 1979

Chapter 6

Opener: © Jane Buron/Bruce Coleman Inc.; **6.3(both):** © Spencer Grant/Photo Edit

Chapter 7

Opener: © Skip Moody/Dembinsky Photo Associates; **7.1(left):** Manfred Kage/Peter Arnold; **7.1(right):** Courtesy of Dr. Kenneth Miller, Brown University; **7.5(both):** © Eric Soder/Tom Stack & Associates

Chapter 8

Opener: © John Gerlach/Animals Animals/Earth Scenes; **8.1:** © Robert A. Caputo/Aurora & Quanta Productions, Inc.; **p. 143:** © Royalty-Free/Corbis; **p. 145(second, third):** © PhotoDisc/Vol. 44; **p. 145(fourth, last):** © Edward S. Ross; 145(first): © Corbis/Vol.145

Chapter 9

Opener: Kindly provided by Prof. Chun-Ming Liu, Chinese Academy of Sciences; **9.1a:** © Lee D. Simon/Photo Researchers; **9.4:** © SPL/Photo Researchers; **9.5:** © Biophoto Associates/Photo Researchers Inc.; **9.7(all):** © Andrew S. Bajer; **9.8a:** © David M. Phillips/Visuals Unlimited; **9.9:** © Cabisco/Visuals Unlimited; **9.12:** © Moredun Animal Health LTD/Science Photo Library/Photo Researchers, Inc.; **p. 166:** Courtesy of Dr. Roland Zell

Chapter 10

Opener: © Andrew S. Bajer; **10.2:** © David Cavagnaro/Peter Arnold; **10.8(all):** © C.A. Hasenkampf/Biological Photo Service; **p. 179:** Sir John Tenniel; **p. 180:** Reprinted with permission from Science Vol. 311, no. 5759, 20 January 2006. Image: Khodjakov. Copyright 2006 AAAS

Chapter 11

Opener: © Richard Gross/Biological Photography; **11.1:** Courtesy of American Museum of Natural History; **11.4:** Courtesy R. W. Van Norman; **11.12a:** From Albert & Blakeslee "Corn and Man," Journal of Heredity V. 5, pg. 511, 1914, Oxford University Press; **11.15(both):** © Fred Bruemmer; **p. 197:** © RubberBall Productions/Getty Images; **11.17a:** © Richard Hutchings/Photo Researchers; **11.17b:** © Cheryl A. Ertelt/Visuals Unlimited; **11.17c:** © William H. Mullins/Photo Researchers; **11.17d:**

© Gerard Lacz/Peter Arnold Inc.; **11.18:** © Heidi Kellams; **11.20(both):** © CBS/Phototake; **11.24a:** Courtesy of Loris McGavran, Denver Children's Hospital; **11.24b:** © Joseph Sohm; ChromoSohm Inc./Corbis; **11.28:** © Bettmann/Corbis; **11.30(both):** © Stanley Flegler/Visuals Unlimited; **11.35:** © Yoav Levy/Phototake; **11.36:** © Pascal Goetgheluck/Photo Researchers, Inc.

Chapter 12
Opener: © Michael Dunning/Getty Images; **12.4(top & a):** From J.D. Watson, *The Double Helix Atheneum*, New York, 1968. Cold Spring Harbor Lab; **12.4b:** © A.C. Barrington Brown/Photo Researchers, Inc.; **12.10:** © David Scharf; **p. 224:** From M. Meselson and F.W. Stahl/*Proceedings of the Nat. Acad. of Sci.* 44 (1958): 671

Chapter 13
Opener: N. Ban, P. Nissen, J. Hansen, P.B. Moore & T.A. Steitz, "The Complete Atomic Structure of the Large Ribosomal Subunit at 2.4 A Resolution," reprinted with permission from *Science* v. 289 #5481, p917,© American Association for the Advancement of Science; **p. 238:** © David M. Phillips/Visuals Unlimited

Chapter 14
Opener: © The Roslin Institute; **14.3(top left):** © Dr. Gopal Murti/ Science Photo Library/Photo Researchers, Inc.; **14.3(top right):** Courtesy of Dr. Ken Culver, Photo by John Crawford, National Institutes of Health; **14.3(bottom left):** © Robert H. Devlin/Fisheries & Oceans Canada; **14.3(bottom right):** Courtesy of Monsanto Corporation; **14.6:** Courtesy of R.L. Brinster, U. of Pennsylvania Sch. of Vet. Med.; **14.9:** Courtesy of Monsanto Corporation; **14.10:** Courtesy of Ingo Potrykus & Peter Beyer, photo by Peter Beyer; **p. 255:** © Jack Hollingsworth/Getty Images; **14.11:** © Getty Images; **p. 257:** © AP/Wide World Photos; **14.13(bottom left):** © Getty Images; **14.13(top left):** © Reuters /Landov; **14.13(bottom second from left):** © AP Wide World; **14.13(top second from left):** © Getty Images; **14.13(bottom third from left):** © AP Photo/Texas A&M University; **14.13(top third from left):** © Giovanna Lazzara/AP Wide World; **14.13(bottom right):** © Reuters/Corbis; **14.13(top right):** © Hwang Woo-suk/AP Wide World; **14.15:** © University of Wisconsin-Madison News & Public Affairs; **14.17:** © SUW-Madison News & Public Affairs, Photo by Jeff Miller; **p. 266 right:** © Seoul National University/Handout/Reuters/Corbis

Chapter 15
Opener: © Breck P. Kent/Animals Animals/Earth Scenes; **15.4:** Courtesy of Michael Richardson and Ronan O'Rahilly; **15.12:** Courtesy of Dr. Victor A. McKusick, Johns Hopkins University; **15.19:** Courtesy of H. Rodd; **p. 293:** © Peter E. Smith/The Natural Sciences Image Library; **15.20a:** © Corbis/ Vol. 6; **15.20b:** © Royalty-Free/Corbis; **15.20c:** © Porterfield/Chickering/Photo Researchers Inc.; **15.21a:** © John Shaw/Tom Stack & Associates; **15.21b:** © Rob & Ann Simpson/Visuals Unlimited; **15.21c:** © Suzanne L. Collins & Joseph T. Collins/ National Audubon Society Collection/Photo Researchers; **15.21d:** © Phil A. Dotson/National Audubon Society Collection/Photo Researchers

Chapter 16
Opener: © Dave Watts/ Tom Stack & Associates; **16.2(top left):** © Dwight R. Kuhn; **16.2(top middle):** © Heather Angel; **16.2(top right):** © S. Maslowski/ Visuals Unlimited; **16.2(bottom left):** © Henry Ausloos/ Animals Animals/Earth Scenes; **16.2(bottom middle):** © John Cancalosi/Peter Arnold Inc.; **16.2(bottom right):** © Manfred Danegger/Peter Arnold, Inc.; **16.4(left):** © Gerard Lacz/Peter Arnold, Inc.; **16.4(middle):** © Grant Heilman/Grant Heilman Photography; **16.4(right):** © Ralph Reinhold/Animals Animals/Earth Scenes; **16.7(top):** © Royalty-Free/Corbis; **16.7(second row left):** © PhotoDisc/Getty Images; **16.7(second row right):** © Corbis/Vol.6; **16.7(third row left):** © PhotoDisc/Getty Images; **16.7(third row right):** © Royalty-Free/Corbis; **16.7(fourth row):** © PhotoDisc/ Getty Images; **16.7(fifth row):** © Getty Images; **16.7(top left):** © PhotoDisc/Vol. 6; **16.7(bottom left):** © Royalty-Free/Corbis; **16.7(bottom first middle):** © Royalty-Free/Corbis; **16.7(bottom second middle):** © Royalty-Free/Corbis; **16.7(bottom second right):** © PhotoDisc/Getty Images; **p. 309:** © PhotoDisc/Vol.44; **16.11b:** © OSF/ Animals Animals/Earth Scenes

Chapter 17
Opener: © Dwight R. Kuhn; **p. 320(top):** NASA/ JPL/Cornell; **p. 320(bottom):** NASA; **17.5:** Courtesy of Dr. Charles Brinton; **17.3:** © Abraham & Beachey/BPS/Tom Stack & Associates; **17.4(top left):** Royalty-Free/Index Stock; **17.4(middle left):** © PhotoDisc/EP073; **17.4(bottom left):** © Royalty-Free/ Corbis; **17.4(top right):** © Gene Ott; **17.4(middle right):** © Eric and David Hosking/ Corbis; **17.4(bottom right):** © Tim Zurowski/ Corbis; **17.8,17.9:** © John D. Cunningham/Visuals Unlimited; **17.11(both):** © Royalty-Free/Corbis; **17.12:** © Bill Keogh/Visuals Unlimited; **17.13:** © L. West/Photo Researchers; **17.14:** © Royalty-Free/Corbis

Chapter 18
Opener: © Royalty-Free/Corbis; **18.1:** © Royalty-Free/Corbis; **18.2:** © Terry Ashley/ Tom Stack & Associates; **18.4a:** © Edward S. Ross; **18.4b:** © Richard Gross/Biological Photography; **18.6:** © Edward S. Ross; **18.7:** Courtesy of Hans Steur, The Netherlands; **18.8:** © E.J. Cable/Tom Stack & Associates; **18.9a:** © Edward S. Ross; **18.9b:** © Royalty-Free/Corbis; **18.9c:** © Kingsley R. Stern; **18.9d:** © Edward S. Ross; **18.9e:** © Rod Planck/ Tom Stack & Associates; **18.11:** © Kingsley R. Stern; **18.13(left):** © Alan and Linda Detrick/ Photo Researchers, Inc.; **18.13(right):** © R.J. Delorit, Agronomy Publications; **18.14:** © Carolina Biological/ Visuals Unlimited; **18.15a:** © Walter H. Hodge/Peter Arnold Inc.; **18.15b:** © Kjell Sandved/ Butterfly Alphabet; **18.15c:** © Runk/Schoenberger/ Grant Heilman Photography; **18.17b:** © Ed Pembleton; **18.18a,b:** © Thomas Eisner; **18.18c:** © OSF/Animals Animals/Earth Scenes; **18.19:** © Michael & Patricia Fogden; **18.22a:** © Patrick Johns/Corbis; **18.22b:** © James L. Castner; **18.22c:** © Kingsley R. Stern

Chapter 19
Opener: © James H. Robinson/Animals Animals/ Earth Scenes; **19.1a:** © Corbis/Volume 86; **19.1b:** © Corbis/Volume 65; **19.1c:** © David M. Phillips/ Visuals Unlimited; **19.1d:** © Royalty-Free/Corbis; **19.1e:** © Edward S. Ross; **19.1f:** © Corbis; **19.1g:** © Cleveland P. Hickman; **19.1h:** © Cabisco/Phototake; **19.1i:** © Ed Reschke; **19.2a:** © Alamy Images; **19.2b:** © Royalty Free/ Corbis; **19.3:** © Gwen Fidler/Tom Stack & Associates; **19.7b:** © Stan Elems/Visuals Unlimited; **19.7a:** © T.E. Adams/Visuals Unlimited; **19.9a:** © Larry Jensen/Visuals Unlimited; **19.9b:** © T.E. Adams/Visuals Unlimited; **19.10a:** © Image Ideas, Inc./ PictureQuest; **19.10b:** © Kjell Sandved/ Butterfly Alphabet; **19.10c:** © Fred Bavendam/Peter Arnold Inc.; **19.11:** © David M. Dennis/Tom Stack & Associates; **19.14a:** © Kjell Sandved/Butterfly Alphabet; **19.14b:** © Edward S. Ross; **19.14c:** © Don Valenti/Tom Stack & Associates; **19.14d:** © John Gerlach/Visuals Unlimited; **19.14e:** © Edward S. Ross; **19.14f:** © J.A. Alcock/Visuals Unlimited; **19.14g:** © Cleveland P. Hickman; **19.16a:** © Randy Morse/Tom Stack & Associates; **19.16b:** © Carl Roessler/Tom Stack & Associates; **19.16c:** © William C. Ober; **19.16d:** © Jeff Rotman; **19.16e:** © Daniel W. Gotshall; **19.16f:** © Alex Kirstitch/ Visuals Unlimited; **19.17:** © Jim & Cathy Church; **19.18:** © Heather Angel; **19.19:** © Eric N. Olsen/The University of Texas MD Anderson Cancer Center; **19.21:** © Stephen Frink Collection/ Alamy Images; **19.23:** © John Shaw/Tom Stack & Associates; **19.28a:** © Erwin & Peggy Bauer/ Tom Stack & Associates; **19.28b:** © Charles Philip Cangialosi/ Corbis; **19.28c:** © Corbis/Vol. 5

Chapter 20
Opener: © Bill Ross/ Corbis; **20.3a:** © Dave G. Houser/ Corbis; **20.3b:** © Royalty-Free/Corbis; **20.3c:** © Corbis/Volume 86; **20.3d,e:** © Edward S. Ross; **20.8:** © Doug Sokell/Tom Stack & Associates; **20.19a:** © Digital Vision/PictureQuest; **20.19b:** © W. Gregory Brown/Animals Animals/Earth Scenes; **20.20a:** © Jim Church; **20.20b:** Courtesy of J. Frederick Grassel, Woods Hole Oceanographic Institution; **20.20c:** © Kenneth L. Smith; **20.21:** © Edward S. Ross; **20.22a:** © Fred Rhode/Visuals Unlimited; **20.22b:** © Dwight Kuhn; **20.26:** © Michael Graybill & Jan Hodder/Biological Photo Service; **20.27:** © E.R. Degginger/Photo Researchers; **20.28:** © S.J. Krasemann/Peter Arnold, Inc.; **20.29:** © J. Weber/Visuals Unlimited; **20.30:** © IFA/Peter Arnold, Inc.; **20.31:** © Charlie Ott/The National Audubon Society Collection/Photo Researchers; **20.32:** © John Shaw/Tom Stack & Associates; **20.33:** © Tom McHugh/Photo Researchers; **20.34:** © Dave Watts/Tom Stack & Associates; **20.35:** © E.R. Degginger/Animals Animals/Earth Scenes; **20.36:** © Gunter Ziesler/Peter Arnold, Inc.; **p. 410:** U.S. Forest Service

Chapter 21
Opener: © ABPL/Gavin Thomson/Animals Animals; **21.1b:** © Manfred Danegger/Peter Arnold, Inc.; **21.6:** Courtesy of National Museum of Natural History, Smithsonian Institution; **21.8a:** © Tim Davis/Photo Researchers; **21.8b:** © Digital Vision;

21.8c: © Royalty-Free/Corbis; 21.13: © Merlin D. Tuttle, Bat Conservation International; 21.14: © N&C Photography/Peter Arnold, Inc.; 21.15: © Michael Fogden/DRK Photo; 21.16a: © Edward S. Ross; 21.16b: Courtesy D. Wixted. Reprinted with permission from Davis, E. L., and A. E. MacGuidwin. 2000. Lesion nematode disease. The Plant Health Instructor. DOI: 10.1094/PHI-I-2000-1030-02; 21.16c: © Roger Wilmshurst/The National Audubon Society Collection/Photo Researchers; 21.17: © Jim Harvey/Visuals Unlimited; 21.18: © Digital Vision/Getty Images; 21.19: Courtesy of Rolf O. Peterson; 21.21a: © Tom J. Ulrich/Visuals Unlimited; 21.22: © Anne Wertheim /Animals Animals/ Earth Scenes; 21.23(both): © Edward S.Ross; 21.24(both): © Lincoln P. Brower, Sweet Briar College; 21.25: © James L. Castner; 21.26: © Royalty Free/PictureQuest; 21.27a: © Edward S. Ross; 21.27b: © Paul A. Opler; 21.28(all): © Edward S. Ross; 21.29: © Peter Chew; 21.31(all): © Tom Bean; p. 436: © Jim Bailey

Chapter 22

Opener: © Renee Lynn/ Photo Researchers; 22.1: © Stone/Getty Images; 22.3 a,b: From J.R. Brown et al, "A defect in nurturing mice lacking…gene for fosB"Cell v. 86, 1996, pp. 297–308, © Cell Press; 22.4: Thomas McAvoy, Life Magazine/© Time, Inc./Getty Images; 22.6: Courtesy of Bernd Heinrich; 22.7: Nina Leen, Life Magazine, © Time Inc/Getty Images.; 22.8: © Bios(C. Thouvenin)/ Peter Arnold, Inc.; p. 451(top left): © Eric & David Hosling/ Corbis; p. 451(top middle): © NGS/ Getty; p. 451(top right): © Alamy Images; p. 451(bottom left): Courtesy of James Traniello; p. 451(bottom middle): © K. Ammann/Bruce Coleman Inc.; p. 451(bottom right): © Mark Moffett / Minden Pictures; 22.11: © Dr. Don W. Fawcett / Visuals Unlimited; 22.12: © Dr. Mark Moffett / Minden Pictures; 22.13: © S. Osolinski/OSF/Animals Animals/Earth Scenes; 22.14: © Nigel Dennis/National Audubon Society Collection/ Photo Researchers; 22.16: © David Hosking/ Photo Researchers; 22.17: © Raymond Mendez/ Animals Animals; 22.18: © Michael Neugebauer; 22.19: © Jim Pickerell/Alamy Images

Chapter 23

Opener: © Steve McCurry/National Geographic Society; 23.2: © Grant Heilman/Grant Heilman Photography; 23.3: © Rob & Ann Simpson/Visuals Unlimited; 23.4: NASA; p. 467: © 1999 Ed Ely/ Biological Photo Service; 23.9: © Byron Augustine/ Tom Stack & Associates; 23.10a: © Royalty-Free/ Corbis; 23.10b: © Creatas/PunchStock; 23.11: © Gary Griffen/Animals Animals/Earth Scenes; 23.12a: © James Blair/National Geographic Society; 23.12b: NASA; 23.12c: © Frans Lanting/Minden Pictures; 23.14: © Stephanie Maze/Woodfin Camp & Associates; 23.20a: Courtesy of University of Wisconsin-Madison Arboretum; 23.20b: Courtesy of University of Wisconsin-Madison Arboretum; 23.21: © Kennan Ward/ Corbis; 23.22: © Merlin D. Tuttle/Bat Conservation International; 23.23(both): Courtesy of Nashua River Watershed Association; 23.24: © T. Henneghan/ImageState

Chapter 24

Opener: © Anthony Bannister/Animals Animals/ Earth Scenes; 24.4: © Jim Merli/Visuals Unlimited; p. 491(top): © Ed Reschke/Peter Arnold Inc.; p. 491(second, third): © Ed Reschke; p. 491(fourth): © Fred Hossler/Visuals Unlimited; p. 491(bottom): © Ed Reschke; p. 493(top): © Biology Media/ Photo Researchers; p. 493(second): © Biophoto Associates/Photo Researchers; p. 493(third): © Chuck Brown/Photo Researchers; p. 493(fourth): © Biophoto Associates/Photo Researchers; p. 493(bottom): © Ken Eward/Science Source/Photo Researchers; 24.6a,b: © Dr. Michael Klein/Peter Arnold, Inc.; p. 496(all): © Ed Reschke; 24.10: © Cleveland P. Hickman, Jr.; 24.11: © David M. Dennis/Tom Stack & Associates; p. 501: © Royalty Free/ Corbis; p. 504(left): © Pete Saloutos/Corbis; p. 504(top): © Digital Vision; p. 504(bottom): © Digital Vision/Getty Images

Chapter 25

Opener: © David M. Phillips/ Visuals Unlimited; 25.5, 25.6: © Ed Reschke; 25.10: © Manfred Kage/ Peter Arnold Inc.; 25.17(all): Courtesy of Frank P. Sloop, Jr.; p. 522(left): © Rob Simpson/Visuals Unlimited; p. 522(right): © Jeremy Woodhouse/ Getty Images

Chapter 26

Opener(top left): © Royalty-Free/Corbis; (bottom left): © Tom Brakefield/Digital Vision/Getty Images; (top right): © Royalty-Free/Corbis; (middle right): © Getty Images; (bottom right): © Royalty-Free/ Corbis; p. 536: © Mark Chivers/Robert Harding World Imagery/Getty Images

Chapter 27

Opener: © Royalty-Free/Corbis; 27.14b: © David M. Phillips/Visuals Unlimited; 27.14c: © Biophoto Associates/Science Source/Photo Researchers

Chapter 28

Opener: © Royalty-Free/Corbis; 28.11: © Larry Brock/Tom Stack & Associates; p. 572: © Michael & Patricia Fogden

Chapter 29

Opener: © CDC/Science Source/Photo Researchers; 29.1: © Jeremy Woodhouse/Getty Images; 29.3: © Rob Simpson/Visuals Unlimited; 29.4: © Manfred Kage/Peter Arnold, Inc.; 29.6: Courtesy of Dr. Gilla Kaplan, Public Health Research Institute, Newark, NJ; 29.9d: From Alan S. Rosenthal, Regulation of the Immune Response-Role of the Macrophage, New England Journal of Medicine 303: 1153, 1980; 29.18: © Visuals Unlimited; 29.19: National Library of Medicine; 29.22: © Oliver Meckes/Photo Researchers; p. 596: © J. Cavallini/ Custom Medical Stock Photo

Chapter 30

Opener: © Dr. Dennis Kunkel/Visuals Unlimited; 30.6: © Dennis Kunkel/Phototake; 30.7b: © John Heuser, Washington University School of Medicine, St. Louis, MO; 30.8: © E.R. Lewis, YY Zeevi, T.E. Everhart, U. of California/Biological Photo Service; 30.11: © Ernest Wilkinson/Animals Animals/Earth Scenes; 30.17: Dr. Marcus E. Rachle, Washington University, McDonnell Center for High Brain Function; 30.18: Photo Lennart Nilsson/Albert Bonnier Forlag AB,Behold Man, Little Brown & Co.; 30.23: © Wendy Shatil/Bob Rozinski/Tom Stack & Associates; 30.31: © Stephen Dalton/Animals Animals/Earth Scenes; 30.33(left): © David M. Dennis/Tom Stack & Associates; 30.33(middle): © Kjell Sandved/Butterfly Alphabet; 30.33(right): © Kjell Sandved/Butterfly Alphabet; 30.35: Courtesy of Beckman Vision Center at UCSF School of Medicine; D. Copenhagen, S. Mittman, M. Maglio; 30.38: Reproduced from Ishihara's Tests for Colour Deficiency, published by Kanehara Trading Inc., Tokyo, Japan; 30.41: © Leonard L. Rue, III

Chapter 31

Openers: © E.R. Degginger/Animals Animals; 31.11b: © John Paul Kay/Peter Arnold, Inc.

Chapter 32

Opener: © Photo Researchers; 32.2a: © Chuck Wise/Animals Animals/Earth Scenes; 32.2b: © Fred McConnaughey/The National Audubon Society Collection/Photo Researchers; 32.4: © David Doubilet; 32.6: © Hans Pfletschinger/Peter Arnold Inc.; 32.8: © Cleveland P. Hickman; 32.9: © Frans Lanting/Minden Pictures; 32.10a: © Jean Phillippe Varin/Jacana/Photo Researchers; 32.10b: © Tom McHugh/The National Audubon Society Collection/ Photo Researchers; 32.10c: © Corbis/Volume 86; 32.13: © David M. Phillips/Photo Researchers; 32.19: © Ed Reschke; 32.20(first three): Photo Lennart Nilsson/Albert Bonniers Forlag AB, A Child is Born, Dell Publishing Co.; 32.20(last): Photo Lennart Nilsson/Albert Bonniers Forlag AB,Behold Man, Little Brown & Co.; 32.23(all): © The McGraw-Hill Companies, Inc./Bob Coyle, photographer

Chapter 33

Opener: © Scott T. Smith/ Corbis; 33.2: © Biophoto Associates/ Photo Researchers; 33.3: © George Wilder/Visuals Unlimited; 33.4: © Lawrence Mellinchamp/Visuals Unlimited; 33.5a: © Dr. Jeremy Burgess/Science Photo Library/ Photo Researchers, Inc.; 33.5b: © Andrew Syred/ Science Photo Library/ Photo Reseachers, Inc.; 33.6d: Courtesy of Wilfred Cote, SUNY College of Environmental Forestry; 33.7a: © Randy Moore/ Visuals Unlimited; 33.8b: © Terry Ashley/Tom Stack & Associates; 33.9a: © Kingsley R. Stern; 33.9b: Photomicrograph by G.S. Ellmore; 33.10: © E.J. Cable/Tom Stack & Associates; 33.12(both): © Ed Reschke; 33.14: © CBS/Phototake; 33.15a: © Michael P. Gadomski/ Dembinsky Photo Associates; 33.15b: © Royalty-Free/Corbis; 33.15c: © Steven J. Baskauf; 33.15d: © Jack Wilburn/ Animals Animals; 33.15e: © Corbis; 33.15f: © Royalty-Free/Corbis; 33.16a: © Kjell Sandved/ Butterfly Alphabet; 33.16b: © Pat Anderson/ Visuals Unlimited; 33.17: © Ed Reschke; 33.21: © Kingsley R. Stern; 33.24(left): © J.A.L. Cooke/ Animals Animals/Earth Scenes; 33.24(right): © Robert Mitchell/Tom Stack & Associates

Chapter 34

Opener: ©Adam Hart-Davis/SPL/PhotoResearchers; **34.1c:** © Jerome Wexler/Photo Researchers; **34.3a:** © Corbis; **34.3b:** Courtesy of www.laspilitas.com; **34.5a:** © Alamy Images; **34.5b:** USDA photo by Jack Dykinga; **34.5c:** © Alamy Images; **34.5d:** © Bill Bonner; **34.5e:** © Royalty-Free/Corbis; **34.5f:** © Ed Reschke/Peter Arnold, Inc.; **34.6(left):** © Holt Studios International Ltd/Alamy Images; **34.6(right):** © Helmut Gritscher/Peter Arnold Inc.; **34.12:** © Runk/Schoenberger/Grant Heilman Photography; **34.13a:** © Malcolm Wilkins; **34.13b:** © Malcolm Wilkins; **34.14:** © Kingsley R. Stern; **34.17(both):** © Runk/Schoenberge/Grant Heilman Photography; **34.17c:** © John D. Cunningham/Visuals Unlimited; **p. 702:** © Runk/Schaenberger/Grant Heilman

Line Art and Text

Chapter 2
2.21: Data from E.J. Heske, et al., *Ecology*, 1994.

Chapter 5
TA 5.13: Data from Darnell, Lodah, Baltimore, *Molecular Cell Biology*, 2/e, Fig 14.3, page 534.

Chapter 16
16.3: From Miles Eldredge, Life in the Balance, in *Natural History*, June 1998. Reprinted by permission of Patricia J. Wynne.

Chapter 21
21.12: Data from Shluter & McPhail, 1993. After Begon, Harper, Townsend, *Ecology: Individuals, Populations, and Communities,* 3/e. Copyright © 1993 Blackwell Publishing. Reprinted by permission.

Chapter 23
23.17: U.S. Census Bureau, International Data Base.

Chapter 24
TA 24.17: Schmidt Nielson, *Animal Physiology* 3/e Cambridge Univ. Press, Fig 6.19, page 215.

Chapter 25
TA 25.2: Schmidt Nielson, *Animal Physiology* 3/e Cambridge Univ. Press, Fig. 4.11, page 110.

Chapter 26
TA 26.3: Schmidt Nielson, *Animal Physiology* 3/e Cambridge Univ. Press 1983, Fig 3.7, page 79.

Chapter 27
27.1: www.mypyramid.gov U.S. Department of Agriculture, U.S. Department of Health and Human Services.

Chapter 29
29.16: From Immunity and the Invertebrates, *Scientific American*, November 1996. Copyright Robert Osti Illustrations. Reprinted with permission.

Chapter 33
33.25: Data from the U.S. Centers for Disease Control and Prevention.

ScienCentral news stories are supported in part by the National Science Foundation (http://www.nsf.gov), Division of Informal Science Education, under Grants No. ESI0206184, ESI021155, and ESI9904457. Any opinions, findings, and conclusions or recommendations expressed in this material are those of the author(s) and do not necessarily reflect the views of the National Science Foundation.

Index

A

AAV (adeno-associated virus), 265–266
ABA (abscisic acid), 695 *table,* 699, 699 *fig.*
Abdominal cavity, 486
ABO blood groups, 199, 592, 592 *fig.*
Abscisic acid (ABA), 695 *table,* 699, 699 *fig.*
Abstinence as birth control method, 665
Acacias, ants and, 425, 425 *fig.*
Accessory pigments, 125, 125 *fig.*
Acetylcholine (ACh), 605
Acetyl-CoA, producing, 144, 144 *fig.*
ACh (acetylcholine), 605
Acid(s), 53
    buffers and, 55, 55 *fig.*
    enzyme activity and pH and, 114, 114 *fig.*
Acid rain, 463, 463 *fig.*
Acini, 554
Acoelomate, 366
Acquired immunodeficiency syndrome (AIDS),
    325, 474, 475 *fig.,* 575, 594 *fig.,* 594–595. *See*
    also Human immunodeficiency virus (HIV)
    vaccine for, 590–591, 595
Acromegaly, 634
ACTH (adrenocorticotropic hormone), 635, 636,
    641 *table*
Actias luna, 373 *fig.*
Actin filaments, 500, 502
Action potentials, 603, 603 *fig.*
Activators, 115, 234–235, 235 *fig.*
Active immunity, 586, 586 *fig.,* 587 *fig.*
Active site of enzyme, 113, 113 *fig.*
Active transport, 103 *fig.,* 103–104, 104 *fig.,*
    105 *table*
Adaptation, evolution and, 269, 288–291
Adaptive radiation, 32, 32 *fig.*
Adeno-associated virus (AAV), 265–266
Adenosine diphosphate (ADP), 116, 116 *fig.*
Adenosine triphosphate (ATP), 103, 116–118
    formation of, 138
    production of, 140–142, 141 *fig.,* 148–150,
        148–150 *fig.*
    structure of, 116, 116 *fig.,* 118
ADH (antidiuretic hormone), 570, 634, 641 *table*
Adhesion, 52
    in plants, 680, 680 *fig.*
Adipose cells, 61 *fig.*
Adipose tissue, 492
ADP (adenosine diphosphate), 116, 116 *fig.*
Adrenal glands, 639
    cortex of, 639, 640 *table*
    medulla of, 639, 640 *table*
Adrenaline, 640 *table*
Adrenocorticotropic hormone (ACTH), 635, 636,
    641 *table*
Adult stem cells, 260
Adventitious plantlets, 688, 688 *fig.*
Adventitious roots, 675
Aerobic respiration, 138, 139 *fig.*
Afferent neurons, 600, 600 *fig.*
African bull elephant, 522
African weaver birds, 456, 456 *fig.*
Agaston, Arthur, 143
Age(s), 377
Age distribution of populations, 418
Age structure of populations, 418
Agglutination, 592, 592 *fig.*
Aggregate organisms, 328

Aging of cells, 664
Agnathans, 378
Agonistic behavior, 450
Agriculture
    chemicals used in, 462, 462 *fig.*
    genetic engineering and, 251 *fig.,* 251–254, 252
        fig., 255
    global warming and, 465
AIDS. *See* Acquired immunodeficiency syndrome
    (AIDS); Human immunodeficiency virus
    (HIV)
Air pollution, 462
Alarm calls, 452
Alarm pheromones, 452
Albert, Prince of England, 212
Albinism, 212
Albumin, serum, 514
Alcohol
    fetal alcohol syndrome and, 661
    production of, 142
Aldosterone, 570, 639, 640 *table*
Alexandra, Empress of Russia, 212
Algae, green, 328, 328 *fig.*
Allantoin, 571
Allantois of reptilian eggs, 381, 381 *fig.*
Alleles, 188, 282
    codominant, 198–199, 199 *fig.*
    frequencies of, 282, 284 *table,* 284–285
Allergies, 593, 593 *fig.*
Allometric growth, 663, 663 *fig.*
Allopatric species, 296, 423
Allopolyploidy, 296–297
Allosteric enzymes, 115
Allosteric sites, 115, 115 *fig.*
Alternation of generations, 339
Alternative hypotheses, 10, 11
Alternative splicing, 232
Altitude, oxygen loading curves and, 436
Altruism, 454, 454 *fig.*
Alveoli, 530, 530 *fig.*
Alzheimer's disease, 612
Amanita mushrooms, 331 *fig.*
Ambulocetus, 273
Amino acids, 61 *fig.,* 62, 62 *fig.,* 541
Ammonia, 571
Amniocentesis, 208 *fig.,* 208–209
Amnion, 381, 381 *fig.,* 658
Amniotic eggs of reptiles, 381, 381 *fig.*
Amoebas, 328, 330 *table*
Amphibians, 380 *fig.,* 380–381
    characteristics of, 380–381
    circulatory system of, 517, 517 *fig.*
    global decline in, 467
    history of, 381
    kidneys of, 567
    lungs of, 528, 528 *fig.*
    reproduction in, 649 *fig.,* 649–650
Amylase, 546
Anabaena, 317
Anabolic steroids, 632
Anaerobic respiration, 142, 142 *fig.*
Analogous structures, 274, 275 *fig.*
Anaphase in meiosis
    anaphase I, 172, 173 *fig.,* 174, 174 *fig.,*
        202 *fig.*
    anaphase II, 173 *fig.,* 174, 175 *fig.,* 202 *fig.*
Anaphase in mitosis, 155 *fig.,* 159
Anastasia, Princess of Russia, 184

Anemia
    pernicious, 549
    sickle-cell, 204 *table,* 206 *fig.,* 206–207, 288
        fig., 288–289
Aneuploidy, 202
Angel dust (PCP), 605
Angina pectoris, 521
Angiogenesis inhibitors, 165
Angiosperms, 346, 350–353
    dicots and monocots, 353, 353 *fig.*
    flowers of, 350 *fig.,* 350–351
    fruits of, 354, 354 *fig.*
    life cycle of, 352 *fig.,* 352–353
    reproduction by, 688–690
Animal(s) (Animalia), 2, 2 *fig.,* 19 *fig.,* 20, 20 *fig.,*
    310, 311 *table,* 357–385
    advent of body cavities in, 368–373
    bilateral symmetry in, 360, 361 *fig.,* 366 *fig.,*
        366–367
    cells of, 86 *fig.,* 87
    cognition in, 445, 445 *fig.*
    defenses of, 430–431, 431 *fig.*
    deuterostome development in, 360, 361 *fig.*
    diversity of form of, 359 *fig.*
    embryonic development in, 359 *table*
    evolutionary trends in, 360, 361 *fig.*
    evolution of body cavities in, 360, 361 *fig.*
    evolution of segmentation in, 360, 361 *fig.*
    evolution of tissues in, 360, 361 *fig.*
    family tree of, 362–363 *table*
    general features of, 358–359 *table*
    habitat diversity of, 359 *table*
    invertebrate, 359 *table*
    learning by, 443
    major phyla of, 362–363 *table*
    movement of, 358 *table*
    radial symmetry in, 360, 361 *fig.*
    simplest, 364–365
    transitions of body plan of, 360, 361 *fig.*
    vertebrate. See Vertebrate entries; specific
        animals
Animal body plans, 484–487
    innovations in, 484–485, 485 *table*
    of vertebrate body, 486–487, 488–489 *fig.*
Annelida, 352 *table,* 371, 371 *fig.*
Anolis, 37, 37 *fig.*
Anomaly histograms, 480
Anopheles gambiae, genome of, 243 *table*
Ant(s), acacias and, 425, 425 *fig.*
Antenna complex, 128
Antennal glands, 562
Anterior, definition of, 366, 366 *fig.*
Anthers, 350
Anthidiine bee, 433
Anthocerophyta, 340 *table,* 342
Anthophyta, 341 *table*
Anthracotheres, 273
Antibodies, 584, 585 *fig.*
    in invertebrates, 588–589
    in medical diagnosis, 592, 592 *fig.*
Anticoagulants, manufacture of, 249, 249 *table*
Anticodon, 230
Antidiuretic hormone (ADH), 570, 634, 641 *table*
Antigen(s), 581
    immunity and, 596
Antigen-presenting cells, 582
Antipollution laws, 468
Anura, 381

Cohesion-adhesion-tension theory, 680, 680 *fig.*
Cohesion in water, 52, 52 *fig.*
Cohorts, 418
Cole, Lamont, 391
Collenchyma cells, 671, 671 *fig.*
Colon, 550
Colon cancer, 164, 165
Colonial organisms, 328, 328 *fig.*
Colony-stimulating factors, manufacture of, 249 *table*
Color blindness, 624, 625 *fig.*
Color vision, 624, 624 *fig.*, 625 *fig.*
Columba livia, 449
Columnar epithelium, 491 *table*
Combination therapy for AIDS, 594
Combustion, 395
Commensalism, 39, 39 *fig.*, 427, 427 *fig.*
Communities, 5, 5 *fig.*, 33, 35, 35 *fig.*, 388, 420–435
    animal defenses and, 430–431, 431 *fig.*
    climax, 434
    coevolution and, 424, 424 *fig.*, 431
    commensalism and, 427, 427 *fig.*
    individualistic concept of, 420
    mimicry and, 432 *fig.*, 432–433, 433
    niche and competition and, 421–422
    parasitism and, 426
    pioneering, 434
    plant defenses and, 430
    predator-prey interactions and, 428 *fig.*, 428–429, 429 *fig.*
    stability of, 434–438
    symbiosis and, 424
    Tanzanian savanna, 420 *fig.*
Community ecologists, 33
Compact bone, 494
Companion cells, 672, 673 *fig.*
Compass sense, 448, 449
Competition, 36, 421
    interspecific, 36, 36 *fig.*, 40, 421
    intraspecific, 36, 36 *fig.*, 421
    reduction by predation, 429, 429 *fig.*
Competitive exclusion, 37, 422
Competitive inhibitors, 115, 115 *fig.*
Complementarity, 218
    of DNA, 69
Complement system, 579, 579 *fig.*
    in invertebrates, 588
Complex carbohydrates, 70, 70 *fig.*, 71 *table*
Compound leaves, 679, 679 *fig.*
Compound microscopes, 80
Compsognathus, 382
Concentration, 74
    measure of, 106
Concentration gradient, 98
Concorde supersonic transport, 449
Condensation, 158
Conditioned stimulus, 443
Conditioning, 443
    classical, 443
    operant, 443
Condoms, 665, 665 *fig.*
Cones
    of conifers, 348
    of eye, 623–624, 624 *fig.*
Confocal microscopes, 81 *table*
Conidiophores, 332
Coniferophyta, 341 *table*
Conifers, 348, 348 *fig.*
Conjugation, bacterial, 322, 322 *fig.*
Connective tissues, 492–494, 493 *table*, 494 *fig.*
    of vertebrates, 486, 486 *fig.*
Connell, J. H., 421
Conscious planning, by animals, 445
Conservative replication, 218, 218 *fig.*
Consumers, 388
Continuous data, 13
Continuous variation, 194, 194 *fig.*
Contraception, 665, 665 *fig.*
Contractile vacuoles, 96, 562 *fig.*
Control experiments, 11
Conus arteriosus of fishes, 516, 516 *fig.*
Convergent evolution, 274, 275 *fig.*

Cooke, Howard, 664
Cooksonia, 338, 343
Cooperation, 6, 7 *table*
Coprophagy, 552
Coral, 39, 39 *fig.*
Cork cambium, 670, 676–677, 677 *fig.*
Cork cells, 677
Corn
    color of, 196
    genetically modified, 252
Corn borer, 253
Cornea, 623, 623 *fig.*
Coronary arteries, blocked, 521
Corpus luteum, 657, 657 *fig.*
Correlation, causation and, 12
Correns, Karl, 200
Cortex of stems, 676
Corticotropin-releasing hormone (CRH), 636, 640 *table*
Cortisol, 639, 640 *table*
Cotton, genetically modified, 252
Cotyledons, 346, 346 *fig.*, 691, 691 *fig.*
Countercurrent flow, 527, 527 *fig.*
Countercurrent heat exchange, 509, 509 *fig.*
Coupled channels, 103, 104 *fig.*
Coupled transport, 105 *table*
Covalent bonds, 48–49, 49 *fig.*
C phase of cell cycle, 155, 155 *fig.*
C_3 photosynthesis, *130 fig.*, 130–131, 131
C_4 photosynthesis, 132, 132 *fig.*, 134, 134 *fig.*
Crabs, 458
Crassulacean acid metabolism (CAM), 134
Creationism science, 277
Crenicichla alta, 290
CRH (corticotropin-releasing hormone), 636, 640 *table*
Crick, Francis, 16, 69, 216, 217 *fig.*
Cristae, 92, 92 *fig.*
Crocodiles, 517
Crops. *See* Agriculture
Crosscurrent flow, 529, 529 *fig.*
Crossing over, 172, 172 *fig.*, 175, 178
Crutzen, Paul, 14
Cryptic coloration, 431, 431 *fig.*
Ctenophora, 365
Cuboidal epithelium, 491 *table*
Cuenot, Lucien, 194
Culture, cells in, 180
Cuscuta, 426 *fig.*
Cutaneous respiration, 381, 517
Cuticle of plants, 338, 672
Cycadophyta, 341 *table*
Cycling, 393
Cyst(s), protist, 327
Cystic fibrosis (CF), 84, 204 *table,* 265
Cystic fibrosis transmembrane conductance regulator (CFTR), 84
Cytokinesis, 155 *fig.*, 160, 160 *fig.*
Cytokinins, 695 *table*, 698, 698 *fig.*
Cytoplasm, 79, 86
Cytoskeleton, 86, 89 *table*, 94 *fig.*, 94–96
Cytotoxic T cells, 581 *table,* 583, 583 *fig.*

D

Daltons, 44
Danaus plexippus, 432 *fig.*
Dance language, 452 *fig.*, 452–453
Darevsky, Ilya, 646
Dark-field microscopes, 81 *table*
Darwin, Charles Robert, 18, 23, 24 *fig.*, 24–25, 270, 273, 277
    Beagle voyage of, 24–26, 25 *fig.*
    Galàpagos Islands research of, 23, 25, 29 *fig.*, 29–30, 30 *fig.*, 37, 37 *fig.*
    natural selection theory of, 24, 27–30 *fig.*, 27–32, 32 *fig.*, 298
    plant studies of, 696, 696 *fig.*
Darwin, Francis, plant studies of, 696, 696 *fig.*
Darwin's Black Box: The Biochemical Challenge in Evolution (Behe), 278
Daughter cells, 154

DDT (dichlorodiphenyltrichloroethane), 477
Deamination, 554
Death(s). *See also* Mortality
    cancer-related, 162
    of cells, 160, 160 *fig.*
Deciduous forests, 408, 408 *fig.*
Decomposers, 333, 390, 390 *fig.*
Deductive reasoning, 8, 8 *fig.*
Deep-sea waters, 402–403, 403 *fig.*
Defense(s), 575–597
    of animals, 430–431, 431 *fig.*
    cellular counterattack as, 576, 578–580, 578–580 *fig.*
    immune system and. See Immune response; Immune system
    of plants, 430
    skin as, 576 *fig.*, 576–577, 577 *fig.*
Defense mechanisms, predation and, 38, 38 *fig.*
Defensive coloration, 430–431, 431 *fig.*
Dehydration synthesis, 60, 60 *fig.*
Deletion, 222, 223 *fig.*
Demography, 418, 418 *fig.*
Denaturation of protein, 65, 65 *fig.*
Dendrites, 497, 497 *fig.*, 602
Dendrobatidae, 38, 38 *fig.*
Density-dependent effects, 416, 416 *fig.*
Density-independent effects, 416
Deoxyribonucleic acid. *See* DNA (deoxyribonucleic acid)
Depression, 606
Derived characters, 306
Dermal tissue of plants, 672, 672 *fig.*
Dermis, 576
Descent of Man, The (Darwin), 28
Deserts, 407, 407 *fig.*
Detrivores, 390, 390 *fig.*
Deuterostomes, 360, 361 *fig.*, 374, 374 *fig.*, 484–485, 485 *table*
Development, 691
    cellular, 328
Devonian period, 377
Diabetes, 249, 556, 593, 637
    embryonic stem cell therapy for, 260
    obesity and, 543, 637
Diaphragm (muscle), 383, 530
Diaphragms (contraceptive), 665, 665 *fig.*
Diastolic pressure, 520
Diatom(s), 330 *table*
Diazepam, 605
Dichlorodiphenyltrichloroethane (DDT), 477
Dicots, 353, 353 *fig.*
    development of, 693, 693 *fig.*
Dictyostelium discoideum, genome of, 243 *table*
Diencephalon, 609
Dieting, 143
Differential-interference-contrast microscopes, 81 *table*
Diffusion, 98, 98 *fig.*, 105 *table*
    facilitated, 102, 102 *fig.*, 105 *table*
    selective, 102
Digestion, 544
    extracellular, 365
Digestive systems, 544–555, 555 *fig.*
    accessory digestive organs and, 554, 554 *fig.*, 555 *fig.*
    defense of, 577
    esophagus and stomach of, 548 *fig.*, 548–549
    intestines of, 550, 551 *fig.*
    mouth and teeth of, 546 *fig.*, 546–547
    types of, 544, 544 *fig.*
    variations in, 552, 552 *fig.*, 553 *fig.*
    of vertebrates, 487, 489 *fig.*, 545 *fig.*, 545–555
Dihybrid individuals, 191, 191 *fig.*
Dileptus, 77
Dilger, William, 442
Dilobderus abderus, 373 *fig.*
Dinoflagellates, 330 *table*
Dioecious plants, 689
Dionaea muscipula, 684, 684 *fig.*
Diploid cells, 170, 170 *fig.*
Dipodomys panamintensis, 567 *fig.*

Nuclear power, 469, 469 *fig.*
Nucleic acids, 60, 67 *fig.*, 67–68, 68 *fig.*
Nucleoid region, 85, 85 *fig.*
Nucleolus, 88, 89 *table*
Nucleosomes, 157
Nucleotides, 16, 17 *fig.*, 61 *fig.*, 67, 67 *fig.*, 216, 216 *fig.*
Nucleus of cells, 86 *fig.*, 86–87, 87 *fig.*, 88, 88 *fig.*, 89 *table*
Nutrition. *See also* Digestive systems; Food(s)
    dieting and, 143
    essential plant nutrients and, 684, 684 *fig.*
    in fish, 378
    food for energy and growth and, 540 *fig.*, 540–542, 541 *fig.*
    of fungi, 332, 332 *fig.*
    link between diabetes and obesity and, 543
    protist, 327
    provided by seeds, 347

O

Obesity, diabetes and, 543, 637
Observation in scientific process, 10
Ocean(s)
    global warming and, 465
    patterns of circulation in, 400 *fig.*, 400–401, 401 *fig.*
Ocean ecosystems, 402 *fig.*, 402–403, 403 *fig.*
Oenone fulgida, 371 *fig.*
Oil spills, 462 *fig.*
Olfactory information, 608
Oligodendrites, 602
Oligotrophic lakes, 405
Omnivores, 390, 390 *fig.*, 544
Oncogenes, 162
On the Origin of Species by Means of Natural Selection (Darwin), 24, 27, 277
Oocytes, 654, 654 *fig.*
Open circulatory systems, 368, 508, 508 *fig.*
Open-sea surface, 402, 402 *fig.*
Operant conditioning, 443
Operculum, 379
Operons, 234
Opiothrix, 375 *fig.*
Opossums, 651
Opposite pairs of leaves, 679
Optic lobes, 608
Optimal foraging theory, 447, 458
Orbitals, 45
Orb-weaving spiders, 447
Orders, 304
Oreaster occidentalis, 375 *fig.*
Organ(s), 4 *fig.*, 5
    fetal development of, 660 *fig.*, 660–661
    of vertebrates, 486–487
Organelles, 4, 4 *fig.*, 85, 86–87, 92–93
    of protists, 327
Organ function, body cavity and, 368
Organic molecules, 60, 60 *fig.*
Organismic water cycle, 393
Organization of life, 4–5, 4–5 *fig.*
Organogenesis, 659 *table*, 660 *fig.*, 660–661
Organ systems, 4 *fig.*, 5
    of vertebrates, 487–489 *fig.*
Origin of life, 318–319
    extraterrestrial life and, 318, 320
    first cells and, 319
    forming of building blocks and, 318 *fig.*, 318–319, 319 *fig.*
Origin of muscles, 500
Ornithorhynchus anatinus, 301
Oryza sativa, genome of, 243 *table*
Osmoregulation, 562 *fig.*, 562–570, 566 *fig.*. *See also* Kidney(s)
Osmosis, 98–99, 99 *fig.*, 105 *table*
Osmotic pressure, 99
Osmotrophs, 327
Osteichthyes, evolution of, 314
Osteoblasts, 494
Osteoporosis, 494, 494 *fig.*

Ostracoderms, 378
Otoliths, 618, 618 *fig.*
Outgroups, 306
Ova, 655. *See also* Eggs
Ovaries, 646, 654, 654 *fig.*
    of flowers, 350
    hormones of, 640 *table*
Overweight, 540–541. *See also* Obesity
Oviducts, 655, 655 *fig.*
Oviparity, 648
Ovoviparity, 648
Ovulation, 654
Ovules, of plants, 346, 690
Oxidation, 138
Oxidation-reduction (redox) reactions, 118, 118 *fig.*
Oxidative metabolism, 92
Oxpeckers, 39, 39 *fig.*
Oxygen
    atomic number and atomic mass of, 45 *table*
    binding of, by hemoglobin, 74, 436
Oxygen loading curves, 74, 536
Oxygen transport, by hemoglobin, 532
Oxytocin, 634, 641 *table*, 657
Oystercatchers, 451 *table*
Oyster fungus, 332, 332 *fig.*
Ozone hole, 9, 9 *fig.*, 10–11, 11 *fig.*, 20, 464, 464 *fig.*

P

Pain perception, 617
Paleozoic era, 377
Paley, William, 278, 279
Palisade mesophyll, 679, 679 *fig.*
Palmately compound leaves, 679, 679 *fig.*
Pancreas, 550, 554, 554 *fig.*, 637, 637 *fig.*, 640 *table*
Pancreatic duct, 554, 554 *fig.*
Panthera leo, 384 *fig.*
Pan troglodytes, genome of, 243 *table*
Papuan kingfisher, 296
Paramecium, 422, 422 *fig.*
Parasitism, 39, 39 *fig.*, 426, 426 *fig.*
Parasitoids, 426
Parastichopus parvimensis, 375 *fig.*
Parasympathetic nervous system, 615, 615 *fig.*
Parathyroid glands, 638–639, 640 *table*
Parathyroid hormone (PTH), 638–639, 639 *fig.*, 640 *table*
Parazoa, 364
Parenchyma cells, 671, 671 *fig.*
Parietal cells, 548
Parkinson's disease, embryonic stem cell therapy for, 260
Parthenogenesis, 179, 646
Passive transport, 105 *table*
Pavlov, Ivan, 443
PCP (angel dust), 605
Pea(s), Mendel's experiments on, 184–187, 185 *fig.*, 186 *table*, 187 *fig.*
Peacock, 450, 450 *fig.*
Pectoral girdle, 499
Pedigree(s), 204 *fig.*, 204–205, 205 *fig.*
Pedigree analysis, 208
Pelvic girdle, 499
Penicillium, 156
Penis, 653, 653 *fig.*
Peptic ulcers, 549
Peptide bonds, 62, 62 *fig.*
Peptide hormones, 632–633, 633 *fig.*
Percents, 74
Peregrine falcons, 477
Pericycle, 674
Periderm, 677
Periods, 377
Peripheral nervous system (PNS), 600, 600 *fig.*
Peristalsis, 548, 548 *fig.*
Permafrost, 409
Permian period, 377
Pernicious anemia, 549
Perognathus, 567

Peroxisomes, 89 *table*, 91
Petals, 350
Petioles, 678
P generation, 185
pH
    of blood, 150
    enzyme activity and, 114, 114 *fig.*
    protein functions and, 74
Phagocytes in invertebrates, 588
Phagocytosis, 100–101, 105 *table*
Phagosomes, 327
Phagotrophs, 327
Pharyngeal pouches in chordates, 376
Pharynx, 369
Phase-contrast microscopes, 81 *table*
Phenotype, 188
    determination of, 192
    mutations and, 192
Phenylketonuria, 204 *table*
Pheromones
    alarm, 452
    trail, 452, 452 *fig.*
Phloem, 672, 673 *fig.*, 684
Phospholipids, 72, 73, 73 *fig.*
    in plasma membrane, 82, 82 *fig.*, 83, 83 *fig.*
Phosphorus
    atomic number and atomic mass of, 45 *table*
    as plant nutrient, 684
Phosphorus cycle, 396–397, 397 *fig.*
Phosphorylation, substrate-level, 140
Photoautotrophs, 323
Photoheterotrophs, 323
Photons, 124, 124 *fig.*
Photoperiodism, 700 *fig.*, 700–701
Photorespiration, 132, 132 *fig.*, 134
Photosynthesis, 110, 116, 118, 122 *fig.*, 122–134
    C_3, 130 *fig.*, 130–131, 131
    C_4, 132, 132 *fig.*, 134, 134 *fig.*
    capture of energy from sunlight and, 124 *fig.*, 124–125
    chloroplasts and, 123
    levels of light and, 134
    photosystems and, 123, 126–129, 126–129 *fig.*
Photosystems, 123, 126–129, 126–129 *fig.*
    architecture of, 126, 127 *fig.*
    conversion of light to chemical energy by, 128, 128 *fig.*, 129 *fig.*
    dual, in plants, 127, 127 *fig.*
Phototrophs, 327
Phototropism, 696, 696 *fig.*
pH scale, 53, 53 *fig.*, 55, 55 *fig.*
pH-tolerant archaea, 312
Phyla, 304
Phylogenetic trees, 306
Phylogenies, 306–307
Phytochromes, 701
Phytoplankton, 392
Pie charts, 13, 298
Pieris rapae, 38, 38 *fig.*, 430, 430 *fig.*
Piggyback vaccines, 249–250, 250 *fig.*, 590
Pigments, 124–125, 125 *fig.*
PIH (prolactin-inhibiting hormone), 636, 640 *table*
Pike cichlid, 290
Pili, 85
    bacterial, 322, 322 *fig.*
Pineal gland, 641 *table*
Pinnately compound leaves, 679, 679 *fig.*
Pinocytosis, 101, 105 *table*
Pioneering communities, 434
Pisaster, 429 *fig.*
Pistol shrimp, 39, 39 *fig.*
Pith, 676
Pituitary gland, 634–636
    anterior, 634–635, 635 *fig.*, 641 *table*
    control by hypothalamus, 635–636, 636 *fig.*
    posterior, 634, 634 *fig.*, 641 *table*
Pit vipers, 626, 626 *fig.*
Pius IX, Pope, 264
Placenta, 383, 383 *fig.*, 658
Placental mammals, 383, 384, 384 *fig.*, 651, 651 *fig.*
Planaria, 367 *fig.*, 508, 508 *fig.*. *See also* Flatworms

R

Radial cleavage, 374
Radial symmetry, 360, 361 *fig.*, 365, 484, 485 *table*
Radiata, 365
Radioactive decay, 46, 56
Radioactive isotopes, 46–47, 47 *fig.*
Radioactive tracers, 46, 47 *fig.*
Radioisotopic dating, 47, 47 *fig.*
Radiolarians, 330 *table*
Radula, 370
Rain
  acid, 463, 463 *fig.*
  global warming and, 465
Rain forests, 407, 407 *fig.*
  tropical, 471, 471 *fig.*
Rain shadows, 399, 399 *fig.*
Rana muscosa, 467
Rana pipiens, 295, 295 *fig.*
Ras protein, 164
Rate, 106
Ratios, 56
Ravens, 445, 445 *fig.*
Ray, John, 305
Reactants, 112, 112 *fig.*
Reaction center, 126
Realized niches, 421
Reasoning, 8, 8 *fig.*
Receptacles of flowers, 350
Receptor(s), sensory, 616
Receptor-mediated endocytosis, 101, 101 *fig.*, 105 *table*
Reciprocity, 454–455
Recombinant DNA, 248, 248 *fig.*
Recombination, 222
Rectum, 550
Recycling, 479
Red algae, 330 *table*
Red blood cells, 514–515, 515 *fig.*
  gas exchange and, 532, 533 *fig.*
  osmotic pressure in, 99, 99 *fig.*
Redox reactions, 118
Red Queen Hypothesis, 179
Redwood sorrel, 35, 35 *fig.*
Redwood trees, 35, 35 *fig.*, 348
Reflexes, 600–601, 601 *fig.*, 614, 614 *fig.*
Regression line, 13, 298
Renal cortex, 568, 569 *fig.*
Renal medulla, 568, 569 *fig.*
Renal pelvis, 568
Replication fork, 220, 221 *fig.*
Repressors, 115, 234, 234 *fig.*
Reproduction, 3
  asexual. See Asexual reproduction
  evolution among vertebrates, 648–651
  fungal, 332, 332 *fig.*
  in mammals, 651, 651 *fig.* See also Human reproduction
  by plants, 339, 688–693. See also Fruit(s); Seeds
  prokaryotic, 322, 322 *fig.*
  sexual. See Sexual reproduction
Reproductive behavior, 450, 450 *fig.*, 451 *table*
Reproductive cloning, 256–258 *fig.*, 256–259
Reproductive isolating mechanisms, 292, 292 *table*
Reproductive isolation, 292
Reproductive strategies, 450
Reproductive system of vertebrates, 487, 489 *fig.*
Reptiles
  characteristics of, 381
  circulatory system of, 517, 517 *fig.*
  kidneys of, 567
  lungs of, 528, 528 *fig.*
  reproduction in, 650 *fig.*, 650–651
Residual volume, 531
Resolution, 80
Resource partitioning, 37, 37 *fig.*, 422–423, 423 *fig.*
Respiration, 526
  in carbon cycle, 395
  cutaneous, 381, 517

Respiratory system, 508, 525–537
  in aquatic vertebrates, 526, 526 *fig.*, 527, 527 *fig.*
  defense of, 577
  gas exchange and, 532 *fig.*, 532–533
  lung cancer and, 153, 163, 164, 224, 534–535
  mammalian, 530 *fig.*, 530–531
  of mammals, 528, 528 *fig.*, 530 *fig.*, 530–531
  severe acute respiratory syndrome and, 325
  in terrestrial vertebrates, 528–529
  types of, 526, 526 *fig.*
  of vertebrates, 487, 488 *fig.*
Resting potential, 603
Restriction endonucleases, 588
Restriction enzymes, 246–247, 247 *fig.*
Reticular formation, 611
Retina, 623, 623 *fig.*
Retinal, 124
Retinoblastoma protein, in lung cancer, 534
Reznick, David, 291
Rheumatoid arthritis, 593
Rh factor, 592
Rhizomes, 688
Rhynia, 338
Ribonucleic acid (RNA). *See* RNA (ribonucleic acid)
Ribosomal RNA (rRNA), 230
Ribosomes, 85, 85 *fig.*, 88, 89 *table*, 227, 230
Rice, genetically modified, 252, 252 *fig.*
Rickets, 639
Right atrium, 518 *fig.*, 519
Rivulus hartii, 290
RNA (ribonucleic acid), 16, 67, 67 *fig.*, 68 *fig.*
  in chromosomes, 156
  formation of, 319
  messenger, 228, 228 *fig.*
  regulation of gene expression by, 236, 237 *fig.*
  ribosomal, 230
  transcription and, 228, 228 *fig.*, 234 *fig.*, 234–235, 235 *fig.*
  transfer, 230 *fig.*, 230–231
  translating message into proteins, 230 *fig.*, 230–231, 231 *fig.*
  translation and, 228, 229–231, 229–231 *fig.*
RNA interference, 236
RNA polymerase, 228
Rodhocetus, 273
Rod(s) of eye, 623–624, 624 *fig.*
Rodrigues fruit bat, 478
Roland, Sherwood, 14
Root(s), 674 *fig.*, 674–675, 675 *fig.*
  adventitious, 675
  lateral, 675, 675 *fig.*
  of vascular plants, 670, 670 *fig.*
  water absorption by, 682, 682 *fig.*
Root cap, 674
Root hairs, 672
Root pressure, 680
Rotifers (Rotifera), 353 *table*, 369, 369 *fig.*
Rough ER, 90, 90 *fig.*
Roundworms, 352 *table*, 369
  body cavities of, 368, 368 *fig.*
rRNA (ribosomal RNA), 230
r-selected adaptations, 417, 417 *table*
RU486, 665
Rubber, natural, 73 *fig.*
Ruminants, 545, 552, 552 *fig.*
Runners, 688

S

Saccharomyces cerevisae, genome of, 243 *table*
St. John's wort, 421
Saliva, 546
Salt
  dissolution in water, 52, 52 *fig.*
  ionic bond in, 48, 48 *fig.*
  plant growth and, 354
Sand dollar, 375 *fig.*
Sand wasp, 433
Sanger, Frederick, 242
Saprozoic feeders, 327

Sarcocystis, 426 *fig.*
Sarcoplasmic reticulum, 503
SARS (severe acute respiratory syndrome), 325
Saturated fats, 72 *fig.*, 73
Savannas, 407, 407 *fig.*
  savanna community in Tanzania and, 420 *fig.*
Scanning electron microscopes, 80, 81 *table*
Scaphinotus velutinus, 35, 35 *fig.*
Scarab beetle, 373 *fig.*
Schleiden, Matthias, 16, 79
Schwann, Theodor, 79
Schwann cells, 602
Science, limitations of, 14
Scientific creationism movement, 277
Scientific method, 14
Scientific names, 303
Scientific process, 10 *fig.*, 10–13
  analyzing relationships between variables and, 12–13
  case study in, 9, 9 *fig.*
  graphs and, 12–13
  reasoning in, 8, 8 *fig.*
  stages of, 10–12
  variables and, 12
Sclereids, 671
Sclerenchyma cells, 671, 671 *fig.*
Scopes, John, 277
Screening
  for genetic disorders, 209, 209 *fig.*
  in genetic engineering, 248, 248 *fig.*
Scrotum, 652, 652 *fig.*
Sea cucumbers, 375 *fig.*
Sears, Barry, 143
Sea stars, 375 *fig.*
Sea urchins, 375 *fig.*
Secondary cell walls, 671, 671 *fig.*
Secondary growth, 343, 670, 676–677, 677 *fig.*
Secondary immune response, 586, 586 *fig.*
Secondary structure of proteins, 64, 64 *fig.*
Secondary succession, 434
Secondary walls, 97
Second law of thermodynamics, 111, 111 *fig.*
Second messengers, 632–633, 633 *fig.*
Seeds, 346, 691, 691 *fig.*
  development of, 691
  dispersal of, 347, 692, 692 *fig.*
  dormancy of, 347
  endosperm of, 352
  germination of, 347, 693, 693 *fig.*
  nourishment provided by, 347
  structure of, 346 *fig.*, 346–347, 347 *fig.*
Segmentation, 484, 485 *table*
  in annelids, 371, 371 *fig.*
  evolution of, 360, 361 *fig.*
  in insects, 372, 372 *fig.*
Segregation, law of, 191
Selection, 285–287
  artificial, 6, 7 *table*, 27–28, 285
  changes in allele frequencies due to, 284 *table*, 285
  directional, 286 *fig.*, 287, 287 *fig.*
  disruptive, 286 *fig.*, 287, 287 *fig.*
  natural. See Natural selection
  stabilizing, 286 *fig.*, 286–287, 287 *fig.*, 289, 289 *fig.*
Selective diffusion, 102
Selective permeability, 102–105
Self, distinguishing from nonself, 588
Self-fertilization, 689
Self-incompatibility, 689
Self mimicry, 433
Self-pollination, 689
Semibalanus balanoides, 421, 421 *fig.*
Semicircular canals, 618, 618 *fig.*
Semiconservative replication, 218, 218 *fig.*
Semidesert, 410, 410 *fig.*
Seminiferous tubules, 652, 652 *fig.*
Sensitization, 443
Sensors, homeostasis and, 560
Sensory nervous system, 616
Sensory neurons, 497 *table*, 600, 600 *fig.*
Sensory organs, 616, 616 *fig.*

mutualism and, 39, 39 *fig.*
parasitism and, 39, 39 *fig.*
Symmetry
bilateral, 360, 361 *fig.*
radial, 360, 361 *fig.*
Sympathetic nervous system, 615, 615 *fig.*
Sympatric speciation, 296–297
Sympatric species, 423
Synapses, 497, 604 *fig.*, 604–607
addictive drugs and, 606–607, 607 *fig.*
kinds of, 604–605, 605 *fig.*
neurotransmitters and, 604, 604 *fig.*, 05
Synapsis in meiosis, 176, 176 *fig.*
Syngamy, 170
Syphilis, 666
Systematics, 306
Systemic circulation, 517
Systems ecologists, 33
Systolic pressure, 520

T

t1/2, 180
T3 (triiodothyronine), 641 *table*
T4 (thyroxine), 638, 641 *table*
Table salt. *See* Salt
Tachyglossus aculeatus, 384 *fig.*
Taiga, 408, 408 *fig.*
Taller, Herman, 143
Tanysiptera hydrocharis, 296
Tanzania, savanna in, 420 *fig.*
Taricha granulosa, 169
Taste sense, 619, 619 *fig.*
tau protein, 612
Taxon(s), 303
Taxonomy, 303, 306
traditional, 307
Tay-Sachs disease, 204 *table,* 07, 207 *fig.*, 602
T cells, 515, 515 *fig.*, 581, 581 *able*, 583, 583 *fig.*
cytotoxic, 583, 583 *fig.*
helper, 582
Teeth
of mammals, 383
vertebrate, 546, 547 *fig.*
Telencephalon, 609
Telomerase, 166
Telomeres, 165, 166
Telophase of meiosis
telophase I, 172, 173 *fi.*, 174, 174 *fig.*
telophase II, 173 *fig.*, 74–175, 175 *fig.*
Telophase of mitosis, 155 *fig.*, 159
Temperate evergreen forests, 410
Temperature
body, regulation of, 51
change in, perception of, 617
enzyme activity and 114, 114 *fig.*
Temporal isolation, 292 *table*, 294
Tendons, 500
Tensil strength in plants, 680, 680 *fig.*
Territorial behavior, 447, 447 *fig.*
Tertiary structure of proteins, 64, 64 *fig.*
Testcross, 190, 190 *fig.*
Testes, 646, 652 *fig.*, 652–653, 653 *fig.*
Testosterone, 641 *table,* 653
Tetrahymena thermophila, 236
Textbooks, evolution controversy and, 277, 280
Thalamus, 609, 611
Thalassoma bifasciatum, 647 *fig.*
Theory, 11, 14
Theory of evolution, 18–20 *fig.*, 20
Theory of heredity, 18, 18 *fig.*
Therapeutic cloning, 262, 263 *fig.*
controversy over, 262
Thermal stratification, 404, 405 *fig.*
Thermal vent system, 403
Thermoclines, 404–405
Thermodynamics, 110
first law of, 111, 111 *fig.*
second law of, 111, 111 *fig.*
Thermophiles, 312
Thigmotropism, 701, 701 *fig.*

Thomson, James, 260
Thoracic breathing of reptiles, 381
Thoracic cavity, 486, 530
Thylakoids, 92, 93 *fig.*, 123
Thymosin, 641 *table*
Thymus gland, 641 *table*
Thyroid gland, 638, 638 *fig.*, 641 *table*
Thyroid-stimulating hormone (thyrotropin or TSH), 630, 635, 636, 641 *table*
Thyrotropin-releasing hormone (TRH), 630, 636, 640 *table*
Thyroxine (T4), 638, 641 *table*
Tidal volume, 531
Tinbergen, Niko, 441, 446, 446 *fig.*
Tissues, 4 *fig.*, 5, 359 *table*
distinct, evolution of, 360, 361 *fig.*
of vertebrates, 486, 486 *fig.*
T lymphocytes, 515, 515 *fig.*
TMV (tobacco mosaic virus), 324, 324 *fig.*
Toads, reproduction in, 649
Tobacco mosaic virus (TMV), 324, 324 *fig.*
Topsoil, 470
Torpor, 572
Totipotent embryonic stem cells, 260
Touch perception, 617, 617 *fig.*
Trace elements, 541–542
Tracers, 46, 47 *fig.*
Trachea(e), 526, 526 *fig.*, 530, 530 *fig.*
Tracheids, 672, 673 *fig.*
Tracts, 610–611
Traditional taxonomy, 307
Tragedy of the commons, 470 *fig.*
Trail pheromones, 452, 452 *fig.*
Traits, 184
gene influence on, 192, 193 *fig.*
Transcription, 228, 228 *fig.*
controlling, 234 *fig.*, 234–235, 235 *fig.*
Transduction of sensory information, 616
Transfer RNA (tRNA), 230 *fig.*, 230–231
Transformation, 214, 214 *fig.*
Transforming principle, 215
Translation, 228, 229–231, 229–231 *fig.*
Translocation, 683, 683 *fig.*
Transmembrane proteins, 83, 83 *fig.*
Transmission electron microscopes, 80, 81 *table*
Transmission of sensory information, 616
Transpiration, 393
in plants, 680–682, 681 *fig.*, 682 *fig.*
Transposable sequences (transposons), 232
Transposition, 223
trans-retinal, 624
Tree of life, 18, 19 *fig.*
TRH (thyrotropin-releasing hormone), 630, 636, 640 *table*
Triacylglycerol, 72 *fig.*, 73
Trichinella, 369
Trichinosis, 369
Trichomes, 672, 672 *fig.*
Tricuspid valve, 518 *fig.*, 519
Triglochin maritima, 354
Triglycerides, 72 *fig.*, 73
Triiodothyronine (T3), 641 *table*
Triple bonds, 49
Trisomy, 156
Trivers, Robert, 454
tRNA (transfer RNA), 230 *fig.*, 230–231
Trophic levels, 388, 388 *fig.*, 391, 391 *fig.*
Trophoblast, 658
Tropical hummingbird, 572
Tropical monsoon (upland) forests, 409, 409 *fig.*
Tropical rain forests, 471, 471 *fig.*
Tropisms, 701, 701 *fig.*
Tropomyosin, 503
Troponin, 503
True-breeding peas, 185
TSH (thyroid-stimulating hormone or thyrotropin), 630, 635, 636, 641 *table*
Tube feet, 375
Tumors, 162
benign, 165
malignant. See Cancer

Tumor-suppressor genes, 162
in lung cancer, 534
Tundra, 409, 409 *fig.*
Turgidity in plants, 681
Turner syndrome, 203
Twin studies, of behavior, 442
Tyrosine kinases, 164–165

U

Ulcers, peptic, 549
Ultimate causation, 440
Ultrasound, prenatal, 208, 208 *fig.*
Unconditioned stimulus, 443
Unsaturated fats, 72 *fig.*, 73
Upland forests, tropical, 409, 409 *fig.*
Urea, 571
Ureters, 568, 568 *fig.*, 569 *fig.*
Urethra, 653
Urey, Harold, 318–319
Uric acid, 571
Uricase, 571
Urinary bladder, 568, 568 *fig.*
Urinary system of vertebrates, 487, 489 *fig.*
Urination, 634
Urodela, 381
Uterine contractions, 634
Uterus, 655

V

Vaccines, 590 *fig.*, 590–591, 591 *fig.*, 596
for AIDS, 590–591, 595
piggyback (subunit), 249–250, 250 *fig.*, 590
Vacuoles, 96, 96 *fig.*
Van der Waals forces, 50
Van Valen, Leo, 384
Variables, 11, 12
analyzing relationships between, 12–13
Variance, 197
Varmus, Harold, 265
Vascular cambium, 670, 676, 677 *fig.*
Vascular endothelial growth factor (VEGF), 165
Vascular plants
angiosperms. See Angiosperms
evolution of, 343, 343 *fig.*, 346 *fig.*, 346–347
organization of, 670, 670 *fig.*
seed, 341 *table.* See also Angiosperms; Gymnosperms
seedless, 340 *table,* 344 *fig.*, 344–345
Vascular tissue of plants, 340, 343, 343 *fig.*, 672, 673 *fig.*
Vas deferens, 653
Vasopressin, 634, 641 *table*
Vectors, 248
for gene therapy, 265–266
Vegetative reproduction, 688, 688 *fig.*
VEGF (vascular endothelial growth factor), 165
Veins, 508, 510, 512, 512 *fig.*
Vena cava, 518 *fig.*, 519
Venules, 510
Ventral, definition of, 366, 366 *fig.*
Ventricles, of fishes, 516, 516 *fig.*
Venules, 510
Venus's-flytrap, 684, 684 *fig.*
Vertebral column in fish, 378
Vertebrates, 19 *fig.*, 359 *table,* 377–384, 486–487. *See also specific animals*
amphibious, respiratory system in, 526, 526 *fig.*, 527, 527 *fig.*
brain of, 608 *fig.*, 608–609
chemical defenses of, 431, 431 *fig.*
chordate, 376
circulatory systems in, 509–512, 510 *fig.*
digestive systems of, 545 *fig.*, 545–555
evolution of, 377
eyes of, 623, 623 *fig.*
immune system of, 589, 589 *fig.*
organs of, 486–487